Studies in Systems, Decision and Control

Volume 296

The series "Studies in Systems, Decision and Control" (SSDC) covers both new developments and advances, as well as the state of the art, in the various areas of broadly perceived systems, decision making and control–quickly, up to date and with a high quality. The intent is to cover the theory, applications, and perspectives on the state of the art and future developments relevant to systems, decision making, control, complex processes and related areas, as embedded in the fields of engineering, computer science, physics, economics, social and life sciences, as well as the paradigms and methodologies behind them. The series contains monographs, textbooks, lecture notes and edited volumes in systems, decision making and control spanning the areas of Cyber-Physical Systems, Autonomous Systems, Sensor Networks, Control Systems, Energy Systems, Automotive Systems, Biological Systems, Vehicular Networking and Connected Vehicles, Aerospace Systems, Automation, Manufacturing, Smart Grids, Nonlinear Systems, Power Systems, Robotics, Social Systems, Economic Systems and other. Of particular value to both the contributors and the readership are the short publication timeframe and the world-wide distribution and exposure which enable both a wide and rapid dissemination of research output.

** Indexing: The books of this series are submitted to ISI, SCOPUS, DBLP, Ulrichs, MathSciNet, Current Mathematical Publications, Mathematical Reviews, Zentralblatt Math: MetaPress and Springerlink.

More information about this series at http://www.springer.com/series/13304

Piotr Kulczycki · Józef Korbicz ·
Janusz Kacprzyk
Editors

Automatic Control, Robotics, and Information Processing

Springer

Editors
Piotr Kulczycki
Systems Research Institute
Polish Academy of Sciences
Warsaw, Poland

Faculty of Physics and Applied
Computer Science
AGH University of Science
and Technology
Kraków, Poland

Janusz Kacprzyk
Systems Research Institute
Polish Academy of Sciences
Warsaw, Poland

Józef Korbicz
Institute of Control
and Computation Engineering
University of Zielona Góra
Zielona Góra, Poland

ISSN 2198-4182 ISSN 2198-4190 (electronic)
Studies in Systems, Decision and Control
ISBN 978-3-030-48586-3 ISBN 978-3-030-48587-0 (eBook)
https://doi.org/10.1007/978-3-030-48587-0

This Springer imprint is published by the registered company Springer Nature Switzerland AG
The registered company address is: Gewerbestrasse 11, 6330 Cham, Switzerland

To Nestors of Polish automatic control and robotics,
our Masters and Mentors,
we dedicate this work

Preface

The history of mankind could be viewed as the history of difficult challenges which had to be solved by human beings for, first, simple survival, and then for the necessity of effective action in increasingly complex situations, often in the face of natural disasters, the necessity of existence and operation in adverse environments, and also self-generated processes in the presence of successive development of technology and war art. As intelligent beings, humans have quickly realized that their specific physical, mental, and cognitive limitations are difficult to overcome and, as a consequence, they prevented them from reaching an effective and efficient performance of many activities requiring physical strength or constant, long-term concentration, and/or focus attention or reaction speed. An obvious conclusion has readily been that these basic human abilities, capabilities, and skills should somehow be supported (augmented) by using appropriate technological tools and techniques available at the current development level of knowledge and technology. A natural example can be the use of the first draught (working, usually pulling) animals, followed by steam engines, and electric motors, up to the ubiquitous use of computation technology today.

In our context, all these activities can be considered as consecutive steps toward an increasingly common automation of all activities, from simple solutions requiring practically only physical strength, through more complex, requiring a high accuracy and speed of action, to the most complex and sophisticated acts, the execution of which needs intelligence, learning skills, planning, effective, and efficient group cooperation or collaboration, to just mention a few. This human quest for developing more effective and efficient tools and techniques for dealing with complex processes and systems has been a natural inspiration for the emergence of numerous fields of science and technology, among which control and automation and, recently, robotics, which are the subjects of papers included in this volume, are of a particular relevance.

The purpose of this book is to present a wide and comprehensive spectrum of issues and problems related to these fields of science and engineering, both from a theoretical and applied perspective.

To be more specific, the works included involve the development of methods and algorithms to determine the best practices, most often regarding decisions or controls, which would ensure the most effective and efficient attainment of goals assumed, under the satisfaction of some constraints.

From an engineering point of view, many papers deal with how to automate a specific process or even a complex system. From a perspective of tools, there are numerous contributions which are concerned with the development of analytic and algorithmic methods and techniques, and then devices and systems which make it possible to attain automation and robotization.

For these issues outlined above as examples of various aspects of the wide field of knowledge that is control and automation, and its related robotics, there are also a number of very important, detailed and more specific elements, in particular those connected with the analysis of developed methods and systems.

Virtually all of the above aspects are the subjects of consideration in the individual chapters of this monograph written by outstanding Polish researchers, scholars, and engineers who are well known in the fields of automatic control, automation, robotics, computer science, IT/ITC, as well as in some other fields such as measurement and sensor technology, reliability and damage detection. The first authors of the individual chapters are members of the Committee for Automatic Control and Robotics, Polish Academy of Sciences, selected by the Polish scientific community in recognition of their outstanding publication and application records, as well as their high stature in international science.

For many years, Polish control and automation has enjoyed a great recognition in the world's scientific community. Immediately after World War II, despite the destruction of the country, higher education was reinstated by resuming the operation of many renowned universities, and then establishing new ones. Already in the early 1950s, the first research teams, usually within University departments, working on automatic control and automation were formed. These centers quickly obtained valuable scientific results and as a consequence the Polish school in these fields began to be highly valued in the world. At that time, top scientists from around the world decided to set up the IFAC (International Federation of Automatic Control), which is today the largest and most prestigious, opinion-forming organization bringing together top experts, both researchers, scholars and practitioners, in automatic control, automation and robotics. Poland was one of the founders of IFAC. Moreover, as both a great distinction for a country so severely damaged by the war, and as a proof of a high recognition for the scientific and technological achievements, Poland was entrusted with the organization of the Fourth IFAC World Congress in 1969. This Congress, held in Warsaw, gathered practically all leading theoreticians and practitioners in the fields from around the world, as well as many young researchers and students. The participation of the Polish authors was particularly significant. To this day, the IFAC World Congresses are held every 3 years and are the most prestigious scientific events in the fields of broadly perceived automatic control and automation, and related areas, at which outstanding innovative theoretical achievements and applications are presented. In addition, representatives of the Polish science organize numerous conferences devoted to various aspects of automation and robotics, both under the auspices of IFAC and other prestigious international organizations: IEEE (Institute of Electrical and Electronics Engineers), IMECO (International Measurement Confederation), IFIP (International Federation of Information Processing), to just mention a few.

The purpose of this volume is to present the current state of knowledge as well as new trends and application potential in the field of broadly understood automatic control automation and robotics in a comprehensible way.

The first part, our starting point, is devoted to various basic issues of mathematical modeling. Currently, without knowing the model, which properly captures the examined fragment of reality, it is difficult, if not impossible, to perform any analysis of a process or system, and in particular to develop an appropriate control algorithm. Various types of mathematical models, both classic and specialized are presented here. As a more specific example, the modeling of physical flow processes is discussed. The issues of noise and disturbance and identification of processes in their presence are widely considered which in modern engineering is a matter of primary importance.

The next part of the monograph is devoted to various aspects of control and regulation. Both linear and non-linear systems are presented, along with a detailed analysis of their fundamental properties, especially controllability and stability. Special attention is paid to the design of control systems, using various concepts and methods, both formal and based on computer simulation.

One of the most important tools used in solving problems in the area of automatic control and automation are optimization methods that make it possible to derive the best, or at least sufficiently good solutions. Some important methods from this area are included in the next part of this book. The use of optimization methods to identify systems with space-time dynamics is shown, the use of discrete optimization methods in industrial automation is presented, and finally, the use of dynamic programming models for multi-stage control tasks under the conditions of uncertain and imprecise (fuzzy) information is discussed.

Moreover, some relevant applications of the tools and techniques derived in the area of control, automation and robotics are discussed, exemplified by the use of advanced methods for the modeling and control of walking and humanoid robots. New approaches and methods for the control of robot teams and the use of multi-agent paradigms are also presented.

Widely understood intelligent methods, notably, tools and techniques stemming from artificial intelligence, are also discussed. Notably, using the decision support paradigm, new approaches to technical diagnostics, development of fault-tolerant, and damage-tolerant control systems are presented. The application of computational intelligence methods, mainly fuzzy systems, evolutionary computations, and neural networks, to control specific technical objects, in particular vessels, as well as selected industrial processes are also dealt with.

A very important part of this volume is devoted to a detailed presentation of specific applications which concern here, to just mention a few, the analysis and control of important issues affecting human comfort and health, such as a reduction of noise or acoustic disturbances, as well as the emission of pollutants into the atmosphere. Innovative application of control methods in gene-cell networks is also presented.

As the editors of this collective work, we hope that various aspects, models, concepts, and methods of modern control, automation and robotics discussed in this volume will be of interest and use for many researchers and practitioners, and also students of various levels, seeking information on both the new trends as well as state-of-the-art concepts in the areas concerned. The material included can also be used for scientific, professional, and teaching purposes.

We hope that this volume will be of a particular interest to the younger generation, both students and graduates, opening to them the fascinating world of these disciplines of science and technology which are widely associated with hopes for a better future. Namely, virtually all forecasts, which primarily capture the fundamental importance of broadly perceived information processing and artificial intelligence for the competitiveness of economies, as well as the ability to solve basic socioeconomic problems, emphasize the key importance of ubiquitous automation and robotization of virtually all systems and activities.

The roots of this volume can be traced to the related book "Automatyka, robotyka i przetwarzanie informacji", eds. P. Kulczycki, J. Korbicz, J. Kacprzyk, issued in Polish by the PWN scientific publisher. Particular topics have been modified and updated in varying degrees. We would like to thank first of all the authors of the chapters for their extremely valuable works, presenting in an insightful but accessible way both the state of the art of modern knowledge as well as the latest trends and tendencies, sometimes too innovative and too early to enter the standard canon of applications. Our gratitude is also due to the peer reviewers whose insightful and constructive remarks and suggestions have helped the authors improve their contributions.

And last but not least, we wish to thank Dr. Tom Ditzinger, Dr. Leontina di Cecco, and Mr. Holger Schaepe from Springer Nature for their dedication and help to implement and finish this important publication project on time, while maintaining the highest publication standards.

Warsaw/Kraków, Poland — Piotr Kulczycki
Zielona Góra, Poland — Józef Korbicz
Warsaw, Poland — Janusz Kacprzyk
March 2020

Contents

Optimization

Robotics

Computational Intelligence and Decision Support

Applications

Mathematical Modeling

Parametric Identification for Robust Control

Piotr Kulczycki

Abstract Many contemporary automatic control applications require parametric identification, taking into account results of estimation errors unavoidable in practice. The subject of this chapter constitutes a procedure enabling effective calculation of optimal, in the sense of minimal expectation of losses, estimator of a parameter in the case when losses resulting from negative and positive errors have significantly different influence on system operation, which forces the use of an asymmetric loss function. The conditional version of the issue is also considered in detail, as the process is significantly dependent on factors available metrologically during the system operation. This material has been provided in ready-to-use form, not requiring additional studies and bibliographical exploration. The presented concept is directed towards robust control problems. As an appendix, the analysis of solutions of differential equations with discontinuous right-hand side, subordinated for this purpose, has been presented, which may be useful for describing the dynamics of such resultant systems.

1 Introduction

In the early period of the development of control engineering, a fundamental discipline of automatic control, concepts were dominated by the intuition and experience of designers, without any dedicated mathematical apparatus. A significant change occurred in the interwar period, thanks to the application of basic elements of operational calculus: Laplace transformation and the transfer function concept [47]. The fundamental requirement formulated towards the constructed regulation systems was stability [4]. The research was facilitated by rich and advantageous mathematical

P. Kulczycki (✉)
Systems Research Institute, Polish Academy of Sciences, Newelska 6, 01447 Warsaw, Poland
e-mail: kulczycki@ibspan.waw.pl; kulczycki@agh.edu.pl

Faculty of Physics and Applied Computer Science, AGH University of Science and Technology, Mickiewicza 30, 30059 Cracow, Poland

P. Kulczycki et al. (eds.), *Automatic Control, Robotics, and Information Processing*, Studies in Systems, Decision and Control 296,
https://doi.org/10.1007/978-3-030-48587-0_1

apparatus provided for that purpose. The main achievement of that period was the regulator PID, exceptional in its simplicity and convenient in interpretation, and simultaneously surprisingly effective [2]. To this day it constitutes a fundamental tool in industrial applications, undergoing permanently various modifications and modernizations.

However, with the growth in requirements formulated towards automatic control systems the natural question arose, which among the stable systems should be recognized and applied as the best (optimal)? Using operational calculus there was no possible way to solve such a defined problem. The conditions of the second World War and the later period of the Cold War, when the best possible arms quality transferred into military success and human life, gradually magnified the growth in importance of optimization and research conducted in this field. In the fifties, operational calculus was replaced by differential equations, which enabled rapid development in the optimal control theory [3, 38]. At the end of this decade two fundamental theories allowing the establishment of control of this kind were published: Pontryagin Maximum Principle and Bellman Dynamical Programming, at this time engineering practice unambiguously distinguished two types of optimal control: time-optimal (minimum-time), relying on reaching the target in the shortest time ([3], Chapter 7), also with a quadratic performance index, directed primarily towards minimalizing energy used ([3], Chapter 9).

Although, in many applications of optimal control, sublime and extreme in nature, it only worked well in the assumed conditions or those not much different from them. As a consequence this led to a transfer in interest of practitioners in the direction of so-called robust control, which also worked correctly when real conditions significantly differed from those assumed [5, 50]. In practice this often means that the values of the parameter or parameters, used in a given model, can significantly differ from the real ones occurring in the object. An illustration of this may be for example the mass of an aircraft, changing due to fuel use, that is impossible to measure directly, which is more precise. Robust control systems were frequently derived from the concept of optimal control; the robustness of the former was obtained at the cost of a small worsening of quality in the sense of the performance index used in the baseline optimal structure being accepted in advance.

In this chapter the ready-to-use procedure for calculation of the parameter value is presented, which minimalizes the expected losses resulting from imprecise estimation, different for negative and positive errors. It will provide the solution for the introduced issue of the robust control problem. Thus, in the next subchapter, the motivation for the identification task considered here will be presented. The third subchapter covers the mathematical preliminaries, the methodology of statistical kernel estimators is described here, which will be used in the fourth main subchapter for the creation of a procedure for calculation of the optimal value of the studied parameter. In the first two points of this subchapter, the subject of detailed considerations are basic estimation issues with asymmetric linear and then quadratic loss function, after which in the third point, the above material will be submitted in the conditional approach. These issues may be generalized into polynomials of higher orders and the multidimensional case, where the subject of investigations are parameters with the correlated impact on estimation errors; these problems will be briefly outlined in the the fifth subchapter. Finally at the end, the appendix presents considerations concerning the solutions of differential equations with discontinuous left-hand side, which maybe helpful for the formal description of a robust control system, arising as a result of the application of the procedure previously worked out in this chapter.

2 Motivation and Problem Formulation

As grounds for the concept outlined here, three illustrative examples from the area of robust control are presented.

Assume the classic dynamic system

$$\begin{bmatrix} \dot{x}_1(t) \\ \dot{x}_2(t) \end{bmatrix} = \begin{bmatrix} 0 & 1 \\ 0 & 0 \end{bmatrix} \begin{bmatrix} x_1(t) \\ x_2(t) \end{bmatrix} + \begin{bmatrix} 0 \\ \frac{1}{M} \end{bmatrix} u(t), \tag{1}$$

where the positive parameter M represents a mass submitted to the action of force, according to the second law of Newtonian dynamics. Here x_1, x_2 and u denote the position and velocity of the above mass and the force treated as a control. This system forms the basis of the majority of research in robotics, leading to considerably more complex models with specificity appropriate for the problem under research [35, 37]. Consider the time-optimal control problem, whose basic form relies on bringing the state of the system to the origin of coordinates by the bounded control, in a minimal and finite time ([3], Subchapter 7.2). Fundamental importance for phenomena taking place in the control system has an appropriate identification of the value of the parameter M, introduced in the formula (1). The control algorithm is however defined on the basis the value of its estimator $\widehat{M}$, different in practice from the value M occurring in the object.

In the purely hypothetical case $\widehat{M} = M$, therefore when the estimator and real values are equal, the process is regular in nature. The system state reaches the origin in a minimal and finite time.

In turn, if the estimator is overestimated (i.e. for $\widehat{M} > M$), then over-regulations appear in the system. Its state oscillates around the origin and reaches it in a finite time, however larger than minimal.

Finally, in the case of underestimation (i.e. when $\widehat{M} < M$), the system state moves along a so-called sliding trajectory [8, 48] and in the end reaches the origin in a finite time, again larger than minimal.

Detailed considerations on the topic of the phenomena occurring in the system under investigation can be found in the paper [30]. On the margin of the above considerations, in order to illustrate the formal complexity of the task, one can mark that in the first two cases, the differential equation (1) possesses unique C-solutions (in the Caratheodory sense) and in the third unique F-solutions (in the Filippov sense). For the purpose of unifying all three cases, one can use K-solutions (in the Krasovski sense), more general than C- and F-solutions. In many issues, the family of K-solutions turns out to be too numerous, however in the system considered here it is fortunately unique [13]. Furthermore, in the probabilistic approach, the family of K-solutions occurring for particular values of the estimated parameter constitutes a stochastic process [14], which enables an application of a rich mathematical apparatus dedicated to this issue. These tasks have been specified and illustrated in the appendix to this chapter.

Therefore, in each of the above three cases, the system trajectory reaches the origin in the finite time, however, the phenomena occurring in the system as a result of under- and overestimation of the parameter $\widehat{M}$ are completely different and thus one can expect significantly different times for reaching the target set in the ranges of negative and positive errors $\widehat{M} - M$.

Figure 1 shows the graph of the time to reach the origin by the system (1) state, with exemplary fixed $M = 1$. It can be seen that the growth in the value of this index is approximately directly proportional to the estimation error $\widehat{M} - M$, although with different coefficients for negative and positive errors:

$$l\left(\widehat{M}, M\right) = \begin{cases} -a\left(\widehat{M} - M\right) & \text{when } \widehat{M} - M \le 0 \\ b\left(\widehat{M} - M\right) & \text{when } \widehat{M} - M \ge 0 \end{cases}, \tag{2}$$

while the constants a and b are positive; it is worth underlining they do not necessarily equal.

In the case of the stronger interaction of the parameter identified on the behavior of the system, for example where its impact is "multitrack" (multifaceted), the quadratic form of the performance index may be appropriate. As illustrative example, consider the dynamic system

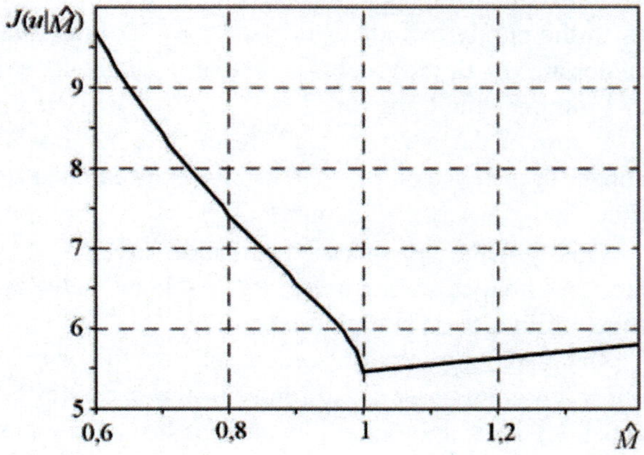

Fig. 1 Value of time-optimal index for different values of estimator $\widehat{M}$, with $M = 1$

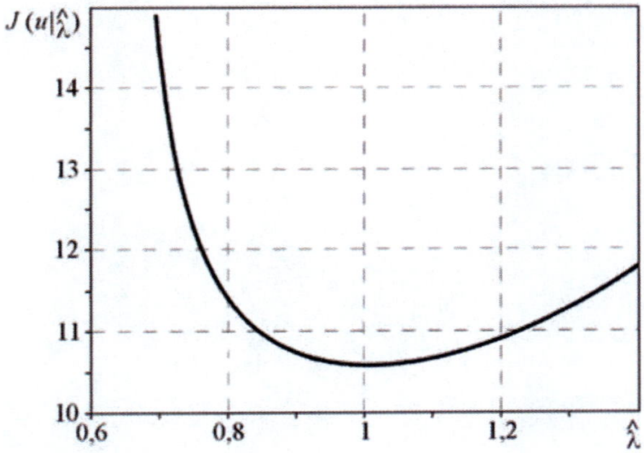

Fig. 2 Value of quadratic index for different values of estimator $\hat{\lambda}$, with $\lambda = 1$

$$\begin{bmatrix} \dot{x}_1(t) \\ \dot{x}_2(t) \end{bmatrix} = \begin{bmatrix} \lambda & 1 \\ 0 & \lambda \end{bmatrix} \begin{bmatrix} x_1(t) \\ x_2(t) \end{bmatrix} + \begin{bmatrix} 0 \\ \lambda \end{bmatrix} u(t), \tag{3}$$

where the constant λ is different from 0. It is worth noting that it occurs both in the state A and in the control B matrixes. The system (3) will be submitted to the classic optimal control with the quadratic performance index ([3], Subchapter 9.5) with unique matrixes and infinite final time. Figure 2 shows the value of the above index as the function of the estimator $\hat{\lambda}$ value, with exemplary fixed $\lambda = 1$. One can see that the the value of the loss function resulting from the estimation error may be successfully approximated by a asymmetric quadratic form:

$$l\left(\hat{\lambda}, \lambda\right) = \begin{cases} a\left(\hat{\lambda} - \lambda\right)^2 & \text{when } \hat{\lambda} - \lambda \leq 0 \\ b\left(\hat{\lambda} - \lambda\right)^2 & \text{when } \hat{\lambda} - \lambda \geq 0 \end{cases}, \tag{4}$$

where again the coefficients a and b are positive and not necessarily equal.

Now consider the two-dimensional case. Assume the dynamic system, originated by the introduction into the system (4) of the drive inertia; therefore, substituting the control u by the third coordinate of the state vector x_3, and so in consequence

$$\begin{bmatrix} \dot{x}_1(t) \\ \dot{x}_2(t) \\ \dot{x}_3(t) \end{bmatrix} = \begin{bmatrix} \lambda & 1 & 0 \\ 0 & \lambda & \lambda \\ 0 & 0 & -\tau \end{bmatrix} \begin{bmatrix} x_1(t) \\ x_2(t) \\ x_3(t) \end{bmatrix} + \begin{bmatrix} 0 \\ 0 \\ 1 \end{bmatrix} u(t), \tag{5}$$

where τ is a positive drive inertia constant. Thus the parameters $\begin{bmatrix} \lambda \\ \tau \end{bmatrix}$ originated from different subsystems (λ from the object itself and τ characterizing the drive) are subjected to estimation, and therefore can be treated as independent. Once again the system will be submitted to control with the quadratic performance index ([3],

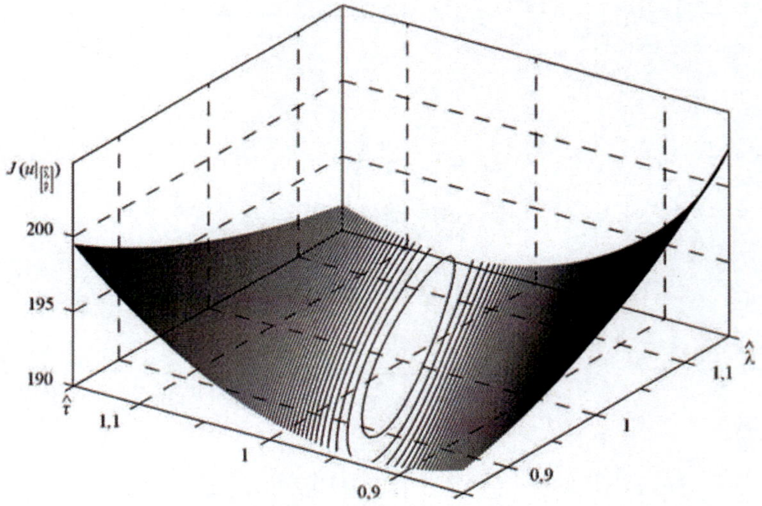

Fig. 3 Value of two-dimensional quadratic index for different values of estimators $\begin{bmatrix} \hat{\lambda} \\ \hat{\tau} \end{bmatrix}$, with $\begin{bmatrix} \lambda \\ \tau \end{bmatrix} = \begin{bmatrix} 1 \\ 1 \end{bmatrix}$

Subchapter 9.5) with unique matrixes and infinite time. Figure 3 shows the spatial graph of this index as the function of the estimators' $\begin{bmatrix} \hat{\lambda} \\ \hat{\tau} \end{bmatrix}$ values, with exemplary fixed $\begin{bmatrix} \lambda \\ \tau \end{bmatrix} = \begin{bmatrix} 1 \\ 1 \end{bmatrix}$. It can be seen that losses resulting from estimation errors can be approximated with high precision by the two-dimensional asymmetric quadratic function:

$$l\left(\begin{bmatrix} \hat{\lambda} \\ \hat{T} \end{bmatrix}, \begin{bmatrix} \lambda \\ T \end{bmatrix}\right) = \begin{cases} a_r\left(\hat{\lambda}-\lambda\right)^2 + a_{ru}\left(\hat{\lambda}-\lambda\right)\left(\hat{T}-T\right) + a_u\left(\hat{T}-T\right)^2 & \text{when } \hat{\lambda}-\lambda \geq 0 \text{ i } \hat{T}-T \geq 0 \\ a_l\left(\hat{\lambda}-\lambda\right)^2 + a_{lu}\left(\hat{\lambda}-\lambda\right)\left(\hat{T}-T\right) + a_u\left(\hat{T}-T\right)^2 & \text{when } \hat{\lambda}-\lambda \leq 0 \text{ i } \hat{T}-T \geq 0 \\ a_l\left(\hat{\lambda}-\lambda\right)^2 + a_{ld}\left(\hat{\lambda}-\lambda\right)\left(\hat{T}-T\right) + a_d\left(\hat{T}-T\right)^2 & \text{when } \hat{\lambda}-\lambda \leq 0 \text{ i } \hat{T}-T \leq 0 \\ a_r\left(\hat{\lambda}-\lambda\right)^2 + a_{rd}\left(\hat{\lambda}-\lambda\right)\left(\hat{T}-T\right) + a_d\left(\hat{T}-T\right)^2 & \text{when } \hat{\lambda}-\lambda \geq 0 \text{ i } \hat{T}-T \leq 0 \end{cases}, \quad (6)$$

where a_l, a_r, a_u, $a_d > 0$, a_{ld}, $a_{ru} \geq 0$ and a_{lu}, $a_{rd} \leq 0$. The mixed coefficients a_{ld}, a_{ru}, a_{lu}, a_{rd} represent mutual correlation of estimation errors of both parameters. In the case when they are equal to zero, the problem de facto reduces to two separate tasks with the asymmetric quadratic loss function (4).

The above examples illustrate the problem of quality of robust control—originating from the conceptions of optimal structures—with respect to the errors of parameters estimation, in the cases where the losses resulting from this reason can be described by the asymmetric linear (2) or quadratic (4), (6) functions. The polynomial form seems to be a suitable compromise between the precision and the complexity and in consequence usefulness of the approach proposed [36].

It is worth furthermore generalizing the classic formulation of the research problem presented above. Thus, the more complex and refined the models currently applied become, the less reasoned is a traditional perception of parameters identification as the task of calculation of a concrete (existing in reality but unknown) value of the parameter under investigation. Currently this task covers establishment of value, which will represent the whole range of multifaceted phenomena, simplified in the model to one, the only formally existing parameter. In such a situation, evaluation of the quality of the parameter identification cannot be made in the classical way by comparing the obtained value to the imagined "real", although unknown, parameter value (because no such value exists), but rather through taking into account the implications of particular values on the functioning of the system under investigation. This differs from the mathematical apparatus existing in the framework of point estimation from the classic mathematical statistic [34], becoming closer to the currently expanding data analysis [31–33]. Fortunately, development of modern, advanced and specialized identification methods enables a rapid expansion of contemporary computer technology, supporting from theoretical point of view IT methods dedicated to it.

In summary: in the method presented here, the value of estimator will be sought, minimizing the expectation value of losses resulting from estimation errors. One assumes availability of a finite number of metrologically obtained values for the estimated parameter. The knowledge of its distribution is not necessary; its establishment constitutes an integral part of the procedure. The conditional approach will also be proposed, designated for the case where the parameter under research is significantly dependent on the factor, which current value can be successively measured during realization of the control process, causing an increase in the models precision. The material is given in the complete form, not requiring additional investigations or bibliographical research.

3 Mathematical Preliminaries: Kernel Estimators

Kernel estimators are counted among the nonparametric methods. Methods of this type do not require initial assumptions concerning the type of the existing distribution. In their basic formula, kernel estimators serve to establish the fundamental functional characteristics of the distribution; its density. First, in Sect. 3.1 the basic one-dimensional case will be presented, generalized in Sect. 3.2 to the multidimensional approach.

3.1 One-Dimensional Case

Let x denote the quantity under investigation. Let the m-elements set of independent measurements

$$x_1, x_2, \ldots, x_m \in \mathbb{R} \tag{7}$$

be given. The kernel estimator of the distribution density of the studied quantity $\hat{f} : \mathbb{R} \to [0, \infty)$ is defined in its standard form by the following formula

$$\hat{f}(x) = \frac{1}{mh} \sum_{i=1}^{m} K\left(\frac{x - x_i}{h}\right), \tag{8}$$

where the positive constant h is called the smoothing parameter, while the function $K : \mathbb{R} \to [0, \infty)$ measurable, symmetrical with respect to zero and having the weak global maximum at this point, with unique integral $\int_{\mathbb{R}} K(y)\mathrm{d}y = 1$, is named as the kernel.

The interpretation of the above definition is illustrated in Fig. 4. In the case of the single element x_i, the function K transposed by the vector x_i and scaled by the coefficient h, represents estimation of the distribution after obtaining the value x_i. For m independent values $x_1, x_2, \ldots, x_m$, it takes the form of the sum of such individual estimations. The coefficient $1/mh$ norms the function obtained to guarantee the condition $\int_{\mathbb{R}} \hat{f}(y)\mathrm{d}y = 1$. Finally, the formula (8) provides the tool for estimation of density of the set (7) distribution; it is illustrative and rather less complex for numerical calculations.

In practice the choice of the kernel form ([20], Section 3.1.3; [49], Subchapters 2.7 and 4.5) has no significant importance, and thanks to this it is possible to primarily take into account the properties of the estimator obtained, e.g. its continuity, differentiability, or bounding of the support; advantageous from the point of view of the

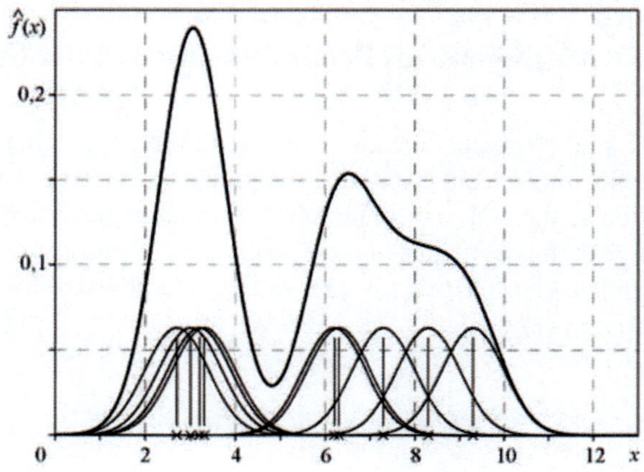

Fig. 4 One-dimensional kernel estimator (8)

specific application. In the method presented here, the Cauchy kernel

$$K(x) = \frac{2}{\pi\left(x^2+1\right)^2} \tag{9}$$

will be applied, because of its convenient integration properties. The constants $W(K) = \int_{-\infty}^{\infty} K(y)^2 \mathrm{d}y$ and $U(K) = \int_{-\infty}^{\infty} y^2 K(y)\mathrm{d}y$, which are necessary for further calculations, equal in this case:

$$W(K) = \frac{5}{4\pi} \tag{10}$$

$$U(K) = 1. \tag{11}$$

Great importance for the estimation quality is, however, fixing of the smoothing parameter h value ([20], Section 3.1.5; [49], Chapter 3 and Subchapter 4.7). Fortunately for practitioners, many convenient calculational procedures, based the set (7), have been investigated. During the initial design phase, as well as in the case of the set size greater than 1000, one can recommend a simple approximate method with linear complexity regarding to the size m. In the remaining cases, the effective *plug-in* procedure, with quadratic complexity, can be implemented.

The first method relies on the direct use of the formula

$$h = \left(\frac{8\sqrt{\pi}}{3}\frac{W(K)}{U(K)^2}\frac{1}{m}\right)^{1/5}\widehat{\sigma}, \tag{12}$$

where $\widehat{\sigma}$ denotes the classic standard deviation estimator:

$$\widehat{\sigma} = \sqrt{\frac{1}{m-1}\sum_{i=1}^{m}(x_i)^2 - \frac{1}{m(m-1)}\left(\sum_{i=1}^{m}x_i\right)^2}. \tag{13}$$

In turn, the second procedure, *plug-in*, is based on the application of the above formula, following several improvements in such obtained value. Thus, in the case of the third degree procedure, one should sequentially apply the following dependencies:

$$d_{10} = \frac{-945}{64\sqrt{\pi}\,\widehat{\sigma}^{11}}, \tag{14}$$

where $\widehat{\sigma}$ is given by the equality (13), and next

$$g_{III} = \left(\frac{-2\widetilde{K}^{(8)}(0)}{d_{10}U\left(\widetilde{K}\right)m}\right)^{1/11} \tag{15}$$

$$g_{II} = \left(\frac{-2\widetilde{K}^{(6)}(0)}{d_{8,j}(g_{III})U(\widetilde{K})m} \right)^{1/9} \tag{16}$$

$$g_{I} = \left(\frac{-2\widetilde{K}^{(4)}(0)}{d_{6,j}(g_{II})U(\widetilde{K})m} \right)^{1/7}; \tag{17}$$

finally

$$h = \left(\frac{1}{d_4(g_I)} \frac{W(K)}{U(K)^2} \frac{1}{m} \right)^{1/5}, \tag{18}$$

while

$$d_p(g) = \frac{1}{m^2 g^{p+1}} \sum_{i=1}^{m} \sum_{k=1}^{m} \widetilde{K}^{(p)}\left(\frac{x_i - x_k}{g} \right) \quad \text{for} \quad p = 4, 6, 8. \tag{19}$$

The basic kernel K, used so far, exists in the above procedure only in the last formula (18). In the remaining formulas one can apply any form convenient for multiple differentiation. Commonly the normal kernel

$$\widetilde{K}(x) = \frac{1}{\sqrt{2\pi}} \exp\left(-\frac{x^2}{2} \right) \tag{20}$$

is accepted here. Then the factors appearing in the dependencies (15)–(17) and (19) are equal respectively:

$$\widetilde{K}^{(8)}(x) = \frac{1}{\sqrt{2\pi}} \left(x^8 - 28x^6 + 210x^4 - 420x^2 - 105 \right) \exp\left(-\frac{x^2}{2} \right), \ \widetilde{K}^{(8)}(0) = -\frac{105}{\sqrt{2\pi}} \tag{21}$$

$$\widetilde{K}^{(6)}(x) = \frac{1}{\sqrt{2\pi}} \left(x^6 - 15x^4 + 45x^2 - 15 \right) \exp\left(-\frac{x^2}{2} \right), \ \widetilde{K}^{(6)}(0) = -\frac{15}{\sqrt{2\pi}} \tag{22}$$

$$\widetilde{K}^{(4)}(x) = \frac{1}{\sqrt{2\pi}} \left(x^4 - 6x^2 + 3 \right) \exp\left(-\frac{x^2}{2} \right), \ \widetilde{K}^{(4)}(0) = \frac{3}{\sqrt{2\pi}} \tag{23}$$

$$U(\widetilde{K}) = 1. \tag{24}$$

3.2 Multidimensional Case

The concept of the kernel estimator will now be extended to the multidimensional case. The role of the measurements (7) is taken then by the m-elements set of n-dimensional vectors:

$$x_1, x_2, \dots, x_m \in \mathbb{R}^n. \tag{25}$$

The kernel estimator of the density of its distribution $\widehat{f} : \mathbb{R}^n \to [0, \infty)$ can now be defined in the form

$$\widehat{f}(x) = \frac{1}{m} \sum_{i=1}^{m} \mathcal{K}(x, x_i, h), \tag{26}$$

where after indication of coefficients

$$x = \begin{bmatrix} x_1 \\ x_2 \\ \vdots \\ x_n \end{bmatrix}, \quad x_i = \begin{bmatrix} x_{i,1} \\ x_{i,2} \\ \vdots \\ x_{i,n} \end{bmatrix} \quad \text{for } i = 1, 2, \dots, m \quad \text{and} \quad h = \begin{bmatrix} h_1 \\ h_2 \\ \vdots \\ h_n \end{bmatrix}, \tag{27}$$

whereas the positive constants h_j plays the role of smoothing parameters for particular coordinates $j = 1, 2, \dots, n$, while with the contemporary applications the multidimensional kernel $\mathcal{K}$ is commonly used in the product form

$$\mathcal{K}(x, x_i, h) = \prod_{j=1}^{n} \frac{1}{h_j} K\left(\frac{x_j - x_{i,j}}{h_j}\right), \tag{28}$$

therefore, as the product of the one-dimensional kernels K (discussed in details in the previous Sect. 3.1) scaled with the coefficient $1/h_j$. Because of such decomposition with respect to particular coordinates, all the previously formulated comments concerning the fixing of the smoothing parameter value and selection of the kernel form are applicable also in the multidimensional case. In particular, the smoothing parameters h_j can be calculated using the approximate or *plug-in* methods, separately for each coordinate, with the one-dimensional kernel K assumed in the Cauchy form (9). The two-dimensional kernel estimator has been illustrated in Fig. 5.

Note also that in the one-dimensional case, after substituting $n = 1$ and omitting the index j redundant here, the definitions (26)–(28) lead to the basic form (8).

In practice, various modifications and generalizations of the presented above standard form of the kernel estimators are possible. One should remember, however, that they increase the complexity of the formulas, and reduce the ease of interpretation and consequently mean that the solution of the problem becomes less convenient for the potential user. Detailed description of the kernel estimators methodology can be

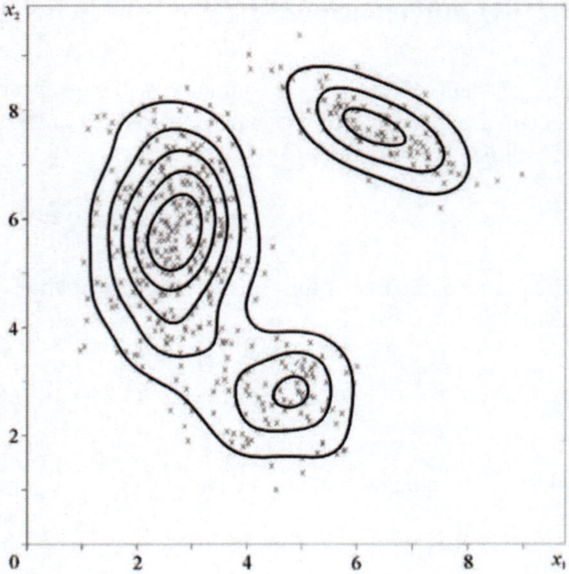

Fig. 5 Two-dimensional kernel estimator (26); elements of set (25) are marked by crosses, lines denote the contours of estimator

found in the monographs [20, 44, 49]. It is also worth recalling the publications [6, 7, 40–42, 45] as well as the new subject book [9], where one can also obtain broad current literature. The sister problem of the kernel regression can be found in [10]. The implicational possibilities to identification of atypical elements (outliers), clustering, and classification, three basic procedures of data analysis and exploration, are summarized in the survey article [21] and also in the work [26] for the conditional approach. In the paper [19] the concept of the sophisticated diagnostic system based on kernel estimators is presented.

4 Procedure for Parametric Identification

4.1 Linear Case

Assume the asymmetric linear loss function (2), presented below using standard notation

$$l(\hat{x}, x) = \begin{cases} -a(\hat{x} - x) & \text{when } \hat{x} - x \leq 0 \\ b(\hat{x} - x) & \text{when } \hat{x} - x \geq 0 \end{cases}, \tag{29}$$

where x represents the investigated parameter, $\hat{x}$ its estimator, and the coefficients a and b are positive (non-necessarily equal to each other). Consider the set

$$x_1, x_2, \ldots, x_m \in \mathbb{R} \tag{30}$$

of values of the parameter under research, obtained metrologically in the framework of independent measurements. If one denotes by $\widehat{f}$ the kernel estimator of its distribution density (calculated following Subchapter 3), then the dependence (29) shows directly that the expected value of losses resulting from taking the estimator $\widehat{x}$ can be given as

$$\Lambda(\widehat{x}) = b \int_{-\infty}^{\widehat{x}} (\widehat{x} - x)\widehat{f}(x)\mathrm{d}x - a \int_{\widehat{x}}^{\infty} (\widehat{x} - x)\widehat{f}(x)\mathrm{d}x. \tag{31}$$

Let the function $\widehat{f}$ be continuous; this condition can be easily obtained while using the continuous kernel K, e.g. Cauchy (9), for the construction of the kernel estimator. It is then not difficult to show, that the function Λ, defined above, reaches the minimum for the value $\widehat{x}$ fulfilling the equation

$$\int_{-\infty}^{\widehat{x}} \widehat{f}(x)\mathrm{d}x = \frac{a}{a+b}. \tag{32}$$

Because $0 < a/(a+b) < 1$, then the above equation has a solution, and if the support of the function $\widehat{f}$ is coherent, it is unique. Such a condition can be obtained using the kernel K with positive values, e.g. Cauchy (9), for the construction of the kernel estimator $\widehat{f}$. And finally, due to equality

$$\frac{a}{a+b} = \frac{\frac{a}{b}}{\frac{a}{b}+1}, \tag{33}$$

it can be seen that the separate fixing of the parameters a and b is not necessary, but only the ratio a/b.

Assume, therefore, that the function $\widehat{f}$ is a kernel estimator constructed using a continuous kernel with positive values. Also, I denotes the primitive function of the kernel K, i.e.

$$I(x) = \int_{-\infty}^{x} K(y)\mathrm{d}y. \tag{34}$$

In the case of the Cauchy kernel (9), this function can be expressed by the analytical formula

$$I(x) = \frac{1}{\pi}\operatorname{arctg}(x) + \frac{x}{\pi\left(x^2+1\right)} + \frac{1}{2}. \tag{35}$$

Let us introduce the auxiliary notation

$$P_i(x) = I\left(\frac{x - x_i}{h}\right) \quad \text{for} \quad i = 1, 2, \ldots, m. \tag{36}$$

The criterion (32) takes the form

$$\frac{1}{m}\sum_{i=1}^{m} P_i\left(\hat{x}\right) - \frac{a}{a+b} = 0. \tag{37}$$

Denoting the left side of the above equation by $L\left(\hat{x}\right)$, i.e.

$$L\left(\hat{x}\right) = \frac{1}{m}\sum_{i=1}^{m} P_i\left(\hat{x}\right) - \frac{a}{a+b}, \tag{38}$$

its derivative is

$$L'\left(\hat{x}\right) = \hat{f}\left(\hat{x}\right). \tag{39}$$

Note that the function I is increasing. This feature, therefore, passes to P_i and in consequence also to L. Because an analytic form of the derivative is accessible, then these properties suggest use of the classic Newton algorithm [12] in order to solve the Eq. (32). Then this solution can be found as a limit of the sequence $\left\{\hat{x}_k\right\}_{k=0}^{\infty}$ defined by the formulas

$$\hat{x}_0 = \frac{1}{m}\sum_{i=1}^{m} x_i \tag{40}$$

$$\hat{x}_{k+1} = \hat{x}_k - \frac{L\left(\hat{x}_k\right)}{L'\left(\hat{x}_k\right)} \quad \text{for } k = 0, 1, \ldots \; . \tag{41}$$

As a stop criterion the following condition is assumed:

$$\left|\hat{x}_k - \hat{x}_{k-1}\right| \leq 0,01\,\hat{\sigma}, \tag{42}$$

where $\hat{\sigma}$ is defined by the formula (13).

Summarizing, having independent measurements of the investigated parameter (30), its optimal (in the sense of the expected loss value, in the case when the loss function is given in the asymmetrical linear form (29)) the estimator can be calculated by the recurrence algorithm (40)–(41) having the stop condition (42) with (13). The functions L and L' occurring there are given by the formulas (38)–(39), and in turn P_i and $\widehat{f}$ are defined by the dependencies (36) and (8). If one uses the Cauchy kernel recommended here, then the functions K and I are denoted by the formulas (9) and (35), respectively. The value of the smoothing parameter h can be calculated directly from the formula (12) with (13), or by the *plug-in* method (14)–(24) and (13). When the Cauchy kernel is used, the values of the functionals U and W existing there are given by the formulas (10)–(11).

It should be noted that if $a = b$, the criterion (32) indicates directly the median.

The first results in the range of estimation with the asymmetrical loss function, concerning the linear case, have been published in the paper [18].

4.2 *Quadratic Case*

The asymmetric quadratic loss function (3), described using standard notation, takes the form

$$l(\widehat{x}, x) = \begin{cases} a(\widehat{x} - x)^2 & \text{for } \widehat{x} - x \le 0 \\ b(\widehat{x} - x)^2 & \text{for } \widehat{x} - x \ge 0 \end{cases}, \tag{43}$$

where x, $\widehat{x}$ and a, b have the same interpretation as in the case of the asymmetric linear loss function (29), similarly as the set of measurements $x_1, x_2, \ldots, x_m$ given there by the formula (30). Let $\widehat{f}$ be established, using the material from Subchapter 3, by the kernel estimator methodology; while the applied kernel K is continuous with positive values, and additionally such that

$$\int_{-\infty}^{\infty} x K(x) \mathrm{d}x < \infty. \tag{44}$$

The Cauchy kernel (9) fulfils the above condition.

The form of the definition (43) constitutes directly that the expectation value of losses, resulting from the accepted value of the estimator $\widehat{x}$, is given by the formula

$$\Lambda(\widehat{x}) = b \int_{-\infty}^{\widehat{x}} (\widehat{x} - x)^2 \widehat{f}(x) \mathrm{d}x + a \int_{\widehat{x}}^{\infty} (\widehat{x} - x)^2 \widehat{f}(x) \mathrm{d}x. \tag{45}$$

It is not difficult to show that this function reaches a minimum for the value $\widehat{x}$, which is the solution of the equation

$$(a-b)\int_{-\infty}^{\widehat{x}}(\widehat{x}-x)\widehat{f}(x)\mathrm{d}x - a\int_{-\infty}^{\infty}(\widehat{x}-x)\widehat{f}(x)\mathrm{d}x = 0, \tag{46}$$

whereas an analysis of its first and second derivatives indicates that this solution exists and is unique. Dividing the above equation by b one can easily note, that it is sufficient to identify the quotient a/b and the separate fixing of these parameter values is not necessary.

Solving the Eq. (46) is not an easy task. However if $\widehat{f}$ is obtained by the kernel estimators methodology with the preoperatively chosen kernel K, then it becomes possible to create an effective numerical algorithm for this purpose. Let us introduce the following function

$$J(x) = \int_{-\infty}^{x} y\,K(y)\mathrm{d}y \tag{47}$$

and the auxiliary notation

$$R_i(x) = \frac{1}{h}J\left(\frac{x-x_i}{h}\right) \quad \text{for} \quad i = 1, 2, \dots, m. \tag{48}$$

In the case of the Cauchy kernel (9), the function J is given in an analytical form as

$$J(x) = -\frac{1}{\pi\left(x^2+1\right)}. \tag{49}$$

If one additionally assumes that the estimator of the expected value is the arithmetic mean, then the criterion (46) may be denoted in the equivalent form

$$\sum_{i=1}^{m}\left[(a-b)\left(\widehat{x}P_i\left(\widehat{x}\right) - R_i\left(\widehat{x}\right)\right) + ax_i\right] - a\widehat{x}m = 0. \tag{50}$$

Denote the left-hand side of the above equation by $L\left(\widehat{x}\right)$:

$$L\left(\widehat{x}\right) = \sum_{i=1}^{m}\left[(a-b)\left(\widehat{x}P_i\left(\widehat{x}\right) - R_i\left(\widehat{x}\right)\right) + ax_i\right] - a\widehat{x}m. \tag{51}$$

Because of the equality $R_i'\left(\widehat{x}\right) = \widehat{x}P_i'\left(\widehat{x}\right)$ resulting directly from the definitions (31) and (43), one obtains

$$L'(\hat{x}) = \sum_{i=1}^{m} \left[(a-b) P_i(\hat{x}) \right] - am. \tag{52}$$

Similarly to the linear case, the solution of the Eq. (46) may be effectively calculated numerically using the Newton method as the limit of the sequence $\{\hat{x}_k\}_{k=0}^{\infty}$ defined recurrently as

$$\hat{x}_0 = \frac{1}{m} \sum_{i=1}^{m} x_i \tag{53}$$

$$\hat{x}_{k+1} = \hat{x}_k - \frac{L(\hat{x}_k)}{L'(\hat{x}_k)} \quad \text{for} \quad k = 0, 1, \ldots \quad , \tag{54}$$

with the stop condition

$$|\hat{x}_k - \hat{x}_{k-1}| \leq 0,01\,\hat{\sigma}, \tag{55}$$

where $\hat{\sigma}$ is given by the dependency (13).

Summarizing, in the case where the loss function takes the asymmetric quadratic form (43), the estimator minimizing the expectation loss value can be calculated using the algorithm (53)–(54) and the stop condition (55) with (13). The functions L and L' are defined by the formulas (51)–(52), while P_i and R_i are given by the dependencies (36) and (48). In cases when the Cauchy kernel is applied, the functions I and J are described by the equalities (35) and (47). The value of the smoothing parameter h can be calculated using the procedures described in Subchapter 3, from the formulas (12)–(13) or by the *plug-in* method (14)–(24) and (13). When using the Cauchy kernel, the values of the functionals U and W occurring there are given by the dependencies (10)–(11).

If $a = b$, then it is easy to observe that the criterion (46) simplifies to the expected value.

Additional aspects of the issue of the parametric estimation with the asymmetric quadratic loss function are found in the paper [27].

4.3 *Conditional Approach*

Both the above presented asymmetric cases, linear and quadratic, can be submitted in the conditional approach, therefore, when the estimated parameter is meaningfully dependent on the factor or factors, available metrologically during system operation. In engineering objects, a typical example of such a factor is the current temperature. Performing the measurement of the actual value of the conditioning factor, i.e. so-called conditioning value, and then introducing it into the algorithm, can contribute to significantly making the model more precise.

Thus, let the set of values of parameter under research x be given

$$x_1, x_2, \ldots, x_m \in \mathbb{R}, \tag{56}$$

obtained during independent measurements, when the conditioning factor took the values respectively

$$w_1, w_2, \ldots, w_m \in \mathbb{R}. \tag{57}$$

Denote the formulas (56)–(57) in the vector form

$$\begin{bmatrix} x_1 \\ w_1 \end{bmatrix}, \begin{bmatrix} x_2 \\ w_2 \end{bmatrix}, \ldots, \begin{bmatrix} x_m \\ w_m \end{bmatrix} \in \mathbb{R}^2. \tag{58}$$

The kernel estimator of distribution density of the parameter x, when the conditioning factor takes the value $w = w_*$ (therefore, for the conditioning value w_*) can be defined as follows:

$$\hat{f}_{w_*}(x) = \frac{\hat{f}_{x,w}\left(\begin{bmatrix} x \\ w_* \end{bmatrix}\right)}{\hat{f}_w(w_*)}, \tag{59}$$

where $\hat{f}_{x,w}$ denotes the joint distribution density of the parameter x and the conditioning factor w, while $\hat{f}_w$ is the distribution density of the conditioning factor w. The first of the above densities can be calculated based on the set (58) applying the material from Sect. 3.2, whereas the second using the set (57) and Sect. 3.1, with a kernel with positive values, e.g. the Cauchy kernel (9). Then

$$\hat{f}_{w_*}(x) = \frac{\frac{1}{mh_x h_w} \sum_{i=1}^m K\left(\frac{x-x_i}{h_x}\right) K\left(\frac{w_*-w_i}{h_w}\right)}{\frac{1}{mh_w} \sum_{i=1}^m K\left(\frac{w_*-w_i}{h_w}\right)}, \tag{60}$$

where h_x and h_w denote smoothing parameters calculated for the sets (56) and (57), respectively. Introducing the positive coefficients, known as conditioning parameters:

$$d_i = K\left(\frac{w_* - w_i}{h_w}\right) \quad \text{for} \quad i = 1, 2, \ldots, m \quad , \tag{61}$$

one can obtain the final form of the kernel estimator of the distribution density of the parameter x for the conditioning value w_*:

$$\hat{f}_{w_*}(x) = \frac{1}{h_x \sum_{i=1}^m d_i} \sum_{i=1}^m d_i K\left(\frac{x - x_i}{h_x}\right). \tag{62}$$

In the case of many conditioning factors, after indication of coefficients

$$w_* = \begin{bmatrix} w_{*,1} \\ w_{*,2} \\ \vdots \\ w_{*,n_w} \end{bmatrix}, \quad w_i = \begin{bmatrix} w_{i,1} \\ w_{i,2} \\ \vdots \\ w_{i,n_w} \end{bmatrix} \quad \text{for } i = 1, 2, \dots, m, \quad \text{and} \quad h = \begin{bmatrix} h_{w_1} \\ h_{w_2} \\ \vdots \\ h_{w_{n_w}} \end{bmatrix}, \tag{63}$$

the above considerations, in particular the formula (62), remain unchanged, except the definition of the conditioning parameters (61) generalized here to

$$d_i = K\left(\frac{w_{*,1} - w_{i,1}}{h_{w_1}}\right) \cdot K\left(\frac{w_{*,2} - w_{i,2}}{h_{w_2}}\right) \cdot \ldots \cdot K\left(\frac{w_{*,n_w} - w_{i,n_w}}{h_{n_w}}\right) \quad \text{for } i = 1, 2, \dots, m. \tag{64}$$

The form of the conditional distribution density estimator (62) together with definitions of the conditioning parameters (61) or (64) allows convenient, illustrative interpretation, especially in comparison with the basic definition (8). Thus, as results from the formulas (61) and (64), the parameter d_i value characterizes the "distance" of the conditioning value w_* from the conditioning factor value w_i, with which the i-th measurement x_i of the parameter under research was obtained. In the basic form (8) each element x_i, and in consequence the related kernel, has the same weight. In turn, the estimator (62) can be interpreted as a linear combination of kernels assigned to the particular measurements x_i, whereas the coefficients of the above combination characterizes a representativeness of these elements for the given conditioning value w_*. In particular, note that if all parameters d_i were equal, then the formula (62) would be simplified to the basic form (8).

If, therefore, in the system under research one can select conditioning factors, then replacing in the issues of estimation with the asymmetric linear and quadratic loss functions, presented in Sects. 3.1 and 3.2, the basic form of the kernel estimator (8) by its conditional form (62) together with the formulas (61) or (64) can effectively make the applied model more precise. In the case of the task with the asymmetric linear loss function, the functions L and L', given by the formulas (38)–(39), take the form

$$L(\hat{x}) = \frac{h}{\sum_{i=1}^{m} d_i} \sum_{i=1}^{m} d_i P_i(\hat{x}) - \frac{a}{a+b} \tag{65}$$

$$L'(\hat{x}) = \hat{f}_{w_*}(\hat{x}). \tag{66}$$

In the case of the asymmetric quadratic loss function, the functions L and L' (51)–(52) are given as

$$L(\hat{x}) = \frac{1}{\sum_{i=1}^{m} d_i} \sum_{i=1}^{m} \{d_i[(a-b)(\hat{x} P_i(\hat{x}) - R_i(\hat{x})) + a x_i]\} - a\hat{x} \tag{67}$$

$$L'(\hat{x}) = \frac{1}{\sum_{i=1}^{m} d_i} \sum_{i=1}^{m} [d_i(a-b)P_i(\hat{x})] - a. \tag{68}$$

The functions P_i and R_i as well as I, J and K, and also the procedures for calculating the smoothing parameters values remain the same.

Broader information on the conditional approach for the single parameter may be found in the paper [24].

5 Generalizations

The convenience of interpretation of the issue itself and the results obtained facilitate various possible generalizations of the asymmetric linear and quadratic cases, presented in Sects. 4.1 and 4.2, similarly as in Sect. 4.3 it was done for the conditional approach. Consider thus the asymmetric loss functions of the 3rd degree or higher:

$$l(\hat{x}, x) = \begin{cases} (-1)^k a(\hat{x} - x)^k & \text{for } \hat{x} - x \leq 0 \\ b(\hat{x} - x)^k & \text{for } \hat{x} - x \geq 0 \end{cases} \quad \text{when} \quad k = 3, 4, \ldots \quad , \tag{69}$$

with the earlier interpretation of the quantities x, $\hat{x}$ and a, b, as previously for the linear (29) and quadratic (43) functions. The expected value of losses resulting from assuming the value of the estimator $\hat{x}$ is then given by the formula

$$\Lambda(\hat{x}) = b \int_{-\infty}^{\hat{x}} (\hat{x} - x)^k \hat{f}(x) \mathrm{d}x + (-1)^k a \int_{\hat{x}}^{\infty} (\hat{x} - x)^k \hat{f}(x) \mathrm{d}x \tag{70}$$

and further reasoning can be conducted as in Sects. 4.1 and 4.2. One should note that in order to ensure the appropriate convergences, the Cauchy kernel (9) must be modified to the form

$$K(x) = \frac{c_k}{(x^2 + 1)^k}, \tag{71}$$

where the constant is given recurrently as:

$$c_2 = \frac{2}{\pi}, \quad c_k = \frac{2(k-1)}{2n-3} c_{k-1} \quad \text{for } k = 3, 4, \ldots \quad . \tag{72}$$

It does not seem, however, that the above case would have significant applicational meaning. In practice, particular parameters do not influence the systems so much in multifaceted ways, such that the cubic form or even higher degrees find rational justifications. Moreover, for such a large k, potentially great identification errors raised to significant power, have multiplied meaning, and in the consequence the optimal; in the sense of minimum value of expected losses; estimator value becomes close to the central values (median or expected value). As a result the application of such advanced methods is rather less effective then. Although incidentally the polynomial form (69) may find application in atypical situations.

The case of asymmetric polynomial loss function has been widely described in the paper [29].

The generalization to the multidimensional case will now be investigated. Consider two independent parameters, for example originating from different subcomponents of a given system, however such that their estimation errors mutually cumulate. In the third example of Subchapter 2, such parameters were λ, included in the description of the object itself, and τ characterizing the actuator. As can be seen in Fig. 3, overestimating one of the parameters linked with underestimation of the second results in considerably greater negative consequences, than when both estimation errors have the same sign.

Designate the investigated parameters by x_1 and x_2, after which denote them in the vector notation by $x = \begin{bmatrix} x_1 \\ x_2 \end{bmatrix}$, similarly as their searching estimators $\widehat{x} = \begin{bmatrix} \widehat{x}_1 \\ \widehat{x}_2 \end{bmatrix}$. The asymmetric two-dimensional quadratic loss function (6) may then be stated using general notations:

$$l\left(\begin{bmatrix} \widehat{x}_1 \\ \widehat{x}_2 \end{bmatrix}, \begin{bmatrix} x_1 \\ x_2 \end{bmatrix}\right) = \begin{cases} a_r(\widehat{x}_1 - x_1)^2 + a_{ru}(\widehat{x}_1 - x_1)(\widehat{x}_2 - x_2) + a_u(\widehat{x}_2 - x_2)^2 & \text{for } \widehat{x}_1 - x_1 \geq 0 \text{ i } \widehat{x}_2 - x_2 \geq 0 \\ a_l(\widehat{x}_1 - x_1)^2 + a_{lu}(\widehat{x}_1 - x_1)(\widehat{x}_2 - x_2) + a_u(\widehat{x}_2 - x_2)^2 & \text{for } \widehat{x}_1 - x_1 \leq 0 \text{ i } \widehat{x}_2 - x_2 \geq 0 \\ a_l(\widehat{x}_1 - x_1)^2 + a_{ld}(\widehat{x}_1 - x_1)(\widehat{x}_2 - x_2) + a_d(\widehat{x}_2 - x_2)^2 & \text{for } \widehat{x}_1 - x_1 \leq 0 \text{ i } \widehat{x}_2 - x_2 \leq 0 \\ a_r(\widehat{x}_1 - x_1)^2 + a_{rd}(\widehat{x}_1 - x_1)(\widehat{x}_2 - x_2) + a_d(\widehat{x}_2 - x_2)^2 & \text{for } \widehat{x}_1 - x_1 \geq 0 \text{ i } \widehat{x}_2 - x_2 \leq 0 \end{cases}, \tag{73}$$

where $a_l, a_r, a_g, a_d > 0$, $a_{ld}, a_{ru} \geq 0$ and $a_{lu}, a_{rd} \leq 0$. The value of expected losses resulting from assuming the estimator value $\widehat{x} = \begin{bmatrix} \widehat{x}_1 \\ \widehat{x}_2 \end{bmatrix}$ is given in this situation by the formula

$$\Lambda\left(\begin{bmatrix} \widehat{x}_1 \\ \widehat{x}_2 \end{bmatrix}\right) = a_r \int_{-\infty}^{\widehat{x}_1} (\widehat{x}_1 - x_1)^2 \widehat{f}_1(x_1) \mathrm{d}x_1 + a_l \int_{\widehat{x}_1}^{\infty} (\widehat{x}_1 - x_1)^2 \widehat{f}_1(x_1) \mathrm{d}x_1$$

$$+ a_{rd} \int_{-\infty}^{\widehat{x}_1} (\widehat{x}_1 - x_1) \widehat{f}_1(x_1) \mathrm{d}x_1 \int_{\widehat{x}_2}^{\infty} (\widehat{x}_2 - x_2) \widehat{f}_2(x_2) \mathrm{d}x_2$$

$$+ a_{ld} \int_{\widehat{x}_1}^{\infty} (\widehat{x}_1 - x_1)\widehat{f}_1(x_1)\mathrm{d}x_1 \int_{\widehat{x}_2}^{\infty} (\widehat{x}_2 - x_2)\widehat{f}_2(x_2)\mathrm{d}x_2$$

$$+ a_{lu} \int_{\widehat{x}_1}^{\infty} (\widehat{x}_1 - x_1)\widehat{f}_1(x_1)\mathrm{d}x_1 \int_{-\infty}^{\widehat{x}_2} (\widehat{x}_2 - x_2)\widehat{f}_2(x_2)\mathrm{d}x_2$$

$$+ a_{ru} \int_{-\infty}^{\widehat{x}_1} (\widehat{x}_1 - x_1)\widehat{f}_1(x_1)\mathrm{d}x_1 \int_{-\infty}^{\widehat{x}_2} (\widehat{x}_2 - x_2)\widehat{f}_2(x_2)\mathrm{d}x_2$$

$$+ a_d \int_{\widehat{x}_2}^{\infty} (\widehat{x}_2 - x_2)^2\widehat{f}_2(x_2)\mathrm{d}x_2 + a_u \int_{-\infty}^{\widehat{x}_2}\int (\widehat{x}_2 - x_2)^2\widehat{f}_2(x_2)\mathrm{d}x_2. \tag{74}$$

The minimum of this function exists and it is unique. It can be established using the two-dimensional Newton algorithm.

Detailed considerations concerning the multidimensional case can be found in the paper [28].

The generalizations presented above may be changed, e.g. by modifying degree of the function (73), although it seems that such an approach will have limited practical significance, as also mutually connected, in particular with the conditional approach. Composition of the multidimensional and conditional approaches has been described in detail in the paper [25].

6 Final Comments

The described methodology has been comprehensively verified numerically, and also checked in control engineering [23] as well as medicine [22] problems; the fuzzy approach was presented in the paper [17]. Below will be given results in an illustrative case, when the distribution of the parameter under investigation is the normal standard. It should be clearly stressed that the application of the concept presented above is not limited to such a trivial case; however, precisely thanks to its simplicity and general familiarity and comprehensibility, the obtained and presented below results become easily interpretable.

Thus, consider first the case of estimating a single parameter, whose measurements (30) origin from a generator with the normal standard distribution $N(0, 1)$, with the asymmetric linear loss function (29), studied in Sect. 3.1. For particular values of the quotient a/b:

$$a/b \qquad 1/10 \qquad 1/3 \qquad 1 \qquad 3 \qquad 10 \tag{75}$$

the following results were obtained

$$\hat{x} \qquad -1.34 \qquad -0.67 \qquad 0 \qquad 0.67 \qquad 1.34. \tag{76}$$

Begin with the case $a/b = 1$, i.e. where $a = b$. Then the losses implied be over- and under-estimation are the same, therefore, the optimal estimator is obviously 0, as the 'center' of the distribution $N(0, 1)$, or more precisely, which was emphasized, the median, which for this distribution equals 0. In the case $a/b = 3$, as results from the function (29) form, the losses connected with underestimation are three times greater than from overestimation; therefore, the optimal estimator takes the value greater than 0, in order to counteract the underestimation more harmful in the case considered. More precisely, as it results from the criterion (32), this constitutes a value such that the probability of underestimation would be three times smaller than overestimation. The above effect is clearer for $a/b = 10$; the optimal value of the estimator is even greater such that the probability of underestimation would be ten times smaller. For $a/b = 1/3$ and $a/b = 1/10$ the calculated values are opposite to those obtained for $a/b = 3$ and $a/b = 10$, respectively, due to the symmetry of the considered distribution $N(0, 1)$ regarding to zero.

In the case of the asymmetric quadratic function considered in Sect. 3.2, for the quotients (75), the following estimator values were respectively calculated

$$\hat{x} \qquad -0.90 \qquad -0.44 \qquad 0 \qquad 0.44 \qquad 0.90, \tag{77}$$

while for the asymmetric cubic function, i.e. of the form (69) with $k = 3$:

$$\hat{x} \qquad -0.72 \qquad -0.34 \qquad 0 \qquad 0.34 \qquad 0.72. \tag{78}$$

The regularity of the results of the linear case (76) were retained in the formulas (77)–(78). It is worth noting that as the degree of the polynomial loss function increases, the calculated estimator values successively approaches zero, which reduces benefits resulting from the application of the methodology given in this chapter for large values of k.

Consider now the conditional approach, presented in Sect. 4.3. For the illustration, the set (58) was obtained from a generator of two-dimensional normal distribution $N\left(\begin{bmatrix} 0 \\ 0 \end{bmatrix}, \begin{bmatrix} 1 & \mathrm{cov} \\ \mathrm{cov} & 1 \end{bmatrix}\right)$, where the parameter $\mathrm{cov} \in (-1, 1)$ describes correlation of the tested parameter x and the conditioning factor w. The estimator value thus achieved for the reference case $a = b$ moved by the value

$$\mathrm{cov} \cdot w_*; \tag{79}$$

therefore, the larger when the stronger is the correlation between the parameter and conditioning factor, and also the more distant the specific conditioning vale w_* is from zero (as the 'central' value of the conditioning factor). Similarly the estimator values $\hat{x}$ changed for other quotients a/b different from unity. The introduction of the conditioning value may thus constitute in practice a valuable supplement to the model leading to the calculation of the proper estimator value, also in the aspect of adaptation to successive changes in the environment, represented by the conditioning factors.

And finally, consider the multidimensional case commented on at the end of Subchapter 5. The values of the estimated parameters $\begin{bmatrix} x_1 \\ x_2 \end{bmatrix}$ were obtained using generator of two-dimensional normal standard distribution $N\left(\begin{bmatrix} 0 \\ 0 \end{bmatrix}, \begin{bmatrix} 1 & 0 \\ 0 & 1 \end{bmatrix}\right)$. If the mixed coefficients a_{ld}, a_{ru}, a_{lu}, a_{rd} are equal to zero, then the problem amounts to two one-dimensional issues and the results are identical to (77), which can be seen in the second column of the presentation below:

$$\begin{array}{lccccc} a_l \;;\; a_r & 10 \;;\; 1 & 1 \;;\; 1 & 1 \;;\; 1 & 1 \;;\; 1 \\ a_{lu} \;;\; a_{ru} & 0 \;;\; 0 & 0 \;;\; 0 & 0 \;;\; 0 & 0 \;;\; 0 \\ a_{ld} \;;\; a_{rd} & 0 \;;\; 0 & 3 \;;\; 0 & 10 \;;\; 0 & 10 \;;\; -3 \\ a_d \;;\; a_u & 1 \;;\; 10 & 1 \;;\; 1 & 1 \;;\; 1 & 1 \;;\; 1 \end{array} \tag{80}$$

$$\begin{bmatrix} \hat{x}_1 \\ \hat{x}_2 \end{bmatrix} \qquad \begin{bmatrix} 0.90 \\ -0.90 \end{bmatrix} \qquad \begin{bmatrix} 0.19 \\ 0.19 \end{bmatrix} \qquad \begin{bmatrix} 0.40 \\ 0.40 \end{bmatrix} \qquad \begin{bmatrix} 0.20 \\ 0.62 \end{bmatrix}. \tag{81}$$

To illustrate the influence of mixed factors, in columns 3–5 the neutral $a_l = a_r = a_d = a_u = 1$ has been assumed. Thus, in the third column the value $a_{ld} = 3$ is introduced, which, accordance with the formula (73), indicates increased losses connected with simultaneous underestimations of both parameters. To reduce the possibility of such an effect, the values of the estimators $\hat{x}_1$ and $\hat{x}_2$ are increased to 0.19. This effect undergoes enlargement in the fourth column, where $a_{ld} = 10$ is assumed, obtaining $\hat{x}_1 = \hat{x}_2 = 0.40$. And finally, in the fifth column, assuming $a_{rd} = -3$ increased losses relating to simultaneous overestimation of the first estimator and underestimation of the second, however, through $a_{ld} = 10$ significantly greater resulting from simultaneous underestimations of both, were declared. As a result the estimator of the first parameter underwent a small increase to $\hat{x}_1 = 0.20$, while as a consequence of cumulation of those assumptions, the estimator of the second was considerably increased to $\hat{x}_2 = 0.62$. The above results indicate the comprehensive possibilities of shaping the values of the estimator obtained, depending on the conditions of the specific application problem under research.

The next element requiring commentary is the postulated, for the needs of the elaborated procedure, size of the measurements set m. In the simplest situation, when one parameter is estimated, its distribution is unimodal, and the coefficients of the loss function are relatively balanced, it is sufficient if $m = 100$. Increasing the dimensionality of the kernel estimator (in the multidimensional of conditional cases), additional modes of distribution, or also unbalanced loss coefficients increase the requirements in this matter; in the majority of practical cases $m = 1{,}000$ is enough, while $m = 10{,}000$ fulfills all needs to excess. With the possibilities of contemporary computer and metrological systems used in modern automated systems, the above conditions do not seem difficult to fulfill in technical applications.

If for calculation of the smoothing parameter the formula (12) is applied, then all procedures have linear calculation complexity, and the calculation time on standard equipment does not exceed 0.01 s. The Newton algorithm performs 5–10 iterations. The calculation time of the order 0.1 s is achieved using the *plug-in* method recommended for sizes not greater than 1000.

Currently investigations are being conducted concerning kernel estimators for streaming data [43], therefore, an infinite sequence; in the issue considered here; of measurements. Others concern categorical attributes characterizing objects by assigning them to one of the finite number of groups on the descriptive, qualitative base [1], both nominal (without an ordered relationship between categories, e.g. colors red, blue, yellow) and ordered (with an order relationship, e.g. primary, secondary, higher education), which in the methodology presented here can broaden the possibilities of the conditional approach. The successive increasing dimensionality of contemporary data forces further research on its reduction ([39; 46, Chapter 3]).

Acknowledgements I am very grateful for the many years of creative effort of my close cooperators, coauthors of the cited publications. Particular words of gratitude are directed towards my past Ph.D.-students Małgorzata Charytanowicz, D.Sc., Piotr A. Kowalski, D.Sc., Szymon Łukasik, Ph.D., and Aleksander Mazgaj, Ph.D.

Appendix (Solutions of Differential Equations with Discontinuous Right-Hand Side)

In the first example of Subchapter 2 the issue of the different types of solutions of differential equations, needed to describe the phenomena appearing in the robust control systems (e.g. sliding trajectories or over-regulations) depending on the relationships between the values of the parameter occurring in the object and the estimator assumed in the control algorithm, was mentioned. The classic solution of differential equations in many such cases does not exist or improperly describes the modeled reality, which frequently leads to theoretical errors and interpretative misunderstandings. In this appendix, the fundamental definitions and relations connected with the issue of solutions of differential equations with discontinuous right-hand side, which can be applied to the description of the dynamics of objects submitted to robust control, will be presented.

Consider the differential equation in the form commonly used in control engineering, i.e. with the control u distinguished:

$$\dot{y}(t) = g(y(t), u(t), t), \tag{82}$$

where $y : T \to \mathbb{R}^n$, $u : T \to \mathbb{R}^m$, $g : \mathbb{R}^n \times \mathbb{R}^m \times T \to \mathbb{R}^n$, and $T \subset \mathbb{R}$ is a time interval with nonempty interior. Equation (82) is inherently connected with the initial condition

$$y(t_0) = y_0, \tag{83}$$

while $t_0 \in T$ and $y_0 \in \mathbb{R}^n$ are fixed.

The classic solution of the differential equation is the function y, differentiable and fulfilling the Eq. (82) for every $t \in T$ (if the boundary of the interval T belongs to it, then a proper one-sided derivative is considered) as well as the condition (83). This solution is unique, if each classic solution is a function identically equal to it. The above definitions are fully consistent with intuition, at least within the scope of the basic problems and aspects concerning dynamic systems.

The solution thus defined can be useful when the functions u and g are continuous. The continuity of the right-hand side of the differential equation constitutes the sufficient condition for the existence of the classic solution, and although it is not a necessary condition, the assumption of continuity is practically inherent in considerations of solutions of this type. In the case of a finite number of discontinuities of the first type (i.e. when in the point of discontinuity there exist bounded one-sided derivatives, not necessarily equal to each other nor the value of the function at this point) of the functions u and g (in the case of the second, with respect to the independent variable t), the above definition can be supplemented by a somewhat informal treatment known as "joining" of solutions. Namely, the interval T is then divided into subintervals, in which the functions u and g are continuous, and in these subintervals classic solutions are found, after which these solutions are linked together ("joined") with maintaining the continuity of the solution y (in the point of these discontinuities of the functions u and g with respect to variable t, the derivative $\dot{y}$ does not exist, due to the lack of equality of the one-sided derivatives). The above concept has a clear physical interpretation, one can then treat that at the moment of discontinuity of the functions u or g, the process described by the differential equation (82) begins again, with the initial condition (83) which is the left-hand limit of the solution from the previous area of continuity. From the mathematical point of view, such a procedure has a very limited scope of applicability to the case of the finite number of discontinuities of the functions u and g with respect to the independent variable.

A typical example illustrating the necessity for the introduction of generalization of the classic solution notion, is the following differential equation, representing in the simplified way the closed-loop control task with the discontinuous function of the feedback controller (e.g. in the classic problem of the minimal-time control):

$$\dot{y}_1(t) = 1, \qquad y_1(0) = y_{01} \tag{84}$$

$$\dot{y}_2(t) = \begin{cases} -1 & \text{gdy } y_2(t) > 0 \\ 1 & \text{gdy } y_2(t) \le 0 \end{cases}, \quad y_2(0) = y_{02}, \tag{85}$$

where $y_{01}, y_{02} \in \mathbb{R}$. Its solutions end after reaching the axis y_1. If the solution could be extended, then as results from the formula (85), it would be "push" above the axis y_1 (because $\dot{y}_2(t) = 1$ when $y_2(t) \leq 0$), but immediately after its "release", "pushed" in this axis again (since $\dot{y}_2(t) = -1$ when $y_2(t) > 0$), which negates the possibility of the existence of such a solution. It is worth observing that in the above example the function g is discontinuous on the axis y_1.

The concepts generalizing the notion of the classic solution for the needs of differential equations with a discontinuous right-hand side (when the function g defined in the formula (85) is discontinuous not only with respect to variable t) did not lead to a uniform approach. Currently the most frequently applied are solutions proposed by Caratheodory, Filippov and Krasovski. Before they have been defined, first the notions of "almost everywhere", the absolutely continuous function, convex closed hull, and compact set will be recalled.

If the measure of the points of the set $A \subset \mathbb{R}$, in which some property is not fulfilled, equals zero, then one can say that this property occurs almost everywhere in A.

Now consider the function $x : [p, q] \to \mathbb{R}$. It is called an absolutely continuous function if for every $\varepsilon > 0$ there exists $\delta > 0$ such that when (p_1, q_1), $(p_2, q_2), \ldots, (p_k, q_k)$ are intervals included in the domain $[p, q]$, separable and such that $\sum_{i=1}^{k} (q_i - p_i) \leq \delta$, then $\sum_{i=1}^{k} \|x(q_i) - x(p_i) \leq \varepsilon\|$ is fulfilled. The absolutely continuous function is also continuous. Furthermore, it is differentiable almost everywhere in $[p, q]$. These two properties particularly predispose the absolutely continuous function to be used for generalization of classic notion of the differential equations solution.

The convex closed hull of the set $C \subset \mathbb{R}^n$, denoted further as $\mathrm{conv}(C)$, is called the smallest (in the sense of an inclusion relation) closed and convex set including the set C. For example, the convex closed hull of the two different points is a segment connecting them, while of three different points, a closed triangle, whose points are the apexes.

The subset of the space $\mathbb{R}$ is compact, if and only if, it is bounded and closed.

Now, the solutions of differential equations in the Caratheodory, Filippov, and Krasovski senses [11] will be defined. The function y, absolutely continuous on every compact subinterval of the set T, is a solution of the differential equations (82)–(83):

- in the Caratheodory sense (C-solution), if the Eq. (82) is fulfilled almost everywhere in T, and also the condition (83) is fulfilled;
- in the Filippov sense (F-solution), if

$$\dot{y}(t) \in F[g](y(t), t) \quad \text{almost everywhere in } T, \tag{86}$$

and also the condition (83) is fulfilled, whereas the Filippov operator F is defined as follows

$$F[g](y(t),t) = \bigcap_{r>0} \bigcap_{\substack{Z \subset \mathbb{R}^n : \\ \mu(Z)=0}} \mathrm{conv}[g((y(t)+rB)\backslash Z, t)]; \qquad (87)$$

– in the Krasovski sense (K-solution), if

$$\dot{y}(t) \in K[g](y(t),t) \quad \text{almost everywhere in } T, \qquad (88)$$

and also the condition (83) is fulfilled, whereas the Krasovski operator K is given by the formula

$$K[g](y(t),t) = \bigcap_{r>0} \mathrm{conv}[g((y(t)+rB),t)]; \qquad (89)$$

where B denotes an open unit ball in the space $\mathbb{R}^n$ and μ is n-dimensional Lebesgue measure. C-, F- or K-solutions are unique, if each, respectively, C-, F- or K-solution is a function identically equal to it. The interpretation of these definitions presented below also provide practical motivations for the forms introduced above.

Thus, in the case of the C-solutions, the derivative $\dot{y}(t)$ is dependent (except the independent variable t) only on the current value of the solution y. This type of solution is in practice a mathematically uniform generalization of the classic solutions "joining".

In the case of the K-solution, the values of the functions g are taken into account not only at the point $y(t)$, but at all points in the ball with the center at this point and positive, although, because of the set-product, of any small radius r. Furthermore, the convex closed hull supplements thus created a set of the function g values with intermediate points. Interpreting the above, one may state that the K-solution, allowing all points from the surrounding of the value $y(t)$, takes into account the measurements errors, unavailable in practice.

Finally, additionally introduced in the definition of the F-solution set-product implies that from the ball with center at point $y(t)$ and radius r, the zero-measure sets, irrelevant from the practical point of view, are eliminated.

Directly from the above definitions it results that:

(A) every C-solution is a K-solution, because $g(y(t),t) \in K[g](y(t),t)$;
(B) every F-solution is K-solution, since $F[g](y(t),t) \subset K[g](y(t),t)$.

The uniqueness of K-solutions thus implies the uniqueness of C- and F-solutions.

Other relations between C-, F- and K-solutions are illustrated below by the example of the following differential equation:

$$\dot{y}_1(t) = \begin{cases} 1 & \text{when } y_2(t) > 0 \\ -1 & \text{when } y_2(t) = 0\,, \\ 1 & \text{when } y_2(t) < 0 \end{cases} \quad y_1(0) = y_{01} \qquad (90)$$

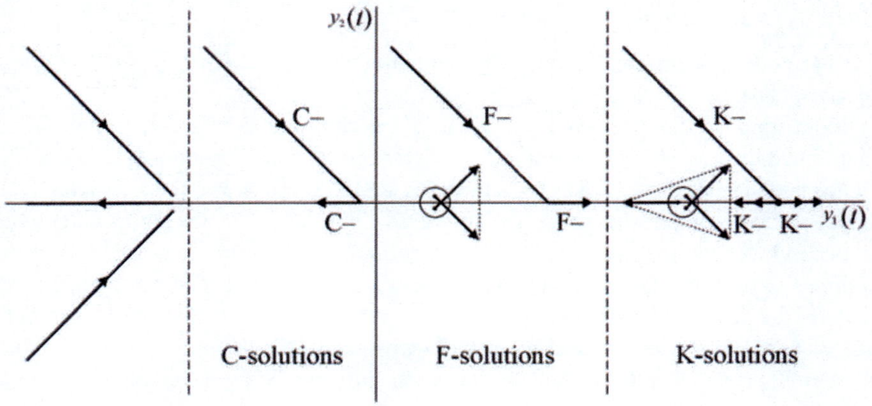

Fig. 6 Solutions of differential equations (90)–(91)

$$\dot{y}_2(t) = \begin{cases} -1 & \text{when } y_2(t) > 0 \\ 0 & \text{when } y_2(t) = 0\,, \\ 1 & \text{when } y_2(t) < 0 \end{cases} \quad y_2(0) = y_{02} \tag{91}$$

where $y_{01}, y_{02} \in \mathbb{R}$ (Fig. 6).

First consider the case, when the initial state is included in the axis y_1. The function $\begin{bmatrix} y_{01} - t \\ 0 \end{bmatrix}$ for $t \in [0, \infty)$ is then a unique C-solution. Because for every $t \in [0, \infty)$ and $y(t)$ contained in the axis y_1, the equality $F[g](y(t), t) = \left\{ \begin{bmatrix} 1 \\ v \end{bmatrix} \text{where} |v| \le 1 \right\}$ is true, then the function $\begin{bmatrix} y_{01} + t \\ 0 \end{bmatrix}$ for $t \in [0, \infty)$ is then a unique F-solution. In turn, for every $t \in [0, \infty)$ and $y(t)$ contained in the axis x_1, the set $K[g](y(t), t)$ is a triangle with apexes $\begin{bmatrix} 1 \\ 1 \end{bmatrix}, \begin{bmatrix} 1 \\ -1 \end{bmatrix}, \begin{bmatrix} -1 \\ 0 \end{bmatrix}$, therefore, the absolute continuous functions of the form $\begin{bmatrix} y_1(t) \\ 0 \end{bmatrix}$, while $y_1(0) = y_{01}$ and $|\dot{y}_1(t)| \le 1$ almost everywhere in $[0, \infty)$, are K-solutions.

In the second case, when the initial state is not included in the axis y_1, the function $\begin{bmatrix} y_{01} + t \\ y_{02} - t \cdot \operatorname{sgn}(y_{02}) \end{bmatrix}$ is, to the moment it crosses this axis, a unique C-, F- and K-solution. When it reaches the axis y_1, the solutions can be extended by the proper solutions described in the previous paragraph.

From the above example it results that in the general case there are no relations between C- and F-solutions, so continuing the comments formulated previously in the form of points (A)–(B), one can also add that:

(C) C-solution need not be F-solution;

(D) F-solution need not be C-solution.

Furthermore, K-solutions frequently rather constitute too rich, from a practical point of view, class.

Now suppose that the functions u and g are continuous, which assumption, as mentioned, is practically inherent to considerations of classic solutions. Because the right-hand side of the Eq. (82) is continuous, so the same is true for a derivative of such a solution, therefore $y \in \mathcal{C}^1$. The function of the class $\mathcal{C}^1$ on the compact interval is a Lipschitz function, so also absolutely continuous; therefore, the classic solution is then a C-solution. Due to point (A) it is also a K-solution. The above conclusions remain true also in the case when functions u and g have a finite number of discontinuities of the first type (the second of them with respect to the independent variable t), so in the case when the classic "joint" procedure can be used. In conclusion:

(E) if the functions u and g are continuous or have a finite number of discontinuities of the first type (the second of them with respect to the independent variable t), then every classic solution is a C-solution.

Finally, in practice C- and in consequence K-solutions constitute a generalization of the classic solutions, also with the possibility of "joining".

The relations between particular types of solutions have been synthetically expressed in the form of points (A)–(E). C-solutions are most frequently used in the case of discontinuities of the function g with respect to the independent variable t, F-solutions for discontinuities with respect to $x(t)$, although K-solutions generalizes the above two types, however often constitute too rich a class. The confusions to which the lack of a proper understanding and using of notions of solutions to differential equations with discontinuous right-hand side, were illustrated in the publications ([20], point 2.3.2; [15]).

In the first example of Subchapter 2 it was mentioned that in the hypothetical case $\widehat{M} = M$, trajectories have regular character and can be even described by classic solutions with "joining" or, more generally, C-solutions. If the parameter $\widehat{M}$ is overestimated, the over-regulations appear in the system; the trajectories are then C-solutions. And finally, when it is underestimated, then the robust control generates the sliding trajectories, described by F-solutions. In order to unify these cases, one can use the concept of K-solutions, more general then both C- and F-solutions. However, as mentioned, in many applications K-solutions constitute too rich a class, in the publications [13, 14] it was shown that in this case they are unique, and furthermore in the probabilistic approach, the family arrived in this way constitutes a stochastic process, which enables the use of advanced methodology available for this purpose.

References

1. Agresti, A.: Categorical Data Analysis. Wiley, Hoboken (2002)
2. Astrom, K.J., Hagglund, T.: Advanced PID Control. ISA, Research Triangle Park (2006)
3. Athans, M., Falb, P.L.: Optimal Control. McGraw-Hill, New York (1966)

4. Bolotin, V.V.: Dynamic stability of structures. In: Kounadis, A.N., Krätzig, W.B. (eds.) Nonlinear Stability of Structures, pp. 3–72. Springer, Vienna (1995)
5. Bubnicki, Z.: Modern Control Theory. Springer, Berlin (2005)
6. Devroye, L., Gyorfi, L.: Nonparametric Density Estimation: The L_1 View. Wiley, New York (1985)
7. Devroye, L., Lugosi, G.: Combinatorial Methods in Density Estimation. Springer, New York (2001)
8. Edwards, C., Spurgeon, S.K.: Sliding Mode Control: Theory and Applications. CRC Press, London (1998)
9. Gramacki, A.: Nonparametric Kernel Density Estimation and Its Computational Aspects. Springer, Cham (2018)
10. Green, P.J., Silverman, B.W.: Nonparametric Regression and Generalized Linear Models: A roughness penalty approach. Chapman and Hall/CRC, New York (1993)
11. Hajek, O.: Discontinuous differential equations, I-II. J. Differ. Equ. **32**, 149–185 (1979)
12. Kincaid, D., Cheney, W.: Numerical Analysis. Brooks/Cole, Pacific Grove (2002)
13. Kulczycki, P.: Time-optimal stabilization of a discontinuous and non-autonomous dynamic object. Control Cybern. **25**, 707–720 (1996)
14. Kulczycki, P.: Almost certain time-optimal positional control. IMA J. Math. Control. Inf. **13**, 63–77 (1996)
15. Kulczycki, P.: Some remarks on solutions of discontinuous differential equations applied in automatic control. Ind. Math. **46**, 119–128 (1996)
16. Kulczycki, P.: A random approach to time-optimal control. J. Dyn. Syst. Meas. Contr. **121**, 542–543 (1999)
17. Kulczycki, P.: Fuzzy controller for mechanical systems. IEEE Trans. Fuzzy Syst. **8**, 645–652 (2000)
18. Kulczycki, P.: An algorithm for Bayes parameter identification. J. Dyn. Syst. Meas. Contr. **123**, 611–614 (2001)
19. Kulczycki, P.: Statistical inference for fault detection: a complete algorithm based on kernel estimators. Kybernetika **38**, 141–168 (2002)
20. Kulczycki, P.: Estymatory jądrowe w analizie systemowej. WNT, Warsaw (2005)
21. Kulczycki, P.: Kernel estimators for data analysis. In: Ram, M., Davim, J.P. (eds.) Advanced Mathematical Techniques in Engineering Sciences, pp. 177–202. CRC/Taylor & Francis, Boca Raton (2018)
22. Kulczycki, P., Charytanowicz, M.: Bayes sharpening of imprecise information. Int. J. Appl. Math. Comput. Sci. **15**, 393–404 (2005)
23. Kulczycki, P., Charytanowicz, M.: Asymmetrical conditional Bayes parameter identification for control engineering. Cybern. Syst. **39**, 229–243 (2008)
24. Kulczycki, P., Charytanowicz, M.: Conditional parameter identification with different losses of under- and overestimation. Appl. Math. Model. **37**, 2166–2177 (2013)
25. Kulczycki, P., Charytanowicz, M.: An algorithm for conditional multidimensional parameter identification with asymmetric and correlated losses of under- and overestimations. J. Stat. Comput. Simul. **86**, 1032–1055 (2016)
26. Kulczycki, P., Franus, K.: Methodically Unified Procedures for Conditional Approach of Outliers Detection, Clustering, and Classification (2020). Inf. Sci. (in press)
27. Kulczycki, P., Mazgaj, A.: An algorithm for Bayes parameter identification with quadratic asymmetrical loss function. Control Cybern. **34**, 1127–1148 (2005)
28. Kulczycki, P., Mazgaj, A.: Bayes identification of parameters' vector with quadratic asymmetrical loss function. In: 19th International Conference on Production Research, Valparaiso (Chile), 29 July–2 August 2007, CD: 250 (2007)
29. Kulczycki, P., Mazgaj, A.: Parameter identification for asymmetrical polynomial loss function. Inf. Technol. Control. **38**(1), 51–60; errata: **38**(2), 167–168 (2009)
30. Kulczycki, P., Wisniewski, R.: Fuzzy controller for a system with uncertain load. Fuzzy Sets Syst. **131**, 185–195 (2002)
31. Larose, T.L.: Discovering Knowledge in Data. Wiley, New York (2005)

32. Larose, T.L.: Data Mining Methods and Models. Wiley, New York (2006)
33. Larose, T.L., Larose, C.D.: Data Mining and Predictive Analytics. Wiley, Hoboken (2015)
34. Lehmann, E.L.: Theory of Point Estimation. Wiley, New York (1983)
35. Liu, S., Chen, G.S.: Dynamics and Control of Robotic Manipulators with Contact and Friction. Wiley, Hoboken (2018)
36. McCullough, B.D.: Optimal prediction with a general loss function. J. Comb. Inf. Syst. Sci. **25**, 207–221 (2000)
37. Murray, R.M., Li, Z., Sastry, S.S.: A Mathematical Introduction to Robotic Manipulation. CRC/Taylor & Francis, Boca Raton (1994)
38. Naidu, D.S.: Optimal Control Systems. CRC, Boca Raton (2002)
39. Pal, K.P., Mitra, P.: Pattern Recognition Algorithms for Data Mining. Chapman & Hall/CRC, Boca Raton (2004)
40. Prakasa Rao, B.L.S.: Nonparametric Functional Estimation. Academic Press, Orlando (1983)
41. Parzen, E.: On estimation of a probability density function and mode. Ann. Math. Stat. **33**, 1065–1076 (1962)
42. Scott, D.W.: Multivariate Density Estimation: Theory, Practice, and Visualization. Wiley, New York (1992)
43. Silva, J., Faria, E., Barros, R., Hruschka, E., de Carvalho, A., Gama, J.: Data stream clustering: a survey. ACM Comput. Surv. **46**(1), paper 13 (2013)
44. Silverman, B.W.: Density Estimation for Statistics and Data Analysis. Chapman and Hall, London (1986)
45. Simonoff, J.S.: Smoothing Methods in Statistics. Springer, New York (1996)
46. Sorzano, C.O.S., Vargas, J., Pascual-Montano, A.: A survey of dimensionality reduction techniques. Cornell University Library, signature (2014). arXiv:1403.2877v1
47. Towill, D.R.: Transfer Function Techniques for Control Engineers. Newnes-Butterworth, London (1970)
48. Utkin, V.I.: Sliding Modes in Control Optimization. Springer, Berlin (1992)
49. Wand, M.P., Jones, M.C.: Kernel Smoothing. Champan and Hall, London (1995)
50. Zhou, K.: Robust and Nonlinear Control. Prentice-Hall, Upper Saddle River (1996)

Flow Process Models for Pipeline Diagnosis

Zdzisław Kowalczuk and Marek Sylwester Tatara

Abstract This chapter examines the problem of modeling and parameterization of the transmission pipeline flow process. First, the base model for discrete time is presented, which is a reference for other developed models. Then, the diagonal approximation (AMDA) method is proposed, in which the tridiagonal sub-matrices of the recombination matrix are approximated by their diagonal counterparts, which allows for a simple determination of the explicit form of the inverse matrix. Another suggestion is the Thomas model (ATM), in which the basic model is reformulated to a form to which the Thomas algorithm applies, at which the computational complexity of the order $\mathcal{O}(N)$ can be obtained. The fourth suggestion is a steady state analytical model (AMSS), characterizing the steady state after transient processes. In addition, the parameterization of the discrete models in space and time is analyzed, proposing a method ensuring the maximum margin of numerical stability. This model is verified by means of simulation tests. Finally, the developed model is compared with the basic model, taking into account the accuracy and time of calculations.

1 Introduction

One of the most efficient ways to transport fluid is to pump it through a transmission pipeline over short and long distances. However, there is a risk of leakage that could lead to financial loss, environmental pollution or, in extreme cases, expose people to danger. That is why it is important to implement Leak Detection and Isolation systems (LDI) to quickly detect each leak and assess its parameters in an effective way.

Z. Kowalczuk (✉) · M. S. Tatara
Department of Robotics and Decision Systems, Faculty of Electronics, Telecommunications and Informatics, Gdańsk University of Technology, ul. Narutowicza 11/12, 80-233 Gdańsk, Poland
e-mail: kova@eti.pg.edu.pl

M. S. Tatara
e-mail: martatar@pg.edu.pl

P. Kulczycki et al. (eds.), *Automatic Control, Robotics, and Information Processing*, Studies in Systems, Decision and Control 296,
https://doi.org/10.1007/978-3-030-48587-0_2

Model-based LDI systems allow continuous assessment of the state of the monitored pipeline. The monitored process can be numerically simulated to compare the results obtained with appropriate measurements. Appropriate processing of residual signals allows us to complete the task of detecting leaks and identifying its parameters (location and size).

Considering the classification of LDI systems taking into account the implemented detection method, we can mention methods based on the state observers [4, 5, 19, 21, 22, 41–43] or Kalman filtration [15], as well as methods using measurements of pressure wave propagation both in the time domain [34, 46] and the frequency domain [29, 30, 36]. Other methods use the rough sets theory and the support vector machine method [32], leak-sensitive cables that change their resistance as a result of a leak reaction [35], georadars to detect leaks from underground pipes [16] or inspection apparatuses flowing through the pipeline [10, 44].

For model-based leak detection systems that are covered in this chapter, the following quality measures can be defined [1]:

- Accuracy—determines the quality of the leak parameter estimates (such as its location or size)
- Sensitivity—shows the minimum detectable leakage
- Reliability—it reflects the probability that the alarm will not be triggered in the event of a leak
- Robustness—it measures the system's ability to continue to function properly under changing conditions.

One of the core elements of a complete LDI system that determines its quality is a mathematical model of the flow process used to conduct analysis or simulation. Such a model must first of all be computable with the conditions of given, real-time constraints. It should also reflect the process being monitored as accurately as possible. Later in this chapter, we will discuss flow process models for transmission pipelines, developed by the authors for their application in diagnostic systems.

2 Base Model of the Flow Process

Let us consider the principal, continuous-time mathematical description of the pressure and mass-flow rate of liquid flowing in a transmission pipeline. Assuming the incompressible isothermal flow, it is expressed by the following two partial differential equations, resulting from the momentum and mass conservation laws [5]:

$$\frac{S}{\nu^2}\frac{\partial p}{\partial t} + \frac{\partial q}{\partial z} = 0 \tag{1}$$

$$\frac{1}{S}\frac{\partial q}{\partial t} + \frac{\partial p}{\partial z} = -\frac{\lambda \nu^2}{2DS^2}\frac{q|q|}{p} - \frac{g \sin\alpha}{\nu^2}p \tag{2}$$

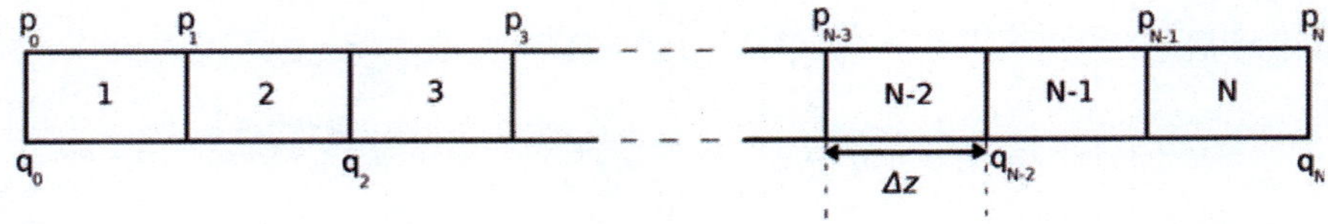

Fig. 1 Discretization scheme of a pipeline with N segments

where S is the cross-sectional area [m^2], v is a surrogate velocity related to the isothermal velocity of sound in the fluid [$\frac{\mathrm{m}}{\mathrm{s}}$], D is the diameter of the pipe [m], q is the mass flow [$\frac{\mathrm{Kg}}{\mathrm{s}}$], p is the pressure [Pa], t is the time [s], z is the spatial coordinate [m], λ is the generalized dimensionless friction factor, α is the inclination angle [rad], and g is the gravitational acceleration [$\frac{\mathrm{m}}{\mathrm{s}^2}$].

Since the operation of a model-based algorithm for pipeline diagnosis usually requires simulation of the behavior of the underlying flow process, the set of Eqs. (1)–(2) is discretized in order to enable computer implementation. In such cases, the pipeline is divided into N segments of equal length Δz, where the pressure at the end of each odd segment, and the flow rate at the end of each even segment, are the main object of calculations. The resulting mass-flow and pressure are computed at the inlet and outlet of the pipeline, as well. A diagram of such a discretized pipeline is represented in Fig. 1.

The discrete-time model is developed by properly inserting low-order central difference schemes [5]:

$$\frac{\partial x}{\partial t} = \frac{3x_d^{k+1} - 4x_d^k + x_d^{k-1}}{2\Delta t} \tag{3}$$

$$\frac{\partial x}{\partial z} = \frac{x_{d+1}^{k+1} - x_{d-1}^{k+1} + x_{d+1}^k - x_{d-1}^k}{4\Delta z} \tag{4}$$

where Δz is a spatial-step size, Δt is a time-step size, subscripts and superscripts stand for the number of the pipeline segments and discrete-time indexes, respectively.

By implementing (3)–(4) in the composed model (1)–(2), one obtains the following equations:

$$ap_d^{k+1} - b\left(q_{d-1}^{k+1} - q_{d+1}^{k+1}\right) = \frac{a}{3}\left(4p_d^k - p_d^{k-1}\right) + b\left(q_{d-1}^k - q_{d+1}^k\right) \tag{5}$$

$$b\left(p_{d+1}^{k+1} - p_{d-1}^{k+1}\right) + cq_d^{k+1} = b\left(p_{d-1}^k - p_{d+1}^k\right) + Y_d p_d^k + \left(\frac{4c}{3} + F_d^k\right) q_d^k - \frac{c}{3} q_d^{k-1} \tag{6}$$

with physical coefficients

$$a = \frac{3S}{2\nu^2 \Delta t}, \quad b = \frac{1}{4\Delta z}, \quad c = \frac{3}{2S\Delta t}, \quad Y_d = -\frac{g \sin \alpha_d}{\nu^2}$$

$$F_d^k \simeq -\frac{\lambda \nu^2}{DS^2} \frac{|q_d^k|}{p_{d-1}^k + p_{d+1}^k}$$

where α_d denotes the inclination angle of the d-th segment.

Rewriting the Eqs. (5)–(6) into the classical form of state-space equations, the model can be represented by the following compact state-space model:

$$\mathbb{A}\hat{\mathbf{x}}^k = \mathbb{B}\hat{\mathbf{x}}^{k-2} + \mathbb{C}\left(\hat{\mathbf{x}}^{k-1}\right)\hat{\mathbf{x}}^{k-1} + \mathbb{D}\mathbf{u}^{k-1} + \mathbb{E}\mathbf{u}^k \tag{7}$$

where the state and input vectors are defined as

$$\hat{\mathbf{x}}^k = \left[\hat{q}_0^k \; \hat{q}_2^k \; \hat{q}_4^k \cdots \hat{q}_N^k \; \hat{p}_1^k \; \hat{p}_3^k \; \hat{p}_5^k \cdots \hat{p}_{N-1}^k\right]^T \in \mathbb{R}^{N+1}$$

and

$$\mathbf{u}^k = \left[p_0^k \; p_N^k\right]^T \in \mathbb{R}^2$$

respectively [19]. Symbols with hats indicate estimates. The form (7), however, is difficult to analyze the numerical stability of the model. Since matrix $\mathbb{A}$ is proved to be invertible [23, 25], Eq. (7) can be rewritten in a nonsingular state-space form:

$$\hat{\mathbf{x}}^k = \mathbb{A}^{-1}\left(\mathbb{B}\hat{\mathbf{x}}^{k-2} + \mathbb{C}\left(\hat{\mathbf{x}}^{k-1}\right)\hat{\mathbf{x}}^{k-1} + \mathbb{D}\mathbf{u}^{k-1} + \mathbb{E}\mathbf{u}^k\right) \tag{8}$$

The recombination matrix $\mathbb{A} \in \mathbb{R}^{(N+1)\times(N+1)}$ [24, 25] can be defined as

$$\mathbb{A} = \left[\begin{array}{c|c} \mathbb{A}_1 & \mathbb{A}_2 \\ \hline \mathbb{A}_3 & \mathbb{A}_4 \end{array}\right] = \left[\begin{array}{ccccccc|cccccc} & & & & & & & \searrow 2b & 0 & \cdots & 0 & 0 \\ & & & & & & & -b & \searrow b & \cdots & 0 & 0 \\ & & & \mathbb{D}_{\frac{N}{2}+1}(c) & & & & \vdots & & \ddots & & \vdots \\ & & & & & & & 0 & 0 & \cdots & \searrow b & 0 \\ & & & & & & & 0 & 0 & \cdots & -b & \searrow b \\ & & & & & & & 0 & 0 & \cdots & 0 & -2b \\ \hline -b & b & 0 & \cdots & 0 & 0 & 0 & & & & & \\ 0 & -b & b & \cdots & 0 & 0 & 0 & & & & & \\ \vdots & & & \ddots & & & \vdots & & & \mathbb{D}_{\frac{N}{2}}(a) & & \\ 0 & 0 & 0 & \cdots & -b & b & 0 & & & & & \\ 0 & 0 & 0 & \cdots & 0 & -b & b & & & & & \end{array}\right] \tag{9}$$

where $\mathbb{D}_W(\theta) \in \mathbb{R}^{W \times W}$ denotes a diagonal matrix with θ on its diagonal. Note that the upper right submatrix is non-square and belongs to $\mathbb{R}^{(\frac{N}{2}+1)\times(\frac{N}{2})}$. For the matrices $\mathbb{B}$, $\mathbb{C}$, $\mathbb{D}$, and $\mathbb{E}$ please refer to [25].

3 Model of Diagonal Approximation

Due to specific structure of the recombination matrix, it is possible to calculate its inverse, and the following matrix inversion lemma can be applied [8]:

$$\mathbb{M}^{-1} = \left[\begin{array}{c|c} \mathbb{M}_1' & \mathbb{M}_2' \\ \hline \mathbb{M}_3' & \mathbb{M}_4' \end{array}\right] = \left[\begin{array}{c|c} (\mathbb{M}_1 - \mathbb{M}_2\mathbb{M}_4^{-1}\mathbb{M}_3)^{-1} & -\mathbb{M}_1^{-1}\mathbb{M}_2(\mathbb{M}_4 - \mathbb{M}_3\mathbb{M}_1^{-1}\mathbb{M}_2)^{-1} \\ \hline -\mathbb{M}_4^{-1}\mathbb{M}_3(\mathbb{M}_1 - \mathbb{M}_2\mathbb{M}_4^{-1}\mathbb{M}_3)^{-1} & (\mathbb{M}_4 - \mathbb{M}_3\mathbb{M}_1^{-1}\mathbb{M}_2)^{-1} \end{array}\right] \tag{10}$$

Since matrix $\mathbb{A}$ consists of two diagonal submatrices and two submatrices having elements only on two diagonals, by (10) the matrix $\mathbb{A}^{-1}$ can be represented as

$$\mathbb{A}^{-1} = \left[\begin{array}{c|c} \mathbb{A}_1 & \mathbb{A}_2 \\ \hline \mathbb{A}_3 & \mathbb{A}_4 \end{array}\right]^{-1} = \left[\begin{array}{c|c} \mathbb{A}_1' & \mathbb{A}_2' \\ \hline \mathbb{A}_3' & \mathbb{A}_4' \end{array}\right] \tag{11}$$

with

$$\mathbb{A}_1' = \left(\mathbb{D}_{\frac{N}{2}+1}(c) - \mathbb{A}_2\mathbb{D}_{\frac{N}{2}}(a^{-1})\mathbb{A}_3\right)^{-1} \tag{12}$$

$$\mathbb{A}_4' = \left(\mathbb{D}_{\frac{N}{2}}(a) - \mathbb{A}_3\mathbb{D}_{\frac{N}{2}+1}\left(c^{-1}\right)\mathbb{A}_2\right)^{-1} \tag{13}$$

$$\mathbb{A}_2' = \mathbb{D}_{\frac{N}{2}+1}\left(-c^{-1}\right)\mathbb{A}_2\mathbb{A}_4' \tag{14}$$

$$\mathbb{A}_3' = \mathbb{D}_{\frac{N}{2}}\left(-a^{-1}\right)\mathbb{A}_3\mathbb{A}_1' \tag{15}$$

From the set of matrices, one can see that $\mathbb{A}_2'$ and $\mathbb{A}_3'$ are dependent on matrices $\mathbb{A}_4'$ and $\mathbb{A}_1'$, which have to be determined first. Thus, (12) and (13) can be directly calculated which results in the following tridiagonal matrices:

$$\mathbb{A}_1' = \begin{bmatrix} c + \frac{2b^2}{a} & -\frac{2b^2}{a} & 0 & \cdots & 0 & 0 & 0 \\ -\frac{b^2}{a} & c + \frac{2b^2}{a} & -\frac{b^2}{a} & \cdots & 0 & 0 & 0 \\ \vdots & & & \ddots & & & \vdots \\ 0 & 0 & 0 & \cdots & -\frac{b^2}{a} & c + \frac{2b^2}{a} & -\frac{b^2}{a} \\ 0 & 0 & 0 & \cdots & 0 & -\frac{2b^2}{a} & c + \frac{2b^2}{a} \end{bmatrix}^{-1} \tag{16}$$

$$A'_4 = \begin{bmatrix} a+\frac{3b^2}{c} & -\frac{b^2}{c} & 0 & \cdots & 0 & 0 & 0 \\ -\frac{b^2}{c} & a+\frac{2b^2}{c} & -\frac{b^2}{c} & \cdots & 0 & 0 & 0 \\ \vdots & & & \ddots & & & \vdots \\ 0 & 0 & 0 & \cdots & -\frac{b^2}{c} & a+\frac{2b^2}{c} & -\frac{b^2}{c} \\ 0 & 0 & 0 & \cdots & 0 & -\frac{b^2}{c} & a+\frac{3b^2}{c} \end{bmatrix}^{-1} \tag{17}$$

We propose in [24] to approximate the tridiagonal matrices (16) and (17) with their diagonal counterparts. Such simplification is feasible only if elements on diagonal are significantly greater than the ones outside main diagonal, which can be given by the following condition:

$$|c| \gg \left|4\frac{b^2}{a}\right| \tag{18}$$

By substituting physical parameters of the pipeline in (18), and incorporating the Courant-Friedrichs-Lewy (CFL, $\Delta t = \mu\frac{\Delta z}{v}$) condition in (18), the following restriction on μ is obtained:

$$\mu < 0.03 \tag{19}$$

By considering (19), the two submatrices of the inverted recombination matrix (16) and (17) can be approximated as

$$A'_1 \approx \breve{A}'_1 = \begin{bmatrix} c+\frac{2b^2}{a} & 0 & 0\cdots 0 & 0 & 0 \\ 0 & c+\frac{2b^2}{a} & 0\cdots 0 & 0 & 0 \\ \vdots & & \ddots & & \vdots \\ 0 & 0 & 0\cdots 0 & c+\frac{2b^2}{a} & 0 \\ 0 & 0 & 0\cdots 0 & 0 & c+\frac{2b^2}{a} \end{bmatrix}^{-1} \tag{20}$$

and

$$A'_4 \approx \breve{A}'_4 = \begin{bmatrix} a+\frac{3b^2}{c} & 0 & \cdots & 0 & 0 & 0 & \\ 0 & a+\frac{2b^2}{c} & 0 & \cdots & 0 & 0 & 0 \\ & & & & & & \\ \vdots & & & \ddots & & & \vdots \\ & & & & & & \\ 0 & 0 & 0 & \cdots & 0 & a+\frac{2b^2}{c} & 0 \\ 0 & 0 & 0 & \cdots & 0 & 0 & a+\frac{3b^2}{c} \end{bmatrix}^{-1} \tag{21}$$

The inverses of these matrices are obtained by substituting their reciprocals on the main diagonal. Therefore, the inverse recombination matrix

$$A^{-1} \approx \left[\begin{array}{cccccc|cccccc} \frac{a}{\sigma} & 0 & & \cdots & 0 & 0 & -\frac{2b}{3b^2+ca} & 0 & & \cdots & 0 & 0 \\ 0 & \frac{a}{\sigma} & & & 0 & 0 & \frac{b}{3b^2+ca} & -\frac{b}{\sigma} & 0 & & 0 & 0 \\ & & & & & & 0 & \frac{b}{\sigma} & -\frac{b}{\sigma} & & 0 & 0 \\ \vdots & & \ddots & & & \vdots & \vdots & & \ddots & & & \vdots \\ & & & & & & 0 & 0 & & \frac{b}{\sigma} & -\frac{b}{\sigma} & 0 \\ 0 & 0 & & & \frac{a}{\sigma} & 0 & 0 & 0 & & 0 & \frac{b}{\sigma} & -\frac{b}{3b^2+ca} \\ 0 & 0 & & \cdots & 0 & \frac{a}{\sigma} & 0 & 0 & & \cdots & 0 & \frac{2b}{3b^2+ca} \\ \hline \frac{b}{\sigma} & -\frac{b}{\sigma} & 0 & \cdots & 0 & 0 & \frac{c}{3b^2+ca} & 0 & & \cdots & 0 & 0 \\ 0 & \frac{b}{\sigma} & -\frac{b}{\sigma} & & 0 & 0 & 0 & \frac{c}{\sigma} & & & 0 & 0 \\ 0 & 0 & \frac{b}{\sigma} & & 0 & 0 & & & & & & \\ \vdots & & \ddots & & & \vdots & \vdots & & \ddots & & & \vdots \\ 0 & 0 & & -\frac{b}{\sigma} & 0 & 0 & & & & & & \\ 0 & 0 & & \frac{b}{\sigma} & -\frac{b}{\sigma} & 0 & 0 & 0 & & & \frac{c}{\sigma} & 0 \\ 0 & 0 & \cdots & 0 & \frac{b}{\sigma} & -\frac{b}{\sigma} & 0 & 0 & & \cdots & 0 & \frac{c}{3b^2+ca} \end{array}\right] \tag{22}$$

is obtained by using (10) with $\sigma = 2b^2 + ca$. The model obtained with this method for the recombination matrix is called the analytic model of diagonal approximation (AMDA).

4 Thomas Model

Similar to the AMDA model, sparsity of matrices in the state-space model (7) allows us to rearrange this model to a form which falls inside the scope of applicability of the Thomas algorithm [12], which reduces the computational complexity of computations.

The procedure starts with the consideration of the structure of submatrices in the recombination matrix. Two of them are square and diagonal, whereas the other two are non-square band matrices with the band width equal to 2:

$$\mathbb{A}_1 = \mathbb{D}_{\frac{N}{2}+1}(c) \in \mathbb{R}^{(\frac{N}{2}+1)\times(\frac{N}{2}+1)} \tag{23}$$

$$\mathbb{A}_4 = \mathbb{D}_{\frac{N}{2}}(a) \in \mathbb{R}^{\frac{N}{2}\times\frac{N}{2}} \tag{24}$$

$$\mathbb{A}_2 = \begin{bmatrix} \searrow 2b & 0 & \cdots & 0 & 0 \\ -b & \searrow b & \cdots & 0 & 0 \\ \vdots & & \ddots & & \vdots \\ 0 & 0 & \cdots & \searrow b & 0 \\ 0 & 0 & \cdots & -b & \searrow b \\ 0 & 0 & \cdots & 0 & -2b \end{bmatrix} \in \mathbb{R}^{(\frac{N}{2}+1)\times\frac{N}{2}} \tag{25}$$

$$\mathbb{A}_3 = \begin{bmatrix} -b & \searrow b & 0 & \cdots & 0 & 0 & 0 \\ 0 & -b & \searrow b & \cdots & 0 & 0 & 0 \\ \vdots & & & \ddots & & & \vdots \\ 0 & 0 & 0 & \cdots & -b & \searrow b & 0 \\ 0 & 0 & 0 & \cdots & 0 & -b & \searrow b \end{bmatrix} \in \mathbb{R}^{\frac{N}{2}\times(\frac{N}{2}+1)} \tag{26}$$

Matrices $\mathbb{A}_2$ and $\mathbb{A}_3$ as a pair have such a property that their multiplication (regardless of its order) results in tridiagonal matrices:

$$\mathbb{A}_2\mathbb{A}_3 = \begin{bmatrix} -2b^2 & 2b^2 & 0 & 0 & 0 & & 0 & 0 & 0 \\ b^2 & -2b^2 & b^2 & 0 & 0 & \cdots & 0 & 0 & 0 \\ 0 & b^2 & -2b^2 & b^2 & 0 & & 0 & 0 & 0 \\ 0 & 0 & b^2 & -2b^2 & b^2 & & 0 & 0 & 0 \\ & \vdots & & & \ddots & & & \vdots & \\ 0 & 0 & 0 & & b^2 & -2b^2 & b^2 & 0 & 0 \\ 0 & 0 & 0 & & 0 & b^2 & -2b^2 & b^2 & 0 \\ 0 & 0 & 0 & \cdots & 0 & 0 & b^2 & -2b^2 & b^2 \\ 0 & 0 & 0 & & 0 & 0 & 0 & 2b^2 & -2b^2 \end{bmatrix} \in \mathbb{R}^{\left(\frac{N}{2}+1\right)\times\left(\frac{N}{2}+1\right)} \tag{27}$$

$$\mathbb{A}_3\mathbb{A}_2 = \begin{bmatrix} -3b^2 & b^2 & 0 & 0 & 0 & & 0 & 0 & 0 \\ b^2 & -2b^2 & b^2 & 0 & 0 & \cdots & 0 & 0 & 0 \\ 0 & b^2 & -2b^2 & b^2 & 0 & & 0 & 0 & 0 \\ 0 & 0 & b^2 & -2b^2 & b^2 & & 0 & 0 & 0 \\ & \vdots & & & \ddots & & & \vdots & \\ 0 & 0 & 0 & & b^2 & -2b^2 & b^2 & 0 & 0 \\ 0 & 0 & 0 & & 0 & b^2 & -2b^2 & b^2 & 0 \\ 0 & 0 & 0 & \cdots & 0 & 0 & b^2 & -2b^2 & b^2 \\ 0 & 0 & 0 & & 0 & 0 & 0 & b^2 & -3b^2 \end{bmatrix} \in \mathbb{R}^{\left(\frac{N}{2}+1\right)\times\frac{N}{2}}. \tag{28}$$

Thus, the following lemma can be formulated:

Lemma 1 *Both matrices obtained from multiplications $\mathbb{A}_2\mathbb{A}_3$ and $\mathbb{A}_3\mathbb{A}_2$ are tridiagonal.*

In order to achieve a simpler form of the model, the right-hand side of (7) is represented by an auxiliary vector:

$$\mathbf{w}\left(\hat{\mathbf{x}}^{k-1}, \hat{\mathbf{x}}^{k-2}, \mathbf{u}^k, \mathbf{u}^{k-1}\right) = \mathbb{B}\hat{\mathbf{x}}^{k-2} + \mathbb{C}\left(\hat{\mathbf{x}}^{k-1}\right)\hat{\mathbf{x}}^{k-1} + \mathbb{D}\mathbf{u}^{k-1} + \mathbb{E}\mathbf{u}^k. \tag{29}$$

Hence, (7) is rewritten as:

$$\mathbb{A}\hat{\mathbf{x}}^k = \mathbf{w}\left(\hat{\mathbf{x}}^{k-1}, \hat{\mathbf{x}}^{k-2}, \mathbf{u}^k, \mathbf{u}^{k-1}\right). \tag{30}$$

Since the matrix $\mathbb{B}$ is diagonal and the matrices $\mathbb{C}$, $\mathbb{D}$ and $\mathbb{E}$ are sparse [24], computation of the vector $\mathbf{w}$ is of complexity $\mathcal{O}(N)$. Hence, one can rewrite (29) taking into account the division of the recombination matrix into four submatrices:

$$\mathbb{A}\hat{\mathbf{x}}^k = \left[\begin{array}{c|c} \mathbb{A}_1 & \mathbb{A}_2 \\ \hline \mathbb{A}_3 & \mathbb{A}_4 \end{array}\right]\hat{\mathbf{x}}^k = \mathbf{w}\left(\hat{\mathbf{x}}^{k-1}, \hat{\mathbf{x}}^{k-2}, \mathbf{u}^k, \mathbf{u}^{k-1}\right) \tag{31}$$

Since

$$\hat{\mathbf{x}}^k = \begin{bmatrix} \mathbf{q}^k \\ \mathbf{p}^k \end{bmatrix} \quad \text{and} \quad \mathbf{w} = \begin{bmatrix} \mathbf{g} \\ \mathbf{h} \end{bmatrix}$$

where $\mathbf{q}^k \in \mathbb{R}^{\frac{N}{2}+1}$ and $\mathbf{p}^k \in \mathbb{R}^{\frac{N}{2}}$ are flow rates and pressures in subsequent segments in k-th computational moment, while $\mathbf{g} \in \mathbb{R}^{\frac{N}{2}+1}$ and $\mathbf{h} \in \mathbb{R}^{\frac{N}{2}}$ are collective results of the right-hand side of (29).

By defining: $\tilde{\mathbf{g}} = \frac{\mathbf{g}}{c}$, $\tilde{\mathbb{A}}_2 = \frac{\mathbb{A}_2}{c}$, $\tilde{\mathbf{h}} = \frac{\mathbf{h}}{a}$ and $\tilde{\mathbb{A}}_3 = \frac{\mathbb{A}_3}{a}$, one eventually obtains:

$$\left(\mathbb{I} - \tilde{\mathbb{A}}_2\tilde{\mathbb{A}}_3\right)\mathbf{q}^k = \tilde{\mathbf{g}} - \tilde{\mathbb{A}}_2\tilde{\mathbf{h}} \tag{32}$$

$$\left(\mathbb{I} - \tilde{\mathbb{A}}_3\tilde{\mathbb{A}}_2\right)\mathbf{p}^k = \tilde{\mathbf{h}} - \tilde{\mathbb{A}}_3\tilde{\mathbf{g}} \tag{33}$$

Taking advantage of the previously defined Lemma 1, the results of the matrix multiplications $\tilde{\mathbb{A}}_3\tilde{\mathbb{A}}_2$ and $\tilde{\mathbb{A}}_2\tilde{\mathbb{A}}_3$ are tridiagonal, therefore the expressions $\left(\mathbb{I} - \tilde{\mathbb{A}}_3\tilde{\mathbb{A}}_2\right)$ and $\left(\mathbb{I} - \tilde{\mathbb{A}}_2\tilde{\mathbb{A}}_3\right)$ are also tridiagonal matrices. Hence, the Thomas algorithm [38] can be applied assuring the computational complexity of order $\mathcal{O}(N)$ [12]. The algorithm is defined for equations given in the following form:

$$\mathbb{T}\mathbf{v} = \mathbf{d} \tag{34}$$

where $\mathbf{v}$ and $\mathbf{d}$ are vectors, and $\mathbb{T}$ is a tridiagonal parameter matrix of the following structure:

$$\mathbb{T} = \begin{bmatrix} \beta_1 & \gamma_1 & & & \mathbf{0} \\ \alpha_2 & \beta_2 & \gamma_2 & & \\ & \alpha_3 & \ddots & \ddots & \\ & & \ddots & \ddots & \gamma_{n-1} \\ \mathbf{0} & & & \alpha_n & \beta_n \end{bmatrix}. \tag{35}$$

Efficient computation of the vector $\mathbf{v}$ is based on LU decomposition of the tridiagonal matrix $\mathbb{T} = \mathbb{L}\mathbb{U}$, giving the following lower and upper triangular matrices [12]:

$$\mathbb{L} = \begin{bmatrix} 1 & & & & \mathbf{0} \\ l_2 & 1 & & & \\ & \ddots & \ddots & & \\ & & l_{n-1} & 1 & \\ \mathbf{0} & & & l_n & 1 \end{bmatrix} \quad \mathbb{U} = \begin{bmatrix} u_1 & r_1 & & & \mathbf{0} \\ & u_2 & r_2 & & \\ & & \ddots & \ddots & \\ & & & u_{n-1} & r_{n-1} \\ \mathbf{0} & & & & u_n \end{bmatrix}. \tag{36}$$

Consequently, equation (34) can be shown as $\mathbb{L}\mathbb{U}\mathbf{v} = \mathbf{d}$. For the purpose of calculating the vector $\mathbf{v}$ for a known $\mathbf{d}$, we can solve subsequently two matrix equations: $\mathbb{L}\mathbf{y} = \mathbf{d}$ and $\mathbb{U}\mathbf{v} = \mathbf{y}$, with initial condition $u_1 = \beta_1$, $y_1 = d_1$, according to the following equations:

$$l = \frac{\alpha_i}{u_{i-1}}, \quad u_i = \beta_i - l\gamma_{i-1}, \quad y_i = d_i - l y_{i-1}, \quad \text{for } i = 1, 2, 3, ..., n \tag{37}$$

$$v_n = \frac{y_n}{u_n}, \quad v_i = \frac{y_i - \gamma_i v_{i+1}}{u_i}, \quad \text{for } i = n-1, n-2, ..., 2, 1. \tag{38}$$

It can be shown that the tridiagonal matrices $\left(\mathbb{I} - \tilde{\mathbb{A}}_3\tilde{\mathbb{A}}_2\right)$ and $\left(\mathbb{I} - \tilde{\mathbb{A}}_2\tilde{\mathbb{A}}_3\right)$ are diagonally dominant, therefore, the Thomas algorithm can successfully solve (32)–(33) for $\mathbf{q}^k$ and $\mathbf{p}^k$, respectively [12]. Then, combining both vectors into a common state vector, a single iteration of the simulation is completed. The simulated model with this state vector, obtained using the algorithm (37)–(38), is called the analytic Thomas model (ATM) [26].

5 Steady-State Solution

In practical model-based diagnostic applications of LDI, it is usually important to accurately determine the steady state flow. Therefore, in this section we will derive a simple analytical model that provides a solution to the partial differential Eqs. (1)–(2) under steady state conditions. We will analyze two cases corresponding to the zero and non-zero angle of inclination. As a starting point for both analyzes, we recall the flow process Eqs. (1)–(2), rewritten using appropriate constants and parameters, which simplifies the form of the model:

$$C_1 \frac{\partial p}{\partial t} + \frac{\partial q}{\partial z} = 0 \tag{39}$$

$$C_2 \frac{\partial q}{\partial t} + \frac{\partial p}{\partial z} = -C_3 \frac{q|q|}{p} - C_4 p \tag{40}$$

where $C_1 = \frac{S}{\nu^2}$, $C_2 = \frac{1}{S}$, $C_3 = \frac{\lambda \nu^2}{2DS^2}$, $C_4 = \frac{g \sin \alpha}{\nu^2}$

Let us assume that the LDI system works for a pipeline with a steady flow, that is, the pressure and flowrate remain unchanged in time. Then, even for small changes in the inlet or outlet pressure measured by sensors (e.g. resulting from sensor outliers or non-stationary work of the pumping system), this assumption means that the characteristics along the pipeline can be calculated according to given boundary conditions (for any computational moment, in this steady state). Therefore, we consider $\frac{\partial q}{\partial t} \to 0$ and $\frac{\partial p}{\partial t} \to 0$ as approximately satisfied for steady flow. In such cases, partial differential equations reduce to ordinary differential equations. We propose here the simplified denotations p and q, hiding their functional dependence on the space, $p(z)$ and $q(z)$, which represent the description of pressure and mass-flow for any spatial coordinate z along the length of the considered pipeline.

5.1 *Zero Inclination Angle, $\alpha = 0$*

In the case of the zero inclination angle, the system of equations describing the flow process, under the previously specified assumptions about steady state ($\frac{\partial q}{\partial t} \to 0$ and $\frac{\partial p}{\partial t} \to 0$), can be, due to $C_4 = 0$, further simplified to:

$$\frac{dq}{dz} = 0 \tag{41}$$

$$\frac{dp}{dz} = -C_3 \frac{q|q|}{p} \tag{42}$$

From the above, after some rearrangements (described in detail in [27]), the following formula for the mass flow rate (independent of z, i.e. constant along the pipe) is obtained:

$$q = \operatorname{sign}\left(p_i^2 - p_o^2\right)\sqrt{\left|\frac{DA^2}{\lambda\nu^2}\frac{p_i^2 - p_o^2}{L}\right|} \tag{43}$$

where $\operatorname{sign}(x)$ is 1 for $x \geq 0$, and -1 otherwise. The appropriate pressure distribution along the pipeline is as follows:

$$p = \sqrt{p_i^2 - \frac{p_i^2 - p_o^2}{L}z} \tag{44}$$

Note that the mass-flow rate sign depends on the sign of the difference $p_i^2 - p_o^2$ (other parameters are positive). If the inlet pressure is higher than the outlet pressure, the fluid flows from the default input to the output. If, however, the inlet pressure is lower, the fluid flows into the opposite direction (this will give you a minus sign).

5.2 *Nonzero Inclination Angle, $\alpha \neq 0$*

Now, taking into account the original set of Eqs. (39)–(40) and a non zero inclination angle, along with the aforementioned steady-state conditions, we have

$$\frac{dq}{dz} = 0 \tag{45}$$

$$\frac{dp}{dz} = -C_3\frac{q|q|}{p} - C_4 p \tag{46}$$

From (45) we have an immediate result of $q(z) = const$. The analysis of (46) leads to the following result for mass flow rate at a non zero angle of inclination:

$$q = \sqrt{\left|\frac{2DS^2}{\lambda\nu^2}\frac{g\sin\alpha}{\nu^2}\left(\frac{p_i^2 - p_o^2 e^{2\frac{g\sin\alpha}{\nu^2}L}}{e^{2\frac{g\sin\alpha}{\nu^2}L} - 1}\right)\right|}\operatorname{sign}\left(p_i^2 - p_o^2 e^{2\frac{g\sin\alpha}{\nu^2}L}\right) \tag{47}$$

and the pressure distribution along the spatial coordinate z:

$$p = \sqrt{e^{-2\frac{g\sin\alpha}{\nu^2}z}p_i^2 + \left(\frac{p_i^2 - p_o^2 e^{2\frac{g\sin\alpha}{\nu^2}L}}{e^{2\frac{g\sin\alpha}{\nu^2}L} - 1}\right)\left(e^{-2\frac{g\sin\alpha}{\nu^2}z} - 1\right)} \tag{48}$$

The two key Eqs. (47)–(48) provide the value of mass-flow rate and pressure distribution for given physical parameters of the pipe, including the non zero inclination angle. Note that in the above formula at $\alpha \to 0$ there is a singularity, in this case, however, it is recommended to use the model derived for the zero angle of inclination ($\alpha = 0$). The convergence of solutions with a non-zero and zero angle of inclination is shown in [27]. As a result, it is also worth noting that the direction of flow depends on the expression sign $\left(p_i^2 - p_o^2 e^{2\frac{g \sin\alpha}{\nu^2}L}\right)$, in which a correction element appears, which takes into account the gravitational force acting on the fluid and absent in the model for the zero angle of inclination.

The dichotomous model developed (for zero and non-zero angle of inclination) describing the steady state flow will be jointly called the Analytical Model of Steady State (AMSS).

5.3 Exemplary Applications of the Model

The derived AMSS model of the flow process through transmission pipelines, represented as a set of two equations (for operating variables q and p), provides an analytical relationship between the most important flow parameters. As shown below, this model is a powerful tool for the pipeline engineer in both the design and operation phases.

At zero angle of inclination, inlet and outlet pressures must be equalized to stop flow. In the case of a pipeline with a non zero angle of inclination, this problem is slightly more complex due to the effect of gravity acting on the medium, which, as can be easily demonstrated [27], leads to the following condition:

$$\frac{g \sin\alpha}{\nu^2} L = \ln\left(\frac{p_i}{p_o}\right) \tag{49}$$

If the above condition is met, the flow is zero ($q = 0$). For example, it can be used to determine the minimum inlet pressure required (at known outlet pressure) to achieve a downward flow ($q \geqslant 0$).

Another application of the AMSS model, in particular (47), is the determination of the resulting flow rate for the determined physical parameters of the pipeline and the given input and output pressure. A 3D plot presenting the dependency of the flowrate on the inlet and outlet pressure for an exemplary pipeline is shown in Fig. 2.

This plot expresses the most essential information about the flow resulting from the selected pair of inlet and outlet pressures. It can be helpful, for instance, when one wants to transport a specific amount of fluid over time, and has to select pumping pressures. You can indicate a curve describing the pressure combination to provide the desired flowrate. In particular, a zero flow curve can be indicated which determines the minimum inlet pressure (for the given outlet pressure) so that the flow occurs. Clearly, negative flowrates mean that the fluid flows in the opposite direction.

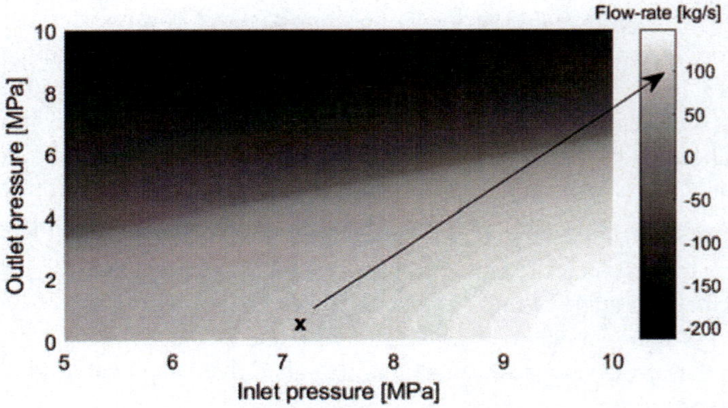

Fig. 2 Distribution of the flowrate, as dependent on the pressure at both ends of the pipeline, with the experimental setting defined by: $\nu = 304\frac{\mathrm{m}}{\mathrm{s}}$, $g = 9.81\frac{\mathrm{m}}{\mathrm{s}^2}$, $L = 75$km, $\lambda = 0.02$, $d = 0.6m$, $\alpha = 3°$

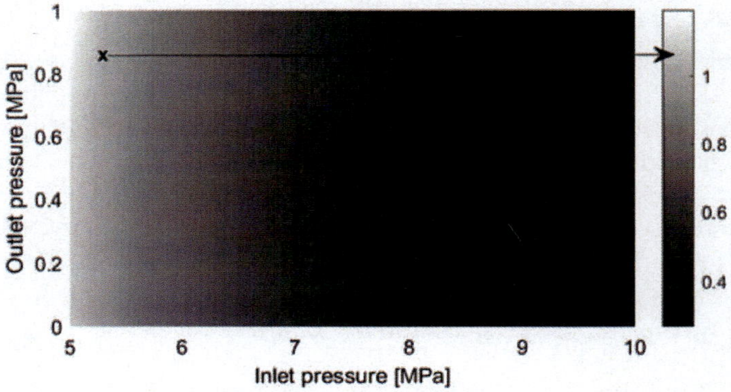

Fig. 3 Distribution of the diameter required to assure mass flowrate $q = 100\frac{\mathrm{kg}}{\mathrm{s}}$ as a function of the pressures on both ends of the pipeline, for the experimental setting defined as: $\nu = 304\frac{\mathrm{m}}{\mathrm{s}}$, $g = 9.81\frac{\mathrm{m}}{\mathrm{s}^2}$, $L = 75$km, $\lambda = 0.02$, $\alpha = 3°$

The AMSS model can also be applied to various other subtasks in pipeline design. For example, it can be used to determine the pipe diameter appropriate to the desired pipeline operation (in terms of pressure and flow velocity). A 3D plot showing an example of the relationship between diameter and inlet and outlet pressure is shown in Fig. 3. The plot precisely shows the necessary diameter D of the pipe for given pressures which assures the desired flowrate ($q = 100\frac{\mathrm{kg}}{\mathrm{s}}$): we also see that as the inlet pressure increases, the diameter required to obtain a given flow rate can be reduced (taking into account the strength of the pipe).

The above results represent only examples of relations and certainly the AMSS model can be the source of other characteristics (2D or 3D) related to the operation of transmission pipelines.

6 Numerical Stability Analysis

Numerical stability is necessary for the proper functioning of any practically implemented or simulated system. Therefore, analyzing this phenomenon in the context of synthesized mathematical models is necessary.

Process stability is directly related to its physical parameters, the selected discretization grid and the implemented discretization scheme. Usually, physical parameters remain constant during analyzed time, while the discretization method is chosen before target software and hardware implementation. However, the discretization grid is often set by the system designer and can even be changed during its operation. There are tools that support the selection of these parameters. Lax's theorem says that the numerical scheme converges, if and only if, it is stable [40]. One of the most important criteria for practice is the Courant-Friedrichs-Lewy (CFL) condition, representing the necessary condition for numerical stability of the difference equation [13]. The importance of this condition is fundamental, which can be seen in numerous references to it [2, 3, 6, 9, 17, 18, 20, 31, 39, 45, 48–50]. Seemingly, there are methods allegedly not requiring compliance with the CFL condition, but you should be aware that they may have other singularities that make it impossible to calculate the result [11, 14, 33, 37, 47].

To perform numerical stability analysis of the base model (as well as related discrete-time AMDA and ATM models), the state-space model (8) will be transformed into a form that allows the application of appropriate methods of control and systems theory.

According to [25], by defining an aggregated state vector $\tilde{\mathbf{x}}^k = \left[\hat{\mathbf{x}}^{k\,T}\ \hat{\mathbf{x}}^{k-1\,T}\right]^T$ and an augmented input vector $\tilde{\mathbf{u}}^k = \left[\mathbf{u}^{k\,T}\ \mathbf{u}^{k-1\,T}\right]^T$, the flow process model can be written as an aggregated state-space dynamic equation given in the affine/linear framework (first stage of linearization):

$$\tilde{\mathbf{x}}^k = \mathbb{A}_c \tilde{\mathbf{x}}^{k-1} + \mathbb{B}_c \tilde{\mathbf{u}}^k \tag{50}$$

where

$$\mathbb{A}_c\left(\tilde{\mathbf{x}}^{k-1}\right) = \left[\begin{array}{c|c} \mathbb{A}^{-1}\mathbb{C}\left(\tilde{\mathbf{x}}^{k-1}\right) & \mathbb{A}^{-1}\mathbb{B} \\ \hline \mathbb{I} & \mathbf{0} \end{array}\right] \tag{51}$$

$$\mathbb{B}_c = \left[\begin{array}{c|c} \mathbb{A}^{-1}\mathbb{E} & \mathbb{A}^{-1}\mathbb{D} \\ \hline \mathbf{0} & \mathbf{0} \end{array}\right] \tag{52}$$

The matrix $\mathbb{A}_c$ is a function of the state vector $\tilde{\mathbf{x}}^{k-1}$, while the matrix $\mathbb{B}_c$ depends on the friction factor λ. During a simulation it is necessary to iteratively calculate the value $\mathbb{A}_c$, which means that the calculations must be non-linear and non-stationary. Nevertheless, (50) has a well-known form of (non-linear) dynamic state-space equations that can be analyzed using convenient theoretical and systemic tools.

To guarantee the stability of the numerical model, it is recommended to make the rate of evaluation of this phenomenon bigger than the analyzed phenomenon itself (which also coincides with the famous sampling theorem) [13]. In consequence (and due to the method of characteristics), the discretization grid $\frac{\Delta z}{\Delta t}$ must be greater or equal to the information exchange rate in the corresponding differential equations for maintaining stability [7]. However, this is only a necessary condition, which may be insufficient in general, because it gives only the lower limit of the expression $\frac{\Delta z}{\Delta t}$.

To enable a stable online simulation, the discretization grid must be properly selected, and because the highest speed present in the flow process is the surrogate sound speed ν, the following (CFL) equality must be met:

$$\Delta t = \mu \frac{\Delta z}{\nu} \tag{53}$$

where μ is a coefficient within the range $(0, 1\rangle$, binding the discretization steps, also referred to as the Courant number.

So, we can now define the mesh parameter (m_h) of the discretization grid as

$$m_h = \frac{\Delta z}{\Delta t} = \frac{\nu}{\mu} \tag{54}$$

Study [24] shows that for a given set of physical parameters, there is a relationship between time and spatial steps that assures the maximum stability margin m. For the considered discrete-time models, the margin is measured as $m = 1 - s_{max}$, where s_{max} is the largest module of the eigenvalues of the state transition matrix $\mathbb{A}_c$.

Simulations were conducted for a number of pipeline diameter values. The results are shown in Fig. 4, where the stability margin is presented as a function of the coefficient μ. By analyzing the graph, it can be seen that for different sets of physical parameter values there is a coefficient μ that clearly provides the maximum stability margin of the numeric scheme.

6.1 Steady-State Analysis and a Linear Model

The results shown in Fig. 4 were collected after 20,000 iterations in each simulation (to obtain a steady state), which required a lot of computational effort. Moreover, it is obvious that the transient processes at the beginning of the simulation are useless from the point of view of the monitoring or diagnostic application. Furthermore, in the initial phase, such processes may require different values of the discretization grid to maintain stability.

In addition, due to the nonlinearity of the model (50), stability during this phase may be temporary (not persistent). Therefore, the model linearization around the steady state (43)–(44) or (47)–(48) was proposed [27], for zero and non-zero incli-

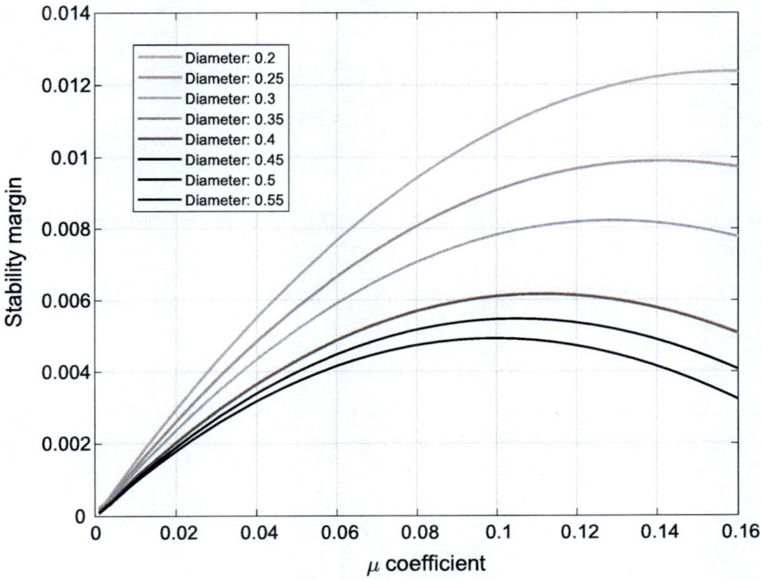

Fig. 4 Stability margin as a function of the μ coefficient for different diameters of a pipeline. The stability margin was measured after 20,000 iterations of the simulation algorithm. Other physical parameters of the pipeline were: N=10, $L = 4$km, λ=0.01, $\nu = 304\frac{\text{m}}{\text{s}}$, $p_i = 3.2$MPa, $p_o = 3$MPa, $\alpha = 0°$ [24]

nation angles respectively. The linearized state transition matrix obtained in this way is calculated as

$$\bar{\mathbb{A}}_c = \mathbb{A}_c \left(\tilde{\mathbf{x}}^{k-1}\right)\Big|_{\tilde{\mathbf{x}}^{k-1}=[\mathbf{q}^{k-1}\mathbf{p}^{k-1}\mathbf{q}^{k-2}\mathbf{p}^{k-2}]} \tag{55}$$

Putting this in (50), we get the following linearized dynamic state equation:

$$\bar{\mathbf{x}}^k = \bar{\mathbb{A}}_c\bar{\mathbf{x}}^{k-1} + \mathbb{B}_c\tilde{\mathbf{u}}^k \tag{56}$$

This approach is a huge methodological simplification of the modeling and computation problem discussed here. It is important that the above steady state of the flow process has been calculated analytically, based on the original partial differential equations of the flow process. Although, in general, numerical (discrete-time) models, equivalent to computer simulations, do not necessarily coincide with analytical solutions, our study [27] shows that the steady-state vector determined for the numerical simulation model converges to the analytical result obtained for the full (continuous in time and space) model of the flow process (the higher the accuracy and complexity of the discrete model, the better). Before linearization, it is beneficial to estimate the value of the friction factor (within a given experimental setting),

Table 1 Range of parameters used for the simulation

Parameter	Min	Max
Length L [m]	1	10^6
Diameter D [m]	0.001	2
Friction factor λ [–]	0.001	0.4
Number of segments N [–]	4	100
Inlet pressure p_i [MPa]	0.1	10
Outlet pressure p_o [MPa]	0.001	Inlet pressure
Surrogate sound velocity of the fluid ν [$\frac{\text{m}}{\text{s}}$]	250	2000

because it allows you to justify the model and its rational use in an unchanged form when operating the pipeline near this work point.

6.2 *Maximum Stability Margin*

To find the optimal discretization grid m_h (or $\frac{\Delta z}{\Delta t}$), the above linearized aggregated model (56) was simulated with a random selection of physical (mechanical) flow parameter values (shown in Table 1), and for each set of the parameters the Courant number μ was optimized to determine μ_{opt}, for which the stability margin is the highest.

To obtain a synthetic description of the utility coefficient μ_{opt}, an optimization algorithm was applied to find such a curve on the $\Delta t - \Delta z$ plane, which provides the maximum margin of stability computed as:

$$s_{max} = 1 - e_{max} \tag{57}$$

where e_{max} is the maximal absolute eigenvalue of the matrix $\mathbb{A}_c$.

For three sample lengths of the pipeline, a graph of the stability margin has been plotted as a function of the Courant number: with the case of a short pipe (100 km) shown in Fig. 5, intermediate (1,000 km)—in Fig. 6, and long (5,000 km)—in Fig. 7.

From the graphs presented, it can be concluded that the CFL condition is sufficient only for one of the pipeline cases (i.e. for intermediate pipes). In other cases, a more restrictive criterion is required.

After a detailed analysis of a wider range of lengths (and other parameters), we can distinguish three types of pipes:

- Short
- Intermediate
- Long.

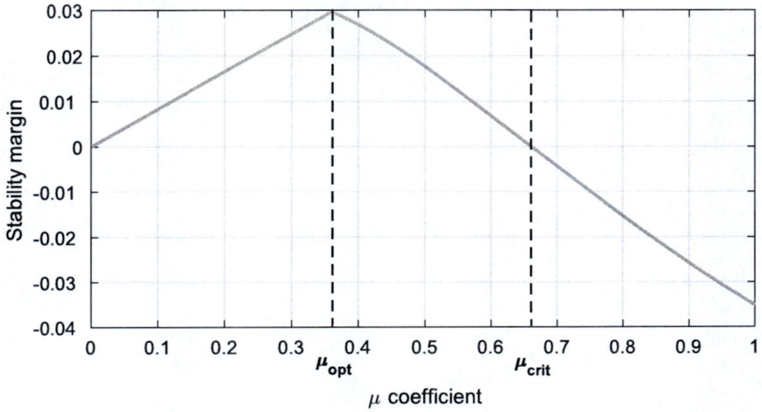

Fig. 5 Dependency of stability margin on μ coefficient for a short pipe ($L = 100$Km). Fixed parameters of the simulation: $v = 1472\frac{\mathrm{m}}{\mathrm{s}}$, $D = 0.52$m, $\lambda = 0.0029$; N = 16; $p_i = 9.4$MPa, $p_o = 7.3$MPa

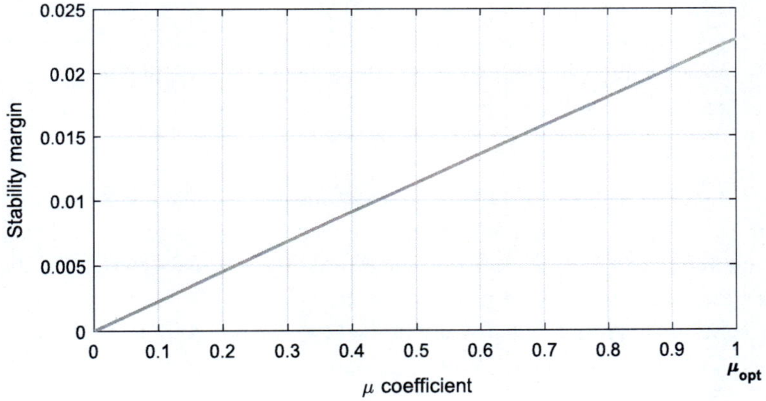

Fig. 6 Dependency of stability margin on μ coefficient for an intermediate pipe ($L = 1000$km). Fixed parameters of the simulation: $v = 1472\frac{\mathrm{m}}{\mathrm{s}}$, $D = 0.52$m, $\lambda = 0.0029$; N = 16; $p_i = 9.4$MPa, $p_o = 7.3$MPa

The research shows that for intermediate pipes the optimal value of the μ coefficient is 1. In the case of short and long pipes, however, its value can be expressed by an approximate formula related to numerical issues, pipe geometry and pressure control. The dual nature of the coefficient μ is revealed here. Basically, this factor is used for parameterization stabilizing the discrete-time model. It now turns out that it is a precise tool in feasibility studies for additional optimization (improvement) of model stability and adaptation to specific, extreme conditions (e.g. short or long

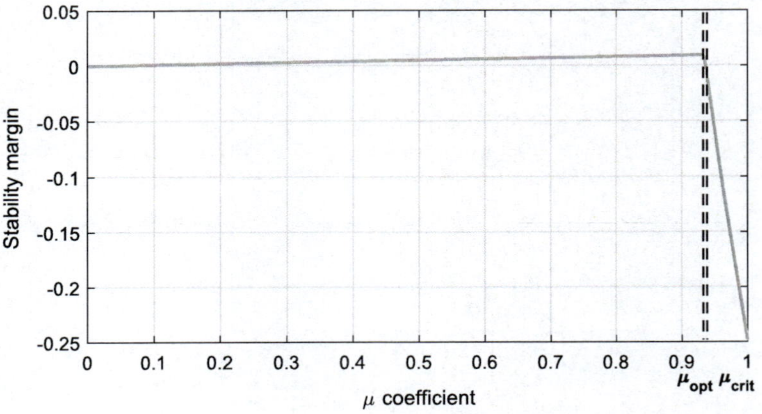

Fig. 7 Dependency of stability margin on μ coefficient for a long pipe (L = 5000km). Fixed parameters of the simulation: $\nu = 1472\frac{\text{m}}{\text{s}}$, $D = 0.52$m, $\lambda = 0.0029$; N = 16; $p_i = 9.4$MPa, $p_o = 7.3$ MPa

pipes). Our research shows that we can express the optimal Courant number for short and long pipes using a common formula, which due to its classification character we will call the pipe factor Π:

$$\Pi = {\Pi_N}^{-i}\vartheta_p \tag{58}$$

with $i = 1$ for long pipes, and $i = -1$ for short pipes, and the coefficients Π_N and ϑ_p defined underneath. The value of the pipe factor, for cases when it is lower than 1, can be associated with μ_{opt}, which means that the optimal Courant number can be calculated for a given set of pipe parameters.

By defining the physical coefficient of the pipe, capturing the known pipe geometry and generalized friction coefficient:

$$\Pi_f = \sqrt{\frac{L\lambda}{D}} \tag{59}$$

the above-shown numerical factor Π_N of a pipe can be described as

$$\Pi_N = \frac{1}{N}\Pi_f \tag{60}$$

The value of μ_{opt} for the varying numerical factor (60) of the pipeline (for fixed inlet and outlet pressures) is presented in Fig. 8, where regions for short, intermediate and long pipes are indicated.

Studying the course of the plot μ_{opt} in Fig. 8 relative to the numerical pipe factor, we see two regions regarding short and long pipes where the CFL condition is insufficient. In addition, two lines with different slopes can be observed in the region of short pipes. For safety reasons, preferring lower values, we will apply a stronger

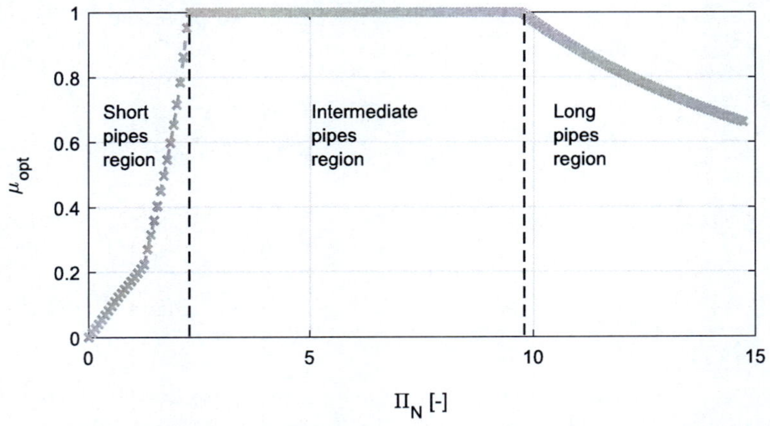

Fig. 8 Exemplary distribution of μ_{opt} in terms of the numerical pipeline factor Π^N controlled by changing the length of the pipe L (with the fixed parameters of the simulation: $v = 1472\,\frac{\mathrm{m}}{\mathrm{s}}$, $D = 0.52$m, λ = 0.0029; N = 16; $p_i = 9.4$MPa, $p_o = 7.3$MPa

restriction on the coefficient μ, which is also proportional to the numerical factor Π_N. Interestingly, and as you can see, for long pipes, the value of μ_{opt} is inversely proportional to the numerical pipe factor.

Unfortunately, the study revealed that the numerical pipe factor is not sufficient to distinguish between short, intermediate and long pipes. Therefore, we have also defined and used in (58) the pressure corrector ϑ_p for long pipes as

$$\vartheta_p^L = 8\sqrt{\frac{p_o^2}{p_i^2 - p_o^2}} \tag{61}$$

and for short pipes as

$$\vartheta_p^S = 0.5\sqrt{\frac{p_i - p_o}{p_i + p_o}} \tag{62}$$

which introduces into the model the effects caused by the pressure difference and the resultant flow.

The report [28] distinguishes between the following four cases:

- Case (i)—The pipeline is long
- Case (ii)—The pipeline is short
- Case (iii)—The pipeline is intermediate, i.e. is neither long nor short
- Case (iv)—The pipeline can be describe as both long and short at the same time.

Cases (i) and (ii) mean that the appropriate integrated pipe factor Π should be used to determine the optimal μ. Therefore, one should use Π^L for long pipes in case (i):

$$\mu_{opt} = \Pi^L = \frac{1}{\Pi_N}\vartheta_p^L = 8N\sqrt{\frac{D}{L\lambda}\frac{p_o^2}{p_i^2 - p_o^2}} \tag{63}$$

and Π^S for short pipelines in case (ii):

$$\mu_{opt} = \Pi^S = \Pi_N\vartheta_p^S = 0.5\frac{1}{N}\sqrt{\frac{L\lambda}{D}\frac{p_i - p_o}{p_i + p_o}} \tag{64}$$

Due to the great practical importance, the integrated pipe factor Π directly related to cases (i) and (ii) will also be individually identified as Π^L and Π^S, for long and short pipes, respectively.

The fundamental, and consistent with the generally recognized theory, is intermediate length, case (iii), in which ('medium') pipelines represent a well-known general arrangement. In this case, the CFL condition is sufficient, and it is recommended to use the Courant number $\mu_{opt} = 1$.

Case (iv) requires more comment. The pipeline in this case can be classified as both short and long. The results in [28] show that there is no intermediate region for a large pressure difference ($p_o \ll p_i$). Exactly for the pressure ratio

$$\frac{p_o}{p_i} < \frac{1}{3} \tag{65}$$

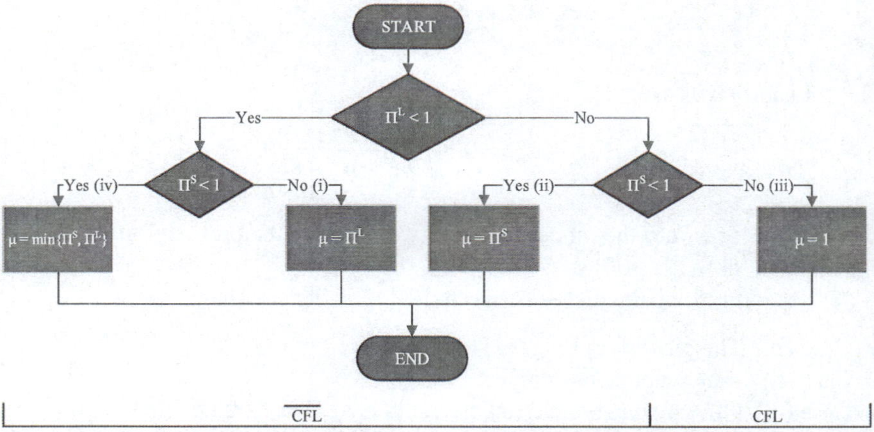

Fig. 9 Procedure for selecting the optimal values of the Courant number. Part denoted as CFL refers to the case, when the CFL condition is sufficient, and $\overline{CFL}$ denotes the cases when the CFL condition is not sufficient

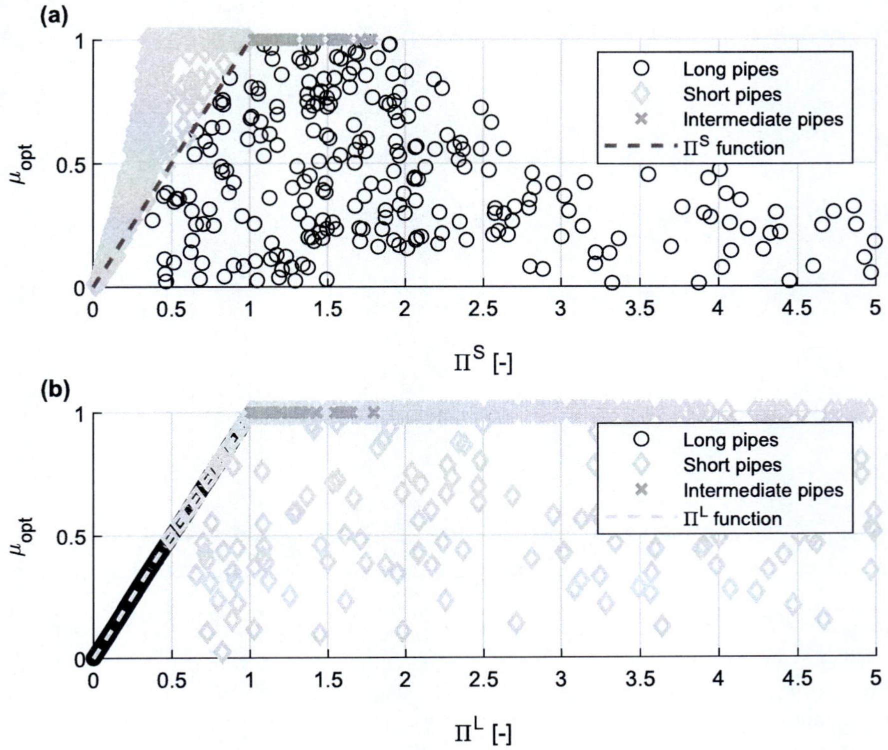

Fig. 10 Result of points classification in terms of functions Π^S (a) and Π^L (b), where the points are classified as long (○), short (◊) or intermediate (×). Note that there are no cases (iv) representing both short and long pipes, because they are (re)classified into the more secure version (short or long)

there is no intermediate (general) region, and the pipe—for specified sets of parameters—is classified as both short and long. In such a case (iv) the optimal Courant number can be safely determined as

$$\mu_{opt} = min\{\Pi^L, \Pi^S\} \tag{66}$$

To select a specific value μ, the procedure presented in Fig. 9 is proposed. First, check whether the considered pipeline can be classified as a long pipe. If so, check that the pipe also falls into the area of short pipes. If so, then you should specify μ as $min\{\Pi^L, \Pi^S\}$, because this is case (iv) and there is no intermediate region. If the pipe is classified as long, but not as a short pipe, it is case (i) when the pipe is long and $\mu = \Pi^L$. A similar situation occurs when the pipe can be classified as short, but not as a long pipe, then we have case (ii) and $\mu = \Pi^S$. If the pipe under consideration can not be classified as long or short, which means general properties (the intermediate region), we have case (iii) when the condition CFL is sufficient ($\mu = 1$). This way of determining the Courant number maximizes the chance that

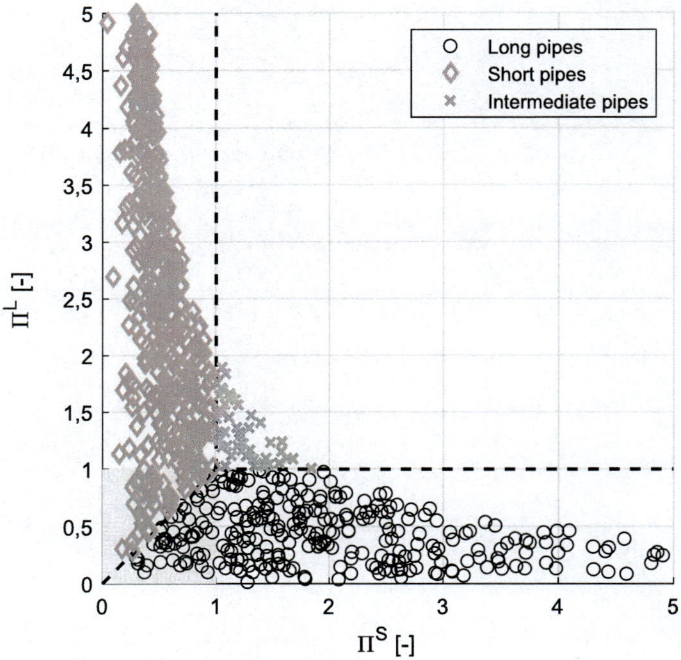

Fig. 11 Results of the pipelines classification on the plane of pipe factors Π^S–Π^L. The cases (iv) representing short-long pipes, placed in the gray square, and were reclassified using the minimum function

the system will be stable, even if the maximum margin of stability is not reached (especially in case (iv), L/S, covering both long and short pipelines).

To verify the correctness of the results, 1000 simulations were carried out with random parameters ((their ranges are listed in Table 1)) [28]. Using the appropriate optimization procedure, the optimal Courant number was determined for each case and compared with the results of the design procedure shown in Fig. 9. The summary of this research is the distribution of the optimal stability margin relative to the indexes Π^S and Π^L, together with the classification results, shown in Fig. 10. In Fig. 11 the classification results are presented in the space of the indicators Π^L and Π^S with the shaded area indicating the case (iv).

One can see in Fig. 10 that most of the points classified as short pipes, are over the Π^S line, and the ones classified as long pipes are exactly on the Π^L line.

It is noteworthy that case (iv) was reclassified as a pipeline either long (i) or short (ii)—depending on which indicator (Π^L or Π^S) had a lower value. This can be seen after analyzing the results in the gray area (iv) in Fig. 11, where due to the minimizing operator applied, the pipelines were assigned the lower coordinate (specifying its 'length type'). Therefore, in the area (iv), the points representing long pipes are under the diagonal, and short ones are above the diagonal.

To verify the obtained results, we conducted simulations of the flow process based on the values of μ calculated in accordance with the procedure presented in Fig. 9 for the values of physical flow parameters selected at random from the ranges defined in Table 1. These experiments showed the convergence of the results of all simulations, and thus the correctness of the proposed method.

7 Comparison of the Models

There are two main criteria for comparing the analyzed models: computation time and accuracy of results, which is a measure of the difference between steady-state vectors. To reduce the computational overhead, the boundary conditions for each model were taken from the AMSS model.

7.1 Validity of the Models

The first indicator of the appropriateness of the model is its reference to the base model which has been verified in [5]. The following ratio of norms was used as a relative criterion for measuring errors during this comparative study:

$$\frac{||\mathbf{x}_b - \mathbf{x}_m||_2}{||\mathbf{x}_b||_2} \tag{67}$$

where $\mathbf{x}_b$ and $\mathbf{x}_m$ are, respectively, the state vector of the base and compared model (steady-state vectors were used in the calculations). The results of this analysis are presented in Fig. 12.

We see that the difference between the base model and the Thomas model remains at the level of numerical noise. We can therefore conclude that both results are consistent. It should also be noted that as the models become more complex (the cardinality increases), the states of both models are asympotically convergent to the state of the AMSS model. We can therefore imply that the AMSS model reflects the real flow and pressure distribution in the steady state with the highest accuracy (assuming perfect measurement data).

Additional analysis and validation of the proposed model-system solutions were obtained by testing the index (67) against the variable value of the Courant coefficient μ, as shown in Fig. 13.

As you can see in this graph, with the decreasing Courant number μ, the AMDA model gives estimates that coincide with the state estimates of the base model. Comparing this with the error graph (67) of Fig. 12 as a function of the cardinality N, it can be concluded that the AMDA model is well suited for discretized systems with

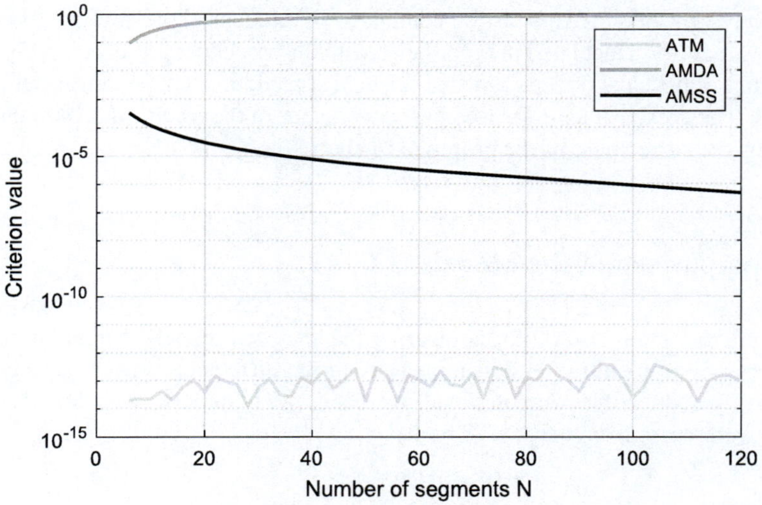

Fig. 12 Comparative criterion of the analyzed models with a variable number of segments, and the experimental setting defined as: $\mu = 0.1$, $L = 75$km, $D = 0.4$m, $\lambda = 0.02$, $v = 300\,\frac{\text{m}}{\text{s}}$, $p_i = 70$ MPa, $p_o = 60$ MPa

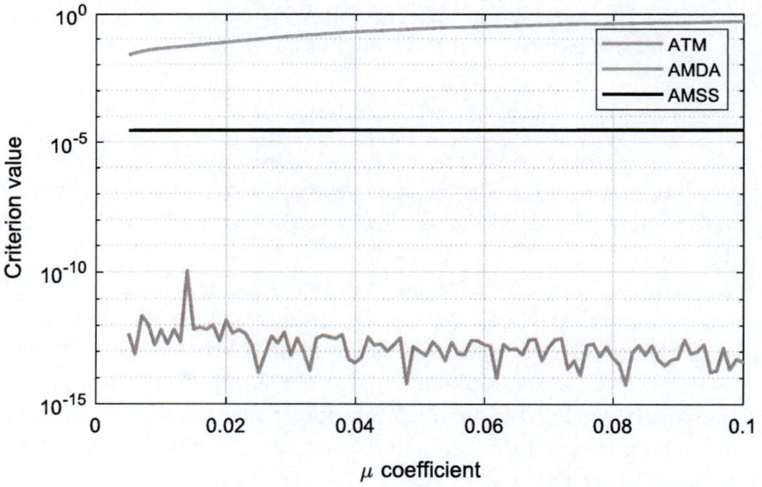

Fig. 13 Comparative criterion for the models analyzed in terms of Courant number, with the experimental setting defined as: $N = 20$, $L = 75$km, $D = 0.4$ rm m, $\lambda = 0.02$, $v = 300\,\frac{\text{m}}{\text{s}}$, $p_i = 70$ MPa, $p_o = 60$ MPa

a small number of segments and a small value of μ. In addition to the AMSS model, which is naturally insensitive to N, the Thomas model (relative to the basic model) also appears to be favorably robust to the Courant number μ (regulating system stability).

7.2 Computation Time

In this part of the comparative study, two aspects were considered: the time required to reach steady state and the average time of a single iteration. It should be noted that all matrices were stored as full matrices (further research into these methods considering sparce matrices is necessary). The times required to reach the steady state by the considered models are shown in Fig. 14.

Because models may require different numbers of iterations to achieve steady states, single iteration times were also analyzed. The results of their comparison are presented in Fig. 15. For the AMSS model, of course, this is only the time required to calculate the flow and pressure once.

The steady state vector is obtained in the fastest way using the one-cycle AMSS model, which results from its specificity (and adaptation to this task). The second fastest is the ATM model. The base and AMDA models require the longest calculations to achieve steady state. However, it should be noted that the AMDA model reaches a slightly different steady state than other models (which can be seen in Fig. 12). That is why properly selected initial conditions, which accelerate calculations for other models, do not give such a good result in the AMDA model. Regarding the time consumption of a single iteration, the results summary shows the computational acceleration of the AMDA model compared to the basic model. In general, however, given the time context, in a wide range of N segments, the Thomas model is the most efficient.

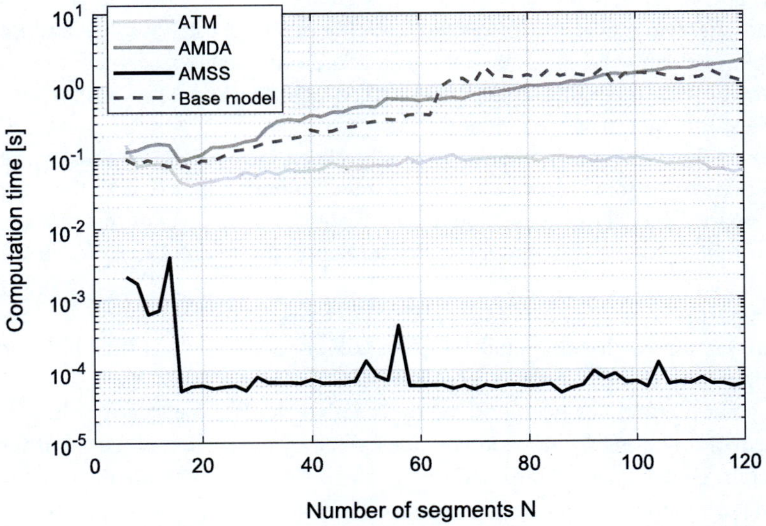

Fig. 14 Time required to reach a steady state by the considered models, with the experimental setting defined by: $\mu = 0.1$, $L = 75\text{km}$, $D = 0.4$ rm m, $\lambda = 0.02$, $\nu = 300\ \frac{\text{m}}{\text{s}}$, $p_i = 70$ MPa, $p_o = 60$ MPa

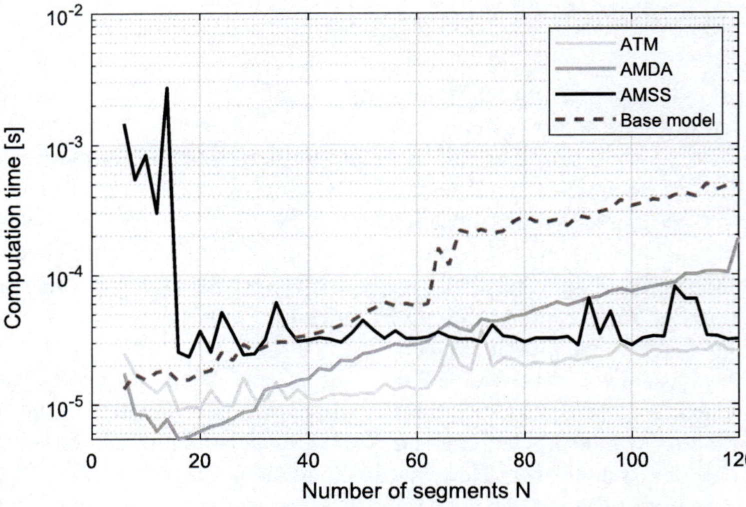

Fig. 15 Time required to calculate a single iteration by the models under consideration, with the experimental setting defined by: $\mu = 0.1$, $L = 75\text{km}$, $D = 0.4$ rm m, $\lambda = 0.02$, $\nu = 300\ \frac{\text{m}}{\text{s}}$, $p_i = 70$ MPa, $p_o = 60$ MPa

8 Conclusions

This chapter discusses the problem of synthesizing an accurate and, at the same time, efficient discrete model of fluid flow through a pipeline. Such models are used, for example, in diagnostic systems. First, a standard (universally accepted and verified) base state-space model has been presented, which then served as a reference point when testing other proposed models.

Next, a diagonal approximation model has been introduced, obtained by approximating the tridiagonal submatrices of the recombination matrix with their diagonal counterparts. This allowed us to obtain an explicit inverse form of this matrix and to simplify the dynamic flow model and ensure appropriate conditions for its application.

The next section describes a model that, by bringing the base model into a system of two equations with tridiagonal matrices, falls under the applicability of the Thomas algorithm with computational complexity $\mathcal{O}(N)$.

Another method has been utilized to directly calculate the steady solution to the problem of modeling the flow process, described by the system of partial differential equations. Under steady-state conditions, the flow rate and pressure distribution have been determined for both a zero and nonzero pipeline inclination angle. Examples of non-diagnostic design applications of the developed model have also been presented.

The next section examines the issue of numerical stability of dynamic discrete-time models (AMDA, ATM and base models). It turned out to be possible to analytically determine the stability margin based on the analysis of the transition matrix of the linearized (standard) base model in steady state. It has been shown that for each set of physical flow parameters (pipeline geometry, fluid parameters, pressures used) there is a point where the stability margin is maximal. Analyzing the location of these points in the considered space of physical parameters, a multidimensional explicit function was formulated showing this relationship. As a result, it was possible to easily determine such a discretization grid that provides the maximum margin of stability, including practical conditions commonly described as the Courant-Friedrichs-Lewy empirical criterion.

Studies have shown that a distinction is needed between short and long pipelines for which the CFL criterion is insufficient, and intermediate pipelines (with a universally recognizable property) for which this criterion is actually a sufficient condition. Consequently, the procedure for determining the optimal Courant number for individual types of pipelines has been presented.

To confirm the validity of the results obtained, comprehensive simulations were carried out for various types of pipelines using the coefficient μ_{opt}, computed with the aid of the proposed design procedure from Fig. 9. In this way, we have demonstrated the stability, convergence and compatibility of simulated flow processes, and thus the correctness of the analysis performed.

Finally, the developed models were compared with the basic model, taking into account the time needed to perform the calculations and the quality of the state vector generated by them.

Analyzing the steady-state calculation time (simulation time), it can be seen that all three models presented require less computation time than the basic model. It is also worth noting that due to the explicit (non-iterative) nature, the AMSS model speeds up calculations up to 1000 times compared to the base model (for 120 segments).

The ATM and base models have also been shown to match, and as the number of segments increases, they strive for 'real' steady-state values, explicitly indicated by the AMSS model. The economical diagonal approximation model AMDA gives better results, a smaller number of segments and a smaller Courant number μ.

The developed models of fluid flow processes enable the planning of further theoretical research on them, as well as in practical implementation and validation of leakage diagnostics procedures in transport pipelines.

References

1. API RP 1130: Computational Pipeline Monitoring for Liquid Pipelines. Recommended Practice from the American Petroleum Institute (2007)
2. Banda, M.K., Seaid, M.: Higher-order relaxation schemes for hyperbolic systems of conservation laws. J. Numer. Math. **13**(3), 171–196 (2005). https://doi.org/10.1515/156939505774286102
3. Bauer, A.L., Loubére, R., Wendroff, B.: On stability of staggered schemes. SIAM J. Numer. Anal. **46**(2), 996–1011 (2008)

4. Besançon, G., Georges, D., Begovich, O., Verde, C., Aldana, C.: Direct observer design for leak detection and estimation in pipelines, In: 2007 European Control Conference (ECC), pp. 5666–5670 (2007)
5. Billmann, L., Isermann, R.: Leak detection methods for pipelines. Automatica **23**(3), 381–385 (1987)
6. Bonzanini, A., Picchi, D., Poesio, P.: Simplified 1D incompressible two-fluid model with artificial diffusion for slug flow capturing in horizontal and nearly horizontal pipes. Energies **10**, 1372 (2017)
7. Bridson, R.: Fluid Simulation for Computer Graphics, 2nd edn. CRC Press (2015)
8. Brogan, W.: Modern Control Theory, 3rd edn. Prentice Hall (1991)
9. Capuano, F., Mastellone, A., Angelis, E.D.: A conservative overlap method for multi-block parallelization of compact finite-volume schemes. Comput. Fluids **159**, 327–337 (2017)
10. Chatzigeorgiou, D., Youcef-Toumi, K., Ben-Mansour, R.: Design of a Novel In-Pipe Reliable Leak Detector. IEEE/ASME Trans. Mechatron. **20**, 824–833 (2015). https://doi.org/10.1109/TMECH.2014.2308145
11. Chen, Z., Zhang, J.: An unconditionally stable 3-D ADI-MRTD method free of the CFL stability condition. IEEE Microwave Wireless Comp. Lett. **11**(8), 349–351 (2001)
12. Conte, S.D., de Boor, C.: Elementary Numerical Analysis: An Algorithmic Approach, 3rd edn. McGraw-Hill (1980)
13. Courant, R., Friedrichs, K., Lewy, H.: On the partial difference equations of mathematical physics. IBM J. Res. Dev. **11**(2), 215–234 (1967)
14. Czernous, W.: Numerical method of characteristics for semilinear partial functional differential systems. J. Num. Math. **16**(1), 1–21 (2008). https://doi.org/10.1515/jnum.2008.001
15. Delgado-Aguiñaga, J.A., Besançon, G., Begovich, O., Carvajal, J.E.: Multi-leak diagnosis in pipelines based on extended Kalman filter. Control Eng. Prac. **49**, 139–148 (2016). https://doi.org/10.1016/j.conengprac.2015.10.008
16. Demirci, S., Yigit, E., Eskidemir, I.H., Ozdemir, C.: Ground penetrating radar imaging of water leaks from buried pipes based on back-projection method. NDT E-Int. **47**, 35–42 (2012). https://doi.org/10.1016/j.ndteint.2011.12.008
17. Duquette, J., Rowe, A., Wild, P.: Thermal performance of a steady state physical pipe model for simulating district heating grids with variable flow. Appl. Energy **178**, 383–393 (2016)
18. Kamga, J.A., Desprs, B.: CFL condition and boundary conditions for DGM approximation of convection diffusion. SIAM J. Num. Anal. **44**(6), 2245–2269 (2006). https://doi.org/10.1137/050633159
19. Gunawickrama, K.: Leak detection methods for transmission pipelines. Ph.D. Thesis, supervised by Z. Kowalczuk. Gdansk University of Technology, Gdańsk (2001)
20. Kornhaas, M., Schäfer, M., Sternel, D.C.: Efficient numerical simulation of aeroacoustics for low mach number flows interacting with structures. Comput. Mech. **55**(6), 1143–1154 (2015). https://doi.org/10.1007/s00466-014-1114-1
21. Kowalczuk, Z., Gunawickrama, K.: Model-based cross-correlation method for leak detection in pipelines. Pomiary Automatyka Kontrola **4**, 140–146 (1998)
22. Kowalczuk, Z., Gunawickrama, K.: Detection and localization of leaks in transmission pipelines. In: Korbicz, J., Kościelny, J.M., Kowalczuk, Z., Cholewa, W. (eds.) Fault Diagnosis. Models, Artificial Intelligence, Applications, pp. 821–864. Springer, Berlin, Heidelberg, New York (2004)
23. Kowalczuk, Z., Tatara, M.: Analytical modeling of flow processes: Analysis of computability of a state-space model. In: XI International Conference on Diagnostics of Processes and Systems, pp. 74.1–12. Łagów Lubuski (2013)
24. Kowalczuk, Z., Tatara, M.: Approximate models and parameter analysis of the flow process in transmission pipelines. In: Kowalczuk, Z. (ed.) Advanced and Intelligent Computations in Diagnosis and Control, vol. 386, pp. 209–220. Springer IP, Switzerland (2016). https://doi.org/10.1007/978-3-319-23180-8_17
25. Kowalczuk, Z., Tatara, M.: Numerical issues and approximated models for the diagnosis of transmission pipelines. In: Advances in the Diagnosis of Faults in Pipeline Networks, pp. 1–24. Springer, Berlin, Heidelberg (2017)

26. Kowalczuk, Z., Tatara, M., Stefański, T.: Reduction of computational complexity in simulations of the flow process in transmission pipelines. In: Kościelny, J.M., Syfert, M., Sztyber, A. (eds.) Advanced Solutions in Diagnostics and Fault Tolerant Control, pp. 241–252. Springer, Cham (2018)
27. Kowalczuk, Z. Tatara, M.: Analytical steady-state model of the pipeline flow process. In 23rd International Conference on Methods and Models in Automation and Robotics, pp. 182–187. Miedzyzdroje, Poland, IEEE Xplore 2018; ISSN: 978-1-5386-4325-9/18 (2018). https://doi.org/10.1109/MMAR.2018.8486066
28. Kowalczuk, Z., Tatara, M.: Analytical 'steady-state'-based derivation and clarification of the Courant-Friedrichs-Lewy condition for pipe flow (internal report prepared for publication) (2019)
29. Li, S., Wen, Y., Li, P., Yang, J., Dong, X., Mu, Y.: Leak location in gas pipelines using cross-time-frequency spectrum of leakage-induced acoustic vibrations. J. Sound Vibration **333**(17), 3889–3903 (2014). https://doi.org/10.1016/j.jsv.2014.04.018
30. Li, S., Zhang, J., Yan, D., Wang, P., Huang, Q., Zhao, X., Cheng, Y., Zhou, Q., Xiang, N., Dong, T.: Leak detection and location in gas pipelines by extraction of cross spectrum of single non-dispersive guided wave modes. J. Loss Prevent. Process Ind. **44**, 255–262 (2016). https://doi.org/10.1016/j.jlp.2016.09.021
31. Lyra, P.R.M., Morgan, K.: A review and comparative study of upwind biased schemes for compressible flow computation. Part I: 1—D first—order schemes. Arch. Comput. Methods Eng. **7**(1), 19–55 (2000). https://doi.org/10.1007/BF02736185
32. Mandal, S.K., Chan, F.T.S., Tiwari, M.K.: Leak detection of pipeline: An integrated approach of rough set theory and artificial bee colony trained SVM. Exp. Syst. Appl. **39**(3), 3071–3080 (2012). https://doi.org/10.1016/j.eswa.2011.08.170
33. Namiki, T., Ito, K.: A new FDTD algorithm free from the CFL condition restraint for a 2D-TE wave. In: IEEE Antennas and Propagation Society International Symposium. 1999 Digest. Held in conjunction with: USNC/URSI National Radio Science Meeting (Cat. No.99CH37010), vol. 1, pp. 192–195 (1999)
34. Ostapkowicz, P., Bratek, A.: Leak detection in liquid transmission pipelines during transient state related to a change of operating point. In: Kowalczuk, Z. (ed.) Advanced and Intelligent Computations in Diagnosis and Control, vol. 386, pp. 253–265. Springer IP, Switzerland (2016). https://doi.org/10.1007/978-3-319-23180-8
35. Sandberg, C., Holmes, J., Mccoy, K., Koppitsch, H.: The application of a continuous leak detection system to pipelines and associated equipment. IEEE Trans. Ind. Appl. **25**(5), 906–909 (1989)
36. Al-Shidhani, I., Beck, S., Staszewski, W.: Leak monitoring in pipeline networks uing wavelet analysis. Key Eng. Mater. **245–246**, 51–58 (2003)
37. Stefański, T., Drysdale, T.D.: Parallel adi-bor-fdtd algorithm. IEEE Microwave Wireless Comp. Lett. **18**(11), 722–724 (2008)
38. Thomas, L.H.: Elliptic Problems in Linear Difference Equations Over a Network. Watson Sci. Comput. Lab. Rept., Columbia University, New York (1949)
39. Thanh, M.D.: Building fast well-balanced two-stage numerical schemes for a model of two-phase flows. Commun. Nonlinear Sci. Num. Simul. **19**(6), 1836–1858 (2014). https://doi.org/10.1016/j.cnsns.2013.10.017
40. Thomas, J.W.: Numerical Partial Differential Equations: Finite Difference Methods, Texts in Applied Mathematics, vol. 22. Springer, New York, NY (1995)
41. Verde, C., Visairo, N., Gentil, S.: Two leaks isolation in a pipeline by transient response. Adv. Water Resourc. **30**, 1711–1721 (2007)
42. Verde, C., Molina, L., Torres, L.: Parameterized transient model of a pipeline for multiple leaks location. J. Loss Preven. Process Ind. **29**, 177–185 (2014)
43. Verde, C., Torres, L.: Referenced model based observers for leaks' location in a branched pipeline In: The 9th International Federation of Automatic Control (IFAC) Symposium SAFEPROCESS-2015. Paris, France (2015)

44. Waleed, D., Mustafa, S.H., Mukhopadhyay, S., Abdel-Hafez, M.F., Jaradat, M.A.K., Dias, K.R., Arif, F., Ahmed, J.I.: An in-pipe leak detection robot with a neural-network-based leak verification system. IEEE Sens. J. **19**(3), 1153–1165 (2019)
45. Wang, P., Ho, M.T., Wu, L., Guo, Z., Zhang, Y.: A comparative study of discrete velocity methods for low-speed rarefied gas flows. Comput. Fluids **161**, 33–46 (2018). https://doi.org/10.1016/j.compfluid.2017.11.006
46. Wang, X.J., Lambert, M.F., Simpson, A.R., Liggett, J.A., Vitkovský, J.P.: Leak detection in pipelines using the damping of fluid transients. J. Hydraulic Eng. **128**(7), 697–711 (2002)
47. Wang, X.Y.: A Summary of the Space-Time Conservation Element and Solution Element. CESE) Method, NASA Report (2015)
48. Woodward, P., Colella, P.: The numerical simulation of two-dimensional fluid flow with strong shocks. J. Comput. Phys. **54**(1), 115–173 (1984)
49. Yabe, T., Ogata, Y.: Conservative semi-lagrangian cip technique for the shallow water equations. Comput. Mech. **46**(1), 125–134 (2010). https://doi.org/10.1007/s00466-009-0438-8
50. Yeung, P.K., Sreenivasan, K.R., Pope, S.B.: Effects of finite spatial and temporal resolution in direct numerical simulations of incompressible isotropic turbulence. Phys. Rev. Fluids **3**(6) (2018)

Output Observers for Linear Infinite-Dimensional Control Systems

Zbigniew Emirsajłow

Abstract This paper studies the general output observation problem for linear infinite-dimensional control systems with bounded input and output operators. Both, the plant and the observer are described by state space models with linear, unbounded system operators, generating strongly continuous semigroups. The plant has two inputs and two outputs. One input is for control, available to the observer, and the other for disturbance. In turn, one output is for the so-called measured signal, available to the observer, and the other for the so-called output to be observed. Using knowledge of the control and the measured output the observer output is to asymptotically follow the output to be observed and that condition is called the output observation. Disturbances are generated by a homogeneous, linear system with an unknown initial condition. Some sufficient conditions for the output observation are derived. These conditions involve the plant, observer and disturbance system operators and consist of two linear operator equations with one of them being an algebraic Sylvester equation. We show that the obtained conditions are constructive and under the assumption on the exponential detectability of the plant allow to derive a design procedure for the output observer. The presented results extend those for linear finite-dimensional control systems.

1 Introduction

The observation problem of the state or output of a dynamical plant is usually the part of a control system design. In the case of finite-dimensional plants the literature provides well developed general theory, see e.g [1–6] and references cited therein. In the case of infinite-dimensional plants the situation is far more complicated and the

---

Z. Emirsajłow (✉)
West Pomeranian University of Technology, Chair of Control and Measurements,
Sikorski st. 37, 70-313 Szczecin, Poland
e-mail: emirsaj@zut.edu.pl

P. Kulczycki et al. (eds.), *Automatic Control, Robotics, and Information Processing*,
Studies in Systems, Decision and Control 296,
https://doi.org/10.1007/978-3-030-48587-0_3

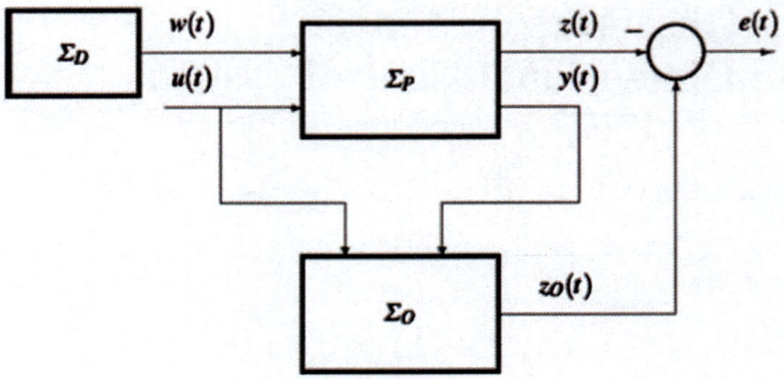

Fig. 1 General interconnection

available theory is rather modest and unsatisfactory. The existing literature mainly concentrates on individual or special cases, see e.g. [7–12] and references cited therein. The general theory for the practically important state space models with unbounded input and output operators, is still missing.

The main goal of the present work is to complete the general theory of the output (sometimes called functional) observation problem for infinite-dimensional plants described by the state space model with bounded input and output operators. Our main effort is to present the results in a way suitable for further development for models with unbounded input and output operators. The unavoidable part of our approach is the infinite-dimensional algebraic Sylvester equation (see [13–16]).

We introduce the following general interconnection as shown in Fig. 1.

Σ_P is the *plant*, Σ_O is the *observer*, Σ_D is the *disturbance source*, $u(t)$ is the *control input*, $d(t)$ is a *disturbance*, $y(t)$ is the *measured output*, $z(t)$ is the *output to be observed*. Both signals $u(t)$ and $y(t)$ are measured and available to the observer and $z_O(t)$ is the *observer output*.

The goal of the interconnection is to observe asymptotically the plant observed output $z(t)$ by the observer output $z_O(t)$, i.e., the *observation error*

$$e(t) := z_O(t) - z(t) \tag{1}$$

is to decay asymptotically

$$\lim_{t\to\infty} e(t) = 0\,, \tag{2}$$

which condition we call the *output observation*.

2 Notation and Assumptions

In order to describe the problem precisely we need some notation and assumptions:

1. We introduce Hibert spaces:

   X—plant state space with the scalar product $\langle \cdot\,,\,\cdot \rangle_X$ and norm $\|\cdot\|_X$
   V—observer state space with the scalar product $\langle \cdot\,,\,\cdot \rangle_V$ and norm $\|\cdot\|_V$
   U—control input space with the scalar product $\langle \cdot\,,\,\cdot \rangle_U$ and norm $\|\cdot\|_U$
   W—disturbance space with scalar product $\langle \cdot\,,\,\cdot \rangle_W$ and norm $\|\cdot\|_W$
   Y—measured output space with the scalar product $\langle \cdot\,,\,\cdot \rangle_Y$ and norm $\|\cdot\|_Y$
   Z—observer output space with the scalar product $\langle \cdot\,,\,\cdot \rangle_Z$ and norm $\|\cdot\|_Z$.

2. The *plant* model is described by

$$\Sigma_P\ :\quad \begin{cases} \dot{x}(t) = Ax(t) + B_1 w(t) + B_2 u(t)\,, \quad x(0) = x_0\,, \\ z(t) = C_1 x(t) + D_{11} w(t) + D_{12} u(t), \\ y(t) = C_2 x(t) + D_{21} w(t) + D_{22} u(t)\,, \end{cases} \tag{3}$$

   where $(x(t))_{t\geq 0} \subset X$, $(u(t))_{t\geq 0} \subset U$, $(w(t))_{t\geq 0} \subset W$, $(y(t))_{t\geq 0} \subset Y$ and $(z(t))_{t\geq 0} \subset Z$. Moreover, for the plant (3) we assume that the system operator $(A, \mathscr{D}(A))$ is linear and unbounded on X and generates a strongly continuous semigroup $(T(t))_{t\geq 0} \subset \mathscr{L}(X)$ with the growth bound $\omega_0(T)$ defined by

$$\omega_0(T) = \inf\{\omega \in \mathbb{R} : \|T(t)\|_{\mathscr{L}(X)} \leq M_\omega \mathrm{e}^{\omega t},\ t \geq 0\}\,. \tag{4}$$

   For $\lambda \in \rho(A)$, where $\rho(A)$ is the resolvent set, $R(\lambda, A) = (\lambda I - A)^{-1} \in \mathscr{L}(X)$ denotes the resolvent. We define (see [17]) the Hilbert space X_1, as $D(A)$ with the scalar product $\langle \cdot, \cdot \rangle_{X_1} := \langle (\lambda I - A)(\cdot), (\lambda I - A)(\cdot) \rangle_X$ and norm $\|\cdot\|_{X_1} = \|(\lambda I - A)(\cdot)\|_X$. The following holds

$$X_1 \hookrightarrow X \tag{5}$$

   with a continuous and dense injection. Moreover, $(T(t))_{t\geq 0} \subset \mathscr{L}(X)$ restricts to $(T_1(t))_{t\geq 0} \subset \mathscr{L}(X_1)$ with generator $(A_1, \mathscr{D}(A_1))$, which is a part of A in $X_1 \subset X$. We have

$$\omega_0(T_1) = \omega_0(T)\,. \tag{6}$$

   For the plant input and output operators we assume

$$\begin{aligned} &B_1 \in \mathscr{L}(W, X)\,,\ B_2 \in \mathscr{L}(U, X)\,, \\ &C_1 \in \mathscr{L}(X, Z)\,,\ \ D_{11} \in \mathscr{L}(W, Z)\,,\ D_{12} \in \mathscr{L}(U, Z)\,, \\ &C_2 \in \mathscr{L}(X, Y)\,,\ \ D_{21} \in \mathscr{L}(W, Y)\,,\ D_{22} \in \mathscr{L}(U, Y)\,. \end{aligned} \tag{7}$$

Moreover, if $(A, \mathscr{D}(A))$ satisfies a stronger assumption and generates a strongly continuous *group* $(T(t))_{t\in\mathbb{R}} \subset \mathscr{L}(X)$ with the growth bound $\omega_0(T)$ [17], given by

$$\omega_0(T) = \inf\{\omega \in \mathbb{R} : \|T(t)\|_{\mathscr{L}(X)} \leq M_\omega \mathrm{e}^{\omega|t|},\ t \in \mathbb{R}\}, \tag{8}$$

then $(T(t))_{t\geq 0} \subset \mathscr{L}(X)$ is a strongly continuous semigroup with a generator $(A, \mathscr{D}(A))$ and growth bound $\omega_0(T)$, and $(T(-t))_{t\geq 0} \subset \mathscr{L}(X)$ is also a strongly continuous semigroup with generator $(-A, \mathscr{D}(A))$ and growth bound $\omega_0(T)$. Moreover $T(-t) = T^{-1}(t)$, i.e. $T(-t)T(t) = T(t)T(-t) = I$ for $t \in \mathbb{R}$.

3. The system generating *disturbances* is described by

$$\Sigma_D : \quad \dot{w}(t) = Qw(t), \quad w(0) = w_0, \tag{9}$$

where $(w(t))_{t\geq 0} \subset W$, $(Q, \mathscr{D}(Q))$ generates a strongly continuous semigroup $(P(t))_{t\geq 0} \subset \mathscr{L}(W)$ with the growth bound $\omega_0(P)$. Sometimes, $(Q, \mathscr{D}(Q))$ satisfies a stronger assumption and generates a strongly continuous *group* $(P(t))_{t\in\mathbb{R}} \subset \mathscr{L}(W)$ with the growth bound $\omega_0(P)$ (see (8)), i.e. $(P(t))_{t\geq 0} \subset \mathscr{L}(W)$ is a strongly continuous semigroup with the generator $(Q, \mathscr{D}(Q))$ and the growth bound $\omega_0(P)$, as well as $(P(-t))_{t\geq 0} \subset \mathscr{L}(W)$ is a strongly continuous semigroup with the generator $(-Q, \mathscr{D}(Q))$ and the growth bound $\omega_0(Q)$. Analogously, as in the case of the plant, we define W_1 and $(P_1(t))_{t\in\mathbb{R}} \subset \mathscr{L}(W_1)$ with the generator $(Q_1, \mathscr{D}(Q_1))$.

4. The *output observer* is described as follows

$$\Sigma_O : \begin{cases} \dot{v}(t) &= Ev(t) + F_1 y(t) + F_2 u(t), \quad v(0) = v_0, \\ z_O(t) &= G_1 v(t) + H_{11} y(t) + H_{12} u(t), \end{cases} \tag{10}$$

where $(v(t))_{t\geq 0} \subset V$ i $(z_O(t))_{t\geq 0} \subset Z$. For the observer (10) we assume that $(E, \mathscr{D}(E))$ is a linear, unbounded operator on V which generates a strongly continuous semigroup $(S(t))_{t\geq 0} \subset \mathscr{L}(V)$ with the growth bound $\omega_0(S)$ defined by

$$\omega_0(S) = \inf\{\omega \in \mathbb{R} : \|S(t)\|_{\mathscr{L}(V)} \leq M_\omega \mathrm{e}^{\omega t},\ t \geq 0\}. \tag{11}$$

Moreover, we assume that the semigroup $(S(t))_{t\geq 0} \subset \mathscr{L}(V)$ is *exponentially stable*, i.e.

$$\omega_0(S) < 0. \tag{12}$$

Analogously, as in the case of the plant, we define V_1 and $(S_1(t))_{t\geq 0} \subset \mathscr{L}(V_1)$ with the generator $(E_1, \mathscr{D}(E_1))$. Obviously, we have

$$V_1 \hookrightarrow V \tag{13}$$

and

$$\omega_0(S_1) = \omega_0(S) < 0. \tag{14}$$

For the observer input and output operators we assume

$$\begin{aligned} &F_1 \in \mathscr{L}(Y, V)\,, \quad F_2 \in \mathscr{L}(U, V)\,, \\ &G_1 \in \mathscr{L}(V, Z)\,, \; H_{11} \in \mathscr{L}(Y, Z)\,, \; H_{12} \in \mathscr{L}(U, Z)\,. \end{aligned} \tag{15}$$

5. The *observation error* is given by

$$e(t) = z_O(t) - z(t)\,, \quad t \geq 0\,, \tag{16}$$

where $(e(t))_{t\geq 0} \subset Z$.

6. *Regularity of functions* describing the disturbance $w(t)$, plant state $x(t)$, observer state $v(t)$, plant outputs $z(t)$, $y(t)$ and observer output $z_O(t)$. Formally, we have

$$\begin{aligned} x(t) &= T(t)x_0 + \textstyle\int_0^t T(t-s)B_1 w(s)\mathrm{d}s + \int_0^t T(t-s)B_2 u(s)\mathrm{d}s\,, \\ z(t) &= C_1 x(t) + D_{11} w(t) + D_{12} u(t)\,, \\ y(t) &= C_2 x(t) + D_{21} w(t) + D_{22} u(t)\,, \end{aligned} \tag{17}$$

where

$$w(t) = P(t)w_0\,, \tag{18}$$

and

$$\begin{aligned} v(t) &= S(t)v_0 + \textstyle\int_0^t S(t-s)F_1 y(s)\mathrm{d}s + \int_0^t S(t-s)F_2 u(s)\mathrm{d}s\,, \\ z_O(t) &= G_1 v(t) + H_{11} y(t) + H_{12} u(t)\,. \end{aligned} \tag{19}$$

Since for $\lambda \in \rho(A)$, $w \in C^1([0,\infty); W)$ and $u \in C^1([0,\infty); U)$ we have

$$\begin{aligned} &\textstyle\int_0^t T(t-s)(B_1 w(s) + B_2 u(s))\mathrm{d}s \\ &= R(\lambda, A)[T(t)(B_1 w(0) + B_2 u(0)) - B_1 w(t) - B_2 u(t) \\ &+\lambda \textstyle\int_0^t T(t-s)(B_1 w(s) + B_2 u(s))\mathrm{d}s + \int_0^t T(t-s)(B_1 w'(s) + B_2 u'(s))\mathrm{d}s] \end{aligned}$$

and for $\mu \in \rho(E)$, $y \in C^1([0,\infty); Y)$ and $u \in C^1([0,\infty); U)$ we have

$$\begin{aligned} &\textstyle\int_0^t S(t-s)(F_1 y(s) + F_2 u(s))\mathrm{d}s \\ &= R(\mu, E)[S(t)(F_1 y(0) + F_2 u(0)) - (F_1 y(t) + F_2 u(t)) \\ &+\mu \textstyle\int_0^t S(t-s)(F_1 y(s) + F_2 u(s))\mathrm{d}s + \int_0^t S(t-s)(F_1 y'(s) + F_2 u'(s))\mathrm{d}s]\,, \end{aligned}$$

then for $w_0 \in W_1$, $x_0 \in X_1$, $v_0 \in V_1$ and $u \in C^1([0,\infty); U)$ we obtain the following regularities for strong solutions

$$\begin{aligned} &w(\cdot) \in C([0,\infty); W_1) \cap C^1([0,\infty); W)\,, \\ &x(\cdot) \in C([0,\infty); X_1) \cap C^1([0,\infty); X)\,, \\ &z(\cdot) \in C^1([0,\infty); Z)\,, \; y(\cdot) \in C^1([0,\infty); Y) \end{aligned} \tag{20}$$

and

$$\begin{aligned} &v(\cdot) \in C([0,\infty); V_1) \cap C^1([0,\infty); V), \\ &z_O(\cdot) \in C^1([0,\infty); Z), \; e(\cdot) \in C^1([0,\infty); Z). \end{aligned} \tag{21}$$

In turn, for $w_0 \in W$, $x_0 \in X$, $v_0 \in V$ and $u \in C([0,\infty); U)$ we obtain lower regularities for mild solutions

$$\begin{aligned} &w(\cdot) \in C([0,\infty); W), \\ &x(\cdot) \in C([0,\infty); X), \\ &z(\cdot) \in C([0,\infty); Z), \; y(\cdot) \in C([0,\infty); Y) \end{aligned} \tag{22}$$

and

$$\begin{aligned} &v(\cdot) \in C([0,\infty); V), \\ &z_O(\cdot) \in C([0,\infty); Z), \; e(\cdot) \in C([0,\infty); Z). \end{aligned} \tag{23}$$

7. *Output observation* condition. For $w_0 \in W$, $x_0 \in X$, $v_0 \in V$ and $u \in C([0,\infty); U)$ the following holds

$$\lim_{t\to\infty} \|e(t)\|_Z := \|z_O(t) - z(t)\|_Z = 0. \tag{24}$$

3 Basic Relations

Combining together descriptions of Σ_P, Σ_D and Σ_O, we get

$$\begin{aligned} \dot{x}(t) &= Ax(t) + B_1 w(t) + B_2 u(t), \quad x(0) = x_0, \\ z(t) &= C_1 x(t) + D_{11} w(t) + D_{12} u(t), \\ y(t) &= C_2 x(t) + D_{21} w(t) + D_{22} u(t), \\ \dot{w}(t) &= Q w(t), \quad w(0) = w_0, \\ \dot{v}(t) &= E v(t) + F_1 y(t) + F_2 u(t), \quad v(0) = v_0, \\ z_O(t) &= G_1 v(t) + H_{11} y(t) + H_{12} u(t), \\ e(t) &= z_O(t) - z(t). \end{aligned} \tag{25}$$

After simple manipulations we obtain an extended system, which is denoted by Σ, i.e.

$$\Sigma : \begin{cases} \dot{v}(t) = Ev(t) + F_1 C_2 x(t) + F_1 D_{21} w(t) + (F_1 D_{22} + F_2) u(t), \; v(0) = v_0, \\ \dot{x}(t) = Ax(t) + B_1 w(t) + B_2 u(t), \quad x(0) = x_0, \\ \dot{w}(t) = Q w(t), \quad w(0) = w_0, \\ e(t) = G_1 v(t) + (H_{11} C_2 - C_1) x(t) + (H_{11} D_{21} - D_{11}) w(t) \\ \qquad + (H_{11} D_{22} - D_{12} + H_{12}) u(t), \end{cases} \tag{26}$$

and in the matrix form

$$\Sigma : \begin{bmatrix} \dot{v} \\ \dot{x} \\ \dot{w} \\ \hline e \end{bmatrix} = \left[\begin{array}{ccc|c} E & F_1C_2 & F_1D_{21} & F_1D_{22} + F_2 \\ 0 & A & B_1 & B_2 \\ 0 & 0 & Q & 0 \\ \hline G_1 & H_{11}C_2 - C_1 & H_{11}D_{21} - D_{11} & H_{11}D_{22} - D_{12} + H_{12} \end{array}\right] \begin{bmatrix} v \\ x \\ w \\ \hline u \end{bmatrix} \tag{27}$$

with the initial conditions

$$\begin{bmatrix} v(0) \\ x(0) \\ w(0) \end{bmatrix} = \begin{bmatrix} v_0 \\ x_0 \\ w_0 \end{bmatrix}. \tag{28}$$

The output observation condition takes the form: for $v(0) = v_0 \in V, x(0) = x_0 \in X$, $w(0) = w_0 \in W$ and $u(\cdot) \in C([0, \infty); U)$ the following convergence holds

$$\begin{aligned} \lim_{t\to\infty} \|e(t)\|_Z = \lim_{t\to\infty} \| & \begin{bmatrix} G_1 & H_{11}C_2 - C_1 & H_{11}D_{21} - D_{11} \end{bmatrix} \begin{bmatrix} v(t) \\ x(t) \\ w(t) \end{bmatrix} \\ & + (H_{11}D_{22} - D_{12} + H_{12})u(t)\|_Z \\ = 0 . & \end{aligned} \tag{29}$$

In order to show that the system Σ is well-posed we introduce simplified notation

$$\mathbb{A} := \begin{bmatrix} E & 0 & 0 \\ 0 & A & B_1 \\ 0 & 0 & Q \end{bmatrix}, \quad \mathbb{P} := \begin{bmatrix} 0 & F_1C_2 & F_1D_{21} \\ 0 & 0 & 0 \\ 0 & 0 & 0 \end{bmatrix}, \quad \mathbb{B} := \begin{bmatrix} F_1D_{22} + F_2 \\ B_2 \\ 0 \end{bmatrix} \tag{30}$$

and

$$\mathbb{C} := \begin{bmatrix} G_1 & H_{11}C_2 - C_1 & H_{11}D_{21} - D_{11} \end{bmatrix}, \quad \mathbb{D} := \begin{bmatrix} H_{11}D_{22} - D_{12} + H_{12} \end{bmatrix}. \tag{31}$$

Defining the state

$$\xi(t) := \begin{bmatrix} v(t) \\ x(t) \\ w(t) \end{bmatrix}, \quad \xi(0) = \xi_0 = \begin{bmatrix} v_0 \\ x_0 \\ w_0 \end{bmatrix}, \tag{32}$$

we can rewrite Σ in the compact form

$$\Sigma : \begin{bmatrix} \dot{\xi}(t) \\ \hline e(t) \end{bmatrix} = \left[\begin{array}{c|c} \mathbb{A} + \mathbb{P} & \mathbb{B} \\ \hline \mathbb{C} & \mathbb{D} \end{array}\right] \begin{bmatrix} \xi(t) \\ \hline u(t) \end{bmatrix}, \quad \xi(0) = \xi_0 . \tag{33}$$

The operator $\mathbb{A}$ is linear and unbounded on the Hilbert space $V \times X \times W$, with the domain $\mathscr{D}(\mathbb{A}) = V_1 \times X_1 \times W_1$. All the remaining operators are linear and bounded. Namely

$$\begin{aligned} &\mathbb{P} \in \mathscr{L}(V \times X \times W), \quad \mathbb{B} \in \mathscr{L}(U, V \times X \times W), \\ &\mathbb{C} \in \mathscr{L}(V \times X \times W, Z), \quad \mathbb{D} \in \mathscr{L}(U, Z). \end{aligned} \tag{34}$$

Recall that for $v(0) = v_0 \in V_1$, $x(0) = x_0 \in X_1$, $w(0) = w_0 \in W_1$ and $u(\cdot) \in C^1([0, \infty); U)$ the solutions $v(t)$, $x(t)$ and $w(t)$ are understood in the strong sense, and for $v(0) = v_0 \in V$, $x(0) = x_0 \in X$, $w(0) = w_0 \in W$ and $u(\cdot) \in C([0, \infty); U)$ are understood in the mild sense.

Assumptions on E, A and Q imply that the operator $\mathbb{A}$ generates a strongly continuous semigroup on $V \times X \times W$ and, analogously as before, we can introduce the space $V_1 \times X_1 \times W_1$. It is known, e.g. [17], that for a bounded perturbation $\mathbb{P}$, the operator $\mathbb{A} + \mathbb{P}$ preserves properties of the operator $\mathbb{A}$ and consequently, for $\xi_0 \in V_1 \times X_1 \times W_1$ and $u(\cdot) \in C^1([0, \infty); U)$ we have

$$\xi(\cdot) \in C([0, \infty); V_1 \times X_1 \times W_1) \cap C^1([0, \infty); V \times X \times W), \quad e(\cdot) \in C^1([0, \infty); Z), \tag{35}$$

i.e. $\xi(t)$ is a strong solution, and for $\xi_0 \in V \times X \times W$ and $u(\cdot) \in C([0, \infty); U)$ we have

$$\xi(\cdot) \in C([0, \infty); V \times X \times W), \quad e(\cdot) \in C([0, \infty); Z), \tag{36}$$

i.e. $\xi(t)$ is a mild solution.

Remark 1 In order to clarify the above considerations notice that the operator

$$\begin{bmatrix} A & B_1 \\ 0 & Q \end{bmatrix} = \begin{bmatrix} A & 0 \\ 0 & Q \end{bmatrix} + \begin{bmatrix} 0 & B_1 \\ 0 & 0 \end{bmatrix}, \tag{37}$$

with the domain $X_1 \times W_1$, generates a strongly continuous semigroup $(\mathcal{U}(t))_{t \geq 0} \subset \mathcal{L}(X \times W)$, explicitly given by

$$\mathcal{U}(t) \begin{bmatrix} x_0 \\ w_0 \end{bmatrix} = \begin{bmatrix} T(t) & \int_0^t T(t-r) B_1 P(r) \mathrm{d}r \\ 0 & P(t) \end{bmatrix} \begin{bmatrix} x_0 \\ w_0 \end{bmatrix}, \quad \begin{bmatrix} x_0 \\ w_0 \end{bmatrix} \in X \times W. \tag{38}$$

This implies that the operator

$$\left[\begin{array}{c|cc} E & F_1 C_2 & F_1 D_{21} \\ \hline 0 & A & B_1 \\ 0 & 0 & Q \end{array}\right] = \left[\begin{array}{c|cc} E & 0 & 0 \\ \hline 0 & A & B_1 \\ 0 & 0 & Q \end{array}\right] + \left[\begin{array}{c|cc} 0 & F_1 C_2 & F_1 D_{21} \\ \hline 0 & 0 & 0 \\ 0 & 0 & 0 \end{array}\right], \tag{39}$$

with the domain $V_1 \times (X_1 \times W_1)$, generates a strongly continuous semigroup $(\mathbb{U}(t))_{t \geq 0} \subset \mathcal{L}(V \times (X \times W))$, explicitly given by

$$\mathbb{U}(t) \begin{bmatrix} v_0 \\ \hline x_0 \\ w_0 \end{bmatrix} = \left[\begin{array}{c|c} S(t) & \int_0^t S(t-r) F_1 \begin{bmatrix} C_2 & D_{21} \end{bmatrix} \mathcal{U}(r) \mathrm{d}r \\ \hline 0 & \mathcal{U}(t) \end{array}\right] \begin{bmatrix} v_0 \\ \hline x_0 \\ w_0 \end{bmatrix}, \tag{40}$$

where $\begin{bmatrix} v_0 \\ \hline x_0 \\ w_0 \end{bmatrix} \in V \times (X \times W)$.

Finally, the output observation condition can be written in the form: for $\xi_0 \in V \times X \times W$ and $u(\cdot) \in C([0, \infty); U)$ the following convergence holds

$$\lim_{t\to\infty} \|e(t)\|_Z = \lim_{t\to\infty} \|\mathbb{C}\xi(t) + \mathbb{D}u(t)\|_Z = 0\,. \tag{41}$$

In the following parts of the paper we will analyze some special cases and then formulate the general result. In all cases we obtain sufficient conditions which involve an operator algebraic Sylvester equation. For this reason we provide in the next section some general results for that equation.

4 Algebraic Sylvester Equation

Below we present some results for the operator algebraic Sylvester equation based on the work [18]. This approach explores the concept of an implemented semigroup which can be found in many papers (see, e.g., [13, 14, 18, 19] and references cited therein).

For the purpose of the Sylvester equation we introduce some new notation and assumptions which remain valid throughout the section.

1. X is a Hilbert space with the scalar product $\langle\cdot\,,\,\cdot\rangle_X$ and the norm $\|\cdot\|_X$ and V is a Hilbert space with the scalar product $\langle\cdot\,,\,\cdot\rangle_V$ with the norm $\|\cdot\|_V$.
2. $(N, \mathscr{D}(N))$ is a linear, unbounded operator on X which generates a strongly continuous semigroup $(T(t))_{t\geq 0} \subset \mathscr{L}(X)$ with the growth bound $\omega_0(T)$. For $\lambda \in \rho(N)$ we define a Hilbert space X_1, $X_1 \hookrightarrow X$, and a strongly continuous semigroup $(T_1(t))_{t\geq 0} \subset \mathscr{L}(X_1)$ with the generator $(N_1, \mathscr{D}(N_1))$, which is part of N in $X_1 \subset X$.
3. $(M, \mathscr{D}(M))$ is a linear, unbounded operator on V which generates a strongly continuous semigroup $(S(t))_{t\geq 0} \subset \mathscr{L}(V)$ with the growth bound $\omega_0(S)$. For $\mu \in \rho(M)$ we define a Hilbert space V_1, $V_1 \hookrightarrow V$, and a strongly continuous semigroup $(S_1(t))_{t\geq 0} \subset \mathscr{L}(V_1)$ with the generator $(M_1, \mathscr{D}(M_1))$, which is part of M in $V_1 \subset V$.
4. $\mathcal{X} := \mathscr{L}(X, V)$ is a Banach space of linear, bounded operators from X to V with norm $\|\cdot\|_{\mathcal{X}}$. $(\mathcal{X}, \|\cdot\|_{\mathcal{X}})$ denotes $\mathscr{L}(X, V)$ with the uniform operator topology (induced by $\|\cdot\|_{\mathcal{X}}$) and $(\mathcal{X}, \tau)$ denotes $\mathscr{L}(X, V)$ with the strong operator topology τ, i.e. topology induced by a family of seminorms $\mathcal{P} = \{p_h\}_{h\in X}$, where $p_h(\Phi) = \|\Phi h\|_V$ for $\Phi \in \mathcal{X}$ and $h \in X$.
5. Combining $(T(t))_{t\geq 0} \subset \mathscr{L}(X)$ and $(S(t))_{t\geq 0} \subset \mathscr{L}(V)$, we define another semigroup.

Definition 1 A family of operators $(\mathcal{U}(t))_{t\geq 0} \subset \mathscr{L}(\mathcal{X})$, defined by the relation

$$\mathcal{U}(t)\Phi := S(t)\Phi T(t), \quad \Phi \in \mathcal{X}, \quad t \geq 0\,, \tag{42}$$

is called an *implemented semigroup*.

It turns out that $(\mathcal{U}(t))_{t\geq 0} \subset \mathscr{L}(\mathcal{X})$ is a semigroup and for every $\Phi \in \mathcal{X}$ enjoys the continuity $\mathcal{U}(\cdot)\Phi \in C([0,\infty);(\mathcal{X},\tau))$. Such a family of functions is said to be τ-continuous. In general this family is not strongly continuous ($\mathcal{U}(\cdot)\Phi \notin C([0,\infty);(\mathcal{X},\|\cdot\|_{\mathcal{X}}))$ or, in other words, it is not $\|\cdot\|_{\mathcal{X}}$-continuous), unless both operators M and N are bounded. However, in the control theory of infinite-dimensional systems [20] the most interesting case is when the operators M and N are unbounded.

Definition 2 A *generator* $\mathcal{A}$ of the semigroup $(\mathcal{U}(t))_{t\geq 0} \subset \mathscr{L}(\mathcal{X})$ is defined by the following limit

$$\mathcal{A}\Phi = \lim_{t\searrow 0} \frac{\mathcal{U}(t)\Phi - \Phi}{t} \text{ in } (\mathcal{X},\tau)\,, \quad X \in \mathscr{D}(\mathcal{A})\,, \tag{43}$$

where $\mathscr{D}(\mathcal{A}) \subset \mathcal{X}$ is the domain of $\mathcal{A}$, defined as follows

$$\mathscr{D}(\mathcal{A}) = \{\Phi \in \mathcal{X} : \lim_{t\searrow 0} \frac{\mathcal{U}(t)\Phi - \Phi}{t} \text{ exists in } (\mathcal{X},\tau)\}\,. \tag{44}$$

6. The domain $\mathscr{D}(\mathcal{A}) \in \mathcal{X}$ and the generator $\mathcal{A}$ have the following properties: $\Phi \in \mathcal{X} = \mathscr{L}(X,V)$ belongs to $\mathscr{D}(\mathcal{A})$ if and only if

$$\Phi \in \mathscr{L}(X,V) \cap \mathscr{L}(X_1,V_1) \tag{45}$$

and continuous extension of $(M\Phi + \Phi N) \in \mathscr{L}(X_1,V)$ to the whole space X belongs to $\mathcal{X}$. The operator $\mathcal{A}$ admits the following representation

$$(\mathcal{A}\Phi)h = M\Phi h + \Phi N h\,,\ \Phi \in \mathscr{D}(\mathcal{A})\,,\ h \in X_1\,, \tag{46}$$

where, by (45), the right hand side of (46) is well defined in V. It is clear that $\mathcal{A}\Phi \in \mathcal{X}$ for all $\Phi \in \mathscr{D}(\mathcal{A})$.
7. The following relations hold:

$$\|\mathcal{U}(t)\|_{\mathscr{L}(\mathcal{X})} = \|S(t)\|_{\mathscr{L}(X)}\|T(t)\|_{\mathscr{L}(V)} \tag{47}$$

and

$$\omega_0(\mathcal{U}) = \omega_0(S) + \omega_0(T)\,, \tag{48}$$

where $\omega_0(\mathcal{U})$ denotes the growth bound of the implemented semigroup.

$$\mathbb{C}_{\omega_0(S)+\omega_0(T)} \subset \rho(\mathcal{A})\,, \tag{49}$$

where we used the notation

$$\mathbb{C}_\omega := \{\lambda \in \mathbb{C} : \Re\lambda > \omega\}\,.$$

$\rho(\mathcal{A})$ is the resolvent set of $\mathcal{A}$ and for $\lambda \in \mathbb{C}_{\omega_0(S)+\omega_0(T)}$ the resolvent operator has the representation

$$\mathcal{R}(\lambda, \mathcal{A})\Phi := (\lambda I - \mathcal{A})^{-1}\Phi \tag{50}$$
$$= \int_0^\infty e^{-\lambda t}\mathcal{U}(t)\Phi \, dt = \int_0^\infty e^{-\lambda t} S(t)\Phi T(t) \, dt \, , \quad \Phi \in \mathcal{X} \, ,$$

where integrals are convergent in $(\mathcal{X}, \tau)$.

8. Basic results for the algebraic Sylvester equation are collected below. Proofs of these results may be found in the work [18].

Lemma 1 *If $\omega > \omega_0(\mathcal{U})$ and $K \in \mathcal{X}$, then the algebraic equation*

$$\omega\Pi - \mathcal{A}\Pi = K \, , \tag{51}$$

where equality is in $\mathcal{X}$, has a unique solution $\Pi \in \mathscr{D}(\mathcal{A})$ and this solution can be expressed in the form

$$\Pi = \mathcal{R}(\omega, \mathcal{A})K = \int_0^\infty e^{-\omega t}\mathcal{U}(t)K dt \, , \tag{52}$$

where the integral is understood in $(\mathcal{X}, \tau)$.

The above lemma implies the following corollaries.

Corollary 1 *If $\omega > \omega_0(S) + \omega_0(T)$ and $K \in \mathcal{L}(X, V)$, then the algebraic Sylvester equation*

$$\omega\Pi h - M\Pi h - \Pi N h = K h \, , \quad h \in X_1 \, , \tag{53}$$

where equality is in V, has a unique solution Π, which satisfy the condition

$$\Pi \in \mathcal{L}(X, V) \cap \mathcal{L}(X_1, V_1) \tag{54}$$

and can be expressed in the form

$$\Pi = \int_0^\infty e^{-\omega t} S(t) K T(t) dt \, , \tag{55}$$

where the integral is convergent in the strong operator topology $\mathscr{L}(X, V)$.

One can see that if $0 > \omega_0(S) + \omega_0(T)$ (the semigroup is exponentially stable), then we can assume $\omega = 0$ and Corollary 1 can be rewritten as follows.

Corollary 2 *If $0 > \omega_0(S) + \omega_0(T)$ and $K \in \mathcal{L}(X, V)$, then the algebraic Sylvester equation*

$$M\Pi h + \Pi N h + K h = 0 \, , \quad h \in X_1 \, , \tag{56}$$

where equality is in V, has a unique solution Π, such that

$$\Pi \in \mathcal{L}(X, V) \cap \mathcal{L}(X_1, V_1) \tag{57}$$

and this solution can be expressed in the form

$$\Pi = \int_0^\infty S(t)KT(t)\mathrm{d}t \,. \tag{58}$$

Corollary 2 provides an easy way to check some sufficient condition for the existence of a unique solution for the operator algebraic Sylvester equation.

5 Plant with No Inputs

Before we start studying a general case we analyze the plant Σ_P with no control and no disturbances, i.e., we assume $u(t) \equiv 0$ and $w(t) \equiv 0$. In this case

$$\Sigma_P \;:\; \begin{cases} \dot{x}(t) = Ax(t)\,, \quad x(0) = x_0\,, \\ z(t) = C_1 x(t), \\ y(t) = C_2 x(t) \end{cases} \tag{59}$$

and

$$\Sigma_O \;:\; \begin{cases} \dot{v}(t) \;= Ev(t) + F_1 y(t)\,, \quad v(0) = v_0\,, \\ z_O(t) = G_1 v(t) + H_{11} y(t)\,. \end{cases} \tag{60}$$

Interconnection of Σ_P and Σ_O, i.e. the extended system Σ, takes the form

$$\Sigma \;:\; \begin{bmatrix} \dot{v} \\ \dot{x} \\ \hline e \end{bmatrix} = \begin{bmatrix} E & F_1 C_2 \\ 0 & A \\ \hline G_1 & H_{11}C_2 - C_1 \end{bmatrix} \begin{bmatrix} v \\ x \end{bmatrix}, \quad \begin{bmatrix} v(0) \\ x(0) \end{bmatrix} = \begin{bmatrix} v_0 \\ x_0 \end{bmatrix}. \tag{61}$$

Lemma 2 *(sufficient conditions) Given the extended system Σ described by the equatuins (61). If the observer (E, F_1, G_1, H_{11}) is exponentially stable, i.e. $\omega_0(S) < 0$, and the following equations*

$$\begin{cases} E\Pi h_1 - \Pi A h_1 + F_1 C_2 h_1 = 0\,, \; h_1 \in X_1\,, \\ G_1 \Pi h + H_{11} C_2 h - C_1 h = 0\,, \; h \in X\,, \end{cases} \tag{62}$$

where equalities are in V and Z, have a solution $\Pi \in \mathcal{L}(X, V) \cap \mathcal{L}(X_1, V_1)$, then for every $x_0 \in X$ and $v_0 \in V$ the output observation condition holds.

Proof In order to prove the lemma we introduce new state variables

$$\begin{bmatrix} p(t) \\ x(t) \end{bmatrix} = \begin{bmatrix} I & -\Pi \\ 0 & I \end{bmatrix} \begin{bmatrix} v(t) \\ x(t) \end{bmatrix}, \tag{63}$$

where $\Pi \in \mathcal{L}(X, V) \cap \mathcal{L}(X_1, V_1)$ is a certain operator. It is easy to check that for every Π there exists an inverse

$$\begin{bmatrix} I & -\Pi \\ 0 & I \end{bmatrix}^{-1} = \begin{bmatrix} I & \Pi \\ 0 & I \end{bmatrix}.$$

By the regularity conditions (35)–(36) and the transformation (63) we get

$$\begin{gathered} \begin{bmatrix} v(\cdot) \\ x(\cdot) \end{bmatrix} \in C([0, \infty); V_1 \times X_1) \cap C^1([0, \infty); V \times X) \\ \Leftrightarrow \\ \begin{bmatrix} p(\cdot) \\ x(\cdot) \end{bmatrix} \in C([0, \infty); V_1 \times X_1) \cap C^1([0, \infty); V \times X) \end{gathered} \tag{64}$$

and

$$\begin{bmatrix} v(\cdot) \\ x(\cdot) \end{bmatrix} \in C([0, \infty); V \times X) \Leftrightarrow \begin{bmatrix} p(\cdot) \\ x(\cdot) \end{bmatrix} \in C([0, \infty); V \times X). \tag{65}$$

Differentiating both sides of the equality (63) (under suitable assumptions) and making use of (61) we easily obtain

$$\begin{aligned} \dot{p}(t) &= \dot{v}(t) - \Pi \dot{x}(t) = Ev(t) + F_1 C_2 x(t) - \Pi A x(t) \\ &= (E\Pi - \Pi A + F_1 C_2) x(t) + Ep(t), \\ e(t) &= (G_1 \Pi + H_{11} C_2 - C_1) x(t) + G_1 p(t). \end{aligned}$$

After substitution of the state equation for $x(t)$, we can write

$$\begin{bmatrix} \dot{p} \\ \dot{x} \\ \hline e \end{bmatrix} = \begin{bmatrix} E & E\Pi - \Pi A + F_1 C_2 \\ 0 & A \\ \hline G_1 & G_1 \Pi + H_{11} C_2 - C_1 \end{bmatrix} \begin{bmatrix} p \\ x \end{bmatrix}, \quad \begin{bmatrix} p(0) \\ x(0) \end{bmatrix} = \begin{bmatrix} v_0 - \Pi x_0 \\ x_0 \end{bmatrix}. \tag{66}$$

Let us now notice that if equations (62) have a solution $\Pi \in \mathcal{L}(X, V) \cap \mathcal{L}(X_1, V_1)$, then (66) simplifies to the form

$$\begin{aligned} \dot{p}(t) &= Ep(t), \quad p(0) = v_0 - \Pi x_0, \\ e(t) &= G_1 p(t), \end{aligned}$$

and since $p(t) = S(t)p(0)$ and $\omega_0(S) < 0$, then for every $p(0) \in V$ ($v_0 \in V$, $x_0 \in X$) we have $\lim_{t\to\infty} \|p(t)\|_V = 0$ and the output observation condition holds, i.e. $\lim_{t\to\infty} \|G_1 p(t)\|_Z = \lim_{t\to\infty} \|e(t)\|_Z = 0$. □

Notice that the first equation in (62) is the algebraic Sylvester equation to which we can apply the theory of Sect. 4. To this end we have to assume $M = E$, $N = -A$ and $K = F_1 C_2$. If E and $-A$ generate strongly continuous semigroups, then by Corollary 2, we can always guarantee the existence of a unique solution of the Sylvester equation by choosing the growth bound $\omega_0(S) < 0$ of the observer small enough so that the inequality $\omega_0(S) + \omega_0(T) < 0$ holds.

6 Plant With Control and Without Disturbances

In this case we assume that in the plant there are no disturbances, i.e. $w(t) \equiv 0$. By combing Σ_P and Σ_O, we obtain the extended system Σ in the form

$$\Sigma : \left[\begin{array}{c} \dot{v} \\ \dot{x} \\ \hline e \end{array}\right] = \left[\begin{array}{cc|c} E & F_1 C_2 & F_1 D_{22} + F_2 \\ 0 & A & B_2 \\ \hline G_1 & H_{11} C_2 - C_1 & H_{11} D_{22} + H_{12} - D_{12} \end{array}\right] \left[\begin{array}{c} v \\ x \\ \hline u \end{array}\right], \quad \begin{bmatrix} v(0) \\ x(0) \end{bmatrix} = \begin{bmatrix} v_0 \\ x_0 \end{bmatrix}. \tag{67}$$

In turn, the output observation condition takes the form: for every $v(0) = v_0 \in V$, $x(0) = x_0 \in X$ and $u(\cdot) \in C([0, \infty); U)$ the following holds

$$\begin{aligned} \lim_{t \to \infty} \|e(t)\|_Z = \lim_{t \to \infty} \| \begin{bmatrix} G_1 & H_{11} C_2 - C_1 \end{bmatrix} \begin{bmatrix} v(t) \\ x(t) \end{bmatrix} \\ + (H_{11} D_{22} + H_{12} - D_{12}) u(t) \|_Z = 0 \,. \end{aligned} \tag{68}$$

In order to derive sufficient conditions which guarantee the convergence (68) we introduce new state variables

$$\begin{bmatrix} p(t) \\ x(t) \end{bmatrix} = \begin{bmatrix} I & -\Pi \\ 0 & I \end{bmatrix} \begin{bmatrix} v(t) \\ x(t) \end{bmatrix}, \tag{69}$$

where $\Pi \in \mathcal{L}(X, V) \cap \mathcal{L}(X_1, V_1)$ is a certain operator. By the regularity of conditions (35)–(36) and by transformation (69) we get

$$\begin{aligned} &\begin{bmatrix} v(\cdot) \\ x(\cdot) \end{bmatrix} \in C([0, \infty); V_1 \times X_1) \cap C^1([0, \infty); V \times X) \\ &\Leftrightarrow \\ &\begin{bmatrix} p(\cdot) \\ x(\cdot) \end{bmatrix} \in C([0, \infty); V_1 \times X_1) \cap C^1([0, \infty); V \times X) \end{aligned} \tag{70}$$

and

$$\begin{bmatrix} v(\cdot) \\ x(\cdot) \end{bmatrix} \in C([0, \infty); V \times X) \Leftrightarrow \begin{bmatrix} p(\cdot) \\ x(\cdot) \end{bmatrix} \in C([0, \infty); V \times X) \,. \tag{71}$$

The extended system Σ with new state variables takes the form

$$\Sigma : \begin{bmatrix} \dot{p} \\ \dot{x} \\ \hline e \end{bmatrix} = \left[\begin{array}{cc|c} E & E\Pi - \Pi A + F_1 C_2 & F_1 D_{22} - \Pi B_2 + F_2 \\ 0 & A & B_2 \\ \hline G_1 & G_1\Pi + H_{11}C_2 - C_1 & H_{11}D_{22} - D_{12} + H_{12} \end{array}\right] \begin{bmatrix} p \\ x \\ \hline u \end{bmatrix} \tag{72}$$

with initial conditions

$$\begin{bmatrix} p(0) \\ x(0) \end{bmatrix} = \begin{bmatrix} v_0 - \Pi x_0 \\ x_0 \end{bmatrix}. \tag{73}$$

The output observation condition converts to: for all $v(0) = v_0 \in V, x(0) = x_0 \in X$ $(p(0) = (v_0 - \Pi x_0) \in V)$ and $u(\cdot) \in C([0, \infty); U)$ the following holds

$$\begin{aligned} &\lim_{t\to\infty} \|e(t)\|_Z \\ &= \lim_{t\to\infty} \| \begin{bmatrix} G_1 & G_1\Pi + H_{11}C_2 - C_1 \end{bmatrix} \begin{bmatrix} p(t) \\ x(t) \end{bmatrix} + (H_{11}D_{22} + H_{12} - D_{12})u(t)\|_Z \\ &= 0 . \end{aligned} \tag{74}$$

Without the loss of generality we can assume that the operators F_2 and H_{12} of the observer are defined by the equations

$$\begin{aligned} F_2 &= \Pi B_2 - F_1 D_{22} , \\ H_{12} &= D_{12} - H_{11} D_{22} . \end{aligned} \tag{75}$$

This assumption is allowed since the choice of operators E, F_1, G_1 and H_{11} of the observer is independent of the choice of F_2 and H_{12}. Under the assumptions (75) the extended system Σ simplifies to the form

$$\Sigma : \begin{bmatrix} \dot{p} \\ \dot{x} \\ \hline e \end{bmatrix} = \left[\begin{array}{cc|c} E & E\Pi - \Pi A + F_1 C_2 & 0 \\ 0 & A & B_2 \\ \hline G_1 & G_1\Pi + H_{11}C_2 - C_1 & 0 \end{array}\right] \begin{bmatrix} p \\ x \\ \hline u \end{bmatrix}, \quad \begin{bmatrix} p(0) \\ x(0) \end{bmatrix} = \begin{bmatrix} v_0 - \Pi x_0 \\ x_0 \end{bmatrix}. \tag{76}$$

As a consequence, the output observation condition takes the form: for all $v(0) = v_0 \in V, x(0) = x_0 \in X$ $(p(0) = (v_0 - \Pi x_0) \in V)$ and $u(\cdot) \in C([0, \infty); U)$ there is

$$\lim_{t\to\infty} \|e(t)\|_Z = \lim_{t\to\infty} \| \begin{bmatrix} G_1 & G_1\Pi + H_{11}C_2 - C_1 \end{bmatrix} \begin{bmatrix} p(t) \\ x(t) \end{bmatrix} \|_Z = 0 . \tag{77}$$

The above considerations can be summed up in the form of the following theorem.

Theorem 1 *(sufficient conditions) Given the extended system Σ described by the equations (72). If the observer $(E, F_1, F_2, G_1, H_{11}, H_{12})$ is exponentially stable, i.e. $\omega_0(S) < 0$, and the equations*

$$\begin{cases} E\Pi h_1 - \Pi A h_1 + F_1 C_2 h_1 = 0 , & h_1 \in X_1 , \\ G_1 \Pi h + H_{11} C_2 - C_1 h = 0 , & h \in X , \end{cases} \tag{78}$$

where equalities are in V and Z, have a solution $\Pi \in \mathcal{L}(X, V) \cap \mathcal{L}(X_1, V_1)$ *and the following equalities are satisfied*

$$\begin{aligned} F_2 &= \Pi B_2 - F_1 D_{22}\,, \\ H_{12} &= D_{12} - H_{11} D_{22}\,, \end{aligned} \tag{79}$$

then

$$\begin{aligned} \dot{p}(t) &= E p(t)\,, \quad p(0) = v_0 - \Pi x_0\,, \\ e(t) &= G_1 p(t) \end{aligned} \tag{80}$$

and for $x_0 \in X$, $v_0 \in V$ *and* $u \in C([0, \infty); U)$, *we obtain*

$$\lim_{t\to\infty} \|p(t)\|_V = \lim_{t\to\infty} \|S(t)p(0)\|_Z = 0\,, \tag{81}$$

which implies that the output observation condition holds.

It is worth noticing that the equations (78) coincide with the corresponding equations (62) of Lemma 2. In Theorem 1 we have the extra conditions (79), which can always be satisfied once the equations (78) hold.

7 General Case

We are ready to consider the output observation problem in the general case, i.e. we assume (see Sect. 3) that the interconnection of the plant Σ_P, the observer Σ_O and the disturbance generator Σ_D, which we called the extended system Σ, takes the form

$$\Sigma : \begin{bmatrix} \dot{v} \\ \dot{x} \\ \dot{w} \\ \hline e \end{bmatrix} = \left[\begin{array}{ccc|c} E & F_1 C_2 & F_1 D_{21} & F_1 D_{22} + F_2 \\ 0 & A & B_1 & B_2 \\ 0 & 0 & Q & 0 \\ \hline G_1 & H_{11} C_2 - C_1 & H_{11} D_{21} - D_{11} & H_{11} D_{22} - D_{12} + H_{12} \end{array}\right] \begin{bmatrix} v \\ x \\ w \\ \hline u \end{bmatrix} \tag{82}$$

with the initial conditions

$$\begin{bmatrix} v(0) \\ x(0) \\ w(0) \end{bmatrix} = \begin{bmatrix} v_0 \\ x_0 \\ w_0 \end{bmatrix}. \tag{83}$$

The output observation condition takes the form: for every $v(0) = v_0 \in V$, $x(0) = x_0 \in X$, $w(0) = w_0 \in W$ and $u(\cdot) \in C([0, \infty); U)$ we have the convergence

$$
\begin{aligned}
\lim_{t\to\infty} \|e(t)\|_Z &= \lim_{t\to\infty} \| \begin{bmatrix} G_1 & H_{11}C_2 - C_1 & H_{11}D_{21} - D_{11} \end{bmatrix} \begin{bmatrix} v(t) \\ x(t) \\ w(t) \end{bmatrix} \\
&\quad + (H_{11}D_{22} - D_{12} + H_{12})u(t)\|_Z \\
&= 0\,.
\end{aligned} \tag{84}
$$

In order to derive sufficient conditions which guarantee (84), we introduce in Σ new sate variables

$$
\begin{bmatrix} p(t) \\ x(t) \\ w(t) \end{bmatrix} = \begin{bmatrix} I & -\Pi_1 & -\Pi_2 \\ 0 & I & 0 \\ 0 & 0 & I \end{bmatrix} \begin{bmatrix} v(t) \\ x(t) \\ w(t) \end{bmatrix}, \tag{85}
$$

where $\Pi_1 \in \mathscr{L}(X, V) \cap \mathscr{L}(X_1, V_1)$ and $\Pi_2 \in \mathscr{L}(W, V) \cap \mathscr{L}(W_1, V_1)$ are certain operators. It is easy to check that for every Π_1 and Π_2 there exits an inverse

$$
\begin{bmatrix} I & -\Pi_1 & -\Pi_2 \\ 0 & I & 0 \\ 0 & 0 & I \end{bmatrix}^{-1} = \begin{bmatrix} I & \Pi_1 & \Pi_2 \\ 0 & I & 0 \\ 0 & 0 & I \end{bmatrix}. \tag{86}
$$

The regularity conditions (35)–(36) combined with the transformation (85) imply

$$
\begin{aligned}
&\begin{bmatrix} v(\cdot) \\ x(\cdot) \\ w(\cdot) \end{bmatrix} \in C([0,\infty); V_1 \times X_1 \times W_1) \cap C^1([0,\infty); V \times X \times W) \\
&\Leftrightarrow \\
&\begin{bmatrix} p(\cdot) \\ x(\cdot) \\ w(\cdot) \end{bmatrix} \in C([0,\infty); V_1 \times X_1 \times W_1) \cap C^1([0,\infty); V \times X \times W)
\end{aligned} \tag{87}
$$

and

$$
\begin{bmatrix} v(\cdot) \\ x(\cdot) \\ w(\cdot) \end{bmatrix} \in C([0,\infty); V \times X \times W) \Leftrightarrow \begin{bmatrix} p(\cdot) \\ x(\cdot) \\ w(\cdot) \end{bmatrix} \in C([0,\infty); V \times X \times W)\,. \tag{88}
$$

The extended system Σ with new state variables takes the form

$$\Sigma : \begin{bmatrix} \dot{p} \\ \dot{x} \\ \dot{w} \\ \hline e \end{bmatrix} = \left[\begin{array}{cccc} E & E\Pi_1 - \Pi_1 A + F_1 C_2 & E\Pi_2 - \Pi_2 Q + F_1 D_{21} - \Pi_1 B_1 & F_1 D_{22} - \Pi_1 B_2 + F_2 \\ 0 & A & B_1 & B_2 \\ 0 & 0 & Q & 0 \\ \hline G_1 & G_1\Pi_1 + H_{11}C_2 - C_1 & G_1\Pi_2 + H_{11}D_{21} - D_{11} & H_{11}D_{22} - D_{12} + H_{12} \end{array}\right] \times \begin{bmatrix} p \\ x \\ w \\ \hline u \end{bmatrix} \tag{89}$$

with the initial conditions

$$\begin{bmatrix} p(0) \\ x(0) \\ w(0) \end{bmatrix} = \begin{bmatrix} v_0 - \Pi_1 x_0 - \Pi_2 w_0 \\ x_0 \\ w_0 \end{bmatrix}. \tag{90}$$

For the output observation condition we have: for all $x(0) = x_0 \in X$, $w(0) = w_0 \in W$, $v(0) = v_0 \in V$ $(p(0) = (v_0 - \Pi_1 x_0 - \Pi_2 w_0) \in V)$ and $u(\cdot) \in C([0, \infty); U)$ the following holds

$$\begin{aligned} &\lim_{t\to\infty} \|e(t)\|_Z \\ &= \lim_{t\to\infty} \|G_1 p(t) + (G_1\Pi_1 + H_{11}C_2 - C_1)x(t) \\ &\qquad +(G_1\Pi_2 + H_{11}D_{21} - D_{11})w(t) + (H_{11}D_{22} - D_{12} + H_{12})u(t)\|_Z \\ &= 0\,. \end{aligned} \tag{91}$$

Extending the results of previous sections we arrive at the following theorem which may be regarded as the most general result of the paper.

Theorem 2 *(sufficient conditions) Given the extended system Σ described by the equation (89)–(90). If the observer $(E, F_1, G_1, H_{11}, F_2, H_{12})$ is exponentially stable, i.e. $\omega_0(S) < 0$, and the equations*

$$\begin{cases} E\begin{bmatrix}\Pi_1 & \Pi_2\end{bmatrix}\begin{bmatrix} h_1 \\ g_1 \end{bmatrix} - \begin{bmatrix}\Pi_1 & \Pi_2\end{bmatrix}\begin{bmatrix} A & B_1 \\ 0 & Q \end{bmatrix}\begin{bmatrix} h_1 \\ g_1 \end{bmatrix} + F_1\begin{bmatrix} C_2 & D_{21}\end{bmatrix}\begin{bmatrix} h_1 \\ g_1 \end{bmatrix} = 0\,, \\ \qquad \begin{bmatrix} h_1 \\ g_1 \end{bmatrix} \in X_1 \times W_1\,, \\ G_1\begin{bmatrix}\Pi_1 & \Pi_2\end{bmatrix}\begin{bmatrix} h \\ g \end{bmatrix} + H_{11}\begin{bmatrix} C_2 & D_{21}\end{bmatrix}\begin{bmatrix} h \\ g \end{bmatrix} - \begin{bmatrix} C_1 & D_{11}\end{bmatrix}\begin{bmatrix} h \\ g \end{bmatrix} = 0\,, \\ \qquad \begin{bmatrix} h \\ g \end{bmatrix} \in X \times W\,, \end{cases} \tag{92}$$

where each of the above equalities is in V*, have solutions*

$$\begin{bmatrix} \Pi_1 \ \Pi_2 \end{bmatrix} \in \mathscr{L}(X_1 \times W_1, V_1) \cap \mathscr{L}(X \times W, V) \tag{93}$$

and the following equalities are satisfied

$$\begin{aligned} F_2 &= \begin{bmatrix} \Pi_1 \ \Pi_2 \end{bmatrix} \begin{bmatrix} B_2 \\ 0 \end{bmatrix} - F_1 D_{22}\,, \\ H_{12} &= D_{12} - H_{11} D_{22}\,, \end{aligned} \tag{94}$$

then for all $x(0) = x_0 \in X$, $w(0) = w_0 \in W$, $v(0) = v_0 \in V$ *(*$p(0) = (v_0 - \Pi_1 x_0 - \Pi_2 w_0) \in V$*) and* $u(\cdot) \in C([0, \infty); U)$ *the following convergence takes place*

$$\lim_{t\to\infty} \|e(t)\|_Z = \lim_{t\to\infty} \|G_1 p(t)\|_Z = \lim_{t\to\infty} \|G_1 S(t) p(0)\|_Z = 0\,, \tag{95}$$

which means that the output observation condition holds.

Proof Let us notice that first of the Eq. (92) can be rewritten in the form

$$\begin{aligned} (E\Pi_1 - \Pi_1 A + F_1 C_2)h_1 + (E\Pi_2 - \Pi_2 Q + F_1 D_{21} - \Pi_1 B_1)g_1 = 0\,, \\ h_1 \in X_1\,,\ g_1 \in W_1\,, \end{aligned} \tag{96}$$

and the second one—in the form

$$(G_1\Pi_1 + H_{11}C_2 - C_1)h + (G_1\Pi_2 + H_{11}D_{21} - D_{11})g = 0\,, \quad h \in X\,,\ g \in W. \tag{97}$$

Taking into account the Eq. (94) written in the form

$$F_1 D_{22} - \Pi_1 B_2 + F_2 = 0\,, \quad H_{11} D_{22} - D_{12} + H_{12} = 0\,, \tag{98}$$

we see that (97) and (98) imply that the extended system (89)–(90) simplifies to the form

$$\Sigma : \begin{bmatrix} \dot p \\ \dot x \\ \dot w \\ \hline e \end{bmatrix} = \left[\begin{array}{ccc|c} E & 0 & 0 & 0 \\ 0 & A & B_1 & B_2 \\ 0 & 0 & Q & 0 \\ \hline G_1 & 0 & 0 & 0 \end{array}\right] \begin{bmatrix} p \\ x \\ w \\ \hline u \end{bmatrix}, \quad \begin{bmatrix} p(0) \\ x(0) \\ w(0) \end{bmatrix} = \begin{bmatrix} v_0 - \Pi_1 x_0 - \Pi_2 w_0 \\ x_0 \\ w_0 \end{bmatrix}. \tag{99}$$

Since $\omega_0(S) < 0$, then for every $x(0) = x_0 \in X$, $w(0) = w_0 \in W$, $v(0) = v_0 \in V$ ($p(0) = (v_0 - \Pi_1 x_0 - \Pi_2 w_0) \in V$) and $u(\cdot) \in C([0, \infty); U)$ we get the convergence (95). □

Let us notice that the first of the equations (92), i.e., the equation

$$E\begin{bmatrix} \Pi_1 \ \Pi_2 \end{bmatrix}\begin{bmatrix} h_1 \\ g_1 \end{bmatrix} - \begin{bmatrix} \Pi_1 \ \Pi_2 \end{bmatrix}\begin{bmatrix} A & B_1 \\ 0 & Q \end{bmatrix}\begin{bmatrix} h_1 \\ g_1 \end{bmatrix} + F_1\begin{bmatrix} C_2 \ D_{21} \end{bmatrix}\begin{bmatrix} h_1 \\ g_1 \end{bmatrix} = 0, \quad \begin{bmatrix} h_1 \\ g_1 \end{bmatrix} \in X_1 \times W_1, \tag{100}$$

where equality is in V, is an operator algebraic Sylvester equation. However it has now a bit more complicated form. Nevertheless, we can complement the above considerations with the following remarks.

Remark 2 If we assume that the operators A and Q generate strongly continuous *groups*, respectively, $(T(t))_{t\in\mathbb{R}} \subset \mathscr{L}(X)$ and $(P(t))_{t\in\mathbb{R}} \subset \mathscr{L}(W)$, then the combined operator $\begin{bmatrix} A & 0 \\ 0 & Q \end{bmatrix}$, generates a strongly continuous *group*

$$(\begin{bmatrix} T(t) & 0 \\ 0 & P(t) \end{bmatrix})_{t\in\mathbb{R}} \subset \mathscr{L}(X \times W).$$

It is clear that, $\mathscr{D}(\begin{bmatrix} A & 0 \\ 0 & Q \end{bmatrix}) = X_1 \times W_1$. It is known (see e.g. [17]) that for every bounded operator $B_1 \in \mathscr{L}(W, X)$ the perturbed operator $\begin{bmatrix} A & B_1 \\ 0 & Q \end{bmatrix}$, also generates a strongly continuous *group*, which we denote by $(\mathcal{U}(t))_{t\in\mathbb{R}} \subset \mathscr{L}(X \times W)$ and its growth bound by $\omega_0(\mathcal{U})$. This group is given explicitly as

$$\mathcal{U}(t)\begin{bmatrix} x_0 \\ w_0 \end{bmatrix} = \begin{bmatrix} T(t) & \int_0^t T(t-r)B_1P(r)\mathrm{d}r \\ 0 & P(t) \end{bmatrix}\begin{bmatrix} x_0 \\ w_0 \end{bmatrix}, \quad \begin{bmatrix} x_0 \\ w_0 \end{bmatrix} \in X \times W. \tag{101}$$

Consequently, the operator

$$-\begin{bmatrix} A & B_1 \\ 0 & Q \end{bmatrix} \tag{102}$$

generates a strongly continuous semigroup $(\mathcal{U}(-t))_{t\geq 0} \subset \mathscr{L}(X \times W)$, with the growth bound $\omega_0(\mathcal{U})$. We have $\mathscr{D}(\begin{bmatrix} A & 0 \\ 0 & Q \end{bmatrix}) = \mathscr{D}(\begin{bmatrix} A & B_1 \\ 0 & Q \end{bmatrix})$. Let us also notice that

$$-F_1\begin{bmatrix} C_2 \ D_{21} \end{bmatrix} \in \mathscr{L}(X \times W, V).$$

We are now allowed to apply Corollary 2 to the Eq. (100). It then follows that for $\omega_0(S) + \omega_0(\mathcal{U}) < 0$ the algebraic Sylvester Eq. (100) has a unique solution

$$\begin{bmatrix} \Pi_1 \ \Pi_2 \end{bmatrix} \in \mathscr{L}(X \times W, V) \cap \mathscr{L}(X_1 \times W_1, V_1). \tag{103}$$

Remark 3 In general case the operators A and Q generate strongly continuous *semigroups* and consequently the operator

$$\begin{bmatrix} A & B_1 \\ 0 & Q \end{bmatrix} = \begin{bmatrix} A & 0 \\ 0 & Q \end{bmatrix} + \begin{bmatrix} 0 & B_1 \\ 0 & 0 \end{bmatrix} \tag{104}$$

generates only a strongly continuous *semigroup* $(\mathcal{U}(t))_{t\geq 0} \subset \mathcal{L}(X \times W)$. It is explicitly given in the form

$$\mathcal{U}(t)\begin{bmatrix} x_0 \\ w_0 \end{bmatrix} = \begin{bmatrix} T(t) & \int_0^t T(t-r)B_1 P(r)\mathrm{d}r \\ 0 & P(t) \end{bmatrix}\begin{bmatrix} x_0 \\ w_0 \end{bmatrix}, \quad \begin{bmatrix} x_0 \\ w_0 \end{bmatrix} \in X \times W\,. \tag{105}$$

We now additionally *assume* that the pair

$$\left(\begin{bmatrix} C_2 & D_{21} \end{bmatrix}, \begin{bmatrix} A & B_1 \\ 0 & Q \end{bmatrix}\right) \tag{106}$$

is *exponentially detectable*, i.e. there exists an *output injection* operator $\begin{bmatrix} L_1 \\ L_2 \end{bmatrix} \in \mathscr{L}(Y, X \times W)$ such that the operator

$$\begin{bmatrix} A & B_1 \\ 0 & Q \end{bmatrix} - \begin{bmatrix} L_1 \\ L_2 \end{bmatrix}\begin{bmatrix} C_2 & D_{21} \end{bmatrix}, \tag{107}$$

withe the domain

$$\mathscr{D}\left(\begin{bmatrix} A & B_1 \\ 0 & Q \end{bmatrix} - \begin{bmatrix} L_1 \\ L_2 \end{bmatrix}\begin{bmatrix} C_2 & D_{21} \end{bmatrix}\right) = X_1 \times W_1\,, \tag{108}$$

generates a strongly continuous semigroup $(\mathcal{T}(t))_{t\geq 0} \subset \mathscr{L}(X \times W)$, which is *exponentially stable*. It means that its growth bound $\omega_0(\mathcal{T})$ satisfies

$$\omega_0(\mathcal{T}) < 0\,. \tag{109}$$

Moreover, let us notice that for every $\begin{bmatrix} \Pi_1 & \Pi_2 \end{bmatrix} \in \mathscr{L}(X \times W, V)$ we always have

$$F_1 := \begin{bmatrix} \Pi_1 & \Pi_2 \end{bmatrix}\begin{bmatrix} L_1 \\ L_2 \end{bmatrix} \in \mathscr{L}(Y, V)$$

and for such F_1 the Eq. (92) can be rewritten in the form

$$E\,[\,\Pi_1\;\Pi_2\,]\begin{bmatrix} h_1 \\ g_1 \end{bmatrix} = [\,\Pi_1\;\Pi_2\,]\Big(\begin{bmatrix} A & B_1 \\ 0 & Q \end{bmatrix} - \begin{bmatrix} L_1 \\ L_2 \end{bmatrix}[\,C_2\;D_{21}\,]\Big)\begin{bmatrix} h_1 \\ g_1 \end{bmatrix},$$

$$\begin{bmatrix} h_1 \\ g_1 \end{bmatrix} \in X_1 \times W_1\,,$$

$$G_1\,[\,\Pi_1\;\Pi_2\,]\begin{bmatrix} h \\ g \end{bmatrix} = (-H_{11}\,[\,C_2\;D_{21}\,] + [\,C_1\;D_{11}\,])\begin{bmatrix} h \\ g \end{bmatrix},$$

$$\begin{bmatrix} h \\ g \end{bmatrix} \in X \times W\,. \tag{110}$$

Under these assumptions we can derive the following characterization of the observer operators $(E, F_1, F_2, G_1, H_{11}, H_{12})$.

1. Choose an *observer state space* V and a *bijective* operator $[\,\Pi_1\;\Pi_2\,] \in \mathscr{L}(X \times W, V)$ (hence there exists $[\,\Pi_1\;\Pi_2\,]^{-1} \in \mathscr{L}(V, X \times W)$) and define the *observer system operator* as follows

$$E := [\,\Pi_1\;\Pi_2\,]\left(\begin{bmatrix} A & B_1R_1 \\ 0 & Q \end{bmatrix} - \begin{bmatrix} L_1 \\ L_2 \end{bmatrix}[\,C_2\;D_{21}R_1\,]\right)[\,\Pi_1\;\Pi_2\,]^{-1}.$$

   This operator generates a strongly continuous semigroup $(\mathcal{S}(t))_{t\geq 0} \subset \mathscr{L}(V)$, given by

$$\mathcal{S}(t) := [\,\Pi_1\;\Pi_2\,]\,\mathcal{T}(t)\,[\,\Pi_1\;\Pi_2\,]^{-1}, \quad t \geq 0\,,$$

   which is similar to $(\mathcal{T}(t))_{t\geq 0} \subset \mathscr{L}(X \times W)$. By similarity and (109) we have

$$\omega_0(\mathcal{S}) = \omega_0([\,\Pi_1\;\Pi_2\,]\,\mathcal{T}\,[\,\Pi_1\;\Pi_2\,]^{-1}) = \omega_0(\mathcal{T}) < 0\,.$$

2. Choose the *observer input operator*

$$F_1 := [\,\Pi_1\;\Pi_2\,]\begin{bmatrix} L_1 \\ L_2 \end{bmatrix} \in \mathscr{L}(Y, V)\,.$$

3. Choose arbitrary H_{11}.
4. Choose the *observer output operator*

$$G_1 := \Big([\,C_1\;D_{11}R_1\,] - H_{11}\,[\,C_2\;D_{21}R_1\,]\Big)[\,\Pi_1\;\Pi_2\,]^{-1}.$$

5. Compute F_2 and H_{12} as follows

$$F_2 := [\,\Pi_1\;\Pi_2\,]\left(\begin{bmatrix} B_2 \\ 0 \end{bmatrix} - \begin{bmatrix} L_1 \\ L_2 \end{bmatrix} D_{22}\right),$$
$$H_{12} := D_{12} - H_{11}D_{22}\,.$$

The above procedure provides a general characterization of the observer parameters $(E, F_1, F_2, G_1, H_{11}, H_{12})$. It is rather clear that the most essential role is played by the choice of the observer state space V. Usually, we try to choose V as small as possible, which leads to a "reduced order" observer. It is always possible to choose $V = X \times W$ and obtain a "full order" observer. How to design a full order observer, which may be called a Luenberger observer [21], is illustrated in the following example.

Example 1 *Luenberger full order observer.* We assume the following:

(A1): The observer state space has the form $V := X \times W$.

(A2): The operator $\left[\, \Pi_1 \; \Pi_2 \,\right]$ is given in the form

$$\left[\, \Pi_1 \; \Pi_2 \,\right] := \begin{bmatrix} I & 0 \\ 0 & I \end{bmatrix} \in \mathscr{L}(X \times W)\,.$$

It is obvious that $\left[\, \Pi_1 \; \Pi_2 \,\right]$ is boundedly invertible.

(A3): There exist bounded output injection operators

$$\begin{bmatrix} L_1 \\ L_2 \end{bmatrix} \in \mathscr{L}(Y, X \times W)$$

such that the operator

$$\begin{bmatrix} A & B_1 \\ 0 & Q \end{bmatrix} - \begin{bmatrix} L_1 \\ L_2 \end{bmatrix} \left[\, C_2 \; D_{21} \,\right] \tag{111}$$

generates a strongly continuous semigroup on $X \times W$ which is exponentially stable, i.e.

$$\omega_0\left(\begin{bmatrix} A & B_1 \\ 0 & Q \end{bmatrix} - \begin{bmatrix} L_1 \\ L_2 \end{bmatrix} \left[\, C_2 \; D_{21} \,\right] \right) < 0\,. \tag{112}$$

The assumption (A1) means that the observer state space V coincides with the extended system state space $X \times W$. In turn, the assumption (A3) is just the *exponential detectability* of the pair $\left(\left[\, C_2 \; D_{21} \,\right], \begin{bmatrix} A & B_1 \\ 0 & Q \end{bmatrix} \right)$. Under the assumptions (A1)-(A3) there always exists an observer which guarantees the output observation condition holds and parameters $(E, F_1, F_2, G_1, H_{11}, H_{12})$ of such observer can be found as follows.

1. Find $\begin{bmatrix} L_1 \\ L_2 \end{bmatrix}$ according to (A3).
2. Choose the observer system operator E in the form

$$E := \begin{bmatrix} A & B_1 \\ 0 & Q \end{bmatrix} - \begin{bmatrix} L_1 \\ L_2 \end{bmatrix} \left[\, C_2 \; D_{21} \,\right].$$

It is clear that

$$\omega_0(E) = \omega_0\left(\begin{bmatrix} A & B_1 \\ 0 & Q \end{bmatrix} - \begin{bmatrix} L_1 \\ L_2 \end{bmatrix}\begin{bmatrix} C_2 & D_{21} \end{bmatrix}\right) < 0 . \tag{113}$$

3. Choose F_1 as follows

$$F_1 := \begin{bmatrix} L_1 \\ L_2 \end{bmatrix}.$$

4. Choose arbitrary H_{11} (e.g. $H_{11} = 0$).
5. Choose the observer output operator G_1 in the form

$$G_1 := \begin{bmatrix} C_1 & D_{11} \end{bmatrix} - H_{11}\begin{bmatrix} C_2 & D_{21} \end{bmatrix}.$$

6. Choose operators F_2 and H_{12} as follows

$$\begin{aligned} F_2 &:= \begin{bmatrix} B_2 \\ 0 \end{bmatrix} - \begin{bmatrix} L_1 \\ L_2 \end{bmatrix} D_{22} , \\ H_{12} &:= D_{12} - H_{11} D_{22} . \end{aligned}$$

Thus we obtain the *full order* Luenberger output observer has the form

$$\Sigma_O : \begin{cases} \dot{v}(t) = \left(\begin{bmatrix} A & B_1 \\ 0 & Q \end{bmatrix} - \begin{bmatrix} L_1 \\ L_2 \end{bmatrix}\begin{bmatrix} C_2 & D_{21} \end{bmatrix}\right)v(t) + \begin{bmatrix} L_1 \\ L_2 \end{bmatrix} y(t) \\ \qquad + \left(\begin{bmatrix} B_2 \\ 0 \end{bmatrix} - \begin{bmatrix} L_1 \\ L_2 \end{bmatrix} D_{22}\right)u(t) , \\ z_O(t) = \left(-H_{11}\begin{bmatrix} C_2 & D_{21} \end{bmatrix} + \begin{bmatrix} C_1 & D_{11} \end{bmatrix}\right)v(t) + H_{11} y(t) \\ \qquad + (D_{12} - H_{11} D_{22})u(t) . \end{cases}$$

8 Conclusions

The results presented in this paper provide sufficient conditions which are to be satisfied by observer parameters in order to guarantee asymptotic tracking of an output of an infinite-dimensional plant in the presence of deterministic input disturbances. The main limitation of the applicability of the results is the assumption that input and output operators of the plant and the observer are bounded which means that e.g. plants and observers with boundary inputs and outputs are not covered [15, 22]. It seems possible to generalize the presented approach to systems with unbounded input and output operators although tough new theoretical problems may arise and have to be solved. We think that the most promising direction of further development is to make use of geometric methods for infinite-dimensional systems (e.g. invariant subspaces [23], generalized inverses [24]) which should allow to obtain effective methods of synthesis of infinite-dimensional output observers in the general setup. Our hope is justified by the existence of well developed theory for parallel problems of output regulation for infinite-dimensional systems (see [25] and references cited therein).

References

1. Aldeen, M., Trinh, H.: Reduced-order linear functional observer for linear systems. IEE Proc. Control Theory Appl. **146**(5), 399–405 (1999)
2. Darouach, M.: Existence and design of functional observers for linear systems. IEEE Trans. Auto. Control **45**(5), 940–943 (2000)
3. Fernando, T., MacDougall, S., Sreeram, V., Trinh, H.: Existence conditions for unknown input functional observers. Int. J. Control **86**(1), 22–28 (2013)
4. Fernando, T., Trinh, H., Jennings, L.: Functional observability and the design of minimum prder linear functional observers. IEEE Trans. Auto. Control **55**(5), 1268–1273 (2010)
5. Trinh, H., Fernando, T.: On the existence and design of functional observers for linear systems. In: Proceedings of the 2007 IEEE International Conference on Mechatronics and Automation, pp. 1974–1979. Harbin, China (2007)
6. Trinh, H., Fernando, T.: Functional Observers for Dynamical Systems. Springer, Berlin (2012)
7. Emirsajłow, Z.: Remarks on functional observers for distributed parameter systems. In: Proceedings of the 23rd International Conference on Methods and Models in Automation and Robotics, pp. 27–30. Miedzyzdroje, Poland (2018)
8. Hidayat, Z., Babuska, R., De Schutter, B., Nunez, A.: Observers for linear distributed-parameter systems: A survey. In: Proceedings of the 2011 IEEE International Symposium on Robotic and Sensors Environments, pp. 166–171. Montreal, Canada (2011)
9. Kohne, M.: Implementation on distributed parameter state observers. Distrib. Param. Syst. Model. Iden. **1**, 310–324 (1978) (Springer-Verlag)
10. Li, Y., Chen, X.: End-point sensing and state observation of a flexible link robot. Mechatronics **6**, 351–356 (2001)
11. Liu, Y.A., Lapidus, L.: Observer theory for distributed-parameter systems. Int. J. Syst. Sci. **7**, 731–742 (1976)
12. Sakawa, Y., Matsushita, T.: Feedback stabilization of a class of distributed systems and construction of a state estimator. IEEE Trans. Auto. Control **20**, 748–753 (1975)
13. Emirsajłow, Z.: Infinite-dimensional Sylvester equations: basic theory and applications to observer design. Int. J. Appl. Math. Comput. Sci. **22**(2), 245–257 (2012)
14. Emirsajłow, Z.: Robustness of solutions of an infinite-dimensional algebraic Sylvester equation under bounded perturbations. Control Cybernet. **42**(1), 27–47 (2013)
15. Emirsajlow, Z., Townley, S.: On application of the implemented semigroup to a problem arising in optimal control. Int. J. Control **78**(4), 298–310 (2005)
16. Vu Quac Phong: The operator equation AX-XB=C with unbounded operators A and B and related Cauchy problems. Mathematische Zeitschrift **208**, 567–588 (1991)
17. Engel, K.-J., Nagel, R.: One-parameter Semigroups for Linear Evolution Equations. Graduate Texts in Mathematics No. 194. Springer-Verlag (2000)
18. Emirsajłow, Z.: Implemented Semigroup in the Analysis of Infinite-Dimensional Sylvester and Lyapunov Equations. Wydawnictwo Uczelniane Politechniki Szczecińskiej, Szczecin (2005). (in polish)
19. Alber, J.: On implemented semigroups. Semigroup Forum **63**(3), 371–386 (2001)
20. Curtain, R., Zwart, H.: An Introduction to Infinite-Dimensional Linear Systems Theory. Springer-Verlag, New York (1995)
21. Luenberger, D.: Observers for multivariable systems. IEEE Trans. Auto. Control **11**, 190–197 (1966)

22. Tucsnak, M., Weiss, G.: Observation and Control for Operator Semigroups. Birhauser, Basel (2009)
23. Zwart, H.: Geometric Theory for Infinite Dimensional Systems. Springer-Verlag, Berlin (1989)
24. Adi Ben-Israel, T.: Greville: Generalized Inverses. Springer-Verlag, Theory and Applications, New York (2003)
25. Paunonen, L.: Controller design for robust output regulation of regular linear systems. IEEE Trans. Auto. Control **61**(10), 2974–2986 (2016)

Non-Gaussian Noise Reduction in Measurement Signal Processing

Jerzy Świątek, Krzysztof Brzostowski, and Jarosław Drapała

Abstract The currently available computational power of machines allows for the modelling and analysis of nonlinear processes measured in the presence of non-Gaussian disturbances. This work gives an overview of methods useful for the analysis and reduction of the noise that can be met when using modern sensors. In order to obtain reliable estimates of the measured values, the noise reduction method should be chosen according to the type of process being measured (linear or nonlinear) and the characteristics of the noise (Gaussian or non-Gaussian). We focused on filters belonging to the Kalman family i.e.: original Kalman filter, extended Kalman filter, unscented Kalman filter, particle filter and ensemble Kalman filter. The key ideas behind the design of these filters were explained and their theoretical properties were described. Importantly, recommendations were made regarding their applicability in various types of measuring systems.

1 Introduction

The problem of removing noise from measurements is one of the key issues in the industrial applications of methods of automatics, robotics, telecommunication, biomedicine, etc. The main purpose of signal denoising is to extract relevant information from measurement data.

Noise present in measurement data may be due to the internal properties of the process (the process noise) or to external disturbances.

J. Świątek (✉) · K. Brzostowski · J. Drapała
Wrocław University of Science and Technology, Faculty of Computer Science and Management, Wyb. Wyspiańskiego 27, 50-370 Wrocław, Polska
e-mail: jerzy.swiatek@pwr.edu.pl

K. Brzostowski
e-mail: krzysztof.brzostowski@pwr.edu.pl

J. Drapała
e-mail: jaroslaw.drapala@pwr.edu.pl

P. Kulczycki et al. (eds.), *Automatic Control, Robotics, and Information Processing*, Studies in Systems, Decision and Control 296,
https://doi.org/10.1007/978-3-030-48587-0_4

Many methods of noise reduction rely on the assumption that the noise has the Gaussian distribution, meaning that the measured quantity is actually a combination of a pure signal with an undesirable value randomly drawn according to the Gaussian probability density function. Unfortunately, this assumption does not hold when working with most available sensors and devices. The most common causes of non-Gaussianity are: thermal effects in electronic devices, sampling frequency fluctuations in these devices and the so-called avalanche effect. Removing noise from a signal subject to non-Gaussian disturbances is a demanding task. It requires the almost from scratch design of a filter for each particular measurement system or at least the tedious adjustment of a carefully selected type of filter. Moreover, the designer cannot take advantage of the many useful properties of noise reduction systems (such as confidence intervals for estimated values or the convergence rate of adaptive algorithms) that are available only for systems affected by Gaussian noise.

This work gives an overview of modern noise reduction methods intended for non-Gaussian disturbances. The work consists of two parts. The first is concerned with mathematical descriptions of measurements and noises and gives some details of the methods of noise analysis. The second part is devoted to methods of noise reduction based on the Kalman filter family.

2 Measurements and Noises

Measurement of something is the process of measuring it in order to obtain a result that is expressed in numbers.

Measurement is the process of measuring in order to obtain the properties of an object. The result of the process is the value, which in real-world conditions has two components. The first of these components is the true value, i.e. the quantity describing the property of the object being observed. The second of these components is noise. Noise is an undesirable component of measurements that makes further analysis and the processing of the results of measuring difficult. The main sources of noise in the measurements are related to:

- Uncertainty in the observed object or measurement process;
- Utilization of measurement devices (i.e. sensors);
- External conditions (environmental);
- Sampling methods;
- The quantization process;
- Errors in the transmission channel.

Generally, noise can be divided into internal and external noise [19]. Internal noise is strictly related to the object, while external noise is caused by environmental conditions and lies outside the measuring system.

2.1 Measurements

Let us consider a typical problem in measurement systems. A certain source (object) generates a signal $s(t)$, which is measured by the sensor unit. The value acquired by the sensor is marked by $y(t)$. In perfect measuring conditions, there are no disturbances affecting the object or the measuring device. The described condition is illustrated in Fig. 1. The mathematical model for this case has the following form:

$$y(k) = h\big(s(k)\big), \tag{1}$$

where k denotes a discrete-time moment and h is a function describing the relation between the measurand and the measurement.

In real-world settings, gathered measurements contain unwanted components. Therefore, more complicated measurement data has to be used. In Fig. 2, we present an extension of the model illustrated in Fig. 1. In this case, external noise is taken into account. Based on Fig. 2, the model can be written as:

$$y(k) = h_v\big(s(k), v(k)\big), \tag{2}$$

where $v(k)$ stands for external noise.

Models (1) and (2) are suited for a static object. In the case of dynamic objects, models are generalized to the following form:

$$\begin{aligned} s(k+1) &= f_n\big(s(k), n(k)\big), \\ y(k) &= h_v\big(s(k), v(k)\big), \end{aligned} \tag{3}$$

where $n(k)$ is a process (dynamical) noise, and f_n represents the model of the process. Model (3) is illustrated in Fig. 3.

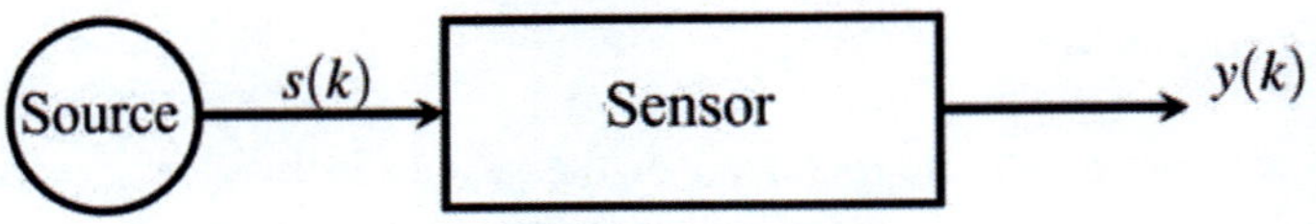

Fig. 1 Scheme of a simple measuring system

Fig. 2 Scheme of a simple measuring system with external noise

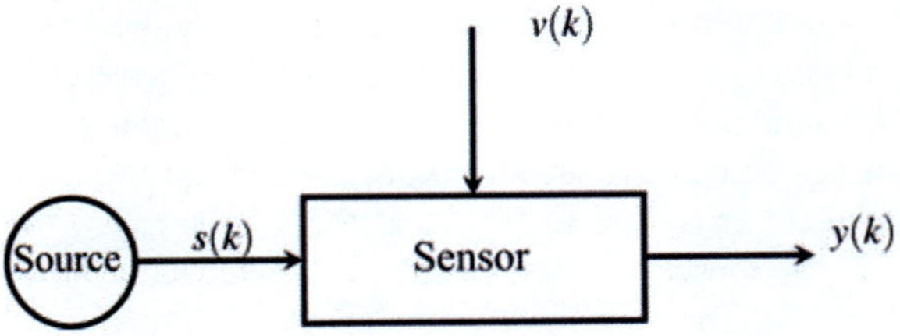

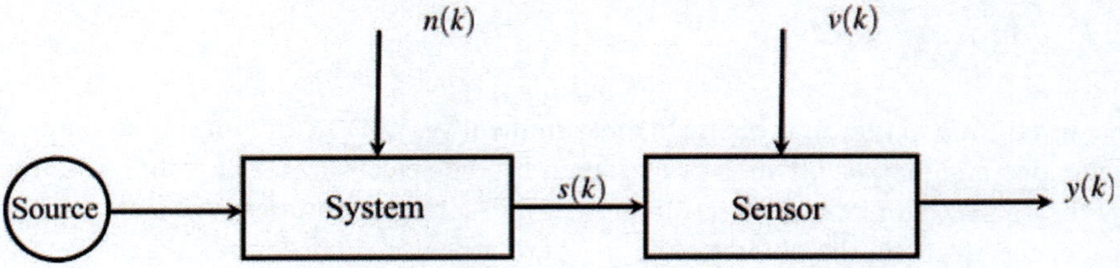

Fig. 3 Scheme of a simple measuring system with process and external noise

Process (dynamical) noise is usually modeled as additive white Gaussian noise (AWGN) [25]. In the case of external noise, application of AWGN to noise modeling is inadequate [20]. In order to improve the overall quality of modeling external noise, it is necessary to use more sophisticated models.

2.2 *Noise*

Process noise is related to the object of interest and its influence is permanent (Eq. 3). On the other hand, external noise is usually connected with environmental conditions (e.g. temperature, humidity, etc.) or with the measuring system.

Since noise affects the results of processing and the analysis of signals, the important issue is the characterization of unwanted components of the acquired measurements. In this case, it is important to comprehend the nature of the noise and its sources. In this section, we discuss the typical problems related to noise modeling. In addition, we spare some part of the work for the analysis of the common mechanisms of noise generation. In the next section, however, we will consider some methods of noise characterization.

2.2.1 Sources of Noise

Generally, the sources of noise generation can be various. In this work, we concentrate on physical causes. Figure 4 presents a taxonomy of noise sources and their types with regards to physical causes. We divide noise into three groups: environmental noise, noise related to electronic devices, and noise related to data transmission.

In environmental causes, we distinguish electromagnetic noise, which is usually related to the radio frequency range (from several kHz to several GHz), and electrostatic noise, which is mainly caused by electric charges. Another source of noise is acoustic noise, which is connected to the movement and vibration of measuring objects, as well as to the phenomenon of sound wave reflection from obstacles.

The sources of noise can also be related to the flow of electric current in electronic devices. The noise in electronic devices is mainly caused by phenomena connected with the flow of electric current and its fluctuations. In electronic devices, we distin-

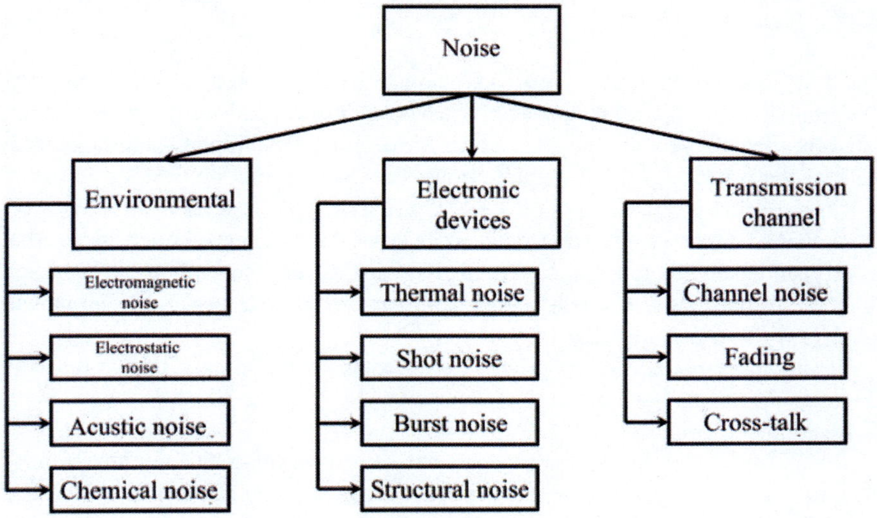

Fig. 4 Taxonomy of noise types

guish, among others, thermal noise, which is connected with chaotic movements of free electrons in the resistance of electronic systems. As a result of chaotic movements of free electrons, we can observe heat emission, which causes the effect of self-heating of the electronic device. Another type of noise, i.e. shot noise, is connected with the flow of electric current in devices and the process of induction electric charges. However, structural noise is related to inaccuracies in the technological process of manufacturing electronic devices. One example of structural noise is avalanche noise, which occurs in semiconductor systems and is associated with the p-n connector [19]. Burst noise is a kind of structural noise which is connected with inaccuracies in the technological process of manufacturing electronic devices. The noise occurs in the frequency range below 100 [Hz]. The characteristic feature of burst noise is that both the moment of its occurrence and its duration are random.

The main source of noise connected with the transmission channel, among others, is channel noise. Another type of noise is related to the phenomenon of signal suppression. It is caused by the overlapping of signals in the transmission channel. Another example of noise is that of cross-talks, which are caused by overlapping of the signal transmitted in one channel with the signal transmitted in another one. Cross-talks may occur, for example, during measurements made with the use of a sensor connected to the energy network. In this case, we can observe in acquired signals unwanted components, in which the frequency is close to the frequency of the electric current of the energy network.

2.2.2 Noise Modelling

The basic model is white noise, which is characterized by a completely flat frequency spectrum. It means that white noise has equal power in any band of a given bandwidth. The other feature of white noise is a lack of memory. This implies that the current state of the process generating white noise does not depend on past states of the process.

A special case of white noise is Gaussian noise. Gaussianity of noise means that its probability distribution function in the time domain has the form of the Gaussian function. Additionally, when the noise is additive, this white Gaussian noise is named *additive white Gaussian noise* (AWGN).

It should be noted that for probability distribution functions different than the Gaussian function, e.g., Cauchy or Poisson functions, the noise is named non-Gaussian white noise.

In real-world problems, noise characterized by a flat frequency spectrum is uncommon. Typically, noise in real-world problems has a more complex nature and it is difficult to capture it by applying, for example, AWGN. It may then be useful to assume that noise in real-world problems is correlated.

Correlated noise (or color noise) is the noise in which the frequency spectrum differs from the spectrum of white noise. It means that the spectral distribution of power is not flat in the whole frequency spectrum. We distinguish the following types of nonwhite noises:

- Red noise (Brownian noise);
- Pink noise;
- Blue noise;
- Violet noise.

Red noise (Brownian noise)

Red noise is generated by Brownian movements. The power of red noise is concentrated in the low-frequency spectrum. It means that the spectral power density decreases with an increasing frequency of the noise. The rate of decrease (3 dB per octave) of the red noise spectrum is inversely proportional to the squared frequency. Red noise can be determined by integrating white noise.

Pink noise

Pink noise has a spectral power density inversely proportional to frequency. Thus, the noise is often named as $\frac{1}{f}$. It is worth mentioning that the relationship between spectral power density and frequency is accomplished in the full frequency range only theoretically. The spectral energy of $\frac{1}{f}$ noise is also concentrated in the low-frequency spectrum.

The pervasiveness of pink noise in measurement signals does not mean that the noise is comprehensively investigated and described. There is also no method to generate pink noise. For example, it is not possible to generate the noise by a simple integration of white noise (as is the case with Brown noise).

Some approximation of pink noise $\frac{1}{f}$ can be obtained by applying a formal generalization of Brown's fractional motion (fBm). Pink noise is an example of non-gaussian noise, [8, 12, 26, 46].

Blue noise

The energy of blue noise agglomerates in the high frequencies. In this case, the spectral density of the power is proportional to f and grows to 3 [dB] per octave. Generally, blue noise can be considered as the inverse of pink noise.

Violet noise

The energy of purple noise is also agglomerated in the high-frequency range similar to blue noise. However, the energy of violet noise is more agglomerated in the high-frequency. The noise can be considered as the inverse of red noise. In this case, the noise energy increases by 6 [dB] per octave and is proportional to f^2. It is worth mentioning that purple noise can be obtained by differentiating white noise.

3 Description of Noise in Sensor Units

A further step in the analysis of noise is related to the determination of its characteristics. Noise characteristics can be used, for example, to design and build generators of the noise with desired properties. It is an important issue in experimental studies concerning, e.g. algorithms of removing undesirable components from measurement signals. To this end, suitable methods should be applied. In this section, two of them will be discussed. The main algorithms widely used to analyze and make a description of the noise in measurements are the Power Spectral Density (PSD) and the Allan Variance (AV).

Power Spectral Density describes the frequency structure of the analyzed signal by estimation of the spectral density of the mean-square value of the signal [7]. Determination of the frequency structure of the signal allows important information on the analyzed signal to be obtained.

The PSD approach has some limitations. These limitations restrict application areas. The approach is only suitable to analyze stationary signals. However, Power Spectral Density is used, among others, to analyze signals acquired from inertial sensors such as accelerometers [33, 36], but also from magnetometers, [27, 31, 38].

An alternative to PSD is the Allan Variance. AV is a method of representing the root mean square of noise as a function of averaging time. One of the applications of the Allan variance method is the determination of the characteristics of the underlying processes that give rise to the noise in measurements [2]. Originally, AV was used to measure the stability of oscillators and clocks [3]. One of the state-of-the-art applications of the Allan Variance is the construction of methods to analyze types of noise and their evaluation for measuring devices (e.g. acceleration sensors). One of the characteristic features of the Allan variance approach is that the sensor data is analyzed in the time domain.

In the AV-based methods for signal analysis, we distinguish two stages. In the first of them, the value of the Allan Variance is calculated. In the second one, the estimation of error values is performed. It is worth noting that the determined value of the Allan variance depends on the length of the time interval. Therefore, the selection of this interval is an important stage. In the next section, we present the main steps of the Allan Variance algorithm.

3.1 The Allan Variance

The description of the Allan Variance algorithm is based on works [14, 21]. Let us assume that there are K consecutive data points, each having a sample time of Deltat. Now let's group together the k consecutive data points (with $k < K/22$). Each element of the formed group is a cluster. Associated with each cluster is a time T, which is equal to $T = k\Delta t$. The cluster average is defined as:

$$\bar{y}_i (T) = \frac{1}{T} \int_{t_I}^{t_i+T} y(\tau)\, d\tau, \tag{4}$$

where T denotes the time associated with each cluster, y is the measurement from the sensor unit, $\bar{y}_i$ represents the cluster average of the output rate for a cluster, and i stands for the index of the cluster. In turn, the definition of the subsequent cluster average is:

$$\bar{y}_{i+1} (T) = \frac{1}{T} \int_{t_{i+1}}^{t_{i+1}+T} y(\tau)\, d\tau, \tag{5}$$

where $t_{i+1} = t_i + T$. Subsequently, based on (4) and (5), the difference between the two adjacent clusters is calculated:

$$\varepsilon_{i+1,i} = \bar{y}_{i+1} (T) - \bar{y}_i (T). \tag{6}$$

Based on these calculations, we can determine the Allan Variance from:

$$\sigma^2_{Allana} (T) = \frac{1}{2(K-2k)} \sum_{i=1}^{K-2k} (\bar{y}_{i+1} (T) - \bar{y}_i (T))^2. \tag{7}$$

3.2 Types of Noise

The noise in the sensor unit can be related to thermal phenomena (e.g. self-heating), the flow of electric current (e.g. fluctuations), or caused by inaccuracies in the technological process of electric circuits. Typical types of noise caused by the mentioned phenomena in measuring devices are [14, 29]:

- Quantization nois;
- Angular random walk and velocity random walk;
- Bias instability;
- Rate random walk;
- Rate ramp.

3.2.1 Quantization Noise

Quantization noise is connected with the operation of the quantization of the measurement signal. This noise can be expressed as a relation between the Power Spectral Density function and the Allan Variance by [14, 29]:

$$\sigma_{qn}^2(T) = \frac{3Q_{qn}^2}{T^2}, \tag{8}$$

where Q_{qn} is a quantization-noise coefficient.

Figure 5 indicates that the quantization noise is represented by a slope of -1 in a $log - logg$ plot of AV versus T. From this plot, we can read off the magnitude of this noise as the slope of the line at $T = 3^{1/2}$.

It is worth noting that quantization noise can only be modeled as additive noise when the quantization resolution is high [22]. Since quantization of the signal is a non-linear operation, it is not possible to obtain small errors at a low quantization resolution. This can only be achieved with a sufficiently high resolution of the A/D converter [22] (Fig. 6).

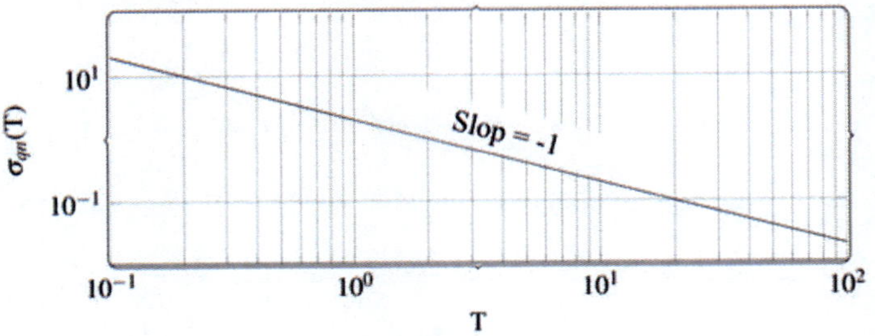

Fig. 5 The Allan Variance plot for quantization noise

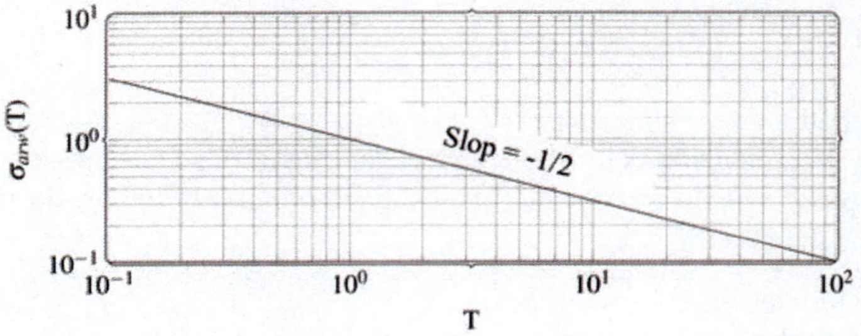

Fig. 6 The Allan variance plot for ARW and VRW

3.2.2 Angular Random Walk and Velocity Random Walk

Angular random walk and velocity random walk noise (ARW and VRW respectively) are noise dominated by high-frequency components. It means that their correlation time is very short. The parameter characterizing ARW and VRW noise, expressed using the Allan variance, [14, 29], has the following form:

$$\sigma_{arw}^2(T) = \frac{Q_{arw}^2}{T}, \tag{9}$$

where Q_{arw} represents a ARW and VRW coefficient.

For ARW and VRW noise, the $log - log$ plot of AV versus T has a slope of $-1/2$. Furthermore, the value of Q_{arw} can be directly obtained by reading the slope line at $T = 1$.

3.2.3 Bias Instability

Bias instability (BI) noise is the result of current flow in electronic circuits and vibrations caused by other components of the measuring sensors [14, 29]. This noise is characterized by a predominance of low-frequency components and is generated by the influence of the magnetic field (flickering noise). BI noise is described by the following equation:

$$\begin{aligned} \sigma_{bi}^2(T) = \frac{2B_{bi}^2}{\pi} \times \Bigg[\ln 2 - \frac{\sin(\pi f_0 T)^3}{2(\pi f_0 T)^2} \Big(\sin(\pi f_0 T) + \\ + 4\pi f_0 T \cos(\pi f_0 T) + c_{bi}(2\pi f_0 T) - c_{bi}(4\pi f_0 T) \Big) \Bigg], \end{aligned} \tag{10}$$

where f_0 represents cutoff frequency, B_{bi} is the bias instability coefficient and c_{bi} represents the cosine-integral function (Fig. 7).

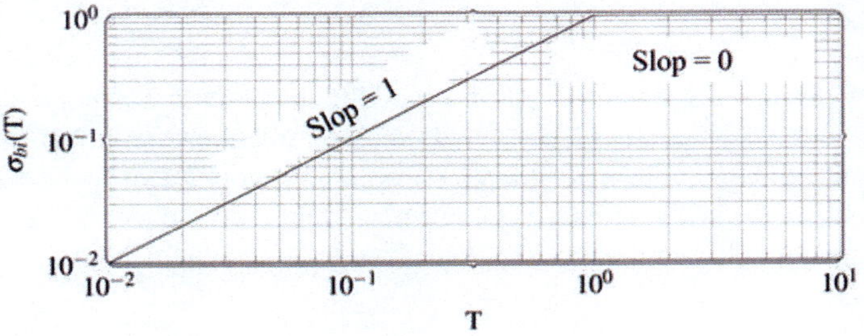

Fig. 7 The Allan variance plot for bias instability

In Fig. 10, a $log - log$ plot of the Allan variance for bias instability is presented. The plot shows that, in this case, AV reaches a plateau for T in a much longer time than the inverse cutoff frequency. The flat region of the $log - log$ plot can be used to estimate the limit of the bias instability.

3.2.4 Rate Random Walk Noise

Rate random walk noise (RRW) is the noise with a predominance of low-frequency components. This is a process that possibly limits an exponentially correlated noise with a very long correlation time. In inertial sensors (e.g. accelerometers), this phenomenon might have a negative effect on the measurement of acceleration (VRRW) and angular rate random walk (ARRW). RRW is determined with:

$$\sigma_{rrw}^2 (T) = \frac{K_{rrw}^2 T}{3}, \tag{11}$$

where K_{rrw} is the RRW coefficient.

Figure 8 indicates that the RRW is represented by a slope of $+1/2$ on a $log - log$ plot of AV versus T. In turn, its magnitude may be read off as the slope line at $T = 3$.

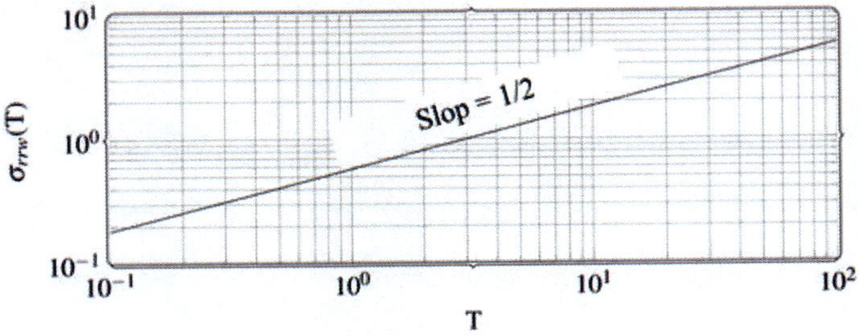

Fig. 8 The Allan variance plot for rate random walk noise

3.2.5 Rate Ramp

The noises discussed previously are of a random character. Rate ramp is the noise associated with a systematic (deterministic) measurement error. One of the sources of systematic errors is the inaccuracy of the measurement sensor. The RR noise expressed using the Allan variance is determined based on the following term:

$$\sigma_{rr}^2(T) = \frac{R_{rr}^2 T^2}{2}, \tag{12}$$

where R_{rr} is the RR coefficient.

Rate ramp noise is illustrated in Fig. 9. It can be seen that RR has a slope of $+1$ in the $log - log$ plot of AV versus T. The amplitude of rate ramp may be determined from the slope line at $T = 2^{1/2}$.

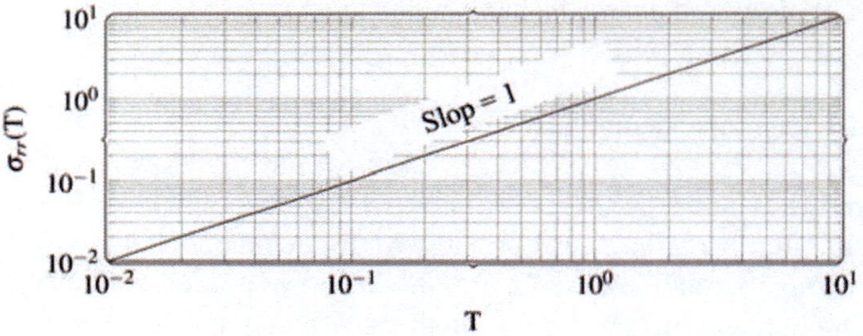

Fig. 9 The Allan variance plot for rate ramp

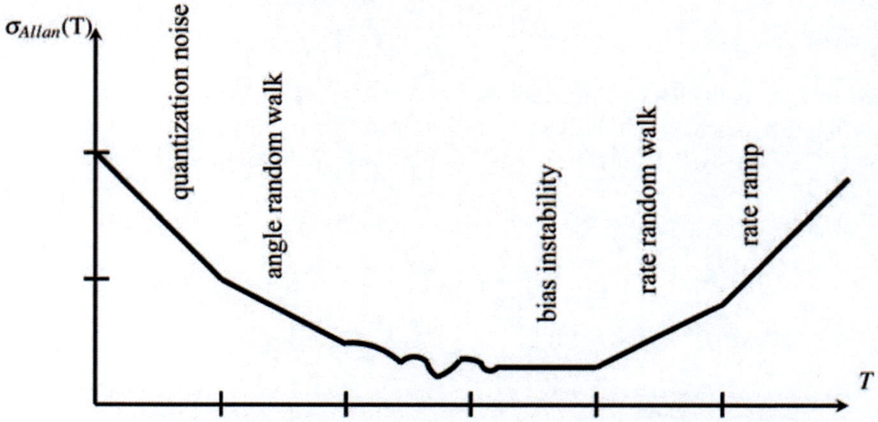

Fig. 10 The Allan variance for different types of noise

3.3 *Summary of the Allan Variance*

Let us summarize our consideration regarding methods of noise in measurement signals based on the Allan variance. If the analyzed data contain only one type of noise, then the appropriate AV plot has a form similar to that previously presented. In real-world problems, measurements may contain various types of noise. As an illustration, in Fig. 10 we present the Allan Variance plot for different types of noise.

The AV plot can be used to analyze measurements in terms of various types of noise. It can help us to assess which noise dominates in the analyzed data. The plot is also useful for evaluating the quality of algorithms to remove undesirable components from the measurement signals (e.g. denoising or detrending methods).

4 Noise Analysis in Acceleration Data

Noise in an electronic circuit (also in acceleration sensors) is associated with phenomena inside the device, as well as with external causes [13, 49]. In this case, the main sources of noise are [6, 10, 50]:

- Thermal effects;
- Fluctuation of electric current;
- Self-heating of sensors and ambient temperature;
- Non-orthogonality of measurement axes.

For example, thermal effects in electronic circuits are the result of chaotic movements of electrons. The flow of electric current causes heat energy in circuits. The result of electric current is noise that has the characteristics of white noise (in accelerometers this noise is called VRW noise). Another factor that has an impact on the quality of measurements, i.e. fluctuations of the electric current, is related to defects in the electric current structure.

Temperature effects in electric circuits are twofold. On the one hand, the temperature effect can be related to ambient heat. And on the other hand, the temperature effect may be caused by self-heating phenomena in the sensor's electric circuits. These phenomena are strongly non-linear. Moreover, the described phenomena cause the non-stationary nature of measurements.

Some of the previously discussed phenomena, which have an impact on the performance of the acceleration sensor, may be modeled. In turn, the results of its modeling may be used to eliminate their negative impact on the measurements. Table 1 summarizes the phenomena affecting the accuracy of the acceleration sensor and the acquired measurement data [50].

Figure 11a illustrates an example of accelerometer's data (in this case the x-axis). In turn, Fig. 11b, c show a representation of the signal applying Power Spectral Density and the Allan variance, respectively. The plots show that AV allows the type of noise that corrupted the analyzed signals to be recognized. For example, analysis of

Table 1 Types of noise in accelerometer data

Type of noise	Description
White noise	Relate to brownian motions
Quantization noise	Relate to signal quantization
RRW noise	Low frequency noise
RR noise	Deterministic noise

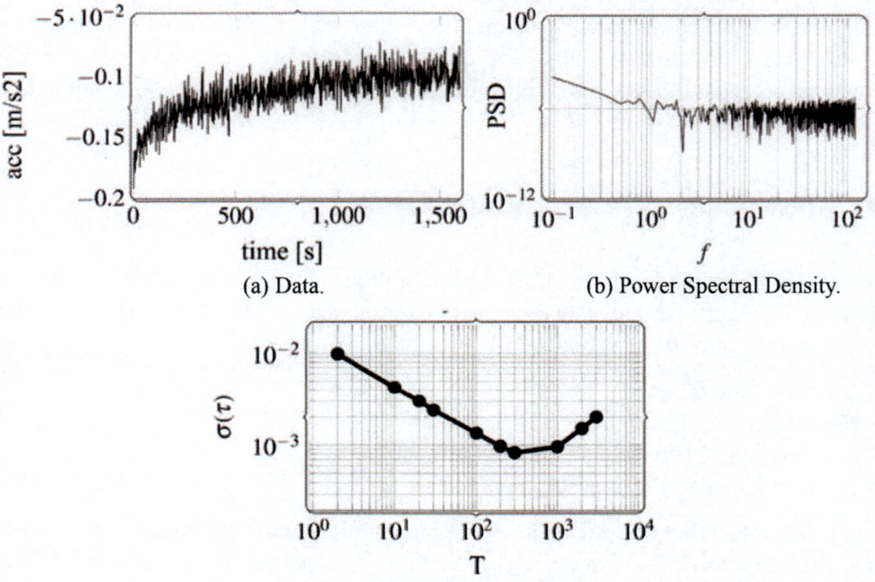

(a) Data.

(b) Power Spectral Density.

(c) The Allan Variance.

Fig. 11 Analysis of acceleration data for x-axis

Fig. 11c leads us to the conclusion that for the short time intervals of T, the dominant noise is quantization noise. On the other hand, for the case of the long-time intervals of T, the dominant noises are RRW and RR.

5 Methods of Noise Reduction

The lesson learned from the previous sections is that a multitude of sources and types of noises make it necessary to expand the toolbox of noise reduction methods with routines that are able to deal with nonlinear and non-Gaussian disturbances.

The second part of the work gives an overview of the Kalman filter family, which is commonly used to state estimation of the dynamic system. Kalman filters and their

Table 2 Classification of filters by type of system and type of noise

	Gaussian noise	Non-Gaussian noise
Linear system	Kalman filter	Unscented Kalman filter
Nonlinear system	Extended, unscented and ensemble Kalman filter	Extended, unscented and particle Kalman filter

varieties are widely used in industry and scientific research and they dominate in the literature of the subject. The filters considered in the work are:

- Kalman filter;
- Extended Kalman filter;
- Unscented Kalman filter;
- Particle filter;
- Ensemble Kalman filter.

Generally speaking, input-output systems may be divided into two types: linear and nonlinear. Moreover, two types of noise may appear: Gaussian and non-Gaussian. Therefore, there are four cases included in Table 2. Different filters are recommended in each of these cases. However, it is not possible to strictly follow these recommendations in practice. They should be treated rather as suggestions of a good start for noise reduction in a particular use case. The user needs to take into account the computational resources available and also requirements regarding operating time. It is very often the case, that a filter operates out of its domain of applicability (when assumptions behind the filter are violated), until it significantly reduces the quality of the signal processing procedure.

It should be noted, that nonlinearity often goes hand in hand with non-Gaussian noise. This is due to the fact, that applying nonlinear state equations to the Gaussian function several times results in significant deformation of its shape.

5.1 Kalman Filter

When dealing with a linear system and when disturbances n, v are additive and have Gaussian distribution, then the original Kalman filter (KF) is the optimal noise reduction method that minimizes the least square error, [40]. In this situation the state equation and measurement equation (3) take a special form:

$$\begin{aligned} s(k+1) &= \vec{A}s(k) + n(k), \\ y(k) &= \vec{C}s(k) + v(k), \end{aligned} \tag{13}$$

where $\vec{A}$ and $\vec{C}$ are the system matrix and the output matrix, respectively. The coefficients of these matrices are determined during the system modelling stage. It is assumed that the noises n and v are mutually independent and may be fully described by the covariance matrices $\vec{Q}_n = E[n^T n]$ and $\vec{Q}_v = E[v^T v]$. As more measurements $y(k)$ are being taken, the system state $s(k)$ estimation is performed and the *Kalman gain* matrix $\vec{K}$ together with the *prediction-error covariance matrix*) $\vec{P}$ are updated, according to formulas:

$$\begin{aligned} \vec{K}(k) &= \vec{A}\vec{P}(k)\vec{C}^T\left(\vec{C}\vec{P}(k)\vec{C}^T + \vec{Q}_v\right)^{-1}, \\ \widehat{s}(k+1) &= \vec{A}\widehat{s}(k) + \vec{K}(k)\big(y(k+1) - \vec{C}\widehat{s}(k)\big), \\ \vec{P}(k+1) &= \vec{A}\vec{P}(k)\vec{A}^T + \vec{Q}_n - \vec{A}\vec{P}(k)\vec{C}^T\vec{Q}_v^{-1}\vec{C}\vec{P}_k\vec{A}^T, \end{aligned} \tag{14}$$

where $\widehat{s}$ is the state estimate. In algorithm (14), the state $\widehat{s}(0)$ is initialized by the expected value of the state vector and the matrix $\vec{P}(0)$ is initialized by the a'priori estimate of the state vector covariance matrix. Full derivation of equations (14) is given in book [41]. A more readable discussion of these equations is carried out in work [16]. In polish literature two books are worth mentioning. The first is monograph [30], where Kalman filter defining equations are explained as originating from recurrent filter equations, which itself has been fully derived from basic principles of signal processing. The second is academic handbook [34], where equations of the discrete Kalman filter are derived and its application to the one-dimensional Wiener process is demonstrated. The reader should keep in mind that the authors use different forms of Kalman filter equations, but all of them are equivalent to the system of equations (14).

5.2 Extended Kalman Filter

If the system dynamics described by the equations f_n and measurement equations h_v are both nonlinear and the noise is additive and Gaussian:

$$\begin{aligned} s(k+1) &= f_n\big(s(k)\big) + n(k), \\ y(k) &= h_v\big(s(k)\big) + v(k), \end{aligned} \tag{15}$$

then one should consider using an Extended Kalman Filter (EKF) for noise reduction. The procedure is analogous to the original Kalman filter. There is a preparation step that relies on transformation of the system model from the form (15) to linear form (13). This linearization is carried out in each iteration k in the vicinity of the current estimate of the state vector $s(k)$. Then, original equations (14) may be applied to remove noise. Perhaps the simplest method of linearization is the Taylor series expansion of functions f_n, h_v and leaving only first-order terms, which leads to matrices:

$$\vec{A}(k+1) = \left.\frac{\partial f_n}{\partial s}\right|_{s=\widehat{s}(k)},$$

$$\vec{C}(k+1) = \left.\frac{\partial h_v}{\partial s}\right|_{s=\widehat{s}(k)}. \tag{16}$$

The extended Kalman filter is relatively easy to implement, but it is necessary to derive analytical formulas for partial derivatives making up the Jacobi matrix. Moreover, state and measurements equations must not by too strongly nonlinear. If that happens, the approximation error that results from cutting off higher-order terms of the Taylor series expansion becomes too high to keep the iterative procedure defined by equations (14) convergent. Of course, higher-order terms of the Taylor series expansion may be included, but at the cost of a significant increase of analytical calculations to be made. Ultimately, this would seriously limit practical applications of the filter. However, derivations of the EKF equations including second-order terms are available in literature, [41]. The unscented Kalman filter, which is described later on, delivers state estimation of nonlinear systems accurate up to the third-order terms, [48].

Full derivation of the EKF with comprehensive analysis of its properties is given in book [41].

5.3 *Unscented Kalman Filter*

The main source of practical difficulties when applying the extended Kalman filter are distortions introduced by nonlinear functions f_n and h_v to the noise characteristics. As time passes, this characteristic deviates more and more from the Gaussian, [43]. Foundation of the unscented Kalman filter (UKF) is laid by the observation that individual points sampled appropriately from noise distributions n and h are not subject to such large deformations. Therefore, the key idea is to take only special points according to the probability distribution and transform them using equations defining the system dynamics, instead of processing the whole Gaussian distribution function. This procedure, termed unscented transform (UT) in literature, is to be applied in each iteration of signal processing. In work [23] the so-called sigma-points are introduced. These points are calculated in a deterministic way using formulas delivered in the aforementioned paper. The most important thing to know is that the sample mean vector and the sample covariance matrix worked out on the basis of the sigma-points match the mean values and covariances of the source Gaussian distribution. This procedure leads to highly accurate estimates of the system state if the noise is Gaussian. If not, estimated values of the state vector are close to those that could be calculated when knowing the analytical form of the noise after transformation by functions f_n and h_v. Thus, the unscented transform applies to every type of noise distribution. However, the closer the distribution to Gaussian, the better quality of noise reduction.

Equations defining the UKF are the same as those for the KF, but covariance matrices and the matrix $\vec{P}$ appearing in equation (14) are estimated using the sigma-points sample. A full description of the UKF algorithm and an explanatory illustration of the unscented transform are given in paper [48].

5.4 Particle Filter

The family of Kalman filters may be considered as a special case of the Bayesian filter, [9]. The Bayesian filter is more of a theoretical construct, in the sense that it cannot be implemented in practice with current computational capabilities. Despite this, it is useful because it unifies the whole family of filters and makes it possible to compare different filters belonging to the family. Filters used in practice may be treated as different ways to approximate the idealized Bayesian filter. From this perspective the original Kalman filter is an attempt to estimate in every iteration k a'posteriori distribution of the current state of the system conditioned by previous measurements. For linear systems the resulting distribution is Gaussian with the expected value equal to $\widehat{s}(k)$. The extended Kalman filter is an attempt to construct a similar statistical characteristic for a nonlinear process using local linear approximation. The unscented Kalman filter uses appropriately chosen sigma-points in place of the function describing the state vector probability distribution. In every k-th iteration new values of sigma-points are determined using nonlinear equations (16) or (3). The values of sigma-points are updated according to new incoming measurements and current state estimates.

Particle Filters (PF) also operate on specific points, termed particles, that are meant to represent the system state. Unlike the UKF, these points are not calculated in a deterministic way but instead are sampled from appropriate probability distribution, [11]. Moreover, there are no assumptions concerning the type of system dynamics, the type of noise and the way it affects the system and measurement devices. This means that the functions f_n and h_v appearing in equations (3) can be of any type. Thanks to this versatility particle filters are excellent tools to remove noise from strongly nonlinear systems subject to non-Gaussian measurement disturbances, [5]. However, the price for this is the high requirements for computational power, which limits their applications to low dimensional systems (having at most a dozen state variables). Every particle maintained by the filter represents an estimate of the whole state vector and the outcome of the filter in the current step k is obtained by taking the mean value of all the particles.

There are two basic types of calculations performed by PF. The first one is the propagation of the a'posteriori distribution of the system state through equations (3), with a note that only the particles are propagated, and not the distribution itself. This subroutine is simple and straightforward, as opposed to the next one. The second operation involves sampling the a'posteriori distribution of the state vector in order to select candidate particles for the next iteration k. This subroutine requires the use of large amount of computational resources due to the potentially high dimensionality of the a'posteriori distribution function and the complexity of its formula. There are efficient sampling routines that work fast with low-dimensional distributions or distributions that have a special analytical form (e.g. Gaussian, Poisson, exponential distributions), but there is no universal, effective sampling procedure that works well with all types of distribution. The issue of sampling high-dimensional probability distribution functions dominates the literature on particle filters. Among the most important examples of methods in the field are: Markov Chain Monte Carlo (MCMC), importance sampling, the Metropolis-Hastings algorithm, and Gibbs sampling, [32, 42]. Researchers interested in the application of particle filters to positioning, navigation and tracking can be recommended an excellent book [18].

Particle filters work fine when dealing with low dimensional state vectors (preferably up to about a dozen variables for real-time systems). This excludes those industrial applications that rely on multidimensional measurements, e.g. performed by sensor grids. Geophysics is a rich source of noise reduction and state estimation use cases. Geophysicists attempt to tackle problems involving tens of millions of state variables. A typical task is weather forecasting in a given area. In this case the system model consists of a large set of strongly nonlinear partial differential equations that describe the atmosphere with high resolution with measurements being provided by irregularly distributed weather stations, [44]. The author of work [45] managed to apply the particle filter to an artificial system imitating weather phenomena which was described by a thousand state variables. This was possible thanks to the appropriate choice of the so-called proposal density, which is a key part of the sampling routine of PF. The authors of paper [1] extended the method's applicability to larger dimensional systems. They demonstrated it on a simplified ocean model containing 65500 state variables.

5.5 *Ensemble Kalman Filter*

The Ensemble Kalman Filter (EnKF) rarely appears in the literature on signal processing. It was developed by geophysicists for practical reasons related to removing noise from meteorological measurements of extremely high dimensional (tens of millions of variables) and nonlinear systems. Guided by experience and intuition, researchers came up with a new filter, which was similar to a particle filter, except that the noise was assumed to be Gaussian but not necessarily additive. EnKF, just like PF, makes use of particles, which here have been called ensemble members. The assumption of gaussianity allowed the sampling subroutine to be significantly simplified and improved the convergence of state vector estimates. Surprisingly, it turned out that EnKF copes well with noise reduction in practical applications, where the noise characteristic deviates from the Gaussian. However, such optimistic results were reported only in the literature regarding applications of geophysics, such as [4, 39]. Still, these reports have been published in reputable magazines, and therefore EnKF should be treated as an interesting and promising approach to noise reduction for researchers from other disciplines. As an introduction to principles of EnKF, we recommend [15, 24]. Relationships of EnKF with other filters belonging to the Kalman family are overviewed in works [28, 37].

6 Final Remarks

The main idea of this review is as follows. In practical applications of noise reduction methods it is typical to assume for convenience that the noise is Gaussian, because this makes the design and theoretical analysis of the filtering system easier. In consequence the filtering systems become vulnerable to non-Gaussianity. The result of this situation is the excellent performance of the filter on idealized benchmark datasets combined with the poor quality of noise removal from real life signals. The design of the noise reduction system should be preceded by thorough analysis of the properties of the processed signal. In this way correct assumptions concerning the nature of the noise would be made. Currently available computational power enables the use of advanced state estimation methods that are designed to filter out non-Gaussian noise from highly nonlinear systems.

References

1. Ades, M., Van Leeuwen, P.J.: The equivalent-weights particle filter in a high-dimensional system. Q. J. R. Meteorolog. Soc. **141**(687), 484–503 (2015)
2. Allan, D.W.: Should the classical variance be used as a basic measure in standards metrology? IEEE Trans. Instrument. Measure. **1001**(2), 646–654 (1987a)
3. Allan, D.W.: Time and frequency(time-domain) characterization, estimation, and prediction of precision clocks and oscillators. IEEE Trans. Ultrason. Erroelect. Freq. Control **34**(6), 647–654 (1987b)
4. Anderson, J.L.: Ensemble Kalman filters for large geophysical applications. IEEE Control Syst. Magazine **29**(3), 66–82 (2009)
5. Arulampalam, M.S., Maskell, S., Gordon, N., Clapp, T.: A tutorial on particle filters for online nonlinear/non-Gaussian Bayesian tracking. IEEE Trans. Signal Process. **50**(2), 174–188 (2002)
6. Aydemir, G.A., Saranlı, A.: Characterization and calibration of mems inertial sensors for state and parameter estimation applications. Measurement **45**(5), 1210–1225 (2012)
7. Bendat, J.S., Piersol, A.G., Dudziewicz, J., Białobrzeski, R.: Metody analizy i pomiaru sygnałów losowych. Państwowe Wydawnictwo Naukowe (1976)
8. Benedetti, C., Buscemi, F., Bordone, P., Paris, M.G.: Quantum probes for the spectral properties of a classical environment. Phys. Rev. A **89**(3), 032114 (2014)
9. Chen, Z.: Bayesian filtering: From Kalman filters to particle filters, and beyond. Statistics **182**(1), 1–69 (2003)
10. Djurić, Z.: Mechanisms of noise sources in microelectromechanical systems. Microelectron. Reliabil. **40**(6), 919–932 (2000)
11. Djurić, P.M., Kotecha, J.H., Zhang, J., Huang, Y., Ghirmai, T., Bugallo, M.F., Miguez, J.: Particle filtering. IEEE Signal Process. Magazine **20**(5), 19–38 (2003)
12. Doyle, J.A., Evans, A.C.: What colour is neural noise? arXiv:1806.03704 (2018)

13. D'Alessandro, A., Vitale, G., Scudero, S., D'Anna, R., Costanza, A., Fagiolini, A., Greco, L.: Characterization of MEMS accelerometer self-noise by means of PSD and allan variance analysis. In: 7th IEEE International Workshop on Advances in Sensors and Interfaces (IWASI), pp. 159–164 (2017)
14. El-Sheimy, N., Hou, H., Niu, X.: Analysis and modeling of inertial sensors using allan variance. IEEE Trans. Instrument. Measure. **57**(1), 140–149 (2008)
15. Evensen, G.: The ensemble Kalman filter: Theoretical formulation and practical implementation. Ocean Dyn. **53**(4), 343–367 (2003)
16. Faragher, R.: Understanding the basis of the Kalman filter via a simple and intuitive derivation. IEEE Signal Process. Magazine **29**(5), 128–132 (2012)
17. Gucma, M., Montewka, J.: Podstawy morskiej nawigacji inercyjnej (in polish). Maritime University of Szczecin (2006)
18. Gustafsson, F.: Particle filter theory and practice with positioning applications. IEEE Aerosp. Elect. Syst. Magazine **25**(7), 53–82 (2010)
19. Hasse, L., Spiralski, L.: Szumy elementów i układów elektronicznych (in polish). Polish Scientific Publishers (1981)
20. Haykin, S.: Signal processing in a nonlinear, nongaussian, and nonstationary world. In: Nonlinear Speech Modeling and Applications, pp. 43–53. Springer (2005)
21. Hou, H.: Modeling inertial sensors errors using Allan variance. University of Calgary, Department of Geomatics Engineering (2004)
22. Jerath, K., Brennan, S., Lagoa, C.: Bridging the gap between sensor noise modeling and sensor characterization. Measurement **116**, 350–366 (2018)
23. Julier, S.J., Uhlmann, J.K.: New extension of the Kalman filter to nonlinear systems. In: International Society for Optics and Photonics Signal processing, sensor fusion, and target recognition VI, vol. 3068, pp. 182–194 (1997)
24. Katzfuss, M., Stroud, J.R., Wikle, C.K.: Understanding the ensemble Kalman filter. Am. Stat. **70**(4), 350–357 (2016)
25. Keesman, K.J.: System Identification: An Introduction. Springer Science & Business Media (2011)
26. Kenfack, L.T., Martin, T., Fai, L.C.: Classification of classical non-gaussian noises with respect to their detrimental effects on the evolution of entanglement using a system of three-qubit as probe (2017)
27. Kiani, M., Pourtakdoust, S.H., Sheikhy, A.A.: Consistent calibration of magnetometers for nonlinear attitude determination. Measurement **73**, 180–190 (2015)
28. Lei, J., Bickel, P., Snyder, C.: Comparison of ensemble Kalman filters under non-Gaussianity. Month. Weather Rev. **138**(4), 1293–1306 (2010)
29. Lv, H., Zhang, L., Wang, D., Wu, J.: An optimization iterative algorithm based on nonnegative constraint with application to allan variance analysis technique. Adv. Space Res. **53**(5), 836–844 (2014)
30. Mańczak, K., Nahorski, Z.: Komputerowa identyfikacja obiektów dynamicznych (in polish). Polish Scientific Publishers (1983)
31. Mateos, I., Patton, B., Zhivun, E., Budker, D., Wurm, D., Ramos-Castro, J.: Noise characterization of an atomic magnetometer at sub-millihertz frequencies. Sensors Actuators A Phys. **224**, 147–155 (2015)
32. Murphy, K.P.: Machine Learning: A Probabilistic Perspective. The MIT Press, Cambridge, MA (2012)
33. Of Electrical, I., Engineers, E.: IEEE Standard Specification Format Guide and Test Procedure for Linear, Single-axis. IEEE Nongyroscopic Accelerometers (1999)
34. Osowski, S.: Modelowanie i symulacja układów i procesów dynamicznych (in polish). A publishing house of Warsaw University of Technology (2007)
35. Paszek, J., Kaniewski, P., Łabowski, M.: Analiza błędów losowych czujników bezwładnościowych przy pomocy metod wariancyjnych (in polish). Przeglqd Elektrotechniczny **92**(1), 62–67 (2016)

36. Petkov, P., Slavov, T.: Stochastic modeling of mems inertial sensors. Cybernet. Inf. Technol. **10**(2), 31–40 (2010)
37. Roth, M., Fritsche, C., Hendeby, G., Gustafison, F.: The ensemble Kalman filter and its relations to other nonlinear filters. In: 23rd European Signal Processing Conference (EUSIPCO), pp. 1236–1240. IEEE (2015)
38. Sheinker, A., Shkalim, A., Salomonski, N., Ginzburg, B., Frumkis, L., Kaplan, B.Z.: Processing of a scalar magnetometer signal contaminated by 1/f α noise. Sensors Actuators A Phys. **138**(1), 105–111 (2007)
39. Shen, Z., Tang, Y.: A modified ensemble Kalman particle filter for non-Gaussian systems with nonlinear measurement functions. J. Adv. Model. Earth Syst. **7**(1), 50–66 (2015)
40. Simon, D.: Kalman filtering. Embed. Syst. Program. **14**(6), 72–79 (2001)
41. Simon, D.: Optimal state estimation: Kalman, H infinity, and nonlinear approaches. John Wiley & Sons (2006)
42. Theodoridis, S.: Machine learning: a Bayesian and optimization perspective. Academic Press, Elsevier (2015)
43. Van Der Merwe, R., Wan, E.A., Julier, S.: Sigma-Point Kalman Filters Nonlinear Estimation and Sensor Fusion- Applications in Integrated Navigation. Proceedings, AIAA Guidance Navigation and Controls Conference (2004)
44. Van Leeuwen, P.J.: Particle filtering in geophysical systems. Month. Weather Rev. **137**(12), 4089–4114 (2009)
45. Van Leeuwen, P.J.: Nonlinear data assimilation in geosciences: an extremely efficient particle filter. Q. J. R. Meteorol. Soc. **136**(653), 1991–1999 (2010)
46. Vasilescu, G.: Electronic noise and interfering signals: principles and applications. Springer Science & Business Media (2006)
47. Vig, J.R.: IEEE standard definitions of physical quantities for fundamental frequency and time metrology-random instabilities (IEEE standard). IEEE, New York (1999)
48. Wan, E.A., Van Der Merwe, R.: The unscented Kalman filter for nonlinear estimation. In: Adaptive Systems for Signal Processing, Communications, and Control Symposium, pp. 53–158 (2000)
49. Wang, L., Zhang, Ch., Gao, S., Wang, T., Lin, T., Li, X.: Application of fast dynamic Allan variance for the characterization of FOGs-Based measurement while drilling. Sensors **16**(12), 1–18 (2016)
50. Woodman, O.J.: An introduction to inertial navigation. University of Cambridge, Computer Laboratory, Tech. Rep. UCAMCL-TR-696 (2007)

Fractional Order Models of Dynamic Systems

Andrzej Dzieliński, Grzegorz Sarwas, and Dominik Sierociuk

Abstract This chapter presents the use of models of non-integer (fractional) order for modeling of dynamic systems. Ultracapacitors have been presented as systems with these types of dynamics. Their internal structure and energy storage method were thoroughly analyzed. The diffusion process used in their operation has in their mathematical description derivatives of non integer order, which confirms the legitimacy of using models of non integer order to describe their dynamics. As part of this chapter, various models of ultracapacitors used to model their dynamics in time and frequency domain are presented. The presented transmittance models have been thoroughly discussed and confirmed by their comparison with properties observed in laboratory experiments with the ultracapacitors. The last chapter presents a fractional order neural network, which, using the differences of the non integer order, despite its simple structure, allows for accurate mapping of the process of charging and discharging ultracapacitors.

1 Introduction

Integral and differential calculus of non integer (fractional) order is a natural generalization of the well-known differential and integral calculus (integer order). Additionally, it turns out that a number of phenomena occurring in the reality that surrounds us can be better modeled using non integer derivatives or integrals (see e.g. [22, 23, 26, 27, 29, 33, 40, 47, 49, 51]). An overview of tools and programs for identifying non

A. Dzieliński (✉) · G. Sarwas · D. Sierociuk
Institute of Control and Industrial Electronics, Warsaw University of Technology, ul. Koszykowa 75, 00-662 Warsaw, Poland
e-mail: andrzej.dzielinski@ee.pw.edu.pl

G. Sarwas
e-mail: grzegorz.sarwas@ee.pw.edu.pl

D. Sierociuk
e-mail: dominik.sierociuk@ee.pw.edu.pl

P. Kulczycki et al. (eds.), *Automatic Control, Robotics, and Information Processing*, Studies in Systems, Decision and Control 296,
https://doi.org/10.1007/978-3-030-48587-0_5

integer order models is provided in [30]. For the case when the order of derivatives or integrals is variable in time, the principle of analog modeling and the properties of such operators are presented in [41–43]. One of the most interesting applications of this type of operators (or their equivalents in the field of the complex variable of Laplace transform, i.e. the fractional powers of s) for modeling real systems are ultracapacitor models. This chapter discusses the basic issues of modeling ultracapacitors with the main focus on models using fractional order calculus. Starting from a detailed description of the internal structure of ultracapacitors, an analytical derivation of a fractional order model was presented along with justification for using this calculus in the modeling process. In addition to models of non integer order, models based on RC systems (RC ladders) were also presented. Next, the properties resulting from various models of fractional order known from literature are discussed. In addition, physical properties that justify the sense of building this type of model class are presented. The next part presents the results of modeling ultracapacitors using selected models of non integer order. It is important to emphasize the non-linearity of ultracapacitors, which justifies the use of artificial neural networks for the process of modeling their dynamics. In order to simplify the neural model, a fractional order neural network model proposed in [44], which allows us to simplify the neural network used to model ultracapacitors.

2 Ultracapacitors

Ultracapacitors, also known as supercapacitors or double-layer capacitors, are electrical devices used to store energy. They offer high energy density, which we are not able to obtain in traditional capacitors. The capacity of ultracapacitors is thousands of times greater than the capacity of electrolytic capacitors and reaches even several thousand farads, however, the possibility of storing energy in them is less than in traditional batteries. This is due to the use of the Helmholtz phenomenon, which implies relatively low voltages obtained in individual cells. The properties of ultracapacitors allow their use in applications such as: energy storage in wind and solar power plants, electric cars or elevators, as well as in devices such as computer network components, photographic equipment, multimedia players and wherever fast storage or release of energy is needed. However, their complicated internal structure causes difficulties in their modeling.

The ultracapacitor can be represented in the form of two porous plates placed in the electrolyte (Fig. 1). These plates are usually made of activated carbon, which is characterized by a high degree of microporosity (1 g of activated carbon has an area exceeding 500 m^2). In the ultracapacitor, the potential of the positive electrode attracts negative ions in the electrolyte, and the potential of the negative electrode attracts positive ions. The semi-permeable separator between the two electrodes prevents them from shorting. Traditional capacitors store electric charge on the electrode. In other devices, such as electrochemical cells or batteries, energy is created by the chemical reaction of the electrodes. In both cases, the ability to store or create an

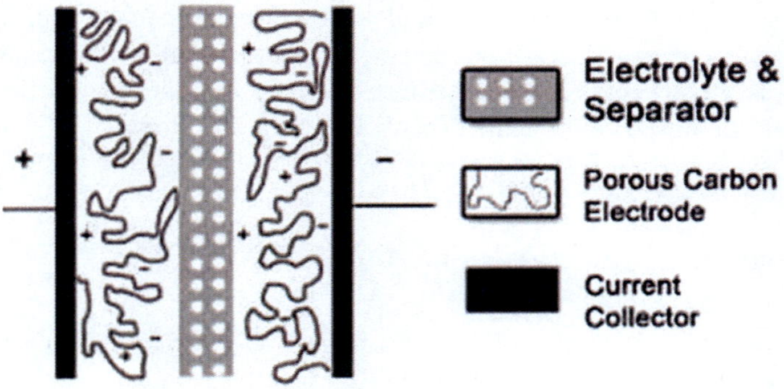

Fig. 1 Ultracapacitor scheme

electric charge is a function of the surface of the electrodes. It is the same with ultracapacitors. Their capacitance is a function of the surface of the electrodes, but unlike electrolytic capacitors, energy is not stored in ion polarization, it is in their movement.

The method of operation of ultracapacitors is based on the storage of energy in a microscopic junction (separator) at the edge of the electrolyte and the conductor. This separator is produced due to the double layer phenomenon (Helmholtz phenomenon). At low voltage, on the border of two substances having electrically charged particles, a thin separating area is created, in which there is an electric field, from where ultracapacitors are also called double-layer capacitors. The use of the Helmholtz phenomenon in supercapacitors forces their specific operating parameters (individual ultracapacitors can only be used in the narrow 0–5 V voltage range). However, this limitation can be bypassed by combining ultracapacitors in series.

As already mentioned, the capacity of ultracapacitors and traditional capacitors is proportional to the surface of the active electrode and inversely proportional to the width of the insulating layer. It is not possible to control the width of the contact region in ultracapacitors, so building ultracapacitors of different capacities (accumulating different amounts of energy that can be stored) involves increasing the surface of the electrodes. To maximize the size of the electrodes, and thus increase the capacity of ultracapacitors, porous materials, such as activated carbon or sintered metal powders, should be used in the construction of the electrodes, but in both cases there is an internal limit of porosity, which defines the upper limit of the surface of the electrodes.

The complex structure of ultracapacitors and the use of ion movement to store energy have a great impact on their dynamics. Modeling and mathematical description of these devices requires the use of more sophisticated methods, of which we can distinguish three main groups:

1. fractional order model,
2. RC cascades,
3. artificial neural network (ANN).

The following sections present examples of models that can be found in the literature. First, however, the model derivation presented in the dissertation of Sarwas [34, 36] will be presented. This derivative justifies the sense of using an fractional order differential calculus to describe the dynamics of ultracapacitors.

3 Derivation of the Ultracapacitor Model

High porosity of the electrodes, different from those observed in conventional capacitors, has a huge impact on their impedance. In order to analytically derive a double-layer capacitor model, it is necessary to focus on the porous structure and electrode-electrolyte connection. Current in the ultracapacitor is a component of the ionic current flowing through the electrolyte, the electrode material and the capacity of the electrode-electrolyte link [4]. Because the conductivity of the electrodes is two orders higher than the insulator layers, the resistance of the electrodes may be neglected, however, in the full ultracapacitor model this resistance should be taken into account due to the high value of currents resulting from their operating conditions. The diffusion of ions at the interface of the electrode-electrolyte layer has the greatest impact on the impedance of a double-layer capacitor. The diffusion process can be modeled using the finite long line model [1] shown in Fig. 2.

In order to derive a simplified model of the capacity of an ultracapacitor based on the transmission line model, it should be assumed that the distribution of resistance R and capacity C are constant, and the cylindrical pores are filled with electrolyte evenly, at the same time d is taken as the length of the line.

Current $i(x, t)$ and voltage $u(x, t)$ are related by partial differential equations resulting from the diffusion equation:

$$\begin{cases} \frac{\partial u(x,t)}{\partial x} = -r \cdot i(x, t), \\ \frac{\partial i(x,t)}{\partial x} = -c \cdot \frac{\partial u(x,t)}{\partial t}. \end{cases} \tag{1}$$

Because the parameters r and c are constant, the Eq. (1) can be written as follows:

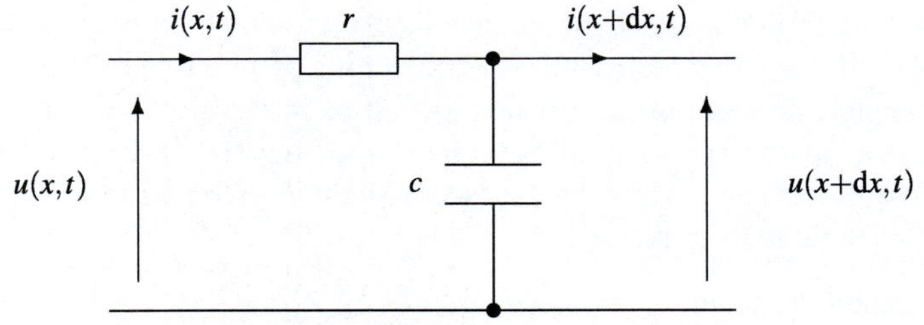

Fig. 2 Scheme of transmission line mathematical model

$$\begin{cases} \frac{\partial^2 u(x,t)}{\partial x^2} = r \cdot c \cdot \frac{\partial u(x,t)}{\partial t}, \\ \frac{\partial^2 i(x,t)}{\partial x^2} = r \cdot c \cdot \frac{\partial i(x,t)}{\partial t}. \end{cases} \tag{2}$$

The solution to the above relation can be obtained with the help of the Laplace transform, analogous to the one-dimensional heat conductivity equation. The Laplace transform of the first equation with (2) versus time looks like this:

$$\frac{\partial^2 U(x,s)}{\partial x^2} = r \cdot c \cdot s \cdot U(x,s) - r \cdot c \cdot U(x,0). \tag{3}$$

For zero initial conditions, the above equation can be transformed into:

$$\frac{\partial^2 U(x,s)}{\partial x^2} - r \cdot c \cdot s \cdot U(x,s) = 0. \tag{4}$$

This equation has the following roots:

$$\omega_{1,2} = \pm\sqrt{r \cdot c \cdot s}, \tag{5}$$

and a solution of the Eq. (4) is:

$$U(x,s) = C_1(s) \sinh\left(x\sqrt{r \cdot c \cdot s}\right) + C_2(s) \cosh\left(x\sqrt{r \cdot c \cdot s}\right), \tag{6}$$

where $C_1(s)$, $C_2(s)$ are the constants of the equation.

To determine the constants from the above equation, it is necessary to use boundary conditions. For $x = 0$ we get:

$$U(0,s) = C_1(s) \sinh(0) + C_2(s) \cosh(0), \tag{7}$$

which in effect gives a relation to the constant $C_2(s)$:

$$C_2(s) = U(0,s). \tag{8}$$

The constant $C_1(s)$ can be calculated from the first equation from the set of Eq. (1):

$$-r \cdot i(x,t) = \frac{\partial u(x,t)}{\partial x}.$$

Using the Laplace transform after time and Eq. (6) we get:

$$\begin{aligned} -r \cdot I(x,s) = {} & C_1(s)\sqrt{r \cdot c \cdot s} \cdot \cosh\left(x\sqrt{r \cdot c \cdot s}\right) \\ & + U(0,s)\sqrt{r \cdot c \cdot s} \cdot \sinh\left(x\sqrt{r \cdot c \cdot s}\right). \end{aligned} \tag{9}$$

In the case of an ultracapacitor, the limit condition at $x = d$ is $i(d,t) = 0$, which implies $I(d,s) = 0$. As a result, the Eq. (9) becomes:

$$0 = C_1(s)\sqrt{r \cdot c \cdot s} \cdot \cosh\left(d\sqrt{r \cdot c \cdot s}\right) + U(0,s)\sqrt{r \cdot c \cdot s} \cdot \sinh\left(d\sqrt{r \cdot c \cdot s}\right), \tag{10}$$

which can be transformed into the form:

$$C_1(s) = -\tanh\left(d\sqrt{r \cdot c \cdot s}\right) U(0,s). \tag{11}$$

Let us return to Eq. (9). For $x = 0$ we obtain:

$$-r \cdot I(0,s) = -\tanh\left(d\sqrt{r \cdot c \cdot s}\right) U(0,s)\sqrt{r \cdot c \cdot s} \cdot 1 + 0, \tag{12}$$

which can be rewritten as:

$$I(0,s) = \frac{\sqrt{r \cdot c \cdot s}}{r} \tanh\left(d\sqrt{r \cdot c \cdot s}\right) U(0,s). \tag{13}$$

Finally, we obtained the transfer function of the transmission line:

$$Z_{imp}(s) = \frac{U(0,s)}{I(0,s)} = \frac{r}{\sqrt{r \cdot c \cdot s}} \coth\left(d\sqrt{r \cdot c \cdot s}\right). \tag{14}$$

This allows us to introduce a total capacity of the line $C = c \cdot d$ and a total resistance $R_l = r \cdot d$. The pulse impedance of transmission line can be rewritten as:

$$\bar{Z}_{imp}(s) = \frac{R_l}{\sqrt{R_l \cdot C \cdot s}} \cdot \coth\left(\sqrt{R_l \cdot C \cdot s}\right). \tag{15}$$

Using Taylor series expansion it is possible to approximate $\coth(x)$ through the following equation [32]:

$$\coth(x) = \frac{\cosh(x)}{\sinh(x)} \cong \frac{1 + \frac{x^2}{2}}{x}. \tag{16}$$

Numerator of the expression $1 + \frac{x^2}{2}$ is an approximation of $\sqrt{1 + x^2}$. For low frequencies, the Eq. (16) becomes:

$$\coth(x) \underset{x \to 0}{\cong} \frac{\sqrt{1 + x^2}}{x}. \tag{17}$$

The development of this equation is also possible for high frequencies:

$$\coth(x) \underset{\infty}{\cong} \frac{e^{\frac{x}{2}}}{e^{\frac{x}{2}}} = 1 \underset{\infty}{\cong} \frac{\sqrt{1 + x^2}}{x}. \tag{18}$$

Putting the above relation to the Eq. (15), we obtain a simplified fractional model of the ultracapacitor's porous impedance:

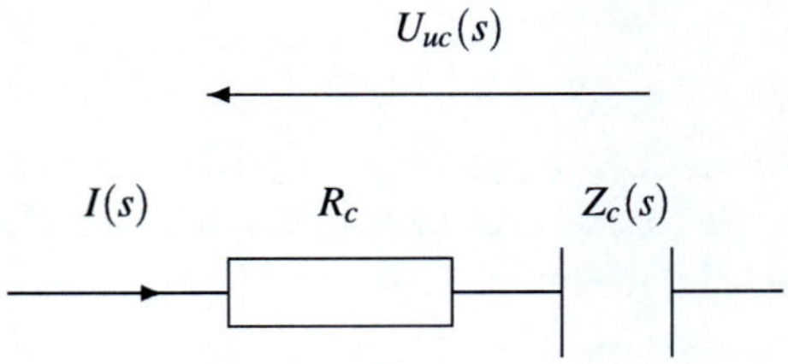

Fig. 3 Ultracapacitor equivalent model

$$Z_{imp}(s) = \frac{\sqrt{1 + R \cdot C \cdot s}}{Cs} = \frac{\sqrt{1 + Ts}}{Cs}. \tag{19}$$

Model (19) was derived using some idealistic assumptions, however, real electrodes should be modeled using more complex methods. Diffusion in liquids [3, 16] suspension in fractional structures like granular substances or a transport of electrons, holes, ions, spin, etc., in disordered solids should be described using anomalous (fractional) diffusion equations [20].

For normal diffusion, the average particle movement is proportional to $t^{0.5}$, however, in the case of subdiffusion it is smaller, proportional to $t^{\frac{\alpha}{2}}$, where $\alpha \in (0, 1)$. As a result, we can formulate the following lemma:

Lemma 1 *The ultracapacitors impedance model based on anomalous diffusion is:*

$$Z_{imp}(s) = \frac{\sqrt{1 + Ts^{\alpha}}}{C_{\alpha}s^{\alpha}}, \tag{20}$$

where C_α is the ultracapacitor fractional capacity and T is the parameter responsible for the changes capacity with frequency.

The complete model of the ultracapacitor including the resistance of the electrodes is presented in Fig. 3. The transfer function of the circuit presented in Fig. 3 is defined as follows:

$$G_{uc}(s) = \frac{U_{uc}(s)}{I(s)}, \tag{21}$$

where $U_{uc}(s)$ is the Laplace transform of the capacitor voltage and $I(s)$ is the Laplace transform of the capacitor current. The whole transfer function of an ultracapacitor presented in Fig. 3 is:

$$G_{uc}(s) = \frac{U_{uc}(s)}{I(s)} = R_c + Z_c(s), \tag{22}$$

where R_c is the resistance of the ultracapacitor and $Z_c(s)$ can be any of the ultracapacitor capacity model. Using the impedance model from (20) can be presented the whole double-layer capacitor model:

$$G_{uc}(s) = R_c + \frac{\sqrt{1 + Ts^\alpha}}{Cs^\alpha}, \tag{23}$$

where $R_c \in \mathbb{R}_+$, $T \in \mathbb{R}_+$, $\alpha \in (0, 1)$.

In [31] J. J. Quintana presented one of the best known fractional order ultracapacitor models:

$$Z_{c_q}(s) = \frac{(1 + Ts)^\alpha}{Cs^\beta}, \tag{24}$$

however, the analytical derivation of this model has never been presented. This model was obtained from model $Z_c = \frac{1}{Cs^\alpha}$ presented in [50]. The Quintana model is the result of fitting the model to the results obtained from impedance spectroscopy and is able to very accurately describe the dynamics of the ultracapacitor, hence in the following chapters this model is deeply analyzed.

The full model of the ultracapacitor in the form proposed by Quintana looks as follows:

$$G_{uc_q}(s) = R_c + \frac{(1 + Ts)^\alpha}{Cs^\beta}, \tag{25}$$

where $R_c \in \mathbb{R}_+$, $T \in \mathbb{R}_+$ and $\alpha, \beta \in (0, 1)$.

Neither model (23), nor model (25) have capacity that can be compared with the capacity of traditional capacitor expressed in farads. The denominator of these expressions is Cs^v, where $v \in \mathbb{R}_+$, which can be called fractional order capacity. To create a more physical model capable of comparing with the traditional model of capacitor, the Quintana capacitance model with $\beta = 1$ is used:

$$G_{DC}(s) = R_c + \frac{(1 + Ts)^\alpha}{Cs}. \tag{26}$$

The model thus obtained was historically proposed by Davidson and Cole in 1951 as a dielectric model immersed in glycerin [7, 8]. Modeling the dynamics of the ultracapacitor using a simplified model of the fractional order was presented in [17–19], and slightly more complex models were presented in [48]. An overview of modeling supercapacitors, accumulators and fuel cells is presented in [14]. The issue of the description of energy stored in ultracapacitors is described in [13].

In the following sections, the properties of the given models are described and discussed. Additionally, the results of identifying the parameters of ultracapacitors using the given models are presented.

4 Other Methods of Ultracapacitors Modeling

Before the introduction of the fractional order model, many scientists tried to describe the dynamics of ultracapacitors using an RC cascade. These models were based on the double-layer equivalent model [21], that constituted a complicated network of

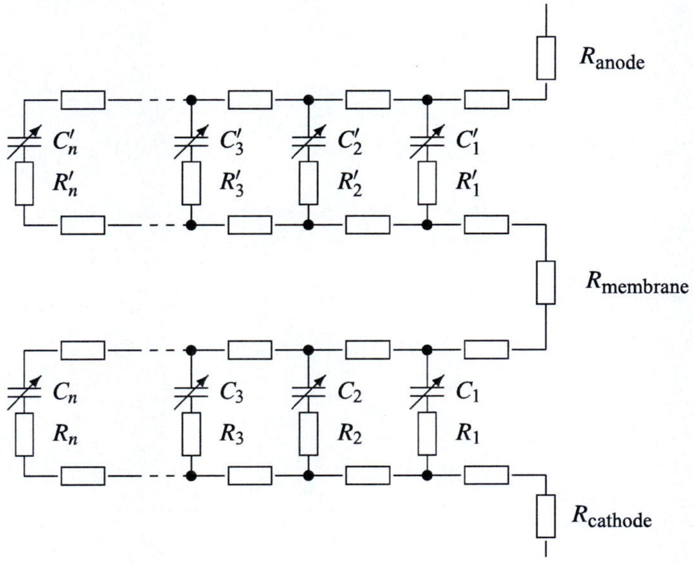

Fig. 4 Ultracapacitor theoretical model

RC branches with non-linear capacitors connected in parallel by resistances Fig. 4. The value of the resistances depends on such parameters as:

- resistance of the electrode material,
- resistance of the electrolytic solvent,
- pores width,
- membrane porosity,
- quality of the connection electrode-collector.

Since the presented model is difficult to use and practically useless, several simplifications have been proposed [1, 5, 12, 15]. Some of these models are shown in Fig. 5. Shortcuts ESR and EPR mean the equivalent series resistance and the equivalent parallel resistance, respectively. The first model (Fig. 5a) is a classic model of an ultracapacitor without non-linear elements. The second one (Fig. 5b) is a variable capacity model. The next models (Fig. 5c, d) are the three, four branches models, respectively. The last one (Fig. 5c) is the transmission line model. Because the presented models are simplified, it is difficult to model the dynamics of the ultracapacitor in a full frequency range.

This problem forced engineers to use nonlinear correction elements. More accurate models can also be obtained by using subsequent branches in the model, however, infinite models are not possible to implement, and the huge number of their elements increases the susceptibility to system damage.

To address the problem of large and non-linear models for modeling ultracapacitors, adaptation of artificial neural networks (ANN) began. This solution also allows us to omit using a precise fractional model of ultracapacitor, which usually is very

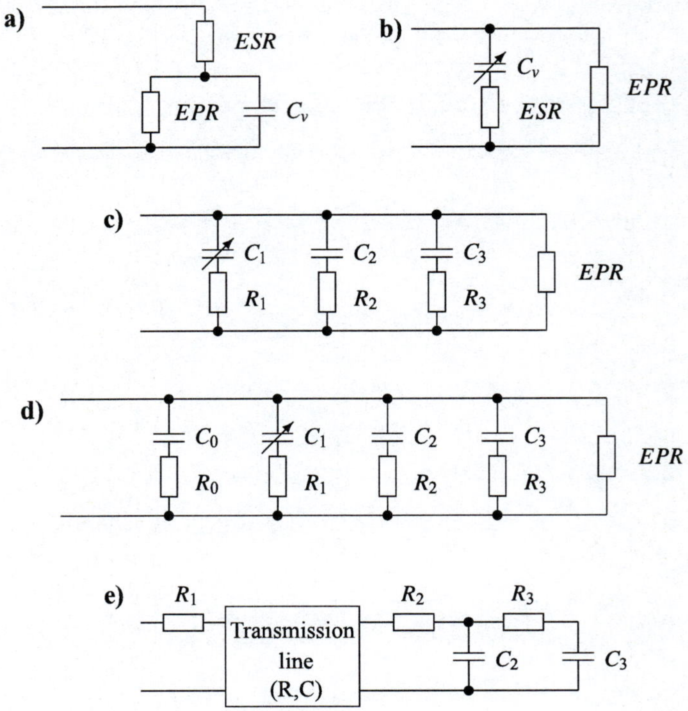

Fig. 5 Examples of ultracapacitors RC models

hard to use. However, using the neural network, you can correct errors resulting from model approximations and solve the problem of non-linearity. Neural networks are a universal approximator and as such can be used to represent dependencies in a designed system, which is why neural networks are often used in applications of systems with ultracapacitors [6, 24]. The neural network for modeling ultracapacitors uses a direct approach (Fig. 6), where parameters can be calculated from measurements. A learning algorithm uses an adaptive mechanism. It updates the network parameters thanks to the Lavenberg-Marquardt method to minimize the error between the y_p system output and its expected y_m output.

One of the simplest network designs used for modeling ultracapacitors is presented by Marie-Francoise et al. [25] which used the neural network (MISO). This model has been trained, tested and checked according to the diagram presented in Fig. 7. For modeling a network the following parameters has been used:

- Inputs: temperature T, current I_{uc}, value of capacity C;
- Output: ultracapacitor voltage U_{uc};
- 3 neurons in Hidden-layer;
- 1 neuron in Output-layer;
- Activation function: tanh function (hidden layer); linear function (output layer);
- Delays output number: $n_1 = 1$, $n_2 = 3$, $n_3 = 3$.

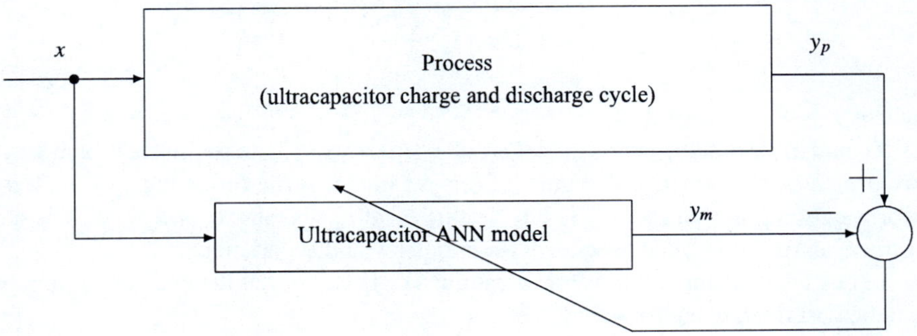

Fig. 6 Ultracapacitor—model direct approach

Fig. 7 Ultracapacitor ANN black-box model

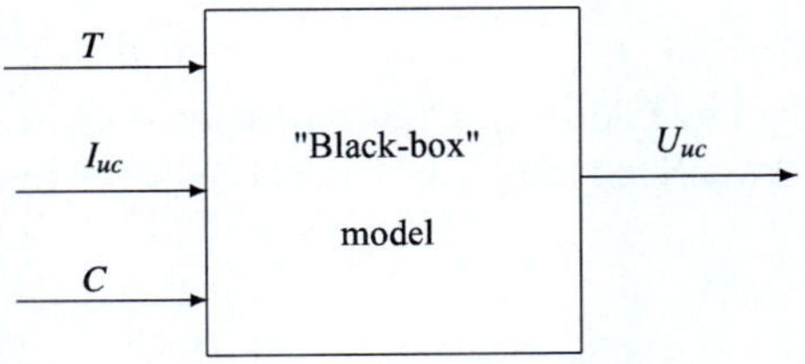

To update the parameters the Levenberg–Marquardt method was performed. This model accurately described the behaviour of the ultracapacitor, not only taking account of electrical, but also thermal influences.

The rest of this chapter presents the properties of selected models of the fractional order. Then, confirming the non-linearity of ultracapacitors, a model based on a fractional order neural networks proposed in the article [35] is described.

5 Properties of Theoretical Models

Section 5.1 presents a discussion on the model derived from anomalous diffusion equations. In Sect. 5.2, a more physical model has been presented. This model has parameters corresponding to the parameters of a traditional capacitor (Quintana model for $\beta = 1$). For this model, the step response in the time domain of the ultracapacitor and the quadripole with the ultracapacitor is shown.

5.1 Full Ultracapacitor Model

Let us focus on the following fractional order capacitance model:

$$Z_c(s) = \frac{\sqrt{1 + Ts^{\alpha}}}{C_{\alpha}s^{\alpha}}, \tag{27}$$

where $T \in \mathbb{R}_{+}$, $\alpha \in (0, 1)$.

To underline that C_{α} in this model is not a capacity of capacitor in traditional way we use an extra index C_{α}. The units of this parameter is the following $\frac{Farads}{(sec)^{1-\alpha}}$. It is worth noticing that in fact C_{α} is not a value of ultracapacitor capacity but it is an ultracapacitor impedance dependent on frequency and α parameter.

Let us take this model into consideration. The behaviour at the ends of its ranges can be described using following relation:

$$G_c(s) = \begin{cases} \frac{\sqrt{T}}{C_{\alpha}s^{\alpha/2}} & \text{for } s \gg \frac{1}{T}, \\ \frac{1}{C_{\alpha}s^{\alpha}} & \text{for } s \ll \frac{1}{T}. \end{cases} \tag{28}$$

At low frequencies, this model tends to the fractional integrator model presented by Westerlund [50]. This model can be rewritten in the following form:

$$\frac{1}{C_{\alpha}s^{\alpha}} = \frac{1}{C_{\alpha}\omega^{\alpha}\left(\sin\frac{\pi}{2}\alpha + j\cos\frac{\pi}{2}\alpha\right)}, \tag{29}$$

which means that for the $\omega \to 0$ the ultracapacitor capacity is equal to the capacity of a traditional capacitor. In Fig. 8 there are presented exemplary Bode diagrams of this model for $T = 1$, $C = 1$ and different $\alpha = 0.6, 0.8, 1$. It can be noticed that a higher value of α (model order) implies faster system dynamics. Although this model is most suitable for describing the dynamics of an ultracapacitor in the frequency domain, it is very difficult to calculate its inverse transformation to determine the step response in the time domain, therefore the Davidson–Cole model is easier to model for ultracapacitors.

5.2 *Davidson–Cole Model Properties*

Recall the Davidson–Cole model (26), which can be obtained from the Quintana model assuming that $\beta = 1$:

$$G_{DC} = \frac{(1 + Ts)^{\alpha}}{Cs}, \tag{30}$$

where C is a capacity of ultracapacitor, $\alpha \in (0, 1)$ is a model order and T is a parameter whose meaning is presented in this subsection.

In this model, parameter C is a traditional capacity with the unit $Farads$, therefore this model is more physically-oriented. The dynamics of this model at the end of the ranges are as follows:

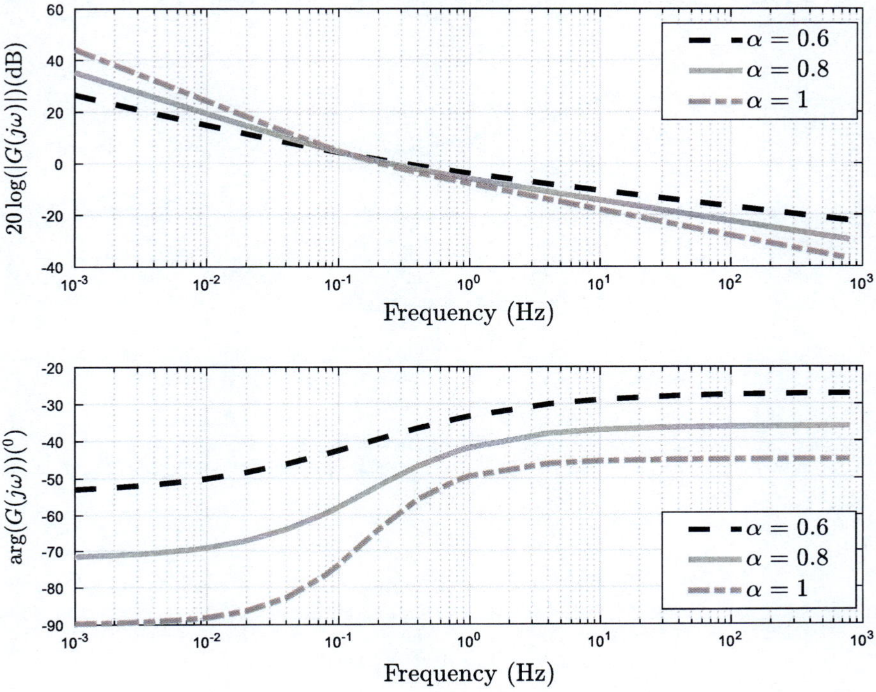

Fig. 8 Bode's diagrams of G_c capacitance model for $\alpha = 0.6, 0.8, 1$

$$G_{DC}(s) = \begin{cases} \frac{T^\alpha}{Cs^{1-\alpha}} & \text{for } s \gg \frac{1}{T}, \\ \frac{1}{Cs} & \text{for } s \ll \frac{1}{T}. \end{cases} \tag{31}$$

It worth noticing, that this model for low frequency ($\omega \to 0$) behaves like model of a traditional capacitor with a decrease of amplitude characteristic, 20 dB per decade. For high values of frequency this model reminds us of a fractional order integrator with order equally $1 - \alpha$. Figure 9 presents Bode diagrams of the Davidson–Cole model for $T = 1$, $C = 1$ and three different values of α (0.4, 0.5, 0.6).

To explain the role of the T parameter, we should consider the spectral transfer function of the Davidson–Cole model given as follows:

$$G_{DC}(j\omega) = \frac{(Tj\omega + 1)^\alpha}{Cj\omega}. \tag{32}$$

The magnitude of this spectral transfer function is:

$$A_{DC}(\omega) = \frac{((T\omega)^2 + 1)^{\frac{\alpha}{2}}}{C\omega}. \tag{33}$$

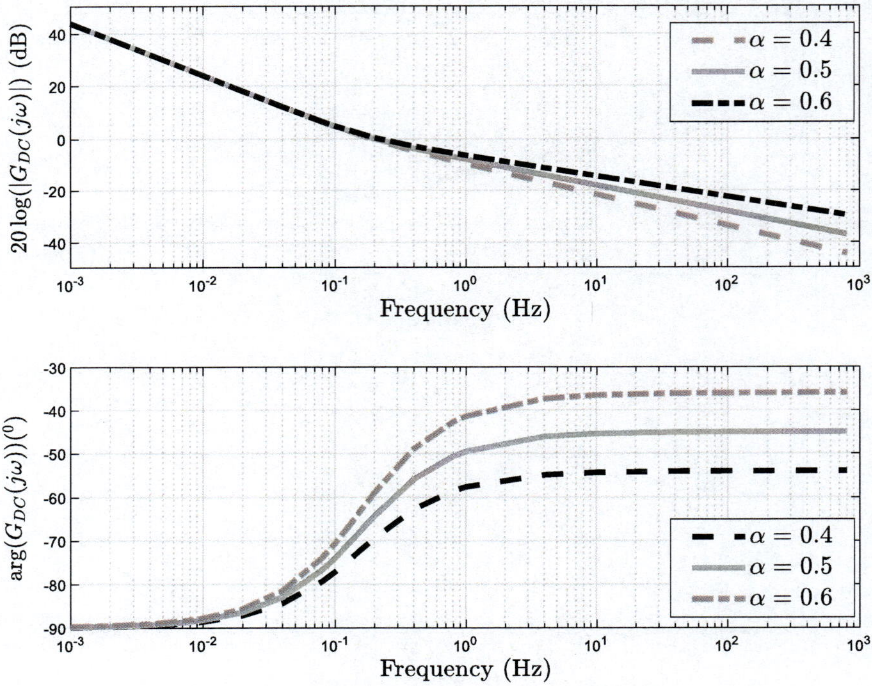

Fig. 9 Bode's diagrams of Davidson–Cole capacitance model for $\alpha = 0.4, 0.5, 0.6$

This magnitude is compared to the magnitude of a traditional capacitor of capacity C' is:

$$\frac{((T\omega)^2 + 1)^{\frac{\alpha}{2}}}{C\omega} = \frac{1}{C'\omega}, \tag{34}$$

which yields:

$$C' = \frac{C}{((T\omega)^2 + 1)^{\frac{\alpha}{2}}}, \tag{35}$$

where C' is the capacity equivalent of the ultracapacitor for the given frequency ω. This equivalent capacity illustrates what capacity the traditional capacitor should have in order to have the same magnitude for the desired value of the frequency. For $\alpha = 0.5$ we have:

$$C' = \begin{cases} C & \text{for } \omega \ll \frac{1}{T}, \\ \frac{C}{\sqrt{T\omega}} & \text{for } \omega \gg \frac{1}{T}. \end{cases} \tag{36}$$

The frequency f_c for which the capacity equivalent decreases by 2 times is given as follows:

$$f_c = \frac{\sqrt{2^{\frac{2}{\alpha}} - 1}}{2\pi T}. \tag{37}$$

Fig. 10 RC quadripole model

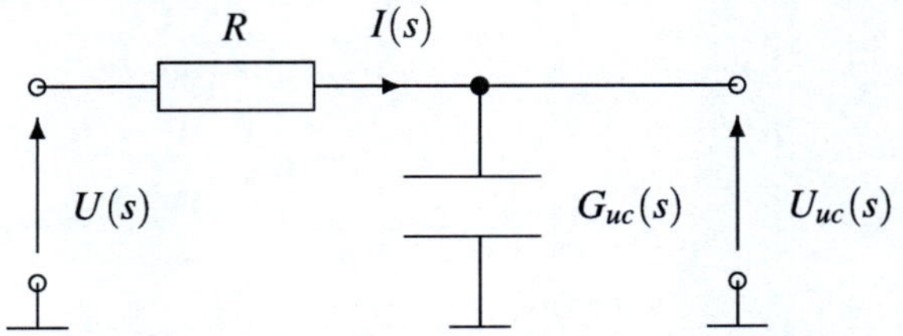

A fact of decrease of the ultracapacitor capacity with the frequency is known by the producers but only the using of a fractional order model of the presented form for the ultracapacitor modeling explains this behaviour.

Now let us take a full ultracapacitor model based on the Davidson–Cole capacitance model:

$$G_c(s) = R_c + \frac{(1+Ts)^\alpha}{Cs}. \tag{38}$$

The step response of this model in the time domain is discussed below:

Lemma 2 *The time domain step response of the ultracapacitor itself is as follows*

$$u_c(t) = \mathscr{L}^{-1}\left[G_c(s)\frac{I}{s}\right] = \left(R_c + \frac{T^\alpha}{C}e^{-\frac{t}{T}}\frac{t^{1-\alpha}}{\Gamma(2-\alpha)}{}_1F_1\left(2; 2-\alpha; \frac{t}{T}\right)\right)I, \tag{39}$$

where the ${}_1F_1\left(2; 2-\alpha; \frac{t}{T}\right)$ *is a confluent hypergeometric function and* $I = \text{const}$.

Proof of this lemma presented in the Sarwas dissertation [34] and in the articles [9, 10].

Analytically, the step response of the ultracapacitor for the linear function of current ($i(t) = It$, where $I = \text{const}$) is described by the following formula:

$$\mathscr{L}^{-1}\left[\frac{G_c(s)I}{s^2}\right] = \left(R_c I + \frac{T^\alpha}{C}e^{-\frac{t}{T}}\frac{t^{2-\alpha}}{\Gamma(3-\alpha)}{}_1F_1\left(3; 3-\alpha; \frac{t}{T}\right)\right)I. \tag{40}$$

Let us take consideration of the RC quadripole presented in Fig. 10.

The transfer function of this system is defined as follows:

$$G_{\text{RC}}(s) = \frac{U_{uc}(s)}{U(s)} = \frac{G_{uc}(s)}{G_{uc}(s)+R}. \tag{41}$$

In the analyzed case, using as an ultracapacitor model $G_c(s)$, the quadripool model $G_{uc}(s)$ is given by the following form:

$$G_{\text{RC}}(s) = \frac{G_c(s)}{G_c(s)+R} = \frac{(Ts+1)^\alpha + R_c Cs}{(Ts+1)^\alpha + (R+R_c)Cs}, \tag{42}$$

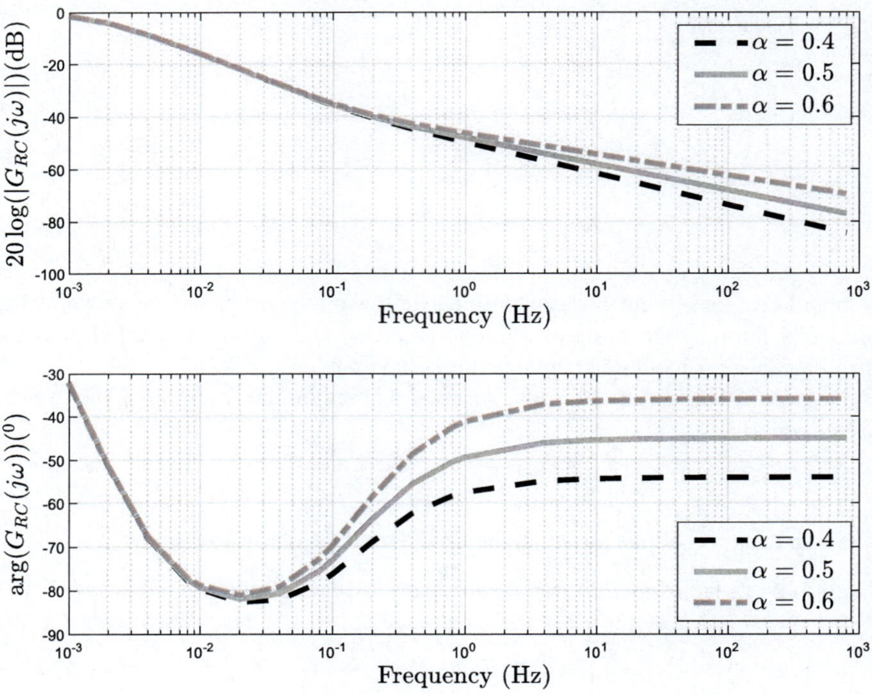

Fig. 11 Fractional order RC quadripole Bode diagrams for $\alpha = 0.4, 0.5, 0.6$

where C, R_c is the capacity and resistance of capacitor, respectively. T is a parameter of capacity decreasing with frequency and R is a system resistance (matching resistance in experimental setup).

Examples of the frequency characteristics of this model for $C = 1, T = 1, R_c = 1$, $R = 100$ and three α orders equal 0.4, 0.5, 0.6 are presented in Fig. 11.

Calculation of the step response of the RC quadripole with the ultracapacitor model $G_c(s)$ for arbitrary α is not easy. Therefore, the step response of this model for $\alpha = 0.5$ is presented. This order can describe the dynamics of the low-capacity ultracapacitor and was confirmed in experiments, presented in Sect. 6.

Lemma 3 *The unit step response of the system with the ultracapacitor for* $\alpha = 0.5$ *is given by the following equation:*

$$u_{uc}(t) = \mathcal{L}^{-1}\left[G_{RC}(s)\frac{1}{s}\right] = e^{-\frac{t}{T}} \sum_{i=1}^{4} \left(\frac{F_i}{t^{0.5}} \cdot E_{1,\,0.5}\left(H_i^2 t\right) + F_i H_i e^{H_i^2 t}\right), \tag{43}$$

where

$$F_1 = \frac{\sqrt{T}}{2}, \quad F_2 = -\frac{\sqrt{T}}{2}, \quad F_3 = \frac{(B-A)T}{\sqrt{T(T^2+4A^2)}}, \quad F_4 = \frac{(A-B)T}{\sqrt{T(T^2+4A^2)}},$$

$$H_1 = \sqrt{\frac{1}{T}}, \quad H_2 = -\sqrt{\frac{1}{T}}, \quad H_3 = \frac{-T^{1.5} + \sqrt{T^3+4A^2T}}{2AT},$$

$$H_4 = \frac{-T^{1.5} - \sqrt{T^3+4A^2T}}{2AT}, \quad A = (R+R_c)C, \quad B = R_cC$$

and $E_{1,0.5}\left(H_i^2 t\right)$ *is a two-parameter Mittag–Leffler function. Parameters* C, R_c, T *are the parameters of the ultracapacitor* $G_{uc}(s)$ *model and* R *is the resistance of RC quadripole's resistor.*

Proof of this lemma was also presented in Sarwas's doctoral dissertation [34] and in the articles [9, 10].

6 Identification

This section presents the results of identification experiments using the presented models. Two types of ultracapacitors were used in the research. The first type is represented by three ultracapacitors with nominal capacities of 0, 047, 0, 1, 0, 33 F, manufactured by Panasonic® [28], called in this chapter low capacity ultracapacitors. The next two ultracapacitors used in the experiments had nominal capacities of 1500 F/2.7 V (BCAP1500) and 3000 F/2.7 V (BCAP3000). These ultra-capacitors are called high-capacity capacitors.

The first experiment presents the identification of parameters of the ultracapacitors with a fractional order. The parameter values were obtained as a result of matching the frequency response charts. The *fmincon* function implemented in the Matlab environment was used to match the graphs, the purpose of which was to minimize the sum of the elements of RMSE (Root Mean Square Error) obtained from the gain and phase graph. It is very important to remember that ultracapacitors are electrolytic capacitors and allow only positive voltages to be used. As a result, when testing low-capacitance ultra-capacitors, a constant voltage value had to be added to the sinusoidal output signal shifting the sinusoidal variable signal above the OX axis:

$$u(t) = 2V + \sin(\omega t). \tag{44}$$

The voltage on the ultracapacitor (in steady state) was expressed by the formula:

$$u_{uc}(t) = A_c(\omega)\sin(\omega t + \varphi_u). \tag{45}$$

In the case of a system with a current transducer, the input signal was a sinusoidal current:

$$i(t) = A_i(\omega)\sin(\omega t + \varphi_i). \tag{46}$$

In this case, the voltage on the ultracapacitor contained an additional voltage constant (u_0) depending on the signal frequency to protect the device under test. As a result, the output voltage was expressed by the relationship:

$$u_{uc}(t) = u_0 + A_c(\omega)\sin(\omega t + \varphi_u). \tag{47}$$

Bode's characteristics were obtained from the following relation:

$$M(\omega) = 20\log\left(\frac{A_c(\omega)}{A_i(\omega)}\right), \varphi(\omega) = \varphi_i(\omega) - \varphi_u(\omega). \tag{48}$$

As a result of the measurements, two graphs were obtained. The first presenting the amplitude of the tested ultracapacitor and the second presenting the phase shift, both as a function of frequency.

6.1 Modeling of Ultracapacitors Using the Davidson–Cole Model

Recall the Davidson–Cole model:

$$G_c = R_c + \frac{(1+Ts)^\alpha}{Cs}, \tag{49}$$

in which R_c is the resistance of the ultracapacitor, C is its capacity defined in the traditional way and T is a parameter describing the decrease capacity with increasing frequency, which has been described and explained in Sect. 5.2.

Model parameters obtained as a result of the identification process are presented in the Table 1.

Table 2 shows the identification error for this model. The values presented in this table emphasize the high compatibility of the identified theoretical model with physical objects. The average error for amplitude and phase is less than 1.5 dB and 1.66°. Examples of identification of ultracapacitors in the frequency domain are

Table 1 Ultracapacitor identified parameters

Capacitor (F)	T	C (F)	R_c (Ω)	α
0.047	5.2261	0.06	32	0.6
0.1	14.7979	0.094	52	0.6
0.33	56.9669	0.27	29	0.57
1500	1.006	1348	0.25m	0.62
3000	0.6369	2410	0.13m	0.7

Table 2 RMSE for Davidson–Cole model

Capacitor (F)	RMSE amplitude	RMSE phase
0.047	0.0985	0.6152
0.1	0.1521	0.6366
0.33	1.3146	0.7857
1500	0.9932	1.6518
3000	0.8154	1.5344

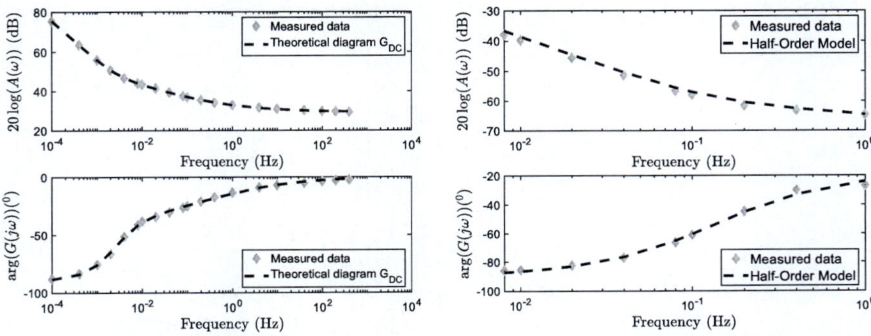

a) Ultracapacitor $0.33F$. **b)** Ultracapacitor $1500F$.

Fig. 12 Measured and theoretical frequency characteristics of D-C model

presented in Fig. 12a, b. The presented results confirm that the Davidson–Cole model is able to very accurately describe the dynamics of the tested objects in the full frequency range. It is worth noting that the order of identified ultracapacitors is approximately $\alpha = 0.6$, which means that it is close to the model derived from the normal diffusion equation, therefore acceptable modeling results can be obtained for $\alpha = 0.5$. The results of modeling using the Half-Order model are presented later in this section.

Let us focus on the T parameter. This parameter is responsible for the decrease in the capacity of the ultracapacitor with an increase in signal frequency. The frequency of f_c, for which the equivalent capacity drops twice, is given by the relationship.

Now, let us focus on the T parameter. This parameter is responsible for the ultracapacitor capacity decreasing with frequency. The f_c, for which the equivalent capacity decreases two times is given by Eq. (37). The value of the frequency of double capacity decrease for individual ultracapacitors is presented in Table 3. To sum up the results it contains, the following conclusions can be drawn. For the High Capacity Ultracapacitors manufactured by Maxwell, the frequency value f_c increases with their capacity, which means that the larger the capacitor, the more stable its capacitance at higher frequencies. In the case of Small Capacity Ultracapacitors, higher capacity means less capacity constant as a function of frequency.

Table 3 Values of frequency for which equivalent decreases capacity two times

Capacitor (F)	T	α	f_c (mHz)
0.047	5.2261	0.6	91.8
0.1	14.7979	0.6	32.4
0.33	56.9669	0.57	9
1500	1.006	0.62	457
3000	0.6369	0.7	625

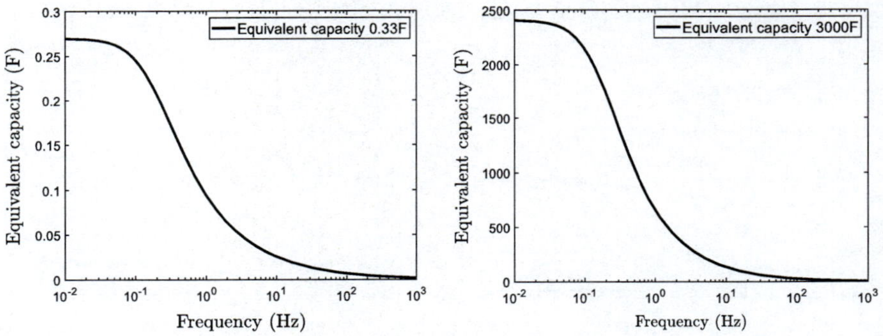

Fig. 13 Equivalent capacity C'

Figure 13a, b represents examples of diagrams of ultracapacitor capacitance decrease as a function of frequency for nominal capacitance ultracapacitors 0.33, 3000 F.

Next, compare the parameters identified with those measured directly from the ultracapacitors. This comparison is possible using the Davidson–Cole model, because unlike other models of the fractional order, this model contains the resistance R_c (Ω) and the capacity C (F) in a traditional form. The actual values of these parameters were measured using the step response of the ultracapacitors tested. The actual value of the resistance R_{real} (Ω) was determined from the voltage across the capacitor immediately after switching on the unit input. The actual value of the capacity C_{real} (F) was determined by dividing the value of the charge supplied to the capacitor by the voltage for it. A comparison of these parameters is provided in Table 4.

Table 4 confirms that using the fractional order model of G_c we can very accurately identify the actual parameters of the ultracapacitor. The identification results are equivalent to the actual values obtained from another method (in this case step response). The differences between the C parameter obtained in the identification process and the actual parameters in the case of modeling high-capacity ultracapacitors are the result of low precision of measurements for fast-changing signals. The lack of high frequency samples has a significant impact on the identification process.

Table 4 Comparison of model and real parameters

Capacitor (F)	C (F)	R_c (Ω)	C_{real} (F)	R_{real} (Ω)
0.047	0.06	32	0.06	32
0.1	0.094	52	0.1	42
0.33	0.27	29	0.27	28
1500	1348	0.25m	1189	0.26m
3000	2446	0.13m	2435.4	0.15m

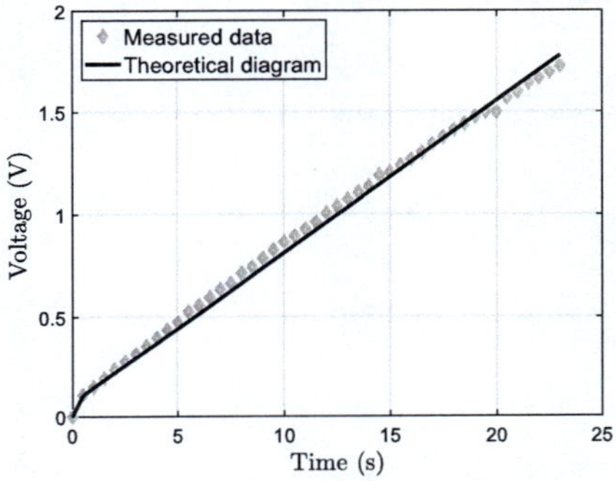

a) Ultracapacitor 1500F and current $I(t) = 100A$.

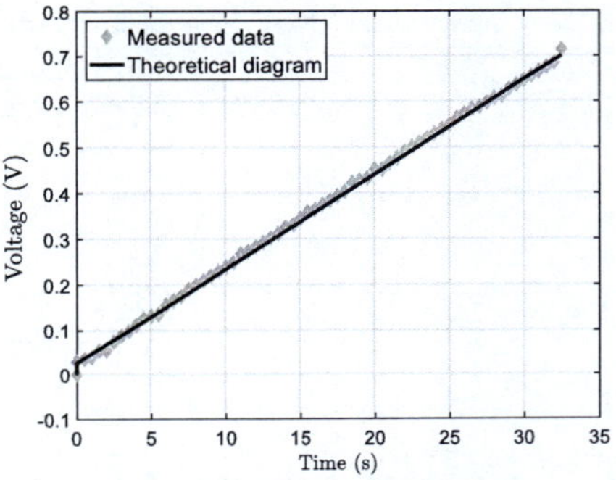

b) Ultracapacitor 3000F and current $I(t) = 50A$.

Fig. 14 Step response of model $G_c(s)$

Having identified the model parameters for all ultracapacitors, we were able to validate the obtained in time domain values by comparing the step responses of the received models (39) with the responses of physical devices. Sample results of this comparison are presented in Fig. 14a, b. First of these figures show the step response of an ultracapacitor with nominal capacity of 1500 F for the current input signal equal to 100 A and second one presents the step response of an ultracapacitor with a nominal capacity of 3000 F for current equal to 50 A. The obtained results confirm the usability of the Davidson–Cole model for modeling the dynamics of ultracapacitors. These results are not perfect, however, the identified parameters correspond to the physical parameters. Let us now examine the model derived from the anomalous diffusion equation. This model should very precisely describe the dynamics of the ultracapacitor in the frequency domain, however, unlike the D-C model, it is non-physically.

Table 5 Full model identified parameters

Capacitor (F)	T	$C_\alpha(\frac{F}{s^{1-\alpha}})$	R_c (Ω)	α	C_{real} (F)	R_{real} (Ω)
0.047	5.9257	0.0533	35	0.9887	0.06	32
0.1	18.0459	0.0941	58	0.9969	0.1	42
0.33	64.4422	0.2512	32	0.9870	0.27	28
1500	1.0132	1490	0.0003	0.9945	1189	0.00026
3000	0.5	2497	0.00018	0.9810	2435	0.00015

Table 6 New model RMSE identification error

Capacitor (F)	RMSE amplitude	RMSE phase
0.047	0.3121	1.4608
0.1	0.3513	1.4979
0.33	0.4190	1.7736
1500	0.3555	1.8486
3000	0.3729	0.7570

6.2 Ultracapacitor Modeling Using a Full Model

An ultracapacitor full model derived from anomalous diffusion equation is:

$$G_{C_\alpha} = R_c + \frac{\sqrt{1 + T_\alpha s^\alpha}}{C_\alpha s^\alpha}, \tag{50}$$

where R_c is an ultracapacitor resistance, C_α is fractional capacity in units $\frac{farad}{(s)^{1-\alpha}}$ and T_α is the parameter responsible for the decrease capacity as a function of frequency.

Comparison of model parameters obtained in the identification process by matching models with actual parameters of ultracapacitors is presented in Table 5. In fact we cannot compare the decrease of ultracapacitor capacity with frequency, because parameter C_α should be interpreted as ultracapacitor impedance rather than capacity like in the case of parameter C for a traditional capacitor (see Sect. 5.1), but we have to notice that the values of the α parameters in Table 5 demonstrate that for this model the best results of ultracapacitor modeling can be achieved for α slightly less than 1, which means also that for $\alpha = 1$ we should achieve similar results. Although the parameters C and C_α have different units, for α approximately equal to 1 we can try to put these parameters together to see if their values are identified correctly. The presented table confirms that the achieved results are proper, because their values are quite close to the real values of ultracapacitor parameters. RMSE of the identification is presented in Table 6.

The obtained values confirm good identification results using the model derived from the anomalous diffusion equation. When using this model, it can be seen that

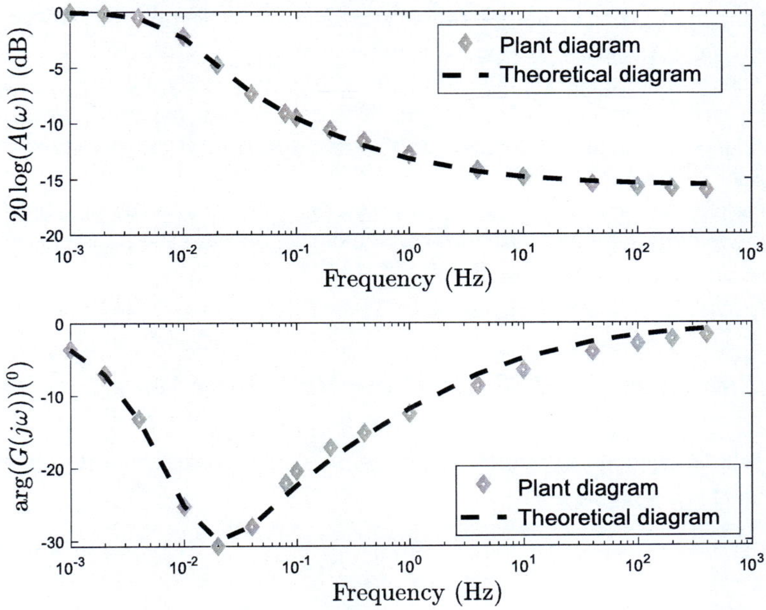

Fig. 15 Measured and new model theoretical Bode's diagrams

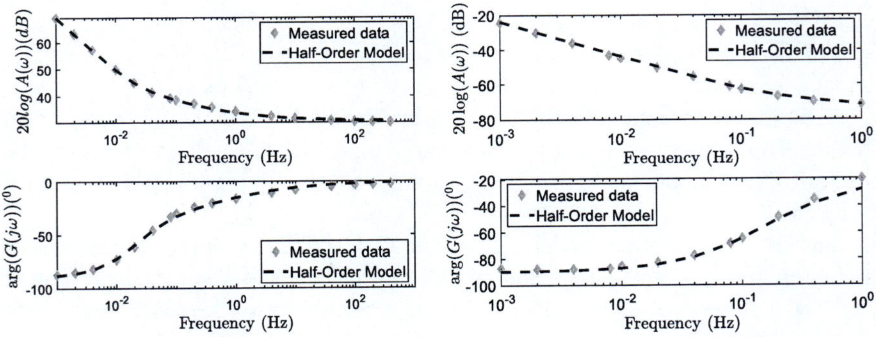

a) Ultracapacitor 0.47F. **b)** Ultracapacitor 3000F.

Fig. 16 Measured and new model theoretical Bode's diagrams of system with ultracapacitor 0.047F

the matching of amplitude characteristics is better than the phase shift characteristics. Sample results of modeling ultracapacitors using the new model are shown in Fig. 15a, b. The maximum amplitude and phase identification error is successively less than 0.8 dB and 3°. Additional confirmation of correct identification is Fig. 16, which shows a comparison of real and theoretical characteristics of the quadripole system with an ultracapacitor with an approximately capacity equal to 0.047 F.

Calculating the step response of this model in the time domain is a difficult issue because it requires the calculation of the Laplace inverse transform from $\sqrt{1 + T_\alpha s^\alpha}$. In the case of this expression, one would have to derive the formula for the Laplace transform for an fractional order. This problem remained an open problem, therefore the validation of the obtained results in the time domain was not presented for the full model.

In the next section we present the results of identification using the Half-Order Model, which through its physical form (the capacity in this model is the capacity in the traditional form) is more useful.

6.3 Ultracapacitor Modeling Using Half-Order Model

A Half-Order Model is an ultracapacitor model derived using the diffusion partial differential equation:

$$G_{0.5} = R_c + \frac{\sqrt{1 + Ts}}{Cs}, \tag{51}$$

where R_c is the ultracapacitor resistance, C is the ultracapacitor capacity in the traditional definition and T is a parameter of capacity decrease with frequency. This model can be also achieved as a particular case of the Davidson–Cole model for $\alpha = 0.5$.

The comparison of the parameters obtained in the identification process with the actual parameters of the ultra-capacitors is presented in the Table 7.

A summary of the identification process is presented in Table 8. It is worth noting that the Root Mean Square Error of this experiment is also small which means that the Half-Order model is precise enough for ultracapacitor modeling. Examples of Bode characteristics are presented in Fig. 17a, b.

Since in the Half order model's parameter C describes the device capacity in the traditional form, it can be compared with traditional capacity. The f_c dependent on T, for which the equivalent capacity decreases two times for the Half Order Model is given as follows:

Table 7 Identification parameters obtained by half-order model

Capacitor (F)	T	C (F)	R_c (Ω)	C_{real} (F)	R_{real} (Ω)
0.047	6.5231	0.056	35	0.06	32
0.1	18.5672	0.1	58	0.1	42
0.33	73.2900	0.27	32	0.27	28
1500	1.3059	1371	0.25m	1189	0.00026
3000	0.9668	2446	0.13m	2435	0.00013

Table 8 Identification error for half-order model

Capacitor (F)	RMSE amplitude	RMSE phase
0.047	0.3175	1.5169
0.1	0.3546	1.4971
0.33	0.5053	2.0026
1500	0.9533	2.7004
3000	0.6875	2.8670

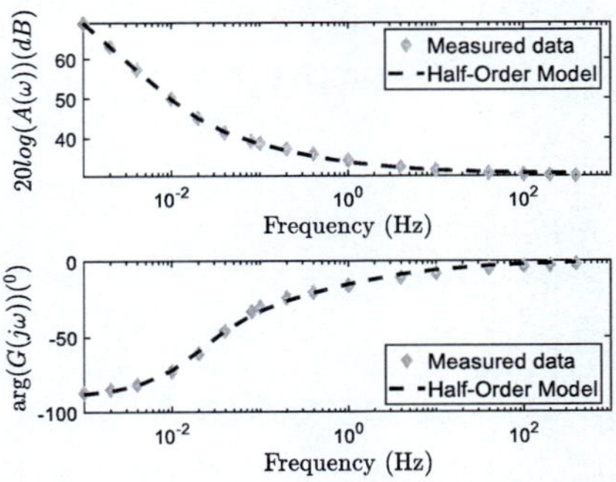

a) Ultracapacitor 0.47F.

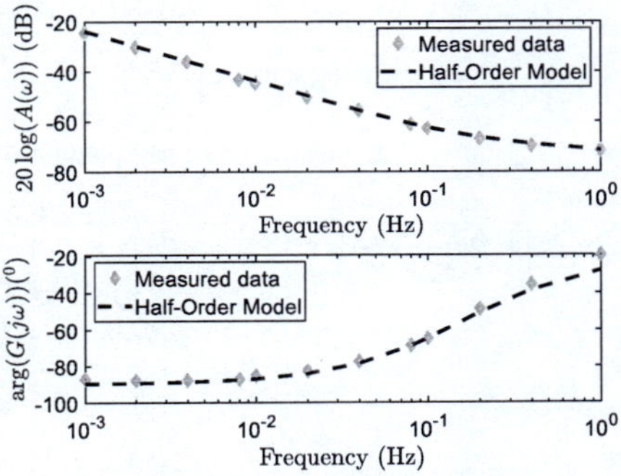

b) Ultracapacitor 3000F.

Fig. 17 Measured and theoretical Bode's diagrams for half-order model

Table 9 Values of frequency for which equivalent capacity decreases two times

Capacitor (F)	T	f_c (mHz)
0.047	6.5231	94.5
0.1	18.5672	33.2
0.33	73.29	8.41
1500	1.3059	472
3000	0.9668	637.6

$$f_c = \frac{\sqrt{2^4 - 1}}{2\pi T}. \tag{52}$$

The calculated frequency values are collected in Table 9. Achieved values are very close to the values obtained for the Davidson–Cole model. Also this model shows that for a low capacity ultracapacitor produced by Panasonic, frequency of two time capacity decrease is smaller for higher capacity and in the case of high capacity supercapacitors produced by Maxwell the value of this parameter is increasing with capacity of the device. Figure 18a, b show ultracapacitor equivalent capacity for ultracapacitors 0.33, 3000 F in function of frequency.

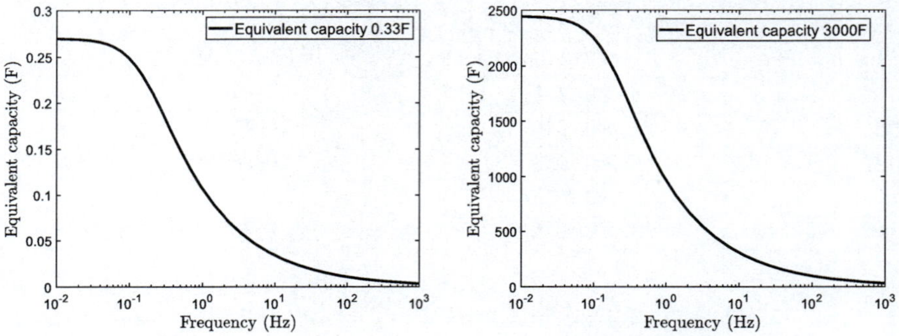

a) Ultracapacitor 0.33F. **b)** Ultracapacitor 3000F.

Fig. 18 Equivalent capacity C' of ultracapacitors

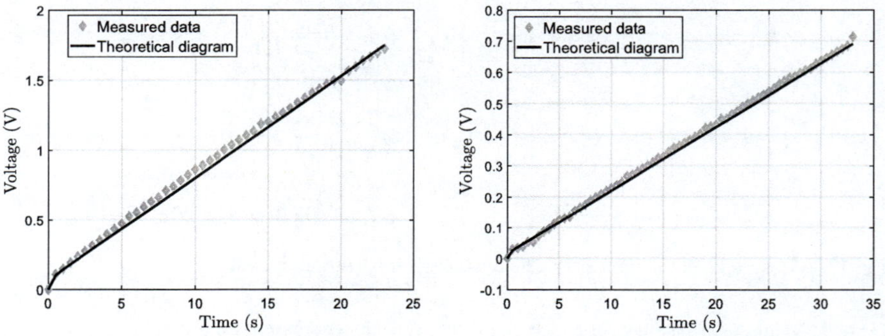

a) Ultracapacitor 1500F and current I(t)=100A. **b)** Ultracapacitor 3000F and current I(t)=50A.

Fig. 19 Step response of model $G_{0.5}$ for current

After receiving the modeling results in the frequency domain, one can check the obtained results based on the step response of ultracapacitors in the time domain and from their theoretical models. Sample results are presented in Fig. 19a, b. The first of them presents a comparison of the step response of an ultracapacitor with a nominal capacity of 1500 F to a current excitation of 100 A. The second graph shows the step response of a 3000 F ultracapacitor to a 50 A current signal.

Presented results confirm good accuracy of the model achieved from a normal diffusion equation with the dynamics of a physical systems, however they are a little worse than results achieved from Davidson–Cola's model. The greatest advantage of a Half-Order model is the integer system order in the denominator which implies the use of physical system parameters able to compare with parameters of a traditional capacitor.

Only linear capacity models are used in this section. It was therefore necessary to check whether the errors and inaccuracies of the measurements were not the result

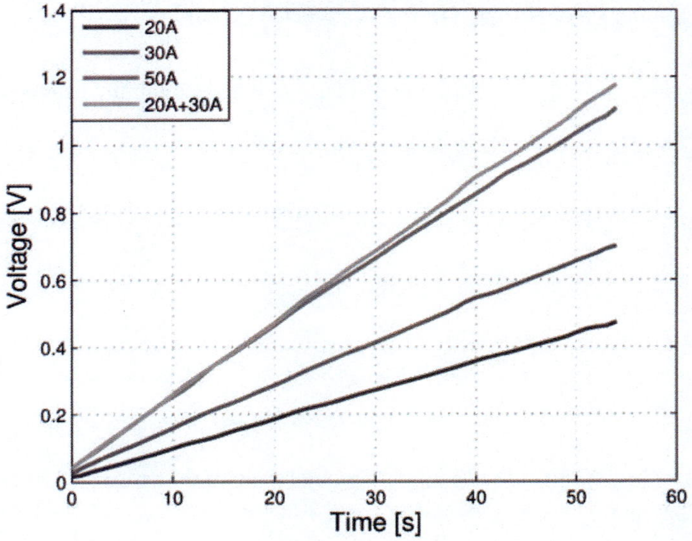

Fig. 20 Test of ultracapacitors linearity

of non-linearity of ultracapacitors. Therefore, tests were carried out to confirm the linearity of these systems. The results of these tests are shown in Fig. 20.

Collected results for the ultracapacitor of a nominal capacity equal to 3000 F show that ultracapacitors have some nonlinearity which leads to the conclusion that the most precise ultracapacitor modeling and control can be achieved only using nonlinear fractional order model. The solution of such a system is not trivial. So for solving the nonlinearity problem and for precise modeling of ultracapacitor dynamics fractional order neural network has been studied. This is described in the next section.

7 Ultracapacitor Modeling by Using a Fractional Order Neural Network

At the end of the previous section, the results of the non-linearity of ultracapacitors were presented (which was also shown in the work [2]). Considering the fractional order of the ultra-capacitor model that best reflects its dynamics, it's worth building a model that takes into account its non-linearity. Since the solution of a non-linear system of fractional order equations is a very complex problem, the most intuitive idea is to use a neural network to model ultracapacitors. This subsection presents the fractional order artificial neural network (FANN) [35] for modeling the dynamics of ultracapacitors.

7.1 Discrete Fractional Order Neural Network

In the beginning let us consider the non-linear system given in the following form:

$$\Delta^{n\alpha} y_{k+n} = g(\Delta^{(n-1)\alpha} y_{k+n-1}, \ldots, \Delta^{\alpha} y_{k+1}, y_k, u_k), \tag{53}$$

which we can rewrite as:

$$\begin{aligned} \Delta^{\alpha} y_{1,k+1} &= x_{2,k}, \\ \Delta^{\alpha} x_{2,k+1} &= x_{3,k}, \\ &\vdots \\ \Delta^{\alpha} x_{n,k+1} &= g\left(y_k, x_{2,k}, \ldots, x_{n,k}, u_k\right). \end{aligned} \tag{54}$$

Very good results of this system modeling can be achieved using the Discrete Fractional Order Neural Network (DFONN) proposed by Sierociuk et al. [45]. This structure is a combination of a standard neural network and the linear discrete fractional state-space system (DFOSS), which was presented in Sierociuk doctoral dissertation [38]. The scheme of this structure useful for simulation of a fractional order neural network is presented in Fig. 21. In this architecture the neural network is a traditional neural network structure in the form dependent on the modeling system. Input signals of the network are system input and output data for k-th sample (u_k, y_k) and the vector differences between previous outputs from $\Delta^{(n-1)\alpha} y_{k+n-1}$ to $\Delta^{\alpha} y_{k+1}$. In the output of the neural network one obtains the prediction of the next step difference $\Delta^{n\alpha} y_{k+n}$. Using this value DFOSS calculates value of the system output and the new differences vector. The size of DFOSS blocks depends on the modelled system structure and their matrices can be obtained in the following way:

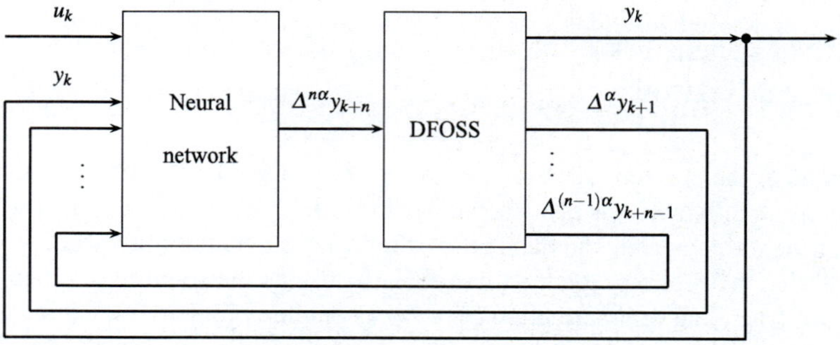

Fig. 21 Discrete fractional order neural network

$$\mathbb{A} = \begin{bmatrix} 0\,1\,0\,0 \cdots 0 \\ 0\,0\,1\,0 \cdots 0 \\ \vdots\;\vdots\;\vdots\;\vdots\;\ddots\;\vdots \\ 0\,0\,0\,0 \cdots 1 \\ 0\,0\,0\,0 \cdots 0 \end{bmatrix}, \quad \mathbb{B} = \begin{bmatrix} 0 \\ \vdots \\ 0 \\ 1 \end{bmatrix}, \quad \mathbb{C} = [I], \quad \mathbb{D} = [0].$$

A properly presented structure can be used to model the dynamics of fractional orders only in off-line mode. In order to use it for control, it was necessary to obtain the difference $\Delta^{n\alpha} y_{k+n}$ at the network output based on earlier signal samples. The results of modeling the non-linear system in the form presented in (53) were presented in the work [45]. The use of DFONN to model the dynamics of non-linear systems of fractional order allows the use of a much simpler neural network structure, however, in the case of ultracapacitors, correct dynamics mapping requires a more complex non-linear system. Sierociuk and Petráš in [44] presented a non-linear system of fractional order in a more general form, which they used to model the process of heat transfer in a rod. For this purpose, they used the following system:

$$\Delta^{n\alpha} y_{k+n} = g\left(\Delta^{(n-1)\alpha} y_{k+n-1}, \ldots, \Delta^{\alpha} y_{k+1}, y_k, \Delta^{(n-1)\alpha} u_{k+n-1}, \ldots, \Delta^{\alpha} u_{k+1}, u_k\right). \tag{55}$$

This system apart from the differences of output signals has the differences of input signals. The calculation of future samples requires only past samples (resulting system). Using the equation time offset shift (55), we get the following relation:

$$\Delta^{n\alpha} y_k = g\left(\Delta^{(n-1)\alpha} y_{k-1}, \ldots, \Delta^{\alpha} y_{k-n+1}, y_{k-n}, \Delta^{(n-1)\alpha} u_{k-1}, \ldots, \Delta^{\alpha} u_{k-n+1}, u_{k-n}\right). \tag{56}$$

Above relation is applied for signal pre-filtering in the neural network training process. The scheme of the DFONN which realizes this relation is presented in Fig. 22. The following configuration system can be implemented in Simulink, using a Neural Network Toolbox and Fractional State–Space Toolkit [37, 39]. The presented diagram can be adapted to the diagram useful for on-line simulations. Such a scheme is presented in Fig. 23. This configuration of DFONN is very useful in modeling commensurate systems, otherwise it is necessary to use two or more system orders independent of each other. Despite the fact that the presented neural network configuration is sufficient for modeling the heat transfer process in the case of the ultracapacitor modeling a non-linear incommensurate system is needed. Description of this system configuration is presented in the next section.

7.2 *Ultracapacitor Modeling and Control Using DFONN*

For ultracapacitors modeling, the following non-linear system is proposed:

$$\Delta^{\alpha+\beta} y_k = g\left(u_{k-2}, \Delta^{\alpha} u_{k-1}, \Delta^{\beta} u_{k-1}, \Delta^{\alpha+\beta} u_k, y_{k-2}, \Delta^{\alpha} y_{k-1}, \Delta^{\beta} y_{k-1}\right). \tag{57}$$

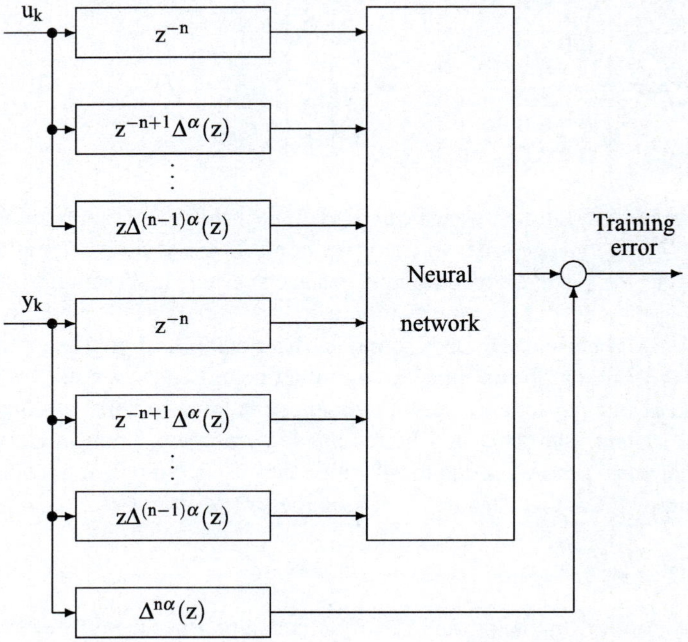

Fig. 22 Signal pre-filtering for neural network learning process

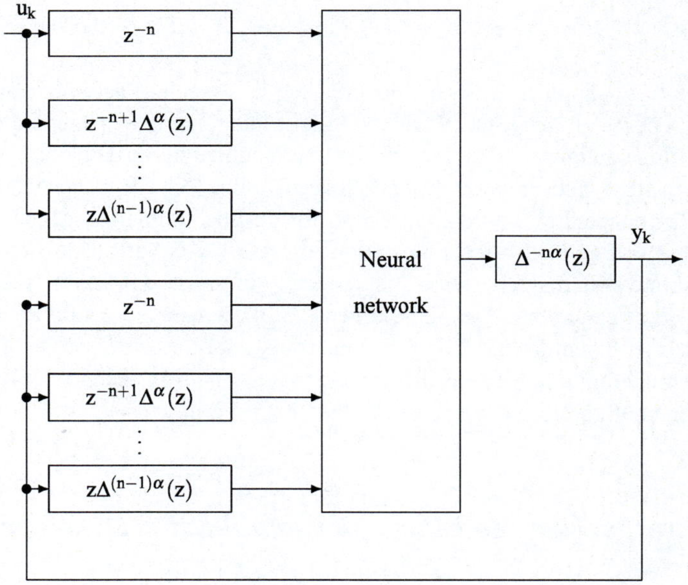

Fig. 23 Scheme for on-line simulation of discrete fractional order neural network

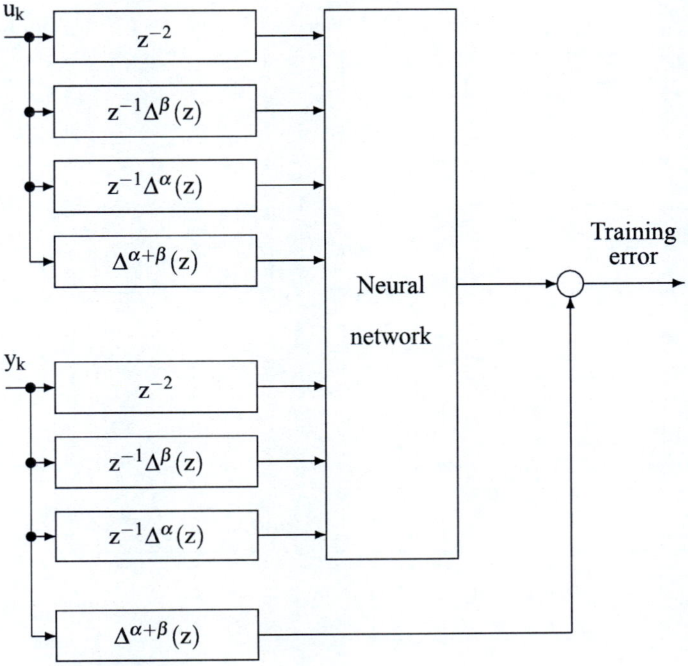

Fig. 24 Signal prefiltering for neural network training process

This system is a particular case of an incommensurate non-linear fractional order system for two system orders (α and β) and the maximum delay amounting two samples. The general formula of this kind of system for any delay is much more complicated. It contains elements depended on higher order differences for α, β and $\alpha + \beta$ orders. Because of the huge and complicated expression for the general incommensurate non-linear system, in this book, we have limited ourselves only to present system in the form useful for ultracapacitor modeling. This system was realized for the model shown in Fig. 24. The presented scheme of signal prefiltering was implemented for the fractional order neural network training process. For on-line simulation of the system, achieved during modeling process, the link configuration presented in Fig. 25 was used.

Experiments for ultracapacitor modeling were carried on of the Maxwell PC5 (4 F, 2.5 V). This ultracapacitor was connected for the experimental setup in the voltage to current converter configuration. For the process of training and testing of the fractional order neural network for ultracapacitors modeling, the responses of

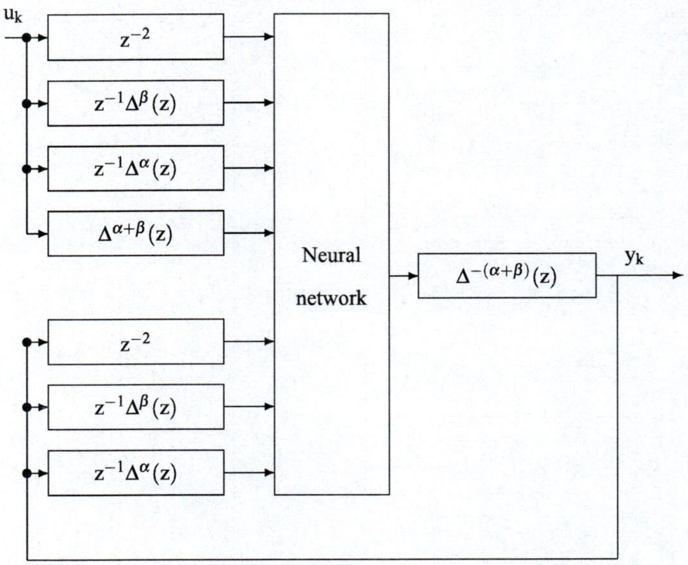

Fig. 25 Scheme of on-line simulation of DFONN ultracapacitor model

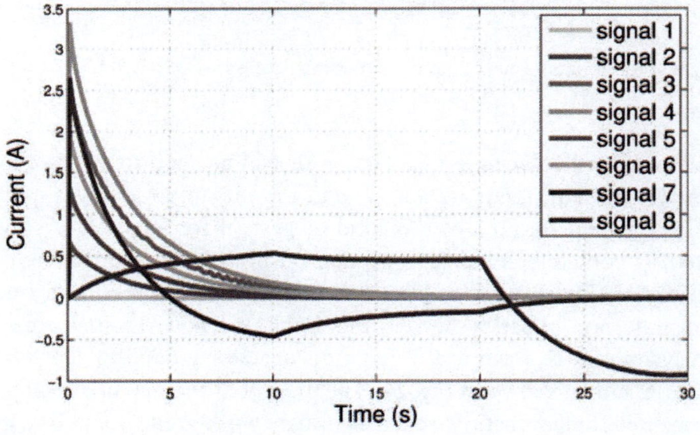

Fig. 26 Training input signals

the system with the ultracapacitor were measured. Using the Matlab and Simulink program and DSpace DS1103 signal acquisition card, the input signal $u_{dac}(t)$ was generated. This signal was converted to input current $i(t)$. The sampling time was $t_s = 0.01$ s. The input signals $i(t)$ are presented in Fig. 26.

A very important action, was to take the zero signal for the neural network training process. Without this signal, the net had a huge problem with generalization in the

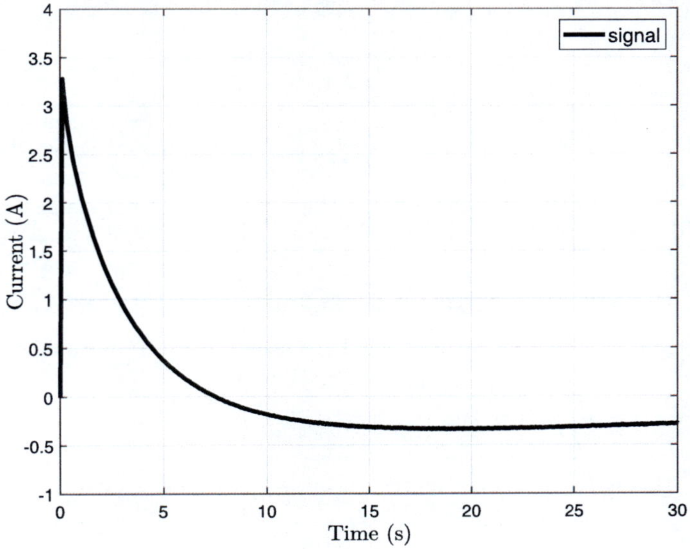

Fig. 27 Test input signals

case of the steady state. To test the network ability for generalization the test signal shown in Fig. 27 was used.

The voltage responses of the ultracapacitor $u_{uc}(t)$ for the presented input signals $i(t)$ measured directly from the ultracapacitor are presented in Fig. 28, where the response called *signal* is the output for the test input signal and another responses are named like input signals presented in Fig. 26 respectively. These signals have been filtered for measurement noise elimination.

At the beginning of the modeling process it was necessary to establish parameters of the neural network. As a result, the following configuration of the fractional order neural network for the ultracapacitor modeling were achieved:

- Seven inputs,
- Two layers,
- Five neurons in Hidden-layer,
- One neuron in Output-layer,
- Activation function: tansig function (hidden layer); linear function (output layer),
- Training algorithm: Levenberg–Marquardt,
- Epochs number: 35,
- $\alpha = 0.15$, $\beta = 0.3$.

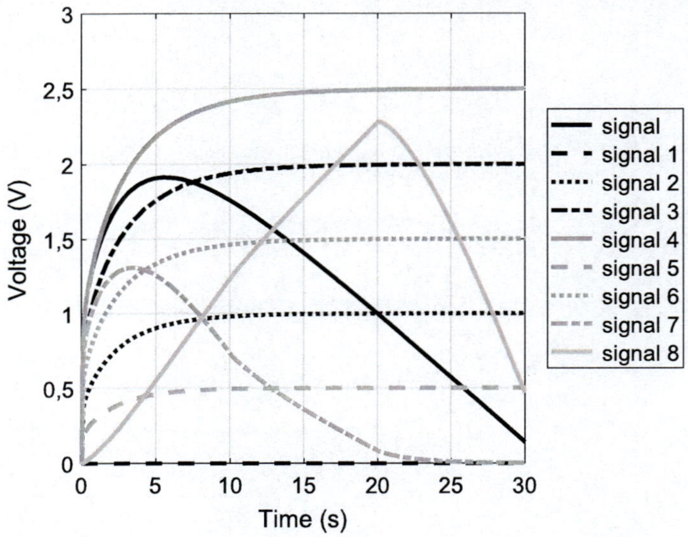

Fig. 28 Ultracapacitor response for input signals

Table 10 Mean error of fractional order neural network model

Signals	Error	Signals	Error	Signals	Error
Signal 1	$1.78e^{-4}$	Signal 4	$2.42e^{-4}$	Signal 7	$1.98e^{-4}$
Signal 2	$3.57e^{-4}$	Signal 5	$8.29e^{-5}$	Signal 8	$2.98e^{-4}$
Signal 3	$1.30e^{-4}$	Signal 6	$9.45e^{-5}$	Signal 9	$3.73e^{-4}$

Configuration of these parameters was established empirically. It is worth to notice that the sum of α and β orders is approximately equal to the order of the ultracapacitor fractional order model presented in Sect. 6. Another important result to highlight is the number of epoch enough for the network training. So, adopted fractional order neural network configuration allows for a very quick learning of the network.

As a result of neural network training, the means of the Euclidean norm of an absolute error vector for all signals have been calculated. Obtained results are presented in Fig. 10.

Exemplary results achieved after the neural network training process are presented in Fig. 29a, b. The generalization result for the test signal is shown in Fig. 30. The presented results confirm the high convergence between the behavior of the modeled system and its obtained model. Figure 30 also confirms the ability of the trained fractional order neural network for generalization.

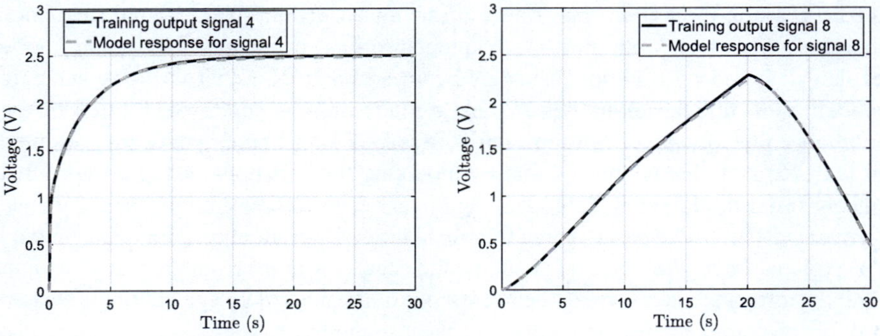

a) Signal 4 **b)** Signal 8

Fig. 29 Comparison of model and system responses

Fig. 30 Model response for test signal

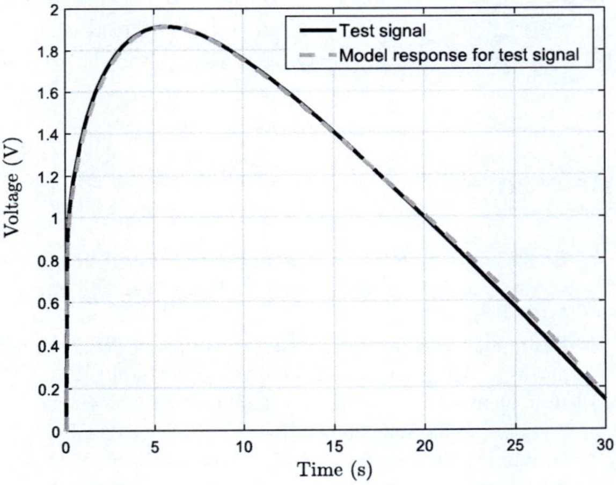

8 Conclusion

In this chapter we presented the different fractional order models for modeling systems with ultracapacitors. At the start, various traditional models used to model the dynamics of ultracapacitors were discussed. Then, based on their internal structure, an analytical derivation of a mathematical model justifying the sense of using an non integer order in the process of modeling ultracapacitors was presented. Subsequently, various dynamic models used for modeling in time and frequency domain, as well as the dynamics of ultracapacitors are described. The presented frequencies models have been thoroughly discussed and confirmed by their comparison with the properties observed in the operation of ultracapacitors. The obtained convergence of parameters confirmed the legitimacy of using models of this class. In addition, the big advantage of the fractional order models for modeling ultracapacitors is to explain the decrease in the usable capacitance of the ultracapacitor as the signal frequency increases. This fact, well known to manufacturers of ultra-capacitors, could not be explained based on traditional models of ultracapacitors. What is more, accurate modeling of ultracapacitors has enabled a much more accurate description of the resonance phenomenon in systems with ultracapacitors. These results are presented in [11, 46] and clearly showed that by modeling these systems using a traditional integer order model we will obtain a significant error in determining the resonance frequency. The model based on the fractional order differential calculus also enabled the explanation of an additional phenomenon occurring in the resonance in the circuit with the ultracapacitor, namely the occurrence of the maximum current of the system not for the resonance frequency. This is because the real part of the impedance also depends on the frequency and above the resonance frequency, in addition, to some extent it decreases faster than the imaginary part increases, which results in the maximum current falling for a different frequency than the resonant one (which cannot be described by an integer order model). In the last part of this chapter, indicating the non-linearity of ultracapacitors, a method of modeling ultracapacitors using a fractional order neural network was presented.

References

1. Belhachemi, F., Raël, S., Davat, B.: A physical based model of power electric double-layer supercapacitors. In: Proceedings of IEEE Industry Applications Conference, vol. 5, pp. 3069–3076, Oct 2000
2. Bertrand, N., Sabatier, J., Briat, O., Vinassa, J.-M.: Fractional non-linear modelling of ultracapacitors. Commun. Nonlinear Sci. Numer. Simul. **15**(5), 1327–1337 (2010)
3. Bouchaud, J.P., Georges, A.: Anomalous diffusion in disordered media: statistical mechanisms, models and physical applications. Phys. Rep. **195**, 127–293 (1990)
4. Brouji, H., Vinassa, J.M., Briat, O., Bertrand, N., Woirgard, E.: Ultracapacitors self discharge modelling using a physical description of porous electrode impedance. In: Proceedings of IEEE Vehicle Power and Propulsion Conference (VPPC), September 2008
5. Buller, S., Karden, E., Kok, D., De Doncker, R.W.: Modeling the dynamic behavior of supercapacitors using impedance spectroscopy. IEEE Trans. Ind. Appl. **38**(6), 1622–1626 (2002)
6. Chu-qi, S., Du-que, Z.: Computer simulation study on ultracapacitor for electric vehicle. In: IEEE International Conference on Intelligent Computing and Intelligent Systems, 2009. ICIS 2009, vol. 1, pp. 199 –203, Nov 2009
7. Davidson, D.W., Cole, R.H.: Dielectric relaxation in glycerine. J. Chem. Phys. **18**, 1417 (1951)
8. Davidson, D.W., Cole, R.H.: Dielectric relaxation in glycerine, propylene glycol, and n-propanol. J. Chem. Phys. **19**, 1484–1490 (1951)
9. Dzieliński, A., Sarwas, G., Sierociuk, D.: Time domain validation of ultracapacitor fractional order model. In: Proceedings of the IEEE Conference on Decision and Control, pp. 3730–3735 (2010)
10. Dzieliński, A., Sarwas, G., Sierociuk, D.: Comparison and validation of integer and fractional order ultracapacitor models. Adv. Differ. Equ. **11** (2011)
11. Dzieliński, A., Sierociuk, D., Sarwas, G.: Fractional-order modeling and control of selected physical systems. In: Petráš, I., (ed.) Handbook of Fractional Calculus with Applications. Applications in Control, vol. 6, pp. 293–320. De Gruyter, Boston (2019)

12. Faranda, R., Gallina, M., Son, D.T.: A new simplified model OS double-layer capacitors. In: Proceedings of International Conference on Clean Electrical Power, ICCEP'07, pp. 706–710 (2007)
13. Fouda, M.E., Elwakil, A.S., Radwan, A.G., Allagui, A.: Power and energy analysis of fractional-order electrical energy storage devices. Energy **111**, 785–792 (2016)
14. Freeborn, T.J., Maundy, B., Elwakil, A.S.: Fractional-order models of supercapacitors, batteries and fuel cells: a survey. Mater. Renew. Sustain. Energy **4**(3) (2015)
15. Górecki, P.: 2700 faradów, czyli super(ultra)kondensatory. Elektronika dla Wszystkich **6**, 21–24 (2001)
16. Haus, J.W., Kehr, K.W.: Diffusionnext term in previous termregular and disordered latticesnext term. Phys. Rep. **150**, 263–406 (1987)
17. Kopka, R., Tarczyński, W.: Measurement system for determination of supercapacitor equivalent parameters. Metrol. Meas. Syst. **20**(4), 581–590 (2013)
18. Kopka, R., Tarczyński, W.: Influence of the operation conditions on the supercapacitors reliability parameters. Pomiary Autom. Robot. 49–54 (2015)
19. Kopka, R., Tarczyński, W.: A fractional model of supercapacitors for use in energy storage systems of next-generation shipboard electrical networks. J. Marine Eng. Technol. **16**(4), 200–208 (2017)
20. Kosztołowicz, T.: Subdiffusion in a system with a thick membrane. J. Membr. Sci. **320**, 492–499 (2008)
21. Lai, J.S., Levy, S., Rose, M.F.: High energy density double-layer capacitors for energy storage applications. IEEE Aerosp. Electron. Syst. Mag. **7** (1992)
22. Lopes, A.M., Tenreiro Machado, J.A., Ramalho, E.: On the fractional-order modeling of wine. Eur. Food Res. Technol. **243**(6), 921–929 (2017)
23. Macias, M., Sierociuk, D.: Modeling of electrical drive system with flexible shaft based on fractional calculus. In: 2013 14th International Carpathian Control Conference (ICCC), pp. 222–227, May 2013
24. Marie-Francoise, J.N., Gualous, H., Berthon, A.: Artificial neural network control and energy management in 42 v dc link. In: European Conference on Power Electronics and Applications, 8 pp. (2005)
25. Marie-Francoise, J.N., Gualous, H., Berthon, A.: Supercapacitor thermal- and electrical-behaviour modelling using ANN. IEE Proc. Electric Power Appl. **153**(2), 255–262 (2006)
26. Neto, J.P., Coelho, R.M., Valério, D., Vinga, S., Sierociuk, D., Malesza, W., Macias, M., Dzieliński, A.: Variable order differential models of bone remodelling. IFAC-Papers OnLine **50**(1), 8066 – 8071 (2017). 20th IFAC World Congress
27. Ostalczyk, P.: Epitome of the fractional calculus: theory and its applications in automatics. Wydawnictwo Politechniki Łódzkiej, Łódź, Poland (2008). (in Polish)
28. Panasonic Corporation: Electric double layer capacitor (gold capacitor), series sd, catalogue. http://industrial.panasonic.com (2006)
29. Podlubny, I.: Fractional Differential Equations. An Introduction to Fractional Derivatives, Fractional Differential Equations, Some Methods of Their Solution and Some of Their Applications. Academic Press, San Diego, USA (1999)
30. Podlubny, I., Petras, I., Skovranek, T., Terpak, J.: Toolboxes and programs for fractional-order system identification, modeling, simulation, and control. In: Petras, I., Podlubny, I., Kacur, J. (eds.) Proceedings of the 2016 17th International Carpathian Control Conference (ICCC), Tatranska Lomnica, Slovakia, May 29–Jun 01, 2016, pp. 608–612 (2016)
31. Quintana, J.J., Ramos, A., Nuez, I.: Identification of the fractional impedance of ultracapacitors. In: Proceedings of the 2nd IFAC Workshop on Fractional Differentiation and its Applications, Porto, Portugal, IFAC FDA'06, 19–21 July 2006
32. Riu, D., Retière, N., Linzen, D.: Half-order modelling of supercapacitors. In: Proceedings of IEEE Industry Applications Conference, 39th IAS Annual Meeting, vol. 4, pp. 2550–2554 (2004)
33. Sakrajda, P., Sierociuk, D.: Modeling heat transfer process in grid-holes structure changed in time using fractional variable order calculus. In: Babiarz, A., Czornik, A., Klamka, J.,

Niezabitowski, M. (eds.) Theory and Applications of Non-integer Order Systems, pp. 297–306. Springer International Publishing, Cham (2017)
34. Sarwas, G.: Modelling and control of systems with ultracapacitors using fractional order calculus. Ph.D. thesis, Faculty of Electrical Engineering, Warsaw University of Technology (2012)
35. Sarwas, G., Sierociuk, D., Dzieliński, A.: Modeling and control with discrete fractional order artificial neural network. In: 2012 13th International Carpathian Control Conference (ICCC), May 2012
36. Sarwas, G., Sierociuk, D., Dzieliński, A.: Ultracapacitor model based on anomalous diffusion. In: The Fifth Symposium on Fractional Differentiation and Its Applications (FDA12), Nanjing, China, May 2012
37. Sierociuk, D.: Fractional order discrete state–space system Simulink Toolkit user guide. http://www.ee.pw.edu.pl/~dsieroci/fsst/fsst.htm
38. Sierociuk, D.: Estymacja i sterowanie dyskretnych układów dynamicznych ułamkowego rzędu opisanych w przestrzeni stanu. Ph.D. thesis, Warsaw University of Technology (Poland) (2007)
39. Sierociuk, D., Dzieliński, A.: Fractional Kalman filter algorithm for states, parameters and order of fractional system estimation. Int. J. Appl. Math. Comput. Sci. **16**, 129–140 (2006)
40. Sierociuk, D., Dzielinski, A., Sarwas, G., Petras, I., Podlubny, I., Skovranek, T.: Modelling heat transfer in heterogeneous media using fractional calculus. Philos. Trans. R. Soc. A: Math. Phys. Eng. Sci. **371**, 2013 (1990)
41. Sierociuk, D., Malesza, W., Macias, M.: Derivation, interpretation, and analog modelling of fractional variable order derivative definition. Appl. Math. Modell. **39**(13), 3876–3888 (2015). http://dx.doi.org/10.1016/j.apm.2014.12.009
42. Sierociuk, D., Malesza, W., Macias, M.: Numerical schemes for initialized constant and variable fractional-order derivatives: matrix approach and its analog verification. J. Vib. Control (2015). https://doi.org/10.1177/1077546314565438
43. Sierociuk, D., Malesza, W., Macias, M.: On the recursive fractional variable-order derivative: equivalent switching strategy, duality, and analog modeling. Circuits, Syst. Signal Process. **34**(4), 1077–1113 (2015)
44. Sierociuk, D., Petráš, I.: Modeling of heat transfer process by using discrete fractional-order neural networks. In: Proceedings of 16th International Conference on Methods and Models in Automation and Robotics, MMAR'11 Międzyzdroje, Aug 2011
45. Sierociuk, D., Sarwas, G., Dzieliński, A.: Discrete fractional order artificial neural network. Acta Mech. Autom. **5**(2), 128–132 (2011)
46. Sierociuk, D., Sarwas, G., Twardy, M.: Resonance phenomena in circuits with ultracapacitors. In: Proceedings of 12th International Conference on Environment and Electrical Engineering (EEEIC), pp. 197–202 (2013)
47. Sierociuk, D., Skovranek, T., Macias, M., Podlubny, I., Petras, I., Dzielinski, A., Ziubinski, P.: Diffusion process modeling by using fractional-order models. Appl. Math. Comput. **257**, 2–11 (2015)
48. Wang, Y., Hartley, T.T., Lorenzo, C.F., Adams, J.L., Carletta, J.E., Veillette, R.J.: Modeling ultracapacitors as fractional-order systems. In: Baleanu, D., Guvenc, Z.B., Machado, J.A.T. (eds.) New Trends in Nanotechnology and Fractional Calculus Applications, pages 257+ (2010). International Workshops on New Trends in Science and Technology (NTST 08)/ Fractional Differentiation and Its Applications (FDA08), Cankaya Univ., Ankara, Turkey, Nov 5–7, 2008
49. Wen, C., Jianjun, Z., Jinyang, Z.: A variable-order time-fractional derivative model for chloride ions sub-diffusion in concrete structures. Fract. Calc. Appl. Anal. **16**(1), 76–92 (2013)
50. Westerlund, S., Ekstam, L.: Capacitor theory. IEEE Trans. Dielectrics Electrical Insul. **1** (1994)
51. Ziubinski, P., Sierociuk, D.: Fractional order noise identification with application to temperature sensor data. In: 2015 IEEE International Symposium on Circuits and Systems (ISCAS), pp. 2333–2336, May 2015

Switched Models of Non-integer Order

Stefan Domek

Abstract In this paper, methods for modeling complex, non-linear dynamic systems of non-integer order, using so-called switched models that are based on the dynamic change of local linear models, depending on the value of an appropriately chosen switching function are discussed. Such a multimodel approach enables a relatively simple description of properties of many complex physical and abstract processes encountered in technology, especially in automation and robotics, but also in nature, biology, medicine and, for example, in economics. Although the theory of switched systems has been developing intensively for over a dozen or so years, many issues and problems have not yet been solved. This is particularly true for systems of a non-integer order. In the paper, basic definitions of fractional switched models, their principal properties and examples of applications in control are presented.

1 Introduction

The vast majority of phenomena, processes and objects surrounding us, both physical and abstract ones, show a non-linear nature. It is often so strong that it is not enough to use linearized models to describe adequately their properties. In turn, determining and subsequent use of non-linear models is usually very difficult. This applies to phenomena and processes from various fields of life, science and technology, for example, biological, social, socio-cognitive, economic, transport, information and many technological processes and systems, not to mention control and monitoring systems, including solutions based on artificial intelligence. One of the methods to overcome these difficulties consists in replacing a complex non-linear model by a set of local linear models valid for small areas around various operating points. The idea of this approach boils down to switching over active models in time so that

S. Domek (✉)
Department of Control Engineering and Robotics, West Pomeranian University of Technology Szczecin, Al. Piastów 17, 70-310 Szczecin, Poland
e-mail: stefan.domek@zut.edu.pl

P. Kulczycki et al. (eds.), *Automatic Control, Robotics, and Information Processing*, Studies in Systems, Decision and Control 296,
https://doi.org/10.1007/978-3-030-48587-0_6

the generalized modeling error does not exceed specified bounds (for example, in terms of a chosen norm) and to obtain a reduction in the computational complexity of the description at the same time. By this means a nonlinear time-invariant process (NLTI) can be treated as a linear time-variant one (LTV). Instantaneous properties of the modeled non-linear process are described then by a quasilinear switched model, composed of batteries (set, network) of local linear submodels [4, 18, 43, 48].

Switched systems have been the subject of intensive studies for decades [1–3, 12, 24, 29, 31, 37, 43, 44, 48, 52, 64, 65]. It has been shown that they can model effectively various complex dynamic systems, including systems with perturbed parameters, chaos, multiple limit cycles and others. They also make it possible to better analyze systems present in modern technology, such as adaptive wide area networks, fault-tolerant systems, systems with multiple sampling periods, etc. It has also been shown that there is a large class of non-linear control processes that can be stabilized by switched local linear controllers, whereas this cannot be done by static state feedback. Yet, despite explaining many theoretical issues, such as stability conditions, controllability, observability and undoubted successes in applications, especially in engineering, chemical, automotive, power industries and in control of autonomous vehicles, a number of theoretical problems still remain to be solved [39, 40].

On the other hand, theory of fractional order systems (FOS) has been intensively developing for the past decades [3, 6, 26, 34, 46, 50, 51, 60, 63]. Following from the research work conducted worldwide, system description by means of non-integer differential equations is one of the more effective modeling methods and opens up new possibilities of modeling real properties of many complex phenomena and industrial processes. In automation and robotics, as in the case of integer-order models, description by means of fractional-order models can be used indirectly for tuning, or directly, for synthesis of linear control algorithms [9, 15, 24, 45, 53, 58]. However, also here, determining and the subsequent use of nonlinear models is very difficult, if not more so than in the case of integer-order processes. Therefore, an idea was conceived to replace complex fractional-order non-linear time-invariant models (FO-NLTI) by switched fractional-order models made up of batteries comprising fractional-order linear time-variant models (FO-LTV) [1, 18, 38].

This Section is structured as follows: in Sect. 2 the concept of switched models, their types and areas of use are discussed; in Sect. 3 the basics of fractional-order differential and difference calculus, as well as linear and non-linear, continuous-time and discrete-time dynamic state space models of non-integer order based on them are recalled; in Sect. 4 selected non-integer order discrete-time switched space state models are defined and their basic properties are discussed, and finally, in Sect. 5 an example of the use of discrete-time non-integer order switched models in the synthesis of nonlinear fractional-order prediction control algorithms is given. The whole is summarized in Conclusions.

2 The Concept of Switched Models

The method of describing a non-linear process by a set of switched linear local models assumes that the entire operating area of the process is divided into sub-areas for which local linear models are to be determined. In switched models a special supervisory system determines an appropriately selected switching signal that defines the instantaneous degree of activity of each local model. That is why they are classified with so-called hybrid systems that combine dynamic systems with logic ones [11, 57, 61, 67]. Elements of the lower layer that model local properties of the process are described by relationships typical of linear dynamic systems (continuous-time or discrete-time), while elements of the higher layer, which governs the operation of the whole model, are described by the laws of logic. In the basic approach these are classic binary logic laws, in the broader approach these are fuzzy logic laws [8, 15, 18, 33]. The decision variable is given by the following two definitions depending on the adopted notation:

Definition 1 Let $\Sigma = \{(t_0, j_0), \ \ldots, (t_k, j_k), \ \ldots\}, t_k \in \mathbb{R}\backslash\mathbb{R}_-, \ j_k \in Z_+$ denote a switching sequence of S local submodels, where $0 \leq t_0 < t_1 < \cdots < t_k < \cdots$, with t_0 being the initial time instant, t_k being the k-th switching time instant, and j_k being the k-th set of active submodels.

The switching signal $\sigma(t) \in \mathbb{R}^N$ is of the form of a piecewise constant vector of weighting variables $\sigma_j(t)$

$$\sigma(t) = \sigma(t_k) = \begin{bmatrix} \sigma_1(t_k) \\ \sigma_2(t_k) \\ \vdots \\ \sigma_S(t_k) \end{bmatrix} \text{ for } t \in [t_k, t_{k+1}), \quad \sum_{j=1}^{S} \sigma_j(t) = 1, \tag{1}$$

with

- $\sigma_j(t) \in \{0, 1\}$ for models switched according to classic binary logic laws,
- $\sigma_j(t) \in [0, 1]$ for models switched according to fuzzy logic laws.

Definition 2 Let $\Sigma = \{(t_0, j_0), \ \ldots, (t_k, j_k), \ \ldots\}, t_k \in \mathbb{R}\backslash\mathbb{R}_-, \ j_k \in Z_+$ denote a switching sequence of S local submodels switched according to laws of binary logic, where $0 \leq t_0 < t_1 < \ldots < t_k < \ldots$, with t_0 being the initial time instant, t_k being k-th switching time instant, and j_k being the index of the submodel activated at the instant t_k.

The switching signal $\sigma(t) \in \mathbb{Z}_+, \ dim\sigma(t) = 1$ is of the form of a piecewise constant selection variable

$$\sigma(t) = \sigma(t_k) = j_k \in \{1, 2, \ldots, S\} \text{ for } t \in [t_k, t_{k+1}). \tag{2}$$

The need for an adequate description of switched systems in automation was recognized a few dozen years ago, when systems containing relays and other elements with hysteresis, important from an engineering point of view, were considered. They were commonly used, for example, in mechanical and power systems, in transport, etc., where discontinuous two-position and three-position controllers and continuous PID controllers with variable structure found a wide application [48, 61]. It turned out then that there exists a large class of non-linear systems that can be stabilized by switching the control algorithm, but they cannot be stabilized through continuous static state feedback [2, 40]. Moreover, it also turned out that switched systems have some specific features. For example, the system as a whole may not be stable for some switching strategies, even when all switched subsystems are stable and vice versa, proper selection of the switching strategy can provide stability to the system in which the switched subsystems are unstable [40, 71]. This indicates that one of the main features of the control process, namely stability, is dependent not only on the dynamics of each subsystem, but also on characteristics of the switching signal. Similarly, it is like this with other important features, like controllability and observability of switched systems [1, 38, 44, 66].

The above properties of switched systems are also the reason that formulating conditions for their stability, controllability or observability, being difficult in general, depends additionally on the scenario of changes and restrictions imposed on the switching signal. From this point of view, switched systems can be divided into two groups:

Definition 3 A switched system where no restrictions are imposed on time variability of the switching signal $\sigma(t)$ is called a system with arbitrary switching.

The second group is when, a switched system is called a system with restricted switching, i.e. the system in which restrictions are imposed on the switching signal $\sigma(t)$ relating to, for example, switching instants, minimum and maximum length of time between switching instants, sequences of changes in value and membership of a specific class of function.

It follows from Definitions 1 and 2 that, in the case of a switching system governed by the laws of fuzzy logic, selected local submodels are activated with various weights at the time instant t_k $\{\mathscr{M}_i, \ldots, \mathscr{M}_j\} \subseteq \{\mathscr{M}_1, \mathscr{M}_2, \ldots, \mathscr{M}_S\}$, $1 \leq i \neq j \leq S$. It also follows that the non-linear system is represented by a fusion of several local submodels from the instant t_k until the next submodel change at the instant t_{k+1}. This is shown diagrammatically in Fig. 1, where $u(t) \in U \subseteq \mathbb{R}^m$ is the input vector, $x(t) \in X \subseteq \mathbb{R}^n$ is the input vector $y(t) \in Y \subseteq \mathbb{R}^p$. In the case of a switching system governed by the laws of binary logic, the t_k the j_k-th local submodel $\mathscr{M}_{j_k}$, $1 \leq j_k \leq S$ is activated at the instant t_k, which means that the non-linear system is represented by the submodel $\mathscr{M}_{j_k}$. t_{k+1} from the instant t_k until the next submodel change at the instant t_{k+1}. This is depicted in Fig. 2.

The switching signal $\sigma(t)$ can be in general a function of time, its past values, state variables, outputs and inputs of the modeled process, as well as a function of an external auxiliary signal. In practice, there can be distinguished various particular classes of switched models depending on the specific form of the switching law [17, 64]:

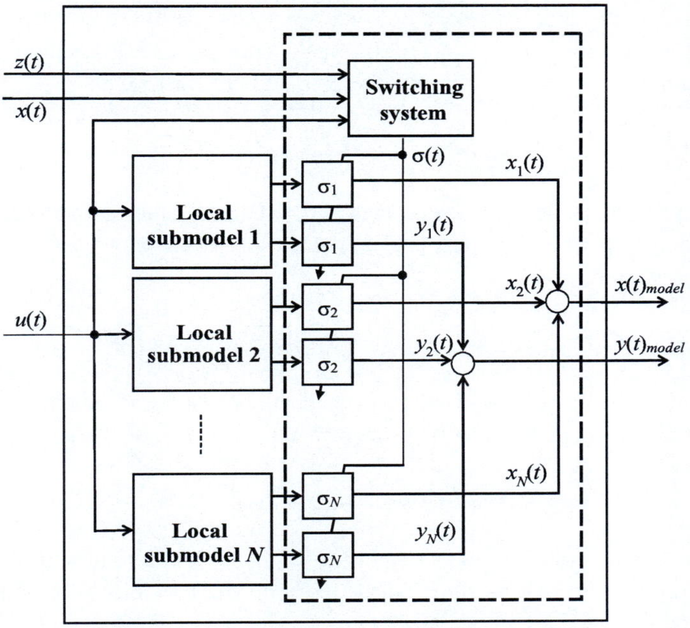

Fig. 1 Block diagram of the model switched according to fuzzy logic laws

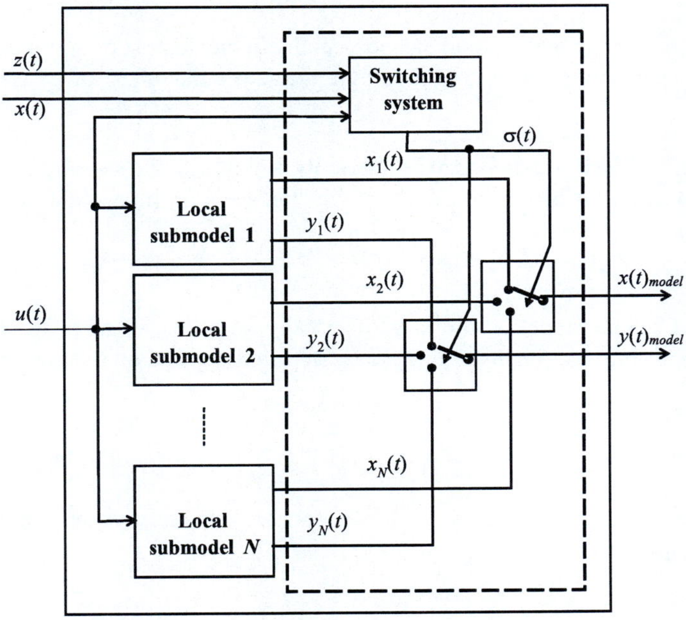

Fig. 2 Block diagram of the model switched according to binary logic laws

Definition 4 The model is said to be switched according to a predetermined plan, if the switching signal $\sigma(t)$ is solely a known function of time that is independent of both the input, state and output process vectors, and its past values

$$\sigma(t) = h(t) \overset{def}{\leftarrow} \Sigma, \quad 0 \leq t_0 \leq t. \tag{3}$$

Definition 5 The model is said to be switched with time, if the switching signal $\sigma(t)$ is a function of time and its past values, but does not depend on the process state vector and other signals

$$\sigma(t+1) = h(t, \sigma(t)), \quad 0 \leq t_0 \leq t, \quad \sigma(t_0) = \sigma_0. \tag{4}$$

Definition 6 The model is said to be switched autonomously, if the switching signal $\sigma(t)$ depends solely on the history of the previous switching scenario

$$\sigma(t+1) = h(\sigma(t)), \quad 0 \leq t_0 \leq t, \quad \sigma(t_0) = \sigma_0. \tag{5}$$

Definition 7 The model is said to be event-switched, if the switching signal $\sigma(t)$ does not depend directly on time, but depends on the previous switching history, state vector and possibly other signals

$$\sigma(t+1) = h(\sigma(t),\ x(t),\ u(t),\ z(t)), \quad 0 \leq t_0 \leq t, \quad \sigma(t_0) = \sigma_0. \tag{6}$$

In practice, of greatest importance, especially in the context of nonlinear process modeling and nonlinear controller design, are particular cases of event-switched models:

Definition 8 The model is said to be internally switched, if the switching signal $\sigma(t)$ depends only on the history of previous switchings and on state and input vectors

$$\sigma(t+1) = h(\sigma(t),\ x(t),\ u(t)), \quad 0 \leq t_0 \leq t, \quad \sigma(t_0) = \sigma_0. \tag{7}$$

Definition 9 The model is said to be externally switched, if the switching signal $\sigma(t)$ depends only on the history of previous switchings and on the auxiliary signal $z(t)$

$$\sigma(t+1) = h(\sigma(t),\ z(t)), \quad 0 \leq t_0 \leq t, \quad \sigma(t_0) = \sigma_0. \tag{8}$$

This switching method is very widely encountered in adaptive systems with auxiliary variable and gain scheduling.

3 Dynamical Models of Non-integer Order

3.1 Differential and Difference Calculus of Non-integer Order

Differential calculus of non-integer order is a generalization of the classic differential calculus. For a derivative of non-integer order $\alpha \in \mathbb{R}$ of a real-valued function $f(t),\ t \in \mathbb{R}$ on the interval $[t_0,\ t]$, denoted by the operator ${}_{t_0}D_t^{\alpha} f(t) = \frac{d^{\alpha} f(t)}{dt^{\alpha}}$, there exist many definitions proposed by various researchers, for example, by Riemann and Liouville, Caputo, Weyl, Fourier, Cauchy and Abel. The definitions differ in their properties and/or area of applicability [35, 50, 51]. In practical applications, especially in digital control systems where discrete values of the function $f(t)$ taken with the sampling interval h are used in a natural way for computations, the most commonly encountered is the definition introduced by Grünwald and Letnikov [50]:

Definition 10 A derivative of order $\alpha \in \mathbb{R}$ of the function $f(t), t \in \mathbb{R}$, according to Grünwald and Letnikov is given by

$$ {}_{t_0}^{GL}D_t^{\alpha} f(t) = \lim_{h \to 0} h^{-\alpha} \sum_{j=0}^{\lfloor \frac{t-t_0}{h} \rfloor} c_j^{\alpha} f(t - jh), \tag{9} $$

where the symbol $\lfloor \cdot \rfloor$ denotes the integer part, $c_j^{\alpha} = (-1)^j \binom{\alpha}{j}$, and the so-called generalized Newton symbol is defined by

$$ \binom{\alpha}{j} = \begin{cases} 1 & \text{for} \quad j = 0 \\ \frac{\alpha(\alpha-1)\ldots(\alpha-j+1)}{j!} & \text{for} \quad j = 1, 2, 3, \ldots \end{cases} \tag{10} $$

Remark 1 It follows from the definition of the generalized derivative that ${}_{t_0}D_t^0 f(t) = f(t)$ and the fractional order derivative of the function $f(t)$ for $\alpha < 0$ is an integral of the order $-\alpha$ frequently symbolized by the operator ${}_{t_0}I_t^{-\alpha} f(t)$.

For discrete-time functions $f(t)$ the difference calculus of non-integer order is the counterpart of the differential calculus of non-integer order. Taking for simplicity the normalized sampling period $h = 1$, by analogy with Definition 10, the following can be formulated [17]:

Definition 11 A discrete difference of fractional order $\alpha \in \mathbb{R}$ of the function $f(t)$, $t \in \mathbb{Z}$ is given by

$$ {}_{t_0}\Delta_t^{\alpha} f(t) = \sum_{j=0}^{t-t_0} c_j^{\alpha} f(t - j), \tag{11} $$

with $t_0 = 0$ being most commonly adopted. If so, (23) takes a simpler form

$$\Delta^{\alpha} f(t) = \sum_{j=0}^{t} c_j^{\alpha} f(t-j). \tag{12}$$

3.2 Discrete-Time Dynamical Models of Non-integer Order

For a nonlinear discrete-time dynamical model of non-integer order, by analogy with the integer order model, we can define the state-space equations

$$x_i(t+1) = f_i(x(t), u(t)), \quad i = 1, 2, \ldots, n, \quad t \in \mathbb{Z}, \tag{13}$$

where f_i is a nonlinear state function for the i-th state variable $x_i(t) \in \mathbb{R}$, $x(t) \in \mathbb{R}^n$, $u(t) \in \mathbb{R}^m$ are state and input vectors, respectively, in the following way:

$$\Delta^{\alpha_i} x_i(t+1) = f_i(x(t), u(t)) - x_i(t), \quad i = 1, 2, \ldots, n, \quad t \in \mathbb{Z}. \tag{14}$$

By introducing matrices of generalized fractional orders

$$\boldsymbol{\Upsilon}_i = diag\left[\binom{\alpha_1}{i} \cdots \binom{\alpha_n}{i}\right] \tag{15}$$

we obtain the generalized nonlinear discrete-time model of non-integer order with the initial condition $x(t_0) = x_0$ in the following form:

$$x(t+1) = \boldsymbol{f}_d(x(t), u(t)) - \sum_{i=1}^{t+1} (-1)^i \boldsymbol{\Upsilon}_i x(t+1-i), \tag{16}$$

$$y(t) = \boldsymbol{g}(x(t), u(t)), \tag{17}$$

where $\boldsymbol{f}_d(x(t), u(t)) = \left[f_1(x(t), u(t)) - x_1(t), \ldots, f_n(x(t), u(t)) - x_n(t)\right]^T$ denotes a modified matrix of nonlinear functions of the state, $g(x(t), u(t))$ is a matrix of nonlinear functions of the output, and $y(t) \in \mathbb{R}^p$ is the output vector.

The discrete-time linear model of non-integer order in state space with the initial condition $x(t_0) = x_0$, in view of (13)–(17), can be written as:

$$x(t+1) = \boldsymbol{A}_d x(t) + \boldsymbol{B} u(t) - \sum_{i=1}^{t+1} (-1)^i \boldsymbol{\Upsilon}_i x(t+1-i), \tag{18}$$

$$y(t) = \boldsymbol{C} x(t), \tag{19}$$

where $\boldsymbol{A} \in \mathbb{R}^{n \times n}$, $\boldsymbol{B} \in \mathbb{R}^{n \times m}$, $\boldsymbol{C} \in \mathbb{R}^{p \times n}$ are state, input and output matrices, respectively $\boldsymbol{A}_d = \boldsymbol{A} - \boldsymbol{I}_n \in \mathbb{R}^{n \times n}$, $\boldsymbol{I}_n$, denotes an identity matrix of dimensions $n \times n$.

3.3 *Finite-Memory Discrete-Time Models of Fractional Order*

In practical applications of models (16), (17) and (18), (19) it is not possible to take for numerical calculations the state vector $x\,(t+1-i)$ samples, the number of which grows rapidly with increasing discrete time t. One of the methods to cope with the problem is to adopt a finite-length memory in which instantaneous values of the state vector are stored. Such an approach, taken from the theory of digital filters with a finite impulse response, leads to so-called finite fractional differences (FFD) [17]. In view of this, the discrete difference of a fractional order $\alpha \in \mathbb{R}$ of the function $f(t)$, $t \in Z$ takes the form

$$\Delta^{\alpha} x\,(t+1) = \sum_{j=0}^{L} c_j^{\alpha} x\,(t+1-j)\,, \tag{20}$$

where L is the adopted memory length. However, in the process of calculations it should be taken into account that the upper limit of summation must be reduced to the value of $t+1$ until enough samples are accumulated. A similar shortening applies to the sum in the generalized models (16) and (18).

On this basis, the generalized discrete-time linear model of fractional order with finite length memory is obtained in the form

$$x\,(t+1) = \boldsymbol{A}_d x(t) + \boldsymbol{B}u(t) - \sum_{i=1}^{L} (-1)^i\, \boldsymbol{\Upsilon}_i x\,(t+1-i)\,. \tag{21}$$

It is worthy of note that for identical orders of discrete differences $\alpha_1 = \cdots = \alpha_n = \alpha$ of individual state variables, the model (21) can be easily approximated by an integer-order model $n \cdot L$ in the following form [22, 55, 56, 62, 63]:

$$\boldsymbol{A} = \begin{bmatrix} \overline{\boldsymbol{A}}_d & -\overline{\boldsymbol{c}}_2 & -\overline{\boldsymbol{c}}_3 & \cdots & -\overline{\boldsymbol{c}}_{L-1} & -\overline{\boldsymbol{c}}_L \\ \boldsymbol{I}_n & \boldsymbol{0}_n & \boldsymbol{0}_n & \cdots & \boldsymbol{0}_n & \boldsymbol{0}_n \\ \boldsymbol{0}_n & \boldsymbol{I}_n & \boldsymbol{0}_n & \cdots & \boldsymbol{0}_n & \boldsymbol{0}_n \\ \vdots & \vdots & \vdots & \ddots & \vdots & \vdots \\ \boldsymbol{0}_n & \boldsymbol{0}_n & \boldsymbol{0}_n & \cdots & \boldsymbol{I}_n & \boldsymbol{0}_n \end{bmatrix} \in \mathbb{R}^{n \cdot L \times n \cdot L}, \tag{22}$$

$$\overline{\boldsymbol{A}}_d = \boldsymbol{A}_d + \alpha \boldsymbol{I}_n, \tag{23}$$

$$\overline{c}_i = c_i^{\alpha} \boldsymbol{I}_n, i = 2, \ldots, L, \tag{24}$$

$$\overline{\boldsymbol{B}} = \left[\boldsymbol{B}^T \ \boldsymbol{0}_{m\times n} \ \cdots \ \boldsymbol{0}_{m\times n} \right]^T \in \mathbb{R}^{n\cdot L\times m}, \tag{25}$$

$$\overline{\boldsymbol{C}} = \left[\boldsymbol{C} \ \boldsymbol{0}_{p\times n} \ \cdots \ \boldsymbol{0}_{p\times n} \right] \in \mathbb{R}^{p\times n\cdot L}. \tag{26}$$

4 Discrete-Time Switched Models of Non-integer Order in State Space

In view of Definition 2 for switched models, all local submodels $\mathscr{M}_j$ that make up the switched model can be written as

$$x(t+1) = \boldsymbol{A}_{d,j} x(t) + \boldsymbol{B}_j u(t) - \sum_{i=1}^{t+1} (-1)^i \boldsymbol{\Upsilon}_{i,j} x(t+1-i), \tag{27}$$

$$y_j(t) = \boldsymbol{C}_j x(t), \qquad j \in \{1, 2, \ldots, S\}, \qquad t \in \mathbb{Z}, \tag{28}$$

$$\boldsymbol{\Upsilon}_{i,j} = diag\left[\binom{\alpha_{1,j}}{i} \cdots \binom{\alpha_{n,j}}{i} \right], \tag{29}$$

where $\boldsymbol{A}_{d,j} \in \mathbb{R}^{n\times n}$ is the complemented state matrix of the j-th local model, $\boldsymbol{\Upsilon}_{i,j} \in \mathbb{R}^{n\times n}$ are matrices of its generalized fractional orders, and $\boldsymbol{B}_j \in \mathbb{R}^{n\times m}$, $\boldsymbol{C}_j \in \mathbb{R}^{p\times n}$ are input and output local matrices of the j-th submodel.

It can be noticed that the multimodel composed of local sub-models (22)–(24) selected on the basis of the switching signal $\sigma(t)$, assumed to describe a non-linear time-invariant process (NLTI), is essentially a piecewise linear time-variant model (PWLTV) defined at the current time instant by one of the so-called fours from S [17]:

$$\mathscr{M}_j \asymp \left(\boldsymbol{A}_{d,j}, \boldsymbol{B}_j, \boldsymbol{C}_j, \left\{ \alpha_{1,j}, \alpha_{2,j}, \ \ldots, \alpha_{n,j} \right\} \right), \quad j = 1, 2, \ldots, S, \tag{30}$$

3 and all local models (27)–(29) taken together form the so-called convex hull

$$(\boldsymbol{A}_d(t), \boldsymbol{B}(t), \boldsymbol{C}(t), \boldsymbol{\Upsilon}_1(t), \boldsymbol{\Upsilon}_2(t), \ldots, \boldsymbol{\Upsilon}_{t+1}(t)) \subset$$

$$\subset conv \left\{ \left(\boldsymbol{A}_{d,1}(t), \boldsymbol{B}_1(t), \boldsymbol{C}_1(t), \boldsymbol{\Upsilon}_{1,1}(t), \ldots, \boldsymbol{\Upsilon}_{t+1,1}(t) \right), \ldots \right.$$

$$\left. \ldots \left(\left(\boldsymbol{A}_{d,S}(t), \boldsymbol{B}_S(t), \boldsymbol{C}_S(t), \boldsymbol{\Upsilon}_{1,S}(t), \ldots, \boldsymbol{\Upsilon}_{t+1,S}(t) \right) \right\} . \tag{31}$$

4.1 Selection of an Active Local Model

For piecewise linear models the linear space $P = U \times X$ made up of the input space $u(t) \in U \subseteq \mathbb{R}^m$ and the state space $x(t) \in X \subseteq \mathbb{R}^n$ of the nonlinear process is divided into S convex polyhedrons $P_1, P_2, \ldots, P_S,\ P_j = U_j \times X_j \subset P$, such that

$$P = \bigcup_{j=1}^{S} P_j \quad \text{and} \quad P_j \bigcap \left(\sum_{i \neq j} P_i \right) = \varnothing \quad \forall j \in \{1, 2, \ldots, S\}, \tag{32}$$

defined by matrices $S_j^x,\ S_j^u$ and S_j^0 and matrix inequalities

$$S_j^x x(t) + S_j^u u(t) \leq S_j^0, \quad j = 1, 2, \ldots, S. \tag{33}$$

It is assumed that the switching signal $\sigma(t) = j$ inside the polyhedron P_j, for which the inequality (33) holds, and the linear submodel $\mathscr{M}_j$ is the best local linearization of the nonlinear process (16), (17).

Example 1 Let there be given four local SISO models with two state variables. Model switching will be determined by the first state variable $x_1(t)$ and the input $u(t)$. Four convex polyhedrons (polygons, in this case) shown in Fig. 3 are defined below:

$$P_1: \qquad 0 \leq u \leq 0.8 \quad \wedge \quad 0 \leq x_1 \leq 0.3u + 0.4, \tag{34}$$

$$P_2: \qquad 0.8 \leq u \leq 1.0 \quad \wedge \quad 0 \leq x_1 \leq 0.3u + 0.4, \tag{35}$$

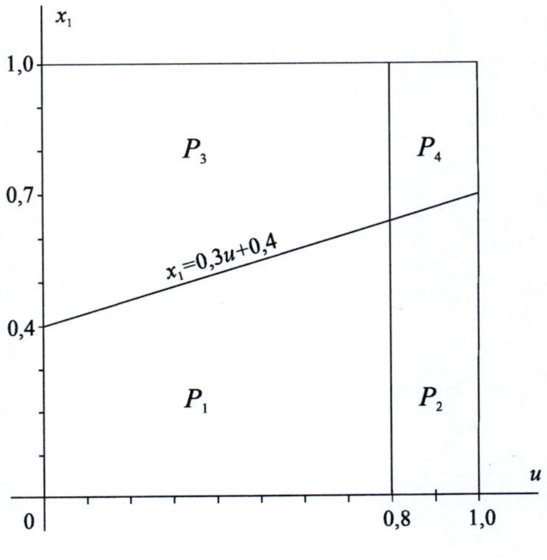

Fig. 3 Division of state and input spaces into convex polyhedrons

$$P_3: \qquad 0 \le u \le 0.8 \quad \wedge \ 0.3u + 0.4 \le x_1 \le 1.0, \tag{36}$$

$$P_4: \qquad 0.8 \le u \le 1.0 \quad \wedge \ 0.3u + 0.4 \le x_1 \le 1.0. \tag{37}$$

Matrices that define polyhedrons in matrix inequalities (28) for individual polyhedrons are the following

$$\boldsymbol{S}_1^x = \boldsymbol{S}_2^x = \boldsymbol{S}_3^x = \boldsymbol{S}_4^x = \begin{bmatrix} -1 & 0 \\ 1 & 0 \\ 0 & 0 \\ 0 & 0 \end{bmatrix}, \tag{38}$$

$$\boldsymbol{S}_1^u = \boldsymbol{S}_2^u = \begin{bmatrix} 0 \\ -0.3 \\ -1 \\ 1 \end{bmatrix}, \qquad \boldsymbol{S}_3^u = \boldsymbol{S}_4^u = \begin{bmatrix} 0.3 \\ 0 \\ -1 \\ 1 \end{bmatrix}, \tag{39}$$

$$\boldsymbol{S}_1^0 = \begin{bmatrix} 0 \\ 0.4 \\ 0 \\ 0.8 \end{bmatrix}, \quad \boldsymbol{S}_2^0 = \begin{bmatrix} 0 \\ 0.4 \\ -0.8 \\ 1 \end{bmatrix}, \quad \boldsymbol{S}_3^0 = \begin{bmatrix} -0.4 \\ 1 \\ 0 \\ 0.8 \end{bmatrix}, \quad \boldsymbol{S}_4^0 = \begin{bmatrix} -0.4 \\ 1 \\ -0.8 \\ 1 \end{bmatrix}. \tag{40}$$

Example 2 Let there be given two discrete-time local models of fractional order $\alpha_1 = \alpha_2 = 0.9$

$$\boldsymbol{A}_{d1} = \begin{bmatrix} 2.7756 & -1.2876 & 0.7985 \\ 2 & 0 & 0 \\ 0 & 0.5 & 0 \end{bmatrix}, \quad \boldsymbol{B}_1 = \begin{bmatrix} 0.0313 \\ 0 \\ 0 \end{bmatrix}, \tag{41}$$

$$\boldsymbol{A}_{d2} = \begin{bmatrix} 2.8474 & -1.3505 & 0.8536 \\ 2 & 0 & 0 \\ 0 & 0.5 & 0 \end{bmatrix}, \quad \boldsymbol{B}_2 = \begin{bmatrix} 0.0078 \\ 0 \\ 0 \end{bmatrix}, \tag{42}$$

$$\boldsymbol{C}_1 = \boldsymbol{C}_2 = \begin{bmatrix} 1 & 0 & 0 \end{bmatrix}, \qquad L = 200. \tag{43}$$

The switched piecewise linear model has the form

$$(\boldsymbol{A}_d, \boldsymbol{B}, \boldsymbol{C}, \boldsymbol{\Upsilon}_i) = \begin{cases} \boldsymbol{A}_{d1}, \ \boldsymbol{B}_1, \boldsymbol{C}_1, \alpha_1 & \text{if} \ \ \boldsymbol{S}_1^x x(t) + \boldsymbol{S}_1^u u(t) \le \boldsymbol{S}_1^0 \\ \boldsymbol{A}_{d2}, \ \boldsymbol{B}_2, \ \boldsymbol{C}_2, \ \alpha_2 & \text{otherwise} \end{cases} \tag{44}$$

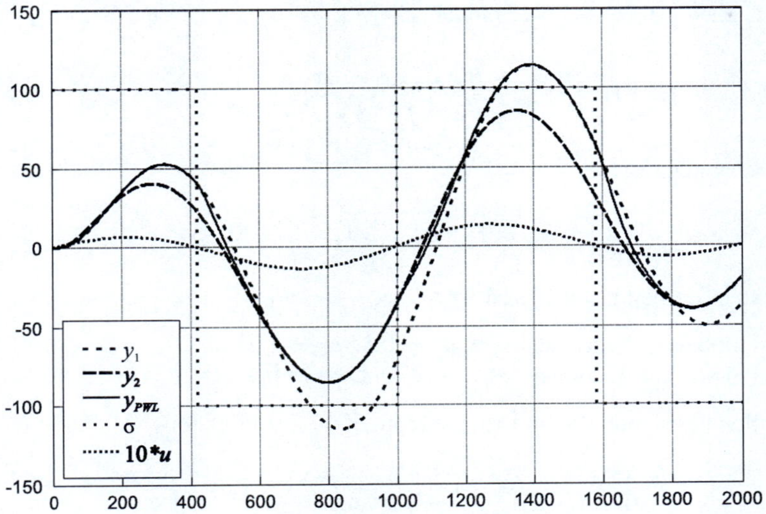

Fig. 4 Time-response characteristics for the switched model of fractional order (α_1 = 0.9, α_2 = 0.9)

with the switching law

$$S_1^x = 0, \quad S_1^u = -1, \quad S_1^0 = 0. \tag{45}$$

The following signal is applied to its input

$$u(t) = \sin\left(\frac{\pi t}{500}\right) - 0.5 \sin\left(\frac{\pi t}{1000}\right), \qquad t \in \mathbb{Z}. \tag{46}$$

Figure 4 depicts time responses of the input and the outputs for the switched model and local models and the switching signal.

Example 3 Let there be given two fractional discrete-time local models with identical matrices

$$\boldsymbol{A}_{d1} = \boldsymbol{A}_{d2} = \begin{bmatrix} 2.7756 & -1.2876 & 0.7985 \\ 2 & 0 & 0 \\ 0 & 0.5 & 0 \end{bmatrix}, \quad \boldsymbol{B}_1 = \boldsymbol{B}_2 = \begin{bmatrix} 0.0313 \\ 0 \\ 0 \end{bmatrix}, \tag{47}$$

$$\boldsymbol{C}_1 = \boldsymbol{C}_2 = \begin{bmatrix} 1 \; 0 \; 0 \end{bmatrix}, \tag{48}$$

$$L = 200. \tag{49}$$

The switched piecewise linear model has the form

$$(\boldsymbol{A}_d, \boldsymbol{B}, \boldsymbol{C}, \boldsymbol{\Upsilon}_i) = \begin{cases} \boldsymbol{A}_{d1},\ \boldsymbol{B}_1, \boldsymbol{C}_1, \alpha_1 & \text{if } \ \boldsymbol{S}_1^x x(t) + \boldsymbol{S}_1^u u(t) \leq \boldsymbol{S}_1^0 \\ \boldsymbol{A}_{d2},\ \boldsymbol{B}_2,\ \boldsymbol{C}_2,\ \alpha_2 & \text{otherwise} \end{cases} \tag{50}$$

with the switching law

$$\boldsymbol{S}_1^x = 0, \quad \boldsymbol{S}_1^u = -1, \quad \boldsymbol{S}_1^0 = 0, \tag{51}$$

with two cases being considered:

- local models of fractional order: $\alpha_1 = 0.8,\ \alpha_2 = 1.2$,
- local models of fractional order: $\alpha_1 = 0.8,\ \alpha_2 = 0.5$.

The following signal is applied to the input of the PWL model

$$u(t) = \sin\left(\frac{\pi t}{500}\right) - 0.5 \sin\left(\frac{\pi t}{1000}\right), \qquad t \in \mathbb{Z}. \tag{52}$$

Figures 5 and 6 show time responses of the input and the outputs for the switched model and local models and the switching signal for the both cases under consideration in the example.

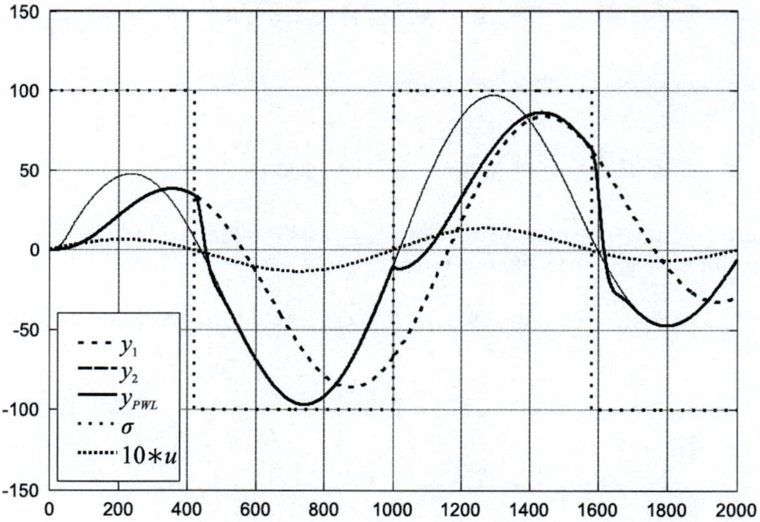

Fig. 5 Time-response characteristics for the switched model of fractional order ($\alpha_1 = 0.8, \alpha_2 = 1.2$)

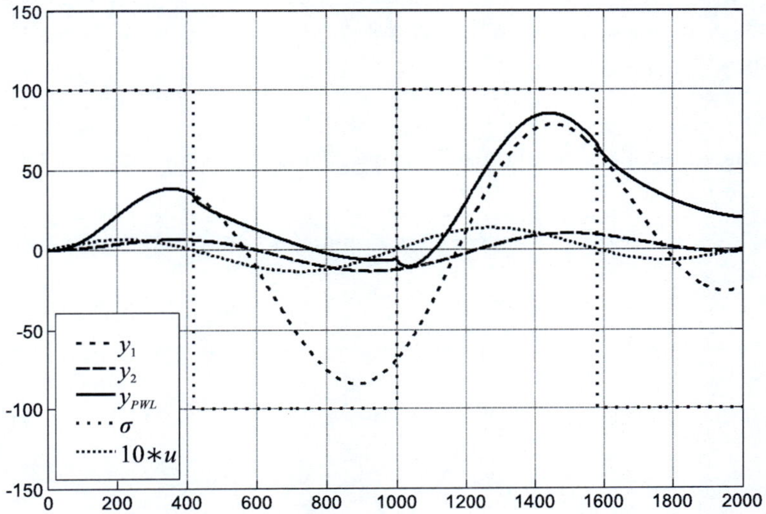

Fig. 6 Time-response characteristics for the switched model of fractional order ($\alpha_1 = 0.8, \alpha_2 = 0.5$)

4.2 Strategies for Switching the Models Order

In switched models, changing the submodel at the k-th instant implies the change of the model matrices $\boldsymbol{A}_{d,j}$, $\boldsymbol{B}_j$, $\boldsymbol{C}_j$ and the change of the orders $\alpha_{i,j}$ or α_j in the case that all state variables are of identical order. The change of the order is hidden in coefficients $c_{i,j}$ determined by Definition 10 or, respectively, in matrices $\boldsymbol{\Upsilon}_i(t)$ given by (15) in the case of the generalized model. For fractional order models, this means that current coefficients apply to historical data that were generated (or sampled) during the validity of the past values of the order. For this reason, alternative proposals for modeling the varying order in switched models may be found in literature [17, 18, 41]. The proposals differ in transient processes when switching local models. The most popular of them are given in [22]:

Definition 12 Discrete finite differences of time-variant fractional order of a discrete-time function $f(t)$, respectively called the type $\mathscr{A}$ to $\mathscr{C}$, are given by:

$$ {}^{\mathscr{A}}\Delta_t^{\alpha(t)} f(t) = \sum_{i=0}^{L} (-1)^i \binom{\alpha(t)}{i} f(t-i), \tag{53} $$

$$ {}^{\mathscr{B}}\Delta_t^{\alpha(t)} f(t) = \sum_{i=0}^{L} (-1)^i \binom{\alpha(t-i)}{i} f(t-i), \tag{54} $$

$$
{}^{\mathscr{C}}\Delta_t^{\alpha(t)} f(t) = \sum_{i=0}^{L} (-1)^i \binom{\alpha(t-L+i)}{i} f(t-i). \tag{55}
$$

Table 1 gives a practical interpretation of individual types of time-variant discrete differences of fractional order. Figures 7, 8 and 9 show switching strategies for particular types of order variability.

Example 4 Let there be given two discrete-time fractional local models with identical matrices

$$
\boldsymbol{A}_{d1} = \boldsymbol{A}_{d2} = \begin{bmatrix} 2.7756 & -1.2876 & 0.7985 \\ 2 & 0 & 0 \\ 0 & 0.5 & 0 \end{bmatrix}, \quad \boldsymbol{B}_1 = \boldsymbol{B}_2 = \begin{bmatrix} 0.0313 \\ 0 \\ 0 \end{bmatrix}, \tag{56}
$$

$$
\boldsymbol{C}_1 = \boldsymbol{C}_2 = \begin{bmatrix} 1 & 0 & 0 \end{bmatrix}, \tag{57}
$$

$$
L = 200, \tag{58}
$$

Table 1 Interpretation of time-variant discrete differences of fractional order

Type	Description
$\mathscr{A}$	Current weighting coefficients apply to all data, including those historical, sampled during the "validity" of previous orders (the changed order affects the whole system memory at one moment)
$\mathscr{B}$	The weighting coefficients "valid" at the time of data sampling are applicable to historical data (the changed order affects faster the newer data)
$\mathscr{C}$	More recent weighting coefficients apply to historical data more distant in time and vice versa, older coefficients apply to newer data (the changed order affects faster the older data)

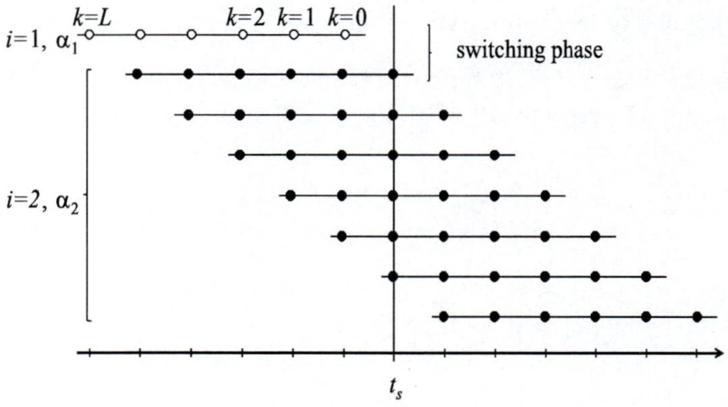

Fig. 7 Switching from submodel 1 to submodel 2 by type $\mathscr{A}$, $L = 5$

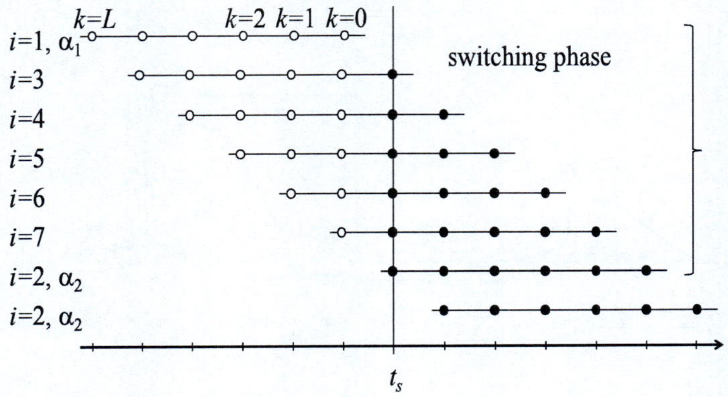

Fig. 8 Switching from submodel 1 to submodel 2 by type $\mathscr{B}$, $L = 5$

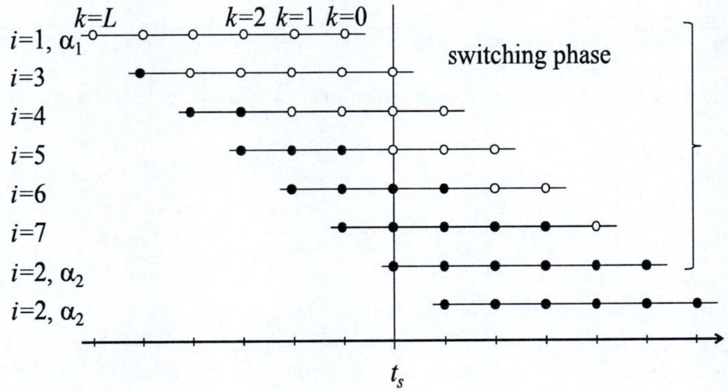

Fig. 9 Switching from submodel 1 to submodel 2 by type $\mathscr{C}$, $L = 5$

$$\alpha_1 = 0.8, \ \alpha_2 = 0.5. \tag{59}$$

The switched piecewise linear model has the form

$$(\boldsymbol{A}_d, \boldsymbol{B}, \boldsymbol{C}, \boldsymbol{\Upsilon}_i) = \begin{cases} \boldsymbol{A}_{d1}, \ \boldsymbol{B}_1, \boldsymbol{C}_1, \alpha_1 & \text{if } \ \boldsymbol{S}_1^x x(t) + \boldsymbol{S}_1^u u(t) \leq \boldsymbol{S}_1^0 \\ \boldsymbol{A}_{d2}, \ \boldsymbol{B}_2, \ \boldsymbol{C}_2, \ \alpha_2 & \text{otherwise} \end{cases} \tag{60}$$

with the switching law

$$\boldsymbol{S}_1^x = 0, \quad \boldsymbol{S}_1^u = -1, \quad \boldsymbol{S}_1^0 = 0. \tag{61}$$

The following signal is applied to the input of the PWL switched model

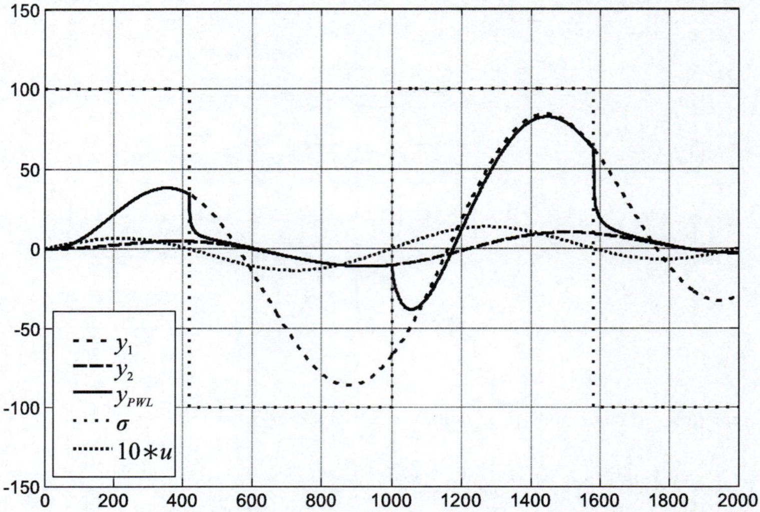

Fig. 10 Time-response characteristics for the fractional switched model of type $\mathscr{B}\alpha_1 = 0.8, \alpha_2 = 0.5)$

$$u(t) = \sin\left(\frac{\pi t}{500}\right) - 0.5\sin\left(\frac{\pi t}{1000}\right), \qquad t \in \mathbb{Z}. \tag{62}$$

The switching of local models will follow the type $\mathscr{B}$.

Figure 10 depicts time responses of the input and the outputs for the switched model, local models, and the switching signal for this kind of fractional order switching. They can be compared with those shown in Fig. 6.

4.3 Switching Due to Memory Impairment

Models of fractional order, as mentioned, can effectively model different complex dynamic systems [27, 28]. Aside from many applications listed in the Introduction, they are particularly suitable for modeling biological processes in which the current state depends in a natural way on historical states, i.e. on system memory [32]. If a memory impairment happens in such processes, the system properties change. Therefore, proper modeling of the system with memory deficiency requires switching the submodel with correct memory to the submodel representing the system with impaired memory. Here, three types of system memory deficiency can be distinguished [23]:

- whole memory deficiency (WMD); all the memory deteriorates at the instant the impairment occurs;

- long-term memory deficiency (LTMD); remembering data distant in time deteriorates at the instant the impairment occurs;
- short-term memory deficiency (STMD); remembering data collected recently deteriorates at the instant the impairment occurs.

For the above-mentioned types of memory deficiency, the following descriptions for switched discrete differences of fractional order may be introduced [23]:

Definition 13 A discrete difference of fractional order $\alpha \in \mathbb{R}$ of a discrete function $f(t),\ t \in \mathbb{Z}$, in the event that long-term memory deficiency occurs to a degree λ $(0 < \lambda \leq 1)$ is given by

$$ {}^{WMD}\Delta^{\alpha,\lambda} f(t) = \begin{cases} \sum_{j=0}^{t} c_j^{\alpha} f(t-j), & \text{for } t < t_s \\ \sum_{j=0}^{t} c_j^{\lambda\alpha} f(t-j), & \text{for } t \geq t_s \end{cases} \tag{63} $$

where $t_s \in \mathbb{Z}$ denotes the instant of impairment (instant of switching), and

$$ c_j^{\alpha} = (-1)^j \binom{\alpha}{j}, \quad c_j^{\lambda\alpha} = (-1)^j \binom{\lambda\alpha}{j}. \tag{64} $$

Definition 14 A discrete difference of fractional order $\alpha \in \mathbb{R}$ of a discrete function $f(t),\ t \in \mathbb{Z}$, in the event that long-term memory deficiency occurs to a degree λ $(0 < \lambda \leq 1)$ is given by

$$ {}^{LTMD}\Delta^{\alpha,\lambda,H} f(t) = \begin{cases} \sum_{j=0}^{t} c_j^{\alpha} f(t-j), & \text{for } t < t_s \\ \sum_{j=0}^{H} c_j^{\alpha} f(t-j) + \sum_{j=H+1}^{t} c_j^{\lambda\alpha} f(t-j), & \text{for } t \geq t_s \end{cases} \tag{65} $$

where $H \in \mathbb{Z}_+$ denotes the adopted short-term memory horizon, $t_s \in \mathbb{Z}$ denotes the instant of impairment (instant of switching), and

$$ c_j^{\alpha} = (-1)^j \binom{\alpha}{j}, \quad c_j^{\lambda\alpha} = (-1)^j \binom{\lambda\alpha}{j}. \tag{66} $$

Definition 15 A discrete difference of fractional order $\alpha \in \mathbb{R}$ of a discrete function $f(t),\ t \in \mathbb{Z}$, in the event that short-term memory deficiency occurs to a degree λ $(0 < \lambda \leq 1)$ is given by

$$ {}^{STMD}\Delta^{\alpha,\lambda,H} f(t) = \begin{cases} \sum_{j=0}^{t} c_j^{\alpha} f(t-j), & \text{for } t < t_s \\ \sum_{j=0}^{H} c_j^{\lambda\alpha} f(t-j) + \sum_{j=H+1}^{t} c_j^{\alpha} f(t-j), & \text{for } t \geq t_s \end{cases} \tag{67} $$

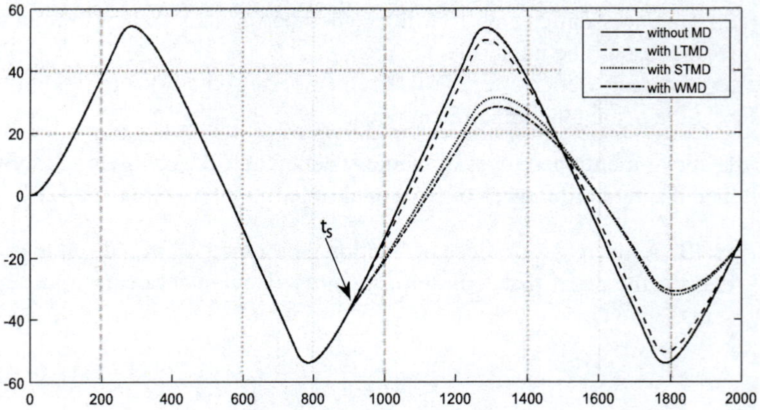

Fig. 11 Time responses of the system of fractional order alpha $\alpha = 0.9$, $L = 1000$ to a rectangular excitation $u = \pm 1$ before and after memory deficiency occurred ($\lambda = 0.75$, $H = 200$, $t_s = 900$)

where $H \in \mathbb{Z}_+$ denotes the adopted short-term memory horizon, $t_s \in \mathbb{Z}$ denotes the instant of impairment (instant of switching), and

$$c_j^\alpha = (-1)^j \binom{\alpha}{j}, \qquad c_j^{\lambda\alpha} = (-1)^j \binom{\lambda\alpha}{j}. \tag{68}$$

Remark 2 The whole memory deficiency corresponds to the discrete difference of time-variant order, as for type $\mathscr{A}$ in Table 1.

Remark 3 In models of fractional order with finite memory (20) the adopted short-term memory horizon H should be significantly shorter than the memory length L.

Figure 11 shows examples of time responses of a fractional system with various types of memory deficiency.

4.4 Basic Properties of Switched Models

Assuming that the switched discrete piecewise linear multimodel (27)–(29) is determined, it is important to know its basic properties, such as stability, observability, controllability, etc. For models of integer order, many works exist, in which these issues are analyzed for both arbitrary and restricted switchings [29, 43, 59, 61, 65, 66, 73]. It has been shown, for example, that a switched model made up of stable local models can be unstable, and vice versa—the model composed of unstable local models can be stable. The situation is similar with respect to observability and controllability of the switched model in relation to the same features of local models.

This depends to a large extent on the form of the switching signal $\sigma(t)$ described by Definitions 4–9.

In survey paper [40] there are basic methods given for assessing stability of linear switched models. For integer-order models, the main tools to assess the stability of PWL multimodels are those derived from the second Lyapunov method, i.e. the Common Quadratic Lyapunov Function (CQLF) and the Switched Quadratic Lyapunov Function (SQLF) [5, 10]. Among others, the following holds true:

Theorem 1 *A sufficient condition for the exponential stability of the switched model of integer order is the existence of CQLF for all local submodels. In the discrete case this can be expressed by a system of matrix inequalities*

$$\boldsymbol{A}_j^T \boldsymbol{P} \boldsymbol{A}_j - \boldsymbol{P} < 0 \qquad \forall j \in \{1, 2, \ldots, S\}, \qquad \boldsymbol{P} \in \mathbb{R}^{n \times n}, \tag{69}$$

where $\boldsymbol{A}_j$ denotes the state matrix of the j-th local model, S is the number of local models, and $\boldsymbol{P}$ is a symmetric positive definite matrix.

The proof of the theorem is given in [40].

Unfortunately, determining the matrix $\boldsymbol{P}$ that would satisfy the system of inequalities (69) for a large number of local S models is not easy, even with numerical methods, and most often quite conservative results are delivered.

The second method is to use SQLF, where the switched matrix $\boldsymbol{P}_{\sigma(t)} \in \{\boldsymbol{P}_1, \boldsymbol{P}_2, \ldots, \boldsymbol{P}_S\}$ joins all matrices corresponding to local models. Among others, the following holds true:

Theorem 2 *If all local models are stable, i.e. if for each submodel with the state matrix $\boldsymbol{A}_j$ there exists a positive definite symmetric matrix $\boldsymbol{P}_j$, such that*

$$\boldsymbol{A}_j^T \boldsymbol{P}_j \boldsymbol{A}_j - \boldsymbol{P}_j < 0 \qquad \forall j \in \{1, 2, \ldots, S\}, \qquad \boldsymbol{P}_j \in \mathbb{R}^{n \times n}, \tag{70}$$

then in order for the integer-order switched model composed of S stable local models with state matrix $\boldsymbol{A}_j$ to be stable exponentially, it is sufficient that there exist positive definite symmetric matrices $\boldsymbol{P}_i$, $\boldsymbol{P}_j$ satisfying the following system of inequalities

$$\begin{bmatrix} \boldsymbol{P}_j & \boldsymbol{A}_j^T \boldsymbol{P}_i \\ \boldsymbol{P}_i \boldsymbol{A}_j & \boldsymbol{P}_i \end{bmatrix} > 0 \qquad \forall i, j \in \{1, 2, \ldots, S\}, \qquad \boldsymbol{P}_i, \boldsymbol{P}_j \in \mathbb{R}^{n \times n}. \tag{71}$$

The proof of the theorem may be found in [17, 40].

Condition (71) is easier to check and gives less conservative results.

The presented briefly methods of testing the stability of integer-order switched models can be used to formulate conditions for the stability of fractional-order switched models. Based on approximation (22)–(26) they find direct application, among others, to testing stability of finite-memory switched systems of fractional orders. This is illustrated by

Theorem 3 *If there exist positive definite symmetric matrices $\boldsymbol{P}_i$, $\boldsymbol{Q}_i \in \mathbb{R}^{n\times n}$, $\boldsymbol{P}_i = \boldsymbol{P}_i^T$, $\boldsymbol{Q}_i = \boldsymbol{Q}_i^T$, auxiliary matrices $\boldsymbol{F}_{ij}$, $\boldsymbol{G}_{ij} \in \mathbb{R}^{n\times n}$ and constants μ_{ij} satisfying the inequalities*

$$\begin{bmatrix} \overline{\boldsymbol{A}}_i^T \boldsymbol{F}_{ij}^T + \boldsymbol{F}_{ij}\overline{\boldsymbol{A}}_i - \boldsymbol{P}_i + \mu_{ij}\boldsymbol{Q}_i & \overline{\boldsymbol{A}}_i^T \boldsymbol{G}_{ij}^T - \boldsymbol{F}_{ij} \\ \boldsymbol{G}_{ij}\overline{\boldsymbol{A}}_i - \boldsymbol{F}_{ij}^T & \boldsymbol{P}_j - \boldsymbol{G}_{ij} - \boldsymbol{G}_{ij}^T + \mu_{ij}\boldsymbol{Q}_j \end{bmatrix} < 0 \tag{72}$$

for all $i, j \in \{1, 2, \ldots, S\}$, where

$$\overline{\boldsymbol{A}}_i = \begin{bmatrix} \overline{\boldsymbol{A}}_{d,i} & -\overline{\boldsymbol{c}}_{2,i} & -\overline{\boldsymbol{c}}_{3,i} & \cdots & -\overline{\boldsymbol{c}}_{L-1,i} & -\overline{\boldsymbol{c}}_{L,i} \\ \boldsymbol{I}_n & \boldsymbol{0}_n & \boldsymbol{0}_n & \cdots & \boldsymbol{0}_n & \boldsymbol{0}_n \\ \boldsymbol{0}_n & \boldsymbol{I}_n & \boldsymbol{0}_n & \cdots & \boldsymbol{0}_n & \boldsymbol{0}_n \\ \vdots & \vdots & \vdots & \ddots & \vdots & \vdots \\ \boldsymbol{0}_n & \boldsymbol{0}_n & \boldsymbol{0}_n & \cdots & \boldsymbol{I}_n & \boldsymbol{0}_n \end{bmatrix}, \tag{73}$$

$$\overline{\boldsymbol{A}}_{d,i} = \boldsymbol{A}_{d,i} + \alpha_i \boldsymbol{I}_n, \tag{74}$$

$$\overline{\boldsymbol{c}}_{k,i} = c_k^{\alpha_i}\boldsymbol{I}_n, \qquad k = 2, \ldots, L, \tag{75}$$

then the discrete fractional system composed of arbitrary switched by type $\mathscr{A}$ finite-memory fractional submodels with matrices $\boldsymbol{A}_i$, $\boldsymbol{B}_i$, $\boldsymbol{C}_i$ and discrete difference Δ^{α_i} is asymptotically stable.

The proof is given in [22].

For infinite-memory non-integer switched systems it is most common to use a transformation of a non-convex stability area, where all eigenvalues of state matrices of stable local non-integer models are to be contained, into a convex stability area for integer-order models [6, 17, 22, 29–31, 64].

5 Use of Discrete Switched Models of Non-integer Order in Predictive Control

One of the natural areas of using fractional switched models in automation is predictive control of complex nonlinear processes. The effectiveness of Model Predictive Control (MPC) depends predominantly on the quality of the controlled process model [7, 14, 21, 47–49, 72]. For complex nonlinear processes of both integer and non-integer order, the use of nonlinear fractional models is an effective and natural modeling method. Unfortunately, this leads to a very complex nonlinear optimization problem that must be solved at each sampling step [13, 17, 42, 54, 68–71]. As for integer-order MPC, one of the methods to mitigate the numerical complexity of a nonlinear algorithm is the use of methods with nonlinear trajectory prediction and cyclic model linearization for the needs of ongoing optimization. There are known several ways to accomplish this general principle of simplifying the implementation of nonlinear predictive control, but the most advanced method seems to be that labeled with acronyms NPL+ or NPLTP (nonlinear prediction of the free response component with model linearization along the predicted trajectory) [13, 36, 68]. For fractional-order MPC it can be formulated as follows [16, 20, 21, 25].

Let us consider a nonlinear discrete-time process model of non-integer order (16), (17) with state $v(t)$ and output $d(t)$ disturbances

$$x\,(t+1) = \boldsymbol{f}_d\,(x(t), u(t)) - \sum_{i=1}^{t+1} (-1)^i\, \boldsymbol{\Upsilon}_i x\,(t+1-i) + v\,(t+1)\,, \tag{76}$$

$$y(t) = \boldsymbol{g}\,(x(t), u(t)) + d\,(t+1)\,. \tag{77}$$

Assuming that the unknown vectors of state and output disturbances include effects induced by unmeasurable disturbances and model inaccuracy, they can be determined indirectly as the difference between the values measured at the instant t (or those estimated in the absence of access to the state vector) and the values calculated for the instant t at the previous instant $t-1$ [17, 35, 36]

$$v(t) = x(t) + \sum_{i=1}^{t} (-1)^i\, \boldsymbol{\Upsilon}_i x\,(t-i) - \boldsymbol{f}_d\,(x\,(t-1)\,, u\,(t-1))\,, \tag{78}$$

$$d(t) = y(t) - \boldsymbol{g}\left(\boldsymbol{f}_d\,(x\,(t-1)\,, u\,(t-1)) + v(t)\right). \tag{79}$$

The task set before the fractional predictive controller FO MPC is to determine the current manipulated variable that would minimize the cost function specified over the prediction N and control N_u horizons

$$\min_{\substack{\Delta U(t)_\rightarrow \\ E^{\min}(t)_\rightarrow \\ E^{\max}(t)_\rightarrow}} J(t) = \Big\{\|Y^r(t)_\rightarrow - Y^p(t)_\rightarrow\|^2_{\boldsymbol{M}} + \|\Delta U(t)_\rightarrow\|^2_{\boldsymbol{L}} \\ + \rho^{\min}\|E^{\min}(t)_\rightarrow\|^2 + \rho^{\max}\|E^{\max}(t)_\rightarrow\|^2\Big\}\,, \tag{80}$$

where $\|V\|^2_{\boldsymbol{W}} = V^T \boldsymbol{W} V$ denotes the vector V norm with weighting matrix $\boldsymbol{W} \geq \boldsymbol{0}$,

$$Y^r(t)_\rightarrow = \begin{bmatrix} y^r\,(t+1|t) \\ \vdots \\ y^r\,(t+N|t) \end{bmatrix} \in \mathbb{R}^{N\cdot p} \tag{81}$$

is the vector of the reference trajectory that starts from the current process output value and approaches the set point in the way intended by the designer,

$$Y^p(t)_{\rightarrow} = \begin{bmatrix} y^p(t+1|t) \\ \vdots \\ y^p(t+N|t) \end{bmatrix} \in \mathbb{R}^{N \cdot p} \tag{82}$$

is the process output prediction vector,

$$\Delta U(t)_{\rightarrow} = \begin{bmatrix} \Delta u(t|t) \\ \vdots \\ \Delta u(t+N_u-1|t) \end{bmatrix} \in \mathbb{R}^{N_u \cdot m} \tag{83}$$

is the sought vector of future increments of the manipulated variable

$$\Delta u(t+j|t) = \begin{cases} u(t|t) - u(t-1), & \text{for } j = 0 \\ u(t+j|t) - u(t+j-1|t), & \text{for } j \geq 1 \\ 0, & \text{for } j \geq N_u \end{cases} \tag{84}$$

with rigid constraints imposed on amplitudes of the manipulated variable and its increments

$$u^{\min} \leq u(t+j|t) \leq u^{\max}, \qquad j = 0, 1, 2, \ldots, N_u - 1, \tag{85}$$

$$\Delta u^{\min} \leq \Delta u(t+j|t) \leq \Delta u^{\max}, \qquad j = 0, 1, 2, \ldots, N_u - 1, \tag{86}$$

and on process outputs, as well

$$y^{\min} - \varepsilon^{\min}(t+j|t) \leq y^p(t+j|t) \leq y^{\max} + \varepsilon^{\max}(t+j|t), \ \ j = 1, 2, \ldots, N. \tag{87}$$

The vectors

$$E^{\min}(t)_{\rightarrow} = \begin{bmatrix} \varepsilon^{\min}(t+1|t) \\ \vdots \\ \varepsilon^{\min}(t+N|t) \end{bmatrix} \in R^{N \cdot p},$$

$$E^{\max}(t)_{\rightarrow} = \begin{bmatrix} \varepsilon^{\max}(t+1|t) \\ \vdots \\ \varepsilon^{\max}(t+N|t) \end{bmatrix} \in R^{N \cdot p} \tag{88}$$

denote additional decision variables that determine the degree of constraints easing in the case that restrictions (87) must be violated to solve the optimization problem. In practical applications of MPC the matrices and weighting coefficients in the cost function (80) usually take the following values $\boldsymbol{M} = \boldsymbol{I}_{p \cdot N}$ or $\boldsymbol{M} = \boldsymbol{0}_{p \cdot N}$, $\boldsymbol{L} = \lambda \cdot \boldsymbol{I}_{m \cdot N_u}$, $\lambda > 0$, $\rho^{\min}$, $\rho^{\max} > 0$.

The prediction vector (82) can be found by determining the process output prediction from (76), (77)

$$\begin{aligned} y^p(t+j|t) = \boldsymbol{g}\Bigg(& \boldsymbol{f}_d\left(x^p(t+j-1|t), u(t+j-1|t)\right) \\ & - \sum_{i=1}^{j-1} (-1)^i \boldsymbol{\Upsilon}_i x^p(t+j-i|t) \\ & - \sum_{i=j}^{t+j} (-1)^i \boldsymbol{\Upsilon}_i x(t+j-i) + v(t+j|t) \Bigg) \\ & + d(t+j|t), \quad j = 1, 2, \ldots, N \end{aligned} \tag{89}$$

and assuming, after the pattern of the well-known DMC predictive algorithm [13, 68], that disturbances within the prediction horizon are constant and equal to disturbances occurring at the current instant (78), (79), that is

$$v(t+j|t) = v(t), \quad j = 1, 2, \ldots, N, \tag{90}$$

$$d(t+j|t) = d(t), \quad j = 1, 2, \ldots, N. \tag{91}$$

From Eq. (89) one can derive the non-linear relationship for the free response component of the process within the prediction horizon

$$y^0(t+j|t) = \boldsymbol{g}\left(x^0(t+j|t)\right) + d(t), \; j = 1, 2, \ldots, N, \tag{92}$$

$$x^0(t+1|t) = \boldsymbol{f}_d(x(t), u(t-1)) - \sum_{i=1}^{t+1} (-1)^i \boldsymbol{\Upsilon}_i x(t+1-i|t) + v(t), \tag{93}$$

$$\begin{aligned} x^0(t+j|t) = {} & \boldsymbol{f}_d\left(x^0(t+j-1|t), u(t-1)\right) - \sum_{i=1}^{j-1} (-1)^i \boldsymbol{\Upsilon}_i x^0(t+j-i|t) \\ & - \sum_{i=j}^{t+j} (-1)^i \boldsymbol{\Upsilon}_i x(t+j-i) + v(t), \; j = 2, \ldots, N. \end{aligned} \tag{94}$$

For the non-linear MPC NPL+ algorithm it is assumed that, as for linear systems, where the superposition principle applies, one can write

$$Y^p(t)_{\rightarrow} = Y^c(t)_{\rightarrow} + Y^0(t)_{\rightarrow} = \boldsymbol{E}_Y \Delta U(t)_{\rightarrow} + Y^0(t)_{\rightarrow} \tag{95}$$

but only the first, forced response component (95) is calculated from the linearized model in the form of the so-called linearized process dynamics matrix $\boldsymbol{E}_Y$, and the second component, depending on the past, is calculated on the basis of a more accurate non-linear model (92) [17, 19, 68]. By this means, the nonlinear optimization problem (80) changes into a quadratic optimization problem with constraints which is much easier to solve

$$\min_{\substack{\Delta U(t)_{\rightarrow} \\ E^{\min}(t)_{\rightarrow} \\ E^{\max}(t)_{\rightarrow}}} J(t) = \|Y^r(t)_{\rightarrow} - \boldsymbol{E}_Y \Delta U(t)_{\rightarrow} - Y^0(t)_{\rightarrow}\|^2_{\boldsymbol{M}} + \|\Delta U(t)_{\rightarrow}\|^2_{\boldsymbol{L}}$$
$$+ \rho^{\min}\|E^{\min}(t)_{\rightarrow}\|^2 + \rho^{\max}\|E^{\max}(t)_{\rightarrow}\|^2, \tag{96}$$

with constraints imposed on amplitudes of the manipulated variable (85), its increments (86) and on outputs (87) given as follows

$$U^{\min} \le \boldsymbol{T}\Delta U(t)_{\rightarrow} + U(t-1) \le U^{\max}, \tag{97}$$

$$\Delta U^{\min} \le \Delta U(t)_{\rightarrow} \le \Delta U^{\max}, \tag{98}$$

$$Y^{\min} - E^{\min}(t)_{\rightarrow} \le \boldsymbol{E}_Y \Delta U(t)_{\rightarrow} \le Y^{\max} + E^{\max}(t)_{\rightarrow}, \tag{99}$$

where

$$U(t-1) = u(t-1)\,1_{N_u \cdot m}, \quad U^{\min} = u_{\min} 1_{N_u \cdot m}, \quad U^{\max} = u_{\max} 1_{N_u \cdot m}, \tag{100}$$

$$\Delta U^{\min} = \Delta u_{\min} 1_{N_u \cdot m}, \quad \Delta U^{\max} = \Delta u_{\max} 1_{N_u \cdot m}, \tag{101}$$

$$Y^{\min} = y_{\min} 1_{N \cdot p}, \quad Y^{\max} = y_{\max} 1_{N \cdot p}, \tag{102}$$

$$\boldsymbol{T} = \begin{bmatrix} \boldsymbol{I}_{N_u} & \boldsymbol{0}_{N_u} & \cdots & \boldsymbol{0}_{N_u} & \boldsymbol{0}_{N_u} \\ \boldsymbol{I}_{N_u} & \boldsymbol{I}_{N_u} & \ddots & \boldsymbol{0}_{N_u} & \boldsymbol{0}_{N_u} \\ \vdots & \ddots & \ddots & \ddots & \vdots \\ \boldsymbol{I}_{N_u} & \ddots & \ddots & \boldsymbol{I}_{N_u} & \boldsymbol{0}_{N_u} \\ \boldsymbol{I}_{N_u} & \boldsymbol{I}_{N_u} & \cdots & \boldsymbol{I}_{N_u} & \boldsymbol{I}_{N_u} \end{bmatrix} \in \mathbb{R}^{n \cdot N_u \times n \cdot N_u}, \tag{103}$$

$$1_k = \begin{bmatrix} 1 \\ \vdots \\ 1 \end{bmatrix} \in \mathbb{R}^k. \tag{104}$$

The dynamics matrix $\boldsymbol{E}_Y$ can be found by considering the linearized model (18), (19) of the nonlinear model (76), (77) with disturbances determined according to (78), (79)

$$x\left(t+1\right) = \boldsymbol{A}_d x(t) + \boldsymbol{B}u(t) - \sum_{i=1}^{t+1} (-1)^i \boldsymbol{\Upsilon}_i x\left(t+1-i\right) + v\left(t+1\right), \tag{105}$$

$$y(t) = \boldsymbol{C}x(t) + d(t). \tag{106}$$

Taking into account that for such a model, the prediction of the state vector is [17, 35]

$$x^p\left(t+j|t\right) = \boldsymbol{\Phi}^{\Upsilon}\left(j\right) x(t) + \sum_{i=0}^{j-1} \boldsymbol{\Phi}^{\Upsilon}\left(j-i-1\right) \boldsymbol{B}u\left(t+i\right) + v\left(t+j|t\right), \tag{107}$$

where the matrix $\boldsymbol{\Phi}^{\Upsilon}\left(j\right)$ is defined by a recurrence relation

$$\boldsymbol{\Phi}^{\Upsilon}\left(j\right) = \left(\boldsymbol{A}+\boldsymbol{\Upsilon}_1\right) \boldsymbol{\Phi}^{\Upsilon}\left(j-1\right) - \sum_{i=2}^{j} \left(-1^i\right) \boldsymbol{\Upsilon}_i \boldsymbol{\Phi}^{\Upsilon}\left(j-i\right), \quad j = 2, 3, \ldots, N, \tag{108}$$

$$\boldsymbol{\Phi}^{\Upsilon}\left(1\right) = \left(\boldsymbol{A}+\boldsymbol{\Upsilon}_1\right), \qquad \boldsymbol{\Phi}^{\Upsilon}\left(0\right) = \boldsymbol{I}_n, \tag{109}$$

we obtain a formula for the forced component of the state vector [17]

$$x^c\left(t+j|t\right) = \sum_{i=0}^{j-1} \boldsymbol{\Phi}^{\Upsilon}\left(j-i-1\right) \boldsymbol{B} \sum_{k=0}^{i} \Delta u\left(t+k|t\right), \; j = 1, 2, \ldots, N, \tag{110}$$

whence it follows directly that

$$\boldsymbol{E}_Y = \underline{\boldsymbol{C}} \cdot \boldsymbol{E} \cdot \boldsymbol{T}, \tag{111}$$

where

$$\boldsymbol{E}=\begin{bmatrix} \boldsymbol{B} & \cdots\cdots & \boldsymbol{0}_n \\ (\boldsymbol{A}+\boldsymbol{\Upsilon}_1)\,\boldsymbol{B} & \boldsymbol{B}\;\vdots & \boldsymbol{0}_n \\ \vdots & \ddots\;\ddots & \vdots \\ \sum_{i=0}^{N_u-1}\boldsymbol{\Phi}^{\Upsilon}(i)\,\boldsymbol{B} & \cdots\cdots & \boldsymbol{B} \\ \sum_{i=0}^{N_u}\boldsymbol{\Phi}^{\Upsilon}(i)\,\boldsymbol{B} & \cdots\cdots & (\boldsymbol{A}+\boldsymbol{\Upsilon}_1)\,\boldsymbol{B} \\ \vdots & \vdots\;\vdots & \vdots \\ \sum_{i=0}^{N-1}\boldsymbol{\Phi}^{\Upsilon}(i)\,\boldsymbol{B} & \cdots\cdots & \sum_{i=0}^{N-N_u}\boldsymbol{\Phi}^{\Upsilon}(i)\,\boldsymbol{B} \end{bmatrix}\in\mathbb{R}^{n\cdot N\times n\cdot N_u}, \tag{112}$$

$$\underline{\boldsymbol{C}}=diag\,(\boldsymbol{C},\ldots,\boldsymbol{C})\in\mathbb{R}^{p\cdot N\times n\cdot N}. \tag{113}$$

Here there is proposed a predictive algorithm of fractional order with nonlinear prediction and linearization along the predicted trajectory (FO MPC NPL+) which works as follows. Once the membership of the state vector $x(t)$ and the input vector $u\,(t-1)$ in the polyhedron $P_k=U_k\times X_k$ has been established at each sampling instant t on the basis of the matrix inequality system (33), the initial model dynamics matrix $\boldsymbol{E}^1_{Y,k}\in\mathbb{R}^{n\cdot N\times n\cdot N_u}$ and then the initial predicted trajectory $X^P(t)^1_{\rightarrow}$ and the initial solution $\Delta U(t)^1_{\rightarrow}$ to the quadratic optimization problem (96) are determined from Eq. (111). Further search for the optimal solution $\Delta U(t)^i_{\rightarrow}$ is done iteratively with the index $i\geq 2$, according to the scheme shown in Table 2. Based on $X^P(t)^{i-1}_{\rightarrow}$ and $\Delta U(t)^{i-1}_{\rightarrow}$ switching sequences for local sub-models, according to the set of matrix inequalities (33), are determined by

$$\mathcal{M}^i_{k,\,N^i_k},\quad k=1,2,3,\ldots,S^i,\quad 1\leq N^i_k\leq N,\ \sum_{k=1}^{S^i}N^i_k=N, \tag{114}$$

i.e. the number of models S^i switched at the i-th iteration step, an ordered list of switched models $\mathcal{M}^i_k$ from the set of all submodels $\{\mathcal{M}_1,\mathcal{M}_2,\ldots,\mathcal{M}_S\}$ and the length $\{N^i_1,N^i_2,\ldots,N^i_{S^i}\}$ of the portion of the prediction horizon where these submodels are active. On this basis, according to (111), the dynamics matrix is created at the i-th iteration step

$$\boldsymbol{E}^i_Y=\underline{\boldsymbol{C}}^i\cdot\begin{bmatrix} \boldsymbol{E}^i_1\in\mathbb{R}^{n\cdot N^i_1\times n\cdot N_u} \\ \hline \boldsymbol{E}^i_2\in\mathbb{R}^{n\cdot N^i_2\times n\cdot N_u} \\ \hline \vdots \\ \hline \boldsymbol{E}^i_{S^i}\in\mathbb{R}^{n\cdot N^1_{S^i}\times n\cdot N_u} \end{bmatrix}\cdot\boldsymbol{T} \tag{115}$$

and the suboptimal vector of future sequences of increments in the manipulated variable is determined at the current sampling instant $\Delta U(t)^i_{\rightarrow}$. Iterations are carried out until one of two conditions that interrupt the iterative search for the suboptimal vector of future sequences of increments in the manipulated variable at the current

Table 2 Computational procedure of the FO MPC NPL+ algorithm

Calculations performed once at each sampling step	
Step 1	Measuring (estimating) the state vector $x(t)$, measuring the output vector $y(t)$
Step 2	Determining the active polyhedron P_k from the set $\{P_1, P_2, \ldots, P_S\}$ according to (33); determining the matrices $\boldsymbol{A}_k, \boldsymbol{B}_k, \boldsymbol{C}_k, \boldsymbol{\Upsilon}_{1,k}, \boldsymbol{\Upsilon}_{2,k}, \ldots$ for the appropriate local model
Step 3	Creating the starting dynamics matrix $\boldsymbol{E}^1_{Y,k}$ from (111)
Step 4	Determining the disturbance vectors $v(t)$ and $d(t)$ from (78), (79)
Step 5	Calculating the starting free trajectory $X^0(t)^1_{\rightarrow}$ and the prediction trajectory $X^P(t)^1_{\rightarrow}$ from (92)–(95)
Step 6	Determining the reference trajectory $X^r(t)_{\rightarrow}$ that starts from the current process output value
Step 7	Solving the quadratic optimization problem (96) with constraints (97)–(99); determining the starting suboptimal vector of increments of the manipulated variable $\Delta U^1_{\rightarrow}(t)$; setting the index of iterative calculations ($i := 2$)
Step 8	**Jump to the iterative procedure, Steps 8.1–8.6**
Step 9	Applying the first m entries $\Delta u\,(t\|t)$ of the vector $\Delta U_{\rightarrow}(t)$ to the process
Calculations repeated iteratively at each sampling step	
Step 8.1	An ordered list of switched models $\mathscr{M}^i_k$ and time instants they are switched $\left\{N^i_1, N^i_2, \ldots, N^i_{S^i}\right\}$ within the prediction horizon is determined according to (33) from $X^P(t)^{i-1}_{\rightarrow}, \Delta U(t)^{i-1}_{\rightarrow}$
Step 8.2	Creating the dynamics matrix $\boldsymbol{E}^i_Y$ at the i-th iteration step; calculating $X^P(t)^i_{\rightarrow}$
Step 8.3	Solving the quadratic optimization problem (96) with constraints; determining the suboptimal vector of increments in the manipulated variable $\Delta U^i_{\rightarrow}(t)$
Step 8.4	For adopted δ_u, δ_i checking whether the conditions (116) for terminating the iterative search for a suboptimal vector of future sequences of increments in the manipulated variable at the current sampling instant t are satisfied or not
Step 8.5	If the conditions for terminating the iterative search (116) are satisfied, then **jump to Step 9**
Step 8.6	The iteration index is increased $i := i + 1$; **jump to Step 8.1**

sampling instant t is fulfilled

$$\|\Delta U(t)^i_{\rightarrow} - \Delta U(t)^{i-1}_{\rightarrow}\| \leq \delta_u \|\Delta U(t)^{i-1}_{\rightarrow}\| \quad \text{lub} \quad i > \delta_i \tag{116}$$

which means a small improvement in the sought vector of increments or exceeding the time limit allotted for iterative searches. The parameters δ_u, δ_i are to be chosen experimentally.

6 Conclusion

This paper discusses modeling of complex, nonlinear dynamic systems of non-integer order using so-called switched models that are based on the dynamic change of local linear models, depending upon an appropriately selected switching function. At first, an insight into the idea of switched models and their types and areas of application is given, and then the basics of differential and difference calculus of fractional order, as well as linear and nonlinear, continuous-time and discrete-time dynamic models of non-integer order based on them in the state space are outlined. In the main part of the work selected discrete-time switched models of non-integer order in the state space are defined, their basic properties are discussed and examples of simulation results are given. Finally, it has been proposed to employ discrete-time switched models of non-integer order for synthesis of nonlinear predictive control algorithms also of non-integer order. Such algorithms have been the subject of numerous theoretical and practical studies for several years.

References

1. Bemporad, A., Ferrari-Trecate, G., Morari, M.: Observability and controllability of piecewise affine and hybrid systems. IEEE Trans. Autom. Control **45**(10), 1864–1876 (2000)
2. Benmessaouda, O., Benzaouia, A., Tadeo, F.: Stabilization of uncertain state constrained discrete-time switched systems. In: Preparation of the 18th IFAC World Congress, Milano, Italy, pp. 5736–5740 (2011)
3. Bin Liu, B., Hill, D.J.: Stabilization for decomposable dissipative discrete-time switched systems. In: Preparation of the 18th IFAC World Congress, Milano, Italy, pp. 5730–5735 (2011)
4. Blesa, J., Puig, V., Bolea, Y.: Fault detection using interval LPV models in an open-flow canal. Contr. Eng. Practice **18**, 460–470 (2010)
5. Branicky, M.S.: Multiple Lyapunov functions and other analysis tools for switched and hybrid systems. IEEE Trans. Autom. Control **43**(3), 475–482 (1998)
6. Busłowicz, M.: Stability of state-space models of linear continuous-time fractional order systems. Acta Mech. Autom. **5**(2), 15–22 (2011)
7. Camacho, E.F., Bordons, C.: Model Predictive Control. Springer, London (2004)
8. Chadli, M.: An LMI approach to design observer for unknown inputs Takagi-Sugeno fuzzy models. Asian J. Control **12**(4), 524–530 (2010)
9. Chen, Y.Q., Petráš, I., Xue, D.: Fractional order control. In: American Control Conference, St. Louis, pp. 1397–1410 (2009)
10. Daafouz, J., Riedinger, P., Iung, C.: Stability analysis and control synthesis for switched systems: a switched Lyapunov function approach. IEEE Trans. Autom. Control **47**(11), 1883–1887 (2002)
11. Deaecto, G.S., Souza, M., Geromel, J.C.: Discrete-time switched linear systems state feedback design with application to networked control. IEEE Trans. Autom. Control **60**(3), 877–888 (2015)
12. Ding, B.: Properties of parameter-dependent open-loop MPC for uncertain systems with polytopic description. Asian J. Control **12**(1), 58–70 (2010)
13. Domek, S.: Robust Model Predictive Control for Nonlinear Processes (In Polish). Technical University of Szczecin Academic Press, Szczecin (2006)

14. Domek, S.: A model based robust control system for a class of MIMO nonlinear systems. In: Malinowski, K., Rutkowski, L. (eds.) Recent Advances in Control and Automation, Challenging Problems of Science, Control and Automation, pp. 1–10. Academic Publishing House EXIT, Warszawa (2008)
15. Domek, S.: Fuzzy predictive control of fractional-order nonlinear discrete-time systems. Acta Mech. Autom. **5**(2), 23–26 (2011)
16. Domek, S.: Switched state model predictive control of fractional-order nonlinear discrete-time systems. In: Pisano, A., Caponetto, R. (eds.) Advances in Fractional Order Control and Estimation, Asian J. Control, Special Issue **15**(3), 658–668 (2013)
17. Domek, S.: Fractional-Order Calculus in Model Predictive Control (In Polish). West Pomeranian University of Technology Academic Press, Szczecin (2013)
18. Domek, S.: Piecewise affine representation of discrete in time, non-integer order systems. In: Mitkowski, W., Kacprzyk, J., Baranowski, J. (eds.) Advances in the Theory and Applications of Non-integer Order Systems. Lecture Notes in Electrical Engineering, vol. 257, pp. 149–160. Springer (2013)
19. Domek, S.: Multiple use of the fractional-order differential calculus in the model predictive control. In: 19th International Conference on Methods and Models in Automation and Robotics, Międzyzdroje, pp. 359–362 (2014). https://doi.org/10.1109/MMAR.2014.6957379
20. Domek, S.: Fractional-order model predictive control with small set of coincidence points. In: Latawiec, K., Łukaniszyn, M., Stanisławski, R. (eds.) Advances in Modelling and Control of Non-integer Order Systems. Lecture Notes in Electrical Engineering, vol. 320, pp. 135–144. Springer (2014)
21. Domek, S.: Model-plant mismatch in fractional order model predictive control. In: Domek, S., Dworak, P. (eds.) Theoretical Developments and Applications of Non-integer Order Systems. Lecture Notes in Electrical Engineering, vol. 357, pp. 281–291. Springer (2016)
22. Domek, S.: Approximation and stability analysis of some kinds of switched fractional linear systems. In: Mitkowski, W., Kacprzyk, J., Oprzędkiewicz, K., Skruch, P. (eds.) Trends in Advanced Intelligent Control, Optimization and Automation. AISC, vol. 577, pp. 442–454. Springer (2017)
23. Domek, S.: Fractional linear systems with memory deficiency and their state-space integer-order approximation. In: Ostalczyk, P., Sankowski, D., Nowakowski, J. (eds.) Non-integer Order Calculus and its Applications. Lecture Notes in Electrical Engineering, vol. 496, pp. 164–179. Springer (2018). https://doi.org/10.1007/978-3-319-78458-8_15
24. Domek, S., Dworak, P., Pietrusewicz, K.: Hybrid model-following control algorithm within the motion control system. In: Proceedings of the IEEE International Symposium on Industrial Electronics, Piscataway, pp. 1476–1481 (2009)
25. Domek, S., Jaroszewski, K.: Model predictive controller for fractional order systems. In: Grzech, A., Świątek, P., Drapała, J. (eds.) Advances in System Science. Computer Science, pp. 9–18. EXIT, Warszawa (2010)
26. Dzieliński, A., Sierociuk, D.: Stability of discrete fractional order state-space systems. J. Vib. Control **14**(9–10), 1543–1556 (2008)
27. Dzieliński, A., Sarwas, G., Sierociuk, D.: Comparison and validation of integer and fractional order ultracapacitor models. Adv. Differ. Equ. (2011). https://doi.org/10.1186/1687-1847-2011-11
28. Dzieliński, A., Sierociuk, D., Sarwas, G.: Some applications of fractional order calculus. Bull. Pol. Acad. Sci. Tech. Sci. **58**(4), 583–592 (2010)
29. Fang, L., Lin, H., Antsaklis, P.J.: Stabilization and performance analysis for a class of switched systems. In: Proceedings of the 43rd IEEE Conference on Decision Control, Atlantis, pp. 1179–1180 (2004)
30. Geromel, J.C., Colaneri, P.: Stability and stabilization of discrete time switched systems. Int. J. Control **79**(7), 719–728 (2006)
31. Geyer, T., Torrisi, F., Morari, M.: Optimal complexity reduction of polyhedral piecewise affine systems. Automatica **44**(7), 1728–1740 (2008)

32. Ionescu, C., Lopes, A., Copot, D., Machado, J.A.T., Bates, J.H.T.: The role of fractional calculus in modeling biological phenomena: a review. Commun. Nonlinear Sci. Numer. Simulat. **51**, 141–159 (2017)
33. Jiménez, A., Al-Hadithi, B.M., Matia, F., Haber-Haber, R.: Improvement of Takagi-Sugeno fuzzy model for the estimation of nonlinear functions. Asian J. Control (2010). https://doi.org/10.1002/asjc.310
34. Kaczorek, T.: Practical stability and asymptotic stability of positive fractional 2D linear systems. Asian J. Control **12**(2), 200–207 (2010)
35. Kaczorek, T.: Selected Problems of Fractional Systems Theory. Springer, Berlin (2011)
36. Ławryńczuk, M.: Nonlinear state-space predictive control with on-line linearisation and state estimation. Int. J. Appl. Math. Comput. Sci. **25**(4), 833–847 (2015)
37. Lian, J., Zhao, J.: Output feedback variable structure control for a class of uncertain switched systems. Asian J. Control **11**(1), 31–39 (2009)
38. Liberzon, D., Morse, A.S.: Basic problems in stability and design of switched systems. IEEE Control Syst. **19**(5), 59–70 (1999)
39. Liberzon, D.: Switching in Systems and Control. Birkhauser, Boston (2003)
40. Lin, H., Antsaklis, P.J.: Stability and stabilizability of switched linear systems: a survey of recent results. IEEE Trans. Autom. Control **54**(2), 308–322 (2009)
41. Macias, M., Sierociuk, D.: An alternative recursive fractional variable-order derivative definition and its analog validation. In: Proceedings of the International Conference on Fractional Differentiation and its Applications (ICFDA), Catania, pp. 1–6 (2014)
42. Maciejowski, J.M.: Predictive Control with Constraints. Prentice Hall, Englewood Cliffs (2002)
43. Mäkilä, P.M., Partington, J.R.: On linear models for nonlinear systems. Automatica **39**, 1–13 (2003)
44. Margaliot, M.: Stability analysis of switched systems using variational principles: an introduction. Automatica **42**, 2059–2077 (2006)
45. Monje, C.A., Chen, Y.Q., Vinagre, B.M., Xue, D., Feliu, V.: Fractional Order Systems and Controls. Springer-Verlag, London (2010)
46. Mozyrska, D., Ostalczyk, P.: Generalized fractional-order discrete-time integrator. Complexity (Hindawi) (2017). https://doi.org/10.1155/2017/3452409
47. Muddu Madakyaru, M., Narang, A., Patwardhan, S.C.: Development of ARX models for predictive control using fractional order and orthonormal basis filter parameterization. Ind. Eng. Chem. Res. **48**(19), 8966–8979 (2009)
48. Murray-Smith, R., Johansen, T.: Multiple Model Approaches to Modeling and Control. Taylor and Francis, London (1997)
49. Nafsun, A.I., Yusoff, N.: Effect of model-plant mismatch on MPC controller performance. J. Appl. Sci. **21**(11), 579–585 (2011)
50. Ostalczyk, P.: The non-integer difference of the discrete-time function and its application to the control system synthesis. Int. J. Syst. Sci. **31**(12), 1551–1561 (2000)
51. Podlubny, I.: Fractional Differential Equations. Academic Press, San Diego (1999)
52. Rodrigues, M., Theilliol, D., Adam-Medina, M., Sauter, D.: A fault detection and isolation scheme for industrial systems based on multiple operating models. Contr. Eng. Practice **16**, 225–239 (2008)
53. Romero, M., De Madrid, Á.P., Mañoso, C., Vinagre, B.M.: Fractional-order generalized predictive control: formulation and some properties. In: Proceedings of the 11th International Conference on Control, Automation, Robotics and Vision, Singapore, pp. 1495–1500 (2010)
54. Romero, M., Vinagre, B.M., De Madrid, Á.P.: GPC control of a fractional–order plant: improving stability and robustness. In: Proceedings of the 17th IFAC World Congress, Seoul, pp. 14266–14271 (2008)
55. Rydel, M., Stanisławski, R., Bialic, G., Latawiec, K.: Modeling of discrete-time fractional-order state space systems using the balanced truncation method. In: Domek, S., Dworak, P. (eds.) Theoretical Developments and Applications of Non-integer Order Systems. Lecture Notes in Electrical Engineering, vol. 357, pp. 119–127. Springer (2016)

56. Safonov, M.G., Chiang, R.Y.: A Schur method for balanced model reduction. IEEE Trans. Autom. Control **AC–2**(7), 729–733 (1989)
57. Savkin, A.V., Evans, R.J.: Hybrid Dynamical Systems. Controller and Sensor Switching Problems. Birkhäuser, Boston (2002)
58. Shantanu, D.: Functional Fractional Calculus for System Identification and Controls. Springer Verlag, Berlin (2008)
59. Shorten, R., Wirth, F., Mason, O., Wulff, K., King, C.: Stability criteria for switched and hybrid systems. SIAM Rev. **49**(4), 545–592 (2007)
60. Sierociuk, D., Dzieliński, A.: Fractional Kalman filter algorithm for the states, parameters and order of fractional system estimation. Int. J. Appl. Math. Comp. Sci. **16**(1), 129–140 (2006)
61. Sontag, E.D.: Interconnected automata and linear systems: a theoretical framework in discrete-time. In: Alur, R., Henzinger, T.A., Sontag, E.D. (eds.) Hybrid Systems III: Verification and Control. Lecture Notes in Computer Science, vol. 1066, pp. 436–448. Springer-Verlag (1996)
62. Stanisławski, R., Latawiec, K.: Normalized finite fractional differences: the computational and accuracy breakthroughs. Int. J. Appl. Math. Comput. Sci. **22**(4), 907–919 (2012)
63. Stanisławski, R., Latawiec, K., Łukaniszyn, M., Gaüek, M.: Time-domain approximations to the Grünwald-Letnikov difference with application to modeling of fractional-order state space systems. In: Proceedings of the 20th International Conference on Methods and Models in Automation and Robotics, Międzyzdroje, Poland, pp. 579–584 (2015)
64. Sun, Z., Ge, S.S.: Switched Linear Systems. Control and Design. Springer, London (2005)
65. Sun, Z., Ge, S.S.: Analysis and synthesis of switched linear control systems. Automatica **41**(2), 181–195 (2005)
66. Sun, Z., Ge, S.S., Lee, H.: Controllability and reachability criteria for switched linear systems. Automatica **38**(5), 775–786 (2002)
67. Szűcs, A., Kvasnica, M., Fikar, M.: Optimal piecewise affine approximations of nonlinear functions obtained from measurements. In: Proceedings of the 4th IFAC Conference on Analysis and Design of Hybrid Systems, Eindhoven, pp. 160–165 (2012)
68. Tatjewski, P.: Advanced Control of Industrial Processes. Springer, London (2007)
69. Tatjewski, P.: Supervisory predictive control and on-line set-point optimization. Int. J. Appl. Math. Comput. Sci. **20**(3), 483–495 (2010)
70. Tatjewski, P.: Offset-free nonlinear predictive control with measured state and unknown asymptotically constant disturbances. In: Malinowski, K., Józefczyk, J., Świ(mohana)tek, J.: (eds.) Actual Problems in Automation and Robotics. Akademicka Oficyna Wydawnicza Exit, Warszawa, pp. 288–299 (2014)
71. Vesely, V., Rosinova, D.: Robust MPC controller design for switched systems using multi-parameter dependent Lyapunov function. Int. J. Innov. Comput. Inform. Control **10**(1), 269–280 (2014)
72. Wang, Y., Wang, X.: Performance diagnosis of MPC with model-plant mismatch. In: Proceedings of the Chinese Control and Decision Conference, Xuzhou, pp. 78–82 (2010)
73. Zhao, X.D., Zhang, L., Shi, P., Liu, M.: Stability and stabilization of switched linear systems with mode-dependent average dwell time. IEEE Trans. Autom. Control **57**(7), 1809–1815 (2012)

Control

Nonlinear Predictive Control

Piotr Tatjewski and Maciej Ławryńczuk

Abstract Model Predictive Control (MPC) algorithms using nonlinear process models are the subject of consideration in this chapter. The applied nonlinear models are in the form of general difference equations or state-space equations. MPC algorithms using directly nonlinear models in the optimization of the trajectory of the manipulated variables are described in the first part of the chapter. This leads to strictly optimal solutions, but is practically restricted to processes with slow dynamics due to difficult, time consuming nonlinear optimization. For the case of state-space models, the original authors' approach to the modeling of disturbances and state estimation is presented. The most extensive part of the paper is devoted to effective, suboptimal MPC algorithms with successive linearizations, which enables us to replace nonlinear optimization by a quadratic one. Several versions of such algorithms are presented, with different linearization structures. This class of algorithms enables us to apply nonlinear modeling to fast dynamical systems, leading generally to suboptimal results, but usually fully acceptable in engineering practice. This is confirmed by the presented results of simulation studies of two processes. Finally, augmentations of MPC algorithms to incorporate current set-point optimization are described, to increase economic efficiency of the control structures.

1 Introduction

Predictive control, well known under the acronym MPC (Model Predictive Control), is currently a widely used advanced control technique [2, 3, 56], accessible as a broad collection of effective control algorithms applied in industrial practice, see,

P. Tatjewski (✉) · M. Ławryńczuk
Warsaw University of Technology, Faculty of Electronics and Information Technology, Institute of Control and Computation Engineering, ul. Nowowiejska 15/19, 00-665 Warsaw, Poland
e-mail: P.Tatjewski@ia.pw.edu.pl

M. Ławryńczuk
e-mail: M.Lawrynczuk@ia.pw.edu.pl

P. Kulczycki et al. (eds.), *Automatic Control, Robotics, and Information Processing*, Studies in Systems, Decision and Control 296,
https://doi.org/10.1007/978-3-030-48587-0_7

e.g., [2, 3, 5, 40, 52–54, 56–58, 61, 64]. Presentation of the predictive control principle and basic algorithms can be found in numerous books and journal papers, see, e.g. [40, 53, 54, 56, 64]. These publications are devoted mainly to algorithms with linear process models of different natures. Nonlinear process models used in MPC algorithms appear mainly in the form of nonlinear state-space equations or difference equations of higher orders, including neural network modeling, see, e.g. [26, 63].

Real industrial processes are in fact always nonlinear and linear feedback control in neighborhoods of working points is not always satisfactory. Especially, nonlinear control is needed for the case of optimal control in a multilayer structure, where the optimizer adjusts on-line working points (set-points) in order to keep the economic efficiency as high as possible, despite changes in the disturbances and product requirements. On the other hand, time-variable trajectories of the process variables are standard in the control of batch processes. Therefore, attempts to formulate nonlinear MPC algorithms can be found in the literature starting from the 1980s.

MPC algorithms directly using the original nonlinear model in MPC optimization problem solved on-line by a nonlinear optimization procedure can be applied, first of all, to strongly nonlinear processes with slow dynamics, and when significant changes of manipulated variables occur. Problems connected with the design and implementation of MPC algorithms with nonlinear and non-quadratic optimization at every sampling instant are presented in the first part of this chapter.

In the design of nonlinear MPC algorithms directed to more practical and more widely applicable solutions, there is a tendency to apply algorithms with model linearization, which can be seen starting from the 1980s [11]. In nonlinear MPC algorithms, linearization is used in such a way that the MPC optimization problem solved on-line is formulated as a convex quadratic programming problem. This problem is usually adapted at every sampling instant of the algorithm, or even several times within one sampling instant, to accommodate the nonlinearity of the process. Structures of nonlinear, suboptimal MPC algorithms using appropriate linearization techniques are presented in the next, dominant part of the chapter.

2 The MPC Principle

We shall briefly recall in this section the MPC formulation needed for further presentation. We shall consider the dynamical process with n_{u} manipulated inputs, n_{y} controlled outputs and n_{x} state variables. Thus $u = \begin{bmatrix} u_1 & \cdots & u_{n_{\mathrm{u}}} \end{bmatrix}^{\mathrm{T}}$ is the vector of manipulated variables (control inputs), $y = \begin{bmatrix} y_1 & \cdots & y_{n_{\mathrm{y}}} \end{bmatrix}^{\mathrm{T}}$ is the vector of controlled outputs (process variables), $x = \begin{bmatrix} x_1 & \cdots & x_{n_{\mathrm{x}}} \end{bmatrix}^{\mathrm{T}}$ is the vector of process states.

The principle of MPC is to evaluate the process control inputs by minimizing, at each sampling instant k, a performance function (cost function) over a predefined future prediction horizon of N samples. The following performance function is one of the most widely used in process control

$$J(k) = \sum_{p=1}^{N} \left\| y^{\mathrm{sp}}(k+p|k) - y(k+p|k) \right\|_{\boldsymbol{\Psi}_p}^2 + \sum_{p=0}^{N_\mathrm{u}-1} \left\| \Delta u(k+p|k) \right\|_{\boldsymbol{\Lambda}_p}^2, \quad (1)$$

where $\|x\|_{\mathbf{R}}^2 = x^T \mathbf{R} x$, $\boldsymbol{\Psi} \geq \mathbf{0}$ and $\boldsymbol{\Lambda} > \mathbf{0}$ are square diagonal scaling matrices of dimensions corresponding to the dimensions n_y and n_u of the process controlled output and control input vectors, respectively (a simpler formulation of Eq. (1) is often used in theoretical considerations, with one scaling scalar λ only, i.e., $\boldsymbol{\Psi} = \mathbf{I}$ and $\boldsymbol{\Lambda} = \lambda\mathbf{I}$). In the above formulation, $N_\mathrm{u} \leq N$ denotes the length of the control horizon, $y^{\mathrm{sp}}(k+p|k)$ and $y(k+p|k)$ are set-point (reference) vector and process output vector, respectively, predicted for the future sample $k+p$, but calculated at the current sample k, $p = 1, \ldots, N$. The first sum in Eq. (1) is the sum of predicted control errors over the prediction horizon, the second sum is the sum of the control increments $\Delta u(k+p|k)$ over the control horizon,

$$\Delta u(k+p|k) = u(k+p|k) - u(k+p-1|k), \quad p = 0, \ldots, N_\mathrm{u} - 1. \quad (2)$$

In MPC algorithms with linear process models, the control increments (2) are usually used as decision variables in the optimization of the function (1). However, when the optimization uses nonlinear process models, it is often more convenient to use the control inputs $u(k+p|k)$, $p = 0, \ldots, N_\mathrm{u} - 1$, as the decision variables. The whole vector of these variables over the control horizon is

$$\mathbf{u}(k) = \begin{bmatrix} u(k|k) \\ \vdots \\ u(k+N_\mathrm{u}-1|k) \end{bmatrix}. \quad (3)$$

Assuming simple inequality constraints on values and on increments of the decision variables (control inputs), and on the values of the predicted process outputs, the MPC optimization problem can be defined in the following form

$$\begin{aligned} &\min_{\mathbf{u}(k)} \left\{ J(k) = \sum_{p=1}^{N} \left\| y^{\mathrm{sp}}(k+p|k) - y(k+p|k) \right\|_{\boldsymbol{\Psi}_p}^2 + \sum_{p=0}^{N_\mathrm{u}-1} \left\| \Delta u(k+p|k) \right\|_{\boldsymbol{\Lambda}_p}^2 \right\} \\ &\text{subject to:} \qquad (4) \\ &u^{\min} \leq u(k+p|k) \leq u^{\max}, \quad p = 0, \ldots, N_\mathrm{u} - 1, \\ &-\Delta u^{\max} \leq \Delta u(k+p|k) \leq \Delta u^{\max}, \quad p = 0, \ldots, N_\mathrm{u} - 1, \\ &y^{\min} \leq y(k+p|k) \leq y^{\max}, \quad p = 1, \ldots, N. \end{aligned}$$

When solving this problem by a numerical optimization procedure, for every consecutive vector of the decision variables $\mathbf{u}(k)$, the trajectory of predicted outputs $y(k+p|k)$, $p = 1, \ldots, N$, is calculated with the use of the nonlinear model (which is not explicitly seen in Eq. (4)). A more general form of the inequality constraints is possible, in the form of any linear functions of process inputs and outputs. But this does not lead to difficulties for the solution of the optimization problem and is therefore omitted in this chapter, to make the formulations simpler.

When the process model is linear, then the MPC optimization problem (4) is a strictly convex quadratic programming (QP) problem, thus having a well defined, unique solution, provided the set defined by the inequality constraints is feasible (non-empty). Thus this problem can be effectively and robustly solved by a standard numerical QP procedure. However, with a nonlinear process model the problem (4) becomes nonlinear and generally nonconvex, thus a nonlinear (and non-quadratic) optimization procedure must be applied or the MPC problem must be redesigned to enable its easier, but generally suboptimal solution.

Once the MPC optimization problem has been solved, only first n_{u} elements of the calculated optimal trajectory of the decision variables (process inputs) $\mathbf{u}^{\mathrm{opt}}(k)$ are used and applied as the process input $u(k)$, $u(k) = u^{\mathrm{opt}}(k|k)$. After the next measurement (at the next sampling instant) the whole algorithm is repeated (receding horizon strategy).

We shall assume $n_{\mathrm{u}} = n_{\mathrm{y}}$ in the paper, which usually (always in the linear case) yields a unique solution of the MPC optimization problem. Nevertheless, the case of $n_{\mathrm{u}} > n_{\mathrm{y}}$, which is not unusual in MPC applications, will be commented on.

3 MPC Algorithms with Nonlinear Optimization (MPC-NO)

The full nonlinear MPC algorithm, with prediction of process output trajectory over the prediction horizon calculated using nonlinear process model, will be denoted by the acronym MPC-NO (MPC with Nonlinear Optimization).

At first glance, the difference between MPC with a linear and MPC with a nonlinear process model, may seem not to be significant, as principle of calculation of the optimal control trajectory remains unchanged. The only difference is that the nonlinear model is used for the prediction of the process outputs over the prediction horizon, instead of the linear one. But this difference is fundamental. The nonlinear dependence of the predicted outputs $y(k + p|k)$ on the decision variables makes the MPC optimization problem (4) non-quadratic and, in general, non-convex. For solving such problems, there are no universal optimization procedures which can guarantee a fast and reliable finding of the solution, over a limited time period and with an assumed accuracy. Moreover, relatively fast gradient procedures find, in general, only local minima of nonlinear functions. These facts imply that MPC algorithms with nonlinear model, used in the optimization problem, solved at every sampling instant by a nonlinear optimization procedure, are usually applied for slow processes, possibly with stronger nonlinearities and when more aggressive changes of control inputs can occur.

The most commonly met forms of nonlinear process models are nonlinear difference equations (presented in the sequel for the case of a SISO process, for the sake of simplicity)

$$y(k) = g(u(k-1), \ldots, u(k-n_{\mathrm{B}}), y(k-1), \ldots, y(k-n_{\mathrm{A}})), \tag{5}$$

or nonlinear state equations

$$x(k+1) = f(x(k), u(k)), \tag{6}$$
$$y(k) = g(x(k)). \tag{7}$$

Calculation of the predicted trajectory of the model outputs is performed in a recurrent way, for every current control trajectory $u(k+p|k)$, $p = 0, \ldots, N-1$, over the prediction horizon – assuming that if $N_{\mathrm{u}} < N$ then for $p = N_{\mathrm{u}}, \ldots, N-1$: $u(k+p|k) = u(k+N_{\mathrm{u}}-1|k)$.

For the nonlinear difference equation (5), calculation of the predicted output trajectory is performed in the following way

$$\begin{aligned}
y(k+1|k) &= g(u(k|k), u(k-1), u(k-2), \ldots, u(k-n_{\mathrm{B}}+1),\\
&\qquad y(k), y(k-1), y(k-2), \ldots, y(k-n_{\mathrm{A}}+1)) + d(k),\\
y(k+2|k) &= g(u(k+1|k), u(k|k), u(k-1), \ldots, u(k-n_{\mathrm{B}}+2),\\
&\qquad y(k+1|k), y(k), y(k-1), \ldots, y(k-n_{\mathrm{A}}+2)) + d(k),\\
&\ \vdots\\
y(k+N|k) &= g(u(k-1+N|k), \ldots, u(k-n_{\mathrm{B}}+N|k),\\
&\qquad y(k-1+N|k), \ldots, y(k-n_{\mathrm{A}}+N|k)) + d(k),
\end{aligned} \tag{8}$$

where $d(k)$ is the estimate of output disturbances (unmeasured) defined in the following way

$$\begin{aligned}
d(k) &= y(k) - y(k|k-1)\\
&= y(k) - g(u(k-1), \ldots, u(k-n_{\mathrm{B}}), y(k-1), \ldots, y(k-n_{\mathrm{A}})).
\end{aligned} \tag{9}$$

For nonlinear state equations (6)–(7), prediction of the state and output values takes the following form, for every trajectory of the control inputs $u(k+p|k)$, $p = 0, \ldots, N-1$

$$\begin{aligned}
x(k+1|k) &= f(x(k), u(k|k)) + v(k),\\
&\qquad\qquad y(k+1|k) = g(x(k+1|k)) + d(k),\\
x(k+2|k) &= f(x(k+1|k), u(k+1|k)) + v(k),\\
&\qquad\qquad y(k+2|k) = g(x(k+2|k)) + d(k),
\end{aligned}$$

$$\vdots$$

$$x(k+N|k) = f(x(k+N-1|k), u(k+N-1|k)) + v(k),$$
$$y(k+N|k) = g(x(k+N|k)) + d(k), \tag{10}$$

where $v(k)$ is the estimate of unmeasured state disturbances defined in the following way

$$\begin{aligned} v(k) &= x(k) - x(k|k-1) \\ &= x(k) - f(x(k-1), u(k-1)). \end{aligned} \tag{11}$$

This formula is used for the estimation of unmeasured state disturbances at the current sampling instant. Next, this estimate is used for prediction of these disturbances over the prediction horizon, because it is used in output prediction equations (10) for every future sample. On the other hand, $d(k)$ is the estimate of the disturbances acting directly on the process output, e.g., disturbances stemming from inaccurate modeling of process output equations or stemming from inaccurate state estimation. The problem of modeling $d(k)$ will be dealt with further on.

Modeling of state disturbances (11) was first introduced for state prediction (as in Eq. (10)) in [56], to get offset-free control for typical in process control cases with deterministic disturbances with non-zero mean values (modeling errors, piecewise step disturbances, etc., see also [60, 62]). Without this model of state disturbances, it is necessary to apply additional modeling and estimation of the disturbance state vector or the augmented process-and-disturbance state vector, both in linear and nonlinear cases, see [45, 46, 50, 51].

For efficient numerical solutions of the nonlinear MPC optimization problem the initial (starting) point is important. The MPC controller solves, at consecutive sampling instants, nonlinear optimization problems which differ only in values of certain parameters: measurements of process outputs (or states), measurements of disturbances, values of process control inputs. The differences in these variables, from step to step, are in most cases not significant. Therefore, the principle of "warm start" here is strongly recommended: the optimal trajectory from the previous sampling instant should be the base of the initial trajectory for the next one. More precisely, this principle applied to the MPC optimization problem means that to define the initial trajectory for the step k, $u^0(k+p|k)$, $p = 0, \ldots, N-1$, we utilize the optimal trajectory from the previous $(k-1)$-th step in the following way: the elements corresponding to the first sample are cut out and the elements corresponding to the last sample are duplicated, to get the required length of the initial trajectory, see [61].

It should be emphasized that at every step (sampling instant) of the MPC algorithm the prediction of the process output trajectory is calculated many times, for every consecutive input trajectory generated by the numerical procedure during the optimization process; starting with the initial trajectory $\mathbf{u}^0(k)$ and terminating with the current optimal trajectory $\mathbf{u}^{\text{opt}}(k)$. Therefore, efficiency is here crucial, as every output prediction is nonlinear, using a nonlinear process model.

If the whole process state vector cannot be measured, the nonlinear state estimation must be applied. We have the following situation here. There is not a precisely known process description, which can be formulated as a set of (unknown) nonlinear state equations

$$\begin{aligned} x(k+1) &= f_{\text{p}}(x(k), u(k)), \\ y(k) &= g_{\text{p}}(x(k)), \end{aligned} \tag{12}$$

and there is a (known) process model, in the form

$$\begin{aligned} x(k+1) &= f(x(k), u(k)), \\ y(k) &= g(x(k)), \end{aligned} \tag{13}$$

possibly augmented by stochastic disturbances (noises) acting on the process states and outputs.

The state is estimated on the basis of the model equations (13), the following methods are most popular

- the extended Luenberger observer (ELO), effective when the noise-to-signal ratios are relatively low,
- the extended Kalman filter (EKF).

The extended Luenberger observer can be defined by the following equation

$$\hat{x}(k+1) = f(\hat{x}(k), u(k)) + \mathbf{K}_{\text{ELO}}[y(k) - g(\hat{x}(k))],$$

see, e.g. [1], where the gain matrix $\mathbf{K}_{\text{ELO}}$ is calculated for the linearized process model, in the standard way as for the Luenberger observer for linear models. For moderate nonlinearities and deviations from a predefined equilibrium point, a constant gain matrix may be sufficient, calculated at this point. However, in general, adaptive ELO may be needed and is more effective, where the gain matrix is recalculated for every consecutive working point (recalculation of the linearized model is also needed). Another possible solution is the use of fuzzy TS (Takagi-Sugeno) ELO: a region of process variability is divided into overlapping subregions, for every subregion the observer gain matrix is calculated corresponding to the process linearization for this region, finally the current gain matrix is calculated by fuzzy reasoning with the subregion gain matrices constituting consequents of the fuzzy rules.

The extended Kalman filter is designed for a model with the following state and output equations

$$\begin{aligned} x(k+1) &= f(x(k), u(k)) + \varepsilon_{\text{x}}(k), \\ y(k) &= g(x(k)) + \varepsilon_{\text{y}}(k), \end{aligned} \tag{14}$$

where the noises ε_{x} and ε_{y} are white and Gaussian, with covariance matrices $\mathbf{Q}$ and $\mathbf{R}$ (in general, these matrices can be nonstationary, $\mathbf{Q}_k$ and $\mathbf{R}_k$).

Initial values for the k-th iteration of the EKF are: the value of the current state estimate $\hat{x}(k-1)$ and the value of the covariance matrix of the estimation error $\mathbf{P}_{k-1}$. Having these values, equations of the EKF are

$$\begin{aligned}
\text{prediction:}\quad & \hat{x}(k|k-1) = f(\hat{x}(k-1), u(k-1)),\\
& \mathbf{P}_{k|k-1} = \mathbf{A}_{k-1}\mathbf{P}_{k-1}\mathbf{A}_{k-1}^{\mathrm{T}} + \mathbf{Q},\\
\text{gain matrix:}\quad & \mathbf{K}_k = \mathbf{P}_{k|k-1}\mathbf{C}_{k|k-1}^{\mathrm{T}}(\mathbf{C}_{k|k-1}\mathbf{P}_{k|k-1}\mathbf{C}_{k|k-1}^{\mathrm{T}} + \mathbf{R})^{-1},\\
\text{correction:}\quad & \hat{x}(k) = \hat{x}(k|k-1) + \mathbf{K}_k[y(k) - g(\hat{x}(k|k-1))],\\
\text{covariance matrix:}\quad & \mathbf{P}_k = (\mathbf{I} - \mathbf{K}_k\mathbf{C}_{k|k-1})\mathbf{P}_{k|k-1},
\end{aligned}$$

where matrices $\mathbf{A}_{k-1}$ and $\mathbf{C}_{k|k-1}$ result from linearizations of the nonlinear model functions at the current state estimate and the process control input values,

$$\mathbf{A}_{k-1} = \left[\frac{\partial f_i}{\partial x_j}(\hat{x}(k-1), u(k-1))\right], \quad \mathbf{C}_{k|k-1} = \left[\frac{\partial g_i}{\partial x_j}(\hat{x}(k|k-1))\right]. \tag{15}$$

Extended Kalman filter is no longer an optimal estimator, it is optimal only if the process model is linear. Therefore, its convergence is no longer guaranteed. It may happen that it becomes divergent, when linearized process matrices loose observability. But such cases occur rather rarely. This is the reason why EKF is popular in practice.

The process is often influenced by deterministic disturbances, then ELO and EKF estimate the process state in the control feedback loop with zero offset only when the disturbances have zero mean values. Under the step disturbance, e.g., acting on the process input, the estimated state will generally differ from the actual process state, the offset will be nonzero. This can be easily shown as a result of a comparison of the disturbed process equations and estimator equations, see [59] in the linear case, or [61, 62] in the nonlinear case. In this situation:

- the state observer, or state filter, estimates the erroneous state value of the process $\hat{x}(k) \neq x(k)$, therefore the controller uses the disturbed estimate of the outputs $\hat{y}(k) = g(\hat{x}(k)) \neq y(k)$ for state and output prediction over the prediction horizon (excluding rare cases of mutual cancelling influences of the disturbances),
- the MPC algorithm with the state-space process model (we shall use the acronym MPCS for brevity) acts to cancel the predicted control errors, calculated as the difference between the reference values and the predicted disturbed (erroneous) model output values, with errors caused by the disturbed output estimates $\hat{y}(k) = g(\hat{x}(k))$, i.e., with the error $g(\hat{x}(k)) - y(k)$. Therefore, at steady-state the error (offset) is obtained, equal to $g(\hat{x}_{\mathrm{ss}}) - y_{\mathrm{ss}}$.

However, the MPCS controller with predictions (10)–(11) and with the observer (or filter) of the process state will generate the correct control action eliminating the offset, provided the errors caused by the incorrect output estimates are eliminated by appropriate corrections introduced to the prediction equations (10). This is done by adding the corrections $d(k)$ calculated as

$$d(k) = y(k) - g(\hat{x}(k)), \quad p = 1, \ldots, N, \tag{16}$$

to these equations, see [62], where strict theoretical proof of this statement is provided.

The value $y(k)$ in the equation of the output error (16) is the measured value, thus the true process output value. Therefore, it accommodates also the influence of all other disturbances acting directly (not through the process state) on the process outputs, e.g., stemming from possible errors in modeling the output equations. This results in the conclusion that the use of the correction (16) in the prediction equations (10) is necessary also in the situation of the state measurement (estimator or filter not needed), if the mentioned disturbances acting directly on the process state-output relations (the output equations) are present.

An alternative to the approach with the estimation of the process state only and appropriate prediction (Eqs. (10), (11), (16)), as presented in this section, is the augmentation of the dynamical process model, by addition of the dynamical model of the deterministic disturbances. Then, the extended process-and-disturbance state is estimated by the observer or filter, see [45, 61, 62] for the nonlinear case. In this approach, estimates of the disturbance states are present in the prediction equations, at appropriate places in the state equations, instead of the state disturbances $v(k)$ and corrections $d(k)$. This leads to offset-free control, provided the disturbance model is correctly chosen with the disturbances appropriately placed in the state equations. However, this more classical approach suffers two important deficiencies when compared to the approach presented in this chapter, namely:

- The number of modeled disturbances (dimensionality of the disturbance state vector) cannot exceed the number of measured process outputs (dimensionality of the output vector). Therefore, when dimensionality of the state vector is larger than dimensionality of the output vector (which is usually the case), the limited number of disturbances must be appropriately placed in the process model. This is an important decision of the control system designer.
- The extended state estimator or filter (ELO or EKF) has increased dimensionality, which results in increased complexity of the control system.

These deficiencies are not present in the approach presented in this chapter: the dimensionality of the vector of disturbances $v(k)$ is always equal to the dimensionality of the process state vector, thus $v(k)$ can influence independently all process state equations. It is simply added to these equations, thus there is no problem of a correct placement of a limited number of disturbances in the model equations. As the estimator or filter of only the process state is applied, the control system is easier to design and apply.

4 Computationally Efficient Nonlinear MPC Algorithms with On-Line Lineariation

Theoretically, the MPC-NO algorithm should give very good control quality because the nonlinear model is able to predict precisely the values of the process output (and state variables when necessary). Furthermore, a nonlinear optimization procedure finds the best possible sequence of manipulated variables. Unfortunately, the use of a nonlinear model means that the predicted trajectory of the output variables over the prediction horizon (8) or the predicted trajectory of state and output variables (10) are nonlinear functions of the calculated vector of the manipulated variables (3). The MPC optimization problem (4) is nonlinear, as the minimized cost-function $J(k)$ and the constraints imposed on the predicted output variables are nonlinear. While it is acceptable to use on-line nonlinear optimization in MPC for slow industrial processes, it is worth noting that MPC algorithms are more and more popular in fast embedded systems, characterized by very short sampling times [6]. In such cases, computationally more efficient alternatives are necessary.

The most straightforward approach is to find off-line a parameter-constant linear approximation of a nonlinear model for a typical operating point (a linearized model) and to next develop a classical MPC algorithm in which for prediction such a linearized model is used. As it is shown in Example 1, this approach does not work when significant changes of the operating point are required.

The following part of this work discusses computationally efficient MPC algorithms with on-line linearization. Two main approaches will be considered: MPC with model linearization and MPC with linearization of the predicted trajectory of the controlled variables. In all cases linearization makes it possible to obtain quadratic optimization problems, which for the weighting matrices ${}_p \geq 0$ and $\boldsymbol{\Lambda}_p > 0$ are strictly convex. Such problems have only one solution, which is the global one.

In all MPC algorithms with on-line linearization the vector of decision variables will be defined as the vector of increments of the manipulated variables $\Delta\mathbf{u}(k)$,

$$\Delta\mathbf{u}(k) = \begin{bmatrix} \Delta u(k|k) \\ \vdots \\ \Delta u(k+N_\mathrm{u}-1|k) \end{bmatrix}. \tag{17}$$

Despite this choice, differing from (3), the following rudimentary optimization problem of the MPC algorithms with on-line linearization is very similar to the optimization problem of the MPC-NO algorithm (Eq. (4))

$$\begin{aligned} &\min_{\Delta\mathbf{u}(k)} \left\{ J(k) = \sum_{p=1}^{N} \left\| y^{\mathrm{sp}}(k+p|k) - y(k+p|k) \right\|^2_{\boldsymbol{\Psi}_p} + \sum_{p=0}^{N_\mathrm{u}-1} \left\| \Delta u(k+p|k) \right\|^2_{\boldsymbol{\Lambda}_p} \right\} \\ &\text{subject to} \\ &u^{\min} \leq u(k+p|k) \leq u^{\max}, \quad p = 0, \ldots, N_\mathrm{u}-1, \end{aligned} \tag{18}$$

$$-\Delta u^{\max} \leq \Delta u(k+p|k) \leq \Delta u^{\max}, \quad p = 0, \ldots, N_{\mathrm{u}} - 1,$$
$$y^{\min} \leq y(k+p|k) \leq y^{\max}, \quad p = 1, \ldots, N,$$

For further transformations the vector-matrix form of the optimization problem (18) is used

$$\begin{aligned} &\min_{\Delta \mathbf{u}(k)} \left\{ J(k) = \|\mathbf{y}^{\mathrm{sp}}(k) - \mathbf{y}(k)\|^2_{\boldsymbol{\Psi}} + \|\Delta \mathbf{u}(k)\|^2_{\boldsymbol{\Lambda}} \right\} \\ &\text{subject to} \\ &\mathbf{u}^{\min} \leq \mathbf{J}\Delta \mathbf{u}(k) + \mathbf{u}(k-1) \leq \mathbf{u}^{\max}, \\ &-\Delta \mathbf{u}^{\max} \leq \Delta \mathbf{u}(k) \leq \Delta \mathbf{u}^{\max}, \\ &\mathbf{y}^{\min} \leq \mathbf{y}(k) \leq \mathbf{y}^{\max}, \end{aligned} \tag{19}$$

where the vector of the-set-point trajectory and the vector of the predicted trajectory

$$\mathbf{y}^{\mathrm{sp}}(k) = \begin{bmatrix} y^{\mathrm{sp}}(k+1|k) \\ \vdots \\ y^{\mathrm{sp}}(k+N|k) \end{bmatrix}, \ \mathbf{y}(k) = \begin{bmatrix} y(k+1|k) \\ \vdots \\ y(k+N|k) \end{bmatrix} \tag{20}$$

have both the length $n_{\mathrm{y}}N$, the vectors related to the manipulated variables

$$\mathbf{u}^{\min} = \begin{bmatrix} u^{\min} \\ \vdots \\ u^{\min} \end{bmatrix}, \ \mathbf{u}^{\max} = \begin{bmatrix} u^{\max} \\ \vdots \\ u^{\max} \end{bmatrix}, \ \Delta \mathbf{u}^{\max} = \begin{bmatrix} \Delta u^{\max} \\ \vdots \\ \Delta u^{\max} \end{bmatrix}, \ \mathbf{u}(k-1) = \begin{bmatrix} u(k-1) \\ \vdots \\ u(k-1) \end{bmatrix} \tag{21}$$

all have the length $n_{\mathrm{u}}N_{\mathrm{u}}$, the vectors concerned with the constraints imposed on the controlled variables

$$\mathbf{y}^{\min} = \begin{bmatrix} y^{\min} \\ \vdots \\ y^{\min} \end{bmatrix}, \ \mathbf{y}^{\max} = \begin{bmatrix} y^{\max} \\ \vdots \\ y^{\max} \end{bmatrix} \tag{22}$$

have both the length $n_{\mathrm{y}}N$ and the auxiliary matrix

$$\mathbf{J} = \begin{bmatrix} \mathbf{I}_{n_{\mathrm{u}} \times n_{\mathrm{u}}} & \mathbf{0}_{n_{\mathrm{u}} \times n_{\mathrm{u}}} & \mathbf{0}_{n_{\mathrm{u}} \times n_{\mathrm{u}}} & \cdots & \mathbf{0}_{n_{\mathrm{u}} \times n_{\mathrm{u}}} \\ \mathbf{I}_{n_{\mathrm{u}} \times n_{\mathrm{u}}} & \mathbf{I}_{n_{\mathrm{u}} \times n_{\mathrm{u}}} & \mathbf{0}_{n_{\mathrm{u}} \times n_{\mathrm{u}}} & \cdots & \mathbf{0}_{n_{\mathrm{u}} \times n_{\mathrm{u}}} \\ \vdots & \vdots & \vdots & \ddots & \vdots \\ \mathbf{I}_{n_{\mathrm{u}} \times n_{\mathrm{u}}} & \mathbf{I}_{n_{\mathrm{u}} \times n_{\mathrm{u}}} & \mathbf{I}_{n_{\mathrm{u}} \times n_{\mathrm{u}}} & \cdots & \mathbf{I}_{n_{\mathrm{u}} \times n_{\mathrm{u}}} \end{bmatrix} \tag{23}$$

is of dimensionality $n_{\mathrm{u}}N_{\mathrm{u}} \times n_{\mathrm{u}}N_{\mathrm{u}}$. The weighting matrices $\boldsymbol{\Psi} = \mathrm{diag}(\boldsymbol{\Psi}_1, \ldots, \boldsymbol{\Psi}_N)$ and $\boldsymbol{\Lambda} = \mathrm{diag}(\boldsymbol{\Lambda}_0, \ldots, \boldsymbol{\Lambda}_{N_{\mathrm{u}}-1})$ are of dimensionality $n_{\mathrm{y}}N \times n_{\mathrm{y}}N$ and $n_{\mathrm{u}}N_{\mathrm{u}} \times n_{\mathrm{u}}N_{\mathrm{u}}$, respectively.

4.1 MPC Algorithms Using Input-Output Models

Let us consider a general formulation of a Multiple Input Multiple Output (MIMO) nonlinear model which is a generalization of the previously discussed SISO model (5). The MIMO model is comprised of n_{y} Multiple Input Single Output (MISO) models. It has the following form

$$\begin{aligned}
y_1(k) = g_1(\mathbf{x}_1(k)) = g_1(&u_1(k-\tau^{1,1}), \ldots, u_1(k-n_{\mathrm{B}}^{1,1}), \ldots, \\
&u_{n_{\mathrm{u}}}(k-\tau^{1,n_{\mathrm{u}}}), \ldots, u_{n_{\mathrm{u}}}(k-n_{\mathrm{B}}^{1,n_{\mathrm{u}}}), \\
&y_1(k-1), \ldots, y_1(k-n_{\mathrm{A}}^{1})), \\
&\vdots \\
y_{n_{\mathrm{y}}}(k) = g_{n_{\mathrm{y}}}(\mathbf{x}_{n_{\mathrm{y}}}(k)) = g_{n_{\mathrm{y}}}(&u_1(k-\tau^{n_{\mathrm{y}},1}), \ldots, u_1(k-n_{\mathrm{B}}^{n_{\mathrm{y}},1}), \ldots, \\
&u_{n_{\mathrm{u}}}(k-\tau^{n_{\mathrm{y}},n_{\mathrm{u}}}), \ldots, u_{n_{\mathrm{u}}}(k-n_{\mathrm{B}}^{n_{\mathrm{y}},n_{\mathrm{u}}}), \\
&y_{n_{\mathrm{y}}}(k-1), \ldots, y_{n_{\mathrm{y}}}(k-n_{\mathrm{A}}^{n_{\mathrm{y}}})). \qquad (24)
\end{aligned}$$

For further transformations we assume that the functions $g_1, \ldots, g_{n_{\mathrm{y}}}$ are differentiable. The integer numbers $n_{\mathrm{A}}^{m}, n_{\mathrm{B}}^{m,n}, \tau^{m,n}$ ($m = 1, \ldots, n_{\mathrm{y}}, n = 1, \ldots, n_{\mathrm{u}}$) determine the order of model dynamics, $\tau^{m,n} \leq n_{\mathrm{B}}^{m,n}$. The vectors $\mathbf{x}_m(k), m = 1, \ldots, n_{\mathrm{y}}$, define arguments of the consecutive MISO models.

The model described by Eq. (24) may have different structures, e.g. polynomials, neural networks or fuzzy systems may be used [48]. Furthermore, in many cases cascade models are used which are comprised of linear dynamical blocks connected in series with nonlinear steady-state blocks [14, 47]. Such models are frequently used because dynamical properties of many processes are linear (or approximately linear) whereas the steady-state characteristics of sensors and/or actuators are nonlinear. As an example of cascade models let us consider the MIMO Wiener model the structure of which is depicted in Fig. 1. The linear dynamical block is described by the equation

$$\mathbf{A}(q^{-1})v(k) = \mathbf{B}(q^{-1})u(k) \qquad (25)$$

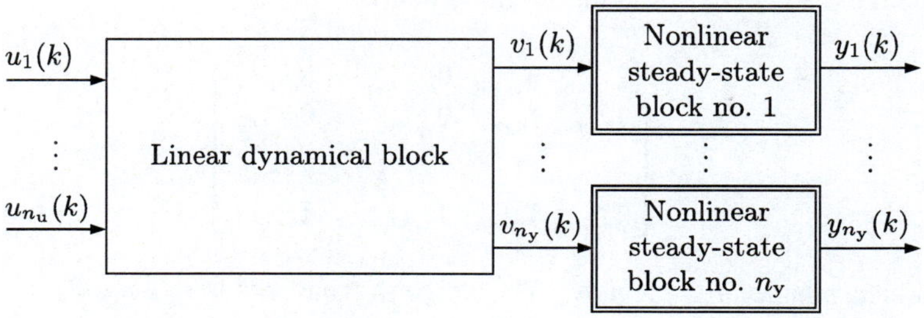

Fig. 1 The structure of the MIMO Wiener model

where q^{-1} denotes a unit discrete-time delay (equal to the sampling time) and

$$\mathbf{A}(q^{-1}) = \begin{bmatrix} 1 + a_1^1 q^{-1} + \cdots + a_{n_A}^1 q^{-n_A} & \ldots & 0 \\ \vdots & \ddots & \vdots \\ 0 & \ldots & 1 + a_1^{n_y} q^{-1} + \cdots + a_{n_A}^{n_y} q^{-n_A} \end{bmatrix},$$

$$\mathbf{B}(q^{-1}) = \begin{bmatrix} b_1^{1,1} q^{-1} + \cdots + b_{n_B}^{1,1} q^{-n_B} & \ldots & b_1^{1,n_u} q^{-1} + \cdots + b_{n_B}^{1,n_u} q^{-n_B} \\ \vdots & \ddots & \vdots \\ b_1^{n_y,1} q^{-1} + \cdots + b_{n_B}^{n_y,1} q^{-n_B} & \ldots & b_1^{n_y,n_u} q^{-1} + \cdots + b_{n_B}^{n_y,n_u} q^{-n_B} \end{bmatrix}. \quad (26)$$

The nonlinear steady-state blocks are characterized by the general equations

$$y_1(k) = g_1(v_1(k)), \quad (27)$$

$$\vdots$$

$$y_{n_y}(k) = g_{n_y}(v_{n_y}(k)). \quad (28)$$

4.1.1 MPC Algorithm with Nonlinear Prediction and Linearization (MPC-NPL)

Using the Taylor series expansion formula, a linear approximation of the nonlinear model (24) may be expressed in the general form

$$\begin{aligned} y_m(k) = g_m(\bar{\mathbf{x}}_m(k)) &+ \sum_{n=1}^{n_u} \sum_{l=1}^{n_B^{m,n}} b_l^{m,n}(\bar{\mathbf{x}}_m(k))(u_n(k-l) - \bar{u}_n(k-l)) \\ &- \sum_{l=1}^{n_A^m} a_l^m(\bar{\mathbf{x}}_m(k))(y_m(k-l) - \bar{y}_m(k-l)), \end{aligned} \quad (29)$$

for $m = 1, \ldots, n_y$. The model linearization is carried out on-line for every successive operating point of the process. The linearization point determined by the vectors $\bar{\mathbf{x}}_1(k), \ldots, \bar{\mathbf{x}}_{n_y}(k)$ depends on the manipulated variables applied to the process at previous sampling instants and on the process controlled variables measured at previous sampling instants (they are denoted as the signals $\bar{u}_n(k-l)$ and $\bar{y}_m(k-l)$). The linearized model (29) may be expressed in the following incremental form

$$\delta y_m(k) = \sum_{n=1}^{n_u} \sum_{l=1}^{n_B^{m,n}} b_l^{m,n}(k) \delta u_n(k-l) - \sum_{l=1}^{n_A^m} a_l^m(k) \delta y_m(k-l), \quad (30)$$

where $\delta y_m(k) = y_m(k) - g_m(\bar{\mathbf{x}}_m(k))$, $\delta u_n(k-l) = u_n(k-l) - \bar{u}_n(k-l)$ and $\delta y_m(k-l) = y_m(k-l) - \bar{y}_m(k-l)$, $a_l^m(k) = a_l^m(\bar{\mathbf{x}}_m(k))$, $b_l^{m,n}(k) = b_l^{m,n}(\bar{\mathbf{x}}_m(k))$. Neglecting the symbol of incrementation (for simplicity), the linearized model becomes

$$y_m(k) = \sum_{n=1}^{n_u} \sum_{l=1}^{n_B^{m,n}} b_l^{m,n}(k) u_n(k-l) - \sum_{l=1}^{n_A^m} a_l^m(k) y_m(k-l). \tag{31}$$

The time-varying coefficients of the linearized model are analytically calculated from

$$a_l^m(k) = - \left. \frac{\partial g_m(\mathbf{x}_m(k))}{\partial y_m(k-l)} \right|_{\mathbf{x}_m(k)=\bar{\mathbf{x}}_m(k)}, \tag{32}$$

for all $l = 1, \ldots, n_A$, $m = 1, \ldots, n_y$ and

$$b_l^{m,n}(k) = \left. \frac{\partial g_m(\mathbf{x}_m(k))}{\partial u_n(k-l)} \right|_{\mathbf{x}_m(k)=\bar{\mathbf{x}}_m(k)}, \tag{33}$$

for all $l = 1, \ldots, n_B$, $m = 1, \ldots, n_y$, $n = 1, \ldots, n_u$.

The linearized model (31) is used for prediction in a recurrent way, in the same way the nonlinear SISO one is used in Eq. (8). This makes it possible to calculate predicted values of the controlled variables (for $m = 1, \ldots, n_y$) over the whole prediction horizon (for $p = 1, \ldots, N$)

$$\begin{aligned}
y_m(k+1|k) = \sum_{n=1}^{n_u} (b_1^{m,n}(k) u_n(k|k) &+ b_2^{m,n}(k) u_n(k-1) \\
&+ b_3^{m,n}(k) u_n(k-2) + \cdots + b_{n_B}^{m,n}(k) u_n(k - n_B^{m,n} + 1)) \\
&- a_1^m(k) y_m(k) - a_2^m(k) y_m(k-1) \\
&- a_3^m(k) y_m(k-2) - \cdots - a_{n_A}^m(k) y_m(k - n_A^m + 1) + d_m(k), \\
y_m(k+2|k) = \sum_{n=1}^{n_u} (b_1^{m,n}(k) u_n(k+1|k) &+ b_2^{m,n}(k) u_n(k|k) \\
&+ b_3^{m,n}(k) u_n(k-1) + \cdots + b_{n_B}^{m,n}(k) u_n(k - n_B^{m,n} + 2)) \\
&- a_1^m(k) y_m(k+1|k) - a_2^m(k) y_m(k) \\
&- a_3^m(k) y_m(k-1) - \cdots - a_{n_A}^m(k) y_m(k - n_A^m + 2) + d_m(k),
\end{aligned} \tag{34}$$

$$\vdots$$

Similarly to Eq. (9), the estimation of an unmeasured disturbance acting on the m-th process output is

$$d_m(k) = y_m(k) - g_m(u_1(k - \tau^{m,1}), \ldots, u_1(k - n_{\mathrm{B}}^{m,1}), \ldots, u_{n_{\mathrm{u}}}(k - \tau^{m,n_{\mathrm{u}}}), \ldots, u_{n_{\mathrm{u}}}(k - n_{\mathrm{B}}^{m,n_{\mathrm{u}}}), y_m(k-1), \ldots, y_m(k - n_{\mathrm{A}}^m)). \tag{35}$$

Using the vector notation, the predicted signals may be expressed as functions of increments of the manipulated variables (the influence of the past process signals is not yet considered)

$$\begin{aligned} y(k+1|k) &= \mathbf{S}_1(k)\Delta u(k|k) + \ldots, \\ y(k+2|k) &= \mathbf{S}_2(k)\Delta u(k|k) + \mathbf{S}_1(k)\Delta u(k+1|k) + \ldots, \\ y(k+3|k) &= \mathbf{S}_3(k)\Delta u(k|k) + \mathbf{S}_2(k)\Delta u(k+1|k) + \mathbf{S}_1(k)\Delta u(k+2|k) + \ldots, \\ &\vdots \end{aligned} \tag{36}$$

where the step response matrices of the MIMO linearized model have dimensionality $n_{\mathrm{y}} \times n_{\mathrm{u}}$ and the following structure

$$\mathbf{S}_p(k) = \begin{bmatrix} s_p^{1,1}(k) & \ldots & s_p^{1,n_{\mathrm{u}}}(k) \\ \vdots & \ddots & \vdots \\ s_p^{n_{\mathrm{y}},1}(k) & \ldots & s_p^{n_{\mathrm{y}},n_{\mathrm{u}}}(k) \end{bmatrix}. \tag{37}$$

The scalar step response coefficients $s_p^{m,n}(k)$ are calculated recurrently using the current values of the coefficients of the linearized models (for the current discrete sampling instant k) over the whole prediction horizon ($p = 1, \ldots, N$) and for all input-output channels ($m = 1, \ldots, n_{\mathrm{y}}$, $n = 1, \ldots, n_{\mathrm{u}}$). Using Eq. (36), the vectors of predicted controlled variables over the prediction horizon (20), of length $n_{\mathrm{y}}N$, may be expressed in the compact form

$$\mathbf{y}(k) = \underbrace{\mathbf{G}(k)\Delta\mathbf{u}(k)}_{\text{future}} + \underbrace{\mathbf{y}^0(k)}_{\text{past}}. \tag{38}$$

As a result of model linearization, the predicted trajectory of the controlled variables $\mathbf{y}(k)$ is a linear function of the future increments of the manipulated variables $\Delta\mathbf{u}(k)$ (decision variables of the optimization problem calculated in the MPC). The so called dynamic matrix

$$\mathbf{G}(k) = \begin{bmatrix} \mathbf{S}_1(k) & \mathbf{0}_{n_{\mathrm{y}}\times n_{\mathrm{u}}} & \ldots & \mathbf{0}_{n_{\mathrm{y}}\times n_{\mathrm{u}}} \\ \mathbf{S}_2(k) & \mathbf{S}_1(k) & \ldots & \mathbf{0}_{n_{\mathrm{y}}\times n_{\mathrm{u}}} \\ \vdots & \vdots & \ddots & \vdots \\ \mathbf{S}_N(k) & \mathbf{S}_{N-1}(k) & \ldots & \mathbf{S}_{N-N_{\mathrm{u}}+1}(k) \end{bmatrix}, \tag{39}$$

has the dimensionality $n_{\mathrm{y}}N \times n_{\mathrm{u}}N_{\mathrm{u}}$ and it consists of step response coefficients of the linear approximation of the nonlinear model calculated at the current operating point of the process. The free trajectory vector

$$\mathbf{y}^0(k) = \begin{bmatrix} y^0(k+1|k) \\ \vdots \\ y^0(k+N|k) \end{bmatrix}, \tag{40}$$

has the length $n_{\mathrm{y}}N$ and it depends only on the past. It is calculated recurrently using the nonlinear model of the process (24) (the linearized model is not used for this purpose) assuming no influence of the future, i.e. $u(k+p|k) = u(k-1)$ for $p \geq 0$.

The suboptimal prediction formula is given by Eq. (38). The linearized model (31) is only an approximation of the rudimentary nonlinear model. It means that the predicted trajectory calculated by means of the linearized model is only an approximation of the truly nonlinear trajectory obtained when the full nonlinear model is used for prediction, as it is done in the MPC-NO scheme. On the other hand, using the suboptimal prediction equation (38), the general MPC optimization task (19) may be transformed to the following quadratic optimization problem

$$\begin{aligned} &\min_{\Delta\mathbf{u}(k)} \left\{ J(k) = \left\| \mathbf{y}^{\mathrm{sp}}(k) - \mathbf{G}(k)\Delta\mathbf{u}(k) - \mathbf{y}^0(k) \right\|^2_{\boldsymbol{\Psi}} + \|\Delta\mathbf{u}(k)\|^2_{\boldsymbol{\Lambda}} \right\} \\ &\text{subject to} \\ &\mathbf{u}^{\min} \leq \mathbf{J}\Delta\mathbf{u}(k) + \mathbf{u}(k-1) \leq \mathbf{u}^{\max}, \\ &-\Delta\mathbf{u}^{\max} \leq \Delta\mathbf{u}(k) \leq \Delta\mathbf{u}^{\max}, \\ &\mathbf{y}^{\min} \leq \mathbf{G}(k)\Delta\mathbf{u}(k) + \mathbf{y}^0(k) \leq \mathbf{y}^{\max}. \end{aligned} \tag{41}$$

4.1.2 MPC Algorithm with Successive Linearization (MPC-SL)

A time-varying linear approximation of the nonlinear model may be used in the MPC, not only to calculate step response coefficients which comprise the dynamic matrix (i.e. to describe the influence of the future), but also to find the free trajectory (i.e. to describe the influence of the past). Such an approach is known as the MPC Algorithm with Successive Linearization (MPC-SL). The quadratic optimization problems solved at each sampling instant of the MPC-SL and MPC-NPL schemes are the same (41), the only difference is in the way the free trajectory is determined. In practice, computational complexity of both algorithms is very similar since it is mainly influenced by complexity of the quadratic optimization. Hence, the MPC-NPL algorithm should be used rather than the MPC-SL one, since the former should always lead to better control quality.

The general formulation of MPC-SL and MPC-NPL algorithms is discussed in [56]. Implementation details for Multi Layer Perceptron (MLP) neural network models are given in [18, 26], details of the MPC-NPL algorithm for Radial Basis Functions

(RBF) neural network models are given in [17]. Applications of the MPC-NPL algorithm include: a low-pressure methanol-water distillation column [18, 22], an insulin administration process for diabetics [36], a yeast fermentation bioreactor [19, 26], a polymerization reactor [17, 18, 22], a combustion engine [55] and a high-pressure high-purity ethylene-ethane distillation column [26, 56].

As mentioned before, the discussed MPC algorithms are universal, the nonlinear model may have different forms. Implementation details of the MPC algorithms with on-line linearization in which Takagi-Sugeno fuzzy models are used for prediction are detailed in [36, 42, 63].

4.1.3 MPC Algorithm with Nonlinear Prediction and Simplified Linearization (MPC-NPSL), MPC Algorithm with Successive Simplified Linearization (MPC-SSL)

The discussed MPC-NPL and MPC-SL algorithms are very general, they may be used with different types of models (the only assumption is that the model (24) is differentiable). In case of cascade models (Hammerstein model, Wiener model, etc.), it is possible to directly take into account their structure in the MPC. More specifically, for model linearization, the Taylor series expansion approach is not used, but a time-varying gain of the nonlinear steady-state block for the current operating point is successively determined and used for correction of the gain of the linear dynamical block. For example, let us consider the Wiener model shown in Fig. 1. For the current operating point the output vector of the model is easily calculated as multiplication of the current gain matrix denoted by $\mathbf{K}^{\mathrm{W}}(k)$ and the vector of auxiliary variables $v(k)$, i.e.

$$y(k) = \mathbf{K}^{\mathrm{W}}(k)v(k). \tag{42}$$

Taking into account that nonlinear steady-state blocks have only one input and one output, the gain matrix is diagonal

$$\mathbf{K}^{\mathrm{W}}(k) = \begin{bmatrix} k^{\mathrm{W}}_{1,1}(k) & \dots & 0 \\ \vdots & \ddots & \vdots \\ 0 & \dots & k^{\mathrm{W}}_{n_{\mathrm{y}},n_{\mathrm{y}}}(k) \end{bmatrix} = \begin{bmatrix} \dfrac{\partial y_1(k)}{\partial v_1(k)} & \dots & 0 \\ \vdots & \ddots & \vdots \\ 0 & \dots & \dfrac{\partial y_{n_{\mathrm{y}}}(k)}{\partial v_{n_{\mathrm{y}}}(k)} \end{bmatrix}. \tag{43}$$

The partial derivatives used in the gain matrix are calculated successively on-line for the current operating point of the process, using the specific form of the nonlinear steady-state blocks. It follows from Eq. (42) that $v(k) = (\mathbf{K}^{\mathrm{W}}(k))^{-1}(k)y(k)$. Using the diference equation of the linear dynamical part of the model given by Eq. (25), we obtain

$$\mathbf{A}(q^{-1})(\mathbf{K}^{\mathrm{W}}(k))^{-1}y(k) = \mathbf{B}(q^{-1})u(k). \tag{44}$$

Since the matrix $\mathbf{K}^{\mathrm{W}}(k)$ is diagonal, we also have

$$\mathbf{A}(q^{-1})y(k) = \mathbf{K}^{\mathrm{W}}(k)\mathbf{B}(q^{-1})u(k). \tag{45}$$

Taking into account the structure of the model polynomial matrices $\mathbf{A}(q^{-1})$ and $\mathbf{B}(q^{-1})$, characterized by Eq. (26), we obtain time-varying coefficients of the linearized model

$$a_l^m(k) = a_l^m, \ l = 1, \ldots, n_{\mathrm{A}}, \ m = 1, \ldots, n_{\mathrm{y}}, \tag{46}$$

$$b_l^{m,n}(k) = k_{m,m}^{\mathrm{W}}(k)b_l^{m,n}, \ l = 1, \ldots, n_{\mathrm{B}}, \ m = 1, \ldots, n_{\mathrm{y}}, \ n = 1, \ldots, n_{\mathrm{u}}. \tag{47}$$

The time-varying step response matrices of the linearized model may be easily calculated as multiplications of the gain matrix $\mathbf{K}^{\mathrm{W}}(k)$ and constant step response coefficients of the linear dynamical blocks $\mathbf{S}_p$, i.e.

$$\mathbf{S}_p(k) = \mathbf{K}^{\mathrm{W}}(k)\mathbf{S}_p = \begin{bmatrix} k_{1,1}^{\mathrm{W}}(k)s_p^{1,1} & \ldots & k_{1,1}^{\mathrm{W}}(k)s_p^{1,n_{\mathrm{u}}} \\ \vdots & \ddots & \vdots \\ k_{n_{\mathrm{y}},n_{\mathrm{y}}}^{\mathrm{W}}(k)s_p^{n_{\mathrm{y}},1} & \ldots & k_{n_{\mathrm{y}},n_{\mathrm{y}}}^{\mathrm{W}}(k)s_p^{n_{\mathrm{y}},n_{\mathrm{y}}} \end{bmatrix}. \tag{48}$$

The full Wiener model is used in the MPC-NPSL algorithm to calculate on-line the nonlinear free trajectory whereas a successively linearized model (45) is used for this purpose in the MPC-SSL. In both cases the same quadratic optimization problem (41) is solved.

The general formulation of the MPC-NPSL and MPC-SSL algorithms is discussed in [26], implementation details for MIMO Hammerstein and Wienera models of different configurations are given, MLP neural networks are used as the nonlinear steady-state part of the model. Implementation of the MPC-NPSL algorithm for a general Hammerstein–Wiener model is presented in [28], the same algorithm for the Wiener-Hammerstein model and its application to a nonlinear heat exchanger are described in [31].

4.1.4 MPC Algorithm with Nonlinear Prediction and Linearization Along the Trajectory (MPC-NPLT)

In all the algorithms with on-line linearization discussed so far, i.e. in the MPC-NPL, MPC-SL, MPC-SSL and MPC-NPSL, a linear approximation of the nonlinear model is calculated on-line for the current operating point of the process, defined using some previous values of process signals. The linearized model is then used for prediction over the whole prediction horizon. When the operating point changes significantly and fast, such a local linearized model is likely to approximate well properties of the original nonlinear model (and properties of the controlled process) only for the beginning of the prediction horizon. When this horizon is long, differences between the calculated suboptimal trajectory and the optimal nonlinear one which could be obtained using the nonlinear model (and thus also the trajectory of the process) are likely to be important. It may results in low control quality.

The MPC-NPLT algorithm uses a more advanced linearization method. In this approach the predicted trajectory of the controlled variables ($\mathbf{y}(k)$) is not calculated recurrently from a successively linearized model, but it is determined directly. Furthermore, linearization is not performed for the current operating point, but along some assumed trajectory of the manipulated variables

$$\mathbf{u}^{\text{traj}}(k) = \begin{bmatrix} u^{\text{traj}}(k|k) \\ \vdots \\ u^{\text{traj}}(k + N_{\text{u}} - 1|k) \end{bmatrix}, \tag{49}$$

which has length $n_{\text{u}}N_{\text{u}}$. Because, typically, the prediction horizon is longer than the control one, $u^{\text{traj}}(k + p|k) = u^{\text{traj}}(k + N_{\text{u}} - 1|k)$ for $p = N_{\text{u}}, \ldots, N - 1$. Using the nonlinear model (24), it is possible to calculate for the assumed trajectory $\mathbf{u}^{\text{traj}}(k)$ the corresponding trajectory of the controlled variables

$$\mathbf{y}^{\text{traj}}(k) = \begin{bmatrix} y^{\text{traj}}(k + 1|k) \\ \vdots \\ y^{\text{traj}}(k + N|k) \end{bmatrix}, \tag{50}$$

which is the vector of length $n_{\text{y}}N$. Estimations of disturbances $d_1(k), \ldots, d_{n_{\text{y}}}(k)$ acting on the process are determined from the nonlinear model, in the same way as it is done in the MPC-NPL algorithm (Eq. (35)). The simplest choice is to define the trajectory $\mathbf{u}^{\text{traj}}(k)$ using over the whole control horizon the manipulated variables calculated and applied to the process at the previous sampling instant $k - 1$, i.e.

$$\mathbf{u}^{\text{traj}}(k) = \mathbf{u}(k - 1) = \begin{bmatrix} u(k - 1) \\ \vdots \\ u(k - 1) \end{bmatrix}. \tag{51}$$

It is also possible to perform linearization for the last $n_{\text{u}}(N_{\text{u}} - 1)$ elements of the optimal trajectory of the manipulated variables determined at the previous sampling instant

$$\mathbf{u}^{\text{traj}}(k) = \mathbf{u}(k|k - 1) = \begin{bmatrix} u(k|k - 1) \\ \vdots \\ u(k + N_{\text{u}} - 3|k - 1) \\ u(k + N_{\text{u}} - 2|k - 1) \\ u(k + N_{\text{u}} - 2|k - 1) \end{bmatrix}. \tag{52}$$

When the trajectory (51) is used for linearization, we obtain the MPC-NPLT1 algorithm, the trajectory (52) results in the MPC-NPLT2 algorithm. Using the Taylor series expansion formula, a linear approximation of the nonlinear predicted trajectory of the controlled variables $\mathbf{y}(k)$ with respect to the trajectory of the future manipulated variables $\mathbf{u}(k)$, i.e. linearization of the function $\mathbf{y}(\mathbf{u}(k)) : \mathbb{R}^{n_\mathrm{u}N_\mathrm{u}} \rightarrow \mathbb{R}^{n_\mathrm{y}N}$, may be expressed in the following form

$$\mathbf{y}(k) = \mathbf{y}^{\mathrm{traj}}(k) + \mathbf{H}(k)(\mathbf{u}(k) - \mathbf{u}^{\mathrm{traj}}(k)), \tag{53}$$

where

$$\mathbf{H}(k) = \left.\frac{\mathrm{d}\mathbf{y}(k)}{\mathrm{d}\mathbf{u}(k)}\right|_{\substack{\mathbf{y}(k) = \mathbf{y}^{\mathrm{traj}}(k) \\ \mathbf{u}(k) = \mathbf{u}^{\mathrm{traj}}(k)}} = \begin{bmatrix} \frac{\partial y^{\mathrm{traj}}(k+1|k)}{\partial u^{\mathrm{traj}}(k|k)} & \cdots & \frac{\partial y^{\mathrm{traj}}(k+1|k)}{\partial u^{\mathrm{traj}}(k+N_\mathrm{u}-1|k)} \\ \vdots & \ddots & \vdots \\ \frac{\partial y^{\mathrm{traj}}(k+N|k)}{\partial u^{\mathrm{traj}}(k|k)} & \cdots & \frac{\partial y^{\mathrm{traj}}(k+N|k)}{\partial u^{\mathrm{traj}}(k+N_\mathrm{u}-1|k)} \end{bmatrix}, \tag{54}$$

is the matrix of the derivatives of the predicted trajectory of the controlled variables $\mathbf{y}^{\mathrm{traj}}(k)$ with respect to the assumed trajectory of the manipulated variables $\mathbf{u}^{\mathrm{traj}}(k)$. The matrix $\mathbf{H}(k)$ has dimensionality $n_\mathrm{y}N \times n_\mathrm{u}N_\mathrm{u}$. Its entries are successively calculated on-line using the nonlinear model of the process (24).

Using Eq. (53), the linear approximation of the nonlinear trajectory of the controlled variables may be expressed in the following form

$$\mathbf{y}(k) = \mathbf{H}(k)\mathbf{J}\triangle\mathbf{u}(k) + \mathbf{y}^{\mathrm{traj}}(k) + \mathbf{H}(k)(\mathbf{u}(k-1) - \mathbf{u}^{\mathrm{traj}}(k)), \tag{55}$$

where the vector $\mathbf{u}(k-1)$ is defined by Eq. (21) and the matrix $\mathbf{J}$ is given by Eq. (23). As a result of linearization, the predicted trajectory (55) is a linear function of the future increments of the maniupulated variables. Therefore, the obtained prediction formula is the equivalent of the superposition rule (38) used in the MPC-NPL algorithm in which the model is linearized on-line for the current operating point.

Using the suboptimal prediction equation of the MPC-NPLT algorithm (55), the rudimentary MPC optimization problem (19) is transformed to the following quadratic optimization task

$$\begin{aligned} &\min_{\triangle\mathbf{u}(k)} \Big\{ J(k) = \big\| \mathbf{y}^{\mathrm{sp}}(k) - \mathbf{H}(k)\mathbf{J}\triangle\mathbf{u}(k) - \mathbf{y}^{\mathrm{traj}}(k) \\ &\qquad\qquad - \mathbf{H}(k)(\mathbf{u}(k-1) - \mathbf{u}^{\mathrm{traj}}(k)) \big\|_{\mathbf{\Psi}}^2 + \|\triangle\mathbf{u}(k)\|_{\mathbf{\Lambda}}^2 \Big\} \\ &\text{subject to} \\ &\mathbf{u}^{\min} \leq \mathbf{J}\triangle\mathbf{u}(k) + \mathbf{u}(k-1) \leq \mathbf{u}^{\max}, \\ &-\triangle\mathbf{u}^{\max} \leq \triangle\mathbf{u}(k) \leq \triangle\mathbf{u}^{\max}, \\ &\mathbf{y}^{\min} \leq \mathbf{H}(k)\mathbf{J}\triangle\mathbf{u}(k) + \mathbf{y}^{\mathrm{traj}}(k) + \mathbf{H}(k)(\mathbf{u}(k-1) - \mathbf{u}^{\mathrm{traj}}(k)) \leq \mathbf{y}^{\max}. \end{aligned} \tag{56}$$

4.1.5 MPC Algorithm with Nonlinear Prediction and Linearization Along the Predicted Trajectory (MPC-NPLPT)

An arbitrarily chosen trajectory of the future values of the manipulated variables over the control horizon, i.e. $\mathbf{u}^{\mathrm{traj}}(k)$, is used in the MPC-NPLT algorithm to calculate a linear approximation of the predicted trajectory of the controlled variables. Because the assumed trajectory $\mathbf{u}^{\mathrm{traj}}(k)$ is a rough estimate of the optimal control strategy calculated from the MPC-NPLT optimization task, a straightforward method to improve accuracy of this approach is to repeat at each sampling instant the following steps: linearization along a trajectory, optimization, update of the trajectory used for linearization, etc.

Let t denote the index of internal iterations ($t = 1, 2, 3, \ldots$). At the internal iteration t the trajectory of the controlled variables is linearized along the trajectory of the manipulated variables $\mathbf{u}^{t-1}(k)$ determined at the previous internal iteration $(t-1)$. For such a trajectory, using a nonlinear model of the process, we may easily calculate the trajectory of the controlled variables $\mathbf{y}^{t-1}(k)$. The initial trajectory $\mathbf{u}^0(k)$ (which is used to initialize the internal loops) may be defined by Eqs. (51) or (52).

If the linearization (using the Taylor series expansion formula) is performed along the vector of the manipulated variables $\mathbf{u}^{t-1}(k)$, then a linear approximation of the nonlinear trajectory of the predicted controlled variables $\mathbf{y}^t(k)$ with respect to the trajectory of the future manipulated variables $\mathbf{u}^t(k)$, i.e. linearization of the function $\mathbf{y}^t(\mathbf{u}^t(k)) : \mathbb{R}^{n_{\mathrm{u}}N_{\mathrm{u}}} \rightarrow \mathbb{R}^{n_{\mathrm{y}}N}$, may be expressed in the following form

$$\begin{aligned}\mathbf{y}^t(k) &= \mathbf{y}^{t-1}(k) + \mathbf{H}^t(k)(\mathbf{u}^t(k) - \mathbf{u}^{t-1}(k)) \\ &= \mathbf{H}^t(k)\mathbf{J}\Delta\mathbf{u}^t(k) + \mathbf{y}^{t-1}(k) + \mathbf{H}^t(k)(\mathbf{u}(k-1) - \mathbf{u}^{t-1}(k)).\end{aligned} \tag{57}$$

Using the suboptimal MPC-NPLPT prediction equation (57), the rudimentary MPC optimization problem (19) can be transformed to the following quadratic task

$$\begin{aligned}&\min_{\Delta\mathbf{u}^t(k)} \Big\{ J(k) = \big\| \mathbf{y}^{\mathrm{sp}}(k) - \mathbf{H}^t(k)\mathbf{J}\Delta\mathbf{u}^t(k) - \mathbf{y}^{t-1}(k) \\ &\qquad\qquad - \mathbf{H}^t(k)(\mathbf{u}(k-1) - \mathbf{u}^{t-1}(k)) \big\|^2_{\boldsymbol{\Psi}} + \big\| \Delta\mathbf{u}^t(k) \big\|^2_{\boldsymbol{\Lambda}} \Big\} \\ &\text{subject to} \\ &\mathbf{u}^{\min} \leq \mathbf{J}\Delta\mathbf{u}^t(k) + \mathbf{u}(k-1) \leq \mathbf{u}^{\max}, \\ &-\Delta\mathbf{u}^{\max} \leq \Delta\mathbf{u}^t(k) \leq \Delta\mathbf{u}^{\max}, \\ &\mathbf{y}^{\min} \leq \mathbf{H}^t(k)\mathbf{J}\Delta\mathbf{u}^t(k) + \mathbf{y}^{t-1}(k) + \mathbf{H}^t(k)(\mathbf{u}(k-1) - \mathbf{u}^{t-1}(k)) \leq \mathbf{y}^{\max}.\end{aligned} \tag{58}$$

If changes of the operating point of the process at the consecutive sampling instants are not significant, we may expect that one internal iteration could be sufficient (which leads to the MPC-NPLT algorithm). Internal iterations are continued if

$$\sum_{p=0}^{N_0}(y^{\mathrm{sp}}(k-p)-y(k-p))^2 \geq \delta_{\mathrm{y}} \tag{59}$$

where the integer number N_0 is a horizon and $\delta_{\mathrm{y}} > 0$ is a threshold parameter. If the difference between the increments of the manipulated variables calculated in two consecutive internal iterations are small, i.e. when

$$\left\|\Delta\mathbf{u}^t(k)-\Delta\mathbf{u}^{t-1}(k)\right\|^2 < \delta_{\mathrm{u}} \tag{60}$$

the internal iterations are terminated, $\delta_{\mathrm{u}} > 0$ is a threshold parameter.

The general formulation of the MPC-NPLT and MPC-NPLPT algorithms is described in [26], implementation details for MLP neural network models are also given. Additionally, algorithms' details for MIMO Hammerstein and Wiener models with different configurations of nonlinear steady-state blocks with MLP neural networks are given. Algorithms' details when Support Vector Machines (SVM) are used in the nonlinear steady-state Wiener model are given in [30]. Example applications of the MPC-NPLT and MPC-NPLPT algorithms include: a high-pressure high-purity ethylene-ethane distillation column [26], a neutralization reactor [25, 30], a polymerization [25, 26], a heat exchanger [31], a proton exchange membrane fuel cell [37], a solid oxide fuel cell [15]. In order to reduce the number of decision variables a parameterisation approach based on Laguerre functions may be used [16]. In that approach the actual increments of the manipulated variables are not calculated directly from the MPC optimization problem, but coefficients of the Laguerre functions. The number of the Laguerre functions can be much lower than the number of decision variables in MPC optimization (i.e. $n_{\mathrm{u}}N_{\mathrm{u}}$).

In addition to the discussed MPC schemes, the MPC algorithm with neural approximations have been also developed. They make it possible to further reduce computational complexity of nonlinear MPC algorithms with on-line linearization. In particular, the analytical MPC algorithms with neural approximation [26, 27] are the most important. In such approaches a specially trained neural network (an approximator) directly calculates on-line time-varying coefficients of the analytical control law. On-line model or trajectory linearization is not necessary. The calculated values of the manipulated variables may be projected onto the admissible set determined by the constraints.

Specialized dynamical models have been also developed for accurate predictions over long prediction horizons. The first approach is to use multi-models [26, 38] which may be used in the case of noise and when the proper process order and thus the order of the dynamical model is not known. The consecutive sub-models do not work recurrently, neural networks have been used for this purpose. Such a model structure significantly simplifies model identification. The second approach is to use structured models which are not used recurrently, also developed with the aim to be used for prediction in MPC [20]. Both model classes may be used in MPC algorithms with on-line linearization. In order to obtain a precise long range prediction, a model identification algorithm for training MLP neural networks has also been developed [21]. The algorithm directly takes into account the role of the model in the MPC, its objective is to minimize model error over the whole prediction horizon.

An important field of research is concerned with fault-tolerant MPC algorithms. When some of the actuators are faulty, the MPC algorithm is reconfigured in order to obtain safe process operation despite the faults. In particular, fault-tolerant versions of fuzzy MPC algorithms have been considered [41, 43]. This approach may also be integrated with the set-point optimization task [44].

Other works are concerned with statistical analysis of control quality [7, 8], in particular when MPC is mis-tuned and/or properties of the process vary in time. Additionally, dependability studies of software implementations of the MPC algorithms have been carried out [12].

Example 1 Let us consider the problem of controlling the value of pH in a neutralization reactor [13]. The base (NaOH) flow stream q_1 (ml/s) is the manipulated variable. Continuous-time differential equations are used for process simulation whereas in the MPC algorithms different discrete-time input-output models are used.

At first, a classical Linear MPC algorithm (LMPC) is developed for which the prediction with a linear model with the second order of dynamics is used. The horizons are: $N = 10$, $N_\mathrm{u} = 3$, the constraints imposed on the magnitude of the manipulated variable are: $q_1^{\min} = 0$, $q_1^{\max} = 30$. The weighting matrices are: $\boldsymbol{\Psi} = \mathbf{I}$ and $\boldsymbol{\Lambda} = \lambda\mathbf{I}$. Simulation results for different values of the penalty factor λ are depicted in Fig. 2.

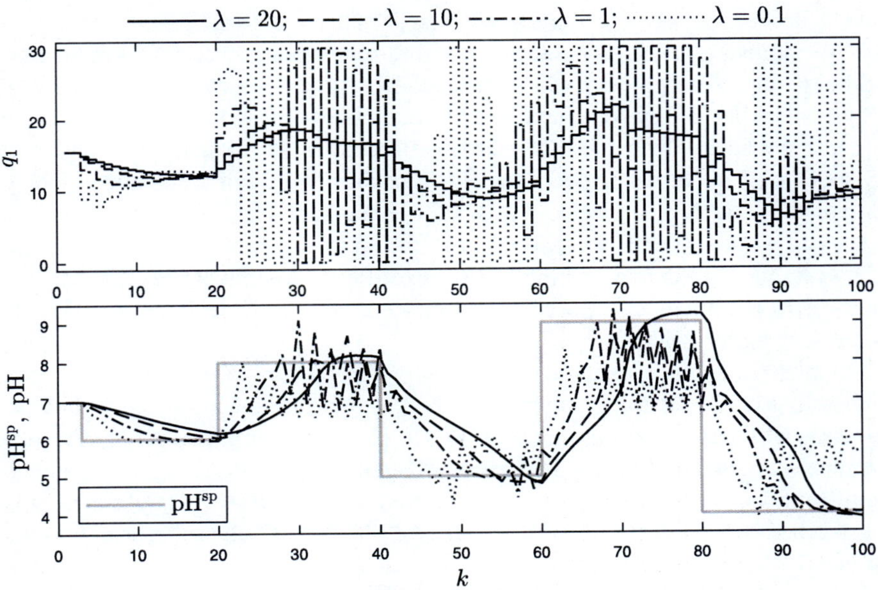

Fig. 2 Simulation results of the control system of the neutralization reactor: the process trajectories for the LMPC algorithm for different values of the penalty factor λ

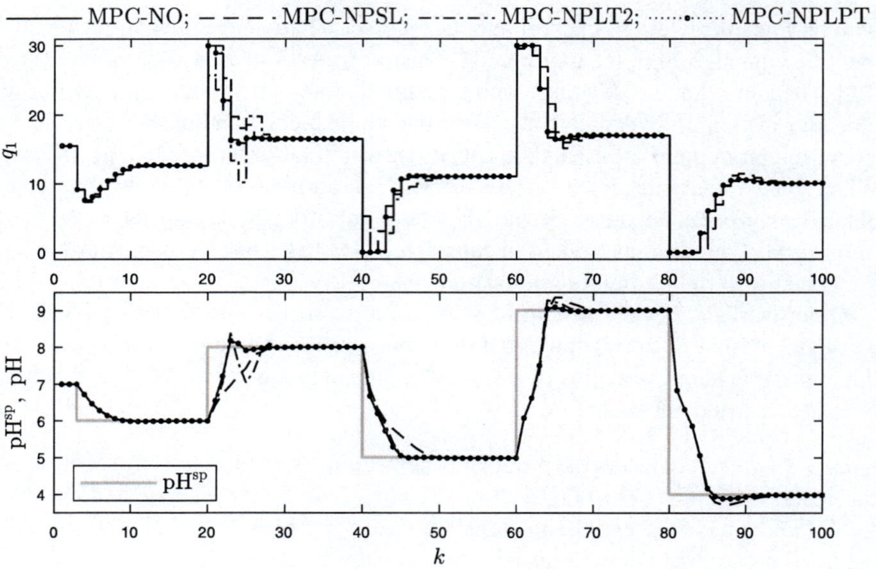

Fig. 3 Simulation results of the control system of the neutralization reactor: the process trajectories for the MPC-NO, MPC-NPSL, MPC-NPLT2 and MPC-NPLPT algorithms, $\lambda = 0.1$

Unfortunately, the LMPC algorithm does not work correctly. For $\lambda = 0.1$, $\lambda = 1$ and $\lambda = 10$ the obtained trajectories are characterized by strong oscillations, which may be eliminated when $\lambda = 20$, but in this case the process trajectories are very slow, the controlled variable never practically reaches the required set-point.

All nonlinear MPC algorithms for the neutralization reactor use, for the prediction, although in a different way, the Wiener model with the structure shown in Fig. 1. Both model blocks have one input and one output. The second-order of dynamics is used in the linear block, an MLP neural network with one hidden layer containing 5 nodes with hyperbolic tangent transfer functions is used in the nonlinear block. Simulation results of the MPC-NO, MPC-NPSL, MPC-NPLT2 and MPC-NPLPT ($\delta_u = \delta_y = 0.1$, $N_0 = 3$) algorithms are depicted in Fig. 3, in all cases the penalty factor $\lambda = 0.1$. In contrast to the LMPC algorithm, all nonlinear MPC algorithms work correctly, there are no oscillations, the controlled variable quickly reaches the time-varying set-point.

Furthermore, it is instructive to point out the link between the ways the nonlinear model is used in the MPC and the obtained quality of the control. Simplified on-line model linearization used in the MPC-NPSL algorithm gives the worst results. A much better control quality is obtained when the MPC-NPLT2 algorithm with an on-line trajectory linearization is used, although the second set-point change results in damped oscillations. The most advanced MPC-NPLPT algorithm using on-line trajectory linearization which is repeated when the process is far from the required set-point, results in process trajectories which are the same as those obtained in

Table 1 Simulation results of the control system of the neutralization reactor: the values of the performance indices E_1 and E_2 for different MPC algorithms, SII denotes the sum of internal iterations of the MPC-NPLPT algorithm

Algorithm	λ	E_1	E_2	SII
LMPC	0.1	1.8558×10^2	1.3116×10^2	–
LMPC	1	1.8865×10^2	5.6437×10^1	–
LMPC	10	2.2991×10^2	2.0226×10^1	–
LMPC	20	3.2619×10^2	8.1189×10^0	–
MPC-NPSL	0.1	1.0561×10^2	4.0473×10^0	–
MPC-NPLT1		1.0244×10^2	2.3263×10^0	–
MPC-NPLT2		1.0245×10^2	1.7022×10^0	–
MPC-NPLPT, $\delta = 10$		1.0245×10^2	1.7022×10^0	98
MPC-NPLPT, $\delta = 1$		1.0245×10^2	1.7022×10^0	102
MPC-NPLPT, $\delta = 0.1$		9.8373×10^1	8.6987×10^{-3}	122
MPC-NPLPT, $\delta = 0.01$		9.8170×10^1	2.6952×10^{-4}	135
MPC-NO		9.8178×10^1	0	–

the "ideal" MPC-NO scheme with nonlinear optimization repeated on-line at each sampling instant.

It is interesting to compare all the considered MPC algorithms in a quantitative way. For this purpose two performance indices are used. Let E_1 denote the sum of squared differences between the set-point and the actual value of the process output over the whole simulation horizon. Let E_2 denote the sum of squared differences between the process output in two cases: when the compared MPC algorithm is used and the "ideal" MPC-NO one. Table 1 compares all MPC algorithms in terms of the values of the performance indices E_1 and E_2. For the linear LMPC algorithm four values of the penalty factor λ are considered (0.1, 1, 10, 20), for all nonlinear MPC algorithms the penalty factor is 0.1, since such a small value results in fast control. The LMPC algorithm is characterized by the biggest values of the indices E_1 and E_2. The values obtained for the nonlinear MPC algorithms confirm the conclusion that the the more advanced the way the model is used, the better the quality of control. For the MPC-NPLPT algorithm, the smaller the value of the parameter $\delta = \delta_\mathrm{u} = \delta_\mathrm{y}$, the better the control quality, but, on the other hand, the higher the sum of internal iterations. For a very big value $\delta = 10$, no internal iterations are present, we obtain the rudimentary MPC-NPLT2 algorithm.

4.2 MPC Algorithms Using State-Space Models

In addition to the input-output process representation, all discussed MPC algorithms may be also developed when the controlled process is described by state-space models (6)–(7). We will shortly discuss the state-space MPC-NPL and MPC-NPLPT schemes.

4.2.1 MPC Algorithm with Nonlinear Prediction and Linearization (MPC-NPL)

Using the Taylor series expansion formula, a linear approximation of the nonlinear state-space model (6)–(7) may be expressed in the following way

$$\begin{aligned} x(k) &= f(\bar{x}(k-1), \bar{u}(k-1)) + \mathbf{A}(k)(x(k-1) - \bar{x}(k-1)), \\ &\quad + \mathbf{B}(k)(u(k-1) - \bar{u}(k-1)), \\ y(k) &= g(\bar{x}(k)) + \mathbf{C}(k)(x(k) - \bar{x}(k)), \end{aligned} \tag{61}$$

where the signals $\bar{x}(k-1)$ (measured or estimated) and $\bar{u}(k-1)$ define the current operating point of the process. Using the incremental variables, the model becomes

$$\begin{aligned} \delta x(k) &= \mathbf{A}(k)\delta x(k-1) + \mathbf{B}(k)\delta u(k-1), \\ \delta y(k) &= \mathbf{C}(k)\delta x(k), \end{aligned} \tag{62}$$

where $\delta x(k) = x(k) - \bar{x}(k) = x(k) - f(\bar{x}(k-1), \bar{u}(k-1))$, $\delta y(k) = y(k) - \bar{y}(k) = y(k) - g(\bar{x}(k))$. Neglecting the symbol of incremental variables, the linearized model becomes

$$\begin{aligned} x(k) &= \mathbf{A}(k)x(k-1) + \mathbf{B}(k)u(k-1), \\ y(k) &= \mathbf{C}(k)x(k). \end{aligned} \tag{63}$$

For the current sampling instant k the matrices of the linearized model, of dimensionality $n_{\text{x}} \times n_{\text{x}}$, $n_{\text{x}} \times n_{\text{u}}$ and $n_{\text{y}} \times n_{\text{x}}$, respectively, are calculated from the general formulae

$$\begin{aligned} \mathbf{A}(k) &= \left. \frac{\mathrm{d}f(x(k-1), u(k-1))}{\mathrm{d}x(k-1)} \right|_{\substack{x(k-1)=\bar{x}(k-1), \\ u(k-1)=\bar{u}(k-1)}} \\ \mathbf{B}(k) &= \left. \frac{\mathrm{d}f(x(k-1), u(k-1))}{\mathrm{d}u(k-1)} \right|_{\substack{x(k-1)=\bar{x}(k-1), \\ u(k-1)=\bar{u}(k-1)}} \\ \mathbf{C}(k) &= \left. \frac{\mathrm{d}g(x(k))}{\mathrm{d}x(k)} \right|_{x(k)=\bar{x}(k)}. \end{aligned} \tag{64}$$

The linearized state equation (63) is used for prediction in a recurrent way. The predicted state vector over the whole prediction horizon ($p = 1, \ldots, N$) is

$$\begin{aligned} x(k+1|k) &= \mathbf{A}(k)x(k) + \mathbf{B}(k)u(k|k) + v(k), \\ x(k+2|k) &= \mathbf{A}(k)x(k+1|k) + \mathbf{B}(k)u(k+1|k) + v(k), \\ \hat{x}(k+3|k) &= \mathbf{A}(k)x(k+2|k) + \mathbf{B}(k)u(k+2|k) + v(k), \\ &\vdots \end{aligned} \tag{65}$$

where $v(k)$ is the estimate of unmeasured state disturbances defined as in Eq. (11).

State predictions may be expressed as some functions of the increments of the manipulated variables (similarly to Eq. (36) the influence of the past is yet not considered)

$$\begin{aligned} x(k+1|k) &= \mathbf{B}(k)\Delta u(k|k) + \ldots, \\ x(k+2|k) &= (\mathbf{A}(k) + \mathbf{I})\mathbf{B}(k)\Delta u(k|k) + \mathbf{B}(k)\Delta u(k+1|k) + \ldots, \\ x(k+3|k) &= (\mathbf{A}^2(k) + \mathbf{A}(k) + \mathbf{I})\mathbf{B}(k)\Delta u(k|k) \\ &\quad + (\mathbf{A}(k) + \mathbf{I})\mathbf{B}(k)\Delta u(k+1|k) + \mathbf{B}(k)\Delta u(k+2|k) + \ldots, \\ &\vdots \end{aligned} \tag{66}$$

The predicted state vector for the whole prediction horizon

$$\mathbf{x}(k) = \begin{bmatrix} x(k+1|k) \\ \vdots \\ x(k+N|k) \end{bmatrix} \tag{67}$$

has the length $n_{\mathrm{x}}N$. Using Eq. (66), it may be expressed in the following way

$$\mathbf{x}(k) = \underbrace{\mathbf{P}(k)\Delta\mathbf{u}(k)}_{\text{future}} + \underbrace{\mathbf{x}^0(k)}_{\text{past}}, \tag{68}$$

where the matrix $\mathbf{P}(k)$ of dimensionality $n_{\mathrm{x}}N \times n_{\mathrm{u}}N_{\mathrm{u}}$ has the following structure

$$\mathbf{P}(k) = \begin{bmatrix} \mathbf{B}(k) & \ldots & \mathbf{0}_{n_{\mathrm{x}} \times n_{\mathrm{u}}} \\ (\mathbf{A}(k) + \mathbf{I})\mathbf{B}(k) & \ldots & \mathbf{0}_{n_{\mathrm{x}} \times n_{\mathrm{u}}} \\ \vdots & \ddots & \vdots \\ (\sum\limits_{i=1}^{N_{\mathrm{u}}-1} \mathbf{A}^i(k) + \mathbf{I})\mathbf{B}(k) & \ldots & \mathbf{B}(k) \\ (\sum\limits_{i=1}^{N_{\mathrm{u}}} \mathbf{A}^i(k) + \mathbf{I})\mathbf{B}(k) & \ldots & (\mathbf{A}(k) + \mathbf{I})\mathbf{B}(k) \\ \vdots & \ddots & \vdots \\ (\sum\limits_{i=1}^{N-1} \mathbf{A}^i(k) + \mathbf{I})\mathbf{B}(k) & \ldots & (\sum\limits_{i=1}^{N-N_{\mathrm{u}}} \mathbf{A}^i(k) + \mathbf{I})\mathbf{B}(k) \end{bmatrix}. \tag{69}$$

The free state trajectory vector of the length $n_x N$ is

$$\mathbf{x}^0(k) = \begin{bmatrix} x^0(k+1|k) \\ \vdots \\ x^0(k+N|k) \end{bmatrix}. \tag{70}$$

Using the linearized output equation (63), we obtain the predicted trajectory of the controlled variables

$$\mathbf{y}(k) = \underbrace{\widetilde{\mathbf{C}}(k)\mathbf{P}(k)\Delta\mathbf{u}(k)}_{\text{future}} + \underbrace{\mathbf{y}^0(k)}_{\text{past}}, \tag{71}$$

where the matrix $\widetilde{\mathbf{C}}(k) = \text{diag}(\mathbf{C}(k), \ldots, \mathbf{C}(k))$ has dimensionality $n_y N \times n_x N$. The free state trajectory $\mathbf{x}^0(k)$ and the free trajectory of the manipulated variables $\mathbf{y}^0(k) = \widetilde{\mathbf{C}}\mathbf{x}^0(k)$ are calculated recurrently using the full (not linearized) nonlinear models (6)–(7), in the same way it is done in the MPC-NO algorithm, using equations (10), but now the trajectory of the manipulated variables over the prediction horizon corresponds to the current linearization point.

Of course, the free trajectories may be calculated using the linearized model, such an approach leads to the MPC-SL algorithm in the state-space.

Using the suboptimal prediction equation of controlled variables (71), the general MPC optimization task (19) may be transformed to the following quadratic state-space MPC-NPL optimization problem

$$\begin{aligned} &\min_{\Delta\mathbf{u}(k)} \left\{ J(k) = \left\| \mathbf{y}^{\text{sp}}(k) - \widetilde{\mathbf{C}}(k)\mathbf{P}(k)\Delta\mathbf{u}(k) - \mathbf{y}^0(k) \right\|^2_{\boldsymbol{\Psi}} + \|\Delta\mathbf{u}(k)\|^2_{\boldsymbol{\Lambda}} \right\} \\ &\text{subject to} \\ &\mathbf{u}^{\min} \leq \mathbf{J}\Delta\mathbf{u}(k) + \mathbf{u}(k-1) \leq \mathbf{u}^{\max}, \\ &-\Delta\mathbf{u}^{\max} \leq \Delta\mathbf{u}(k) \leq \Delta\mathbf{u}^{\max}, \\ &\mathbf{y}^{\min} \leq \widetilde{\mathbf{C}}(k)\mathbf{P}(k)\Delta\mathbf{u}(k) + \mathbf{y}^0(k) \leq \mathbf{y}^{\max}. \end{aligned} \tag{72}$$

The obtained optimization problem is very similar to the task (41) which is used in the MPC-NPL algorithm with input-output models.

The general formulation of state-space MPC-SL and MPC-NPL schemes is discussed in [26], implementation detail for MLP neural network models are also given. As an example application of these algorithms, two processes can be mentioned: a polymerization reactor [26] and a boiler-turbine unit in a power station [32].

4.2.2 MPC Algorithm with Nonlinear Prediction and Linearization Along the Predicted Trajectory (MPC-NPLPT)

It may be proved that for a state-space model of the controlled process, the resulting MPC-NPLPT quadratic optimization problem has the same general formulation as in the input-output case, i.e. Eq. (58) [26]. Of course, the state-space model must be used for calculation of the predicted trajectory of the controlled variables $\mathbf{y}^{t-1}(k)$ which corresponds to the trajectory of the manipulated variables $\mathbf{u}^{t-1}(k)$. The general prediction equations for state and controlled variables are given by Eq. (10). A specific structure of the state-space model must also be taken into account for calculation of the derivatives of the predicted trajectory of the controlled variables with respect to the trajectory of the manipulated variables, i.e. the entries of the matrix $\mathbf{H}^t(k)$. They are determined from differentiation of the prediction equations (10). The MPC-NPLT and MPC-NPLPT algorithms for general state-space models are discussed in [26, 29], implementation details for neural network models are given in [26]. Effectiveness of the discussed MPC algorithms has been demonstrated for a simulated polymerization reactor. Recently, the MPC-NPLPT algorithm has been considered for a general class of state-space MIMO Wiener models [39]. Good control quality of the algorithm is shown for changing set-points and for unmeasured disturbances. In particular, it is demonstrated that the MPC-NPLPT algorithm with the disturbance model used in Eq. (10) gives better compensation of disturbances than the conventional offset-free MPC with state estimation [45].

Example 2 The second considered process is a polymerization reactor [9]. The flow rate F_{I} (m^3/h) of the initiator is the manipulated variable, the Number Average Molecular Weight (NAMW) (kg/kmol) of the product is the controlled variable. A set of continuous-time differential equations are used for the process simulation, discrete-time state-space representation is used in the MPC.

In all compared MPC algorithms the horizons are: $N = 10$, $N_{\mathrm{u}} = 3$, the weighting matrices are: $\mathbf{\Psi} = \mathbf{I}$, $\mathbf{\Lambda} = \lambda\mathbf{I}$, where $\lambda = 5 \times 10^{10}$, and the constraints on the manipulated variable are: $F_{\mathrm{I}}^{\mathrm{min}} = 0.003$, $F_{\mathrm{I}}^{\mathrm{max}} = 0.06$. A Kalman filter is used to estimate process state in the LMPC algorithm, an Extended Kalman Filter is used in all nonlinear MPC algorithms. The parameters of all estimators are the same: the covariance matrices are $\mathbf{Q} = 0.1\mathbf{I}_{4\times4}$ and $\mathbf{R} = 1$. In addition to the time-varying set-point, the process is affected by three unmeasured disturbances: between the sampling instants $k = 20$ and $k = 59$ the additive disturbance of the manipulated variable is -0.005, then from $k = 60$ it changes to -0.01; an unmeasured disturbance of the controlled variable changes its value from 0 to 2000 at the sample $k = 100$.

Figure 4 depicts simulation results of the LMPC algorithm in which a linear state-space model is used. In spite of the fact that the initial condition of the process is known and measurement noise is not present, the LMPC algorithm does not work due to large modeling errors. Figure 5 compares the trajectories obtained when the nonlinear MPC-NO, MPC-NPL and MPC-NPLPT algorithms are used in which a nonlinear model is used for prediction (in different ways). It is assumed that the initial condition of the process is known and the measurement noise is not present. Because the nonlinear model gives a much better prediction quality, all nonlinear algorithms work much better than the LMPC one. We may conclude that the polymerization reactor is simpler to control in comparison with the neutralization process since even the simple MPC-NPL algorithm gives the trajectories very similar to those obtained in the "ideal" MPC-NO scheme. The MPC-NPLPT algorithm gives exactly the same

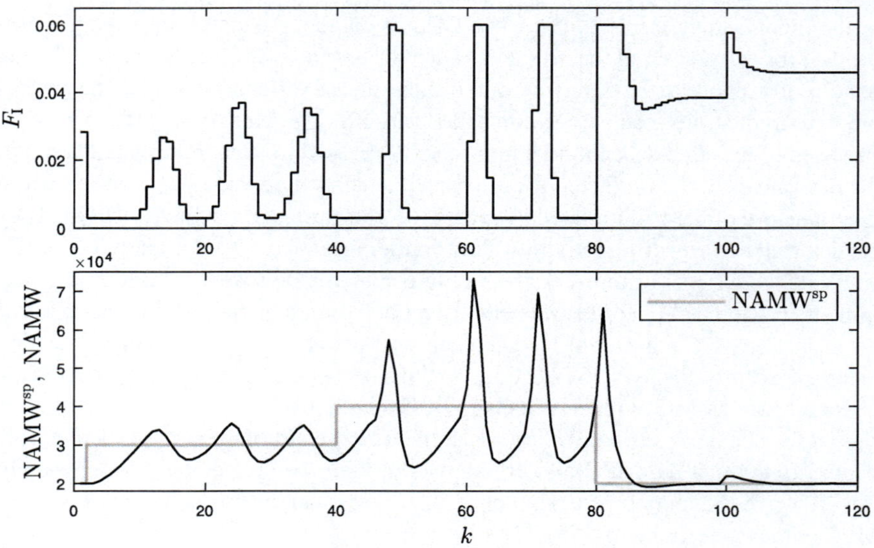

Fig. 4 Simulation results of the control system of the polymerization reactor: process trajectories for the LMPC algorithm, initial condition is known, measurement noise is not present

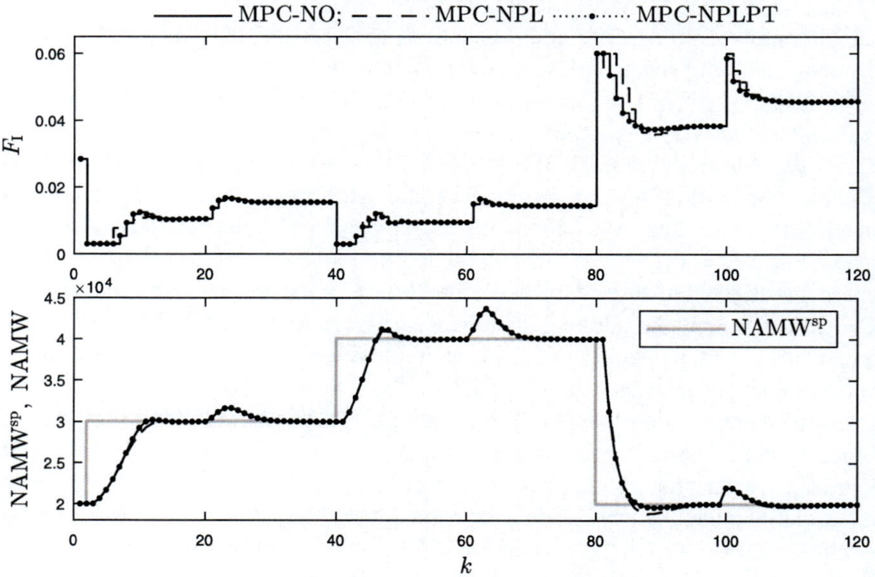

Fig. 5 Simulation results of the control system of the polymerization reactor: the process trajectories for the MPC-NO, MPC-NPL and MPC-NPLPT algorithms, the initial condition is known, measurement noise is not present

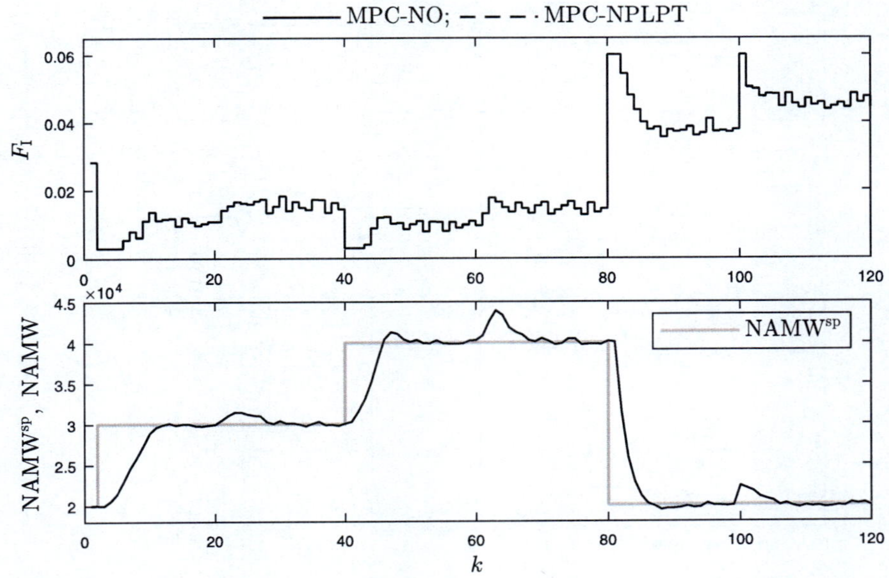

Fig. 6 Simulation results of the control system of the polymerization reactor: the process trajectories for the MPC-NO and MPC-NPLPT algorithms, the initial condition is not known, measurement of the process controlled variable is noised

results as the MPC-NO one, it has the parameters: $N_0 = 3$, $\delta_y = 100$, $\delta_u = 0.00001$. In all nonlinear MPC algorithms, the controlled variable follows changes of the set-point and compensates for the influence of the consecutive disturbance steps which occur at the sampling instants $k = 20, 60, 100$, respectively.

Figure 6 depicts simulation results of the MPC-NO and MPC-NPLPT algorithms when the initial condition is not known (the initial condition of the estimator is $\tilde{x} = [4\ 0.3\ 0.001\ 40]^T$) and measurements of the process controlled variable are noisy (the noise is with normal distribution with zero mean and standard deviation 250). Both MPC algorithms work very well, all changes of the controlled variable are followed correctly and influence of all disturbances is compensated. The trajectories of the MPC-NPLPT algorithm are the same as those obtained in the MPC-NO scheme. A comparison between real state trajectories of the process and the estimated ones, in the MPC-NPLPT algorithm, is given in Fig. 7. Because of the initial wrong condition of the estimator and due to the presence of the unmeasured disturbances, state estimation errors are inevitable. On the other hand, thanks to the fact that adequate models of state and output unmeasured disturbances are used in the MPC ($v(k)$ and $d(k)$, respectively), no steady-state error is present (offset-free control).

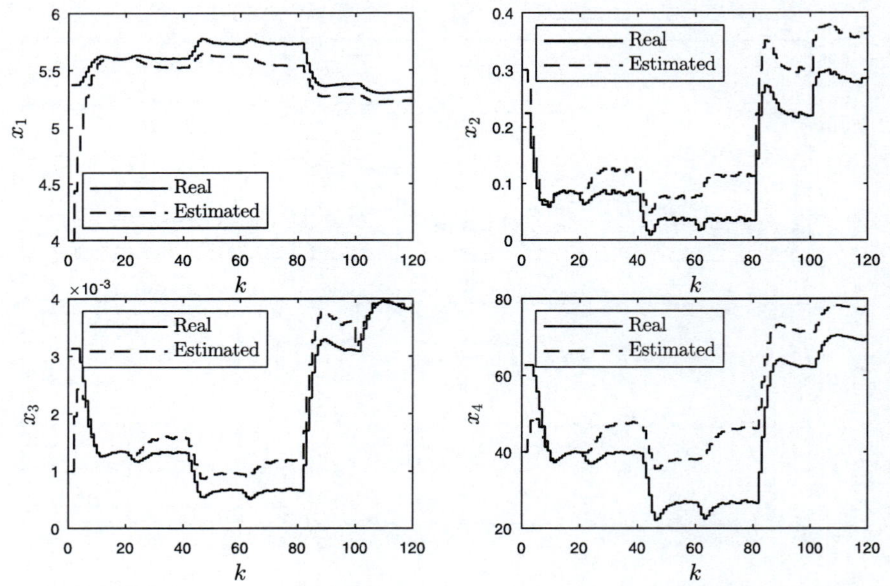

Fig. 7 Simulation results of the control system of the polymerization reactor: real versus estimated process state trajectories for the MPC-NPLPT algorithm, the initial condition is not known, measurement of the process controlled variable is noised

5 Mutual Cooperation of MPC and Set-Point Optimization

The basic goal of industrial control systems is usually of economic nature: maximization of production profit. To achieve this goal, several conditions and constraints must be taken into account, such as the constraints describing quality of the products and ecological requirements. It is also important to take into account conditions assuring safety of the production. Therefore, three main partial goals of the control system can be formulated as follows:

(a) assuring safe operation of the processes in the controlled plant, i.e., lowering to acceptable limits the probability of emergency situations and break-downs,
(b) assuring required quality of the products, i.e., assuring that the process variables do not exceed required limits (e.g., concerning product composition),
(c) the proper economic optimization of the process operation; usually profit maximization, i.e., maximization of the production volume under limitations of raw materials or utilities, or minimization of production costs assuming desired level of production, over longer horizons.

Implementing these goals in a multilayer, hierarchical control structure is common in the industry. This structure is depicted in Fig. 8, see e.g. [4, 10, 56–58, 61].

The main goal of the direct control layer is to assure safe and continuous process operation. It is the only layer having direct access to the process. The systems

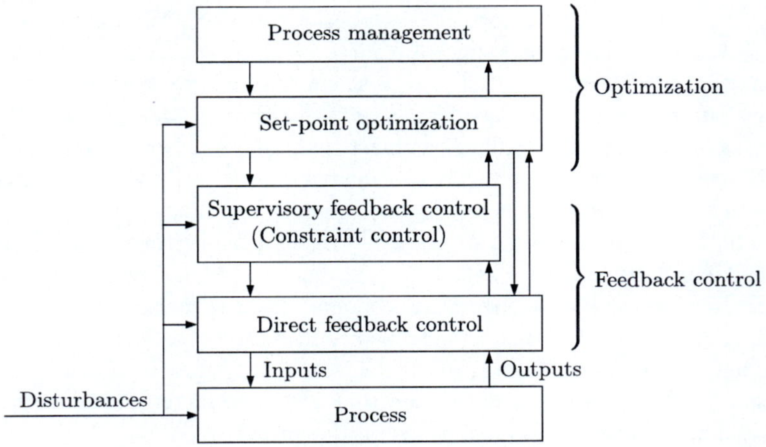

Fig. 8 Multilayer control structure

of blockades, backups and basic diagnostics are an immanent part of this layer. Algorithms of direct control should be safe, robust and relatively easy, that is why classic PID algorithms are still dominant. However, the computing power of DCS systems, modern PLCs or PCs enables more demanding solutions. Thus, in places where classic PID control leads to unsatisfactory control quality, more advanced control algorithms can be implemented. Especially appropriate are augmentations of the PID algorithm and the MPC algorithms, in fast operating realizations such as unconstrained or explicit MPCs. The sampling period of the direct control layer is the shortest, directly adjusted to the process dynamics, in particular to the fast changing variables of the process.

The supervisory control layer, also called the set-point control layer or constraint control layer, is the next one. High quality feedback control of the process variables determining the production quality is the goal of this layer. It determines on-line set-points for certain controllers of the direct control layer. Its intervention frequency is significantly lower, at least by an order of magnitude. Usually, MPC algorithms are implemented here, as high quality multivariable control is needed here, typically also under inequality constraints on manipulated variables (control signals) and often also on process variables. It should be mentioned that this layer does not always occur in the control structure. It should not be designed in cases when there is no need for set-point control (the constraint control) in the sense described above. Moreover, this layer cannot fully separate the direct control layer from the optimization layer. The set-point values for a certain part of direct controllers can be directly transmitted from the optimization layer, as it is shown in Fig. 8.

The local steady-state optimization layer (LSSO) is the third layer in the control structure [56–58, 61]. The objective of its operation is to calculate optimal set-point values for the controllers of the subordinate feedback control layers. These values usually result from the optimization of an economic objective function which defines the profit or running costs of the process operation. The optimization task to be solved is usually a static one (a mathematical programming problem). The optimal values for the set-points should be calculated for current values of the slow varying disturbances influencing optimality of the set-points. Measured or estimated properties of raw materials, utilities, ambient conditions, etc., are examples of such disturbances. Therefore, the frequency of intervention of this layer depends on the dynamics of these disturbances.

The highest layer presented in Fig. 8 is the plant management (or production planning) layer. Its task is to establish operating conditions for the optimization layer, i.e., the form and parameters of the economic objective function and constraints. This layer operates on the brink of the economic environment of the process, reacting accordingly to orders concerning assortment and amount of production, prices, sales, etc., coming directly from the market environment or the larger plant environment. The frequency of intervention of this layer is the lowest.

Taking into account the goal of maximization of production profit and the constraints, the following task is solved by the optimization layer

$$\begin{aligned} &\min_{u^{\mathrm{ss}}} \left\{ J_{\mathrm{E}} = c_{\mathrm{u}}^{\mathrm{T}} u^{\mathrm{ss}} - c_{\mathrm{y}}^{\mathrm{T}} y^{\mathrm{ss}} \right\} \\ &\text{subject to} \\ &u^{\min} \leq u^{\mathrm{ss}} \leq u^{\max}, \\ &y^{\min} \leq y^{\mathrm{ss}} \leq y^{\max}, \\ &y^{\mathrm{ss}} = f^{\mathrm{ss}}(u^{\mathrm{ss}}, h^{\mathrm{ss}}). \end{aligned} \tag{73}$$

The vectors $c_{\mathrm{u}} = \left[c_{\mathrm{u},1} \ldots c_{\mathrm{u},n_{\mathrm{u}}}\right]^{\mathrm{T}}$, $c_{\mathrm{y}} = \left[c_{\mathrm{y},1} \ldots c_{\mathrm{y},n_{\mathrm{y}}}\right]^{\mathrm{T}}$ describe prices, the index "ss" denotes the steady-state. The optimization procedure uses the static model of the process

$$y^{\mathrm{ss}} = f^{\mathrm{ss}}(u^{\mathrm{ss}}, h^{\mathrm{ss}}), \tag{74}$$

where $f^{\mathrm{ss}} : \mathbb{R}^{n_{\mathrm{u}}+n_{\mathrm{h}}} \rightarrow \mathbb{R}^{n_{\mathrm{y}}}$ and $h^{\mathrm{ss}} = h(k)$ denotes measurement (or estimation) of the disturbances influencing the location of the optimal point. Applying the model equations (74), the optimization task (73) can be reformulated to the form

$$\begin{aligned} &\min_{u^{\mathrm{ss}}} \left\{ J_{\mathrm{E}} = c_{\mathrm{u}}^{\mathrm{T}} u^{\mathrm{ss}} - c_{\mathrm{y}}^{\mathrm{T}} (f^{\mathrm{ss}}(u^{\mathrm{ss}}, h^{\mathrm{ss}})) \right\} \\ &\text{subject to} \\ &u^{\min} \leq u^{\mathrm{ss}} \leq u^{\max}, \\ &y^{\min} \leq f^{\mathrm{ss}}(u^{\mathrm{ss}}, h^{\mathrm{ss}}) \leq y^{\max}. \end{aligned} \tag{75}$$

Let us denote the vector resulting from the optimization by $u^{\mathrm{ss}}_{\mathrm{lsso}}(k)$. Using this vector and the static model (74) and substituting $u^{\mathrm{ss}} = u^{\mathrm{ss}}_{\mathrm{lsso}}(k)$, $h^{\mathrm{ss}} = h(k)$, the optimal

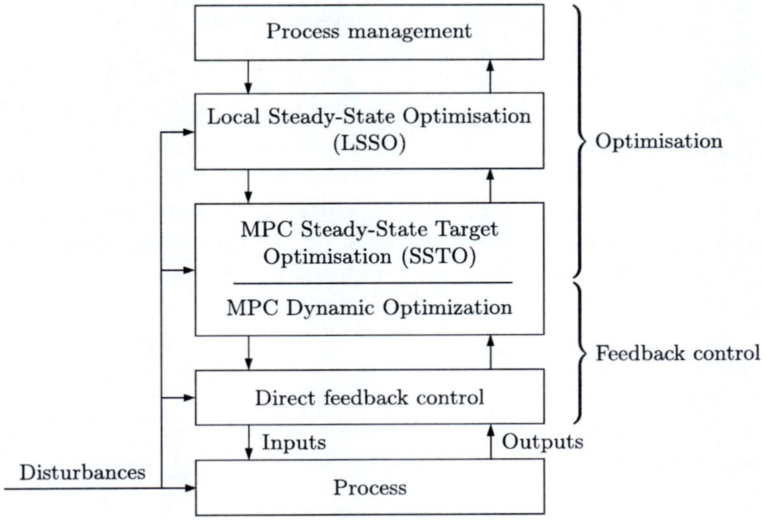

Fig. 9 Multilayer control structure with steady-state target optimization

steady-state point $y^{ss}_{lsso}(k)$ is calculated, which is then passed as the set-point to the supervisory feedback control layer (the MPC control layer).

The static model (74) is usually complex and nonlinear. The steady-state optimization problem (75) is therefore nonlinear. Thus, the algorithms solving this problem suffer from the same drawbacks as those mentioned when describing the MPC-NO algorithm: complexity of calculations, problems with local minima and lack of guarantee in finding the true solution in a prescribed time.

Classical multilayer structures with a low intervention frequency of the LSSO layer can be effective only in cases when dynamics (rates of change) of the disturbances affecting the location of the economically optimal steady-state point is slow, i.e., significantly slower than the dynamics of the feedback controlled process. If both these dynamics are comparable (similarly fast or slow), then it is obvious that the application of significantly less frequent set-point optimization may lead to economic losses. In such cases it is recommended to apply an additional, simpler and more frequent on-line set-point optimization, called the steady-state target optimization (SSTO). It is designed to directly cooperate with the MPC algorithm, i.e. it is activated at the beginning of each step (sampling interval) of the MPC [2, 35, 56–58, 61]. The control structure with the SSTO is presented in Fig. 9.

For MPC algorithms using linear process models, such as the popular DMC algorithm, the simplest way is to apply, in the SSTO, a linear static process model directly resulting from the linear dynamic model used in the MPC algorithm, i.e. the gain matrix **H** of this model. With the linear performance function taken from the LSSO problem (73), the SSTO problem then becomes a simple linear programming problem. It adjusts, at the beginning of every step of the MPC algorithm, the set-point values for the MPC dynamic optimization problem, to make these values better corresponding to actual values of the disturbances, see, e.g.,[2, 56, 61].

For nonlinear MPC, there are two ways of formulation for the local linear static process model for the SSTO problem, assuming we want to preserve the linear structure of this problem, for the sake of computation efficiency required for the application of SSTO at every MPC step. The simplest solution is to exploit the nonlinear dynamic model applied in the nonlinear MPC algorithm. Then, at every MPC step the local gain matrix $\mathbf{H}(k)$ is calculated, i.e., the gain matrix of the linearized model. The SSTO problem can be then formulated in the form

$$\begin{aligned}
&\min_{u^{\mathrm{ss}}} \left\{ J_{\mathrm{E}} = c_{\mathrm{u}}^{\mathrm{T}} u^{\mathrm{ss}} - c_{\mathrm{y}}^{\mathrm{T}} [y^0(k+N|k) + \mathbf{H}(k)(u^{\mathrm{ss}} - u(k-1))] \right\} \\
&\text{subject to} \\
&u^{\min} \leq u^{\mathrm{ss}} \leq u^{\max}, \\
&y^{\min} \leq y^0(k+N|k) + \mathbf{H}(k)(u^{\mathrm{ss}} - u(k-1)) \leq y^{\max}.
\end{aligned} \tag{76}$$

Let us denote the result of the optimization by $u^{\mathrm{ss}}(k)$. Then, the point $y^{\mathrm{ss}}(k) = y^0(k + N|k) + \mathbf{H}(k)(u^{\mathrm{ss}}(k) - u(k-1))$ is passed to the MPC as the current set-point, where $y^0(k+N|k)$ is the steady-state value of the process variables corresponding to the point $u(k-1)$, calculated using the nonlinear dynamic process model. Certainly, the prediction horizon N must be sufficiently long for $y^0(k+N|k)$ to stabilize as the steady-state value. It should be noted that the calculation of this value accommodates the influence of the unmeasured disturbances, due to the use of the current estimate of these disturbances in MPC prediction equations.

The second solution, not always possible but potentially more accurate, is to take the nonlinear static process model (74) from the LSSO and to make use of it in the SSTO problem. It means the linearization of this static model at every sample of the MPC algorithm. Using the Taylor series expansion, linear approximation of the nonlinear static model (74) can be formulated in the form

$$y^{\mathrm{ss}} = f^{\mathrm{ss}}(u^{\mathrm{ss}}, h^{\mathrm{ss}})\Big|_{\substack{u^{\mathrm{ss}} = u(k-1) \\ h^{\mathrm{ss}} = h(k)}} + \mathbf{H}(k)(u^{\mathrm{ss}} - u(k-1)), \tag{77}$$

where the current process point consists of the most actual signals $u(k-1)$ and $h(k)$. The matrix $\mathbf{H}(k)$,

$$\mathbf{H}(k) = \left.\frac{\mathrm{d}f^{\mathrm{ss}}(u^{\mathrm{ss}}, h^{\mathrm{ss}})}{\mathrm{d}u^{\mathrm{ss}}}\right|_{\substack{u^{\mathrm{ss}} = u(k-1) \\ h^{\mathrm{ss}} = h(k)}} = \frac{\mathrm{d}f^{\mathrm{ss}}(u(k-1), h(k))}{\mathrm{d}u(k-1)}, \tag{78}$$

of dimensionality $n_{\mathrm{y}} \times n_{\mathrm{u}}$, contains partial derivatives of the nonlinear function describing the static model. Using this gain matrix, the SSTO problem can be defined as the following linear programming problem

$$\begin{aligned}
&\min_{u^{\mathrm{ss}}} \left\{ J_{\mathrm{E}} = c_{\mathrm{u}}^{\mathrm{T}} u^{\mathrm{ss}} - c_{\mathrm{y}}^{\mathrm{T}} [f^{\mathrm{ss}}(u(k-1), h(k)) + \mathbf{H}(k)(u^{\mathrm{ss}} - u(k-1))] \right\} \\
&\text{subject to} \\
&u^{\min} \leq u^{\mathrm{ss}} \leq u^{\max}, \\
&y^{\min} \leq f^{\mathrm{ss}}(u(k-1), h(k)) + \mathbf{H}(k)(u^{\mathrm{ss}} - u(k-1)) \leq y^{\max}.
\end{aligned} \tag{79}$$

Denoting the optimal solution to the above problem by $u^{\mathrm{ss}}_{\mathrm{assto}}(k)$ and substituting the values $u^{\mathrm{ss}} = u^{\mathrm{ss}}_{\mathrm{assto}}(k)$ and $h^{\mathrm{ss}} = h(k)$ into the linearized model (77), we can calculate the current steady-state optimal value of the process variables $y^{\mathrm{ss}}_{\mathrm{assto}}(k)$. This value is further passed to the MPC algorithm as the current set-point value.

The multilayer control structure presented in this section was applied to several chemical processes: a polymerization reactor and Van de Vusse reactor [35], a yeast fermentation reactor [24] and a neutralization reactor [23, 26].

There is an alternative to the presented multilayer structure. This is an integrated approach, where the problems of steady-state optimization and dynamic MPC optimization are integrated into one, more complex optimization problem [33]. However, this approach may only be effective when both optimization problems are of low or, at most, moderate complexity. There is also another control system design possibility: a set-point optimizing predictive controller [34], where a simple unconstrained MPC is applied and the supervising optimization problem considers both set-point optimality and constraint satisfaction. In all these design alternatives any of the optimal or suboptimal nonlinear MPC algorithms, presented in this chapter, can be applied.

6 Conclusions

Model predictive control (MPC) algorithms using nonlinear process models are presented in this chapter. In the first part, the MPC-NO algorithm is presented, using directly the nonlinear process model, both to the prediction of the process variables (outputs), and to the optimization of the trajectories of the manipulated variables (control inputs). This implies that nonlinear optimization must be used, which is often numerically difficult and not sufficiently effective for faster on-line applications. Such an approach can be successively applied to the control of relatively slow industrial processes, when the sampling interval equals tens of seconds or several minutes [52, 56]. In the main part of this chapter, numerically more effective MPC algorithms with nonlinear prediction and successive model linearizations to get the simplified MPC optimization problem are presented. These algorithms, although suboptimal, can be successively applied to the processes with much faster dynamics and thus with shorter sampling intervals. Due to the possibility to use modern calculation platforms with high numerical potential, such as microcontrollers [6] and logical arrays [65], it is believed that these algorithms can be applied to the feedback control of fast processes requiring short sampling intervals of tens of milliseconds or even several milliseconds, e.g. robots [49], combustion engines [55] or drones [66].

References

1. Birk, J., Zeitz, M.: Extended Luenberger observer for non-linear multivariable systems. Int. J. Control **47**(6), 1823–1835 (1988)
2. Blevins, T.L., McMillan, G.K., Wojsznis, W.K., Brown, M.W.: Advanced Control Unleashed. The ISA Society, Research Triangle Park, NC (2003)
3. Blevins, T.L., Wojsznis, W.K., Nixon, M.: Advanced Control Foundation. The ISA Society, Research Triangle Park, NC (2013)
4. Brdys, M.A., Tatjewski, P.: Iterative Algorithms for Multilayer Optimizing Control. Imperial College Press/World Scientific, London/Singapore (2005)
5. Camacho, E.F., Bordons, C.: Model Predictive Control. Springer, London (1999)
6. Chaber, P., Ławryńczuk, M.: Fast analytical model predictive controllers and their implementation for STM32 ARM microcontroller. IEEE Trans. Ind. Inform. **15**, 4580–4590 (2019)
7. Domański, P., Ławryńczuk, M.: Assessment of predictive control performance using fractal measures. Nonlinear Dyn. **89**, 773–790 (2017)
8. Domański, P., Ławryńczuk, M.: Assessment of the GPC control quality using non-Gaussian statistical measures. Int. J. Math. Comput. Sci. **27**, 291–307 (2017)
9. Doyle III, F.J., Ogunnaike, B.A., Pearson, R.K.: Nonlinear model-based control using second-order Volterra models. Automatica **31**(5), 697–714 (1995)
10. Findeisen, W., Bailey, F.N., Brdyś, M., Malinowski, K., Tatjewski, P., Woźniak, A.: Control and Coordination in Hierarchical Systems. Wiley, Chichester, New York, Brisbane, Toronto (1980)
11. Garcia, C.E.: Quadratic/dynamic martix control of nonlinear processes: an application to a batch reaction process. In: Proceedings of the AIChE Annual Meeting, San Francisco (1984)
12. Gawkowski, P., Ławryńczuk, M., Marusak, P., Sosnowski, J., Tatjewski, P.: Fail-bounded implementations of the numerical model predictive control algorithms. Control Cybern. **39**, 1117–1134 (2010)
13. Gómez, J.C., Jutan, A., Baeyens, E.: Wiener model identification and predictive control of a pH neutralisation process. Proc. IEE Part D Control Theory Appl. **151**, 329–338 (2004)
14. Janczak, A.: Identification of Nonlinear Systems Using Neural Networks and Polynomial Podels: Block Oriented Approach. Lecture Notes in Control and Information Sciences, vol. 310. Springer, Berlin (2004)
15. Ławryńczuk, M.: Constrained computationally efficient nonlinear predictive control of solid oxide fuel cell: tuning, feasibility and performance. ISA Trans. (accepted)
16. Ławryńczuk, M.: Nonlinear model predictive control for processes with complex dynamics: a parameterisation approach using Laguerre functions. Int. J. Appl. Math. Comput. Sci. (accepted)
17. Ławryńczuk, M.: A computationally efficient nonlinear predictive control algorithm with RBF neural models and its application. In: Kryszkiewicz, M., Peters, J.F., Rybiński, H., Skowron, A. (eds.) The International Conference Rough Sets and Emerging Intelligent Systems Paradigms. Lecture Notes in Artificial Intelligence, vol. 4585, pp. 603–612. Springer, Berlin (2007)
18. Ławryńczuk, M.: A family of model predictive control algorithms with artificial neural networks. Int. J. Appl. Math. Comput. Sci. **17**(2), 217–232 (2007)
19. Ławryńczuk, M.: Modelling and nonlinear predictive control of a yeast fermentation biochemical reactor using neural networks. Chem. Eng. J. **145**(2), 290–307 (2008)
20. Ławryńczuk, M.: Efficient nonlinear predictive control based on structured neural models. Int. J. Appl. Math. Comput. Sci. **19**(2), 233–246 (2009)
21. Ławryńczuk, M.: Training of neural models for predictive control. Neurocomputing **73**, 1332–1343 (2010)
22. Ławryńczuk, M.: Accuracy and computational efficiency of suboptimal nonlinear predictive control based on neural models. Appl. Soft Comput. **11**(2), 2202–2215 (2011)
23. Ławryńczuk, M.: On-line set-point optimisation and predictive control using neural Hammerstein models. Chem. Eng. J. **166**, 269–287 (2011)

24. Ławryńczuk, M.: Online set-point optimisation cooperating with predictive control of a yeast fermentation process: a neural network approach. Eng. Appl. Artif. Intell. **24**, 968–982 (2011)
25. Ławryńczuk, M.: Practical nonlinear predictive control algorithms for neural Wiener models. J. Process Control **23**, 696–714 (2013)
26. Ławryńczuk, M.: Computationally Efficient Model Predictive Control Algorithms: A Neural Network Approach. Studies in Systems, Decision and Control, vol. 3. Springer, Cham (2014)
27. Ławryńczuk, M.: Explicit nonlinear predictive control algorithms with neural approximation. Neurocomputing **129**, 570–584 (2014)
28. Ławryńczuk, M.: Nonlinear predictive control for Hammerstein–Wiener systems. ISA Trans. **55**, 49–62 (2015)
29. Ławryńczuk, M.: Nonlinear state-space predictive control with on-line linearisation and state estimation. Int. J. Appl. Math. Comput. Sci. **25**(4), 833–847 (2015)
30. Ławryńczuk, M.: Modelling and predictive control of a neutralisation reactor using sparse support vector machine Wiener models. Neurocomputing **205**, 311–328 (2016)
31. Ławryńczuk, M.: Nonlinear predictive control of dynamic systems represented by Wiener–Hammerstein models. Nonlinear Dyn. **86**, 1193–1214 (2016)
32. Ławryńczuk, M.: Nonlinear predictive control of a boiler-turbine unit: a state-space approach with successive on-line model linearisation and quadratic optimisation. ISA Trans. **67**, 476–495 (2017)
33. Ławryńczuk, M., Marusak, P., Tatjewski, P.: Efficient model predictive control integrated with economic optimisation. In: Proceedings of the 15th Mediterranean Conference on Control and Automation, MED, Ateny (2007)
34. Ławryńczuk, M., Marusak, P., Tatjewski, P.: Set-point optimisation and predictive constrained control for fast feedback controlled processes. In: Proceedings of the 13th IEEE IFAC International Conference on Methods and Models in Automation and Robotics, MMAR, Międzyzdroje, pp. 357–362 (2007)
35. Ławryńczuk, M., Marusak, P., Tatjewski, P.: Cooperation of model predictive control with steady-state economic optimisation. Control Cybern. **37**, 133–158 (2008)
36. Ławryńczuk, M., Marusak, P., Tatjewski, P.: Efficient predictive control algorithms based on soft computing approaches: application to glucose concentration stabilization. In: Iskander, M., Kapila, V., Karim, M.A. (eds.) Novel Algorithms and Techniques in Telecommunications, Automation and Industrial Electronics, pp. 425–430. Springer, Berlin (2011)
37. Ławryńczuk, M., Söffker, D.: Wiener structures for modeling and nonlinear predictive control of proton exchange membrane fuel cell. Nonlinear Dyn. **95**, 1639–1660 (2019)
38. Ławryńczuk, M., Tatjewski, P.: Nonlinear predictive control based on neural multi-models. Int. J. Appl. Math. Comput. Sci. **20**(1), 7–21 (2010)
39. Ławryńczuk, M., Tatjewski, P.: Offset-free state-space nonlinear predictive control for Wiener systems. Inf. Sci. **511**, 127–151 (2020)
40. Maciejowski, J.M.: Predictive Control. Prentice Hall, Harlow, England (2002)
41. Marusak, P.: Easily reconfigurable analytical fuzzy predictive controllers: actuator faults handling. In: Kang, L., Cai, Z., Yan, X., Liu, Y. (eds.) The International Symposium on Intelligence Computation and Applications. Lecture Notes in Computer Science, vol. 5370, pp. 396–405. Springer, Berlin (2008)
42. Marusak, P.: Advantages of an easy to design fuzzy predictive algorithm in control systems of nonlinear chemical reactors. Appl. Soft Comput. **9**, 1111–1125 (2009)
43. Marusak, P.: Easily reconfigurable analytical fuzzy predictive controllers: actuator faults handling. In: Szczuka, M., Kryszkiewicz, M., Ramanna, S., Jensen, R., Hu, Q. (eds.) The International Conference on Rough Sets and Current Trends in Computing. Lecture Notes in Computer Science, vol. 6086, pages 551–559. Springer, Berlin (2010)
44. Marusak, P., Tatjewski, P.: Actuator fault tolerance in control systems with predictive constrained set-point optimizers. Int. J. Appl. Math. Comput. Sci. **18**, 539–551 (2008)
45. Morari, M., Maeder, U.: Nonlinear offset-free model predictive control. Automatica **48**, 2059–2067 (2012)

46. Muske, K.R., Badgwell, T.A.: Disturbance modeling for offset-free linear model predictive control. J. Process Control **12**, 617–632 (2002)
47. Mzyk, G.: Combined Parametric-nonparametric Identification Of Block-Oriented Systems. Lecture Notes in Control and Information Sciences, vol. 454. Springer, Berlin (2014)
48. Nelles, O.: Nonlinear System Identification. From Classical Approaches to Neural Networks and Fuzzy Models. Springer, Berlin (2001)
49. Okulski, M., Ławryńczuk, M.: Development of a model predictive controller for an unstable heavy self-balancing robot. In: Proceedings of the 23th IEEE IFAC International Conference on Methods and Models in Automation and Robotics, MMAR, Międzyzdroje, pp. 503–508 (2018)
50. Pannocchia, G., Bemporad, A.: Combined design of disturbance model and observer for offset-free model predictive control. IEEE Trans. Autom. Control **52**(6), 1048–1053 (2007)
51. Pannocchia, G., Rawlings, J.B.: Disturbance models for offset-free model predictive control. AIChE J. **49**(2), 426–437 (2003)
52. Qin, S.J., Badgwell, T.A.: A survey of industrial model predictive control technology. Control Eng. Pract. **11**, 733–764 (2003)
53. Rawlings, J.B., Mayne, D.Q.: Model Predictive Control: Theory and Design. Nob Hill Publishing, Madison (2009)
54. Rossiter, J.A.: Model-Based Predictive Control. CRC Press, Boca Raton, London, New York, Washington, D.C. (2003)
55. Saraswati, S., Chand, S.: Online linearization-based neural predictive control of air-fuel ratio in SI engines with pid feedback correction scheme. Neural Comput. Appl. **19**(6), 919–933 (2010)
56. Tatjewski, P.: Advanced Control of Industrial Processes. Springer, London (2007)
57. Tatjewski, P.: Advanced control and on-line process optimization in multilayer structures. Annu. Rev. Control **32**, 71–85 (2008)
58. Tatjewski, P.: Supervisory predictive control and on-line set-point optimization. Int. J. Appl. Math. Comput. Sci. **20**(3), 483–496 (2010)
59. Tatjewski, P.: Disturbance modeling and state estimation for offset-free predictive control with state-spaced process models. Int. J. Appl. Math. Comput. Sci. **24**(2), 313–323 (2014)
60. Tatjewski, P.: Offset-free nonlinear predictive control with measured state and unknown asymptotically constant disturbances. In: Malinowski, K., Józefczyk, J., Światek, J. (eds.) Aktualne problemy automatyki i robotyki (Actual Problems in Automation and Robotics), pp. 288–299. Akademicka Oficyna Wydawnicza Exit, Warszawa (2014)
61. Tatjewski, P.: Sterowanie zaawansowane procesów przemysłowych (Advanced Control of Industrial Processes), Second, revised edition (e-book, in Polish). EXIT Academic Publishers, Warszawa (2016)
62. Tatjewski, P.: Offset-free nonlinear model predictive control with state-spaced process models. Arch. Control Sci. **27**(4), 595–615 (2017)
63. Tatjewski, P., Ławryńczuk, M.: Soft computing in model-based predictive control. Int. J. Appl. Math. Comput. Sci. **16**(1), 7–26 (2006)
64. Wang, L.: Model Predictive Control System Design and Implementation Using MATLAB. Springer, London (2009)
65. Wojtulewicz, A., Ławryńczuk, M.: Implementation of multiple-input multiple-output dynamic matrix control algorithm for fast processes using field programmable gate array. IFAC PapersOnLine **51–6**, 324–329 (2018)
66. Yuan, Q., Zhan, J., Li, X.: Outdoor flocking of quadcopter drones with decentralized model predictive control. ISA Trans. **71**, 84–92 (2017)

Positive Linear Control Systems

Tadeusz Kaczorek

Abstract This chapter is devoted to the analysis of the positive dynamical linear continuous-time and discrete-time systems. The basic definitions and theorems concerning positivity and stability of the standard and descriptor (singular) linear systems described by the state equation and transfer matrices are given. The Kharitonov theorem is extended to positive linear systems with interval state matrices. The notions of the convex combinations of Hurwitz polynomials and Schur polynomials and of the state matrices are introduced. The considerations are illustrated by numerical examples of positive linear continuous-time and discrete-time systems.

1 Introduction

In the last 20 years we observe increasing interest in the theory and applications of the positive dynamical systems [1, 3, 8, 11–13, 16, 19, 22, 24, 26, 28–38, 40, 41, 49, 50, 56]. The dynamical system is called positive if its state variables and outputs of the system take nonnegative values for all nonnegative initial conditions and all nonnegative inputs. Variety of models having positive behavior can be found in the engineering, economics, biology medicine and social sciences.

In classical theory of linear dynamical systems the basic tool of analysis is the mathematical theory of linear operators in linear spaces. In analysis of positive systems the linear spaces should be substituted by the spaces of cones. Therefore, the analysis of positive linear systems is more complicated and less advanced.

In this chapter the basic notions and properties of positive linear continuous-time and discrete-time systems will be analyzed. Besides the standard linear systems the descriptor (singular) linear systems will be also analyzed [6, 8–15, 17, 18, 23, 27, 29, 33, 35, 39, 48, 57].

T. Kaczorek (✉)
Faculty of Electrical Engineering, Bialystok University of Technology, Wiejska 45A Street, 15-351 Białystok, Poland
e-mail: kaczorek@ee.pw.edu.pl

P. Kulczycki et al. (eds.), *Automatic Control, Robotics, and Information Processing*, Studies in Systems, Decision and Control 296,
https://doi.org/10.1007/978-3-030-48587-0_8

The considerations will be extended to positive interval linear systems, i.e. to linear systems with parameters belonging to some given intervals [37, 38]. The Kharitonov theorem will be extended to this class positive interval linear systems [20, 21, 45]. The convex linear combination of the state matrices of linear systems will be introduced and its basic properties will be analyzed.

The content of this chapter is a compendium of modern knowledge about positive dynamical linear continuous-time and discrete-time systems. The chapter is organized as follows. In Sect. 1 the basic notions of the analysis of this class of linear systems are given. Section 2 is devoted to analysis of the stability and the Kharitonov theorem is extended to the positive interval linear continuous-time systems. The stability analysis and the Kharitonov theorem for positive linear descriptor systems is given in Sect. 3. Bibliography ends the chapter.

2 Positive Linear Systems

2.1 Positive Continuous-Time Systems

Consider the continuous-time linear system

$$\dot{x}(t) = Ax(t) + Bu(t), \tag{1}$$

$$y(t) = Cx(t) + Du(t), \tag{2}$$

where $x(t) \in \Re^n$, $u(t) \in \Re^m$, $y(t) \in \Re^p$ are the state, input and output vectors and $A \in \Re^{n\times n}$, $B \in \Re^{n\times m}$, $C \in \Re^{p\times n}$, $D \in \Re^{p\times m}$. The transfer matrix of the system (1), (2) is given by

$$T(s) = C[I_n s - A]^{-1}B + D \ \in \Re^{p\times m}(s), \tag{3}$$

where $\Re^{p\times m}(s)$ is the set of rational matrices in s.

Definition 1 The system (1), (2) is called (internally) positive if $x(t) \in \Re^n_+$ and $y(t) \in \Re^p_+$, $t \geq 0$ for any initial conditions $x(0) \in \Re^n_+$ and all inputs $u(t) \in \Re^m_+$, $t \geq 0$.

Theorem 1 ([19, 23]) *The system (1), (2) is positive if and only if*

$$A \in M_n, \quad B \in \Re^{n\times m}_+, \quad C \in \Re^{p\times n}_+, \quad D \in \Re^{p\times m}_+, \qquad \square \tag{4}$$

where M_n is the set of Metzer matrixes, i.e. squire matrixes with nonnegative elements besite its diagonal, while $\Re^{n\times m}_+$ is the set of matrixes with dimension $n \times m$ and nonnegative elements.

Definition 2 The positive system (1), (2) is called asymptotically stable if

$$\lim_{t\to\infty} x(t) = 0. \tag{5}$$

Theorem 2 ([19, 23]) *The positive system* (1) *is asymptotically stable if and only if one of the equivalent conditions is satisfied:*

(1) *all coefficients of the characteristic polynomial*

$$\det[I_n s - A] = s^n + a_{n-1}s^{n-1} + \cdots + a_1 s + a_0 \tag{6}$$

are positive, i.e. $a_k > 0$ *for* $k = 0, 1, \ldots, n-1$,

(2) *all principal minors* $\bar{M}_i$ $(i = 1,\ 2,\ \ldots,\ n)$ *of the matrix* $-A$ *are positive, i.e.*

$$\bar{M}_1 = |-a_{11}| > 0,\ \bar{M}_2 = \begin{vmatrix} -a_{11} & -a_{12} \\ -a_{21} & -a_{22} \end{vmatrix} > 0, \ldots, \bar{M}_n = \det[-A] > 0, \tag{7}$$

(3) *there exists strictly positive vector* $\lambda^T = [\lambda_1,\ \lambda_2, \ldots, \lambda_n]^T$, *i.e.* $\lambda_k > 0$ *for* $k = 1, 2, \ldots, n$, *such that*

$$A\lambda < 0 \text{ or } A^T\lambda < 0. \tag{8}$$

If $\det A \neq 0$ *then we may choose* $\lambda = -A^{-1}c$, *where* $c \in \Re^n$ *is any strictly positive vector.* □

2.2 Positive Discrete-Time Linear Systems

Now let us consider the discrete-time linear system

$$x_{i+1} = Ax_i + Bu_i, \quad i \in Z_+ = \{0, 1, \ldots\}, \tag{9}$$

$$y_i = Cx_i + Du_i, \tag{10}$$

where $x_i \in \Re^n$, $u_i \in \Re^m$, $y_i \in \Re^p$ are the state, input and output vectors and $A \in \Re^{n\times n}$, $B \in \Re^{n\times m}$, $C \in \Re^{p\times n}$, $D \in \Re^{p\times m}$.

Definition 3 The system (9), (10) is called (internally) positive if $x_i \in \Re^n_+$ and $y_i \in \Re^p_+$, $i \in Z_+$ for any initial conditions $x_0 \in \Re^n_+$ and all inputs $u_i \in \Re^m_+$, $i \in Z_+$.

The transfer matrix of the system (9), (10) is given by

$$T(z) = C[Iz - A]^{-1}B \quad \in \Re^{p\times m}(z), \tag{10a}$$

where $\Re^{p\times m}(z)$ is the set of rational matrices in z.

Theorem 3 ([19, 23]) *The system (9), (10) is positive if and only if*

$$A \in \Re_{+}^{n\times n},\quad B \in \Re_{+}^{n\times m},\quad C \in \Re_{+}^{p\times n},\quad D \in \Re_{+}^{p\times m}, \qquad \square \tag{11}$$

where $\Re_{+}^{n\times n}$ is the set of matrixes with nonnegative elements.

Definition 4 The positive system (9), (10) is called asymptotically stable if

$$\lim_{i\to\infty} x_i = 0. \tag{12}$$

Theorem 4 ([19, 23]) *The positive system (9) is asymptotically stable if and only if one of the equivalent conditions is satisfied:*

(1) *all coefficients of the characteristic polynomial*

$$\det[I_n(z+1) - A] = z^n + a_{n-1}z^{n-1} + \cdots + a_1 z + a_0 \tag{13}$$

are positive, i.e. $a_k > 0$ for $k = 0, 1, \ldots, n-1$,

(2) *all principal minors of the matrix*

$$\bar{A} = I_n - A = \begin{bmatrix} \bar{a}_{11} & \ldots & \bar{a}_{1n} \\ \vdots & \ldots & \vdots \\ \bar{a}_{n1} & \ldots & \bar{a}_{nn} \end{bmatrix} \tag{14}$$

are positive, i.e.

$$|a_{11}| > 0, \begin{vmatrix} \bar{a}_{11} & \bar{a}_{12} \\ \bar{a}_{21} & \bar{a}_{22} \end{vmatrix} > 0, \ldots, \det \bar{A} > 0, \tag{15}$$

(3) *there exists strictly positive vector $\lambda^T = [\lambda_1,\ \lambda_2,\ \ldots,\ \lambda_n]^T$, i.e. $\lambda_k > 0$ for $k = 1,\ 2,\ \ldots,\ n$ such that*

$$[A - I_n]\lambda < 0. \tag{16}$$

2.3 Descriptor Positive Continuous-Time Linear System

Consider the descriptor (singular) continuous-time linear system

$$E\dot{x}(t) = Ax(t) + Bu(t), \tag{17}$$

$$y(t) = Cx(t), \tag{18}$$

where $x(t) \in \Re^n$, $u(t) \in \Re^m$, $y(t) \in \Re^p$ are the state, input and output vectors and $E, A \in \Re^{n\times n}$, $B \in \Re^{n\times m}$, $C \in \Re^{p\times n}$. It is assumed that the matrix pencil (E, A) of (17) is regular [15, 20], i.e.

$$\det[Es - A] \neq 0 \quad \text{for some} \quad s. \tag{19}$$

Definition 5 The descriptor system (17), (18) is called (internally) positive if $x(t) \in \Re^n_+$, $y(t) \in \Re^p_+$, $t \geq 0$ for every consistent nonnegative initial conditions $x(0) \in \Re^n_+$ and all inputs such that $u^{(k)}(t) = \frac{d^k u(t)}{dt^k} \in \Re^m_+$ for $t \geq 0$, $k = 0, 1, \ldots, q$, where q is the index of E.

The transfer matrix of the system (17), (18) has the form

$$T(s) = C[Es - A]^{-1}B \quad \in \Re^{p\times m}(s). \tag{20}$$

The above matrix can always be decomposed into the strictly proper matrix

$$T_{sp}(s) = C_1[I_{n_1}s - A_1]^{-1}B_1 \tag{21}$$

and the polynomial matrix

$$P(s) = D_0 + D_1 s + \cdots + D_q s^q \quad \in \Re^{p\times m}[s] \tag{22}$$

Theorem 5 The descriptor system (17), (18) is positive if and only if

$$A_1 \in M_{n_1}, \quad B_1 \in \Re^{n_1\times m}_+, \quad C_1 \in \Re^{p\times n_1}_+ \tag{23}$$

and

$$D_k \in \Re^{p\times m}_+ \quad \text{dla} \quad k = 0, 1, \ldots, q. \tag{24}$$

Proof By Theorem 3 the descriptor system with (21) is positive if and only if the conditions (23) are satisfied. Taking into account that $T(s) = T_{sp}(s) + P(s)$ the descriptor system (17), (18) is positive if and only if the conditions (23) and (24) are met. □

Example 1 Consider the descriptor system (17), (18) with

$$E = \begin{bmatrix} 0 & 2 & 0 & 12 \\ 4 & 4 & 0 & 0 \\ 4 & 12 & 0 & 8 \\ 2 & 0 & 0 & 16 \end{bmatrix}, \quad A = \begin{bmatrix} 0 & -4 & 1 & 2 \\ 1 & 2 & 0 & -4 \\ 2 & -8 & 0 & 4 \\ 0 & 2 & 2 & -4 \end{bmatrix},$$

$$B = \begin{bmatrix} 3 & 4.5 \\ 3 & 7 \\ -2 & -4.5 \\ -3 & -5 \end{bmatrix},$$

$$C = \begin{bmatrix} 3 & 2 & 1 & 0 \end{bmatrix}. \tag{25}$$

In this case the transfer matrix (20) has the form

$$T(s) = \begin{bmatrix} \frac{s^3+7s^2+16s+11}{s^2+4s+3} & \frac{2s^3+15s^2+36s+25}{s^2+4s+3} \end{bmatrix} = P(s) + T_{sp}(s) \tag{26}$$

which can be decomposed into the strictly proper part

$$T_{sp}(s) = \begin{bmatrix} \frac{s+2}{s^2+4s+3} & \frac{2s+4}{s^2+4s+3} \end{bmatrix} \tag{27}$$

and the polynomial part

$$P(s) = D_0 + D_1 s = \begin{bmatrix} s+3 & 2s+7 \end{bmatrix}. \tag{28}$$

The transfer matrix (27) represents the state space description with

$$A_1 = \begin{bmatrix} -2 & 1 \\ 1 & -2 \end{bmatrix}, \quad B_1 = \begin{bmatrix} 1 & 2 \\ 0 & 2 \end{bmatrix}, \quad C_1 = \begin{bmatrix} 1 & 0 \end{bmatrix}. \tag{29}$$

The matrices (29) satisfy the conditions (23). For the polynomial part (28) we have

$$D_0 = \begin{bmatrix} 3 & 7 \end{bmatrix}, \quad D_1 = \begin{bmatrix} 1 & 2 \end{bmatrix}. \tag{30}$$

Therefore, by Theorem 5 the descriptor system (17), (18) with (25) is positive.
Now let us consider the descriptor discrete-time linear system

$$Ex_{i+1} = Ax_i + Bu_i, \quad i \in Z_+ \tag{31}$$

$$y_i = Cx_i, \tag{32}$$

where $x_i \in \Re^n$, $u_i \in \Re^m$, $y_i \in \Re^p$ are the state, input and output vectors and $E, A \in \Re^{n\times n}$, $B \in \Re^{n\times m}$, $C \in \Re^{p\times n}$. It is assumed that the matrix pencil (E, A) of (31) is regular, i.e.

$$\det[Ez - A] \neq 0. \tag{33}$$

Definition 6 The descriptor system (31), (32) is called (internally) positive if $x_i \in \Re^n_+$ and $y_i \in \Re^p_+$, $i \in Z_+$ for every consistent nonnegative initial conditions $x_0 \in \Re^n_+$ and all inputs $u_i \in \Re^m_+$ dla $i \in Z_+$.

The transfer matrix of the system (31), (32) is given by

$$T(z) = C[Ez - A]^{-1}B \in \Re^{p\times m}(z) \tag{34}$$

and can be decomposed into the strictly proper part

$$T_{sp}(z) = C_1[I_{n_1}z - A_1]^{-1}B_1 \tag{35}$$

and the polynomial part

$$P(z) = D_0 + D_1 z + \cdots + D_q z^q \quad \in \Re^{p\times m}[z]. \tag{36}$$

Theorem 6 *The descriptor system* (31), (32) *is positive if and only if*

$$A_1 \in \Re^{n\times n}_+, \quad B_1 \in \Re^{n_1\times m}_+, \quad C_1 \in \Re^{p\times n_1}_+ \tag{37}$$

and

$$D_k \in \Re^{p\times n_1}_+ \quad \text{dla} \quad k = 0,\ 1,\ \ldots,\ q. \tag{38}$$

Proof is similar to the proof of Theorem 5. □

It is assumed that the singular matrix E has only $n_1 < n$ linearly independent columns and the matrix pencil (E, A) is regular. In this case there exist nonsingular matrices $P \in \Re^{n\times n}$ and $Q \in \Re^{n\times n}$ such that

$$PEQ = \begin{bmatrix} I_{n_1} & 0 \\ 0 & N \end{bmatrix}, \quad PAQ = \begin{bmatrix} A_1 & 0 \\ 0 & I_{n_2} \end{bmatrix}, \quad n = n_1 + n_2, \tag{39}$$

where $N \in \Re^{n_2\times n_2}$ is the nilpotent matrix such that $N^\mu = 0$, $N^{\mu-1} \neq 0$, where μ is the nilpotency index, $A_1 \in \Re^{n_1\times n_1}$ and $n_1 = \deg(\det[Es - A])$. Premultiplying the Eq. (17) by the matrix $P \in \Re^{n\times n}$ and defining new state vector

$$\begin{bmatrix} x_1(t) \\ x_2(t) \end{bmatrix} = Q^{-1}x(t), \quad x_1(t) \in \Re^{n_1}, \quad x_2(t) \in \Re^{n_2} \tag{40}$$

we obtain

$$\dot{x}_1(t) = A_1 x_1(t) + B_1 u(t), \tag{41}$$

$$N\dot{x}_2(t) = x_2(t) + B_2 u(t), \tag{42}$$

where $A_1 \in \Re^{n_1 \times n_1}$, $B_1 \in \Re^{n_1 \times m}$, $B_2 \in \Re^{n_2 \times m}$ and $\begin{bmatrix} B_1 \\ B_2 \end{bmatrix} = PB$. Note that if $Q \in \Re_+^{n \times n}$ is a monomial matrix (i.e. in each row and in each column only one entry is positive and the remaining entries are zero), then $Q^{-1} \in \Re_+^{n \times n}$, $x_1(t) \in \Re_+^{n_1}$ and $x_2(t) \in \Re_+^{n_2}$, $t \geq 0$ if $x(t) \in \Re_+^n$, $t \geq 0$. Defining $CQ = [C_1 \; C_2]$, $C_1 \in \Re_+^{p \times n_1}$, $C_2 \in \Re^{p \times n_2}$ for any $C \in \Re_+^{p \times n}$ from (18) we have

$$y(t) = C_1 x_1(t) + C_2 x_2(t). \tag{43}$$

It is easy to verify that

$$\begin{aligned} T(s) &= C[Es - A]^{-1}B = CQ[P(Es - A)Q]^{-1}PB \\ &= \begin{bmatrix} C_1 & C_2 \end{bmatrix} \begin{bmatrix} I_{n_1}s - A_1 & 0 \\ 0 & Ns - I_{n_2} \end{bmatrix}^{-1} \begin{bmatrix} B_1 \\ B_2 \end{bmatrix} \\ &= C_1[I_{n_1}s - A_1]^{-1}B_1 - C_2[I_{n_2} + Ns + \cdots + N^{\mu-1}s^{\mu-1}]B_2. \end{aligned} \tag{44}$$

From (44) we have the following theorem.

Theorem 7 *The descriptor continuous-time system* (17), (18) *is positive if and only if*

$$A_1 \in M_{n_1}, \quad B_1 \in \Re_+^{n_1 \times m}, \quad -B_2 \in \Re_+^{n_2 \times m}, \quad C_1 \in \Re_+^{p \times n_1}, \quad C_2 \in \Re_+^{p \times n_2}. \tag{45}$$

Example 2 (Continuation of Example 1). It is easy to check that the pencil (E, A) for (25) is regular since

$$\det[Es - A] = \begin{vmatrix} 0 & 2s + 4 & -1 & 12s - 2 \\ 4s - 1 & 4s - 2 & 0 & 4 \\ 4s - 2 & 12s + 8 & 0 & 8s - 4 \\ 2s & -2 & -2 & 16s + 4 \end{vmatrix} = 64s^3 + 32s^2 + 16s - 24. \tag{46}$$

In this case the matrices P and Q have the forms

$$P = \begin{bmatrix} 1 & 0 & 1 & 0 \\ 0 & 1 & 0 & 1 \\ 2 & 0 & 0 & 2 \\ 0 & 1 & 2 & 0 \end{bmatrix}, \quad Q = \begin{bmatrix} 0 & 2 & 0 & 0 \\ 0 & 0 & 0 & 2 \\ 0 & 0 & 1 & 0 \\ 1 & 0 & 0 & 0 \end{bmatrix} \tag{47}$$

and

$$PEQ = \begin{bmatrix} 1\,0\,0\,0 \\ 0\,1\,0\,0 \\ 0\,0\,0\,1 \\ 0\,0\,0\,0 \end{bmatrix}, \quad PAQ = \begin{bmatrix} -2 & 1 & 0\,0 \\ 1 & -2\,0 & 0 \\ 0 & 0 & 1\,0 \\ 0 & 0 & 0\,1 \end{bmatrix},$$

$$PB = \begin{bmatrix} 1 & 0 \\ 0 & 2 \\ 0 & -1 \\ -1 & -2 \end{bmatrix},$$

$$CQ = \begin{bmatrix} 1\,0\,1\,3 \end{bmatrix}, \tag{48}$$

And the Eqs. (41), (42) take the forms:

$$\begin{aligned} \dot{x}_1(t) &= \begin{bmatrix} -2 & 1 \\ 1 & -2 \end{bmatrix} x_1(t) + \begin{bmatrix} 1\,2 \\ 0\,2 \end{bmatrix} u(t), \\ \begin{bmatrix} 0\,1 \\ 0\,0 \end{bmatrix} \dot{x}_2(t) &= x_2(t) + \begin{bmatrix} 0 & -1 \\ -1 & -2 \end{bmatrix} u(t), \\ y(t) &= \begin{bmatrix} 1 & 0 \end{bmatrix} x_1(t) + \begin{bmatrix} 1 & 3 \end{bmatrix} x_2(t). \end{aligned} \tag{49}$$

By Theorem 7 the descriptor system (17), (18) and (25) is positive since

$$\begin{aligned} &A_1 = \begin{bmatrix} -2 & 1 \\ 1 & -2 \end{bmatrix} \in M_2, \quad B_1 = \begin{bmatrix} 1\,0 \\ 0\,2 \end{bmatrix} \in \Re_+^{2\times 2}, \quad -B_2 = \begin{bmatrix} 0\,1 \\ 1\,2 \end{bmatrix} \in \Re_+^{2\times 2}, \\ &C_1 = \begin{bmatrix} 1\,0 \end{bmatrix} \in \Re_+^{1\times 2}, \quad C_2 = \begin{bmatrix} 1\,3 \end{bmatrix} \in \Re_+^{1\times 2}. \end{aligned} \tag{50}$$

Repeating the considerations for the descriptor discrete-time system (17), (18) we obtain the following theorem.

Theorem 8 *The descriptor discrete-time system* (31), (32) *is positive if and only if*

$$A_1 \in \Re_+^{n_1\times n_1}, \quad B_1 \in \Re_+^{n_1\times m}, \quad -B_2 \in \Re_+^{n_2\times m}, \quad C_1 \in \Re_+^{p\times n_1}, \quad C_2 \in \Re_+^{p\times n_2}. \tag{51}$$

2.4 Transfer Matrices of Positive Descriptor Linear Systems

Consider the standard positive continuous-time linear system (1), (2).

Theorem 9 *If the matrix $A \in M_n$ is Hurwitz and $B \in \Re_+^{n\times m}$, $C \in \Re_+^{p\times n}$, $D \in \Re_+^{p\times m}$ of the linear positive system* (1), (2)*, then all coefficients of the transfer matrix* (3) *are positive.*

Proof First by induction with respect to n we shall show that the matrix

$$[I_n s - A]^{-1} \quad \in \Re^{n\times n}(s) \tag{52}$$

has positive coefficients.

The hypothesis is true for $n = 1$ since

$$[s + a]^{-1} = \frac{1}{s + a}, \tag{53}$$

and for $n = 2$ we have

$$[I_2 s - A_2]^{-1} = \begin{bmatrix} s + a_{11} & -a_{12} \\ -a_{21} & s + a_{22} \end{bmatrix}^{-1} = \frac{1}{s^2 + a_1 s + a_0}\begin{bmatrix} s + a_{22} & a_{12} \\ a_{21} & s + a_{11} \end{bmatrix}, \tag{54}$$

where $a_0 = a_{11}a_{22} - a_{12}a_{21} \geq 0$.

Assuming that the hypothesis is valid for $n - 1$ we shall show that it is also true for n, i.e. all coefficients of the matrix $[I_n s - A_n]^{-1}$ are positive. It is easy to check that the inverse matrix of the matrix

$$[I_n s - A_n] = \begin{bmatrix} I_{n-1}s - A_{n-1} & u_n \\ v_n & s + a_{nn} \end{bmatrix},$$
$$u_n = -\begin{bmatrix} a_{1n} \\ \vdots \\ a_{n-1,n} \end{bmatrix}, \quad v_n = -\begin{bmatrix} a_{n1} \dots a_{n,n-1} \end{bmatrix} \tag{55}$$

has the form

$$[I_n s - A_n]^{-1} = \begin{bmatrix} \bar{A}_{n-1}^{-1} + \frac{\bar{A}_{n-1}^{-1}u_n v_n \bar{A}_{n-1}^{-1}}{a_n} & -\frac{\bar{A}_{n-1}^{-1}u_n}{a_n} \\ -\frac{v_n \bar{A}_{n-1}^{-1}}{a_n} & \frac{1}{a_n} \end{bmatrix}, \tag{56}$$

where $\bar{A}_{n-1} = [I_{n-1}s - A_{n-1}]$ and

$$a_n = (s + a_{nn}) - v_n[I_{n-1}s - A_{n-1}]^{-1}u_n. \tag{57}$$

By assumption all coefficients of the matrix $[I_{n-1}s - A_{n-1}]^{-1}$ are positive. We shall show that coefficients of the matrix (56) are also positive. Taking into account that u_n and v_n have nonnegative entries, we conclude that $-\frac{[I_{n-1}s - A_{n-1}]^{-1}u_n}{a_n}$ and $-\frac{v_n[I_{n-1}s - A_{n-1}]^{-1}}{a_n}$ are column and row rational vectors with positive coefficients. By the same arguments the matrix

$$\frac{[I_{n-1}s - A_{n-1}]^{-1}u_n v_n [I_{n-1}s - A_{n-1}]^{-1}}{a_n} \tag{58}$$

has also all rational entries in s with positive coefficients and the matrix (56) has positive coefficients. Therefore, if $B \in \Re_+^{n\times m}$, $C \in \Re_+^{p\times n}$ and $D \in \Re_+^{p\times m}$ then all coefficients of the transfer matrix (3) are positive. □

Theorem 10 *If the positive descriptor system* (17), (18) *is asymptotically stable then all coefficients of its transfer matrix are positive.*

Proof By Theorem 5 the descriptor system (17), (18) is positive if the conditions (45) are satisfied. If the conditions (37) are satisfied then by Theorem 9 the transfer matrix of the strictly proper part has positive coefficients. Note that by Theorem 5 if the descriptor system is positive, then the conditions (23) and (24) are satisfied and the transfer matrix of the positive descriptor system (17), (18) has positive coefficients. □

Example 3 (Continuation of Example 2). The transfer matrix of the positive descriptor linear system (17), (18) and (25) has the form (26) and its coefficients are positive. Note that the transfer matrix (27) of its strictly proper part has also positive coefficients.

Theorem 11 *If the matrix $A \in \Re_+^{n\times n}$ of the descriptor discrete-time system* (31), (32) *is a Schur matrix, then the matrix*

$$[I_n(z+1) - A]^{-1} \quad \in \Re^{n\times n}(z) \tag{59}$$

has positive coefficients.

Proof By Theorem 4 if $A \in \Re_+^{n\times n}$ of the discrete-time system is a Schur matrix then the matrix $A - I_n \in M_n$ is the Hurwitz matrix. The remaining elements of the proof are similar as in the proof of Theorem 9. □

Example 4 The matrix

$$A = \begin{bmatrix} 0.6 & 0 & 0.2 \\ 0.1 & 0.4 & 0.2 \\ 0.2 & 0.1 & 0.5 \end{bmatrix} \tag{60}$$

of the discrete-time linear system is asymptotically stable (Schur) since the polynomial

$$\det[I_3(z+1) - A] = \begin{vmatrix} z+0.4 & 0 & -0.2 \\ -0.1 & z+0.6 & -0.2 \\ -0.2 & -0.1 & z+0.5 \end{vmatrix} = z^3 + 1.5z^2 + 0.68z + 0.082 \tag{61}$$

has positive coefficients. The inverse matrix (59) for (60) has the form

$$\det[I_3(z+1)-A]^{-1} = \begin{bmatrix} z+0.4 & 0 & -0.2 \\ -0.1 & z+0.6 & -0.2 \\ -0.2 & -0.1 & z+0.5 \end{bmatrix}^{-1}$$

$$= \frac{1}{z^3+1.5z^2+0.68z+0.082} \begin{bmatrix} z^2+1.1z+0.28 & 0.02 & 0.2z+0.12 \\ 0.1z+0.01 & z^2+0.9z+0.16 & 0.2z+0.06 \\ 0.2z+0.11 & 0.1z+0.04 & z^2+z+0.24 \end{bmatrix}. \tag{62}$$

Note that all coefficients of the matrix (62) are positive. The transfer function of the positive asymptotically stable system with (60) and

$$B = \begin{bmatrix} 1 \\ 0 \\ 0 \end{bmatrix}, \quad C = \begin{bmatrix} 1 & 0 & 2 \end{bmatrix} \tag{63}$$

has the form

$$T(z) = C[I_n z - A]^{-1} B = \frac{z^2 - 0.5z + 0.5}{z^3 - 1.5z^2 + 0.68z - 0.094}. \tag{64}$$

Note that (64) has some negative coefficients. Therefore, Theorem 9 is not true for positive discrete-time linear systems.

3 Extension of Kharitonov Theorems to Interval Linear Systems

3.1 Kharitonov Theorem and Convex Combination of Hurwitz Polynomials

Consider the set (family) of the n-th degree polynomials

$$p_n(s) = a_n s^n + a_{n-1} s^{n-1} + \cdots + a_1 s + a_0 \tag{65}$$

with the interval coefficients

$$\underline{a_i} \le a_i \le \overline{a_i}, \quad i = 0, 1, \ldots, n. \tag{66}$$

Using (65), (66) we define the following four polynomials:

$$p_{1n}(s) := \underline{a_0} + \underline{a_1} s + \overline{a_2} s^2 + \overline{a_3} s^3 + \underline{a_4} s^4 + \underline{a_5} s^5 + \cdots$$

$$
\begin{aligned}
p_{2n}(s) &:= \underline{a_0} + \overline{a_1}s + \overline{a_2}s^2 + \underline{a_3}s^3 + \underline{a_4}s^4 + \overline{a_5}s^5 + \cdots \\
p_{3n}(s) &:= \overline{a_0} + \underline{a_1}s + \underline{a_2}s^2 + \overline{a_3}s^3 + \overline{a_4}s^4 + \underline{a_5}s^5 + \cdots \\
p_{4n}(s) &:= \overline{a_0} + \overline{a_1}s + \underline{a_2}s^2 + \underline{a_3}s^3 + \overline{a_4}s^4 + \overline{a_5}s^5 + \cdots
\end{aligned} \tag{67}
$$

Kharitonov Theorem [21, 45]. The set of polynomials (65) is asymptotically stable if and only if the four polynomials (67) are asymptotically stable. □

The polynomial

$$
p(s) = s^n + \overline{a_{n-1}}s^{n-1} + \cdots + \overline{a_1}s + \overline{a_0} \tag{68}
$$

is called Hurwitz if its zeros s_i satisfy the condition $\operatorname{Re} s_i < 0$ for $i = 1, 2, \ldots, n$.

Definition 7 The polynomial

$$
p(s) = (1-k)p_1(s) + kp_2(s) \quad \text{for } k \in [0, 1] \tag{69}
$$

is called convex linear combination of the polynomials

$$
\begin{aligned}
p_1(s) &= s^n + a_{n-1}s^{n-1} + \cdots + a_1 s + a_0, \\
p_2(s) &= s^n + b_{n-1}s^{n-1} + \cdots + b_1 s + b_0.
\end{aligned} \tag{70}
$$

Theorem 12 *The convex linear combination (69) of the Hurwitz polynomials (70) of the positive linear system is also a Hurwitz polynomial.*

Proof By Theorem 2 the polynomials (70) are Hurwitz if and only if

$$
a_i > 0 \text{ and } b_i > 0 \quad \text{for } i = 1, 2, \ldots, n. \tag{71}
$$

The convex linear combination (69) of the Hurwitz polynomials (70) is a Hurwitz polynomial if and only if

$$
(1-k)a_i + kb_i > 0 \quad \text{for } k \in [0, 1] \text{ and } i = 1, 2, \ldots, n. \tag{72}
$$

Note that the conditions (72) are always satisfied if (71) holds. Therefore, the convex linear combination (69) of the Hurwitz polynomials (70) of the positive linear system is always the Hurwitz polynomial. □

Example 5 Consider the convex linear combination (69) of the Hurwitz polynomials

$$
p_1(s) = s^2 + 5s + 2, \ \ p_2(s) = s^2 + 3s + 4. \tag{73}
$$

The convex linear combination (69) of the polynomials (73) is a Hurwitz polynomial since

$$(1-k)5+3k=5-2k>0 \text{ and } (1-k)2+k4=2+2k>0 \quad \text{for } k\in[0,1]. \tag{74}$$

The above considerations for two polynomials (70) of the same order n can be extended to two polynomials of different orders [37].

Consider the set of positive interval linear continuous-time systems with the characteristic polynomials

$$p(s)=p_n s^n+p_{n-1}s^{n-1}+\cdots+p_1 s+p_0, \tag{75}$$

where the coefficients are defined on the intervals, i.e.

$$p_i\in[\underline{p_i},\overline{p_i}] \text{ (or equivalently } 0<\underline{p_i}\le p_i\le\overline{p_i}) \quad \text{for } i=0,1,\ldots,n. \tag{76}$$

Theorem 13 *The positive interval linear system with the set of characteristic polynomials (75) is asymptotically stable if and only if the conditions (76) are satisfied.*

Proof By Kharitonov theorem the set of polynomials (75), (76) is asymptotically stable if and only if the polynomials (67) are asymptotically stable. Note that the coefficients of polynomials (75) are positive if the conditions (76) are satisfied. Therefore, by Theorem 12 the positive interval linear system with the characteristic polynomials (75) is asymptotically stable if and only if the conditions (76) are satisfied. □

Example 6 Consider the positive linear system with the characteristic polynomial

$$p(s)=a_3s^3+a_2s^2a_1s+a_0 \tag{77}$$

with the interval coefficients

$$0.5\le a_3\le 2,\quad 1\le a_2\le 3,\quad 0.4\le a_1\le 1.5,\quad 0.3\le a_0\le 4. \tag{78}$$

By Theorem 13 the interval positive linear system with (77) is asymptotically stable since all coefficients of the polynomial (77) are positive.

3.2 *Extension of Kharitonov Theorem to Positive Interval Continuous-Time Linear Systems*

Consider the interval positive linear continuous-time system

$$\dot{x}=Ax \tag{79}$$

where $x=x(t)\in\Re^n$ is the state vector and the matrix $A\in M_n$ is defined by

$$A_1 \leq A \leq A_2 \text{ (or equivalently } A \in [A_1, A_2]). \tag{80}$$

Definition 8 The interval positive system (79) is called asymptotically stable if the system is asymptotically stable for all matrices $A \in M_n$ satisfying the condition (80).

By Theorem 2 the positive system (79) is asymptotically stable if there exists strictly positive vector $\lambda > 0$ such that the condition (8) is satisfied. For the following two positive linear systems

$$\dot{x}_1 = A_1 x_1, \quad A_1 \in M_n, \tag{81}$$

$$\dot{x}_2 = A_2 x_2, \quad A_2 \in M_n, \tag{82}$$

there exists a strictly positive vector $\lambda \in \Re_+^n$ such that

$$A_1 \lambda < 0 \text{ and } A_2 \lambda < 0 \tag{83}$$

if and only if the systems (81), (82) are asymptotically stable.

Example 7 Consider the positive linear continuous-time systems (81), (82) with the matrices

$$A_1 = \begin{bmatrix} -0.6 & 0.3 \\ 0.4 & -0.4 \end{bmatrix}, \quad A_2 = \begin{bmatrix} -0.6 & 0.3 \\ 0.3 & -0.4 \end{bmatrix}. \tag{84}$$

It is easy to verify that for $\lambda^T = [0.8 \;\; 1]$ we have

$$\begin{aligned} A_1 \lambda &= \begin{bmatrix} -0.6 & 0.3 \\ 0.4 & -0.4 \end{bmatrix} \begin{bmatrix} 0.8 \\ 1 \end{bmatrix} = \begin{bmatrix} -0.18 \\ -0.08 \end{bmatrix} < 0, \\ A_2 \lambda &= \begin{bmatrix} -0.6 & 0.3 \\ 0.3 & -0.4 \end{bmatrix} \begin{bmatrix} 0.8 \\ 1 \end{bmatrix} = \begin{bmatrix} -0.18 \\ -0.16 \end{bmatrix} < 0. \end{aligned} \tag{85}$$

Therefore, by the condition (8) of Theorem the positive systems are asymptotically stable.

Theorem 14 *If the matrices A_1 and A_2 of positive systems* (81), (82) *are asymptotically stable then their convex linear combination*

$$A = (1 - k)A_1 + kA_2 \quad \text{for } k \in [0, 1] \tag{86}$$

is also asymptotically stable.

Proof By condition (8) of Theorem 2 if the positive linear systems (81), (82) are asymptotically stable then there exists strictly positive vector $\lambda \in \Re_+^n$ such that

$$A_1 \lambda < 0 \text{ and } A_2 \lambda < 0. \tag{87}$$

Using (86) and (87) we obtain $A\lambda = [(1-k)A_1+kA_2]\lambda = (1-k)A_1\lambda+kA_2\lambda < 0$ for $0 \leq k \leq 1$. Therefore, if the positive linear systems (81), (82) are asymptotically stable and (87) holds, then their convex linear combination is also asymptotically stable. □

Theorem 15 *The interval positive systems (79) are asymptotically stable if and only if the positive linear systems (81), (82) are asymptotically stable.*

Proof By condition (8) of Theorem 2 if the matrices $A_1 \in M_n$, are asymptotically stable then there exists a strictly positive vector $\lambda \in \Re_+^n$ such that (87) holds. The convex linear combination (86) satisfies the condition $A\lambda < 0$ if and only if (87) holds. Therefore, the interval system (79) is asymptotically stable if and only if the positive linear system is asymptotically stable. □

Example 8 Consider the interval positive linear continuous-time system (79) with the matrices

$$A_1 = \begin{bmatrix} -2 & 1 \\ 2 & -3 \end{bmatrix}, \quad A_2 = \begin{bmatrix} -3 & 2 \\ 4 & -4 \end{bmatrix}. \tag{88}$$

Using the condition (8) of Theorem 2 we choose for A_1 the vector $\lambda_1 = [1 \quad 1]^T$ and we obtain

$$A_1\lambda_1 = \begin{bmatrix} -2 & 1 \\ 2 & -3 \end{bmatrix}\begin{bmatrix} 1 \\ 1 \end{bmatrix} = \begin{bmatrix} -1 \\ -1 \end{bmatrix} < 0; \tag{89}$$

and for A_2 with $\lambda_2 = [0.8 \quad 1]^T$:

$$A_2\lambda_2 = \begin{bmatrix} -3 & 2 \\ 4 & -4 \end{bmatrix}\begin{bmatrix} 0.8 \\ 1 \end{bmatrix} = \begin{bmatrix} -0.4 \\ -0.8 \end{bmatrix} < 0, \tag{90}$$

Therefore, the matrices (88) are Hurwitz.

Note that

$$A_1\lambda_2 = \begin{bmatrix} -2 & 1 \\ 2 & -3 \end{bmatrix}\begin{bmatrix} 0.8 \\ 1 \end{bmatrix} = \begin{bmatrix} -0.6 \\ -1.4 \end{bmatrix} < 0. \tag{91}$$

Therefore, for both matrices (88) we may choose $\lambda = \lambda_1 = \lambda_2 = [0.8 \quad 1]^T$ and by Theorem 15 the interval positive system (79) with (88) is asymptotically stable.

3.3 Extensions of Kharitonov Theorem to Positive Interval Discrete-Time Linear Systems

Consider the interval positive discrete-time linear system with the interval matrix $A \in \Re_+^{n\times n}$ defined by

$$A_1 \leq A \leq A_2 \text{ (or equivalently } A \in [A_1, A_2]). \tag{92}$$

Definition 9 The interval positive system with (92) is called asymptotically stable if the system is asymptotically stable for all matrices $A \in \Re_+^{n\times n}$ belonging to the interval $[A_1, A_2]$.

Definition 10 The matrix

$$A = (1-k)A_1 + kA_2, \tag{93}$$

where $0 \leq k \leq 1$, $A_1 \in \Re^{n\times n}$, $A_2 \in \Re^{n\times n}$ is called the convex linear combination of the matrices A_1 and A_2.

Theorem 16 *The convex linear combination (93) is asymptotically stable if and only if the matrices $A_1 \in \Re^{n\times n}$ and $A_2 \in \Re^{n\times n}$ are asymptotically stable.*

Proof If the matrices $A_1 \in \Re^{n\times n}$ and $A_2 \in \Re^{n\times n}$ are asymptotically stable then there exists strictly positive vector $\lambda \in \Re_+^n$ such that

$$A_1\lambda < \lambda \text{ and } A_2\lambda < \lambda. \tag{94}$$

In this case using (93) and (94) we obtain

$$A\lambda = [(1-k)A_1 + kA_2]\lambda = (1-k)A_1\lambda + kA_2\lambda < \lambda \quad \text{for } 0 \leq k \leq 1. \tag{95}$$

Therefore, if the matrices A_1 and A_2 are asymptotically stable, then the convex linear combination (93) is also asymptotically stable. Necessity follows immediately from the fact that k can be equal to 0 and 1. □

Theorem 17 The interval discrete-time positive system with (94) is asymptotically stable if and only if the matrices $A_1 \in \Re^{n\times n}$ and $A_2 \in \Re^{n\times n}$ are Schur matrices.

Proof By condition (16) of Theorem 4 the matrices $A_1 \in \Re^{n\times n}$ and $A_2 \in \Re^{n\times n}$ are Schur matrices if and only if there exists strictly positive vector $\lambda \in \Re_+^n$ such that (94) holds. The convex linear combination (95) satisfies the condition $A\lambda < \lambda$ if and only if (94) holds. Therefore, the interval fractional positive systems with (94) is asymptotically stable if and only if $A_1 \in \Re^{n\times n}$ and $A_2 \in \Re^{n\times n}$ are Schur matrices. □

Example 9 Consider the interval positive linear system with the matrices

$$A_1 = \begin{bmatrix} 0.3 & 0.1 \\ 0.05 & 0.4 \end{bmatrix}, \quad A_2 = \begin{bmatrix} 0.5 & 0.3 \\ 0.2 & 0.6 \end{bmatrix}. \tag{96}$$

It is easy to check that for $\lambda^T = [1 \quad 1]$ we have

$$A_1\lambda = \begin{bmatrix} 0.3 & 0.1 \\ 0.05 & 0.4 \end{bmatrix}\begin{bmatrix} 1 \\ 1 \end{bmatrix} = \begin{bmatrix} 0.4 \\ 0.45 \end{bmatrix} < \begin{bmatrix} 1 \\ 1 \end{bmatrix},$$
$$A_2\lambda = \begin{bmatrix} 0.5 & 0.3 \\ 0.2 & 0.6 \end{bmatrix}\begin{bmatrix} 1 \\ 1 \end{bmatrix} = \begin{bmatrix} 0.3 \\ 0.8 \end{bmatrix} < \begin{bmatrix} 1 \\ 1 \end{bmatrix}. \tag{97}$$

Therefore, by Theorem 17 the interval positive system with (79) with (96) is asymptotically stable.

3.4 Convex Linear Combination of Schur Polynomials and Stability of Interval Positive Linear Systems

Definition 11 The polynomial

$$p(z) = b_n z^n + b_{n-1} z^{n-1} + \cdots + b_1 z + b_0 \tag{98}$$

is called Schur polynomial if its zeros satisfy the condition

$$|z_l| < 1 \quad \text{for } l = 1, 2, \ldots, n. \tag{99}$$

Definition 12 The polynomial

$$(1-k)p_1(z) + kp_2(z) \quad \text{dla } k \in [0, 1] \tag{100}$$

is called convex linear combination of the polynomials

$$p_1(z) = b_{1,n} z^n + b_{1,n-1} z^{n-1} + \cdots + b_{1,1} z + b_{1,0}$$
$$p_2(z) = b_{2,n} z^n + b_{2,n-1} z^{n-1} + \cdots + b_{2,1} z + b_{2,0}. \tag{101}$$

Theorem 18 *The convex linear combination of the Hurwitz polynomials is also a Hurwitz polynomial.*

Proof is similar to the proof of Theorem 14. □

For positive linear systems we have the following relationship between Hurwitz and Schur polynomials.

Theorem 19 The polynomial

$$p(s) = a_n s^n + a_{n-1} s^{n-1} + \cdots + a_1 s + a_0 \tag{102}$$

is Hurwitz and the polynomial

$$p(z) = b_n z^n + b_{n-1} z^{n-1} + \cdots + b_1 z + b_0 \tag{103}$$

is Schur polynomial if and only if their coefficients a_i and b_i $i = 0, 1, \ldots, n$ are related by

$$\begin{aligned} a_0 &= b_0 + b_1 + \cdots + b_n, \\ a_1 &= b_1 + 2b_2 + \cdots + nb_n, \\ &\vdots \\ a_{n-1} &= b_{n-1} + nb_n, \\ a_n &= b_n. \end{aligned} \tag{104}$$

Proof It is well-known [21] that for positive linear discrete-time and continuous-time systems the zeros z_l of the polynomial (103) and the zeros s_l of the polynomial (102) are related by

$$z_l = s_l + 1 \quad \text{for } l = 1, 2, \ldots, n. \tag{105}$$

Substituting $z = s + 1$ into the polynomial (103) we obtain

$$\begin{aligned} &b_n(s+1)^n + b_{n-1}(s+1)^{n-1} + \cdots + b_1(s+1) + b_0 \\ &= a_n s^n + a_{n-1} s^{n-1} + \cdots + a_1 s + a_0. \end{aligned} \tag{106}$$

It is easy to verify that the coefficients a_i and b_i $(i = 0, 1, \ldots, n)$ are related by (104).

The polynomial (102) is Hurwitz if and only if $a_i > 0$ for $i = 0,\ 1,\ \ldots,\ n$ and the polynomial (103) is Schur if and only if $b_i > 0$ for $i = 0,\ 1,\ \ldots,\ n$. From (104) it follows that if $b_i > 0$, then $a_i > 0$ for $i = 0,\ 1,\ \ldots,\ n$. □

Example 10 The polynomial

$$p(z) = z^2 + 0.6z + 0.08 \tag{107}$$

of positive discrete-time linear system is Schur polynomial since its zeros are $z_1 = -0.2$ and $z_2 = -0.4$. Substituting $z = s + 1$ into (107) we obtain

$$p(s) = (s+1)^2 + 0.6(s+1) + 0.08 = s^2 + 2.6s + 1.68, \tag{108}$$

with the zeros $s_1 = -1.2$ and $s_2 = -1.4$. Therefore, the polynomial (108) is Hurwitz.

Theorem 20 *The interval positive discrete-time linear system with the characteristic polynomial* (103) *with interval coefficients* $\underline{b}_i \leq b_i \leq \overline{b}_i$, $i = 0, 1, \ldots, n$ *is asymptotically stable if and only if the lower bound* $\underline{b}_i > 0$ *of its coefficients are positive for* $i = 0, 1, \ldots, n$.

Proof From (104) it follows that $\underline{b}_i > 0$ implies $a_i > 0$ for $i = 0, 1, \ldots, n$ and the characteristic polynomial (102) is Hurwitz. By Theorem 2 the continuous-time system is asymptotically stable. Similar result we obtain for the upper bound. Therefore, the interval positive discrete-time system with (103) is asymptotically stable if the lower and upper bound of the coefficients are positive. □

Remark 1 The equalities (104) can be used to compute the lower and upper bounds of the coefficients a_i of polynomial (101) knowing the lower and upper bounds of the coefficients b_i of polynomial (102) for $i = 0, 1, \ldots, n$.

Example 11 Consider the characteristic polynomial

$$p(z) = b_2 z^2 + b_1 z + b_0 \tag{109}$$

of positive discrete-time systems with the interval coefficients

$$1 \leq b_2 \leq 3, \quad 2 \leq b_1 \leq 3, \quad 1 \leq b_0 \leq 4. \tag{110}$$

The equivalent characteristic polynomial of continuous-time system has the form

$$p(s) = b_2(s+1)^2 + b_1(s+1) + b_0 = a_2 s^2 + a_1 s + a_0 \tag{111}$$

where

$$a_2 = b_2, \; a_1 = b_1 + 2b_2, \; a_0 = b_0 + b_1 + b_2. \tag{112}$$

Therefore, the interval coefficients of characteristic polynomial of continuous-time system are

$$1 \leq a_2 \leq 3, \quad 4 \leq a_1 \leq 9, \quad 4 \leq a_0 \leq 10. \tag{113}$$

By Theorem 20 the interval positive discrete-time linear system with (109) is asymptotically stable since the lower bounds (113) are positive.

4 Descriptor Linear Systems

4.1 Positivity of Descriptor Continuous-Time Linear Systems

Consider the autonomous descriptor continuous-time linear system

$$E\dot{x} = Ax, \tag{114}$$

where $x = x(t) \in \Re^n$ is the state vector and $E, A \in \Re^{n\times n}$. It is assumed that

$$\det[Es - A] \neq 0 \quad \text{for some } s \in C \tag{115}$$

In this case the system (114) has unique solution for admissible initial conditions $x_0 = x(0) \in \Re^n$.

It is well-known [20, 21] that if (115) holds then there exists a pair of nonsingular matrices $P, Q \in \Re^{n\times n}$ such that

$$P[Es - A]Q = \begin{bmatrix} I_{n_1}s - A_1 & 0 \\ 0 & Ns - I_{n_2} \end{bmatrix}, A_1 \in \Re^{n_1\times n_1}, N \in \Re^{n_2\times n_2}, \tag{116}$$

where $n_1 = \deg(\det[Es - A])$ and N is the nilpotent matrix, i.e. $N^\mu = 0$, $N^{\mu-1} \neq 0$ (μ is the nilpotency index).

To simplify the considerations it is assumed that the matrix N has only one block. The nonsingular matrices P and Q can be found for example by the use of elementary row and column operations [20, 21]:

(1) multiplication of any i-th row (column) by the number $c \neq 0$; this operation will be denoted by $L[i \times c]$ ($R[i \times c]$);
(2) addition to any i-th row (column) of the j-th row (column) multiplied by any number $c \neq 0$; this operation will be denoted by $L[i + j \times c]$ ($R[i + j \times c]$);
(3) interchange of any two rows (columns); this operation will be denoted by $L[i, j]$ ($R[i, j]$).

Definition 13 The descriptor system (114) is called (internally) positive if $x(t) \in \Re^n_+$, $t \geq 0$ for all admissible nonnegative initial conditions $x(0) \in \Re^n_+$.

Definition 14 A real matrix $A = [a_{ij}] \in \Re^{n\times n}$ is called Metzler matrix if its off-diagonal entries are nonnegative, i.e. $a_{ij} \geq 0$ for $i \neq j$. The set of $n \times n$ Metzler matrices will be denoted by M_n.

Theorem 21 *The descriptor system* (114) *is positive if and only if the matrix E has only linearly independent columns and the matrix* $A_1 \in M_{n_1}$.

Proof Knowing $n_1 = \deg(\det[Es - A])$ and rank(E) we may find the nilpotency index $\mu = rank(E) - n_1 + 1$ of the matrix N. Using column permutation of E we choose its n_1 linearly independent columns as its first columns. Next using elementary row operations we transform the matrix E to the form $\begin{bmatrix} I_{n_1} & 0 \\ 0 & N \end{bmatrix}$ and the matrix A to the form $\begin{bmatrix} A_1 & 0 \\ 0 & I_{n_2} \end{bmatrix}$.

The system (114) has been decomposed into two independent subsystems:

$$\dot{x}_1 = Ax_1, \;\; x_1 \in \Re^{n_1} \tag{117}$$

and

$$N\dot{x}_2 = x_2, \;\; x_2 \in \Re^{n_2}, \tag{118}$$

where

$$Q^{-1}x = \begin{bmatrix} x_1 \\ x_2 \end{bmatrix}, \tag{119}$$

and Q, Q^{-1} are permutation matrices.

It is well-known [20, 21] that the solution $x_1 = e^{A_1 t}x_1(0)$ of (117) is not negative if and only if $A_1 \in M_{n_1}$ and the solution x_2 of (118) is zero for $t > 0$. □

Definition 15 The positive system (117) is called asymptotically stable if

$$\lim_{t \to \infty} x_1(t) = 0 \;\; \text{for all admissible } x_1(0) \in \Re_+^{n_1}. \tag{120}$$

Theorem 22 ([23, 40]) *The positive system* (117) *is asymptotically stable if and only if one of the equivalent conditions is satisfied:*

(1) *all coefficients of the polynomial*

$$\det[I_{n_1}s - A_1] = s^{n_1} + a_{n_1-1}s^{n_1-1} + \cdots + a_1 s + a_0 \tag{121}$$

are positive, i.e. $a_k > 0$ *for* $k = 0, 1, \ldots, n_1 - 1$;

(2) *all principal minors* $\bar{M}_i$ $(i = 1, 2, \ldots, n_1)$ *of the matrix* $-A_1$ *are positive, i.e.*

$$\bar{M}_1 = |-a_{11}| > 0, \bar{M}_2 = \begin{vmatrix} -a_{11} & -a_{12} \\ -a_{21} & -a_{22} \end{vmatrix} > 0, \ldots, \bar{M}_{n_1} = \det[-A_1] > 0; \tag{122}$$

(3) *there exists a strictly positive vector* $\lambda = [\lambda_1, \; \lambda_2, \; \cdots, \; \lambda_{n_1}]^T$, *i.e.* $\lambda_k > 0$ *for* $k = 1, 2, \ldots, n_1$ *such that*

$$A_1\lambda < 0 \text{ or } A_1^T\lambda < 0. \tag{123}$$

If $\det A_1 \neq 0$ *then we may choose* $\lambda = -A_1^{-1}c$, *where* $c \in \Re^{n_1}$ *is any strictly positive vector.* □

Example 12 Consider the descriptor system (114) with the matrices

$$E = \begin{bmatrix} 0 & 0 & 0 & 2 \\ 0 & 1 & 0 & -2 \\ 1 & -2 & 0 & 0 \\ 0 & 0 & 0 & -2 \end{bmatrix}, \quad A = \begin{bmatrix} 0 & 1 & 0 & -4 \\ 1 & -4 & 0 & 4 \\ 0 & 6 & 1 & 0 \\ 1 & -1 & 0 & 4 \end{bmatrix}. \tag{124}$$

The condition (115) for (124) is satisfied since

$$\det[Es - A] = \begin{vmatrix} 0 & -1 & 0 & 2s+4 \\ -1 & s+4 & 0 & -2s-4 \\ s & -2s-6 & -1 & 0 \\ -1 & 1 & 0 & -2s-4 \end{vmatrix} = -2s^2 - 10s - 12, \tag{125}$$

and $n_1 = 2$. In this case $rank(E) = 3$ and $\mu = rank(E) - n_1 + 1 = 2$.

To transform the matrix $Es - A$ with (124) to the desired form

$$\begin{bmatrix} I_2 s - A_1 & 0 \\ 0 & Ns - I_2 \end{bmatrix} \text{ with } A_1 = \begin{bmatrix} -2 & 1 \\ 0 & -3 \end{bmatrix}, N = \begin{bmatrix} 0 & 1 \\ 0 & 0 \end{bmatrix}, \tag{126}$$

the following elementary column operations $R[4 \times \frac{1}{2}]$, $R[4, 1]$ and elementary row operations $L[2 + 4 \times (-1)]$, $L[4 + 1 \times 1]$, $L[3 + 2 \times 2]$ have been performed. In this case the matrices Q and P have the form

$$Q = \begin{bmatrix} 0 & 0 & 0 & 1 \\ 0 & 1 & 0 & 0 \\ 0 & 0 & 1 & 0 \\ \frac{1}{2} & 0 & 0 & 0 \end{bmatrix}, \quad P = \begin{bmatrix} 1 & 0 & 0 & 0 \\ 0 & 1 & 0 & -1 \\ 0 & 2 & 1 & -2 \\ 1 & 0 & 0 & 1 \end{bmatrix}. \tag{127}$$

Note that the matrix A_1 defined by (126) is the stable Metzler matrix and the descriptor system with (124) is positive and asymptotically stable.

4.2 *Convex Linear Combination of Hurwitz Polynomials and Extension of Kharitonov Theorem*

The polynomial

$$p(s) = s^n + \overline{a_{n-1}} s^{n-1} + \cdots + \overline{a_1} s + \overline{a_0} \tag{128}$$

is called Hurwitz if its zeros s_i satisfy the condition $\operatorname{Re} s_i < 0$ for $i = 1, 2, \ldots, n$.

Definition 16 The polynomial

$$(1-k)p_1(s) + kp_2(s) \ \text{ dla } \ k \in [0, 1] \tag{129}$$

is called convex linear combination of the polynomials

$$p_1(s) = s^n + a_{n-1}s^{n-1} + \cdots + a_1 s + a_0,$$
$$p_2(s) = s^n + b_{n-1}s^{n-1} + \cdots + b_1 s + b_0. \tag{130}$$

Theorem 23 *The convex linear combination (129) of the Hurwitz polynomials (130) of the positive linear system is also a Hurwitz polynomial.*

Proof By Theorem 2 the polynomials (130) are Hurwitz if and only if

$$a_i > 0 \text{ and } b_i > 0 \quad \text{for } i = 0, 1, \ldots, n-1. \tag{131}$$

The convex linear combination (129) of the Hurwitz polynomials (130) is a Hurwitz polynomial if and only if

$$(1-k)a_i + kb_i > 0 \quad \text{for } k \in [0, 1] \text{ and } i = 0, 1, \ldots, n-1. \tag{132}$$

Note that the conditions (132) are always satisfied if (131) holds. Therefore, the convex linear combination (129) of the Hurwitz polynomials (130) of the positive linear system is always the Hurwitz polynomial. □

Example 13 Consider the convex linear combination (129) of the Hurwitz polynomials

$$p_1(s) = s^2 + 5s + 2,$$
$$p_2(s) = s^2 + 3s + 4. \tag{133}$$

The convex linear combination (129) of the polynomials (133) is a Hurwitz polynomial since

$$(1-k)5 + 3k = 5 - 2k > 0 \text{ and } (1-k)2 + k4 = 2 + 2k > 0 \quad \text{for } k \in [0, 1]. \tag{134}$$

The above considerations for two polynomials (130) of the same order n can be extended to two polynomials of different orders [37].

Consider the set of positive interval linear continuous-time systems with the characteristic polynomials

$$p(s) = p_n s^n + p_{n-1}s^{n-1} + \cdots + p_1 s + p_0, \tag{135}$$

where

$$0 < \underline{p_i} \leq p_i \leq \overline{p_i}, \quad i = 0, 1, \ldots, n. \tag{136}$$

Theorem 24 *The positive interval linear system with the characteristic polynomial (135) is asymptotically stable if and only if the conditions (136) are satisfied.*

Proof By Kharitonov Theorem the set of polynomials (135) is asymptotically stable if and only if the polynomials (67) are asymptotically stable. From (136) it follows that the coefficients of polynomials (135) are positive. Therefore, the positive interval linear system with the characteristic polynomials (135) is asymptotically stable if and only if the conditions (136) are satisfied. □

Example 14 Consider the positive linear system with the characteristic polynomial

$$p(s) = a_3 s^3 + a_2 s^2 + a_1 s + a_0 \tag{137}$$

with the interval coefficients

$$0.5 \le a_3 \le 2, \quad 1 \le a_2 \le 3, \quad 0.4 \le a_1 \le 1.5, \quad 0.3 \le a_0 \le 4. \tag{138}$$

By Theorem 24 the interval positive linear system with the polynomial (137), (138) is asymptotically stable since the coefficients a_k, $k = 0, 1, 2, 3$ of the polynomial (137) are positive.

4.3 Stability of Descriptor Positive Linear Systems with Interval State Matrices

Consider the autonomous descriptor positive linear system

$$E\dot{x} = Ax, \tag{139}$$

where $x = x(t) \in \Re^n$ is the state vector, $E \in \Re^{n\times n}$ is constant (exactly known) and $A \in \Re^{n\times n}$ is an interval matrix defined by

$$\underline{A} \le A \le \overline{A} \quad \text{(or equivalently } A \in [\underline{A}, \overline{A}]). \tag{140}$$

It is assumed that

$$\det[Es - \underline{A}] \ne 0 \text{ and } \det[Es - \overline{A}] \ne 0, \tag{141}$$

and the matrix E has only linearly independent columns.

If these assumptions are satisfied then there exist two pairs of nonsingular matrices (P_1, Q_1), (P_2, Q_2) such that

$$P_1[Es - \underline{A}]Q_1 = \begin{bmatrix} I_{\underline{n}_1}s - \underline{A}_1 & 0 \\ 0 & \underline{N}s - I_{\underline{n}_2} \end{bmatrix}, \underline{A}_1 \in \Re^{\underline{n}_1 \times \underline{n}_1}, \underline{N} \in \Re^{\underline{n}_2 \times \underline{n}_2}, \underline{n}_1 + \underline{n}_2 = n, \tag{142}$$

and

$$P_2[Es - \overline{A}]Q_2 = \begin{bmatrix} I_{\bar{n}_1}s - \overline{A}_1 & 0 \\ 0 & \overline{N}s - I_{\bar{n}_2} \end{bmatrix}, \overline{A}_1 \in \Re^{\bar{n}_1 \times \bar{n}_1}, \overline{N} \in \Re^{\bar{n}_2 \times \bar{n}_2}, \bar{n}_1 + \bar{n}_2 = n, \tag{143}$$

where $\underline{n}_1 = \deg(\det[Es - \underline{A}])$ and $\bar{n}_1 = \deg(\det[Es - \overline{A}])$.

Theorem 25 *If the above assumptions are satisfied then the interval descriptor system (139) is positive if and only if*

$$\underline{A}_1 \in M_{\underline{n}_1} \text{ and } \overline{A}_1 \in M_{\bar{n}_1}. \tag{144}$$

Proof is similar to the proof of Theorem 12. □

Definition 17 The descriptor interval positive system (114) is called asymptotically stable (Hurwitz) if the system is asymptotically stable for all matrices $A \in [\underline{A}, \overline{A}]$.

Theorem 26 *If the matrices $\underline{A}$ and $\overline{A}$ of the positive system (114) are asymptotically stable then their convex linear combination*

$$A = (1-k)\underline{A} + k\overline{A} \quad \text{for } k \in [0, 1] \tag{145}$$

is also asymptotically stable.

Proof By Theorem 14 if the positive systems are asymptotically stable then there exists strictly positive vector $\lambda \in \Re_+^n$ such that

$$\underline{A}\lambda < 0 \text{ and } \overline{A}\lambda < 0. \tag{146}$$

Using (145) and (146) we obtain

$$A\lambda = [(1-k)\underline{A} + k\overline{A}]\lambda = (1-k)\underline{A}\lambda + k\overline{A}\lambda < 0 \quad \text{for } k \in [0, 1]. \tag{147}$$

Therefore, if the matrices $\underline{A}$ and $\overline{A}$ are asymptotically stable and (145) hold then the convex linear combination is also asymptotically stable. □

Theorem 27 *The interval descriptor positive system (139), (140) and matrix E with only linearly independent columns is asymptotically stable if and only if there exists a strictly positive vector $\lambda \in \Re_+^n$ such that*

$$P_n\underline{A}\lambda < 0 \text{ and } P_n\overline{A}\lambda < 0, \tag{148}$$

where P_n is the submatrix of P consisting of its first n rows.

Proof If by assumption the matrix E has only linearly independent columns, then $\lambda = Q\lambda_q \in \Re_+^n$ with all positive components for any $\lambda_q \in \Re_+^n$ with all positive components. By Theorems 12 and 13 the interval descriptor positive system (139), (140) is asymptotically stable if and only if the conditions (148) are satisfied. □

Example 15 (Continuation of Example 12). Consider the descriptor positive system (139) with the matrix E of the form (124) and the interval matrix A with

$$\underline{A} = \begin{bmatrix} 0 & -1 & 0 & -2 \\ 1 & -3 & 0 & 2 \\ 0 & 4 & 1 & 0 \\ 1 & -1 & 0 & 2 \end{bmatrix}, \quad \overline{A} = \begin{bmatrix} 0 & -1 & 0 & -6 \\ 1 & -5 & 0 & 6 \\ 0 & 8 & 1 & 0 \\ 1 & -1 & 0 & 6 \end{bmatrix}. \tag{149}$$

The matrices $(E, \underline{A})$ and $(E, \overline{A})$ satisfy the assumptions (141), and the matrix E given by (124) has only linearly independent columns.

In this case

$$n = \deg(\det[Es - \underline{A}]) = \deg\begin{vmatrix} 0 & -1 & 0 & 2s+2 \\ -1 & s+3 & 0 & -2s-2 \\ s & -2s-4 & -1 & 0 \\ -1 & 1 & 0 & -2s-2 \end{vmatrix} = \deg(-2s^2 - 6s - 4) = 2, \tag{150}$$

$$n = \deg(\det[Es - \overline{A}]) = \deg\begin{vmatrix} 0 & -1 & 0 & 2s+6 \\ -1 & s+5 & 0 & -2s-6 \\ s & -2s-8 & -1 & 0 \\ -1 & 1 & 0 & -2s-6 \end{vmatrix} = \deg(-2s^2 - 14s - 24) = 2, \tag{151}$$

and from (127) we have

$$P_2 = \begin{bmatrix} 1 & 0 & 0 & 0 \\ 0 & 1 & 0 & -1 \end{bmatrix}. \tag{152}$$

Using (148) and (149) for $\lambda = [1\ 1\ 1\ 1]^T$ we obtain

$$P_2\underline{A}\lambda = \begin{bmatrix} 1 & 0 & 0 & 0 \\ 0 & 1 & 0 & -1 \end{bmatrix} \begin{bmatrix} 0 & -1 & 0 & -2 \\ 1 & -3 & 0 & 2 \\ 0 & 4 & 1 & 0 \\ 1 & -1 & 0 & 2 \end{bmatrix} \begin{bmatrix} 1 \\ 1 \\ 1 \\ 1 \end{bmatrix} = \begin{bmatrix} -3 \\ -2 \end{bmatrix} < \begin{bmatrix} 0 \\ 0 \end{bmatrix} \tag{153}$$

and

$$P_2\overline{A}\lambda = \begin{bmatrix} 1 & 0 & 0 & 0 \\ 0 & 1 & 0 & -1 \end{bmatrix} \begin{bmatrix} 0 & -1 & 0 & -6 \\ 1 & -5 & 0 & 6 \\ 0 & 8 & 1 & 0 \\ 1 & -1 & 0 & 6 \end{bmatrix} \begin{bmatrix} 1 \\ 1 \\ 1 \\ 1 \end{bmatrix} = \begin{bmatrix} -7 \\ -4 \end{bmatrix} < \begin{bmatrix} 0 \\ 0 \end{bmatrix}. \quad (154)$$

By Theorem 27 the interval positive descriptor system is asymptotically stable.

4.4 Positive Descriptor Discrete-Time Linear Systems

Consider the autonomous descriptor discrete-time linear system

$$Ex_{i+1} = Ax_i, \ \ i \in Z_+, \quad (155)$$

where $x_i \in \Re^n$ is the state vector and $E, A \in \Re^{n\times n}$. It is assumed that

$$\det[Ez - A] \neq 0 \ \ \text{for some } z. \quad (156)$$

In this case the system (155) has unique solution for admissible initial conditions $x_0 \in \Re^n_+$.

It is well-known [20, 21] that if (156) holds, then there exists a pair of nonsingular matrices $P, Q \in \Re^{n\times n}$ such that

$$P[Ez - A]Q = \begin{bmatrix} I_{n_1}z - A_1 & 0 \\ 0 & Nz - I_{n_2} \end{bmatrix}, A_1 \in \Re^{n_1\times n_1}, N \in \Re^{n_2\times n_2}, \quad (157)$$

where $n_1 = \deg(\det[Ez - A])$ and N is the nilpotent matrix, i.e. $N^\mu = 0, N^{\mu-1} \neq 0$ (μ is the nilpotency index). To simplify the considerations it is assumed that the matrix N has only one block. The nonsingular matrices P and Q can be found for example by the use of the elementary row and column operations.

Definition 18 ([19, 23]) The autonomous discrete-time linear system

$$x_{i+1} = Ax_i, \quad (158)$$

where $A \in \Re^{n\times n}$ is called (internally) positive if $x_i \in \Re^n_+$, $i \in Z_+$ for all $x_0 \in \Re^n_+$.

Theorem 28 ([19, 23]) *The system* (158) *is positive if and only if*

$$A \in \Re^{n\times n}_+. \quad (159)$$

Definition 19 The positive system (158) is called asymptotically stable (Schur) if

$$\lim_{i\to\infty} x_i = 0 \ \ \text{for all } x_0 \in \Re^{n_1}_+. \quad (160)$$

Theorem 29 ([32]) *The positive system (158) is asymptotically stable if and only if one of the equivalent conditions is satisfied:*

(1) *all coefficients of the characteristic polynomial*

$$\det[I_n(z+1)-A]=z^n+a_{n-1}z^{n-1}+\cdots+a_1z+a_0 \tag{161}$$

are positive, i.e. $a_k>0$ *for* $k=0,\ 1,\ \ldots,\ n-1$;

(2) *there exists a strictly positive vector* $\lambda=[\lambda_1,\ \lambda_2,\cdots,\lambda_n]^T$, *i.e.* $\lambda_k>0$ *for* $k=1,\ 2,\ \ldots,\ n$ *such that*

$$A\lambda<\lambda. \tag{162}$$

Definition 20 The descriptor system (155) is called (internally) positive if $x_i\in\Re_+^n$ for $i\in Z_+$ and all admissible nonnegative initial conditions $x_0\in\Re_+^n$.

Theorem 30 *The descriptor system (155) j is positive if and only if the matrix E has only linearly independent columns and the matrix* $A_1\in\Re_+^{n_1\times n_1}$.

Proof Using the column permutation (the matrix Q) we choose n_1 linearly independent columns of the matrix E as its first columns. Next using elementary row operations (the matrix P) we transform the matrix E to the form $\begin{bmatrix} I_{n_1} & 0 \\ 0 & N \end{bmatrix}$ and the matrix A to the form $\begin{bmatrix} A_1 & 0 \\ 0 & I_{n_2} \end{bmatrix}$. From (157) it follows that the system (155) has been decomposed into the following two independent subsystems

$$x_{1,i+1}=A_1x_{1,i},\ x_{1,i}\in\Re^{n_1},\ i\in Z_+ \tag{163}$$

and

$$Nx_{2,i}=x_{2,i},\ x_{2,i}\in\Re^{n_2},\ i\in Z_+, \tag{164}$$

where

$$Q^{-1}x_i=\begin{bmatrix} x_{1,i} \\ x_{2,i} \end{bmatrix},\ i\in Z_+, \tag{165}$$

and Q, Q^{-1} are permutation matrices. Note that the solution $x_{1,i}=A_1^ix_{10}$ of the Eq. (163) is nonnegative if and only if $A_1\in\Re_+^{n_1\times n_1}$ and the solution $x_{2,i}$ of the Eq. (164) is zero for $i=1,2,\ldots$. □

Example 16 Consider the descriptor system (155) with the matrices

$$E = \begin{bmatrix} 0 & 0 & 0 & 2 \\ 0 & 1 & 0 & -2 \\ 1 & -2 & 0 & 0 \\ 0 & 0 & 0 & -2 \end{bmatrix}, \quad A = \begin{bmatrix} 0 & 1 & 0 & 1 \\ 1 & -\frac{2}{3} & 0 & -1 \\ 0 & \frac{2}{3} & 1 & 0 \\ 1 & -1 & 0 & -1 \end{bmatrix}. \tag{166}$$

The condition (156) is satisfied since

$$\det[Ez - A] = \begin{vmatrix} 0 & -1 & 0 & 2z-1 \\ -1 & z+\frac{2}{3} & 0 & -2z+1 \\ z & -2z-\frac{2}{3} & -1 & 0 \\ -1 & 1 & 0 & -2z+1 \end{vmatrix} = -2z^2 + \frac{5}{3}z - \frac{1}{3} \tag{167}$$

and $n_1 = 2$. In this case $rank(E) = 3$ and $\mu = rank(E) - n_1 + 1 = 2$. Performing on the matrix

$$Ez - A = \begin{bmatrix} 0 & -1 & 0 & 2z-1 \\ -1 & z+\frac{2}{3} & 0 & -2z+1 \\ z & -2z-\frac{2}{3} & -1 & 0 \\ -1 & 1 & 0 & -2z+1 \end{bmatrix} \tag{168}$$

the following column elementary operations $R[4 \times \frac{1}{2}]$, $R[4, 1]$ and the row operations $L[2 + 4 \times (-1)]$, $L[4 + 1 \times 1]$, $L[3 + 2 \times 2]$ we obtain

$$A_1 = \begin{bmatrix} \frac{1}{2} & 1 \\ 0 & \frac{1}{3} \end{bmatrix}, \quad N = \begin{bmatrix} 0 & 1 \\ 0 & 0 \end{bmatrix}. \tag{169}$$

In this case the matrices Q and P have the forms

$$Q = \begin{bmatrix} 0 & 0 & 0 & 1 \\ 0 & 1 & 0 & 0 \\ 0 & 0 & 1 & 0 \\ \frac{1}{2} & 0 & 0 & 0 \end{bmatrix}, \quad P = \begin{bmatrix} 1 & 0 & 0 & 0 \\ 0 & 1 & 0 & -1 \\ 0 & 2 & 1 & -2 \\ 1 & 0 & 0 & 1 \end{bmatrix}. \tag{170}$$

By Theorem 30 the descriptor system (155) with (166) is positive since $A_1 \in \Re_+^{2\times 2}$ and the matrix Q is monomial.

4.5 *Stability of Positive Descriptor Discrete-Time Linear Systems*

Consider the descriptor system (155) satisfying the condition (156).

Lemma 1 The characteristic polynomial of the system (155) and of the matrix $A_1 \in \Re^{n_1 \times n_1}$ are related by

$$\det[I_{n_1} z - A_1] = c \det[Ez - A], \tag{171}$$

where $c = (-1)^{n_2} \det(P) \det(Q)$.

Proof From (157) we have

$$\begin{aligned}\det[I_{n_1} z - A_1] &= (-1)^{n_2} \det \begin{bmatrix} I_{n_1} z - A_1 & 0 \\ 0 & Nz - I_{n_2} \end{bmatrix} \\ &= (-1)^{n_2} \det P \det[Ez - A] \det Q = c \det[Ez - A],\end{aligned} \tag{172}$$

which completes the proof. □

Theorem 31 *The positive descriptor system (155) is asymptotically stable if and only if one of the following equivalent conditions is satisfied:*

(1) *all coefficients of the characteristic polynomial*

$$\det[I_{n_1}(z+1) - A_1] = z^{n_1} + a_{n_1-1} z^{n_1-1} + \cdots + a_1 z + a_0 \tag{173}$$

are positive, i.e. $a_k > 0$ *for* $k = 0, 1, \ldots, n_1 - 1$;

(2) *all coefficients of the characteristic equation of the matrix* $Ez - A$

$$\det[E(z+1) - A] = \bar{a}_{n_1} z^{n_1} + \bar{a}_{n_1-1} z^{n_1-1} + \cdots + \bar{a}_1 z + \bar{a}_0 = 0 \tag{174}$$

are positive, i.e. $a_k > 0$ *for* $k = 0, 1, \ldots, n_1$;

(3) *there exists a strictly positive vector* $\lambda = [\lambda_1, \lambda_2, \ldots, \lambda_{n_1}]^T$, *i.e.* $\lambda_k > 0$ *for* $k = 1, 2, \ldots, n_1$ *such that*

$$A_1 \lambda < \lambda; \tag{175}$$

(4) *there exists a strictly positive vector* $\lambda = [\lambda_1, \lambda_2, \ldots, \lambda_{n_1}]^T$, *i.e.* $\bar{\lambda}_k > 0$ *for* $k = 1, 2, \ldots, n_1$ *such that*

$$\bar{P} \tilde{A} \bar{\lambda} < \bar{\lambda}, \tag{176}$$

where

$$\bar{P} = \bar{Q}_{n_1} P_{n_1}, \tag{177}$$

$\bar{Q}_{n_1} \in \Re_+^{n_1 \times n_1}$ *consists of* n_1 *nonzero rows of* $Q_{n_1} \in \Re_+^{n \times n_1}$ *which is built of first* n_1 *columns of the matrix* Q *defined by (116),* $P_{n_1} \in \Re^{n_1 \times n}$ *consists of* n_1 *rows of the matrix* P *defined by (116) and* $\tilde{A} \in \Re^{n \times n_1}$ *consists of* n_1 *columns of* $A \in \Re^{n \times n}$ *corresponding to the nonzero rows of* Q_{n_1}.

Proof Proof of condition 1 follows immediately from condition 1 of Theorem 29. By Lemma 1 $\det[I_{n_1}(z+1) - A_1] = 0$ if and only if $\det[E(z+1) - A] = 0$. Therefore, the positive descriptor system (155) is asymptotically stable if and only if all coefficients of (174) are positive.

From (142) we have

$$A_1 = P_{n_1} A Q_{n_1} \tag{178}$$

and using (162) we obtain

$$A_1 \lambda = P_{n_1} A Q_{n_1} \lambda < \lambda \tag{179}$$

for some strictly positive vector $\lambda \in \Re_+^{n_1}$. Premultiplying the Eq. (178) by the matrix $\bar{Q}_{n_1}$ and taking into account that $\bar{Q}_{n_1}\lambda = \bar{\lambda}$ and eliminating from A all columns corresponding to zero rows of Q_{n_1} we obtain (176), (177). □

Example 17 (Continuation of Example 16). Using Theorem 31 check the asymptotic stability of the positive descriptor system (155) with the matrices (166). The matrix A_1 of the system is given by (169) and its characteristic polynomial

$$\det[I_2(z+1) - A_1] = \begin{bmatrix} z + \frac{1}{2} & -1 \\ 0 & z + \frac{2}{3} \end{bmatrix} = z^2 + \frac{7}{6}z + \frac{1}{3} \tag{180}$$

has positive coefficients. Therefore, by condition 1 of Theorem 31 the matrix A_1 is asymptotically stable. The characteristic Eq. (174) of the matrices (166) is given by

$$\det[E(z+1) - A] = \begin{vmatrix} 0 & -1 & 0 & 2z+1 \\ -1 & z + \frac{5}{3} & 0 & -2z-1 \\ z+1 & -2z - \frac{8}{3} & -1 & 0 \\ -1 & 1 & 0 & -2z-1 \end{vmatrix} = 2z^2 + \frac{7}{3}z + \frac{2}{3} = 0, \tag{181}$$

so it has positive coefficients and by condition 2 of Theorem 31 the positive system is asymptotically stable. In this case we have

$$\bar{P} = \bar{Q}_{n_1} P_{n_1} = \begin{bmatrix} 0 & 1 \\ \frac{1}{2} & 0 \end{bmatrix} \begin{bmatrix} 1 & 0 & 0 & 0 \\ 0 & 1 & 0 & -1 \end{bmatrix} = \begin{bmatrix} 0 & 1 & 0 & -1 \\ \frac{1}{2} & 0 & 0 & 0 \end{bmatrix}, \quad \tilde{A} = \begin{bmatrix} 1 & -1 \\ -\frac{2}{3} & 1 \\ \frac{2}{3} & 0 \\ -1 & 1 \end{bmatrix} \tag{182}$$

and

$$\bar{P}\tilde{A} = \begin{bmatrix} 0 & 1 & 0 & -1 \\ \frac{1}{2} & 0 & 0 & 0 \end{bmatrix} \begin{bmatrix} 1 & -1 \\ -\frac{2}{3} & 1 \\ \frac{2}{3} & 0 \\ -1 & 1 \end{bmatrix} = \begin{bmatrix} \frac{1}{3} & 0 \\ \frac{1}{2} & -\frac{1}{2} \end{bmatrix}. \tag{183}$$

Therefore, using (176), (182) and (183) we obtain

$$\bar{P}\tilde{A}\bar{\lambda} = \begin{bmatrix} \frac{1}{3} & 0 \\ \frac{1}{2} & -\frac{1}{2} \end{bmatrix} \begin{bmatrix} 1 \\ 1 \end{bmatrix} < \begin{bmatrix} 1 \\ 1 \end{bmatrix} \tag{184}$$

and by condition 4 of Theorem 31 the positive system is asymptotically stable.

4.6 *Stability of Positive Descriptor Discrete-Time Linear Systems with Interval State Matrices*

Consider the autonomous descriptor positive linear system

$$Ex_{i+1} = Ax_i, i \in Z_+, \tag{185}$$

where $x_i \in \Re^n$ is the state vector, $E \in \Re^{n\times n}$ is constant (exactly known) and $A \in \Re^{n\times n}$ is an interval matrix defined by

$$\underline{A} \le A \le \overline{A} \text{ (or equivalently } A \in [\underline{A}, \overline{A}]). \tag{186}$$

It is assumed that

$$\det[Ez - \underline{A}] \ne 0 \text{ and } \det[Ez - \overline{A}] \ne 0, \tag{187}$$

and the matrix E has only linearly independent columns. If these assumptions are satisfied then there exist two pairs of nonsingular matrices (P_1, Q_1), (P_2, Q_2) such that

$$P_1[Es - \underline{A}]Q_1 = \begin{bmatrix} I_{\underline{n}_1}z - \underline{A}_1 & 0 \\ 0 & \underline{N}z - I_{\underline{n}_2} \end{bmatrix}, \underline{A}_1 \in \Re^{\underline{n}_1\times\underline{n}_1}, \underline{N} \in \Re^{\underline{n}_2\times\underline{n}_2}, \underline{n}_1 + \underline{n}_2 = n \tag{188}$$

and

$$P_2[Ez - \overline{A}]Q_2 = \begin{bmatrix} I_{\bar{n}_1}z - \overline{A}_1 & 0 \\ 0 & \overline{N}z - I_{\bar{n}_2} \end{bmatrix}, \overline{A}_1 \in \Re^{\bar{n}_1\times\bar{n}_1}, \overline{N} \in \Re^{\bar{n}_2\times\bar{n}_2}, \bar{n}_1 + \bar{n}_2 = n, \tag{189}$$

where $\underline{n}_1 = \deg(\det[Ez - \underline{A}])$ and $\bar{n}_1 = \deg(\det[Ez - \overline{A}])$.

Theorem 32 *If the assumptions are satisfied then the interval descriptor system* (185), (186) *is positive if and only if*

$$\underline{A}_1 \in \Re_+^{\underline{n}_1 \times \underline{n}_1} \text{ and } \overline{A_1} \in \Re_+^{\bar{n}_1 \times \bar{n}_1}. \tag{190}$$

Proof is similar to the proof of Theorem 30. □

Definition 21 The positive descriptor interval system (185) is called asymptotically stable (Schur) if the system is asymptotically stable for all matrices $A \in [\underline{A}, \overline{A}]$.

Theorem 33 *If the matrices $\underline{A}$ and $\overline{A}$ of the positive system* (185) *are asymptotically stable then their convex linear combination*

$$A = (1-k)\underline{A} + k\overline{A} \quad \text{for } 0 \le k \le 1 \tag{191}$$

is also asymptotically stable.

Proof By condition 2 of Theorem 29 if the positive systems are asymptotically stable then there exists strictly positive vector $\lambda \in \Re_+^n$ such that

$$\underline{A}\lambda < \lambda \text{ and } \overline{A}\lambda < \lambda. \tag{192}$$

Using (191) and (192) we obtain

$$A\lambda = [(1-k)\underline{A} + k\overline{A}]\lambda = (1-k)\underline{A}\lambda + k\overline{A}\lambda < (1-k)\lambda + k\lambda = \lambda \quad \text{for } k \in [0, 1]. \tag{193}$$

Therefore, if the matrices $\underline{A}$ and $\overline{A}$ are asymptotically stable and (192) holds, then the convex linear combination is also asymptotically stable. □

Theorem 34 *The positive descriptor system* (185) *with the matrix E with only linearly independent columns and interval matrix A is asymptotically stable if and only if there exists a strictly positive vector $\lambda \in \Re_+^n$ such that*

$$\bar{P}\underline{A}\lambda < \lambda \text{ and } \bar{P}\bar{A}\lambda < \lambda, \tag{194}$$

where $\bar{P}$ is defined by (189).

Proof By assumption the matrix E has only linearly independent columns and $\lambda = Q\lambda_q \in \Re_+^n$ is strictly positive for any vector $\lambda_q \in \Re_+^n$ with all positive components. By condition 2 of Theorem 29 and by Theorem 14 the positive descriptor system with interval (186) is asymptotically stable if and only if the conditions (194) are satisfied. □

Example 18 (Continuation of Example 12). Consider the positive descriptor system (185) with the matrix E given by (124) and the interval matrix A with

$$\underline{A} = \begin{bmatrix} 0 & 1 & 0 & 0.4 \\ 0 & -0.7 & 0 & -0.4 \\ 1 & -0.6 & 1 & 0 \\ 0 & -1 & 0 & -0.4 \end{bmatrix}, \overline{A} = \begin{bmatrix} 0 & 1 & 0 & 0.8 \\ 0 & -0.4 & 0 & -0.8 \\ 1 & -1.2 & 1 & 0 \\ 0 & -1 & 0 & -0.8 \end{bmatrix}. \tag{195}$$

We shall check the stability of the system by the use of Theorem 34. The matrices Q and P have the same forms (127) as in Example 12. Therefore, the matrix $\bar{P}$ in (194) is given by (182). Taking into account that in this case

$$\underline{\tilde{A}} = \begin{bmatrix} 1 & 0.4 \\ -0.7 & -0.4 \\ -0.6 & 0 \\ -1 & -0.4 \end{bmatrix} \text{ and } \overline{\tilde{A}} = \begin{bmatrix} 1 & 0.8 \\ -0.4 & -0.8 \\ -1.2 & 0 \\ -1 & -0.8 \end{bmatrix} \tag{196}$$

and using (195) we obtain

$$\bar{P}\underline{\tilde{A}}\bar{\lambda} = \begin{bmatrix} 0 & 1 & 0 & -1 \\ 0.5 & 0 & 0 & 0 \end{bmatrix} \begin{bmatrix} 1 & 0.4 \\ -0.7 & -0.4 \\ -0.6 & 0 \\ -1 & -0.4 \end{bmatrix} \begin{bmatrix} 1 \\ 1 \end{bmatrix} = \begin{bmatrix} 0.3 & 0 \\ 0.5 & 0.2 \end{bmatrix} \begin{bmatrix} 1 \\ 1 \end{bmatrix} < \begin{bmatrix} 1 \\ 1 \end{bmatrix} \tag{197}$$

$$\bar{P}\overline{\tilde{A}}\bar{\lambda} = \begin{bmatrix} 0 & 1 & 0 & -1 \\ 0.5 & 0 & 0 & 0 \end{bmatrix} \begin{bmatrix} 1 & 0.8 \\ -0.4 & -0.8 \\ -1.2 & 0 \\ -1 & -0.8 \end{bmatrix} \begin{bmatrix} 1 \\ 1 \end{bmatrix} = \begin{bmatrix} 0.6 & 0 \\ 0.5 & 0.4 \end{bmatrix} \begin{bmatrix} 1 \\ 1 \end{bmatrix} < \begin{bmatrix} 1 \\ 1 \end{bmatrix}. \tag{198}$$

Therefore, by Theorem 34 the positive descriptor system is asymptotically stable.

5 Summary

Basic definitions and theorems for positive linear continuous-time and discrete-time systems have been presented. Theorems concerning positivity and stability of standard and descriptor (nonsingular) linear systems described by state equations or transfer matrices have been established. The Kharitonov theorem has been extended to positive linear systems with interval state matrices. The notion of convex linear combination of interval state matrices and interval polynomials of continuous-time and discrete-time linear systems has been introduced. New results on positivity and stability of linear systems standard and descriptor systems have been presented.

The dynamical positive systems are currently at the beginning of its development, which makes them interesting for research and for potential of future applications. In particular, the results presented in this chapter can be extended to nonlinear systems (see e.g. [26, 32, 40]) and to fractional-order systems.

Acknowledgements This work was supported by National Science Centre in Poland under work No. 2017/27/B/ST7/02443.

I wish to thank very much to Dr. Łukasz Sajewski and Dr. Kamil Borawski for their essential help in preparation of the finale version of this chapter.

References

1. Ait Rami, M., Tadeo, F.: Controller synthesis for positive linear systems with bounded controls. IEEE Trans. Circuits Syst. **54**(2), 151–155 (2007)
2. Antsaklis, E., Michel, A.: Linear Systems. Birkhauser, Boston (2006)
3. Berman, A., Plemmons, R.J.: Nonnegative Matrices in the Mathematical Sciences. SIAM (1994)
4. Bose, N.K.: Applied Multidimensional Systems Theory. Van Nostrand Reinhold Co., New York (1982)
5. Bose, N.K.: Multidimensional Systems Theory Progress, Directions and Open Problems. D. Reidel Publishing Company, Dordrecht, Holland (1985)
6. Borawski, K.: Characteristic equations for descriptor linear electrical circuits. In: Proceedings of the 22nd International Conference on Methods and Models in Automation and Robotics (MMAR 2017), Międzyzdroje (2017)
7. Borawski, K.: Comparison of two different methods of observer synthesis for descriptor discrete-time linear systems. In: Advances in Intelligent Systems and Computing, vol. 550, pp. 151–164. Springer (2017)
8. Borawski, K.: Modification of the stability and positivity of standard and descriptor linear electrical circuits by state feedbacks. Electr. Rev. **93**(11), 176–180 (2017)
9. Borawski, K.: State vector estimation in descriptor electrical circuits using the shuffle algorithm. In: Advances in Intelligent Systems and Computing, vol. 577. Springer (2017)
10. Borawski, K.: Analysis of the positivity of descriptor continuous-time linear systems by the use of Drazin inverse matrix method. In: Szewczyk, R., Zieliński, C., Kaliczyńska, M. (eds.) Automation 2018. Advances in Intelligent Systems and Computing, vol. 743, pp. 172–182. Springer, Cham (2018)
11. Bru, R., Coll, C., Romero-Vivo, S., Sanchez, E.: Some problems about structural properties of positive descriptor systems. In: Positive Systems (Rome, 2003). Lecture Notes in Control and Information Sciences, vol. 294, pp. 233–240. Springer, Berlin (2003)
12. Bru, R., Coll, C., Sanchez, E.: About positively discrete-time singular systems, system and control: theory and applications. Electr. Comput. Eng. Ser. World Sci. Eng. Soc. Press, Athens **2000**, 44–48 (2000)
13. Bru, R., Coll, C., Sanchez, E.: Structural properties of positive linear time-invariant difference-algebraic equations. Linear Algebra Appl. **349**(1–3), 1–10 (2002)
14. Campbell, S.L.: Singular Systems of Differential Equations. Research Notes in Mathematics, vol. 40, Pitman (Advanced Publishing Program), Boston, Mass (1980)
15. Campbell, S.L.: Singular Systems of Differential Equations, vol. 2. Pitman, San Franciso (1982)
16. Commalut, C., Marchand, N.: Positive Systems. Lecture Notes in Control and Information Sciences, vol. 341. Springer, Berlin (2006)
17. Dail, L.: Singular Control Systems. Lectures Notes in Control and Information Sciences. Springer, Berlin (1989)

18. Dodig, M., Stosic, M.: Singular systems state feedbacks problems. Linear Algebra Appl. **431**(8), 1267–1292 (2009)
19. Farina, L., Rinaldi, S.: Positive Linear Systems; Theory and Applications. Wiley, New York (2000)
20. Kaczorek, T.: Linear Control Systems, vols. 1 and 2. Wiley, New York (1993)
21. Kaczorek, T.: Theory of Control and Systems. PWN, Warszawa (1993)
22. Kaczorek, T.: Positive singular discrete-time linear systems. Bull. Pol. Ac. Tech. **45**(4), 619–631 (1997)
23. Kaczorek, T.: The singular general model of 2-D systems and its solution. IEEE Trans. Autom. Control AC **33**(11), 1060–1061 (1988)
24. Kaczorek, T.: Polynomial and Rational Matrices, Applications in Dynamical Systems Theory. Springer, London (2007)
25. Kaczorek, T.: Decomposition of the pairs (A, B) and (A, C) of the positive discrete-time linear systems. Arch. Control Sci. **20**(3), 341–361 (2010)
26. Kaczorek, T.: Selected Problems of Fractional Systems Theory. Springer, Berlin (2012)
27. Kaczorek, T.: Descriptor positive discrete-time and continuous-time nonlinear systems. Proc. SPIE **9290** (2014). https://doi.org/10.1117/12.2074558
28. Kaczorek, T.: Analysis of positivity and stability of discrete-time and continuous-time nonlinear systems. Comput. Prob. Electr. Eng. **5**(1), 127–130 (2015)
29. Kaczorek, T.: Descriptor standard and positive discrete-time nonlinear systems. Arch. Control Sci. **25**(2), 227–235 (2015)
30. Kaczorek, T.: Positivity and linearization of a class of nonlinear continuous-time systems by state feedbacks. Int. J. Appl. Math. Comput. Sci. **25**(4), 827–831 (2015)
31. Kaczorek, T.: Positivity and stability of discrete-time and continuous-time nonlinear systems. Electr. Rev. **91**(8), 127–130 (2015)
32. Kaczorek, T.: Positivity and stability of discrete-time nonlinear systems. In: IEEE 2nd International Conference on Cybernetics, pp. 156–159 (2015)
33. Kaczorek, T.: Positive descriptor time-varying discrete-time linear systems and their asymptotic stability. TransNav, Int. J. Mar. Navig. Saf. Sea Trans. **9**(1), 83–83 (2015)
34. Kaczorek, T.: Standard and positive electrical circuits with zero transfer matrices. Poznan Univ. Technol. Acad. J. Electr. Eng. **85**, 11–28 (2016)
35. Kaczorek, T.: Descriptor positive nonlinear systems. In: Advances in Intelligent Systems and Computing, vol. 550, pp. 34–44. Springer (2017)
36. Kaczorek, T.: Analysis of positive linear continuous-time systems using the conformable derivative. Int. J. Appl. Math. Comput. Sci. **28**(2), 335–340 (2018)
37. Kaczorek, T.: Stability of interval positive continuous-time linear systems. Bull. Pol. Ac. Tech. **66**(1) (2018)
38. Kaczorek, T.: Stability of interval positive discrete-time linear systems. Int. J. Appl. Math. Comput. Sci. **28**(3) (2018)
39. Kaczorek, T., Borawski, K.: Minimum energy control of descriptor discrete-time linear systems by the use of Weierstrass-Kronecker decomposition. Arch. Control Sci. **26**(2), 177–187 (2016)
40. Kaczorek, T., Rogowski, K.: Fractional Linear Systems and Electrical Circuits. Springer, Cham, Heidelberg (2015)
41. Kaczorek, T., Sajewski, Ł.: The Realization Problem for Positive and Fractional Systems. Springer, Switzerland (2014)
42. Kailath, T.: Linear Systems. Prentice Hall, Englewood Cliffs, New York (1980)
43. Kalman, R.: On the general theory of control systems. In: Proceedings of the First International Congress on Automatic Control, London, UK, pp. 481–493. Butterworth (1960)
44. Kalman, R.: Mathematical description of linear systems. SIAM J. Control **1**(2), 152–192 (1963)
45. Kharitonov, V.L.: Asymptotic stability of an equilibrium position of a family of systems of differential equations. Differentsialnye uravneniya **14**, 2086–2088 (1978)
46. Kilbas, A.A., Srivastava, H.M., Trujilio, J.J.: Theory an Applications of Fractional Differential Equations. Elsevier, Amsterdam (2006)
47. Klamka, J.: Controllability of Dynamical Systems. Kluwer Academic Press, Dordrecht (1991)

48. Kucera, V., Zagalak, P.: Fundamental theorem of state feedback for singular systems. Automatica **24**(5), 653–658 (1988)
49. Mitkowski, W.: Remarks on stability of positive linear systems. Control Cybern. **29**(1), 205–304 (2000)
50. Mitkowski, W.: Dynamical properties of Metzler systems. Bull. Pol. Ac. Tech. **56**(4), 309–312 (2008)
51. Ortigueira, M.D.: Fractional Calculus for Scientists and Engineers. Springer, Dordrecht (2011)
52. Ostalczyk, P.: Discrete Fractional Calculus, Applications in Control and Image Processing. World Scientific, New Jersey (2016)
53. Polyak, B.T., Shcherbakov, P.S.: Superstable linear control systems. I. Analysis. Autom. Remote Control **63**(8), 1239–1254 (2002)
54. Polyak, B.T., Shcherbakov, P.S.: Superstable linear control systems. II. Design. Autom. Remote Control **63**(11), 1745–1763 (2002)
55. Rosenbrock, H.: State-Space and Multivariable Theory. Wiley, New York (1970)
56. Sajewski, Ł.: Positive stable realization of fractional discrete-time linear systems. Asian J. Control **16**(3), 922–927 (2014)
57. Sajewski, Ł.: Descriptor fractional discrete-time linear system with two different fractional orders and its solution, Bull. Pol. Ac. Tech. **64**(1), 15–20 (2016)
58. Wolovich, W.A.: Linear Multivariable Systems. Springer, New York (1974)
59. Zhang, H., Xie, D., Zhang, H., Wang, G.: Stability analysis for discrete-time switched systems with unstable subsystems by a mode-dependent average dwell time approach. ISA Trans. **53**, 1081–1086 (2014)
60. Zhang, J., Han, Z., Wu, H., Hung, J.: Robust stabilization of discrete-time positive switched systems with uncertainties and average dwell time switching. Circuits Syst. Signal Process. **33**, 71–95 (2014)

Controllability and Stability of Semilinear Fractional Order Systems

Jerzy Klamka, Artur Babiarz, Adam Czornik, and Michał Niezabitowski

Abstract In the chapter two the most important properties of fractional order dynamical systems, namely, controllability and stability are presented. At the beginning the basic notations and the fundamental definitions are recalled. The first part of the chapter is devoted to controllability and contains the formulation of the problem, main hypotheses and theorems about controllability of semilinear fractional order systems with distributed and point multiplicities constant or variable delays in the state variables and controls. Next, using fixed point theorems approximate controllability problem in infinite dimensional spaces, in particular Banach or Hilbert space, are discussed. Second part of the chapter is devoted to the stability problem of fractional order systems. The problem of stability and the problem of the existence of solutions for linear and nonlinear fractional order systems are also presented.

1 Introduction

Controllability plays a very important role in various areas of engineering and science. In particular in control systems many fundamental problems of control theory, such as optimal control, stabilizability or pole placement can be solved with assumption that the system is controllable [68, 89]. Controllability in general means that there exists a control function which steers the solution of the system from its initial state

J. Klamka (✉) · A. Babiarz · A. Czornik · M. Niezabitowski
Silesian University of Technology, Faculty of Automatic Control, Electronics and Computer Science, Akademicka 16, 44-100 Gliwice, Poland
e-mail: Jerzy.Klamka@polsl.pl

A. Babiarz
e-mail: Artur.Babiarz@polsl.pl

A. Czornik
e-mail: Adam.Czornik@polsl.pl

M. Niezabitowski
e-mail: Michal.Niezabitowski@polsl.pl

P. Kulczycki et al. (eds.), *Automatic Control, Robotics, and Information Processing*, Studies in Systems, Decision and Control 296,
https://doi.org/10.1007/978-3-030-48587-0_9

to a final state using a set of admissible controls, where the initial and final states may vary over the entire space. A standard approach for nonlinear and semilinear systems is to transform the controllability problem into a fixed point problem for an appropriate operator in a functional space. There are many papers devoted to the controllability problem, in which authors used the theory of fractional calculus [7, 8, 25, 41, 46, 57, 60, 67, 78, 79, 82] and a fixed point approach [5, 28, 29, 32, 33, 38, 45, 74, 75, 91].

The subject of fractional calculus and its applications has gained a lot of importance during the past four decades. This was mainly because it has become a powerful tool in modeling several complex phenomena in numerous seemingly diverse and widespread fields such as engineering, chemistry, mechanics, aerodynamics, physics, etc. [10, 24, 36, 39, 40, 43, 49, 65, 69].

For infinite-dimensional systems two basic concepts of controllability can be distinguished: approximate (weak) and exact (strong), as in infinite-dimensional spaces there exist linear subspaces which are not closed. means that there is possible to steer the system to an arbitrarily small neighbourhood of final state. The second one, i.e. exact controllability means that system can be steered to arbitrary final state. From these definitions it is obvious that approximate controllability is essentially weaker notion than exact. In the case of finite-dimensional systems notions of approximate and exact controllability coincide. It implies from the fact, that in finite-dimensional spaces all linear subspaces are closed.

Many control systems arising from real models can be described have the form of ordinary or partial fractional differential or integrodifferential equations [6, 21, 86, 87]. In [95] authors present a new approach to the problem of existence of mild solutions and controllability of nonlinear systems. For this purpose they avoid hypotheses of compactness on the semigroup generated by the linear part and any conditions on the multivalued nonlinearity expressed in terms of measures of noncompactness. Author of [93] focuses on fractional evolution equations and differential inclusions, and presents their applications to control theory. The existence of solutions for fractional semilinear differential or integrodifferential equations has been studied by many authors [20, 22, 58, 85, 94].

The impulsive differential systems can be used to model processes which are subject to sudden changes and which cannot be described by classical differential systems [59]. The controllability problem for impulsive differential and integrodifferential systems in Banach spaces has been discussed in [90]. Papers [83, 92] are devoted to the controllability of fractional evolution systems. The problem of controllability and optimal controls for functional differential systems has been extensively studied in many papers [1, 2, 27].

In addition to controllability, very important issue is the stability of the fractional order systems, that has crucial role for delay systems, what caused that it has been investigated for many years. Delay differential equations constitute basic mathematical models for real phenomena, for instance in engineering, mechanics, and economics, [31]. Delay is very often encountered in different technical systems, such as electric, pneumatic and hydraulic networks, chemical processes, long transmission lines, etc. [62]. Delays are inherent in many physical and engineering systems. In particular, pure delays are often used to ideally represent the effects of transmission and transportation. This is because these systems have only limited time to receive information and react accordingly. Such a system cannot be described by purely differential equations, but has to be treated with differential difference equations or the so-called differential equations with difference variables. The basic theory concerning the stability of systems described by equations of this type was developed by Pontryagin in 1942. Also, important works have been written by Bellman and Cooke in 1963, [11]. The presence of time delays in a feedback control system leads to a closed-loop characteristic equation which involves the exponential type transcendental terms. The exponential transcendentally brings infinitely many isolated roots, and hence it makes the stability analysis of time-delay systems a challenging task. It is well recognized that there is no simple and universally applicable practical algebraic criterion, like the Routh–Hurwitz criterion for stability of delay-free systems, for assessing the stability of linear time-invariant time-delayed systems. On the other side, the existence of pure time delay, regardless if it is present in the control or/and state, may cause an undesirable system transient response, or generally, even an instability. Numerous reports have been published on this matter, with a particular emphasis on the application of Lyapunov‘s second method, or on using the idea of matrix measure [54]. The analysis of time-delay systems can be classified such that the stability or stabilization criteria involve the delay element or not. In other words, delay independent criteria guarantee global asymptotic stability for any time-delay that may change from zero to infinity.

Recently there have been many papers in the control theory of fractional dynamic systems for stability such as robust stability, bounded input–bounded output stability, internal stability, finite time stability, practical stability, root-locus, robust controllability and observability, etc. For example, regarding linear fractional differential systems of finite dimensions in a state-space form, both internal and external stabilities are investigated in [64]. For the fractional order system is not possible to use algebraic tools, e.g. Routh-Hurwitz criteria, because instead of a characteristic polynomial there is a rational function. An analytical approach was suggested by Chen and Moore [18], who considered the analytical stability bound using Lambert function W. Further, analysis of stability and stabilization of fractional delay systems of retarded/neutral type are considered in [13, 14] as well as BIBO stability [37]. Whereas Lyapunov methods have been developed for the stability analysis and the control law synthesis of integer linear systems and have been extended to stability of fractional systems, only few studies deal with non-Lyapunov stability of fractional systems. For the first time, the finite time stability analysis of fractional time delay systems has been presented and reported in papers [50, 53]. Here, a Gronwall-Bellman‘s approach is proposed, using a "classical" Gronwall-Bellman inequality as well as a recently obtained generalized Gronwall inequality reported in [88] as a starting point. In the above papers for a particular class of (non)linear (non)autonomous fractional order time-delay systems has been examined the problem of sufficient conditions that enable system trajectories to stay within the a priori given sets was investigated.

In spite of intensive research, the stability problem of time delay fractional order systems remains an open problem. As for linear time invariant integer order systems, it is now well known that the stability of a linear fractional order system depends on the location of the system poles in the complex plane. However, the poles location analysis remains a difficult task in the general case. For commensurating fractional order systems, powerful criteria have been proposed. The most well-known is Matignon's stability theorem [64]. It permits us to check the system stability through the location in the complex plane of the matrix eigenvalues of the state space dynamic system representation. Matignon's theorem is the starting point of several results devoted to the stability problem. As it is known, due to the presence of the exponential function, the characteristic equation has an infinite number of roots, which makes the analytical stability analysis of a time-delay system extremely difficult. In the literature few theorems are available for stability testing of fractional-delay systems. Almost all of these theorems are based on the locations of the transfer function poles [14, 63] and since there is no universally applicable analytical method for solving fractional-delay equations in Laplace transformation domain, the numerical approach is commonly used.

2 Basic Notations

In this section, we introduce some definitions. Let $(X, \|\cdot\|)$ be a Banach space, $J = [0, t_1]$, $\alpha \in (0, 1)$ and $f : J \to X$ be a given function.

Definition 1 ([43]) The Caputo fractional derivative of order α is given as follows

$$^{C}D^{\alpha} f(t) = \frac{1}{\Gamma(1-\alpha)} \int_0^t \frac{f'(s)ds}{(t-s)^{\alpha}},$$

where: f is the function which has absolutely continuous derivative, Γ is the Gamma function and f' is the derivative of function f.

For completeness of presentation, below the definition of the measure noncompactness is shown. It is the generalization of the Schauder's fixed point theorem [44].

Definition 2 Let $(X, \|\cdot\|)$ be a Banach space and E be a bounded subset of X. Then the measure noncompactness of the set E is defined as

$$\mu(E) = \inf\{r > 0 : E \text{ can be covered by}$$

$$\text{a finite number of balls whose radii are smaller than r}\}.$$

Theorem 1 ([76]) *Let Q be a nonempty, bounded, convex and closed subset of the space X and let $F : Q \to Q$ be a continuous function such that*

$$\mu(F(S)) \leq k\mu(S),$$

for all nonempty subset S of Q, where $k \in [0, 1)$ is a constant. Then F has a fixed point in the set Q.

Theorem 2 ([48]) *Let Q be a closed convex subset of X and $F : Q \to Q$ be a continuous function. Then F has at least one fixed point in the set Q.*

Let us introduce the following necessary notation:

- $0 < \alpha < 1$ is order of derivative,
- ϕ is a continuous function on $[-h, 0]$, $h \in [0, \infty)$, $\phi : [-h, 0] \to \mathbb{R}^n$,
- A, B are $n \times n$ dimensional matrices and C is $n \times m$ dimensional matrix,
- u is the control function $u : [-h, \infty) \to \mathbb{R}^m$,
- $\mathscr{L}$ is the Laplace transform,
- $X_\alpha(t) = \mathscr{L}^{-1}\Big[[s^\alpha \cdot I - A - Be^{-s}]^{-1}s^{\alpha-1}\Big](t)$,
- $X_{\alpha,\alpha}(t) = t^{t-\alpha} \int_0^t \frac{(t-s)^{\alpha-2}}{\Gamma(\alpha-1)} X_\alpha(s)ds$,
- $x_L(t; \phi) = X_\alpha(t)\phi(0) + \int_{-h}^0 (t-s-h)^{\alpha-1} X_{\alpha,\alpha}(t-s-h)\phi(s)ds$,
- $H(t, s)$ is an $n \times m$ matrix, continuous in t for fixed s, $H : J \times [-h, 0] \to \mathbb{R}^{n\times m}$,
- $\int_{-h}^0 d_s H(t, s)$ denotes the integrals in the Lebesgue-Stieltjes sense with respect to s,
- exist positive real constants K and k with $0 \leq k < 1$ such that

$$|f(t, x, y, z, u)| \leq K, \tag{1}$$

$$|f(t, x, y, z, u) - f(t, x, y, \overline{z}, u)| \leq k(|z - \overline{z}|) \tag{2}$$

for all $x, y, z, \overline{z} \in \mathbb{R}^n$ and $u \in \mathbb{R}^m$, where f is nonlinear function.

3 Fractional System with Distributed Delays in Control

In this section mathematical models of fractional systems with different delays in state and control will be presented.

The fractional delay dynamical system with distributed delays in control can be presented by the following equation:

$$^C D^\alpha x(t) = Ax(t) + Bx(t-h) + \int_{-h}^0 d_s H(t, s)u(t+s) + \tag{3}$$

$$+ f\big(t, x(t), x(t-h), {}^C D^\alpha x(t), u(t)\big)$$

$$x(t) = \phi(t), \quad t \in [-h, 0].$$

Moreover the relative and approximate controllability definitions are recall below.

Definition 3 ([71]) The dynamical system (3) is said to be relatively controllable on interval $[0, t_1]$, if for every initial function ϕ and every final state $x_1 \in \mathbb{R}^n$ there exists a control function u defined on $[0, t_1]$ such that the solution of dynamical system satisfies $x(t_1) = x_1$.

Definition 4 ([71]) The dynamical system (3) is said to be approximately controllable on interval $[0, t_1]$, if for every desired final state x_1 and $\varepsilon > 0$ there exists a control function u such that the solution of dynamical system satisfies

$$\|x(t_1) - x_1\| < \varepsilon.$$

Using the well-known result of the unsymmetric Fubini theorem [30], the solution of (3) can be expressed by the following form:

$$x(t) = x_L(t; \phi)+$$

$$+\int_{-h}^{0} dH_\tau \int_{\tau}^{0} \big(t-(s-\tau)\big)^{\alpha-1} X_{\alpha,\alpha}\big(t-(s-\tau)\big) H(s-\tau, \tau) u_0(s) ds+$$

$$+\int_{0}^{t}\int_{-h}^{0} \big(t-(s-\tau)\big)^{\alpha-1} X_{\alpha,\alpha}\big(t-(s-\tau)\big) d_\tau H_t(s-\tau, \tau) u(s) ds+$$

$$+\int_{0}^{t} (t-s)^{\alpha-1} X_{\alpha,\alpha}(t-s) f\big(s, x(s), x(s-h), {}^C D^\alpha x(s), u(s)\big) ds,$$

where

$$H_t(s, \tau) = \begin{cases} H(s, \tau), & s \le t, \\ 0, & s > t, \end{cases}.$$

The below, we present the main result of relative controllability for the system (3).

Theorem 3 *Assume that the nonlinear function f satisfies the conditions* (1) *and* (2) *and suppose that the controllability Gramian*

$$W = \int_{0}^{t_1} S(t_1, s) S^*(t_1, s) ds$$

where:

$$S(t, s) = \int_{-h}^{0} \big(t-(s-\tau)\big)^{\alpha-1} X_{\alpha,\alpha}\big(t-(s-\tau)\big) d_\tau H_t(s-\tau, \tau).$$

is positive definite. Then the nonlinear system (3) *is relatively controllable on* J.

In order to prove of Theorem 3 authors using Darbo's fixed point theorem. Specifying the matrix function $H(t, s)$, it is possible to obtain systems with different lumped delays in control.

3.1 Fractional Systems with Multiple Delays in Control

In the same article [71] the authors focus on the implicit fractional delay dynamical system with time varying multiple delays in control given by the equation

$$^{C}D^{\alpha}x(t) = Ax(t) + Bx(t-h) + \sum_{i=0}^{M} C_i u(\sigma_i(t)) + \tag{4}$$

$$+ f\big(t, x(t), x(t-h), {}^{C}D^{\alpha}x(t), u(t)\big),$$

$$x(t) = \phi(t), \quad t \in [-h, 0],$$

where C_i for $i = 0, 1, \ldots, M$ are $n \times l$ matrices.

In order to examine relative controllability of (4), the following assumptions should be introduced.

Hypothesis 1 *The functions* $\sigma_i : J \to \mathbb{R}$, $i = 0, 1, \ldots, M$, *are twice continuously differentiable and strictly increasing in* J. *Moreover* $\sigma_i(t) \le t$, $i = 0, 1, \ldots, M$, *for* $t \in J$.

Hypothesis 2 *Introduce the time lead functions* $r_i(t) : [\sigma_i(0), \sigma_i(t_1)] \to [0, t_1]$, $i = 0, 1, \ldots, M$, *such that* $r_i(\sigma_i(t)) = t$ *for* $t \in J$. *Further,* $\sigma_0(t) = t$ *and for* $t = t_1$ *the following inequality holds:*

$$\begin{aligned} &\sigma_M(t_1) \le \cdots \le \sigma_{l+1}(t_1) \le 0 = \sigma_l(t_1) \\ &< \sigma_{l-1}(t_1) = \cdots = \sigma_1(t_1) = \sigma_0(t_1) = t_1. \end{aligned} \tag{5}$$

Hypothesis 3 *Given* $\sigma > 0$, *for functions* $u : [-\sigma, t_1] \to \mathbb{R}^l$ *and* $t \in [0, t_1]$, *we use the symbol* u_t *to denote the function on* $[-\sigma, 0]$ *defined by* $u_t(s) = u(t+s)$ *for* $s \in [-\sigma, 0)$.

Using (5), we can write solution of (4):

$$x(t) = x_L(t;\phi) + H(t) + \sum_{i=0}^{l} \int_0^t (t - r_i(s))^{\alpha-1} X_{\alpha,\alpha}(t - r_i(s)) C_i \dot{r}_i(s) u(s) ds + \tag{6}$$

$$+ \int_0^t (t-s)^{\alpha-1} X_{\alpha,\alpha}(t-s) f\big(s, x(s), x(s-h), {}^C D^\alpha x(s), u(s)\big) ds,$$

where

$$H(t) = \sum_{i=0}^{l} \int_{\sigma_i(0)}^{0} (t - r_i(s))^{\alpha-1} X_{\alpha,\alpha}(t - r_i(s)) C_i \dot{r}_i(s) u_0(s) ds +$$

$$+ \sum_{i=l+1}^{M} \int_{\sigma_i(0)}^{\sigma_i(t)} (t - r_i(s))^{\alpha-1} X_{\alpha,\alpha}(t - r_i(s)) C_i \dot{r}_i(s) u_0(s) ds.$$

Under above-mentioned hypotheses, the main theorem can be recalled.

Theorem 4 *Assume that the Hypotheses 1–3 hold. Further assume that the nonlinear function satisfies the condition* (1) *and* (2) *and suppose that determinant of Gramian matrix*

$$W = \sum_{i=0}^{l} \int_0^{t_1} \Big(X_{\alpha,\alpha}(t_1 - r_i(s)) C_i \dot{r}_i(s)\Big)\Big(X_{\alpha,\alpha}(t_1 - r_i(s)) C_i \dot{r}_i(s)\Big)^T ds$$

is positive definite. Then the nonlinear system (4) *is relatively controllable on* J.

As before, the proof was obtained using Darbo's fixed point theorem.

3.2 Controllability of Nonlinear Fractional Delay Systems

In this subsection, the results for relative controllability of nonlinear fractional delay systems are presented. The fractional delay systems described as follows [72]:

$$^C D^\alpha x(t) = Ax(t) + Bx(t-h) + Cu(t) + f(t, x(t), x(t-h), u(t)), \tag{7}$$

$$x(t) = \phi(t), \quad t \in [-h, 0]$$

are the special case of dynamical systems (3) when we use one delay instead distributed delays in control signal.

Then, the solution of (7) can be expressed by:

$$x(t) = X_\alpha(t)\phi(0) + B \int_{-h}^{0} (t-s-h)^{\alpha-1} X_{\alpha,\alpha}(t-s-h)\phi(s) ds +$$

$$+ \int_0^t (t-s)^{\alpha-1} X_{\alpha,\alpha}(t-s) f(s)\big[Cu(s) + f(s, x(s), x(s-h), u(s))\big] ds.$$

To obtain the main result on the controllability of the system (7), we have to introduce $p = (x, y, u) \in \mathbb{R}^n \times \mathbb{R}^n \times \mathbb{R}^m$ and let $|p| = |x| + |y| + |u|$. After that, the main theorem has the form.

Theorem 5 ([72]) *Let the function f satisfy the condition:*

$$\|f(t, p)\| \leq \sum_{i=1}^{k} q_i(t)\phi_i(p), \tag{8}$$

where: $\phi_i : \mathbb{R}^n \times \mathbb{R}^n \times \mathbb{R}^m \to \mathbb{R}_+$ are measurable functions and $q_i : J \to \mathbb{R}_+$ are L_1 functions for $i = 1, 2, 3, \ldots, k$. Suppose that the linear system

$$^{C}D^{\alpha}x(t) = Ax(t) + Bx(t-h) + Cu(t) \tag{9}$$

is relative controllable and the following condition holds

$$\limsup_{r\to\infty} \left(r - \sum_{i=1}^{p} c_i \sup \phi_i(p) : \|p\| \leq r \right) = +\infty$$

for constants $c_i = \max\{\gamma_i, \upsilon_i\}$, where:

- $E_{\alpha,\beta}(z) = \sum_{k=0}^{\infty} \frac{z^k}{\Gamma(k\alpha+\beta)}$, $\alpha, \beta > 0$ *- is the Mittag-Leffler function,*
- $a_1 = \sup \|E_{\alpha,\alpha}\big(A(t_1 - t)^{\alpha}\big)\|$,
- $W = \int_0^{t_1} (t_1 - s)^{\alpha-1}\big[E_{\alpha,\alpha}\big(A(t_1 - s)^{\alpha}\big)C\big]\big[E_{\alpha,\alpha}\big(A(t_1 - s)^{\alpha}\big)C\big]^* ds$,
- $\gamma_i = 4a_1^2(t_1)^{\alpha}\|q_i\|_{L_1}\|C^*\|\|W^{-1}\|\alpha^{-1}$,
- $\upsilon_i = 4a_1(t_1)^{\alpha}\|q_i\|_{L_1}\alpha^{-1}$.

Then the nonlinear fractional delay dynamical system (7) is relative controllable on J.

The proof of Theorem 5 is presented in [72] using fixed point technique.

3.3 The Controllability of Nonlinear Implicit Fractional Delay systems

The nonlinear implicit fractional delay systems are given by the form [71]:

$$^{C}D^{\alpha}x(t) = Ax(t) + Bx(t-h) + Cu(t) + \tag{10}$$

$$+ f\big(t, x(t), x(t-h), {}^{C}D^{\alpha}x(t), u(t)\big).$$

As we can see, the nonlinear part of the Eq. (10) is dependent not only on time, state and control (see Eq. 7), but also on derivative in Caputo sense.

For such an equation, the solution, with the initial function $x(t) = \phi(t)$, can be expressed as follows:

$$x(t) = x_L(t;\phi) + \int_0^t (t-s)^{\alpha-1} X_{\alpha,\alpha}(t-s)Cu(s)ds +$$

$$+ \int_0^t (t-s)^{\alpha-1} X_{\alpha,\alpha}(t-s) f\big(s, x(s), x(s-h), {}^C D^\alpha x(s), u(s)\big)ds.$$

By Gramian matrix and some conditions, the main result can be formulated as follows.

Theorem 6 *If the controllability Gramian matrix*

$$W = \int_0^{t_1} [X_{\alpha,\alpha}(t_1 - s)C][X_{\alpha,\alpha}(t_1 - s)C]^T ds$$

is positive definite and the nonlinear function f satisfies the condition (1) *and* (2), *then the nonlinear system* (10) *is relative controllable on J.*

The proof of the Theorem 6 is obtained with Darbo's fixed point theorem.

4 Controllability Problem of Fractional Semilinear Systems In infinite Space

This section considers the controllability of fractional semilinear systems in Banach and Hilbert space without and with delay. Precisely speaking, we will present results for approximate controllability which are proved with fixed point theorem.

4.1 Approximate Controllability of Fractional Order Semilinear systems

For dynamical systems defined on infinite dimensional state space it is necessary to introduce two main concepts of controllability: approximate (weak) controllability and exact (strong) controllability. Generally speaking approximate controllability means the possibility of control the dynamical system to the final states that form a dense subspace of final states. In practice, approximate controllability means that we are be able to approach the final state with any given a priori accuracy. It is clear that from exact controllability the approximate controllability follows which in term has a great practical meaning. In real systems exact controllability is rather an exception then a rule. But approximate controllability is a good tool to the develop properties of the dynamical systems.

In paper [80], authors investigate approximate controllability of semilinear delay systems in the Banach space. Roughly speaking, the approximate controllability gives the possibility of steering the dynamical system to the states which form the dense subspace in the state space.

Let $\mathscr{X}$ and $\widehat{\mathscr{X}}$ be Banach spaces and $\mathscr{C}\big([-h, 0]; \mathscr{X}\big)$ be the Banach space of all continuous function from $[-h, 0]$ to $\mathscr{X}$. The norm on $\mathscr{C}\big([-h, 0]; \mathscr{X}\big)$ is defined as

$$\|x_t\|_{\mathscr{C}} := \sup_{-h\le\theta\le 0} \|x_t(\theta)\|_{\mathscr{X}} \qquad \text{for } t \in]0, t_1].$$

The semilinear delay system is given by the following equation

$$\begin{aligned} {}^C D^{\alpha} x(t) &= Ax(t) + Bu(t) + f(t, x_t, u(t)), \\ t &\in]0, t_1], x_0(\theta) = \phi(\theta), \qquad \theta \in [-h, 0], \end{aligned} \tag{11}$$

where:

- $\frac{1}{2} < \alpha < 1$ is an order of the fractional derivative,
- $A : D(A) \subseteq \mathscr{X} \to \mathscr{X}$ is a closed linear operator with dense domain $D(A)$ generating a C_0-semigroup $S(t), t \ge 0$,
- x is a state taking values in the Banach space $\mathscr{X}$,
- u is control function taking values in $\widehat{\mathscr{X}}$,
- B is a bounded linear operator from $L_2\big([0, t_1]; \widehat{\mathscr{X}}\big)$,
- $f : [0, t_1] \times \mathscr{C}\big([-h, 0]; \mathscr{X}\big) \times \widehat{\mathscr{X}} \to \mathscr{X}$ is nonlinear operator.

If $x : [-h, t_1] \to \mathscr{X}$ is a continuous function, then $x_t : [-h, 0] \to \mathscr{X}$ is defined as $x_t(\theta) = x(t + \theta)$ for $\theta \in [-h, 0]$ and $\phi \in \mathscr{C}\big([-h, 0]; \mathscr{X}\big)$.

We introduce the fundamental solution $T(t)$ of the linear equation (11) (when $B = 0$ and $f = 0$), which is an operator valued function in the form $T(t) = S(t)$, $t \in]0, t_1]$, $T(0) = I$ and $T(\theta) = 0$, for $\theta \in [-h, 0]$ [84]. It is well known that C_0-semigroup $S(t)$ is bounded on $[0, t_1]$ and $T(t)$ is also bounded on $[0, t_1]$. Now we can define the solution of (11).

Lemma 1 ([80]) *A function $x(\cdot, \phi, u) \in \mathscr{C}\big([-h, t_1]; \mathscr{X}\big)$ is said to be the mild solution of* (11) *iff it satisfies*

$$x_t(\theta) = T(t)\phi(0) +$$

$$+ \frac{1}{\Gamma(\alpha)} \int_0^t (t-s)^{\alpha-1} T(t-s)\Big[Bu(s) + f(s, x_s, u(s))\Big] ds$$

on $[-h, t_1]$.

Let us impose some assumptions.

Hypothesis 4 *The nonlinear operator $f(t, x_t, u(t))$ satisfies the Lipschitz condition, i.e. there exists a constant $\ell > 0$ such that*

$$\|f\big(t, x_t, u(t)\big) - f\big(t, y_t, \vartheta(t)\big)\|_{\mathscr{X}} \leq \ell\Big[\|x_t - y_t\|_{\mathscr{C}} + \|u(t) - \vartheta(t)\|_{\widehat{\mathscr{X}}}\Big]$$

for all $x_t, y_t \in \mathscr{C}\big([-h, 0]; \mathscr{X}\big)$; $u, \vartheta \in \widehat{\mathscr{X}}$ and $t \in]0, t_1]$.

Hypothesis 5 *Range of f is the subset of the range of B i.e. $R(f) \subseteq R(B)$.*

Hypothesis 6 *The linear system*

$$^{C}D^{\alpha}x(t) = Ax(t) + Bu(t) \tag{12}$$

is approximately controllable.

Hypothesis 7 *There exists a positive constant β such that $\|B\vartheta\| \geq \beta\|\vartheta\|$ for all $\vartheta \in \widehat{\mathscr{X}}$.*

The main result is presented as follows

Theorem 7 *Under Hypotheses 4–7 the semilinear control system (11) is approximately controllable if $\ell < \beta$.*

The authors using fixed point technique to prove above-mentioned theorem.

The article [47] is related to approximate controllability of fractional order seminlinear systems with bounded delay. Author focus on dynamical systems which is given by the equation:

$$^{C}D^{\alpha}x(t) = Ax(t) + Bu(t) + f\big(t, x(t - h)\big), \tag{13}$$

where:

- V and $\widehat{V}$ are the Hilbert spaces,
- x is state and takes values in the space V,
- u is a control and takes values in the space $\widehat{V}$,
- $A : D(A) \subseteq V \to V$ is a closed linear operator with dense domain $D(A)$ and generates a C_0-semigroup $T(t)$,
- B is a bounded linear operator from Y to Z,
- $f : [0, t_1] \times V \to V$ is nonlinear function,
- $\phi \in C\big([-h, 0]; V\big)$,
- $Z = L_2\big([0, t_1]; V\big)$ and $Z_h = L_2\big([-h, t_1]; V\big)$ are the function spaces corresponding to V,
- $Y = L_2\big([0, t_1]; \widehat{V}\big)$ is the function space corresponding to $\widehat{V}$.

Lemma 2 *([94]) A function $x(\cdot) \in Z_h$ is said to be the mild solution of (13) if it satisfies*

$$x(t) = S_\alpha(t)\phi(0)+$$

$$+\int_0^t (t-s)^{\alpha-1} T_\alpha(t-s)\big[Bu(s) + f(s, x(s-h))\big]ds,$$

where:

$$S_\alpha(t)x = \int_0^\infty \varphi_\alpha(\theta) T(t^\alpha \theta) x d\theta,$$

$$T_\alpha(t)x = \alpha \int_0^\infty \theta \varphi_\alpha(\theta) T(t^\alpha \theta) x d\theta,$$

$$\varphi_\alpha(\theta) = \frac{1}{\alpha}\theta^{-1-\frac{1}{\alpha}} \psi_\alpha(\theta^{-\frac{1}{\alpha}}).$$

It should be noticed that $\varphi_\alpha(\theta)$ satisfies the conditions of a probability density function defined on $(0, \infty)$, that is $\varphi_\alpha \geq 0$, and $\int_0^\infty \varphi_\alpha(\theta) d\theta = 1$. Moreover, $\psi_\alpha(\theta)$ is given by the formula

$$\psi_\alpha(\theta) = \frac{1}{\pi}\sum_{n=1}^{\infty}(-1)^{n-1}\theta^{-n\alpha-1}\frac{\Gamma(n\alpha+1)}{n!}\sin(n\pi\alpha), \quad \theta \in (0, \infty).$$

Let us define the linear operator $\mathscr{L}$ from Z to V by the following formula

$$\mathscr{L}p = \int_0^\tau (\tau - s)^{\alpha-1} T_\alpha(\tau - s) p(s) ds$$

and the range of operator B by $R(B)$, and its closure by $\overline{R(B)}$.

Next, to formulate the main theorem we have to impose some assumptions.

Hypothesis 8 *The nonlinear function $f(t, x)$ satisfies the Lipschitz condition, i.e. there exists a positive constant ℓ such that*

$$\|f(t, x) - f(t, y)\| \leq \ell\|x - y\|$$

for all $x, y \in V$, $0 < t \leq \tau$.

Hypothesis 9 *The C_0-semigroup is compact.*

Hypothesis 10 *For each $p \in Z$ there exists a function $q \in \overline{R(B)}$ such that $\mathscr{L}p = \mathscr{L}q$.*

The main result of paper [47] is contained in the following theorem.

Theorem 8 *The semilinear control systems* (13) *is approximately controllable under Hypotheses 8–10.*

All the results in [47] are obtained using contraction principle and the Schauder's fixed point theorem.

5 Stability of Fractional Systems

This section focuses on stability problem of fractional dynamical systems. The main topics covered in the following subsections concern the existence of a solution, linear systems and nonlinear dynamical systems with delays in the state variables.

5.1 Solution Existence

The simplest form of delay fractional differential equation is

$$\begin{cases} {}^{C}D^{\alpha}x(t) = f(t, x(t), x(t-h)), \quad t \in [0, T], \\ x(t) = \phi(t), \quad \forall t \in [-h, 0], \end{cases} \tag{14}$$

where $\alpha \in (0, 1)$ is the order of the Caputo fractional derivative ${}^{C}D^{\alpha}$, and the initial condition ϕ is a continuous function on the interval $[-h, 0]$, $h, T > 0$ are fixed real parameters. For this equation, the first basic and important problem is to show the existence and uniqueness of solutions under some reasonable conditions. It is well known that in the case of ordinary differential equations (α is an integer), under some Lipschitz conditions a delay equation has an unique local solution [34, Sect. 2.2]; furthermore, by using continuation property (see [34, Sect. 2.3]) one can derive global solutions as well. However, in the fractional case (non-integer α) the problem of existence and uniqueness of (local and global) solutions is more complex because of the fractional order feature of the equation which implies history dependence of the solutions, hence, among others, the continuation property is not applicable. In the investigation of long term behavior of the DFDEs, as in the classical theory of dynamic systems, the understanding of growth rate of the solutions is of basic importance. One needs to know whether the solutions are exponentially bounded so that the theory of Lyapunov exponents as well as the tools of the Laplace transform are applicable to the study of the qualitative behavior of the systems.

Let h be an arbitrary positive constant, and $\phi \in C := C([-h, 0]; \mathbb{R}^d)$ be a given continuous function, we study the delay Caputo fractional differential equations

$$ {}^{C}D^{\alpha}_{0+}x(t) = f(t, x(t), x(t-h)), \quad t \in [0, T], \tag{15}$$

with the initial condition

$$x(t) = \phi(t), \quad \forall t \in [-h, 0], \tag{16}$$

where $x \in \mathbb{R}^d$, $T > 0$ and $f : [0, T] \times \mathbb{R}^d \times \mathbb{R}^d \to \mathbb{R}^d$ is continuous, $f(\cdot, 0, 0) \equiv 0$.

We also consider the initial condition problem (15) and (16) on the infinite time interval $[-h, \infty)$ as well with the obvious change from finite T to ∞.

A function $\varphi(\cdot, \phi) \in C([-r, T]; \mathbb{R}^d)$ is called a solution of the initial condition problem (15) and (16) over the interval $[-h, T]$ if

$$\begin{cases} {}^C D^{\alpha}_{0+}\varphi(t, \phi) = f(t, \varphi(t, \phi), \varphi(t - r, \phi)), & t \in [0, T], \\ \varphi(t, \phi) = \phi(t), \quad \forall t \in [-r, 0]. \end{cases}$$

Most of the terminologies for the theory of delay fractional differential equations used in this article are inherited from [24, 26]. For an interval $J = [a, b]$, we mean by $C[a, b]$ the space of all continuous functions defined on J with values in the d-Euclidean space $\mathbb{R}^d$. This is a Banach space with the norm $\|x\|_{\max} = \max\{\|x(t)\| : t \in J\}$, $x \in C[a, b]$, where $\|\cdot\|$ is the Euclidean norm on $\mathbb{R}^d$. If z is a function defined at least on $[t - h, t] \to \mathbb{R}^d$, then the new function $z_t : [-h, 0] \to \mathbb{R}^d$, is defined by

$$z_t(s) = z(t + s) \text{ for } -h \le s \le 0.$$

The next theorem from [19] presents a fundamental theorem about the uniqueness of the global solution of the initial condition problem (15) and (16).

Theorem 9 *Assume that $f : [0, T] \times \mathbb{R}^d \times \mathbb{R}^d \to \mathbb{R}^d$ is continuous and satisfies the following Lipschitz condition with respect to the second variable: there exists a non-negative continuous function $L : [0, T] \times \mathbb{R}^d \to \mathbb{R}_{\ge 0}$ such that*

$$\|f(t, x, y) - f(t, \hat{x}, y)\| \le L(t, y)\|x - \hat{x}\| \tag{17}$$

for all $t \in [0, T]$, $x, y, \hat{x} \in \mathbb{R}^d$. Then, the initial condition problem (15) and (16) has a unique global solution $\varphi(\cdot, \phi)$ on the interval $[-r, T]$.

We will use the following definitions of [23].

Definition 5 System (14) is said to be stable if for any $\varepsilon > 0$, there is a $\delta = \delta(\varepsilon)$, such that $\|\phi\|_{\max} < \delta$ implies that $\|\varphi_t(\phi)\|_{\max} < \varepsilon$, asymptotically stable if it is stable and there is $\delta_a > 0$ such that $\|\phi\|_{\sup} < \delta_a$ implies that

$$\lim_{t\to\infty} \varphi(t, \phi) = 0.$$

5.2 Linear Equation

Consider problem (15) and (16) with

$$f(t, x, y) = Ax + By$$

with $d \times d$ real matrices A, B, i.e. we consider linear equation of the form

$$^{C}D^{\alpha}x(t) = Ax(t) + Bx(t-h). \tag{18}$$

It should be noted that the problem of necessary and sufficient conditions for (asymptotic) stability of the zero solution of (18) seems to be a very complicated matter which is not answered neither in the integer-order case $\alpha = 1$ (for a particular result on this problem we refer to [42]).

To the best of our knowledge, Chen and Moore [17, 18] maybe firstly considered the analytical stability bound by using Lambert function for a class of fractional differential equations with time-delay. In [18], the authors considered a simple class of fractional delay equations with a constant time delay and Riemann-Liouville derivative.

The next theorem from [55] presents a sufficient conditions for of (18).

Theorem 10 *If all the roots of the characteristic equation*

$$det(s^{\alpha}I - A - Be^{-s\tau}) = 0$$

have negative real parts, then the zero solution of system (18) is asymptotically stable.

In [70] the following systems of linear equation

$$^{C}D^{\alpha_i}x(t) = \sum_{j=1}^{n} a_{ij}x_i\left(t - \tau_{ij}\right), \tag{19}$$

where $\alpha_i \in (0, 1)$, $\tau_{ij} \in (0, \infty)$, $i, j = 1, ..., n$, is considered. The multiple fractional order of (19) is denoted by $\alpha = (\alpha_1, \alpha_2, ..., \alpha_n)$.

The next two theorems give necessary and sufficient condition for particular cases of equation (18). The initial value is given by

$$x_i(t) = \varphi_i(t) \in C[-\tau_{\max}, 0], \ i = 1, ..., n,$$

where

$$\tau_{\max} = \max_{i,j=1,...,n} \tau_{ij}.$$

For subsequent discussion, we shall set

$$\Delta(s) = \left[s^{\alpha_{i1}} - a_{ij}e^{-s\tau_{ij}}\right]_{i,j=1,...,n}.$$

Theorem 11 *If all the roots of the characteristic equation*

$$\det \Delta(s) = 0$$

have negative real parts, then the zero solution of (19) is asymptotically stable.

The next theorem deals with equation (18) with $B = 0$.

Theorem 12 [64]. *Equation (18) with $B = 0$ is asymptotically stable if and only if*

$$|arg(\lambda)| > \alpha\pi$$

for all eigenvalues λ of matrix A. In this case the components of the state decay towards 0 like $t^{-\alpha}$.

In the one dimensional case of Eq. (18) we are able to provide necessary and sufficient conditions for asymptotic stability.

Theorem 13 [15]. *The one dimensional ($d = 1$) Eq. (18) is asymptotically stable if and only if the pair (a, b) is an interior point of the area bounded by the line*

$$a + b = 0$$

from above and by the parametric curve

$$a = \frac{\phi^{\alpha} sin(\tau\phi + \alpha\pi/2)}{sin(\tau\phi)},$$

$$b = -\frac{\phi^{\alpha} sin(\alpha\pi/2)}{sin(\tau\phi)}, \ \phi \in \left(\frac{(1-\alpha)\pi}{\tau}, \frac{\pi}{\tau}\right).$$

from below. In this case the components of the state decay towards 0 like $t^{-\alpha}$.

5.3 Nonlinear System

To the best of our knowledge, there have been only very few contributions to the problem of stability of nonlinear delay equation until now. In [61, 66], the authors considered the stability of some particular types of fractional differential equations with constant delays. In [16], the authors discussed stability and asymptotic properties of linear fractional-order differential systems involving both delayed and non-delayed terms. The stability and bifurcation analysis of a generalized scalar DFDE was discussed in [12]. The stability and performance analysis for positive fractional-order systems with time-varying delays was reported in [77]. In this paragraph we will present two approaches to this problem: one of them is based on the Lyapunov function, whereas the second one on the linearization method.

Consider first the following version of (15)

$$ {}^{C}D_{0+}^{\alpha}x(t) = f(t, x(t-h)), \quad t \in [0, T]. \tag{20} $$

The next theorem from [9] provides sufficient condition for stability and asymptotic stability of (20) expressed in terms of Lyapunov function.

Theorem 14 *Suppose that $f : \mathbb{R} \times C \to \mathbb{R}^d$ is such that for any subset D of C the set $f(\mathbb{R} \times D)$ is bounded, and $\alpha_1, \alpha_2, \alpha_3 : \mathbb{R}_+ \to \mathbb{R}_+$ are continuous nondecreasing functions, $\alpha_1(s), \alpha_2(s)$ are positive for $s > 0$, and $\alpha_1(0) = \alpha_2(0) = 0$, α_2 strictly increasing. If there exists a continuously differentiable function $V : \mathbb{R} \times \mathbb{R}^d \to \mathbb{R}$ such that*

$$ \alpha_1(\|x\|) \le V(t, x) \le \alpha_1(\|x\|), $$

for all $t \in \mathbb{R}$, $x \in \mathbb{R}^d$ and

$$ {}^{C}D_{0+}^{\alpha}V(t, x(t)) \le -\alpha_3\left(\|x(t)\|\right), $$

whenever

$$ V\left(t+\theta, x\left(t+\theta\right)\right) \le V\left(t, x\left(t\right)\right) $$

for $\theta \in [-h, 0]$, then system (20) is stable.

If, in addition, $\alpha_3\left(s\right) > 0$ for $s > 0$ and there exists a continuous nondecreasing function $p(s) > s$ for $s > 0$ such that

$$ {}^{C}D_{0+}^{\alpha}V(t, x(t)) \le -\alpha_3\left(\|x(t)\|\right), $$

whenever

$$ V\left(t+\theta, x\left(t+\theta\right)\right) \le p(V\left(t, x\left(t\right)\right)) $$

for $\theta \in [-h, 0]$, then system (20) is asymptotically stable.

An alternative approach to the problem of stability in nonlinear case is based on linearization method. Consider Eq. (15) with the function $f : \mathbb{R}^d \times \mathbb{R}^d \to \mathbb{R}^d$ of the following form

$$ f(x, y) = Ay + g(x, y). $$

Here, A is a square d by d real matrix and $g : \mathbb{R}^d \times \mathbb{R}^d \to \mathbb{R}^d$.

Theorem 15 *Suppose that*

1. *function g satisfies the following condition*

$$ g(0, 0) = 0; \tag{21} $$

2. *g is locally Lipschitz continuous in a neighborhood of the origin*
3.

$$\lim_{\rho \to 0} l(\rho) = 0,$$

with

$$l(\rho) = \sup_{x, y, \overline{x}, \overline{y} \in B(\rho)} \frac{\|g(x, y) - g(\overline{x}, \overline{y})\|}{\|x - \overline{x}\| + \|y - \overline{y}\|},$$

where

$$B(\rho) = \left\{x \in \mathbb{R}^d : \|x\| \le \rho\right\}.$$

4.

$$|arg(\lambda)| > \alpha\pi$$

for all eigenvalues λ *of matrix* A.

Then, the system (14) is asymptotically stable. In the context on the above theorem is worth to mention the following three papers [3, 4, 81].

A system could be stable but still completely useless because it possesses undesirable transient performances. Thus, it may be useful to consider the stability of such systems with respect to certain subsets of state-space which are defined a priori in a given problem. Besides that, it is of particular significance to concern the behaviour of dynamic systems only over a finite time interval. These boundedness properties of system responses, i.e. the solution of system models, are very important from the engineering point of view. Realizing this fact, numerous definitions of the so-called technical and practical stability were introduced. Roughly speaking, these definitions are essentially based on the predefined boundaries for the perturbation of initial conditions and allowable perturbation of system response. Thus, the analysis of these particular boundedness properties of solutions is an important step, which precedes the design of control signals, when finite time or practical stability control is concern. This controllability problem for fractional systems has been also investigated in [35, 51–53, 56, 73].

Consider the following definition from [52].

Definition 6 System (14) is said to be finite time stable with respect to $\{\delta, \varepsilon, J\}$, $\delta < \varepsilon$, $J = [0, T]$, $T > 0$ if and only if:

$$\|\phi\|_{\max} < \delta$$

implies that

$$\|\varphi_t(\phi)\| < \varepsilon$$

for $t \in J$.

Consider the following linear system

$$^{C}D^{\alpha}x(t) = A_0 x(t) + \sum_{i=1}^{n} A_i x(t - h_i), \tag{22}$$

$$0 \leq \tau_1 < \tau_2 < \ldots < \tau_n = \Delta.$$

In [51] the following sufficient condition for finite time stability of (22) was proved.

Theorem 16 *System (22) is finite time stable with respect to* $\{\delta, \varepsilon, J\}$, $\delta < \varepsilon$, $J = [0, T]$, $T > 0$ *if the following condition is satisfied*

$$\left[1 + \frac{t^{\alpha} \sum_{i=1}^{n} \sigma_{\max}(A_i)}{\Gamma(\alpha + 1)}\right] \exp\left(\frac{t^{\alpha} \sum_{i=1}^{n} \sigma_{\max}(A_i)}{\Gamma(\alpha + 1)}\right) \leq \frac{\varepsilon}{\delta}, \; t \in J$$

where $\sigma = t^{\alpha} \sum_{i=1}^{n} \sigma_{\max}(A_i)$ *and* $\sigma_{\max}(A)$ *is the largest singular value of matrix* A.

6 Summary

In this chapter we made literature review of the most important results concerning controllability and stability of semilinear fractional order systems. In the first part we presented results describing relation between the controllability of semilinear system with controllability of its linear part. Next certain results where the Lipschitz continuity plays the main rule are presented. In the second part of this chapter we included a review of the recent results concerning the problem of stability of fractional order Caputo derivative systems with delay. In the base of the review we concluded that the stability problem for the linear system is completely solved. However, the same problem for nonlinear system is still open. Recently, a fractional order Lyapunov function [49] has been proposed to deal with this problem. Similarly, the problem of stability of nonlinear fractional order system with delay is far from being completely solved. We believe that in the near future the open questions will be solved.

Acknowledgments The research presented here was done as parts of the projects funded by the National Science Centre in Poland granted according to decisions UMO-2017/25/B/ST7/02236 (JK) and DEC-2017/25/B/ST7/02888 (AB, MN). The work of Adam Czornik was supported by Polish Ministry for Science and Higher Education funding for statutory activities 02/990/BK_19/0121.

References

1. Agarwal, R.P., de Andrade, B., Siracusa, G.: On fractional integro-differential equations with state-dependent delay. Comput. Math. Appl., **62**(3), 1143–1149 (2011)
2. de Andrade, B., dos Santos, J.P.C. , et al.: Existence of solutions for a fractional neutral integro-differential equation with unbounded delay. Electron. J. Differ. Equat. **2012**(90), 1–13 (2012)
3. Anh, P.T., Babiarz, A., Czornik, A., Niezabitowski, M., Siegmund, S.: Asymptotic properties of discrete linear fractional equations. Bull. Pol. Acad. Sci.: Tech. Sci. **67**(4), 749–759 (2019)

4. Anh, P.T., Babiarz, A., Czornik, A., Niezabitowski, M., Siegmund, S.: Variation of constant formulas for fractional difference equations. Arch. Control Sci. **28** (2018)
5. Babiarz, A., Klamka, J., Niezabitowski, M.: Schauder's fixed-point theorem in approximate controllability problems. Int. J. Appl. Math. Comput. Sci. **26**(2), 263–275 (2016)
6. Bachelier, O., Dąbkowski, P., Gałkowski, K., Kummert, A.: Fractional and nd systems: a continuous case. Multidimension. Syst. Signal Process. **23**(3), 329–347 (2012)
7. Balasubramaniam, P., Tamilalagan, P.: Approximate controllability of a class of fractional neutral stochastic integro-differential inclusions with infinite delay by using mainardis function. Appl. Math. Comput. **256**, 232–246 (2015)
8. Balasubramaniam, P., Vembarasan, V., Senthilkumar, T.: Approximate controllability of impulsive fractional integro-differential systems with nonlocal conditions in hilbert space. Numer. Funct. Anal. Optim. **35**(2), 177–197 (2014)
9. Baleanu, D., Sadati, S.J., Ghaderi, R., Ranjbar, A., Abdeljawad Maraaba, T., Jarad, F.: Razumikhin stability theorem for fractional systems with delay. In: Abstract and Applied Analysis, vol. 2010. Hindawi (2010)
10. Baleanu, D., Diethelm, K., Scalas, E.: Fractional Calculus: Models And Numerical Methods. Nonlinearity And Chaos. World Scientific Publishing Company, Series on Complexity (2012)
11. Bellman, R., Cooke, K.L.: Differential Difference Equations. New York (1963)
12. Bhalekar, S.: Stability and bifurcation analysis of a generalized scalar delay differential equation. Chaos: An Interdisc. J. Nonlinear Sci. **26**(8), 084306 (2016)
13. Catherine Bonnet and Jonathan R Partington. Stabilization of fractional exponential systems including delays. Kybernetika **37**(3), 345–353 (2001)
14. Bonnet, C., Partington, J.R.: Analysis of fractional delay systems of retarded and neutral type. Automatica **38**(7), 1133–1138 (2002)
15. Čermák, J., Došlá, Z., Kisela, T.: Fractional differential equations with a constant delay: stability and asymptotics of solutions. Appl. Math. Comput. **298**, 336–350 (2017)
16. Čermák, J., Horníček, J., Kisela, T.: Stability regions for fractional differential systems with a time delay. Commun. Nonlinear Sci. Numer. Simul. **31**(1–3), 108–123 (2016)
17. Chen, Y.Q., Moore, K.L.: Analytical stability bound for a class of delayed fractional-order dynamic systems. In: Proceedings of the 40th IEEE Conference on Decision and Control, 2001, vol. 2, pp. 1421–1426. IEEE (2001)
18. Chen, Y.Q., Moore, K.L.: Analytical stability bound for a class of delayed fractional-order dynamic systems. Nonlinear Dyn. **29**(1–4), 191–200 (2002)
19. Nguyen Dinh Cong and Hoang The Tuan: Existence, uniqueness, and exponential boundedness of global solutions to delay fractional differential equations. Mediterr. J. Math. **14**(5), 193 (2017)
20. Debbouche, A., Nieto, J.J.: Sobolev type fractional abstract evolution equations with nonlocal conditions and optimal multi-controls. Appl. Math. Comput. **245**, 74–85 (2014)
21. Debbouche, A., Torres, D.F.M.: Approximate controllability of fractional delay dynamic inclusions with nonlocal control conditions. Appl. Math. Comput. **243**, 161–175 (2014)
22. Debbouche, A., Torres, D.F.M.: Sobolev type fractional dynamic equations and optimal multi-integral controls with fractional nonlocal conditions. Fract. Calc. Appl. Anal. **18**(1), 95–121 (2015)
23. Deng, W., Li, C., Lü, J.: Stability analysis of linear fractional differential system with multiple time delays. Nonlinear Dyn. **48**(4), 409–416 (2007)
24. Diethelm, K.: The Analysis of Fractional Differential Equations: An Application-Oriented Exposition Using Differential Operators of Caputo Type. Springer Science & Business Media (2010)
25. Ding, X., Nieto, J.J.: Controllability and optimality of linear time-invariant neutral control systems with different fractional orders. Acta Math. Sci. **35**(5), 1003–1013 (2015)
26. Driver, R.D.: Ordinary and delay differential equations. In: Applied Mathematical Sciences. Springer New York (2012)
27. Feckan, M., Wang, J.R., Zhou, Y.: Controllability of fractional functional evolution equations of sobolev type via characteristic solution operators. J. Optim. Theory Appl. **156**(1), 79–95 (2013)

28. Xianlong, F.: Controllability of non-densely defined functional differential systems in abstract space. Appl. Math. Lett. **19**(4), 369–377 (2006)
29. Xianlong, F., Liu, X.: Controllability of non-densely defined neutral functional differential systems in abstract space. Chin. Ann. Math., Ser. B **28**(2), 243–252 (2007)
30. Fubini, G.: Sugli integrali multipli. Rend. Acc. Naz. Lincei **16**, 608–614 (1907)
31. Henryk, G.: Analysis and Synthesis of Time Delay Systems. Wiley (1989)
32. Górniewicz, L., Ntouyas, S.K., O'regan, D.: Controllability of semilinear differential equations and inclusions via semigroup theory in banach spaces. Rep. Math. Phys. **3**(56), 437–470 (2005)
33. Guendouzi, T., Farahi, S.: Approximate controllability of sobolev-type fractional functional stochastic integro-differential systems. Boletín de la Sociedad Matemática Mexicana **21**(2), 289–308 (2015)
34. Hale, J.K., Verduyn Lunel, S.M.: Introduction to Functional Differential Equations, vol. 99. Springer Science & Business Media (2013)
35. Hei, X., Ranchao, W.: Finite-time stability of impulsive fractional-order systems with time-delay. Appl. Math. Model. **40**(7–8), 4285–4290 (2016)
36. Richard,H.: Fractional Calculus: An Introduction for Physicists. World Scientific (2011)
37. Hotzel, R., Fliess, M.: On linear systems with a fractional derivation: introductory theory and examples. Math. Comput. Simul. **45**(3–4), 385–395 (1998)
38. Ji, S., Li, G., Wang, M.: Controllability of impulsive differential systems with nonlocal conditions. Appl. Math. Comput. **217**(16), 6981–6989 (2011)
39. Kaczorek, T.: Selected Problems of Fractional Systems Theory. Lecture Notes in Control and Information Sciences. Springer, Berlin Heidelberg (2011)
40. Kaczorek, T., Sajewski, L.: The Realization Problem for Positive and Fractional Systems. Studies in Systems, Decision and Control. Springer International Publishing (2014)
41. Karthikeyan, S., Balachandran, K., Sathya, M.: Controllability of nonlinear stochastic systems with multiple time-varying delays in control. Int. J. Appl. Math. Comput. Sci. **25**(2), 207–215 (2015)
42. Khokhlova, T., Kipnis, M., Malygina, V.: The stability cone for a delay differential matrix equation. Appl. Math. Lett. **24**(5), 742–745 (2011)
43. Kilbas, A.A., Srivastava, H.M., Trujillo, J.J.: Theory and Applications of Fractional Differential Equations. Number t. 13. Elsevier Science (2006)
44. Klamka, J.: Schauder's fixed-point theorem in nonlinear controllability problems. Control and Cybern. **29**, 153–165 (2000)
45. Klamka, J., Babiarz, A., Niezabitowski, M.: Banach fixed-point theorem in semilinear controllability problems-a survey. Bull. Pol. Acad. Sci. Tech. Sci. **64**(1), 21–35 (2016)
46. Klamka, J., Sikora, B.: New controllability criteria for fractional systems with varying delays. In: Theory and Applications of Non-integer Order Systems, pp. 333–344. Springer (2017)
47. Kumar, S., Sukavanam, N.: Approximate controllability of fractional order semilinear systems with bounded delay. J. Differ. Equat. **252**(11), 6163–6174 (2012)
48. Kumlin, P.: A note on fixed point theory. In: Functional Analysis Lecture (2004)
49. Lakshmikantham, V., Leela, S., Devi, J.V.: Theory of Fractional Dynamic Systems. CSP (2009)
50. Lazarević, M.P.: Finite time stability analysis of pdα fractional control of robotic time-delay systems. Mech. Res. Commun. **33**(2), 269–279 (2006)
51. Lazarević, M.P., Debeljković, D.L.: Finite time stability analysis of linear autonomous fractional order systems with delayed state. Asian J. Control **7**(4):440–447 (2005)
52. Lazarević, M.P., Debeljković, D.L., Nenadić, Z., Milinković, S.: Finite-time stability of delayed systems. IMA J. Math. Control Inf. **17**(2), 101–109 (2000)
53. Lazarević, M.P., Spasić, A.M.: Finite-time stability analysis of fractional order time-delay systems: Gronwalls approach. Math. Comput. Model. **49**(3–4), 475–481 (2009)
54. Lee, T.N., Dianat, S.: Stability of time-delay systems. IEEE Trans. Autom. Control **26**(4), 951–953 (1981)
55. Li, H., Zhong, S.-M., Li, H.-B.: Stability analysis of fractional order systems with time delay. Int. J. Math., Comput. Sci. Eng. **8**(4), 400–403 (2014)

56. Li, M., Wang, J.: Finite time stability of fractional delay differential equations. Appl. Math. Lett. **64**, 170–176 (2017)
57. Liang, J., Yang, H.: Controllability of fractional integro-differential evolution equations with nonlocal conditions. Appl. Math. Comput. **254**, 20–29 (2015)
58. Liang, S., Mei, R.: Existence of mild solutions for fractional impulsive neutral evolution equations with nonlocal conditions. Adv. Differ. Equat. **2014**(1), 101 (2014)
59. Liu, Z., Li, X.: On the controllability of impulsive fractional evolution inclusions in banach spaces. J. Optim. Theory Appl. **156**(1), 167–182 (2013)
60. Liu, Z., Li, X.: On the exact controllability of impulsive fractional semilinear functional differential inclusions. Asian J. Control **17**(5), 1857–1865 (2015)
61. Luo, Y., Chen, Y.Q.: Stabilizing and robust fractional order PI controller synthesis for first order plus time delay systems. Automatica **48**(9), 2159–2167 (2012)
62. Malek-Zavarei, M., Jamshidi, M.: Time-Delay Systems: Analysis, Optimization and Applications. Elsevier Science Inc. (1987)
63. Matignon, D.: *Représentations en variables d'état de modèles de guides d'ondes avec dérivation fractionnaire*. Ph.D. thesis, Paris 11 (1994)
64. Denis, M.: Stability results for fractional differential equations with applications to control processing. In: Computational Engineering in Systems Applications, vol. 2, pp. 963–968. IMACS, IEEE-SMC Lille, France (1996)
65. Monje, C.A., Chen, Y.Q., Vinagre, B.M., Xue, D., Feliu-Batlle, V.: Fractional-order Systems and Controls: Fundamentals and Applications. Advances in Industrial Control. Springer, London (2010)
66. Kamran Akbari Moornani and Mohammad Haeri: Necessary and sufficient conditions for bibo-stability of some fractional delay systems of neutral type. IEEE Trans. Autom. Control **56**(1), 125–128 (2011)
67. Mur, T., Henriquez, H.R.: Relative controllability of linear systems of fractional order with delay. Math. Control Relat. Fields **5**(4), 845–858 (2015)
68. Paszke, W., Dąbkowski, P., Rogers, E., Gałkowski, K.: New results on strong practical stability and stabilization of discrete linear repetitive processes. Syst. Control Lett. **77**, 22–29 (2015)
69. Podlubny, I.: Fractional differential equations: an introduction to fractional derivatives. In: Fractional Differential Equations, to Methods of Their Solution and Some of Their Applications, vol. 198. Elsevier (1998)
70. Qian, D., Li, C., Agarwal, R.P., Wong, P.J.Y.: Stability analysis of fractional differential system with Riemann–Liouville derivative. Math. Comput. Model.**52**(5–6), 862–874 (2010)
71. Joice Nirmala Rajagopal and Krishnan Balachandran: The controllability of nonlinear implicit fractional delay dynamical systems. Int. J. Appl. Math. Comput. Sci. **27**(3), 501–513 (2017)
72. Nirmala Rajagopal, J., Balachandran, K., Rodrguez-Germa, L., Trujillo, J.J.: Controllability of nonlinear fractional delay dynamical systems. Rep. Math. Phys.**77**(1), 87–104 (2016)
73. Rakkiyappan, R., Velmurugan, G., Cao, J.: Finite-time stability analysis of fractional-order complex-valued memristor-based neural networks with time delays. Nonlinear Dyn. **78**(4), 2823–2836 (2014)
74. Sakthivel, R., Nieto, J.J., Idrisoglu Mahmudov, N.: Approximate controllability of nonlinear deterministic and stochastic systems with unbounded delay. Taiwanese J. Math. **14**(5), 1777–1797 (2010)
75. Sakthivel, R., Ren, Y., Idrisoglu Mahmudov, N.: Approximate controllability of second-order stochastic differential equations with impulsive effects. Mod. Phys. Lett. B **24**(14), 1559–1572 (2010)
76. Schmeidel, E., Zbąszyniak, Z.: An application of darbos fixed point theorem in the investigation of periodicity of solutions of difference equations. Comput. Math. Appl. **64**(7), 2185–2191 (2012)
77. Shen, J., Lam, J.: Stability and performance analysis for positive fractional-order systems with time-varying delays. IEEE Trans. Autom. Control **61**(9), 2676–2681 (2016)
78. Sikora, B.: Controllability criteria for time-delay fractional systems with a retarded state. Int. J. Appl. Math. Comput. Sci. **26**(3), 521–531 (2016)

79. Sikora, B.: Controllability of time-delay fractional systems with and without constraints. IET Control Theory Appl. **10**(3), 320–327 (2016)
80. Sukavanam, N., Kumar, S.: Approximate controllability of fractional order semilinear delay systems. J. Optim. Theory Appl. **151**(2), 373–384 (2011)
81. Tuan, H.T., Czornik, A., Nieto, J.J., Niezabitowski, M., et al.: Global attractivity for some classes of Riemann–Liouville fractional differential systems. J. Integr. Equat. Appl. **31**(2), 265–282 (2019)
82. Wang, J.R., Fan, Z., Zhou, Y.: Nonlocal controllability of semilinear dynamic systems with fractional derivative in banach spaces. J. Optim. Theory Appl. **154**(1), 292–302 (2012)
83. Wang, J.R., Zhou, Y.: Complete controllability of fractional evolution systems. Commun. Nonlinear Sci. Numer. Simul. **17**(11), 4346–4355 (2012)
84. Wang, L.: Approximate controllability for integrodifferential equations with multiple delays. J. Optim. Theory Appl. **143**(1), 185–206 (2009)
85. Yan, Z.: Existence results for fractional functional integrodifferential equations with nonlocal conditions in banach spaces. Ann. Polonici Math. **3**, 285–299 (2010)
86. Yan, Z.: Controllability of fractional-order partial neutral functional integrodifferential inclusions with infinite delay. J. Franklin Inst. **348**(8), 2156–2173 (2011)
87. Yan, Z.: Approximate controllability of partial neutral functional differential systems of fractional order with state-dependent delay. Int. J. Control **85**(8), 1051–1062 (2012)
88. Ye, H., Gao, J., Ding, Y.: A generalized Gronwall inequality and its application to a fractional differential equation. J. Math. Anal. Appl. **328**(2), 1075–1081 (2007)
89. Zawiski, R.: On controllability and measures of noncompactness. **1637**, 1241–1246 (2014)
90. Zhang, X., Huang, X., Liu, Z.: The existence and uniqueness of mild solutions for impulsive fractional equations with nonlocal conditions and infinite delay. Nonlinear Anal.: Hybrid Syst. **4**(4), 775–781 (2010)
91. Zhang, X., Zhu, C., Yuan, C.: Approximate controllability of fractional impulsive evolution systems involving nonlocal initial conditions. Adv. Differ. Equat. **2015**(1), 244 (2015)
92. Zhou, X.-F., Wei, J., Liang-Gen, H.: Controllability of a fractional linear time-invariant neutral dynamical system. Appl. Math. Lett. **26**(4), 418–424 (2013)
93. Yong, Z.: Fractional Evolution Equations and Inclusions: Analysis and Control. Academic Press (2016)
94. Zhou, Y., Jiao, F.: Existence of mild solutions for fractional neutral evolution equations. Comput. Math. Appl. **59**(3), 1063–1077 (2010)
95. Zhou, Y., Vijayakumar, V., Murugesu, R.: Controllability for fractional evolution inclusions without compactness. Evol. Equat. Control Theory **4**(4), 507–524 (2015)

Computer Simulation in Analysis and Design of Control Systems

Ewa Niewiadomska-Szynkiewicz and Krzysztof Malinowski

Abstract This chapter is concerned with computational research for complex systems. It presents a summary of the authors' experience with the development of methods for optimization, control and simulation of real-life large scale systems. In the presented approach, a complex system is one that consists of many components described by a significant number of state variables or interconnected subsystems, equipped with local decision units. The principal objective is to present basic ideas related to computer-aided analysis and design of these types of systems. A computer simulation experiment is seen here as a basic tool to support the design process and/or a component of the designed control system to be implemented.

1 Introduction

Nowadays, design of control structures and decision mechanisms for real-life systems requires a combination of a good knowledge of the controlled process, understanding of relevant theoretical issues, and the use of computer simulation tools. Modelling and simulation play a crucial role in science and engineering. Natural complexity, uncertainty and scale of modern systems limit the applicability of purely analytic approaches. Thus, computer simulation is a standard tool for understanding and predicting the behavior of a system under the influence of realistic and stochastic input scenarios. It enables us to investigate the proposed control structure and decision mechanisms prior to their implementation in the real world.

Simulation covers a very wide range of problems, from modeling to optimization, the design of control systems and decision support systems in operational mode. The works of various authors usually concern selected issues and focuses on certain

E. Niewiadomska-Szynkiewicz (✉) · K. Malinowski
Research and Academic Computer Network (NASK), Kolska 12, 01-045 Warsaw, Poland
e-mail: ewan@nask.pl

K. Malinowski
e-mail: Krzysztof.Mmalinowski@nask.pl

P. Kulczycki et al. (eds.), *Automatic Control, Robotics, and Information Processing*, Studies in Systems, Decision and Control 296,
https://doi.org/10.1007/978-3-030-48587-0_10

aspects. Many of them are devoted to modeling and computer implementation of system models from selected domains [9, 12, 23]. Model design, sensitivity assessment techniques, numerical methods, various simulation techniques, experiment design principles, and results verification methods are described and discussed [5, 6, 8, 26, 37, 44]. A significant part of the literature refers to the design of control systems, whose task is to determine the decisions that guarantee the desired operation of the system under the conditions of constant changes occurring in its environment [36, 44, 56, 63]. Simulation is one of the basic tools used in their construction, especially when we are dealing with complex problems, inspired by reality, which cannot be solved by using formal methods.

Furthermore, the current trends and needs in complex systems optimization involve the application of a simulation-based optimization scheme, which merges optimization and simulation technologies [2, 3]. The volume of mathematical and practical knowledge input into the optimization model often results in numerous formulas, the solution of which can be obtained only numerically. Thus, in many cases, simulation is the only viable method to calculate the given objective functions and to obtain crucial performance characteristics. This type of approach requires the use of effective and robust optimization methods. In most cases, these are simple techniques that work via so-called "brute force" or more sophisticated solutions involving the imitation of certain phenomena occurring in biological, chemical, physical and other systems. Recently, intensive use of computer simulation has been observed to create new numerical algorithms that solve difficult, poorly conditioned tasks [15, 19, 34]. Other issues addressed are concerned with implementation of simulation and design of software environments which provide frameworks for simulation experiments conducted on single or parallel computers and computer networks. In recent years numerous specialized simulation languages, software packages and frameworks have been developed [53]. Simulations are experiments performed with a mathematical model over time. Because of the complexity of systems taken into consideration, simulations require significant execution time. The spread of multiprocessor and multi-core machines, clusters, grids and clouds, as well as the use of graphics and signal processors for calculations have resulted in the development of parallel processing technology. The goal of parallelization of simulations is to speed up calculations and enable a solution of tasks that are too large for a single computer [17, 44, 58].

The authors' attention focuses on presenting the methodology of designing and analyzing control structures and mechanisms for complex systems, in which simulation plays the major role. The recommendations for the designer are preceded in sections devoted to discussing various architectures of complex systems and control structures. A unified description of simulators of continuous and discrete time systems and discrete events is presented and the stages of creating simulation models are discussed. Particular attention is devoted to methods of solving tasks, in which a simulation experiment is used to evaluate the performance of the developed control structure or decision mechanism. The discussed methodology was verified

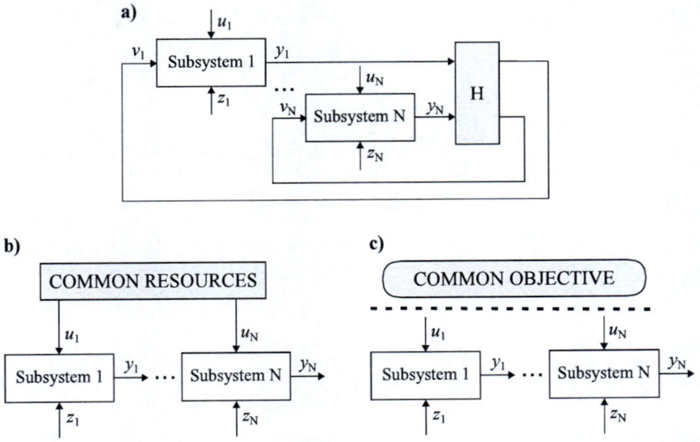

Fig. 1 Complex systems; types of subsystems interconnection: **a** through physical links, **b** through common resources, **c** through the common goal

via the development of methods for controlling different complex physical systems. The design process of control structures and decision mechanisms for examples are described in this chapter.

2 Complex Systems and Control

2.1 Complex Systems

The subjects for considerations are issues of analysis, design and computer verification of complex control systems, also called large-scale systems [16, 23, 32]. Some common features of such types of systems can be distinguished. In general, complex systems are often comprised of a set of components, i.e. subsystems. However, by controlling the components, we are forced to consider the task set for the whole system. Subsystems, as shown in Fig. 1, can be interconnected in different ways, i.e. directly, when they interact each other (for example there are physical links between them, namely the outputs of one subsystem are the inputs of others) or indirectly through the common objective of the system operation or global constraints on their common resources. Some physical systems have several types of interconnections, while others have one. We can distinguish two types of systems. They differ in the nature and strength of internal couplings.

Networks of components: systems in which many components can be separated by dividing the state vector. Thus, the components have their own state subvectors and state transition functions, but inputs and outputs are usually common. The interactions are modelled through state transition functions, namely a change of the state of one component affects the state of the others. Components cannot operate autonomously. A generic model of such a complex system composed of N components can be defined as follows:

$$SK = \langle T, U, Y, P_N, \{Sk_i\}_{i \in P_N} \rangle, \tag{1}$$

where SK denotes the system comprised of N components, T is the observation horizon, U and Y are sets of system inputs and outputs, P_N is a set of indices of all components and Sk_i denotes the i-th component. Each component is defined as follows:

$$Sk_i = \langle X_i, I_i, O_i, f_i, g_i \rangle, \tag{2}$$

where X_i is a state vector of the i-th component, $I_i \subset P_N$ is the set of indices of components that affect Sk_i ($i \notin I_i$), $O_i \subset P_N$ is the set of indices of components that are affected by Sk_i, $f_i : \times X_{m \in I_i} \times X_i \times U \rightarrow X_i$ denotes the state transition function and $g_i : \times X_{m \in I_i} \times X_i \times U \rightarrow Y$ the output function. This group may contain systems with components physically linked or connected indirectly through the common objective of the system operation.

Networks of subsystems: a composite system consisting of multiple interconnected, often spatially deployed subsystems, in which control units are located in different places. Subsystems can operate independently. They have their own state vectors, inputs and outputs. The generic model of a network of subsystems is as follows:

$$SP = \langle T, U_{sp}, Y_{sp}, P_N, \{Sp_i\}_{i \in P_N}, \{I_i\}_{i \in P_N \cup \{sp\}}, \{d_i\}_{i \in P_N \cup \{sp\}} \rangle, \tag{3}$$

where SP denotes the system composed of N subsystems, T is the observation horizon, U_{sp} the environmental impact on the system, Y_{sp} a vector of outputs that affect the environment, sp index that denotes the environment of a system, P_N a set of indices of all subsystems. Sp_i denotes the i-th subsystem, $I_i \subset P_N \cup \{sp\}$ a set of indices of subsystems affecting the subsystem Sp_i, d_i the function that describe interconnections of subsystems, $d_i : \times Q_{j \in I_i} \rightarrow Q_i$, where $Q_j = U_j$ for $j = sp$, $Q_j = Y_j$ for $j \neq sp$ and $Q_i = Y_i$ for $i = sp$ and $Q_i = X_i$ for $i \neq sp$. Each subsystem can be modelled as follows.

$$Sp_i = \langle T, U_i, Y_i, X_i, \{Y_m\}_{m \in I_i}, f_i, g_i \rangle, \tag{4}$$

where U_i, Y_i i X_i are vectors of inputs, outputs and states of the i-th subsystem, $\{Y_m\}_{m \in I_i}$ outputs affecting Sp_i (links between Sp_i and other subsystems, $m \neq i$), $f_i : \times Y_{m \in I_i} \times U_i \times X_i \rightarrow X_i$ is the state transition function and $g_i : \times Y_{m \in I_i} \times U_i \times X_i \rightarrow Y_i$ the output function.

Let us focus on modelling different types of interconnections between subsystems Sp_i, $i = 1, \ldots, N$. We can divide the input variable vector of all subsystems into two parts $[u, v]$, where $u = [u_1, u_2, \ldots, u_N]$, $u_i \in U_i$ denotes a local input variables of Sp_i and v other inputs that bind subsystems and cannot be decomposed, so-called interactive inputs. The interconnection of subsystems through physical links (Fig. 1a) can be modelled by the following equations $v_i = \sum_{j=1}^{N} H_{ij} y_j$, $i = 1, \ldots, N$. $v_i \in \{Y_m\}_{m \in I_i}$ is the subvector of v. H_{ij} is the connection matrix, and denotes the local interaction operator. The output subvector y_j is given by $y_j = g_j(u_j, v_j, x_j)$.

Indirect interconnection can be expressed through the following constraints $\sum_{i=1}^{N} z_i(u_i, v) \leq w$ or $\sum_{i=1}^{N} h_i(u_i, v) = s$. In the case of systems with only indirect links between subsystems—shared resources (Fig. 1b), the above formula is reduced to the following one $\sum_{i=1}^{N} z_i(u_i) \leq w$ and $\sum_{i=1}^{N} h_i(u_i) = s$. Furthermore, the following additional constraints for v may occur $r_0(v) = 0$ i $q_0(v) \leq 0$ and for $u_i, v, r_i(u_i, v) = 0, q_i(u_i, v) \leq 0$.

2.2 *Control Structures*

While proposing control structures for composite systems, consisting of multiple interconnected subsystems endowed with their local decision units, one can consider three basic configurations presented in Fig. 2: structure with one centralized controller replacing local control units, decentralized structure with local units in full charge of the system, and a hierarchical structure involving both local units and a coordinator. In the centralized case all decisions $u_1, \ldots, u_N$ are made by one overall decision mechanism. The basic advantage of a centralized solution (Fig. 2a) is its structural simplicity and provision to introduce a control mechanism providing for operational objectives defined with a reference to the whole system. Such an approach, however, possesses several disadvantages. A control center may have to make all decisions while using the currently available complete information about and from all subsystems. This requires concentration of this information and of decision computing. In the case of complex systems composed of subsystems located in distant places, data transmissions may lead to delays or even suffer failures [41]. Also, decisions taken by the central control unit are not being immediately worked out, especially in situations when the decision mechanism, e.g. based on large scale optimization, is complicated. Another major disadvantage of using a centralized decision mechanism is that the local decision units are overridden. They do not participate in working out the operational decisions and are possibly responsible only for implementation of those decisions. In engineering practice, control of composite systems, in particular regulatory control, is therefore, if possible, realized in a decentralized mode (Fig. 2b), with local decision units in charge. This involves division of competence and, usually, division of available information; it is assumed that each decision unit is in possession of partial information about the state of the system and its environment. The decision process is decentralized and all decision makers take an active role in this process. Full decentralization, however, may not allow for the meeting all objectives related to the desired system operation, either due to local decision makers being concerned only with their local goals or due to locally restricted available information. In such a situation a supervisory unit (coordinator) may be introduced to monitor and, if needed, to modify operations of the local decision makers. In this way

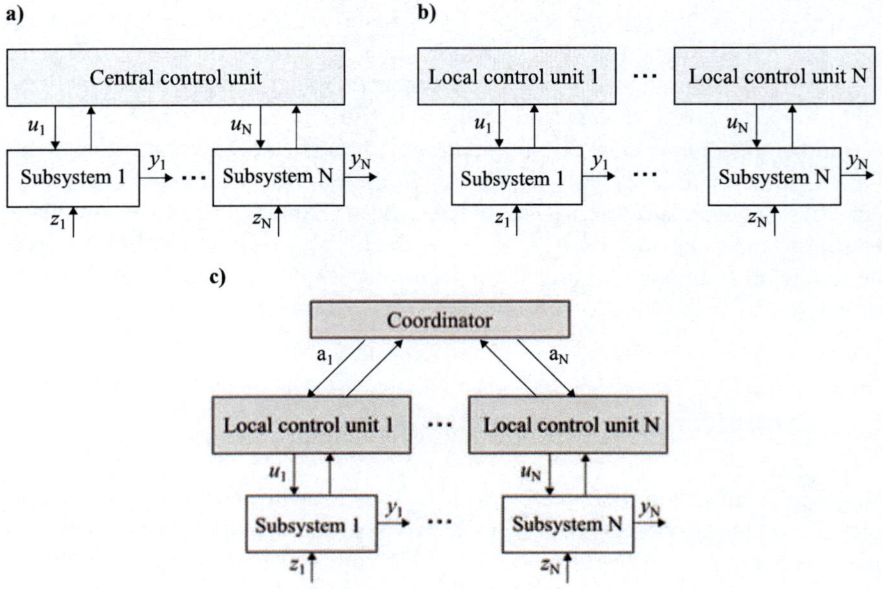

Fig. 2 Control structures for complex systems

a hierarchical control structure is created. Of course, a basic two-level structure may be, if needed, developed further into a multiple-level system with sub-coordinators responsible for supervising and harmonizing groups of lower level units. Figure 2c depicts a basic two-level structure with N local decision makers and a coordinator. Similarly, as in a decentralized case, both the available information and the decision process are distributed between the local units. Yet, actions of these units are influenced by the coordinator through coordinating controls $\mathbf{a}_i$, $i = 1, \ldots, N$ modifying local decision mechanisms $u_i(\mathbf{a}_i)$. In this way the coordinator provides, assuming that the whole control system is properly designed, for all objectives being fulfilled. Coordinating instruments $\mathbf{a}_i$ depend on the mode of coordination and on other problem specific factors. Hierarchical structures allow us to combine the advantages of both centralized and fully decentralized controllers and in numerous cases offer a reasonable solution to control design [22, 45, 47].

It is typical that the actions taken by the coordinator, leading to fulfillment of overall system objectives, result in the degrading of some control goals as perceived by the local decision makers, which of course may make those decision makers, if they are human operators, not willing to comply with the coordinator's actions. Then it becomes very important to precisely define the scopes of competence and responsibility of all decision units. In particular, it is possible to consider, such a coordination mode in which the coordinator is monitoring decentralized decision processes and reacts only in the case when realization of the overall objectives are endangered, or, in the opposite situation, such a mode in which the coordinator can, in specified situations, override local decisions.

3 Computer Simulation

There are many definitions of the term "simulation". Simulation is defined by Bratley et. al. as a process of driving a model of a system with suitable inputs and observing the corresponding outputs [11], by Shannon as the process of using the model of a conceptual system to conduct experiments for the purpose of understanding the performance of the system and/or evaluating alternative management strategies and decision-making processes using simulation results [54]. A general definition was formulated by Korn and Wait [25]: *Simulation is an experiment performed on a model*. Based on these definitions, a simulation is implemented in two stages. In the first, a simulation model is created. The second involves developing an experiment plan for the prepared test data set and establishing a method for analyzing the results as well as conducting experimental tests and evaluating the results. The experimenter is meant to be a modeler, user and decision maker. Therefore, the simulation process encompasses the activities included in model building as well as its analytic use.

To sum up, simulation is a standard tool for understanding and predicting the behavior of a system under the influence of realistic and stochastic input scenarios. It can be used to show the real effects of alternative conditions, decisions and courses of action. Simulation is also used when the real system cannot be engaged, because it is not accessible, dangerous or unacceptable to engage, or it does not exist. Key issues in simulation include the acquisition of source of information about the simulated systems, relevant selection of key characteristics and behaviors, the use of simplifying approximations and assumptions within the simulation, and fidelity and validity of the simulation outcomes. Procedures and protocols for model verification and validation are an ongoing field of academic study, refinement, research and development in simulations technology or practice, particularly in the field of computer simulation.

3.1 Simulation Model

A simulation model, referred to as the simulator, is a physical entity or a computer program imitating, for a defined purpose, behavior of a given device or plant, say system, in terms of processes that one is interested in. Presently, computer based simulators are predominantly used and so the attention here is focused on such solutions. Design of a computer simulator begins with formulation of an appropriate formal or intuitive base model representing, in accordance with needs and subjective understanding of the user, behavior of the considered system, in case of a composite system behavior of its elements and their interactions. When the objective is to

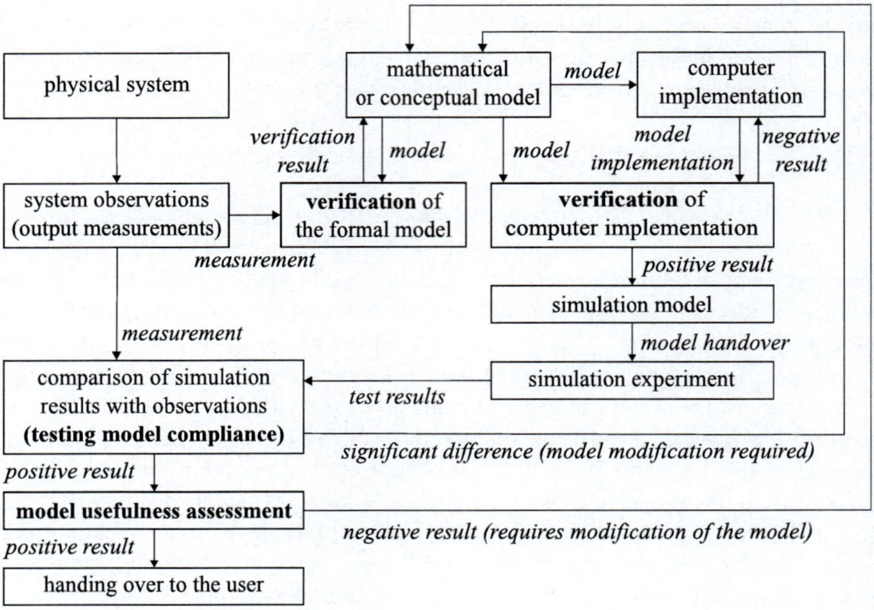

Fig. 3 The stages of the simulation model development

imitate only the selected processes in the system, design of the base model may not require full knowledge about the system components and properties. The role of the model is to enable analysis of the impact of system inputs, including both decisions being made by the controller and uncontrolled inputs, on specified outputs and outcomes like profits and losses, environmental consequences, etc. The next step requires program implementation of the base model. The computer model consists of active components (procedures, functions, classes, etc.) and passive components (data). This model should be consistent with the formal (intuitive) model. Comparing simulation results with those predicted from the initial model is necessary to verify whether the computer model is properly constructed. Verification of a computer model is a complex endeavor, covering a broad range of issues, from numerical analysis to checking correctness of programming. The result of a positive verification is the statement that the computer model is complete and consistent with the base model. Then it is necessary to check whether the results of computer simulation are consistent with observations and measurements related to real system behavior. Of course as far as the processes in which one is interested in are concerned. This is called computer model validation. If there are significant, i.e. greater than allowed, discrepancies between real and simulated quantities, in particular relevant system outputs, then the computer model has to be modified. The final form of the simulation model is an outcome of an iterative process involving modifications being made until the results of the simulation experiments match sufficiently well the observed real system behavior (Fig. 3).

Summarizing the above, it is useful to stress that the simulation model represents selected aspects of real system behavior, related to required model usage; design of the simulator does not require taking into account all available knowledge concerning laws and internal properties governing all real system processes [5, 8, 9].

One can formulate basic principles to be observed by the model builder: postulate of usefulness, relevance and simplicity of the model. According to these postulates a simulation model is developed for specified applications, it has to represent processes relevant for those applications and should be as simple as possible. Following these rules is particularly important in the case when computer simulation is to be used multiple times, in particular for on-line decision making, and has to be computationally very efficient. According to the terminology of structural programming one can consider two approaches to modeling:

bottom-up approach—Modeling process begins with a precise, detailed, description of the considered system, its components and properties. Broad and detailed knowledge about the underlying laws governing system processes is required. Such a model is usually complicated, involving large number of equations and describes well internal system structures and properties. Then, taking into account the desired model usage, this model may eventually, or must, be simplified.

top-down approach—Modeling begins with establishing what the model is needed for. Then a sufficient description of system processes is formulated. It is not necessary to take care of all internal system properties and therefore a less detailed knowledge is needed. Such a model contains a smaller number of equations when compared to precise description resulting from a bottom-up approach. Therefore a top-down approach follows the model simplicity postulate; yet it may result with a large number of parameters that are unknown and require identification based on experiments on the real system. Also, such a model may fail the validation procedure and need to introduce modifications leading to a more detailed and sufficiently accurate description of the relevant processes.

In engineering both modeling approaches are used, if possible, the bottom-up approach is preferred when one wants to understand well system behavior. When modeling complex biological systems, the top-down approach is usually the only possible approach. Also, when building models for the purpose of designing control systems, the top-down approach is typical. The reason this is that, even if one is able to create and use a detailed model, the fulfillment of control objectives may not require such a model to be used. In particular this is the case of regulation where it is needed to provide for desired patterns of specified system quantities to be followed. In the case of operational optimizing control, it situation may be different. When dealing with composite systems modeling, it is usually realized on several layers. One may use modular and hierarchical approaches. A modular approach follows the system structure. In the case of a hierarchical control modeling, it becomes more detailed as decision mechanisms at lower layers or levels are considered. At the upper layer, an aggregated system description is used.

3.2 Simulation Techniques

A computer simulation is an experiment conducted on a computer. The objective is to investigate a given system and observe time-based behaviors within a system for the determined time horizon, referred to as the period of a simulated time. On the other hand, the experiment takes a lot of time for computing processing. Thus, we can define two time scales: simulated time and simulation time.

Definition 1 Simulation time t is the time of the simulation experiment conducted on the computer (computation time).

Definition 2 Simulated time T is the execution time of processes running in a simulated physical system.

Physical systems, and at the same time their simulators can be categorized as continuous-time, discrete-time and discrete-events. Let us present the unified description of these three types of simulators derived in [67].

In the continuous-time simulation, simulated time $T \in \mathbb{R}$ is continuous and state, input and output, change continuously over time. Differential equations are used to model the simulated system. A general specification—DESS (*Differential Equation System Specification*) is as follows

$$DESS = (U, X, Y, f, g), \tag{5}$$

where $U \subset \mathbb{R}^m$ denotes a set of inputs, $X \subset \mathbb{R}^n$ a set of states, $Y \subset \mathbb{R}^p$ a set of outputs, $f : X \times U \to X$ the state transition function, $g : X \times U \to Y$ the output function.

A discrete-time simulation is one in which the state variables change only at a discrete set of points in time. Inputs, outputs and state variables are specified only for discrete time stamps $T_1, \ldots, T_k, \ldots$. Similarly to DESS all variables are continuous. A general specification—$DTSS$ (*Discrete Time System Specification*) is as follows

$$DTSS = (U, X, Y, f, g, h), \tag{6}$$

where $U \subset \mathbb{R}^m, X \subset \mathbb{R}^n, Y \subset \mathbb{R}^p, f : X \times U \to X, g : X \times U \to Y, h$ is the fixed time step, $T_{k+1} = T_k + h$. The state variables change at fixed intervals, so we know exactly when these changes will occur.

A discrete-event simulation is a special case of a discrete simulation with a variable time increment. Let us define a sequence of events. Each event occurs at a particular instant in time and changes the values of state variables. No change in the system is assumed to occur between the consecutive events. Thus the simulation can directly jump to the occurrence time of the next event, which is called next-event time progression. A general specification—DEVS (*Discrete Event System Specification*) is as follows

$$DEVS = (U, X, Y, f_{ext}, f_{int}, g, f_T), \qquad (7)$$

where U denotes the series of input events, $Y \subset \mathbb{R}^p$ a set of outputs, $X \subset \mathbb{R}^n$ a set of state variables, τ, $0 \le \tau \le f_T(x)$ the time period between the consecutive event that will change the current state $x \in X$. Thus, we can define an extended state vector as $q = (x, \tau)$. In this specification two types of events, i.e. "internal event" and "external event" and two state transition functions are defined. External events are caused by changes to the inputs u. We have no influence on these types of events and we usually do not know a prior to when they will occur. $f_{ext} : Q \times U \to X$ denotes the "external transition function". Internal events are caused by internal processes and are not caused by environmental impact. We know times of their occurrence and we can make a list of such events. $f_{int} : X \to X$ denotes the "internal transition function". $g : X \to Y$ is the output function. $f_T : X \to \mathbb{R}^+ \cup \{0\}$ computes the time interval $\Delta T \in [0, \infty]$ to the next external event. ΔT is calculated based on the internal events list. $f_T(x) = 0$ denotes that x is the transient state, $f_T(x) = \infty$ denotes that x is the passive state that can be changed only by the external event.

Thus, state variables may be changed by internal and external events. New system inputs cause a change of the system state, $x' = f_{ext}(x, u)$ and $\Delta T = f_T(x')$, $\tau = 0$. In the case when no external events have occurred in the time interval $\Delta T = f_T(x')$, new outputs and states of the system are calculated $y = g(x')$ and $x'' = f_{int}(x')$, $\Delta T = f_T(x'')$, $\tau = 0$. In the presented specification the calculation of new system outputs have to precedes the execution of the internal transition function. A new external event for $\tau < f_T(x')$, involves only the change of the system state.

Thus, the discrete-event simulator maintains a list of external and internal simulation events that will occur in the future. These events are the results of previously simulated events. Each event is described by the time at which it will occur and indicates the computing process that will be used to simulate that event. Simulation is implemented by maintaining a global clock (GVT—Global Virtual Time). The event with the smallest time-stamp is removed from the event list. When events are instantaneous, activities that extend over time are modeled as sequences of events. The simulation can be implemented in two ways, the so-called *event-driven simulation*, when events occurring in a physical processes are chronologically simulated, and in each iteration, the global clock assumes a value corresponding to the time stamp of the event being performed. Another implementation is a *time-driven simulation*. In this variant, after each iteration, the clock (GVT) increases by a fixed value (one clock cycle) and all events are carried out in turn with the corresponding time stamp.

In the case of discrete event simulators of composite real-life systems, it is natural to model such a simulator as a set of computing processes which then can be handled by distributed machines or processors. For the last few decades, parallel and distributed simulation has been an active research area. Distributed simulations not only reduce the computation time and permit it to execute large programs that cannot be executed on a single machine, but they first of all reflect better the structure of the physical system to be simulated, which usually consists of a set of components. Parallel and distributed discrete-event simulations can be described in terms of logical processes LP_i, $i = 1, \ldots, N$, that communicate with each other through message passing. LP_i simulates the i-th real life physical process FP_i. Each logical process starts processing as a result of an event occurrence (from the event list or having received a new message). It performs a few calculations and usually generates some local events and messages for other processes. In parallel simulators, each logical process maintains its own local clock (LVT—Local Virtual Time). Local times of different processes may advance asynchronously. Events arriving at the local input message queue of a logical process are executed according to the local clock and the local schedule scheme. Synchronization mechanisms fall into three categories: conservative, optimistic and lookback-based. They differ in their approach to time management [13, 17, 39, 46, 55, 67]. Conservative schemes avoid the possibility of causality error occurrence. These protocols determine safe events that can be executed. Optimistic schemes allow for the occurrence of causality errors. They detect such errors and provide mechanisms for their removal. The calculations are rolled back to a consistent state by sending out antimessages. It is obvious that, in order to allow rollback, all results of previous calculations have to be recorded. Lookback schemes permit us to consider only causality errors, which require rollback but no antimessages, so they fall in between conservative and optimistic schemes.

4 Computer Simulation in Complex Control Systems Design

Last years given us a revolution in computing, both in terms of hardware and software. It also became apparent that in systems engineering as in physics, chemistry and many other areas, traditional theoretical and experimental research has been supplemented by computational research. Computer experiments should be performed at the pre-implementation phase and precede real-life experiments. In view of this, the analysis and design of systems, based on computer simulation become the only reasonable solution. A simulation is used both at the stage of building the system model, for its identification, testing compliance with the physical system, and when designing structures and control mechanisms or at the stage of planning activities. It allows: (1) to learn the characteristics of a system and understand its operation, (2) to estimate the values of indices describing the quality of the system operation, (3) to evaluate the correctness of the control system design and to indicate directions of its improvement.

The simulators are used indirectly for: (1) studies covering investigating the internal structure of the system, i.e. information connections and rules to which processes in the system are subject, (2) forecasting, i.e. predicting the future behavior of the system, (3) design, the purpose of which is to develop an optimal, in the sense of given criteria, control structure and decision mechanisms. The simulator can also be used explicitly, in decision determination procedures in an operational mode [44, 45]. The role of computational research is particularly and increasingly important in the field of large scale systems [16]. Here, the size of the considered problem, partitioning of processes into several interconnected units, multiple time scales and multiple decision units make it impossible to assess system behavior, under realistic conditions, through a theoretical analysis.

Computational research could be possibly thought of as a form of experimental research but, in fact, it is not that. It requires different methodology, different tools and involves a need to solve many additional problems. Design of control structures for complex systems requires a combination of good knowledge of the controlled process, understanding of relevant theoretical issues and the use of computer simulation and optimisation tools. There is a need to develop specialised software systems for different classes of applications that can become essential tools for evaluating alternative solutions prior to actual implementation. In view of this, we can talk about a new system analysis tool, namely computer analysis of systems that combines technical fields, e.g. automatic control that raises technical problems and gives methods to solve them with computer science, enabling software and hardware implementation. It is currently difficult to imagine a decision-making process without a computer simulation, both at the planning, design and operational stages.

Definition 3 Computer-aided analysis and simulation (CAS) of control systems is type of formal and explicit research based on a computer simulation supporting the design of control systems for systems operating under uncertainty. Its purpose is to verify and evaluate the performance of developed control structures and decision mechanisms and to assess the possible effects of their implementation, taking into account the complexity of phenomena occurring in the real system.

In view of the above definition, the computer together with the appropriate software, performs an essential function in the design process. It replaces the real system, simulates the operation of the control system, and creates the environment in which the experiment is performed. CAS allows the complexity of phenomena in a real system, system-environment interaction and its mathematical model to be scoped in.

Control system design stages. During design of a complex control system one can distinguish two main stages:

1. Preparatory stage:
   - Problem understanding: objectives specification, process modeling, computer process simulation.
   - Preliminary design of a control structure and its mechanisms: theoretical analysis of simplified cases, algorithms for identification, forecasting and decision making.
   - Formulation of assumptions for the entire system simulation, i.e. determining the time scale, decision and transmission delays, the possibility of failure in the operation of the system, etc.

2. Interactive design stage with CAS:
   - Computer implementation of the structure and control mechanisms.
   - Data preparation and creation of databases for which the tests will be carried out.
   - Input scenarios definition and design of experiments.
   - Computer-aided analysis: computer experiments to check reliability of control mechanisms, timeliness and robustness of operation, quantitative results of operation under different scenarios of external inputs, tuning and modification of control algorithms.
   - Evaluation of results and preparation of the report.
   - Real-life, practical, implementation of the control structure: assess the system behavior, under realistic condition.

At the preparatory stage the description (model) and the properties of the process to be controlled are obtained and investigated in broad terms. The objectives are specified and expressed in the form required. Then, the layout of the control structure is proposed, together with algorithms to be used during on-line operation for: identification (filtering), forecasting and decision making. It is obvious that the complexity of those algorithms is growing. As we consider more complex problems and structures and as more powerful computing facilities are available for the on-line usage. This last factor is in fact extremely important as it somehow changes the basic paradigm of control design. For many years the control and system engineers were used to the idea that the control rule had to be simple and computationally low demanding. Many sophisticated techniques were proposed to design such control algorithms and, due to rather simple, mostly linear form, it was possible to investigate the properties of those algorithms by the theoretical analysis. Now, we see the major change. Control algorithms can be made and are made quite complex and computationally demanding. In particular model-based predictive control for nonlinear systems involving repetitive, on-line, usage of nonlinear models, real-time forecasting tools and optimisation techniques, is considered for many applications, e.g. [16, 21, 29]. During the preparatory stage, good understanding of underlying theory is needed as well as the knowledge about the process to be controlled. Computer simulation may be required to get better understanding of the process properties.

The interactive design stage begins when the structural decisions regarding the control structure and algorithms are made. These algorithms then have to be made operational, tuned, checked for reliability and timeliness and validated in terms of meeting the design objectives. Eventually they can be modified or even replaced with others. At this stage, computer-aided analysis and simulation (CAS) is necessary if we want to have the control system tested and verified prior to the actual real implementation. The objectives of CAS are:

- to test the functioning of all the algorithms involved: their correctness, accuracy, timeliness and reliability,

- to investigate system performance under various possible scenarios of uncontrolled inputs or disturbances, and under variations of the process parameters (parameters of the process model),
- to seek a trade-off—where necessary—between the use of more complicated decision techniques, involving larger decision delays, and the use of less complicated algorithms—allowing for prompt control actions but offering less in terms of control objectives.

CAS: components and tools. In order to perform CAS efficiently and in a proper manner it is required to have:

- Process simulator, consisting—if required—of several modules related to the process components.
- Simulator of the process environment; in particular the module, or modules, allowing for generation of the uncontrolled inputs to the process and other data required by the control algorithms.
- Data related to the process and the process environment; parameters, historical data, probability distributions etc., time scale, decision and transmission delays.
- Control structures and algorithms for forecasting and decision making (e.g. optimisation routines)—these are chosen within the preparatory stage but have to be available as software modules with well define interfaces.
- Software environment (shell) allowing for convenient organization of the simulation experiments—with: friendly user interface, facilities for timing and communication between participating tasks and data base for storing historical data, input scenarios and the results of simulation.

Both process, and process environment simulators are necessary to create a "virtual reality". CAS will provide useful, meaningful, results only in the case when we can put our trust in this reality. One cannot also overestimate the importance of having sufficient and reliable data related to the process and its environment. Computer-aided analysis cannot be done without such data.

Various algorithms forming control mechanisms should be available prior to a simulation experiment. It may happen, however, that some of these algorithms will be very complicated and it will be useful to replace them, for the purpose of the simulation, with much simpler ones. In particular, assume that a real-time control system involves the use of complicated forecasting procedures. As far as the results of the computer simulation are concerned the use of forecast dummies could be possible. The forecast dummy would be obtained by using a much simpler algorithm. It is only essential to make sure that on average the forecast dummies have the same probabilistic error characteristics as the real forecasts operationally generated on-line.

Software frameworks. While proper process and process environment simulators are required to obtain meaningful results from computer-based analysis, good software frameworks are needed to make this analysis fast and effective. There are two basic directions to take when developing such software frameworks for CAS [5, 23, 46]:

1. Development of a problem dedicated (specialised) system, specific for a given type of process and control problem,
2. Creation of general purpose (universal) systems, allowing for setting up simulation experiments and analysis for different types of processes.

The advantage of a specialised system is that one can have in-build in the software framework, both typical algorithms for identification, and control as well as process simulators. The disadvantage is that such framework has a restricted usage and it is difficult for the user to modify it when new features need to be introduced.

General purpose systems for CAS have a much wider range of applications. This class of systems allows to set up simulation experiments and to analyse different types of processes. However, one must pay for this universal applicability with having to prepare, for each particular case study, those software modules which are specific to this study. Very well known integrated software packages like Matlab, Simulink belong to this category of systems. These computer tools provide libraries of tools and framework organization for different classes of applications.

5 Computer Simulation in Analysis and Operational Control

In the previous sections attention was focused on the role of computer simulation in designing and performance evaluation of control systems. However, simulation models are intensively used not only to investigate physical systems [12, 60], or to design and evaluate control structures and decision mechanisms [9, 30, 42]. They are important components of many decision mechanisms used in operational controls.

5.1 Simulation-Based Optimization Problem Formulation

The current trends and needs in complex systems optimization involve the application of a simulation-based optimization scheme, which merges optimization and simulation technologies. Traditionally, the complex optimization problems are solved using linear and nonlinear techniques, which normally assume that the performance function and the set of admissible solutions are known in analytical form. However, the load of mathematical and practical knowledge input into the optimization model often results in numerous formulas, the solution of which can be obtained only numerically. Thus, in many cases, simulation is the only viable method to calculate given objective functions and to obtain crucial performance characteristics. In this section we will focus our attention on control systems in which the final decision is determined by solving the optimization problem with the help of the repeatedly running simulator. In recent years we have observed a rapid growth of applications of simulation-based optimization [2, 3, 15].

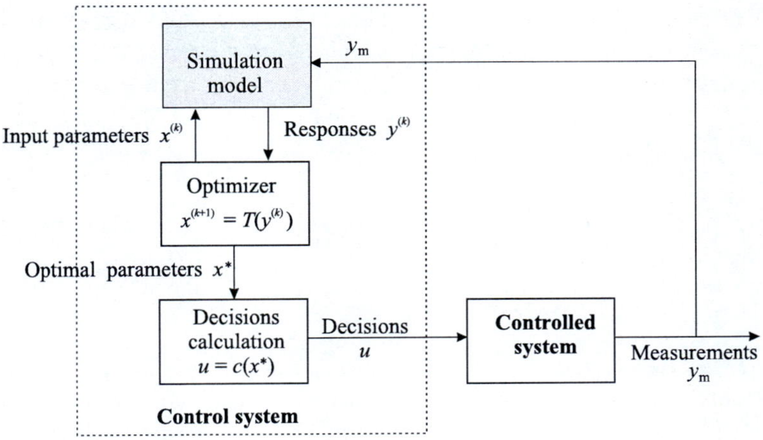

Fig. 4 Simulation optimization scheme for control decisions calculations in operational control

Let us formulate a definition.

Definition 4 Simulation-based optimization (or simulation-optimization) is a search or optimization technique that uses the simulation experiment to evaluate the expected performance of the system for each set of decision variables (system inputs).

Figure 4 shows the control system in which control decisions are calculated based on the results of a simulation experiment.

In the context of simulation optimization, a simulation model is a function, $S : \mathbb{R}^n \to \mathbb{R}^m$, mapping input parameters, $x \in \mathbb{R}^n$, to outputs, $y \in \mathbb{R}^m$. The explicit form of the function S is usually unknown. Its numerical representations (sets of y) that are the results of simulation experiments performed for various input parameters x are called *response surfaces*. We can formulate the general problem of optimization through embedded simulation:

$$\min_x [f(x) = \mathcal{J}(x, S(x))], \qquad x \in D_x, \tag{8}$$

where $\mathcal{J} : \mathbb{R}^{n+m} \to \mathbb{R}$ denotes a performance measure calculated for simulator responses $y = S(x)$ for input variables x, and D_x specifies the constraints on the input variables. So, in this context the role of the simulator is to transform input parameters x into outputs y used to calculate the performance measure $\mathcal{J}$.

As an abstraction of real-life systems operation, simulators have limits of credibility. Such limits may result from simplifications in mathematical modelling and its computer implementation, as well as omission of some possible values of inputs and the random effects of the considered system (environment interactions). Assume that vector $V = [\vartheta_1, \ldots, \vartheta_m]$ represents the amalgamation of many individual random effects in the simulator. The simulator response for decision variables x and random effects V are calculated as $y = \tilde{S}(x, V)$, where $\tilde{S}$ denotes a simulation model considering the random effects in the simulated system. Let us denote the sample realization of the performance function calculated from running the simulation by q. Hence, we can formulate the stochastic simulation-based optimization problem:

$$\min_x \left[f(x) = E\left\{ q(x, V) = \mathcal{J}(x, \tilde{S}(x, V)) \right\} \right], \qquad x \in D_x, \tag{9}$$

where f denotes an average performance measure over all possible values of V at the specified x.

Two situations concerned with the type of influence on the system by the inputs x can be distinguished [57]:

1. Inputs x effect directly the outputs y and does not influence the random vector V (*structural parameters*). The performance function in (9) is calculated as follows:

$$f(x) = E\{q(x, V)\} = \int_{D_v} q(x, v)p(v)dv, \tag{10}$$

   where $p(v)$ is the density of probability distribution of V, and D_v the domain of V. Moreover, for independent and uniform distribution of vectors $V_1, V_2, \ldots, V_N$, the form of the indicator is simplified to

$$f(x) = \frac{1}{N} \sum_{i=1}^{N} q(x, V_i). \tag{11}$$

2. Inputs x influence probability distribution of V (*distributional parameters*). In this case the performance function in (9) is calculated as follows:

$$f(x) = E\{q(x, V)\} = \int_{D_v} q(x, v)p(v|x)dv. \tag{12}$$

   where $p(v|x)$ is the density of probability distribution of V.

In general, most of the optimization solvers assume case (10), therefore, if possible, performance (12) is often modified and transformed to the form (10).

To sum up, simulation optimization is generally complex but seems to be the only viable option to solve many practical engineering problems. The restrictions are caused by the computational complexity demands on memory and CPU. The efficient optimization process requires a large number of iterations and system performance evaluations. For every iteration of the optimization algorithm we have to perform, time-consuming simulation experiment (see Fig. 4), and it is not guaranteed that it will be successful for each set of inputs.

5.2 Numerical Methods

The optimization problems (8) and (9) are usually nonlinear, often non-convex and gradient information is rarely available. The features that complicate the task are the demand that the solution is globally optimal and that constraints may be put both on the decision variables (simulator inputs x) and the responses y. If the constraints depend only on input variables x, a feasible trial solution can be checked before the simulation execution, otherwise the feasibility of new trials is not known before running the simulation experiment. Additionally, the constraints provided explicitly are accompanied by a number of implicit constraints, defined on the internal variables of the modelled system. Hence, it is difficult or even impossible to assess whether the domain D_x is convex and compact.

Many optimization techniques that could be employed for solving complex optimization problems (8) and (9) have been reported in the literature [2, 4, 15, 34, 38, 57]. The most popular are: (1) stochastic approximation (gradient based approaches), (2) sample path optimization and response surface methodology, (3) deterministic search methods, (4) random search methods, (5) metaheuristics, i.e., genetic algorithms, evolutionary strategies, simulated annealing, etc. Most of these methods need numerous simulation runs for performance measure calculations and slowly converge to the solution. A direction, which could bring benefits, is parallel computing. Another approach to reduce the computational load, is the use of a metamodel. Popular techniques used to build metamodels are: linear and non-linear regression, neural networks, etc. The metamodel is employed to estimate performance in (8) or (9) instead of a simulation model (see Fig. 5). It approximates the response surface of S function and the performance measure $\mathcal{J}$, corresponding to the simulation model. It is often used as a filter, the goals of which are to predict the performances of new trial solutions, compare them with the current best known one and eliminate low quality solutions from further consideration. In such an approach the decision concerned with trial points rejection is a trade off between the speed and accuracy of the search.

The techniques for solving simulation-based optimization tasks are briefly described below.

Stochastic approximation. In many applications that require a solution of the optimization problem (9) we are unable to determine the value of the performance measure f because we do not know the probability distribution of V. We can, however, calculate the values of the simulator outputs and calculate q in (9). Let $\tilde{S}$ and q be convex, continuous and differentiable in relation to x for all V. We can calculate partial derivatives. The vector of partial derivatives, $\nabla q = \frac{\partial q}{\partial x}$, is called a stochastic gradient because it depends on random V disturbances. Thus, the optimization task (9) can be solved using the stochastic approximation algorithm. It is a well known family of iterative gradient search methods [57].

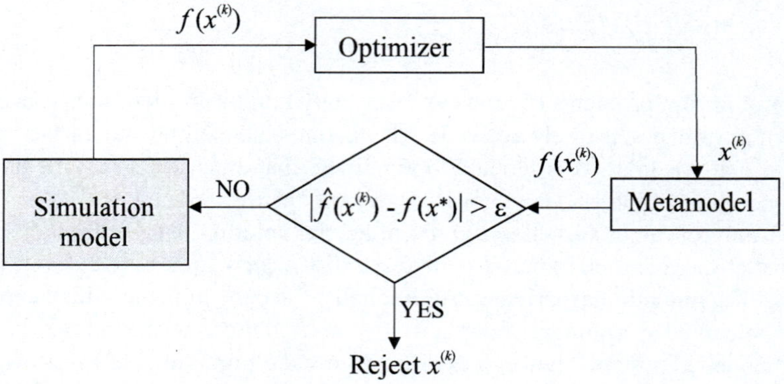

Fig. 5 Metamodel application to optimization; x^*—the best (current) solution, $f(x)$—the performance obtained based on simulation, $\hat{f}(x)$—the performance obtained by evaluating the metamodel, ε—assumed accuracy

1. Set $k = 0$. Choose the initial values $x^{(k)} = x_0, x_0 \in D_x$ and a sequence of positive step sizes $\{\alpha^{(k)}\}$ that satisfies this restriction: $\sum_{k=0}^{\infty} \alpha^{(k)} = \infty$ and $\sum_{k=0}^{\infty} \left[\alpha^{(k)}\right]^2 < \infty$.
2. Run the simulation and compute the estimated value of a gradient $\nabla f(x^{(k)})$, i.e. $\nabla \hat{f}(x^{(k)}) = \frac{\partial q(x, V^{(k)})}{\partial x}|_{x=x^{(k)}}$.
3. Compute a new potential solution $x^{(k+1)} = \Pi \left(x^{(k)} - \alpha^{(k)} \nabla \hat{f}(x^{(k)})\right)$, where Π denotes a projection operator on the set D_x.
4. Terminate the algorithm if the stopping criterion is met; else set $k = k + 1$, $x^{(k)} = x^{(k+1)}$ and return to step 2.

The proof of convergence of the algorithm is presented in [57]. Two basic problems that need to be solved are: determination of an estimate of the stochastic gradient and determination of the sequence of step sizes that ensure the convergence of the algorithm. Various methods of matching the step are proposed and described in [65, 66]. The following sequences: $\alpha^{(k)} = a/k$ (k—denotes iteration step), $a > 0$ and $\alpha^{(k)} = \frac{b}{1+k/k_0}$ ($b > 0$, $k_0 >> 1$) satisfy the conditions defined in step 1 of the algorithm.

Gradient estimation. Assume that q and p are differentiable with respect to x. A key issue in the gradient calculation is the question of validity of the interchange of the order of integration and differentiation proposed in the following theorem proven in [62].

Theorem 1 *Let $h(x, v)$ and $\partial h(x, v)/\partial x$ be continuous on $D_x \times D_v$. Suppose that there exist nonnegative functions $c_0(v)$ and $c_1(v)$ such that*

$$\forall (x, v) \in D_x \times D_v \quad |h(x, v)| \leq c_0(v), \quad \left|\frac{\partial h(x, v)}{\partial x}\right| \leq c_1(v) \tag{13}$$

and $\int_{D_v} c_0(v)\mathrm{d}v < \infty$ *and* $\int_{D_v} c_1(v)\mathrm{d}v < \infty$, *then*

$$\frac{\partial}{\partial x}\int_{D_v} h(x,v)\mathrm{d}v = \int_{D_v}\frac{\partial h(x,v)}{\partial x}\mathrm{d}v. \tag{14}$$

Let us show how the proposed interchange is used to compute the estimate of the gradient ∇f under the condition that $q(x,v)p(v|x)$ satisfies the condition (14)

$$\frac{\partial f}{\partial x} = \frac{\partial}{\partial x}\int_{D_v} q(x,v)p(v|x)\mathrm{d}v = \int_{D_v}\left[q(x,v)\frac{\partial p(v|x)}{\partial x} + \frac{\partial q(x,v)}{\partial x}p(v|x)\right]\mathrm{d}v \tag{15}$$

Assuming that $p(v|x) > 0$ and multiply (15) by $p(v|x)/p(v|x)$ we obtain

$$\begin{aligned}\frac{\partial f}{\partial x} &= \int_{D_v}\left[\frac{q(x,v)}{p(v|x)}\frac{\partial p(v|x)}{\partial x} + \frac{\partial q(x,v)}{\partial x}\right]p(v|x)\mathrm{d}v \\ &= E\left\{q(x,V)\frac{\partial \log p(V|x)}{\partial x} + \frac{\partial q(x,V)}{\partial x}\right\}.\end{aligned} \tag{16}$$

From (16) we can simply produce unbiased estimates of f at any $x^{(k)}$. In practice, at each k-th step we randomly select with the probability $p(v|x = x^{(k)})$ N independent variables $V^{(k)} = [V_1^{(k)}, V_2^{(k)}, \ldots, V_N^{(k)}]$ and compute the estimate $\nabla f(x^{(k)})$

$$\nabla\hat{f}(x^{(k)}) = \frac{1}{N}\sum_{i=1}^{N}\left[q(x^{(k)},V_i^{(k)})\frac{\partial \log p(V_i^{(k)}|x)}{\partial x}\Big|_{x=x^{(k)}} + \frac{\partial q(x,V_i^{(k)})}{\partial x}\Big|_{x=x^{(k)}}\right] \tag{17}$$

It is useful to distinguish types of input parameters presented in Sect. 5.1 when gradient optimization methods are considered. It is proved in [57] that in the case of the first variant (*structural parameters*) the infinitesimal perturbation analysis (IPA) method is used to estimate the stochastic gradient. In the case of the second variant (*distributional parameters*) the likelihood ratio method (LR) is applied, for both types of inputs IPA/LR is used.

IPA: $\frac{\partial q(x,V)}{\partial x}$,

LR: $q(V)\frac{\partial \log p(V|x)}{\partial x} = \frac{q(V)}{p(V|x)}\frac{\partial p(V|x)}{\partial x}$, $p(V|x) > 0$,

IPA/LR: $q(x,V)\frac{\partial \log p(V|x)}{\partial x} + \frac{\partial q(x,V)}{\partial x}$.

The detailed considerations on IPA and LR, including examples of applications can be found by in [52, 57]. It should be pointed out that the main advantage of these techniques is that both IPA and LR require only a single simulation run for one set of inputs and involve differentiating in equations (10) or (12) directly. Is seems that the simulation outcome has to contain gradients evaluations.

Practical methods for gradient estimation. In general, both IPA and LR are difficult to implement. They involve modifications in a simulation model. Another approach is to use finite differences and multiple simulations to calculate the derivatives of the performance function f.

$$\frac{\partial \hat{f}(x)}{\partial x_i} = \frac{\hat{q}(x + he_i, V_i^{(+)}) - \hat{q}(x, V_i^{(-)})}{h}, \tag{18}$$

where e_i denotes i-th unit vector with 1 at the i-th position, $\hat{q}$ is the estimate of the function f in (8) computed, based on the simulations conducted for $x + he_i$, x and random disturbances $V_i^{(+,-)}$. The derivative can be estimated more precisely using a central difference.

$$\frac{\partial \hat{f}(x,)}{\partial x_i} = \frac{\hat{q}(x + he_i, V_i^{(+)}) - \hat{q}(x - he_i, V_i^{(-)})}{2h}. \tag{19}$$

The central difference method assumes simultaneous disruption of function in all directions is used.

$$\frac{\partial \hat{f}(x)}{\partial x_i} = \frac{\hat{q}(x + hd, V^{(+)}) - \hat{q}(x - hd, V^{(-)})}{2hd_i}, \tag{20}$$

where elements of $d = [d_1, d_2, \ldots, d_n]^T$ are randomly selected values.

Sample path method. In the sample path method the original optimization problem is converted into an approximated deterministic problem. The approximation $\hat{f}$ of the performance function f is calculated based on at least K simulations performed for a random generated set of independent observations V, i.e., $V_1, V_2, \ldots V_M$, $M \neq K$ referred to as the *sample path*. Then standard optimization algorithms are used to locate the optimal solution. In the case when the cumulative distribution function of the random vector V does not depend on inputs x, the optimization problem (9) is transformed to

$$\min_x \left[\hat{f}(x) = \frac{1}{M} \sum_{i=1}^{M} q(x, V_i) \right]. \tag{21}$$

A fundamental issue in the sample path method is the determination of the distribution for generating the fixed sample consider ensemble $V_1, V_2, \ldots V_M$. A problem arises when the distribution of V_i depends on x, which involves the generation of new sets of V for all new values of x. The solution of this problem is presented in [57] and briefly described below. Suppose that there exists a fixed value $x' \in D_x$ such that for all $x \in D_x$ and $v \in D_v$ the condition $p(v|x) = 0$ is implied by the condition $p(v|x') = 0$. Using the Eq. (12) we get

$$f(x) = \int_{D_v} q(x, v)p(v|x)dv = \int_{D_v} \left[q(x, v)\frac{p(v|x)}{p(v|x')} \right] p(v|x')dv$$
$$= E\left\{ q(x, V)\frac{p(V|x)}{p(V|x')} \right\}. \tag{22}$$

Let $V_1, V_2, \ldots, V_M$ be a sequence of M values of V generated according to $p(v|x')$, then

$$\hat{f}(x) = \frac{1}{M}\sum_{i=1}^{M} q(x, V_i)\frac{p(V_i|x)}{p(V_i|x')}. \tag{23}$$

Response surface methodology. RSM is a sequential strategy based on local approximation of the performance f in the current neighborhood $D_x^{(k)}$ of x. A function $F(x, \alpha)$ is used to approximate f. The parameters α are calculated using simulations and k denotes the iteration number. The following operations are repeated unless the acceptable solution is found.

1. Set $k = 0$. Select M values of input data $x_1^{(k)}, \ldots, x_M^{(k)} \in D_x^{(k)}$.
2. Perform M simulation experiments at $x_1^{(k)}, \ldots, x_M^{(k)}$. Compute set of parameters $\alpha^{(k)}$ for which the approximation error $F(x, \alpha)$ is minimized.
3. Calculate the minimal value of $F(x, \alpha^{(k)})$, i.e. $x^B = arg\min_x F(x, \alpha^{(k)})$.
4. Run the simulator at x^B and compute $\hat{f}(x^B)$. If $\hat{f}(x^B) < \hat{f}(x^{(k)})$, then set $k = k + 1$ and $x^{(k)} = x^B$, otherwise augment $D_x^{(k)}$ or modify F. Return to step 2.

Various functions F are proposed in the literature [2, 52, 57]. Widely used are linear or quadratic regression. While RMS has a long history of success, there is no guarantee of greater efficiency in computing an optimal solution than other stochastic optimization methods.

Direct search techniques. Standard deterministic search techniques ([48, 50]) for nonlinear optimization, such as algorithms developed by Hook and Jeeves, Rosenbrock or Nelder and Mead can be applied to solve non-differentiable simulation-based optimization problems. In the Hook-Jeeves and Rosenbrock algorithms the neighborhood of the current solution is explored to find a new one. New points determined in directions that are calculated as the convex combinations of coordinate vectors are evaluated. Another commonly used direct search is the downhill simplex algorithm (Nelder and Mead). It is based on the concept of a simplex, a convex figure of $n + 1$ vertices in $\mathbb{R}^n$. At each iteration a new point is calculated and replaces the worst of the vertices of the current simplex. Although the downhill simplex method was originally a deterministic technique, it has frequently been used in a stochastic setting with noisy performance function measurements.

All optimization algorithms discussed so far can be successfully used for calculating the local optima. In the case of non-convex optimization problems, other techniques are recommended. As a result of the growing possibilities of modern computers, we can observe increasing interest in the development of the global algorithms that are addressed to compute the global optima of non-convex functions.

During the last decades many theoretical and computational contributions can helped to solve multimodal problems. Global optimization methods are widely used in many industrial and scientific applications. Their approach to simulation optimization is based on viewing the simulation model as a black box function evaluator. These approaches are flexible, robust and less demanding when regarding properties of the problem. Deterministic techniques are often based on branch-and-bound approaches that adaptively perform partition, search and bounding [19]. Numerous global solvers use the stochastic algorithms typically based on random search [57]. The idea of pure random search is very simple. New points are randomly selected from uniform distribution. Simulation experiments are conducted for these points and objective functions are estimated. The best one is taken as the problem solution $\hat{f}$. Instead of random generators the predetermined sequences of n-tuples can be used for new points selecting. They fill n-space more uniformly, than uncorrelated random points and are referred as quasi-random sequences [48]. This term is somewhat of a misnomer, since there is nothing "random" about quasi-random sequences. Three commonly known such sequences were developed by Halton, Sobol and Faure. Multi-start local search is another proposition. The current solution is shifted by randomly generated vector $d^{(k)} \in \mathbb{R}^n$, $x^{(k+1)} = x^{(k)} + d^{(k)}$ or in a modified version $x^{(k+1)} = x^{(k)} + d^{(k)} + b^{(k)}$, where $b^{(k)}$ is a deviation vector determined according to the rule $b^{(k+1)} = \alpha b^{(k)} + \beta d^{(k)}$, α and β are coefficients. Searching the neighbourhood of the current point can also be carried out using the previously mentioned simple deterministic techniques. Adaptive search methods, like *Controlled Random Search* CRS is another option. In principle, they are simple, general purpose algorithms which are based on a combination of random sampling and direct search methods, e.g. downhill simplex method. The CRS2 algorithm starts from the creation of a set P of N_P points. Then, simulations are run for all these points and values of the estimate of the performance measure are calculated $\hat{f}(x_i), i = 1, \ldots, N_P$. Two points are selected $x_l = arg\min_{x \in P} \hat{f}(x)$ and $x_h = arg\max_{x \in P} \hat{f}(x)$. The P set is transformed over time. Currently, many versions of this algorithm are described in [1, 31, 35, 49]. Different methods implement different algorithms for set P transformation. They are robust and can be easy to speed up through parallel implementation.

Methods based on heuristics and metaheuristics. Most global techniques utilize heuristics and do not guarantee the convergence to the optimum solution, but rather provide a reasonable solution in a reasonable time. *Genetic Algorithms* GA [7, 18], *Evolutionary Strategies* ES [7, 34], *Simulated Annealing* SA [14, 24, 48, 57, 64] are all of a heuristic nature. Nowadays application of heuristics and metaheuristics to simulation-based optimization problems is very popular. SA is a sequential search technique that avoids getting trapped in local minima. In addition to transitions corresponding to a decrease in function value, transitions corresponding to an increase in function value are accepted. In the course of the minimization process, the probability of accepting deterioration decreases slowly towards zero and the procedure leads to a global minimum. Simulated annealing originated from the analogy with the physical annealing process of finding low energy states of a solid in a heat bath [14]. Firstly, this idea was adopted to solve discrete combinatorial optimization problems [24, 48]. Formulations of the simulated annealing for continuous optimization have been addressed by a number of authors (see [14, 48, 57]). In general, the simulated annealing algorithm operates as follows. The initial values of x and the control parameter T are determined, and $f(x)$ is evaluated. Next, the new points are selected from the domain and values of the performance function are compared. Better solutions, satisfying the condition $f(x^{(k)}) \leq f(x^{(k-1)})$ are always accepted. Acceptance probability of worse solutions depends on the value of the following expression $exp\left(\frac{\hat{f}(x^{(k-1)})-\hat{f}(x^{(k)})}{T^{(k)}}\right)$. The main problem in the practical implementation of SA is the choice of the *cooling schedule* for decreasing the control parameter of the algorithm, that is an analogue of temperature T. This parameter influences the probability acceptance of inferior solutions, namely the level of acceptance decreases as T decreases. Different cooling schemes are introduced and described in [14, 48]. The widely known ones are presented below.

1. $T^{(k+1)} = (1-\epsilon)T^{(k)}$, where $\epsilon \in (0, 1)$; the parameter T is decreased after every k iterations
2. $T^{(k+1)} = T_0(1-\frac{k}{K})^{\alpha}$, where k denotes the iteration number, K maximum number of iterations, α the constant parameter.
3. $T^{(k+1)} = \beta(f(x^{(k)}) - f(\hat{x}))$, where $\hat{x}$ denotes the best solution calculated so far.
4. $T^{(k+1)} = T^{(k)}\left(1 + \frac{T^{(k)}ln(1+\rho)}{3\sigma(T^{(k)})}\right)^{-1}$, where ρ is a parameter experimentally chosen, $\sigma(T^{(k)})$ standard deviation of f calculated for the current value of T.

The SA algorithm can be described using Markov chains. The proof of asymptotic convergence of this algorithm is presented in [14].

Genetic algorithms and evolutionary strategies are stochastic search methods inspired by the principles of biological evolution [18, 33]. The general algorithm is to sample a number of independent points (individuals) from a given distribution P, evaluate these points using values of the performance measures f "fitness". Then, the selection, mutation and (sometimes) recombination operations are performed. Finally, the population evolves towards a global minimum target solution. Two groups of algorithms can be distinguished: classical genetic algorithms using fixed-length binary strings for individuals representation and evolutionary strategies using real-valued individuals. In evolutionary strategies each individual is usually represented by a pair of vectors, i.e. (x, σ), where x denotes a given solution and σ is the standard deviation.

Various variants of evolution techniques are described in the literature [10, 34]. To speed up convergence to the optimum, the popular approach is to employ parallel computation and combination of global methods with fast local or global search [27, 48]. Recently *Covariance Matrix Adaptation Evolution Strategy* CMA-ES has been considered to be one of the best choices against ill-conditioned, non-convex simulation-based optimization problems [10, 59]. CMA-ES [28] is an evolutionary algorithm based on Gaussian mutation and deterministic selection. In the CMA-ES algorithm, P is a multivariate normal distribution that is a generalization of the one-dimensional (univariate) normal distribution to higher dimensions. Hence, a random vector is said to be n-variate normally distributed if every linear combination of its n components has a univariate normal distribution. Normal distribution is a good candidate for a randomized search, it has the largest entropy for given mean values, variances and covariances of all distributions in $\mathbb{R}^n$ and coordinate directions are not distinguished in any way. Therefore, in CMA-ES, a population of new search points is generated by sampling a multivariate normal distribution. At every iteration k new points $x_i^{(k)} \in \mathbb{R}^n$ are calculated as follows

$$x_i^{(k+1)} = m^{(k)} + \sigma^{(k)} \times \mathcal{N}_i^{(k)}(0, C^{(k)}), \quad i = 1, \ldots, I, \tag{24}$$

where $m^{(k)}$ and $C^{(k)}$ denote the approximated mean value and the $n \times n$ covariance matrix of the search distribution at the k-th iteration, $\sigma^k > 0$ is the standard deviation, step-size at the k-th iteration, $\mathcal{N}_i^{(k)}(0, C^{(k)})$ a normal distribution with the mean equal 0 and I is a the population size. Hence, the mutation is performed by a perturbation with a covariance matrix which is iteratively updated to guide the search towards areas with expected lower objective values. After a generation of the population of individuals they are evaluated on f (8), sorted and transformed according to (24). Every iteration of all distribution parameters $(m^{(k)}, C^{(k)}, \sigma^{(k)})$ are updated.

6 Simulation in the Design of Control Systems and Operational Control

Computer-aided analysis and simulation methodology was verified via the development of decision mechanisms for controlling two large scale water systems. The first one is a water supply system. The system's complexity is determined by the dimensions of the input, state and control vectors. Referring to the Sect. 2.1, it is a system of a component network type (1) and (2). The second example shows the process of the development of control structures for a multi-reservoir system operating under flood conditions. A system composed of multiple water storage reservoirs with local decision units is an example of a network of subsystems defined according to the formulas (3) and (4).

6.1 Optimizing Control for Municipal Water Distribution System

A Water Distribution System (WDS) for a large town consists of multiple components: water treatment plants, clean water wells reservoirs including underground reservoirs and elevated tanks, pumping stations and transmission utilities including pipes and various types of control valves. A network of water mains transports water from pumping stations to storage facilities and to demand nodes. Multiple valves, including pressure regulating valves (PRV), are important components of the distribution network. In general, the objective of operational optimizing control of WDS is to meet service delivery levels and other requirements (pressures, reservoir levels, water quality) while minimizing the electrical cost of water pumping. Scheduling pump operations, which is main task of operational control, makes a very complex problem, in particular because of its large dimension. Consider, for example, the City of Toronto Water Supply System (TWS), one of the largest in North America, consisting of treated water procurement at four filtration plants, 29 pumping stations, floating storage at 19 reservoirs and 9 elevated tanks, and approximately 520 km of water mains that transport treated water from four water treatment plants situated at the Lake Ontario level up through the system. The system has over 3500 main pipes, 4000 junction nodes, and over 1000 demand nodes. Water is pumped through a hierarchy of pressure districts with elevated storage facilities (reservoirs and tanks). Within each district, there are a number of water supply connections from the transmission water mains to the local water distribution systems. The primary objective of the Transmission Operations Optimizer (TOO) is to ensure that the required water delivery standards are met, while minimizing electrical cost of water pumping, by scheduling operations of 153 pumps. TOO works as an on-line tool 'above' the SCADA control system. From the point of view of operational control dynamics of the system, it is determined by working out capacities of drinking water reservoirs and by quasi-periodicity of demands and electrical tariffs. In the case of TWS it appeared necessary to consider planning of pumping operations over a weekly planning period divided into fifteen minutes time steps. Pumping scheduling has to be repeated at least once a day; more often if so required. A basic role in developing and using the automatic tool for this purpose is played by the simulation model of the transmission network, which must be very precise. It has to be stressed that in the case of optimizing operational control it is of essential importance to use a precise model of the controlled system to be able to assess both feasibility and quality, in particular in terms of monetary costs, of the decisions. Having a good, reliable, hydraulic model of WDS one can propose a decision process consisting of a model-based repetitive planning of pumping schedules. In order to determine such a schedule one may formulate and then solve a dynamic optimization problem. It must be noted that energy costs are based on a complex structure with commodity cost depending on spot market pricing as well as peak power demand charges. Further, since the optimization works towards producing a pumping schedule from the present time into future hours and days, predicted information is required for water demand/consumption and energy rates.

In the case of a large water distribution system, like TWS, the computational complexity makes it impossible to solve such optimization problems directly with pumping schedules as there are decision variables. To find a good sub-optimal solution one may use a two-stage approach. In case of TWS, in the first stage a large scale nonlinear optimization problem, involving the full hydraulic model of WDS, which results in a few hundred thousand, to over a million variables, depending upon the level of accepted model simplification and the number of hourly intervals considered, was formulated concerning aggregated hourly flows and average hourly pressure gains for each pumping station as decision variables. In the second stage, after solving this problem [51, 61], the Pump Scheduler is to translate the optimal solution (aggregated hourly flows and average hourly pressure gains for each pumping station) into an ON-OFF pump operation schedule for all required pumping stations. To achieve this, the EPANET hydraulic simulation is used at each scheduling iteration, and the actual pump schedule is adjusted and corrected, taking into account:

- precise hourly flows and operating pressures for each pump delivered by EPANET,
- pressure and pump warnings issued during hydraulic simulation,
- minimum ON and OFF times specified for each individual pump.

Furthermore, the actual pump schedule is validated and corrected (as required) for minimum and maximum pressure limit violations at pumping stations discharge nodes. This is done by reducing the number of pumps working in parallel (to decrease pressure) or forcing a pump ON (to increase pressure). At the same time, for a given pumping station, the scheduler is trying to minimize an accumulated difference between aggregated optimal flow found in the first stage and disaggregated flow (resulting from the actual pump schedule simulated by EPANET), by shifting some flow volume between adjacent hourly intervals and adjusting properly the duration of pump switches. Moreover, complex rule-based heuristics are implemented in the scheduler for merging adjacent pump ON switches in hourly intervals and minimizing the number of used identical or similar pumps over the considered horizon. The starting point for the scheduler is obtained by means of the solution of an auxiliary nonlinear optimization problem for each pumping station with full objective functions, taking into account all the terms of energy cost. It should be noted that hydraulic system simulation plays essential role both in the first and in the second stage. It is thus useful to present here shortly the main software tool, the EPANET.

EPANET is a public domain software developed by the Water Supply and Water Resources Division of the U.S. Environmental Protection Agency's National Risk Management Research Laboratory [51]. EPANET provides an integrated environment for editing network input data, running hydraulic and water quality simulations, and viewing the results in a variety of formats. The hydraulic simulation performed by EPANET delivers information such as flows and head losses in links (pipes, pumps and valves), heads, pressures and demands at junctions, levels and volumes for water storage. This allows computing of pumping energy and cost. EPANET's computational engine is available also as a separate library (called the EPANET Toolkit) for incorporation into other applications. The network hydraulics solver employed by EPANET uses the gradient method, first proposed by Todini and Pilati [61], which is a variant of Newton-Raphson method.

The EPANET Toolkit could be used for developing specialized applications, such as optimization or automated calibration models that require running network analyses as selected input parameters are iteratively modified. It was, therefore, decided that EPANET will be used as the hydraulic model component of the TOO system, since the EPANET Toolkit is capable of providing all of the hydraulic modelling functionality.

The on-line TOO Optimizer uses EPANET for several purposes. The modified and extended versions of the EPANET Toolkit and OOTEN library (C++ wrapper for EPANET Toolkit) have been built into the TOO Optimizer and are used to build the optimization models, to generate the starting point for the optimizer and to check the aggregated optimal results provided by solving the nonlinear optimization problem. The EPANET Toolkit is also used during iterations of the TOO Pump Scheduler to adjust the optimized pump schedule by feedback from hydraulic simulations of the current iteration of the pumping schedule. Furthermore, it also provides the hydraulic and quality simulation results for the final optimized pump schedule, such as reservoir profiles for volume level and water quality, pressure profiles for pumping stations discharges and node pressures and energy and power calculations for all pumping stations.

A further possible use of simulation would be to carry the computer based analysis of the WDS, with repetitive scheduling of pump operations and various scenarios of water withdrawals and energy tariffs, over representative time intervals. Such analysis, similar to the described below concerning operation of a flood control system, would in case of a large WDS lead to very complex and time consuming computational task. As far as the Toronto Water System was concerned it has been decided to carry out extensive on-line experiments using the proposed repetitive two-stage planning of pumping schedules. The experiments appeared successful and, after extensive testing, the described mechanism was introduced to operational practice.

6.2 *Control System for Flood Operation in a Multiple Reservoirs Water System*

Let us consider a complex system that consists of a set of N water storage reservoirs located on tributaries of the main river, each with a local dispatcher. Our goal is to develop an optimal control structure and decision mechanisms for flood operation in such a system. It is obvious that the main goal pursued by floodplain communities is to minimize flood induced losses. Increased protection of municipal, industrial and agricultural developments can be achieved by employment of off-line and on-line activities. The control of retention reservoirs is the basic on-line activity, which can purposely change flow discharges at important damage centers. It should be pointed that all operators of reservoirs have to be forced to achieve the common goal for the whole system (a type of the complex system depicted in Fig. 1c).

The flood protection system and its main components are shown in Fig. 6. Measurement data and forecasts of all inflows are provided by hydrological stations. We can see multiple decision units, reservoir operators and the central dispatcher.

Development of an optimal flood operation in the system presented in Fig. 6 is not a simple task and often requires a trade-off. Design works began with determining the aim of the control. Then it was decided that the whole system will be decomposed into several subsystems. The mathematical models of these subsystems

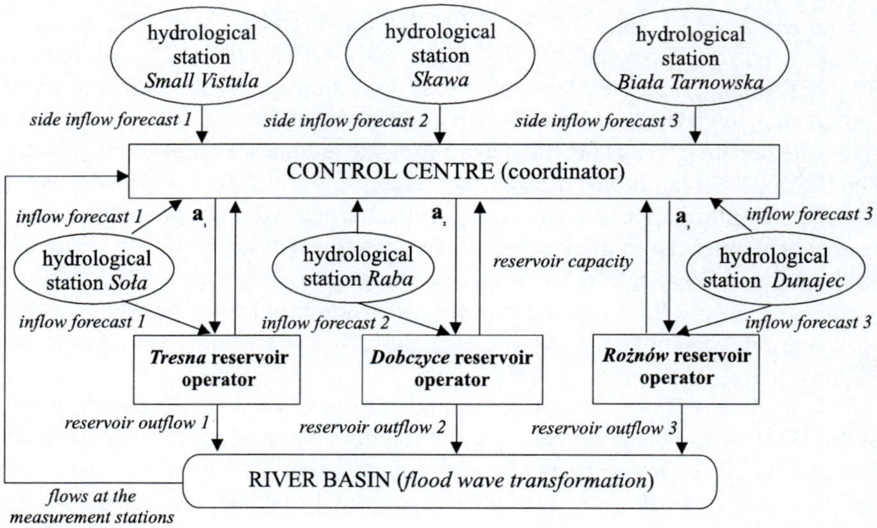

Fig. 6 Flood protection system; Vistula river basin

were formulated. Then, three control structures, i.e. decentralized (DCS), centralized (CCS) and hierarchical (HCS) depicted in Fig. 2 were proposed and investigated.

Flood damages mitigation schemes are usually evaluated by means of multicriteria analysis. The main purpose of flood control is obviously to minimize damages caused by inundations. These can be related to the peak water levels as well as to the duration of flooding and to the current use of the floodplain. In most cases it is assumed that flood damages J can be expressed as a function of high flow levels Q downstream the dams, at the K important cross sections. This is basically due to the fact that the impact of other flow attributes on flood damages is very complicated and not easily identifiable. Thus, the following performance measure can be considered

$$J(Q_{[t_0,t_f]}) = \sum_{k=1}^{K} \alpha_k \max(Q_k^{cul} - Q_k^{limit}, 0), \quad Q_k^{cul} = \max_{t\in[t_0,t_f]} Q_k(t), \tag{25}$$

where $[t_0, t_f]$ is the control horizon, Q_k denotes the flow at the k-th damage center ($k = 1, \ldots, K$), Q_k^{limit} the highest safe discharge, Q_k^{cul} the peak flow and α_k denotes the weighting factor related to the flow at the k-th damage center (different points have different importance). An additional control criterion is to fill i-th reservoir to the desired volume $\overline{w}_i$ at the end of the flood, i.e. $w_i(t_f) = \overline{w}_i$, $i = 1, \ldots, N$.

The natural objective in managing of the i-th retention reservoir is minimization of damages created by high water levels directly downstream to the reservoir (26) and filling up the reservoir to the desired volume $\overline{w}_i$ at the end of the flood, i.e. $w_i(t_f) = \overline{w}_i$, $i = 1, \ldots, N$, where w_i denotes the capacity of the i-th reservoir.

$$q(u(\cdot)) = \max_{[t_0, t_f]} u(t). \tag{26}$$

The next stage is to develop models of all components of the system, namely retention reservoirs and river channels. Each reservoir is described by the dynamic of a simple tank $\dot{w}(t) = d(t) - u(t)$, with one input $d(t)$ (inflow) and one controlled output $u(t)$ (outflow) from the reservoir, w denotes the reservoir capacity at time t. The volume of water stored in the reservoir is obviously constrained $w_{min} \leq w(t) \leq w_{max}$. The construction of spillways in the dam results in constraints on the outflows $u_{min}(w(t)) \leq u(t) \leq u_{max}(w(t))$.

Propagation of flood waves along the river reaches can be described in many ways at different levels of accuracy. The possible range varies from nonlinear partial differential equations proposed by B. de Saint Venant, to simple (discrete in time and space) linear models like Kalininin-Miliukov or Muskingum ones. For the practical application, the river reach is often divided into a number of sub-sections. Each of those sections is then treated as one of a series of a few storage elements described by identical linear or nonlinear equations. Simple models play an important role in decision support systems and are recommended for operational flood control and flood forecasting systems. An example can be the flood control system for the Vistula river basin described in [45].

Control of dynamic systems by repetitive, on-line use of optimization requires forecasts of the future inflows to the system [21, 29]. The deterministic optimal release scheduling problem is formulated under the idealistic assumption that the future external inflows can be exactly predicted. Flood control methods that use stochastic models of inflows (e.g. Markov chains, ARIMA models) are presented in [20]. Another possibility allowing for directly taking into account the stochastic nature of inflows is to use the repetitive optimization based on multiple inflow forecasts [45].

Assume that a reasonable i-th inflow forecasting beyond the next ΔT_i hours (a long-term forecast) is possible at times $t_l, t_{l+1}, \ldots$ in the form of inflow wave pattern $\overline{d}_i(t)$, $t \in [t_l, t_f]$, where t_f is the estimated control termination time. The central dispatcher gathers data from the whole river basin and calculates a new decision every $t_c, t_{c+1}, \ldots$ and $t_{l+1} - t_l \leq t_{c+1} - t_c$. The proposed design process was as follows. First, the mathematical models of all components of the water system were formulated. Then, three control structures as described in Sect. 2 and deterministic and stochastic decision mechanisms were developed. All models and control mechanisms were implemented and the simulator of the whole system was developed. The dataset of historical and generated inputs was collected. Finally, numerous simulation experiments and evaluated the performance of proposed solutions were performed. Let us focus on the design process in more details.

In the case of a decentralised control the operation of each reservoir is independent of the others, hence the common goal, i.e. flood mitigation in the whole river basin can not be achieved. It is obvious that the objectives of local dispatchers are often in contradiction to the global objectives. Various rules can be used to determine water outflows, including predictive control with repetitive minimization of q (26). Open loop feedback and closed loop feedback control can be implemented, which are described in [21, 29].

In the centralised case one has to design a multidimensional rule for establishing simultaneously the releases from all reservoirs. One possibility of computing the releases is to formulate and solve the dynamic optimization problem, using the models of reservoirs, dynamic models of river reaches and forecasted inflows. Another approach is to calculate suboptimal outflows from the reservoirs under the assumption that for the i-th reservoir release, it is related to a step function (switching between two constant values): $u_i(t) = \overline{u}_i^1$ for $t \in [t_c, T_i^\star)$ and $u_i(t) = \overline{u}_i^2$ for $t \in [T_i^\star, t_{w_i}]$, where t_{w_i} denotes the time when the reservoir was filled. Thus, the objective of an operator in charge of the entire water system is to determine at times $t_c, t_{c+1}, \ldots$ release trajectories $\hat{u}_i$ ($i = 1, \ldots, N$; N is the number of reservoirs) minimizing the performance measure (25), satisfying the constraints on the reservoirs storages and releases, corresponding to the optimal switching functions described by $\hat{\overline{u}}_i^1$, $\hat{\overline{u}}_i^2$ and the time of switching $T_i^\star$. By an appropriate choice of parameters $\overline{u}_i^1$, $\overline{u}_i^2$ and $T_i^\star$ one can try to desynchronize the peaks of the flows on various rivers and prevent the culmination from different reservoirs from overlaping.

Finally, in the case of a hierarchical control structure the local release rules are designed in such way that a central authority may adjust them in the process of periodic coordination so as to achieve the cooperation of reservoirs in minimising global damages. Thus, it incorporates two decision levels: the upper level with the central authority (coordinator) and the local level formed by the operators of the reservoirs. Within this structure, the central dispatcher performs an analysis of possible future scenarios of the flood and determines the optimal vector of coordinating parameters $\mathbf{a}$ influencing local operator decisions about the outflows from the reservoirs solving the optimization problem: $\min_{\mathbf{a} \in A} J(\mathbf{a} \in A)$, with J defined in (25). The local decision rules were designed in such a way that a central authority could adjust them in the process of periodic coordination, so as to achieve the coordination of reservoirs in minimizing global damages. It was assumed that the vector $\mathbf{a}_i$ of coordinating parameters for the i-th reservoir, $i = 1, \ldots, N$, considered at time t_l was related to the weighting function $\alpha_i(t)$ defined as follows: $\alpha_i(t) = 1 + (c_i - 1) \cdot \mathbf{1}(t - T_i^*)$, i.e., $\alpha_i(t) = 1$ for $t \in [t_l, T_i^*)$ and $\alpha_i(t) = c_i$ for $t \in [T_i^*, t_f]$. Finally, the vector $\mathbf{a}_i$ was given as $\mathbf{a}_i = [c_i, T_i^*]$. Coordinating parameters were used to modify the local performance measures (26). Thus, the task of the i-th operator at every time step t_l is to calculate the outflow minimizing the following performance measure $q(u(\cdot), \mathbf{a}_i) = \max_{[t_l, t_f]}(u(t) \cdot \alpha_i(t))$. The decision making process of the central authority is presented in detail in Fig. 7.

Thus, the preparatory design stage was completed. The next step was to develop the simulators of all components of the controlled systems and tools for forecasts generation. Then, the database containing possible values of inputs was built. The test scenarios were defined and simulation experiments were conducted. The goal of

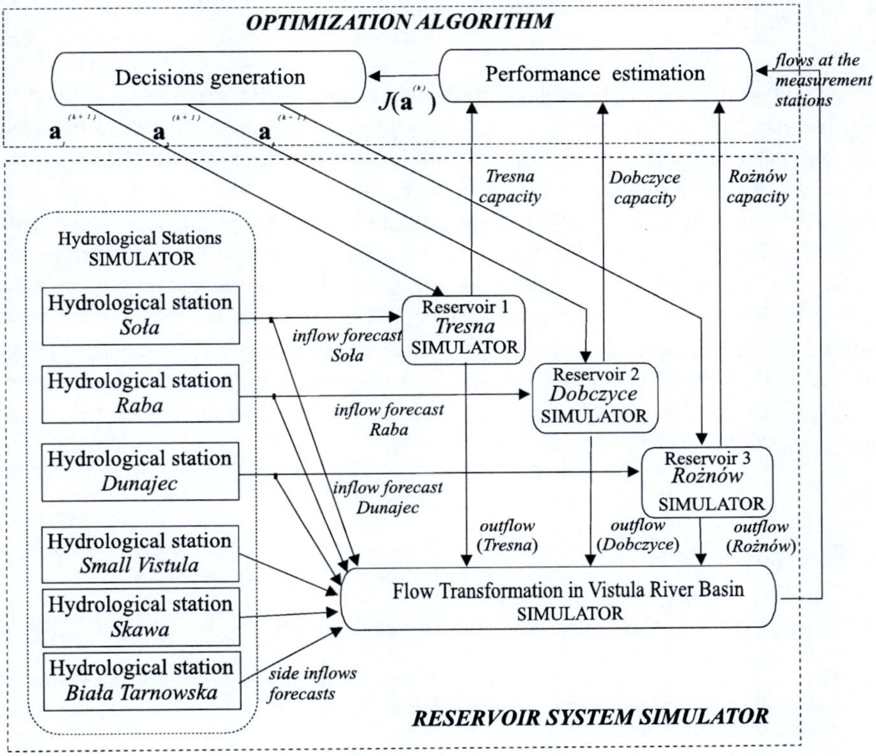

Fig. 7 Two-level structure for flood control

simulations was to evaluate and compare the proposed solutions. Finally, the control system that gave the best results (minimal flood damages) was recommended for real life implementation. The process of development of the control system for flood operation in the Upper Vistula river basin system in the Southern part of Poland is described in [21, 40, 42, 45].

It should be noted that the presented example also shows the use of simulations to determine controls in real-time decision making processes. Both in centralized and hierarchical structures (Fig. 2), the estimation of expected flood losses J requires simulation of the operation of all decision makers and transformation of the proposed discharges by the river channel system. This is a typical example of the use of a simulation-based optimization scheme described in Sect. 5.1. The paper [43] describes the decision support system for the central authority of the Upper Vistula reservoir system, in which decisions are made on the basis of repeated simulations.

7 Summary

This chapter addresses issues related to the design and analysis of control structures and mechanisms for complex systems. The goal is to draw attention to the role of computer simulation in the process of designing structures and decision-making mechanisms, and in making decisions in operational mode. The chapter begins with a synthetic description of complex systems and an overview of existing control structures. Next, computer simulation techniques and stages of building simulation models are discussed. The next sections are devoted to the presentation of control system design methodology using the concept of computer-aided analysis in which simulation is used to determine controls in an operational mode. Various approaches and methods of solving formulated optimization tasks are discussed. The chapter ends with the presentation of two examples showing the application of the presented design methodology in practical control tasks.

References

1. Ali, M.M., Törn, A., Viitanen, S.: A numerical comparison of some modified controlled random search algorithms. J. Global Optim. **11**(4), 377–385 (1997)
2. Amaran, Satyajith., Sahinidis, Nikolaos V., Sharda, Bikram, Bury, Scott J.: Simulation optimization: a review of algorithms and applications. J. Oper. Res. **12**(4), 301–333 (2014)
3. April, J., Glover, F., Kelley, J.P., Laguna M.: Practical introduction to simulation optimization. In *2003 Winter Simulation Conference*, pp. 71–78 (2003)
4. Arabas, J.: Wykłady z algorytmów ewolucyjnych. WNT (2001)
5. Banks, J.: Handbook of Simulation. Wiley, New York (1998)
6. Banks, J., Carson, J.S., Nelson, B.L., Nicol, D.M.: Discrete-Event Systems Simulation. PEARSON (2000)
7. Bartz-Beielstein, T.: A survey of model-based methods for global optimization. In: Conference Bioinspired Optimization Methods and their Applications, pp. 1–18 (2016)
8. Beisbart, C., Saam, N.J.: Computer Simulation Validation: Fundamental Concepts, Methodological Frameworks, and Philosophical Perspectives. Springer-Verlag (2018)
9. Berry, S., Lowndes, V., Trovati, M.: Guide to Computational Modelling for Decision Processes: Theory, Algorithms. Springer, Techniques and Applications (2017)
10. Beyer, H., Sendhoff, B.: Simplify your covariance matrix adaptation evolution strategy. IEEE Trans. Evol. Comput. **21**(5), 746–759 (2017)
11. Bratley, P., Fox, B.L., Schrage, L.E.: A guide to simulation. Springer, New York (1987)
12. Cellier, F.E.: Continuous System Modeling. Springer (1991)
13. Chen, G., Szymanski, B.K.: Four types of lookback. In: Proceedings of 17th Workshop on Parallel and Distributed Simulation (PADS 2003), pp. 3–10 (2003)
14. Dekkers, A., Aarts, E.: Global optimization and simulated annealing. Math. Program. **50**, 367–393 (1991)
15. Figueira, C., Almada-Lobo, B.: Hybrid simulation-optimization methods: a taxonomy and discussion. Simul. Model. Pract. Theory **46**, 118–134 (2014)
16. Findeisen, W., Bailey, F.N., Brdyś, M., Malinowski, K., Woźniak, A.: Control and Coordination in Hierarchical Systems. Wiley, London (1990)
17. Ghos, S., Lee, T.S.: Modelling and Asynchronous Distributed Simulation. IEEE Press (2000)
18. Goldberg, D.E.: Genetic Algorithms in Search and Optimization and Machine Learning. Addison-Wesley and Reading (1989)

19. Horst, R., Pardalos, P.M.: Handbook of Global Optimization. Kluwer Academic Publishers (1995)
20. Jasper, K., Gurtz, J., Lang, H.: Advanced flood forecasting in alpine watersheds by coupling meterological observations and forecasts with a distributed hydrological model. J. Hydrol. **267**, 40–52 (2002)
21. Karbowski, A., Malinowski, K., Niewiadomska-Szynkiewicz, E.: A hybrid analytic rule-based approach to real-time flood-control in a reservoir. Decis. Support Syst. **38**, 599–610 (2005)
22. Karpowicz, M., Arabas, P., Niewiadomska-Szynkiewicz, E.: Energy-aware multi-level control system for a network of linux software routers: design and implementation. IEEE Syst. J. **12**, 571–582 (2018)
23. Kheir, N.A.: Systems Modeling and Computer Simulation. Marcel Dekker Inc, New Yorks (1996)
24. Kirkpatrick, S., Gelatt, C.D., Vecchi, M.P.: Optimization by simulated annealing. Science, New Series **220**(4598), 671–680 (1983)
25. Korn, G.A., Wait, G.V.: Digital Continuous Simulation. Prentice-Hall, Englewood Cliffs, N.J. (1978)
26. Lenhard, J.: The Oxford Handbook of Philosophy of Science, Chapter Modern Simulation and Modeling. Oxford University Press (2016)
27. Li, X.G., Wei, X.: An improved genetic algorithm-simulated annealing hybrid algorithm for the optimization of multiple reservoirs. Water Resour. Manage. **22**, 1031–1049 (2008)
28. Lozano, J.A., Larranaga, P., Inza, I., Bengoetxea, E.: Towards a New Evolutionary Computation. Advances on Estimation of Distribution Algorithms. Springer, Berlin (2006)
29. Malinowski, K.: Control Applications of Optimization, vol. 115, Chapter Repetitive Optimization for Predictive Control of Dynamic Systems under Uncertainty, pp. 163–175. Birkhäuser Publishers (1994)
30. Malinowski, K., Błaszczyk, J.B, Allidina, A.: Optimizing control for large scale dynamic systems; general issues and case study results: transmission operations optimizer for toronto water system. In: 2017 International Conference on Engineering, Technology and Innovation (ICE/ITMC), pp. 161–168 (2017)
31. Manzanares-Filho, N., Albuquerque, R.B.F., Sousa, B.S., Santos, L.G.C.: A comparative study of controlled random search algorithms with application to inverse aerofoil design. Eng. Optim. **50**(6), 996–1015 (2018)
32. Mesarovic, M.D.: Theory of Hierarchical Multilevel Systems. Academic Press (1970)
33. Michalewicz, Z.: Algorytmy Genetyczne + Struktury Danych = Programy Ewolucyjne. WNT, Warszawa (1996)
34. Michalewicz, Z. , Fogel, D.B.: How to Solve It: Modern Heuristics. Springer (2000)
35. Mohan, C., Shanker, K.: A controlled random search technique for global optimization using quadratic approximation. J. Oper. Res. **11**, 93–101 (1994)
36. Murray-Smith, D.J.: Continuous System Simulation. Chapman & Hall (1995)
37. Napiórkowski, J.J.: Modelling and Control of Floods, Monografia E-3 (365). Oficyna Wydawnicza Instytutu Geofizyki PAN (2003)
38. Nguyen, A.T.: A review on simulation-based optimization methods applied to building performance analysis. Appl. Energy **113**, 1043–1058 (2014)
39. Nicol, D.M., Fujimoto, R.: Parllel simulation today. Ann. Oper. Res. **53**, 249–285 (1994)
40. Niewiadomska-Szynkiewicz, E.: Development of flood control methodologies based on computer simulation. In: Advanced Simulation Technologies Conference (ASTC 2000), pp. 1–7. SCS Series (2000)
41. Niewiadomska-Szynkiewicz, E.: Decision delays in predictive control of retention reservoir under uncertainty. Acta Geol. Pol. **50**(3), 479–496 (2002)
42. Niewiadomska-Szynkiewicz, E.: Computer simulation of flood operation in multireservoir systems. Simulation **80**(2), 101–116 (2004)
43. Niewiadomska-Szynkiewicz, E.: Fc-mws: a software environment for flood operation in multiple reservoir systems. Acta Geol. Pol. **50**(3), 479–496 (2004)

44. Niewiadomska-Szynkiewicz, E., Blaszczyk, J.: Simulation-based optimization methods applied to large scale water systems control. In: 2016 Int. IEEE Conferences on Ubiquitous Intelligence Computing, Advanced and Trusted Computing, Scalable Computing and Communications, Cloud and Big Data Computing, Internet of People, and Smart World Congress (UIC/ATC/ScalCom/CBDCom/IoP/SmartWorld), pp. 649–656 (July 2016)
45. Niewiadomska-Szynkiewicz, E., Malinowski, K., Karbowski, A.: Predictive methods for real time control of flood operation of a multireservoir system—methodology and comparative study. Water Resour. Res. **32**(9), 2885–2895 (1996)
46. Niewiadomska-Szynkiewicz, E., Żmuda, M., Malinowski, K.: Application of java-based framework to parallel simulation of large-scale systems. Int. J. Appl. Math. Comput. Sci. **13**(4), 537–547 (2003)
47. Niewiadomska-Szynkiewicz, E., Sikora, A., Kołodziej, J.: Modeling mobility in cooperative ad hoc networks. Mob. Netw. Appl. **18**(5), 610–621 (2013)
48. Press, W.H., Teukolsky, T.W., Vettering, S.A., Flannery, B.P.: Numerical Recipes in C. Cambridge University Press (1997)
49. Price, W.L.: Global optimization algorithms for a cad workstation. J. Optim. Theory Appl. **55**(1), 133–146 (1987)
50. Rios, L.M., Sahinidis, N.V.: Derivative-free optimization: a review of algorithms and comparison of software implementations. J. Glob. Optim. **56**, 1247–1293 (2013)
51. Rossman, L.A.: Epanet 2 users manual. Technical Report EPA/600/R-00/057, U.S. States Environmental Protection Agency, National Risk Management Research Laboratory, Office of Research and Development, Cincinnati, Ohio, USA (2000)
52. Rubinstein, R.Y., Melamed, B.: Modern Simulation and Modeling. Wiley (1998)
53. Sanders, J., Kandtor, E.: CUDA w przykładach. Wprowadzenie do ogólnego programowania procesorów GPU. Helion S.A. (2012)
54. Shannon, R.E.: Introduction to simulation. In: Proceedings of the 24th Conference on Winter Simulation, WSC 1992, pp. 65–73. ACM, New York, NY, USA (1992)
55. Sikora, A., Niewiadomska-Szynkiewicz, E.: A federated approach to parallel and distributed simulation of complex system. Int. J. Appl. Math. Comput. Sci. **17**(1), 99–106 (2007)
56. Sokolowski, J., Banks, C.M.: Modeling and Simulation for Analysing Global Events. Wiley, New Jersey (2009)
57. Spall, J.C.: Introduction to Stochastic Search and Optimization. Wiley, New Jersey (2003)
58. Szynkiewicz, P.: A novel gpu-enabled simulator for large scale spiking neural networks. J. Telecommun. Inf. Technol. **2**, 34–42 (2016)
59. Szynkiewicz, P.: A comparative study of pso and cma-es algorithms on black-box optimization benchmarks. J. Telecommun. Inf. Technol. **4**, 5–17 (2018)
60. Szynkiewicz, P., Kozakiewicz, A.: Design and evaluation of a system for network threat signatures generation. J. Comput. Sci. **22**, 187–197 (2017)
61. Todini, E., Pilati, S.: Computer Applications in Water Supply: Systems Analysis and Simulation, vol. 1, Chapter A Gradient Algorithm for the Analysis of Pipe Networks, pp. 1–20. Research Studies Press Ltd., Letchworth, Hertfordshire, England (1988)
62. Fleming, W.: Functions of Several Variables. Springer (1977)
63. Wainer, G.A.: Discrete-Event Modeling and Simulation. A Practitioner's Approach. CRS Press and Taylor & Francis Group (2009)
64. Wang, Z., Zhao, Y., Liu, Y., Lv, C.: A speculative parallel simulated annealing algorithm based on apache spark. Concurr. Comput. Pract. Exp. **30**(14) (2018)
65. Wardi, Y.: Stochastic algorithms with armijo stepsizes for minimization of functions. J. Optim. Theory Appl. **64**(2), 399–417 (1990)
66. Yan, D., Mukai, H.: Optimization algorithm with probabilistic estimation. J. Optim. Theory Appl. **79**(2), 345–371 (1993)
67. Zeigler, B.P., Praehofer, H., Kim, T.G.: Theory of Modeling and Simulation. Academic Press (2000)

Optimization

Optimal Sensor Selection for Estimation of Distributed Parameter Systems

Dariusz Uciński and Maciej Patan

Abstract This chapter is focused on spatial sensor location for maximization of identification accuracy of distributed parameter systems. The following problem formulation, in terms of optimum experimental design, is adopted, in which the optimality criterion is a convex function defined on the Fisher information matrix associated with estimated parameters. The locations of a given number of sensors are selected from among a fixed, but possibly very large, set of candidate locations. A systematic procedure is presented for finding optimal solutions through a relaxation of the original combinatorial problem, which boils down to determining an optimum spatial density of sensors instead of the actual locations. Thus a convex optimization problem with linear constraints is obtained. Its form makes it suitable for a solution using an extremely efficient algorithm of simplicial decomposition yielding an additional reduction in problem dimensionality. A technique is also discussed to post-process the computed optimal sensor density in order to get an equivalent form with minimal spatial support. The work is complemented with a numerical example and a discussion of various variants of the basic problem, which illustrates the generality and flexibility of the proposed approach.

1 Introduction

Distributed parameter systems (DPSs) are a class of dynamic systems whose states depend not only on time, but also on the spatial variable. Numerous examples of their common occurence in practice are provided by thermodynamics (heat exchangers, industrial furnaces), mechanics and construction (vibrations of structural elements),

D. Uciński (✉) · M. Patan
University of Zielona Góra, Institute of Control and Computation Engineering, ul. Szafrana 2, 65-516 Zielona Góra, Poland
e-mail: D.Ucinski@issi.uz.zgora.pl

M. Patan
e-mail: M.Patan@issi.uz.zgora.pl

P. Kulczycki et al. (eds.), *Automatic Control, Robotics, and Information Processing*, Studies in Systems, Decision and Control 296,
https://doi.org/10.1007/978-3-030-48587-0_11

environmental engineering (dispersion of pollutants in the atmosphere or groundwater), electrical engineering (long transmission lines), nuclear power (energy transformations in reactors) or meteorology (atmospheric phenomena and their forecasting).

An adequate description of DPSs are partial differential equations (PDEs). On the one hand, their application largely complicates the analysis and synthesis of control systems owing to the necessity of using functional analysis tools and large-scale numerical computations. On the other hand, the formalism of this type increases the modeling accuracy and, as a result, has a strong impact on the control quality. The approach consisting of replacing the more accurate description in the form of PDEs by an approximation in the form of a system of ordinary differential equations, which is preferred by engineers, most often leads to an unacceptable loss of information about the spatial aspects of system dynamics.

The control theory of DPSs has been developed since the 1960s [35] attracting more and more interest not only among theoreticians, but also among practitioners. Numerous monographs [15, 17, 33, 60], invited sessions at top conferences (CDC, ACC, ECC) or specialized workshops which are now devoted to this area. What is more, the increasing availability of highly efficient numerical PDE solvers (COMSOL, FEniCS, FreeFem++) makes the appropriate control algorithms accessible to engineers [28, 59].

Despite achieving some maturity by the system approach to DPSs, there still exist problems not fully resolved and constituting challenges strongly motivated by practice. One of them is the observation problem for estimation of unknown values of the parameters underlying the relevant mathematical model (unfortunately, most often, not all physical parameters can be directly measured and they have to be estimated via their tuning yielding the best fit of the model output to the observations of the actual system; the latter are provided by measurement sensors). Although the problems of identifiability and parameter estimation are already well examined and described for both deterministic [7] and stochastic uncertainty descriptions (from the frequentionist [5] and Bayesian points of view [16]), this regards the situations in which the observations have already been collected using the sensors located in the spatial domain. Since the number of sensors is finite, and frequently quite low due to economic limitations, the appropriate strategy of their location so as to get the most valuable information about the estimated parameters becomes of paramount importance. In order to illustrate this issue, consider the following example.

1.1 Motivating Example

Heat is propagated in a rectangular metal block with a rectangular crack, which occupies the domain displayed in Fig. 1. The appropriate mathematical model of this process is the heat equation

$$\frac{\partial y}{\partial t}(\mathbf{x}, t) = \theta \Delta y(\mathbf{x}, t), \quad \mathbf{x} \in \Omega, \quad t \in (0, t_f], \tag{1}$$

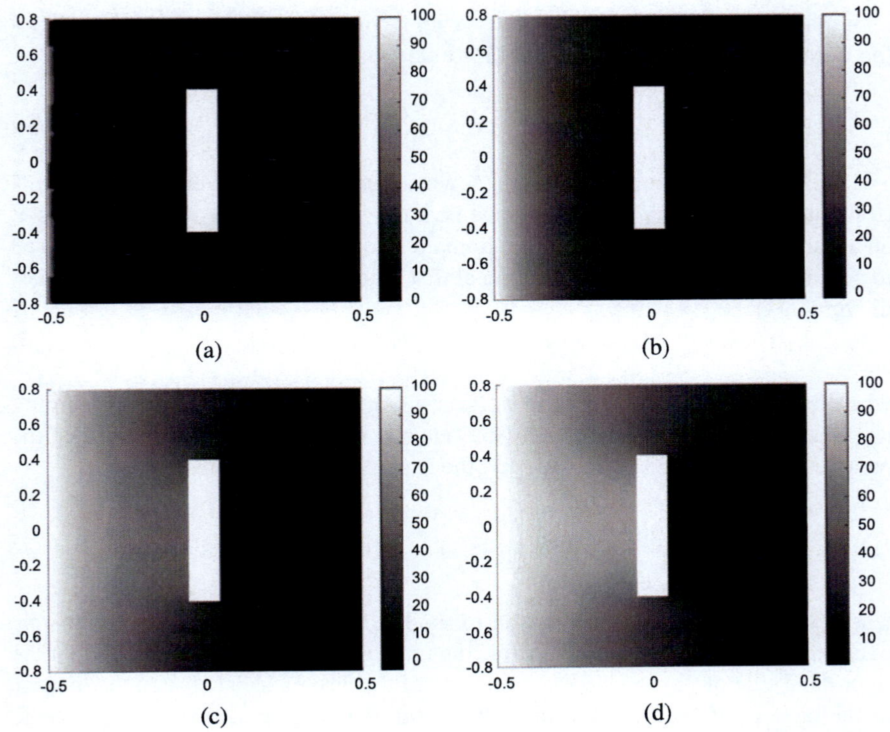

Fig. 1 Temperature evolution in a block with a crack, which is heated on the left-hand boundary

where $y(\mathbf{x}, t)$ is the temperature at spatial point $\mathbf{x} = (x_1, x_2)$ and at time instant t, $\theta > 0$ stands for a constant heat conduction coefficient, t_f signifies a finite observation horizon, $\Delta(\cdot) = \partial(\cdot)/\partial x_1^2 + \partial(\cdot)/\partial x_2^2$ means the Laplacian with respect to spatial variables.

The left-hand side of the block (i.e., the left-hand boundary Ω) is heated by a thermostat forcing its temperature to the level of 100 °C. On the right-hand side of the block, heat flows to the surrounding air with a known speed. The remaining parts of the boundary are insulated, which means lack of heat flow. In this way, we get the following system of boundary conditions:

$$y(\mathbf{x}, t) = 100 \text{ on the left boundary of } \Omega \text{ (Dirichlet condition)}, \tag{2}$$

$$\frac{\partial y}{\partial \mathbf{n}} = -10 \text{ on the right boundary of } \Omega \text{ (von Neumann condition)}, \tag{3}$$

$$\frac{\partial y}{\partial \mathbf{n}} = 0 \quad \text{on the remainder of the boundary (von Neumann condition)}, \tag{4}$$

where $\mathbf{n}$ stands for the outward normal of the boundary of Ω, and $\partial(\cdot)/\partial\mathbf{n}$ signifies the derivative in the direction of $\mathbf{n}$.

Equations (1)–(4) are complemented with the initial condition which determines the starting temperature of the block. Here it is assumed to be zero, i.e.,

$$y(\mathbf{x}, 0) = 0, \quad \mathbf{x} \in \Omega. \tag{5}$$

If the value of the coefficient θ is unknown, it can be estimated using measurements from a temperature sensor (a thermoelement or a semiconductor sensor) located at an arbitrarily selected point $\mathbf{x}^1$ of the domain Ω or its boundary $\partial\Omega$. It is supposed to continuously record the temperature at that point. Then the output equation takes the form

$$z(t) = y(\mathbf{x}^1, t) + \varepsilon(\mathbf{x}^1, t), \quad t \in [0, t_f], \tag{6}$$

where $\varepsilon(\mathbf{x}^1, \cdot)$ is the measurement noise at point $\mathbf{x}^1$ (some assumptions regarding its properties as a stochastic process are made in the seuqel). The estimate $\widehat{\theta}$ of the parameter θ is obtained by minimizing the least-squares criterion, i.e.,

$$\widehat{\theta} = \arg\min_{\vartheta} \int_0^{t_f} (z(t) - y(\mathbf{x}^1, t; \vartheta))^2 \, \mathrm{d}t, \tag{7}$$

where $y(\cdot, \cdot; \vartheta)$ signifies the solution to the PDE (1) complemented with (2)–(5), obtained for the value ϑ substituted for the unknown value of θ. The integral on the right-hand side of (7) constitutes a measure of the discrepancy between the outputs of the model and the actual system. It should be emphasized that, in spite of the linearity of Eq. (1), the dependence of its solution on the parameter θ is strongly nonlinear.

The presence of the output z in the integrand on the right-hand side of (7) implies the dependence of the estimate $\widehat{\theta}$ on the realization of the measurement process. This means that $\widehat{\theta}$ becomes a random variable and repetition of the measurements in the same conditions, followed by estimation in accordance with (7), will yield different estimates $\widehat{\theta}$ owing to different realizations of noise.

The appearance of the sensor location $\mathbf{x}^1$ in the criterion (7) suggests that it may have an impact on the variation in $\widehat{\theta}$. The discussion presented in monograph [62, p. 17] confirms that this is true. What is more, this influence can be high, which constitutes a motivation behind the problem of selecting an "optimal" sensor location. In the case considered, the notion of optimality is made specific in quite a natural manner. Assume first that for each arbitrarily selected sensor location $\mathbf{x}^1$ the least-squares estimator (7) is unbiased, which means that repetition of the measurements at point $\mathbf{x}^1$ would lead to a variation in the estimates $\widehat{\theta}$, but nonetheless they would concentrate around the sought value θ. Clearly, this is possible provided that some set of assumptions are satisfied [5, 44], but they are not excessively restrictive. As the optimality criterion for the sensor location the variance is then selected,

$$\mathscr{J}(\mathbf{x}^1) = \operatorname{var}(\widehat{\theta}) = \mathrm{E}\big[(\widehat{\theta} - \theta)^2\big], \tag{8}$$

where E stands for the expectation. For a sensor located at a point minimizing the criterion (8) the estimates $\widehat{\theta}$ will concentrate around the value of θ, but their dispersion will be the lowest. Conducting only one experiment, as is the case in practice, only one estimate $\widehat{\theta}$ is obtained, but we can have quite a firm guarantee of its proximity to θ.

This example reflects the idea of the traditional approach to optimal sensor placement, which orignates in optimum experimental design [4, 19, 44]. It is very intuitive, but putting it into practice is not that obvious. Optimal observation points ought to be determined prior to taking measurements, which implies difficulties in computing the value of the criterion (8). The value of θ is not known, nor is an economic procedure of estimating the expected value in (8) which would not resort to time-consuming Monte Carlo simulations. The approach also has to be generalized for the case of multiple estimated parameters. As the problem has been reduced to a nonlinear optimization problem, adequate numerical algorithms should be proposed to solve it, especially in the case of locating hundreds or even thousands of sensors, which is typical in modern sensor networks.

1.2 Literature Review

The traditional approach to optimal sensor location for parameter estimation in spatiotemporal systems is based on minimizing various optimality criteria defined in the Fisher information matrix (FIM) associated with these parameters. The inverse of the FIM constitutes an estimate of the covariance matrix of the maximum likelihood estimator of the unknown parameters (the least-squares estimator is one of its forms). Its history dates back to monograph [74] and the work [46, 47]. A comprehensive overview of the works published within the first three decades since that period is contained in the classical work [32] and comprehensive monographs [37, 62]. The remainder of this work, which is introductory in nature, is devoted just to this approach. It is also worth mentioning multi-aspect monographs [55, 56] which are directed towards applications in model calibration for underground water.

In the past decade, the interest in this problem has increased rapidly due to the growing popularity of sensor networks [53, 58, 77], which form an ideal platform for the application of theoretical solutions, and the great progress in Bayesian data assimilation techniques [11, 34, 51] which are used in numerical models of weather or ocean circulation forecasting. Data assimiliation consists in integrating the mathematical model of a stochastic dynamic system with measurement data incoming on-line (from satellites, radars, lidars or direct measurements taken at various locations) so as to obtain a very precise forecasting model which most often employs large-scale numerical computations.

A severe difficulty in parameter estimation of spatiotemporal systems is the infinite-dimensional character of the state space, which is pronounced when the estimated parameters depend on spatial varables or when one of the objectives is an estimation of the initial state. The correct formulation of some of the theoretical problems, such as the existence of the least-squares estimator, the continuous dependence on data or the convergence of the employed approximations, require the compactness of the parameter space. Disregarding these subtle aspects may make the estimation process an ill-posed problem, which is most often manifested through the instability of the estimates excited even by very low measurement noise or simply slight numerical errors [26, 75]. This adverse phenomenon caused a rapid development of regularization techniques, among which, Tikhonov's regularization is best known [75]. However, most often these techniques abstract away from stochastic aspects of the estimation problem. In recent years an alternative has emerged, which makes it possible to naturally include prior statistical information about the distribution of the unknown parameters into the estimation process. This has been made possible through solving a number of theoretical problems related to the Bayes formula for infinite dimensional parameter spaces [16, 54]. Works [1–3, 21–25] are inspired by these advances and very prospective, but a formidable obstacle at the current stage of their development is the large scale of the attendant numerical computations.

The number of located sensors is usually conditioned by economic limitations. Most techniques are reduced to the selection of optimal sensor locations from a finite (but possibly very large) set of candidate locations. The problem of assigning sensors to specific spatial locations can equivalently be treated in terms of activating an optimal subset of all the available sensors deployed in the spatial area (the non-activated sensors remain dormant). The latter framework corresponds directly to the strategy of taking measurements in modern sensor networks. This interpretation will dominate remainder of this work.

A great difficulty underlying selection of an optimal subset from among a given set of candidate locations is the combinatorial nature of the optimization problem. When increasing the numbers of sensors and candidate locations the exhaustive search of the search space quickly consumes all the available computational resources. This stimulated attempts to solve this problem in a more constructive manner. For problems with low or moderate dimensionalities, in works [38, 64, 71] a branch-and-bound method was proposed, which most often drastically reduces the search space. For large-scale sensor networks, various approaches dominate, which replace the original NP-hard combinatorial problem with its convex relaxation in the form of a convex programming problem. This paves the way for application of efficient convex optimization algorithms with polynomial complexity.

In this trend, in works [39–41, 63, 65, 67] optimal densities of active sensors are sought instead of the direct search for their locations. But most works relax the binary variables indicating whether or not the corresponding sensors are active and allow them to take any real values from the interval [0, 1]. In this way, the optimization problem takes the form suitable for application of interior-point methods [30], which was exploited in works [12, 13], or polyhedral approximation methods [38, 64, 71]. The latter, although less known, turn out to substantially reduce the problem of dimensionality while fully exploiting the problem specificity. They also remain functional in the case of large-scale sensor networks, which may be problematic in the case of much more general interior point methods. What is more, they make it possible to include very effective algorithms of optimum experimental design as components [4, 44]. Therefore, the remainder of this work is devoted to them.

1.3 Looking for Optimal Sensor Activation Strategies

The intention of this chapter is to present the algorithmic aspects of selecting an optimal subset of activated sensors for the case of a large-scale sensor network and a convex optimality criterion specified on the FIM. Due to space limitations, the attention was limited to the original approach developed by the authors of this chapter over the last decade, whose elements are described in detail in papers [39–41, 63, 65, 67]. The characteristic features of the methodology are as follows:

- dimensionality reduction by operating on the spatial density of activated sensors and not directly on sensor positions;
- solution of the continuous convex optimization task using the simplicial decomposition algorithm belonging to the class of polyhedral approximation methods; this allows solving relatively small-scale tasks at every step;
- post-optimal analysis leading to the positions of the activated sensors concentrated on a minimum support spatial area.

Notation In this chapter we adopt the convention that $\mathbb{R}_+$ and $\mathbb{R}_{++}$ mean the sets of nonnegative and positive reals, respectively. The set of all $m \times n$ matrices with real elements is denoted by $\mathbb{R}^{m \times n}$. The notation $\mathbb{S}^m$ signifies the set of symmetric $m \times m$ matrices, $\mathbb{S}^m_+$ the set of symmetric and nonnegative definite $m \times m$ matrices, $\mathbb{S}^m_{++}$ the set of symmetric and positive definite $m \times m$ matrices. Given a set H, $|H|$ and $\bar{H}$ denotes its cardinality and closure, respectively. The curly inequalities $\succeq$ (respectively, $\succ$) are employed to denote generalized inequalities. More precisely, between vectors, the symbols represent ordinary inequalities between the corresponding components, and between matrices they represent the Löwner ordering: given $\mathbf{A}, \mathbf{B} \in \mathbb{S}^m$, the notation $\mathbf{A} \succeq \mathbf{B}$ means that the matrix difference $\mathbf{A} - \mathbf{B}$ is nonnegative (resp., positive). Symbols $\mathbf{1}$ and $\mathbf{0}$ denote all vectors whose elements are zeros or ones, respectively. Their lengths results from context. For vectors $\mathbf{a}$ and $\mathbf{b}$ of the same length, the notation $\mathbf{a} \cdot \mathbf{b}$ means the vector whose consecutive components are the products of the corresponding components of $\mathbf{a}$ and $\mathbf{b}$.

For a given set of points F, $\mathrm{conv}(F)$ stands for its convex hull. The standard (or canonical) simplex in $\mathbb{R}^n$ is defined as

$$S_n = \mathrm{conv}\big(\{\mathbf{e}_1, \dots, \mathbf{e}_n\}\big) = \{\mathbf{p} \in \mathbb{R}^n_+ \mid \mathbf{1}^\top \mathbf{p} = 1\}, \tag{9}$$

where $\mathbf{e}_j$ is the versor (unit vector) corresponding to the j-th coordinate in $\mathbb{R}^n$.

For $\mathbf{A} \in \mathbb{S}^m$ we define the mapping

$$\mathrm{svec}_0(\mathbf{A}) = (a_{11}, a_{21}, \dots, a_{m1}, a_{22}, a_{32}, \dots, a_{m2}, \dots, a_{mm})^\top, \tag{10}$$

which yields a column vector with length $m(m+1)/2$, obtained by stacking the consecutive elements of the lower triangle of $\mathbf{A}$.

2 Optimal Sensor Location Problem

2.1 Quantification of Estimation Accuracy

Consider a DPS defined in a bounded spatial domain $\Omega \subset \mathbb{R}^d$ with a sufficiently smooth boundary $\partial\Omega$, evolving in a bounded time interval $T = (0, t_f]$. Denote by $y(\mathbf{x}, t; \boldsymbol{\theta})$ its scalar state at spatial point $\mathbf{x} \in \bar{\Omega} \subset \mathbb{R}^d$ and at time moment $t \in \bar{T}$. It is a solution of a PDE (here we do not specify this equation explictly) whose known parameters are represented by the vector $\boldsymbol{\theta} \in \mathbb{R}^m$ to be estimated based on continuous-time observations of the state variable. We assume that these observations over the time interval T are supposed to be provided by N pointwise sensors. The estimation process is formally represented by the equation

$$z^j(t) = y(\mathbf{x}^j, t; \boldsymbol{\theta}) + \varepsilon(\mathbf{x}^j, t), \quad t \in T \tag{11}$$

for $j = 1, \dots, N$, where $z^j(t)$ is the scalar output, $\mathbf{x}^j \in X$ stands for the location of the j-th sensor, X signifies the part of Ω, in which the measurements are allowed (we assume that X is a compact subset of $\bar{\Omega}$ with nonempty interior), and ε signifies measurement noise. It is usually assumed (and we shall also do it in the sequel) that ε is temporally and spatially zero-mean white noise with a constant variance [32].

The estimate $\widehat{\boldsymbol{\theta}}$ of the parameter vector $\boldsymbol{\theta}$ is obtained through minimization of the relevant least-squares criterion [6, 7]. Determination of the sensor location that would guarantee the best accuracy of the least-squares estimates of $\boldsymbol{\theta}$ reduces to the choice of $\mathbf{x}^j$, $j = 1, \dots, N$, so as to minimize some scalar performance measure Ψ defined on the Fisher information matrix (FIM) [49]

$$\mathbf{M}(\mathbf{x}^1, \dots, \mathbf{x}^N) = \sum_{j=1}^{N} \int_T \mathbf{g}(\mathbf{x}^j, t)\mathbf{g}^\top(\mathbf{x}^j, t)\, \mathrm{d}t, \tag{12}$$

where $\mathbf{g}(\mathbf{x}, t) = \nabla_{\boldsymbol{\vartheta}} y(\mathbf{x}, t; \boldsymbol{\vartheta})\big|_{\boldsymbol{\vartheta}=\boldsymbol{\theta}^0}$ means the vector of sensitivity coefficients, which defines the sensitivity of the state variable to changes in the parameters (the gradient of the state variable with respect to the vector of the unknown parameters), $\boldsymbol{\theta}^0$ signifies a preliminary estimate of the unknown parameter $\boldsymbol{\theta}$ [4]. Discussion and comparison of numerical techniques for determining sensitivity coefficients is included in [62, Chap. 2.6]. Up to a constant, the inverse of the FIM constitutes an approximation to the covariance matrix of the least-squares estimator of the vector $\boldsymbol{\theta}$ [4, 44]. This matrix is a generalization of the notion of the variance from Eq. (8).

Basically, the derivatives determining sensitivity coefficients should be determined at point $\boldsymbol{\vartheta} = \boldsymbol{\theta}$ [4, 44]. However, a key problem is that the value of $\boldsymbol{\theta}$ is unknown (the sensors are to be positioned prior to identification of $\boldsymbol{\theta}$). It is undoubtedly the disadvantage of optimal experiment design methods for nonlinear systems, which is to some extent eliminated by substituting the above estimate $\boldsymbol{\theta}^0$ for $\boldsymbol{\theta}$ (if no such estimate is available, it can be obtained, e.g., by conducting a preliminary experiment). More advanced (and, unfortunately, much more complex) techniques are discussed in monograph [62, Chap. 6].

In what follows, we assume that the criterion Ψ is convex on $\mathbb{S}_+^m$ and strictly convex on $\mathbb{S}_{++}^m$. Moreover, we require Ψ to be differentiable on $\mathbb{S}_{++}^m$, where $\Psi'(\mathbf{A}) \equiv \frac{\mathrm{d}}{\mathrm{d}\mathbf{A}}\Psi(\mathbf{A})$ stands for the derivative $\Psi(\mathbf{A})$ with respect to the matrix argument $\mathbf{A} \in \mathbb{R}^{m\times m}$. $\Psi'(\mathbf{A})$ is an $m \times m$ matrix whose element at the intersection of the i-th row and the j-th column has the form $\partial\Psi(\mathbf{A})/\partial a_{ji}$, cf. [8, p. 410].

As for the form of Ψ, optimal experimental design theory offers a number of options [4, 44]. The most common choice is the D-optimality criterion

$$\Psi_D(\mathbf{M}) = \begin{cases} \log\det(\mathbf{M}^{-1}) & \text{if } \mathbf{M} \text{ is invertible,} \\ +\infty & \text{otherwise.} \end{cases} \tag{13}$$

Its minimization leads to minimization of the volume of the uncertainty ellipsoid for the estimates [44]. Another frequent choice is the A-optimality criterion

$$\Psi_A(\mathbf{M}) = \begin{cases} \operatorname{trace}(\mathbf{M}^{-1}) & \text{if } \mathbf{M} \text{ is invertible,} \\ +\infty & \text{otherwise.} \end{cases} \tag{14}$$

Its minimization forces minimization of the sum of the variances of the estimates.

For the above criteria we have [8, Chapter 10]

$$\Psi_D'(\mathbf{M}) = -\mathbf{M}^{-1}, \quad \Psi_A'(\mathbf{M}) = -\mathbf{M}^{-2} \tag{15}$$

whenever $\mathbf{M}$ is nonsingular.

The introduction of the optimality criterion enables the formulation of the optimal sensor location problem in terms of the following optimization problem: Minimize $\Psi(\mathbf{M}(\mathbf{x}^1, \dots, \mathbf{x}^N))$ with respect to the locations $\mathbf{x}^j$, $j = 1, \dots, N$ subject to the constraint that they remain in the feasible set X.

2.2 *Problem Conversion to the Search for an Optimal Sensor Density*

It turns out that the direct minimization of the criterion $\Psi(\mathbf{M}(\mathbf{x}^1, \dots, \mathbf{x}^N))$ with respect to the sensor locations most often leads to the phenomenon of clustering [62, p. 20], which is manifested by the location of multiple sensors at a single spatial point. This is a consequence of the simplifying assumption about the independence of measurements performed by various sensors even in a situation where their positions coincide. Admittedly, this undesirable phenomenon could be avoided by taking account of the correlation between measurements made by adjacent sensors, but this would disproportionately complicate the optimization task, see [78]. Instead of this, sensor positions are selected from a finite set of arbitrarily specified feasible positions, but the number of elements in this set is possibly very large [32]. Note that this technique is compatible with the way sensor networks operate: network nodes are already located in space, and the decision problem consists in the determination of sensors best suited for activation and collection of valuable information through making measurements.

Unfortunately, when the number of feasible locations gets large, which is quite often the case in applications involving sensor networks, computations related to the determination of the optimal sensor locations based on the aforementioned optimization approach quickly become extremely hard from the viewpoint of computational complexity. An especially efficient method of avoiding these difficulties consists in giving up the idea of treating the individual locations of active sensors as decision variables and adopting spatial sensor density (i.e., the number of sensors per unit area) in this role [18, 62].

Mathematically, the density of sensor locations can be represented by a probability measure ξ defined on the σ-field $\mathscr{B}(X)$ of Borel subsets of the set X (i.e., the σ-field generated by all open subsets of the set X treated as a topological subspace of $\mathbb{R}^d$). It turns out that, despite appearances of abstractness, this form of feasible solutions makes it possible to employ convenient and effective mathematical tools of convex programming theory. As for the usefulness of the potential results, it becomes visible after arbitrarily partitioning the set X into subsets $X_i, i = 1, \dots, n$ of relatively small areas. It is assumed that these sets are compact and apart from their boundaries they are pairwise disjoint. This type of partition is, e.g., a by-product of partitioning the spatial domain into finite elements formed by the finite element method. In each subset X_i we allocate as many as

$$N_i = \left\lceil N \int_{X_i} \xi(\mathrm{d}\mathbf{x}) \right\rceil \tag{16}$$

sensors (the symbol $\lceil \rho \rceil$ means the least integer no less than ρ, and the integral should be understood as the one with respect to the measure ξ).

In this way, the class of sensor locations is identified with the set $\Xi(X)$ of all probability measures ξ defined on the measurable space $(X, \mathscr{B}(X))$, which are absolutely continuous with respect to the Lebesgue measure (absolute continuity excludes the concentration of the measure at single spatial points and implies the existence of a probability density function with respect to the Lebesgue measure [43]).

In Eq. (12) the FIM then replaced with the expression

$$\mathbf{M}(\xi) = \int_X \mathbf{G}(\mathbf{x})\, \xi(\mathrm{d}\mathbf{x}), \tag{17}$$

where $\mathbf{G}(\mathbf{x}) = \int_T \mathbf{g}(\mathbf{x}, t)\mathbf{g}^\top(\mathbf{x}, t)\,\mathrm{d}t$.

In order to avoid solutions that might lead to an excessive concentration of sensors in specific spatial subdomains, a maximal allowable sensor density is introduced, which is represented by a measure ω ($N\omega(\mathrm{d}\mathbf{x})$ is to be interpreted as a maximal possible "number" of sensors per segment $\mathrm{d}\mathbf{x}$ [18]). This leads to the condition $\xi \le \omega$, cf. [14, 18], which is a shorthand notation for the requirement $\xi(B) \le \omega(B),\ \forall B \in \mathscr{B}(X)$. Since $\int_X \xi(\mathrm{d}\mathbf{x}) = 1$, we must have $\int_X \omega(\mathrm{d}\mathbf{x}) \ge 1$.

As a result, we deal with the following optimization problem:

Problem 1 Find

$$\xi^\star = \arg \min_{\xi \in \Xi(X)} \Psi[\mathbf{M}(\xi)] \tag{18}$$

subject to $\xi \le \omega$.

The measure $\xi^\star$ defined above is said to be a (Ψ, ω)-optimum design by analogy with the definition introduced in [14] in the context of directly constrained design measures.

3 Conversion to a Weight Optimization Problem

In order to make Problem 1 suitable for the solution using numerical tools, it has to be properly discretized. The partition of the set X into a sum of small nonempty sets X_i, $i = 1, \dots, n$ with disjount interiors turns to be very helpful here. In this way, we have the representation $X = \bigcup_{i=1}^n X_i$, as was already mentioned in Sect. 2.2. Observe that the measure $\xi \in \Xi(X)$ assigns each subset X_i a weight $p_i = \xi(X_i)$. In view of (16), the knowledge of the coefficients p_i is sufficient to determine an optimal distribution of active nodes of the sensor network among the subsets X_i. Assuming that the variation in the the values of $\mathbf{G}(\,\cdot\,)$ on each subset X_i is negligible (we can achieve this by a sufficiently fine partition of the set X), we get

$$\begin{aligned} \mathbf{M}(\xi) &= \int_X \mathbf{G}(\mathbf{x})\,\xi(\mathrm{d}\mathbf{x}) \\ &\approx \int_X \Big(\sum_{i=1}^n \mathbf{G}(\mathbf{x}_i)\mathbb{1}_{X_i}(\mathbf{x})\Big)\,\xi(\mathrm{d}\mathbf{x}) \\ &= \sum_{i=1}^n \mathbf{G}(\mathbf{x}_i) \int_X \mathbb{1}_{X_i}(\mathbf{x})\,\xi(\mathrm{d}\mathbf{x}) = \sum_{i=1}^n p_i \mathbf{M}_i =: \mathbf{M}(\mathbf{p}), \end{aligned} \tag{19}$$

where $\mathbf{p} = (p_1, \dots, p_n)$, $\mathbf{M}_i = \mathbf{G}(\mathbf{x}_i)$, $\mathbf{x}_i$ is an arbitrarily fixed point in X_i (e.g., the center of mass), $i = 1, \dots, n$, $\mathbb{1}_A$ stands for the indicator function of the set A. Thus, fixing $\mathbf{b} = (\omega(X_1), \dots, \omega(X_n))$, we get the following approximation to Problem 1:

Problem 2 Find a vector of weights $\mathbf{p} \in \mathbb{R}^n$ to minimize

$$\Phi(\mathbf{p}) = \Psi\big(\mathbf{M}(\mathbf{p})\big) \tag{20}$$

subject to

$$\mathbf{1}^\top \mathbf{p} = 1, \tag{21}$$

$$\mathbf{0} \preceq \mathbf{p} \preceq \mathbf{b}. \tag{22}$$

With no loss of generality, we shall further assume that $\mathbf{b} \succ \mathbf{0}$.

The set of feasible solutions P defined by (21) and (22) is the intersection of the standard simplex S_n and the hyperbox $B = \big\{\mathbf{q} \in \mathbb{R}^n \mid \mathbf{0} \preceq \mathbf{q} \preceq \mathbf{b}\big\}$, i.e., a relatively simple convex polyhedron, although with a large number of vertices whenever n is large. This property is going to be exploited below in the numerical procedure of weight optimization.

Another specific feature of the above formulation is the convexity of typical optimality criteria (including (13) and (14)) on the cone $\mathbb{S}^m_+$ and their strict convexity on the cone $\mathbb{S}^m_{++}$ [45]. What is more, they are differentiable at points S_n yielding a nonsingular FIM, and its gradient $\phi(\mathbf{p}) := \nabla\Phi(\mathbf{p})$ is

$$\boldsymbol{\phi}(\mathbf{p}) = \Big(\operatorname{trace}\big(\Psi'(\mathbf{M}(\mathbf{p}))\mathbf{M}_1\big), \ldots, \operatorname{trace}\big(\Psi'(\mathbf{M}(\mathbf{p}))\mathbf{M}_n\big)\Big)^\top. \tag{23}$$

4 Optimality Conditions and Linear Separability of the Gradient Components

Problem 2 can be easily solved using the simplicial decomposition algorithm [39, 67] or Newton's method in the manner usually employed in interior-point methods [30]. Regardless of which method we use, any minimizer $\mathbf{p}^\star$ is characterized by the following necessary and sufficient optimality conditions [71]:

Theorem 1 *Assume that the matrix $\mathbf{M}(\mathbf{p}^\star)$ is nonsingular for some value $\mathbf{p}^\star \in P$. A vector $\mathbf{p}^\star$ is a global minimizer of the function Φ over P if, and only if, there exists a real $\lambda^\star$ satisfyling*

$$\phi_i(\mathbf{p}^\star) \begin{cases} \le \lambda^\star & \text{if } p_i^\star = b_i, \\ = \lambda^\star & \text{if } 0 < p_i^\star < b_i, \\ \ge \lambda^\star & \text{if } p_i^\star = 0. \end{cases} \tag{24}$$

The property (24) is a linear separability condition for the components of the gradient $\nabla\Phi(\mathbf{p}^\star)$. Indeed, fix $g_i = \phi_i(\mathbf{p}^\star)$ and imagine a plane with n points with coordinates $\big(i, g_i\big)$, $i = 1, \ldots, n$. Optimality means that it is possible to draw a straight horizontal line described by the equation $g = \lambda^\star$ such that the points corresponding to the zero weights lie on this line or above it, the points corresponding to the weights in between zero and their upper bounds lie exactly on this line, and the points corresponding to the weights equal to their upper bounds lie on this line or below it.

5 Numerical Algorithm for Weight Optimization

5.1 *Simplicial Decomposition for Problem 2*

The structure of Problem 2 makes it especially suitable for the application of the optimization technique known as simplicial decomposition (SD). This method is an extremely efficient tool of solving large-scale minimization problems of pseudoconvex functions on polyhedral sets of feasible solutions [10, 42]. In its basic form, SD alternates between linear and nonlinear optimization problems, called the column generation problem (CGP) and the restricted master problem (RMP). In the RMP we deal with the relaxation of the original problem consisting of replacing the polyhedral set of feasible solutions P by its approximation being the convex hull of a finite set $\tilde{P}$, whose number of elements is usually drastically lower than the number of vertices of P (in this way a substantial reduction in the problem dimensionality is achieved). In the CGP this interior approximation of the set P is improved by including a vertex of the set P, which lies furthest in the direction of the antigradient computed at the point being the current solution to the RMP. A distinctive feature of SD is that the sequence of solutions to the RMP tends to the solution of the original problem in such a manner that the consecutive values of of the objective funtion decrease and systematically approach its optimal value.

SD can be interpreted as a kind of modular nonlinear programming provided that an efficient code for solving the RMP and a code exploitng the linearity of the CGP are available. In the sequel, we show that this is the case for Problem 2. What is more, since we deal with minimization of a differentiable convex function Φ on the polyhedral set P, convergence of the algorithm in a finite number of steps is guaranteed.

Adaptation of the SD scheme to the needs of the problem considered leads to Algorithm 1. In what follows, its consecutive steps are detailed.

5.2 *Termination of Algorithm 1*

Originally, in the SD scheme the computations are terminated when the current point $\mathbf{p}^{(k)}$ fulfills the basic condition of the impossibility of a decrease (up to first-order terms) in the value of the optimality criterion over the entire set of feasible solutions, i.e., if

$$\min_{\mathbf{p}\in P} \boldsymbol{\phi}(\mathbf{p}^{(k)})^\top (\mathbf{p} - \mathbf{p}^{(k)}) \geq 0. \tag{32}$$

Algorithm 1 Simplicial decomposition for Problem 2.

Step 0: (Initalization)
Input an initial approximated solution $\mathbf{p}^{(0)} \in P$ such that $\mathbf{M}(\mathbf{p}^{(0)})$ is nonsingular. Set $I = \{1, \ldots, n\}$, $Q^{(0)} = \{\mathbf{p}^{(0)}\}$ and $k = 0$.

Step 1: (Checking the stopping condition)
Set

$$I_g^{(k)} = \{i \in I \mid p_i^{(k)} = b_i\}, \tag{25}$$

$$I_m^{(k)} = \{i \in I \mid 0 < p_i^{(k)} < b_i\}, \tag{26}$$

$$I_d^{(k)} = \{i \in I \mid p_i^{(k)} = 0\}. \tag{27}$$

If there is a value $\lambda \in \mathbb{R}$ such that

$$\phi_i(\mathbf{p}^{(k)}) \begin{cases} \leq \lambda & \text{jeli } i \in I_g^{(k)}, \\ = \lambda & \text{jeli } i \in I_g^{(k)}, \\ \geq \lambda & \text{jeli } i \in I_g^{(k)}, \end{cases} \tag{28}$$

then $\mathbf{p}^{(k)}$ is optimal and terminate.

Step 2: (Solving the column generation problem)
Compute

$$\mathbf{q}^{(k+1)} = \arg \min_{\mathbf{p} \in P} \boldsymbol{\phi}(\mathbf{p}^{(k)})^\top \mathbf{p} \tag{29}$$

and set

$$Q^{(k+1)} = Q^{(k)} \cup \{\mathbf{q}^{(k+1)}\}. \tag{30}$$

Step 3: (Solving the restricted master problem)
Find

$$\mathbf{p}^{(k+1)} = \arg \min_{\mathbf{p} \in \operatorname{conv}(Q^{(k+1)})} \Phi(\mathbf{p}) \tag{31}$$

and purge $Q^{(k+1)}$ of all the elements with associated zero weights in the expression determining $\mathbf{p}^{(k+1)}$ as a convex combination of the elements of the set $Q^{(k+1)}$. Increment k by one and return to Step 1.

The alternative condition (28) results from the characterization of the point $\mathbf{p}^\star$ being a global minimizer satisfying $\Phi(\mathbf{p}^\star) = \min_{\mathbf{p} \in P} \Phi(\mathbf{p})$, which is discussed in Sect. 4.

5.3 Solving the Column Generation Problem

In Step 2 of Algorithm 1 we deal with the linear programming problem

$$\begin{aligned} &\text{minimize} \quad \mathbf{c}^\top \mathbf{p} \\ &\text{subject to} \quad \mathbf{p} \in P, \end{aligned} \tag{33}$$

where $\mathbf{c} = \boldsymbol{\phi}(\mathbf{p}^{(k)})$. The set of feasible solutions is defined through $2n$ interval constraints (22) i one equality constraint (21).

For such a simple problem it is possible to set forth an extremely simple algorithm to determine its solution. Its idea is based on the following characterization resulting from Karush-Kuhn-Tucker optimality conditions, cf. [65].

Theorem 2 *A vector $\mathbf{q} \in P$ constitutes a global solution to Problem (33) if, and only if, there exists a scalar ρ satisfying*

$$c_i \begin{cases} \leq \rho & \text{if } q_i = b_i, \\ = \rho & \text{if } 0 < q_i < b_i, \\ \geq \rho & \text{if } q_i = 0 \end{cases} \tag{34}$$

for $i = 1, \dots, n$.

From the above result it follows that in order to solve (33) it is enough to select the consecutive least elements c_i of vector $\mathbf{c}$ and to fix the corresponding weights q_i at their maximal allowable values b_i. The process is repeated until the sum of the assigned weights exceeds the value of one. Then the last assigned weight is corrected so as to satisfy the constraint $\mathbf{1}^\top \mathbf{p} = 1$, and all the remaining weights (i.e., the ones with no value assigned) are simply zeroed.

5.4 Solving the Restricted Master Problem

Suppose that in the $(k+1)$-th iteration of Algorithm 1 we have

$$Q^{(k+1)} = \{\mathbf{q}_1, \dots, \mathbf{q}_r\}, \tag{35}$$

possibly for $r < k + 1$ owing to the built-in mechanism of purging $Q^{(j)}, 1 \leq j \leq k$ of points which contribute nothing to the convex combinations yielding the consecutive approximations $\mathbf{p}^{(j)}$. Step 3 of Algorithm 1 comprises minimization of the criterion (20) over the set

$$\operatorname{conv}(Q^{(k+1)}) = \left\{ \sum_{j=1}^{r} w_j \mathbf{q}_j \,\middle|\, \mathbf{w} \succeq \mathbf{0},\ \mathbf{1}^\top \mathbf{w} = 1 \right\}. \tag{36}$$

From the representation of any $\mathbf{p} \in \operatorname{conv}(Q^{(k+1)})$ as

$$\mathbf{p} = \sum_{j=1}^{r} w_j \mathbf{q}_j, \tag{37}$$

it follows that

$$\mathbf{M}(\mathbf{p}) = \sum_{i=1}^{n} p_i \mathbf{M}_i = \sum_{j=1}^{r} w_j \mathbf{M}(\mathbf{q}_j). \tag{38}$$

As can be easily seen, the RMP can equivalently be formulated as follows:

Problem 3 Find a sequence of weights $\mathbf{w} \in \mathbb{R}^r$ minimizing

$$\Pi(\mathbf{w}) = \Psi\big(\mathbf{H}(\mathbf{w})\big) \tag{39}$$

subject to

$$\mathbf{1}^\top \mathbf{w} = 1, \tag{40}$$

$$\mathbf{w} \succeq \mathbf{0}, \tag{41}$$

where

$$\mathbf{H}(\mathbf{w}) = \sum_{j=1}^{r} w_j \mathbf{H}_j, \qquad \mathbf{H}_j = \mathbf{M}(\mathbf{q}_j), \quad j = 1, \dots, r. \tag{42}$$

The above problem constitutes one of the classical computational problems considered in optimum experimental design theory (the search for a convex combination of fixed nonnegative definite matrices so as to minimize a convex optimality criterion). Its solution can be obtained using one of the following effective tools, e.g., the multiplicative algorithm [76], or alternatively, the interior point method [36] (the latter work is accompanied with efficient Matlab codes, including the one for the multiplicative algorithm).

6 Sparsity Promoting Postprocessing of Solutions

In most cases the solution $\mathbf{p}^\star$ to Problem 2 is not unique and the same value of the optimality criterion may be attained at infinitely many values of the weight vector. Observe, however, that the corresponding FIM $\mathbf{M}^\star = \mathbf{M}(\mathbf{p}^\star)$ is unique, i.e., it does not depend on a specific minimizer. In fact, Problem 2 can be interpreted as minimization of the criterion Ψ on a convex and compact set of matrices $\mathfrak{M} = \big\{\mathbf{M}(\mathbf{p}) : \mathbf{1}^\top\mathbf{p} = 1,\ \mathbf{0} \preceq \mathbf{p} \preceq \mathbf{b}\big\}$, cf. [62, Lemma 3.2, p. 39]. But the criterion Ψ is strictly convex on the cone $\mathbb{S}^m_{++}$, which constitutes the interior of $\mathbb{S}^m_{+}$ with respect to $\mathbb{S}^m$ [45, s. 10], and this very fact implies the uniqueness of the optimal FIM.

Thus there may be infinitely many optimal weight vectors. What is more, they form a convex set. This offers an additional degree of freedom, which can be exploited to satisfy some extra requirements. One of them, which practitioners often care about, is the following: Given a minimizer $\mathbf{p}^\star$ (owing to the convexity of Ψ, any local minimizer constitues a global one, too), thus, it is required to find another minimizer $\widehat{\mathbf{p}}$, which would concentrate the experimental effort on a possibly least area. Mathematically, $\widehat{\mathbf{p}}$ would be a discrete probability distribution with minimal support, i.e., a low number of nonzero components.

First of all, observe that the optimality conditions imply that only some components of $\widehat{\mathbf{p}}$ can differ from the corresponding components of $\mathbf{p}^\star$. This observation can drastically reduce the search space. From Eq. (23) its is clear that the gradients of $\phi(\mathbf{p}^\star)$ and $\phi(\widehat{\mathbf{p}})$ must be the same, as they implicitly depend on their arguments through the optimal FIM, which is unambiguous, i.e., $\mathbf{M}^\star = \mathbf{M}(\mathbf{p}^\star) = \mathbf{M}(\widehat{\mathbf{p}})$. Thus, partitioning the set of indices $I = \{1, \dots, n\}$ into the following subsets:

$$I_< = \{i \in I : \phi_i(\mathbf{p}^\star) < \lambda^\star\} = \{i \in I : \phi_i(\widehat{\mathbf{p}}) < \lambda^\star\}, \tag{43}$$

$$I_= = \{i \in I : \phi_i(\mathbf{p}^\star) = \lambda^\star\} = \{i \in I : \phi_i(\widehat{\mathbf{p}}) = \lambda^\star\}, \tag{44}$$

$$I_> = \{i \in I : \phi_i(\mathbf{p}^\star) > \lambda^\star\} = \{i \in I : \phi_i(\widehat{\mathbf{p}}) > \lambda^\star\}, \tag{45}$$

we notice that these subsets are the same for all minimizers. In consequence, the optimality of $\phi(\widehat{\mathbf{p}})$ and Theorem 1 force

$$\widehat{p}_i = \begin{cases} b_i & \text{for } i \in I_<, \\ \text{undefined} & \text{for } i \in I_=, \\ 0 & \text{for } i \in I_>. \end{cases} \tag{46}$$

As can be seen, we can focus solely on determining the components $\widehat{p}_i$, $i \in I_=$.

For notational convenience, replace the variables $\widehat{p}_i$, $i \in I_=$ by w_j, $j \in J$, where $J = \{1, \dots, r\}$, $r = |I_=|$. This is possible because there is a bijection π of the set J onto $I_=$ such that $w_j = \widehat{p}_{\pi(j)}$, $j \in J$. As a result, we get the following formulation:

Problem 4 Find a vector $\mathbf{w} \in \mathbb{R}^r$ minimizing

$$\mathscr{L}_0(\mathbf{w}) = \|\mathbf{w}\|_0 \tag{47}$$

subject to

$$\mathbf{A}\mathbf{w} = \mathbf{c}, \tag{48}$$

$$\mathbf{1}^\top \mathbf{w} = \alpha, \tag{49}$$

$$\mathbf{0} \preceq \mathbf{w} \preceq \mathbf{d}, \tag{50}$$

where $\|\mathbf{w}\|_0$ is the number of nonzero components of the vector $\mathbf{w}$, i.e., $\|\mathbf{w}\|_0 = |\text{supp}(\mathbf{w})|$, $\text{supp}(\mathbf{w}) = \{j \in J : w_j \neq 0\}$, and

$$\mathbf{A} = \Big[\, \mathrm{svec}_0(\mathbf{M}_{\pi(1)}) \Big| \dots \Big| \mathrm{svec}_0(\mathbf{M}_{\pi(r)}) \,\Big], \tag{51}$$

$$\mathbf{c} = \mathrm{svec}_0\Big(\mathbf{M}^\star - \sum\nolimits_{i \in I_<} b_i \mathbf{M}_i\Big), \tag{52}$$

$$\mathbf{d} = (b_{\pi(1)}, \dots, b_{\pi(r)})^\top, \tag{53}$$

$$\alpha = 1 - \sum\nolimits_{i \in I_<} b_i. \tag{54}$$

The constraint (48) expresses the requirement of a constant FIM, i.e., we must have $\mathbf{M}^\star = \mathbf{M}(\widehat{\mathbf{p}})$. In turn, (49) follows from the condition $\mathbf{1}^\top \widehat{\mathbf{p}} = 1$ and from (46). Finally, (50) ensures that the condition $\mathbf{0} \preceq \widehat{\mathbf{p}} \preceq \mathbf{b}$ is met.

In the literature, $\|\cdot\|_0$ is termed the ℓ_0-norm, although it is not a norm, as it does not satisfies the condition of absolute homogeneity. In recent years, the problem of minimizing the ℓ_0-'norm' subject to linear constraints has received much attention owing to its presence in many domains, e.g., machine learning, data mining or compressed sensing [57]. It is one of the NP-complete problems [29], which mean that there is no algorithm for its solution that has polynomial time complexity. In spite of this negative result, which is rather discouraging, the problem has met with great interest and a number of efficient algorithms have been invented to produce approximate solutions. One of them consists of replacing the ℓ_0-'norm' by the ℓ_q-'norm'

$$\|\mathbf{w}\|_q = \Big(\sum_{j=1}^{r} |w_j|^q\Big)^{\frac{1}{q}}, \tag{55}$$

which is justified by the property $\|\mathbf{w}\|_0 = \lim_{q \downarrow 0} \|\mathbf{w}\|_q$ valid for each fixed $\mathbf{w}$. Observe that given $0 < q < 1$, $\|\cdot\|_q$ it is not a norm, either, as it does not satisfy the triangle inequality.

In work [20] it was shown that for a sufficiently small value of $q > 0$ the problems of minimizing 'norms' ℓ_0 and ℓ_q under the same linear constraints are equivalent in the sense of possessing identical optimal solutions. Making use of this result, Problem 4 is replaced by its counterpart for a fixed $q \in (0, 1)$:

Problem 5 Find a vector $\mathbf{w} \in \mathbb{R}^r$ minimizing

$$\mathscr{L}_q(\mathbf{w}) = \|\mathbf{w}\|_q^q = \sum_{j=1}^{r} w_j^q \tag{56}$$

subject to the constraints (48)–(50).

Note, that this is the problem of finding a global minimum of a separable convex function (in this case, separability means that the objective function is the sum of convex functions of one variable) on a polyhedron. One of its characteristics is that its local minima (and thus a global minimum as well) are reached at the extreme points of this polyhedron. From the point of view of computational complexity, we thus get an NP-hard problem. Its special features, however, make it more tractable than other NP-hard problems.

To solve this type of problem, in work [20] an algorithm of consecutive linearizations based on linear programming has been proposed. However, this method cannot guarantee that the calculations will end once the global minimum is reached. An interesting alternative, based on a relatively simple branch and bound algorithm, is contained in the work [52]. It finds a global minimum in Problem 5 in a finite number of iterations. In each of these iterations, the algorithm determines the lower bounds of the optimality criterion values for each convex subproblem through a solution of a relaxed formulation based on linear programming. It partitions the solution space using rectangular divisions and branches computations based on the best currently available approximation to the optimal solution. Numerical experiments confirm a very good behavior of this approach for the problems of sensor location with hundreds of weights, even in a low-budget personal computer.

7 Computational Example

In order to illustrate the described approach for optimal sensor subset selection, consider the process of estimating pollution sources using a large-scale sensor network with nodes capable of measuring pollution concentration. In the interior of the monitored domain Ω, which has the form of a square with length of 1 km, there are two active sources emitting carbon dioxide to the atmosphere (see Fig. 2). The changes in the spatial concentration $y(\mathbf{x}, t)$ of SO_2 in the time interval $T = [0, 1000]$ (in seconds) can be mathematically described by the following advection-diffusion equation:

$$\begin{aligned} \frac{\partial y(\mathbf{x}, t)}{\partial t} &+ \nabla \cdot \big(\mathbf{v}(\mathbf{x}, t) y(\mathbf{x}, t)\big) + \varrho y(\mathbf{x}, t) \\ &= \nabla \cdot \big(\kappa \nabla y(\mathbf{x}, t)\big) + f_1(\mathbf{x}) + f_2(\mathbf{x}) \quad \text{in } \Omega \times T, \end{aligned} \tag{57}$$

complemented with the following homogeneous boundary and initial conditions:

$$\frac{\partial y(\mathbf{x}, t)}{\partial \mathbf{n}} = 0 \quad \text{on } \partial\Omega \times T, \quad y(\mathbf{x}, 0) = 0 \quad \text{in } \Omega, \tag{58}$$

where $f_\ell(\mathbf{x}) = \mu_\ell \exp\big(-100\|\mathbf{x} - \boldsymbol{\chi}^\ell\|^2\big)$ represents the emission model of a source with intensity μ_ℓ located at point $\boldsymbol{\chi}^\ell = (\chi_1^\ell, \chi_2^\ell)$, $\ell = 1, 2$. Averaged changes in the wind velocity in the domain Ω are approximated by the following model (scaled in [km/h])

$$\mathbf{v}(\mathbf{x}, t) = \begin{bmatrix} (2 - 8x_1 + 2x_2)\cos(\pi t \cdot 10^{-3}) \\ (2 + 2x_1 - 8x_2)\sin(\pi t \cdot 10^{-3}) \end{bmatrix}. \tag{59}$$

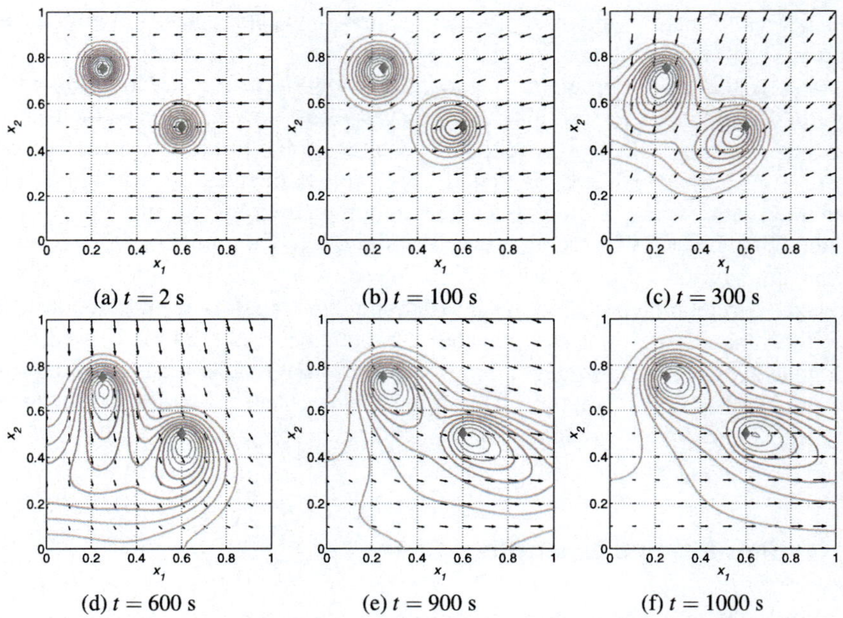

Fig. 2 Contour plots of concentration changes at consecutive time instants. The arrows represent the varying wind velocity field, and the diamonds denote the locations of the pollution sources

Moreover, κ stands for the unknown diffusion coefficient, and $\varrho = 0.005\ \mathrm{s}^{-1}$ represents the rate of the disappearance of the toxic substance as a result of the reaction of sulfur dioxide with air components. Figure 2 illustrates the complicated dynamics of the spatiotemporal emission process.

In the considered computational experiment we look for a D-optimal spatial configuration of active sensor network nodes in order to identify as accurately as possible the positions of the pollution sources and estimate their intensities along with the unknown diffusivity coefficient, i.e., to estimate the parameter vector

$$\boldsymbol{\theta} = (\mu_1, \chi_1^1, \chi_2^1, \mu_2, \chi_1^2, \chi_2^2, \kappa). \tag{60}$$

It is assumed that the design region X coincides with $\bar{\Omega} = \Omega \cup \partial\Omega$. The measure determining the upper bound to the density of active sensor network nodes has the form $\omega(A) = \gamma\, \mathfrak{m}(A)/\mathfrak{m}(X)$ for each Borel subset A of set X, where $\mathfrak{m}(\,\cdot\,)$ signifies the area (i.e., Lebesgue measure) of its argument. For the two considered scenarios we have $\gamma = 5$ and $\gamma = 4$. This means a coverage of approximately $1/\gamma$ of the whole area of Ω, i.e., 20% and 25%, respectively. At the same time, owing to the large scale of the sensor network, we are interested in a solution reducing a large number of active sensor networks. This can be achieved through postoptimal analysis and a possibly large reduction in the domains with assigned sensor densities not achieving the constraint ω.

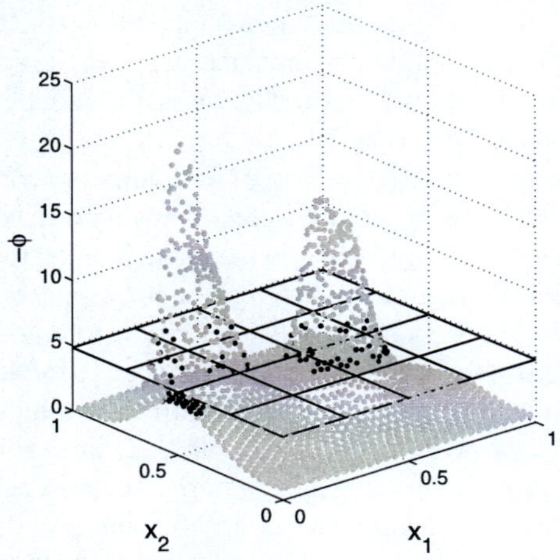

Fig. 3 Visualization of the optimality condition (24) for the target 20% coverage of the domain Ω: the gray points represent the values of the anitgradient components (i.e., $-\phi_i(\mathbf{p}^\star)$), computed at points $\mathbf{x}^i$ being the centers of mass of the mesh triangles; they are separated using a horizontal plane on a level of $-\lambda^\star = 4.85$. The black points denote the subdomains at which the the gradient components are approximately equal to $\lambda^\star$. The corresponding weights will form the set of decision variables during the generation of a sparse solution

Both simplicial decomposition and the sparsity promoting postoptimal procedure were implemented in MATLAB 2015b. Problem (57)–(58) was solved using effective numerical procedures of COMSOL software using the finite element method. Computations were performed on a triangular spatial mesh composed of 2976 triangles, 1549 nodes, and an even partition of the time interval T into 100 subintervals. The triangles of the spatial mesh were simultaneously used as subdomains X_i, cf. Fig. 4. As preliminary estimates of the unknown parameters the following nominal values were used:

$$\boldsymbol{\theta}^0 = (75\ \text{kg/s}, 0.25\ \text{km}, 0.75\ \text{km}, 60\ \text{kg/s}, 0.6\ \text{km}, 0.5\ \text{km}, 50\ \text{m}^2/). \tag{61}$$

All tests were made on a low-budget personal computer equipped with an Intel Core i7 processor (2.53GHz, 12 GB RAM) working on Windows 10.

In Fig. 4a,d D-optimal configurations of active sensors are shown. They were produced by the simplicial decomposition algorithm for two test scenarios corresponding to 20% and 25% coverage of set X. It is clearly seen that the active sensors concentrate in domains of the largest changes in the polluton concentrations emitted by the sources. In the first scenario (20% coverage) the support of optimal configuration consists of 573 triangles, including 359 with the spatial density of active sensors attaining the maximal allowable value (marked in black) and 214 with the density taking nonzero values below the maximum value (marked in gray). In the second scenario (25% coverage) the support of the D-optimal spatial measure is composed of 766 triangles (597 of which attained the maximal allowable density and 169 correspond to a positive density but below the maximal one). It is clear that in both the cases a margin exists for a substantial reduction in the number of elements in the support of the optimal configuration.

The key role of the gradient $\phi(\mathbf{p}) = \nabla\Phi(\mathbf{p})$ for optimality conditions is illustrated in Fig. 3 for the first scenario, where the subdomains with maximal density can be separated from those with zero intensity via the plane parallel to the the plane x_1–x_2. The black points represent the domains in which the corresponding components of the gradient of ϕ_i equal $\lambda^\star = -4.85$ (obviously, up to some tolerance introduced to take account of numerical errors). They simultaneously identify subdomains with positive density values lower than the upper bound. The corresponding weights form the set of decision variables in the next step, which produces a sparse solution. The results of solving this postoptimal problem using the ℓ_q-'norm' are presented in Fig. 4b,c for the 20% coverage and the corresponding values $q = 0.5$ and $q = 0.05$. A black color is used to mark the domains selected from the candidate set which are assigned maximal allowable density values while the other subdomains, which were not selected, are

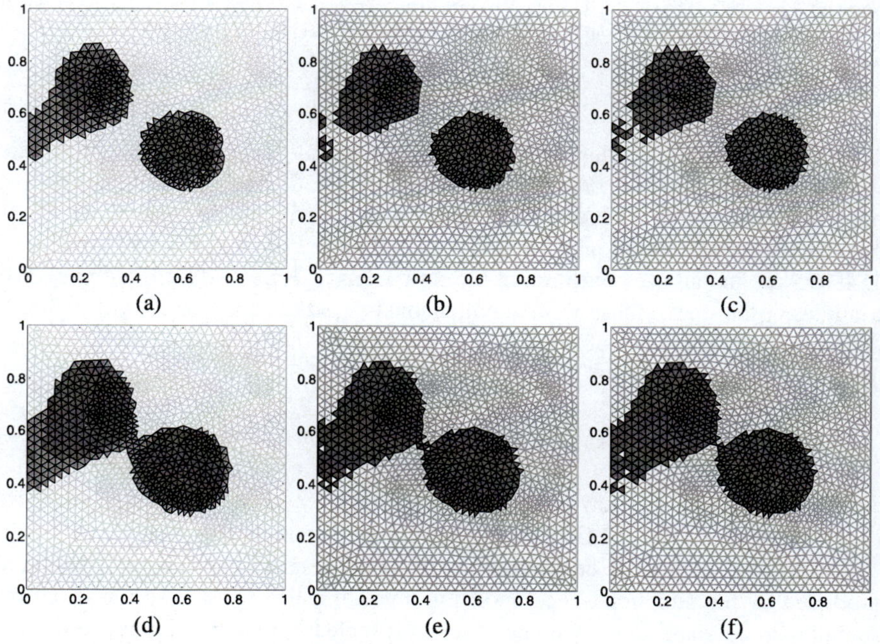

Fig. 4 Optimal spatial densities of active sensor locations: the solution produced by simplicial decomposition (subdomains with the density equal to the upper bound or a positive density below it are marked in black and gray, respectively) for 20% coverage (**a**) and 25% coverage (**d**); the solution after postoptimal analysis using the $\ell_{0.5}$ 'norm' (black triangles denote the subdomains selected from the candidate set marked in gray) for 20% coverage (**b**) and 25% coverage (**e**); the solution determined in the same way, but using the $\ell_{0.05}$ 'norm' for 20% coverage (**c**) and 25% coverage (**f**)

assigned zero density. Both the configurations in Fig. 4b,c look similar. The number of subdomains with active sensor network nodes was reduced to 506 (a reduction of 67) and 506 (a reduction of 68), respectively. This means a significant (over 30%) reduction in the number of triangles among these, which were assigned a density lower than the maximal one. For the second scenario (25% coverage), the results are displayed in Fig. 4e,f for $q = 0.5$ and $q = 0.05$, respectively. In this case the number of subdomains with active sensor network nodes was reduced to 711 (a reduction of 55) and 709 (a reduction of 57), respectively.

As for the efficiency of the algorithms, simplicial decomposition returned each time a result in a time not exceeding 30 s (the number of iterations did not exceed 20). The postoptimal analysis in both the cases lead to a sparse solution after exploring no more than 2000 nodes of the branch-and-bound tree in a time not exceeding 2 min.

8 Comments and Final Remarks

This chapter presents an original approach to the sensor location aiming at maximizing the parameter estimation accuracy of spatiotemporal systems. The problem formulation makes the proposed strategy suitable for sensor networks with an extremely large number of nodes, of which only those providing the most valuable information on estimated parameters should be activated. The approach is a type of alternative to effective methods described in the works [12, 13, 30], drawing on the advantages of interior-point and semi-definite programming methods, which, however, have some significant limitations in the case of a large number of decision variables, which results primarily from their high generality. Attention should be paid to the fact that in the proposed simplicial decomposition algorithm the restricted master problem has the form of a classical task of finding an optimum experimental design on a finite set of design points. This opens up the possibility of using not only a plethora of effective experimental design algorithms [44, Chap. 9], developed over the last five decades, but also interior-point methods as described in the work [36]. Simplicial decomposition is not so much a computational algorithm, but a way of organizing calculations focused primarily on dimensionality reduction. The components of this scheme (CGP and RMP) can be specified in various ways, including algorithms most suitable for the specificity of the problem being solved.

It is to be noticed that simplicial decomposition can also be successfully used in the case of direct search for optimal sensor locations, and not their density. Then the problem of choosing N optimal measurement points from among given candidate positions $\mathbf{x}^1, \dots, \mathbf{x}^L$ formulated in Sect. 2.1 is treated as the problem of finding a weight vector $\mathbf{v} = (v_1, \dots, v_L)$ minimizing the performance index

$$\mathscr{J}(\mathbf{v}) = \Psi\Big(\sum_{j=1}^{L} v_j \mathbf{G}(\mathbf{x}^j)\Big) \tag{62}$$

subject to

$$\mathbf{1}^\top \mathbf{v} = N, \tag{63}$$

$$v_j \in \{0, 1\}, \quad j = 1, \ldots, L, \tag{64}$$

where

$$w_j = \begin{cases} 1 & \text{if an active sensor is placed at } \mathbf{x}^j, \\ 0 & \text{otherwise.} \end{cases} \tag{65}$$

Of course, the combinatorial nature of the problem formulated in this way constitutes a huge barrier in determining an optimal solution for a large number of feasible positions L. The approach recommended in the literature [12, 13, 30] is a relaxation, consisting in replacing binary weights with the ones that can also take fractional values, and the constraints (64) with the restrictions $w_j \in [0, 1]$, $j = 1, \ldots, L$. Clearly, this task is practically identical with Problem 2 (weight normalization so as they add up to the number of active sensors, as in (63), not to unity, as in (21), does not change anything; indeed, one task can be reduced to another by scaling the weights). For its solution, simplicial decomposition can be directly applied. What is more, a technique that promotes a sparse form of the solution, described in Sect. 6, can be used here without significant changes. In this way, an approximate solution is obtained, which results from the relaxation of the original problem. If this is not satisfactory, exact solutions can be sought based on the branch-and-bound method described in the work [71], in which the simplicial decomposition is a component playing a key role in determining a lower bound to the optimal value of the objective function.

Due to space limitations, the methodology presented here should be treated only as an introduction to the problem of determining optimal locations of measurement sensors. Some modifications of the approach extend its applicability to the following cases:

1. Deploying sensors to maximize prediction accuracy.
   In work [27] the problem of estimation of the initial temperature distribution and thermal parameters of a prototype machine was considered. The machine was constructed for extremely precise positioning of a mechanical tool, which is achieved by computing the machine deformation due to heating and making appropriate corrections to the nominal position of the tool. Therefore, the goal is not parameter estimation in itself, but prediction of the tool position obtained based on them. A significant difference from the approach presented here lies in a slightly different form of the criterion, which can be treated as some modification of the D-optimality criterion. However, the SD scheme remains unchanged.
2. Sensor location taking into account restrictions on the maximal costs of making measurements.
   A usually overlooked, but very important aspect of taking measurements is its cost, which may depend on the spatial variable (in some areas these costs may be much higher than in others). In work [39] additional linear constraints were introduced in the formulation for not exceeding an arbitrarily set total budget

for experiments. This slightly changes the optimality conditions, but they still retain the form of linear separability of gradient elements.

3. Location of scanning sensors.
Spatiotemporal system dynamics may significantly change the location of points offering the most valuable information about the estimated parameters. In this situation, sensors with mobility capability are more desirable [31]. In works [66, 67], an arbitrary partition of the time horizon into subintervals was introduced and optimal sensor locations were sought in each of them. The overall optimization task has a much larger dimension, but the convenient form of the optimality criterion allows application of successive block coordinate descent methods [9, Chap. 2.7], which boils down to repetition of solving convex programming tasks with respect to sensor positions for consecutive time intervals using simplicial decomposition.
4. Sensor location for the purposes of fault detection in distributed parameters systems.
Experimental design for fault detection is a relatively rarely explored problem in the context of distributed parameter systems. In works [38, 65] a method was proposed which is based on parameter estimation and testing the null hypothesis about the invariability of the subset of the parameters crucial for diagnosis. A symptom of a fault is the rejection of the null hypothesis, and the sensor locations are selected so as to maximize the power of the above test. It turns out that such a task corresponds to the criterion being a modified version of the D-optimality criterion, referred to as the D_s-optimality criterion [4]. Again, a key in getting numerical results turns out to be SD.

Of course, this work only slightly touches on the complexity of the optimal sensor location problem, focusing primarily on the placement of static sensors, that is, the ones that have no mobility capabilities, and performing measurements not correlated in space. A fascinating challenge is the use of measurement sensors carried by vehicles (mobile robots or aircraft). To design optimal trajectories of such vehicles is extremely difficult, but as a result we can track points providing the most valuable information about the estimated parameters. The work [49] was pioneering in this respect, focusing on the characterization of optimal solutions. This approach was continued in works [61, 68–70, 78] and the monograph [62] whose significant part was dedicated to it. They proposed a number of algorithms derived from optimal control. In [72], this approach was used to identify trajectories and intensities of mobile contamination sources. In work [73] a method of determination of optimal measurement strategies for a hybrid sensor network (i.e., the one consisting of both static and mobile sensors) was proposed. However, the development of an approach enabling real-time vehicle control still remains an open problem.

A prospective research direction seems to be the joint design of optimal sensor locations and input signals affecting the distributed parameter system at hand. The second of these issues is considered in many aspects in monographs [48, 50], and both in time and frequency domains. Integrating both problems is not trivial, and has been a huge challenge for a number of years. However, it should be expected that in the future it will be one of the most interesting application-oriented research directions.

References

1. Alexanderian, A., Gloor, P.J., Ghattas, O.: On Bayesian A- and D-optimal experimental designs in infinite dimensions. Bayesian Anal. **11**(3), 671–695 (2016). https://doi.org/10.1214/15-BA969
2. Alexanderian, A., Petra, N., Stadler, G., Ghattas, O.: A-optimal design of experiments for infinite-dimensional Bayesian linear inverse problems with regularized ℓ_0-sparsification. SIAM J. Sci. Comput. **36**(5), A2122–A2148 (2014). https://doi.org/10.1137/130933381
3. Alexanderian, A., Petra, N., Stadler, G., Ghattas, O.: A fast and scalable method for A-optimal design of experiments for infinite-dimensional Bayesian nonlinear inverse problems. SIAM J. Sci. Comput. **38**(1), A243–A272 (2016). https://doi.org/10.1137/140992564
4. Atkinson, A.C., Donev, A.N., Tobias, R.D.: Optimum Experimental Designs, with SAS. Oxford University Press, Oxford (2007)
5. Banks, H.T., Fitzpatrick, B.G.: Statistical methods for model comparison in parameter estimation problems for distributed systems. J. Math. Biol. **28**, 501–527 (1990)
6. Banks, H.T., Hu, S., Clayton Thompson, W.: Modeling and Inverse Problems in the Presence of Uncertainty. Monographs and Research Notes in Mathematics. CRC Press, Boca Raton, FL (2014)
7. Banks, H.T., Kunisch, K.: Estimation Techniques for Distributed Parameter Systems. Systems & Control: Foundations & Applications. Birkhäuser, Boston (1989)
8. Bernstein, D.S.: Matrix Mathematics. Theory, Facts, and Formulas with Application to Linear Systems Theory. Princeton University Press, Princeton, NJ (2005)
9. Bertsekas, D.P.: Nonlinear Programming, 2nd edn. Optimization and Computation Series. Athena Scientific, Belmont, MA (1999)
10. Bertsekas, D.P.: Convex Optimization Algorithms. Athena Scientific, Belmont, MA (2015)
11. Cacuci, D.G., Navon, I.M., Ionescu-Bujor, M.: Computational Methods for Data Evaluation and Assimilation. CRC Press, Boca Raton, FL (2014)
12. Chepuri, S.P.: Sparse sensing for statistical inference: Theory, algorithms, and applications. Ph.D. thesis, Technische Universiteit Delft, Delft, The Netherlands (2016)
13. Chepuri, S.P., Leus, G.: Sparsity-promoting sensor selection for non-linear measurement models. IEEE Transactions on Signal Processing **63**(3), 684–698 (2015)
14. Cook, D., Fedorov, V.: Constrained optimization of experimental design. Statistics **26**, 129–178 (1995)
15. Curtain, R.F., Zwart, H.: An Introduction to Infinite-Dimensional Linear Systems Theory. Texts in Applied Mathematics. Springer, New York (1995)
16. Dashti, M., Stuart, A.M.: The Bayesian approach to inverse problems. In: Ghanem, R., Higdon, D., Owhadi, H. (eds.) Handbook of Uncertainty Quantification, pp. 311–428. Springer, Cham (2017)
17. Fattorini, H.O.: Infinite Dimensional Optimization and Control Theory. Cambridge University Press, Cambridge (1999)
18. Fedorov, V.V.: Optimal design with bounded density: optimization algorithms of the exchange type. J. Stat. Plann. Infer. **22**, 1–13 (1989)
19. Fedorov, V.V., Leonov, S.L.: Optimal Design for Nonlinear Response Models. CRC Press, Boca Raton, FL (2014)
20. Fung, G.M., Mangasarian, O.L.: Equivalence of minimal ℓ_0- and ℓ_p-norm solutions of linear equalities, inequalities and linear programs for sufficiently small p. J. Optim. Theory Appl. **151**(1), 1–10 (2011). https://doi.org/10.1007/s10957-011-9871-x

21. Gejadze, I.Y., Le Dimet, F.X., Shutyaev, V.: On optimal solution error covariances in variational data assimilation problems. J. Comput. Phys. **229**(6), 2159–2178 (2010). https://doi.org/10.1016/j.jcp.2009.11.028
22. Gejadze, I.Y., Shutyaev, V.: On computation of the design function gradient for the sensor-location problem in variational data assimilation. SIAM J. Sci. Comput. **34**(2), B127–B147 (2012). https://doi.org/10.1137/110825121
23. Haber, E., Horesh, L., Tenorio, L.: Numerical methods for experimental design of large-scale linear ill-posed inverse problems. Inverse Prob.**24**(5), 055,012 (2008). https://doi.org/10.1088/0266-5611/24/5/055012
24. Haber, E., Horesh, L., Tenorio, L.: Numerical methods for the design of large-scale nonlinear discrete ill-posed inverse problems. Inverse Prob. **26**(2), 025,002 (2010). https://doi.org/10.1088/0266-5611/26/2/025002
25. Haber, E., Magnant, Z., Lucero, C., Tenorio, L.: Numerical methods for A-optimal designs with a sparsity constraint for ill-posed inverse problems. Comput. Optim. Appl. **52**(1), 293–314 (2012). https://doi.org/10.1007/s10589-011-9404-4
26. Hansen, P.C.: Discrete Inverse Problems: Insight and Algorithms. Fundamentals of Algorithms. Society for Industrial and Applied Mathematics, Philadelphia (2010)
27. Herzog, R., Riedel, I., Uciński, D.: Optimal sensor placement for joint parameter and state estimation problems in large-scale dynamical systems with applications to thermo-mechanics. Optim. Eng., pp. 591–627. (2018). https://doi.org/10.1007/s11081-018-9391-8
28. Hinze, M., Pinnau, R., Ulbrich, M., Ulbrich, S.: Optimization with PDE Constraints. Springer (2009)
29. Jokar, S., Pfetsch, M.E.: Exact and approximate sparse solutions of underdetermined linear equations. SIAM J. Sci. Comput. **31**(1), 23–44 (2008). https://doi.org/10.1137/070686676
30. Joshi, S., Boyd, S.: Sensor selection via convex optimization. IEEE Trans. Signal Process. **57**(2), 451–462 (2009)
31. Khapalov, A.Y.: Mobile Point Sensors and Actuators in the Controllability Theory of Partial Differential Equations. Springer, Cham (2017)
32. Kubrusly, C.S., Malebranche, H.: Sensors and controllers location in distributed systems—a survey. Automatica **21**(2), 117–128 (1985)
33. Lasiecka, I., Triggiani, R.: Control Theory for Partial Differential Equations: Continuous and Approximation Theories, Encyclopedia of Mathematics and Its Applications, vol. I and II. Cambridge University Press, Cambridge (2000)
34. Law, K., Stuart, A., Zygalakis, K.: Data Assimilation: A Mathematical Introduction. Springer, Cham (2015)
35. Lions, J.L.: Contrôle optimal de systèmes gouvernés par des équations aux dérivées partielles. Études mathématiques. Dunod, Paris (1968)
36. Lu, Z., Pong, T.K.: Computing optimal experimental designs via interior point method. SIAM J. Matrix Anal. Appl. **34**(4), 1556–1580 (2013). https://doi.org/10.1137/120895093
37. Patan, M.: Optimal Sensor Networks Scheduling in Identification of Distributed Parameter Systems. Springer, Berlin (2012)
38. Patan, M., Uciński, D.: Configuring a sensor network for fault detection in distributed parameter systems. Int. J. Appl. Math. Comput. Sci. **18**(4), 513–524 (2008)
39. Patan, M., Uciński, D.: Cost-constrained D-optimum node activation for large-scale monitoring networks. In: 2016 American Control Conference (ACC), pp. 1643–1648 (2016). https://doi.org/10.1109/ACC.2016.7525152
40. Patan, M., Uciński, D.: D-optimal spatio-temporal sampling design for identification of distributed parameter systems. In: 2016 IEEE 55th Conference on Decision and Control (CDC), pp. 3985–3990 (2016). https://doi.org/10.1109/CDC.2016.7798872
41. Patan, M., Uciński, D.: A sparsity-enforcing method for optimal node activation in parameter estimation of spatiotemporal processes. In: 2017 IEEE 56th Annual Conference on Decision and Control (CDC), pp. 3089–3094 (2017). https://doi.org/10.1109/CDC.2017.8264110
42. Patriksson, M.: Simplicial decomposition algorithms. In: Floudas, C.A., Pardalos, P.M. (eds.) Encyclopedia of Optimization, vol. 5, pp. 205–212. Kluwer Academic Publishers, Dordrecht, The Netherlands (2001)

43. Pollard, D.: A User's Guide to Measure Theoretic Probability. Cambridge University Press, Cambridge (2002)
44. Pronzato, L., Pàzman, A.: Design of Experiments in Nonlinear Models. Optimality Criteria amd Small-Sample Properties. Springer, New York, Asymptotic Normality (2013)
45. Pukelsheim, F.: Optimal Design of Experiments. Probability and Mathematical Statistics. Wiley, New York (1993)
46. Quereshi, Z.H., Ng, T.S., Goodwin, G.C.: Optimum experimental design for identification of distributed parameter systems. Int. J. Control **31**(1), 21–29 (1980)
47. Rafajłowicz, E.: Design of experiments for eigenvalue identification in distributed-parameter systems. Int. J. Control **34**(6), 1079–1094 (1981)
48. Rafajłowicz, E.: Choice of Optimum Input Signals in Linear Distributed-Parameter Systems Identification Monographs. Technical University Press, Wrocław (1986). (In Polish)
49. Rafajłowicz, E.: Optimum choice of moving sensor trajectories for distributed parameter system identification. Int. J. Control **43**(5), 1441–1451 (1986)
50. Rafajłowicz, E.: Optimal Input Signal Design for Parameter Estimation in Linear Systems with Spatiotemporal Dynamics. De Gruyter, Berlin (2020). (in preparation)
51. Reich, S., Cotter, C.: Probabilistic Forecasting and Bayesian Data Assimilation. Cambridge University Press, Cambridge (2015)
52. Shectman, J.P., Sahinidis, N.V.: A finite algorithm for global minimization of separable concave programs. J. Global Optim. **12**(1), 1–36 (1998). https://doi.org/10.1023/A:1008241411395
53. Song, Z., Chen, Y., Sastry, C.R., Tas, N.C.: Optimal Observation for Cyber-physical Systems: A Fisher-information-matrix-based Approach. Springer, London (2009)
54. Stuart, A.M.: Inverse problems: aBayesian perspective. Acta Numerica **19**, 451–559 (2010). https://doi.org/10.1017/S0962492910000061
55. Sun, N.Z.: Inverse Problems in Groundwater Modeling. Theory and Applications of Transport in Porous Media. Kluwer Academic Publishers, Dordrecht, The Netherlands (1994)
56. Sun, N.Z., Sun, A.: Model Calibration and Parameter Estimation For Environmental and Water Resource Systems. Springer, New York (2015)
57. Sun, X., Zheng, X., Li, D.: Recent advances in mathematical programming with semi-continuous variables and cardinality constraint. J. Oper. Res. Soc. China **1**(1), 55–77 (2013). https://doi.org/10.1007/s40305-013-0004-0
58. Tricaud, C., Chen, Y.: Optimal Mobile Sensing and Actuation Policies in Cyber-physical Systems. Springer, London (2012)
59. Tröltzsch, F.: Optimal Control of Partial Differential Equations: Theory Methods and Applications. American Mathematical Society, Providence, Rhode Island (2010)
60. Tucsnak, M., Weiss, G.: Observation and Control for Operator Semigroups. Birkhäuser, Basel (2009)
61. Uciński, D.: Optimal sensor location for parameter estimation of distributed processes. Int. J. Control **73**(13), 1235–1248 (2000)
62. Uciński, D.: Optimal Measurement Methods for Distributed-Parameter System Identification. CRC Press, Boca Raton, FL (2005)
63. Uciński, D.: An algorithm to configure a large-scale monitoring network for parameter estimation of distributed systems. In: Proceedings of the European Control Conference 2007, Kos, Greece, July 2–5 (2007). Published on CD-ROM
64. Uciński, D.: D-optimum sensor activity scheduling for distributed parameter systems. In: Preprints of the 15th IFAC Symposium on System Identification, Saint-Malo, France, July 6–8, (2009). Published on CD-ROM
65. Uciński, D.: Sensor network scheduling for identification of spatially distributed processes. Int. J. Appl. Math. Comput. Sci. **21**(1), 25–40 (2012)
66. Uciński, D.: An optimal scanning sensor activation policy for parameter estimation of distributed systems. In: H.G. Bock, T. Carraro, W. Jäger, S. Körkel, R. Rannacher, J.P. Schlöder (eds.) Model Based Parameter Estimation: Theory and Applications, pp. 89–124. Springer, Berlin (2013). https://doi.org/10.1007/978-3-642-30367-8_4

67. Uciński, D.: D-optimum scanning node activation for large-scale monitoring networks. In: A. van der Schaft, H. Trentelman, J. Scherpen, C. De Persis (eds.) Proceedings of the 21st International Symposium on Mathematical Theory of Networks and Systems, Groningen, The Netherlands, 7–11 July 2014, pp. 116–123 (2014). Paper MoA05.1, ISBN 978-90-367-6321-9
68. Uciński, D., Chen, Y.: Time-optimal path planning of moving sensors for parameter estimation of distributed systems. In: Proc. 44th IEEE Conference on Decision and Control, and the European Control Conference 2005, Seville, Spain (2005). Published on CD-ROM
69. Uciński, D., Chen, Y.: Sensor motion planning in distributed parameter systems using Turings measure of conditioning. In: Proceedings of the 45th IEEE Conference on Decision and Control, San Diego, CA (2006). Published on CD-ROM
70. Uciński, D., Korbicz, J.: Optimal sensor allocation for parameter estimation in distributed systems. J. Inverse Ill-Posed Prob. **9**(3), 301–317 (2001)
71. Uciński, D., Patan, M.: D-Optimal design of a monitoring network for parameter estimation of distributed systems. J. Global Optim. **39**, 291–322 (2007)
72. Uciński, D., Patan, M.: Sensor network design for the estimation of spatially distributed processes. Int. J. Appl. Math. Comput. Sci. **20**(3), 459–481 (2010)
73. Uciński, D., Patan, M.: Constrained mobile sensor routing for parameter estimation of spatiotemporal processes. IFAC Proceedings Volumes **45**(16), 1317–1322 (2012). https://doi.org/10.3182/20120711-3-BE-2027.00363. 16th IFAC Symposium on System Identification
74. Uspenskii, A.B., Fedorov, V.V.: Computational Aspects of the Least-Squares Method in the Analysis and Design of Regression Experiments. Moscow University Press, Moscow (1975). (In Russian)
75. Vogel, C.R.: Computational Methods for Inverse Problems. Frontiers in Applied Mathematics. Society for Industrial and Applied Mathematics, Philadelphia (2002)
76. Yu, Y.: Monotonic convergence of a general algorithm for computing optimal designs. Ann. Stat. **38**(3), 1593–1606 (2010)
77. Zhao, F., Guibas, L.J.: Wireless Sensor Networks: An Information Processing Approach. Morgan Kaufmann Publishers, Amsterdam (2004)
78. Ziba, T., Uciński, D.: Mobile sensor routing for parameter estimation of distributed systems using the parallel tunneling method. Int. J. Appl. Math. Comput. Sci. **18**(3), 307–318 (2008)

Discrete Optimization in the Industrial Computer Science

Czesław Smutnicki

Abstract In the work there is presented a comprehensive overview of scientific and research directions accomplished by the Author and cooperating team in recent years. The team is working in the area of modeling and optimization dedicated for the implementation in discrete manufacturing systems, transport systems, warehousing systems etc. The immediate result of the these research is the original uniform approach to design of various algorithms for problems derived from the industry. Foundations of the approach combine not only graphs, operations research, combinatorial optimization, mathematical programming, but also modern solution technologies and powerful computing tools. In the review there are discussed general technologies of discrete processes modeling as well as the evolution of solving methods in the recent fifty years. There are some selected fundamental discrete optimization problems discussed in relation to the landscape of the solution space. There are also collected the main theoretical results obtained by the Author, his leading publications as well as the spectacular algorithms developed so far. The future research directions are also outlined. The presentation is illustrated by references to papers of the Author and the bibliography of collaborators.

1 Introduction

Design and implementation of the control system dedicated for discrete event systems, object and/or process require a sequence of standard actions: (1) collecting, processing and analyzing of the measurement data, (2) modeling and identification of the mathematical model, (3) formulating the problem of control or planning, (4) solving the problem of optimal control or planning, (5) implementation of the control device (hardware) or control algorithm. Mathematical and technological tools applied in these steps strongly depend on the characteristics of the object and pro-

C. Smutnicki (✉)
Faculty of Electronics, Wrocław University of Science and Technology, 50-370 Wrocław, Wybrzeże Wyspiańskiego 27, 50-370 Wrocław, Poland
e-mail: czeslaw.smutnicki@pwr.edu.pl

P. Kulczycki et al. (eds.), *Automatic Control, Robotics, and Information Processing*, Studies in Systems, Decision and Control 296,
https://doi.org/10.1007/978-3-030-48587-0_12

cess, while their quality is the subject of a permanent investigation of scientists and practitioners. In recent years the hardware implementation of the control system loses the importance because of the technological development. Hardware solutions are commonly replaced by specialized software on universal, cheap, programmable controllers, software on a PC, software on mobile devices, services available in computing clouds. Such trends imply, among others, the increasing role of theoretical findings, thus enlarges the importance of the research dealing with mathematical methods of optimization and control problems inspired by the industry.

Advanced methods of modeling and analysis (fuzzy, stochastic, ...) have good practical interpretation and numerous implementations. Permeation of domains in widely perceived modern technologies of computer science and communication, i.e.: electronics, telecommunication, control, robotics, mechatronics, computer science and engineering change the perception of technology innovations. In consequence, design and implementation of suitable algorithms (in fact, this is a piece of software) can be perceived as an expression of the area called shortly *industrial computer science*. In particular, the flexible scheduling of tasks appears as a part of Industry 4.0.

Presented in this section research areas comprise optimization problems in discrete industrial processes with the use of continuous and discrete optimization technologies. Research initiated by the Author in the seventies of the previous age, was carried out at first in the Institute of Engineering Cybernetics and then in the Department of Automation, Mechatronics and Control Systems, which are part of the Faculty of Electronics in Wroclaw University of Science and Technology. The research is currently multi-directional, and is performed by a small team and deals with at least:

(a) methodology of solving hard combinatorial optimization problems and control of discrete production processes, [35, 36, 39, 42, 44–46, 48];
(b) implementation of combinatorial algorithms in parallel computing environments, [5, 6];
(c) modeling and solving complex scheduling problems, [30–33, 37, 43];
(d) optimization of transport in manufacturing and service systems, [7, 8];
(e) modeling and optimization of warehouse systems, [49];
(f) cyclical manufacturing systems, [51, 56–59];
(g) some industrial applications.

The enclosed bibliography list does not exhaust completely Author's own publications concerning these subjects. It provides only the reference to the most significant/representative papers in suitable research directions.

Problems of planning and control in discrete manufacturing systems (widely understood control and robotics), problems of placement, packing, cutting, order picking, transport, distribution, delivery, timetabling, personnel scheduling, planning of robot tracks, etc. can be perceived as a process based on a sequence of dependant events which appear in some fixed time moments (an event causes next event). The process can be described, among others, by the finite sets of events occurring on the input, output and inside the system. Events have assigned some attributes, and obligatory—the time attribute. Events inside the system can be represented by the system state, so set of states is finite, discrete and can be switched either by an outer event (system input) or by an inner event (a lapse of the time or a composition of inner events to fire transition). Positive strand of having finite sets is in fact illusory, because the number of events grows usually exponentially with the problem size and may exceed quickly the amount of atoms in the Universe. This implies commonly known numerical troubles with problems, their analysis and solution methods. Dependence between input, output and states is usually described by certain rules of transitions, with the significant role of time. Appropriately to the character of dependence between events, one can distinguish a few models: deterministic, stochastic, fuzzy one. Regarding the way of modeling, one can find in the literature: dynamic discrete event system (DDES), Petri networks (PN), combinatorial optimization (CO), Markov chains (MC), computer simulation (CS). As the theoretical methods of solving control and planning optimization problems, there are used: mathematical programming (MP), graph theory (GT), max-plus algebra (MPA) and constraint satisfaction (CS). General methodologies, as an example DDES, MC or PN, provide an uniform, quite general approach applicable for many various practical problems. Alternatively, models dedicated for relatively narrow class of problems, usually by using special problem properties, are able to improve numerical properties of the solution/optimization method.

The dilemma "which is the most useful modeling and solving tool" remains a problem of philosophic nature. Author of this chapter constantly admits the rule: "if something is for all, it is for nothing". This is why I prefer dedicated approaches for each narrow class of problems, not neglecting significance of any other approaches.

Although majority of works in the literature refer to asynchronous event systems (events cause successive events, see e.g. classical batch scheduling or PN), modeling and solution technologies used for asynchronous non-cyclic processes are distinct from those for cyclic processes, synchronous as well as asynchronous. Variety of problems and solution methods for cyclic/non-cyclic systems follows from the philosophy of mass/dedicated/flexible production, for mass customer or an individual client, equally in the static and dynamic context. Therefore, the domain of operation research embraced a large number of various problems: batch, cyclical, multi-criteria. Modern large manufacturing or transport systems performed main process and/or component sub-processes in the repetitive manner, ensuring their proper synchronization.

Generally, as the cyclical synchronous processes there are considered schedules, in which the same type of events is repeated infinite number of times shifted constantly, called the cycle time. Thus, our aim is to provide periodical schedule. In the case of cooperating processes (e.g. multi-modal transport, supply chains, distribution of goods) not only efficiency of particular cycle has meaning, but also the synchronization of sub-processes has an influence on the quality and cost of system activity.

2 Modeling

The creation of the mathematical model for a discrete process allow us to observe the behavior of the system in the laboratory conditions instead of expensive and risky experiment on the real object. Additionally, using the mathematical model one can formulate the task of optimal control/planning, usually hard from the computational point of view. There exist a few 'routine' approaches implying different tools for modeling and solving. The former two have been discussed herein as the fundamental research field of the Author. The third—is the natural consequence of the second. The next five are quoted for completeness.

2.1 Operations Research

From the theoretical point of view, problems of analysis activity networks linked with the allocation of discrete renewable resources usually lead us to discrete or discrete-continuous optimization tasks. Classical problems of this type refer to service stages, personnel team, machines, processors, and so on. Separate class comprises problems known in the literature as Resource Constrained Project Scheduling, (RCPS), see, among others, Sect. 3.4. RCPS has an application in computer science (software project management), as well as in widely understood automation (production schedule). All problems belong to the class unusually troublesome from the computational point of view. So far, there have been several factors responsible for the hardness of solution algorithms for the discrete optimization tasks identified, [49]. There are, among others, huge (growing exponentially) number of local extremes distributed irregularly, curse of dimension, strong NP-hardness, roughness of landscape of solution space. Pessimism in this domain decreases with the development of solution methods. Frequently, instead of expensive searching of the optimal solution, one tries to find quickly an approximate solution. Quality of the approximation has the opposing tendency to computing time, which means that obtaining a better solution needs longer running time of the algorithm. Unfortunately, this dependence has strongly nonlinear and complex character. Therefore, the domain of discrete manufacturing processes, planning, transport and so on, owns considerable variety of models, as well as solution methods.

Limitation of the generality of models aims to detect such special properties of the problem which may substantially improve numerical properties of the algorithm, we mention here the running time, speed of convergence, quality of sub-optimal solution. That is why, for a selected NP-hard problem there exists in the literature a few or a dozen various algorithms with essentially different numerical features. Knowledge of the domain allow one for each newly formulated problem to fit proper algorithm satisfying the user. We should pinpoint here that in the considered domain the goal is not to find any sort of model and solution method, but the overall aim to find the simple model and the solution method reasonable for computer implementation and acceptable by the user of decision support system. This is overall aim of the research leading from years on the Faculty of Electronics, Wroclaw University of Science and Technology.

As it is expressed constantly in our papers, the exploitation of special problem features usually allow us to design better solution algorithm. There exists common belief, that for strongly NP-hard problems it implies the increase of the instance size which can be solved only by a few or a dozen units, so insignificantly. It is true, but only for the exact methods of overview (e.g. branch-and-bound schemes), but not true for many modern approximate methods. Moreover, recent technologies, such as parallel calculations (multi-core processors, multi-processor computers, clusters), distributed computer networks, computing clouds, quantum computers, strongly supports the development of optimization methods, through the increase of available (virtual) computing power. Thus, in turn, enlarges the search intensification, which has the direct influence on the quality of the provided approximate solutions.

2.2 *Deterministic Models*

Most of the problems in the classical deterministic scheduling theory are formulated as so called batch scheduling cases. They require knowledge in advance about the set of jobs and resources in the planning horizon with full information/data about them in order to create the master schedule. This assumption is fully justified in manufacturing and transport systems without human workers (see ideas of Industry 4.0), but somewhat controversial in the presence of humans. The basic philosophical problem still remains: whether we would like either to model and solve harder optimization problems with uncertain data (i.e. fuzzy, stochastic), or to enforce repairing process in order to stabilize data fluctuation and then to approximate the system by a deterministic model.

Taking into account immensity of papers in the literature for each particular scheduling case, the mentioned diversity of models and algorithms implies a natural question about the level of generality and real applicability of scientific research results. The valid contribution to this subject introduces the taxonomy presented in [18]. The fundamentals of this taxonomy refer to given definitions of structures of the systems, additional constraints, criteria functions, solution classes, computational complexity. Unfortunately, the taxonomy does not provide any relation to real manufacturing, transporting or warehousing systems, allowing for creation of a number of artificial problems interesting for scientist from the theoretical point of view, but not for users. Closer to practice taxonomy and classes of solutions were discussed in detail in our books [44, 55]. Cases with rational practical applicability begin from the problem known in the literature as job shop scheduling. Simpler cases are treated as auxiliary for the mentioned class. More advanced—take into account complex operation networks, additional technological constraints, enhanced optimization criteria. Already classical job shop scheduling problems belong to the class of the hardest discrete optimization problems. There were considered up to now, mainly for regular scheduling, criteria with the significant dominance of the makespan, [3, 20]. In turn, cyclic scheduling cases were formulated and analyzed rather for simple problems, e.g. for single machine or flow shop.

Strong NP-hardness already for the simplest versions of the job shop problem limits the application of exact algorithms finally to instances with relatively small size, i.e. 200–250 operations, 10–20 jobs, 10–15 machines. That is why in order to find satisfying solutions in real industrial applications of the greater size we use fast and accurate approximate algorithms based on various local search technologies. These methods usually perform two-level decomposition of the problem: on the upper level—finding the optimal configuration (solution $x \in \mathcal{Y} \subseteq \mathcal{X}$ in equation (1)) and on the lower level—multiple finding of the minimal criterion value $K(x)$ for the given x. Ideas shown in [52] allow us to use any function $K(x)$ defined by the user for the evaluation of the solution x, namely: approximation, algorithm, fuzzy measure, moment of distribution and so on. By using the skilful decomposition and aggregation of calculations for the solution x, in the Wroclaw center there were designed the so called computational accelerators for some well-known scheduling problems, which can significantly speed up the process of local neighborhood overlooking. Providing that for classical scheduling problems the solution of the lower level can be obtained in an efficient way, the solution of the lower level problem can be obtained in the timely efficient way by analyzing the specific graph, then in cyclic scheduling the solution of the lower level problem is relatively time-consuming, since in general needs to solve certain linear programming (LP) task. Therefore, any particular problem features, allowing us on more efficient calculation of the criterion value, cycle time, schedule, reduction of the size of the of the local neighborhood and acceleration of the speed of neighborhood overlooking, are especially welcome. Majority of results obtained for classical manufacturing structures (flow shop, hybrid flow shop, job shop) can be easily extended to cases with additional technological constraints, e.g. limited waiting time, limited buffers, setups, pallets etc. These systems still remain the object of researchers interest because of practical needs and troubles with finding appropriately good solution algorithms, see [9, 43, 44, 57].

2.3 Models with Uncertain Data

There exist common conviction about dispensable stiffness of the deterministic models following from the numerous technological constraints and arbitrary *a priori* data settings. Thus, finding in this way optimal schedules are poorly resistant to the potential disturbance of the input data. Moreover, the problem of existence of any feasible solution in many practical cases is NP-complete. Uncertain data are possible to implement in the fuzzy sets theory and in queening models (see next). Replacing in the activity networks deterministic data by their equivalent fuzzy/random variable, we obtain other methods of evaluating enumerated in the review [52]. Introducing proper operators on fuzzy numbers (sum, min, max,...), we get the fuzzy evaluation of the schedule, and then we can select the robust schedule.

2.4 Petri Networks

Petri networks (PN) were introduced about fifty years ago primarily for the modeling communication with automates. This tool was next developed and modified through years and currently they have been used for modeling, design and analysis of concurrent systems, cyclical processes, synchronization processes, communication, flexible manufacturing, [65]. Graphical representation of PN is the bipartite graph with two types of nodes: first places (drawn as small circles) are called and second transitions (drawn as short line segments) are called. Enumerated elements of the graph are linked by arcs in such a way, that any two places and any two transitions are not linked directly. Equally, places and transitions (both perceived as graph nodes) may have different inter and outer degree. Places in the network may contain tokens, marked graphically by dots. Working of the network consist of iterative relocation of tokens, according to some rules, performed by the fire transitions. Conditions to fire are set differently for various Petri networks. There may take the form of requirements imposing on the number and type of tokens on the input, necessary to fire a selected transition, and also for the number of tokens on the output.

The fundamental concept of the Petri networks were developed through years in order to extend as much as possible the modeling capability. Finally, there were introduced a few new concepts to the basic model, namely: (1) labeled networks, (2) network of places and transitions, (3) time-dependant networks, (4) colored networks. From the beginning, two essential properties attract attention of theoreticians: aliveness and limitation. Aliveness means whether network using certain starting tagging (tokens distribution) are able to realize activity expressing by the continuous relocation of tokens. Tagging is limited if in any place the number of tokens is limited. In the case of limited tagging we are sure that the number of reachable stages is finite, although can be very large (surprisingly, even for the relatively small networks can be greater than the number of atoms in the Universe). Network is correctly defined, if alive and limited tagging exist. Referring to cyclical activity of the network, one expects that there are no network parts permanently switched off. On the other hand, we trace the buffers capacity and scheduling policy to prevent deadlocks.

Petri network constitutes the comfortable and universal tool for the description of various type systems. It finds particular applications in software engineering, where it is used for description and analysis of concurrent processes, in automation of business processes (workflow), task are passed in the chain user-by-user. In computer science, PN are used for modeling and communication between concurrent processes inside programs. In this case places of the network are interpreted as conditions (or stages), but transitions—as actions (instructions or functions). Structure of the Petri networks reflects the structure of the program, while relocation of tokens models the running processes. Application of Petri networks is not limited to this single domain.

Petri networks can be also perceived as a composition of events and conditions. Single event has certain initial conditions, which have to be fulfilled, and output conditions—which will appear if the event occur. In order to run, each event should be activated, so on some the input should be tokens, whereas output should be empty.

Networks created by the places of such events are called conditional-event networks. The networks represent dependency of changing logical conditions (local stages). Token in the place corresponding to the condition means fulfillment of this condition, lack of token means falsification. The number of configurations reachable from the initial tagging is finite and constitutes the space of the system stages.

Petri networks is the tool dedicated chiefly for modeling, but not for optimization. It means that we can observe the system behavior accordingly to the rule "if ... then" by performing simulation of network behavior for a few selected scenarios through the suitable software package. Clearly, by making the test for several scenarios, we can optimize locally (in a limited scope). Thus, the optimization possibilities are rather poor.

2.5 *Service Systems and Queuing Networks*

Systems with elements of mass services policy may be described as an open/close queuing systems and networks [13]. The flow of components or final products can be perceived as the stream of random entries, to formulate a queue of items waiting for the service before each stage. Usually, there is assumed that inflow prompts, and service time are random variables. Inflow prompts are described by some stochastic process, whereas the queuing service rule is fixed (one of the a few commonly available). Basic models of systems refer to the single-channel service stage and homogeneous stream of entries coming in the infinite way or multi-channel service system, it is not necessary homogeneous stage but of the same type of entry. Alternatively, cyclic complex system of his type require the schedule considering sequence of the jobs in the cycle. Regarding to cyclic manufacturing systems we require fixed set of nonuniform entries. A cyclical manufacturing system uses a finite set of prompts (of different size) and job are flown by different technology routes, so the more adequate name of this system will be the queuing network. For very simple systems of this type (chiefly single stage) we are able to find analytical expressions on quality indicators. Unfortunately, for more complex systems, analytical evaluations cannot be obtained, which means that the computer simulation is the only method to analyze queuing networks.

2.6 *Event Systems*

The proper name is finite deterministic (dynamic) event system (DES or DDES). This formalism is relatively young since known from about thirty years. It allows to model asynchronous discrete processes, but does not offer any solution method to solve the optimization tasks. It operates on the finite set of deterministic time–dependent events characterizing the input of the system, output of the system and states of the system. Input events influence on the internal state creating the schedule of changes of the system state. Although the schedule is a continuous function of time we save only events in a compact form (state, time).

Actually, the origin of DES one can find in old well known automata theory, since DES definition is very close to finite state automate. In DES description we have: (1) finite set of input events, (2) finite set of output events, (3) finite set of states, (4) distinguished initial state, (5) function of time flow, (6) function of changing the state spontaneously in time or caused by an input event, (7) function determining the output event.

For systems modeled with the use of DES formalism there are analyzed several fundamental features: state reaching, deadlock avoidance, aliveness.

2.7 Multi-modal Systems

This quite original methodology of modeling and solving cyclical processes (manufacturing, transport, communication and so on) allow us to solve various optimization tasks by using the consistent methodology CSP (Constraint Satisfaction Problem) [2]. Commonly known difficulties in this area imply small number of papers on this subject. Therefore we should pinpoint a paper providing a distinct approach even for problems from the classical scheduling [4].

In the multi-modal approach we assume that global cyclical process can be decomposed on cyclical manufacturing and transportation sub-processes. Decomposition, features of sub-processes and their synchronization influence on indicator of efficiency of the global process. There are formulated necessary conditions of the process and the schedule expressed in terms of CSP constraints. Next by using standard software package we seek for any feasible solution satisfying constraints instead of optimal one. This approach is the second (beside the deterministic models) suitable for solving optimization problems in cyclical systems.

2.8 Simulations

Difficulties in modeling and analysis of real complex manufacturing systems combined with uncertainty of data incline users to apply simulation tools. This technology takes the form of specialized programming languages, object oriented libraries, integrated environments with 2D and 3D graphics, complete software packages with all mentioned possibilities. Although simulators are designed for many various processes and objects, [11], we are interested only these dedicated for discrete manufacturing system. Simulator generates and analyses events occurring in the system in discrete time moments. One can virtually trace material flow processes as well as can measure quality of task service policy and resource allocation. Optimization can be made by testing a small number of different scenarios and variant analysis in the mode “what ... if”. Simulator provides statistical information about system behavior and tools for processing these data.

While using simulation tools we should take into account several accompanying activities. It is evident that credibility of the simulation depends on the reliability of input data. In case of uncertain data (random or fuzzy) we have to perform measures accordingly to statistical rules. Thus, the statistician may support us in estimation, planning experiments, identification, hypotheses testing and so on. Examination of many alternative scenarios linked with the variability analysis of system parameters (moments, distributions) need to perform huge number of simulator runs, usually computationally very expensive. As we said before, reliable probabilistic characteristics need a sufficient number of measurements (means runs), whereas cost of calculation rise very quickly with the increasing instance size. Finally, the obtained results have a statistical meaning in the stable work, so evaluation of the transitional states can be imprecise and rough. In spite of numerous critical comments, simulation packages still remain the most frequently used tool for modeling and analysis of discrete processes, due to convenience and implementation time.

3 Solution Methods

Accordingly to the operations research domain we refer to the general optimization problem, formulated as follows

$$K(x^*) \stackrel{\text{def}}{=} \min_{x \in \mathcal{X}} K(x) \stackrel{\text{def}}{=} \min_{x \in \mathcal{X}} [K_1(x), K_2(x), \ldots, K_s(x)], \tag{1}$$

where $K(x)$ is the vector ($s > 1$) or scalar ($s = 1$) goal function, x is the solution, x^* is the optimal solution, $\mathcal{X}$ is the finite set of feasible solutions defined by a set of constraints. Referring to real applications, x is discrete (binary, integer, combinatorial object, and so on), $\mathcal{X}$ is the set of discrete objects, $K_i(x)$ is non-linear, non-differentiated, optimization problem (1) is strongly NP-hard.

Plentitude of approaches applied to solve optimization problems changes through years. Tasks with single extreme, convex $K(x)$, differentiated and scalar functions $K(x)$ disappeared from the research laboratories, because efficient solving methods and software packages for them were already developed. Currently, scientists focus on particularly hard cases: with multiple extreme, vector optimization, non-differentiate $K(x)$, NP-hard, discrete x, huge dimension of $\mathcal{X}$. These are problems deriving from the practice of control, planning and scheduling, transport, design, management, and so on. These problems, generated by the industry, market and business, cause serious troubles in the process of searching solutions satisfying practitioners expectations. Great effort has been done in recent years in order to increase the power of solution algorithms. The obtained development in the solution methodology linked with the increase of the computer power causes greater expectations of practitioners. Therefore there constantly exist the need of making the research

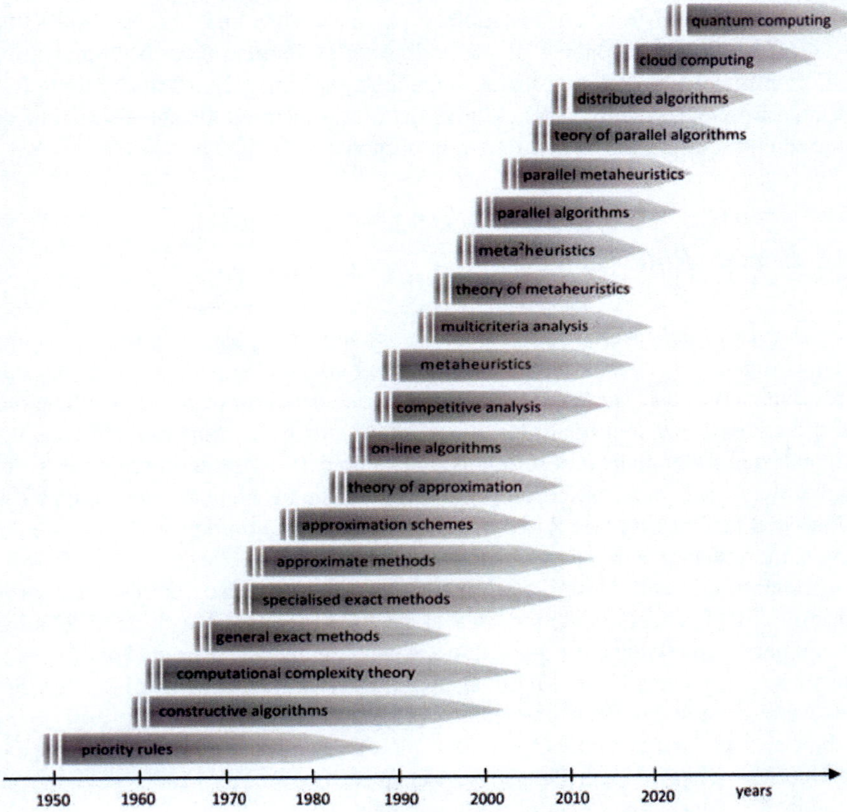

Fig. 1 Evolution of the solution approaches

dealing with the theory and solving methodology of mentioned classes of problems. From the fifties of XX century, the philosophy of approaches has evolved, see Fig. 1. As it happens in science, new successive research directions are created as an answer to exhausted solving capability of the known up to now approaches. Such natural evolution of research has been observed also in the research group in Wroclaw.

For NP-hard problems, for which there have been recommended chiefly the approximate methods, the variety of approaches observed in the literature is fully justified by „no free lunch" theorem, [64]. The theorem confirms inability of finding universal good approximate algorithm (from quality point of view) for the whole spectrum of instances. Negation of the mentioned theorem allows us to improve the quality evaluation on some approximate algorithms by elimination of extreme instances [40, 61].

That is why one accepts commonly numerous alternative approximate algorithms dedicated for relatively narrow classes of instances. Variety of solution methods (over forty of chiefly used metaheuristics, over eighty by adding parallel metaheuristics) is the expression of trend in this subject, see the review and conclusions in [44, 46, 48, 55]. Nobody wonders of methods such as Particle Swarm Optimization (PSO), Artificial Bee Colony (ABC), Genetic Algorithms (GA), Tabu Search (TS), Artificial Immune System (AIS). A huge list of bio-inspired metaheuristics known in the scientific literature [12] seems to be surprising.

3.1 Priority Rules

The oldest and simplest approach to problems of workload balancing, route planning and queuing service is to assign to each object a rank number and to specify the rule of resolving conflicts. There are plenty of rules dedicated for particular applications and indicators of system quality. Priority algorithms offer short running time but own poor quality of generated solutions. In example, from the empirical analysis of distribution of schedule makespan (see Fig. 2) for the benchmark ta45 from [60] it follows that fast priority rule RANDOM (i.e. take any solution) provides solution in average more than two times worse ($RE = (RANDOM - C^*)/C^* \approx 120\%$) from the optimal solution C^*. Next, it was verified experimentally that Shortest Processing Time rule (SPT) generates solutions in average 30–50% worse than optimal solution. For comparison, metaheuristic algorithm TSAB mentioned in the sequel of this paper generates solution very close to optimal, approximately 1-2% in sense of RE. Priority rules were discussed in detail in the book [44]. Currently, research of priority rules has been reduced. Known rules are used in some *ad hoc* constructive algorithms, on-line algorithms, hybrid algorithms, in service queues, as auxiliary for more advanced methods, in an ensemble of cooperating methods.

3.2 Exact Methods

In the sixties and seventies of the XX century, researchers were convinced that they could create a universal exact method for finding x^* in Eq. (1) for at least

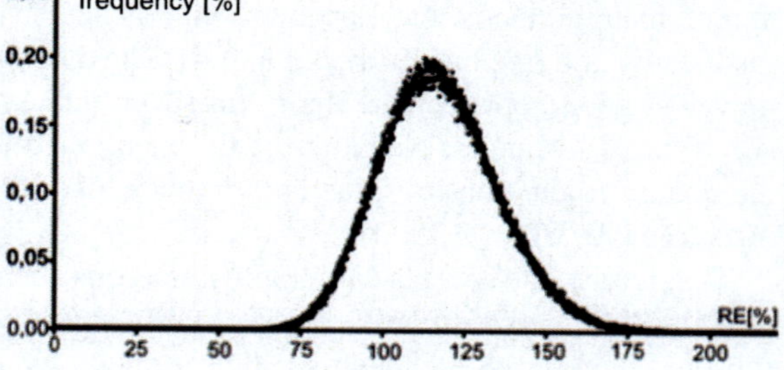

Fig. 2 Frequency of makespan deviation RE = (C-OPT)/OPT of feasible solutions C from optimal solution OPT. A sample of 500,000 random solutions for the benchmark ta45 of the job shop scheduling problem

$s = 1$. For NP-hard problems they put hope to Branch-and-Bound scheme (B&B), Dynamic Programming scheme (DP), Integer Linear Programming (ILP) or Binary Programming (BP) methods. In these years we observed zestful development of discrete mathematical programming. Finally, the computational complexity theory confirmed limited scope of usability of these methods. Computer tests show that the explosion of calculations always appears beginning from a relatively small size of the instance. Further development toward dedicated exact methods were continued several next years due to application of special properties of particular problems. They either increase the power of bounds (see [16] and own chapter in the book [17]), and/or offer additional elimination properties (critical path in a graph and so called block method, [16]). Block method appeared finally more useful in approximation algorithms than in B&B scheme, see [30–32].

After many unfortunate experiences and experiments with B&B scheme, we observe the retreat from exact methods toward approximate approaches. On the battle ground there are remained general PLC and PB methods built into standard solvers (e.g. software packages known under commercial names Gurobi, LINDO, CPLEX), surprisingly efficient in some applications despite the long time of calculations.

3.3 Approximation Algorithms

Primal theoretical results dealing with approximation algorithms (with guaranteed in advance quality of solution) and approximation schemes (family of algorithms with flexible control of the compromise between quality and running time) appeared in the end of the seventies and the beginning of the eighties of the XX century. These research not only introduce independent measures of the approximation, but also order the domain and fix directions of the development.

Conclusions from Sect. 3.2 directed our research also on approximation algorithms. We start analysis from a few algorithms known in the literature such as: Earliest Ready Time (ERT), Earliest Due Date (EDD), Most Work Remaining (MWR) and other for critical stages, see Fig. 3. These algorithms are interesting in the context

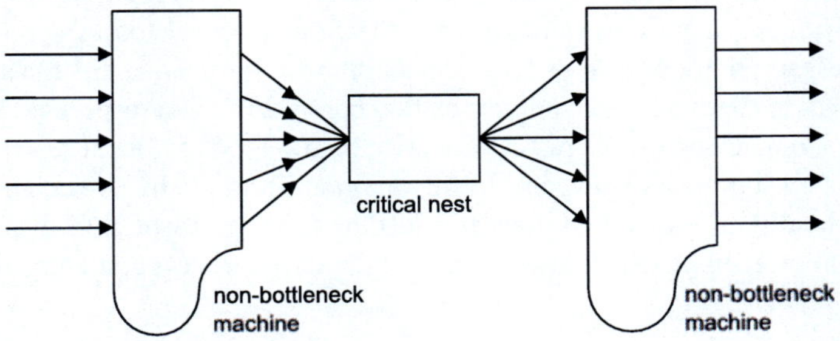

Fig. 3 Optimization of the critical nest

Table 1 Worst case ratio η of the algorithm A for the problem $1|r_j, q_j|C_{\max}$

Algorithm	η	Computational complexity	Comments
R,Q	2	$O(n \log n)$	Rules ERT, EDD
S	2	$O(n \log n)$	Rule MWR
P	3/2	$O(n^2 \log n)$	Iterated S
HS	4/3	$O(n^2 \log n)$	P+inwerse P
NS	3/2	$O(n \log n)$	Own
A-1	$1+\epsilon$	$O(16^{1/\epsilon}(n/\epsilon)^{3+4/\epsilon})$	PTAS
A-2	$1+\epsilon$	$O(n \log n + n(4/\epsilon)^{8/\epsilon^2+8/\epsilon+2})$	PTAS

PTAS polynomial-time approximation scheme, FPTAS fully polynomial-time approximation scheme;

competitive to B&B schemes based on blocks [17], as well as in methods supporting calculation of the lower bound in the specialized B&B scheme for the job shop scheduling problem [16, 17]. In this area there were proposed new own algorithm NS with competitive computational complexity and quality, see Table 1, [27].

Another stream of the research focus on the analysis of the worst-case ratio for numerous approximate algorithms for the flow shop scheduling problems with typical optimization criteria, namely the makespan, weighted sum of completion times (see listing in Table 2, from own papers [24–26, 29, 42]). All obtained results fill in existed gap in the theoretical research. The findings are rewarded doubly. Firstly, there are given quality evaluations for algorithms known for years in the literature as an example CDS, RA, P, HR, NEH. Secondly, quality evaluations are provided as an exact function (but not the order of function which is a common trick) and moreover algorithm has small computational complexity. Finally, despite the time lapse, nobody proposes any competitive solution method. Other published algorithms and approximation schemes offer better quality already for huge size of the instance, or under high cost of calculations.

The development of approximation algorithms for some NP-hard cases is still continuing, despite some pessimistic circumstances. The fundamental argument follows from the elementary result of non-approximation "if P$\neq$ NP then polynomial-time ϵ-approximation algorithm (already!) for the flow shop scheduling problem with the makespan criterion for $\epsilon < 1.25$" does not exist. In practice this means that we cannot ensure relative error of approximation below 25% with respect to the optimal solution. Comparison of this result with error below 1–5% for local search methods seems to be a significant drawback. More important shortcoming is a consequence of great generality of scheduling problems leading to rough theoretical evaluations in contrast to modern metahueristics running without theoretical guarantee of quality.

Table 2 Lower $\underline{\eta}^A$ and upper $\overline{\eta}^A$ bounds on the worst-case ratio η^A for the algorithm A and criteria K

K	A	$\underline{\eta}^A$	$\overline{\eta}^A$
C_{max}	any	m	m
	R	$\lceil m/2 \rceil$	$\lceil m/2 \rceil$
	CDS	$\lceil m/2 \rceil$	$\lceil m/2 \rceil$
	RA	$m/\sqrt{2} + O(1/m)$	$m/\sqrt{2} + O(1/m)$
	RACS,RAES	$m/\sqrt{2} + O(1/m)$	$m/\sqrt{2} + O(1/m)$
	P	$m/\sqrt{2} + O(1/m)$	$m/\sqrt{2} + O(1/m)$
	NEH	$m/\sqrt{2} + O(1/m)$	$(m + 1)/2$
	HR	$m/\sqrt{2} + O(1/m)$	$m/\sqrt{2} + O(1/m)$
	G	$m - 1$	$m - 1$
	TG	$(m + 1)/2$	$(m + 1)/2$
	IE,M	m	m
	KS1,KS2	m	m
	CDS+HC	$m/2$	$\lceil m/2 \rceil$
	T/Yk	$\lceil m/k \rceil \eta^{Yk}(C_{max})$	$\lceil m/k \rceil \eta^{Yk}(C_{max})$
$\overline{C}$	any	n	n
	SPT	m	m
	RCo	$2m/3 + 1/3$	m
	RCo,m=2	1.908	2
	RC1,RC2	n	n
	RC3	$2m/3 + 1/(3m)$	n
	RC3,m=2	1.577	n
	HK,m=2	$2b/(a + b)$	$2b/(a + b)$
	KS1,...,KS5	$n - 1$	n
	R,CDS+API	$n - (2 - 4/(n + 2))$	n
	CDS+NPI	$n/3$	n
C	any	$1 + (n - 1)(\overline{w}/\underline{w})$	$1 + (n - 1)(\overline{w}/\underline{w})$
	CDS,G,P,RA, CDS+HC,G+HC, P+HC,RA+HC	$1 + (n - 1)(\overline{w}/\underline{w})$	$1 + (n - 1)(\overline{w}/\underline{w})$
	F	m	m
	Q/X	$\lceil m/2 \rceil \eta^{X}(C_{sum})$	$\lceil m/2 \rceil \eta^{X}(C_{sum})$
	T/Yk	$\lceil m/k \rceil \eta^{Yk}(C_{sum})$	$\lceil m/k \rceil \eta^{Yk}(C_{sum})$

3.4 Decision Support Systems

This short-time, subsidiary research direction is associated with Author's participation in the international project named "A DSS for Resource Constrained Project Scheduling Problem", realized under auspices of the International Institute of Applied System Analysis (IIASA), Luxemburg, Austria, in years 1989–1991. The research were conducted next a few years within the research agreement IASA-PAN (Poland) and within an independent project founded by KBN (Poland).

Decision Support System (DSS) can be perceived as the complex information system, equipped with at least advanced database management tools, HMI (Human-Machine Interface), model checker, multipurpose solving engine oriented on potential applications, manager of solving process, solution evaluation module. Current software technologies are relevant to realize advanced database and HMI. However, we still observe some lacks in methods of defining the model, model consistency evaluation, verifying correctness, solution algorithms, methods of evaluation solutions (including multi-criteria case), management of the solving process, methods of guiding the dialog with the user. The goal of the international project was to formulate new ideas to attack mentioned challenges. Among a few international research groups, the system designed and implemented by the team with the Author obtained the highest rank. Results are published among others in [28].

3.5 Algorithms with Adaptive Memory

First the most spectacular achievement with the use of the search memory in a meta-heuristic algorithm (called TSAB) was the paper with Author's contribution on the job shop scheduling problem, issued in 1994 and published in *Management Science* in 1996, [30]. It showed that the problem commonly considered from years as extremely hard optimization case can be solved quickly by relatively simple method. In this paper there were formulated and applied some theoretical features of so called block properties, developed and used next by Author in successive papers [31–33, 35–37, 39], in the combination with the tabu search technology of Glover [14]. The paper [30] also firstly introduced new original combinatorial model called next in the literature as permutation-and-graph, in fact completely different than disjunctive graphs commonly used in the world's literature. The proposed model has the smallest redundancy, smallest computational complexity of finding the critical path, and ensures feasibility of all solutions generated on the search trajectory. To illustrate the problem of feasibility we recall a well-known benchmark FT10 as one of the smallest. In FT10 only one solution per 10^{17} is feasible, whereas cost of testing feasibility is the same as calculating single goal function value. Then, in the mentioned paper there were formulated sufficient conditions for optimal solution. Surprisingly good numerical results, despite the lack of connectivity property, provided appreciation to Authors (expressed by the number of citations in SCI, SCIE, Scopus), defining the promising research directions for several successive years.

The approach based on search memory introduces unusual progress among solution methods for some NP-hard discrete optimization problems. Algorithms allow to find solutions very close to optimal, below 5% for instances from practice, already on the ordinary PC. Indirect appreciation of our income to the domain are vignettes in monographs: F. Glover, M. Laguna, Tabu search, Kluwer 1997 [14], and M. Laguna, R. Marti, Scatter Search, Kluwer 2003, and also publications in foreign books. Notice, numerically excellent tabu search algorithm TSAB, despite the lack of theoretical guarantee of quality is more attractive for scientists and users than theoretically excellent but not exciting experimentally approximation algorithms and schemes.

Incredible features of TSAB algorithm can be observed through publications of foreign authors. Firstly, there are papers [21, 62, 63] examining TSAB phenomenon explicitly. Second, features of the critical path and blocks have been used in many other algorithms based on other local search technologies and evolutionary methods. These confirm the strong influence of mentioned elements on the domain development.

The approach has been used next in a few successive algorithms for the flow shop, flow shop with additional technological constraints, open shop, hybrid system, see e.g. [32, 33, 37, 39]. Beside the efficient application of the extended critical path and block notions, there were formulated original theoretical methods of algorithm acceleration by skilful decomposition and aggregation of calculations, called from this moment in the literature software accelerator.

3.6 Parallel Algorithms

In the sixties and seventies of the 20th century, optimization methods were implemented solely on single-processor computers. From the beginning of the nineties, thanks to the progress of the technology, parallel calculations become the standard. In order to fully employ capabilities of parallel computing environments, we should re-design solution algorithms, passing from sequential to parallel versions, which is quite non-trivial. To this aim a completely new research domain was created and called respectively parallel algorithms and parallel metaheuristcs [54]. In fact, parallel methods are considered as the powerful medicine for observed troubles of the discrete optimization. Because the quality of approximate algorithms depends on the number of checked solutions, it is obvious that parallel algorithms offer better quality in a shorter time. A passage from sequential to parallel algorithm needs a broad theoretical knowledge: architecture of processors (commonly denoted by abbreviation SISD, SIMD, MIMD, MIST), structure of the memory (shared, distributed), memory access (commonly denoted by abbreviations EREW, CREW, ERCW, CRCW PRAM), programming languages, concurrency, decomposition strategy (small, medium and coarse grain calculations), Amdahl and Gustafson laws, [1, 19], data structures, algorithms, measures of acceleration. Extensive and wide research are carried out on this subject in the team from the University, let I mention some papers [5, 6, 50, 54].

3.7 Vector Optimization

Vector optimization problems (case $s > 1$ in formula (1)) are classified accordingly the expression of user preferences:

(1) preferences are defined *a priori* as a combination of component criteria or an order of validity of component criteria,
(2) user selects preferred solution *a posteriori* from the set of non-dominated solutions,
(3) user preferences are not defined,
(4) preferences are set in the dialog between system and user in order to approximate user preferences and to find satisfactory solution.

Further details with the list of known and recommended approaches one can find in [52]. Currently, in the literature dominate approaches based on approximation of Pareto front (partially or completely). Indeed, sequential or parallel metaheuristics may provide only certain approximation of the front, so comparison of approximate algorithm in the scope of front is an interesting research topic, [23]. Since cost of calculations necessary to find the set of non-dominated solutions is usually quite large, most of approaches uses parallel genetic algorithms to this aim. Nevertheless, as was shown in own papers [38, 50, 53], there exist other sufficiently efficient approaches. Further research is carried out in direction (4) to reduce cost of calculations.

3.8 Batch and Cyclic Manufacturing

Developed through years the philosophy of conventional manufacturing systems reached finally the fairly acceptable level of knowledge and experience. Currently, manufacturing tends to systems without humans, with exiguous human personnel in the robotized production units, see foundations of the Industry 4.0. Production systems realizing multi-assortment, multi-series manufacturing under slow changes of the production assortment provide final products only in a cyclic, almost deterministic way. Modern systems of such type ensure requirements by: (1) uniform, homogeneous in time production with an some changes from-time-time the assortment in order to fill the stocks, (2) by providing cyclically a mix of assortments according to clients' demands. Assuming that both approaches are acceptable, the former is more attractive since eliminates or limits the warehouse of final products.

Cyclical systems constitute rather poorly examined class of the discrete manufacturing systems, chiefly because of troubles with modeling and solving technology. Quite recently, there appeared new results for more complex cyclical systems, see [10, 22]. These papers refer to disjunctive graphs, block approach and TS metaheuristics, which have been, in fact, crucial methodologies created and developed in Wroclaw University of Science and Technology. The block approach and TS were also applied in earlier Author's own papers for batch scheduling case [30–33, 35–37, 39]. The

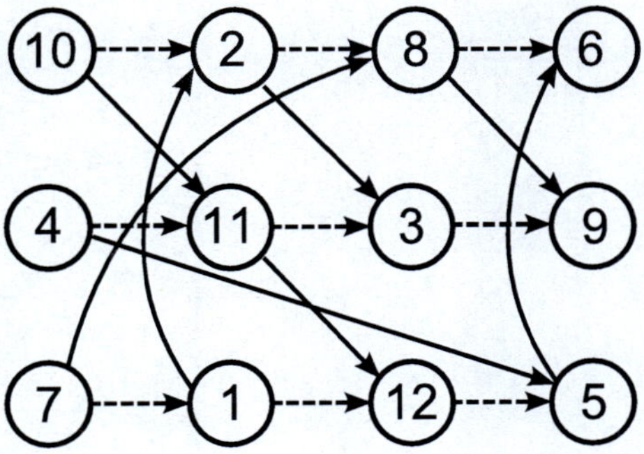

Fig. 4 Graph $G(\pi)$ for the instance from Table 3 and processing order (2). Batch case

papers provided the most known in the world algorithms: TSAB and i-TSAB; see [30, 36]. Because of continuing success of the whole family of TSAB algorithms, in recent years there have been carried out research to extend the approach on cyclic systems [39]. This proposal constitutes an alternative to Kampmayer's model, [22]. The approach is based on the extended permutation-and-graph model, extended notion of the critical path, extended blocks and procedure similar to TASB algorithm.

Excellent results obtained for the family of algorithms TSAB in [30, 36] incline us to recommend their usage in cyclic scheduling as well. Unfortunately, the current success of TSAB is strictly associated with the makespan criterion and is a consequence of elimination features of the critical path and blocks used there. Research made in recent years allow one to extend notion of the critical path on classes of problems with $f_{\max}$ and $h_{\max}$ criterion. The basic dilemma is how to re-define the problem of minimizing cycle time by another scheduling problem with regular criteria of mini-max type. If we can do so, all useful features of TSAB-like algorithms can be applied here, [56].

In order to illustrate the transformation of the job shop scheduling problem from batch scheduling to cyclic scheduling case we analyze the Instance given in Table 3 and processing order defined by the vector of permutations

$$\pi = ((10, 2, 8, 6), (4, 11, 3, 9), (7, 1, 12, 5)) \tag{2}$$

The suitable graph $G(\pi)$ for batch scheduling problem is shown in Fig. 4. Technological arcs from $\mathcal{R}$ are drawn by solid line, sequential arcs Ĺ,uki $\mathcal{E}(\pi)$ are drawn by broken lines. Graph model is very useful for the modeling of the examination of the solution feasibility, for the testing of the feasibility of the solution represented by π, as well as for finding the detailed schedule and the goal function value.

Passage from the batch to cyclic scheduling representation (Fig. 5) generates the cyclic graph with an unknown parameter T, for which there are defined different analytical arguments and different feasibility conditions,[56]. There were several features proved, and what is more not only special properties but also a set of numerically special features given. Cyclic problems with some additional constraints were described also in Author's own papers [47, 51, 58].

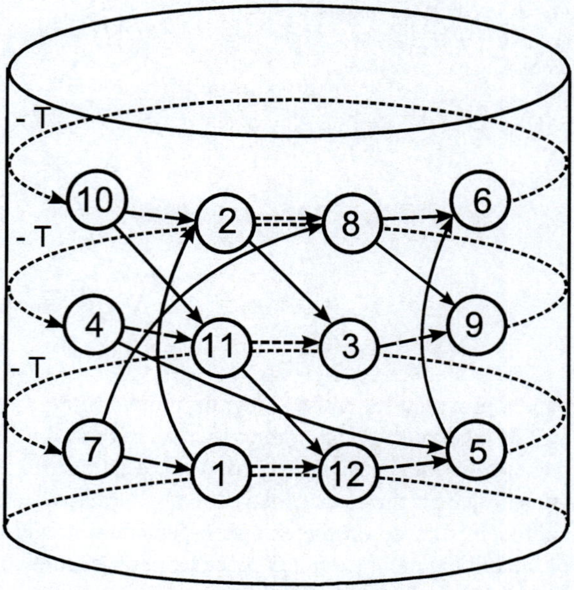

Fig. 5 Graph $H(\pi, T)$ for the instance from Table 3 and processing order (2). Cyclic case

Table 3 Data for the instance

Job	Operation	Machine	Duration
1	1	3	59
1	2	1	65
1	3	2	94
2	4	2	86
2	5	3	60
2	6	1	10
3	7	3	49
3	8	1	43
3	9	2	8
4	10	1	71
4	11	2	25
4	12	3	98

3.9 *Storage Systems*

Storage is the immanent part of manufacturing, logistics, supply, delivering and distribution of goods. We store materials, components, final products, tools and so on. The overall aim of the storage is to polish up the natural fluctuation in the demand of these elements and to achieve the stability and to ensure continuity of the production process and/or sales. Size of the storage and storage cost depend on fluctuation of the flow (expressed e.g. by standard deviation) observed before the service stage. Therefore the primal optimization goal is to minimize the variance of providing/retrieving components.

There are many various types of the storage, depending the level of automation, strategy of management, size and variety of stored elements, strategy of their distribution inside. The simplest storage units are single-level or contains only a few stiles with a small assortment of products. They were serviced by humans through any method with large logistic warehouses, occupying the size of a small locality, using the automated transport with small number of human workers, but chiefly using robots equipped with advanced information and control systems. Independently, see other large commercial Internet sales networks.

There exists many mediate solutions between small and large system. In each of these systems we detect specific optimization problems. They need a definition of system quality indicator, see the taxonomy in [49]). The main component of the completion of a client order is called the order picking process. Although warehouses tend definitely to humanless systems, there still exist systems with human personnel in the process of commissioning. The worker usually gets the support from the advanced computer science infrastructure, transport, communication and so on to increase the efficiency of the worker's job. Nevertheless there still remain order picking criteria such as the minimization of the commission time or minimization of the route of the order picker worker.

The overall aim of the order picking system is to fill the set of bins by the set of various types of items, whereas filling the bin is carried out according to the individual order. As an instance we consider the line of preparing the electric bundle for the car. The number and variety of elements involving the bundle depends on the features of the car ordered by a client (ABS, central lock, etc,). The number of parts collected in the bin is small 20–30, however there are chosen from the set of the set of 200–500 various types of parts located in the storage. The commissioner task is to choose the given set of items from the given locations in the store. There are either given start place and end place of the worker called depots. Depots can be centered or distributed. The order of retrieving elements defines in an unique way the route of the worker only if the product has unique location in the warehouse. More complex problem is when we can chose the item from among various storage places. In case when alternative paths between points of retrieving exists, the shortest path are included to the route.

Optimization of the order picking systems is possible with the use of models and algorithms from the transport and scheduling problems, see own results, [49]. Depending the type of the system, the optimization problem can be transformed to TSP problem or the problem of scheduling on the single machine with setup times. Enhanced optimization can be formulated by introducing multiple locations of items in the store or alteration of location of item in the store.

4 Space Landscape

After the series of applications for various combinatorial problems, the question about the source and determinants leading to the success of the tabu search method in mentioned cases was stated. Taking a broader context, we would like to identify the factors of the success/failure of newly designed discrete optimization algorithms. One of the Author's own research direction was to examine the structure of the search space and search trajectories performed by different algorithms, [15, 34, 45, 59], see also Fig. 6. It has been concluded that solution algorithm should adapt itself in the static or dynamic way do the landscape of the solution space. Identification of particular problem features allow us to design solution algorithm the most fitted to the problem, thus the most efficient.

The evaluation of the search space landscape requires the notion of the distance between solutions in the space $\mathcal{X}$. There are known distance measures in euclidean spaces, but this is not the subject. In discrete solution spaces the problem of setting the measure is much more complex. We need to take into account not only features of the measure but also the computational complexity of its calculation. In Table 4 there are alternative measures between permutations chosen (fundamental for TSP, QAP, as well as for many other scheduling problems). This is a subsequent example on a link between control science and computer science.

The distribution of solutions and local minima in the discrete solution space is usually irregular. These can be observed tracing random or goal-oriented search trajectories passing through neighboring solutions in the solution space. The distance between neighboring solution is one unit in the sense of a chosen measure); see Table 4 and Fig. 6. To test the space and the landscape we use random sampling of the space. It has at least two goals:

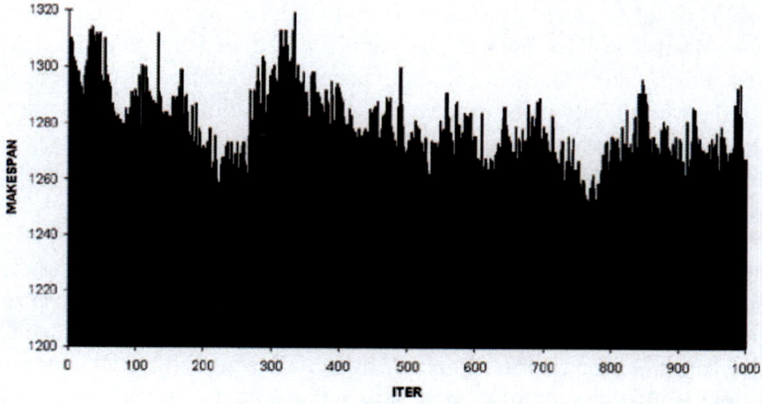

Fig. 6 Search trajectory in TS algorithm. Neighbouring solutions in the space for flexible flow line scheduling

Table 4 Distance measures $D(\alpha, \beta)$ between permutations α and β

	Technology of transformation $\alpha \to \beta$		
	Adjacent swap	Swap	Insert
Recipt	Number of inversions in $\alpha^{-1} \circ \beta$	n minus number of cycles in $\alpha^{-1} \circ \beta$	n minus the length of maximal increasing subsequence in $\alpha^{-1} \circ \beta$
Maximum	$n(n-1)/2$	$n-1$	$n-1$
Mean	$n(n-1)/4$	$n - \sum_{i=1}^{n} 1/i$	$n - O(\sqrt{n})$
Variance	$n(n-1)(2n+5)/72$	$\sum_{i=1}^{n}(i-1)/i^2$	$\theta(n^{1/3})$
Complexity	$O(n^2)$	$O(n^2)$	$O(n \log n)$

(1) identification of regions of the solution space having the feasible solutions from $\mathcal{X}$,
(2) identification the most promising regions of the solution space in sense of the criterion $K(x)$.

Referring to the already mentioned instance [60] of the job shop scheduling problem with the makespan criterion we conclude that the distance of any solution to optimal solution has distribution close to normal with the mean 50% of the space diameter.[1] Feasible solutions are located slightly closer, approximately 20% of space diameter. The distribution of feasible solutions taking account of criterion $K(x)$ is shown in Figure 2. This distribution is also close to normal. Approximation of the least square method provides the mean 115,8 and standard deviation 17,8. It means that random instance is more than twice worse than optimal one. Notice, the probability of finding the solution with $RE \leq 1\%$ by random sampling of the space $\mathcal{X}$ is approximately equal $5 \cdot 10^{-11}$, so infinitesimally small even we consider 10^6 solutions. On the other hand, in $\mathcal{X}$ we have $2 \cdot 10^{39}$ solutions with $RE \leq 1\%$. There is no possibility of overlooking them completely or even partially. Such conclusions from the landscape analysis are very useful for the algorithm design.

Research of the solution space comprise problems of big valley detection [35, 37], roughness examination [45], visualization of the space in 2D and 3D [34], visualization of the search trajectories [48]. We define the big valley phenomenon if there exists positive correlation between goal function value and distance to optimal solution (best found solution). In the big valley we observed the density of local extremes, so the searching process can be directed there. Big valley were detected in many discrete problems, let we mention TSP, scheduling. Roughness is the mathematical measure characterizing the dispersion of the close (neighboring) solutions. Greater diversity means sharp and unpredictable changes of $K(x)$, see Fig. 6. Smaller roughness means a flat landscape. Visualization of the trajectories and the space maps can be perceived as a transformation from n-dimensional space to R^2 or R^3. Some known methods in the pattern analysis [41] can be useful for this aim.

[1] Space diameter is the maximal distance between any two solutions.

5 Conclusion

The domain of discrete production processes is characterized by a significant variety of problems generated by practice. Mutual constant contacts of scientists with practitioners lead to the compromise between expectations and possibilities of both sides. Coexistence of various modeling and solving technologies of event systems is quite natural and each time are verified by practical application. Strong competition on the market forces applications with evident economical profits and technological innovations. This in order implies preferences to solutions and efficient algorithms created with the use of computer science. This in order implies that there are preferred specialized approaches instead of general modeling technology. Implementation of systems without humans provides dominance of deterministic models. Referring to the development of the solution methods, we expect the passage to parallel, distributed and cloud algorithms. The next step is the use of quantum computers.

References

1. Amdahl, G.M.: Validity of the single processor approach to achieving large-scale computing capabilities. AFIPS Conf. Proc. **483–485** (1967). https://doi.org/10.1145/1465482.1465560
2. Barták, R., Salido, M.A., Rossi, F.: Constraint satisfaction techniques in planning and scheduling. J. Intell. Manuf. **21**(1), 5–15 (2010)
3. Błażewicz, J., Domschke, W., Pesch, E.: The job shop scheduling problem: conventional and new solution techniques. Eur. J. Oper. Res. **93**(1), 1–33 (1996)
4. Bocewicz, G., Wójcik, R., Banaszak, Z.: Cyclic scheduling of multimodal concurrently flowing processes. In: Świątek J. et al. (eds.) Advances in Systems Science, pp. 587–598. Springer, Cham (2014). https://doi.org/10.1007/978-3-319-01857-7_57
5. Bożejko W., Pempera J., Smutnicki C.: Parallel simulated annealing for the job shop scheduling problem. In: Allen, G. et al. (eds.) ICCS 2009, Part I, Lecture Notes in Computer Science 5544, pp. 631–40 (2009)
6. Bożejko, W., Pempera, J., Smutnicki, C.: Parallel tabu search algorithm for the hybrid flow shop problem. Comput. Ind. Eng. **65**(3), 466–474 (2013)
7. Bożejko, W., Jagiełło, S., Lower, M., Smutnicki, C.: On underwater vehicle routing problem. In: Moreno-Diaz, R., Pichler, F., Quesada-Arencibia, A. (eds.) Computer Aided Systems Theory EUROCAST 2015. Lecture Notes in Computer Science, vol. 9520, pp. 861–868. Springer, Cham (2015)
8. Bożejko, W., Smutnicki, C., Wodecki, M.: Metropolitan delivery with time windows as a scheduling problem. In: Koukol, M. (ed.) Smart Cities Symposium, Prague (SCSP), pp. 1–5. Piscataway, NJ, IEEE (2016)
9. Brucker, P., Kampmeyer, T.: Cyclic job shop scheduling problems with blocking. Ann. Oper. Res. **159**, 161–181 (2008)
10. Brucker, P., Jurish, B., Sieviers, B.: A fast branch and bound algorithm for the job shop problem. Discret. Appl. Math. **49**, 107–127 (1994)
11. Buzzone A., Laboratories of Simulation, http://www.itim.unige.it/
12. EC Bestiary http://conclave.cs.tsukuba.ac.jp/research/bestiary/ (2017)
13. Filipowicz, B.: Modele stochastyczne w badaniach operacyjnych (Polish). WNT, Warszawa (1996)
14. Glover, F., Laguna, M.: Tabu Search. Kluwer Academic Publishers, Norwell (1997)

15. Gnatowski, A., Smutnicki, C., Wodecki, M.: Wykorzystanie własności krajobrazu przestrzeni rozwiązań w konstrukcji algorytmów etaheurystycznych (Polish). In: 1st International Conference on Decision Making in Manufacturing and Services & XX Jubilee Symposium on Applications of Systems Theory, DMMS 2017 & ZTS XX, Conference Proceedings, Zakopane 2017, pp. 209–214 (2017)
16. Grabowski, J., Nowicki, E., Smutnicki, C.: Algorytm blokowy szeregowania operacji w systemie gniazdowym (Polish), Przegląd Statystyczny **35**(1), 67–80 (1988)
17. Grabowski, J., Nowicki, E., Smutnicki, C.: Metoda blokowa w zagadnieniach szeregowania zadań (Polish). Akademicka Oficyna Wydawnicza EXIT, Warszawa (2003)
18. Graham, R.L., Lawler, E.L., Lenstra, J.K., Kan, A.H.G.R.: Optimization and approximation in deterministic sequencing and scheduling: a survey. Ann. Discret. Math. **5**, 87–326 (1979)
19. Gustafson, J.L.: Reevaluating Amdahl's Law. Commun. ACM **31**(5), 532–3 (1988). https://doi.org/10.1145/42411.42415
20. Jain, A.S., Meeran, S.: Deterministic job-shop scheduling: past, present and future. Eur. J. Oper. Res. **113**, 390–434 (1999)
21. Jain, A.S., Rangaswamy, B., Meeran, S.: New and stronger job-shop neighbourhoods: a focus on the method of Nowicki and Smutnicki (1996). J. Heuristics **6**(4), 457–480 (2000)
22. Kampmeyer, T.: Cyclic scheduling problems, Ph.D. thesis, Osnabrück University (2006)
23. Kowalczuk, Z., Białaszewski, T.: Approximate criteria for the evaluation of truly multi-dimensional optimization problems. In: MMAR Conference (2018)
24. Nowicki, E., Smutnicki, C.: Worst-case analysis of an approximation algorithm for flow-shop scheduling. Oper. Res. Lett. **8**(3), 171–177 (1989)
25. Nowicki, E., Smutnicki, C.: Worst-case analysis of Dannenbring's algorithm for flow-shop scheduling. Oper. Res. Lett. **10**(8), 473–480 (1991)
26. Nowicki, E., Smutnicki, C.: New results in the worst-case analysis for flow-shop scheduling. Discret. Appl. Math. **46**(1), 21–41 (1993)
27. Nowicki, E., Smutnicki, C.: An approximation algorithm for a single-machine scheduling problem with release times and delivery times. Discret. Appl. Math. **48**(1), 69–79 (1994)
28. Nowicki, E., Smutnicki, C.: A decision support system for the resource constrained project scheduling problem. Eur. J. Oper. Res. **79**(2), 183–195 (1994)
29. Nowicki, E., Smutnicki, C.: A note on worst-case analysis of approximation algorithms for a scheduling problem. Eur. J. Oper. Res. **74**(1), 128–134 (1994)
30. Nowicki, E., Smutnicki, C.: A fast taboo search algorithm for the job shop problem. Manag. Sci. **42**(6), 797–813 (1996)
31. Nowicki, E., Smutnicki, C.: A fast tabu search algorithm for the permutation flow-shop problem. Eur. J. Oper. Res. **91**(1), 160–175 (1996)
32. Nowicki, E., Smutnicki, C.: The flow shop with parallel machines: a tabu search approach. Eur. J. Oper. Res. **106**(2–3), 226–253 (1998)
33. Nowicki, E., Smutnicki, C.: Flow line scheduling by tabu search. In: Meta-Heuristics, pp. 175–189. Springer, Boston, MA (1999)
34. Nowicki, E., Smutnicki, C.: 2D and 3D representations of solution spaces for CO problems. In: International Conference on Computational Science, pp. 483–490. Springer, Berlin, Heidelberg (2004)
35. Nowicki, E., Smutnicki, C.: Some new ideas in TS for job shop scheduling. In: Metaheuristic Optimization via Memory and Evolution, pp. 165–190. Springer, Boston, MA (2005)
36. Nowicki, E., Smutnicki, C.: An advanced tabu search algorithm for the job shop problem. J. Sched. **8**(2), 145–159 (2005)
37. Nowicki, E., Smutnicki, C.: Some aspects of scatter search in the flow-shop problem. Eur. J. Oper. Res. **169**(2), 654–666 (2006)
38. Pempera, J., Smutnicki, C., Żelazny, D.: Optimizing bicriteria flow shop scheduling problem by simulated annealing algorithm. Procedia Comput. Sci. **18**, 936–945 (2013)
39. Pempera, J., Smutnicki, C.: Open shop cyclic scheduling. Eur. J. Oper. Res. **269**(2), 773–781 (2018)

40. Rudy, J., Bożejko, W.: (2018) Reversed Amdahl's Law for Hybrid Parallel Computing. In: Moreno-Diaz, R., Pichler, F. Quesada-Arencibia, A. (eds) Computer Aided Systems Theory, EUROCAST, : Lecture Notes in Computer Science, 10671. Springer, Cham (2017)
41. Siedlecki, W., Siedlecka, K., Sklansky, J.: An overview of mapping techniques for exploratory pattern analysis. Pattern Recogn. **21**(5), 411–429 (1988)
42. Smutnicki, C.: Some results of the worst-case analysis for flow shop scheduling. Eur. J. Oper. Res. **109**(1), 66–87 (1998)
43. Smutnicki, C.: A two-machine permutation flow shop scheduling problem with buffers. Oper. Res. Spekt. **20**(4), 229–235 (1998)
44. Smutnicki, C.: Algorytmy szeregowania (Polish). Akademicka Oficyna Wydawnicza EXIT, Warszawa (2002)
45. Smutnicki, C.: Efektywność algorytmów poszukiwań lokalnych a struktura przestrzeni rozwiązań (Polish). In: Greblicki, W., Smutnicki, C. (eds.) Systemy sterowania. WKiŁ, Warszawa (2005)
46. Smutnicki C.: Optimization technologies for hard problems. In: Janos, F., Klempous, R., Paz Suarez Araujo, C. (eds.) Recent Advances in Intelligent Engineering Systems, pp. 79–104. Springer, Berlin; Heidelberg (2011)
47. Smutnicki, C.: An efficient algorithm for finding minimal cycle time in cyclic job shop scheduling problem. IEEE **2012**, 381–386 (2011)
48. Smutnicki, C.: Metody optymalizacji dyskretnej (Polish). In: Bożejko, W., Pempera, J. (eds.) Optymalizacja dyskretna w informatyce, automatyce i robotyce. Oficyna Wydawnicza PWr, Wrocław (2012)
49. Smutnicki, C.: Problemy kolekcjonowania (Polish). In: Bożejko, W., Pempera, J. (eds.) Optymalizacja dyskretna w informatyce, automatyce i robotyce. Oficyna Wydawnicza PWr, Wrocław (2012)
50. Smutnicki, C., Rudy, J., Żelazny, D.: Very fast non-dominated sorting. Decis. Making Manuf. Serv. **8**(1–2), 13–23 (2014)
51. Smutnicki, C.: New features of the cyclic job shop scheduling problem. In: MMAR 2015, pp. 1000–1005. IEEE (2015)
52. Smutnicki, C.: Optimization in CIS systems. In: Zamojski, W., Sugier, J. (eds.) Dependability Problems of Complex Information Systems, Advances in Intelligent Systems and Computing, Springer, vol. 307, pp. 111–128 (2015)
53. Smutnicki, C., Pempera, J., Rudy, J., Żelazny, D.: A new approach for multi-criteria scheduling. Comput. Ind. Eng. **90**, 212–220 (2015)
54. Smutnicki, C., Bożejko, W.: Parallel and distributed metaheuristics. In: Moreno-Diaz, R., Pichler, F., Quesada-Arencibia, A. (eds.) Computer Aided Systems Theory, EUROCAST 2015. Lecture Notes in Computer Science, vol. 9520, pp. 72–79. Springer, Cham (2015)
55. Smutnicki, C.: Algorytmy szeregowania zadań (Polish). Oficyna Wydawnicza PWr, Wrocław (2017)
56. Smutnicki, C.: Cykliczny problem gniazdowy (Polish). In: Bożejko, W., Pempera, J., Smutnicki, C., Wodecki, M. (eds.) Zaawansowane modele i algorytmy optymalizacji w systemach cyklicznych. EXIT, Warszawa (2017)
57. Smutnicki, C.: Minimizing cycle time in manufacturing systems with additional technological constraints. In: MMAR 2017, pp. 463–470. IEEE (2017)
58. Smutnicki, C.: Cyclic Scheduling in interlaced and non-interlaced mode. In: MMAR (2018)
59. Smutnicki, C., Bożejko, W.: Tabu Search and solution space analyses. The job shop case. In: Moreno-Diaz, R., Pichler, F., Quesada-Arencibia, A. (eds.) Computer Aided Systems Theory, EUROCAST 2017. Lecture Notes in Computer Science, vol. 10671, pp. 383–391. Springer, Cham (2017)
60. Taillard, E.: Benchmarks for basic scheduling problems. Eur. J. Oper. Res. **64**, 278–285 (1993)
61. Valenzuela, M.L., Rozenblit, J.W.: Learning using anti-training with sacrificial dat. J. Mach. Learn. Res. 1–42 (2016)
62. Watson, J.P., Beck, J.C., Howe, A.E., Whitley, L.D.: Problem difficulty for tabu search in job-shop scheduling. Artif. Intell. **143**, 189–217 (2003)

63. Watson, J.P., Howe, A.E., Whitley, L.D.: Deconstructing Nowicki and Smutnicki's i-TSAB tabu search algorithm for the job-shop scheduling problem. Comput. Oper. Res. **33**(9), 2623–2644 (2006)
64. Wolpert, D.H., Macready, W.G.: No free lunch theorems for optimization. IEEE Trans. Evol. Comput. **1**(1), 67–82 (1997)
65. Zhou, M., Venkatesh, K.: Modeling, Simulation, and Control of Flexible Manufacturing Systems: A Petri Net Approach. World Scientific (1999)

Dynamic Programming with Imprecise and Uncertain Information

Janusz Kacprzyk

Abstract Basic models of multistage fuzzy control are presented in which goals on the consecutively attained states (outputs) and constraints on the consecutively applied controls (inputs) are specified in an imprecise form as fuzzy sets defined in the space of states and controls, respectively. In such a setting, first, the classic problem of the multistage control of a deterministic system, given as a state transition equation, is considered followed by a discussion of the control of a stochastic system given as a Markov chain. For both cases the problems with the finite, fixed and specified termination time and infinite termination time are discussed. They and solved by using fuzzy dynamic programming in the case of the finite, fixed and specified termination time. In the case of the infinite termination time they are solved by using a policy iteration type algorithm. Some well known and successful applications of the models proposed are discussed.

1 Introduction

Decision making is one of fundamental and widely occurring acts in all kinds of human activities. Its essence is basically the choice or selection of some best option (variant, alternative, choice ...) from a set of possible or feasible ones. There can be various criteria for the choice, for instance a financial outcome, or we can only look for a sufficiently good option, not the best one. The underlying information on the basis of which decisions are made can be characterized by certainty, uncertainty, risk, imprecision or even ignorance, or by other forms of information imperfection. Moreover, we can have one or more criteria for the choice and also dynamics in the sense that the decision process proceeds over time, that is, as it is most often assumed, that decisions are made at consecutive time moments over a control horizon which can be finite or infinite, precisely specified or not.

J. Kacprzyk (✉)
Systems Research Institute, Polish Academy of Sciences, ul. Newelska 6, 01447 Warsaw, Poland
e-mail: kacprzyk@ibspan.waw.pl

P. Kulczycki et al. (eds.), *Automatic Control, Robotics, and Information Processing*,
Studies in Systems, Decision and Control 296,
https://doi.org/10.1007/978-3-030-48587-0_13

Moreover, the decision making process can be considered from both a cognitive and psychological point of view, but also in the context of economics, organization and management, etc. In all these situations we can imagine that the purpose is to investigate how real sciences decision making processes proceed or how to determine the best possible decisions. This corresponds to descriptive (behavioral) and prescriptive approaches, respectively. More information on decision making and decision processes can be found in many books which are available from virtually all publishers. One can also use, which is becoming increasingly popular, respected Internet encyclopedia, for instance Stanford Encyclopedia of Philosophy—cf. Steele and Stefansson [84].

In this work we concentrate on the problem of decision making which is characterized by dynamics, that is, decisions are made sequentially, at consecutive stages, and the termination time of the process can be both finite and infinite. For a conceptual simplicity and clarity we will assume that our decision making process is a control process so that we will look for controls (decisions) at the consecutive control stages which will best satisfy goals imposed on states (equated with outputs, for simplicity) attained and constraints on controls (equated with inputs) applied. What is important is that information about the goals, constraints, etc. is specified in an imprecise (fuzzy) way, for instance as natural language statements like "the value of control should be possibly low". We will use elements of Zadeh's [97] theory of fuzzy sets (cf. [49, 52]). We will also assume that in addition to the imprecision there will be a joint occurence of the stochastic uncertainty as to the dynamics of the system under control, more specifically as to the probabilities of state transitions. A joint occurence of imprecision (fuzziness) and uncertainty (probabilities) is clearly very interesting though it complicates the analysis.

As we have already mentioned, a convenient paradigm for considering such processes is that of multistage control. For instance, suppose that we have a problem of inventory control, exemplified by the classic model. Suppose that we have a demand for a product which is time distributed over a planning horizon and assume that we do not allow for an unfulfilled demand. In the beginning we do not have anything in the inventory so that we have to manufacture goods for which there is a demand. We should therefore bear some fixed costs related to the initialization of manufacturing, which does not depend on the size of manufacturing, and then some variable costs which depends on the number of items manufactured. Moreover, there are some holding costs related to the keeping of products in the inventory which depends on the number of products as well. The problem is to find the number of products to be manufactured at consecutive stages to fulfill the demand at the lowest total cost over the planning horizon. This is a simple example of a multistage control (decision making) problem, that is, decisions are made sequentially.

In this paper we will present the application of dynamic programming to solve such classes of multistage control problems. Dynamic programming is a general approach for the formulation and solution of difficult problems of such a type via their decomposition into simpler problems. It is based on the famous so–called Bellman's optimality principle which states that in each sequence of decisions (controls) from stage $t = 0$ to stage $t = N$, where N is a planning (control) horizon, its subsequence from stage $t = k$, $0 \geq k \geq N$, to stage $t = N$ must itself be optimal. The roots of dynamic programming date back to the 1950s and Richard E. Bellman has been considered its founder. More information can be found, e.g., in the books by Bellman [4], Bertsekas [6, 8], Bertsekas and Shreve [7], Denardo [9], Sniedovich [82, 83], etc.

In real control processes most elements of problems considered are uncertain, for instance, costs, prices, state transitions under certain controls. Traditionally, this uncertainty has been equated with randomness and probabilistic tools and techniques have been used (cf. [6, 8, 25]). It has however been clear since the beginning that in addition to uncertainty related to randomness, there are some other facets of imperfect information exemplified by imprecision, vagueness, and even ignorance. In this work we assume that we deal with imprecision of meanings which is a consequence of the use of natural language which is the only fully natural means of articulation and communication for human beings. This imprecision can be expressed and processed in a simple and elegant way using Zadeh's [97] theory of fuzzy sets.

The application of fuzzy sets in fuzzy dynamic programming has been proposed in the pioneering work by Bellmana and Zadeh [5], followed by many works, including many books, mainly by Kacprzyk [37, 38, 49, 52].

First we will present main elements of the theory of fuzzy dynamic programming as a tool for the formulation, analysis and solution of such multistage control problems. We will start with a brief exposition of the theory of fuzzy sets, followed by fuzzy dynamic systems, and then present basic aspects of the use of fuzzy dynamic programming, in the case of a finite termination time, and a slightly related approach based on a function equation, in the case of an infinite termination time. We will finish with a brief presentation of some relevant applications of the models presented.

2 Fuzzy Sets, and Deterministic, Stochastic and Fuzzy Dynamic Systems

First we will present main concepts and properties of fuzzy sets theory, and then of the deterministic, stochastic and fuzzy dynamic systems.

2.1 *Main Elements of Fuzzy Sets Theory*

Fuzzy sets theory [97] is a simple yet strong, effective and efficient tool for the representation and processing of imprecise information (concepts) exemplified by "*high houses*", "*large numbers*" ... Tools and techniques of fuzzy sets theory are often used for the representation and processing of terms used in natural language which is the only fully natural means of articulation and communication for human beings.

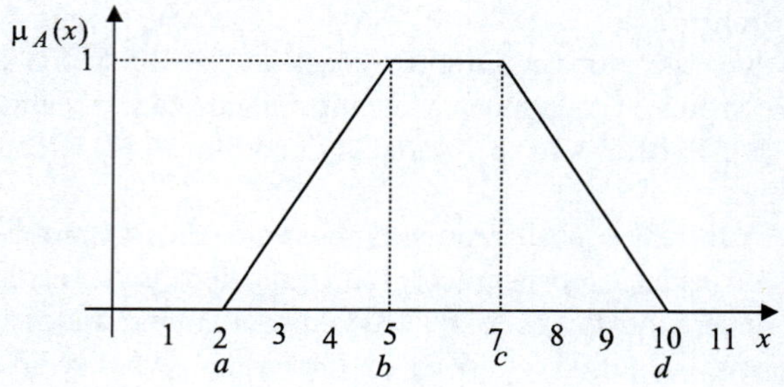

Fig. 1 Piecewise linear membership function of the fuzzy set "integer numbers *more or less equal* 6"

A *fuzzy set* (or a *fuzzy subset*) A in a universe of discourse $X = \{x\}$, written A in X, is defined as a set of pairs

$$A = \{(\mu_A(x), x)\} \tag{1}$$

where $\mu_A : X \longrightarrow [0, 1]$ is a *membership function* of fuzzy set A, and $\mu_A(x) \in [0, 1]$ is a *degree of membership, or membership degree* of element $x \in X$ in fuzzy set A.

Notice that in a fuzzy set the transition from the membership to non-membership of an element of a universe of discourse to a fuzzy set is gradual (from 0 to 1, through all intermediate values being real numbers from (0,1)), and not abrupt like in traditional sets to which elements can either belong, with the degree of membership 0, or not belong, with the degree of membership 1. This is what we do need for the representation of imprecise concepts and properties. In practice, the fuzzy sets are often equated with their membership functions.

For instance, in a fuzzy set (labeled, which is often omitted) "integer numbers *more or less equal* 6", we can say that $x = 6$ for sure belongs to such a fuzzy set so that $\mu_A(6) = 1$, the numbers 5 and 7 belong "almost for sure" so that $\mu_A(5)$ and $\mu_A(7)$ are close to 1, and the more a number differs from 6 the the lower is the value of $\mu_A(.)$. The numbers below 1 and above 10 do not belong to such a fuzzy set so that for them $\mu_A(.) = 0$). This is certainly subjective which results from the very essence of such imprecise terms This can be depicted as in Fig. 1 (in a simpler, often used form of piecewise linear membership function $\mu_A(.)$ which can now be defined by the specification of four points only: $a = 2, b = 5, c = 7$ and $d = 10$; this simplicity is very relevant in practice).

In the sequel we assume that all universes of discourse are finite and then a fuzzy set A in X is written as $A = \{(\mu_A(x), x)\} = \{\mu_A(x)/x\} = \mu_A(x_1)/x_1 + \cdots + \mu_A(x_n)/x_n = \sum_{i=1}^{n} \mu_A(x_i)/x_i$ where "+" and "$\sum$" are meant in the set-theoretic sense.

Below we will briefly present basic definitions and properties related to fuzzy sets which will be of relevance for our considerations.

Fuzzy set A in X is empty, $A = \emptyset$, if and only if $\mu_A(x) = 0$, for all $x \in X$.

Two fuzzy sets A and B in X are equal, $A = B$, if and only if $\mu_A(x) = \mu_B(x)$, for all $x \in X$.

Fuzzy set A in X contained in (or is a subset of) fuzzy set B in X, $A \subseteq B$, if and only if $\mu_A(x) \leq \mu_B(x)$, for all $x \in X$.

Fuzzy set A in X is normal if and only if $\max_{x \in X} \mu_A(x) = 1$; otherwise, it is subnormal.

Now, we will present some important non-fuzzy sets related to fuzzy sets.

The support of fuzzy set A in X, supp$A$, is the (non-fuzzy) set: supp$A = \{x \in X : \mu_A(x) > 0\}$, and—clearly—$\emptyset \subseteq$ supp$A \subseteq X$.

The α-cut, or the α-level set, of fuzzy set A in X, A_α, is the following (non-fuzzy) set: $A_\alpha = \{x \in X : \mu_A(x) \geq \alpha\}$, for all $\alpha \in (0, 1]$.

The α-cuts play an extremely important role in both the theory and applications of fuzzy sets because they make it possible to uniquely represent a fuzzy set by a sequence of non-fuzzy sets, due to the so called representation theorem (cf. [49, 52]). It states that each fuzzy set A in X can be represented as $A = \sum_{\alpha \in (0,1]} \alpha A_\alpha$, where A_α is an α-cut of fuzzy set A, "$\sum$" is meant in the set theoretic sense, and αA_α stands for a fuzzy set the membership degrees of which are equal to $\mu_{\alpha A_\alpha}(x) = \begin{cases} \alpha \text{ dla } x \in A_\alpha \\ 0 \text{ otherwise} \end{cases}$.

An important concept is the cardinality of a fuzzy set. Wygralak's [91] book provides here a rich source of information. The simplest non-fuzzy cardinality of fuzzy set $A = \mu_A(x_1)/x_1 + \cdots + \mu_A(x_n)/x_n$, is called the *sigma-count* denoted $\sum$Count(A) and is defined as (Zadeh 1978, [101]): $\sum\text{Count}(A) = \sum_{i=1}^{n} \mu_A(x_i)$.

Now, we will proceed to definitions of some relevant set theoretic and algebraic operations on fuzzy sets.

The complement of fuzzy set A in X, $\neg A$, is

$$\mu_{\neg A}(x) = 1 - \mu_A(x), \qquad \text{for all } x \in X \tag{2}$$

and it semantically corresponds to the negation "not".

The intersection of two fuzzy sets A and B, both in X, $A \cap B$, is

$$\mu_{A \cap B}(x) = \mu_A(x) \wedge \mu_B(x), \qquad \text{for all } x \in X \tag{3}$$

where "$\wedge$" is the minimum operation, i.e. $a \wedge b = \min(a, b)$; the intersection semantically corresponds to the connective "and".

The union of two fuzzy sets A and B, both in X, $A + B$, is

$$\mu_{A+B}(x) = \mu_A(x) \vee \mu_B(x), \qquad \text{for all } x \in X \tag{4}$$

where "$\vee$" is the maximum operation, i.e. $a \vee b = \max(a, b)$; semantically, the union corresponds to the connective "or".

These traditional definitions of the intersection and union, dating back to Zadeh's [97] source paper, can be generalized by using, respectively, the t-norms and t-konorms (s-norms), defined as:

The t-norm (triangular norm) is defined as a function $t : [0, 1] \times [0, 1] \longrightarrow [0, 1]$, such that for all $a, b, c \in [0, 1]$: (1) $t(a, 1) = a$, (2) $a \leq b \Longrightarrow t(a, c) \leq t(b, c)$, (3)

$t(a, b) = t(b, a)$, (4) $t[a, t(b, c)] = t[t(a, b), c]$. The t-norm is obviously nondecreasing with respect to both arguments, and $t(a, 0) = 0$.

Some important t-norms are: (1) $t(a, b) = a \wedge b = \min(a, b)$, which is the most widely used, (2) $t(a, b) = a \cdot b$, (3) the Łukasiewicz t-norm, i.e. $t(a, b) = \max(0, a + b - 1)$.

The t-conorm (s-norm) is defined as the function $s : [0, 1,] \times [0, 1] \longrightarrow [0, 1]$, such that for all $a, b, c \in [0, 1]$: (1) $s(a, 0) = a$, (2) $a \leq b \Longrightarrow s(a, c) \leq s(b, c)$, (3) $s(a, b) = s(b, a)$, (4) $s[a, s(b, c)] = s[s(a, b), c]$.

Some important s-norms are: (1) $s(a, b) = a \vee b = \max(a, b)$, which is the most widely used, (2) $s(a, b) = a + b - ab$, (3) the Łukasiewicz s-norm, i.e. $s(a, b) = \min(a + b, 1)$.

Obviously, the t-norm is dual to the s-norm, that is, $s(a, b) = 1 - t(1 - a, 1 - b)$. More information on some newer aggregation operators can be found in Kacprzyk et al. [67].

Among many other operations on fuzzy sets, the following ones can be useful for our next discussion.

The product of scalar $a \in R$ and fuzzy set A in X, aA, is defined as $\mu_{aA}(x) = a\mu_A(x)$, for all $x \in X$, and—obviously—$0 \leq a \leq \frac{1}{\mu_A(x)}$, for all $x \in X$.

The k-th power of fuzzy set A in X, A^k, is defined as $\mu_{A^k}(x) = [\mu_A(x)]^k$, for all $x \in X; k \in R$ and, obviously, $0 \leq [\mu_A(x)]^k \leq 1$.

The concept of a fuzzy relation is clearly crucial for the theory and applications of fuzzy sets. The fuzzy relation R between two (non-fuzzy) sets $X = \{x\}$ and $Y = \{y\}$ is defined as a fuzzy set in the Cartesian product $X \times Y$, that is,

$$\begin{aligned} R &= \{(\mu_R(x, y), (x, y))\} = \\ &= \{\mu_R(x, y)/(x, y)\}, \qquad \text{for all } (x, y) \in X \times Y \end{aligned} \tag{5}$$

where $\mu_R(x, y) : X \times Y \longrightarrow [0, 1]$ is the membership function of the fuzzy relation R, and $\mu_R(x, y) \in [0, 1]$ is the degree to which the elements $x \in X$ and $y \in Y$ are one and another in this relation R. Obviously, this definition be extended to k-nary fuzzy relations, $k \geq 3$. In this context a fuzzy set is a unary fuzzy relation.

Since fuzzy relations are fuzzy sets, then all definitions and properties of fuzzy sets given so far can be extended to the case of fuzzy relations. Therefore, we will briefly mention below only those of them which are more specific for fuzzy relations and are relevant for our discussion.

The max-min composition of two fuzzy relations R in $X \times Y$ and S in $Y \times Z$, $R \circ_{\max-\min} S$, is defined as the following fuzzy relation in $X \times Z$: $\mu_{R \circ_{\max-\min} S}(x, y) = \max_{y \in Y}[\mu_R(x, y) \wedge \mu_S(y, z)]$, for all $x \in X, z \in Z$.

The max-min composition is the original Zadeh's [97] definition and is still most often used. However, if we assume that the . "min" ($\wedge$) and "max" ($\vee$) are just special cases of the t-norm and s-norm (t-conorm), respectively, then we can easily define a more general s–t composition. We will use in this work the max–min composition which will be denoted as "$\circ$".

The Cartesian product of two fuzzy sets, A in X and B in Y, $A \times B$, is defined as the following fuzzy relation in $X \times Y$: $\mu_{A \times B}(x, y) = [\mu_A(x) \wedge \mu_B(y)]$, for all $x \in X, y \in Y$.

In applications an important role is played by Zadeh's [100] linguistic approach to the analysis of complex systems and decision processes. This approach is basically a foundation of the traditional fuzzy (logic) control which is presumably the most famous example of real applications of fuzzy sets/logic. It has recently evolved towards computing with words (and perceptions) proposed by Zadeh in the mid-1990s (cf [104]).

The point of departure is here a linguistic variable, e.g. "temperature", which takes on its values not as numeric values, e.g. 150°, but as linguistic values exemplified by "high", "low", etc. represented semantically as some fuzzy sets in the universe of discourse of temperature values which is implied by an inherent imprecision (fuzziness) of natural languages (cf. [49]).

For the representations of relationships (dependencies) between linguistic variables fuzzy conditional statements are employed. For instance, if we have two linguistic variables, L and K, such that the value of L is fuzzy set A in X, and the value of K is fuzzy set B in Y, then the relationship between L and K can be described as

$$\text{IF } (L = A) \text{ THEN } (K = B) = \text{IF A THEN } B = A \times B \tag{6}$$

or, in a shorter form, as IF A THEN B, where "$A \times B$" is the Cartesian product of fuzzy sets A and B which is in turn a fuzzy relation in $X \times Y$.

More information on Zadeh's linguistic approach can be found, for instance, in the books by Kacprzyk [49] or by Zadeh and Kacprzyk [104].

For further discussion it will be very important to be in a position to determine the probability of fuzzy events, for instance the probability that there will be a *good weather tomorrow* or the probability of *high inflation in the next year*. In these cases we have therefore a joint occurrence of uncertainty and imprecision.

The first approach is due to Zadeh [98]. Its point of departure is the concept if a fuzzy event which is a fuzzy set A in $X = \{x\} = \{x_1, \ldots, x_n\}$ with a Borel measurable membership function. It is assumed that the probabilities of (non-fuzzy) elementary events $x_1, \ldots, x_n \in X$ are known and equal to $p(x_1)$, ..., $p(x_n) \in [0, 1]$, with $p(x_1) + \cdots + p(x_n) = 1$.

Among more important concepts and properties related to fuzzy events, the following ones can be mentioned: (1) fuzzy events A and B in X are independent if and only if $p(AB) = p(A) \cdot p(B)$, (2) the conditional probability of a fuzzy event A in X given another fuzzy event B in X occurs, $p(A \mid B)$, is defined as $p(A \mid B) = \frac{p(AB)}{p(B)}$, $p(B) > 0$, and if the fuzzy events A and A are independent, then $p(A \mid B) = p(A)$. Notice that these above concepts are analogous to their non-fuzzy counterparts.

The (non-fuzzy) probability of fuzzy event A in $X = \{x_1, \ldots, x_n\}$, $p(A)$, is defined as

$$p(A) = \sum_{i=1}^{n} \mu_A(x_i) p(x_i) \tag{7}$$

that is, as the expected value of the membership function of fuzzy set A, $\mu_A(x)$.

It is easy to notice that: (1) $p(\emptyset) = 0$, (2) $p(\neg A) = 1 - p(A)$, (3) $p(A + B) = p(A) + p(B) - p(A \cap B)$, (4) $p(\sum_{i=1}^r A_i) = \sum_{i=1}^r p(A_i) - \sum_{j=1}^r \sum_{k=1,k<j}^r p(A_j \cap A_k) + \sum_{j=1}^r \sum_{k=1,k<j}^r \sum_{l=1,l<k}^r p(A_j \cap A_k \cap A_l) + \cdots + (-1)^{r+1} p(A_1 \cap A_2 \cap \ldots \cap A_r)$, so that the use of the term "probability" is justified.

Example 1 Suppose that $X = \{1, 2, \ldots, 5\}$, $p(x_1) = 0.1$, $p(x_2) = 0.1$, $p(x_3) = 0.1$, $p(x_4) = 0.3$, $p(x_5) = 0.4$, and $A = 0.1/2 + 0.5/3 + 0.7/4 + 0.9/5$.

Then

$$p(A) = 0.1 \times 0.1 + 0.1 \times 0.5 + 0.3 \times 0.7 + 0.4 \times 0.9 = 0.73$$

•

The conditional probability of a fuzzy event A in X given a fuzzy event B in X occurs is denoted $p(A \mid B)$ and is defined as

$$p(A \mid B) = \frac{p(AB)}{p(B)}, \qquad p(B) > 0 \tag{8}$$

and, if the fuzzy events are independent, then

$$p(A \mid B) = p(A) \tag{9}$$

and here, again, these concepts are equivalent to their non-fuzzy counterparts,

The above classic definition of the (non-fuzzy) probability of a fuzzy event in Zadeh's [98] sense is the most widely used, also in this work. However, notice that in this definition the event is fuzzy but its probability is non-fuzzy, equal to a real number from [0, 1] which may be viewed counterintuitive. Therefore, some other definitions have been proposed in which the probability of a fuzzy event is defined as a fuzzy set in [0, 1].

One of the early definitions, and still used, is the classic Yager's [92] definition. We assume, as before, that the fuzzy event is a fuzzy set A in $X = \{x_1, \ldots, x_n\}$ with a Borel measurable membership function, and the probabilities of elementary events are known and equal to $p(x_1), \ldots, p(x_n) \in [0, 1]$, and $p(x_1) + \cdots + p(x_n) = 1$.

The (fuzzy) probability of a fuzzy event A in $X = \{x_1, \ldots, x_n\}$ is denoted as $P(A)$ and defined as the following fuzzy set in [0, 1]

$$P(A) = \sum_{\alpha \in (0,1]} \alpha / p(A_\alpha) \tag{10}$$

or equivalently, in terms of the membership functions,

$$\mu_{P(A)}[p(A_\alpha)] = \alpha, \qquad \text{for all } \alpha \in (0, 1] \tag{11}$$

where A_α is the α-cut of fuzzy set A.

Example 2 Suppose that: $X = \{1, 2, 3, 4\}$, $p(1) = 0.1$, $p(2) = 0.3$, $p(3) = 0.5$ i $p(4) = 0.1$, and $A = 0.2/1 + 0.5/2 + 0.8/3 + 1/4$. Then, $\alpha \in \{0.3, 0.5, 0.8, 1\}$ and

$$A_1 = \{4\} \; A_{0.8} = \{3, 4\} \; A_{0.5} = \{2, 3, 4\} \; A_{0.3} = \{1, 2, 3, 4\}$$

so that

$$p(A_1) = 0.1 \; p(A_{0.8}) = 0.6 \; p(A_{0.5}) = 0.9 \; p(A_{0.3}) = 1$$

•

Yager's [92] definition is intuitively appealing but does not fulfill all formal requirements for the probability which are satisfied by Zadeh's [98] definition. We will not consider these issues and refer the interested reader to literature, e.g. Kacprzyk's [49] book.

2.2 *Deterministic, Stochastic and Fuzzy Dynamic Systems*

One of the crucial elements in our model of multistage fuzzy decision making (control) is a dynamic system under control which is meant here as the determination of state transitions under controls. It is assumed that this is known for all control stages. In the simplest case this relationship can be deterministic, that is, if we are at a specified control stage at some specified (non-fuzzy) state, apply a specified (non-fuzzy) control then we proceed to a specified next (non-fuzzy) state. Next, this can be a stochastic relationship in the sense that if we are at a specified control stage at a specified (non-fuzzy) state, apply a specified (non-fuzzy) control, then we only know probabilities of attaining next (non-fuzzy) states. Finally, this relationship can be fuzzy in that if we are at a specified control stage at a specified fuzzy state, apply a specified fuzzy (or non-fuzzy) control, then we proceed to a specified fuzzy next state. In this work we will deal with the cases of a deterministic and stochastic dynamic system under control.

2.2.1 Deterministic Dynamic System Under Control

Suppose that the (non-fuzzy) states are equated with outputs, for simplicity, and the state space is $X = \{s_1, \ldots, s_n\}$, and the (non-fuzzy) controls are equated with inputs, for simplicity, and the control space is $U = \{c_1, \ldots, c_m\}$; both the state and control spaces are here assumed to be finite.

The state transitions of the deterministic dynamic system under control are given by the function $f : X \times U \longrightarrow X$ such that

$$x_{t+1} = f(x_t, u_t), \qquad t = 0, 1, \ldots \tag{12}$$

where $x_t, x_{t+1} \in X$ are states at the control stages, t and $t+1$, respectively, and $u_t \in U$ is a control at stage t.

In a more general case one can assume that the state transitions can be given by a function $f_t : X \times U \longrightarrow X, t = 0, 1, \ldots$ such that $x_{t+1} = f_t(x_t, u_t), t = 0, 1, \ldots$, that is, for each particular control stage there is a different state transition function.

2.2.2 Stochastic Dynamic System Under Control

In this case we assume that the state and control spaces are finite and given as, respectively, $X = \{s_1, \ldots, s_n\}$ and $U = \{c_1, \ldots, c_m\}$.

The state transitions of a stochastic dynamic system under control are determined by a function $p : X \times U \times X \longrightarrow [0, 1]$, such that

$$p(x_{t+1} \mid x_t, u_t) \in [0, 1], \qquad t = 0, 1, \ldots \tag{13}$$

is the conditional probability which specified the probability of reaching state $x_{t+1} \in X$ from state $x_t \in X$ under control $u_t \in U$.

Here, too, the stochastic state transition function can be different for each control stage, that is, defined as $p_t : X \times U \times X \longrightarrow [0, 1]$ such that $p_t(x_{t+1} \mid x_t, u_t) \in [0, 1], t = 0, 1, \ldots$ is the conditional probability determining the probability of reaching state $x_{t+1} \in X$ from state $x_t \in X$ under control $u_t \in U$ at a specified control stage t.

2.2.3 Fuzzy Dynamic System Under Control

Also here it is assumed that the state and control spaces are finite and given as, respectively, $X = \{s_1, \ldots, s_n\}$ i $U = \{c_1, \ldots, c_m\}$.

A fuzzy state at control stage t is now specified as a fuzzy set X_t in the state space X with its membership function $\mu_{X_t}(x_t)$. The control at control stage t can either be non-fuzzy, that is $u_t \in U$, or fuzzy, that is given as a fuzzy set U_t defines in the control space U with its membership function $\mu_{U_t}(u_t)$.

The state transitions of the fuzzy dynamic system under control is determined by a function $F : \mathscr{L}(X) \times \mathscr{L}(U) \longrightarrow \mathscr{L}(X)$ such that

$$X_{t+1} = F(X_t, U_t), \qquad t = 0, 1, \ldots \tag{14}$$

where $\mathscr{L}(.)$ is a family of all fuzzy sets defined in, respectively, the state or control space.

Of course, again, the state transitions can be determined differently for the particular control stages, i,e. by the function $F_t : \mathscr{L}(X) \times \mathscr{L}(U) \longrightarrow \mathscr{L}(X)$ such that $X_{t+1} = F_t(X_t, U_t), t = 0, 1, \ldots$.

This completes our brief review of basic properties related to fuzzy sets theory and various dynamic systems under control.

3 Multistage Decision Making and Control Under Fuzziness

We will briefly present now main elements and problem classes of multistage decision making (or control) under fuzziness, to be called multistage fuzzy control which is more appropriate for our context. For solving the particular control problems we will basically employ fuzzy dynamic programming, a straightforward version of the classic dynamic programming reflecting imprecise (fuzzy) information about elements of the problem formulation. We will start with a convenient conceptual framework, which will be the point of departure for all our considerations, namely Bellman and Zadeh's [5] general approach to decision making in a fuzzy environment.

3.1 Decision Making in a Fuzzy Environment—The Classic Bellman and Zadeh's Approach

In this model imprecision (imprecise information) is represented by the introduction of the so called fuzzy environment which is comprised of fuzzy constraints and fuzzy goals, and a fuzzy decision.

Suppose that we have a set of possible (feasible, admissible, ...) options (alternatives, variants, decisions, ...), $X = \{x\}$.

A fuzzy goal is defined as a fuzzy set G in X characterized by its membership function $\mu_G : X \longrightarrow [0, 1]$ such that $\mu_G(x) \in [0, 1]$ specifies a degree of membership of option $x \in X$ in G.

A fuzzy constraints is defined in a similar way as a fuzzy set C in X characterized by its membership function $\mu_C : X \longrightarrow [0, 1]$ such that $\mu_C(x) \in [0, 1]$ specifies a degree of membership of option $x \in X$ in C.

For instance, if $X = R$, where R is the real axis, then the fuzzy goal "x should be *much greater than* 5" and can be defined as the fuzzy set $\mu_G(x)$, the fuzzy constraint 'x should be *more or less* equal 6" – as $\mu_C(x)$ as shown in Fig. 2.

Notice that if we assume that $f : X \longrightarrow R$ is a conventional performance (quality) function which assigns to each $x \in X$ a real number $f(x) \in R$, and $f(x) \leq M < \infty$, for all $x \in X$, $M = \max_{x \in X} f(x)$, then $\mu_G(x)$ can be meant as a normalized performance function f, that is, $\mu_G(x) = \frac{f(x)}{M} = \frac{f(x)}{\max_{x \in X} f(x)}$, for all $x \in X$.

The fuzzy goal can be interpreted in terms of satisfaction levels which is particularly clear in the case of its piecewise linear membership function as shown in Fig. 2 which should be understood as: if the value attained x is at least equal $\overline{x}_G$ (= 8), which is the satisfaction level for x, that is, for $x \geq \overline{x}_G$, then $\mu_G(x) = 1$ which means that we are fully satisfied of the value of x attained. If the value of x attained does not exceed $\underline{x}_G$ (= 5), which is the lowest possible values of x, then $\mu_G(x) = 0$ which means that we are fully dissatisfied of such a value of x, and all this through intermediate levels of satisfaction.

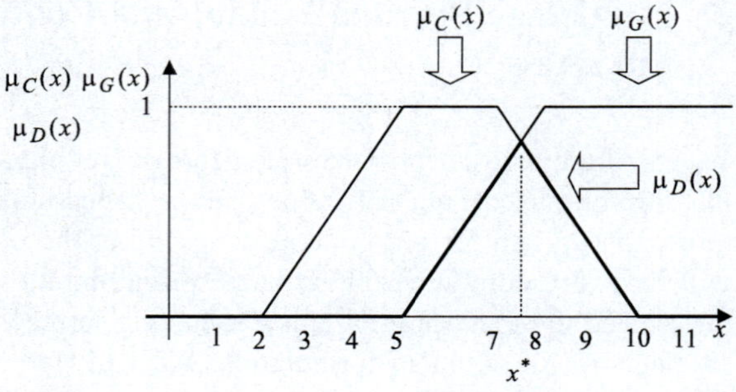

Fig. 2 Fuzzy goal, fuzzy constraint, fuzzy decision and optimal (maximizing) decision

The fuzzy goal G and fuzzy constraint C are, in this approach, treated analogously which implies the following general formulation of the problem of decision making in a fuzzy environment

$$\text{“Attain } G \text{ \underline{and} satisfy } C\text{''} \tag{15}$$

and the above decision, called a fuzzy decision, should be a fuzzy set defined in the set of options which results from an aggregation of the fuzzy goal and fuzzy constraint through the intersection of two fuzzy sets (3) which corresponds to the connective "and", i.e.

$$\mu_D(x) = \mu_G(x) \wedge \mu_C(x) = \min[\mu_G(x), \mu_C(x)], \qquad \text{for all } x \in X \tag{16}$$

where "$\wedge$" is usually, also in this work, the minimum. In such a case this fuzzy decision is called a minimum type fuzzy decision. It will be used throughout this paper.

For instance, suppose that the fuzzy goal G is, "x should be much greater than 5", and the fuzzy constraint C—"x should be more or less equal 6" as in Fig. 2.

The fuzzy decision, shown by a heavy line, should be understood as follows: the set of possible options (for the solution) is the interval [5, 10] because $\mu_D(x) > 0$, for $5 \leq x \leq 10$. Other options are infeasible because for them $\mu_D(x) = 0$. The value of $\mu_D(x) \in [0, 1]$ may be viewed as a degree of satisfaction of the value $x \in X$ as a solution to the problem. Notice that if $\mu_D(x) < 1$, for all $x \in X$, then there is no option which would fully satisfy both the fuzzy goals and constraints.

Obviously, to implement the solution obtained in practice, we should find a non-fuzzy solution. The optimal (maximizing) decision, which is natural and commonly used, is defined as such an $x^* \in X$ that

$$\mu_D(x^*) = \max_{x \in X} \mu_D(x) \tag{17}$$

which is illustrated in Fig. 2 where $x^* = 7.5$. In general, the optimal (maximizing) decision can also be found by a defuzzification of $\mu_D(x)$ (cf. [49]).

In many cases, both in theory and practice, it can be important to be in a position to reflect on the different importance of fuzzy goals and constraints. In this respect there are many possible solutions, for instance, in the popular Yager's (1992b) approach if $w_G, w_C \in [0, 1]$ are importances (importance coefficients) of G i C, respectively, then the following basic forms of the fuzzy decision are possible:

$$\mu_D(x) = [(1 - w_C) \vee \mu_C(x)] \wedge [(1 - w_G) \vee \mu_G(x)] \quad (18)$$

$$\mu_D(x) = [(1 - w_C) + w_c\mu_C(x)] \wedge [(1 - w_G) + w_g\mu_G(x)] \quad (19)$$

$$\mu_D(x) = [\mu_C(x)]^{w_C} \wedge [\mu_G(x)]^{w_G} \quad (20)$$

and, as we will see later, only the simplest form (20) can be of relevance due to the use of optimization, and then the multistage type of problems.

In virtually all nontrivial cases there are many fuzzy goals and many fuzzy constraints, which is clearly implied by the complexity of real problems. If we have, for instance, $n > 1$ fuzzy goals, $G_1, \ldots, G_n$, and $m > 1$ fuzzy constraints, $C_1, \ldots, C_m$, defined in X, then the fuzzy decision can be defined analogously as

$$\mu_D(x) = \mu_{G_1}(x) \wedge \ldots \mu_{G_n}(x) \wedge \wedge \mu_{C_1}(x) \wedge \ldots \wedge \mu_{C_m}(x), \qquad \text{for all } x \in X \quad (21)$$

In practice, however, we need an extension of this approach to the case of fuzzy goals and fuzzy constraints defined in different spaces. Suppose that C is a fuzzy constraint defined as a fuzzy set in $X = \{x\}$, G is a fuzzy goal defined as a fuzzy set in $Y = \{y\}$ and a function $f : X \longrightarrow Y$, $y = f(x)$, is known. For instance, X and Y may be, respectively, the sets of causes and effects, inputs (controls) and outputs (states), etc. Then the induced fuzzy goal G' in X generated by the given fuzzy goal G in Y is defined as $\mu_{G'}(x) = \mu_G[f(x)]$, for all $x \in X$, and the fuzzy decision is defined as

$$\mu_D(x) = \mu_{G'}(x) \wedge \mu_C(x), \qquad \text{for all } x \in X \quad (22)$$

Finally, for $n > 1$ fuzzy goals $G_1, \ldots, G_n$ in Y, $m > 1$ fuzzy constraints $C_1, \ldots, C_m$ in X, and function $f : X \longrightarrow Y$, $y = f(x)$, we obtain

$$\mu_D(x) = \mu_{G'_1}(x) \wedge \cdots \wedge \mu_{G'_n}(x) \wedge \wedge \mu_{C_1}(x) \wedge \cdots \wedge \mu_{C_n}(x), \qquad \text{for all } x \in X \quad (23)$$

and we can also include, similarly as before, different importance degrees assigned to particular fuzzy goals and fuzzy constraints.

In all cases we seek an optimal (maximizing) decision, $\mu_D(x^*) = \max_{x \in X} \mu_D(x)$.

3.2 Multistage Control in a Fuzzy Environment

As we have already mentioned, the multistage decision making in a fuzzy environment (under fuzziness) has a dynamic character and in our context it is to be best discussed as a problem of multistage control in a fuzzy environment the essence of which is shown in Fig. 3.

Suppose that the control (input) space is $U = \{u\} = \{c_1, \ldots, c_m\}$ and the state (output) space is $X = \{x\} = \{s_1, \ldots, s_n\}$. We are in an initial state $x_0 \in X$. We apply a control $u_0 \in U$ on which there is imposed a fuzzy constraint $\mu_{C^0}(u_0)$. We arrive at a state $x_1 \in X$ according to a known state transition equation of the system under control, S. On x_1 a fuzzy goals is imposed $\mu_{G^1}(x_1)$. Next, we apply u_1, on which $\mu_{C^1}(u_1)$ is imposed, and arrive at x_2 on which $\mu_{G^2}(x_2)$ is imposed, etc.

Suppose first that the systems under control is deterministic and its dynamics is described by the state transition equation

$$x_{t+1} = f(x_t, u_t); t = 0, 1, \ldots \tag{24}$$

where $x_t, x_{t+1} \in X = \{s_1, \ldots, s_n\}$ are states at the control stages t i $t + 1$, respectively, and $u_t \in U = \{c_1, \ldots, c_m\}$ is a control at stage t. One can assume, more generally, that $x_{t+1} = f_t(x_t, u_t)$, $t = 0, 1, \ldots$.

At each control stage t, on $u_t \in U$ a fuzzy constraint $\mu_{C^t}(u_t)$ is imposed, and on the state attained $x_{t+1} \in X$ a fuzzy goal $\mu_{G^{t+1}}(x_{t+1})$; $t = 0, 1, \ldots$, is imposed. The termination time (control horizon, planning horizon, etc.) of the control process, that is, the maximum number of control stages, denoted as $N \in \{1, 2, \ldots\}$, and can be finite or infinite.

The performance (quality) criterion is the fuzzy decision:

$$\mu_D(u_0, \ldots, u_{N-1} \mid x_0) = \mu_{C^0}(u_0) \wedge \mu_{G^1}(x_1) \wedge \ldots \wedge \mu_{C^{N-1}}(u_{N-1}) \wedge \mu_{G^N}(x_N) \tag{25}$$

where, through the consecutive application of the state transition equation (24), we consecutively obtain

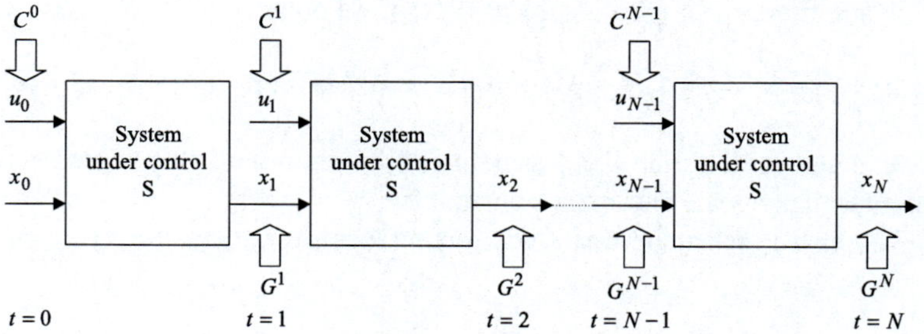

Fig. 3 Multistage control in a fuzzy environment

$$\begin{cases} x_1 = f(x_0, u_0) \\ x_2 = f(x_1, u_1) = f(f(x_0, u_0), u_1) \\ \dots\dots\dots\dots\dots\dots\dots\dots\dots\dots \\ x_N = f(x_{N-1}, u_{N-1}) = f(f(\dots(f(x_0, u_0), u_1), \dots, u_{N-2}), u_{N-1} \end{cases} \tag{26}$$

Quite often a simplified form of the fuzzy decision is applied by assuming that on all consecutively applied fuzzy controls, $u_0, u_1, \dots, u_{N-1}$, fuzzy constraints are imposed, $\mu_{C^0}(u_0)$, $\mu_{C^1}(u_1)$, ..., $\mu_{C^{N-1}}(u_{N-1})$, respectively, but a fuzzy goal is only imposed on the final state attained, x_N, $\mu_{G^N}(x_N)$, so that

$$\mu_D(u_0, \dots, u_{N-1} \mid x_0) = \mu_{C^0}(u_0) \wedge \dots \wedge \mu_{C^{N-1}}(u_{N-1}) \wedge \mu_{G^N}(x_N) \tag{27}$$

The assignment of importance (weights) to the particular control stages usually takes the form of so called *discounting*, the essence of which is that the importance (weight) of what happens earlier is higher that of what happens later. Discounting can be added to the fuzzy decision in a simple way (with the discounting factor $b > 1$):

$$\mu_D(u_0, \dots, u_{N-1} \mid x_0) = b^0[\mu_{C^0}(u_0) \wedge \mu_{G^1}(x_1)] \wedge \dots$$
$$\dots \wedge b^{N-1}[\mu_{C^{N-1}}(u_{N-1}) \wedge \mu_{G^N}(x_N)] = \bigwedge_{t=0}^{N-1} b^t[\mu_{C^t}(u_t) \wedge \mu_{G^{t+1}}(x_{t+1})] \tag{28}$$

The problem of multistage fuzzy control is then stated as the determination of an optimal sequence of controls $u_0^*, \dots, u_{N-1}^*, u_t^* \in U, t = 0, 1, \dots, N-1$, such that:

$$\mu_D(u_0^*, \dots, u_{N-1}^* \mid x_0) = \max_{u_0, \dots, u_{N-1} \in U} \mu_D(u_0, \dots, u_{N-1} \mid x_0) \tag{29}$$

It is often more convenient to express the solution, i.e. the controls, as a *control policy* (or policy) defined, for stage t, as

$$a_t : X \longrightarrow U; t = 0, 1, \dots,$$

such that $u_t = a_t(x_t)$, $t = 0, 1, \dots$, that is, the control to be applied at stage t is expressed as a function of state at t.

However, it is not always possible to assume such a simple definition of the policy function, and then it should be defined as:

$$a_t : X \times U \times X \times \dots \times U \times X \longrightarrow U; t = 0, 1, \dots,$$

such that $u_t = a_t(x_0, u_0, x_1, \dots, u_{t-1}, x_t)$, $t = 0, 1, \dots$, that is, the control at the current stage t depends not only on x_t but also on the entire past trajectory, i.e. the consecutive past states and controls.

In practice the above general definition usually take a simpler form:

$$a_t : [0, 1] \times X \longrightarrow U; t = 0, 1, \ldots,$$

such that $u_t = a_t[w_t(x_0, u_0, x_1, \ldots, u_{t-1}), x_t]$, $t = 0, 1, \ldots$, where $w_t : X \times U \times X \times \cdots \times X \times U \longrightarrow [0, 1]$, $t = 0, 1, \ldots$, is a function subsuming the past trajectory. Therefore, the control at the current stage t depends on the current state and some "summary" of the past (trajectory up to $t - 1$).

We have a very important concept of a stationary policy, $a : X \longrightarrow U$ such that $u_t = a(x_t)$, $t = 0, 1 \ldots$, that is, the current control always depends on the current state in the same way.

Sometimes we need the stationary policy defined as $a : X \times U \times X \times \cdots \times U \times X \longrightarrow U$ such that

$$u_t = a(x_0, u_0, x_1, \ldots, u_{t-1}, x_t); t = 0, 1, \ldots,$$

and we can often assume its simplified form $a : [0, 1] \times X \longrightarrow U$ such that

$$u_t = a[w(x_0, u_0, x_1, \ldots, u_{t-1}), x_t]; t = 0, 1, \ldots,$$

where $w : X \times U \times X \times \cdots \times X \times U \longrightarrow [0, 1]$, $t = 0, 1, \ldots$, is some function subsuming the past trajectory.

The control strategy, or strategy, for short, is defined as a sequence of control policies $A = (a_0, a_1, \ldots, a_{N-1})$, and the stationary strategy is defined as $a_N = a = \underbrace{(a, a, \ldots, a)}_{N}$. Of course, in the case of an infinite termination time the strategies (in fact, the stationary strategies because it makes little sense to consider other ones) are denoted as $a_\infty = (a, a, \ldots)$.

Therefore, in terms of strategies and policies the fuzzy decision takes the form:

$$\mu_D(A \mid x_0) = \mu_{C^0}(a_0(x_0)) \wedge \cdots \wedge \mu_{C^{N-1}}(a_{N-1}(x_{N-1})) \wedge \mu_{G^N}(x_N) \quad (30)$$

and the problem of multistage control in a fuzzy environment is to find an *optimal strategy* $A^* = (a_0^*, \ldots, a_{N-1}^*)$ such that

$$\mu_D(A^* \mid x_0) = \max_A \mu_D(A \mid x_0) \quad (31)$$

with the following ordering "$\succeq$" among the strategies: $A' \succeq A'' \Longleftrightarrow \mu_D(A' \mid x_0) \geq \mu_D(A'' \mid x_0)$, for all $x_0 \in X$. That is, we say that a strategy A' is better than strategy A'' if and only if the above inequality is satisfied for any initial stage. An optimal strategy A^* clearly satisfies the above property for each strategy A, i.e. $A^* \succeq A$, for all A.

It is easy to notice that many extensions of the multistage fuzzy control problem considered above are possible. The most important aspects are obviously:

– the type of the termination time:

1. fixed and specified in advance,
2. implicitly specified by the moment when a set of termination states is reached,
3. fuzzy, i.e. imprecisely specified,
4. infinite (of course, the former types concern the finite termination times),

– the type of the system under control:

1. deterministic,
2. stochastic,
3. fuzzy,

– type of the fuzzy decision (aggregation operation) exemplified by the minimum or product type.

The above aspects determine, to the highest extent, the specifics of the problem considered and its solution which, in our context, will proceed by using fuzzy dynamic programming and its derivations. In the next parts of the paper we will therefore discuss some cases mentioned above. We will concentrate on the deterministic and stochastic systems under control, and on the fixed and specified in advance and infinite termination times, and in all these case the aggregation of partial scores will be by using the minimum operation, i.e. using the minimum type fuzzy decision.

4 Multistage Fuzzy Control with a Fixed and Specified Termination Time

We will now present the basic class of multistage fuzzy control problem with a fixed and specified in advance, finite termination time. We will consider the cases of control of two systems: the deterministic and stochastic. For the analysis of control of the fuzzy system we refer the readers to Kacprzyk's [49, 52] books or Kacprzyk's [53] paper.

4.1 Control of a Deterministic System

The dynamics of the deterministic system under control is given as a state transition equation $x_{t+1} = f(x_t, u_t), t = 0, 1, \ldots$, where: $x_t, x_{t+1} \in X = \{s_1, \ldots, s_n\}$ is a state (output) at stage t i $t + 1$, and $u_t \in U = \{c_1, \ldots, c_m\}$ is a control at stage t. The initial state is $x_0 \in X$, and the termination time (finite!) N is fixed and specified in advance, and given as a positive integer number, the highest number of stages to complete in the control process.

At each t, on control $u_t \in U$ a fuzzy constraint, $\mu_{C^t}(u_t)$, is imposed, and on the final state $x_N \in X$, a fuzzy goal, $\mu_{G^N}(x_N)$, is imposed.

The fuzzy decision is

$$\mu_D(u_0, \ldots, u_{N-1} \mid x_0) = \mu_{C^0}(u_0) \wedge \ldots \wedge \mu_{C^{N-1}}(u_{N-1}) \wedge \mu_{G^N}(x_N) \tag{32}$$

where $x_N = f(x_{N-1}, u_{N-1}) = f[f(x_{N-2}, u_{N-2}), u_{N-1}] = \ldots$.

The problem is to find an optimal sequence of controls, $u_0^*, \ldots, u_{N-1}^*$; $u_t^* \in U$, $t = 0, \ldots, N-1$, such that

$$\mu_D(u_0^*, \ldots, u_{N-1}^* \mid x_0) = \max_{u_0, \ldots, u_{N-1}} \mu_D(u_0, \ldots, u_{N-1} \mid x_0) = \\ = \max_{u_0, \ldots, u_{N-1}} [\mu_{C^0}(u_0) \wedge \ldots \wedge \mu_{C^{N-1}}(u_{N-1}) \wedge \mu_{G^N}(x_N)] \tag{33}$$

This problem can be solved using the following two traditional methods: (fuzzy) dynamic programming and branch and bound, and two new methods: an evolutionary (genetic) algorithm, and a neural network. In this work the application of dynamic programming will be discussed, and the use of other methods will only be mentioned.

The application of dynamic programming for the solution of (33), which results in fuzzy dynamic programming, dates back to the source Bellman and Zadeh's [5] paper.. The problem is to find $u_0^*, \ldots, u_{N-1}^*$ such that

$$\mu_D(u_0^*, \ldots, u_{N-1} \mid x_0) = \max_{u_0, \ldots, u_{N-1}} [\mu_{C^0}(u_0) \wedge \ldots \wedge \mu_{C^{N-1}}(u_{N-1}) \wedge \mu_{G^N}(f(x_{N-1}, u_{N-1}))] \tag{34}$$

It is easy to notice that the structure of this problem, (34), makes it possible to solve it by using dynamic programming. Namely, two right hand side last terms, i.e. $\mu_{C^{N-1}}(u_{N-1}) \wedge \mu_{G^N}(f(x_{N-1}, u_{N-1}))$, depend on control u_{N-1} only and do not depend on other controls.

The maximization in (34) proceeds therefore in two phases: (1) over the sequence of controls $u_0, \ldots, u_{N-2}$, and (2) the maximization over u_{N-1} which can be written as follows:

$$\mu_D(u_0^*, \ldots, u_{N-1}^* \mid x_0) = \max_{u_0, \ldots, u_{N-2}} \{\mu_{C^0}(u_0) \wedge \ldots \wedge \mu_{C^{N-2}}(u_{N-2}) \wedge \\ \wedge \max_{u_{N-1}} [\mu_{C^{N-1}}(u_{N-1}) \wedge \mu_{G^N}(f(x_{N-1}, u_{N-1}))]\} \tag{35}$$

And further, since $\mu_{C^{N-2}}(u_{N-2})$ depends only on u_{N-2}, then (35) can be described as

$$\mu_D(u_0^*, \ldots, u_{N-1}^* \mid x_0) = \\ = \max_{u_0, \ldots, u_{N-3}} (\mu_{C^0}(u_0) \wedge \ldots \wedge \mu_{C^{N-3}}(u_{N-3}) \wedge \max_{u_{N-2}} \{\mu_{C^{N-2}}(u_{N-2}) \wedge \\ \wedge \max_{u_{N-1}} [\mu_{C^{N-1}}(u_{N-1}) \wedge \mu_{G^N}(f(x_{N-1}, u_{N-1}))]\}) \tag{36}$$

This backward iteration, which reflects the very essence of dynamic programming, can be repeated for $u_{N-3}, u_{N-4}, \ldots, u_0$, which leads to the following set of recurrence dynamic programming equations

$$\begin{cases} \mu_{G^{N-i}}(x_{N-i}) = \max_{u_{N-i}}[\mu_{C^{N-i}}(u_{N-i}) \wedge \mu_{G^{N-i+1}}(x_{N-i+1})] \\ x_{N-i+1} = \mathrm{f}(x_{N-i}, u_{N-i}), \qquad i = 0, 1, \ldots, N \end{cases} \tag{37}$$

where $\mu_{G^{N-i}}(x_{N-i})$ can be viewed as a fuzzy goal at $t = N - i$ induced by a fuzzy goal at $t = N - i + 1, i = 0, 1, \ldots, N$.

An optimal sequence of control $u_0^*, \ldots, u_{N-1}^*$ sought is given by the consecutive maximum values of controls $c_{N-i}, i = 1, \ldots, N$ in (37), and each u_{N-i}^* is obtained as a function of state, x_{N-i}, that is, we clearly obtain an optimal policy $\mathrm{a}_{N-i}^* : X \longrightarrow U, i = 1, \ldots, N$, such that $u_{N-i}^* = \mathrm{a}_{n-i}^*(x_{N-i}), i = 1, \ldots, N$. An optimal solution, $u_0^*, \ldots, u_{N-1}^*$, exists if there is at least one sequence of controls, $u_0, \ldots, u_{N-1}$, for which $\mu_D(u_0, \ldots, u_{N-1} \mid x_0) > 0$.

Example 3 Let: $X = \{s_1, s_2, s_3\}$, $U = \{c_1, c_2\}$, $N = 2$, the fuzzy constraints and fuzzy goal are given as, respectively:

$$C^0 = 0.7/c_1 + 1/c_2$$
$$C^1 = 1/c_1 + 0.8/c_2$$
$$G^2 = 0.3/s_1 + 1/s_2 + 0.8/s_3$$

and the dynamics of the deterministic system under control be given as

$$x_{t+1} = \begin{array}{r|ccc} & x_t = s_1 & s_2 & s_3 \\ \hline u_t = c_1 & s_1 & s_3 & s_1 \\ c_2 & s_2 & s_1 & s_3 \end{array} \tag{38}$$

First, from (37), for $i = 1$ we obtain $G^1 = 0.6/s_1 + 0.8/s_2 + 0.6/s_3$, and the optimal policy is

$$a_1^*(s_1) = c_2 \; a_1^*(s_2) = c_1 \; a_1^*(s_3) = c_2$$

Then, the next iteration of (37), for $i = 2$, yields $G^0 = 0.8/s_1 + 0.6/s_2 + 0.6/s_3$ and the optimal policy

$$a_0^*(s_1) = c_2 \; a_0^*(s_2) \in \{c_1, c_2\} \; a_1^*(s_3) \in \{c_1, c_2\}$$

Therefore, if we start at stage $t = 0$ from state $x_0 = s_1$, then we apply control $u_0^* = a_0^*(s_1) = c_2$ and we attain $x_1 = s_2$. Then, st stage $t = 1$, $u_1^* = a_1^*(s_2) = c_1$ and $\mu_D(u_0^*, u_1^* \mid s_1) = \mu_D(c_2, c_1 \mid s_1) = 0.8$. •

It is easy to notice that a similar procedure can be applied to other types of fuzzy decision, notably the min-type fuzzy decision, in which the aggregation of partial scores (satisfaction of a fuzzy constraint and fuzzy goal at a particular control stage is performed using the minimum operation. Moreover, the traditional fuzzy decision with the (weighted) average of partial scores can be employed.

There are some alternative methods for solving problem (37) which try to alleviate the so called curse of dimensionality of dynamic programming which can make its use difficult for larger problems. First, one should mention the solution by using the branch and bound method proposed by Kacprzyk [32, 35], with a different version of the branch and bound procedure proposed by Esogbue et al. [19]. Second, the use of various types of genetic algorithms are proposed in Kacprzyk's ([46–49, 51, 54] papers. Third, the use of some non-traditional, feed forward neural network emulating fuzzy dynamic programming is proposed by Francelin, Gomide and Kacprzyk [22], and Francelin, Kacprzyk and Gomide [21].

Moreover, Kacprzyk and Iwański (1983) and Kacprzyk [57] have proposed a further extension of the multistage fuzzy control model in which a fuzzy linguistic quantifier driven aggregation of partial (stage) scores is used so that the following problem formulation is obtained: find an optimal sequence of control over N; $N < \infty$, which best satisfies the satisfaction of fuzzy constraints on controls applied and fuzzy goals on states obtained at Q (e.g., most, almost all, much more than a half, ...) control stages. This formulation is, however, beyond the scope of this paper.

4.2 Control of a Stochastic System

The stochastic system under control is the Markov chain, dynamics of which, i.e. the state transitions, is determined by the conditional probabilities $p(x_{t+1} \mid x_t, u_t)$, $t = 0, 1, \ldots$. The value of the fuzzy decision $\mu_D(u_0, \ldots, n_{N-1} \mid x_0)$ is now clearly a random variable.

Two basic formulations of the control problem are known:

- Bellman and Zadeh's [5] formulation: find an optimal sequence of controls $u_0^*, \ldots, u_{N-1}^*$ maximizing the probability of attainment of the fuzzy goal subject to the fuzzy constraints, i.e.

$$\mu_D(u_0^*, \ldots, u_{N-1}^* \mid x_0) = \max_{u_0,\ldots,u_{N-1}} [\mu_{C^0}(u_0) \wedge \ldots \wedge \mu_{C^{N-1}}(u_{N-1}) \wedge E\mu_{G^N}(x_N)] \tag{39}$$

- Kacprzyk and Staniewski's [60] approach: find an optimal sequence of controls $u_0^*, \ldots, u_{N-1}^*$ maximizing the expected value of the fuzzy decision, i.e.

$$\begin{aligned}\mu_D(u_0^*, \ldots, u_{N-1}^* \mid x_0) &= \\ &= \max_{u_0,\ldots,u_{N-1}} E\mu_D(u_0, \ldots, u_{N-1} \mid x_0) = \\ &= \max_{u_0,\ldots,u_{N-1}} E[\mu_{C^0}(u_0) \wedge \ldots \wedge \mu_{C^{N-1}}(u_{N-1}) \wedge \mu_{G^N}(x_N)]\end{aligned} \tag{40}$$

In both formulations we look for an optimal strategy $A^* = (a_0^*, \ldots, a_{N-1}^*)$, where $a_t^* : X \longrightarrow U$ such that"re $u_t^* = a_0^*(x_t)$, $t = 0, 1, \ldots, N-1$, is an optimal policy at stage t.

Notice that these two formulations are not equivalent because, in general, $E\mu_D(u_0, \ldots, u_{N-1} \mid x_0) \neq \mu_{C^0}(u_0) \wedge \ldots \wedge \mu_{C^{N-1}}(u_{N-1}) \wedge E\mu_{G^N}(x_N)$.

We assume that in both formulations the probability of a fuzzy event is non-fuzzy and is defined in Zadeh's [98] sense, that is, as a real number in [0, 1].

4.2.1 Bellman and Zadeh's Formulation

We look here for an optimal sequence of controls $u_0^*, \ldots, u_{N-1}^*$ such that

$$\mu_D(u_0^*, \ldots, u_{N-1}^* \mid x_0) = \max_{u_0,\ldots,u_{N-1}} [\mu_{C^0}(u_0) \wedge \ldots \wedge \mu_{C^{N-1}}(u_{N-1}) \wedge E\mu_{G^N}(x_N)] \tag{41}$$

The fuzzy goal is considered to be a fuzzy event in X and the conditional probability of this event, given x_{N-1} and u_{N-1}, is

$$E\mu_{G^N}(x_N) = E\mu_{G^N}(x_N \mid x_{N-1}, u_{N-1}) = \sum_{x_N \in X} p(x_N \mid x_{N-1}, u_{N-1})\, \mu_{G^N}(x_N) \tag{42}$$

The structure of problem (41) is clearly the same as that of problem (33) for the deterministic system under control. We obtain the following set of dynamic programming recurrence equations:

$$\begin{cases} \mu_{G^{N-1}}(x_{N-1}) = \max_{u_{N-1}}[\mu_{C^{N-i}}(u_{N-i}) \wedge E\mu_{G^{N-i+1}}(x_{N-i+1})] \\ E\mu_{G^{N-1+1}}(x_{N-i+1}) = \sum_{x_{N-i} \in X} p(x_{N-i+1} \mid x_{N-i}, u_{N-i})\mu_{G^{N-i+1}}(x_{N-i+1}) \\ i = 1, \ldots, N \end{cases} \tag{43}$$

where $\mu_{G^{N-i}}(x_{N-i})$ can be viewed as a fuzzy goal at stage $t = N - i$ induced by the fuzzy goal at stage $t = N - i + 1$.

The consecutive maximizing values u_{N-i}, u_{N-i}^*, $i = 1, 2, \ldots, N$, determine the optimal sequence of controls sought. We obtain in fact the optimal policies a_{N-i}^* such that $u_{N-i}^* = a_{N-i}^*(x_{N-i})$, $i = 1, \ldots, N$. Notice that the optimal policies are here Markovian, that is, they express the current control only as a function of the current state.

For illustration we consider a simple example [5].

Example 4 Let: $X = \{s_1, s_2, s_3\}$, $U = \{c_1, c_2\}$, $N = 2$, and:

$$C^0 = 0.7/c_1 + 1/c_2$$
$$C^1 = 1/c_1 + 0.6/c_2$$
$$G^2 = 0.3/s_1 + 1/s_2 + 0.8/s_3$$

The dynamics of the stochastic system under control is defined as:

$$p(x_{t+1} \mid x_t, u_t) = \begin{array}{r|ccc} & x_{t+1} = s_1 & s_2 & s_3 \\ \hline u_t = c_1\ x_t = s_1 & 0.8 & 0.1 & 0.1 \\ s_2 & 0 & 0.1 & 0.9 \\ s_3 & 0.8 & 0.1 & 0.1 \end{array}$$

$$\begin{array}{r|ccc} & x_{t+1} = s_1 & s_2 & s_3 \\ \hline u_t = c_2\ x_t = s_1 & 0.1 & 0.9 & 0 \\ s_2 & 0.8 & 0.1 & 0.1 \\ s_3 & 0.1 & 0 & 0.9 \end{array}$$

In the first iteration of (43), for $i = 1$, we first obtain

$$E\mu_{G^2}(x_2) = \begin{array}{r|ccc} & x_1 = s_1 & s_2 & s_3 \\ \hline u_1 = c_1 & 0.42 & 0.82 & 0.42 \\ c_2 & 0.93 & 0.42 & 0.75 \end{array}$$

and then $G^1 = 0.6/s_1 + 0.82/s_2 + 0.6/s_3$, which corresponds to the following optimal policy:

$$a_1^*(s_1) = c_2\ a_1^*(s_2) = c_1\ a_1^*(s_3) = c_3$$

Next, the second iteration of (43), for $i = 2$, yields first

$$E\mu_{G^1}(x_1) = \begin{array}{r|ccc} & x_0 = s_1 & s_2 & s_3 \\ \hline u_0 = c_1 & 0.62 & 0.62 & 0.62 \\ c_2 & 0.8 & 0.62 & 0.6 \end{array}$$

and then $G^0 = 0.8/s_1 + 0.62/s_2 + 0.62/s_3$, which corresponds to the following optimal policy:

$$a_0^*(s_1) = c_1\ a_0^*(s_2) \in \{c_1, c_2\}\ a_0^*(s_3) = c_1$$

•

An extension of this model on the fuzzy probabilities of fuzzy events is proposed in Kacprzyk's [36, 40, 52] works and interested readers are referred thereto.

Finally, notice that Bellman and Zadeh's formulation (41) is not "natural" because much more natural approach would have been to look for a sequence of optimal controls maximizing the expected value of the membership function of the fuzzy decision which would have paralleled the usual approach in the non-fuzzy stochastic control case. Unfortunately, for the case of a performance function (here the fuzzy decision) which is other than the ordinary additive one (possibly with weights) it is impossible to obtain the solution in the form of Markovian optimal policy (cf. [25, 26]). A novel, more "natural" formulation was proposed later by Kacprzyk and Staniewski [60] and will be presented below.

4.2.2 Kacprzyk and Staniewski's Formulation

In Kacprzyk and Staniewski's [60] formulation we look for an optimal sequence of controls $u_0^*, \ldots, u_{N-1}^*$ such that

$$\mu_D(u_0^*, \ldots, u_{N-1}^* \mid x_0) = \max_{u_0, \ldots, u_{N-1}} E[\mu_{C^0}(u_0) \wedge \ldots \wedge \mu_{C^{N-1}}(u_{N-1}) \wedge \mu_{G^N}(x_N)] \tag{44}$$

The probabilities of attaining state x_N from state x_0 through a sequence of control $u_0, \ldots, u_{N-1}$ is obviously equal to $p(x_1 \mid x_0, u_0) \cdot p(x_2 \mid x_1, u_1) \cdot \ldots \cdot p(x_N \mid x_{N-1}, u_{N-1})$, and the expected value of the random variable $\mu_D(u_0, \ldots, u_{N-1} \mid x_0)$ is equal to

$$\begin{aligned} &E\mu_D(u_0, \ldots, u_{N-1} \mid x_0) = \\ &= \sum_{(x_0, \ldots, x_N) \in X^{N+1}} \mu_D(u_0, \ldots, u_{N-1} \mid x_0) \times \\ &\times p(x_1 \mid x_0, u_0) \cdot p(x_2 \mid x_1, u_1) \cdot \ldots \cdot p(x_N \mid x_{N-1}, u_{N-1}) \end{aligned} \tag{45}$$

We introduce a sequence of functions $\hat{h}_i, h_j, i = 0, 1, \ldots, N, j = 1, \ldots, N-1$, such that:

$$\begin{cases} \hat{h}_N(x_N, u_0, \ldots, u_{N-1}) = \mu_D(u_0, \ldots, u_{N-1} \mid x_0) \\ \ldots\ldots\ldots\ldots\ldots\ldots\ldots\ldots\ldots\ldots \\ \hat{h}_k(x_k, u_0, \ldots, u_{k-1}) = \max_{u_k} h_k(x_k, u_0, \ldots, u_k) \\ h_{k-1}(x_{k-1}, u_0, \ldots, u_{k-1}) = \sum_{x_k \in X} \hat{h}_k(x_k, u_0, \ldots, u_{k-1}) \cdot p(x_k \mid x_{k-1}, u_{k-1}) \\ \ldots\ldots\ldots\ldots\ldots\ldots\ldots\ldots\ldots\ldots \\ \hat{h}_0(x_0) = \max_{u_0} h_0(x_0, u_0) \end{cases} \tag{46}$$

that is, it is the expected value $E\mu_D(. \mid x_0)$ assuming the optimal continuation of controls from stage $t = k$ until the end stage ($t = N$), i.e. $u_k^*, \ldots, u_{N-1}^*$. Then, $\hat{h}_k$ denotes the highest expected value of the fuzzy decision which can be attained if we are at the current stage.

It can be shown that since X and U are finite, then there exists such functions $w_k : X \times \underbrace{U \times \cdots \times U}_{k} \longrightarrow U, k = 0, 1, \ldots, N-1,$ that $\hat{h}_k(x_k, u_0, \ldots, u_{k-1})$ $= h_k[x_k, u_0, \ldots, u_{k-1}, w_k(x_k, u_0, \ldots, u_{k-1})]$.

If we introduce now a sequence of functions $g_k^* : \underbrace{X \times \cdots \times X}_{k} \longrightarrow U$, $k = 0, 1, \ldots, N-1$, such that

$$g_k^*(x_0, \ldots, x_k) = w_k[x_k, g_0^*(x_0), \ldots, g_{k-1}^*(x_0, \ldots, x_{k-1})] \tag{47}$$

where $g_0^*(x_0) = w_0(x_0)$, then we have the following important property (Kacprzyk and Staniewski 1980a, [49, 52]): if h_k, $\hat{h}_k$, w_k and g_k^* are given above, then the functions g_k^* are the optimal policies sought, that is,

$$u_k^* = g_k^*(x_0, \ldots, x_k); k = 0, \ldots, N-1 \tag{48}$$

and also $h_0(x_0) = \max_{u_0,\ldots,u_{N-1}} E\mu_D(u_0, \ldots, u_{N-1} \mid x_0)$, where the optimal policy relates the optimal control not only to the current state but to a summary of the past trajectory, too.

The above property implies an algorithm for the determination of the optimal strategy sought (cf. [49, 52]) the essence of which is the consecutive use of the set of equations (46) for controls represented by optimal policies (47). Obviously, the optimal policies obtained are not Markovian!

Now we will solve a simple example, the same as for the Bellman and Zadeh [5] approach.

Example 5 Let: $X = \{s_1, s_2, s_3\}$, $U = \{c_1, c_2\}$, $N = 2$,:

$$\begin{aligned} C^0 &= 0.7/c_1 + 1/c_2 \\ C^1 &= 1/c_1 + 0.6/c_2 \\ G^2 &= 0.3/s_1 + 1/s_2 + 0.8/s_3 \end{aligned}$$

and

$p(x_{t+1} \mid x_t, u_t) =$

	$x_{t+1} = s_1$	s_2	s_3
$u_t = c_1$ $x_t = s_1$	0.8	0.1	0.1
s_2	0	0.1	0.9
s_3	0.8	0.1	0.1

	$x_{t+1} = s_1$	s_2	s_3
$u_t = c_2$ $x_t = s_1$	0.1	0.9	0
s_2	0.8	0.1	0.1
s_3	0.1	0	0.9

We obtain the following optimal policies:

$$\begin{aligned} &a_1^*(1, s_1) = c_2 \quad a_1^*(1, s_2) = c_1 \quad a_1^*(1, s_3) = c_2 \\ &a_1^*(0.7, s_1) = c_2 \; a_1^*(0.7, s_2) = c_1 \; a_1^*(0.7, s_3) = c_2 \end{aligned}$$

$$a_0^*(s_1) = c_2 \; a_0^*(s_3) = c_2 \; a_0^*(s_3) = c_1$$

and, obviously, the results are different than those obtained for the Bellman and Zadeh's [5] formulation. •

Notice that the optimal policies obtained are not Markovian, i.e. they relate optimal control, not only to the current state but also to a summary (history) of the past trajectory which is represented by the value of $v_0(u_0) \in [0, 1]$.

Many years later Iwamoto [28], Iwamoto and Sniedovich [29], and Iwamoto et al. [30] have reconsidered Kacprzyk and Staniewski's [60] formulation and have presented a deeper analysis of the algorithm for the determination of Markovian and non-Markovian policies.

It is easy to notice that the solution of the multistage fuzzy control problem with the stochastic system under control—even in the setting of Bellman and Zadeh [5] but more so in the setting of Kacprzyk and Staniewski [60]—is difficult even in the basic case of a fixed and specified, finite termination time. Therefore, Kacprzyk (1995, [48, 49, 51, 54]) has proposed the application to the solution of the problem of evolutionary algorithms, first a genetic algorithm and then other evolutionary type algorithms, notably bacterial evolutionary algorithms (cf. [23, 74]). This has proved to be an interesting, effective and efficient approach which yields good results. This is, however, beyond the scope of this work and we refer interested readers to the above references cited.

Now we will proceed to the case of multistage fuzzy control with an infinite termination time and will consider the deterministic and stochastic systems under control. Readers interested in the control of a fuzzy system, as well as in the control with an implicitly specified and a fuzzy termination times, are referred to Kacprzyk's [49, 52] books.

5 Multistage Fuzzy Control with an Infinite Termination time

So far we have discussed the case of a fixed and specified, finite termination time, that is, with iterations over $t = 0, 1, \ldots, N$, $N < \infty$. It is clear that the iteration over the subsequent control stages, from 0 to N, makes sense when the number of control stages is not too high, and above all, when the control process exhibits some variability over time. There exist however many control processes which proceed over very long control horizons (termination times), the essence of which is just to maintain the current level of activities. There are many examples in technological, economic, social, etc. systems when, for instance, the problems involve industry branches which cannot expect any development because the demand for their products is saturated, many regions (e.g. rural) for which a stagnation can only be expected due to outmigration of younger population and aging of the society, etc. (cf. [64]).

In such problems a reasonable approach would be to assume an infinite termination time and to look for an optimal stationary strategy, with the iterations over the consecutive control stages to be replaced by iteration schemes with a finite number of steps using a policy iteration scheme (cf., for instance, the classic books by Howard [25, 26]; Bertsekas [6]; Bertsekas and Shreve [7] or Puterman [80].

The problem of multistage fuzzy control with an infinite termination time was first formulated and solved by Kacprzyk and Staniewski (1980b, [37, 62], and then extended in, for instance, Kacprzyk, Safteruk and Staniewski (1981a, b); cf. also

Kacprzyk's [37, 38, 49, 52] books. It has been shown that for the solution a policy iteration type algorithm can be developed.

The problem considered is as follows. At each control stage $t, t = 0, 1 \ldots$, a fuzzy constraint is imposed on control $u_t \in U = \{c_1, \ldots, c_m\}$, $\mu_C(u_t \mid x_t)$, which depends on the state. The fuzzy goal, $\mu_G(x_t)$, is clearly the same for all $t, t = 1, 2, \ldots, \infty$.

The fuzzy decision is

$$\begin{aligned} &\mu_D(u_0, u_1, \ldots \mid x_0) = \\ &= [\mu_C(u_0 \mid x_o) \wedge \mu_G(x_1)] \wedge [\mu_C(u_1 \mid x_1) \wedge \mu_G(x_2)] \wedge \ldots = \\ &= \lim_{N\to\infty} \bigwedge_{t=0}^{N} [\mu_C(u_t \mid x_t) \wedge \mu_G(x_{t+1})] \end{aligned} \tag{49}$$

or, in the case with the discount factor, $b > 1$:

$$\begin{aligned} &\mu_D(u_0, u_1, \ldots \mid x_0, b) = \\ &= [\mu_C(u_0 \mid x_0) \wedge \mu_G(x_1)] \wedge b[\mu_C(u_1 \mid x_1) \wedge \mu_G(x_2)] \wedge \wedge b^2[\mu_C(u_2 \mid x_2) \wedge \mu_G(x_3)] \wedge \ldots = \\ &= \lim_{N\to\infty} \bigwedge_{t=0}^{N} b^t [\mu_C(u_t \mid x_t) \wedge \mu_G(x_{t+1})] \end{aligned} \tag{50}$$

The discount factor can express a higher importance of earlier control stages, i.e. what happens earlier is more important than what happens later. This can diminish the influence of later outcomes, constraints and goals, which are clearly less certain and known with a lesser certainty than earlier ones. Moreover, such a discounting based attitude is usually common in human judgment and behavior. Technically, the use of discounting can have a positive effect in the efficiency of problem solving, and even for the possibility to solve the problem.

The problem considered is now to find an optimal stationary strategy $a^*_\infty = (a^*, a^*, \ldots)$ such that

$$\mu_D(a^*_\infty \mid x_0) = \max_{a_\infty} \mu_D(a_\infty \mid x_0) = \max_{a_\infty} \lim_{N\to\infty} \bigwedge_{t=0}^{N} [\mu_C(a(x_t) \mid x_t) \wedge \mu_G(x_{t+1})] \tag{51}$$

and, in the case of using a discount factor $b > 1$:

$$\mu_D(a^*_\infty \mid x_0, b) = \max_{a_\infty} \mu_D(a_\infty \mid x_0) = \max_{a_\infty} \lim_{N\to\infty} \bigwedge_{t=0}^{N} b^t [\mu_C(a(x_t) \mid x_t) \wedge \mu_G(x_{t+1})] \tag{52}$$

with the ordering "$\succeq$" in the set of strategies given as: $a'_\infty \succeq a''_\infty \iff \mu_D(a_\infty{}' \mid x_0) \geq \mu_D(a_\infty{}'' \mid x_0)$, for all $x_0 \in X$, and, obviously, $a^*_\infty \succeq a_\infty$, for all a_∞.

We will proceed now to the analysis of the cases with the deterministic and stochastic system under control.

5.1 Control of a Deterministic System

The fuzzy decision with the discount factor $b > 1$ is:

$$\mu_D(a_\infty \mid x_0) = \lim_{N\to\infty} \bigwedge_{t=0}^{N} b^t[\mu_C(a(x_t) \mid x_t) \wedge \mu_G(f(x_t, a(x_t)))] \tag{53}$$

and the problem is to find an optimal stationary strategy $a^*_\infty = (a^*, a^*, \ldots)$ such that

$$\mu_D(a^*_\infty \mid x_0) = f \max_{a_\infty} \mu_D(a_\infty \mid x_0) =$$

$$= \lim_{N\to\infty} \bigwedge_{t=0}^{N} b^t[\mu_C(a(x_t) \mid x_t) \wedge \mu_G(f(x_t, a(x_t)))] \tag{54}$$

In this case the use of the discount factor simplifies the analysis and solution of the problem as shown by Kacprzyk and Staniewski [60, 62, 63]; cf. also Kacprzyk's [49, 52] books.

First we have to determine a functional equation which refers the value of the fuzzy decision from a control stage, t, until the end of the control process (i.e. ∞), to the value of the fuzzy decision from the control stage $t + 1$ until the end of the control process; $t = 0, 1 \ldots$.

Thje fuzzy decision (53) can be written as

$$\mu_D(A \mid x_0) = \mu_C(a_0(x_0) \mid x_0) \wedge \mu_G[f(x_0, a_0(x_0))] \wedge b^t \mu_D[TA \mid f(x_0, a_0(x_0))] \tag{55}$$

where $TA = T(a_0, a_1, \ldots) = (a_1, a_2, \ldots)$, i.e. T is a one step forward shifting operator (over the sequence of policies). For the stationary strategy a_∞ we clearly have $Ta_\infty = a_\infty$. Therefore, (55) plays the role of a functional equation.

More generally, we have

$$\mu_D(T^t A \mid x_t) = \mu_C(a_t(x_t) \mid x_t) \wedge \mu_G[f(x_t, a_t(x_t))] \wedge b\mu_D[T^{t+1} A \mid f(x_t, a_t(x_t))] \tag{56}$$

where $T^t A = T^t(a_0, a_1, \ldots) = (a_t, a_{t+1}, \ldots)$; $t = 0, 1, \ldots$.

It is easy to notice that since the sets of states and controls, X i U, are finite, by assumption, then $\mu_D[TA \mid f(x_0, a_0(x_0))] = H(a_0)\, b\, \mu_D(A \mid x_0)$, where $H(a_0)$ is a 0–1 $m \times n$ transition matrix which transforms the values (from a finite set) of $\mu_D(A \mid .)$ into the values of $\mu_D(TA \mid .)$, also from a finite set.

We introduce now an operator $Z : [0, 1]^n \longrightarrow [0, 1]^n$ such that $Z(a_t)w = \mu_C(a_t (x_t) \mid x_t) \wedge \mu_G[f(x_t, a_t(a_t(x_t))]$. Therefore, if we denote $(a, A) = (a, a_1, a_2, \ldots)$, then $\mu_D((a, A) \mid x_0) = Z(a)\mu_D(A \mid x_0)$.

It can be shown that: (1) if $\mu_D(A^* \mid x_0) \geq \mu_D((v, A^*) \mid x_0)$, for each policy $v : X \longrightarrow U$, then strategy A^* is optimal, (2) if $\mu_D((a, A) \mid x_0) > \mu_D(A \mid x_0)$, for each policy a, then $\mu_D(a_\infty \mid x_0) > \mu_D(A \mid x_0)$, that is, if we can improve the strategy, then we can find a better stationary strategy. Moreover., we have the following important theorem which is a basis for an algorithm for the determination of an optimal stationary strategy (cf. [49, 52]).

Theorem 1 *Denote first by $B(i, a)$, for all $i \in \{1, \ldots, n\}$ and for some stationary policy $a : X \longrightarrow U$, the set of all such policies $v : X \longrightarrow U$ that*

$$\mu_D(a_\infty \mid s_i) < \\ < \mu_C(v(s_i) \mid s_i) \wedge \mu_G[f(s_i, v(s_i))] \wedge b\, \mu_D[a_\infty \mid f(s_i, a(s_i))] \tag{57}$$

If:

1. *$B(i, a) = \emptyset$, for all $i \in \{1, \ldots, n\}$, then a_∞ is optimal;*
2. *$B(i, a) \neq \emptyset$, then for all policies $z : X \longrightarrow U$ such that:*
   a. *$z(s_i) \in B(i, a)$, for some $i \in \{1, \ldots, m\}$, and*
   b. *$z(s_i) = a(s_i)$, for all $z(s_i) \notin B(i, a)$,*

holds

$$\mu_D(z_\infty \mid x_0) > \mu_D(a_\infty \mid x_0) \tag{58}$$

Theorem 1 says that any strategy is either optimal, which corresponds to the case $B(i, a) = \emptyset$, or can be improved which corresponds to the case $B(i, a) \neq \emptyset$. Since the number of stationary strategies is finite, which is clearly implied by the finiteness of the sets of states and controls, X and U, then an optimal stationary strategy obviously exists though it need not be unique. The optimal stationary strategy relates the current control to the current state only.

It can easily be shown that the above properties, notably Theorem 1, lead to a policy iteration algorithm for the determination of an optimal stationary strategy (cf. [49, 52]).

Now we will proceed to a more interesting and challenging case of a stochastic system under control.

5.2 *Control of a Stochastic System*

We assume first that the fuzzy decision is now without a discount factor, that is

$$\mu_D(u_0, u_1, \ldots \mid x_0) = \\ = [\mu_C(u_0 \mid x_0) \wedge \mu_G(x_1)] \wedge [\mu_C(u_1 \mid x_1) \wedge \mu_G(x_2)] \wedge \ldots = \\ = \lim_{N \longrightarrow \infty} \bigwedge_{t=0}^{N} [\mu_C(u_t \mid x_t) \wedge \mu_G(x_{t+1})] \tag{59}$$

We look for an optimal stationary strategy $a^*_\infty = (a^*, a^*, \ldots)$ such that

$$E\mu_D(a^*_\infty \mid x_0) = \max_{a_\infty} E\mu_D(a_\infty \mid x_0) \tag{60}$$

with a natural ordering in the set of strategies, "$\succeq$", given as

$$a'_\infty \succeq a''_\infty \Longleftrightarrow E\,\mu_D(a'_\infty \mid x_0) \geq E\,\mu_D(a'' \mid x_0), \qquad \text{dlaka“rdegox}_0 \in \mathrm{X} \tag{61}$$

and, of course, $a^*_\infty \succeq a_\infty$, for all a_∞; $E\mu_D(. \mid .)$ is the expected value of the membership function of the fuzzy decision.

We assume therefore Kacprzyk and Staniewski's (1980a) formulation to the control of a stochastic system. Bellman and Zadeh's [5] approach makes no sense in the case of an infinite termination time.

It easy to notice that the process of solving problem (60) involves two phases:

1. the determination of the value of $E\,\mu_D(a_\infty \mid x_0)$, and
2. the determination of an optimal stationary strategy a^*_∞.

The determination of the expected value of the membership function of the fuzzy decision of the minimum type for an infinite number of control stages is not easy and involves a couple of analytic and algorithmic steps. Due to a lack of space in this work, we cannot discuss this problem in detail and refer the interested reader to Kacprzyk's [49, 52] books.

Moreover, the solution of the problem considered (60), i.e. the determination of an optimal stationary strategy is analytically and algorithmically difficult. We will just present here some main elements of problem analysis and an idea of the algorithm, and for details we will again refer the interested reader to Kacprzyk's [49, 52] books.

First, we introduce the notation $h_t = (x_0, u_0, x_1, u_1, \ldots, x_{t-1}, u_{t-1}, x_t)$, $t = 0, 1, \ldots$, where, by definition, $h_0 = (x_0)$ and $h = (x_0, u_0, x_1, u_1, x_2, \ldots)$. Therefore, h_t represents a trajectory from $t = 0$ to t and h represents the full (total) trajectory, from the beginning to the end of the control process. Moreover, denote by $H_t = \{h_t\}$, for all t, the set of trajectories h_t.

A policy is generally defined as $a_t : H_t \longrightarrow U, t = 0, 1, \ldots$, such that $u_t = a_t(h_t) = a_t(x_0, u_0, x_1, \ldots, x_{t-1}, u_{t-1}, x_t)$, $t = 0, 1, \ldots$, where $u_0 = a_0(h_0) = a_0(x_0)$.

For simplicity, we denote $r_{t-1} = r(x_{t-1}, a_{t-1}(h_{t-1}), x_t) = \mu_C(a_{t-1}(h_{t-1}) \mid x_{t-1}) \wedge \mu_G(x_t)$ and $R(h_t) = r_0 \wedge \ldots \wedge r_{t-1}$ i $R(h) = r_0 \wedge r_1 \wedge \ldots = \lim_{t\longrightarrow\infty} R(h_t)$. By $v^T(h_t, A)$ we denote the expected value of $R(h_t)$, for some $T < \infty$, and for some h_t i $A = (a_0, a_1, \ldots, a_t)$, $t < T$. This is obviously equal to

$$\begin{aligned} v^T(h_t, A) = & \sum_{(x_{t+1}, x_{t+2} \ldots, x_T)} [(R(h_t) \wedge r_t \wedge \ldots \wedge r_{T-1}) \cdot p(x_{t+1} \mid x_t, a(h_t)) \times \cdots \\ & \cdots \times p(x_T \mid x_{T-1}, a_{T-1}(h_{T-1}))] = \\ = & \sum_{(x_{t+1}, x_{t+2}, \ldots, x_T)} [(R(h_t) \wedge \bigwedge_{k=t}^{T-1} r_k) \cdot \prod_{q=t}^{T-1} p(x_{q+1} \mid x_q, a_q(h_q))] \end{aligned} \tag{62}$$

Next, we denote by: $v(h_T, A) = \lim_{T \longrightarrow \infty} v^T(h_T, A)$, $f^T(h_t) = \max_A v^T(h_t, A)$ i $f(h_t) = \lim_{T \longrightarrow \infty} f^T(h_t)$, and the maximization over A proceeds only over the set of policies $\{a_t, a_{t+1}, ..., a_{T-1}\}$, because the policies $a_0, a_1, \ldots, a_{t-1}$ are fixed since h_t is given, by assumption. As, which is a result of using the minimum operation, "$\wedge$", the sequences $(v^T(h_t, A))_{t=0,1,\ldots}$ and $(f^T(h_t))_{t=0,1,\ldots}$ are non-increasing, then the above limits exist.

The strategies A^* are called optimal if $v(h_t, A^*) = f(h_t), t = 0, 1, \ldots$.

The functional equation is

$$v(h_t, A) = \sum_{x_{t+1}} v(x_{t+1}, A) \cdot p(x_{t+1} \mid x_t, a_t(h_t)) \tag{63}$$

for all $t = 0, 1, \ldots, h_t$ and A.

The optimality condition is: for all t and h_t, there holds

$$f(h_t) = \max_{a_t} \sum_{x_{t+1}} f(h_{t+1}) \cdot p(x_{t+1} \mid x_t, a_t(h_t)) \tag{64}$$

Now, $\hat{A} = (\hat{a}_0, \hat{a}_1, \ldots)$ is called a conserving strategy if $f(h_t) = \sum_{x_{t+1}} f(h_{t+1}) \cdot p(x_{t+1} \mid x_t, \hat{a}_t(h_t))$. It can be shown that: (1) there always exists a conserving strategy, and (2) each conserving strategy is optimal, so that there exists an optimal strategy.

The strategy $\breve{A} = (\breve{a}_0, \breve{a}_1, \ldots)$ is called non-improving (in one step!) if

$$v(h_t, \breve{A}) = \max_{A} \sum_{x_{t+1}} v(h_{t+1}, \breve{A}) \cdot p(x_{t+1} \mid x_t, \breve{a}_t(h_t)) \tag{65}$$

so that a non-improving strategy is the one which satisfies the optimality condition (64). It is also easy to notice that each optimal strategy is non-improving but not vice versa.

If the strategy $A = (a_0, a_1, \ldots)$ is not non-improving, then—obviously—a strategy $\overline{A} = (\overline{a}_0, \overline{a}_1, \ldots)$ can be devised such that $v(h_t, A) \leq \sum_{x_{t+1}} v(h_{t+1}, A) \cdot p(x_{t+1} \mid x_t, \overline{a}_t(h_t))$, which is said to improve (in one step!) the strategy A. Therefore, if the strategy $\overline{A}$ improves the strategy A, then $v(h_t, A) \leq v(h_t, \overline{A})$, for all t and h_t, with the strict equality "$<$" for at least one t and h_t.

Therefore, the iteration scheme for the determination of the consecutive strategies $A_0, A_1, \ldots$ such that A_{i+1} improves A_i, for $i = 0, 1, \ldots$, should terminate when a non-improving strategy is found. Such a non-improving strategy will certainly be a very good solution though, strictly speaking, a non-improving strategy need not be optimal.

Since only the stationary strategies are of interest to us, then we will consider first the notion of stationarity. Denote first $w(h_t) = (x_t, R(h_t))$, where $R(h_t)$ is a summary of the trajectory h_t, that is, from the beginning of the control process (i.e. $t = 0$) to stage t. Both x_t and $R(h_t)$ can obviously take on a finite number of values, so that $w(h_t)$ can also take on a finite number of values (i.e. the pairs of values from

$X \times [0, 1]$). Assume that $w(h_t) \in W = \{y_1, \ldots, y_q\}$, where W is the set of different values $w(h_t)$.

Strategy $A = (a_0, a_1, \ldots)$ is called to be stationary, $a_\infty = (a, a, \ldots)$, if $a_t(h_t) = a_{t'}(h_{t'})$, for all h_t and $h_{t'}$ such that $w(h_t) = w(h_{t'})$.

We have now the following important property: for each stationary strategy a_∞, if $w(h_t) = w(h_{t'})$, then $v(h_t, a_\infty) = v(h_{t'}, a_\infty)$ and $f(h_t) = f(h_{t'})$.

Therefore, there exists a stationary conserving strategy, and the (one step) improvement yields better and better strategies and an optimal stationary strategy is to be sought from among the equivalent strategies obtained as a result of strategy improvement, that is, among non-improving strategies.

This result implies a (sub) optimal algorithm of a policy iteration type for the solution of the problem considered. It yields the non-improving strategies in the sense of (65). An optimal strategy is among them. If the number of them is not too high, then via an exhaustive search the optimal strategies can be found. Otherwise, some heuristic search method should rather be employed.

This completes our short presentation of the problem of multistage fuzzy control with the infinite termination time and with a stochastic system. For details we refer the interested readers to Kacprzyk's [49, 52] books [1–3, 12–17, 20, 24, 31, 33, 34, 39, 41–45, 50, 55, 56, 58, 59, 61, 65, 66, 68, 70–73, 75, 76, 78, 79, 81, 85, 87, 89, 93–96, 99, 102, 103, 105].

6 Some Applications of Fuzzy Dynamic Programming

In the previous sections we have discussed various models of fuzzy multistage control. In this part we will briefly mention some examples of practical applications of those models for solving so-called "soft" problems. These examples of applications concern:

- socio-economic regional development (e.g. [54]; Kacprzyk, Francelin and Gomide,[64]),
- flooding control [19],
- R&D planning (e.g. [10]),
- determination of switching schedule for power generators (e.g. [27, 77, 86]),
- control of anaesthesia administration during surgery (e.g. [11]),
- resource allocation (e.g. [1, 18]),
- inventory control (e.g. [60]),
- marketing decision support [90],
- design of a destillation column ([69]),
- control of autonomous vehicles in elastic manufacturing systems (e.g. [53]),
- optimization of spare part inventories in power stations (e.g. [88]), etc.

This completes our short survey of some well known applications of fuzzy dynamic programming. We have mainly concentrated on some better known works.

7 Concluding Remarks

We have presented main models of multistage fuzzy control for the deterministic and stochastic systems under control, and with the fixed and specified (finite) and infinite termination times. For the solution of the resulting problems, we have shown the use of (fuzzy) dynamic programming for the case of the finite, fixed and specified termination time. For the case of an infinite termination time a functional equation is derived which is conceptually similar to the dynamic programming recurrence equations. Finally, we have mentioned some important applications.

References

1. Abo-Sinna, M.A.: Multiple objective (fuzzy) dynamic programming problems: a survey and some applications. Appl. Math. Comput. **157**(3), 861–888 (2004)
2. Baldwin, J.F., Pilsworth, B.W.: Dynamic programming for fuzzy systems with fuzzy environment. J. Math. Anal. Appl. **85**, 1–23 (1982)
3. Belarbi, K.H., Megri, F.: A stable model-based fuzzy predictive control based on fuzzy dynamic programming. IEEE Trans. Fuzzy Syst. **15**(4), 746–754 (2007)
4. Bellman, R.E.: Dynamic Programming. Princeton University Press (1957)
5. Bellman, R.E., Zadeh, L.A.: Decision making in a fuzzy environment. Manag. Sci. **17**, 141–164 (1970)
6. Bertsekas, D.P.: Dynamic Programming and Stochastic Control. Academic Press, New York (1976)
7. Bertsekas, D.P., Shreve, S.E.: Stochastic Optimal Control: The Discrete Time Case. Academic Press, New York (1978)
8. Bertsekas, D.P.: Dynamic Programming: Deterministic and Stochastic Models. Prentice-Hall (1987)
9. Denardo, E.V.: Dynamic Programming: Models and Applications. Dover, Mineola, NY (2003)
10. Esogbue A.O.: Dynamic programming, fuzzy sets and the modeling of R&D management control systems. IEEE Trans. Syst. Man Cybern. SMC-13, 18–30 (1983)
11. Esogbue A.O.: A fuzzy dynamic programming model of intra-operative anesthesia administration. In: Kacprzyk, J., Yager, R.R. (eds.) Management Decision Support Systems Using Fuzzy Sets and Possibility Theory. Verlag TÜV Rheinland, Cologne, str. 155–161 (1985)
12. Esogbue, A.O.: Optimal clustering of fuzzy data via fuzzy dynamic programming. Fuzzy Sets Syst. **18**, 283–298 (1986)
13. Esogbue, A.O.: Dynamic Programming for Optimal Water Resources Systems Analysis. Prentice-Hall, Englewood Cliffs, NJ (1989)
14. Esogbue, A.O., Ahipo, Z.M.: A fuzzy sets model for measuring the effectiveness of public participation in water resources planning. Water Resour. Bull. **18**, 451–456 (1982)
15. Esogbue, A.O., Bellman, R.E.: Fuzzy dynamic programming and its extensions. TIMS/Stud. Manag. Sci. **20**, 147–167 (1984)
16. Esogbue, A.O., Fedrizzi, M., Kacprzyk, J.: Fuzzy dynamic programming with stochastic systems. Kacprzyk, J., Fedrizzi, M. (eds.) Combining Fuzzy Imprecision with Probabilistic Uncertainty in Decision Making. Springer, Berlin/New York, str. 266–285 (1988)
17. Esogbue, A.O., Kacprzyk, J.: Fuzzy dynamic programming: a survey of main developments and applications. Fuzzy Sets Syst. **81**, 31–45 (1996)
18. Esogbue, A.O., Ramesh, V.: Dynamic programming and fuzzy allocation processes. Technical Memo No. 202, Department of Operations Research, Case Western University, Cleveland, OH (1970)

19. Esogbue, A.O., Theologidu, M., Guo, K.: On the application of fuzzy sets theory to the optimal flood control problem arising in water resources systems. Fuzzy Sets Syst. **48**, 155–172 (1992)
20. Filev, D., Angelov, P.: Fuzzy optimal control. Fuzzy Sets Syst. **47**, 151–156 (1992)
21. Francelin, R.A., Kacprzyk, J., Gomide, F.A.C.: Neural network based algorithm for dynamic system optimization. Asian J. Control. **3**(2), 131–142 (2001)
22. Francelin, R.A., Gomide, F.A.C., Kacprzyk, J.: A biologically inspired neural network for dynamic programming. Int. J. Neural Syst. **11**(6), 561–572 (2001)
23. Gál, L., Kóczy, L.T.: Advanced bacterial memetic algorithms. Acta Technica Jauriniensis, Series Intelligentia Combinatorica **1**(3), 481–498 (2008)
24. Gluss, B.: Fuzzy multistage decision-making, fuzzy state and terminal regulators and their relationship to non-fuzzy quadratic state and terminal regulators. Int. J. Control **17**, 177–192 (1972)
25. Howard, R.A.: Dynamic Programming and Markov Processes. MIT Press, Cambridge, MA (1960)
26. Howard, R.A.: Dynamic Probabilistic Systems, vol. 1 and 2. Wiley, New York (1971)
27. Huang, C.J., Lin, C.E., Huang, C.L.: Fuzzy approach for generator maintenance scheduling. Electr. Power Syst. Res. **24**, 31–38 (1992)
28. Iwamoto, S.: Fuzzy dynamic programming in stochastic environment. In: Yoshida, Y. (ed.) Dynamic Aspects in Fuzzy Decision Making, pp. 27–51. Physica-Verlag, Heidelberg and New York (2001)
29. Iwamoto, S., Sniedovich, M.: Sequential decision making in fuzzy environment. J. Math. Anal. Appl. **222**, 208–224 (1998)
30. Iwamoto, S., Tsurusaki, K., Fujita, T.: On Markov policies for minimax decision processes. J. Math. Anal. Appl. **253**, 58–78 (2001)
31. Kacprzyk, J.: Control of a non-fuzzy system in a fuzzy environment with a fuzzy termination time. Syst. Sci. **3**, 320–334 (1977)
32. Kacprzyk, J.: A branch-and-bound algorithm for the multistage control of a non-fuzzy system in a fuzzy environment. Control Cybern. **7**, 51–64 (1978)
33. Kacprzyk, J.: Control of a stochastic system in a fuzzy environment with a fuzzy termination time. Syst. Sci. **4**, 291–300 (1978)
34. Kacprzyk, J.: Decision-making in a fuzzy environment with fuzzy termination time. Fuzzy Sets Syst. **1**, 169–179 (1978)
35. Kacprzyk, J.: A branch-and-bound algorithm for the multistage control of a fuzzy system in a fuzzy environment. Kybernetes **8**, 139–147 (1979)
36. Kacprzyk, J.: Control of a stochastic system in a fuzzy environment with Yager's probability of a fuzzy event. Busefal **12**, 77–88 (1982)
37. Kacprzyk, J.: Wieloetapowe Podejmowanie Decyzji w Warunkach Rozmyto"sci (in Polish). PWN, Warszawa–"L"od"z (1983)
38. Kacprzyk, J.: Multistage Decision Making under Fuzziness. Verlag TÜV Rheinland, Cologne (1983)
39. Kacprzyk, J.: A generalization of fuzzy multistage decision making and control via linguistic quantifiers. Int. J. Control. **38**, 1249–1270 (1983)
40. Kacprzyk, J.: Yager's probability of a fuzzy event in stochastic control under fuzziness. In: Gupta, M.M., Sanchez, M. (eds.) Fuzzy Information, Knowledge Representation and Decision Analysis, pp. 379–384. Pergamon Press, Oxford (1984)
41. Kacprzyk, J.: Towards human-consistent multistage decision making and control models via fuzzy sets and fuzzy logic. Bellman Memorial Issue (A.O. Esogbue, ed.). Fuzzy Sets Syst. **18**, 299–314 (1986)
42. Kacprzyk, J.: Stochastic systems in fuzzy environments: control. In: Singh, M.G. (ed.) Systems and Control Encyclopedia, pp. 4657–4661. Pergamon Press, Oxford (1987)
43. Kacprzyk, J.: Interpolative reasoning in optimal fuzzy control. In: Proceedings of Second IEEE International Conference on Fuzzy Systems—FUZZ–IEEE'93, vol. II, pp. 1259–1263 (1993)

44. Kacprzyk, J.: Interpolative reasoning for computationally efficient optimal fuzzy control. In: Proceedings of Fifth International Fuzzy Systems Association World Congress '93 (Seoul, Korea), vol. II, str. 1270–1273 (1993)
45. Kacprzyk, J.: Fuzzy dynamic programming—battling against the curse of dimensionality via interpolative reasoning. In: Proceedings of Third International Conference on Fuzzy Logic, Neural Nets and Soft Computing—IIZUKA'94, pp. 245–246 (1994)
46. Kacprzyk, J.: Multistage fuzzy control using a genetic algorithm. Proc. Sixth World IFSA Congr. (Sao Paolo, Brazil) **II**, 225–228 (1995)
47. Kacprzyk, J.: A modified genetic algorithm for multistage control of a fuzzy system. In: Proceedings of Third European Congress on Intelligent Techniques and Soft Computing—EUFIT'95 (Aachen, Germany), vol. 1, pp. 463–466 (1995)
48. Kacprzyk, J.: Multistage control under fuzziness using genetic algorithms. Control Cybern. **25**, 1181–1215 (1996)
49. Kacprzyk, J.: A genetic algorithm for the multistage control of a fuzzy system in a fuzzy environment. Mathw. Soft Comput. **I**(3), 219–232 (1997)
50. Kacprzyk, J.: Multistage Fuzzy Control. Wiley, Chichester (1997)
51. Kacprzyk, J.: Multistage control of a stochastic system in a fuzzy environment using a genetic algorithm. Int. J. Intell. Syst. **13**, 1011–1023 (1998)
52. Kacprzyk, J.: Wieloetapowe Sterowanie Rozmyte (in Polish). WNT, Warsaw (2001)
53. Kacprzyk, J.: Fuzzy dynamic programming: interpolative reasoning for an efficient derivation of optimal control policies. Control Cybern. **42**(1), 63–84 (2013)
54. Kacprzyk, J.: Multistage fuzzy control of a stochastic system using a bacterial genetic algorithm. In: Grzegorzewski, P., G"agolewski, M., Hryniewicz, O., Gil, M.A. (eds.) Strengthening Links Between Data Analysis and Soft Computing, pp. 273–284. Springer (2015)
55. Kacprzyk, J., Esogbue, A.O.: Fuzzy dynamic programming: main developments and applications. Fuzzy Sets Syst. **81**, 31–45 (1996)
56. Kacprzyk, J., Iwański, C.: A generalization of discounted multistage decision making and control through fuzzy linguistic quantifiers: an attempt to introduce commonsense knowledge. Int. J. Control **45**, 1909–1930 (1987)
57. Kacprzyk, J., Orlovski, S.A. (eds.): Optimization Models Using Fuzzy Sets and Possibility Theory. Reidel, Dordrecht (1987)
58. Kacprzyk, J., Orlovski, S.A.: Fuzzy optimization and mathematical programming: a brief introduction and survey. In: Kacprzyk, J., Orlovski, S.A. (pod red.) Optimization Models Using Fuzzy Sets and Possibility Theory. Reidel, Dordrecht, str. 50–72 (1987)
59. Kacprzyk, J., Romero, R.A., Gomide, F.A.C.: Involving objective and subjective aspects in multistage decision making and control under fuzziness: dynamic programming and neural networks. Int. J. Intell. Syst. **14**(1), 79–104 (1999)
60. Kacprzyk, J., Staniewski, P.: A new approach to the control of stochastic systems in a fuzzy environment, pp. 433–443. Archiwum Automatyki i Telemechaniki, XXV (1980)
61. Kacprzyk, J., Safteruk, K., Staniewski, P.: On the control of stochastic systems in a fuzzy environment over infinite horizon. Syst. Sci. **7**, 121–131 (1981)
62. Kacprzyk, J., Staniewski, P.: Long-term inventory policy-making through fuzzy decision-making models. Fuzzy Sets Syst. **8**, 117–132 (1982)
63. Kacprzyk, J., Staniewski, P.: Control of a deterministic system in a fuzzy environment over an infinite planning horizon. Fuzzy Sets Syst. **10**, 291–298 (1983)
64. Kacprzyk, J., Straszak, A.: Determination of stable trajectories for integrated regional development using fuzzy decision models. IEEE Trans. Syst. Man Cybern. SMC-14, 310–313 (1984)
65. Kacprzyk, J., Yager, R.R.: Linguistic quantifiers and belief qualification in fuzzy multicriteria and multistage decision making. Control Cybern. **13**, 155–173 (1984)
66. Kacprzyk, J., Yager, R.R.: "Softer" optimization and control models via fuzzy linguistic quantifiers. Inf. Sci. **34**, 157–178 (1984)
67. Kacprzyk, J., Yager, R.R., Merigó, J.M.: Towards human-centric aggregation via ordered weighted aggregation operators and linguistic data summaries: a new perspective on Zadeh's inspirations. IEEE Comput. Intell. Mag. **14**(1), 16–30 (2019)

68. Komolov, S.V., Makeev, S.P., G.P. Serov, Shakhnov, I.F.: On the problem of optimal control of a finite automaton with fuzzy constraints and fuzzy goal (in Russian). Kybernetika (Kiev) **6**, 30–34 (1979)
69. Kras"lawski A., G"orak, A., Vogelpohl, A.: Fuzzy dynamic programming in the synthesis of distillation column systems. Comput. Chem. Eng. **13**, 611–618 (1989)
70. Li, L., Lai, K.K.: Fuzzy dynamic programming approach to hybrid multiobjective multistage decision-making problems. Fuzzy Sets Syst. **117**(1), 13–25 (2001)
71. Mamdani, E.H., Assilian, S.: An experiment in linguistic synthesis with a fuzzy logic controller. Int. J. Man-Mach. Stud. **7**, 1–13 (1975)
72. Morin, T.L., Esogbue, A.O.: Some efficient dynamic programming algorithms for the optimal sequencing and scheduling in water supply projects. Water Resour. **7**, 479–484 (1971)
73. Narshima, S.V., Tiwari, R.N., Sastry, K.S.: Dynamic programming approach to multiple objective control problems having deterministic or fuzzy goals. Fuzzy Sets Syst. **57**, 195–202 (1993)
74. Nawa, N.E., Furuhashi, T., Hashiyama, T., Uchikawa, Y.: A Study on the discovery of relevant fuzzy rules using pseudo-bacterial genetic algorithm. IEEE Trans. Ind. Electron. **46**(6), 1080–1089 (1999)
75. Nawa, N.E., Furuhashi, T.: Fuzzy system parameters discovery by bacterial evolutionary algorithm. IEEE Trans. Fuzzy Syst. **7**(5), 608–616 (1999)
76. Nguyen, D.P., Phan, V.T.: The fuzzy dynamic programming problems. J. Intell. Fuzzy Syst. **30**(3), 1663–1674 (2016)
77. Patra, S., Goswami, S.K., Goswami, B.: Fuzzy based fast dynamic programming solution of unit commitment with ramp constraints. Expert Syst. **26**(4), 307–319 (2009)
78. Pedrycz, W.: Fuzzy Control and Fuzzy Systems. Research Studies, 2nd edn. Press/Wiley, Taunton/New York (1993)
79. Pedrycz, W. (ed.): Fuzzy Modelling: Paradigms and Practice. Kluwer, Boston (1996)
80. Puterman, M.L.: Markov Decision Processes. Discrete Stochastic Dynamic Programming, Wiley, New York (1994)
81. Sirbiladze, G.: Fuzzy dynamic programming problem for extremal fuzzy dynamic dystem. In: Lodwick, W.E., Kacprzyk, J. (eds.) Fuzzy Optimization—Recent Advances and Applications, pp. 231–270, Springer, Heidelberg (2010)
82. Sniedovich, M.: Dynamic Programming. Marcel Dekker, New York (1992)
83. Sniedovich, M.: Dynamic Programming: Foundations and Principles. Taylor and Francis (2010)
84. Steele, K, Stefansson, H.O.: Decision theory. In: Stanford Encyclopedia of Philosophy (2015). ISSN 1095-5054 (https://plato.stanford.edu/entries/decision-theory/)
85. Stein, W.E.: Optimal stopping in a fuzzy environment. Fuzzy Sets Syst. **3**, 253–259 (1980)
86. Su, C.C, Hsu, Y.Y.: Fuzzy dynamic programming: An application to unit commitment. IEEE Trans. Power Syst. PS-6, 1231–1237 (1991)
87. Sugeno, M., Yasukawa, T.: A fuzzy-logic-based approach to qualitative modeling. IEEE Trans. Fuzzy Syst. **1**, 7–31 (1993)
88. Sugianto, L.F., Mielczarski, W.: Dynamic programming application to optimize spare parts inventory. Control Cyberne. **26**(1), 111–130 (1997)
89. Tang, Y., He, W., Ni, Z., Zhong, X., Xu, X.: Fuzzy-based goal representation adaptive dynamic programming. IEEE Trans. Fuzzy Syst. **24**(5), 1159–1175 (2016)
90. Weber, K., Sun, Z.: Fuzzy stochastic dynamic programming for marketing decision support. Int. J. Intell. Syst. **15**(8), 763–783 (2000)
91. Wygralak, M.: Intelligent Counting Under Information Imprecision: Applications to Intelligent Systems and Decision Support. Springer, Heidelberg (2013)
92. Yager, R.R.: A note on probabilities of fuzzy events. Inf. Sci. **18**, 113–129 (1979)
93. Yager, R.R., Filev, D.P.: Foundations of Fuzzy Control. Wiley, New York (1992)
94. Yager, R.R., Kacprzyk, J. (eds.): The Ordered Weighted Averaging Operators: Theory. Methodology and Applications, Kluwer, Boston, MA (1997)
95. Yoshida, Y.: Markov chains with a transition possibility measure and fuzzy dynamic programming. Fuzzy Sets Syst. **66**, 39–57 (1994)

96. Yoshida, Y. (ed.): Dynamical Aspects in Fuzzy Decision Making. Physica-Verlag, Heidelberg and New York (2001)
97. Zadeh, L.A.: Fuzzy sets. Inf. Control **8**, 338–353 (1965)
98. Zadeh, L.A.: Probability measures of fuzzy events. J. Math. Anal. Appl. **23**, 421–427 (1968)
99. Zadeh, L.A.: A rationale for fuzzy control. Meas. Control **34**, 3–4 (1972)
100. Zadeh, L.A.: Outline of a new approach to the analysis of complex systems and decision processes. IEEE Trans. Syst. Man Cybern. SMC-2: 28–44 (1973)
101. Zadeh, L.A.: A computational approach to fuzzy quantifiers in natural languages. Comput. Math. Appl. **9**, 149–184 (1983)
102. Zadeh, L.A.: Fuzzy probabilities. Inf. Process. Manag. **20**, 363–372 (1986)
103. Zadeh, L.A., Kacprzyk, J. (eds.): Fuzzy Logic for the Management of Uncertainty. Wiley, New York (1992)
104. Zadeh, L.A., Kacprzyk, J. (eds.): Computing with Words in Information/Intelligent Systems. 1. Foundations, 2. Applications. Physica-Verlag, Heidelberg and New York (1999)
105. Zimmermann, H.-J.: Fuzzy Set Theory and its Applications, 2nd edn. Kluwer, Boston (1991)

Robotics

Endogenous Configuration Space Approach in Robotics Research

Krzysztof Tchoń

Abstract The Endogenous Configuration Space Approach (ECSA) was invented as a robotics research methodology providing a unified conceptual framework for dealing with robotic manipulators (holonomic robotic systems) and mobile robots (non-holonomic robotic systems). Conceptually, this approach has been derived from control theory. Taking as a point of departure a model of kinematics or dynamics of a mobile robot represented in the form of a non-linear control system with output function, the ECSA builds on an analogy between the kinematics map of a robotic manipulator and the input-output (end-point) map transforming control functions of the control system into the task space. Within this perspective, the derivative of the input-output map with respect to control defines the mobile robot's Jacobian. Consequently, all Jacobian-oriented concepts and instruments are introduced, like regular and singular configurations, dexterity measures and, last but not least, Jacobian motion planning algorithms. Numerical computations within the ECSA are enabled by either parametric or non-parametric implementations of the Jacobian motion planning algorithms.

1 Introduction

Endogenous Configuration Space Approach (hereafter referred to as ECSA) is a methodology for robotics research that offers a uniform theory of both robotic manipulators (holonomic) as well as mobile (non-holonomic) robots. The ECSA relies on concepts of control theory. Specifically, a control system is used as a basic representation of the kinematics or the dynamics of a mobile robot. A cornerstone of the ECSA is an analogy between the kinematics map of a holonomic robot and the input-output (or the end-point) map of the control system representation of the

K. Tchoń (✉)
Department of Cybernetics and Robotics, Wrocław University of Science and Technology, Wybrzeże Wyspiańskiego street 27, 50-370 Wrocław, Poland
e-mail: krzysztof.tchon@pwr.edu.pl

P. Kulczycki et al. (eds.), *Automatic Control, Robotics, and Information Processing*, Studies in Systems, Decision and Control 296,
https://doi.org/10.1007/978-3-030-48587-0_14

non-holonomic robot. This input-output map transforms control functions into the task space. Taking the derivative of the input-output map with respect to the control function one can define the robot's Jacobian, and adapt to non-holonomic robots all Jacobian-originated concepts and tools like regular and singular configurations, dexterity measures, and Jacobian motion planning algorithms. A feature of ECSA is that the Jacobian motion planning algorithms are obtained by means of the method of homotopy (continuation).

Essential ideas of the ECSA, even before the name ECSA had been given, were formulated almost 20 years ago in [1], and then advanced in a number of publications whose list precedes the list of references placed at the end of this chapter. In order to accentuate the analogy between the manipulator and the mobile robot the ideas of the ECSA have originally been exposed in the context of mobile manipulators being robotic devices composed of a mobile platform and a holonomic manipulator, consult [2–7, 10–15, 18, 19, 22, 25, 39] and [42] or a non-holonomic on-board manipulator: serial [9] and parallel [33]. In the works dealing with mobile manipulators we have assumed that the configuration of a mobile manipulator consists of the control function of the platform and the position of the joints of the on-board manipulator. The terms "endogenous configuration" and "endogenous configuration space" were coined later, see [5], as a sort of terminological compromise alleviating possible objections against the identification of a control function with a sort of configuration. The adjective *endogenous* has been borrowed from the Greek language ($\varepsilon\nu\delta o\nu = \varepsilon\nu\ \delta o\mu o\varsigma$i) and refers to something of "intrinsic origin"; it has been used to stress that the control function comprised in the configuration has an intrinsic character that does not manifest directly, but, e.g. by means of the position and orientation of the mobile robot.

Non-holonomic robotic systems subject to the Pfaffian constraints have the endogenous configuration space composed solely of the control functions; such systems have been dealt with in [20, 24, 26–32, 35, 36, 40, 41, 43–46] as well as in [47, 48, 50]. Primarily, the ECSA serves as a source of motion planning algorithms, however its first applications for robot trajectory tracking have already been described in [38, 53]. The task of motion planning by the ECSA of underactuated systems has been addressed in [23, 27].

Mathematical foundations of the ECSA can be tracked to the homotopy or continuation method introduced into mobile robotics as a tool for Jacobian motion planning by Sussmann in [91], and developed further in [55, 60–62, 67, 90]. The same homotopy method regarded as a method of solving non-linear equations was originated by Poincaré and Klein [54], whereas the basic formula defining the Jacobian motion planning is called the Ważewski-Davidenko equation [65, 94]. The input-output map of a control system without outputs is referred to as the end-point map; its differentiability that guarantees the existence of the mobile robot Jacobian has been shown in [89]. Formal existence conditions of a global Jacobian inverse have been studied in [21, 42]. The analysis of singular configurations of a mobile robot relies on the concept of singular control, well established in the theory of optimal control [58, 63, 75, 80]. Detection of singular configurations of a mobile robot that makes use of the conditions for local controllability of its control system representation has been considered in [19, 85].

In this chapter we focus on the theoretical aspects of ECSA. More specifically, we assume that the motion of a mobile robot is subordinated to affine Pfaffian constraints, and obtain the mathematical model of the mobile robot in the form of a control-affine system, possibly with output function. The control functions entering this system are referred to as the endogenous configurations of the mobile robot. The space of admissible control functions defines the endogenous configuration space. With reference to this space we introduce the input-output map and its Jacobian. Properties of the Jacobian allow us to distinguish regular and singular configurations of the mobile robot. Further on we expose a connection between the singular configurations and the singular controls as well as foster the idea of using normal forms of control-affine systems under feedback as a description of singular configurations of mobile robots. Moreover, we state the motion planning problem and define several classes of Jacobian motion planning algorithms. Within the class of algorithms based on inversion of the Jacobian, special attention is paid to the Lagrangian Jacobian algorithm [35, 43, 45, 47], the dynamically consistent algorithm [34, 39–41] and the extended Jacobian algorithm [10, 11, 13, 14, 34]. As an alternative to the Jacobian inverse based algorithms we present the Jacobian transpose algorithm [2, 5, 42]. We also describe the idea of a multiple task planning algorithm, with different task priorities [28, 31, 44, 46]. An analog of the manipulability matrix of a manipulator is the mobility matrix of the mobile robot. Its eigenvalues enable one to define performance measures of the mobile robot, corresponding to the manipulability and the condition number of a manipulator [7, 12, 15]. Another fundamental property of a manipulator whose analog for mobile robots is described in this chapter is the property of repeatability [3]. Finally, for the reason that the equation defining Jacobian motion planning algorithms for mobile robots operates in an infinite-dimensional space of control functions, we discuss some computational methods of finding out its solution [5, 27, 31, 37, 43].

In the following text we purposely concentrate on theoretical-conceptual issues; applications of ECSA to the motion planning of concrete mobile robots, including a ball-shape robot, the trident snake, a balancing robot, and a space manipulator can be found in [5, 17, 27–30, 32, 36, 46, 50–53].

ECSA relies on very specific methods of control theory. A kind of counterpoint for them is constituted by the Lie algebraic methods of robot motion planning, developed in Poland by the scientific school of I. Duleba, [68]. A reader interested in more classical approaches to the control of mobile robots is directed to monographs [81, 93]. Achievements in this subject can be found in the publications of Galicki [71, 72] and Mazur [82]. The computational scheme of the majority of Jacobian motion planning algorithms exploited in ECSA relies on the classic Newton algorithm [66, 70]. Furthermore, these algorithms can be embedded in the context of predictive control [16, 76], learning iterative control [22], as well as sequential quadratic programming [45].

A composition of this chapter is the following. In Sect. 2 we introduce basic concepts of the ECSA, such as the affine Pffafian constraints, the control-affine system as a mathematical model of a mobile robot, the endogenous configuration space, the input-output map, the Jacobian, regular and singular configurations, singular control, and the feedback equivalence of control systems, concluding with a statement of the motion planning problem for the mobile robot. Sect. 3 is devoted to Jacobian motion planning algorithms of non-holonomic robots, with focus on the algorithms involving the inverse Jacobian, the dynamically consistent Jacobian, the Jacobian transpose, and the extended Jacobian. That subsection is closed with a description of a multiple tasks Jacobian motion planning algorithm with priorities. Referring to the Jacobian and to the mobility matrix, in Sect. 4 we define performance indices of a mobile robot understood as certain functions of eigenvalues of the mobility matrix. We also present a re-construction of the concept of repeatability for mobile robots. The next Sect. 5, is concerned with computational methods for solving the motion planning problems. A parametric and non-parametric approaches have been described, revealing advantages of the latter. Section 6, contains final remarks. Subsection References has been preceded by a list of selected publications related to ECSA.

2 Basic Concepts

2.1 Pfaffian Constraints

We shall study mobile robots described by generalized coordinates $q \in \mathbb{R}^n$ and corresponding generalized velocities $\dot{q} \in \mathbb{R}^n$, subject to $l < n$ phase constraints in the affine Pfaffian form

$$\mathscr{A}(q)\dot{q} = b(q). \tag{1}$$

It will be assumed that the Pfaffian matrix $\mathscr{A}(q)$ refers to independent constraints, so it is a matrix of dimension $l \times n$ and of full rank l, whereas $b(q) \in \mathbb{R}^l$ is a vector; both $\mathscr{A}(q)$ as well as $b(q)$ depend smoothly (C^∞) on the coordinates. It follows that the motion of a system obeying the affine constraints can be represented by a control-affine system equipped with an output function

$$\begin{cases} \dot{q} = f(q) + G(q)u = f(q) + \sum_{i=1}^{m} g_i(q)u_i \\ y = h(q) = (h_1(q), \ldots, h_r(q))^\top, \end{cases} \tag{2}$$

with a state variable $q = (q_1, q_2, \ldots, q_n)^\top \in \mathbb{R}^n$, control variable $u = (u_1, u_2, \ldots, u_m)^\top \in \mathbb{R}^m$, $m = n - l$, and output variable $y = (y_1, y_2, \ldots, y_r)^\top \in \mathbb{R}^r$, $r \leq n$. Control system (2) is defined in the following way. Suppose that $\mathscr{A}^{\#}(q)$ denotes a right inverse of the Pfaffian matrix $\mathscr{A}(q)$, i.e. there holds $\mathscr{A}(q)\mathscr{A}^{\#}(q) = \mathbb{I}_l$. This being so, the drift vector field will be taken as $f(q) = \mathscr{A}^{\#}(q)b(q)$. Control matrix $G(q)$ has a dimension $n \times m$, and its columns $g_i(q)$, $i = 1, 2, \ldots, m$, span the null space of $\mathscr{A}(q)$, so that $\mathscr{A}(q)g_i(q) = 0$. By definition, the drift vector field and the control vector fields depend on q in a smooth way. Additionally, we shall assume that all the components of the output function are smooth and independent, the last means that the rank of the Jacobian matrix, rank $\frac{\partial h(q)}{\partial q} = r$.

It may be noticed that representation (2) has not been defined uniquely. Namely, if $f(q)$ fulfills condition $\mathscr{A}(q)f(q) = b(q)$ then any other $\tilde{f}(q) = f(q) + G(q)\alpha(q)$, where $\alpha(q) \in \mathbb{R}^m$, will also fulfill this condition. Similarly, for control vector fields if $\mathscr{A}(q)G(q) = 0$ then matrix $\tilde{G}(q) = G(q)\beta(q)$, for a non-singular $m \times m$ matrix $\beta(q)$, also satisfies condition $\mathscr{A}(q)\tilde{G}(q) = 0$. For the reason that vector function $\alpha(q)$ and matrix function $\beta(q)$ define the feedback, it is said that control-affine system (2) is defined "modulo" a feedback transformation.

Let in control system (2) $T > 0$ denote a control time horizon. We shall choose the control functions $u(\cdot)$ in the system as square Lebesgue integrable functions of time defined on interval $[0, T]$ and taking values in $\mathbb{R}^m$. Control functions with inner product $\langle u(\cdot), \tilde{u}(\cdot)\rangle = \int_0^T u^\top(t)\tilde{u}(t)dt$ form a Hilbert space $\mathscr{U} = L^2_m[0, T]$, that will be referred to as the endogenous configuration space of the mobile robot. Consequently, a control function $u(\cdot) \in \mathscr{U}$ will be called the endogenous configuration of the robot. For a given initial state q_0 the state trajectory of system (2) acted on by the control function $u(\cdot)$ is denoted by $q(t) = \varphi_{q_0,t}(u(\cdot))$, while the corresponding output trajectory $y(t) = h(q(t))$. It is assumed that trajectory $q(t)$ (so also $y(t)$) exists for every time instant $t \in [0, T]$.

2.2 *Provenance of the Affine Representation*

Control-affine system (2) emerges in robotics in three circumstances. Firstly, the best known is the case of modeling the kinematics of wheeled mobile robots moving without the slip of the wheels. Conditions excluding the slip are given the form of linear Pfaffian constraints, with vector $b(q) = 0$, and further, converted to a driftless control system

$$\dot{q} = G(q)u, \quad y = h(q). \tag{3}$$

As an example of a mobile robot subordinated to linear Pfaffian constraints there may serve the trident snake robot introduced in [73], shown in Fig. 1, and analyzed by ECSA in [29, 32]. Secondly, let us consider the dynamics of a robotic system characterized by positions and velocities $q, \dot{q}$, whose Lagrangian does not depend on a certain subvector q^1 of the position coordinates, i.e. takes the form

$$L(q, \dot{q}) = \frac{1}{2}\begin{pmatrix}\dot{q}^1\\ \dot{q}^2\end{pmatrix}^\top \begin{bmatrix} Q_{11}(q^2) & Q_{12}(q^2) \\ Q_{12}^\top(q^2) & Q_{22}(q^2 \end{bmatrix}\begin{pmatrix}\dot{q}^1\\ \dot{q}^2\end{pmatrix} - V(q^2). \tag{4}$$

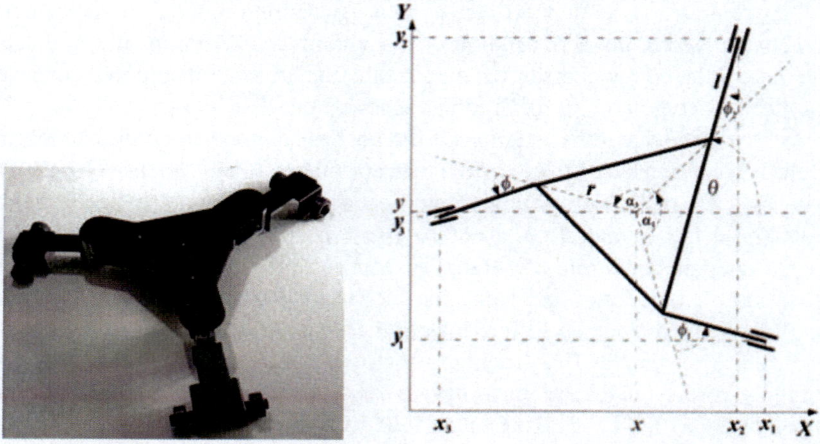

Fig. 1 Trident snake

A system of this kind is sometimes called the Caplygin system. The Euler-Lagrange equations of this system obey the momentum conservation principle

$$\frac{d}{dt}\frac{\partial L}{\partial \dot{q}^1} = 0, \tag{5}$$

which means that

$$Q_{11}(q^2)\dot{q}^1 + Q_{12}(q^2)\dot{q}^2 = \text{const}, \tag{6}$$

resulting in the affine Pffafian constraints and the control-affine representation of the robot's dynamics. It can be demonstrated that this type of behavior is characteristic of free-floating space robots. An example of such a robot is the space manipulator designed in the Space Research Center (SRC) of the Polish Academy of Science, see Fig. 2, described in [87], and used for a performance study of motion planning algorithms in [47]. The third source of representation (2) is the motion of robotic systems whose kinematics obey the linear Pfaffian constraints, whereas the dynamics result from the application of d'Alembert's Principle. As we have already mentioned, the kinematic constraints transform into a driftless control system

$$\dot{q} = G(q)\eta, \quad \eta \in \mathbb{R}^m. \tag{7}$$

Suppose that $L(q, \dot{q}) = \frac{1}{2}\dot{q}^\top Q(q)\dot{q} - V(q)$ stands for the Lagrangian of the system. The standard form of Euler-Lagrange equations is

$$Q(q)\ddot{q} + \mathsf{C}(q, \dot{q})\dot{q} + D(q) = F + \mathsf{B}(q)u, \tag{8}$$

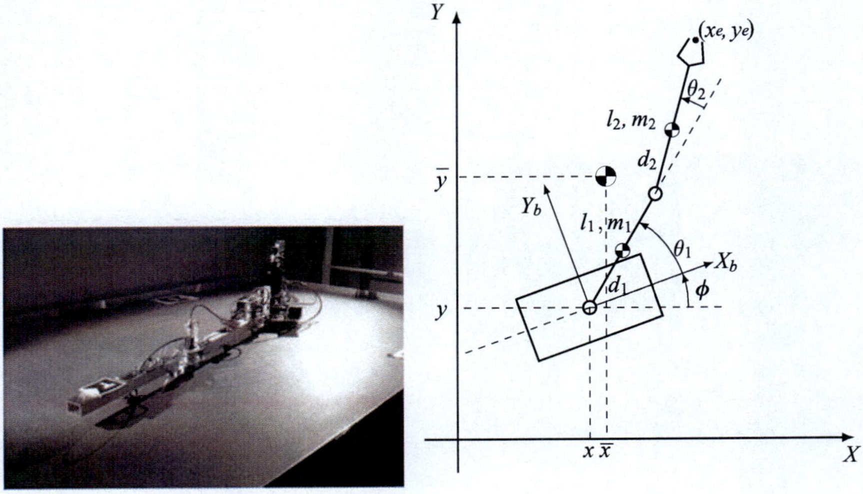

Fig. 2 SRC space manipulator

where matrix C and vector D refer to, respectively, the Coriolis and centripetal forces, and the potential forces, symbol F denotes traction forces preventing the slip, and the component $\mathsf{B}(q)u$ represents the action of controls $u \in \mathbb{R}^p$ on individual generalized coordinates. After invoking d'Alembert's Principle we obtain for a certain Lagrange multiplier $\lambda \in \mathbb{R}^l$

$$F = \mathscr{A}^\top(q)\lambda, \tag{9}$$

and further, having substituted to the dynamics equations and multiplied them from the left by matrix $G^\top(q)$ in order to eliminate the traction forces ($G^\top \mathscr{A}^\top = 0$), we arrive at formula

$$G^\top Q\ddot{q} + G^\top(\mathsf{C}\dot{q} + D) = G^\top \mathsf{B}u. \tag{10}$$

Exploiting once again the form of driftless system $\dot{q} = G(q)\eta$ we compute $\ddot{q} = \dot{G}\eta + G\dot{\eta}$, and conclude with the following equations of motion

$$\begin{cases} \dot{q} = G(q)\eta \\ \dot{\eta} = \left(G^\top QG\right)^{-1} G^\top(-Q\dot{G}\eta - \mathsf{C}G\eta - D) + \left(G^\top QG\right)^{-1} G^\top \mathsf{B}u. \end{cases} \tag{11}$$

It is easily observed that these equations describe a control-affine system with state vector $x = (q, \eta) \in \mathbb{R}^{n+m}$, drift $f(x) = (G\eta, \left(G^\top QG\right)^{-1} G^\top(-Q\dot{G}\eta - \mathsf{C}G\eta - D)$, and control vector fields $g_i(x) = (0, \left(G^\top QG\right)^{-1} G^\top B_i)$, where B_i denotes the i-th column of matrix B. Such a system is called fully actuated when the number of components of control vector u is the same as that of vector η, i.e. $m = p$, and matrix B has full rank. In case when $p < m$ we say that the system is underactuated. An example of the system with dynamics model (11), either fully actuated or

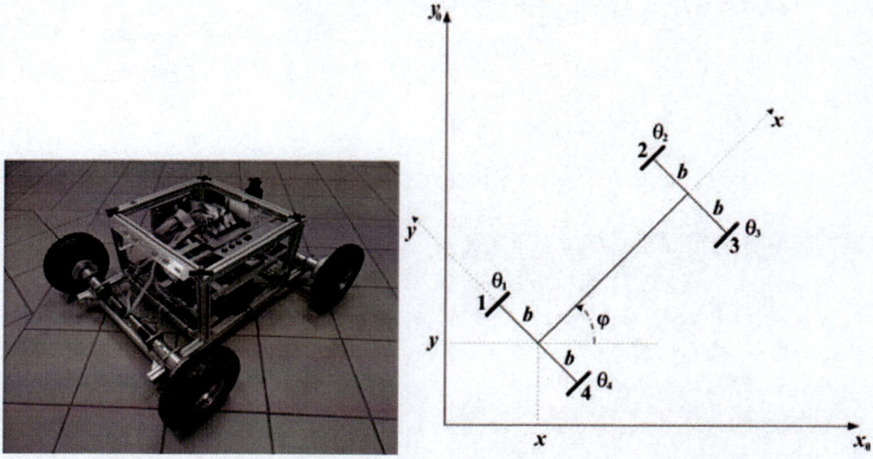

Fig. 3 Mobile robot Rex

underactuated, is the mobile robot Rex designed at the Department of Cybernetics and Robotics of Wrocław University of Science and Technology, displayed in Fig. 3. Several models of kinematics and dynamics of the Rex robot, adapted to diverse motion conditions, can be found in [36].

2.3 Input-Output Map

Given the output trajectory computed for a fixed initial state q_0 we define the input-output map

$$\mathscr{H}_{q_0,T} : \mathscr{U} \longrightarrow \mathbb{R}^r, \tag{12}$$

of system (2) that determines the value of the output variable at time instant T of the system in endogenous configuration $u(\cdot) \in \mathscr{U}$. This means that

$$\mathscr{H}_{q_0,T}(u(\cdot)) = h(\varphi_{q_0,T}(u(\cdot))). \tag{13}$$

It has been proved that the map (13) is continuously differentiable [89]. Its derivative with respect to the control function,

$$J_{q_0,T}(u(\cdot)) = \mathsf{D}\mathscr{H}_{q_0,T}(u(\cdot)), \tag{14}$$

will be called the Jacobian of the mobile robot (2) (symbol D refers to the derivative in a function space). It is easily noticed that, for fixed q_0 and $u(\cdot)$, the Jacobian is a transformation

$$J_{q_0,T}(u(\cdot)) : \mathscr{U} \longrightarrow \mathbb{R}^r \tag{15}$$

from the endogenous configuration space to the task space of the mobile robot. The value taken by this transformation for the given control function $v(\cdot)$ can be computed in the following way. Let $q(t) = \varphi_{q0,t}(u(\cdot))$. For a current time instant $t \in [0, T]$ and function $v(\cdot) \in \mathscr{U}$ we get the derivative

$$\mathsf{D}h(\varphi_{q0,t}(u(\cdot)))v(\cdot) = \frac{d}{d\alpha}|_{\alpha=0}h(\varphi_{q0,t}(u(\cdot) + \alpha v(\cdot))) =$$
$$\frac{\partial h(q(t))}{\partial q}\mathsf{D}\varphi_{q0,t}(u(\cdot))v(\cdot). \tag{16}$$

Now, we introduce an auxiliary variable $\xi(t) = \mathsf{D}\varphi_{q0,t}(u(\cdot))v(\cdot) = \frac{d}{d\alpha}|_{\alpha=0}\varphi_{q0,t}$ $(u(\cdot) + \alpha v(\cdot))$. Changing the order of differentiation and making use of the definition of system (2) we arrive at the following expression

$$\dot{\xi}(t) = \frac{d}{dt}\frac{d}{d\alpha}|_{\alpha=0}\varphi_{q0,t}(u(\cdot) + \alpha v(\cdot))) = \frac{d}{d\alpha}|_{\alpha=0}(f(\varphi_{q0,t}(u(\cdot) + \alpha v(\cdot)))+ \tag{17}$$
$$G(\varphi_{q0,t}(u(\cdot) + \alpha v(\cdot)))(u(t) + \alpha v(t))) = \frac{\partial(f(q(t)) + G(q(t))u(t))}{\partial q}\xi(t) + G(q(t))v(t).$$

Finally, the Jacobian

$$J_{q_0,T}(u(\cdot))v(\cdot) = \eta(T) = C(T)\int_0^T \Phi(T,s)B(s)v(s)ds, \tag{18}$$

is computed by means of the linear approximation of the affine control system

$$\dot{\xi} = A(t)\xi + B(t)v, \quad \eta(t) = C(t)\xi, \tag{19}$$

along the control and state trajectory $(u(t), q(t))$, whose matrices are defined as

$$A(t) = \frac{\partial(f(q(t)) + G(q(t))u(t))}{\partial q}, \; B(t) = G(q(t)), \; C(t) = \frac{\partial h(q(t))}{\partial q}, \tag{20}$$

and the fundamental matrix $\Phi(t, s)$ satisfies equation

$$\frac{\partial \Phi(t,s)}{\partial t} = A(t)\Phi(t,s), \; \Phi(s,s) = \mathbb{I}_n. \tag{21}$$

2.4 *Regular and Singular Configurations*

A control function (endogenous configuration) $u(\cdot) \in \mathscr{U}$ for which the Jacobian is surjective (the onto map) is called a regular configuration of system (2) (i.e. of the mobile robot), otherwise $u(\cdot)$ is named a singular configuration. By definition, in a regular configuration for any task space point $w \in \mathbb{R}^r$ there exists a solution $v(t)$ of the Jacobian equation

$$J_{q_0,T}(u(\cdot))v(\cdot) = w. \tag{22}$$

It is well known that configuration $u(\cdot)$ is regular if and only if matrix

$$\mathscr{M}_{q_0,T}(u(\cdot)) = C(T)\int_0^T \Phi(T,s)B(s)B^\top(s)\Phi^\top(T,s)ds\, C^\top(T) \tag{23}$$

associated with the mobile robot has full rank r. Having in mind an analogy with the manipulability matrix of a manipulator, we shall call matrix $\mathscr{M}_{q_0,T}(u(\cdot))$ the mobility matrix of the mobile robot in configuration $u(\cdot)$. Observe that in singular configurations the mobility matrix loses rank. One can show that the Gram matrix

$$M(T) = \int_0^T \Phi(T,s)B(s)B^\top(s)\Phi^\top(T,s)ds \tag{24}$$

appearing in (23) solves the Lyapunov equation

$$\dot{M}(t) = B(t)B^\top(t) + A(t)M(t) + M(t)A^\top(t) \tag{25}$$

with initial condition $M(0) = 0$. It follows from the theory of linear, time dependent control systems that condition rank $M(T) = n$ is a necessary and sufficient condition for controllability of linear approximation (19) over interval $[0, T]$, [86]. Furthermore, it is also a sufficient condition for the local controllability of system (2).

2.5 Singular Controls

The notion of singular control, borrowed from control theory, proves to be applicable in the analysis of singular configurations of mobile robots. We recall that the singular control in system (2) is defined in the following way. For adjoint variable $p \in \mathbb{R}^n$ we introduce the Hamiltonian

$$\tilde{H}(q, p, u) = p^\top(f(q) + G(q)u). \tag{26}$$

In accordance with the Pontryagin's Maximum Principle, we associate with Hamiltonian (26) the Hamilton canonical equations

$$\dot{q} = \frac{\partial \tilde{H}}{\partial p} = f(q) + G(q)u, \quad \dot{p}^\top = -\left(\frac{\partial \tilde{H}}{\partial q}\right)^\top = -p^\top\frac{\partial(f(q)+G(q)u)}{\partial q}, \tag{27}$$

and the optimality condition

$$\frac{\partial \tilde{H}(q,p,u)}{\partial u} = 0, \quad \text{i.e} \quad p^\top(t)G(q(t)) = 0. \tag{28}$$

Additionally, $p^\top(t)f(q(t)) = \text{const}$. The triplet of functions $(u(t), q(t), p(t))$, $p(t) \neq 0$, satisfying the conditions given above bears the name of the singular extremal of the Hamiltonian, whereas control $u(t)$ appearing in the singular extremal is called the singular control. Invoking the definition of matrices $A(t)$ i $B(t)$ of linear approximation (19) we conclude that at every time instant $t \in [0, T]$ the singular extremal fulfills the system of equations

$$\dot{q}(t) = f(q(t)) + G(q(t))u(t), \quad \dot{p}^\top(t) = -p^\top(t)A(t), \quad p^\top(t)B(t) = 0, \tag{29}$$

and that along the singular extremal the Hamiltonian is constant.

As a consequence of definition of system (2), matrix $C(T)$ has full rank r. Suppose that $r = n$. Then, the rank of mobility matrix (23) is the same as the rank of the Gram matrix $M(T) = \int_0^T \Phi(T,s)B(s)B^\top(s)\Phi^\top(T,s)ds$, and therefore configuration $u(\cdot)$ is singular if and only if the rank of matrix $M(T)$ is less than n. We shall show that the last property holds if and only if control $u(\cdot)$ is singular.

We begin by assuming that $u(\cdot)$ denotes a singular control in system (2). If so, the singular extremal $(u(t), q(t), p(t))$ satisfies equations (29) for $p(t) \neq 0$ and matrices $A(t)$ and $B(t)$ computed in accordance with (20). Using (25) and (29) we find

$$\frac{d}{dt}\left(p^\top M\right) = (p^\top M)A^\top(t), \quad (p^\top M)(0) = 0. \tag{30}$$

Now, because $p^\top M$ is a solution of a linear differential equations with zero initial conditions, we obtain $(p^\top M)(t) = 0$ for every $t \in [0, T]$. This implies the existence of a non-zero vector $p(T)$ such that $p^\top(T)M(T) = 0$, thus matrix $M(T)$ must be singular and, eventually, control $u(\cdot)$ is a singular configuration. In turn, let $u(\cdot)$ be a singular configuration. After solving for $u(t)$ the Hamilton canonical equations we get a trajectory $(q(t), p(t))$. Our objective is to show that $(u(t), q(t), p(t))$ is a singular extremal of system (2). Since matrix $M(T)$ is singular, there exists a non-zero vector $\eta \in \mathbb{R}^n$ such that $\eta^\top M(T)\eta = 0$, so taking into account the definition of matrix $M(T)$, it follows that for every $t \in [0, T]$ there must be $B^\top(t)\Phi^\top(T,t)\eta = 0$. Next, the identity $\dot{p}^\top = -p^\top A(t)$ yields $p(t) = -\Phi^\top(T,t)p(T)$, therefore setting $p(T) = \eta$ results in $B^\top(t)p(t) = 0$. In this way we have demonstrated that functions $(u(t), q(t), p(t))$ satisfy conditions (27) and (28). From the constancy of the maximum value of the Hamiltonian we deduce $p^\top(t)f(q(t)) = \text{const}$. Summarizing, $(u(t), q(t), p(t))$ is a singular extremal and $u(\cdot)$ is a singular control in system (2). We have shown that if $r = n$ then singular configurations of the mobile robot coincide with singular controls of control-affine system (2).

2.6 Equivalence and Normal Forms

Consider two control-affine systems Σ and $\tilde{\Sigma}$ of form (2)

$$\Sigma : \dot{q} = f(q) + G(q)u, \quad \tilde{\Sigma} : \dot{\tilde{q}} = \tilde{f}(\tilde{q}) + \tilde{G}(\tilde{q})\tilde{u}, \tag{31}$$

with state variables $q, \tilde{q} \in \mathbb{R}^n$ and control variables $u, \tilde{u} \in \mathbb{R}^m$. We recall that control systems (31) are referred to as feedback equivalent [74, 84] if there exists a smooth coordinate transformation (a diffeomorphism) $\tilde{q} = \varphi(q)$ and a feedback $u = \alpha(q) + \beta(q)\tilde{u}$ determined by a smooth function $\alpha(q)$ and a smooth non-singular matrix $\beta(q)$ such that

$$\frac{\partial\varphi(q)}{\partial q}(f(q) + G(q)\alpha(q)) = \tilde{f}(\varphi(q)) \quad \text{and} \quad \frac{\partial\varphi(q)}{\partial q}G(q)\beta(q) = \tilde{G}(\varphi(q)). \tag{32}$$

The feedback equivalence is called local if the coordinate transformation is defined locally (it is a local diffeomorphism). The feedback establishes an equivalence relation between a given control-affine system and an equivalence class of systems comprising more or less complex systems than the given system. A system Σ_0 of possible simplest form that is feedback equivalent to given system Σ is named a normal form of system Σ. In the paper [50] we have proved that the feedback preserves singular controls. By the definition of the normal form it can be expected that singular controls for the normal form are easily computed. This property will allow us to exploit the normal forms as a tool for characterizing singular configurations of mobile robots, in a similar way as the characterization of kinematic singularities of manipulators by the normal forms of their kinematics, see [92].

2.7 Motion Planning Problem

Let an initial state q_0 and a control horizon $T > 0$ of system (2) be fixed. Then, the motion planning problem of the mobile robot consists in determining a control function (endogenous configuration) $u_d(\cdot) \in \mathscr{U}$ of the mobile robot, such that the output variable at T assumes a prescribed value y_d, i.e.

$$\mathscr{K}_{q_0,T}(u_d(\cdot)) = y_d. \tag{33}$$

3 Jacobian Motion Planning Algorithms

Motion planning algorithms that use the Jacobian are called Jacobian motion planning algorithms. In this subsection we shall focus on three kinds of the Jacobian algorithms: Jacobian inverse algorithms with special attention paid to the dynamically consistent Jacobian algorithm, Jacobian transpose algorithms, and extended Jacobian algorithms. The subsection will be concluded with a description of a multitask Jacobian algorithm with priorities. We want to mention that all the presented algorithms do not take into account any constraints imposed on the state or the controls of system (2). Jacobian algorithms including such constraints, called imbalanced, have been studied in [18, 26].

3.1 Inverse Jacobian Algorithms

Observe that a solution of the motion planning problem is reduced to finding out an inverse map to the input-output map $\mathscr{H}_{q_0,T}(u(\cdot))$. This inverse map can be determined by inverting Jacobian $J_{q_0,T}(u(\cdot))v(\cdot)$ in a given endogenous configuration $u(\cdot)$. In order to define a Jacobian motion planning algorithm we resort to the following reasoning characteristic to the method of homotopy. We start from an arbitrary control function $u_0(\cdot)$. If, accidentally, this function solves the motion planning problem, we stop. Otherwise, we have $\mathscr{H}_{q_0,T}(u_0(\cdot)) \neq y_d$, an we choose in the endogenous configuration space a smooth curve $u_\theta(\cdot) \in \mathscr{U}$ passing through $u_0(\cdot)$, parametrized by $\theta \in \mathbb{R}$. Along this curve we compute the planning error $e(\theta) = \mathscr{H}_{q_0,T}(u_\theta(\cdot)) - y_d$. The essential step of the algorithm comes from the requirement that along the curve $u_\theta(\cdot)$ the error $e(\theta)$ decreases exponentially, which will be guaranteed if the derivative of the error with respect to θ obeys equation $e'(\theta) = -\gamma e(\theta)$, for a certain convergence ratio $\gamma > 0$. Bearing in mind the definitions of the error and of the Jacobian we derive an implicit differential equation for the curve $u_\theta(\cdot)$,

$$J_{q_0,T}(u_\theta(\cdot))u'_\theta(\cdot) = -\gamma e(\theta) \tag{34}$$

called sometimes the Ważewski-Davidenko equation [65, 94]. In order to make this equation explicit we need to use a certain right inverse of the Jacobian

$$J^{\#}_{q_0,T}(u(\cdot)) : \mathbb{R}^r \longrightarrow \mathscr{U}, \tag{35}$$

such that $J_{q_0,T}(u_\theta(\cdot))J^{\#}_{q_0,T}(u(\cdot)) = \mathbb{I}_r$, providing us with an explicit differential equation for $u_\theta(\cdot)$, of the form

$$u'_\theta(\cdot) = -\gamma J^{\#}_{q_0,T}(u_\theta(\cdot))(\mathscr{H}_{q_0,T}(u_\theta(\cdot)) - y_d), \quad u_{\theta=0}(\cdot) = u_0(\cdot). \tag{36}$$

A solution of the motion planning problem is obtained as a limit of the trajectory of equation (36),

$$u_d(t) = \lim_{\theta \to +\infty} u_\theta(t). \tag{37}$$

Equation (36) along with the limit (37) defines a motion planning algorithm associated with right inverse $J^{\#}_{q_0,T}(u(\cdot))$ of the Jacobian.

To be more specific, we shall present a family of motion planning algorithms based on the Lagrangian inverse of the Jacobian. We begin by recalling the observation that, by definition of regular configuration $u(\cdot) \in \mathscr{U}$ of the mobile robot, it follows that for every $w \in \mathbb{R}^r$ there exists solution $v(\cdot)$ of the Jacobian equation

$$J_{q_0,T}(u(\cdot))v(\cdot) = w. \tag{38}$$

This solution can be found by means of a right inverse of the Jacobian

$$J^{\#}_{q_0,T}(u(\cdot)) : \mathbb{R}^r \to \mathscr{U}, \tag{39}$$

so that $v(t) = (J^{\#}_{q_0,T}(u(\cdot))w)(t)$. For this objective we shall regard equation (38) as an equality constraints $C(T)\xi(T) = w$ in an optimal control problem for system (19), with objective function of the Lagrange form,

$$\min_{v(\cdot)} \frac{1}{2}\int_0^T \begin{pmatrix}\xi(t)\\ v(t)\end{pmatrix}^\top \begin{bmatrix} Q(t) & S(t)\\ S^\top(t) & R(t)\end{bmatrix} \begin{pmatrix}\xi(t)\\ v(t)\end{pmatrix} dt. \tag{40}$$

Matrices contained in objective function (40) need to satisfy conditions $Q(t) = Q^\top(t) \geq 0$, $R(t) = R^\top(t) > 0$ and $Q(t) - S(t)R^{-1}(t)S^\top(t) \geq 0$, thanks to which the matrix under the integral gets positive semi-definite. The Jacobian inverse obtained as the solution of this optimal control problem will be denoted by $J^{GLJ\#}_{q_0,T}(u(\cdot))$ and called the Lagrangian inverse of the Jacobian (Lagrangian Jacobian inverse). By an application of the Pontryagin's Maximum Principle it can be shown that the Lagrangian Jacobian inverse takes the following form, see [47],

$$\begin{aligned}&\left(J^{GLJI\#}_{q_0,T}(u(\cdot))w\right)(t) =\\ &R^{-1}(t)\left(B^\top(t)\phi_{22}(t) - S^\top(t)\phi_{12}(t)\right)\phi_{22}^{-1}(T)C^\top(T)\mathscr{F}^{-1}_{q_0,T}(u(\cdot))w,\end{aligned} \tag{41}$$

where

$$\mathscr{F}_{q_0,T}(u(\cdot)) = C(T)\phi_{12}(T)\phi_{22}^{-1}(T)C^\top(T). \tag{42}$$

The fundamental block matrix $\Phi(t) = \left[\phi_{ij}(t)\right]$, $i, j = 1, 2$ should solve the system of differential equations

$$\dot{\Phi}(t) = \begin{bmatrix} A(t) - B(t)R^{-1}(t)S^\top(t) & B(t)R^{-1}(t)B^\top(t)\\ Q(t) - S(t)R^{-1}(t)S^\top(t) & -A^\top(t) + S(t)R^{-1}(t)B^\top(t)\end{bmatrix}\Phi(t), \tag{43}$$

with the initial condition $\phi_{ij}(0) = \delta_{ij}I_n$, where δ_{ij} $i, j = 1, 2$, stands for the Kronecker delta function. By formula (41), the Lagrangian Jacobian inverse is well defined on condition that matrix $\mathscr{F}_{q_0,T}(u(\cdot))$ is non-singular. This matrix can be re-written in the form

$$\mathscr{F}_{q_0,T}(u(\cdot)) = C(T)F(T)C^\top(T), \tag{44}$$

where $F(t) = \phi_{12}(t)\phi_{22}^{-1}(t)$ solves the Riccati differential equation

$$\dot{F}(t) = B(t)R^{-1}(t)B^\top(t) + \left(A(t) - B(t)R^{-1}(t)S^\top(t)\right)F(t) \\ + F(t)\left(A^\top(t) - S(t)R^{-1}(t)B^\top(t)\right) - F(t)\left(Q(t) - S(t)R^{-1}(t)S^\top(t)\right)F(t), \tag{45}$$

with initial condition $F(0) = 0$.

In the particular case when matrices $S(t)$ and $Q(t)$ are equal to zero and $R(t) = \mathbb{I}_m$, expression (41) boils down to the pseudoinverse of the Jacobian (Jacobian pseudoinverse)

$$\left(J^{JP\#}_{q_0,T}(u(\cdot))w\right)(t) = B^\top(t)\Phi^\top(T,t)C^\top(T)\mathscr{M}^{-1}_{q_0,T}(u(\cdot))w, \tag{46}$$

where $\Phi(t, s)$ denotes the Gram matrix (21). Correspondingly, it is easily seen that matrix $\mathscr{M}_{q_0,T}(u(\cdot)) = C(T)M(T)C^\top(T)$ contains matrix $M(t)$ being a solution of Lyapunov equation (25).

3.2 *Dynamically Consistent Jacobian Algorithm*

Although, formally speaking, this algorithm belongs to the class of inverse Jacobian algorithms, it deserves a separate description due to its robotic interpretation. The idea of defining a Jacobian inverse that would respect certain natural properties of forces acting in the configuration space as well as in the tasks space of a manipulator should be credited to Khatib [77, 78]. Its extension to mobile robots has been presented in [39–41, 49], and can be tracked along the following lines. Let $J_{q_0,T}$ denote the Jacobian of a mobile robot and $J^*_{q_0,T}$ be the dual Jacobian. Relying on the interpretation of the Jacobian as a transformation of velocities from the endogenous configuration space to the task space, we understand the dual Jacobian as a transformation of forces from the task space to the configuration space. As a consequence of the definition, a right inverse of the Jacobian is a velocity transformation

$$J^\#_{q_0,T} : \mathbb{R}^r \longrightarrow \mathscr{U}, \tag{47}$$

while the dual Jacobian inverse

$$J^{\#*}_{q_0,T} : \mathscr{U}^* \longrightarrow (\mathbb{R}^r)^* \tag{48}$$

transforms forces acting in the endogenous configuration space into forces that act in the task space of the mobile robot.

Having defined the dual Jacobian and the dual Jacobian inverse at a certain configuration $u(\cdot)$ we decompose the dual endogenous configuration space into a direct sum of two subspaces

$$\mathscr{U}^* = \operatorname{Im} J^*_{q_0,T}(u(\cdot)) \oplus \operatorname{Ker} J^{\#*}_{q_0,T}(u(\cdot)), \tag{49}$$

that allows us to represent a force acting in the endogenous configuration space as a sum of two components (the symbol of transposition means that forces are elements of the dual space)

$$f^\top = f_1^\top + f_2^\top, \tag{50}$$

such that the former force $f_1^\top = J^*_{q_0,T}(u(\cdot))\Gamma$, Γ refers to the force acting in the task space, while the latter force $f_2^\top$ acts solely in the endogenous configuration space. Due to identity $J^{\#*}_{q_0,T}(u(\cdot))\, f_2^\top = 0$, this last force is not exerted in the task space.

Now, let us represent the subsystem describing the dynamics of a fully actuated controllable model (11) of the mobile robot in the form

$$P(q)\dot{\eta} = f^\top, \tag{51}$$

where $P(q) = G^\top Q G$, $f^\top = G^\top(-Q\dot{G}\eta - \mathsf{C}G\eta - D + \mathsf{B}u)$. Assume that force $J^{\#*}_{q_0,T}(u(\cdot)) f^\top = 0$. Then, the inverse $J^{\#}_{q_0,T} = J^{DC\#}_{q_0,T}$ of the Jacobian will be called dynamically consistent if the following condition holds

$$J_{q_0,T} P^{-1} f^\top = 0, \tag{52}$$

i.e. force $f^\top$ belonging to Ker $J^{\#*}_{q_0,T}(u(\cdot))$ does not cause any motion in the task space.

The analysis performed in [40] yields that the dynamically consistent Jacobian inverse is a special kind of the Lagrangian Jacobian inverse, if one plugs into the latter inverse matrix $R(t) = P(q(t)) = P(\varphi_{q_0,t}(u(\cdot))$, i.e.

$$\left(J^{DC\#}_{q_0,T}(u(\cdot))w\right)(t) = P^{-1}(q(t))B^\top(t)\Phi^\top(T,t)C^\top(T)(\mathscr{M}^{DC}_{q_0,T})^{-1}(u(\cdot))w, \tag{53}$$

where

$$\mathscr{M}^{DC}_{q_0,T}(u(\cdot)) = C(T)\int_0^T \Phi(T,t)B(t)P^{-1}(q(t))B^\top(t)\Phi^\top(T,t)dt\, C^\top(T). \tag{54}$$

3.3 Jacobian Transpose Algorithm

For robotic manipulators the Jacobian transpose algorithm was introduced by Sciavicco and Siciliano [88]. Its equivalent for mobile robots and mobile manipulators can be found in [2, 5, 90]. It follows from the definition that the Jacobian of the mobile robot is a linear map

$$J_{q_0,T}(u(\cdot)) : \mathscr{U} \longrightarrow \mathbb{R}^r \tag{55}$$

of the endogenous configuration space into the task space. The dual map to the Jacobian

$$J^*_{q_0,T}(u(\cdot)) : \mathbb{R}^{r*} \longrightarrow \mathscr{U}^*, \tag{56}$$

acting between respective dual spaces, will be called the dual (or transpose) Jacobian.

In order to provide an operational formula for the Jacobian transpose we choose a transpose vector $w^\top \in \mathbb{R}^{r*}$ and an endogenous configuration $v(\cdot) \in \mathscr{U}$, and compute

$$\left(J^*_{q_0,T}(u(\cdot))w^\top\right)v(\cdot) = w^\top J_{q_0,T}(u(\cdot))v(\cdot) = w^\top C(T)\int_0^T \Phi(T,t)B(t)v(t)dt =$$
$$\langle B^\top(\cdot)\Phi^\top(T,\cdot)C^\top(T)w, v(\cdot)\rangle, \tag{57}$$

where $\langle\ \rangle$ denotes the inner product in the endogenous configuration space. It is now easily deduced that the Jacobian transpose is determined by function

$$\left(\mathsf{J}^*_{q_0,T}(u(\cdot))w^\top\right)(t) = B^\top(t)\Phi^\top(T,t)C^\top(T)w, \tag{58}$$

For a fixed endogenous configuration $u(\cdot)$, let us define the squared motion planning error

$$E(u(\cdot)) = \frac{1}{2}(H_{q_0,T}(u(\cdot) - y_d)^\top(H_{q_0,T}(u(\cdot) - y_d). \tag{59}$$

Recall that the gradient of a function E is a transformation grad $E : \mathscr{U} \longrightarrow \mathscr{U}$, such that

$$\langle(\text{grad } E)(u(\cdot)), v(\cdot)\rangle = \mathsf{D}E(u(\cdot))v(\cdot). \tag{60}$$

Making use of this definition and of the definition of the Jacobian transpose we obtain

$$(\text{grad } E)(u(\cdot))(t) = \left(\mathsf{J}^*_{q_0,T}(u(\cdot))(H_{q_0,T}(u(\cdot)) - y_d)^\top\right)(t) =$$
$$B^\top(t)\Phi^\top(T,t)C^\top(T)(H_{q_0,T}(u(\cdot) - y_d). \tag{61}$$

Suppose that $u_\theta(\cdot)$ is a smooth curve in the endogenous configuration space. We shall request that along this curve the motion planning error decrease as fast as possible (exhibits the steepest descent), which implies that the direction of motion along curve $u_\theta(\cdot)$ needs to be opposite to the gradient grad E of the error. In this way we have produced the Jacobian transpose motion planning algorithm defined by differential equation

$$u'_\theta(t) = -\gamma\left(\mathsf{J}^*_{q_0,T}(u_\theta(\cdot))(H_{q_0,T}(u_\theta(\cdot)) - y_d)^\top\right)(t), \tag{62}$$

where $\gamma > 0$. A solution of the motion planning problem is obtained as the limit

$$u_d(t) = \lim_{\theta\to+\infty} u_\theta(t). \tag{63}$$

3.4 Extended Jacobian Algorithm

The extended Jacobian algorithm for motion planning is a Jacobian inverse type algorithm, nevertheless is will be discussed separately due to a specific definition of the inverse, [4, 11, 13, 14], that generalizes the concept of extended Jacobian inverse for manipulators [56]. Let the input-output map

$$H_{q_0,T} : \mathcal{U} \longrightarrow \mathbb{R}^r \tag{64}$$

be given, defined by expression (13). We introduce an extension of this map to a map of the endogenous configuration space into itself. To this objective we identify the task space with a subspace of the endogenous configuration space, decompose the latter space into the direct sum of two linear subspaces

$$\mathcal{U} \cong \mathbb{R}^r \oplus \mathcal{U}/\mathbb{R}^r, \tag{65}$$

and define an augmenting map

$$K_{q_0,T} : \mathcal{U} \longrightarrow \mathcal{U}/\mathbb{R}^r, \tag{66}$$

that results in the extended input-output map

$$L_{q_0,T} = (H_{q_0,T}, K_{q_0,T}) : \mathcal{U} \to \mathcal{U} \tag{67}$$

of the mobile robot. We assume that the extended map is continuously differentiable. Its derivative

$$\mathsf{D}\, L_{q_0,T}(u(\cdot)) = (J_{q_0,T}, \mathsf{D}\, K_{q_0,T})(u(\cdot)) = \bar{J}_{q_0,T}(u(\cdot)) \tag{68}$$

will be called the extended Jacobian of the mobile robot. It would be most desirable that Jacobian $\bar{J}_{q_0,T}(u(\cdot))$ be a linear isomorphism of the endogenous configuration space, however this expectation is not very realistic as the extended Jacobian not only inherits the singularities of the original Jacobian, but also may bring in additional singularities called algorithmic, that result from the extension procedure.

Restricted to regular endogenous configurations we define the extended Jacobian inverse

$$J^{\#E}_{q_0,T}(u(\cdot)) : \mathbb{R}^r \longrightarrow \mathcal{U} \tag{69}$$

as

$$J^{\#E}_{q_0,T}(u(\cdot))w = \bar{J}^{-1}_{q_0,T}(u(\cdot))(w, 0(\cdot)), \tag{70}$$

where $w \in \mathbb{R}^r$ and $0(\cdot) \in \mathcal{U}/\mathbb{R}^r$ denotes the zero element of the quotient space. It is straightforward to demonstrate the following two properties of Jacobian inverse (70):

1. Right inverse property

$$J_{q_0,T}(u(\cdot))J^{\#E}_{q_0,T}(u(\cdot)) = \mathbb{I}_r, \tag{71}$$

2. Annihilation property

$$\mathsf{D}K_{q_0,T}(u(\cdot))J^{\#E}_{q_0,T}(u(\cdot)) = 0(\cdot). \tag{72}$$

The extended Jacobian motion planning algorithm assumes the standard form

$$u'_\theta(t) = -\gamma\left(J^{\#E}_{q_0,T}(u_\theta(\cdot))(H_{q_0,T}(u_\theta(\cdot)) - y_d)\right)(t). \tag{73}$$

As we have already mentioned, algorithm (73) is well defined in regular configurations. As usual, a solution of the motion planning problem is obtained as the limit

$$u_d(t) = \lim_{\theta\to+\infty} u_\theta(t). \tag{74}$$

It should be noticed that, thanks to the annihilation property, the augmenting map is constant on trajectories of system (73).

3.5 Prioritarian Jacobian Algorithm

This type of motion planning algorithm is applied in the case of multiple task planning. In the field of robotic manipulators the planning of many tasks was studied a.o. in [64, 83]. An extension of these ideas to mobile robots will be presented for the case of two tasks, with reference to a number of relevant publications [23, 27, 28, 31, 44]. Suppose that a mobile robot should execute two tasks. The first of them is a classic motion planning task stated in terms of the input-output map $H_{q_0,T}$, and consists in reaching a prescribed point $y_d \in \mathbb{R}^r$ in the task space, while the second task refers to reaching the zero value by a certain task map $K_{q_0,T} : \mathcal{U} \longrightarrow \mathbb{R}^s$. For both these maps we compute Jacobians denoted, respectively, as $J^H_{q_0,T}$ and $J^K_{q_0,T}$. It will be assumed that the motion planning task is superior, with a higher priority, and will be solved by a Jacobian inverse algorithm employing the Jacobian inverse $J^{H\#}_{q_0,T}$. The inferior task, with lower priority, will also be solved by means of a Jacobian inverse algorithm, with inverse $J^{K\#}_{q_0,T}$. Given endogenous configuration $u(\cdot) \in \mathcal{U}$ we define a map

$$P^H_{q_0,T}(u(\cdot)) : \mathcal{U} \longrightarrow \operatorname{Ker} J^H_{q_0,T}(u(\cdot)) \tag{75}$$

associated with the Jacobian of the superior task, in the following way

$$P^H_{q_0,T}(u(\cdot)) = \mathrm{id}_{\mathcal{U}} - J^{H\#}_{q_0,T}(u(\cdot))J^H_{q_0,T}(u(\cdot)), \tag{76}$$

where $\mathrm{id}_{\mathcal{U}}$ denotes the identity map of the endogenous configuration space into itself. By definition, we obtain:

1. $P^H_{q_0,T}(u(\cdot))$ is idempotent, i.e. $P^H_{q_0,T}(u(\cdot))P^H_{q_0,T}(u(\cdot)) = P^H_{q_0,T}(u(\cdot))$ ($P^H_{q_0,T}(u(\cdot))$ is a projection),
2. $J^H_{q_0,T}(u(\cdot))P^H_{q_0,T}(u(\cdot)) = 0$,
3. $P^H_{q_0,T}(u(\cdot))J^{H\#}_{q_0,T}(u(\cdot)) = 0$,
4. If $J^H_{q_0,T}(u(\cdot))v(\cdot) = 0$ then $P^H_{q_0,T}(u(\cdot))v(\cdot) = v(\cdot)$.

Let $u_\theta(\cdot)$ be a certain curve in the endogenous configuration space. Under the assumed establishments a prioritarian Jacobian motion planning algorithm can be defined by the following differential equation

$$u'_\theta(\cdot) = -\gamma_H J^{H\#}_{q_0,T}(u_\theta(\cdot))e_H(\theta) - \gamma_K P^H_{q_0,T}(u(\cdot))J^{K\#}_{q_0,T}(u_\theta(\cdot))e_K(\theta), \tag{77}$$

$\gamma_H, \gamma_K > 0$, where symbols $e_H = H_{q_0,T}(u(\cdot)) - y_d$ and $e_K = K_{q_0,T}(u(\cdot))$ refer to the errors of solving the superior and the inferior task. A solution of the prioritarian motion planning problem is given as the limit

$$u_d(t) = \lim_{\theta\to+\infty} u_\theta(t). \tag{78}$$

In order to provide some insight into the operation of (77) let us observe that the left multiplication of both sides of equation (77) by $J^H_{q_0,T}(u_\theta(\cdot))$ results in equality

$$J^H_{q_0,T}(u_\theta(\cdot))u'_\theta = \frac{d}{d\theta}H_{q_0,T}(u_\theta(\cdot)) = e'_H(\theta) = -\gamma_H e_H(\theta), \tag{79}$$

that guarantees the accomplishment of the superior task. In turn, after multiplying both sides of this equation by $P^H_{q_0,T}(u_\theta(\cdot))$, and taking into account properties of the projection we deduce

$$P^H_{q_0,T}(u_\theta(\cdot))u'_\theta = -\gamma_K P^H_{q_0,T}(u_\theta(\cdot)))J^{K\#}_{q_0,T}(u_\theta(\cdot))e_K(\theta). \tag{80}$$

Finally, using the definition of error e_K we compute $u'_\theta = J^{K\#}_{q_0,T}(u_\theta(\cdot))e'_K(\theta)$ and re-write the result in the form

$$P^H_{q_0,T}(u_\theta(\cdot))J^{K\#}_{q_0,T}(u_\theta(\cdot))\left(e'_K(\theta) + \gamma_K e_K(\theta)\right) = 0. \tag{81}$$

The inverse $J^{K\#}_{q_0,T}(u(\cdot))$ transfers the planning error to the endogenous configuration space. By the last of the properties of projection $P^H_{q_0,T}$ we draw conclusion that if expression $J^{K\#}_{q_0,T}(u_\theta(\cdot))\left(e'_K(\theta) + \gamma_K e_K(\theta)\right)$ belongs to the null space of Jacobian $J^H_{q_0,T}$ of the superior task then the solution of the inferior task fulfills condition

$$P^H_{q_0,T}(u_\theta(\cdot))J^{K\#}_{q_0,T}(u_\theta(\cdot))\left(e'_K(\theta) + \gamma_K e_K(\theta)\right) = J^{K\#}_{q_0,T}(u_\theta(\cdot))\left(e'_K(\theta) + \gamma_K e_K(\theta)\right) = 0. \tag{82}$$

Since the right Jacobian inverse is an injection, the error of solving the inferior task satisfies equation

$$e'_K(\theta) = -\gamma_K e_K(\theta). \tag{83}$$

The presented reasoning can be adopted in a natural way to an arbitrary number of lower priority tasks [31].

4 Performance Evaluation

Performance quality measures of robotic manipulators are well known since the publication of [79, 95]. Similarly as in the case of the manipulator, the performance of the mobile robot, its mobility, depends on properties of the Jacobian. Local performance evaluation can be based on certain numeric indices characterizing the endogenous configuration, defined analogously to the dexterity measures of manipulators [1, 7, 12, 15]. Another, more traditional approach, has been proposed in [57].

The contribution of the ECSA to the performance evaluation of mobile robots will be sketched below. For a fixed endogenous configuration $u(\cdot) \in \mathscr{U}$ we define the unit sphere

$$S_{q_0,T}(u(\cdot)) = \{v(\cdot) \in \mathscr{U} \mid ||v(\cdot)|| = 1\} \tag{84}$$

of directions of motion in the endogenous configuration space, centered at $u(\cdot)$. Performance indices of the mobile robot depend on the directions of motion in the task space that correspond to this sphere. Let $u(\cdot)$ denote a regular, endogenous configuration and let $J^{JP\#}_{q_0,T}$ be pseudoinverse (46) of the Jacobian. Then the image of the unit sphere in the task space has the form

$$\begin{aligned} &J_{q_0,T}(u(\cdot))S_{q_0,T} = \{w = J_{q_0,T}(u(\cdot))v(\cdot) \mid ||v(\cdot)|| = 1\} \qquad (85)\\ &= \{w \in \mathbb{R}^r \mid ||J^{JP\#}_{q_0,T}(u(\cdot))w|| = 1\} = \{w \in \mathbb{R}^r \mid w^T \mathscr{M}^{-1}_{q_0,T}(u(\cdot))w = 1\} = E_{q_0,T}(u(\cdot)), \end{aligned}$$

and is an ellipsoid determined by the mobility matrix of the robot. This ellipsoid will be referred to as the mobility ellipsoid of the robot in configuration $u(\cdot)$. It is easy to check that the mobility ellipsoid is inscribed into a sphere in $\mathbb{R}^r$ centered at point $0 \in \mathbb{R}^r$ and radius equal to the square root $\overline{\lambda}^{1/2}_{\mathscr{M}_{q_0,T}(u(\cdot))}$ of the biggest eigenvalue of the mobility matrix, and circumscribed on a sphere whose radius equals the square root $\underline{\lambda}^{1/2}_{\mathscr{M}_{q_0,T}(u(\cdot))}$ of the smallest eigenvalues of this matrix. As a singular configuration is approached, the smallest eigenvalue tends to zero and the mobility ellipsoid degenerates.

As local performance indices of the mobile robot we choose certain non-negative, vanishing at singular configurations functions of eigenvalues of the mobility matrix. A typical example is the square root of the determinant,

$$m_{q_0,T}(u(\cdot)) = \sqrt{\det \mathscr{M}_{q_0,T}(u(\cdot))}, \tag{86}$$

which is called the mobility of endogenous configuration $u(\cdot)$. The mobility is proportional to the volume of the mobility ellipsoid, therefore the mobility of a singular configuration is zero. Another example of a local performance index is the condition number

$$\kappa_{q_0,T}(u(\cdot)) = \overline{\lambda}^{-1/2}_{\mathscr{M}_{q_0,T}(u(\cdot))} \underline{\lambda}^{-1/2}_{\mathscr{M}_{q_0,T}(u(\cdot))} \tag{87}$$

of the mobility matrix at a given configuration. The condition number computes a degree of "anisotropy" od configuration $u(\cdot)$; if equals 1 then the mobility ellipsoid converts to a sphere, this means that the motion in any direction of the endogenous configuration space is "amplified" by the Jacobian in the same way. For this reason a configuration whose condition number is equal to 1 is called isotropic. Note that in a singular configuration the condition number grows up to infinity.

Besides the local performance indices one can define global ones, understood as the average value of the local index over a certain region of the endogenous configuration space. Determination of these global indices needs to be preceded by the introduction of a finite-dimensional parametrization of endogenous configurations, for example by approximating control functions by a finite number of functions forming a basis of the endogenous configuration space.

By optimizing the local performance indices it is possible to determine endogenous configurations (control functions) that define so-called patterns of motion of mobile robots. The optimization of global performance indices aids the design process of mobile robots. More details can be found in [7, 15].

4.1 Repeatability

Repeatability is an important property of robot motion planning algorithms. In the field of robotic manipulators repeatability can be explained in the following way. Suppose that we have a sequence of motion planning problems characterized by points $y_{d1}, y_{d2}, \ldots, y_{dk}$ in the task space. Let the initial configuration of the manipulator be q_{01}. The solution of subsequent problems amounts to finding out configurations of the manipulator that allow the manipulator to reach the prescribed points in the task space. There is a natural assumption that in computation of solution of a given problem the algorithm uses as the initial configuration (the starting point) the solution of the preceding problem. Let a certain configuration q_k be a solution of the problem of reaching y_{dk}. Repeatability of the motion planning algorithms means that if $y_{dk} = y_{d1}$ then $q_k = q_1$. In a somewhat condense manner we say that in a repeatable motion planning algorithm, a cycle of problems results in a cycle of solutions. The property of repeatability plays a vital role in the control of manipulators executing cyclic tasks. A geometric condition for repeatability of manipulators has been formulated by Brockett [59].

An adaptation of the concept of repeatability to mobile robots and mobile manipulators is not straightforward [3, 10] and goes along the following lines. Assume that a motion planning algorithm for the mobile robot should solve a sequence of motion planning problems defined by task space points $y_{d1}, \ldots, y_{dk}$, each problem having the same initial state q_0 and the same control horizon T. A solution for each motion planning problem is a certain endogenous configuration $u(\cdot)$, and we assume that the initial configuration for the planning algorithm of a given problem results from solving the preceding problem. In this context, let $u_1(\cdot)$ denote the solution of problem y_{d1} delivered by the motion planning algorithm initialized at a certain $u_0(\cdot)$. Repeatability of the motion planning algorithm means that if the solution of task y_{dk} is $u_k(\cdot)$ and $y_{dk} = y_{d1}$ then $u_k(\cdot) = u_1(\cdot)$. As in the case of the manipulator we shall say that to a cycle of planning problems (all with the same q_0 and T) the repeatable motion planning algorithm assigns a cycle of solutions in $\mathscr{U}$.

To be more instructive, suppose that the motion planning algorithm applied to the mobile robot uses a certain Jacobian inverse $J^{\#}_{q_0,T}(u(\cdot))$, so that

$$u'_\theta(\cdot) = -\gamma J^{\#}_{q_0,T}(u_\theta(\cdot))(\mathscr{K}_{q_0,T}(u_\theta(\cdot)) - y_d), \quad u_{\theta=0}(\cdot) = u_0(\cdot). \tag{88}$$

In the geometric sense, the right hand side of this identity contains r vector fields on $\mathscr{U}$ that span the distribution

$$D^{\#}_{q_0,T} = span_{C^\infty(\mathscr{U})}\left\{J^{\#}_{q_0,T}e_1, \ldots, J^{\#}_{q_0,T}e_r\right\}, \tag{89}$$

where e_i denotes the i-th unit vector in $\mathbb{R}^r$. Similarly as for the repeatability of motion planning algorithms for manipulators, the motion planning algorithm defined by an inverse Jacobian $J^{\#}_{q_0,T}(u(\cdot))$ is repeatable if distribution $D^{\#}_{q_0,T}$ is integrable, i.e. the Lie bracket of vector fields belonging to this distribution remains in the distribution,

$$[J^{\#}_{q_0,T}e_i, J^{\#}_{q_0,T}e_j] \in D^{\#}_{q_0,T}, \tag{90}$$

for every $i, j = 1, 2 \ldots, r$. Integrability of the distribution yields that endogenous configurations which solve a sequence of motion planning problems belong to the integral manifold of distribution $D^{\#}_{q_0,T}$ passing through $u_0(\cdot)$ and that there exists only one configuration on this manifold that corresponds to a given point in the task space.

5 Computational Methods

As we have shown, Jacobian motion planning algorithms of mobile robots rely on solving numerically functional differential equations, either (36), (62) or (77), to get control function trajectory $u_\theta(\cdot)$ in the endogenous configuration space. Two computational methods apply to this case: a parametric method and a non-parametric method, which will be presented below following the exposition in [31]. We shall consider motion planning algorithms based on a Jacobian inverse, defined by equation

(36). In order to facilitate the computations required by the motion planning within the ECSA, a dedicated, free accessible software package has been developed, reported in [37].

5.1 Parametric Method

The parametric approach consists of expanding the control functions of system (2) into an orthogonal series, and truncating the series after a finite number of terms (control parameters) [5, 8, 76]. Thanks to this, computations involved in the solution of the motion planning problem become finite-dimensional, and reduce to the determination of a trajectory in a finite-dimensional space of control parameters. More formally, let $u_i(t)$ denote the value at time instant $t \in [0, T]$ of the i-th component of control function $u(\cdot) \in \mathscr{U}$. Function $u_i(t)$ can be represented in a certain orthogonal basis $\varphi_0(t), \varphi_1(t), \ldots, \varphi_j(t) \ldots$ of endogenous configuration space $L^2_m[0, T]$ as an infinite series, but we retain only the terms up to a number p, so that

$$u_i(t) = \sum_{k=0}^{p} \lambda_{ik}\varphi_k(t). \tag{91}$$

Continuing in this way for $i = 1, 2, \ldots m$ we arrive at a representation of control function $u(\cdot)$ in the form

$$u_\lambda(t) = P_s(t)\lambda, \tag{92}$$

where $s = m(p+1)$, and matrix $P_s(\lambda)$ is a block-diagonal matrix composed of m blocks of basic function

$$P_s(t) = \begin{bmatrix} P(t) & 0 & \ldots & 0 \\ 0 & P(t) & \ldots & 0 \\ \vdots & & & \vdots \\ 0 & 0 & \ldots & P(t) \end{bmatrix}, \tag{93}$$

each of the row of the form $P(t) = \left[\varphi_0(t), \varphi_1(t), \ldots, \varphi_p(t)\right]$. The vector of control parameters $\lambda \in \mathbb{R}^s$ consists of parameters of subsequent components $u_i(t)$ of the control function. The basic functions are orthogonal in the sense that

$$\int_0^T P_s^\top(t) P_s(t) dt = \mathbb{I}_s. \tag{94}$$

After passing to parametrized control functions the endogenous configuration space becomes finite dimensional, $\mathscr{U} \cong \mathbb{R}^s$, and control-affine system (2) takes the form

$$\dot{q} = f(q) + G(q)u_\lambda, \quad y = h(q). \tag{95}$$

Now, let $q_\lambda(t) = \varphi_{q_0,t}(u_\lambda(\cdot))$ be a parametrized trajectory of the system and let $H_{q_0,T}(\lambda) = h(q_\lambda(T))$ denote the parametrized input-output map. For the pair of control and state trajectories $(u_\lambda(t), q_\lambda(t))$ we compute matrices $A_\lambda(t)$, $B_\lambda(t)$, $C_\lambda(t)$ of linear approximation (20) and fundamental matrix $\Phi_\lambda(t,s)$. Then, the parametrized Jacobian expresses as matrix

$$J_{q_0,T}(\lambda) = \frac{\partial H_{q_0,T}(\lambda)}{\partial \lambda} = C_\lambda(T)\int_0^T \Phi_\lambda(T,t)B_\lambda(t)P_s(t)dt, \tag{96}$$

of dimension $r \times s$, where $H_{q_0,T}(\lambda) = \mathscr{H}_{q_0,T}(u_\lambda(\cdot))$. It is easy to notice that a necessary condition for regularity of configuration λ is $s \geq r$, which means that the number of control parameters should be greater or equal to the dimension of the task space (i.e. of the output space of system (2)).

Suppose that $J^{\#}_{q_0,T}(\lambda)$ denotes a right inverse of matrix $J_{q_0,T}(\lambda)$. Consider a curve $\lambda(\theta) \in \mathbb{R}^s$, $\theta \in \mathbb{R}$. Then, the parametric inverse Jacobian motion planning algorithm is defined by the differential equation

$$\lambda'(\theta) = -\gamma J^{\#}_{q_0,T}(\lambda(\theta))(H_{q_0,T}(\lambda(\theta)) - y_d), \quad u_{\lambda(0)}(t) = u_0(t), \tag{97}$$

along with the limit condition $\lambda_d = \lim_{\theta\to+\infty}\lambda(\theta)$. The parametric algorithm is well defined in regular configurations at which the rank of parametrized mobility matrix $\mathscr{M}_{q_0,T}(\lambda) = J_{q_0,T}(\lambda)J^{\top}_{q_0,T}(\lambda)$ equals r. Differential equation (97) needs to be solved numerically, by means of suitable computational procedures. The simplest one, the Euler method, relies on the constant step discretizing of variable θ that leads to the difference equation

$$\lambda_{i+1} = \lambda_i - \gamma J^{\#}_{q_0,T}(\lambda_i)(H_{q_0,T}(\lambda_i) - y_d). \tag{98}$$

5.2 *Non-parametric Method*

In contrast to the parametric method, the non-parametric method consists of a direct solving numerically of equation (36), without resorting to any parametrization of control functions. The procedure is the following. We are searching for a certain curve $u_\theta(\cdot)$ in the endogenous configuration space. It is assumed that the control horizon T is given, and that for every $\theta \in \mathbb{R}$ the initial state is the same, equal to $q_\theta(0) = q_0 \in \mathbb{R}^n$. The initial condition for the control takes the form $u_{\theta=0}(t) = u_0(t)$. The procedure consists of two stages. At the first stage, for given θ we solve equation. (2) of system

$$\dot{q}_\theta(t) = f(q_\theta(t)) + G(q_\theta(t))u_\theta(t) \tag{99}$$

and get trajectory $q_\theta(t)$. Next, for given $u_\theta(t)$, $q_\theta(t)$, we compute the motion planning error

$$e(\theta) = h(q_\theta(T)) - y_d \tag{100}$$

and matrices

$$A_\theta(t) = \frac{\partial\,(f(q_\theta(t)) + G(q_\theta(t))u_\theta(t))}{\partial q}, \quad B_\theta(t) = G(q_\theta(t)),$$
$$C_\theta(t) = \frac{\partial h(q_\theta(t))}{\partial q} \tag{101}$$

of the linear approximation. The first stage ends up with a determination of Jacobian $J_{q_0,T}(u_\theta(\cdot))$, and its inverse $J^{\#}_{q_0,T}(u_\theta(\cdot))$. At the second stage we perform one step of integration of the differential equation (36)

$$u'_\theta(t) = -\gamma J^{\#}_{q_0,T}(u_\theta(\cdot))e(\theta), \tag{102}$$

and, with a new value θ return to the stage no 1. After sufficient repetitions of this procedure we obtain solution $u_d(t)$ for the motion planning problem. In order to solve equations (99) and (102) two solvers of differential equations need to be employed, e.g. chosen from among the solvers available in MATLAB. For more details the interested reader is directed to [31].

A comparative analysis of both these methods confirms an intuition that when dimension s of the parametrized endogenous configuration space is low, the parametric methods is computationally more efficient than the non-parametric one. However, it has been observed that as the number of control parameters increases (e.g. $s = 30$ considered in [31]), the computation times required by the two methods become comparable. For higher dimensional parametrized endogenous configuration spaces advantages of the non-parametric method prevail, such as a bigger accuracy of computations and a better quality of obtained trajectories.

6 Conclusion

This chapter has been devoted to an overview of concepts, ideas, and algorithms that characterize ECSA as a unifying robotics methodology. Connections of ECSA and the contemporary control theory have been highlighted. Specific results obtained within ECSA and its diverse applications are described in the Bibliography of ECSA preceding the list of references.

Acknowledgements This research has been supported by Wrocław University of Science and Technology under a statutory research project.

References

Bibliography of selected works devoted to the Endogenous Configuaration Space Approach, set in a chronological order

1. Tchoń K, Muszyński, R.: Instantaneous kinematics and dexterity of mobile manipulators. In: Proc 2000 IEEE Int Conf Robot Automat, pp. 2493–2498, San Francisco, CA (2000)
2. Tchoń, K., Jakubiak, J., Muszyński, R.: Kinematics of mobile manipulators: a control theoretic perspective. Arch. Control Sci. **11**, 195–221 (2001)
3. Tchoń, K.: Repeatability of inverse kinematics algorithms for mobile manipulators. IEEE Trans. Autom. Control **47**, 1376–1380 (2002)
4. Tchoń, K., Jakubiak, J.: Extended Jacobian inverse kinematics algorithms for mobile manipulators. J. Robotic Syst. **19**, 443–454 (2002)
5. Tchoń, K., Jakubiak, J.: Endogenous configuration space approach to mobile manipulators: A derivation and performance assessment of Jacobian inverse kinematics algorithms. Int. J. Control **76**, 1387–1419 (2003)
6. Tchoń, K., Jakubiak, J., Muszyński, R.: Regular Jacobian motion planning algorithms for mobile manipulators. In: Proc 15th IFAC World Congress, pp. 121–126, Barcelona Spain (2003)
7. Tchoń, K., Zadarnowska, K.: Kinematic dexterity of mobile manipulators: An endogenous configuration space approach. Robotica **21**, 521–530 (2003)
8. Tchoń, K., Jakubiak, J.: Fourier vs. non-Fourier band-limited Jacobian inverse kinematics algorithms for mobile manipulators. In: Proc 10th IEEE Int Conf MMAR, Międzyzdroje Poland, pp. 1005–1010 (2004)
9. Tchoń, K., Jakubiak, J., Zadarnowska, K.: Doubly non-holonomic mobile manipulators. In: Proc 2004 IEEE ICRA, pp. 4590– 4595, New Orleans LO (2004)
10. Tchoń, K., Jakubiak, J.: A repeatable inverse kinematics algorithm with linear invariant subspaces for mobile manipulators. IEEE Trans. Syst. Man Cybernet. Part B Cybernet. **35**, 1051–1057 (2005)
11. Tchoń, K., Jakubiak, J.: A hyperbolic, extended Jacobian inverse kinematics algorithm for mobile manipulators. In: Proc 16th IFAC World Congress, pp. 43–48, Prague, Czechia (2005)
12. Zadarnowska, K.: Dexterity Measures of Mobile Manipulators. Doctoral Dissertation, Wrocław University of Technology (in Polish) (2005)
13. Tchoń, K.: Repeatable, extended Jacobian inverse kinematics algorithm for mobile manipulators. Syst. Control Lett. **55**, 87–93 (2006)
14. Tchoń, K., Jakubiak, J.: Extended Jacobian inverse kinematics algorithm for nonholonomic mobile robots. Int. J. Control **79**, 895–909 (2006)
15. Zadarnowska, K., Tchoń, K.: A control theory framework for performance evaluation of mobile manipulators. Robotica **25**, 703–715 (2007)
16. Muszyński, R., Jakubiak, J.: On predictive approach to inverse kinematics of mobile manipulators. In: Proc IEEE Int Conf Control Automation, pp. 2423–2428, Guangzhou, China (2007)
17. Kabała, M.: Practical Realization of Robot Control Algorithms Based on the Model of Dynamics: Sperical Robot RoBall—Design, Modeling, Motion Planning. Doctoral Dissertation, Wrocław University of Technology (in Polish) (2008)
18. Janiak, M.: Jacobian Inverse Kinematics Algorithms for Mobile Manipulators with Constraints on State, Control, and Performance. Doctoral Dissertation, Wrocław University of Technology (in Polish) (2009)
19. Małek, Ł.: Convergence of Jacobian Inverse Kinematics Algorithms Based on the Method of Homotopy. Doctoral Dissertation, Wrocław University of Technology (in Polish) (2009)
20. Tchoń, K., Jakubiak, J., Małek, Ł.: Motion planning of nonholonomic systems with dynamics. Computational Kinematics, Springer-Verlag, pp.125–132 (2009)

21. Tchoń, K., Małek, Ł.: On dynamic properties of singularity robust Jacobian inverse kinematics. IEEE Trans. Autom. Contr. **54**, 1402–1406 (2009)
22. Tchoń, K.: Iterative learning control and the singularity robust Jacobian inverse for mobile manipulators. Int. J. Control. **83**, 2253–2260 (2010)
23. Karpińska, J., Ratajczak, A., Tchoń, K.: Task-priority motion planning of underactuated systems: An endogenous configuration space approach. Robotica **28**, 885–892 (2010)
24. Jakubiak, J., Tchoń, K., Magiera, W.: Motion planning in velocity affine mechanical systems. Int. J. Control **83**, 1965–1974 (2010)
25. Janiak, M., Tchoń, K.: Towards constrained motion planning of mobile manipulators. In: Proc 2010 IEEE ICRA, pp. 4990–4995, Anchorage, Alaska (2010)
26. Janiak, M., Tchoń, K.: Constrained motion planning of nonholonomic systems. Syst. Control Lett. **60**, 625–631 (2011)
27. Ratajczak, A.: Motion Planning of Underactuated Robotic Systems. Doctoral Dissertation, Wrocław University of Science and Technology (in Polish) (2011)
28. Ratajczak, A., Tchoń, K.: Motion planning of a balancing robot with threefold sub-tasks: An endogenous configuration space approach. In: Proc 2011 ICRA, pp. 6096–6101, Shanghai, China (2011)
29. Paszuk, D., Tchoń, K., Pietrowska, Z.: Motion planning of the trident snake robot equipped with passive or active wheels. Bull. Polish Ac. Sci. Ser. Tech. Sci. **60**, 547–554 (2012)
30. Kędzierski, K.: Control System of a Social Robot. Doctoral Dissertation, Wrocław University of Science and Technology (in Polish) (2013)
31. Ratajczak, A., Tchoń, K.: Multiple-task motion planning of non-holonomic systems with dynamics. Mech. Sci. **4**, 153–166 (2013)
32. Pietrowska, Z., Tchoń, K.: Dynamics and motion planning of trident snake robot. J. Intell. Robotic Syst. **75**, 17–28 (2014)
33. Jakubiak, J., Magiera, W., Tchoń, K.: Control and motion planning of a non-holonomic parallel orienting platform. J. Mech. Robotics **7** (2015). https://doi.org/10.1115/1.4029891
34. Ratajczak, J.: Design of inverse kinematics algorithms: extended Jacobian approximation of the dynamically consistent Jacobian inverse. Arch. Control Sci. **25**, 35–50 (2015)
35. Tchoń, K., Ratajczak, A., Góral, I.: Lagrangian Jacobian inverse for nonholonomic robotic systems. Nonlinear Dyn. **82**, 1923–1932 (2015)
36. Tchoń, K., et al.: Modeling and control of a skid-steering mobile platform with coupled side wheels. Bull. Polish Ac. Sci. Ser. Tech. Sci. **63**, 807–818 (2015)
37. Chojnacki, Ł.: Framework for ECSA Algorithms. Research report, Department of Cybernetics and Robotics, Wrocław University of Science and Technology (in Polish) (2016)
38. Ratajczak, A.: Trajectory reproduction and trajectory tracking problem for the nonholonomic systems. Bull. Polish Ac. Sci. Ser. Tech. Sci. **64**, 63–70 (2016)
39. Ratajczak, J., Tchoń, K.: Dynamically consistent Jacobian inverse for mobile manipulators. Int. J. Control **89**, 1159–1168 (2016)
40. Tchoń, K., Ratajczak, J.: Dynamically consistent Jacobian inverse for nonholonomic systems. Nonlinear Dyn. **85**, 107–122 (2016)
41. Ratajczak, J., Tchoń, K.: On dynamically consistent Jacobian inverse for non-holonomic robotic systems. Arch. Control Sci. **27**, 555–573 (2017)
42. Tchoń, K.: Endogenous configuration space approach: An intersection of robotics and control theory. In: Nonlinear Systems, pp. 209–233, Springer (2017)
43. Góral, I., Tchoń, K.: Lagrangian Jacobian motion planning: A parametric approach. J. Intell. Robotic Syst. **85**, 511–522 (2017)
44. Ratajczak, A.: Egalitarian vs. prioritarian approach in multiple task motion planning for nonholonomic systems. Nonlinear Dyn. **88**, 1733–1747 (2017)
45. Tchoń, K., Góral, I.: Optimal motion planning for non-holonomic robotic systems. In: Proc 20th IFAC World Congress, pp. 1946–1951 Toulouse, France (2017)
46. Zadarnowska, K.: Switched modeling and task-priority motion planning of wheeled mobile robots subject to slipping. J. Intell. Robotic Syst. **85**, 449–469 (2017)

47. Tchoń, K., Ratajczak, J.: General Lagrange-type Jacobian inverse for nonholonomic robotic systems. IEEE Trans. Robotics **34**, 256–263 (2018)
48. Ratajczak, A.: Motion planning for nonholonomic systems with earlier destination reaching. Arch. Control Sci. **27**, 269–283 (2018)
49. Ratajczak, J., Tchoń, K.: Dynamic nonholonomic motion planning by means of dynamically consistent Jacobian inverse. IMA J. Math. Control Inf. **35**, 479–489 (2018)
50. Tchoń, K., Respondek, W., Ratajczak, J.: Normal forms and configuration singularities of a space manipulator. J. Intell. Robotic Syst. **93**, 621–634 (2019)
51. Tchoń, K., Ratajczak, J.: Singularities, Normal Forms, and Motion Planning for Non-holonomic Robotic System. In: Proc. 6th Int. Conf. Control, Dynamic Systems, Robotics (CDSR'19), Ottawa, Canada, CDSR 127-1–cdsr-127-8 (2019)
52. Tchoń, K., Ratajczak, J.: Feedback equivalence and motion planning of a space manipulator. In: Advances in Mechanism and Machine Science, Mechanisms and Machine Science, vol. 73, T. Uhl (ed.), pp. 1691–1700, Springer Nature Switzerland AG (2019)
53. Ratajczak, A., Ratajczak, J.: Trajectory Reproduction Algorithm in Application to an On–Orbit Docking Maneuver with Tumbling Target. In: Proc. 12th Int. Workshop RoMoCo, vol. 8–10, pp. 172–177, Poznań University of Technology, Poznań, Poland (2019)

References

54. Allgower, E.L., Georg, K.: Numerical Continuation Methods. Springer-Verlag, Berlin (1990)
55. Alouges, F., Chitour, Y., Long, R.: A motion-planning algorithm for the rolling-body problem. IEEE Trans. Robot. **26**, 827–836 (2010)
56. Baillieul, J.: Kinematic programming alternatives for redundant manipulators, pp. 722–728. Proc IEEE ICRA, St. Luois, MO (1985)
57. Bayle, B., Fourquet, J.Y., Renaud, M.: Manipulability of wheeled mobile manipulators: application to motion generation. Int. J. Robot. Res. **22**, 565–581 (2003)
58. Bonnard, B., Chyba, M.: Singular Trajectories and Their Role in Control Theory Springer. Paris (2003)
59. Brockett, R.W.: Robotic manipulators and the product of exponentials formula. W: Mathematical Theory of Networks and Systems. Springer-Berlin, pp. 120–129 (1984)
60. Chelouah, A., Chitour, Y.: On the motion planning of rolling surfaces. Forum Math. **15**, 727–758 (2003)
61. Chitour, Y.: A homotopy continuation method for trajectories generation of nonholonomic systems. ESAIM: Control Optim. Calc. Var. **12**, 139–168 (2006)
62. Chitour, Y., Sussmann, H.J.: Motion planning using the continuation method. In: W. Baillieul, J., Sastry, S.S., Sussmann, H.J. (eds.), 91–125, Essays on Mathematical Robotics, Springer-Verlag, New York (1998)
63. Chitour, Y., Jean, F., Trélat, E.: Singular trajectories of control-affine systems. SIAM J. Contr. Opt. **47**, 1078–1095 (2008)
64. Choi, Y., et al.: Multiple task manipulation for a robotic manipulator. Adv. Robot. **18**, 637–653 (2004)
65. Davidenko, D.: On a new method of numerically integrating a system of nonlinear equations. Dokl Akad Nauk SSSR **88**, 601–604 (1953)
66. Deuflhard, P.: Newton Methods for Nonlinear Problems. Springer, Berlin (2004)
67. Divelbiss, A.W., Seereeram, S., Wen, J.T.: Kinematic path planning for robots with holonomic and nonholonomic constraints. In: Sastry, S.S., Sussmann, H.J. (eds.) Baillieul J, pp. 127–150. Essays on Mathematical Robotics, Springer-Verlag, New York (1998)
68. Dulęba, I.: Algorithms of Motion Plannng for Nonholonomic Robots. Oficyna Wydawnicza PWr, Wrocław (1998)

69. Dulęba, I., Khefifi, W., Karcz-Dulęba, I.: Layer, Lie algebraic method of motion planning for nonholonomic systems. J. Franklin Inst. **349**, 201–215 (2012)
70. Dulęba, I., Sąsiadek, J.Z.: Non-holonomic motion planning based on Newton algorithm with energy optimization. IEEE Trans. Contr. Syst. Technol. **11**, 355–363 (2003)
71. Galicki, M.: Inverse kinematics solution to mobile manipulators. Int. J. Robot. Res. **22**, 1041–1064 (2003)
72. Galicki, M.: Tracking the kinematically optimal trajectories by mobile manipulators. J. Intell. Robot. Syst. (2018). https://doi.org/10.1007/s10846-018-0868-7
73. Ishikawa, M., Minati, Y., Sugie, T.: Development and control experiment of the trident snake robot. IEEE/ASME Trans. Mech. **15**, 9–15 (2010)
74. Jakubczyk, B.: Equivalence and invariants of nonlinear control systems. W: Nonlinear Controllability and Optimal Control, pp. 177–218, M. Dekker, New York (1990)
75. Johnson, C.D., Gibson, J.E.: Singular solutions in problems of optimal control. IEEE Trans. Autom. Contr. **8**, 4–15 (1963)
76. Jung, S., Wen, J.T.: Nonlinear model predictive control for the swing-up of a rotary inverted pendulum. Trans. ASME **126**, 666–673 (2004)
77. Khatib, O.: Motion/Force redundancy of manipulators. W: Proc Japan-USA Symposium on Flexible Automation: A Pacific Rim Conference, pp. 337–342, The Institute, Kyoto (1990)
78. Khatib, O.: Inertial properties in robotics manipulation: An object-level framework. Int. J. Robot. Res. **14**, 19–36 (1995)
79. Klein, ChA, Blaho, B.E.: Dexterity measures for the design and control of kinematically redundant manipulators. Int. J. Robot. Res. **6**, 72–82 (1987)
80. L'Affito, A., Haddad, W.M.: Abnormal optimal trajectory planning of multi-body systems in the presence of holonomic and nonholonomic constraints. J. Intell. Robot. Syst. **89**, 51–67 (2018)
81. Li, Z., Ge, S.: Fundamentals in modelling and control of mobile manipulators. CRC Press, Boca Raton (2013)
82. Mazur, A.: Model-based Control of Non-holonomic Mobile Manipulators. Oficyna Wydawnicza PWr, Wrocław (in Polish) (2009)
83. Nakamura, Y., Hanafusa, H., Yoshikawa, T.: Task-priority based redundancy control of robot manipulators. Int. J. Robot. Res. **6**, 3–15 (1987)
84. Nijmeijer, H., van der Schaft, A.J.: Nonlinear Dynamical Control Systems. Springer, New York (1990)
85. Popa, D.O., Wen, J.T.: Singularity computation for iterative control of nonlinear affine systems. Asian J. Control **2**, 57–75 (2000)
86. Rugh, W.J.: Linear System Theory. Prentice Hall, Englewood Cliffs (1993)
87. Rybus, T., Seweryn, K.: Planar air-bearing microgravity simulators: review of applications, existing solutions and design parameters. Acta Astronaut. **120**, 239–259 (2016)
88. Sciavicco, L., Siciliano, B.: Coordinate transformation: A solution algorithm for one class of robots. IEEE Trans. Syst. Man Cybernet. **16**, 550–559 (1986)
89. Sontag, E.D.: Mathematical Control Theory. Springer-Verlag, New York (1990)
90. Sontag, E.D.: A general approach to path planning for systems without drift. In: Baillieul, J., Sastry, S.S., Sussmann, H.J. (eds.) Essays on Mathematical Robotics, pp. 151–168. Springer-Verlag, New York (1998)
91. Sussmann, H.J.: A continuation method for nonholonomic path finding problems. W: Proc 32nd IEEE CDC, pp. 2718–2723, San Antonio, TX (1993)
92. Tchoń, K., Muszyński, R.: Singular inverse kinematic problem for robotic manipulators: A normal form approach. IEEE Trans. Robot. Automat. **14**, 93–104 (1998)
93. Tzafestas, S.G.: Introduction to Mobile Robot Control. Elsevier, Amsterdam (2014)
94. Ważewski, T.: Sur l'évaluation du domaine d'existence des fonctions implicites réelles ou complexes. Ann. Soc. Pol. Math. **20**, 81–120 (1947)
95. Yoshikawa, T.: Manipulability of robotic mechanisms. Int. J. Robot. Res. **4**, 3–9 (1985)

Control of a Mobile Robot Formation Using Artificial Potential Functions

Krzysztof Kozłowski and Wojciech Kowalczyk

Abstract This chapter presents three control algorithms for a group of differentially-driven mobile robots: based on linearization, persistent excitation and vector field orientation method. All the presented algorithms have been extended by the authors with a collision avoidance mechanism based on the artificial potentials functions. For each of them, stability analysis was conducted using Lyapunov method. Both stability analysis and numerical verification of the effectiveness of the control methods were carried out in two stages: first, the case of a single robot moving in the environment with static circle-shaped obstacles was considered, then the same algorithm was extended to multi-robot version. The robots were executing synchronous motion while avoiding inter-agent collisions.

1 Introduction

Until the mid-eighties of the last century the majority of researchers believed that the problem of collision avoidance in robotics can be solved by planning collision-free motion trajectories. In such an approach, the task of the control system is to track the reference trajectory as precisely as possible by a manipulator or mobile platform. Planning, as a process usually requiring a greater computing power, was carried out once, before the start of the control process. Such solution can be effective in static environment or when the workspace is fully controlled. A practical example is a robotic nest on a production line where the safety cage prevents human from entering the vicinity of the robot, and the place and time of delivery of the parts to be processed are known.

K. Kozłowski (✉) · W. Kowalczyk
Poznan University of Technology, Piotrowo 3A, Poznań, Poland
e-mail: krzysztof.kozlowski@put.poznan.pl

W. Kowalczyk
e-mail: wojciech.kowalczyk@put.poznan.pl

P. Kulczycki et al. (eds.), *Automatic Control, Robotics, and Information Processing*, Studies in Systems, Decision and Control 296,
https://doi.org/10.1007/978-3-030-48587-0_15

In the rapidly expanding area of service robotics, where the presence of human close to the robot is a natural situation, and in multi-robot autonomous systems, where independent mobile platforms perform a common task, the implementation of collision avoidance at the trajectory planning level may not be sufficient due to inability to guarantee a real-time response of the system.

The first researcher who proposed a solution to this problem was Khatib. First, in his Ph.D. thesis [1] and later in a widely known paper [2] proposed to use the artificial potential functions (APFs) to avoid collisions (the robot was pushed away from the obstacles) while simultaneously executing motion to the goal (by attracting). Khatib presented not only theoretical concepts but also solution of a practical problem using the Puma 560 robot, popular in the 1980s. In the verification he used innovative at that time simulator of the manipulator arm and the environment. It is worth noting that much earlier, in 1977, Laitmann and Skowronski [3] have published a paper in which they considered the control of two autonomous agents, one of whose goal was to avoid collisions. This work was purely theoretical. The authors continued their research in this field in the following years [4].

A separate class of collision avoidance algorithms are methods based on the harmonic potential functions [5, 6].

The APFs have proved to be an effective tool solving collision avoidance problem, although this approach is not free of weaknesses. Due to the local nature of the potential functions proposed by Khatib, local minima may occur in some cases, which are a crucial problem from a point of view of robotic applications.

The answer to the problem of local minima in APFs was the method proposed by Rimon and Koditschk in 1988. In the paper [7] they proposed global potential function, which, if properly 'fine-tuned', guarantees that local minima do not appear. This 'tuned' function is called a navigation function. When used, almost the entire workspace is an attraction set of global minimum. Exceptions are the single points located close to the obstacles. A set of these points is of measure zero. The first article [7] presents the navigation function for a space in which only spherical obstacles occur (a circle in the case of a two-dimensional space). In the next work [8], this method was extended to star type obstacles (obstacles that have a point from which, leading a radius in any direction, the edge of the obstacle will be crossed only once). The core of the method is an innovative idea to transform star-shaped obstacles into a spheres. The surroundings of the obstacles are also transformed. Resulting auxiliary space is the space of spheres. Due to the fact that the transforming function is a diffeomorphism, it is possible to design the direction of motion in every point of the auxiliary space of spheres and transform the result into the space of stars. This solves the problem of finding the direction to the goal in the space of stars. Notice that star type obstacles include a sub-class of non-convex obstacles for which there was no solution in a classic approach based on the local APFs.

In 2004 Urakubo [9] proposed to extend Rimon and Koditschk's algorithm to non-holonomic mobile platforms. Later, Urakubo has published the paper [10] presenting an in-depth analysis of the stability and convergence of the algorithm. The use of the navigation functions to control multiple robots has been investigated in many works [11–16], and [17].

Articles presenting the practical use of navigation functions have been published relatively recently [18–22]. There are also publications on behaviour of robots close to saddle points and their control [23, 24]. Some of the most recent papers concern improvement of the convergence of the proposed algorithms [25–27]. These works provide solutions to the situation when in some cases the navigation function can be almost flat, and consequently the gradient vector used in the control can have a very small module.

Local artificial potential functions have also been used in multi-robot systems [28–31], and [32]. Recently, an important research problem is collision avoidance that addresses limited motion curvature executed by mobile robots [33, 34].

Further in this chapter, local APFs will be considered, which are mainly used to avoid collisions in real time, especially in a highly dynamic environment. A navigation-function approach will not be considered.

Section 2 introduces a kinematic model of a single differentially-driven mobile robot is presented. Section 3 presents an APF that is used to avoid collisions with static obstacles and between robots. Section 4 introduces an extension of the linearization algorithm from [35] to include a collision avoidance mechanism. This subsection contains stability analysis of the algorithm and numerical simulations that illustrate the effectiveness of the proposed method. Section 5 presents model of the system of N mobile robots, while the Sect. 6 presents an APF for this system. In Sect. 7 the algorithm from the work [35] is extended to the case of N robots forming a chain. Stability analysis and simulation results are also given. Section 8 presents an algorithm based on the article [36], in which two modifications were made: it is used to control single robot and extended to collision avoidance functionality. It includes stability analysis and numerical verification. In Sect. 9 the same algorithm is applied to a group of robots forming a chain. The authors expanded the algorithm to include collision avoidance and conducted stability analysis. The results of numerical simulations are also presented. Section 10 presents the vector field orientation algorithm from the paper [37] applied to a single robot moving in an environment with obstacles. Stability analysis and simulation results are provided. The results of the experimental verification of this algorithm are presented in papers [38–40], and [41]. Section 11 introduces a vector field orientation algorithm for a multiple robots performing synchronized motion while avoiding collisions. Stability analysis and numerical simulation results illustrating the effectiveness of the method are presented. In the last subsection a summary of the chapter is outlined.

2 Kinematic Model of the Robot

The kinematic model of the two-wheeled mobile robot is given by the following equation:

$$\dot{\mathbf{q}} = \begin{bmatrix} \cos\theta & 0 \\ \sin\theta & 0 \\ 0 & 1 \end{bmatrix} \mathbf{u} \tag{1}$$

where vector $q = [x \ \ y \ \ \theta]^\top$ denotes the pose and x, y, θ are position coordinates and orientation of the robot with respect to a global, fixed coordinate frame. Vector $u = [v \ \omega]^\top$ is the control vector with v denoting linear velocity control and ω denoting angular velocity control of the platform.

3 Artificial Potential Function

Collision avoidance behaviour is based on the APF. All obstacles are surrounded by APFs that raise to infinity near objects border r_j (j—number of the obstacle) and decreases to zero at some distance R_j, $R_j > r_j$.

One can introduce the following function [38]:

$$B_{aj}(l_j) = \begin{cases} 0 & for \quad l_j < r_j \\ e^{\frac{l_j - r_j}{l_j - R_j}} & for \ r_j \le l_j < R_j \\ 0 & for \quad l_j \ge R_j \end{cases}, \tag{2}$$

that gives output $B_{aj}(l_j) \in \langle 0, 1)$. Distance between the robot and the j-th obstacle is defined as the Euclidean length $l_j = \left\| [x_j \ y_j]^\top - [x \ y]^\top \right\|$.

Function given by Eq. (2) is scaled within the range $\langle 0, \infty)$ as follows:

$$V_{aj}(l_j) = \frac{B_{aj}(l_j)}{1 - B_{aj}(l_j)}. \tag{3}$$

that will be used later to avoid collisions. Function $V_{aj}(l_j)$ and its derivatives are bounded for $l_j > r_j$.

In further description terms 'collision area' or 'collision region' is used for locations fulfilling conditions $l_j < r_j$. The range $r_j < l_j < R_j$ is called 'collision avoidance area' or 'collision avoidance region'.

4 Control Based on Linearization

This subsection presents extension of the control algorithm from [35] allowing collision avoidance. At first, a version of the algorithm for a single robot in the environment with static obstacles is presented. Later, this method is used to control a group of robots moving one after another. The coordinates of the preceding robot after adding a certain displacement (which is a design parameter) are used as a reference signal.

4.1 Control Algorithm

The purpose of the control is to track a desired trajectory by a group of robots while avoiding collisions. It is assumed that the planned trajectory is collision-free, i.e. it does not pass through the collision avoidance region of any obstacle.

Tracking the desired trajectory is equivalent to bring the following quantities to zero:

$$\begin{aligned} p_x &= x_d - x \\ p_y &= y_d - y \\ p_\theta &= \theta_d - \theta, \end{aligned} \tag{4}$$

where x_d and y_d are the desired position coordinates, and θ_d is the desired orientation. The tracking error expressed with respect to the local coordinate frame fixed to the robot is as follows:

$$\begin{bmatrix} e_x \\ e_y \\ e_\theta \end{bmatrix} = \begin{bmatrix} \cos(\theta) & \sin(\theta) & 0 \\ -\sin(\theta) & \cos(\theta) & 0 \\ 0 & 0 & 1 \end{bmatrix} \begin{bmatrix} p_x \\ p_y \\ p_\theta \end{bmatrix}. \tag{5}$$

Using the above equations and nonholonomic constraint equation $\dot{y}\cos(\theta) - \dot{x}\sin(\theta) = 0$ the error dynamics can be described as follows:

$$\begin{aligned} \dot{e}_x &= e_y\omega - v + v_d \cos e_\theta \\ \dot{e}_y &= -e_x\omega + v_d \sin e_\theta \\ \dot{e}_\theta &= \omega_d - \omega \end{aligned}. \tag{6}$$

Position correction variables are introduced to combine position error and collision avoidance terms:

$$\begin{aligned} P_x &= p_x - \textstyle\sum_{j=1}^{M} \frac{\partial V_{aj}}{\partial x} \\ P_y &= p_y - \textstyle\sum_{j=1}^{M} \frac{\partial V_{aj}}{\partial y} \end{aligned}, \tag{7}$$

where M—number of obstacles. V_{aj} is function of x and y according to Eq. (3). Correction variables can be transformed to the local coordinate frame fixed to the robot as follows:

$$\begin{bmatrix} E_x \\ E_y \\ e_\theta \end{bmatrix} = \begin{bmatrix} \cos(\theta) & \sin(\theta) & 0 \\ -\sin(\theta) & \cos(\theta) & 0 \\ 0 & 0 & 1 \end{bmatrix} \begin{bmatrix} P_x \\ P_y \\ p_\theta \end{bmatrix}. \tag{8}$$

Differentiating first two equations in (4) with respect to p_x and p_y one obtains:

$$\frac{\partial x}{\partial p_x} = -1, \quad \frac{\partial y}{\partial p_y} = -1. \tag{9}$$

Using (9) one can write:

$$\frac{\partial V_{aj}}{\partial p_x} = \frac{\partial V_{aj}}{\partial x}\frac{\partial x}{\partial p_x} = -\frac{\partial V_{aj}}{\partial x}, \qquad \frac{\partial V_{aj}}{\partial p_y} = \frac{\partial V_{aj}}{\partial y}\frac{\partial y}{\partial p_y} = -\frac{\partial V_{aj}}{\partial y}. \tag{10}$$

Using Eqs. (7) and (8) the gradient of the APF can be expressed with respect to the local coordinate frame fixed to the robot:

$$\begin{bmatrix} \frac{\partial V_{aj}}{\partial e_x} \\ \frac{\partial V_{aj}}{\partial e_y} \end{bmatrix} = \begin{bmatrix} \cos\theta & \sin\theta \\ -\sin\theta & \cos\theta \end{bmatrix} \begin{bmatrix} \frac{\partial V_{aj}}{\partial p_x} \\ \frac{\partial V_{aj}}{\partial p_y} \end{bmatrix}. \tag{11}$$

Equation (11) can be verified easily by taking partial derivatives of $V_{aj}(d_x - p_x, d_y - p_y) = V_{aj}(d_x - p_x(e_x, e_y), d_y - p_y(e_x, e_y))$ with respect to e_x, e_y and taking into account inverse transformation of the first two equations of (5).

Using (10) above equation can be written as follows:

$$\begin{bmatrix} \frac{\partial V_{aj}}{\partial e_x} \\ \frac{\partial V_{aj}}{\partial e_y} \end{bmatrix} = \begin{bmatrix} -\cos\theta & -\sin\theta \\ \sin\theta & -\cos\theta \end{bmatrix} \begin{bmatrix} \frac{\partial V_{aj}}{\partial x} \\ \frac{\partial V_{aj}}{\partial y} \end{bmatrix}. \tag{12}$$

Equations (8) using (11) can be transformed to the following form:

$$\begin{aligned} E_x &= p_x \cos(\theta) + p_y \sin(\theta) + \textstyle\sum_{j=1}^{M} \frac{\partial V_{aj}}{\partial e_x} \\ E_y &= -p_x \sin(\theta) + p_y \cos(\theta) + \textstyle\sum_{j=1}^{M} \frac{\partial V_{aj}}{\partial e_y} \\ e_\theta &= p_\theta \end{aligned} \tag{13}$$

where each derivative of the APF is transformed form the global coordinate frame to the local coordinate frame fixed to the robot. Finally, correction variables expressed with respect to the local coordinate frame are as follows:

$$E_x = e_x + \sum_{j=1}^{M} \frac{\partial V_{aj}}{\partial e_x}, \qquad E_y = e_y + \sum_{j=1}^{M} \frac{\partial V_{aj}}{\partial e_y}. \tag{14}$$

For $l_j > R_j$ components of the gradient vector $\frac{\partial V_{aj}}{\partial e_x} = 0$ and $\frac{\partial V_{aj}}{\partial e_y} = 0$. In this case $E_x = e_x$ and $E_y = e_y$.

The algorithm borrowed from [35] is extended to collision avoidance as follows:

$$\begin{aligned} v &= v_d + k_1 E_x \\ \omega &= \omega_d + k_2 \text{sign}(v_d) E_y + k_3 e_\theta \end{aligned}. \tag{15}$$

Assumption 4A

When robot gets into collision avoidance region of the obstacle, its reference signals are frozen, i.e. $\dot{x}_d = 0$, $\dot{y}_d = 0$, reference velocities v_d and ω_d are substituted as 0.

When robot leaves collision avoidance region its reference trajectory is updated to the new value.

Regardless of the definition of sign(•) function for $v_d = 0$ we propose to keep the second term in second Eq. (15) as $k_2 E_y$ in order to avoid possible deadlocks.

Assumption 4B
If the value of the linear control signal is less then considered threshold value v_t, i.e. $|v| < v_t$ (v_t—positive constant), it is replaced by a new scalar function $\tilde{v} = S(v)v_t$, where

$$S(v) = \begin{cases} -1 & \text{for } v < 0 \\ 1 & \text{for } v \geq 0 \end{cases}. \tag{16}$$

Substituting (15) into (6) error dynamics can be expressed a follows:

$$\begin{aligned} \dot{e}_x &= e_y\omega - k_1 E_x + v_d(\cos e_\theta - 1) \\ \dot{e}_y &= -e_x\omega + v_d \sin e_\theta \\ \dot{e}_\theta &= -k_2 \text{sign}(v_d) E_y - k_3 e_\theta \end{aligned} . \tag{17}$$

Transforming (17) using (15) and taking into account *Assumption 4A* (when robot gets into collision avoidance region, velocities v_d and ω_d are substituted as 0) error dynamics can be expressed in the following form:

$$\begin{aligned} \dot{e}_x &= k_3 e_y e_\theta + k_2 e_y E_y - k_1 E_x \\ \dot{e}_y &= -k_3 e_x e_\theta - k_2 e_x E_y \\ \dot{e}_\theta &= -k_2 E_y - k_3 e_\theta \end{aligned} . \tag{18}$$

4.2 Stability of the System

In this subsection stability analysis of the closed-loop system is presented. When the robot is outside of the collision avoidance region of the obstacle (APF takes the value zero) the analysis given in [35] is valid and will be not repeated here. Further analysis is presented for the situation in which the robot is in the collision avoidance region of the obstacle.

Consider the following Lyapunov-like function:

$$V = \frac{1}{2}(e_x^2 + e_y^2 + e_\theta^2) + \sum_{j=1}^{M} V_{aj}. \tag{19}$$

If the robot is in the collision avoidance region of the obstacle, time derivative of the Lyapunov-like function is calculated as follows:

$$\frac{dV}{dt} = e_x\dot{e}_x + e_y\dot{e}_y + e_\theta\dot{e}_\theta + \sum_{j=1}^{M}\left(\frac{\partial V_{aj}}{\partial e_x}\dot{e}_x + \frac{\partial V_{aj}}{\partial e_y}\dot{e}_y\right). \tag{20}$$

Taking into account Eqs. (14) the above formula can be transformed to the following form:

$$\frac{dV}{dt} = E_x\dot{e}_x + E_y\dot{e}_y + e_\theta\dot{e}_\theta. \tag{21}$$

Next, using Eq. (18) one obtains:

$$\dot{V} = k_3E_xe_ye_\theta + k_2E_xe_yE_y - k_1E_x^2 - k_3E_ye_xe_\theta - k_2e_xE_y^2 - k_2E_ye_\theta - k_3e_\theta^2. \tag{22}$$

For further analysis a new variable is introduced: $\theta_E = Atan2(E_y, E_x)$[1]—auxiliary orientation variable. Substituting $E_x = D\cos\theta_E$, $E_y = D\sin\theta_E$, where $D = \sqrt{E_x^2 + E_y^2}$ in the above equation one obtains:

$$\begin{aligned}\dot{V} = {} & k_3D\cos\theta_E e_ye_\theta + k_2D^2e_y\cos\theta_E\sin\theta_E - k_1D^2\cos^2\theta_E - k_3D\sin\theta_E e_xe_\theta \\ & - k_2e_xD^2\sin^2\theta_E - k_2e_\theta D\sin\theta_E - k_3e_\theta^2.\end{aligned} \tag{23}$$

Using an identity substitution

$$-k_3e_\theta^2 = -\frac{1}{4}k_3e_\theta^2 - \frac{1}{4}k_3e_\theta^2 - \frac{1}{2}k_3e_\theta^2 \tag{24}$$

equation (23) can be written as follows:

$$\begin{aligned}\dot{V} = {} & (k_3D\cos\theta_E e_ye_\theta - \frac{1}{4}k_3e_\theta^2) + (-k_3D\sin\theta_E e_xe_\theta - \frac{1}{4}k_3e_\theta^2) \\ & + (-k_2De_\theta\sin\theta_E - \frac{1}{2}k_3e_\theta^2) + k_2D^2e_y\cos\theta_E\sin\theta_E \\ & - k_1D^2\cos^2\theta_E - k_2e_xD^2\sin^2\theta_E \\ = {} & -\left[k_3\left(\frac{1}{2}e_\theta - D\cos\theta_E e_y\right)^2 - k_3D^2\cos^2\theta_E e_y^2\right] \\ & - k_3\left[\left(\frac{1}{2}e_\theta + D\sin\theta_E e_x\right)^2 - D^2\sin^2\theta_E e_x^2\right] \\ & - \left[\left(\sqrt{\frac{k_3}{2}}e_\theta + k_2\frac{1}{\sqrt{2k_3}}D\sin\theta_E\right)^2 - \frac{k_2^2}{2k_3}D^2\sin^2\theta_E\right] \\ & + k_2D^2e_y\cos\theta_E\sin\theta_E - k_1D^2\cos^2\theta_E - k_2e_xD^2\sin^2\theta_E.\end{aligned} \tag{25}$$

[1] $Atan2(\bullet, \bullet)$ is a version of the $Atan(\bullet)$ function covering all four quarters of the Euclidean plane.

To simplify further calculations new scalar functions are introduced:

$$
\begin{aligned}
a &= \frac{1}{2}e_\theta - D\cos\theta_E e_y, \\
b &= \frac{1}{2}e_\theta + D\sin\theta_E e_x, \\
c &= \sqrt{\frac{k_3}{2}}e_\theta + \frac{k_2}{\sqrt{2k_3}}D\sin\theta_E .
\end{aligned}
\tag{26}
$$

Taking into account these functions and after some algebra $\dot{V}$ can be written as follows:

$$
\begin{aligned}
\dot{V} &= -k_3a^2 - k_3b^2 - c^2 - k_1D^2\cos^2\theta_E + D^2\sin^2\theta_E\left(e_x\sqrt{k_3} - \frac{k_2}{2\sqrt{k_3}}\right)^2 \\
&+ \left(\sqrt{k_3}D\cos\theta_E e_y + Dk_2\sin\theta_E\frac{1}{2\sqrt{k_3}}\right)^2 \\
&\leq -k_3a^2 - k_3b^2 - c^2 - k_1D^2\cos^2\theta_E + D^2\left(e_x\sqrt{k_3} - \frac{k_2}{2\sqrt{k_3}}\right)^2 \\
&+ D^2\left(\sqrt{k_3}\cos\theta_E e_y + k_2\sin\theta_E\frac{1}{2\sqrt{k_3}}\right)^2 \\
&= -k_3a^2 - k_3b^2 - c^2 - k_1D^2\cos^2\theta_E + D^2k_3\left(e_x - \frac{k_2}{2k_3}\right)^2 \\
&+ D^2k_3\left(\cos\theta_E e_y + \frac{k_2}{2k_3}\sin\theta_E\right)^2 .
\end{aligned}
$$

The closed-loop system is stable ($\dot{V} \leq 0$) if the following condition is fulfilled:

$$
k_1D^2\cos^2\theta_E - D^2k_3\left(e_x - \frac{k_2}{2k_3}\right)^2 - D^2k_3\left(\cos\theta_E e_y + \frac{k_2}{2k_3}\sin\theta_E\right)^2 \geq 0. \tag{27}
$$

If $D^2 \neq 0$ the condition (27) can be rewritten in te following form:

$$
k_1\cos^2\theta_E \geq k_3\left(e_x - \frac{k_2}{2k_3}\right)^2 + k_3\left(\cos\theta_E e_y + \frac{k_2}{2k_3}\sin\theta_E\right)^2 . \tag{28}
$$

If the term $\cos^2\theta_E$ is close to zero, special procedure described in *Assumption 4B* is applied. It pushes the robot away from this state assuring that $\theta_E \neq \pm\frac{\pi}{2} \pm \pi d$ ($d = 0, \pm1, \pm2, ...$)). In other cases the condition (28) can be met by setting sufficiently high value of k_1.

As shown in [42] collision avoidance is guaranteed if $\dot{V}_{aj} \leq 0$ and $\lim_{||[x\ y]^\top - [x_{aj}\ y_{aj}]^\top|| \to r^+} V_{aj} = +\infty$, where vector $[x_{aj}\ y_{aj}]^\top$ represents location of the centre of the j-th obstacle.

4.3 Numerical Simulation

This subsection presents numerical simulation for the algorithm based on linearization in which the robot avoids collisions with static obstacles. The task is to move along circular desired trajectory with center point (0, 6 m) and radius is 6 m. The motion is counterclockwise with a linear velocity of 0.3 m/s and an angular velocity 0.05 rd/s. The robot is initially at the point (10 m, −10 m) and its orientation is 0 rd. In points (11 m, −8 m) and (6 m, −2 m) two static obstacles are located and their parameters are $r = 0.5$ m, the range of the potential field is $R = 2$ m. The position of the obstacles were chosen in such a way that the robot has to bypass them in the transient state.

The coefficients of the algorithm were set as follows: $k_1 = 1$, $k_2 = 1$, $k_3 = 1$. Figure 1a presents the robot's path in the (x, y)-plane (solid line) and its reference trajectory (dashed line). The robot enters the collision avoidance regions one after another, avoids collisions and finally reaches the reference trajectory. Figure 1b, c show time graphs of the robot coordinates and reference trajectory. The robot reaches reference trajectory after 10 s. Figure 1d, e presents time graphs of the control signals. Dashed line represents the velocity of the reference robot. Initially, control signals are very high, not physically achievable, but this state is very short. Based on the authors' experience on carried out experiments on a real robot, scaling of the wheel velocities may be required, which would allow to maintain the direction of motion of the platform. This would cause a certain extension of the transient state. Figure 1f presents the freeze signal (1—the freeze state, 0—outside the collision avoidance area of the obstacle). The freeze is temporarily activated when the robot is close to the obstacle, but it takes a very short time.

5 Model of the System of N Robots

In the case of the system of N differentially-driven mobile robots model of the kinematics is as follows:

$$\dot{\mathbf{q}}_i = \begin{bmatrix} \cos\theta_i & 0 \\ \sin\theta_i & 0 \\ 0 & 1 \end{bmatrix} \mathbf{u}_i \tag{29}$$

where vectors $\mathbf{q}_i = [x_i\ \ y_i\ \ \theta_i]^\top$ and $\mathbf{u}_i = [v_i\ \ \omega_i]^\top$ are counterparts of the vectors in Eq. (1), where index i denotes number of the robot, N is number of robots.

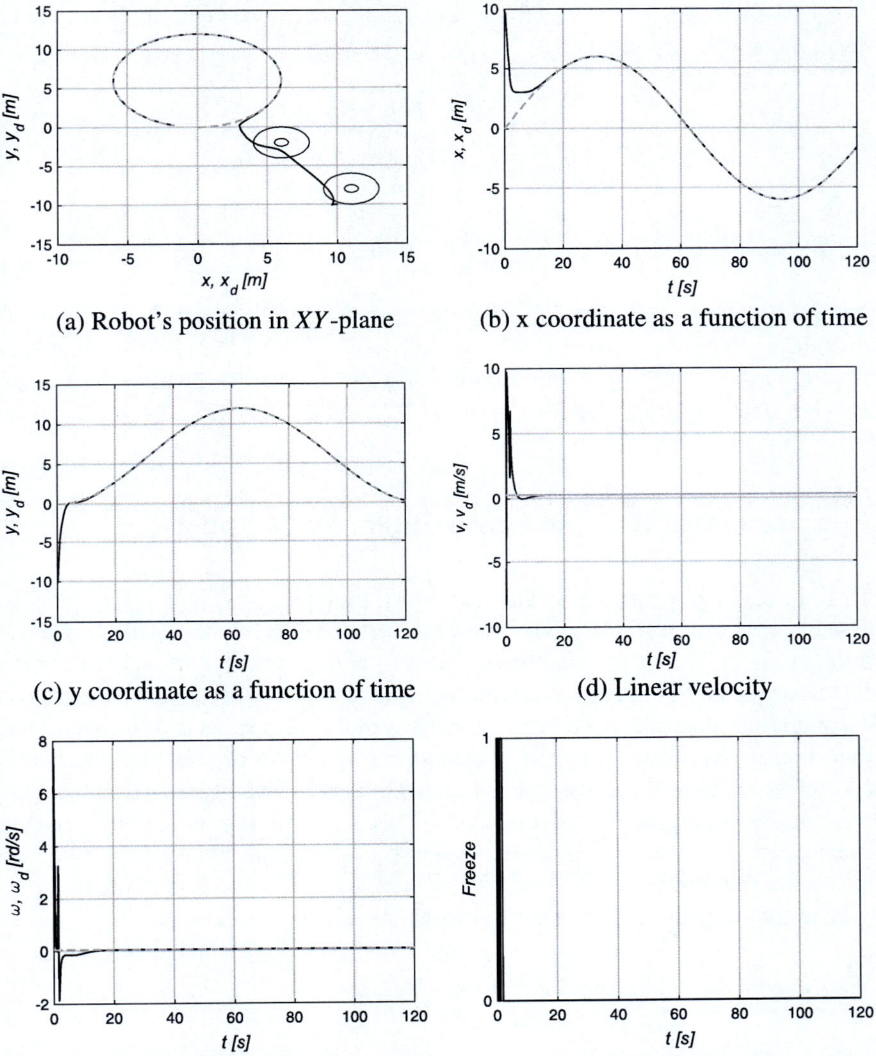

(a) Robot's position in XY-plane

(b) x coordinate as a function of time

(c) y coordinate as a function of time

(d) Linear velocity

(e) Angular velocity

(f) Freeze signal

Fig. 1 Numerical simulation 1: algorithm based on linearization—one robot

6 Artificil Potential Function for Interacting Robots

APFs can be used also to avoid collisions between robots. Each of N robots is surrounded by artificial potential field having the same properties like the one presented in Sect. 3.

In the multi-robot case functions (2) i (3) are extended with additional index denoting number of the robot:

$$B_{aij}(l_{ij}) = \begin{cases} 0 & for \quad l_{ij} < r_j \\ e^{\frac{l_{ij}-r_j}{l_j-R_j}} & for\ r_j \le l_j < R_j \\ 0 & for \quad l_{ij} \ge R_j \end{cases}, \tag{30}$$

$$V_{aij}(l_{ij}) = \frac{B_{aij}(l_{ij})}{1 - B_{aij}(l_{ij})}, \tag{31}$$

Euclidean distance between robots i-th and j-th is given by equation: $l_{ij} = \left\| [x_j \ y_j]^\top - [x_i \ y_i]^\top \right\|$. Function $V_{aij}(l_{ij})$ and its derivative with respect to spatial coordinates are bounded for $l_{ij} > r_j$.

7 Control Based on the Linearization for N Robots

This subsection presents control algorithm for N mobile robots, which is an extension of the method for a single robot from [35]. The robot called the leader (index 1), imitates the motion of a virtual robot (index 0) with a certain spatial displacement. Virtual robot moves along the reference trajectory. The subsequent robots form a chain in which the i-th robot treats the trajectory of the $i - 1$ robot as its own reference trajectory. In the control process, robots use not only the position and orientation coordinates of their predecessor, but also its both linear and angular velocities. As a tracking algorithm, the method presented in Sect. 4 is used. In addition, the algorithm is extended to a collision avoidance mechanism, which is activated if the robots are close to each other.

The control goal is to bring the following quantities to zero:

$$\begin{aligned} p_{ix} &= x_{i-1} - x_i - d_{x\,i-1,i} \\ p_{iy} &= y_{i-1} - y_i - d_{y\,i-1,i} \\ p_{i\theta} &= \theta_{i-1} - \theta_i \end{aligned} \tag{32}$$

where $d_{x\,i-1,i}$ i $d_{y\,i-1,i}$ are design parameters representing components of the spatial displacement between $i - 1$ and i-th robot.

Error expressed in the coordinate frame fixed to the i-th robot is as follows:

$$\begin{bmatrix} e_{ix} \\ e_{iy} \\ e_{i\theta} \end{bmatrix} = \begin{bmatrix} \cos(\theta_i) & \sin(\theta_i) & 0 \\ -\sin(\theta_i) & \cos(\theta_i) & 0 \\ 0 & 0 & 1 \end{bmatrix} \begin{bmatrix} p_{ix} \\ p_{iy} \\ p_{i\theta} \end{bmatrix}. \tag{33}$$

Taking into account above equalities and nonholonomic constraints $\dot{y}_i \cos(\theta_i) - \dot{x}_i \sin(\theta_i) = 0$ error dynamics between the leader and the follower is:

$$\begin{aligned}
\dot{e}_{ix} &= e_{iy}\omega_i - v_i + v_{i-1}\cos e_{i\theta} \\
\dot{e}_{iy} &= -e_{ix}\omega_i + v_{i-1}\sin e_{i\theta} \\
\dot{e}_{i\theta} &= \omega_{i-1} - \omega_i .
\end{aligned} \tag{34}$$

One can introduce the position correction variables that consist of position error of the i-th robot and collision avoidance terms:

$$P_{ix} = p_{ix} - \sum_{j=1, j\neq i}^{N} \frac{\partial V_{aij}}{\partial x_i}, \qquad P_{iy} = p_{iy} - \sum_{j=1, j\neq i}^{N} \frac{\partial V_{aij}}{\partial y_i}. \tag{35}$$

V_{aij} is function of x_i and y_i according to Eq. (32). Variables can me transformed to the local coordinate frame fixed to the robot i-th:

$$\begin{bmatrix} E_{ix} \\ E_{iy} \\ e_{i\theta} \end{bmatrix} = \begin{bmatrix} \cos(\theta_i) & \sin(\theta_i) & 0 \\ -\sin(\theta_i) & \cos(\theta_i) & 0 \\ 0 & 0 & 1 \end{bmatrix} \begin{bmatrix} P_{ix} \\ P_{iy} \\ p_{i\theta} \end{bmatrix}. \tag{36}$$

Following the same procedure as in Sect. 5 one can show that:

$$\begin{bmatrix} \frac{\partial V_{aij}}{\partial e_{ix}} \\ \frac{\partial V_{aij}}{\partial e_{iy}} \end{bmatrix} = \begin{bmatrix} -\cos\theta_i & -\sin\theta_i \\ \sin\theta_i & -\cos\theta_i \end{bmatrix} \begin{bmatrix} \frac{\partial V_{aij}}{\partial x_i} \\ \frac{\partial V_{aij}}{\partial y_i} \end{bmatrix}. \tag{37}$$

and

$$E_{ix} = e_{ix} + \sum_{j=1, j\neq i}^{N} \frac{\partial V_{aij}}{\partial e_{ix}}, \qquad E_{iy} = e_{iy} + \sum_{j=1, j\neq i}^{N} \frac{\partial V_{aij}}{\partial e_{iy}}. \tag{38}$$

The details of this derivation are presented in [43].

The algorithm [35] for N robots is extended to collision avoidance as follows:

$$\begin{aligned}
v_i &= v_{i-1} + k_1 E_{ix} \\
\omega_i &= \omega_{i-1} + k_2 \mathrm{sign}(v_{i-1}) E_{iy} + k_3 e_{i\theta}
\end{aligned} . \tag{39}$$

In each of N controllers the same procedure as described under Eq. (15) is applied.

Substituting Eq. (39) to Eq. (34) error dynamics is calculated as follows:

$$\begin{aligned}
\dot{e}_{ix} &= e_{iy}\omega_i - k_1 E_{ix} + v_{i-1}(\cos e_{i\theta} - 1) \\
\dot{e}_{iy} &= -e_{ix}\omega_i + v_{i-1}\sin e_{\theta} \\
\dot{e}_{i\theta} &= -k_2 \mathrm{sign}(v_{i-1}) E_{iy} - k_3 e_{i\theta}
\end{aligned} . \tag{40}$$

If the i-th robot is in the collision avoidance region of the other robot its reference signals are temporarily frozen, i.e. v_{i-1} and ω_{i-1} are substituted as 0. The tracking of the leader is suspended because collision avoidance has a higher priority. When the robot leaves collision avoidance region of the other robot its reference signal is updated to the current value of the reference signal (that is position of the robot $i-1$).

Error dynamics in a freeze state is as follows:

$$\begin{aligned} \dot{e}_{ix} &= k_3 e_{iy} e_{i\theta} + k_2 e_{iy} E_{iy} - k_1 E_{ix} \\ \dot{e}_{iy} &= -k_3 e_{i\theta} e_{ix} - k_2 e_{ix} E_{iy} \\ \dot{e}_{i\theta} &= -k_2 E_{iy} - k_3 e_{i\theta} \end{aligned} . \tag{41}$$

7.1 Stability Analysis for N Robots

In this subsection, the stability analysis of a system of N mobile robots is presented. In the area where APF is zero the analysis presented in [35] remains valid, and it is not repeated here. The subsection presents the analysis of the case if the robots are in the APFs of other robots.

Consider the following Lyapunov-like function:

$$V = \sum_{i=1}^{N} \left[\frac{1}{2}(e_{ix}^2 + e_{iy}^2 + e_{i\theta}^2) + \sum_{j=1, j \neq i}^{N} V_{aij} \right]. \tag{42}$$

If the robot avoids collision ($l_j < R_j$) time derivative of the Lyapunov-like function is as follows:

$$\frac{dV}{dt} = \sum_{i=1}^{N} \left[e_{ix}\dot{e}_{ix} + e_{iy}\dot{e}_{iy} + e_{i\theta}\dot{e}_{i\theta} + \sum_{j=1, j \neq i}^{N} \left(\frac{\partial V_{aij}}{\partial e_{ix}} \dot{e}_{ix} + \frac{\partial V_{aij}}{\partial e_{iy}} \dot{e}_{iy} \right) \right]. \tag{43}$$

Taking into account (38) the above equation can be written as follows:

$$\frac{dV}{dt} = \sum_{i=1}^{N} \left[E_{ix}\dot{e}_{ix} + E_{iy}\dot{e}_{iy} + e_{i\theta}\dot{e}_{i\theta} \right]. \tag{44}$$

Next steps of the analysis are analogous to that presented in Sect. 4.2 for a single robot. Auxiliary orientation variable θ_{iE}, correction variables for individual robots E_{ix}, E_{iy} and scalar functions a_{i1}, b_{i1} and c_{i1} are introduced, which are equivalent to their counterparts presented in Sect. 4.2.

Time derivative of the Lyapunov-like function is less then zero ($\dot{V} \leq 0$) if the following condition is fulfilled:

$$\sum_{i=1}^{N}\left[k_1\cos^2\theta_{iE}\right] \geq \sum_{i=1}^{N}\left[k_3\left(e_{ix}-\frac{k_2}{2k_3}\right)^2+k_3\left(\cos\theta_{iE}e_{iy}+\frac{k_2}{2k_3}\sin\theta_{iE}\right)^2\right]. \tag{45}$$

If the term $\cos^2\theta_{iE}$ is close to zero, special procedure described in *Assumption 4B* is applied. It pushes i-th robot away from this state assuring that $\theta_{iE} \neq \pm\frac{\pi}{2} \pm \pi d$ ($d = 0, \pm 1, \pm 2, ...$)). In other cases the condition (45) can be met by setting sufficiently high value of k_1.

As shown in [42] collision avoidance is guaranteed if $\dot{V}_{aij} \leq 0$ and $\lim_{\|[x_i\ y_i]^\top - [x_{aj}\ y_{aj}]^\top\| \to r^+} V_{aij} = +\infty$, where vector $[x_{aj}\ y_{aj}]^\top$ represents location of the centre of the j-th obstacle.

7.2 Numerical Simulation

This subsection presents numerical simulation of the algorithm based on linearization applied to a group of N mobile robots moving in a formation. The robots executes circular reference trajectories with a radius of 5 m. The desired displacements between the subsequent robots is 2.5 m in the y axis. There are no displacements in the x axis. The following settings of the algorithm were used: $k_1 = 1$, $k_2 = 1$ and $k_3 = 1$. The graph of the robot paths on the (x, y)-plane are shown in Fig. 2a. The dashed line represents the path of the virtual leader, the solid line are the paths of the robots ($N = 3$). Figure 2b, c show the time graphs of the position coordinates. All mobile platforms achieve locations close to reference in about 10 s. The time graphs of the control signals are shown in Fig. 2d, e. After a short transient state they stabilize at constant values. Figure 2f presents time graph of the freeze signal.

8 Control Algorithm with Persistent Excitation

This subsection presents the trajectory tracking algorithm proposed in [36]. It is extended by collision avoidance. Since the structure of the algorithm is similar to the method presented in Sect. 4, there are references to some of the formulas presented here to avoid repetition.

8.1 Control Algorithm

Equations (4)–(14) presented in Sect. 4 remain actual.

Control algorithm from [36] is extended to collision avoidance and can be written in the following form:

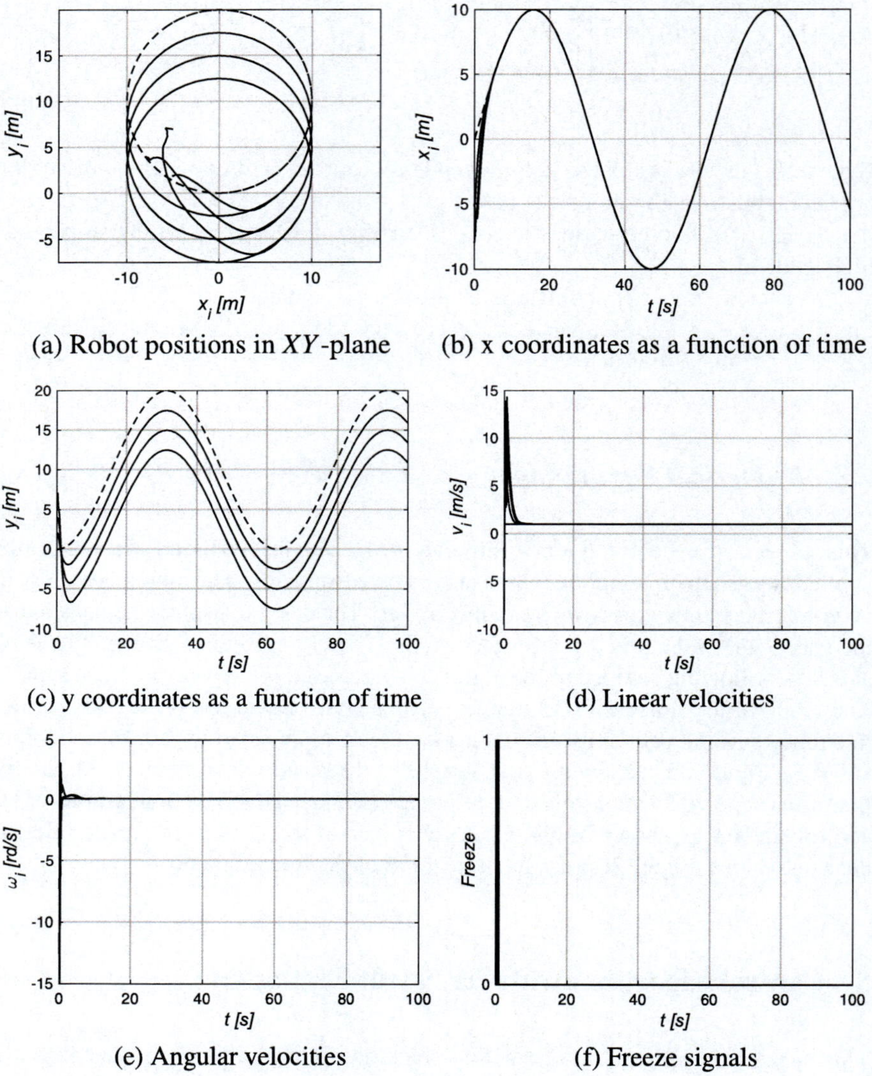

(a) Robot positions in XY-plane (b) x coordinates as a function of time

(c) y coordinates as a function of time (d) Linear velocities

(e) Angular velocities (f) Freeze signals

Fig. 2 Numerical simulation 2: algorithm based on the linearization

$$\begin{aligned} v &= v_d + c_2 E_x \\ \omega &= \omega_d + h(t, E_y) + c_1 e_\theta \end{aligned}, \tag{46}$$

where $h(t, E_y)$ is bounded, depends linearly on E_y, and continuously differentiable function. It must be properly chosen to ensure persistent excitation of the reference angular velocity [44]. Positive constants c_1 and c_2 are design parameters.

Substituting (46) into (6) error dynamics can be written as follows:

$$\begin{aligned} \dot{e}_x &= e_y\omega - c_2 E_x + v_d(\cos e_\theta - 1) \\ \dot{e}_y &= -e_x\omega + v_d \sin e_\theta \\ \dot{e}_\theta &= -h(t, E_y) - c_1 e_\theta \end{aligned} \quad . \tag{47}$$

If the robot gets into avoidance region of the obstacle its desired trajectory is temporarily frozen: $\dot{x}_d = 0$, $\dot{y}_d = 0$ and error dynamics are:

$$\begin{aligned} \dot{e}_x &= h(t, E_y)e_y + c_1 e_y e_\theta - c_2 E_x \\ \dot{e}_y &= -h(t, E_y)e_x - c_1 e_\theta e_x \\ \dot{e}_\theta &= -h(t, E_y) - c_1 e_\theta \end{aligned} \quad . \tag{48}$$

8.2 *Stability Analysis*

Consider the following Lyapunov-like function:

$$V = \frac{1}{2}(e_x^2 + e_y^2 + e_\theta^2) + \sum_{j=1}^{M} V_{aj}. \tag{49}$$

If the robot is outside the collision avoidance region, i.e. $l_j \geq R_j$, $j = 1, ..., M$, the system is equivalent to the one presented in [36] and the analysis given in this paper is valid.

If robot is in the collision avoidance region of the obstacle, time derivative of the Lyapunov-like function is calculated as follows:

$$\frac{dV}{dt} = e_x\dot{e}_x + e_y\dot{e}_y + e_\theta\dot{e}_\theta + \sum_{j=1}^{M}\left(\frac{\partial V_{aj}}{\partial e_x}\dot{e}_x + \frac{\partial V_{aj}}{\partial e_y}\dot{e}_y\right). \tag{50}$$

Taking into account (14) above equality can be transformed as follows:

$$\frac{dV}{dt} = E_x\dot{e}_x + E_y\dot{e}_y + e_\theta\dot{e}_\theta. \tag{51}$$

Next, using Eq. (48) one obtains:

$$\dot{V} = c_1 E_x e_y e_\theta - c_1 E_y e_x e_\theta - E_{iy} e_x h(t, E_y) + E_x e_y h(t, E_y) - e_\theta h(t, E_y) - c_1 e_\theta^2 - c_2 E_x^2. \tag{52}$$

Substituting $E_x = D\cos\theta_E$ i $E_y = D\sin\theta_E$, $D = \sqrt{E_x^2 + E_y^2}$ in the above equation one obtains:

$$\begin{aligned} \dot{V} = {} & c_1 D\cos\theta_E e_y e_\theta - c_1 D\sin\theta_E e_x e_\theta - D\sin\theta_E e_x h(t, E_y) \\ & + D\cos\theta_E e_y h(t, E_y) - e_\theta h(t, E_y) - c_1 e_\theta^2 - c_2 D^2\cos^2\theta_E. \end{aligned} \tag{53}$$

Equation (53) can be rewritten after some algebra follows:

$$\begin{aligned}\dot{V} = -c_1\left(\tfrac{e_\theta}{\sqrt{2}} + \tfrac{1}{\sqrt{2}}D\sin\theta_E e_x\right)^2 - c_1\left(\tfrac{e_\theta}{\sqrt{2}} - \tfrac{1}{\sqrt{2}}D\cos\theta_E e_y\right)^2 - c_2 D^2\cos^2\theta_E \\ +\left(\sqrt{\tfrac{c_1}{2}}D\sin\theta_E e_x - \tfrac{h(t,E_y)}{\sqrt{2c_1}}\right)^2 + \left(\sqrt{\tfrac{c_1}{2}}D\cos\theta_E e_y + \tfrac{h(t,E_y)}{\sqrt{2c_1}}\right)^2 - \tfrac{h^2(t,E_y)}{c_1} - e_\theta h(t,E_y).\end{aligned} \tag{54}$$

The first three terms on the right hand side of Eq. (54) are always less or equal to zero. In the next steps remaining terms will be analysed. One can write their sum as follows:

$$\dot{V}_r = \left(\sqrt{\tfrac{c_1}{2}}D\sin\theta_E e_x - \tfrac{h(t,E_y)}{\sqrt{2c_1}}\right)^2 + \left(\sqrt{\tfrac{c_1}{2}}D\cos\theta_E e_y + \tfrac{h(t,E_y)}{\sqrt{2c_1}}\right)^2 - \tfrac{h^2(t,E_y)}{c_1} - e_\theta h(t,E_y). \tag{55}$$

Equation (55) can be transformed to the following form:

$$\begin{aligned}\dot{V}_r = {} & D^2\frac{c_1}{2}\sin^2\theta_E e_x^2 + D^2\frac{c_1}{2}\cos^2\theta_E e_y^2 \\ & - D\sin\theta_E e_x h(t,E_y) + D\cos\theta_E e_y h(t,E_y) - e_\theta h(t,E_y).\end{aligned} \tag{56}$$

The condition $\dot{V}_r \leq 0$ is fulfilled if the following inequalities hold true:

$$\begin{aligned}& D^2\frac{c_1}{2}\sin^2\theta_E e_x^2 + D^2\frac{c_1}{2}\cos^2\theta_E e_y^2 \\ & \leq \left|-D\sin\theta_E e_x h(t,E_y) + D\cos\theta_E e_y h(t,E_y) - e_\theta h(t,E_y)\right| \\ & \leq D|\sin\theta_E||e_x||h(t,E_y)| + D|\cos\theta_E||e_y||h(t,E_y)| + |e_\theta||h(t,E_y)| \\ & \leq D|e_x||h(t,E_y)| + D|e_y||h(t,E_y)| + |e_\theta||h(t,E_y)|\end{aligned} \tag{57}$$

that can be rewritten in the compact form as follows:

$$D||e^*|||h(t,E_y)| + |e_\theta||h(t,E_y)| \geq D^2\frac{c_1}{2}(\sin^2\theta_E e_x^2 + \cos^2\theta_E e_y^2) \tag{58}$$

where $e^* = [e_x \;\; e_y]^T$.

Reduction of the parameter c_1 increases the chances of satisfying inequality (58) that supports stability of the closed-loop system ($\dot{V} \leq 0$). Notice, however, that satisfying $\dot{V}_r \leq 0$ may not always be possible. In such case stability of the system can be achieved by increasing the value of parameter c_2 (refer to Eq. (46)). Note that the procedure described in the *Assumption 4B* pushes the robot away from the state where auxiliary orientation variable $\theta_E \cong \frac{\pi}{2} + \pi d$. Note that $\cos^2\theta_E$ in Eq. (54) cannot be arbitrarily close to zero.

The error dynamics (48) with frozen reference signals may be decomposed into two subsystems (Fig. 3). Properties of these subsystems are inherited from the non-collision case described in [36]: the system Σ_1 is uniformly asymptotically stable at the origin, provided that $c_2 > 0$ and ω_i is persistently exciting, globally Lipschitz, and bounded. The origin of the system Σ_2 is exponentially stable if $c_1 > 0$. As a matter of

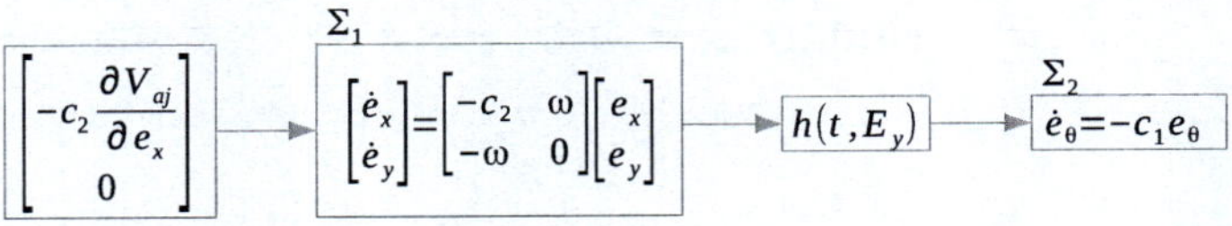

Fig. 3 Diagram of the control system in the collision avoidance mode

fact, it may also be established that each of these subsystems is input to state stable (ISS). The subsystem $h(t, E_y)$ is also uniformly bounded and satisfy $h(t, 0) \equiv 0$. Stability of the origin may be concluded invoking the small-gain theorem for ISS systems [45].

If the robot is in the collision avoidance area of the j-th obstacle, but close to its outer edge (in this case $|| \begin{bmatrix} \frac{\partial V_{aj}}{\partial e_x} & \frac{\partial V_{aj}}{\partial e_y} \end{bmatrix} || \ll ||[e_x \;\; e_y]^T||$), collision avoidance terms can be neglected. Correction variables (14) are approximated as follows:

$$E_x \cong e_x, \qquad E_y \cong e_y, \tag{59}$$

and the algorithm becomes the same as in [36] with the stability analysis presented there.

In the opposite situation, i.e. for $|| \begin{bmatrix} \frac{\partial V_{aj}}{\partial e_x} & \frac{\partial V_{aj}}{\partial e_y} \end{bmatrix} || \gg ||[e_x \;\; e_y]^T||$ (i-th robot is close to the boundary of the j-th obstacle, error terms and terms related to collision avoidance with other obstacles are neglected) the correction variables can be approximated in the following way (refer to Eq. (14)):

$$E_x \cong \frac{\partial V_{aj}}{\partial e_x}, \qquad E_y \cong \frac{\partial V_{aj}}{\partial e_y}. \tag{60}$$

When the robot is close to the j-th obstacle, in most cases the condition $|E_x| \gg 0$ is fulfilled. The exception is situation when θ_E is in the neighbourhood of $\frac{\pi}{2} + \pi d$. This state is non-attracting, but procedure given in *Assumption 4B* is applied to push the robot away from this region (if E_x is close to zero linear velocity control v is close to zero, refer to Eq. (46)). Notice that in this case the centre of the obstacle lays close to the axis of the robot wheels, and thus, application of some arbitrary linear velocity control $\tilde{v}$ to the robot does not carry the risk of collision with the obstacle.

When the robot is away from this state $\frac{\partial V_{aj}}{\partial e_x}$ has a large value and the boundedness of the output of the collision avoidance subsystem is necessary to prove stability. Substituting first equation in (14) into first equation in (48) one can write the following approximation:

$$\dot{e}_x = h(t, E_y)e_y + c_1 e_y e_\theta - c_2 e_x - c_2 \frac{\partial V_{aj}}{\partial e_x}. \tag{61}$$

Note that the above approximation assumes that the robot is located close to the boundary of a single obstacle j, and terms related to the other obstacles are neglected.

If $\frac{\partial V_{aj}}{\partial e_x}$ is sufficiently high (that happens if the robot is very close to the obstacle), i.e. $\frac{\partial V_{aj}}{\partial e_x} \gg e_x$, $\frac{\partial V_{aj}}{\partial e_x} \gg e_y$, $\frac{\partial V_{aj}}{\partial e_x} \gg e_\theta$, and $\frac{\partial V_{aj}}{\partial e_x} \gg h(t, E_y)$ Eq. (61) can be approximated as follows:

$$\dot{e}_x \cong -c_2 \frac{\partial V_{aj}}{\partial e_x}. \tag{62}$$

From the Eq. (62) it is clear that $\dot{e}_x$ and $\frac{\partial V_{aj}}{\partial e_x}$ have different signs and as a result $\frac{\partial V_{aj}}{\partial e_x}\dot{e}_x < 0$. To fulfill the condition that $\dot{V}_{aj} = \frac{\partial V_{aj}}{\partial e_x}\dot{e}_x + \frac{\partial V_{aj}}{\partial e_y}\dot{e}_y$ is less then zero the second term on the right hand side must be less then the first one taking their absolute values. This can be obtained by reducing c_1 parameter (refer to Eq. (48)). The property $\dot{V}_{aj} \leq 0$ guarantees boundedness of both V_{aj} and $\frac{\partial V_{aj}}{\partial e_x}$. Finally, one can state that collision avoidance block, that is input to the system shown in Fig. 3, has also bounded output and both error components e_x and e_y in Σ_1 are bounded.

The third considered case is when modules of the position error of the robot and gradient of the collision avoidance function of the j-th obstacle are similar: $||[e_x \ \ e_y]^T|| \cong ||\left[\frac{\partial V_{aj}}{\partial e_x} \ \ \frac{\partial V_{aj}}{\partial e_y}\right]||$ (in this case collision avoidance terms of other obstacles are neglected). Both vectors can point in arbitrary directions but one situation is special, if both of them point in exactly opposite directions, robot is in the saddle point. This results in $E_x = 0$ and $E_y = 0$ and, finally, $v = 0$ that activates procedure described in *Assumption 4B*. This pushes the robot out of the saddle point usually. The only exception is when the auxiliary orientation variables is $0 + \pi d$ (this is the worst case, robot has the obstacle in the front or at the back and its goal is exactly on the other side of the obstacle; notice that this state is set of measure zero) which can lead to oscillations around the saddle point. In [23] authors investigated other method of leaving the saddle point. The paper includes extensive tests on real non-holonomic mobile robot.

The above considerations assume that the robot is not located near the boundaries of two or more obstacles at the same time. Taking into account that collision avoidance component of the control is large (close to the obstacle) the robot is driven away quickly and these considerations are correct in most cases. Other situations can be considered exceptionally and they are not investigated here.

8.3 Numerical Simulation

This subsection presents numerical simulation results of the algorithm with persistent excitation. The reference trajectory, obstacle positions, and initial conditions are the same as presented in Sect. 4.3. The settings of the algorithm were as follows: $c_1 = 1$, $c_2 = 1$. Persistent excitation signal was given by the function $h(t, E_y) = \phi(t) tanh(E_y)$. The $\phi(t)$ function is a non-differentiable pulse function with an amplitude of 0.5, a period of 4 s and a duty cycle of 80%. This signal intro-

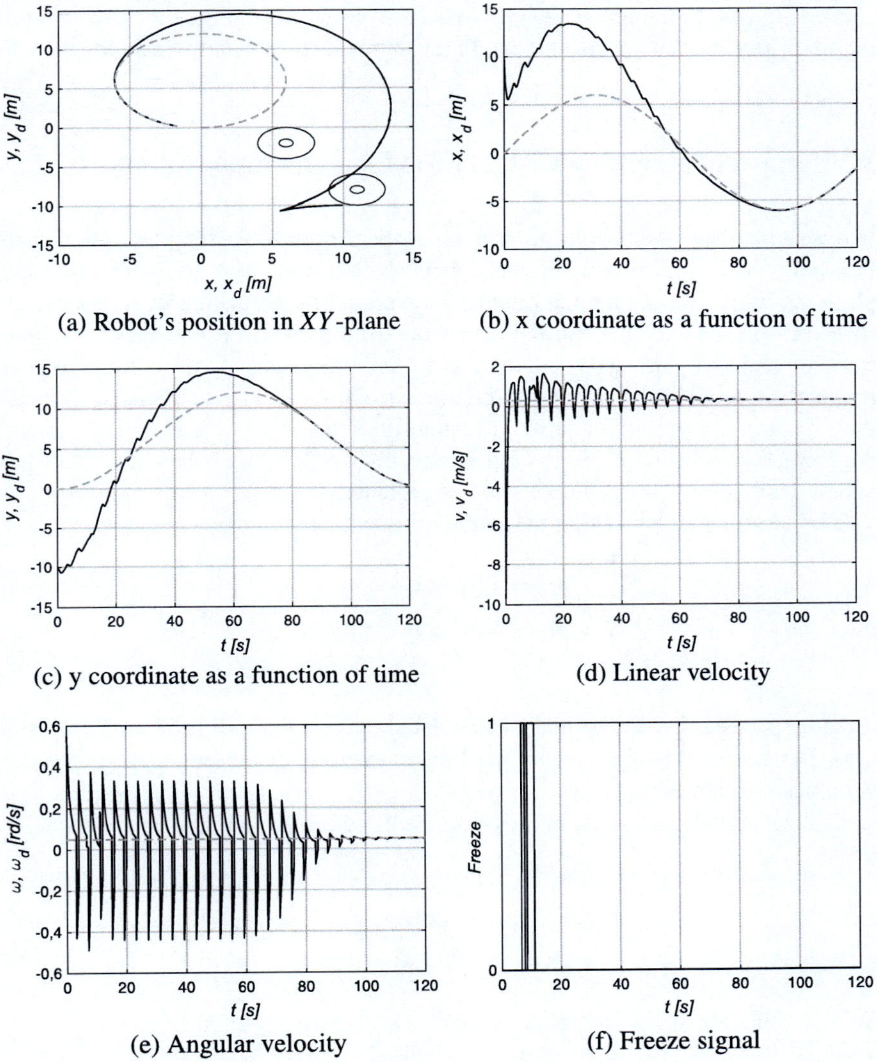

(a) Robot's position in XY-plane

(b) x coordinate as a function of time

(c) y coordinate as a function of time

(d) Linear velocity

(e) Angular velocity

(f) Freeze signal

Fig. 4 Numerical simulation 3: persistent excitation algorithm—single robot

duces persistent excitation necessary to stabilize the mobile platform in the y axis of the local frame.

Figure 4 presents robots path in the (x, y)-plane (solid line). Dashed line represents the reference path. Figure 4b, c show time graphs of the position coordinates. Comparing them with the results obtained for the linearization algorithm, one can see a much slower convergence. The path's shape on the (x, y) plane is entirely different. Signals converge to reference values in about 100 s. Figures 4d, e present control signals. In the transient state peaks are observed with the magnitude that is not realizable by most of real robots. Figure 4f presents time graph o freeze signal.

9 Persistent Excitation Algorithm for *N* Mobile Robots

In this subsection the control algorithm for N mobile robots is presented. The control task is defined as in Sect. 7, i.e. the robots create a formation in which one robot plays the role of the leader and mimics the motion of a virtual leader with a certain spatial displacement. The subsequent robots form a chain in which the i-th robot treats the trajectory of the $i-1$th robot as its reference signal. In the control process, robots use not only the position and orientation coordinates of their leaders, but also their linear and angular velocities. The algorithm presented here is an extension of the method described in [36] to collision avoidance. The case where a single robot avoids a collision with a static obstacle is presented in the previous subsection.

The control goal is to bring the following quantities to zero:

$$\begin{aligned} p_{ix} &= x_{i-1} - x_i - d_{x\,i-1,i} \\ p_{iy} &= y_{i-1} - y_i - d_{y\,i-1,i} \\ p_{i\theta} &= \theta_{i-1} - \theta_i . \end{aligned} \tag{63}$$

Error expressed in the coordinate frame fixed to the i-th robot is given by formula (33). Equations (34)–(38) describing transformations of the erors and correction variables also remain actual.

Control signals described in [36] extended to the collision avoidance are as follows:

$$\begin{aligned} v_i &= v_{i-1} + c_2 E_{ix} \\ \omega_i &= \omega_{i-1} + h(t, E_{iy}) + c_1 e_{i\theta} . \end{aligned} \tag{64}$$

Substituting (64) to (34) one obtains the following error dynamics formula:

$$\begin{aligned} \dot{e}_{ix} &= e_{iy}\omega_i - c_2 E_{ix} + v_{i-1}(\cos e_{i\theta} - 1) \\ \dot{e}_{iy} &= -e_{ix}\omega_i + v_{i-1}\sin e_{i\theta} \\ \dot{e}_{i\theta} &= \omega_{i-1} - h_i(t, E_{iy}) - c_1 e_{i\theta} . \end{aligned} \tag{65}$$

If the robot detects the obstacle its trajectory is temporarily frozen and procedure described in Sect. 4 after Eq. (17) is applied.

Error dynamics in the freeze state is as follows:

$$\begin{aligned}\dot{e}_{ix} &= h_i(t, E_{iy})e_{iy} + c_1 e_{iy} e_{i\theta} - c_2 E_{ix} \\ \dot{e}_{iy} &= -h_i(t, E_{iy})e_{ix} - c_1 e_{i\theta} e_{ix} \\ \dot{e}_{i\theta} &= -h_i(t, E_{iy}) - c_1 e_{i\theta}.\end{aligned} \tag{66}$$

9.1 Stability Analysis

Consider the following Lyapunov-like function:

$$V = \sum_{i=1}^{N} \left[\frac{1}{2}(e_{ix}^2 + e_{iy}^2 + e_{i\theta}^2) + \sum_{j=1, j\neq i}^{N} V_{aij} \right]. \tag{67}$$

If the robot is outside the collision avoidance regions, i.e. $l_{ij} \geq R_j$, $i = 1, ..., N$, $i \neq j$, the system is equivalent to the one presented in [36] and the analysis given in this paper is actual.

If the robot is in collision avoidance mode $(l_j < R_j)$ then the time derivative of the Lyapunov-like function takes the same form as in the previously presented method—the formula (43), which can be converted to (44) using formula (38).

Further stability analysis is analogous to the one presented in Sect. 8.2. For each of the N robots the auxiliary orientation variable θ_{iE} and the correction variables E_{ix}, E_{iy} are introduced.

Conducting further considerations just like in Sect. 8.2 one can write time derivative of the Lyapunov-like function as follows (refer to Eq. (54)):

$$\begin{aligned}\dot{V} = \sum_{i=1}^{N} \Bigg[&-c_1 \left(\frac{e_{i\theta}}{\sqrt{2}} + \frac{1}{\sqrt{2}} D_i \sin\theta_{iE} e_{ix} \right)^2 - c_1 \left(\frac{e_{i\theta}}{\sqrt{2}} - \frac{1}{\sqrt{2}} D_i \cos\theta_{iE} e_{iy} \right)^2 \\ &- c_2 D_i^2 \cos^2\theta_{iE} + \left(\sqrt{\frac{c_1}{2}} D_i \sin\theta_{iE} e_{ix} - \frac{h(t, E_{iy})}{\sqrt{2c_1}} \right) \\ &+ \left(\sqrt{\frac{c_1}{2}} D_i \cos\theta_{iE} e_{iy} + \frac{h(t, E_{iy})}{\sqrt{2c_1}} \right)^2 - \frac{h^2(t, E_{iy})}{c_1} - e_{i\theta} h(t, E_{iy}) \Bigg]\end{aligned} \tag{68}$$

and its part $\dot{V}_r$ analogous to Eq. (55) in the form:

$$\begin{aligned}\dot{V}_r = \sum_{i=1}^{N} \Bigg[&\left(\sqrt{\frac{c_1}{2}} D_i \sin\theta_{iE} e_{ix} - \frac{h(t, E_{iy})}{\sqrt{2c_1}} \right)^2 + \left(\sqrt{\frac{c_1}{2}} D_i \cos\theta_{iE} e_{iy} + \frac{h(t, E_{iy})}{\sqrt{2c_1}} \right)^2 \\ &- \frac{h^2(t, E_{iy})}{c_1} - e_{i\theta} h(t, E_{iy}) \Bigg].\end{aligned} \tag{69}$$

The condition $\dot{V}_r \leq 0$ is fulfilled if the following inequality (obtained using transformations equivalent to (56) and (57)) holds true:

$$\sum_{i=1}^{N}\left[D_i||e_i^*|||h(t, E_{iy})| + |e_{i\theta}||h(t, E_{iy})|\right] \geq \sum_{i=1}^{N}\left[D_i^2\frac{c_1}{2}(\sin^2\theta_{iE}e_{ix}^2 + \cos^2\theta_{iE}e_{iy}^2)\right] \tag{70}$$

where $e_i^* = [e_{ix} \;\; e_{iy}]^T$.

Note that Eq. (70) can be decomposed into N separate inequalities:

$$D_i||e_i^*|||h(t, E_{iy})| + |e_{i\theta}||h(t, E_{iy})| \geq D_i^2\frac{c_1}{2}(\sin^2\theta_{iE}e_{ix}^2 + \cos^2\theta_{iE}e_{iy}^2). \tag{71}$$

This approach allows to consider the stability of each robot independently.

Reduction of the parameter c_1 increases the chances of satisfying inequalities (70) and (71) that supports stability of the closed-loop system ($\dot{V} \leq 0$). Notice, however, that satisfying $\dot{V}_r \leq 0$ may not always be possible. In such case stability of the system can be achieved by increasing the value of parameter c_2 (refer to Eq. (68)). Note that the procedure described in *Assumption 4B* pushes the robot away from the state where auxiliary orientation variable $\theta_{iE} \cong \frac{\pi}{2} + \pi d$ and $\cos^2\theta_{iE}$ in Eq. (68) cannot be arbitrarily close to zero.

The stability analysis presented in Sect. 8.2 after Fig. 3 is applicable.

9.2 Numerical Simulation

This subsection presents numerical simulation that illustrates effectiveness of the persistent excitation algorithm used to control a group of robots moving in formation. The robots execute motion along the reference trajectory with the same parameters as in Sect. 7.2. Also their initial positions are the same.

Figure 5a presents robots paths in the (x, y)-plane. The dashed line represents the virtual leader path (the center of the reference trajectory circle is at (0, 10 m)), the solid lines represent the real robot paths ($N = 3$). Graph 5b, c show the time graphs of the position coordinates. Note that the convergence to values close to the reference signals occurred after almost three times longer time than in the case of the algorithm based on linearization (Sect. 7.2). The control signals (Fig. 5d, e) also stabilize much longer. In Fig. 5f shows the freeze signal as a function of time.

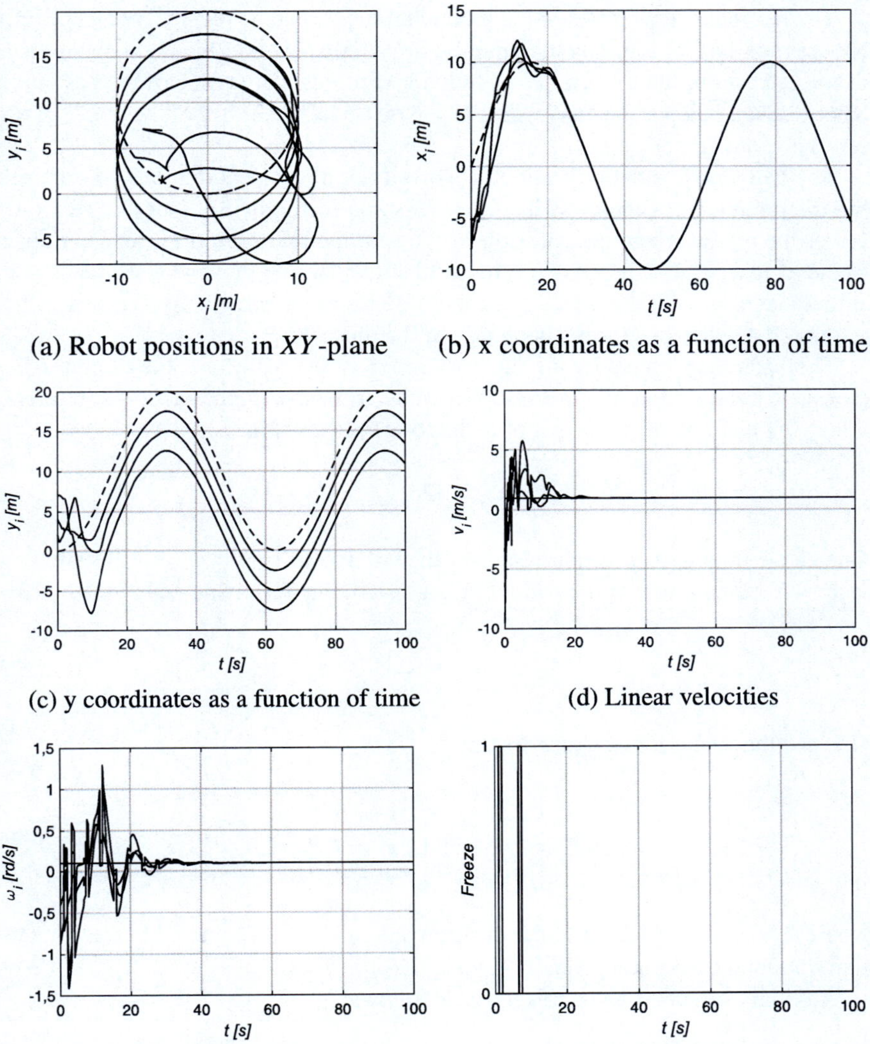

(a) Robot positions in XY-plane (b) x coordinates as a function of time

(c) y coordinates as a function of time (d) Linear velocities

(e) Angular velocities (f) Freeze signals

Fig. 5 Numerical simulation 4: persistent excitation algorithm

10 Vector Field Orientation Method

Vector field orientation method (VFO) for the single mobile platform was proposed in paper [37]. This subsection describes VFO method applied to control N robots executing parallel trajectory tracking task. In order to solve the collision avoidance problem APFs were used.

The reference trajectories of the individual robots can be planned independently. However, if the task is to execute a motion by a group of robots maintaining the shape of the formation, the trajectories of individual robots can be determined as curves having desired displacements relative to the reference trajectory of the whole group, e.g. the geometric center.

The collision avoidance problem is solved not by suspending the task of tracking the reference trajectory as it is done in some methods [31], but until the robot leaves the region where the collision may occur, the reference trajectory is stopped (reduced to a point). The robot has to avoid collisions and move to a reference point simultaneously, with the priority of each of these tasks being a non-linear function of the robot's distance from a boundary of the obstacle.

The reference trajectory of the i-th robot is described by the vector $\mathbf{q}_{di} = [x_{di}\ y_{di}]^T$. The reference orientation results from the direction of the reference linear velocity vector and is given by the following equation:

$$\theta_{di} = atan2c(\dot{y}_{di}, \dot{x}_{di}). \tag{72}$$

One can denote $\mathbf{q}_i$ as position vector of the i-th robot: $\mathbf{q}_i = \begin{bmatrix} x_i\ y_i \end{bmatrix}^T$. Desired trajectory tracking error for the i-th robot is given by the following equation:

$$\mathbf{e}_i = \begin{bmatrix} e_{xi} \\ e_{yi} \end{bmatrix} = \mathbf{q}_{di} - \mathbf{q}_i. \tag{73}$$

One can introduce convergence vector:

$$\mathbf{h}_i = \begin{bmatrix} h_{xi} \\ h_{yi} \\ h_{\theta i} \end{bmatrix} = \begin{bmatrix} k_p e_{xi} + \dot{x}_{di} \\ k_p e_{yi} + \dot{y}_{di} \\ k_\theta e_{ai} + \dot{\theta}_{ai} \end{bmatrix}, \tag{74}$$

where $k_p,\ k_\theta > 0$ are gains of the position and orientation control blocks, and auxiliary orientation error is given by Eq. $e_{ai} = \theta_{ai} - \theta_i$.

Auxiliary orientation variable θ_{ai} is as follows:

$$\theta_{ai} = atan2c(h_{yi}, h_{xi}), \tag{75}$$

and $\dot{\theta}_{ai}(t)$ is obtained by differentiating Eq. (75) with respect to time:

$$\dot{\theta}_{ai} = \frac{\dot{h}_{yi} h_{xi} - h_{yi} \dot{h}_{xi}}{h_{xi}^2 + h_{yi}^2}. \tag{76}$$

Control law $\mathbf{u}_i = \begin{bmatrix} u_{vi}\ u_{\omega i} \end{bmatrix}^T$ for the mobile robot is designed as follows:

$$\begin{matrix} u_{vi} = h_{xi}\cos\theta_i + h_{yi}\sin\theta_i \\ u_{\omega i} = h_{\theta i} \end{matrix}, \tag{77}$$

where u_{vi} is named pushing control, and $u_{\omega i}$ orientation control, respectively.

Formulae $\dot{h}_{xi}$ and $\dot{h}_{yi}$ result form differentiation of (74) with respect to time and are as follows:

$$\begin{matrix} \dot{h}_{xi} = k_p\dot{e}_{xi} + \ddot{x}_{di} \\ \dot{h}_{yi} = k_p\dot{e}_{yi} + \ddot{y}_{di} \end{matrix}, \tag{78}$$

where $\left[\,\dot{e}_{xi}\ \dot{e}_{yi}\,\right]^T$ represents time derivative of the position error:

$$\dot{\mathbf{e}}_i = \left[\,\dot{e}_{xi}\ \dot{e}_{yi}\,\right]^T = \dot{\mathbf{q}}_{di} - \dot{\mathbf{q}}_i. \tag{79}$$

Extension of the VFO algorithm to the collision avoidance is obtained by replacement of $\mathbf{e}_i$ in Eq. (74) by $\mathbf{E}_i$ given by the following equation:

$$\mathbf{E}_i = \begin{bmatrix} E_{xi} \\ E_{yi} \end{bmatrix} = \mathbf{e}_i - \sum_{j=1, j\neq i}^{N} \left[\frac{\partial V_{arij}(l_{ij})}{\partial \mathbf{q}_i}\right]^T - \sum_{k=1}^{M} \left[\frac{\partial V_{aoik}(l_{ik})}{\partial \mathbf{q}_i}\right]^T, \tag{80}$$

where i and j are indexes of the robots, $V_{arij}(l_{ij})$ is APF of the j-th robot acting on the i-th robot, $V_{aoik}(l_{ik})$ is APF of the k-th static obstacle acting on the i-th robot, $l_{ij} = \left\|\mathbf{q}_i - \mathbf{q}_j\right\|$ for $i,\ j = 1, \ldots, N, i \neq j$ is distance between i-th and j-th robot, and $l_{ik} = \|\mathbf{q}_i - \mathbf{q}_k\|$ for $i = 1, \ldots, N$ and $k = 1, \ldots, M$ is distance between i-th robot and k-th obstacle.

If the i-th robot is outside of the detection region of other robots and static obstacles the following vector equality is fulfilled: $\sum_{j=1, j\neq i}^{N} \left[\frac{\partial V_{arij}(l_{ij})}{\partial \mathbf{q}_i}\right]^T + \sum_{k=1}^{M} \left[\frac{\partial V_{aoik}(l_{ik})}{\partial \mathbf{q}_i}\right]^T = \begin{bmatrix} 0 \\ 0 \end{bmatrix}$, and it results in $\mathbf{E}_i = \mathbf{e}_i$.

Taking into account obstacles requires to make some changes in the controller. Convergence vector Eq. (74) takes the form:

$$\mathbf{h}_i = \begin{bmatrix} h_{xi} \\ h_{yi} \\ h_{\theta i} \end{bmatrix} = \begin{bmatrix} k_p E_{xi} + \dot{x}_{di} \\ k_p E_{yi} + \dot{y}_{di} \\ k_\theta e_{ai} + \dot{\theta}_{ai} \end{bmatrix}. \tag{81}$$

Time derivative of the convergence vector is obtained by differentiating (81):

$$\dot{\mathbf{h}}_i = \begin{bmatrix} k_p \dot{E}_{xi} + \ddot{x}_{di} \\ k_p \dot{E}_{yi} + \ddot{y}_{di} \\ k_\theta \dot{e}_{ai} + \ddot{\theta}_{ai} \end{bmatrix}, \tag{82}$$

where an auxiliary angular velocity error is as follows:

$$\dot{e}_{ai} = \dot{\theta}_{ai} - \omega_i. \tag{83}$$

$$\dot{\mathbf{E}}_i(t) = \begin{bmatrix} \dot{E}_{xi} \\ \dot{E}_{yi} \end{bmatrix} = \dot{\mathbf{e}}_i - \sum_{j=1, j\neq i}^{N} \left[\frac{\partial^2 V_{arij}(l_{ij})}{\partial \mathbf{q}_i^2} \right] \dot{\mathbf{q}}_i - \sum_{k=1}^{M} \left[\frac{\partial^2 V_{aoik}(l_{ik})}{\partial \mathbf{q}_i^2} \right] \dot{\mathbf{q}}_i. \tag{84}$$

Assumption 10A
Reference trajectories are planned in such a way that, if accurately traced (zero position errors), robots APFs do not affect other robots.

Assumption 10B
If the robot is in the interaction region its reference trajectory is frozen, i.e.:

$$\mathbf{q}_{di}(t) = \mathbf{q}_{di}(t^-), \tag{85}$$

where t^- is the moment before the interaction takes place. The state described by Eq. (85) is maintained until the robot leaves the interaction region.

During interaction, the reference velocities and the higher derivatives of the reference position with respect to time remain zero:

$$\dot{\mathbf{q}}_{di}(t) = \mathbf{0}, \tag{86}$$

$$\ddot{\mathbf{q}}_{di}(t) = \mathbf{0}. \tag{87}$$

In case of collision avoidance the same procedure was used as in [42].

Assumption 10C
If $e_{ai} \in (\frac{\pi}{2} + \pi d - \delta, \frac{\pi}{2} + \pi d + \delta)$, where δ is small positive constant, $d = 0, \pm 1, \pm 2, ...$ auxiliary orientation variable θ_{ai} is replaced with $\tilde{\theta}_{ai} = \theta_{ai} + \text{sign}\left(e_{ai} - \left(\frac{\pi}{2} + \pi d\right)\right) \varepsilon$, where ε is small value fulfilling condition $\varepsilon > 0$.

Assumption 10D
If the robot is in a saddle point, the reference trajectory is disturbed to push the robot out of the local equilibrium point. At this point the following condition is fulfilled:

$$\left\| \mathbf{h}_\mathbf{i}^* \right\| = 0, \tag{88}$$

where $\mathbf{h}_\mathbf{i}^* = \left[h_{xi} \; h_{yi} \right]^T$, there is no solution of the Eq. (75). In such a case $\theta_{ai}(t) = \theta_{ai}(t^-)$, where t^- is the time just before the situation described by the formula (88) occurs.

Assumption 10A ensures that, in case of collision avoidance, the reference trajectory will not be frozen in the collision avoidance area. In this case, the robot's task would be to achieve two conflicting objectives: avoid collision and move to a point in the collision avoidance area. In practice, such a situation leads to a deadlock.

Assumption 10B (or more precisely, fulfillment of Eq. (86)) causes the solution of the Eq. (72) cannot be calculated, and therefore the reference orientation is not specified. In this case, it is assumed that:

$$\theta_{di}(t) = \theta_{di}(t^{-}). \tag{89}$$

This solution can be interpreted as follows: when an interaction occurs, the vector of linear reference velocity is replaced by the vector of virtual reference velocity, which has no direct effect on position control, but allows to determine the reference orientation.

Assumption 10C refers to a situation where tracking the reference trajectory requires to move in direction parallel to the robot's wheel axis and therefore unrealizable due to non-holomonic constraints.

The auxiliary orientation variable θ_{ai} is disturbed in a small vicinity of the point $e_{ai} = \frac{\pi}{2} \pm \pi d$.

Assumption 10D addresses the problems associated with the existence of local, unstable equilibrium points (saddle points). There is one saddle point close to each obstacle. In the environment with M obstacles there are also M local equilibrium points [46]. Since the set of such points is of measure zero and they are not attracting, in practice they should not cause any problems. A good solution, however, is to use a mechanism for detecting and pushing robots from such points. The paper [9] proposes a mechanism based on time-dependent trigonometric functions. These functions create a disturbance vector that is added to the control signal.

In addition the condition (88) may occur outside of detection region, when the following conditions are fulfilled: $k_p e_{xi} = -\dot{x}_{di}$ and $k_p e_{yi} = -\dot{y}_{di}$.

In such case solution of (75) does not exist, but it is not needed, zero control shall be used $u_{vi} = 0$, $u_{\omega i} = 0$.

According to *Assumption 10B*, control (77) has a finite number of discontinuity points. Discontinuity points occur when robot leaves the interaction area. This is due reference signal update $\dot{\mathbf{q}}_{di}$.

10.1 Stability Analysis

Stability and convergence analysis for the controller (77) is carried out in three steps:

- proof of stability and non-asymptotic convergence of robot position $\begin{bmatrix} x_i & y_i \end{bmatrix}^T$ to reference trajectory $\begin{bmatrix} x_{di} & y_{di} \end{bmatrix}^T$,
- $\lim_{t\to\infty} \theta_i = \theta_{ai}$—proof of the convergence of robots orientation θ_i to the auxiliary orientation variable θ_{ai},

- proof of non-asymptotic convergence of the auxiliary orientation variable θ_{ai} to reference orientation θ_{di}.

This subsection presents proof of stability and position convergence for a single robot in the environment with a single obstacle. The center of the obstacle is at the point $\left[\, x_a \; y_a \,\right]^T$. For simplicity, the indexes indicating the number of the robot and obstacles were omitted.

Consider the following Lyapunov-like function:

$$V_l = V_{tl} + V_{al}, \tag{90}$$

where V_{tl} is the trajectory tracking component and takes the following form:

$$V_{tl} = \frac{1}{2}\mathbf{e}^T\mathbf{e} = \frac{1}{2}\left(e_x^2 + e_y^2\right), \tag{91}$$

and V_{al} is the collision avoidance component given by the following equation:

$$V_{al} = V_a(l), \tag{92}$$

where V_a is the APF given by Eq. (3).

Euclidean distance l between the robot and the obstacle is given by the following equation $l = \left\| [x \; y]^T - [x_a \; y_a]^T \right\|$.

Differentiating (90) with respect to time one obtains:

$$\frac{dV_l}{dt} = e_x\dot{e}_x + e_y\dot{e}_y + \frac{\partial V_a}{\partial x}\dot{x} + \frac{\partial V_a}{\partial y}\dot{y}. \tag{93}$$

Below we show that $\frac{dV_l}{dt} \leqslant 0$. First, the pushing control u_v will be expressed as a function of the convergence vector components h_x, h_y and auxiliary orientation error e_θ.

Substituting $\theta = \theta_a - e_a$ to the first equation in (77) one obtains:

$$u_v = h_x\cos(\theta_a - e_a) + h_y\sin(\theta_a - e_a). \tag{94}$$

Transforming (94) using trigonometric formula one can write:

$$u_v = h_x(\cos(\theta_a)\cos(e_a) + \sin(\theta_a)\sin(e_a)) + h_y(\sin(\theta_a)\cos(e_a) - \cos(\theta_a)\sin(e_a)). \tag{95}$$

Substituting to Eq. (95):

$$\sin(\theta_a) = \frac{h_y}{\|\mathbf{h}^*\|}, \quad \cos(\theta_a) = \frac{h_x}{\|\mathbf{h}^*\|}, \tag{96}$$

results in the following equation:

$$u_v = \left\|\mathbf{h}^*\right\| \cos(e_a). \tag{97}$$

Components of the velocity vector of the robot $\dot{x}_i$, $\dot{y}_i$ will be expressed as a function of h_x, h_y and e_θ.

Substituting (97) to the first two equations of the robot's model (1) one obtains:

$$\begin{aligned} \dot{x} &= \|\mathbf{h}^*\| \cos(e_a)\cos(\theta) \\ \dot{y} &= \|\mathbf{h}^*\| \cos(e_a)\sin(\theta). \end{aligned} \tag{98}$$

Then, plugging $\theta = \theta_a - e_a$ to the above equations and using trigonometric formula one can write:

$$\dot{x} = \left\|\mathbf{h}^*\right\| \cos(e_a)\left[\cos(\theta_a)\cos(e_a) + \sin(\theta_a)\sin(e_a)\right] \tag{99}$$

$$\dot{y} = \left\|\mathbf{h}^*\right\| \cos(e_a)\left[\sin(\theta_a)\cos(e_a) - \cos(\theta_a)\sin(e_a)\right]. \tag{100}$$

Substituting Eq. (96) to Eqs. (99) and (100) and simplifying, the following formulae are obtained:

$$\dot{x} = h_x \cos^2(e_a) + h_y \sin(e_a)\cos(e_a) \tag{101}$$

$$\dot{y} = -h_x \sin(e_a)\cos(e_a) + h_y \cos^2(e_a). \tag{102}$$

Substituting into Eq. (93) the corresponding Eqs. of the right hand side of the vector Eq. (79) one can write:

$$\frac{dV_l}{dt} = e_x(\dot{x}_d - \dot{x}) + e_y(\dot{y}_d - \dot{y}) + \frac{\partial V_a}{\partial x}\dot{x} + \frac{\partial V_a}{\partial y}\dot{y}. \tag{103}$$

The above equation can be transformed into a form:

$$\frac{dV_l}{dt} = e_x\dot{x}_d + e_y\dot{y}_d + \dot{x}\left(\frac{\partial V_a}{\partial x} - e_x\right) + \dot{y}\left(\frac{\partial V_a}{\partial y} - e_y\right). \tag{104}$$

Substituting $e_x = E_x + \frac{\partial V_a}{\partial x}$ and $e_y = E_y + \frac{\partial V_a}{\partial y}$ to Eq. (104) one obtains:

$$\frac{dV_l}{dt} = e_x\dot{x}_d + e_y\dot{y}_d - E_x\dot{x} - E_y\dot{y}. \tag{105}$$

Next, substituting (101) and (102) into Eq. (105) and simplifying, one obtains:

$$\begin{aligned} \frac{dV_l}{dt} &= e_x\dot{x}_d + e_y\dot{y}_d + E_x(-h_x\cos^2(e_a) - h_y\sin(e_a)\cos(e_a)) \\ &\quad + E_y(h_x\sin(e_a)\cos(e_a) - h_y\cos^2(e_a)). \end{aligned} \tag{106}$$

Substituting in the above equation for h_x and h_y the corresponding rows of the Eq. (81) and making simplifications, one obtains a derivative of the Lyapunov-like function with respect to time in the form:

$$\frac{dV_l}{dt} = e_x\dot{x}_d + e_y\dot{y}_d - k_p E_x^2\cos^2(e_a) - k_p E_y^2\cos^2(e_a) - E_x\dot{x}_d\cos^2(e_a)$$

$$- E_y\dot{y}_d\cos^2(e_a) - E_x\dot{y}_d\sin(e_a)\cos(e_a) + E_y\dot{x}_d\sin(e_a)\cos(e_a). \tag{107}$$

Two cases will be considered separately:

1. The robot is in the interaction area of the other robot,
2. The robot is outside the interaction area.

In case 1, according to the *Assumption 10B*, Eq. (107) takes the form ($\dot{x}_d = 0, \dot{x}_d = 0$):

$$\frac{dV_l}{dt} = -k_p E_x^2\cos^2(e_a) - k_p E_y^2\cos^2(e_a), \tag{108}$$

that meets the condition $\dot{V}_l \leqslant 0$.

Remark 10A
According to the theorem presented in [47] fulfillment of the condition $\dot{V} \leqslant 0$, while at the same time ensuring

$$\lim_{\|\mathbf{q}-\mathbf{q}_a\|\to r^+} V_a = +\infty, \tag{109}$$

where $\mathbf{q}_a = [\, x_a \;\, y_a \,]^T$, guarantees the collision avoidance.

The APF presented in Sect. 3 meets the condition (109) because from Eq. (2):

$$\lim_{l\to r^+} B_a(l) = 1 \tag{110}$$

which implies that the Eq. (3) meets the condition:

$$\lim_{l\to r^+} V_a(l) = \infty. \tag{111}$$

In case 2, the robot is outside of the collision avoidance region. Then $E_x = e_x$ and $E_y = e_y$ according to $\frac{\partial V_a}{\partial x} = \frac{\partial V_a}{\partial y} = 0$.

Time derivative of the Lyapunov-like function is as follows:

$$\frac{dV_l}{dt} = e_x\dot{x}_d + e_y\dot{y}_d - k_p e_x^2\cos^2(e_a) - k_p e_y^2\cos^2(e_a) - e_x\dot{x}_d\cos^2(e_a)$$

$$- e_y\dot{y}_d\cos^2(e_a) - e_x\dot{y}_d\sin(e_a)\cos(e_a) + e_y\dot{x}_d\sin(e_a)\cos(e_a). \tag{112}$$

Using Pythagorean identity and substituting into Eq. (112) $e_x\dot{y}_d - e_y\dot{x}_d = \|\mathbf{e} \times \dot{\mathbf{q}}_\mathbf{d}\|$, where $\times$ denotes the cross product:

$$\frac{dV_l}{dt} = e_x \dot{x}_d \sin^2(e_a) + e_y \dot{y}_d \sin^2(e_a)$$

$$- k_p e_x^2 \cos^2(e_a) - k_p e_y^2 \cos^2(e_a) - \|\mathbf{e} \times \dot{\mathbf{q}}_\mathbf{d}\| \sin(e_a)\cos(e_a). \tag{113}$$

Considering that $\|\mathbf{e} \times \dot{\mathbf{q}}_\mathbf{d}\| = \|\mathbf{e}\| \, \|\dot{\mathbf{q}}_d\| \sin\alpha$, where $\alpha = \angle(\mathbf{e}, \dot{\mathbf{q}}_\mathbf{d})$ and introducing new matrix:

$$\mathbf{Q} = \begin{bmatrix} k_p \cos^2(e_a) & 0 \\ 0 & k_p \cos^2(e_a) \end{bmatrix}, \tag{114}$$

one can write:

$$\frac{dV_l}{dt} \leqslant -\mathbf{e}^T \mathbf{Q} \mathbf{e} + \|\mathbf{e}\| \, \|\dot{\mathbf{q}}_d\| \sin^2(e_a) - \|\mathbf{e}\| \, \|\dot{\mathbf{q}}_d\| \sin\alpha \sin(e_a)\cos(e_a) \tag{115}$$

$$\leqslant -\mathbf{e}^T \mathbf{Q} \mathbf{e} + \|\mathbf{e}\| \, \|\dot{\mathbf{q}}_d\| \sin^2(e_a) + \|\mathbf{e}\| \, \|\dot{\mathbf{q}}_d\| \, |\sin\alpha| \, |\sin(e_a)| \, |\cos(e_a)| \tag{116}$$

$$\leqslant -\mathbf{e}^T \mathbf{Q} \mathbf{e} + 2 \, \|\mathbf{e}\| \, \|\dot{\mathbf{q}}_d\| \tag{117}$$

$$= - \|\mathbf{e}\| \left(\|\mathbf{e}\| \, \lambda_{min}(\mathbf{Q}) - 2 \, \|\dot{\mathbf{q}}_d\| \right), \tag{118}$$

where λ_{min} is the smallest eigenvalue of the matrix $\mathbf{Q}$.

The condition $\dot{V}_l < 0$ is fulfilled if the following condition is met:

$$\|\mathbf{e}\| > \frac{2 \, \|\dot{\mathbf{q}}_d\|}{\lambda_{min}(\mathbf{Q})}. \tag{119}$$

Remark 10B
The term $\cos^2(e_a)$ remains greater then zero according to *Assumption 10C*.

By changing the coefficient k_p in the matrix $\mathbf{Q}$ one can influence the value of $\lambda_{min}(\mathbf{Q})$ and thus ensures that the condition (119) is fulfilled. Meeting this condition guarantees the stability of the robot's position $\mathbf{q} = \begin{bmatrix} x & y \end{bmatrix}^T$ if it is outside the collision avoidance region of the obstacle. Fulfillment of inequality (119) guarantees that the robot's position converge to the reference value, but this is not an asymptotic convergence. The condition $\dot{V} < 0$ can be fulfilled by an appropriate setting the k_p coefficient.

In the next step, proof of the convergence of the robot orientation θ_i to an auxiliary orientation θ_{ai} i.e. $\lim_{t\to\infty} \theta_i = \theta_{ai}$ is shown. Substituting the last row of Eq. (81) to the second equation (77) and then, the result to the last row in (29) one obtains equality $\dot{\theta}_i = k_\theta e_{ai} + \dot{\theta}_{ai}$. Substituting $\dot{e}_{ai} = \dot{\theta}_{ai} - \dot{\theta}_i$ one obtains $\dot{e}_{ai} = -k_\theta e_{ai}$.

From the above equation it follows that e_{ai} decreases exponentially to zero and the orientation of the robot θ_i converges exponentially to the auxiliary orientation variable θ_{ai}.

The convergence of the auxiliary orientation variable θ_{ai} to the reference orientation θ_{di} results from a comparison of Eq. (72) and Eq. (75), in which h_{xi} and

h_{yi} are given by Eq. (81). The convergence of the robot position to the reference value is connected with the convergence of the components $k_p E_{xi}$ and $k_p E_{yi}$ to zero and it results in convergence of the auxiliary orientation variable to the reference orientation. This convergence is not asymptotic.

From the above considerations it is concluded that the robot's orientation converge non-asymptotically to the reference orientation.

10.2 Numerical Simulation

Figure 6a presents robot's path in (x, y)-plane (solid line). Dashed line represents desired trajectory. In the transient state, the robot avoids collision with a static obstacle located at point (12 m, 0) and with a radius of $r = 0.5$ m. The range of the APF was $R = 1.8$ m. Figure 6b, c show time graphs of the position coordinates. In Fig. 6d, e control signals of the mobile platform are shown. Figure 6f presents the time graph of the position error components. They reach values close to zero in about 10 s. Figure 6g shows the time graph of the freeze signal.

11 Vector Field Orientation Algorithm for *N* Robots

This subsection presents VFO control for N mobile robots. Their goal is to follow reference trajectories $\mathbf{q_{di}} = [\, x_{di} \;\, y_{di} \,]^T, i = 1, \ldots, N$ while ensuring inter-agent collision avoidance. The positions of robots are described by vectors: $\mathbf{q_i} = [\, x_i \;\, y_i \,]^T$.

11.1 Stability Analysis for N Robots

Consider the following Lyapunov-like function:

$$V_l = \sum_{i=1}^{N} \left[V_{tli} + \sum_{j=1, j \neq i}^{N} V_{alij} \right], \tag{120}$$

where V_{tli} is trajectory tracking component of the control of the i-th robot, V_{aij} represents APFs acting on the i-th robot, given by Eq. (3).

The Lypunov-like function is presented in more detailed form:

$$V_l = \sum_{i=1}^{N} \left[\frac{1}{2} \mathbf{e_i}^T \mathbf{e_i} + \sum_{j=1, j \neq i}^{N} \frac{1}{2} V_{aij} \right], \tag{121}$$

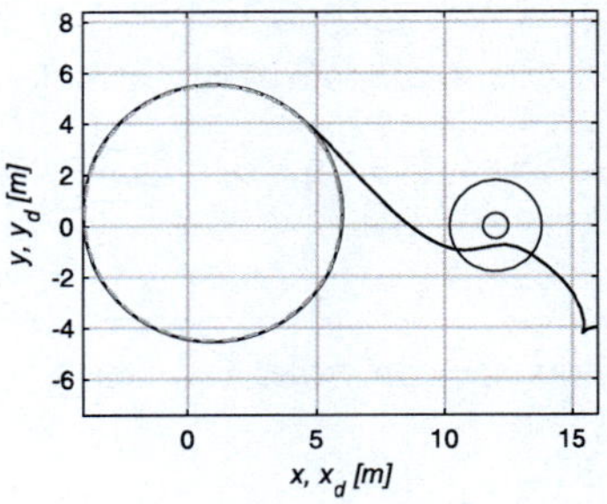

(a) Robot's position in XY-plane

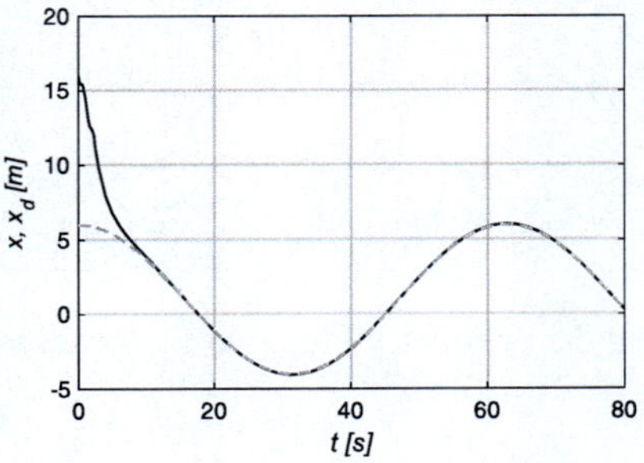

(b) x coordinate as a function of time

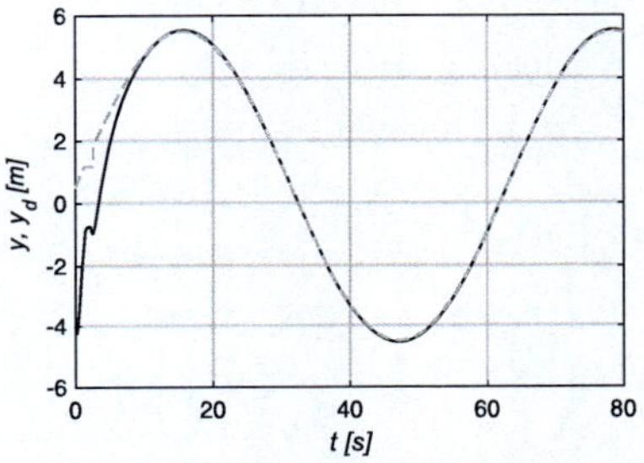

(c) y coordinate as a function of time

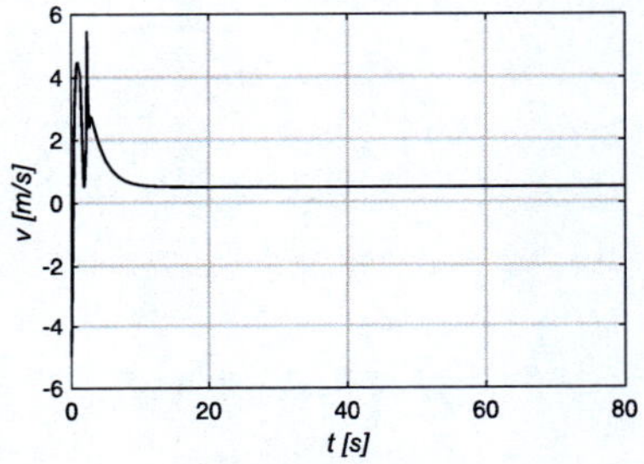

(d) Linear velocity

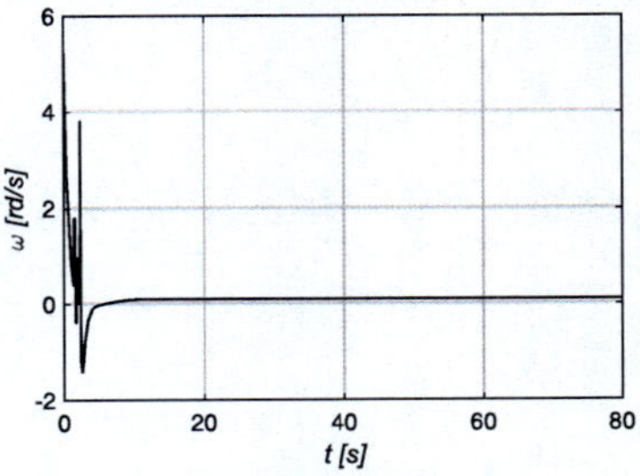

(e) Angular velocity

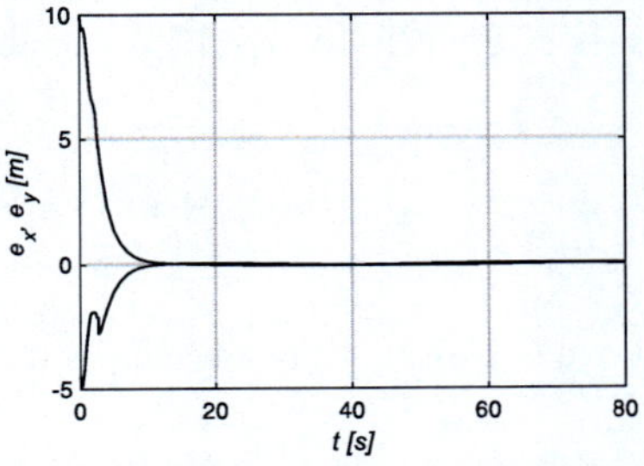

(f) Spatial components of position error

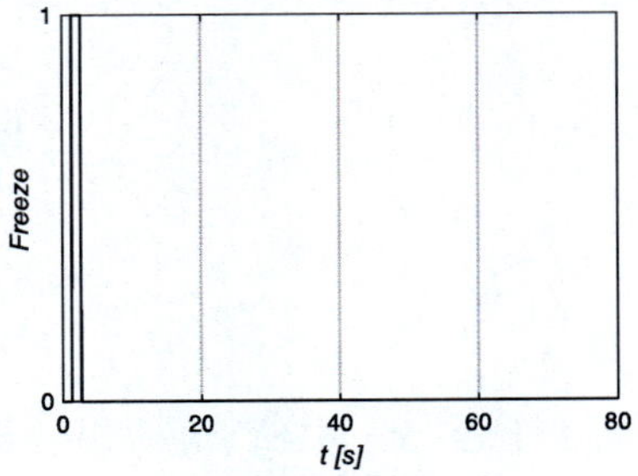

(g) Freeze signal

Fig. 6 Numerical simulation 5: VFO algorithm

where $V_{alij} = \frac{1}{2} V_{aij}(l_{ij})$, $l_{ij} = \left\| [\, x_i \; y_i \,]^T - [\, x_j \; y_j \,]^T \right\|$ and $i, \; j = 1, \ldots, N, i \neq j$.
Differentiating (121) with respect to time results in the following formula:

$$\frac{dV_l}{dt} = \sum_{i=1}^{N} \left(e_{xi}\dot{e}_{xi} + e_{yi}\dot{e}_{yi} + \sum_{j=1, j\neq i}^{N} \frac{1}{2} \left(\frac{\partial V_{aij}}{\partial x_i}\dot{x}_i + \frac{\partial V_{aij}}{\partial x_j}\dot{x}_j + \frac{\partial V_{aij}}{\partial y_i}\dot{y}_i + \frac{\partial V_{aij}}{\partial y_j}\dot{y}_j \right) \right). \tag{122}$$

Using the symmetry of the APFs between robots one can write:

$$\frac{\partial V_{aij}}{\partial x_i} = -\frac{\partial V_{aij}}{\partial x_j} = -\frac{\partial V_{aji}}{\partial x_j} = \frac{\partial V_{aji}}{\partial x_i} \tag{123}$$

and

$$\frac{\partial V_{aij}}{\partial y_i} = -\frac{\partial V_{aij}}{\partial y_j} = -\frac{\partial V_{aij}}{\partial y_j} = \frac{\partial V_{aij}}{\partial y_j}. \tag{124}$$

For $i = 1, \ldots, N$ equality (122) can be transformed as follows:

$$\frac{dV_l}{dt} = \sum_{i=1}^{N} \left(e_{xi}\dot{e}_{xi} + e_{yi}\dot{e}_{yi} + \sum_{j=1, j\neq i}^{N} \left(\frac{\partial V_{aij}}{\partial x_i}\dot{x}_i + \frac{\partial V_{aij}}{\partial y_i}\dot{y}_i \right) \right). \tag{125}$$

It will be shown that $\frac{dV_l}{dt} \leqslant 0$. Substituting corresponding equations from the right hand side of Eq. (79) to Eq. (125) one obtains:

$$\frac{dV_l}{dt} = \sum_{i=1}^{N} \left(e_{xi}(\dot{x}_{di} - \dot{x}_i) + e_{yi}(\dot{y}_{di} - \dot{y}_i) + \sum_{j=1, j\neq i}^{N} \left(\frac{\partial V_{aij}}{\partial x_i}\dot{x}_i + \frac{\partial V_{aij}}{\partial y_i}\dot{y}_i \right) \right). \tag{126}$$

The above equation can be transformed as follows:

$$\frac{dV_l}{dt} = \sum_{i=1}^{N} \left(e_{xi}\dot{x}_{di} + e_{yi}\dot{y}_{di} + \dot{x}_i \left(\sum_{j=1, j\neq i}^{N} \frac{\partial V_{aij}}{\partial x_i} - e_{xi} \right) + \dot{y}_i \left(\sum_{j=1, j\neq i}^{N} \frac{\partial V_{aij}}{\partial y_i} - e_{yi} \right) \right). \tag{127}$$

Substituting $e_{xi} = E_{xi} + \sum_{j=1, j\neq i}^{N} \frac{\partial V_{aij}}{\partial x_i}$ and $e_{yi} = E_{yi} + \sum_{j=1, j\neq i}^{N} \frac{\partial V_{aij}}{\partial y_i}$ to Eq. (127) one can write:

$$\frac{dV_l}{dt} = \sum_{i=1}^{N} \left(e_{xi}\dot{x}_{di} + e_{yi}\dot{y}_{di} - E_{xi}\dot{x}_i - E_{yi}\dot{y}_i \right). \tag{128}$$

Substituting Eqs. (99) and (100) into (128) (extended by the index i representing number of the robot) and simplifying one obtains:

$$\frac{dV_l}{dt} = \sum_{i=1}^{N} (e_{xi}\dot{x}_{di} + e_{yi}\dot{y}_{di} + E_{xi}(-h_{xi}\cos^2(e_{ai}) - h_{yi}\sin(e_{ai})\cos(e_{ai}))$$

$$+E_{yi}(+h_{xi}\sin(e_{ai})\cos(e_{ai}) - h_{yi}\cos^2(e_{ai}))\big). \tag{129}$$

Substituting in the above equation for h_{xi} and h_{yi} the corresponding rows of the equation (81) and making simplifications, one obtains a time derivative of the Lyapunov-like function:

$$\frac{dV_l}{dt} = \sum_{i=1}^{N}\big(e_{xi}\dot{x}_{di} + e_{yi}\dot{y}_{di} - k_pE_{xi}^2\cos^2(e_{ai}) - k_pE_{yi}^2\cos^2(e_{ai}) - E_{xi}\dot{x}_{di}\cos^2(e_{ai})$$

$$-E_{yi}\dot{y}_{di}\cos^2(e_{ai}) - E_{xi}\dot{y}_{di}\sin(e_{ai})\cos(e_{ai}) + E_{yi}\dot{x}_{di}\sin(e_{ai})\cos(e_{ai})\big). \tag{130}$$

Two cases will be considered:

1. All robots are located in the collision avoidance regions of other robots.
2. No robot is located in the collision avoidance regions of other robots.

In case 1, according to *Assumption 10B*, Eq. (130) takes the form:

$$\frac{dV_l}{dt} = \sum_{i=1}^{N}\big(-k_pE_{xi}^2\cos^2(e_{ai}) - k_pE_{yi}^2\cos^2(e_{ai})\big), \tag{131}$$

that fulfills the condition $\dot{V}_l \leqslant 0$.

The statements given in *Remark 10A* remain actual in this case.

In case 2 the following equalities are true: $\sum_{j=1,j\neq i}^{N}\frac{\partial V_{aij}}{\partial x_i} = \sum_{j=1,j\neq i}^{N}\frac{\partial V_{aij}}{\partial y_i} = 0$. Time derivative of the Lyapunov-like function is as follows:

$$\frac{dV_l}{dt} = \sum_{i=1}^{N}\big(e_{xi}\dot{x}_{di} + e_{yi}\dot{y}_{di} - k_pe_{xi}^2\cos^2(e_{ai}) - k_pe_{yi}^2\cos^2(e_{ai}) - e_{xi}\dot{x}_{di}\cos^2(e_{ai})$$

$$-e_{yi}\dot{y}_{di}\cos^2(e_{ai}) - e_{xi}\dot{y}_{di}\sin(e_{ai})\cos(e_{ai}) + e_{yi}\dot{x}_{di}\sin(e_{ai})\cos(e_{ai})\big). \tag{132}$$

Using Pythagorean identity and substituting into Eq. (132) $e_{xi}\dot{y}_{di} - e_{yi}\dot{x}_{di} = \|\mathbf{e_i} \times \dot{\mathbf{q}}_{\mathbf{di}}\|$ one obtains:

$$\frac{dV_l}{dt} = \sum_{i=1}^{N}\big(e_{xi}\dot{x}_{di}\sin^2(e_{ai}) + e_{yi}\dot{y}_{di}\sin^2(e_{ai})$$

$$-k_pe_{xi}^2\cos^2(e_{ai}) - k_pe_{yi}^2\cos^2(e_{ai}) - \|\mathbf{e_i} \times \dot{\mathbf{q}}_{\mathbf{di}}\|\sin(e_{ai})\cos(e_{ai})\big). \tag{133}$$

Taking into account $\|\mathbf{e_i} \times \dot{\mathbf{q}}_{\mathbf{di}}\| = \|\mathbf{e_i}\|\,\|\dot{\mathbf{q}}_{di}\|\sin\alpha_i$, where $\alpha_i = \angle(\mathbf{e_i}, \dot{\mathbf{q}}_{\mathbf{di}})$ and introducing a new matrix:

$$\mathbf{Q_i} = \begin{bmatrix} k_p \cos^2(e_{ai}) & 0 \\ 0 & k_p \cos^2(e_{ai}) \end{bmatrix}, \tag{134}$$

one can rewrite Eq. (133) as follows:

$$\frac{dV_l}{dt} \leqslant \sum_{i=1}^{N} \left(-\mathbf{e_i}^T \mathbf{Q_i} \mathbf{e_i} + \|\mathbf{e_i}\| \, \|\dot{\mathbf{q}}_{di}\| \sin^2(e_{ai}) - \|\mathbf{e_i}\| \, \|\dot{\mathbf{q}}_{di}\| \sin\alpha_i \sin(e_{ai}) \cos(e_{ai})\right) \tag{135}$$

$$\leqslant \sum_{i=1}^{N} \left(-\mathbf{e_i}^T \mathbf{Q_i} \mathbf{e_i} + \|\mathbf{e_i}\| \, \|\dot{\mathbf{q}}_{di}\| \sin^2(e_{ai}) + \|\mathbf{e_i}\| \, \|\dot{\mathbf{q}}_{di}\| \, |\sin\alpha_i| \, |\sin(e_{ai})| \, |\cos(e_{ai})|\right) \tag{136}$$

$$\leqslant - \left\| \begin{bmatrix} \mathbf{e}_1 \\ \vdots \\ \mathbf{e}_N \end{bmatrix} \right\| \left(\left\| \begin{bmatrix} \mathbf{e}_1 \\ \vdots \\ \mathbf{e}_N \end{bmatrix} \right\| \min_i \{\lambda_{min\,i}(\mathbf{Q_i})\} - 2 \left\| \begin{bmatrix} \dot{\mathbf{q}}_{d1} \\ \vdots \\ \dot{\mathbf{q}}_{dN} \end{bmatrix} \right\| \right). \tag{137}$$

The condition $\dot{V}_l < 0$ is fulfilled if:

$$\left\| \begin{bmatrix} \mathbf{e}_1 \\ \vdots \\ \mathbf{e}_N \end{bmatrix} \right\| > \frac{2 \left\| \begin{bmatrix} \dot{\mathbf{q}}_{d1} \\ \vdots \\ \dot{\mathbf{q}}_{dN} \end{bmatrix} \right\|}{\min_i \{\lambda_{min\,i}(\mathbf{Q_i})\}}. \tag{138}$$

The statements made in *Remark 10B* concerning the Eq. (119) remain actual for (138)—the robots positions are stable and non-asymptotically converge to the reference values.

If some robots are in the collision avoidance regions of other robots, the corresponding rows in the vectors (138) disappear. In this case, the corresponding components of the Lyapunov-like function will appear in the Eq. (131).

11.2 Numerical Simulation

Figure 7a presents robots paths in the (x, y)-plane (solid lines) and their reference trajectories (dashed lines), the number of robots: $N = 4$. Figure 7b, c show time graphs of position coordinates. They converge to the values close to the reference signals in about 10 s. Figure 7d, e present time graphs of control signals. In the transient state they reach high values, which in practice are not realizable. In the authors opinion, the implementation on the real robots should use scaling procedure that changes wheel velocities to be achievable by the actuators. Figure 7f presents time graph of the position error components. They are reduced to values close to zero values after about 10 s. Figure 7g shows time graph of the freeze signal.

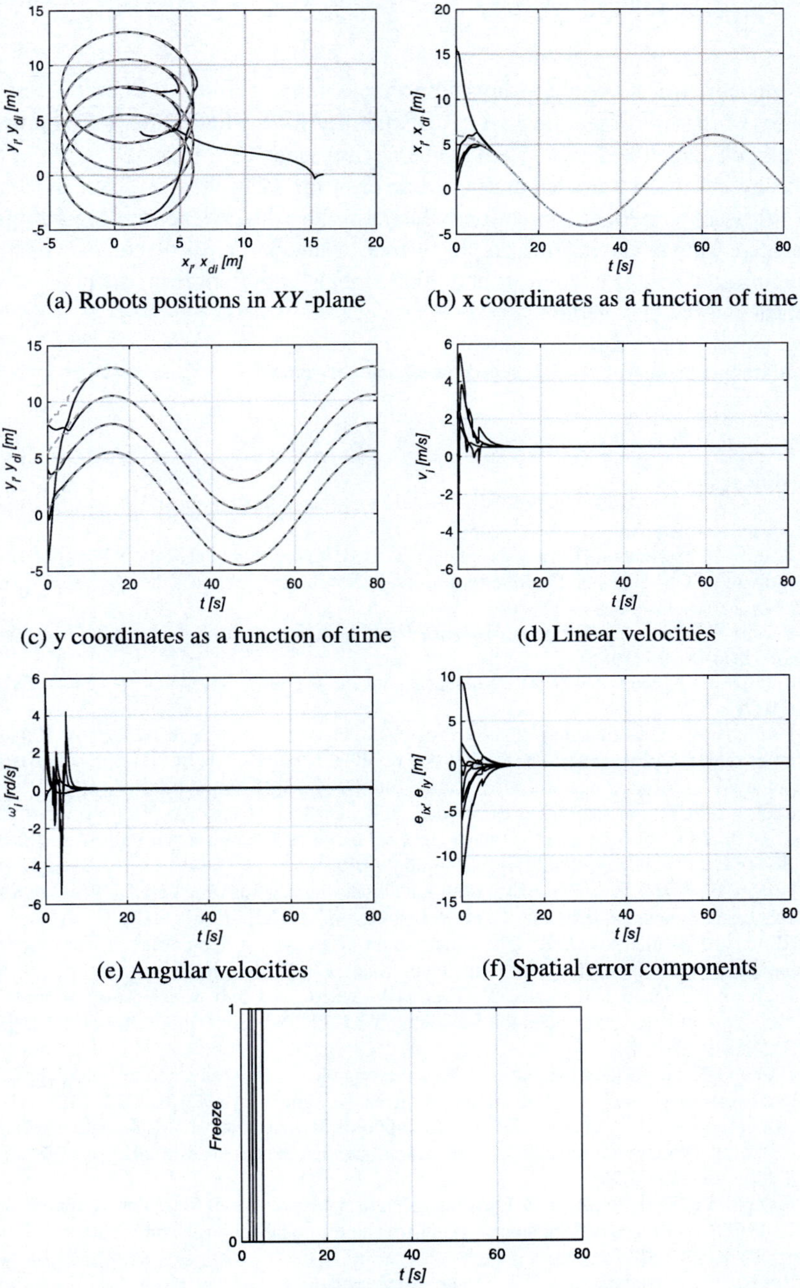

(a) Robots positions in XY-plane

(b) x coordinates as a function of time

(c) y coordinates as a function of time

(d) Linear velocities

(e) Angular velocities

(f) Spatial error components

(g) Freeze signals

Fig. 7 Numerical simulation 6: VFO for N robots

12 Summary

The paper presents three algorithms for nonholonomic mobile robot control. For each of them in the first stage the analysis of stability for a single robot in the presence of static obstacles is shown. Then the same process is carried out for a group of N robots executing coordinated motion while ensuring inter-agent collision avoidance. For all cases, numerical simulations illustrating the effectiveness of algorithms are presented. The results include graphs of robot paths on (x, y)-plane, time graphs of configuration variables, controls and other signals important from the point of view of particular control method.

Acknowledgements This work is supported by statutory grant 09/93/SBAD/0911.

References

1. Khatib, O.: Commande Dynamique dans l'Espace Opérationnel des Robots Manipulateurs en Présence d'Obstacles, Ph.D. dissertation, Ecole Nationale Superieure de l'Aeronautique et de l'Espace, Touluse, France (1980)
2. Khatib, O.: Real-time obstacle avoidance for manipulators and mobile robots. Int. J. Robot. Res. **5**(1), 90–98 (1986)
3. Leitmann, G., Skowronski, J.: Avoidance control. J. Optim. Theory Appl. **23**(4), 581–591 (1977)
4. Leitmann, G.: Guaranteed avoidance strategies. J. Optim. Theory Appl. **32**(4), 569–576 (1980)
5. Szulczyński, P., Pazderski, D., Kozłowski, K.: Real-time obstacle avoidance using harmonic potential functions. J. Autom. Mob. Robot. Intell. Syst. **5**(3), 59–66 (2011)
6. Szulczyński, P., Pazderski, D., Kozłowski, K.R.: Obstacle avoidance and trajectory tracking using fluid-based approach in 2D space. In: Lecture Notes in Control and Information Sciences Book Series (LNCIS, vol. 422), pp. 293–304 (2012)
7. Rimon, E., Koditschek, D.E.: Exact robot navigation using cost functions: the case of distinct spherical boundaries. IEEE Int. Conf. Robot. Autom. **3**, 1791–1796 (1988)
8. Rimon, E., Koditschek, D.E.: The construction of analytic diffeomorphisms for exact robot navigation on star worlds. Trans. Am. Math. Soc. **327**(1), 71–116 (1991)
9. Urakubo, T., Okuma, K., Tada, Y.: Feedback control of a two wheeled mobile robot with obstacle avoidance using potential functions. IEEE/RSJ Int. Conf. Intell. Robot. Syst. (IROS) **3**, 2428–2433 (2004)
10. Urakubo, T.: Feedback stabilization of a nonholonomic system with potential fields: application to a two-wheeled mobile robot among obstacles. Nonlinear Dyn. **81**(3), 1475–1487 (2015)
11. Dimarogonasa, D.V., Loizoua, S.G., Kyriakopoulos, K.J., Zavlanosb, M.M.: A feedback stabilization and collision avoidance scheme for multiple independent non-point agents. Automatica **42**(2), 229–243 (2005)
12. Filippidis, I., Kyriakopoulos, K.J.: Adjustable navigation functions for unknown sphere worlds. In: IEEE Conference on Decision and Control and European Control Conference (CDC-ECC), pp. 4276–4281 (2011)
13. Roussos, G., Kyriakopoulos, K.J.: Decentralized and prioritized navigation and collision avoidance for multiple mobile robots. Distrib. Auton. Robot. Syst.-Springer Tracts Adv. Robot. **83**, 189–202 (2013)
14. Roussos, G., Kyriakopoulos, K.J.: Completely decentralised navigation of multiple unicycle agents with prioritisation and fault tolerance. In: IEEE Conference on Decision and Control (CDC), pp. 1372–1377 (2010)

15. Roussos, G., Kyriakopoulos, K.J.: Decentralized navigation and conflict avoidance for aircraft in 3-D space. IEEE Trans. Control. Syst. Technol. **20**(6), 1622–1629 (2012)
16. Roussos, G., Chaloulos, G., Kyriakopoulos, K., Lygeros, J.: Control of multiple non-holonomic air vehicles under wind uncertainty using model predictive control and decentralized navigation functions. In: 47th IEEE Conference on Decision and Control, pp. 225–1230, Cancun (2008)
17. Roussos, G., Dimarogonas, D., Kyriakopoulos, K.: Distributed 3D navigation and collision avoidance for nonholonomic aircraft-like vehicles. In: Control Conference (ECC), 2009 European, pp. 1830–1835 (2009)
18. Przybyła, M., Kowalczyk, W., Kozłowski, K.: Navigation function used for parallel parking in restricted area by a differentially-driven mobile robot. IFAC-PapersOnLine **50**(1), 4312–4317 (2017)
19. Kowalczyk, W., Przybyła, M., Kozłowski, K.: Set-point control of mobile robot with obstacle detection and avoidance using navigation function—experimental verification. J. Intell. Robot. Syst. **85**(3-4), 539–552
20. Kowalczyk, W., Kozłowski, K.: Control of the differentially-driven mobile robot in the environment with a non-convex star-shape obstacle: simulation and experiments. Acta Polytechnica Hungarica **13**(1), 123–135
21. Kowalczyk, W., Przybyła, M., Kozłowski, K.: Control of a mobile robot and collision avoidance using navigation function-experimental verification. In: 10th International Workshop on Robot Motion and Control (RoMoCo), pp. 148–152, July 6–8, Poznan, Poland (2015)
22. Kowalczyk, W., Kozłowski, K.: Feedback linearization of the differentially driven mobile robot: an experimental verification. In: Advances in Soft Computing, Intelligent Robotics and Control, pp. 41–54
23. Kowalczyk, W., Przybyła, M., Kozłowski, K.: Saddle point detection of the navigation function in nonholonomic mobile robot control. In: 21st International Conference on Methods and Models in Automation and Robotics, pp. 936–941, Miedzyzdroje (2016)
24. Kowalczyk, W., Przybyła, M., Kozłowski, K.: Saddle point detection of the navigation function in nonholonomic mobile robot control. In: 21st International Conference on Methods and Models in Automation and Robotics (MMAR), pp. 936–941, August 29–September 1, Miedzyzdroje, Poland (2016)
25. Kowalczyk, W.: Rapid navigation function control for two-wheeled mobile robots. J. Intell. Robot. Syst., 1–11 (2018)
26. Kowalczyk, W., Przybyła, M., Kozłowski, K.: Rapid navigation function control for omnidirectional mobile platform. In: 22nd International Conference on Methods and Models in Automation and Robotics (MMAR), pp. 137–140, August 28–31, 2017 Miedzyzdroje, Poland
27. Kowalczyk, W.: Rapidly converging navigation function control for differentially driven mobile robots. In: 2017 11th International Workshop on Robot Motion and Control (RoMoCo), pp. 244–250, 3–5 July 2017, Wasowo, Poland (2017)
28. Do, K.: Formation tracking control of unicycle-type mobile robots with limited sensing ranges. IEEE Trans. Control. Syst. Technol. **16**(3), 527–538 (2008)
29. Kozłowski, K., Kowalczyk, W.: Formation control and vision based localization of a system of mobile robots. In: Lecture Notes in Electrical Engineering, Informatics in Control, Automation and Robotics, pp. 3–27 (2016)
30. Kowalczyk, W., Kozłowski, K., Tar, J.: Trajectory tracking for multiple unicycles in the environment with obstacles. In: IEEE 19th International Workshop on Robotics in Alpe-Adria-Danube Region (RAAD), pp. 451–456 (2010)
31. Cruz, D., McClintok, J., Perteet, B., Orqueda, O., Cao, Y., Fierro, R.: Decentralized cooperative control—a multivehicle platform for research in networked embedded systems. IEEE Control. Syst. Mag. **27**(3), 58–78 (2007)
32. Kowalczyk, W., Kozłowski, K.: Artificial Potential Based Control for a Large Scale Formation of Mobile Robots, Climbing and Walking Robots, pp. 191–199. Springer, Berlin, Heidelberg (2004)
33. Gawron, T., Michałek, M.: Procedura Predykująca VFO dla Algorytmów Planowania Ruchu Robotów Mobilnych z Ograniczoną Krzywizną, XIV Krajowa Konferencja Robotyki, pp. 499-508, Polanica-Zdrój, Poland (2016) (in Polish)

34. Gawron, T., Michałek, M.: Algorithmization of constrained motion for car-like robots using the VFO control strategy with parallelized planning of admissible funnels. In: 2018 IEEE/RSJ International Conference on Intelligent Robots and Systems, pp. 6945–6951, At Madrid, Spain (2018)
35. Canudas de Wit, C., Khennouf, H., Samson, C., Sordalen, O.J.: Nonlinear Control Design for Mobile Robots, Recent Trends in Mobile Robots, pp. 121–156 (1994)
36. Loria, A., Dasdemir, J., Alvarez Jarquin, N.: Leader-follower formation and tracking control of mobile robots along straight paths. IEEE Trans. Control. Syst. Technol. **24**(2), 727–732 (2016). https://doi.org/10.1109/TCST.2015.2437328
37. Michałek, M., Kozłowski, K.: Vector-field-orientation feedback control method for a differentially driven vehicle. IEEE Trans. Control. Syst. Technol. **18**(1), 45–65 (2010)
38. Kowalczyk, W., Michałek, M., Kozłowski, K.: Trajectory tracking control with obstacle avoidance capability for unicycle-like mobile robot. Bull. Pol. Acad. Sci. Tech. Sci. **60**(3), 537–546 (2012)
39. Kowalczyk, W., Michałek, M., Kozłowski, K.: Trajectory tracking control and obstacle avoidance for a differentially driven mobile robot. IFAC Proc. Vol. **44**(1), 1058–1063 (2011)
40. Michałek, M., Kowalczyk, W., Kozłowski, K.: Strategia śledzenia trajektorii z unikaniem kolizji dla robota mobilnego klasy (2, 0), Prace Naukowe Politechniki Warszawskiej. Elektronika, pp. 381–390 (2010)
41. Kozłowski, K., Kowalczyk, W.: Motion control for formation of mobile robots in environment with obstacles. In: Towards Intelligent Engineering and Information Technology, pp. 203–219 (2009)
42. Mastellone, S., Stipanovic, D., Spong, M.: Formation control and collision avoidance for multi-agent non-holonomic systems: theory and experiments. In: The International Journal of Robotics Research, pp. 1–14 (2008)
43. Kowalczyk, W., Kozłowski, K.: Leader-follower control and collision avoidance for the formation of differentially-driven mobile robots. In: MMAR 2018—23rd International Conference on Methods and Models in Automation and Robotics, August 27–30, pp. 132–137, Miedzyzdroje, Poland (2018)
44. Loria, A., Panteley, E., Teel, A.: Relaxed persistency of excitation for uniform asymptotic stability. IEEE Trans. Autom. Control **46**(12), 1363–1368 (2001). December
45. Khalil, H.K.: Nonlinear Systems, 3rd edn. NY, USA, Prentice-Hall, New York (2002)
46. Rimon, E., Koditschek, D.: Exact robot navigation using artificial potential functions. IEEE Trans. Robot. Autom. **8**(5), 501–518 (1992)
47. Stipanovic, D.M., Hokayem, P.F., Spong, M.W., Siljak, D.D.: Cooperative avoidance control for multiagent systems. J. Dyn. Syst. Meas. Control. **127**, 699–707 (2007)

Biologically Inspired Motion Design Approaches for Humanoids and Walking Machines

Teresa Zielińska

Abstract On the basis of our research the paper discusses how certain biological patterns are used for motion synthesis of multi-legged walking machines and humanoid robots. In particular, by focusing on humanoid (anthropomorphic) robots, it discusses how proper foot design allows to achieve body inclinations coordinated with legs movement for supporting the dynamic stability of the walking posture. The first part of the chapter summarizes methods of gait synthesis for multi-legged walking machines and humanoid robots. The criterion of dynamic gait stability (ZMP criterion) widely used for bipedal gait synthesis is discussed. This criterion was our starting point for determining the forces acting on the foot of a humanoid robot. Considering a compliant foot, a method for determining the force components acting on the foot was developed. Confirming the usefulness of the method, a calculation example for human gait was considered. The force components exerted by the foot on the ground were obtained for different foot dimensions. The method shows how a compliant foot supports the dynamic stability of the gait.

Keywords Walking machines · Humanoids · Motion synthesis · Postural stability

1 Introduction

The advantages of having a human-like robot are numerous: typical environment does not have to be adapted, having characteristics similar to humans robots can execute daily tasks using human tools. Human-like robots can work as care givers without stressing the patients by strange appearance or by unusual equipment. To achieve human like motion, not only the body must be adequately shaped, but also and adequate motion generation strategy it needed. Robots are expected to act in an environment typical for a human, cooperating with a human being and using tools

---

T. Zielińska (✉)
Faculty of Power and Aeronautical Engineering, Institute of Aeronautics and Applied Mechanics, Warsaw University of Technology, Nowowiejska 24, 00665 Warsaw, Poland
e-mail: teresaz@meil.pw.edu.pl

P. Kulczycki et al. (eds.), *Automatic Control, Robotics, and Information Processing*, Studies in Systems, Decision and Control 296,
https://doi.org/10.1007/978-3-030-48587-0_16

designed for humans, therefore not only leg-end trajectories, but the movements of all body parts are important. In the simplest approach the human motion data is recorded and the robot follows the captured trajectories of joints or body points. Obviously the motion adaptation to the external conditions is not possible in this case. The human motion has been studied for decades, from the point of view of robots. M. Vukobratovic investigated dynamic stability of human gait. For that purpose he defined the Zero Moment Point (ZMP) criterion [1] now commonly used for motion synthesis of humanoids [2].

Many different methods of humanoids motion synthesis are used, but still a robust method of planning postural adjustments, when performing more complex activities, such as carrying, lifting and manipulating objects, especially if they are heavy and big, is missing.

Currently, we observe a growing interest in developing service and personal robots. These robots must operate in a human-oriented environment, that is full of objects, where the situation changes dynamically. Personal robots have to help in home activities, especially taking care of children, the elderly, or people with disabilities.. Often such robots have an anthropomorphic shape, which facilitates their acceptance. Anthropomorphism enables functioning in an environment designed for a human being, thus facilitates the use of everyday objects. Despite examples of some efficient human-like robots, the problem of their motion synthesis still remains open.

There are no publicly known methods that enable real-time human like motion generation for complex activities such as placing or lifting objects, or recovering balance lost due to external push or due to slippery surface. Returning to the seemingly simplest example, i.e. walk, it should be emphasized that not only the legs but the other parts of the body have part in ascertaining the postural stability during locomotion [3]. While walking, the upper part of the body performs rhythmic tilts forward and backward, as well as sideways, the pelvis rotates, the hips move alternately forward and backward relative to the central axis of the body, and the hands swing. All these movements create appropriate inertial forces and force moments contributing to the dynamic stabilization of the posture. The researchers try to achieve a similar synergy of movements in humanoid robots, but this task is obstructed by the non-human specificity of the body structure and different than in human, features of the actuating system.

The first humanoid robots moved on bent legs, keeping the body in a vertical position, and without moving the hands. In this case bent legs helped in postural stabilization, making the movements performed by other parts of the body unnecessary. Modern robots in this respect do not resemble those clumsy precursors. Awkward walking bipedal robots with few degrees of freedom have evolved into complex, efficiently moving humanoid robots. These robots wave their arms and balance the upper body like a human. There are also attempts to create a humanoid robot with a spine. The attention of researchers is also focused on the design of feet. The feet play an important role in achieving smooth human like movements.

In this chapter we show how certain biological patterns are transferred to the area of multi-legged walking machines and humanoid robots. Focusing on anthropomorphic robots, we show how the correct foot structure facilitates robot motion.

Walking is a manner of moving by using legs. Gait types (gaits) were classified and described taking into account the observation of the animal world. The type of walk is the order in which legs are transferred , but it is not associated, as it might seem, with the speed of movement. Certain leg transfer sequences are, nevertheless, typical for a given group of animals moving at a certain speed. The features of the insect gait were identified over forty years ago in the works of Hughes and Wendler, and then summarized by Wilson [4]. They are described by five simple rules:

- the legs on one side of the body are moved one after another, and such order goes from the back to the front of the body, no leg will be transferred before the leg behind is placed on the ground,
- the transfer time is constant for every leg and does not depend on the walking speed,
- opposite legs in the same body segment are never moved simultaneously,
- time of the leg support phase (when the leg is in contact with the ground) decreases when the walking speed increases, what means that with the increase of the walking speed the walking frequency increases,
- the time intervals between lifting of two adjacent legs on the same side of the body are similar, but those intervals change when the walking frequency changes.

In 1981 Delcomyn [5] observed one more rule:

- the leg is raised only when the leg preceding it in a sequence of displacements is already on the ground.

This rule is a natural consequence of the first and last principle formulated by Wilson.

The above features are used for gait synthesis of six-legged machines. Such machines move like insects. Their gait is statically stable. This means that at any instant static stability of their posture is preserved, i.e. the projection of the center of gravity of the entire machine is located inside the support polygon spread on the footprints of the supporting legs. Gait synthesis is not difficult. It is not difficult to select the back stance and stride such that the static stability of a rhythmic gait is maintained.

However, a more difficult problem is to plan a sequence of leg movements required to avoid obstacles, as well as choose the shape of the trajectory of the leg-ends. It is still the current goal of research. The effective solutions that combine methods of environment perception and motion planning are still needed. In our works, we described the methods for generating statically stable gaits for six-legged machines and designed their control systems [6–9] enabling autonomous navigation matching the type of gait to the terrain properties. Figure 1 shows the six-legged machines built by the team. They were used to test the mentioned solutions. We also dealt with the specification of design requirements, selection of actuators considering the energy costs [10], and the method of selecting geometric dimensions of the legs which ensure the minimization of energy expenditure [11].

Many animals, including humans, move maintaining dynamic, not static postural stability. Dynamic stability is the result of the balance of forces—forces and torques acting on body parts of a moving animal or robot. Due to this equilibrium

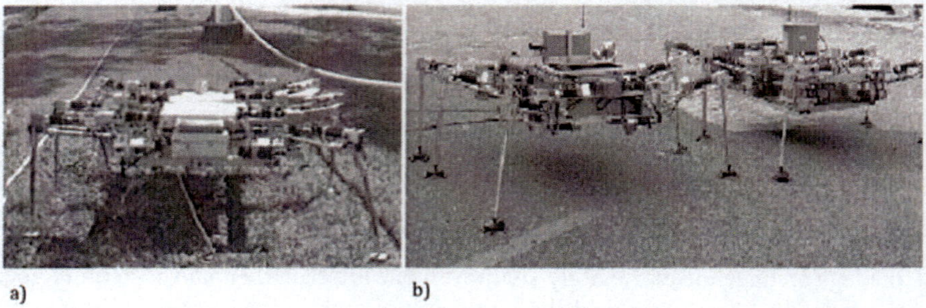

Fig. 1 Navigation of six-legged machines: **a** machine following autonomously a path defined using geographical coordinates, **b** two autonomous machines performing an exploration task

the body reproduces the desired trajectory of motion, while maintaining the desired configuration. At higher movement speeds and when avoiding obstacles, the insects also retain dynamic, not static motion stability. Unfortunately, there are no synthetic descriptions which, as in the case of static stability, would allow to adopt dynamical motion patterns to the world of robotics.

It is worth to add that also the slowest gaits of four-legged animals are statically stable, but unfortunately their gaits can not be described as simply as the insect gaits, due to the many different sequences of leg transfers, different support time for each leg, and different time intervals between leg transfers. Our works devoted to locomotion of quadrupeds include the synthesis of statically stable gait [9, 12–14], design [15, 16], control [17] and generation of quasi-static gaits, where the stabilizing effect is achieved due to feet with built-in compliance [18, 19]. In Fig. 2 a four-legged machine built by our team is shown [12]. The machine was used to test the proposed solutions. Photographs illustrate that the large leg work spaces enable extensive configuration changes. The right hand side of that figure shows the sequence of postures realized when the machine rolls over by 180°. Such motion is useful in walking machines that perform exploratory tasks in difficult, unusual conditions [20], because this enables to return to the normal position after tumbling over. We can give examples of four-legged machines that move quickly, maintaining dynamic stability. The precursor in this type of research was and is M. Raibert, building one-legged, then bipedal, and then four-legged jumping machines [21]. Now, under his supervision, the most impressive four-legged machines such as Big Dog, Spot Mini and others are made. The team led by this scientist developed also a very efficient humanoid robots: PETMAN, ASTROMAN and ATLAS [22]. Unfortunately the method of generating coordinated movements of the body parts for these spectacular prototypes, with impressive motor capabilities, has not been disclosed. Thus still many researchers look for patterns of human movement using the models of double or single inverted pendulum [23–25]. Our works in this area, proposes a reconfigurable pendulum with a moving point mass [26].

Some researchers use models of biological motion generators in robotics. Periodic animal gaits (wave gaits) are usually fast, and due to highly coordinated leg

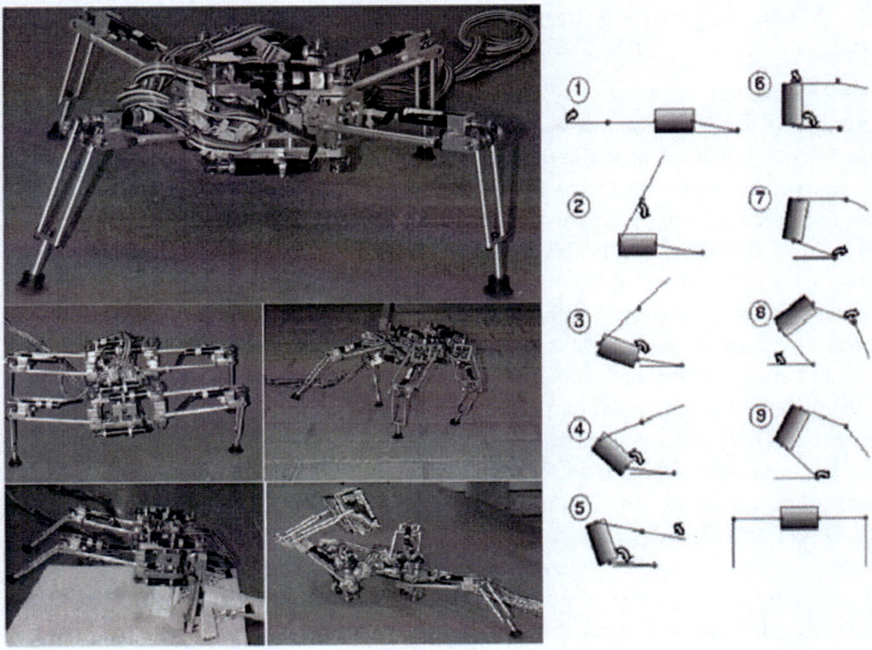

Fig. 2 Reconfiguration of four-legged walking machine

motions they are energy efficient. Such gaits are controlled by Central Pattern Generator (CPG) rhythms. The hypothesis that CPG is responsible for periodic processes, including body part movement was formulated at the beginning of the XX century. Later research confirmed its existence. Nowadays the CPG inspired methods are used in robots for motion generation. The oscillators, such as the gait pattern generators for multi-legged and two-legged locomotion have attracted attention of researchers for many years. In the investigations of CPG inspired locomotion models the concept of distributed self-organized generators is often used [27]. Due to the simplicity of Matsuoka oscillators [28], that drive the two opposing phases of a rhythmic movement (i.e., muscles flexion and extension), they are often used for the implementation of animal-like motion design. Another CPG inspired approach uses coupled oscillators, for which the closed-form solution does not exist. Due to the couplings the oscillators enable fast phase locking as well as higher-order locking— therefore not only periodicity of their solutions, but also of their derivatives can be obtained. Unfortunately due to bifurcation and significant number of unknown parameters the analysis of those oscillators is difficult. Van der Pol oscillators which attracted our attention [29, 30] are of that type.

2 Motion Synthesis for Humanoid Robots

Many different methods are used for humanoid motion synthesis. In [31] the humanoid was described as a multi-body system. Similarly in [32] a whole body model was considered, while dealing with obstacles. Even for a typical gait of the robot, several factors must be considered simultaneously: the leg-end trajectories, foot placement strategy, a plan of upper body motions to assure the postural stability. A motion planner which is able to deal with all those aspects simultaneously is still sought for [33–35] are the examples of works where the whole body model is involved in the motion generation. The body parameters such as: geometry, mass, inertia, and each link CoM (Center of Mass) location influence the motion dynamics. Unfortunately detailed models produce computationally intensive solutions. Berenson et al. [36] proposed an universal motion planning strategy using the so called Task Space Regions (TSR). It is a probabilistic, sample-based approach, that nominally can avoid the heavy computational load. Three main components of the method are: constraint manifold, constraint-satisfaction strategy and planning algorithm called Constrained Bi-Directional Rapidly exploring Random Tree (CBiRRT). The constraint manifold contains all configurations that do not violate the imposed constraints. Unfortunately this method has some limitations concerning motion generation for load lifting tasks. Real-time based motion generation methods often use preview control utilizing state equations of the inverted pendulum. The body is reduced to a single point mass on a massless rod. The paper [24] proposes a simple concept of gait generation using a 3D linear inverted pendulum. The primary objective of the model preview control is to obtain the CoM motion trajectory resembling that of a human CoM. Using the simplified model of a humanoid, the predictive controller generates the trajectory of the center of mass. The control system of the real robot tracks it the best it can. A model of an inverted pendulum on a cart moving on a table was proposed as a model of human motion [37]. The trajectory of the ZMP was given and the CoM motion, assuring close tracking of predefined ZMP trajectory was obtained using the preview control method. The paper [38] presents an inverted pendulum model combined with the models of flexible joints achieving fast, real-time motion generation. Lanari et al. [39] studied linear inverted pendulum (LIP) dynamics to identifying stable and unstable walking states. They obtained the so called *Bounded Solution* for avoiding the unstable states. Unfortunately the LIP model is not able to represent the behavior of the upper part of the body. Therefore Double Inverted Pendulum models (DIP) were used, e.g. [23] presented human-like balancing motion using double inverted pendulum. There two actions performed by a human being were analyzed. The human motion capture data was used to develop a hip balancing strategy providing human-like postural adjustments while keeping the ZMP inside the support polygon. In [26] a DIP with moving masses was used to study the motion coordination of upper and lower parts of the human body. The human motion capture data for four different movements was analyzed and a strong coincidence with DIP mass displacement was noted. A double pendulum with moving masses was also used by [40] to analyze the postural recovery after unexpected push. Figure 3 is showing the double inverted

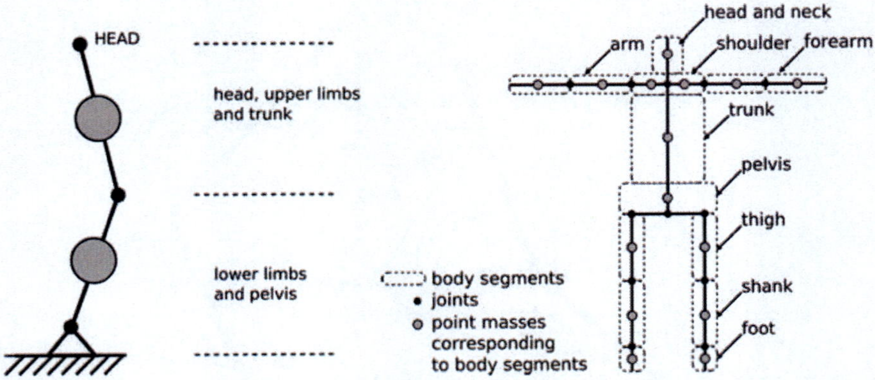

Fig. 3 The double inverted pendulum representing the human body (left), and the segmentation of a human body considered for definition of pendulum parameters (right)

pendulum representing the human body, and the segmentation of a human body considered for definition of pendulum parameters. The Spring Loaded Inverted Pendulum model (SLIP) is often applied to study fast locomotion of humans and animals having legs. Such a model represents well the dynamic behavior of the CoM during running [41]. SLIP models are also used in dynamic controllers of hopping robots, e.g. [42] applied the SLIP model and a simple planner generating walking, running and jumping motions of bipeds. The model was successfully exploited simulating the locomotion on uneven terrain. The single or double inverted pendulum models are also commonly used to study human balancing [43–45]. They enable human-like postural recovery when the normal motion pattern is disturbed.

In opposition to model based methods, the mentioned already CPG approach refers to the work of a neural structure located in the spinal cord, which among others, generate the rhythm of locomotion.

In this section we summarize the concept of CPG inspired motion pattern generation method using the Van der Pol oscillators.

The equations describing the dynamical properties of van der Pol oscillators have the following form:

$$\begin{aligned} \ddot{\alpha}_1 - \mu_1 \cdot (p_1^2 - x_a^2) \cdot \dot{\alpha}_1 + g_1^2 \cdot x_a &= q_1 \\ \ddot{\alpha}_2 - \mu_2 \cdot (p_2^2 - x_b^2) \cdot \dot{\alpha}_2 + g_2^2 \cdot x_b &= q_2 \\ \ddot{\alpha}_3 - \mu_3 \cdot (p_3^2 - x_c^2) \cdot \dot{\alpha}_3 + g_3^2 \cdot x_c &= q_3 \\ \ddot{\alpha}_4 - \mu_4 \cdot (p_4^2 - x_d^2) \cdot \dot{\alpha}_4 + g_4^2 \cdot x_d &= q_4 \end{aligned} \tag{1}$$

where

$$\begin{aligned} x_a &= \alpha_1 - \lambda_{21} \cdot \alpha_2 - \lambda_{31} \cdot \alpha_3 \\ x_b &= \alpha_2 - \lambda_{12} \cdot \alpha_1 - \lambda_{42} \cdot \alpha_4 \\ x_c &= \alpha_3 - \lambda_{13} \cdot \alpha_1 - \lambda_{43} \cdot \alpha_4 \\ x_d &= \alpha_4 - \lambda_{24} \cdot \alpha_2 - \lambda_{34} \cdot \alpha_3 \end{aligned}$$

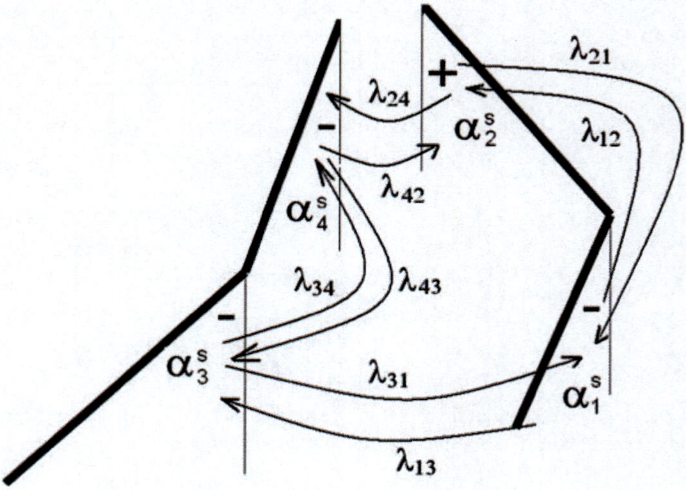

Fig. 4 The joint angles and the couplings between them

In the equation (1) there are 24 unknown parameters: μ_1, μ_2, μ_3, μ_4, p_1^2, p_2^2, p_3^2, p_4^2, g_1^2, g_2^2, g_3^2, g_4^2, q_1, q_2, q_3, q_4, λ_{13}, λ_{31}, λ_{12}, λ_{21}, λ_{24}, λ_{42}, λ_{43}, λ_{34}.

$$\ddot{x}_{osc} - \mu \cdot (p^2 - x_{osc}^2) \cdot \dot{x}_{osc} + g^2 \cdot x_{osc} = q \tag{2}$$

The variables μ, p^2, g^2, q influence the phase shift in oscillators.

α_1^s, α_2^s, α_3^s, α_4^s represent the re-scaled angles α_1 (in degrees) for the lower limb joints. In our notation they are positive if the considered link is in the frontal position with respect to the vertical line and negative in the opposite situation (Fig. 4). First the unknown parameters were selected on the basis of heuristic search, but in in our more recent research a genetic algorithm was used for for final tuning of oscillator parameters [46].

Finally the selected parameters are [46]: $\mu_1 = \mu_3 = 3.59375, \mu_2 = \mu_4 = 2, p_1^2 = p_3^2 = 2, p_2^2 = p_4^2 = 1, g_1^2 = g_3^2 = 28.0039, g_2^2 = g_4^2 = 17.7031, q_1 = q_3 = 15.8516, q_2 = q_4 = -7.04492$, $\lambda_{12} = \lambda_{21} = \lambda_{34} = \lambda_{43} = -0.451172$, $\lambda_{24} = \lambda_{42} = \lambda_{31} = \lambda_{13} = 0.417969$. The search for parameter values using a genetic algorithm can be compared to the phase of learning basic locomotion rhythms by humans, as described by Bernstein [47]. He identified three stages in the development of human motion abilities. In the first stage (early infancy) the number of DOFs is reduced by locking some of them and highly coupled rhythmic movements are learned. This stage in-prints in the CPG the body parts couplings for motion rhythms. In the second stage (young child) the locked DOFs are freed and thus the human gains the ability to produce complex movements. In the third stage the person attains the ability of using the advantages of passive dynamics when producing complex movements. Motion efficiency is enhanced. The genetic algorithm applied by us took on the function of

Fig. 5 The robot using the gait generated by coupled oscillators

tuning the oscillating rhythms for more complex motions, what is the second stage in Bernstein's description.

Our research proved that adequate changes of the oscillator parameters cause the required modifications of the gait. The oscillator equations were modified for real-time implementation in the biped prototype. Results are discussed in [46, 48]. Figure 5 illustrates our robot walking.

3 The Role of the Foot During Locomotion

Now we will focus on dynamic stability, and on the role of the foot. The simplest way for specifying the gait of anthropomorphic robots is recording and then playing back the human movement, as was the case in the first Honda humanoids. Another method is to use the motion trajectory of a chosen degree of freedom, e.g. a human knee joint, and mathematically matching to it the trajectories of the other joints. In still another approach, the trajectories of movement of individual degrees of freedom are defined as polynomials of unknown parameters, the choice of parameters takes place with regard to postural stability. We proposed a method that uses the model of the robot's motion dynamics and the joint controllers. Here motion control synthesis is made using the dynamics model and demanding the reproduction of given trajectories of leg-end forces [49]. This method leads to stabilizing the repositioning of the upper body in a similar manner as in human motion. Despite the existence of particular solutions, still the synthesis of motion is a difficult task, waiting for a universal solution that allows the generation of the posture of the whole body.

Before implementation, it must be checked whether the developed motion pattern ensures dynamic stability, i.e. whether the moving robot will not fall over. For this purpose, the Zero Moment Point (ZMP) criterion, formulated in 1968 by Vukobratovic [1, 2], is most often used. This criterion, in the modified version, is also used to determine stabilizing posture corrections during real robot movement. The ZMP criterion takes into account the forces, force moments and torques equilibrium conditions in the support phase. For defining this criterion a simplified kinematic model of the body of a human or robot is used, with the distributed masses replaced by point masses. These masses are subjected to accelerations during walking, so inertial forces and torques arise. As a result of the existence of these forces, the gravitational forces, the torques are created that cause the overall body rotation. In particular there are rotations around the axes of the reference system with the origin in O. In the single support phase this point is located in the ankle joint of the supporting leg. In the double support phase, those revolutions are around axes of the reference frame with the origin in the projection of the total body mass onto the supporting surface. For simplicity of illustration, we will discuss the ZMP formula for a single-support phase.

Lets $\mathbf{F_r}$ denote the reaction forces vector (Fig. 6b). Forces equilibrium condition is expressed by:

$$\sum_i (\mathbf{r_i} \times \mathbf{F_i}) + \mathbf{I_i}\dot{\omega_i} + \omega_\mathbf{i} \times \mathbf{I_i}\ \omega_\mathbf{i} = -\mathbf{r_P} \times \mathbf{F_r} \tag{3}$$

where $\mathbf{r_i}$ is the vector starting in O and ending in i-th point mass (mass m_i), $\mathbf{F_i}$ is the vector of inertia and gravity forces which act on the i-th mass, $\dot{\omega_i}$ denotes the vector of revolute velocities, and $\mathbf{I_i}$ is an inertia tensor, r_P is a vector from O to point P where the force $\mathbf{F_r}$ is applied. Point P is called the Zero Moment Point (ZMP). Considering the equilibrium condition (3) we conclude that P denotes such point in the support area where the reaction i $\mathbf{F_r}$ must be applied for creating the force moment equilibrating the moments due to the motion of the body parts. The reaction force concerns the supporting foot therefore point P must stay within the foot print of the supporting foot [2]. This is the postural stability condition.

Now the position of P will be obtained using (3). In the considerations only M_x causing the side sway of the body and M_y causing the front-back sway will be used. The small moment M_z, producing the rotation around vertical axis, is neglected, because during the gait it is equilibrated by the moment due to friction forces.

M_x and M_y is obtained:

$$\begin{aligned} M_x = &\sum_i [y_i(m_i\ddot{z}_i + m_i \cdot g)] - \sum_i (z_i \cdot m_i \ddot{y}_i) + \\ &\sum_i (I^i_{xx} \cdot \dot{\omega}^i_x + I^i_{xy} \cdot \dot{\omega}^i_y + I^i_{xz} \cdot \dot{\omega}^i_z) + \\ &\sum_i [\omega^i_y (I^i_{zx}\, \omega^i_x + I^i_{zy}\, \omega^i_y + I^i_{zz}\, \omega^i_z) - \\ &\omega^i_z (I^i_{yx}\, \omega^i_x + I^i_{yy}\, \omega^i_y + I^i_{yz}\, \omega^i_z)] + p_y F^r_z + h\, F^y_R \end{aligned} \tag{4}$$

$$\begin{aligned} M_y = &-\sum_i [x_i(m_i\ddot{z}_i + m_i \cdot g)] + \sum_i (z_i \cdot m_i \ddot{x}_i) + \\ &\sum_i (I^i_{yx} \cdot \dot{\omega}^i_x + I^i_{yy} \cdot \dot{\omega}^i_y + I^i_{yz} \cdot \dot{\omega}^i_z) + \\ &\sum_i [-\omega^i_x (I^i_{zx}\, \omega^i_x + I^i_{zy}\, \omega^i_y + I^i_{zz}\, \omega^i_z) + \\ &\omega^i_z (I^i_{xx}\, \omega^i_x + I^i_{xy}\, \omega^i_y + I^i_{xz}\, \omega^i_z)] - p_x F^r_z + h\, F^x_R \end{aligned} \tag{5}$$

x_i, y_i, z_i are coordinates of point masses m_i with respect to the reference frame $OXYZ$, p_x, p_y are coordinates of P, h is the distance from O (ankle joint) to the support plane. This distance, being of small value, is neglected in farther considerations. O marks the vertical projection of the ankle joint onto the supporting plane (Fig. 6). In equations (4) and (5) one small term vanishes. The F^r_z component of reaction force $\mathbf{F_r}$ is expressed by $F^r_z = -\sum_i (m_i\ddot{z}_i + m_i \cdot g)$. Therefore:

$$\begin{aligned} M_x = &\sum_i [y_i(m_i\ddot{z}_i + m_i \cdot g)] - \sum_i (z_i \cdot m_i \ddot{y}_i) + \\ &\sum_i (I^i_{xx} \cdot \dot{\omega}^i_x + I^i_{xy} \cdot \dot{\omega}^i_y + I^i_{xz} \cdot \dot{\omega}^i_z) + \\ &\sum_i [\omega^i_y (I^i_{zx}\, \omega^i_x + I^i_{zy}\, \omega^i_y + I^i_{zz}\, \omega^i_z) - \\ &\omega^i_z (I^i_{yx}\, \omega^i_x + I^i_{yy}\, \omega^i_y + I^i_{yz}\, \omega^i_z)] - p_y \sum_i (m_i\ddot{z}_i + m_i \cdot g) \end{aligned} \tag{6}$$

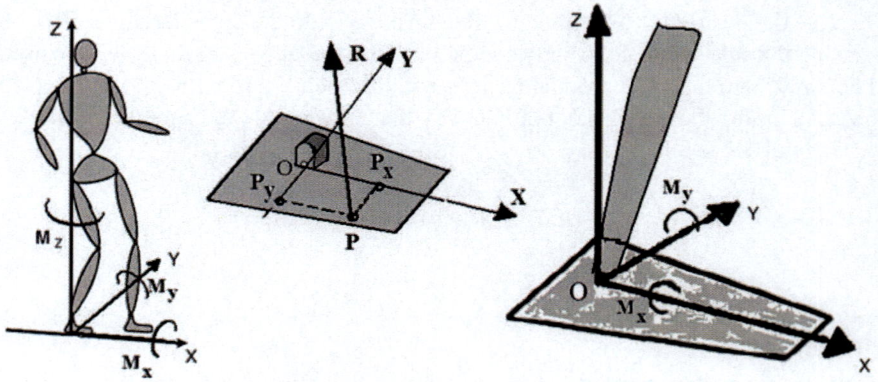

Fig. 6 Zero Moment Point—illustration

$$M_y = -\sum_i [x_i(m_i\ddot{z}_i + m_i \cdot g)] + \sum_i (z_i \cdot m_i\ddot{x}_i) + \sum_i (I^i_{yx} \cdot \dot{\omega}^i_x + I^i_{yy} \cdot \dot{\omega}^i_y + I^i_{yz} \cdot \dot{\omega}^i_z) + \sum_i [-\omega^i_x(I^i_{zx}\,\omega^i_x + I^i_{zy}\,\omega^i_y + I^i_{zz}\,\omega^i_z) + \omega^i_z(I^i_{xx}\,\omega^i_x + I^i_{xy}\,\omega^i_y + I^i_{xz}\,\omega^i_z)] + p_x \sum_i (m_i\ddot{z}_i + m_i \cdot g) \tag{7}$$

The inertia tensor $\mathbf{I_i}$ and angular velocities and acceleration are relatively small, therefore two small terms (6), (7), but more difficult to calculate, are neglected what results in:

$$M_x = \sum_i [y_i(m_i\ddot{z}_i + m_i \cdot g)] - \sum_i (z_i \cdot m_i\ddot{y}_i) - p_y \sum_i (m_i\ddot{z}_i + m_i \cdot g) = 0 \tag{8}$$

$$M_y = -\sum_i [x_i(m_i\ddot{z}_i + m_i \cdot g)] + \sum_i (z_i \cdot m_i\ddot{x}_i) + p_x \sum_i (m_i\ddot{z}_i + m_i \cdot g) = 0 \tag{9}$$

Regrouping the terms of (8) and (9) we obtain coordinates p_y, p_x of point P (ZMP):

$$p_y = \frac{\sum_i [y_i(m_i\ddot{z}_i + m_i \cdot g)] - \sum_i (z_i \cdot m_i\ddot{y}_i)}{\sum_i (m_i\ddot{z}_i + m_i \cdot g)} \tag{10}$$

$$p_x = \frac{\sum_i [x_i(m_i\ddot{z}_i + m_i \cdot g)] - \sum_i (z_i \cdot m_i\ddot{x}_i)}{\sum_i (m_i\ddot{z}_i + m_i \cdot g)} \tag{11}$$

The small error introduced by neglecting the two terms is compensated by the safety margin; the requirement that P must be located within the foot print reduced by a certain distance from its edges.

Designing the locomotion of humanoids we use the knowledge about human motion properties. The sway of upper part of the body (torso) supports the postural stability. This can be easily proved. Lets us assume that m_1 denotes the torso mass which is the biggest one. In (8), (9) we have products: $y_1 \cdot m_1 \cdot g$ in (8) and $x_1 \cdot m_1 \cdot g$ in (9). Because m_1 is large in those products, the small change of y_1 (side sway of upper part of the body) and x_1 (front-back sway) significantly influences resulting p_y i p_x. Therefore by changing the position of the upper part of the body we can adjust the position of P without changing the motion pattern of the legs.

4 Aiding Postural Stability During Locomotion

In the standing position of a human and some mammals the whole foot sole touches the ground. With such starting posture not only the hip and knee joint, but also the ankle joint contributes significantly to the stable walk [50, 51]. The animals with the above posture are called plantigrade.

The angular motion of the ankle joint aids the foot take-off in the beginning of the transfer phase and helps in impact absorption during the touch-down. The change of the foot orientation during the support phase is compared to the rolling wheel [52]. After the touch-down moment only the heel is on the ground, next the full contact occurs, ending with the toe contact during take-off (Fig. 7). Besides orientation changes also the foot compliance is important in locomotion. The earlier robotic feet were built of rigid elements, but currently compliant elements are more often used. Such feet facilitate locomotion similarly as human feet do [53–55]. The robot foot compliance can be active, when the actuators are used, or passive, when elastic materials or mechanical elements such as springs and dampers are used. Passive compliance results in simpler and lighter mechanical design, however it requires studies and careful parameter selection.

In our earlier works we proposed a foot with one compliant element (spring) located near to the ankle joint [56–58]. Theoretical and practical studies confirmed that such an element aids postural stability.

Fig. 7 The sequence of foot configurations during the support phase

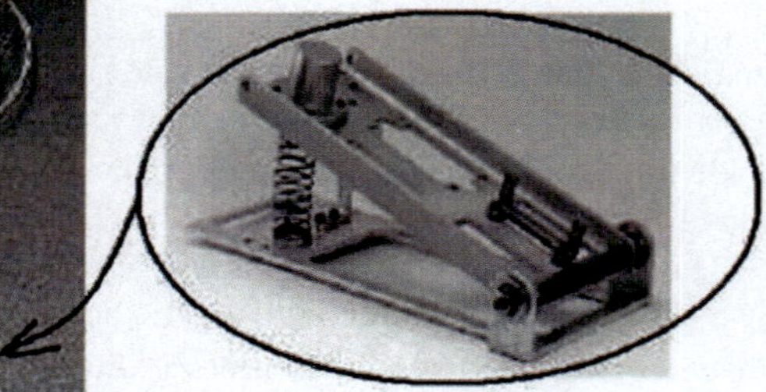

a b

Fig. 8 Our prototype used in the research: **a** view of the robot, **b** the foot with the compliant element

Figure 8 presents designed by us biped and its compliant feet. In each foot is installed small motor which, using the cable, compresses the spring during the leg-end transfer phase and releases it delivering the take-off impulse when the foot starts its transfer. In is interesting that similar solution is used in the human shoes, however the work of springs is passive. Figure 9 shows the rear view of the robot feet. Pictures were taken during the walk generated by coupled oscillators. Frame 1 shows the moment after touch down by the left leg-end, the spring of the right foot is released supporting the leg-end take off, in frame 2—right leg is in transfer phase, spring is compressed by the cable attached to the small motor, the spring in left foot is fully compressed, frame 3—right foot touched the ground, the spring in the left foot is released, frame 4—take off moment of the left foot, its spring is fully released producing the take off impulse, frame 5—left leg is in transfer, its spring is under compression by the motor, frame 6—end of the whole cycle (situation from frame

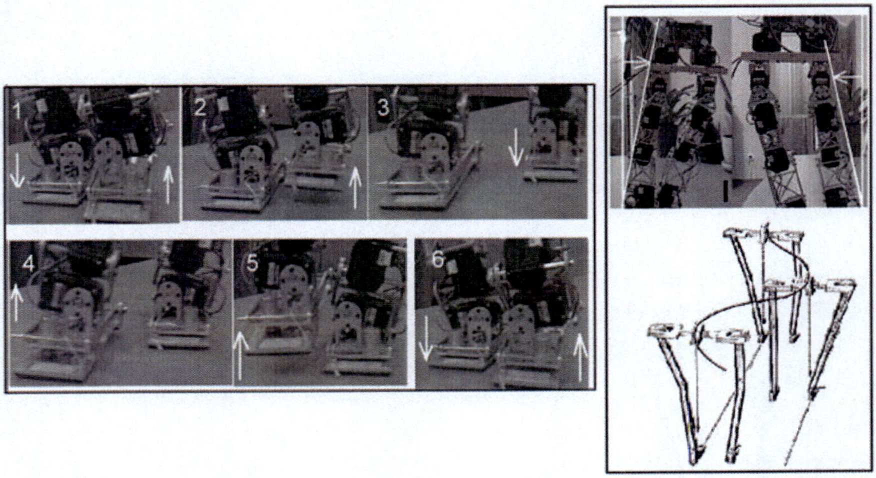

Fig. 9 The view of the feet during the walk (left), side sway during the walk (right)

1 repeats). The robot sway illustrated in right side of Fig. 9 is similar to the human body sway, which is especially visible during the gait of small children.

The concept of a compliant foot inspired by a human foot will be discussed in more detail in the next section. It will be explained how the compliance aids the postural stability during walking. Similar considerations were presented in another our work [59], however a simplified design concept and calculation method were used.

5 Forces Acting on a Foot

The human foot arch works as a spatial spring [61] introducing vertical compliance of the foot arch located along transverse and longitudinal axes of the foot. Figure 10 illustrates the structure of a human foot and its mechanical model [60]. Basing on

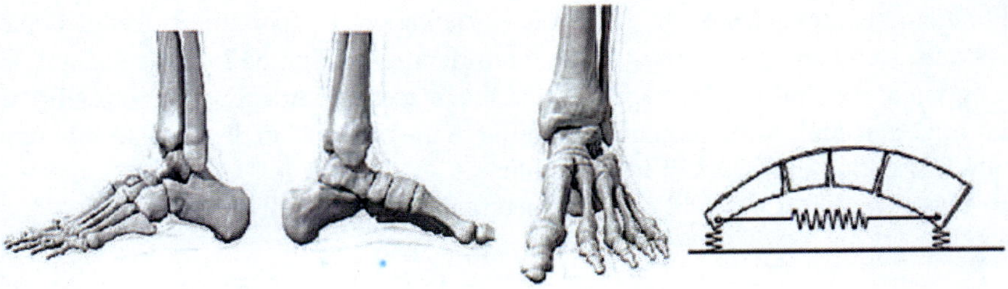

Fig. 10 Human foot and its simplified model [60]

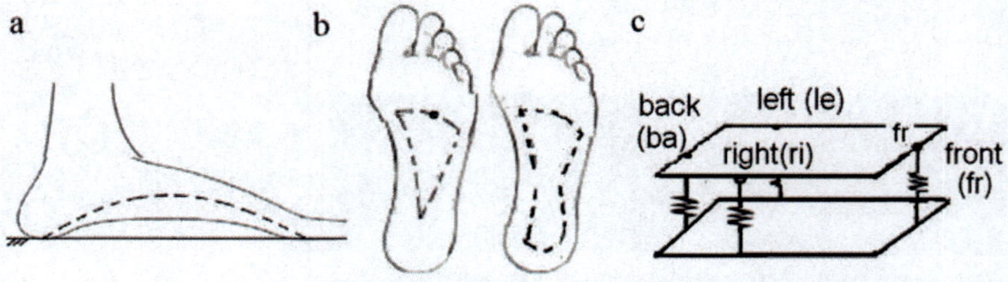

Fig. 11 Foot compliance **a** side view, longitudinal foot arch is shown, **b** top view illustrating the transverse and the whole arch, **c** the model of compliant foot

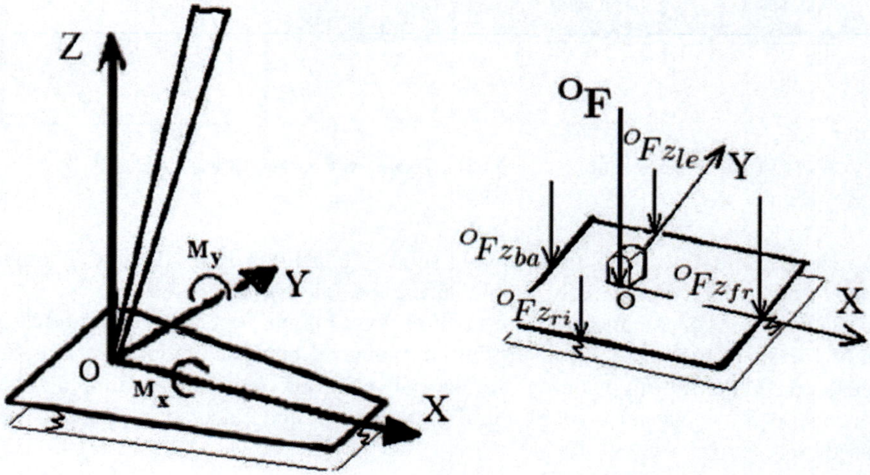

Fig. 12 Forces acting on a foot, the origin of $OXYZ$ reference frame is located in the ankle joint

the above we proposed a virtual model of the robotic foot shown in Fig. 11. The model consists of two rigid layers connected with vertical springs. The springs are called virtual, because it is assumed that the vertical force is split into the two side springs and next the same force is split to the frontal and rear spring. Such situation is possible when using spatial springs [62].

Our virtual spring is compressed proportionally to the acting vertical force. The horizontal components of the force vectors acing on the foot are neglected as they are very small and not relevant to our considerations. Figure 12 shows a simplified diagram of the foot, the acting forces and the reference frame. For such structure we introduce an equilibrium conditions similar to the one used in the ZMP formulation, however taking into account foot compliance.

Force equilibrium condition expressed in the frame $OXYZ$ is given by:

$$\sum_i m_i({}^{\mathbf{O}}\ddot{\mathbf{r}}_{\mathbf{i}} + {}^{\mathbf{O}}\mathbf{g}) = {}^{\mathbf{O}}\mathbf{F} = [{}^{O}Fx,\ {}^{O}Fy,\ {}^{O}Fz]^T \tag{12}$$

where m_i is the mass of the i-th body segment, $^{\mathbf{O}}\mathbf{r_i} = [^O x_i, ^O y_i, ^O z_i]^T$ is the vector between point O and point mass m_i, $^{\mathbf{O}}\mathbf{g} =^O [0, 0, -g]^T$ is the gravitational acceleration, $m_i(^{\mathbf{O}}\ddot{\mathbf{r}}_{\mathbf{i}} + ^{\mathbf{O}}\mathbf{g}) = ^{\mathbf{O}}\mathbf{F_i}$ represents the gravitational force acting on the mass m_i and the inertial force developed due to the motion of the i-th mass, $^{\mathbf{O}}\mathbf{F}$ is the resultant force vector applied to point O:

$$^{\mathbf{O}}\mathbf{F} = \sum_j {^{\mathbf{O}}\mathbf{F_j}} \tag{13}$$

where $j = \{fr, ba\}$ or $j = \{le, ri\}$, and $^{\mathbf{O}}\mathbf{F_{fr}}$, $^{\mathbf{O}}\mathbf{F_{ba}}$ and $^{\mathbf{O}}\mathbf{F_{le}}$, $^{\mathbf{O}}\mathbf{F_{ri}}$ are the forces acting on the front, back, left and right side of the foot respectively. The overall ground reaction force is equal to $(-^{\mathbf{O}}\mathbf{F})$. In stable posture the moments from this force with respect to the $OXYZ$ axes equilibrate the moments due to the body dynamics (in other words, the moments $sum_i {^{\mathbf{O}}\mathbf{r_i}} \times {^{\mathbf{O}}\mathbf{F_i}}$) from all forces $^O F_i$ are equilibrated):

$$\sum_i {^{\mathbf{O}}\mathbf{r_i}} \times {^{\mathbf{O}}\mathbf{F_i}} = \sum_j {^{\mathbf{O}}\mathbf{r_j}} \times {^{\mathbf{O}}\mathbf{F_j}} \tag{14}$$

Using the left side of equation (14) we evaluate M_x^B M_y^B produced by the moving system (body of the robot or a human):

$$\begin{aligned} M_x^B &= \sum_i ({^O y_i} \, {^O F z_i} - {^O z_i} \, {^O F y_i}) \\ M_y^B &= \sum_i (-{^O x_i} \, {^O F z_i} + {^O z_i} \, {^O F x_i}) \end{aligned} \tag{15}$$

Moment M_z, producing the rotation around the vertical axis, is equilibrated by the moment from shear friction forces.

Basing on (14) and (15):

$$\begin{gathered} \sum_i ({^O y_i} \, {^O F z_i} - {^O z_i} \, {^O F y_i}) = \\ \sum_j ({^O y_j} \, {^O F z_j} - {^O z_j} \, {^O F y_j}), \quad for \quad j = \{ri, le\} \\ \sum_i (-{^O x_i} \, {^O F z_i} + {^O z_i} \, {^O F x_i}) = \\ \sum_j (-{^O x_j} \, {^O F z_j} + {^O z_j} \, {^O F x_j}), \quad for \quad j = \{fr, ba\} \end{gathered} \tag{16}$$

First we assume that the ankle position is at constant height over the ground, equal to h ($^O z_j = h$). After regrouping the terms in (16) we get:

$$
\begin{aligned}
{}^{O}Fz_{ri} &= \frac{{}^{O}y_{le}\,{}^{O}Fz - h\,{}^{O}Fy - M_x^B}{{}^{O}y_{le} - {}^{O}y_{ri}} \\
{}^{O}Fz_{le} &= {}^{O}Fz - {}^{O}Fz_{ri} \\
{}^{O}Fz_{fr} &= \frac{{}^{O}x_{ba}\,{}^{O}Fz - h\,{}^{O}Fx + M_y^B}{{}^{O}x_{ba} - {}^{O}x_{fr}} \\
{}^{O}Fz_{ba} &= {}^{O}Fz - {}^{O}Fz_{fr}
\end{aligned} \tag{17}
$$

${}^{O}Fz$ is derived from (13), where M_x^B, M_y^B are expressed by (15).

6 Postural Stability Aided by Feet Compliance

6.1 Force Distribution During Human Gait

By studying the distribution of leg-end forces depending on the foot dimensions we conclude, how the compliant feet can aid the postural stability during locomotion. We consider a simplified model of the human body (illustrated in Fig. 13) [56]. The model concerns 50-centile adult man 1.75 m tall with 75 kg body mass. The gait pattern (joint trajectories) recorded using a human motion tracking system provide motion animation. A typical slow walk with a speed of about 3.5 km/h and the data for

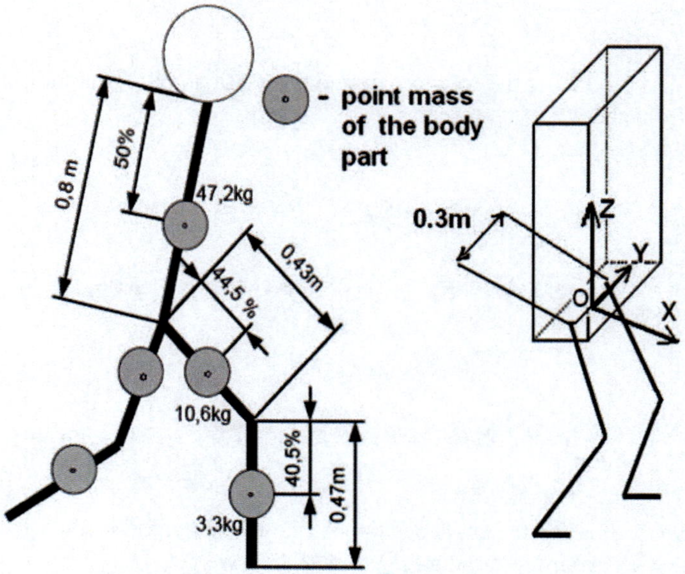

Fig. 13 Considered model of the body

right foot are considered. The displacements, velocities and finally the accelerations of the body point masses were calculated first and then the inertia forces and the torques being the components of (17) were obtained.

Equation (17) presents the forces acting on a foot. By changing ${}^{O}y_{le}, {}^{O}y_{ri}$ and ${}^{O}x_{ba}, {}^{O}x_{fr}$ this equation was used to study the relation between foot dimensions and the force distribution. Results are presented in Fig. 14. In the legends to parts a and c, the first listed data is the distance from the ankle joint to the right edge of the foot, and second – is the distance from the ankle joint to the left edge. In the legend to part b, the first data is the distance from the foot front to the ankle joint, second – is the distance from the ankle joint to the rear of the foot. The words *left*, *right*, *back*, *front* denote the vertical forces acting on the left, right, back and front virtual spring respectively. If the foot is wider the difference between the forces acting on side edges is smaller, and for the longer foot the difference between the forces acting at the front and back edges is smaller. The last part of Fig. 14 illustrates how the lack of foot symmetry, with respect to the Y axis, results in an asymmetrical force distribution. For wider outer part of the foot (from ankle joint to the right edge) the difference between the side force vectors increases with the progress of the support phase. If the inner dimension is bigger the situation is opposite. The presented drawings illustrate how the foot dimensions influence the distribution of the edge forces during the single support phase.

6.2 *Foot Compliance Aiding Postural Stability in a Gait*

Correctly selected foot compliance aids postural stability. We assume that the foot construction assures different stiffness of front-back and left-right edges. The stiffness must be such that the resultant foot sway will aid postural stability during the gait.

Compressions of each spring conforms to the linear spring model; the compression is equal to $\Delta h_j = \frac{{}^{O}F_{zj}}{k_j}$ $(j = \{ri, le\}$ or $j = \{fr, ba\})$, where k_j is the stiffness coefficient of $j - th$ spring.

Let us consider the foot size typical for an adult man. The frontal dimension is 20 cm (from ankle to the end of toes), the rear is 8 cm. The foot longitudinal axis of symmetry is located 3 cm from each edge of the foot. Figure 15 illustrates foot sway during the support phase. The stiffness of the virtual springs was selected by trial and error in such a way that the resultant body sway is similar to that of a walking human and aids the postural stability. The obtained model of the human body for the considered stiffness of the front edge of the foot was equal to 0.05 MN/m, rear stiffness was equal to 2 MN/m, left and right stiffness were equal to 0.9 MN/m. The results are illustrated in Fig. 15. The side sway is illustrated in Fig. 15a, and front-back sway is shown in Fig. 15b. To prove that the compliant foot aids the postural stability the upper body sway was especially considered as smaller than that during the human gait. The upper body (torso) front-back sway for a rigid foot was in the range $\langle 0^o, 1^o \rangle$,

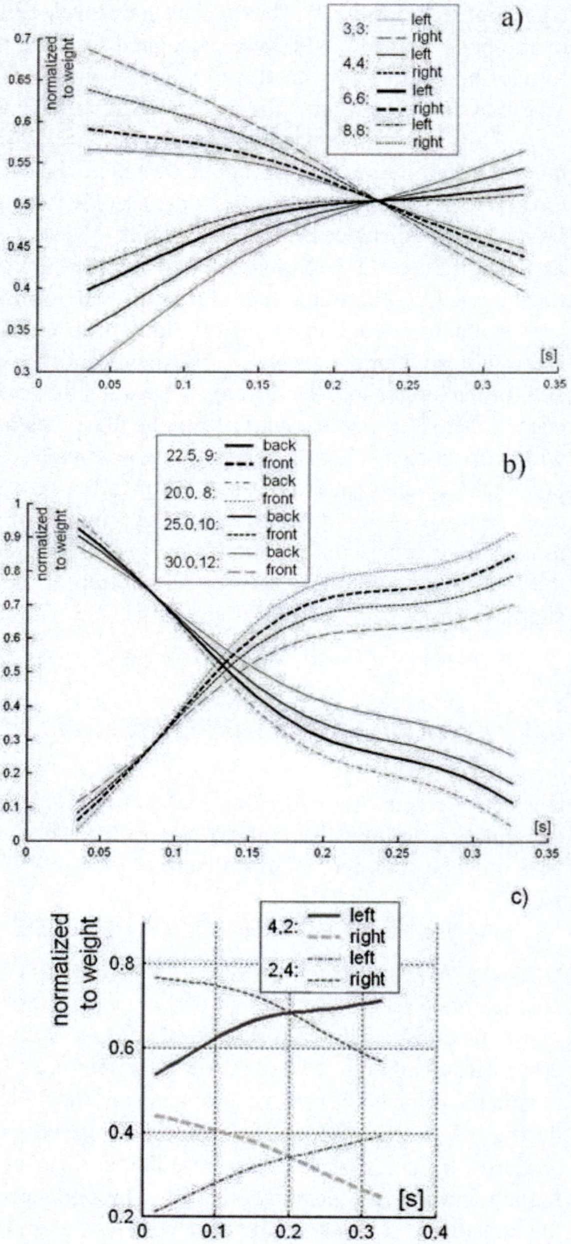

Fig. 14 Force distribution during the single support phase: **a** side forces depend on the foot dimension, **b** front-back forces depend on the foot dimension, **c** side forces for non-symmetrical foot. All forces are normalized to the body weight. In the legend to a and c the first listed value is the distance from the middle of the ankle joint to the right edge, the second value is the distance to the left foot edge, and in the legend to b the first listed value is the distance from the ankle joint to the front edge of the foot and so on

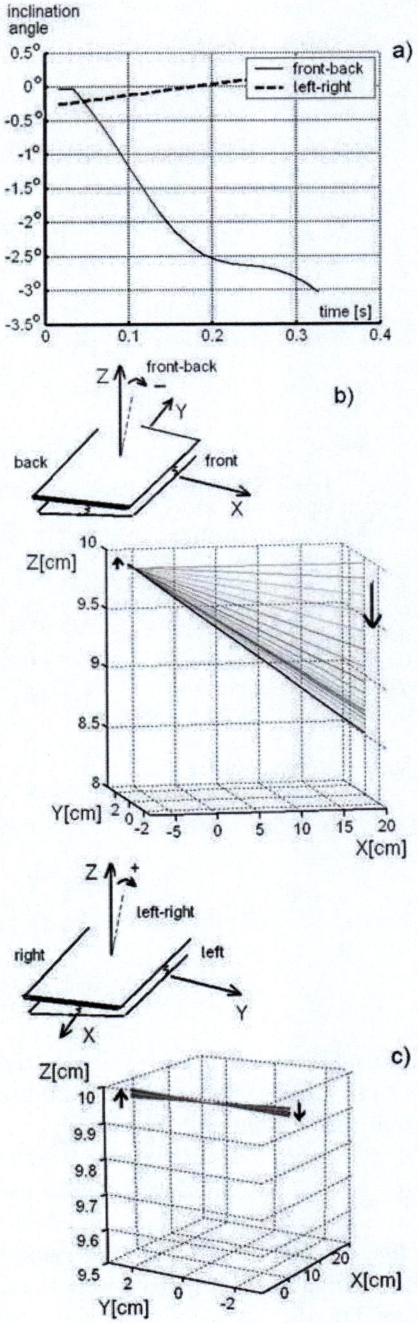

Fig. 15 Sway of the upper layer of the foot: **a** the sway angles; negative sign indicates sway to the front or sway to the right side, **b** front-back sway during support phase, **c** side sway. Arrows mark the time flow

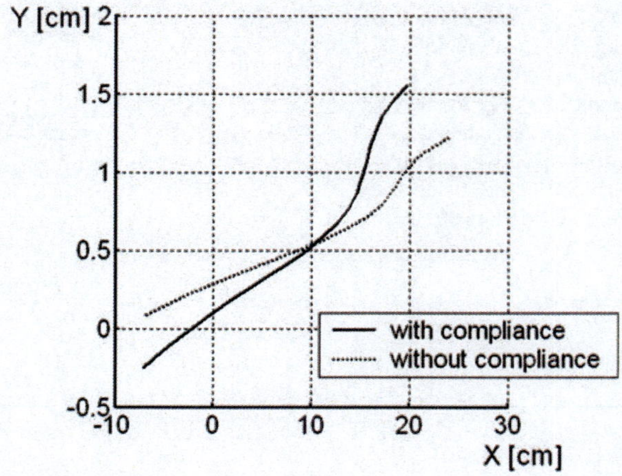

Fig. 16 ZMP trajectories for a compliant and a rigid foot

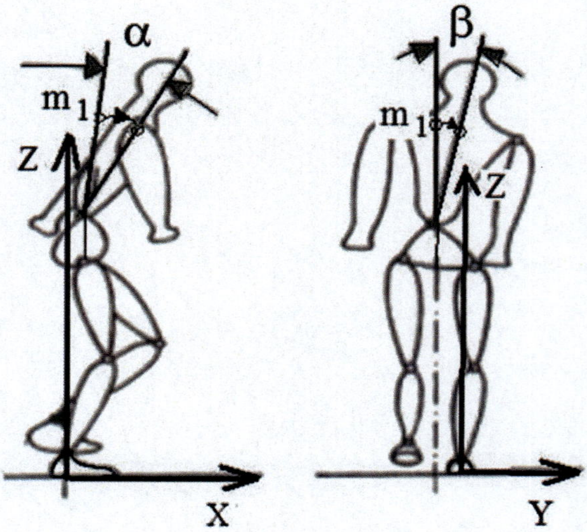

Fig. 17 Illustration of the sway of the human body, which changes the position of the dominant mass (m_1). The biggest sway to the front occurs in the middle of the transfer phase (left side of the picture), the side sway is towards the supporting leg, this sway maximum occurs in the middle of transfer phase of the opposite leg

and side sway in range $\langle -0.2^o, 0.2^o \rangle$, but during the human gait the sway is in the ranges $\langle 8^o, 8^o \rangle$, and $\langle -4^o, 4^o \rangle$ respectively. Under such conditions the posture for the rigid feet was unstable, because p_x coordinate of the ZMP exceeded 20 cm by the end of the support phase. The robot would fall over to the front. With the compliant foot the p_x coordinate is always smaller than 20 cm, what is shown in Fig. 16. Moreover, with the compliant feet we observed bigger shift of p_y coordinate in the positive direction by the end of the single support phase. This creates appropriate conditions for the double support phase when the ZMP moves from one foot print to the other one (see Fig. 18). This confirms that the postural stability can be aided by the compliant feet assuming that their compliance is appropriate (Figs. 17 and 18).

Fig. 18 Upper part of the figure illustrates the change of posture obtained due to the compliant foot. The right part illustrates the sketch of the ZMP trajectory during double support phase, which follows after the single support phase

7 Summary and Conclusions

This chapter presents a synthetic overview of the motion generation approaches used in walking machines and humanoid robots. Particular attention was paid to the role of compliance in bipedal gait. This is a very important issue, because introduction of compliant feet, as well as a suitably compliant whole body is relevant to the development of efficient robot assistants. Hence we devoted more attention to it.

Correctly designed compliance of a passive foot aids dynamic stability of the gait. The advantage is the simplification of both the design and the synthesis of motion of the robot, as no active degrees of freedom have to be included in the leg and the degrees of freedom providing the upper body tilt. Another option is to combine the controlled body tilt and compliant feet, that is, to build a robot to a greater extent imitating the features of the human body. Contemporary robotic feet are more similar to the human feet, but there is no method for determining their compliance. The feet are usually built of rigid components supported by embedded compressible elements. Solutions ensuring spatial compliance similar to that of a human foot are still not used. The proposition of a spatial springs makes a step towards a compliant foot.

As it was shown, by the content of this chapter, the use of biological patterns is important for modern robots. Knowledge of the locomotion of animals and knowledge about the structure of their bodies is useful when building the robots aiding humans, i.e. humanoid robots and walking machines. The discussed problems concern only the motion, but the use of biological patterns in robotics goes much further and concerns, among others, the recognition of human emotions by a robot, expressing emotions, predicting human actions, [63] and intelligent cooperation with a human being [64, 65].

References

1. Vukobratovic, M.: Szagajuszczije Masziny i Antropomoficzeskij Mechanism. MIR Moccov (1978)
2. Vukobratovic, M., Borovac, B.: Zero—moment point—thirty five years of its life. Int. J. Hum.Oid Robot. **1**(1), 157–173 (2004)
3. Winter, D.A.: Human balance and posture control during standing and walking. Gait Posture **3**(4), 193–214 (1995)
4. Wilson, D.M.: Insect walking. Annu. Rev. Entomol. **11**, 103–122 (1966)
5. Delcomyn, F.: Insect Locomotion on Land. Locomotion and Excercise in Antropods. In: Herreid, C.F., Fourtner, C.R. (ed.), pp. 103–125. New York, Plenum (1981)
6. Zielinska, T.: Self-navigation principles for a team of walking robots. Electro-Mech. Rev. **82**(1), 53–57 (2006)
7. Zielinska, T.: Control and navigation aspects of a group of walking robots. Robotica **24**, 23–29 (2006)
8. Zielinska, T., Heng, J.: Real-time control system for a group of autonomous walking robots. Adv. Robot. **20**(5), 543–561 (2006)
9. Zielinska, T.: Walking Machines (in Polish), II edn. PWN, Warsaw (2014)
10. Zielinska, T.: Design issues and robots autonomy. new trends in mechanisms and machine science. In: Viadero, F., Ceccarelli, M. (eds.) Theory and Applications in Engineering, pp. 691–699. Springer (2012)
11. Zielińska, T.: Minimizing energy cost in multi-legged walning machines. J. Intell. Robot. Syst. **85**, 431–447 (2017)
12. Shaoping, B., Low, K.H., Zielinska, T.: Quadruped free gait generation based on the primary/secondary gait. Robotica **17**, 405–412 (1999)
13. Shaoping, B., Low, K.H., Zielinska, T.: Quadruped free gait generation for straight-line and circular motion. Adv. Robot. J. **13**(5), 513–538 (1999)
14. Zhou, D., Low, K.H., Zielinska, T.: An efficient foot-force distribution algorithm for quadruped walking robots. Robotica **18**, 403–413 (2000)
15. Zielińska, T., Heng, J.: Multifunctional quadruped. Robotica **20**, 585–593 (2002)
16. Zielinska, T., Heng, J.: Mechanical design of multifunctional quadruped. Mech. Mach. Theory (38), 463–478 (2003)
17. Zielinska, T., Heng, J.: Development of walking machine: mechanical design and control problems. Mechatronics (12), 737–754 (2002)
18. Zielinska, T., Trojnacki, M.: Dynamical approach to the diagonal gait synthesis: theory and experiments. J. Autom. Mob. Robot. Intell. Syst. **3**(2), 3–7 (2009)
19. Zielinska, T., Trojnacki, M.: Analiza rozkładu sił reakcji podłoża podczas dynamicznie stabilnego chodu robota dwunożnego. Pomiary Automatyka Robotyka. Nr **7–8**, 6–10 (2009)
20. Zielinska, T: Robots for space exploration: barriers, perspecives and implementations. In: Sasiadek, J. (ed.) Aerospace Robotics II, pp. 1–11. Springer (2015)
21. Raibert, M.H.: Legged Robots that Balance. MIT Press, Cambridge, MA (1986)
22. Feng, S., Whitman, E., Xinjilefu, X., Atkeson, C.G.: Optimization based full body control for the atlas robot. In: EEE-RAS International Conference on Humanoid Robots, pp. 120–127 (2014)
23. Hwang, J., Suh, I.H., Park, G., Kwon, T.: Human character balancing motion generation based on a double inverted pendulum model. In: 10-th International Conference on Motion in Games (ACM) pp. 1–11 (2017)
24. Kajita, S., Kanehiro, F., Kaneko, K., Yokoi, K., Hirukawa, H.: The 3d linear inverted pendulum model: a simple modeling for a biped walking pattern generation. IEEE/RSJ Int. Conf. Intell. Robot. Syst. **1**, 239–246 (2013)
25. Omran, S., Sakka, S., Aoustin, Y.: Using the generalized inverted pendulum to generate less energy-consuming trajectories for humanoid walking. Arch. Mech. Eng. **63**(2), 245–262 (2016)

26. Żurawska, M., Szumowski, M., Zielińska, T.: Reconfigurable double inverted pendulum applied to the modelling of humanoid robot motion. J. Autom. Mob. Robot. Intell. Syst. **11**(2), 12–20 (2017)
27. Collins, J.J., Richmond, S.A.: Hard-wired central pattern generators for quadrupedal locomotion. Biol. Cybern. **71**, 375–385 (1994)
28. Matsuoka, K.: Sustained oscillations generated by mutually inhibiting neurons with adaptation. Biol. Cybern. **52**, 367–376 (1985)
29. Zielinska, T.: Coupled oscillators utilised as gait rhythm generators of two legged walking machine. Int. J. Biol. Cybern. **4**(3), 263–273 (1996)
30. Zielinska, T.: Utilisation of biological patterns in biped gait generation. J. Biocybern. Biomed. Eng. **18**(102), 79–88 (1998)
31. Vukobratovic, M., Potkonjak, V., Babkovic, K., Borovac, B.: Simulation model of general human and humanoid motion'. Multibody Syst. Dyn. **17**(1), 71–96 (2007)
32. Harada, K., Yoshida, E., Yokoi, K.: Motion Planning for Humanoid Robots. Springer Science and Business Media (2010)
33. Hirai, K., Hirose, M., Haikawa, Y., Takenaka, T.: The development of honda humanoid robot. IEEE Int. Conf. Robot. Autom. Proc. **2**, 1321–1326 (1998)
34. Yamaguchi, J., Soga, E., Inoue, S., Takanishi, A.: Development of a bipedal humanoid robot control method of whole body cooperative dynamic biped walking. In: IEEE International Conference on Robotics and Automation Proceedings, vol.1, pp. 368–374. IEEE (1999)
35. Wang, T., Chevallereau, C.: A new control law for a 3d biped robot based on regulation of the zero moment point and joint path. In: 10th IEEE-RAS International Conference on Humanoid Robots (Humanoids), pp. 27–32. IEEE (2010)
36. Berenson, D., Srinivasa, S., Kuffner, J.: Task space regions: a framework for pose-constrained manipulation planning. Int. J. Robot. Res. **12**(30), 1435–1460 (2011)
37. Kajita S., Hirukawa H., Harada K., Yokoi K.: Introduction to Humanoid Robotics, vol. 101. Springer (2014)
38. Xue, F., Chen, X., Liu, J., Nardi, D.: Real time biped walking gait pattern generator for a real robot. In: Robot Soccer World Cup, pp. 210–221. Springer (2011); Journal of Neurophysiology, vol. 107(5), pp. 1513–1521 (2012)
39. Lanari L., Hutchinson, S., Marchionni L.: Boundedness issues in planning of locomotion trajectories for biped robots'. In: 14th IEEE-RAS International Conference on Humanoid Robots (Humanoids), pp. 951–958 (2014)
40. Villalobos, J., Zielinska, T.: Study of postural adjustments for humanoidal helpmates. J. Autom. Mob. Robot. Intell. Syst. **11**(04), 15–25 (2017)
41. Full, R.J., Koditschek, D.E.: Templates and anchors: neuromechanical hypotheses of legged locomotion on land. J. Exp. Biol. **202**(23), 3325–3332 (1999)
42. Mordatch, I., De Lasa, M., Hertzmann, A.: Robust physics-based locomotion using low-dimensional planning. ACM Trans. Graph. (TOG) **29**, 71 (2010)
43. Alexandrov, A.V., Lippi, V., Mergner, T., Frolov, A.A., Hettich, G., Husek, D.: Human-inspired eigenmovement concept provides coupling-free sensorimotor control in humanoid robot. Front. Neurorobot. **11**, 1–13 (2017)
44. Mohammadreza, K., Nuno, L., Pereira, A.: Comparison study of well-known inverted pendulum models for balance recovery in humanoid robot'. In: MAPiS 2019—First MAP-i Seminar Proceedings, pp.1–6 (2019)
45. Zielinska, T., Gao, Z., Zurawska, M., Zheng, Q., Mergner, T., Lippi, V.: Postural balance using a disturbance rejection method. IN: 11th IEEE Workshop on Robot Motion and Control (RoMoCo), pp. 23–28 (2017)
46. Zielinska, T., Chew, C.-M., Kryczka, P., Jargiło, T.: Robot gait synthesis using the scheme of human motion skills development. Mech. Mach. Theory **44**(3), 541–558 (2009)
47. Bernstein, N.A.: The Coordination and Regulation of Movements. Pergamon, Great Britain (1967)
48. Zielinska, T.: Biological inspiration used for robots motion synthesis. J. Physiol.-Paris **103**(3–5), 133–140 (2009)

49. Ramamoorthy, L., Zielińska, T.: Robot motion synthesis using ground reaction forces pattern. Int. J. Adv. Robot. Syst. **14**(4), 1–12 (2017)
50. Lark, S., et al.: Adequate joint stiffness is critical during the single support phase to control forward and downward body momentum. Clin. Biomech. **18**, 848–855 (2003)
51. Ludvig, D., et al.: Time-varying system identification for understanding the control of human knee impedance. Syst. Identif. **16**, 1306–1310 (2012)
52. Adamczyk, P.G., Collins, S.H., Kuo, A.D.: The advantages of a rolling foot in human walking. J. Exp. Biol. **209**, 3953–3963 (2006)
53. Manoonpong, P., et al.: Compliant ankles and flat feet for improved self-stabilization and passive dynamics of the biped robot "RunBot". In: International Conference on Humanoid Robots, pp. 276–281 (2011)
54. Meyer, F., et al.: Simple and low-cost compliant leg-foot system. In: Intelligent Robots and Systems. IEEE/RSJ IROS International Conference, vol. 1, pp. 515–520 (2004)
55. Zhang, P., et al.: Adaptive compliant control of humanoid biped foot with elastic energy storage. In: IEEE/ASME International Conference on Advanced Intelligent Mechatronics, pp. 928–933 (2009)
56. Zielinska, T., Chmielniak, A.: Biologically inspired motion synthesis method of two-legged robot with compliant feet. Robotica **29**, 1049–1057 (2011)
57. González de Alba, A., Zielińska, T.: Postural equlibrium criteria concerning feet properties for biped robots. J. Autom. Mob. Robot. Intell. Syst. **6**(1), 22–27 (2012)
58. Zurawska, M., Zielinska, T.: Feet compliance in two legged locomotion. In: Flores, P., Viadero, F. (eds.) New Trends in Mechanism and Machine Science, pp. 595–602. Springer (2015)
59. Zurawska, M., Zielinska, T., Szumowski, M.: The role of compliant elements in two legged robot foot's model. J. Autom. Mob. Robot. Intell. Syst. **9**(1), 68–76 (2015)
60. Wright, W.G., Ivanenko, Y.P., Gurfinkel, V.S.: Foot Anatomy Specialization for Postural Sensation and Control
61. Elftman, H.: Dynamic structure of the human foot. Artif. Limbs **13**(1), 49–58 (1969)
62. Choi, K., Jiang, S., Li, Z.: Spatial stiffness realization with parallel springs using geometric parameters. IEEE Trans. Robot. Autom. **18**(3), 274–284 (2002)
63. Vibekananda, D., Zielinska, T.: Predicting human actions taking into account object affordances. J. Intell. Robot. Syst. **90**, 1–17 (2018)
64. Zielińska, T.: Professional and personal service robots. Int. J. Robot. Appl. Technol. **4**(1), 63–82 (2016)
65. Zielinski, C., Zielinska, T.: Digital Care (in Polish), no. 2/2017, pp. 43–45 (2017)

Robotic System Design Methodology Utilising Embodied Agents

Cezary Zieliński

Abstract The paper presents a holistic robotic system specification methodology taking into account both the system structure and its activities. It is based on the concept of an embodied agent. Each agent is decomposed into cooperating subsystems. Subsystem activities are defined by a hierarchical finite state automaton and subsystem transition functions. Diverse robotic system architectures produced by the postulated design methodology are presented. Classification of robotic systems facilitating the presentation of their architectures is proposed.

Keywords Robotic system specification methodology · Hierarchical finite state automaton · Transition function · Robotic system architecture

1 Motivation

A robotics system is a system composed of one or more robots and possibly some cooperating devices. Such systems are composed of hardware devices and software running on the control computer or a network of such computers. This work primarily concentrates on the robot control software.

Following the separation of concerns principle formulated by Edsger Dijkstra [1], that is at the foundation of software engineering, the stages of system specification and its implementation should be clearly distinguished. Thus a system designer should first answer the question what is to be attained and then how it is to be attained. This chapter concentrates on the former problem, as implementation aspects of robotics software are no different than those of any real-time system software produced by computer scientists or engineers.

C. Zieliński (✉)
Warsaw University of Technology, Faculty of Electronics and Information Technology, Institute of Control and Computation Engineering, Nowowiejska 15/19, 00-665 Warsaw, Poland
e-mail: C.Zielinski@elka.pw.edu.pl

P. Kulczycki et al. (eds.), *Automatic Control, Robotics, and Information Processing*, Studies in Systems, Decision and Control 296,
https://doi.org/10.1007/978-3-030-48587-0_17

Both the number of robot types and the tasks that they execute is huge. Besides very rudimentary devices, each of those robots is computer controlled, thus runs specific control software. The number of robot applications will be growing exponentially, thus the problem of designing their control software is of paramount importance. Software engineering focuses on software reuse to facilitate the implementation of software systems. The reason for that is that the already developed modules are cheeper to reuse than to write from scratch, and even more importantly, they have been thoroughly tested, thus are much more reliable than the newly produced ones. However, this approach pertains to the implementation phase of system creation. The question is whether there exist software patterns that can be reused at the specification stage? Does robotic system software and the design methodology conform to patterns that can be used in the design of any such system?

Three aspects of software creation tend to be singled out as of importance: understanding the domain that the software is designed for, drafting the architecture of the system and selecting the programming language for the implementation of the system [2]. Especially knowledge of the domain and architecture are worth underscoring. Both point at the necessity of elaborating a domain specific structural model of a system. Even better, the aspiration should be to create a meta-model that will guide the designer in the creation of any robotic system. Such a meta-model should be based on concepts stemming from robotics and use the principles established in software engineering. Robotic systems are inherently complex, thus must be decomposed, and for that purpose it is prudent to use the Dijkstra's separation of concerns principle.

Any informal discussion of a problem lacks precision and does not provide a deeper insight into that problem. Thus some form of formalisation of the problem at hand is of benefit. For that purpose it is reasonable to rely on the concepts rooted in the discussed domain. However, for the purpose of specification it is important to single out such concepts that at the later stage will be easy to transform into implementation. The discussion of the mentioned problem is based on the concept of an embodied agent, which is well established both in robotics and computer science.

2 Historical Perspective

The stage has to be set for the discussion of the proposed meta-model. This will be done from a historical perspective. Software and architecture, as well as languages, are tightly intertwined with each other, thus all of those subjects are relevant to this discussion.

2.1 Industrial Robot Programming Methods

The history of robotics gained momentum in the beginning of the 60-ties of the XX century, when the first industrial robots appeared. At that time microprocessors did not exist, thus electro-mechanical means, i.e. diode matrices and limit switches, were used to program industrial robots [3]. With the advent of microprocessors the capabilities of the control equipment increased rapidly, thus programming by teaching could be introduced. Two methods of on-line programming were developed: Point to Point (PTP) and Continuous Path (CP). In the former case the operator guides the robot arm to consecutive points on the required path and memorises them by pressing a button on the teach-pendant. Once the path points are stored in memory they can be replayed using interpolation between those points. In the case of Continuous Path as the operator guides the manipulator the internal timer causes the robot to memorise the points at a fast rate, thus during playback no interpolation is needed. The advantage of both methods is that the operator does not need extensive training. The disadvantage was that the robot had to participate in the programming proces, thus was not productive, every modification of the path required reprogramming from scratch, no program documentation existed and incorporation of sensor data was next to impossible. Hence the attention of the robotics community turned to textual means of programming. Off-line programming did away with the disadvantages of on-line methods, but required higher skills from the programmer, and above all needed program calibration. Eventually hybrid forms of programming emerged. In this case the program is created off-line in a textual form, however some characteristic points are taught-in, thus calibration can be avoided. Current industrial robot controllers, besides controlling the components of the robot, have an in-built interpreter of a specialised robot programming language, in which a specific task is coded. Each industrial robot vendor has his own language, e.g. ABB RAPID [4], Kuka KRL [5], Comau PDL2 [6], Mitsubishi Melfa-Basic [7], Kawasaki AS [8], Staübli VAL3 [9], Universal Robots URscript [10], Toshiba SCOL [11], Festo FTL [12]. All those languages, although having differing syntax, have common features. Program control flow utilises standard conditional and looping statements derived either from BASIC or Pascal. Elementary data types and mathematical functions are the same as those that exist in general programming languages. Robot specific data types include: vectors, rotations, as well as coordinate frames and transforms. The arm motion is described in two parts. Parameters such as: velocity, duration of motion, arm configuration or load are specified separately from the command initiating the motion and defining the used method of interpolation. Interpolation is defined either in configuration or operational space and is either linear or circular. Moreover tool control instructions are included in those languages. Although a dominant number of commands of those languages pertains to motion specification, the limited interpolation capabilities are surprising. Those languages could be significantly simplified if parametric trajectory generators would be introduced [13], however the conservatism of the industry prevented the adoption of that solution. Current research of industrial robot programming focuses on programming by demonstration. In this case the robot should observe the actions of a human demonstrating what the robot should do [14, 15]. For that purpose the robot uses cameras and microphones. The latter are used to provide comments on what the robot can observe that is being done. Such a form of programming is very difficult to implement, thus simplifications are being introduced. Teaching-in is assisted by cameras and force-torque sensors [16]. Enhanced reality is also tested. Another approach describes the task by a set of parameterised skills [17] defined by standard operational procedures for production workers [18].

2.2 Robot Programming Languages

Industrial robot programming languages inherited many of their features from languages developed in research laboratories. One of the first such languages was WAVE [19, 20] implemented in Stanford Artificial Intelligence Laboratory in the 70-ties of the XX century. It was a precursor of many specialised robot programming languages. It introduced homogeneous transforms, to represent coordinate frames, and used cameras and force sensing to gather data from the environment. A better developed form of this language was AL [21]. The arguments of the motion instructions of the WAVE language pertained to the manipulator end-effector, while in the case of AL they referred to transferable objects, what necessitated the representation of the environment, i.e. world modelling. This modification rendered necessary the reflection in the world model of the changes occurring in the environment due to the actions of the robot. AL in turn became the inspiration for the SRL language [22]. Both AL and SRL concentrated on the incorporation of sensing into languages, what was very important from the point of view of robot control. However they ignored the fact that separate motion instructions and commands modifying the world model, in the case of a programming error, may lead to a discrepancy between the model and reality, leading to potentially catastrophic consequences. This flaw was eliminated by TORBOL [23]. All of the mentioned languages represented the manipulator and objects by using homogeneous transforms. The RAPT language [24, 25] took an alternative approach. It used for that purpose elementary shapes, such as: straight lines, circles, planes, cylinders and spheres. Such geometric shapes simplified graphic simulation of the robot and the environment. World modelling was necessary to automatically plan robot activities. However it requires the solution of the fundamental problem, how to determine the current state of the environment being represented. This can be done only through sensors, but addition of new sensors entailed the modification of the language. As a consequence specialised languages became less favoured and the attention of the research community shifted to the use of general purpose programming languages and robot programming libraries.

2.3 Robot Programming Libraries

Libraries contain functions, procedures, objects or components created in a universal programming language such as C, C++ or Pascal, controlling the motion of the effector, modifying the world model and aggregating sensor data. Program control flow and mathematical transformations are being coded in a universal programming language. The advantages of libraries is that they do not require the definition of a new language and thus the implementation of its compiler. Modification or addition of new functions is done in the universal programming language.

Early libraries employed such languages as Pascal and C. The former was used by PASRO (Pascal for Robots) [22, 26] and the latter by RCCL (Robot Control C Library) [27]. RCCL subsequently inspired ARCL (Advanced Robot Control Library) [28], RCI (Robot Control Interface) [29] and KALI [30–33]. Libraries facilitated robotics research, however they did not provide guidance to their users what should be the structure of the designed system. This paved the way for the search of patterns. The advantages of code reuse and pursuit of patterns turned libraries into robot programming frameworks.

2.4 Robot Programming Frameworks

The monograph [2] distinguishes three classes of patterns: architectural, design and idioms. Architectural patterns pertain to the general structure of the designed system. They dictate system decomposition into subsystems and define how those subsystems interact with each other. Design patterns define the inner structure of the subsystems and interaction of their components. Idioms are patterns resulting from the use of a specific programming language. Majority of frameworks focuses on idioms. There is an abundance of robot programming frameworks—to name just a few: Player [34–36], OROCOS (Open Robot Control Software) [37, 38], ROS (Robot Operating System) [39], YARP (Yet Another Robot Platform) [40, 41], ORCA [42, 43], MRROC++ (Multi-Robot Research Oriented Controller based on C++) [44–48]. The majority of frameworks tries to resolve at least some of the following problems [49]:

- facilitate device (i.e. sensor, actuator) handling,
- create a multi-layered communication pattern between subsystems,
- handling specific equipment and providing specific services, such as vision systems, e.g. DisCODe [50] or OpenCV [51]),
- enable wrapping legacy software.

Some solve all of the enumerated problems, and some only some of them, however all of them deal with the first two. Currently frameworks evolve in the direction of toolchains, providing sets of tools for developing robotics systems [52, 53]).

2.5 Domain Specific Languages

Domain Specific Languages (DSLs) are derived from Model Driven Engineering (MDE) [54], which fosters the creation of a formal model of the system being created and then transforming it, preferably automatically, into a real system. The model represents the overall structure of the created system. It abstracts away the implementation details. Those are added at the transformation stage. In the case of software systems MDE involves automatic code generation. In robotics both general purpose modelling languages, such as UML [55], and specialised languages dealing with robots [54] are used.

An early example of automatic code generator for robotic purposes is $\mathsf{G^{en}oM}$ (Generator of Modules) [56, 57]. This tool uses the system specification to create the reactive layer of that system. Automatic code generation requires the standardisation of architecture. This is also one of the requirements of MDE. Unfortunately the establishment of one standard architecture in robotics is very difficult, because robots are used for many purposes and their mechanical structures are very different from each other [58]. To create an architecture conforming to an application domain one should focus on those elements of the system that do not change when the hardware or the tasks of the system are modified [59]. Moreover, the software controlling the hardware should be separated from the one that is responsible for the interaction of the robot with the environment. Another important problem is the abstraction level (ontological level) that the DSL uses [60]. It is suggested that the meta-model should be expressed by UML diagrams [55] and subsequently an instance of the meta-model should be used to generate the code.

The paper [60] presents domain specific languages (DSL) at several levels of abstraction, facilitating programming of robots by employing the task frame formalism (TFF). In general, internal DSLs employ existing general purpose languages for the representation of the concepts of a particular domain, while external ones rely on languages developed especially for that purpose. In that paper a high level Ecore metamodel is defined for the representation of actions employing the TFF. Those actions are subsequently integrated into an FSM invoking them. This paper postulates the expression of a metamodel in terms of UML type diagrams and generation of code on the basis of an instance of such a metamodel. Work also progresses in the area of software language engineering, i.e. creation of Domain-Specific Languages by using metamodels [61]. Software languages, in contrast to computer or programming languages which target the creation of code for computers, are languages for modelling and creating software, thus they encompass a much wider field of knowledge. Model-driven development focuses on automatic creation of standard portions of code, thus higher-level instances of abstract syntax have to be automatically transformed into lower-level target languages. As the software applications are getting more complex the languages used for their creation must be at an appropriately high level of abstraction. Ideas can be expressed clearly only if the unnecessary details are filtered out. Those details can be added at a later stage, preferably automatically.

The use of DSLs in the case of industrial robots has also attracted some attention of researchers. First of all, industrial robot programs are intrinsically hard to modify (especially when such a program interacts with other programmable equipment) and have to adhere to timing constraints (scheduling actions). Besides, the optimization of execution times is a tedious task, especially when dealing with systems containing robots from different vendors, thus coding in different programming languages is time consuming. Those are the main reasons why a modelling, planning and code generation tool suite called Automax was proposed [62]. It is a Domain-Specific Modelling Language (DSML) and as such provides notation and constructs relevant to the problem domain, i.e. robotics. The Automax planner focuses on task timing requirements, generating the base robot model automatically. Key modelling concepts were derived by reverse-engineering the code of industrial robot programs, thus a manipulator level language emerged. The models are expressed in UML [55]. Such models are subsequently used to generate program code for robots from different vendors. The model includes all the devices present in the assembly line. Implementation of Automax is based on the Eclipse modelling environment.

2.6 *Robotic System Architectures*

The most popular architecture employed by robotic systems is the layered architecture. The decomposition into layers is formed either on the basis of the frequency of system component behaviour repetition [63] or on the basis of the decomposition of the task into subtasks. Two [64] or three layered structures dominate, e.g. Sense–Plan–Act (SPA), subsumption [65–67], hybrid planning–reactive [68], hierarchic [57, 69], biologically inspired [70–72], using belief–desire–intension (BDI) approach [73, 74]..

Although a multitude of robotic systems has been created there does not exist a single method of presenting the system architecture or classification of such structures, e.g. [75–83]. Some papers describing robotic system architectures, e.g. [84, 85], distinguish:

- architectural structure, i.e. presentation of subsystems and their connections,
- architectural style, i.e. the description of computational communication concepts utilised.

Nevertheless [84] points out that for the majority of created systems it is difficult to determine both their structure and style. Although some work has been done on formalisation of robotic system control software specification [70, 71, 86–88] and its formal verification [89, 90] unfortunately this approach has not gained a wider acceptance in the robotics community. Usually architectures are presented by using informal textual descriptions and block diagrams with a very varied level of detail. This usually leads to problems during implementation and documentation of such systems.

3 Specification

The above historic overview of robotics software engineering shows that the research effort has concentrated on implementation aspects of design. The current trend is to provide a repository of components solving specific robot computational or control problems and the middleware enabling the interaction of those components, e.g. ROS [39]. This leads to the design of systems composed of a set of components connected together in such a way that the required data is passed between them. Such systems frequently do not conform to any particular architecture. This is the consequence of providing the designer with a multitude of freely available building blocks and virtually unlimited freedom of choice. The provided components do something useful for the system, however not always in the most efficient way and not always exactly what is required, thus wrapping is used as a remedy. Lack of architectural guidance was already evident in 2008 [91], however the situation has not significantly improved till now [84]. Moreover, the implementation aspects of system design dominate, thus domain specific information has been, at the most, used in the design of components. However robotic systems have a specific structure, stemming from the fact that a loop is formed: environment–perception–decision–actuation–environment. This strongly influences the system architecture. As those facts have been noticed, currently the attitude is changing, thus freedom from choice is fostered [92]. With the advent of DSLs supported by toolchains some guidance is provided to the designer. However much more stress should be placed on the specification phase than the implementation phase of robotics software development.

How detailed should the system specification be depends on the designer's competence and the documentation requirements. With respect to this, the methodology presented herein is flexible. A very detailed specification is usually huge, while a concise one is superficial, hence ambiguous. Optimum is somewhere between those extremes. Nevertheless the creation of a specification, besides conforming to the Dijkstra's postulate of separation of concerns, enables the discussion of the adequacy of the assumed system structure, but also of subsystem behaviours. Such a discussion can be conducted prior to writing even the first line of code, and thus the cost of its modification is negligible compared to the cost of modification of the system code. If the system is created out of earlier created modules, which have been thoroughly testes and thus reliable, then the level of detail of the specification can be decreased. Such modules can be treated as black boxes delivering the required functions. Their interface is an important part of their specification. It is of paramount importance that guidance is provided in the specification of architecture. The optimal solution is to provide a meta-model that can be fine-tuned to the problem at hand.

4 Embodied Agent Meta-Model

The concept of an embodied agent was popularised by Rodney Brooks [67], while debating the problem whether robots need world models and planning to act intelligently in a complex environment. Brooks claimed that agents should be embodied, i.e. have a material body, and be situated, i.e. act in a real environment, not in a simplified model of it [66, 93, 94]. They will be perceived as intelligent, as a result of emergence of intelligence due to the interaction of the agent with the real environment. Neither the proponents of GOFAI (Good Old Fashioned Artificial Intelligence) nor Brooks proved that their approach is the only one to be followed, thus hybrid agents having the properties of both reactive and deliberative ones appeared [68]. Nevertheless this debate spurred many authors to center their texts around the concept of an agent. The classic text on AI [95] from the most general perspective treats an agent as something that acts. Such an all encompassing definition is too general, thus this textbook endows agents with specific properties: autonomous control, i.e. acting without the participation of a human, ability to perceive the environment, persistence, adaptation to changes occurring in the environment, ability to undertake the tasks of other agents. Moreover, rational agents produce an optimal result of their activities, or an optimal expected result, if they act in an uncertain environment. Learning agents improve their capabilities with the gained experience.

The concept of an agent penetrated also into computer science. According to [96] software agents can be classified with respect to diverse criteria. They can be either static or mobile, depending on whether they reside on a single computer or they roam the network. Moreover agents may be: reactive, deliberative or hybrid. A specific class of agents is distinguished—interface agents, that enable users to interact with diverse systems. Wooldridge [97] defines an agent as a computer system situated in a certain environment in which it can act autonomously to attain a specific goal. This definition is the basis for the text [74], which enumerates the desirable properties of an agent. An agent should be reactive, i.e. react to changes occurring in the environment, proactive, i.e. strive to attain its goal, social, i.e. capable of interacting with other agents, flexible, i.e. able to reach the goal in diverse ways, robust, i.e. able to cope with possible errors.

4.1 *Specification of Structure*

All of the above texts on agents concentrate on their general properties, not their structure. Obviously those properties have to result from the structure, but a robotic system designer requires some guidance as to how to create agents and how to integrate them into a complete system? The first part of the question pertains to the internal structure of an agent. Hence, here an embodied agent is defined as anything that can be viewed as perceiving its environment through receptors (sensors) and acting upon that environment through effectors, having an internal urge to attain a

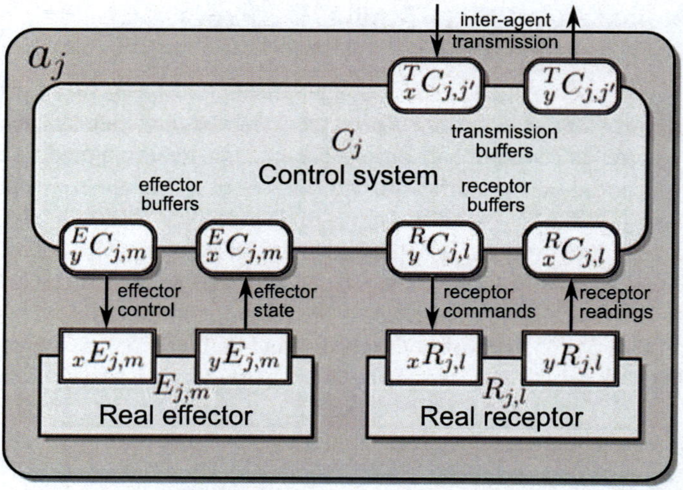

Fig. 1 General structure of an embodied agent a_j

certain goal. The general structure of an agent presented in Fig. 1 reflects this definition. The receptors gather the information from the environment. This information is used by the control system, endowed with the knowledge about the task to be executed, to produce signals controlling the effectors, which in turn influence the environment. The loop closes through the environment.

A systematic notation is used here for the description of the structure and operation of an embodied agent. As the above definition suggests, an embodied agent a_j, where j is its designator, consists of real effectors E_j, i.e. devices affecting the state of the environment, real receptors R_j, i.e. sensors gathering the information about the state of the environment, and the control system R_j i.e. the part of the system that is responsible for the execution of the task. As the task can be executed only if the state of the environment is known, a connection between receptors and the control system is necessary. Moreover, the task can only be accomplished when the agent can influence the environment, thus a connection between the control system and the effectors has to be created. Thus the structure presented in Fig. 1 results. An agent may need many receptors $R_{j,l}$ and effectors $E_{j,m}$ to perform its activities, thus those must be distinguished, by their designators: l and m respectively. In the case of complex systems a monolithic control system R_j is not feasible—it must be decomposed. As processing receptor data and effector commands usually is laborious, it is reasonable to distinguish subsystems responsible for that purpose, thus virtual receptors $r_{j,k}$ and effectors $e_{j,n}$ (where k and n are their designators respectively) appear in the structure presented in Fig. 2 [87, 98–100]. The virtual subsystems transform the ontology form rudimentary low level concepts into the high level concepts used by the control subsystem c_j, which is primarily responsible for the execution of the task.

The subsystems of an embodied agent process data. This data is either contained in the internal memory of the discussed subsystem or originates in other subsystems. For the purposes of communication with other subsystems buffers are used. Because many such buffers appear in any complex system, a systematic notation for designating them facilitates the system specification. The central symbol describes the type of the discussed subsystem, i.e. c, e, r, E, R. If reference is being made to a generic subsystem of an embodied agent, then s is used. The left superscript denotes the type of the subsystem that the considered subsystem communicates with, i.e. c, e, r, E, R, T. The last symbol, i.e. T, refers to a transfer buffer used by the control subsystem for communication with other agents. The subscript to the right of the central symbol is the designator of a particular agent and subsystem separated by commas. The left subscript can be nonexistent, then the subsystem internal memory is referred to, or it can be an x or y, to distinguish input and output buffers respectively. If particular component of a buffer has to be referenced, then its name is placed between square brackets, which are located to the right of the central symbol. The right superscript refers to the discrete time instant at which the contents of the buffer or memory is considered. Generally i is a current instant and $i + 1$ is the next instant. Each subsystem operates at its own sampling rate, however not to unnecessarily multiply symbols, in all cases i is used as a time stamp regardless of the considered subsystem. An example of the use of this notation is: ${}^{e}_{y}c^{i}_{\mathrm{robot,rm}}$. It denotes an output buffer of the control subsystem of an agent named robot, connected to a virtual effector named rm (abbreviation for right manipulator). The contents are considered at an instant i.

The contents of the memory and the buffers represents the concepts the subsystem uses to exercise control over a part of the system. Those concepts form an ontology. Thus through those concepts the designer defines the ontological level at which the task of the subsystem will be expressed. As the subsystems must interact with other subsystems the respective input and output buffer contents must match. Hence the subsystems also transform concepts from lower level ontology to a higher level ones and vice versa. Virtual receptors and effectors usually use concepts associated with the controlled robot, while the control system may use concepts related to the environment in which the robot acts. In the latter case a world model needs to be created in the control subsystem, e.g. [101].

4.2 *Specification of Activities*

Transition function The activities of each subsystem can be described by a pattern. Any subsystem s, named e.g. v, of an agent a_j, acquires data from its input buffers at instant i, computes its transition function ${}^{s}f_{j,v,\omega}$, to despatch the results to the other subsystems and its own internal memory at instant $i + 1$. Thus the transition function ${}^{s}f_{j,v,\omega}$, where ω is its designator, is defined as:

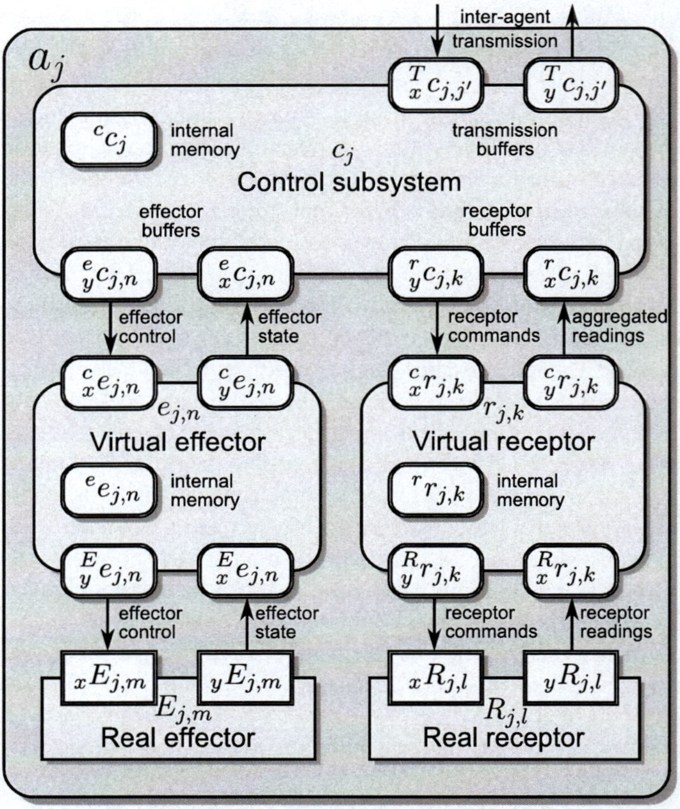

Fig. 2 Internal structure of an embodied agent a_j

$$\left[{}^s s^{i+1}_{j,v},\ {}_y s^{i+1}_{j,v}\right] := {}^s f_{j,v,\omega}({}^s s^i_{j,v},\ {}_x s^i_{j,v}) \tag{1}$$

A transition function performs only data processing, and only for a single discrete time instant. Diverse methods of defining transition functions are presented in [48, 98, 99, 102–104]. The most exact method of specifying a transition function is by providing a mathematical formula transforming the contents of input buffers into the contents of output buffers for one sampling period [98, 99, 104]. However, data flow diagrams (DFD) [99] and pseudo-code [105] have been used too.

Behaviour An elementary action ${}^s\mathcal{A}_{j,v,\omega}$ of a subsystem $s_{j,v}$ is composed of the following subactions: computation of the transition function ${}^s f_{j,v,\omega}$, dispatching the results contained in ${}_y s_{j,v}$ to the other subsystems, incrementation of time i, acquiring new data into the input buffers ${}_x s_{j,v}$. Behaviour ${}^s\mathcal{B}_{j,v,\omega}$ is an iterative execution of an elementary action ${}^s\mathcal{A}_{j,v,\omega}$. Elementary action ${}^s\mathcal{A}_{j,v,\omega}$ is repeated by the behaviour ${}^s\mathcal{B}_{j,v,\omega}$ as long as neither its terminal condition ${}^s f^{\tau}_{j,v,\xi}$ nor its error condition ${}^s f^{\varepsilon}_{j,v,\beta}$ are fulfilled. Both of those conditions are predicates taking as arguments the the

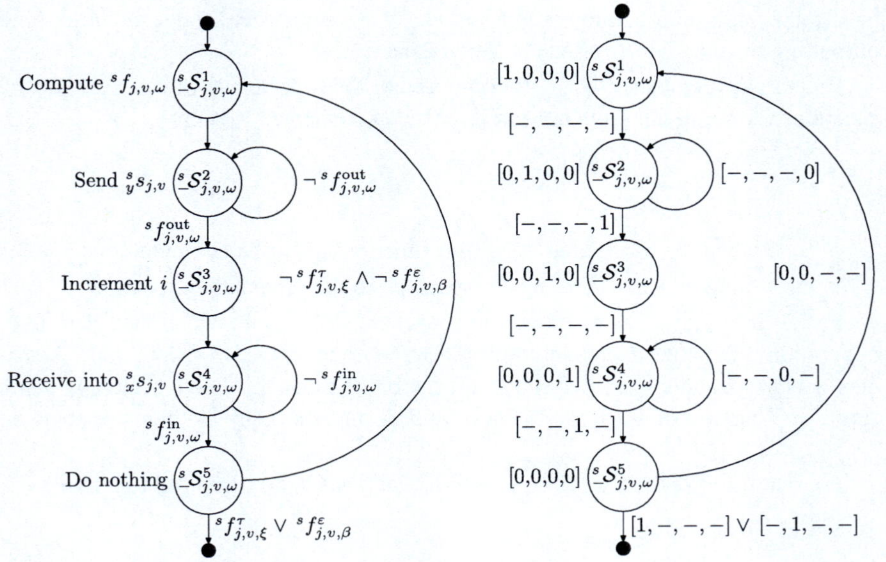

Fig. 3 Graph of the FSA ${}^s\mathcal{F}^{\mathcal{B}}_{j,v,\omega}$ executing the behaviour ${}^s\mathcal{B}_{j,v,\omega}$ of the v-th subsystem of agent a_j; left: explicitly defined subactions and predicates; right: binary input/output version (dash represents a don't care value)

contents of internal memory ${}^s s^i_{j,v}$ and the input buffers ${}_x s^i_{j,v}$ at instant i. The sampling time is the period from i to $i+1$. The sample counter i can also be treated as the behaviour iteration counter.

A behaviour ${}^s\mathcal{B}_{j,v,\omega}$ is represented by a Finite State Automaton (FSA) ${}^s\mathcal{F}^{\mathcal{B}}_{j,v,\omega}$ (Fig. 3):

$$ {}^s\mathcal{B}_{j,v,\omega} \equiv {}^s\mathcal{F}^{\mathcal{B}}_{j,v,\omega} = \langle {}^s_{-}\hat{\mathcal{S}}_{j,v,\omega}, {}^s_{-}\hat{\mathcal{I}}_{j,v,\omega}, {}^s_{-}\hat{\mathcal{O}}_{j,v,\omega}, {}^s_{-}\Theta_{j,v,\omega}, {}^s_{-}\Lambda_{j,v,\omega} \rangle \quad (2) $$

In general sets are distinguished from other concepts by having a hat over the central symbol. The set of substates ${}^s_{-}\hat{\mathcal{S}}_{j,v,\omega}$ contains five members, i.e. substates:

$$ {}^s_{-}\hat{\mathcal{S}}_{j,v,\omega} = \{ {}^s_{-}\mathcal{S}^1_{j,v,\omega}, {}^s_{-}\mathcal{S}^2_{j,v,\omega}, {}^s_{-}\mathcal{S}^3_{j,v,\omega}, {}^s_{-}\mathcal{S}^4_{j,v,\omega}, {}^s_{-}\mathcal{S}^5_{j,v,\omega} \} \quad (3) $$

The input vector set ${}^s_{-}\hat{\mathcal{I}}_{j,v,\omega}$ of the FSA of behaviour ω of subsystem $s_{j,v}$ is composed of four predicates of the form:

$$ {}^s_{-}\mathcal{I}_{j,v,\omega} = [{}^s f^{\tau}_{j,v,\xi}, {}^s f^{\varepsilon}_{j,v,\beta}, {}^s f^{\text{in}}_{j,v,\omega}, {}^s f^{\text{out}}_{j,v,\omega}], \qquad {}^s_{-}\mathcal{I}_{j,v,\omega} \in {}^s_{-}\hat{\mathcal{I}}_{j,v,\omega} \quad (4) $$

Thus the set of inputs is composed of 4-element binary vectors. The first two components are the values of the terminal and error conditions, while the latter two predicates define the communication mode of the subsystem while inputting and outputting the data into/from the buffers respectively.

The output vector set ${}^{s}_{-}\hat{\mathcal{O}}_{j,v,\omega}$ is composed of the four binary values representing subactions forming the sequence executed as an elementary action:

$$ {}^{s}_{-}\mathcal{O}_{j,v,\omega} = [{}^{s}\mathfrak{a}^{1}_{j,v,\omega}, {}^{s}\mathfrak{a}^{2}_{j,v,\omega}, {}^{s}\mathfrak{a}^{3}_{j,v,\omega}, {}^{s}\mathfrak{a}^{4}_{j,v,\omega}], \qquad {}^{s}_{-}\mathcal{O}_{j,v,\omega} \in {}^{s}_{-}\hat{\mathcal{O}}_{j,v,\omega} \tag{5} $$

${}^{s}\mathfrak{a}^{1}_{j,v,\omega} = 1$ when the associated transition function has to be computed, ${}^{s}\mathfrak{a}^{2}_{j,v,\omega} = 1$ when the computed values are to be dispatched to the other subsystems, ${}^{s}\mathfrak{a}^{3}_{j,v,\omega} = 1$ when the time index i is to be incremented, and ${}^{s}\mathfrak{a}^{4}_{j,v,\omega} = 1$ when new data is to be input into the buffers. When nothing is to be done ${}^{s}_{-}\mathcal{O}_{j,v,\omega} = [0, 0, 0, 0]$. As the elementary subactions are disjoint in all the other cases this vector contains just a single 1. Thus out of 16 possibilities only 5 occur—as many as there are states in ${}^{s}\mathcal{F}^{\mathcal{B}}_{j,v,\omega}$.

The definition (2) requires two functions, i.e. FSA state transition function[1]:

$$ {}^{s}_{-}\Theta_{j,v,\omega} : {}^{s}_{-}\hat{\mathcal{S}}_{j,v,\omega} \times {}^{s}_{-}\hat{\mathcal{I}}_{j,v,\omega} \rightarrow {}^{s}_{-}\hat{\mathcal{S}}_{j,v,\omega} \tag{6} $$

and FSA output function:

$$ {}^{s}_{-}\Lambda_{j,v,\omega} : {}^{s}_{-}\hat{\mathcal{S}}_{j,v,\omega} \rightarrow {}^{s}_{-}\hat{\mathcal{O}}_{j,v,\omega} \tag{7} $$

Thus a Moore type deterministic automaton has been defined. It represents the execution pattern of a behaviour.

A behaviour ${}^{s}\mathcal{B}_{j,v,\omega}$ is parameterised by an associated transition function (1) and four predicates (4). The terminal condition ${}^{s}f^{\tau}_{j,v,\xi}$ uses the input buffers to establish whether the behaviour can continue its current activities. The error condition ${}^{s}f^{\varepsilon}_{j,v,\beta}$ uses the same buffers to detect abnormal situations, resulting in the interruption of the current behaviour. The structure of the predicates ${}^{s}f^{\,\mathrm{out}}_{j,v,\omega}$ and ${}^{s}f^{\,\mathrm{in}}_{j,v,\omega}$ defines the communication mode, i.e. states whether the communication of subsystem $s_{j,v}$ with other subsystems is blocking or non-blocking. Receiving data in the blocking mode requires $s_{j,v}$ to wait for new data, while in the non-blocking mode, if new data is not available, $s_{j,v}$ continues its activities without waiting. In the latter case the definition of the transition function must take into account the possible lack of new data. Sending data in the blocking mode the subsystem $s_{j,v}$ has to wait for the acknowledgement produced by the receiving subsystem, while in the non-blocking mode such an acknowledgement is not necessary.

One way to organise the above described modes of communication is to introduce into the input buffers ${}_{x}s_{j,v}$ an additional Boolean variable for each subsystem with

[1]FSA state transition function should not be associated with the transition function. Both have different arguments and values. The former dictates how the FSA switches states, while the latter transforms the data contained in memory and input buffers into the data inserted into output buffers and the memory.

which $s_{j,v}$ communicates. Those variables turn True upon insertion of new data into ${}_{x}s_{j,v}$. When the data in ${}_{x}s_{j,v}$ is accessed by $s_{j,v}$ those variables are switched to False. In the case of output buffers ${}_{y}s_{j,v}$, if blocking mode of communication is used, extra Boolean variables are necessary to receive acknowledgement of the fact that data from ${}_{y}s_{j,v}$ has been transferred to the communicating subsystem. When the communicating subsystem acknowledges receiving the data from $s_{j,v}$ the associated Boolean variables in ${}_{y}s_{j,v}$ are switched to True. When $s_{j,v}$ detects this fact it switches those variables back to False and resumes its normal activities. It is assumed that all data from a data producing subsystem is conveyed atomically. If not, each data item will need its own Boolean variable.

In the non-blocking mode of operation $s_{j,v}$ only checks whether the input data in ${}_{x}s_{j,v}$ is fresh and reacts appropriately. However in the blocking mode it needs an extra predicate ${}^{s}f^{\text{in}}_{j,v,\omega}$. For instance, when $s_{j,v}$ receives data from $s_{j,v'}$ and $s_{j,v''}$, then ${}_{x}s_{j,v}$ will contain two Boolean variables, e.g. ${}_{x}s_{j,v}[b']$ and ${}_{x}s_{j,v}[b'']$, which become True, when $s_{j,v'}$ and $s_{j,v''}$ provide new data. Only in the blocking mode ${}_{x}s_{j,v}$ has to wait until both of those variables become True.

$$
{}^{s}f^{\text{in}}_{j,v,\omega} = {}_{x}s_{j,v}[b'] \wedge {}_{x}s_{j,v}[b''] \tag{8}
$$

For the sake of consistency, for the non-blocking mode predicate ${}^{s}f^{\text{in}}_{j,v,\omega}$ can be treated as always True, i.e. ${}^{s}f^{\text{in}}_{j,v,\omega} = \text{True}$.

If $s_{j,v}$ communicates with more systems in blocking mode the conjunction (8) is extended by adding extra Boolean variables b associated with those subsystems. If indefinite waiting is anticipated a timeout variable $b^{\bullet}$ can be upended to (8):

$$
{}^{s}f^{\text{in}}_{j,v,\omega} = {}_{x}s_{j,v}[b'] \wedge {}_{x}s_{j,v}[b''] \vee {}^{s}s_{j,v}[b^{\bullet}] \tag{9}
$$

It turns True when the prescribed time elapses.

When $s_{j,v}$ dispatches data in the blocking mode the Boolean variables in the output buffer ${}_{y}s_{j,v}$, acknowledging the fact that the communicating subsystem has received the data, are used in a similar way to construct ${}^{s}f^{\text{out}}_{j,v,\omega}$, i.e. in (8) or (9) ${}_{x}s_{j,v}$ is substituted by ${}_{y}s_{j,v}$. It should be noted that the use of blocking mode is potentially dangerous. If a loop is formed, two subsystems can both wait for the fulfilment of their respective ${}^{s}f^{\text{out}}$ and ${}^{s}f^{\text{in}}$ predicates, thus producing a deadlock. It is up to the system designer to avoid such loops by, e.g., using the blocking mode only by one subsystem and only when, e.g., dispatching data.

Hierarchic Finite State Automaton Behaviours ${}^{s}\mathcal{B}_{j,v,\omega}$, being themselves FSAs, form the lowest level of a Hierarchic Finite State Automaton (HFSA), thus they can be treated as elementary building blocks of a HFSA. A HFSA of a subsystem v of an agent a_j is defined recursively as follows:

$$
{}^{s}\mathcal{F}_{j,v} = \langle {}^{s}\hat{\mathcal{S}}_{j,v}, {}^{s}_{+}\hat{\mathcal{S}}_{j,v}, {}^{s}\hat{\mathcal{I}}_{j,v}, {}^{s}\hat{\mathcal{B}}_{j,v}, {}^{s}\Theta_{j,v}, {}^{s}\Lambda_{j,v}, {}^{s}_{+}\Lambda_{j,v} \rangle \tag{10}
$$

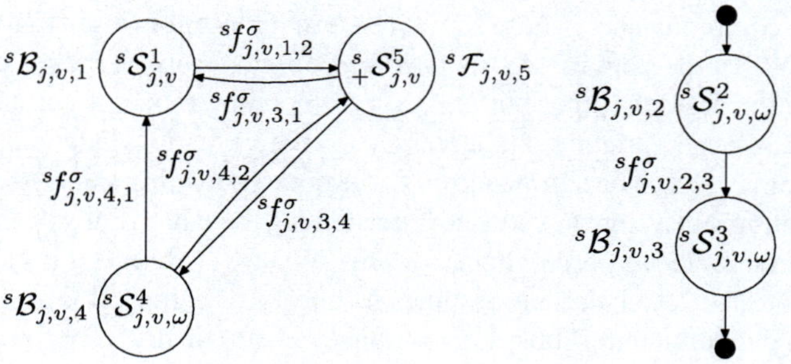

Fig. 4 Graph of an exemplary HFSA ${}^s\mathcal{F}_{j,v}$ and an FSA ${}^s\mathcal{F}_{j,v,5}$ associated with its superstate ${}^s_+\mathcal{S}^5_{j,v}$

where ${}^s\hat{\mathcal{S}}_{j,v}$ is a finite set of states of the HFSA defining the activities of the v-th subsystem of the agent a_j, ${}^s_+\hat{\mathcal{S}}_{j,v}$ is a finite set of superstates of that HFSA, ${}^s\hat{\mathcal{I}}_{j,v}$ is the set of input vectors composed of initial conditions, which are predicates, ${}^s\hat{\mathcal{B}}_{j,v}$ is a set of behaviours ${}^s\mathcal{B}_{j,v,\omega} \in {}^s\hat{\mathcal{B}}_{j,v}$. Moreover three functions are defined: HFSA state transition function:

$$ {}^s\Theta_{j,v} : ({}^s\hat{\mathcal{S}}_{j,v} \cup {}^s_+\hat{\mathcal{S}}_{j,v}) \times {}^s\hat{\mathcal{I}}_{j,v} \rightarrow ({}^s\hat{\mathcal{S}}_{j,v} \cup {}^s_+\hat{\mathcal{S}}_{j,v}) \tag{11} $$

HFSA output function:

$$ {}^s\Lambda_{j,v} : {}^s\hat{\mathcal{S}}_{j,v} \rightarrow {}^s\hat{\mathcal{B}}_{j,v} \tag{12} $$

and HFSA superoutput function:

$$ {}^s_+\Lambda_{j,v} : {}^s_+\hat{\mathcal{S}}_{j,v} \rightarrow {}^s\mathcal{F}_{j,v} \tag{13} $$

It is assumed that no direct or indirect self-recursion of an automaton occurs in the composition of the HFSA. The HFSA starts its activities either in an initial state or an initial superstate—depending on the structure of the HFSA graph. This is a single entry point for the execution of the task of the subsystem. It should be noted that after substitution of all HFSAs for superstates and behaviours for states a single flat FSA results, thus this is not a composition of separately acting parallel FSAs. The resulting single flat FSA is always in a single substate of an active behaviour. A single definition of HFSA would suffice, however here a two level one is presented intentionally. The reason is that a behaviour has a structure distinct from all the higher layers of the HFSA. As the behaviour structure implies a software pattern it is beneficial to define this lowest layer of the HFSA separately.

A simple example of composition of HFSAs is presented in Fig. 4. The result of substitution of ${}^s\mathcal{F}_{j,v,5}$ for superstate ${}^s_+\mathcal{S}^5_{j,v}$ is presented in Fig. 5. The general rule is that all the directed arcs going into a superstate (e.g. ${}^s_+\mathcal{S}^5_{j,v}$) go to the state connected to the entrance point of the subautomaton (here ${}^s\mathcal{S}^2_{j,v}$). All the directed

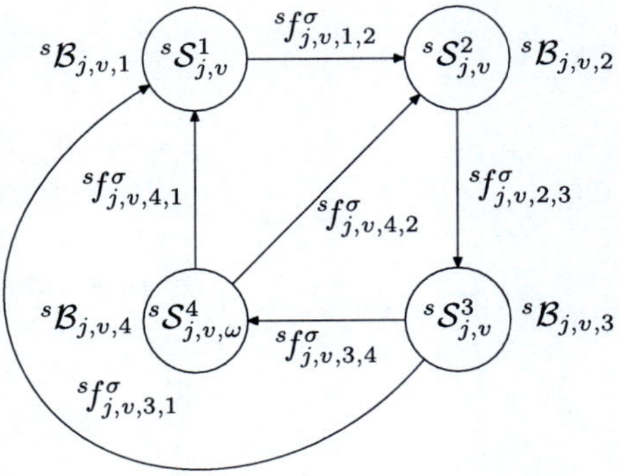

Fig. 5 Flattened out graph of an exemplary HFSA ${}^{s}\mathcal{F}_{j,v}$ from Fig. 4

arcs emerging from the superstate go out from the state connected to the exit point of the subautomaton (here ${}^{s}\mathcal{S}^{3}_{j,v}$). Further on the flattened version from Fig. 5 will be discussed.

Once a behaviour terminates its activities, due to either its terminal condition ${}^{s}f^{\tau}_{j,v,\xi}$ or error condition ${}^{s}f^{\varepsilon}_{j,v,\beta}$ being fulfilled, the next behaviour has to be chosen. The choice results from the state transition function (11). Each behaviour is associated with one of the ${}^{s}\mathcal{F}_{j,v}$ states: ${}^{s}\mathcal{S}^{\delta}_{j,v}$, where δ is the state designator. In general, any behaviour ${}^{s}\mathcal{B}_{j,v,\omega}$, and thus its FSM ${}^{s}\mathcal{F}^{\mathcal{B}}_{j,v,\omega}$, can be associated with any state ${}^{s}\mathcal{S}^{\delta}_{j,v}$, however, to increase the readability of the example, it was assumed that $\delta = \omega$ for ${}^{s}\mathcal{F}_{j,v}$. Initial conditions labeling the directed arcs of the graph are used for deciding which is the next state that the automaton will assume and thus which behaviour will be executed next. Initial conditions are predicates. If the behaviour ${}^{s}\mathcal{B}_{j,v,\omega}$, executed in state ${}^{s}\mathcal{S}^{\omega}_{j,v}$, terminates due to the satisfaction of its terminal condition ${}^{s}f^{\tau}_{j,v,\xi}$ or error condition ${}^{s}f^{\varepsilon}_{j,v,\beta}$, the next state is chosen based on the initial conditions ${}^{s}f^{\sigma}_{j,v,\omega,\gamma}({}^{s}s^{i}_{j,v}, {}_{x}s^{i}_{j,v})$ labeling the directed arcs starting with ${}^{s}\mathcal{S}^{\omega}_{j,v}$ and ending in any ${}^{s}\mathcal{S}^{\gamma}_{j,v}$, thus ${}^{s}\mathcal{S}^{\gamma}_{j,v} \in {}^{s}\hat{\mathcal{S}}^{\omega}_{j,v}$, where ${}^{s}\hat{\mathcal{S}}^{\omega}_{j,v}$ is the set of states pointed at by the arcs emerging from ${}^{s}\mathcal{S}^{\omega}_{j,v}$. As deterministic automatons are considered here, assuming that an automaton ${}^{s}\mathcal{F}_{j,v}$ is in the state ${}^{s}\mathcal{S}^{\omega}_{j,v}$, the initial conditions at instant i, labeling the arcs emerging from state ${}^{s}\mathcal{S}^{\omega}_{j,v}$, must fulfil the following conditions:

$$\begin{aligned}
&\forall_{\gamma \neq \gamma'}\ {}^{s}f^{\sigma}_{j,v,\omega,\gamma}({}^{s}s^{i}_{j,v}, {}_{x}s^{i}_{j,v}) \wedge {}^{s}f^{\sigma}_{j,v,\omega,\gamma'}({}^{s}s^{i}_{j,v}, {}_{x}s^{i}_{j,v}) = \text{False and} \\
&\bigvee_{\gamma}\ {}^{s}f^{\sigma}_{j,v,\omega,\gamma}({}^{s}s^{i}_{j,v}, {}_{x}s^{i}_{j,v}) = \text{True}, \qquad \text{where } {}^{s}\mathcal{S}^{\gamma}_{j,v}, {}^{s}\mathcal{S}^{\gamma'}_{j,v} \in {}^{s}\hat{\mathcal{S}}^{\omega}_{j,v}
\end{aligned} \tag{14}$$

For example, assuming that the automaton ${}^{s}\mathcal{F}_{j,v}$ (Fig. 5) is in the state ${}^{s}\mathcal{S}^{4}_{j,v}$ (i.e. $\omega = 4$), and behaviour ${}^{s}\mathcal{B}_{j,v,4}$ has terminated its activities in the instant i, taking into account that ${}^{s}\hat{\mathcal{S}}^{4}_{j,v} = \{{}^{s}\mathcal{S}^{1}_{j,v}, {}^{s}\mathcal{S}^{2}_{j,v}\}$, the initial conditions must fulfil the following conditions: ${}^{s}f^{\sigma}_{j,v,4,1}({}^{s}s^{i}_{j,v}, {}_{x}s^{i}_{j,v}) \wedge {}^{s}f^{\sigma}_{j,v,4,2}({}^{s}s^{i}_{j,v}, {}_{x}s^{i}_{j,v}) = \text{False}$ and ${}^{s}f^{\sigma}_{j,v,4,1}({}^{s}s^{i}_{j,v}, {}_{x}s^{i}_{j,v}) \vee {}^{s}f^{\sigma}_{j,v,4,2}({}^{s}s^{i}_{j,v}, {}_{x}s^{i}_{j,v}) = \text{True}$.

Table 1 Types of agents

Type	Standard function	Composition
C	Zombie	$C, \bullet, \bullet, \bullet$
CE	Blind agent	$C, E, \bullet, \bullet$
CR	Monitoring agent	$C, \bullet, R, \bullet$
CT	Computational agent	$C, \bullet, \bullet, T$
CER	Autonomous agent	$C, E, R, \bullet$
CRT	Remote sensor	$C, \bullet, R, T$
CET	Teleoperated agent	$C, E, \bullet, T$
CERT	Embodied agent	C, E, R, T

5 Types of Agents and Robotic Systems

Taking into account the most general structure of an embodied agent, as presented in Fig. 1, four distinct components can be distinguished: C—the control system, R—receptors, E—effectors and T—inter agent communication resources. An agent cannot exist without a control system, however the three other components are optional. Based on the composition of an agent eight types can be distinguished (Table 1) [106]. A bullet • represents a missing component.

A C-type agent is a purely computational entity without access to the environment and no contact with the other agents, so it can neither influence its surroundings nor do any useful work for the other agents, hence it is useless—it is a zombie. CE-type agents can influence the environment, however they do not get any information from it, e.g., a positioner works like that—it repeats blindly preprogrammed motions. A CR-type agent only monitors the state of the environment. However, it is not able to convey the gathered information to the other agents—it acts as a black box in an airplane. In the case of a black box, the gathered data is extracted by extraordinary means once a disaster occurs. All of the above agent types are of minor utility. However a CT-type agent is very useful. It can either execute computations for the other agents or can coordinate them. The CER-type agents are fully autonomous as they do not have the means to contact other agents. However they can use non-technical means of communication, i.e. stigmergy [106, 107]—leave signs in the environment. The agents of CRT-type can act in the same way that CT-type agents act, but besides that they can also be utilised as remote sensors. The CET-type agents can influence the environment, but they do not get sensor feedback, thus they must be teleoperated by using the communication link. The CERT-type agents are full-blown embodied agents. As all the other types are subtypes of embodied agents it suffices to discuss just this one type of agents. However, when composing a system it is beneficial to explicitly define the types of integrated agents. If templates of all useful agent types are created they can be reused when creating a new systems, so knowledge of the agent types included in the system is relevant. It is also reasonable to note that CT-type agents can act in cyberspace, thus a system can span both a

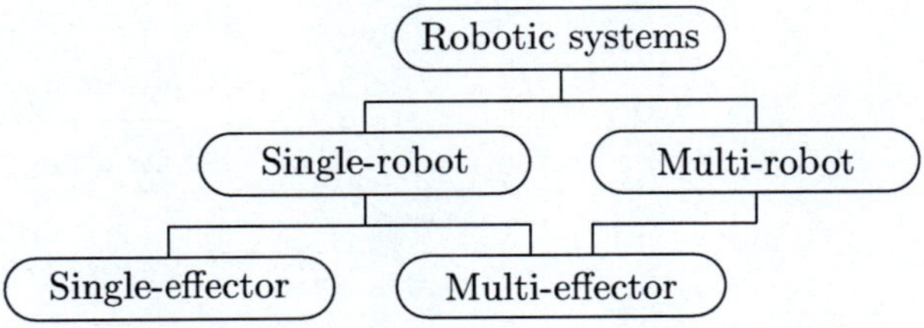

Fig. 6 Classification of robotic systems with respect to the number of effectors

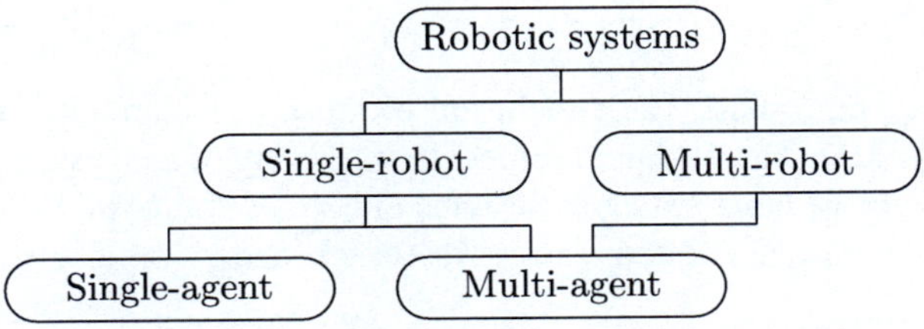

Fig. 7 Classification of robotic systems with respect to the number of robots and agents

physical environment and cyberspace [108]. Usually the task, that the system is to execute, dictates the composition of that system.

The presented classification of agent types enables the distinction between an effector, agent and robot. An effector (E) has no knowledge about the executed task, thus it can only be a component of an agent or a robot. A robot must have a control system (C) and at least one effector (E). Thus C, CR, CT and CRT types of agents should not be treated as robots. An effector can be composed of many devices, e.g. a manipulator having many actuators. A multi-effector robot results from including many effectors in one mechanical structure. It can be assumed that each effector is controlled by a single virtual effector. As virtual effectors can exist in one or many agents, both single-agent and multi-agent robots occur, but also single-effector multi-agent robots can be created (in this case some agents will be of the type not containing E). Multi-agent multi-effector robots are also possible, but single-agent multi-robot systems are not, as they would have to be represented by an agent containing many control systems and that possibility has been excluded. Thus multi-robot systems have to be multi-agent. This discussion leads to classifications with respect to the number of robots, agents and effectors (Figs. 6 and 7).

The structure of the created system can be fixed at the design stage, so during operation of the system it does not change. However, when the tasks that the system has to execute change, sometimes it is useful to design variable structure systems. Nevertheless fixed structure systems can also cope with the variability of tasks. In that case the control system has to have an interpreter of the language in which new tasks are delegated to it. Usually industrial robots employ such a scheme. Designing a variable structure system for a fixed task, taking into account that variable structure systems are much more complex than fixed structure systems performing the same

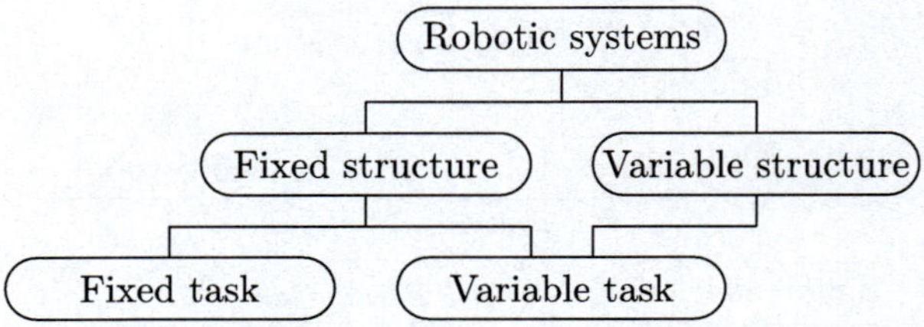

Fig. 8 Classification of robotic systems with respect to the variability of the system structure and tasks

task, is not recommended. The conclusion of those considerations is that robotic systems can be also classified with respect to variability of the system structure and tasks (Fig. 8). In variable structure systems either the composition of subsystems changes or the connections between subsystems change.

6 Methodology

The meta-model of an embodied agent introduced in Sect. 4 is the basis for the formulation of a general robotic system design procedure. Design of a complex system requires an iterative stepwise refinement procedure. Thus the presented design method provides only a set of necessary steps [109]—their order is subject to the designer's preference. Some designers prefer a bottom-up approach, others like the top-down method, still others might use any mixture of the two. It is assumed that the requirements have been formulated and thus the task, or the class of tasks, that the system will have to execute, is defined. The required steps are as follows:

- Choose the necessary real effectors and receptors required for the execution of the task,
- Decompose the designed system into agents, i.e. define the system architecture, deciding whether single or multi-robot system, as well as whether fixed/variable task/structure has to be employed,
- Assign real effectors and receptors to respective agents,
- Allocate real effectors and receptors to virtual effectors and receptors,
- Define sampling rates of each subsystem (sampling rate should not be too high as not to overload the control computer with unnecessary computations and not too low to enable the system to react appropriately),
- Specify the contents of the input and output buffers as well as the internal memory of subsystems, i.e. define the ontologies that the subsystems use,
- Define the necessary behaviours for each subsystem,
- Elaborate the graphs of HFSAs for subsystems, including the formulation of initial conditions,

– Define the subsystem transition functions, terminal and error conditions as well as the communication predicates,
– Validate the model, i.e. assure that, e.g., no communication deadlocks or violations of conditions (14) occur.

7 Structures of Robotic Systems

Following the classification presented in Sect. 5 some examples of robotic systems that have been created will be presented below. Fixed structure systems [110] prevail over the variable structure ones [111], as task variability in research oriented systems does not occur often.

7.1 *Fixed Structure Systems*

Single-robot single-agent systems Such systems are most abundant as robotics research frequently focuses on a single proof-of-the-concept task. A robotic system created for the investigation of position-force manipulator control is an example of such a system [112]. A modified IRp-6 robot manipulator, having 6 degrees of freedom (d.o.f.), was used as an effector of that system. Moreover, it was equipped with a six-component force-torque sensor mounted in its wrist. It was treated as a proprioceptor, thus it was used by the virtual effector, hence a single CE type agent resulted. The task of the robot was either to turn a crank or follow an unknown contour.

Opening a door is usually effortless for humans, but poses a considerable problem to a robot. Robots, if they are to assist humans, have to share with them house or office environment. As people do not like to change their ambiance, its adaptation to the capabilities of robots is out of question. The robot capable of opening doors consisted of: impedance controlled Kuka LWR4+ manipulator, three-fingered position controlled BarretHand and a RGB camera [113]. The first two devices were controlled by their own virtual effectors, thus this was a multi-effector system. Visual data was processed by a virtual receptor. The system was represented by a single CER type agent. The camera was used for detecting handles, so a trajectory leading to the vicinity of the handgrip could be planned. Touch sensors integrated with the gripper, treated as proprioceptors, detected contact with the doors. Subsequently the gripper holding the handle, was moved by the manipulator, exerting a force in such a way that the doors were gradually opened. The force-torque sensors, also treated as proprioceptors, enabled the estimation of the distance of the handle from the hinges, and in consequence the estimation of the radius of the arc that the gripper should follow.

To investigate the communication between subsystems, and thus between the FSAs controlling those subsystems, a simple mobile robot collecting table tennis

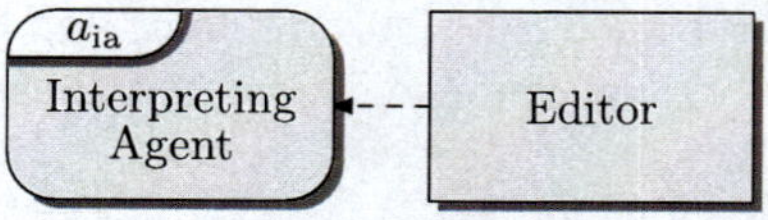

Fig. 9 Structure of the system interpreting user programs

balls from the floor of a room was used [114–116]. A single CER type agent contained two virtual effectors and two virtual receptors. One of the virtual effectors controlled the mobile platform, while the other one activated a vacuum producing device that sucked the detected ball into a container mounted on its back. One of the receptors processed visual data to locate balls, while the other serviced ultrasonic sensors enabling the robot to avoid obstacles. The aggregated data from the camera and the sonars was conveyed to the control subsystem making control decisions forwarded subsequently to the effectors. The fact that each of the subsystems operated at its own sampling rate and the fact that five subsystems operated in parallel made this test-bed a good example for testing the inter-subsystem communication.

The above single-agent systems executed only a single task that they had been designed for. The tasks dictated the structure of those systems. However a universal agent can also be created. Its task does not have to be known at the design stage. In that case the control subsystem has to contain an interpreter of the programming language. The task is conveyed to the robot by downloading a text file created by a programmer using an editor. The file containing commands is being processed by the control subsystem of the Interpreting Agent a_{ia} (Fig. 9). The change of the task consists in the agent a_{ia} downloading a new file. One of the tasks of the robot was to sort apples brought by a conveyer to the manipulator workspace. In this system the tasks were coded in Python [117]. Interpretation of the commands consisted in the invocation of library of functions executing them and enabling the communication of the subsystems of the agent.

Single-robot multi-agent systems Wrocław University of Technology designed the Rex robot (Fig. 10a) to study motion control of skid steering platforms with four independently actuated wheels [105]. Rex uses three agents: a CERT type Locomotion Agent a_{loc}, a CRT type Map Agent a_{map} and a CT type Ontology Agent a_{ont} (Fig. 10b). The Locomotion Agent a_{loc} controls the only effector of the robot, i.e. the 4-wheel mobile platform. Its virtual receptor aggregates data from a camera stereo-pair used for visual odometry and for avoiding obstacles that are not represented on the map. Using this information the control subsystem c_{loc} modifies the planned trajectory subsequently delivered to the virtual effector. The Ontology Agent a_{ont} uses its knowledge base to generate motion plans dispatched to the Locomotion Agent a_{loc} for execution. The plan is represented as a list of points that the mobile platform should pass through. The Map Agent a_{map} possesses the map of the environment and a receptor enabling the localisation of the robot in relation to this map. The virtual receptor of this agent uses the data obtained from the motion capture system. The agent a_{loc} obtains an estimate of the position of the robot from the agent a_{map}. Using this estimate and the motion plan obtained from a_{ont} the agent a_{loc} commands its virtual effector to drive the four motors propelling the wheels. As this agent has

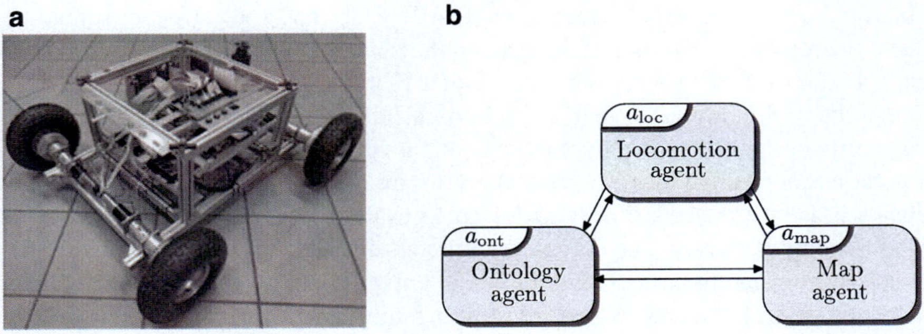

Fig. 10 **a** REX robot, **b** system structure

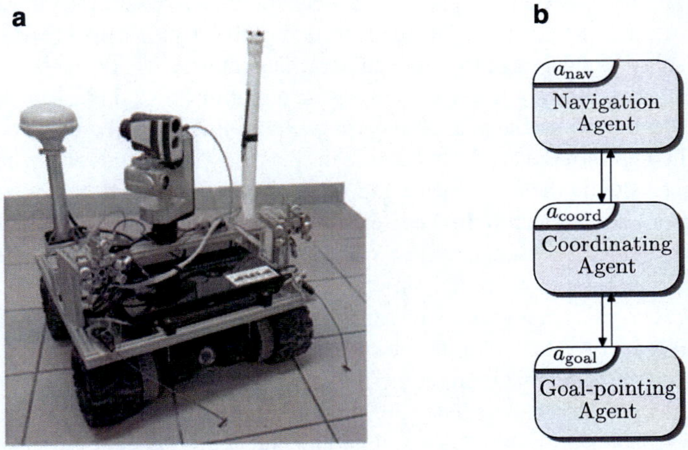

Fig. 11 **a** SCOUT robot (PIAP), **b** system structure

to react quickly the virtual effector treats the Inertial Measurement Unit, the touch sensors located in the bumpers and the four encoders integrated with the motors as proprioceptors. The motor control signals are generated on the basis of the platform model. For that purpose platform state estimator is used. Finally the trajectory is generated in the endogenous space [118].

The three agent system structure resulted from the following causes. Motion planning is a computationally intensive task, thus the software executing it has to reside on a separate computer, hence a separate agent is necessary. The motion capture system can localises robots only inside itself, but mobile robots often operate outside buildings, thus another sensor has to be used. A separate agent facilitates such a substitution. Due to that motion planning and platform localisation were extracted from a_{loc}.

Another example of a robotic system composed of many agents is an exploration mobile robot [119]. Its mobile base was constructed by the Industrial Research Institute for Automation and Measurements (PIAP) (Fig. 11a). Special forces sometimes have to ascertain that at a specific location there are no explosives. For that purpose an operator uses a joystick to steer the robot to that location. Usually it is far away, as the probable striking distance tends to be high. Taking into account the low velocity of the robot the time taken to reach the goal is high, what results in fatigue of the operator, and thus reduces his concentration on the task to be carried out at that location. Hence the task of driving the robot to the exploration site should be carried out autonomously. The task of autonomous driving is divided into two phases: goal-pointing and motion execution. The prototype robot was equipped with a pan-tilt head on which a camera and a laser rangefinder were mounted. The head was actuated by two motors integrated with encoders. Moreover the robot had ultrasonic sensors and touch sensors mounted in the bumpers, inclinometers, compass, GPS and an operator's panel. The operator's panel contained a monitor screen and a keybord substituting the joystick. During operation of the system a cross appears on the screen displaying the image from the camera. The intersection of the horizontal and the vertical lines forming the cross defines the target location. The laser rangefinder provides the distance to the goal. The other two coordinates can be deduced from the pan and tilt angles of the head. By moving the head any visible target can be set. GPS and compass define the coordinates and orientation of the robot so the coordinates of the target with respect to the head can be transformed into geographical coordinates. The keyboard is used to move the head. Once the target is defined the operator commands the robot to move to it autonomously. During that motion the robot uses ultrasonic sensors, camera and touch sensors to detect obstacles and avoid them. GPS and compass are used to determine the current position of the robot. Inclinometers are used to warn the robot that it might tip over. Besides that direct control of the vehicle by the operator was retained. The control system of the robot performs three tasks: establishes the goal of motion, drives the robot and communicates with the operator. Each of those tasks is realised by one agent (Fig. 11b): Goal-pointing Agent a_{goal} determines the target, Navigation Agent a_{nav} autonomously steers the robot to the goal and Coordinating Agent a_{coord} communicates with the operator and coordinates the system. Such a decomposition results from the observation that agents a_{nav} and a_{goal} never act simultaneously. Determination of the motion target and the execution of the motion are disjoint. Thus a_{coord} alternately cooperates with those agents. All the three agents are of the CERT type. Each one of them has its own effectors and receptors. Agent a_{coord} communicates either with a_{goal} or a_{nav}. Agent a_{nav} uses wheel motor encoders for odometry, GPS and compass to self-localise itself, ultrasonic and touch sensors to avoid obstacles and inclinometers to prevent overturning. Agent a_{coord} provides to a_{nav} the goal of motion. That goal is obtained from a_{goal}. Once the goal is reached manual control is resumed, thus a_{coord} transforms the operator commands acquired from the keyboard (treated as a receptor) and sends them to a_{nav}. Agent a_{coord} obtains images from the camera, which is its receptor, and displays them on the screen of the monitor, which is treated as an effector. While searching for the goal, the camera images as well as the pan-tilt head motion commands obtained from the keyboard are supplied to a_{goal}. As a_{coord} is the sole owner of the camera, during motion to the goal it provides images to a_{nav} to assist with obstacle detection.

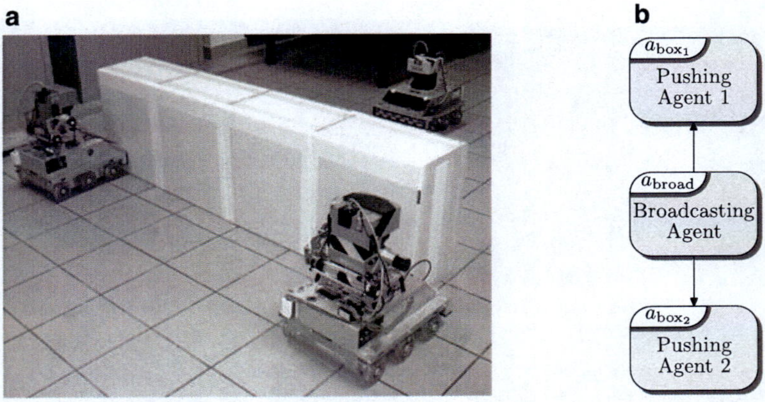

Fig. 12 **a** Box pushing robots, **b** system structure

Actually a_{coord} owns the camera because both a_{goal} and a_{nav} need the images for their operation, so to resolve this contention the camera was assigned to a_{coord}. As a side effect of this the communication link between a_{goal} and a_{nav} became unnecessary.

Multi-robot multi-agent systems Robot communication is an important aspect of research. They can either communicate directly, using a technical communication channel, or indirectly, observing the changes induced into the environment by the activities of other robots, i.e. using stigmergy [107]. The advantage of stigmergy is that its organisation does not change with the number of communicating robots. Thus this method is immune to the failure of some robots. The loss of some robots reduces the effectiveness of task execution, but does not affect the possibility of attaining the goal. This shifted the research interests of many scientists to the investigation of such systems.

One of the benchmarks for the investigation of such systems is the box pushing task. It imitates the cooperation of social insects, e.g. ants [120], especially when transporting objects of considerable size [107]. The investigated system contained three mobile robots: two pushing a box and one deciding where to push it to (Fig. 12b). The last of the mentioned robots also determined the current orientation of the box (Fig. 12a) [106]. Each of the robots was represented by a single CERT type agent. The structure of both box pushing agents is the same. They contain the control subsystem, virtual effector controlling the motion of the mobile platform, and virtual receptor aggregating data obtained from a laser scanner. The Broadcasting Agent a_{broad} determines the current goal of motion and the current orientation of the box, sending this information to the Pushing Agents a_{box_1} and a_{box_2}. Pushing Agents passively receive this information from a_{broad}. They do not actively contact a_{broad}, neither by replying nor by seeking information. Pushing Agents do not communicate with each other. Neither the inner structure of the agents nor their behaviour changes when new Pushing Agents are added to the system. Virtual effectors of the agents controlled two motors, each one of them actuating coupled wheels on one side of the

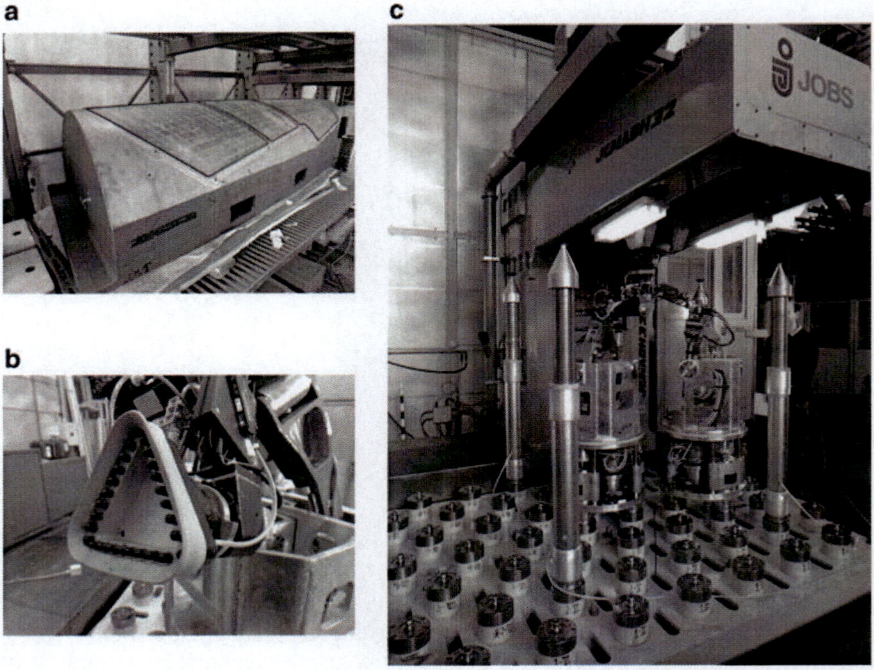

Fig. 13 **a** A fixture, **b** magnetorheological head with pins, **c** SwarmItFIX system

platform. It solved the direct and inverse velocity kinematics. Virtual receptors of the agents aggregated data obtained from the laser scanner, determining the current orientation of the box in relation to the robot and the angle and the distance to the nearest corner of the box. Control subsystems of the Pushing Agents a_{box} computed the values of four feedback functions determining: the error resulting from the difference between the desired and current direction of platform motion, the discrepancy of orientation of the robot from the desired direction of pushing, the distance and the angle of the robot in relation to the nearest corner of the box and the desired velocity of the platform. Each of those partial transition functions determined the suggested rotational and translational velocity of the body of the robot. The resultant velocities were computed as a weighted superposition of the suggested partial velocities. The resultant velocities were conveyed by the control subsystem to the virtual effector for execution. Thus the control subsystem transition function was computed as a superposition of the mentioned partial transition functions. As a result a behaviour emerged that caused the transfer of the box to the desired location, although the Pushing Agents did not communicate with each other directly. They only observed the behaviour of the box and adjusted their activities, by taking into account the feedback, in such a way as to attain the desired motion. As the box is an element of the environment, a system using stigmergy was resulted. A multi-robot, multi-agenty, multi-effector system was constructed.

A different multi-robot multi-agent system was created within an EU project *Self Reconfigurable Intelligent Swarm Fixtures* (SwarmItFIX). The fuselages and wings of airplanes and bodies of cars are made of thin metal sheets pressed into complex shapes. They have to be machined, e.g. holes are drilled in them or the edges are milled. Those operations can be carried out only when the workpiece has been formed into the final shape. The reverse order of operations would cause the holes to be elliptical instead of circular. Due to the flimsiness of the sheets during machining they must be supported. Each shape needs a different fixture (Fig. 13a). This necessitates the creation of a large storage area for the big, rigid and heavy fixtures. They have to be transported from the store to the CNC machine and back. The cost of the fixtures and the store as well as the cumbersome exploitation initiated a search for alternative solutions. One of promising remedies is the utilisation of robots to support the workpiece. During machining the workpiece has to be rigid only in the vicinity of the place that the CNC tool operates [121, 122]. Robots can translocate themselves under a sagging workpiece fixed to four poles. When they are near the machining location they push up the workpiece and attach themselves to it rigidly. As the trajectory of the tool is known the motions of robots can be planned ahead in such a way that the supporting heads make the workpiece rigid in the vicinity of the tool. The designed system assumed the cooperation of several robots, but the prototype contained just two (Fig. 13c).

Cooperating robots translocate themselves using their three legs. An elastic head [123] is pushed up by a parallel-serial structure manipulator mounted on the three-legged platform. When the head is in its required location, supporting the workpiece, vacuum is created sucking the sheet to the head. At this stage the head is solidified and the workpiece becomes rigid in the vicinity of the head. Forces acting on the workpiece during machining are not able to pull it off the head. As the tool pierces the workpiece during machining the head cannot be placed directly underneath. However the head cannot be too far from the location of machining as this would cause the workpiece to vibrate. Thus the head cannot be located neither too near not too far form the tool. If the robots could translocate themselves freely underneath the workpiece the localisation of the head would not be precise enough. This is why the robots translocate themselves over a bench. The bench contains a mesh of docking elements placed in the vertices of equilateral triangles (Fig. 13c) [124]. The three legs protruding from the mobile base match any three docking elements protruding from the bench. Thus the robot can be fixed rigidly and accurately to the bench by its three legs. In this position the manipulator composed of a parallel kinematic machine (PKM) with a serial structure wrist can rise the head to the desired location. If the desired location of the head is too far for the manipulator to reach it the mobile base has to translocate itself. For that it raises two legs and rotates around the third one by a multiple of 60°. The two legs are lowered onto the docking devices in the new position, so the robot again becomes rigidly attached to the platform. Precise localisation of the robot base and the rigidity of the manipulator (PKM) [125] assure very exact placement of the head. The docking devices not only hold the robot rigidly in a precisely defined position, but also deliver compressed air and electric power to the robot. Compressed air is fed into pneumatic actuators

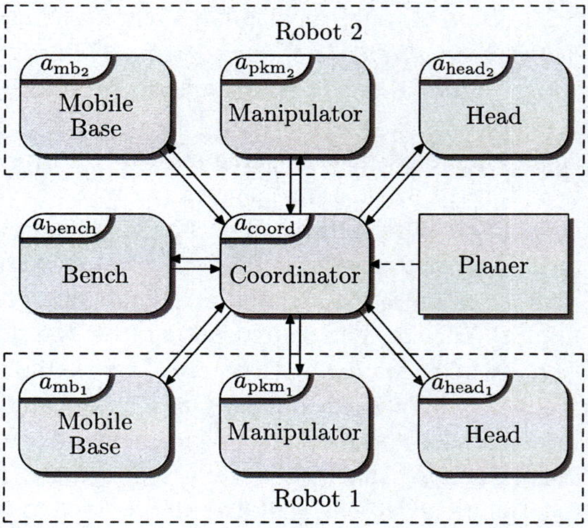

Fig. 14 Structure of the multi-robot reconfigurable fixture

raising or lowering the legs, and creating vacuum in the head. The pneumatic valves located in the docking elements are opened and closed mechanically when the leg attached/disangaged itself to/from the docking element, but the electric power supply had to be computer controlled. Two types of heads were tested. The first was a sac filled with sand. Its shape adjusted to the shape of the workpiece when the head was pushed against it. When the air from between the sand grains was pumped out the head solidified providing firm support for the workpiece. The second head had a ring of pins (Fig. 13b). One end of each pin was submerged in a magneto-rheological fluid. Before a magnetic field produced by an electromagnet was applied the pins could move freely adjusting to the shape of the workpiece. Once the magnetic field appeared the pins could not move, thus again rigid support was created.

The described system is a multi-robot multi-effector and multi-agent one. Each robot has three distinct parts: mobile base (mb), manipulator containing a parallel kinematic machine (pkm) and the head (head). Each of those parts is represented by a separate agent: a_{mb}, a_{pkm} and a_{head} (Fig. 14) [126, 127]. Each robot was divided into three agents, because each of the enumerated devices was controlled independently, and moreover each one of them was designed and tested separately. Each of those agents controlled a single effector: legs, manipulator or head. Each was of a CET type—because of the precision of motion of those devices no receptors, except proprioceptors, were needed. Besides agents controlling the robots an agent responsible for controlling the bench was needed, i.e. a_{bench}. It turned on/off the power supply delivered to the robots through the docking devices and also operated the clamping devices in the docking elements. Robots execute a predefined sequence: translocate the mobile base with a folded manipulator, raise the manipulator, activate the head, deactivate the head, lower the manipulator, move the base. Thus the plan of actions must take into account this sequence of disjoint actions, hence each agent is stimulated by the coordinator separately. The agents do not act simultaneously. Plan of actions was generated off-line using the CAD data describing the workpiece and the CAM data specifying the motions of the tool [128, 129]. The planer delivered the plan to the coordinator a_{coord}, a CT type agent. The coordinator orchestrated the actions of all other agents on the basis of this plan. As the prototype system contained only two robots it consisted of eight agents.

7.2 *Variable Structure Robotic Systems*

In variable structure systems usually the communication network between agents changes its structure. The connections between agents are created and subsequently disappear. Less often the agents forming the nodes of the network are generated and annihilated. As at any given moment the network is fixed, it can be described by methods described in Sect. 4. The system described here required both the changes in the network of connections and the creation/destruction of agents. This system is a single-robot multi-effector multi-agent one.

In many countries the proportion of the elderly in relation to the younger population constantly increases [130]. With age people tend to be impaired both physically and mentally, what leads to social exclusion. In such cases either the family or social workers have to take care of such persons. Unfortunately the number of employed people in relation to the number of people needing support constantly decreases. Hence alternative solutions are being sought. One of them is the creation of robot companions [131–134].

A robot companion should be capable of executing any tasks allotted to it by its owner, thus both the mechanical structure and its control system should be prepared to execute a vast variety of operations. As the needs of the owner are potentially unbounded and the resources of traditional fixed structure control systems are limited, this type of systems cannot be used in this case. This problem can be solved by employing virtually unlimited resources of a computational cloud and a variable structure control system in the robot. The cloud can provide computational services to the robot, and moreover it can contain a repository of software, from which the robot can pick and download whatever is needed for the execution of the task at hand. Information contained in the downloaded software is the only source of knowledge about how to control the robot so that it will execute the required task. As a result the acquired software has to assume command of the system. This implies that to realise new tasks not only the system composition has to change, but also supervisory responsibilities must be shifted between its components. Due to that the robot companion control systems designed within the EU project *Robotic Applications for Delivering Smart User Empowering Applications* (RAPP) had a variable structure with transferred supervision responsibilities [111, 135, 136].

The structure of the RAPP system evolves in time (Fig. 15). Initially it is composed only of the RAPP Platform residing in the cloud and one or more robots (here a one robot version is presented for brevity). The RAPP Platform initially contains two agents: the RAPP Store Agent a_{store} and the Platform Agent a_{plat} connected

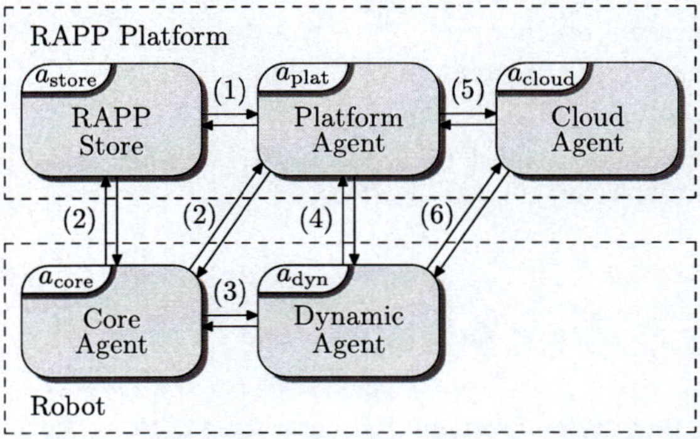

Fig. 15 Structure and phases of creation of a RAPP system

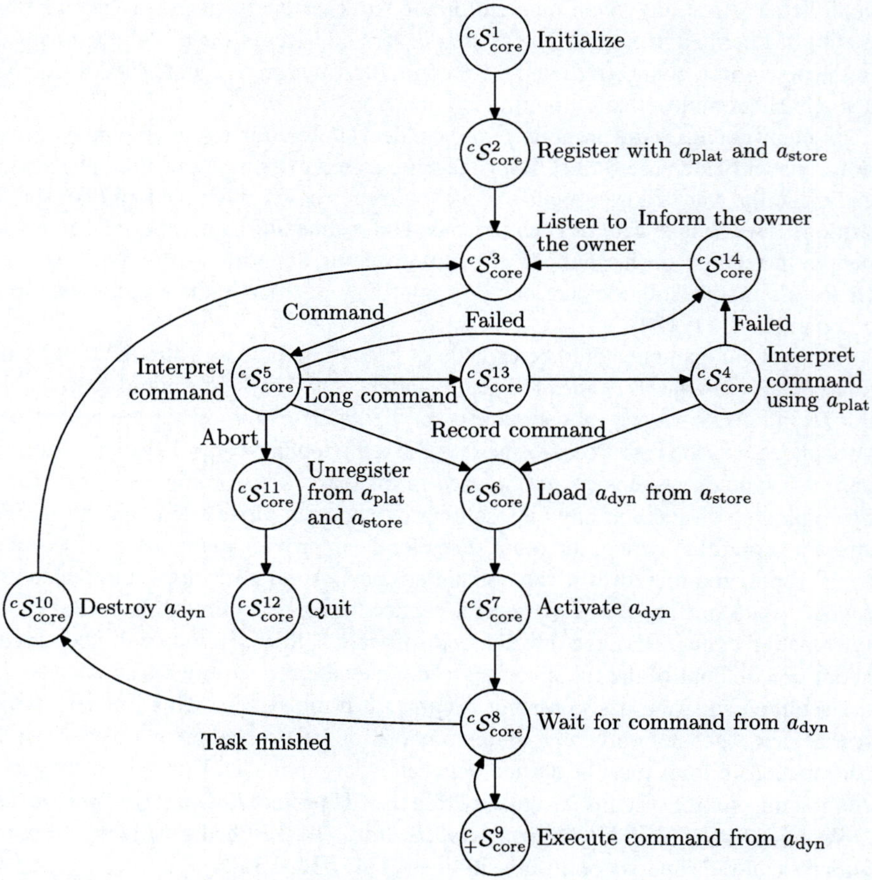

Fig. 16 Graph of the FSA of the control system of the Core Agent a_{core}

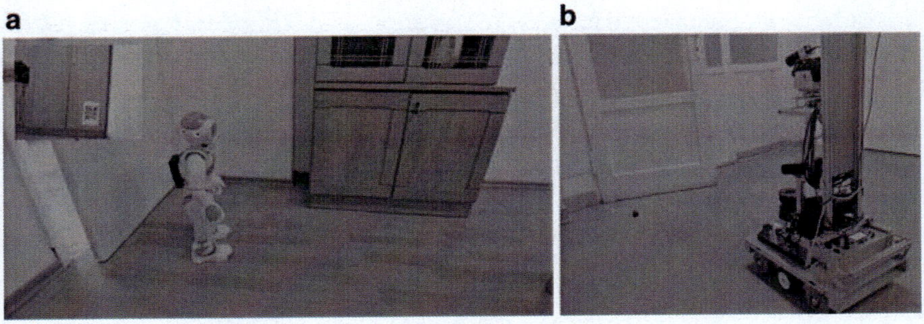

Fig. 17 **a** Nao robot, **b** Elektron robot

by a communication link (1). Both agents are of CT type. Each robot is initially represented by a single Core Agent a_{core}. It is of a CERT type. All the effectors and receptors of the robot belong to this agent. Once the robot is powered up the connections (2) are established between a_{core} and the two agents residing in the RAPP Platform: a_{store} and a_{plat}. From that moment onwards a_{core} expects to get voice commands from the owner. Short commands are processed by a_{core}, but longer ones are sent to a_{plat} for interpretation. In both cases a Dynamic Agent a_{dyn}, of a CT type, is acquired by a_{core} from a_{store}. The Dynamic Agent is initiated by a_{core} and thus communication link (3) is created. Usually a_{dyn} is able to execute the task without extra resources, besides the ones composing the system as it is at this stage. However, if this is not possible it must use the services of a_{plat}. For that purpose a communication link (4) is created between a_{core} and a_{plat}. If a certain sequence of services is required then a_{plat} may create a CT type Cloud Agent a_{cloud} executing this sequence. For that purpose a communication link (5) is established. In this case the communication between a_{dyn} and a_{cloud} is indirect, through a_{plat}. The benefit of this solution is that a_{dyn} instead of communicating with a_{cloud} to get each service of the sequence does this only once for all of those services. If indirect communication through a_{plat} introduces an excessive delay a direct communication link (6) can be created between a_{dyn} and a_{cloud}. Figure 16 presents the case of a single a_{cloud}, however many such agents can be brought to life. Once the task is accomplished the unnecessary agents and communication links are destroyed in the reverse order to the one described above. In this explanation agents a_{dyn} and a_{cloud} are just representatives of their respective classes. Many different a_{dyn} and a_{cloud} are necessary to carry out all possible tasks. They differ by the code that processes the execution of a particular task.

Until agent a_{dyn} is brought to existence the system is supervised by the agent a_{core}. Once a_{dyn} is activated it supervises the system. Only it knows how to realise the task commanded by the owner. When a_{dyn} informs a_{core} that it has executed the task, a_{core} destroys a_{dyn}, and once more assumes the supervisory responsibilities. Thus not only the number of agents and communication links varies throughout the life of the system, but also the supervisory responsibilities are transferred from one agent to the other [111]. Figure 17b presents the FSA of the control system of the Core Agent a_{core}

interacting with the owner (in states: ${}^{c}\mathcal{S}^{3}_{\text{core}}$, ${}^{c}\mathcal{S}^{13}_{\text{core}}$ and ${}^{c}\mathcal{S}^{14}_{\text{core}}$), interpreting his/her commands (${}^{c}\mathcal{S}^{4}_{\text{core}}$ and ${}^{c}\mathcal{S}^{5}_{\text{core}}$), requesting and downloading (${}^{c}\mathcal{S}^{6}_{\text{core}}$) and activating (${}^{c}\mathcal{S}^{7}_{\text{core}}$), destroying (${}^{c}\mathcal{S}^{10}_{\text{core}}$) a Dynamic Agent a_{dyn}, and making itself subordinate to it (${}^{c}\mathcal{S}^{8}_{\text{core}}$ and ${}^{c}_{+}\mathcal{S}^{9}_{\text{core}}$). The system implementation included two different robots : Nao (rys. 17a) and Elektron (rys. 17b), hence two different Core Agents a_{core} had to be created.

8 Conclusions

A procedure of designing robotics systems based on the embodied agent meta-model has been presented. It follows the separation of concerns principle fostered by software engineering, yet it is well embedded in the robotics domain. It is based on such concepts as effectors, receptors, control, transition function or HFSA. At its foundation is the loop starting from the environment and proceeding through the receptors, control system and effectors back to he environment. This loop is reflected in the proposed general agent architecture. The activities of the agent are described by a hierarchic finite state automatom, which invokes behaviours that are represented by a pattern parameterised by transition function and several predicates forming conditions. This hierarchic structure again follows the separation of concerns principle. Larger systems can be formed by composing agents into networks. To facilitate this process classifications of robotics systems with respect to different criteria has been presented. Thus robotics systems can be single/multi-robot, single/multi-agent, while robots can be single/multi-effector as well as single/multi-agent. The designed systems can have a fixed or variable structure. The agents themselves can be categorised into eight types, however the CERT type is the generalisation of all the other ones. System design is an art, but the proposed methodology provides significant guidance, especially to unexperienced designers.

The included examples show how such systems can be structured. The methodology focuses on the specification phase of the system design process, which is often neglected in the design of prototype robots. Formalisation of the methodology enables the discussion of the features of the system before its implementation. Taking into account that the concepts appearing in the specification reflect those used in imperative programming languages its transformation into code does not pose significant problems. This can be done either manually or automatically [109]. The latter is the current subject of interest. Embodied agents can be implemented as processes composed of threads representing subsystems or agents can be compositions of processes. Buffers and internal memory can contain data of any type thus their implementation is straightforward. Transition functions and conditions are simply programming language functions. Behaviours are FSAs and so are the FSAs governing the actions of subsystems. The code executing an FSA can either reflect its structure or specific code operating on a state transition table treated as data can be written. The meta-model is used to specify the model of a particular system. Subsequently all the elements of this model are transformed into its implementation using the mentioned programming concepts. The presented design methodology has been used to specify and implement many systems. Even variable structure controllers can be specified that way.

The presented methodology was conceived to conform to the needs of the robotics domain, however it can be employed beyond that. An interface to the National Cybersecurity Platform was designed using it. The platform needed a voice and gesture command interface. Both of those modes of input are treated as receptors. The output is the presentation of the analysis of the network activity on the screen of the monitor. The data is presented in windows appearing on the screen. The voice and gesture commands affect those windows. The windows influence the operator being the part of the environment, thus they are treated as effectors.

This methodology is currently enhanced to simplify the representation of inter-subsystem communication. To do that the FSAs have been substituted by a hierarchic Petri net [137, 138]. This approach is currently undergoing testing on exemplary prototype systems.

References

1. Dijkstra, E.: On the role of scientific thought. In: Selected Writings on Computing: A Personal Perspective, pp. 60–66. Springer (1982)
2. Kaisler, S.: Software Paradigms. Wiley Interscience, Hoboken (2005)
3. Parent, M., Laurgeau, C.: Logic and Programming. Volume 5 of Series on Robot Technology. Kogan Page, London (1984)
4. RAPID Reference Manual. ABB Flexible Automation, ABB Robotics Products AB, Sweden (2012)
5. Kuka: KUKA System Software 5.5. Operating and Programming Instructions for System Integrators. KUKA Robot Group (2010)
6. Comau: Comau Robotics Instruction Handbook. PDL2—Programming Language Manual. COMAU S.p.A (2012)
7. Mitsibishi: Melfa. Mitsibishi Industrial Robot Controller. Instruction Manual (Detailed Explanation of Functions and Operations). Mitsibishi Electric. Industrial Automation (2007)
8. Kawasaki: AS Language Reference Manual. Kawasaki Heavy Industries, Ltd. (2007)
9. Staübli: VAL3 Reference Manual. Staübli (2006)
10. Robots, U.: The URScript Programming Language. Universal Robots (2012)
11. Toshiba: TS 3000 Series Robot Controller. Instruction Manual. Robot Language Manual. Toshiba Machine Co., Ltd. (2009)
12. Festo: Festo Controller: Programming Instructions CMXR Based on FTL. Festo AG & Co. (2010)
13. Zieliński, C.: By how much should a general purpose programming language be extended to become a multi-robot system programming language? Adv. Robot. **15**(1), 71–96 (2001)
14. Billard, A., Calinon, S., Dillmann, R., Schaal, S.: Robot programming by demonstration. In: Siciliano, B., Khatib, O. (eds.) Springer Handbook of Robotics, pp. 1371–1394. Springer (2008)
15. Billard, A., Calinon, S., Dillmann, R.: Learning from humans. In: Siciliano, B., Khatib, O. (eds.) Springer Handbook of Robotics, pp. 1995–2014. Springer (2016)
16. Pan, Z., Polden, J., Larkin, N., Duin, S.V., Norrish, J.: Recent progress on programming methods for industrial robots. Robot. Comput.-Integr. Manuf. **28**(2), 87–94 (2012)

17. Stenmark, M., Haage, M., Topp, E.A.: Simplified programming of re-usable skills on a safe industrial robot—prototype and evaluation. In: Proceedings of the IEEE/ACM Conference on Human-Robot Interaction (HRI), Vienna, Austria. (March, 6–9 2017)
18. Pedersen, M.R., Nalpantidis, L., Andersen, R.S., Schou, C., Bøgh, S., Krüger, V., Madsen, O.: Robot skills for manufacturing: from concept to industrial deployment. Robot. Comput.-Integr. Manuf. **37**, 282–291 (2016)
19. Paul, R.: WAVE: a model based language for manipulator control. Industr. Robot **4**(1), 10–17 (1977)
20. Paul, R.: The early stages of robotics. In: IFAC Real Time Digital Control Applications, Guadalajara, Mexico, pp. 19–32 (1983)
21. Mujtaba, S., Goldman, R.: AL users' manual. Stanford Artificial Intelligence Laboratory (January 1979)
22. Blume, C., Jakob, W.: Programming Languages for Industrial Robots. Springer (1986)
23. Zieliński, C.: TORBOL: an object level robot programming language. Mechatronics **1**(4), 469–485 (1991)
24. Popplestone, R.J., Ambler, A.P., Bellos, I.: RAPT: a language for describing assemblies. Industr. Robot **5**(3), 131–137 (1978)
25. Ambler, A.P., Corner, D.F.: RAPT1 User's Manual. University of Edinburgh, Department of Artificial Intelligence (1984)
26. Blume, C., Jakob, W.: PASRO: Pascal for Robots. Springer, Berlin (1985)
27. Hayward, V., Paul, R.P.: Robot manipulator control under Unix RCCL: a robot control C library. Int. J. Robot. Res.**5**(4), 94–111 (Winter 1986)
28. Corke, P., Kirkham, R.: The ARCL robot programming system. In: International Conference on Robots for Competitive Industries, pp. 484–493 (14–16 July 1993)
29. Lloyd, J., Parker, M., McClain, R.: Extending the RCCL programming environment to multiple robots and processors. In: IEEE International Conference on Robotics and Automation, Philadelphia, USA, pp. 465–469, 24–29 April (1988)
30. Hayward, V., Hayati, S.: Kali: An environment for the programming and control of cooperative manipulators. In: 7th American Control Conference, pp. 473–478 (1988)
31. Backes, P., Hayati, S., Hayward, V., Tso, K.: The kali multi-arm robot programming and control environment. In: NASA Conference on Space Telerobotics, USA, pp. 173–182 (1989)
32. Hayward, V., Daneshmend, L., Hayati, S.: An overview of KALI: a system to program and control cooperative manipulators. In: Waldron, K. (ed.) Advanced Robotics, pp. 547–558. Springer, Berlin (1989)
33. Nilakantan, A., Hayward, V.: The synchronisation of multiple manipulators in Kali. Robot. Auton. Syst. **5**(4), 345–358 (1989)
34. Gerkey, B.P., Vaughan, R.T., Støy, K., Howard, A., Sukhatme, G.S., Mataric, M.J.: Most valuable player: a robot device server for distributed control. In: Proceedings of the IEEE/RSJ International Conference on Intelligent Robots and Systems (IROS), pp. 1226–1231 (2001)
35. Collett, T., MacDonald, B., Gerkey, B.: Player 2.0: toward a practical robot programming framework. In: Proceedings of the Australasian Conference on Robotics and Automation (ACRA) (December 2005)
36. Vaughan, R.T., Gerkey, B.P.: Reusable robot software and the Player/Stage project. In Brugali, D. (ed.) Software Engineering for Experimental Robotics. Volume 30 of Springer Tracts in Advanced Robotics, pp. 267–289. Springer (2007)
37. Bruyninckx, H.: Open robot control software: the OROCOS project. In: International Conference on Robotics and Automation (ICRA), vol. 3, pp. 2523–2528. IEEE (2001)
38. Bruyninckx, H.: The real-time motion control core of the OROCOS project. In: Proceedings of the IEEE International Conference on Robotics and Automation, pp. 2766–2771. IEEE (September 2003)
39. Quigley, M., Gerkey, B., Conley, K., Faust, J., Foote, T., Leibs, J., Berger, E., Wheeler, R., Ng, A.: ROS: an open-source Robot Operating System. In: Proceedings of the Open-Source Software Workshop at the International Conference on Robotics and Automation (ICRA) (2009)

40. Fitzpatrick, P., Metta, G., Natale, L.: Towards long-lived robot genes. Robot. Auton. Syst. **56**(1), 29–45 (2008)
41. Metta, G., Fitzpatrick, P., Natale, L.: YARP: yet another robot platform. Int. J. Adv. Robot. Syst. **3**(1), 43–48 (2006)
42. Brooks, A., Kaupp, T., Makarenko, A., Williams, S., Orebäck, A.: Orca: A component model and repository. In: Brugali, D. (ed.) Software Engineering for Experimental Robotics. Volume 30 of Springer Tracts in Advanced Robotics, pp. 231–251. Springer (2007)
43. Brooks, A., Kaupp, T., Makarenko, A., Williams, S., Orebäck, A.: Towards component-based robotics. In: IEEE/RSJ International Conference on Intelligent Robots and Systems (IROS 2005), pp. 163–168. (August 2005)
44. Zieliński, C.: The MRROC++ system. In: Proceedings of the First Workshop on Robot Motion and Control, RoMoCo 1999, pp. 147–152. (June 1999)
45. Zieliński, C., Szynkiewicz, W., Mianowski, K., Nazarczuk, K.: Mechatronic design of open-structure multi-robot controllers. Mechatronics **11**(8), 987–1000 (2001)
46. Zieliński, C.: Formal approach to the design of robot programming frameworks: the behavioural control case. Bull. Pol. Acad. Sci. Tech. Sci. **53**(1), 57–67 (2005)
47. Zieliński, C.: Systematic approach to the design of robot programming frameworks. In: Proceedings of the 11th IEEE International Conference on Methods and Models in Automation and Robotics (on CD), Technical University of Szczecin, pp. 639–646 (29 August–1 September 2005)
48. Zieliński, C.: Transition-function based approach to structuring robot control software. In: Kozłowski, K. (ed.) Robot Motion and Control. Volume 335 of Lecture Notes in Control and Information Sciences, pp. 265–286. Springer (2006)
49. Brugali, D., Broten, G.S., Cisternino, A., Colombo, D., Fritsch, J., Gerkey, B., Kraetzschmar, G., Vaughan, R., Utz, H.: Trends in robotic software frameworks. In: Brugali, D. (ed.) Software Engineering for Experimental Robotics, pp. 259–266. Springer (2007)
50. Stefańczyk, M., Kornuta, T.: Handling of asynchronous data flow in robot perception subsystems. In: Simulation, Modeling, and Programming for Autonomous Robots. Volume 8810 of Lecture Notes in Computer Science, pp. 509–520. Springer (2014)
51. Bradski, G., Kaehler, A.: Learning OpenCV: Computer Vision with the OpenCV Library, 1st edn. O'Reilly (September 2008)
52. Fritsch, J., Wrede, S.: An integration framework for developing interactive robots. In: Brugali, D. (ed.) Software Engineering for Experimental Robotics, pp. 291–305. Springer (2007)
53. Bonarini, A., Matteucci, M., Restelli, M.: MRT: Robotics off-the-shelf with the modular robotic toolkit. In: Brugali, D. (ed.) Software Engineering for Experimental Robotics, pp. 345–364. Springer (2007)
54. Brugali, D.: Model-driven software engineering in robotics. IEEE Robot. Autom. Mag. **22**(3), 155–166 (2015)
55. Pilone, D., Pitman, N.: UML 2.0 in a Nutshell. O'Reilly (2005)
56. Fleury, S., Herrb, M., Chatila, R.: GenoM: a tool for the specification and the implementation of operating modules in a distributed robot architecture. In: Proceedings of the 1997 IEEE/RSJ International Conference on Intelligent Robots and Systems (IROS 1997), vol. 2, pp. 842–849. (September 1997)
57. Alami, R., Chatila, R., Fleury, S., M. Ghallab, M., Ingrand, F.: An architecture for autonomy. Int. J. Robot. Res. **17**(4), 315–337 (1998)
58. Brugali, D., Agah, A., MacDonald, B., Nesnas, I., Smart, W.: Trends in robot software domain engineering. In: Brugali, D. (ed.) Software Engineering for Experimental Robotics, pp. 3–8. Springer (2007)
59. Brugali, D.: Stable analysis patterns for robot mobility. In: Brugali, D. (ed.) Software Engineering for Experimental Robotics, pp. 9–30. Springer (2007)
60. Klotzbücher, M., Smits, R., Bruyninckx, H., De Schutter, J.: Reusable hybrid force-velocity controlled motion specifications with executable domain specific languages. In: IEEE/RSJ International Conference on Intelligent Robots and Systems, September 25–30, 2011, San Francisco, USA, pp. 4684–4689 (2011)

61. Kleppe, A.: Software Language Engineering: Creating Domain-Specific Languages Using Metamodels. Addison-Wesley (2009)
62. Sun, Y., Gray, J., Bulheller, K., von Baillou, N.: A model-driven approach to support engineering changes in industrial robotics software. In et al., R.F. (eds.) Model Driven Engineering Languages and Systems: 15th International Conference MODELS 2012, LNCS 7590, pp. 368–382. Springer, Berlin, Heidelberg (October 2012)
63. Dudek, W., Szynkiewicz, W., Winiarski, T.: Nao Robot Navigation system structure development in an agent-based architecture of the RAPP Platform. In: Szewczyk, R., Zieliński, C., Kaliczyńska, M. (eds.) Recent Advances in Automation, Robotics and Measuring Techniques. Volume 440 of Advances in Intelligent Systems and Computing (AISC), pp. 623–633. Springer (2016)
64. Nesnas, I.: The CLARAty project: coping with hardware and software heterogenity. In: Brugali, D. (ed.) Software Engineering for Experimental Robotics, pp. 31–70. Springer (2007)
65. Brooks, R.A.: A robust layered control system for a mobile robot. IEEE J. Robot. Autom. **2**(1), 14–23 (1986)
66. Brooks, R.A.: Intelligence without representation. Artif. Intell. **47**(1–3), 139–159 (1991)
67. Brooks, R.A.: Intelligence without reason. Artif. Intell. Crit. Concepts **3**, 107–163 (1991)
68. Arkin, R.C.: Behavior-Based Robotics. MIT Press (1998)
69. Gat, E.: On three-layer architectures. In: Kortenkamp, D., Bonnasso, R.P., Murphy, R. (eds.) Artificial Intelligence and Mobile Robots, pp. 195–210. AAAI Press Cambridge, MA (1998)
70. Lyons, D.M., Arbib, M.A.: A formal model of computation for sensory-based robotics. IEEE Trans. Robot. Autom. **5**(3), 280–293 (1989)
71. Lyons, D.M.: A Schema-theory approach to specifying and analysing the behavior of robotic systems. In: Prerational Intelligence. Volume 2: Adaptive Behavior and Intelligent Systems without Symbols and Logic of Studies in Cognitive Systems, pp. 51–70. Kluwer Academic (2001)
72. Tang, F., Parker, L.: A Complete Methodology for Generating Multi-Robot Task Solutions using ASyMTRe-D and Market-Based Task Allocation. In: 2007 IEEE International Conference on Robotics and Automation, pp. 3351–3358. IEEE (2007)
73. Shoham, Y.: Agent-oriented programming. Artif. Intell. **60**(1), 51–92 (1993)
74. Padgham, L., Winikoff, M.: Developing Intelligent Agent Systems: A Practical Guide. Wiley (2004)
75. Matarić, M.J., Michaud, F.: Behavior-Based Systems. In: The Handbook of Robotics, pp. 891–909. Springer (2008)
76. Bulter, Z., Rizzi, A.: Distributed and cellular robots. In: Khatib, O., Siciliano, B. (eds.) Springer Handbook of Robotics, pp. 911–920. Springer (June 2008)
77. Parker, L.E.: Multiple mobile robot systems. In: Khatib, O., Siciliano, B. (eds.) Springer Handbook of Robotics, pp. 921–941. Springer (June 2008)
78. Yim, M., Shen, W.M., Salemi, B., Rus, D., Moll, M., Lipson, H., Klavins, E., Chirikjian, G.S.: Modular self-reconfigurable robot systems [grand challenges of robotics]. IEEE Robot. Autom. Mag. **14**(1), 43–52 (2007)
79. Matarić, M.J.: Issues and approaches in the design of collective autonomous agents. Robot. Auton. Syst. **16**(2), 321–331 (1995)
80. Farinelli, A., Iocchi, L., Nardi, D.: Multirobot systems: a classification focused on coordination. IEEE Trans. Syst. Man Cybern. B Cybern., Part B (Cybern.) **34**(5) (Oct 2004) 2015–2028
81. Dudek, G., Jenkin, M.R.M., Milios, E., Wilkes, D.: A taxonomy for multi-agent robotics. Auton. Robot. **3**(4), 375–397 (1996)
82. Doriya, R., Mishra, S., Gupta, S.: A brief survey and analysis of multi-robot communication and coordination. In: 2015 International Conference on Computing, Communication Automation (ICCCA), pp. 1014–1021 (May 2015)
83. Chibani, A., Amirat, Y., Mohammed, S., Matson, E., Hagita, N., Barreto, M.: Ubiquitous robotics: recent challenges and future trends. Robot. Auton. Syst. **61**(11), 1162–1172 (2013)
84. Kortenkamp, D., Simmons, R., Brugali, D.: Robotic systems architectures and programming. In: Siciliano, B., Khatib, O. (eds.) Springer Handbook of Robotics, 2nd edn, pp. 283–306. Springer (2016)

85. Coste-Maniere, E., Simmons, R.: Architecture, the backbone of robotic systems. In: Proceeding of the IEEE International Conference on Robotics and Automation ICRA 2000, vol. 1, pp. 67–72 (2000)
86. Zieliński, C.: A quasi-formal approach to structuring multi-robot system controllers. In: Second International Workshop on Robot Motion and Control, RoMoCo 2001, pp. 121–128 (18–20 October 2001)
87. Zieliński, C., Kornuta, T.: Diagnostic requirements in multi-robot systems. In: Korbicz, J., Kowal, M. (eds.) Intelligent Systems in Technical and Medical Diagnostics. Volume 230 of Advances in Intelligent Systems and Computing, pp. 345–356. Springer (2014)
88. Armbrust, C., Kiekbusch, L., Ropertz, T., Berns, K.: Soft robot control with a behaviour-based architecture. In: Soft Robotics, pp. 81–91. Springer (2015)
89. Kiekbusch, L., Armbrust, C., Berns, K.: Formal verification of behaviour networks including sensor failures. Robot. Auton. Syst. **74**, 331–339 (2015)
90. Kiekbusch, L., Armbrust, C., Berns, K.: Formal verification of behaviour networks including hardware failures. **302**, 1571–1582 (2016)
91. Kortenkamp, D., Simmons, R.: Robotic systems architectures and programming. In: Khatib, O., Siciliano, B. (eds) Springer Handbook of Robotics, pp. 187–206. Springer (2008)
92. Stampfer, D., Lotz, A., Lutz, M., Schlegel, C.: The SmartMDSD toolchain: an integrated MDSD workflow and integrated development environment (IDE) for robotics software. J. Soft. Eng. Robot. (JOSER) **7**(1), 3–19 (2016)
93. Brooks, R.: Elephants don't play chess. Robot. Auton. Syst. **6**(1–2), 3–15 (1990)
94. Brooks, R.A.: New approaches to robotics. Science **253**, 1227–1232 (1991)
95. Russell, S., Norvig, P.: Artificial Intelligence: A Modern Approach. Prentice Hall, Upper Saddle River, N.J. (1995)
96. Nwana, H.S., Ndumu, D.T.: A Brief Introduction to Software Agent Technology, pp. 29–47. Springer, Berlin Heidelberg, Berlin, Heidelberg (1998)
97. Wooldridge, M.: Multiagent Systems, pp. 27–77. MIT Press, Cambridge, MA, USA (1999)
98. Zieliński, C., Winiarski, T.: General specification of multi-robot control system structures. Bull. Pol. Acad. Sci. Tech. Sci. **58**(1), 15–28 (2010)
99. Kornuta, T., Zieliński, C.: Robot control system design exemplified by multi-camera visual servoing. J. Intell. Robot. Syst. **77**(3–4), 499–524 (2013)
100. Zieliński, C., Kornuta, T., Winiarski, T.: A systematic method of designing control systems for service and field robots. In: 19-th IEEE International Conference on Methods and Models in Automation and Robotics, MMAR, IEEE, pp. 1–14 (2014)
101. Zieliński, C., Kornuta, T.: An object-based robot ontology. Volume 323 of Advances in Intelligent Systems and Computing (AISC), pp. 3–14. Springer (2015)
102. Zieliński, C.: A unified formal description of behavioural and deliberative robotic multi-agent systems. In: 7th International IFAC Symposium on Robot Control (SYROCO), vol. 7, pp. 479–486 (2003)
103. Zieliński, C.: Specification of behavioural embodied agents. In: Kozłowski, K., ed.: Fourth International Workshop on Robot Motion and Control (RoMoCo 2004), pp. 79–84. (June 17–20 2004)
104. Zieliński, C., Winiarski, T.: Motion generation in the MRROC++ robot programming framework. Int. J. Robot. Res. **29**(4), 386–413 (2010)
105. Janiak, M., Zieliński, C.: Control system architecture for the investigation of motion control algorithms on an example of the mobile platform Rex. Bull. Pol. Acad. Sci. Tech. Sci. **63**(3), 667–678 (2015)
106. Zieliński, C., Trojanek, P.: Stigmergic cooperation of autonomous robots. J. Mech. Mach. Theory **44**, 656–670 (2009)
107. Bonabeau, E., Dorigo, M., Theraulaz, G.: Swarm Intelligence: From Natural to Artificial Systems. Oxford University Press, New York, Oxford (1999)
108. Ambroszkiewicz, S., Cetnarowicz, K.: On the concept of agent in multi-robot environment. In: Innovative Concepts for Autonomic and Agent-Based Systems. Volume 3825 of Lecture Notes in Computer Science, pp. 135–146. Springer (2006)

109. Zieliński, C., Figat, M.: Robot system design procedure based on a formal specification. In: Szewczyk, R., Zieliński, C., Kaliczyńska, M. (eds.) Recent Advances in Automation, Robotics and Measuring Techniques. Volume 440 of Advances in Intelligent Systems and Computing (AISC), pp. 511–522. Springer (2016)
110. Zieliński, C., Winiarski, T., Kornuta, T.: Agent-based structures of robot systems. In: Kacprzyk, J., et al. (eds.) Trends in Advanced Intelligent Control, Optimization and Automation. Volume 577 of Advances in Intelligent Systems and Computing, pp. 493–502. (2017)
111. Zieliński, C., Stefańczyk, M., Kornuta, T., Figat, M., Dudek, W., Szynkiewicz, W., Kasprzak, W., Figat, J., Szlenk, M., Winiarski, T., Banachowicz, K., Zielińska, T., Tsardoulias, E.G., Symeonidis, A.L., Psomopoulos, F.E., Kintsakis, A.M., Mitkas, P.A., Thallas, A., Reppou, S.E., Karagiannis, G.T., Panayiotou, K., Prunet, V., Serrano, M., Merlet, J.P., Arampatzis, S., Giokas, A., Penteridis, L., Trochidis, I., Daney, D., Iturburu, M.: Variable structure robot control systems: the RAPP approach. Robot. Auton. Syst. **94**, 226–244 (2017)
112. Zieliński, C., Szynkiewicz, W., Winiarski, T.: Applications of MRROC++ robot programming framework. In Kozłowski, K. (ed.) Proceedings of the 5th International Workshop on Robot Motion and Control, RoMoCo 2005, Dymaczewo, Poland, pp. 251–257 (June, 23–25 2005)
113. Winiarski, T., Banachowicz, K., Seredyński, D.: Multi-sensory feedback control in door approaching and opening. In: Filev, D., Jabłkowski, J., Kacprzyk, J., Krawczak, M., Popchev, I., Rutkowski, L., Sgurev, V., Sotirova, E., Szynkarczyk, P., Zadrozny, S. (eds.) Intelligent Systems' 2014. Volume 323 of Advances in Intelligent Systems and Computing, pp. 57–70. Springer (2015)
114. Zieliński, C., Figat, M., Hexel, R.: Communication within multi-fsm based robotic systems. J. Intell. Robot. Syst. **93**(3), 787–805 (2019)
115. Zieliński, C., Figat, M., Hexel, R.: Robotic systems implementation based on fsms. In: Szewczyk, R., Zieliński, C., Kaliczyńska, M. (eds.) Automation 2018: Advances in Automation, pp. 441–452. Springer, Robotics and Measurement Techniques (2018)
116. Figat, M., Zieliński, C., Hexel, R.: Fsm based specification of robot control system activities. In: 2017 11th International Workshop on Robot Motion and Control (RoMoCo), pp. 193–198. (July 2017)
117. Lutz, M.: Programming Python. O'Reilly Media, Inc. (2010)
118. Tchoń, K., Jakubiak, J.: Endogenous configuration space approach to mobile manipulators: a derivation and performance assessment of Jacobian inverse kinematics algorithms. Int. J. Control **76**(14), 1387–1419 (2003)
119. Zieliński, C., Kornuta, T., Trojanek, P., Winiarski, T.: Metoda projektowania układów sterowania autonomicznych robotów mobilnych. Część 2. Przykład zastosowania. Pomiary Automatyka Robotyka (10), pp. 84–91 (2011)
120. Gordon, D.: Ant Encounters: Interaction, Networks and Colony Behavior. Primers in Complex Systems. Princeton University Press (2010)
121. Molfino, R., Zoppi, M., Zlatanov, D.: Reconfigurable swarm fixtures. In: ASME/IFToMM International Conference on Reconfigurable Mechanisms and Robots, pp. 730–735. (June 22–24 2009)
122. Leonardo, L., Zoppi, M., Xiong, L., Gagliardi, S., Molfino, R.: Developing a New Concept of Self Reconfigurable Intelligent Swarm Fixtures, pp. 321–331 (2012)
123. Gagliardi, S., Li, X., Zoppi, M., de Leonardo, L., Molfino, R.: Adaptable fixturing heads for Swarm fixtures: discussion of two designs. In: ASME: 11th Biennial Conference on Engineering Systems Design and Analysis (ESDA2012), p. 2012. France, Nantes (2012)
124. de Leonardo, L., Zoppi, M., Li, X., Zlatanov, D., Molfino, R.: Swarmitfix: a multi-robot-based reconfigurable fixture. Industr. Robot., 320–328 (2013)
125. Neumann, K.: US patent number 4732525 (1988)
126. Zieliński, C., Kornuta, T., Trojanek, P., Winiarski, T., Walcki, M.: Specification of a multi-agent robot-based reconfigurable fixture control system. In: Robot Motion & Control 2011 (Lecture Notes in Control & Information Sciences), vol. 422, pp. 171–182 (2012)
127. Zieliński, C., Kasprzak, W., Kornuta, T., Szynkiewicz, W., Trojanek, P., Walcki, M., Winiarski, T., Zielińska, T.: Control and programming of a multi-robot-based reconfigurable fixture. Industr. Robot: An Int. J. **40**(4), 329–336 (2013)

128. Szynkiewicz, W., Zielińska, T., Kasprzak, W.: Robotized machining of big work pieces: localization of supporting heads. Front. Mech. Eng. China **5**(4), 357–369 (2010)
129. Zielińska, T., Kasprzak, W., Szynkiewicz, W., Zieliński, C.: Path planning for robotized mobile supports. J. Mech. Mach. Theory **78**, 51–64 (2014)
130. Lutz, W., Sanderson, W., Scherbov, S.: The coming acceleration of global population ageing. Nature **451**(7179), 716–719 (2008)
131. Mitkas, P.: Assistive robots as future caregivers: the rapp approach. In Szewczyk, R., Zieliński, C., Kaliczyńska, M. (eds.) Progress in Automation, Robotics and Measuring Techniques. Vol. 2 Robotics. Volume 351 of Advances in Intelligent Systems and Computing (AISC), pp. 171–179. Springer (2015)
132. Reppou, S., Karagiannis, G.: Social inclusion with robots: A RAPP case study using NAO for technology illiterate elderly at Ormylia foundation. In: Szewczyk, R., Zieliński, C., Kaliczyńska, M. (eds.) Progress in Automation, Robotics and Measuring Techniques. Vol. 2 Robotics. Volume 351 of Advances in Intelligent Systems and Computing (AISC), pp. 233–241. Springer (2015)
133. Reppou, S., Karagiannis, G., Tsardoulias, E., Kintsakis, A., Symeonidis, A., Mitkas, P., Psomopoulos, F., Zieliński, C., Prunet, V., Iturburu, M., Arampatzis, S.: RAPP: a robotic-oriented ecosystem for delivering smart user empowering applications for older people. Int. J. Soc. Robot. (June 2016)
134. Tsardoulias, E.G., Kintsakis, A.M., Panayiotou, K., Thallas, A.G., Reppou, S.E., Karagiannis, G.G., Iturburu, M., Arampatzis, S., Zieliński, C., Prunet, V., Psomopoulos, F.E., Symeonidis, A.L., Mitkas, P.A.: Towards an integrated robotics architecture for social inclusion-the rapp paradigm. Cogn. Syst. Res. (2016)
135. Psomopoulos, F., Tsardoulias, E., Giokas, A., Zieliński, C., Prunet, V., Trochidis, I., Daney, D., Serrano, M., Courtes, L., Arampatzis, S., Mitkas, P.: Rapp system architecture. In: IROS 2014—Assistance and Service Robotics in a Human Environment, Workshop in conjunction with IEEE/RSJ International Conference on Intelligent Robots and Systems, Chicago, Illinois, pp. 14–18, September 14 (2014)
136. Tsardoulias, E., Zieliński, C., Kasprzak, W., Reppou, S., Symeonidis, A., Mitkas, P., Karagiannis, G.: Merging robotics and aal ontologies: the rapp methodology. In: Szewczyk, R., Zieliński, C., Kaliczyńska, M. (eds.) Progress in Automation, Robotics and Measuring Techniques. Vol. 2 Robotics. Volume 351 of Advances in Intelligent Systems and Computing (AISC), pp. 285–298. Springer (2015)
137. Figat, M., Zieliński, C.: Hierarchical petri net representation of robot systems. In: Szewczyk, R., Zieliński, C., Kaliczyńska, M. (eds.) Automation 2019, pp. 492–501. Springer International Publishing, Cham (2019)
138. Figat, M., Zieliński, C.: Methodology of designing multi-agent robot control systems utilizing hierarchical petri nets. In: 2019 International Conference on Robotics and Automation (ICRA), pp. 3363–3369, (May 2019)

Computational Intelligence and Decision Support

Fault-Tolerant Control: Analytical and Soft Computing Solutions

Józef Korbicz, Krzysztof Patan, and Marcin Witczak

Abstract The complexity of modern systems and industrial installations, along with continuously growing requirements regarding their operation and control quality, is a serious challenge in the development of control theory as well as process and system diagnostics. The dynamic evolution of fault tolerant control theory witnessed in recent years is a partial solution if this problem. In his chapter, selected issues in fault tolerant control system design using analytical and soft computing methods and approaches are presented. Different structures of fault tolerant control systems are considered, including those with virtual sensors and actuators. Fault estimation and compensation are also discussed. In the second part of the chapter, neural predictive control is considered, with the neural model of processes implemented using feedforward networks with delays and recurrent networks. Included is an example of process control in a two tank system, which well illustrates three implemented control strategies: robust control, robust control with fault compensation, and fault predictive fault tolerant control. Another example of process control in a flow tank system shows the possibilities and efficiency of neural modelling and control.

1 Fault-Tolerant Control

Permanent growth of complexity, efficiency and reliability demands of modern industrial systems raises the need for new control fault diagnosis (FD)) [3, 5, 16, 22, 56] techniques both at the theoretical and practical levels. A balanced synergy of the

J. Korbicz (✉) · K. Patan · M. Witczak
Institute of Control and Computation Engineering, University of Zielona Góra,
ul. Szafrana 2, 65-516 Zielona Góra, Poland
e-mail: j.korbicz@issi.uz.zgora.pl

K. Patan
e-mail: k.patan@issi.uz.zgora.pl

M. Witczak
e-mail: m.witczak@issi.uz.zgora.pl

P. Kulczycki et al. (eds.), *Automatic Control, Robotics, and Information Processing*,
Studies in Systems, Decision and Control 296,
https://doi.org/10.1007/978-3-030-48587-0_18

above paradigms is intensively investigated under the name of fault-tolerant control (FTC) [9, 17, 31, 35, 38, 57]. It leads to an optimized integration of advanced fault diagnosis [21, 22, 56] and control strategies [3]. A general FTC scheme is portrayed in Fig. 1 [55, 56].

The system being controlled and diagnosed constitutes its main component. It is divided into three crucial parts: actuators, process and sensors. Each can be affected by unknown inputs, which are formed with external exogenous disturbances as well as measurement noise. When using analytic redundancy [3, 5, 9, 17, 22, 31, 35, 56], the unknown input can be extended by modelling uncertainty.

The diagnosed system can also be affected by faults. These are defined as an unacceptable change of a characteristic parameter of the system from its nominal value. On the other hand, failure stands for a permanent system ability of performing a desired mission. Figure 2 portrays possible system performance regions along with suitable FTC-based recovery actions. Such a strategy makes it possible to bring the system back to the required performance level without using its structural hardware modification.

According to above-introduced nomenclature (cf. Fig. 1), the faults are divided into [5, 57]

- *actuator faults*,
- *sensor faults*,
- *process faults*.

The objective of the subsequent part of this chapter is to introduce the reader into FTC, along with indicated issues and challenges in this important area of research and science.

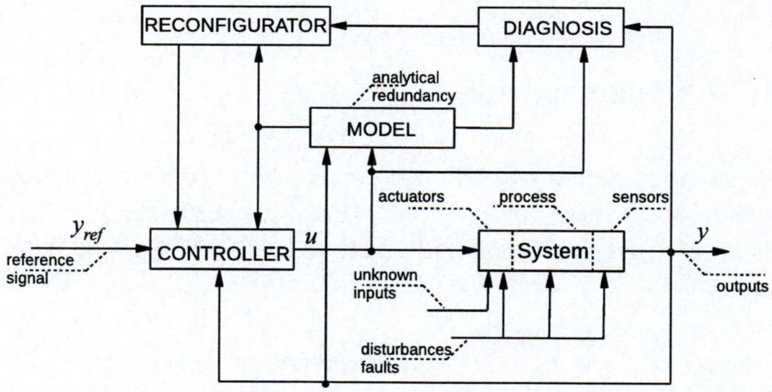

Fig. 1 Modern control system

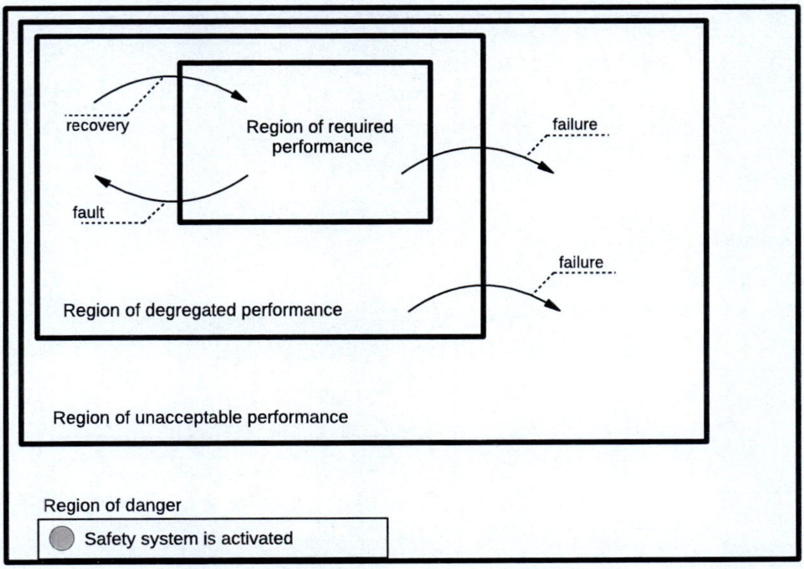

Fig. 2 System operation regions

2 Review of Fault-Tolerant Control

2.1 *Passive FTC*

FTC systems can be divided into two main classes [57, 64], i.e., passive and active ones. In the case of passive FTC, [27, 28, 46, 63], the controller is designed in such a way as to be robust against a set of predefined faults. Such an approach eliminates the necessity of fault diagnosis as it treats the fault as an exogenous disturbance acting on the system. However, its obvious limitation pertains that fact that it can tolerate a significantly restricted set of possible faults. Another disadvantage of such an approach is associated with the fact that the controller performs sub-optimally in a fault-free environment. Indeed, it must be capable of dealing with faults, irrespective of their presence or absence. A general structure of passive FTC is portrayed in Fig. 3.

2.2 *Active FTC: Restructurable and Reconfigurable*

Contrarily to passive FTC, active one reacts actively to the faults of system components. This is realized by changing the control law, which is determined by the diagnostic knowledge about the faults. This means that the level of fault tolerance is strongly related with the quality of fault diagnosis [22, 26, 56]. Such an approach results in a control architecture capable of attaining system stability under prede-

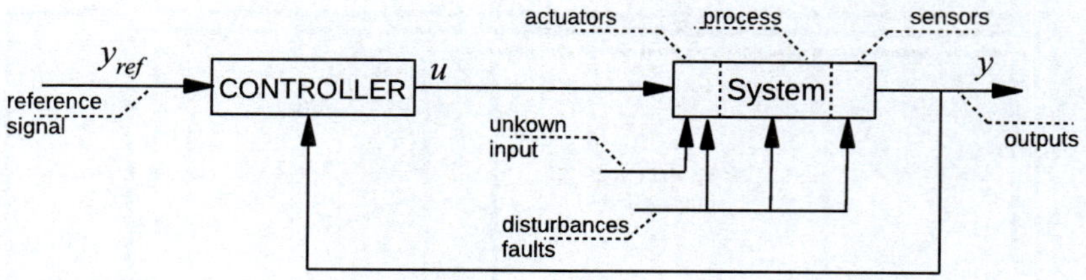

Fig. 3 Passive FTC

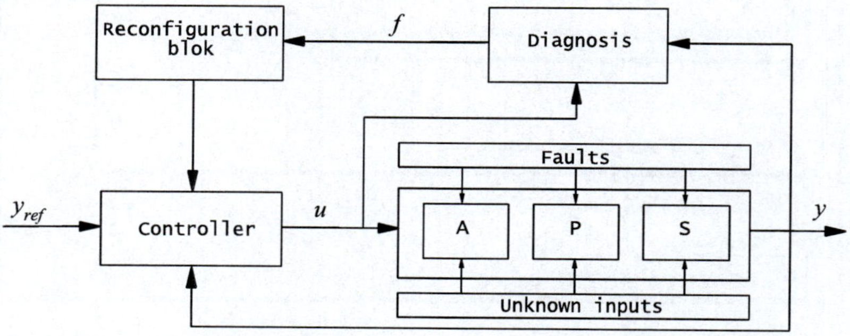

Fig. 4 Reconfigurable FTC (A: actuators, P: process, S: sensors)

fined quality requirements both in fault-free and faulty conditions. According to Fig. 4, FTC is composed of three elements: [64]:

- reconfigurable controller,
- fault diagnosis,
- controller reconfiguration.

Introduction of the above elements provides a clear distinction between active and passive FTC. Thus, a key feature is to design a controller which can be easily reconfigured while keeping high quality control. It should also be indicated that, in some case, controller reconfiguration can be insufficient to stabilize a faulty system. Then, the controller structure must be changed. This is associated with the use of an alternative input/output structure for a new controller. In other words, it is necessary to develop a new control law. Such FTC is called restructurable and its overall scheme is shown in Fig. 5.

2.3 *Virtual Sensors and Actuators*

Apart from the well-known problems associated with fault diagnosis, the following research directions shape the development of FTC [57]:

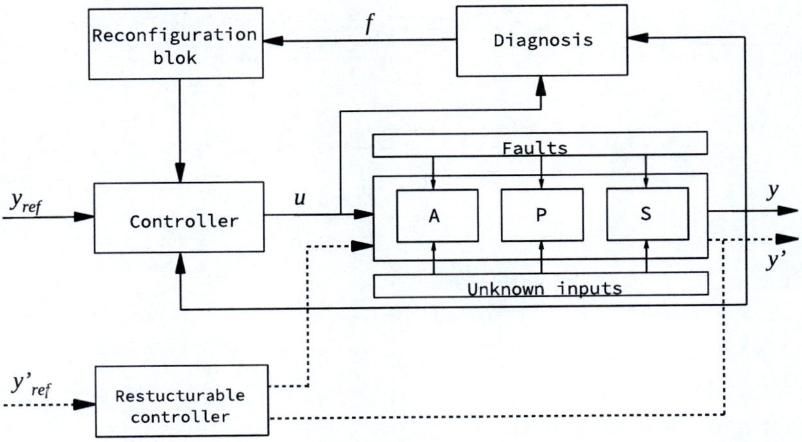

Fig. 5 Restructurable FTC (A: actuator, P: process, S: sensors)

- advanced methods of system and/or controller reconfiguration,
- advanced fault estimation methods for nonlinear systems,
- integration of fault estimation and reconfiguration mechanisms,
- on-line implementation of reconfiguration mechanisms.

The above-listed problems can be partially solved with the application of the so-called virtual sensors and actuators [3, 10, 30]. In the case of the latter, the design problem boils down to removing the controller reconfigurator and placing a virtual actuator instead. It links the output of the controller with the input to all actuators available in the system being diagnosed and controller.

This means that the virtual actuators (Fig. 6) is responsible for control allocation among all available actuators. This is realized in such a way as to minimize the impact of the fault acting on the system [10, 41, 44, 48, 49].

A similar strategy can be realized using the so-called virtual sensors [3, 51, 57]. After fault detection and isolation of a given sensor fault, FTC is reconfigured in such a way as to exclude the measurements from the faulty sensor and replace them with the data from its virtual counterpart.

3 Fault Estimation

Knowledge about the size of a fault constitutes a crucial aspect for controller reconfiguration. Such a task can be solved with various estimations techniques, e.g., by extending the state vector, or applying a two-stage Kalman filter [18], a minimum input and state variance filter [12, 19], an adaptive estimation [62], sliding mode [54], as well as robust $\mathcal{H}_\infty$ [33] estimation.

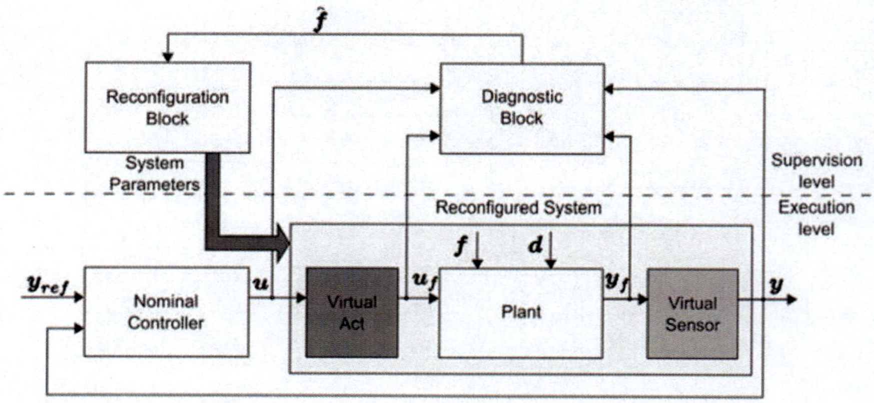

Fig. 6 FTC with virtual sensors and actuators

The problem of fault estimation can be also tackled using parameter estimation [47]. In the case of nonlinear systems, most solutions rely on state estimators [2, 43]. For example, in [13], an estimation for state affine systems is considered, while in [14, 42] the authors consider input affine ones.

Taking into account the above observations, let us consider the following nonlinear system:

$$\boldsymbol{x}_{k+1} = \boldsymbol{A}\boldsymbol{x}_k + \boldsymbol{B}\boldsymbol{u}_k + \boldsymbol{D}\boldsymbol{d}_k + \boldsymbol{g}\left(\boldsymbol{x}_k, \boldsymbol{u}_k\right) + \boldsymbol{B}\boldsymbol{f}_{a,k} + \boldsymbol{W}_1\boldsymbol{w}_k, \tag{1}$$

$$\boldsymbol{y}_k = \boldsymbol{C}\boldsymbol{x}_k + \boldsymbol{f}_{s,k} + \boldsymbol{W}_2\boldsymbol{w}_k, \tag{2}$$

where $\boldsymbol{x}_k \in \mathbb{X} \subset \mathbb{R}^n$ is the state, $\boldsymbol{u}_k \in \mathbb{U} \in \mathbb{R}^r$ is the input, $\boldsymbol{y}_k \in \mathbb{R}^m$ is the output, $\boldsymbol{f}_{a,k} \in \mathbb{R}^r$ is the actuator fault, $\boldsymbol{f}_{s,k} \in \mathbb{R}^m$ denotes a sensor fault, $\boldsymbol{d}_k \in \mathbb{R}^q$ is an unknown input, while a $\boldsymbol{w}_k$ denotes exogenous external disturbance acting on the system.

Considering (1)–(2), the problem is to design an estimator allowing simultaneous state $\boldsymbol{x}_k$ and fault estimation $\boldsymbol{f}_{a,k}$ and $\boldsymbol{f}_{s,k}$, while decoupling $\boldsymbol{d}_k$ and minimizing the influence of $\boldsymbol{w}_k$.

In order to solve such a problem, the following estimator was proposed [59]:

$$\boldsymbol{z}_{k+1} = \boldsymbol{N}\boldsymbol{z}_k + \boldsymbol{G}\boldsymbol{u}_k + \boldsymbol{L}\boldsymbol{y}_k + \boldsymbol{T}\boldsymbol{B}\hat{\boldsymbol{f}}_k + \boldsymbol{T}\boldsymbol{g}\left(\hat{\boldsymbol{x}}_k, \boldsymbol{u}_k\right), \tag{3}$$

$$\hat{\boldsymbol{x}}_k = \boldsymbol{z}_k - \boldsymbol{E}\boldsymbol{y}_k, \tag{4}$$

$$\hat{\boldsymbol{f}}_{k+1} = \hat{\boldsymbol{f}}_k + \boldsymbol{F}(\boldsymbol{y}_k - \boldsymbol{C}\hat{\boldsymbol{x}}_k), \tag{5}$$

where $\hat{\boldsymbol{x}}_k$ and $\hat{\boldsymbol{f}}_k$ stand for the state and fault estimates, respectively.

It should be pointed out that the above scheme can also be used for sensor fault estimation. However, it requires appropriate transformations. For the sake of simplicity, they are presented for linear systems only.

Let us consider the following system:

$$\bar{\boldsymbol{x}}_{k+1} = \bar{\boldsymbol{A}}\bar{\boldsymbol{x}}_k + \bar{\boldsymbol{B}}\boldsymbol{u}_k + \bar{\boldsymbol{B}}\boldsymbol{f}_{a,k} + \bar{\boldsymbol{W}}_1\boldsymbol{w}_k, \tag{6}$$

$$\bar{\boldsymbol{y}}_k = \bar{\boldsymbol{C}}\bar{\boldsymbol{x}}_k + \bar{\boldsymbol{C}}_f\boldsymbol{f}_{s,k} + \bar{\boldsymbol{W}}_2\boldsymbol{w}_k, \tag{7}$$

where $\bar{\boldsymbol{x}}_k \in \mathbb{R}^n$, $\boldsymbol{u}_k \in \mathbb{R}^r$, $\bar{\boldsymbol{y}}_k \in \mathbb{R}^m$ stand for the state, input and output of the system, respectively.

Let us introduce the following filter:

$$\boldsymbol{s}_{k+1} = \boldsymbol{D}(\bar{\boldsymbol{y}}_k - \boldsymbol{s}_k), \tag{8}$$

where $\boldsymbol{D} \in \mathbb{R}^{m\times m}$ is a matrix with eigenvalues located in a unit circle. Substituting (7) into (8) yields

$$\boldsymbol{s}_{k+1} = -\boldsymbol{D}\boldsymbol{s}_k + \boldsymbol{D}\bar{\boldsymbol{y}}_k = -\boldsymbol{D}\boldsymbol{s}_k + \boldsymbol{D}\bar{\boldsymbol{C}}\bar{\boldsymbol{x}}_k + \boldsymbol{D}\bar{\boldsymbol{C}}_f\boldsymbol{f}_{s,k} + \boldsymbol{D}\bar{\boldsymbol{W}}_2\boldsymbol{w}_k. \tag{9}$$

As a result, the following extended state vector is obtained:

$$\boldsymbol{x}_{k+1} = \begin{bmatrix}\bar{\boldsymbol{x}}_{k+1}\\ \boldsymbol{s}_{k+1}\end{bmatrix} = \boldsymbol{A}\boldsymbol{x}_k + \boldsymbol{B}\boldsymbol{u}_k + \boldsymbol{L}\boldsymbol{f}_k + \boldsymbol{W}\boldsymbol{w}_k, \tag{10}$$

where

$$\boldsymbol{f}_k = \begin{bmatrix}\boldsymbol{f}_{a,k}\\ \boldsymbol{f}_{s,k}\end{bmatrix}, \quad \boldsymbol{A} = \begin{bmatrix}\bar{\boldsymbol{A}} & \boldsymbol{0}\\ \boldsymbol{D}\bar{\boldsymbol{C}} & -\boldsymbol{D}\end{bmatrix}, \quad \boldsymbol{B} = \begin{bmatrix}\bar{\boldsymbol{B}}\\ \boldsymbol{0}\end{bmatrix}, \quad \boldsymbol{L} = \begin{bmatrix}\bar{\boldsymbol{B}} & \boldsymbol{0}\\ \boldsymbol{0} & \boldsymbol{D}\bar{\boldsymbol{C}}_f\end{bmatrix},$$

$$\boldsymbol{W} = \begin{bmatrix}\bar{\boldsymbol{W}}_1\\ \boldsymbol{D}\bar{\boldsymbol{W}}_2\end{bmatrix},$$

while

$$\boldsymbol{y}_k = \boldsymbol{C}\boldsymbol{x}_k \tag{11}$$

with

$$\boldsymbol{C} = [\boldsymbol{0} \quad \boldsymbol{I}].$$

Thus, it can be observed that $\bar{\boldsymbol{x}}_k$, $\boldsymbol{f}_{a,k}$ and $\boldsymbol{f}_{s,k}$ reduces to estimating $\boldsymbol{x}_k$ and $\boldsymbol{f}_k$ described by (10)–(11). This allows a direct application of (3)–(5) to simultaneous estimation of sensor and actuator faults.

Unfortunately, the above solution cannot be used to estimate process faults, which can be introduced as follows:

$$\boldsymbol{x}_{k+1} = \left(\boldsymbol{A} + \sum_{i=1}^{n_p} \boldsymbol{A}_{i,f} f_{p,i}\right)\boldsymbol{x}_k + \boldsymbol{B}\boldsymbol{u}_k + \boldsymbol{B}\boldsymbol{f}_{a,k} + \boldsymbol{W}_1\boldsymbol{w}_k, \tag{12}$$

$$\boldsymbol{y}_k = \boldsymbol{C}\bar{\boldsymbol{x}}_k + \boldsymbol{C}_f \boldsymbol{f}_{s,k} + \boldsymbol{W}_2\boldsymbol{w}_k, \tag{13}$$

where $f_{p,i}, i = 1, \ldots, n_p$, stands for the i-th process fault. While analysing the above system description, it can be concluded that even for linear systems the problem of estimating a process fault belongs to the class of nonlinear ones. Thus is because of the product $f_{p,i}\boldsymbol{x}_k$.

The existing approaches which can cope with the above problem are based on sliding mode estimation [52], adaptive observers [6, 11, 62] as well as observers for time-varying parameter systems [50].

4 Fault Accommodation

Knowledge about the source and size of faults makes it possible to undertake their accommodation.

Similarly as before, let us start the discussion with actuator faults. The idea of integrating fault estimation and control for a class of nonlinear systems (1)–(2) was introduced in [60].

The above solution is portrayed in Fig. 7 and obeys the following control strategy:

$$\boldsymbol{u}_{f,k} = -\hat{\boldsymbol{f}}_k + \boldsymbol{K}_1(\boldsymbol{x}_k - \hat{\boldsymbol{x}}_{f,k}) + \boldsymbol{K}_2(\boldsymbol{g}(\boldsymbol{x}_k) - \boldsymbol{g}(\hat{\boldsymbol{x}}_{f,k})) + \boldsymbol{u}_k, \tag{14}$$

where $\boldsymbol{x}_k$ and $\boldsymbol{u}_k$ stand for the reference state and input.

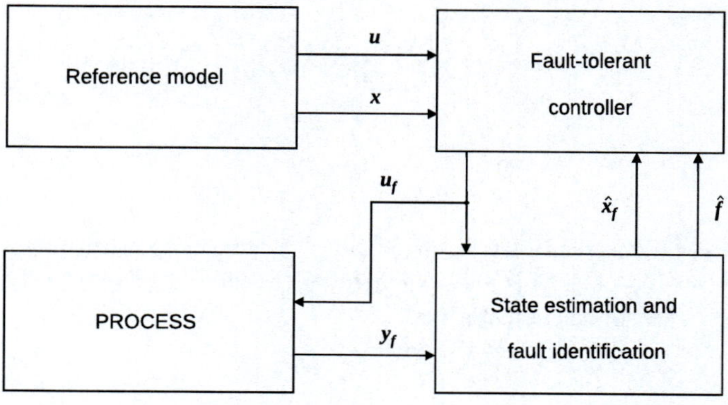

Fig. 7 Fault accommodation by means of a virtual actuator

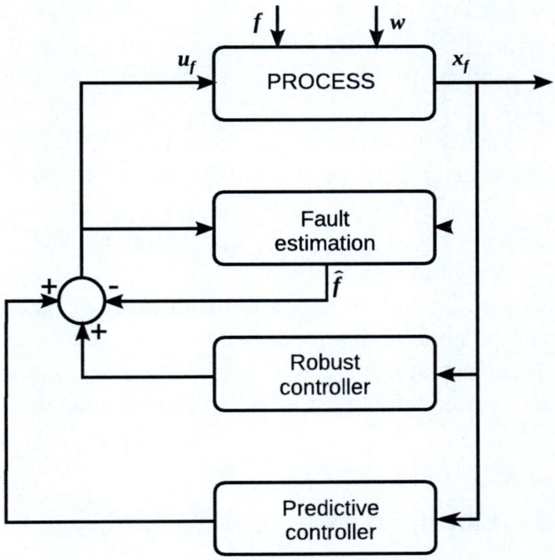

Fig. 8 Fault compensation with virtual actuators and predictive control

The scheme works in such a way that the faulty system tracks the reference state using suitable fault compensation. The control law is based on the nominal controller (characterized by $\boldsymbol{K}_1$ and $\boldsymbol{K}_2$) as well as fault compensation $-\hat{\boldsymbol{f}}_k$. In the light of the above discussion, such a solution can be classified as a virtual actuator.

The main disadvantage of the above solution pertains to the fact that fault compensation is the only reconfigurability mechanism. If it fails, then a divergence between the reference and actual state should be expected.

To tackle such a difficulty, a control allocation-based strategy was proposed in [58]. This results in a triple-stage strategy presented in Fig. 8.

It starts with fault estimation and then the fault is compensated with a robust controller. If the above approach does not provide acceptable results, then a predictive controller is started, which aims at allocating the control effort among all actuators. This is realized in such a way as to attain a desired fault accommodation effect on a given prediction horizon.

The above strategy is formed by combining the results of the work [60] with a predictive control paradigm proposed in [24]. As a result, the control law is

$$\boldsymbol{u}_{f,k} = -\boldsymbol{K}\boldsymbol{x}_k - \hat{\boldsymbol{f}}_{k-1} + \boldsymbol{c}_k, \tag{15}$$

which can be detailed by

$$\boldsymbol{u}_{f,j} = \begin{cases} -\boldsymbol{K}\boldsymbol{x}_j - \hat{\boldsymbol{f}}_{k-1} + \boldsymbol{c}_j, \; j = k, \ldots, k + n_c - 1, \\ -\boldsymbol{K}\boldsymbol{x}_j - \hat{\boldsymbol{f}}_{k-1}, \qquad\qquad j \geq k + n_c, \end{cases} \tag{16}$$

where n_c stands for the prediction horizon, $\boldsymbol{K}$ is the robust $\mathcal{H}_\infty$ controller, $\hat{\boldsymbol{f}}_{k-1}$ is an actuator fault estimate, $\boldsymbol{c}_j$ is a variable responsible for a control allocation.

An extension of the above technique for Takagi–Sugeno systems was proposed in [61].

Finally, it should be pointed out that sensor fault-tolerant control is significantly easier to realise than its actuator-like counter part. Indeed, the control law reduces to

$$\boldsymbol{u}_{f,k} = -\boldsymbol{K}\hat{\boldsymbol{x}}_k, \tag{17}$$

where $\hat{\boldsymbol{x}}_k$ is a fault estimate obtained by excluding a fault sensor. Summarizing, it can be noticed that process fault-tolerant control can be attained in a similar way as the actuator one. However, fault compensation depends of the type of the fault considered, and hence it cannot be described using a universal framework.

5 Example: Two-Tank System

The objective of this section is to illustrate the FTC performance described in Sects. 3 and 4 with a two-tank system example detailed by [58]:

$$\dot{\boldsymbol{x}} = \boldsymbol{A}_c\boldsymbol{x} + \boldsymbol{B}_c\boldsymbol{u} + \boldsymbol{W}_c\boldsymbol{w} \tag{18}$$

with

$$\boldsymbol{A}_c = \begin{bmatrix} -\frac{k}{2F\sqrt{h_1^s}} & 0 \\ \frac{k}{2F\sqrt{h_1^s}} & \frac{k}{2F\sqrt{h_2^s}} \end{bmatrix}, \quad \boldsymbol{B}_c = \begin{bmatrix} \frac{1}{F} & 0 \\ 0 & \frac{1}{F} \end{bmatrix}, \tag{19}$$

where $\boldsymbol{x} = [(h_1 - h_1^s), (h_2 - h_2^s)]^T$, h_1 and h_2 stand for the liquid levels, $F = 12.566\,cm^2$ is the tank cross-section area, and $k = 3.667\,cm^2 s^{-1}$ is a respective flow constants. The above system was linearized under $h_1^s = h_2^s = 0.10\,m$, and discretized with the sampling time equal to $1s$.

The objective of the subsequent part of this section is to compare the approach presented in Sects. 3 and 4:

C1: Robust control,

$$\boldsymbol{u}_{f,k} = -\boldsymbol{K}\boldsymbol{x}_{f,k}. \tag{20}$$

C2: Robust control with fault compensation,

$$\boldsymbol{u}_{f,k} = -\hat{\boldsymbol{f}}_{k-1} - \boldsymbol{K}\boldsymbol{x}_{f,k}. \tag{21}$$

C3: Predictive FTC,

$$\boldsymbol{u}_{f,k} = -\boldsymbol{K}\boldsymbol{x}_k - \hat{\boldsymbol{f}}_{k-1} + \boldsymbol{c}_k. \tag{22}$$

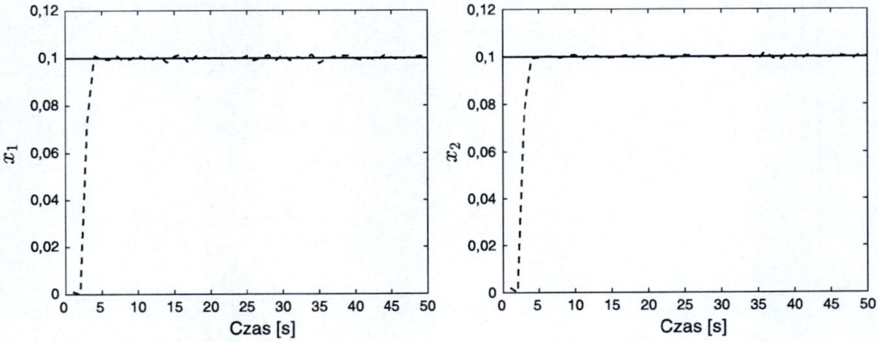

Fig. 9 State of the two-tank system ($k = 0, \ldots, 20$)

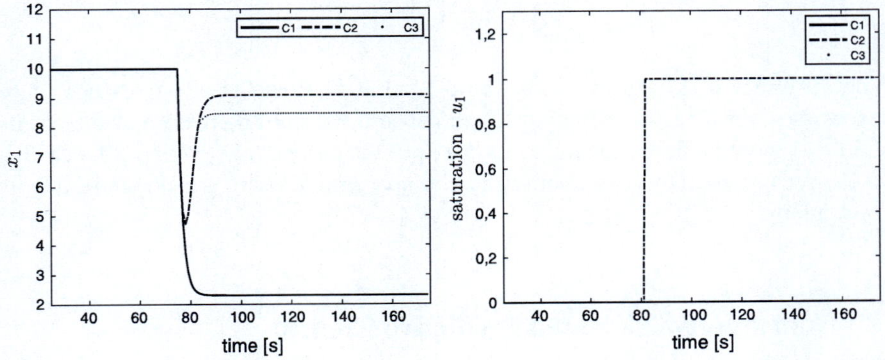

Fig. 10 State $x_{f,1}$ (left) for C1 (solid line), C2 (dashed line), C3 (dotted line) and the saturation indicator for u_1 (right)

Figure 9 presents the performance of the above strategies in the fault-free mode. In this case, neither fault compensation nor predictive control is activated, and hence C2 and C3 perform in the same fashion as C1. In all cases, a desired set-point $h_1^s = h_2^s = 0.10m$ was achieved, starting from an initial condition $\boldsymbol{x}_{f,0} = \mathbf{0}$.

Let us consider a first pump fault pertaining 90% loss of its effectiveness:

$$\boldsymbol{f}_{1,k} = \begin{cases} -0.9\boldsymbol{u}_{f,1,k} & 75 \leqslant k \leqslant 175, \\ 0 & \text{otherwise} \end{cases}$$
$$\boldsymbol{f}_{2,k} = 0.$$

The obtained results are presented in Figs. 10 and 11. For such a serious fault, robust control C1 makes the water levels change significantly due to the fault. Strategy C2 leads the fault pump to the saturation mode (cf. Fig. 10 (right)), while C3 activates the predictive control, which overcomes this unappealing phenomenon by allocating suitable control effort using the second pump. Note that C2 and C3 lead to a similar

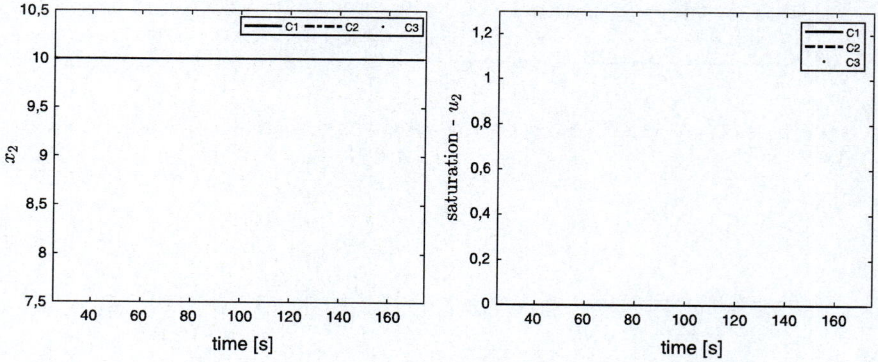

Fig. 11 State $x_{f,2}$ (left) for C1 (solid line), C2 (dashed line), C3 (dotted line) as well as the saturation indicator for u_2 (right)

control effect with respect to the liquid level $x_{f,1}$. However, C3 does not cause excessive over-actuation of pump 1, and hence it does not increase a chance of its failure. Figure 11 shows the liquid level $x_{f,2}$ in the second tank, which is fed with a fault-free pump. In this case, both strategies give similar result and do not lead to its saturation.

6 Neural Network Based Predictive Control

Nowadays, predictive control is a very popular control technique, receiving serious attention among researchers and engineers [4, 25, 29, 53]. The intensively carried out research succeeded in a wide range of applications [1, 20, 45]. Popularity of predictive control follows from the fact that different constraints imposed on the state, control and process variables can be easily and naturally implemented. Moreover, principles of operation are intuitive and easy to understand. Predictive control uses a model to predict the behavior of the process on the finite prediction horizon and by solving suitable formulated optimization problem derives the optimal control sequence. A model predicts the future process outputs based on the actual and future values of the control signal. Taking into account that industrial processes are characterized by nonlinear behavior, artificial neural networks can be effectively employed as a model. A block scheme of neural-network-based predictive control is presented in Fig. 12. This scheme can be easily equipped with a fault detection and isolation (FDI) module, which can be used to develop fault-tolerant predictive control.

Predictive control relies on optimization of a certain cost function. Let us consider minimization of the following cost function:

$$J(k) = \sum_{i=N_1}^{N_2} e^2(k+i) + \rho \sum_{i=1}^{N_u} \Delta u^2(k+i-1) \tag{23}$$

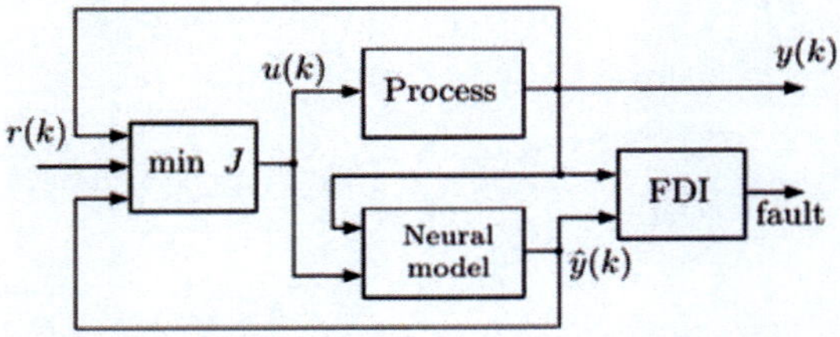

Fig. 12 Block scheme of neural-network-based predictive control

with respect to N_u future controls,

$$\boldsymbol{u}(k) = \left[\, u(k) \ \ldots \ u(k+N_u-1) \,\right]^T, \tag{24}$$

and subject to the constraints

$$\Delta u(k+i) = 0, \quad N_u \leqslant i \leqslant N_2 - 1, \tag{25}$$

where $e(k+i) = r(k) - \hat{y}(k+i)$ is the tracking error, $r(k)$ is the reference signal, $\hat{y}(k+i)$ represents the estimated process output, $\Delta u(k+i-1) = u(k+i-1) - u(k+i-2)$, $u(k)$ is the control signal, N_1 stands for the minimal prediction horizon, N_2 is the prediction horizon, N_u represents the control horizon, ρ is a factor penalizing changes in the control signal. It is assumed that $N_u < N_2$. The optimization procedure derives N_u future controls only. Beyond the value of N_u the control is assumed to be constant. In the criterion (23), $\hat{y}(k+i)$ denotes the minimum variance i-step ahead prediction. This predictor can be achieved by successive recursion of a one-step ahead nonlinear model:

$$\hat{y}(k+1) = f(y(k), \ldots, y(k-n_a+1), u(k), \ldots, u(k-n_b+1)), \tag{26}$$

where n_a and n_b stand for numbers representing past outputs and inputs, respectively. Based on the model (26), the i-step ahead prediction of the plant output is calculated as follows:

$$\hat{y}(k+i) = f(y(k+i-1), \ldots, y(k+i-n_a), u(k+i-1), \ldots, u(k+i-n_b)). \tag{27}$$

Taking into account that measurements of the output are available up to time k, one should substitute predictions for actual measurements since these do not exist,

$$y(k+i) = \hat{y}(k+i), \quad \forall i > 1. \tag{28}$$

A nonlinear function f in (26) can be realized by means of an artificial neural network. It is obvious that neural networks provide an excellent modeling tool for dealing with nonlinear problems [15, 36]. The crucial property of neural networks is their powerful approximation abilities [7]. Another attractive property is the self-learning ability using data recorded in real industrial processes.

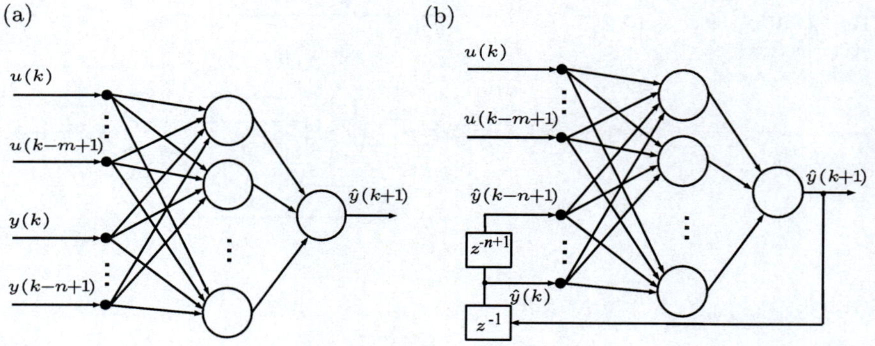

Fig. 13 Model representation: using a feed-forward network (**a**), using a recurrent network (**b**)

The most frequently used neural network in the framework of modeling dynamic systems is the multi-layer feed-forward network with external dynamics. Such a model is able to realize the mapping (26), as presented in Fig. 13a. The training of such a model leads to the so-called serial-parallel identification model [32, 36]. The application of real plant outputs guarantees stability of the neural model and simplifies the training procedure. Unfortunately, in order to design the i-step ahead predictor, it is required to feed the network with past predicted outputs according to (27) and (28). In that case, the series-parallel identification model may be insufficient.

Figure 13b presents neural-network realization of (27). Here the neural network is a recurrent one and has to be trained using the parallel identification scheme [32, 36]. Unfortunately, recurrent networks suffer from stability problems as well as time-consuming and complex training algorithms. In general, designing a model with acceptable performance is a hard task.

In the presented approach we used a feed-forward network with external dynamics described by the following formula:

$$\hat{y}(k+1) = f(\boldsymbol{x}) = \sigma_o(\boldsymbol{W}_2\sigma_h(\boldsymbol{W}_1\boldsymbol{x} + \boldsymbol{b}_1) + \boldsymbol{b}_2), \tag{29}$$

where $\boldsymbol{x} = [y(k), \ldots, y(k-n_a+1), u(k), \ldots, u(k-n_b+1)]^T$, $\boldsymbol{W}_1 \in \mathbb{R}^{n_a+n_b\times v}$ and $\boldsymbol{W}_2 \in \mathbb{R}^{v\times 1}$ are weight matrices, $\boldsymbol{b}_1 \in \mathbb{R}^v$ and $\boldsymbol{b}_2 \in \mathbb{R}^1$ are bias vectors, $\sigma_h : \mathbb{R}^v \to \mathbb{R}^v$ is a nonlinear activation function of the hidden layer, $\sigma_o : \mathbb{R}^1 \to \mathbb{R}^1$ denotes the activation function of the output layer, frequently selected as the linear one. Parameters of the neural network (29) need to be determined using the series-parallel identification scheme.

With the cost function (23) and the neural predictor (26), nonlinear predictive control may be formulated in the form of the following optimization problem:

$$u(k) \triangleq \arg\min_{u} \; J(k) \tag{30a}$$
$$\text{s.t.} \;\; \underline{u} \leqslant u(k+j) \leqslant \overline{u}, \quad \forall j \in [0, N_u - 1], \tag{30b}$$
$$\underline{y} \leqslant y(k+j) \leqslant \overline{y}, \quad \forall j \in [0, N_2], \tag{30c}$$
$$\Delta u(k + N_u + j) = 0, \quad \forall j \geqslant 0, \tag{30d}$$

where $\underline{u}$ and $\overline{u}$ are the lower and upper limits of the control signal, and $\underline{y}$ and $\overline{y}$ are the lower and upper limits of the process output.

When using a neural network model, process output predictions are nonlinear in control inputs. In consequence, the optimization becomes a complex nonlinear programming problem, which should be solved in real time. Moreover, a solver should assure fast convergence and numerical robustness. That is the reason for using second-order optimization algorithms. In this work, a combination of the Newton–Raphson and Levenberg–Marquardt methods suggested in [34, 39, 40] is employed.

$$\boldsymbol{u}^{(i+1)} = \boldsymbol{u}^{(i)} - \left(\boldsymbol{H}^{(i)} + \lambda^{(i)}\boldsymbol{I}\right)^{-1} \boldsymbol{G}^{(i)}, \tag{31}$$

where $\boldsymbol{u}^{(i)}$ is the current iterate of the future control sequence, $\boldsymbol{H}^{(i)}$ is the Hessian derived at the i-th iteration, $\boldsymbol{G}^{(i)}$ represents the gradient derived at the i-th iteration and $\lambda^{(i)}$ is the parameter used to assure positive definiteness of the Hessian.

6.1 Diagnostic Module

A very popular fault diagnosis method is the binary diagnostic matrix (BDM) [22, 23], which defines relation between symptoms and faults. An extension of BDM is the idea of the multi-valued diagnostic matrix (MDM), developed to increase fault distinguishability [23]. Symptoms can be derived as residuals calculated on the basis of the so-called partial models of the controlled process. Partial models can be designed for the smallest possible parts of the system using available measurements. With a set of partial models, we are able to observe fault influence on the diagnostic signal s:

$$s(r_i) = \begin{cases} 0 & \text{if } r_i \in [T_l^i, T_u^i], \\ +1 & \text{if } r_i > T_u^i, \\ -1 & \text{if } r_i < T_l^i, \end{cases} \tag{32}$$

where T_l^i and T_u^i represent the lower and upper diagnostic threshold assigned to the i-th residual r_i. The set of partial models should be designed in such a way as to cover the whole controlled system.

For example, for the tank system considered in Sect. 7, a set of five single-input single-output models is proposed:

- pump model $\hat{P} = f_{m1}(F_1)$,
- tank model $\hat{L}_1 = f_{m2}(F_1)$,
- valve model $\hat{F}_1 = f_{m3}(dP)$,
- positioner model $\hat{F}_2 = f_{m4}(CV)$,
- process model $\hat{L}_2 = f_{m5}(CV)$,

where f_{m1}, f_{m2}, f_{m3}, f_{m4} and f_{m5} represent nonlinear functions, $\hat{L}_1$ and $\hat{L}_2$ are estimated water levels of the process L_1 and model L_2, respectively, $\hat{F}_1$ and $\hat{F}_2$ are estimated water flows determined using the valve model and positioner model, respectively, and $\hat{P}$ stands for the pressure estimate determined using the pump model. The residual is calculated as the difference between a measured value and an estimated one, e.g., for the pump model it is $r_1 = P - \hat{P}$. Each residual is sensitive to a certain set of faults. Diagnostic matrices are constructed on the basis of the expert knowledge. An example of the multi-valued diagnostic matrix is shown in Table 1. Analysing the table it is clear that residuals r_2 and r_5 provide the same data. However, they still can be useful taking into account their rate of changes. Moreover, we can point out faults which are not distinguishable (the pairs of faults $\{f_3, f_8\}$ and $\{f_6, f_7\}$). Therefore, to definitely improve fault isolation quality, the rate of changes of residuals can be taken into account. One of the possible solutions is to investigate the order in which that symptoms occur [22]. In our case, however, for both pairs of faults, the order of symptoms is the same. Hopefully, we observe that the intensity of a fault may be a key solution of this problem:

$$S(f_k) = \frac{1}{N} \sum_{i:r_i \in R(f_k)} \sum_{j=1}^{N} \frac{r_i(j)}{T_i(j)}, \quad T_i(j) = \begin{cases} T_u^i & \text{for } r_i(j) > T_u^i \\ T_l^i & \text{for } r_i(j) < T_l^i \end{cases}, \tag{33}$$

where $S(f_k)$ is the size of the fault f_k, $R(f_k)$ represents the set of residuals sensitive to the occurrence of f_k, N is the length of the moving window used to perform fault diagnosis and to calculate fault intensity (33). Summarizing, fault diagnosis is composed of two phases. The purpose of the first phase is to analyze the diagnostic matrix, e.g., Table 1. The second phase is carried out only if certain faults are not distinguishable.

Table 1 Multi-valued diagnostic matrix

r	f							
	f_1	f_2	f_3	f_4	f_5	f_6	f_7	f_8
r_1	0	0	−1	−1	0	0	0	−1
r_2	−1	+1	+1	−1	−1	+1	+1	+1
r_3	0	0	−1	−1	0	0	0	−1
r_4	0	0	−1	−1	−1	−1	−1	−1
r_5	−1	+1	+1	−1	−1	+1	+1	+1

6.2 Sensor Fault Estimation

It is obvious that any automatic control system can hide faults from being observed, especially those of a small size. As for actuator faults, predictive control may operate in a right way, sensor faults should be treated in a different way. Let us consider a tank system equipped with the water level sensor. If a fault occurs in the sensor, a wrongly measured water level is transferred to the controller, which tries to change the system output while the real water level is correct. In such cases, the control signal should not be changed. Thus, sensor fault isolation as well as estimation of its size is very important in the fault-tolerant predictive control design process in order to preserve the assumed control performance.

The first step is to point out the set of residuals sensitive to that fault ($R(f_k)$). This set can be further reduced to a subset containing residuals determined by means of isolated sensors only ($\bar{R}(f_k)$). If only one residual is sensitive to fault f_k, its size can be determined with the value of this residual, otherwise the fault size is calculated as the mean value of all residuals belonging to $\bar{R}(f_k)$.

After detecting and isolating a fault, it is time for reconstructing the value measured by the faulty sensor. To this end, a suitable partial model representing the signal under consideration is used. Since the fault is detected and isolated, the output of this partial model replaces the value measured by the faulty sensor.

7 Example: Tank System

Fault-tolerant predictive control presented in the previous section was experimentally verified using the example of a tank system [39, 40]. The whole system consists of two tanks (one of them plays the role of a storage tank), a valve with positioner V_1, a pump and a set of sensors (LT, FT_1, FT_2, PDT, PT). The main tank is in the form of horizontally placed cylindrical vessel. Due to that the static characteristic of the system is highly nonlinear. The tank system was implemented in MATLAB©/Simulink©environment. Figure 14 presents the block scheme of the system considered with measurably available process variables marked. The abbreviation LRC denotes the level regulation controller. The place where the fault was introduced is marked with the cross sign. Table 2 summarizes available process variables.

7.1 Modeling

To design a model of the process, the neural network (29) was employed. The input was the control signal (CV), and the output was the water level in the tank (L). Training data was recorded using open-loop control feeding the process a sequence of random steps with the levels from the interval (0, 100). Additionally, the output

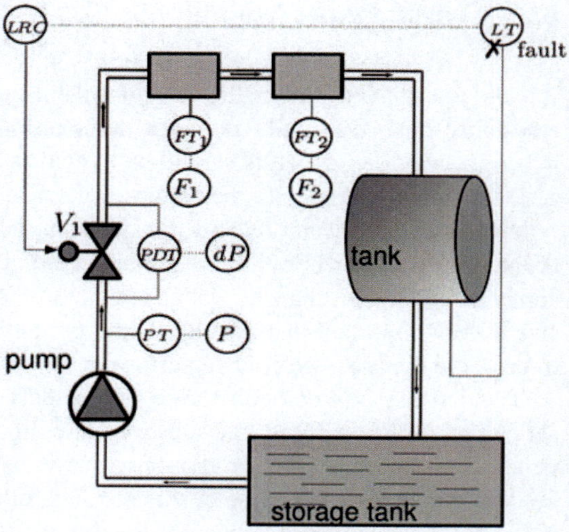

Fig. 14 Tank system block scheme

Table 2 Process variables specification

Variable	Description	Range
CV	Control	0–100 %
dP	Pressure difference V_1	0–275 kPa
P	Pressure before the valve V_1	0–500 kPa
F_1	Flow (electromagnetic flowmeter)	0–5 m^3/h
F_2	Flow (Vortex flowmeter)	0–5 m^3/h
L	Water level	0–0.5 m

data was contaminated with white noise. The structure of the neural model was selected experimentally as a compromise between model complexity and its quality [8]. Eventually, the model had seven hidden neurons ($v = 7$), and the number of delayed inputs and outputs was $n_a = 2$ and $n_b = 2$, respectively. Hidden neurons had the hyperbolic tangent activation function while the output neuron was the linear one. The performance of the model was checked using the sum of squared errors Q_{SSE}. The designed model was used to develop k-step ahead predictors. The results of evaluation of the selected predictors are listed in Table 3. Although it is well known that the approximation error of the one-step ahead predictor affects recursively the prediction quality of k-step ahead predictors, it is clear that the developed predictors mimic the behavior of the process pretty well.

Table 3 Quality of derived k-step ahead predictors

Quality index	k-step ahead predictor					
	$k = 1$	$k = 3$	$k = 5$	$k = 7$	$k = 10$	$k = 15$
Q_{SSE}	10.84	11.44	12.68	14.64	18.61	26.46

7.2 Control

Control system settings were as follows: $N_1 = 1$, $N_2 = 15$, $N_u = 2$, $\rho = 2 \cdot 10^{-5}$. Additionally, there were imposed constraints on the control signal: $\overline{u} = 100$ (maximum admissible value) and $\underline{u} = 20$ (to avoid operation in the dead zone). The control results are shown in Fig. 15. The upper graphic presents reference tracking, and the lower illustrates the derived control signal. The achieved performance of the reference tracking was pretty well ($Q_{SSE} = 73.56$). Comparing these results to the performance of the control system employing the classical PID controller, tracking control performance is about 7% better. What is more, predictive control guarantees a zero error in the steady-state.

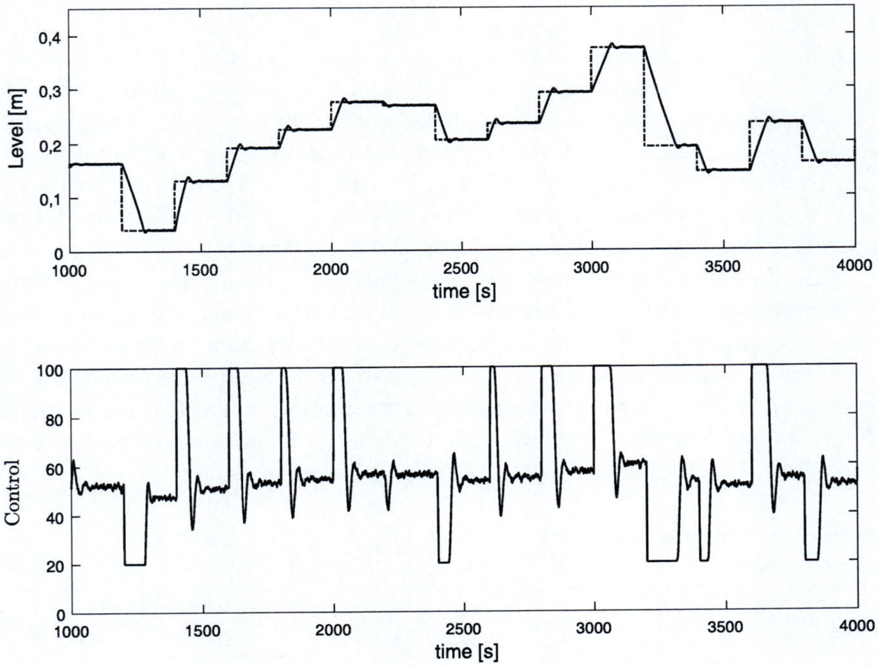

Fig. 15 Water level control results

7.3 Fault Tolerance

The diagnostic module was developed by means of partial models introduced in Sect. 6.1. Each model was designed using the recurrent neural network (29) with a topology illustrated in Fig. 13b). The regression vector was

$$\boldsymbol{x} = [\hat{y}(k), \ldots, \hat{y}(k - n_a + 1), u(k), \ldots, u(k - n_b + 1)]^T. \tag{34}$$

Each model was trained using the parallel identification scheme with data recorded in closed-loop control during the normal work of predictive control. The main characteristics of the partial model are included in Table 4.

The output of each partial model is used to derive the residual defined as a difference between the model output and the value provided by the sensor. Fault detection was carried out by means of thresholding, described in Sect. 6.1.

Thresholds were determined using the rule-of-thumb (three-sigma method) [37]. The purpose of the experiment was detection and isolation of the fault in the water level sensor LT. The fault was simulated as an additive one. Simply, the measured water level was increased by a value equal to −0.05. Based on the expert knowledge, a fault signature in this case is

$$s = [0 \quad -1 \quad 0 \quad -1]. \tag{35}$$

The fault was simulated at the 300-h second of simulation. The detection time was $t_d = 1$ s, while the isolation time was $t_i = 2.85$ s. In the case considered, the work of the control system should not change as the sensor fault does not influence the system itself. Then, reconstruction of the measured value of the water level in the tank is carried out beginning with fault isolation. In order to do this, the estimated water level $\hat{L}$ derived by means of a partial tank model is used. The difference between measured L and estimated $\hat{L}$ water levels is $dL = -0.0459$.

Tank control with and without fault accommodation is illustrated in Fig. 16a and b, respectively. Clearly, the absence of fault accommodation leads to ill functioning of the system, see Fig. 16b. The real water level may be too high and in consequence may lead to serious problems, e.g., to overflow of the water in open tanks. On the other hand, the fault accommodation mechanism makes it possible to maintain proper functioning of the system (Fig. 16a). As a matter of fact, it was not possible to achieve

Table 4 Partial models specification

Model	v	n_a	n_b	σ_h	σ_o	t_s [s]	T_l	T_u
Pump	3	1	1	tanh	Linear	0.05	−0.0072	0.0178
Tank	10	2	2	tanh	Linear	0.05	−0.0023	−0.000195
Valve	5	2	1	tanh	Linear	0.05	−0.0166	0.007
Positioner	5	1	1	tanh	Linear	0.05	−0.0245	0.0093

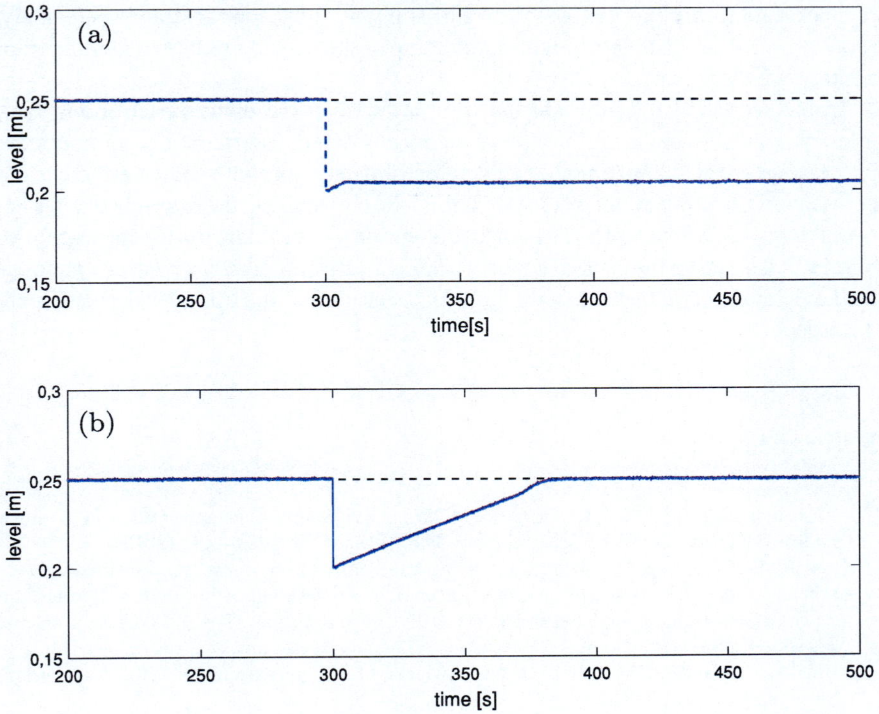

Fig. 16 Control in the presence of a water level sensor fault. The value measured by the sensor: with (**a**) and without (**b**) fault accommodation

the ideal compensation ($dL \neq -0.05$), because fault accommodation was turned on after about 3 s. This time-lag was caused by the processing time of the diagnostic module.

8 Concluding Remarks

This chapter gave an overview of selected problems and issues encountered in the design of fault tolerant control systems using analytical and soft computing methods and approaches. The basic classes of such systems were considered, namely, active and passive ones. Within these classes, control system structures with virtual actuators and sensors were discussed. An important issue of controller reconfiguration using an estimator for a discrete nonlinear system was studied. The problem of fault compensation with the knowledge about the location and size of a fault was also discussed. The fault tolerant control approach was illustrated with an example of a two tank laboratory system. The results of applying three control strategies, namely, robust control, robust control with fault compensation, and predictive fault tolerant control, were compared.

In the second part of the chapter, the problems of predictive control and fault tolerant control using neural networks were considered. Neural models of nonlinear processes were also implemented by feedforward networks with delays and recurrent networks. The problem of fault localization was shown using the example of a partial neural model of a flow tank system, and its solution was developed with the diagnostic matrix. Also, using the tank system example, a neural predictive control system as well as a fault tolerant system with fault compensation of the water level sensor were discussed.

References

1. Alessandri, A., Cervellera, C., Gaggero, M.: Predictive control of container flows in maritime intermodal terminals. IEEE Trans. Control Syst. Technol. **21**, 1423–1431 (2013)
2. Amato, F., Cosentino, C., Mattei, M., Paviglianiti, G.: A mixed direct/functional redundancy scheme for the FDI on a small commercial aircraft. In: Proceedings of the 5th IFAC Symposium on Fault Detection, Supervision and Safety of Technical Processes, SAFEPROCESS 2003, pp. 331–339 (2003)
3. Blanke, M., Kinnaert, M., Lunze, J., Staroswiecki, M.: Diagnosis and Fault-Tolerant Control. Springer, New York (2016)
4. Camacho, E.F., Bordóns, C.: Model Predictive Control, 2nd edn. Springer, London (2007)
5. Chen, J., Patton, R.J.: Robust Model Based Fault Diagnosis for Dynamic Systems. Kluwer Academic Publishers, London (1999)
6. Chunyan, G., Qing, Z., Guang-Ren, D.: Multiplicative fault estimation for a type of nonlinear uncertain system. In: Chinese Control Conference, pp. 4355–4360 (2011)
7. Cybenko, G.: Approximation by superpositions of a sigmoidal function. Math. Control Signals Syst. **2**, 303–314 (1989)
8. Czajkowski, A., Patan, K., Szymański, M.: Application of the state space neural network to the fault tolerant control system of the PLC-controlled laboratory stand. Eng. Appl. Artif. Intell. **30**, 168–178 (2014)
9. Ducard, G.J.J.: Fault-tolerant Flight Control and Guidance Systems: Practical Methods for Small Unmanned Aerial Vehicles. Advances in Industrial Control. Springer, London (2009)
10. Dziekan, L., Witczak, M., Korbicz, J.: Active fault-tolerant control design for Takagi-Sugeno fuzzy systems. Bull. Pol. Acad. Sci. Tech. Sci. **59**(1), 93–102 (2011)
11. Gao, C., Duan, G.: Robust adaptive fault estimation for a class of nonlinear systems subject to multiplicative faults. Circuits Syst. Signal Process. **31**(6), 2035–2046 (2012)
12. Gillijns, S., De Moor, B.: Unbiased minimum-variance input and state estimation for linear discrete-time systems. Automatica **43**, 111–116 (2007)
13. Hammouri, H., Kabore, P., Othman, S., Biston, J.: Observer-based approach to fault detection and isolation for nonlinear systems. IEEE Trans. Autom. Control **44**(10), 1879–1884 (1999)
14. Hammouri, H., Kinnaert, M., Yaagoubi, E.H.E.: Observer-based approach to fault detection and isolation for nonlinear systems. IEEE Trans. Autom. Control **44**, 1879–1884 (1999)
15. Haykin, S.: Neural Networks: A Comprehensive Foundation, 2nd edn. Prentice-Hall, New Jersey (1999)
16. Isermann, R.: Fault Diagnosis Systems. An Introduction from Fault Detection to Fault Tolerance. Springer, New York (2006)

17. Isermann, R.: Fault Diagnosis Applications: Model Based Condition Monitoring, Actuators, Drives, Machinery, Plants, Sensors, and Fault-tolerant Systems. Springer, Berlin (2011)
18. Keller, J.Y., Darouach, M.: Two-stage Kalman estimator with unknown exogenous inputs. Automatica **35**(2), 339–342 (1999)
19. Kemir, K., Ben Hmida, F., Ragot, J., Gossa, M.: Novel optimal recursive filter for state and fault estimation of linear systems with unknown disturbances. Int. J. Appl. Math. Comput. Sci. **21**(4), 629–638 (2011)
20. Koerber, A., King, R.: Combined feedback-feedforward control of wind turbines using state-constrained model predictive control. IEEE Trans. Control Syst. Technol. **21**, 1117–1128 (2013)
21. Korbicz, J., Kościelny, J. (eds.): Modeling, Diagnostics and Process Control: Implementation in the DiaSter System. Springer, Berlin (2010)
22. Korbicz, J., Kościelny, J., Kowalczuk, Z., Cholewa, W. (eds.): Fault Diagnosis. Models, Artificial Intelligence, Applications. Springer, Berlin, Heidelberg (2004)
23. Kościelny, J.M., Bartyś, M., Syfert, M.: Method of multiple fault isolation in large scale systems. IEEE Trans. Control Syst. Technol. **20**, 1302–1310 (2012)
24. Kouvaritakis, B., Rossiter, J.A., Schuurmans, J.: Efficient robust predictive control. IEEE Trans. Autom. Control **45**(8), 1545–1549 (2000)
25. Ławryńczuk, M.: Computationally Efficient Model Predictive Control Algorithms. A Neural Network Approach. Studies in Systems, Decision and Control, vol. 3. Springer, Switzerland (2014)
26. Li, H., Zhao, Q., Yang, Z.: Reliability modeling of fault tolerant control systems. Int. J. Appl. Math. Comput. Sci. **17**(4), 491–504 (2007)
27. Liang, Y., Liaw, D., Lee, T.: Reliable control of nonlinear systems. IEEE Trans. Autom. Control **45**(4), 706–710 (2000)
28. Liao, F., Wang, J., Yang, G.: Reliable robust flight tracking control: a LMI approach. IEEE Trans. Control Syst. Technol. **10**(1), 76–89 (2000)
29. Maciejowski, J.: Predictive Control with Constraints. Prentice-Hall, Harlow (2002)
30. Madria, S., Kumar, V., Dalvi, R.: Sensor cloud: a cloud of virtual sensors. IEEE Softw. **31**(2), 70–77 (2014)
31. Mahmoud, M., Jiang, J., Zhang, Y.: Active Fault Tolerant Control Systems: Stochastic Analysis and Synthesis. Springer, Berlin (2003)
32. Narendra, K.S., Parthasarathy, K.: Identification and control of dynamical systems using neural networks. IEEE Trans. Neural Netw. **1**, 12–18 (1990)
33. Nobrega, E., Abdalla, M., Grigoriadis, K.: Robust fault estimation of unceratain systems using an LMI-based approach. Int. J. Robust Nonlinear Control **18**(7), 1657–1680 (2008)
34. Nørgaard, M., Ravn, O., Poulsen, N., Hansen, L.: Networks for Modelling and Control of Dynamic Systems. Springer, London (2000)
35. Noura, H., Theilliol, D., Ponsart, J., Chamseddine, A.: Fault-Tolerant Control Systems: Design and Practical Applications. Springer, Berlin (2003)
36. Patan, K.: Artificial Neural Networks for the Modelling and Fault Diagnosis of Technical Processes. Lecture Notes in Control and Information Sciences. Springer, Berlin (2008)
37. Patan, K.: Neural network based model predictive control: fault tolerance and stability. IEEE Trans. Control Syst. Technol. **23**, 1147–1155 (2015)
38. Patan, K.: Robust and Fault-Tolerant Control. Neural-Network-Based Solutions. Studies in Systems, Decision and Control. Springer, London (2019)
39. Patan, K., Korbicz, J.: Nonlinear model predictive control of a boiler unit: a fault tolerant control study. Int. J. Appl. Math. Comput. Sci. **22**(1), 225–237 (2012)
40. Patan, K., Korbicz, J.: Sensor fault estimation in the framework of model predictive control. Boiler case study. In: Proceedings of the 8th IFAC Symposium on Fault Detection, Supervision and Safety of Technical Processes, SAFEPROCESS 2012 Mexico City, Mexico (2012) (CD-ROM)
41. Pawlowski, A., Cervin, A., Guzmán, J.L., Berenguel, M.: Generalized predictive control with actuator deadband for event-based approaches. IEEE Trans. Ind. Inform. **10**(1), 523–537 (2014)

42. Persis, C.D., Santis, R.D., Isidori, A.: An $\mathcal{H}_\infty$-suboptimal fault detection filter for bilinear systems. In: Proceedings of the Nonlinear Control in the Year 2000, pp. 331–339 (2000)
43. Persis, C.D., Santis, R.D., Isidori, A.: Nonlinear actuator fault detection and isolation for a VTOL aircraft. In: Proceedings of the 2001 American Control Conference, 25–27 June 2001
44. Pratt, J., Torres, A., Dilworth, P., Pratt, G.: Virtual actuator control. In: Proceedings of the 1996 IEEE/RSJ International Conference on Intelligent Robots and Systems, vol. 3, pp. 1219–1226 (1996)
45. Prodan, I., Olaru, S., Stoica, C., Niculescu, S.: Predictive control for trajectory tracking and decentralized navigation of multi-agent formations. Int. J. Appl. Math. Comput. Sci. **23**(1), 91–102 (2013)
46. Qu, Z., Ihlefeld, C., Yufang, J., Saengdeejing, A.: Robust fault-tolerant self-recovering control of nonlinear uncertain systems. Automatica **39**(10), 1763–1771 (2003)
47. Rotondo, D., Nejjari, F., Puig, V.: A virtual actuator and sensor approach for fault tolerant control of LPV systems. J. Process Control **24**(3), 203–222 (2014)
48. Rotondo, D., Puig, V., Nejjari, F., Romera, J.: Fault-hiding approach for the switching quasi-LPV fault tolerant control of a four-wheeled omnidirectional mobile robot. IEEE Trans. Ind. Electron. **62**(6), 3932–3944 (2015)
49. Seron, M.M., De Doná, J., et al.: Fault tolerant control using virtual actuators and invariant-set based fault detection and identification. In: Proceedings of the 48th IEEE Conference on Decision and Control, 2009 held Jointly with the 2009 28th Chinese Control Conference. CDC/CCC 2009, pp. 7801–7806. IEEE (2009)
50. Shi, F., Patton, R.: A robust adaptive approach to wind turbine pitch actuator component fault estimation. In: International Conference on Control (UKACC), pp. 468–473. IEEE (2014)
51. Shin, J.Y., Kurdahi, F., Dutt, N.: Cooperative on-chip temperature estimation using multiple virtual sensors. IEEE Embed. Syst. Lett. **7**(2), 37–40 (2015)
52. Tan, Y.: Advances in computational intelligence for fault diagnosis systems. Int. J. Appl. Math. Comput. Sci. **14**(1), 63–68 (2004)
53. Tatjewski, P.: Advanced Control of Industrial Processes: Structures and Algorithms. Advances in Industrial Control. Springer, London (2007)
54. Veluvolu, K., Kim, M., Lee, D.: Nonlinear sliding mode high-gain observers for fault estimation. Int. J. Syst. Sci. **42**(7), 1065–1074 (2011)
55. Witczak, M.: Advances in model-based fault diagnosis with evolutionary algorithms and neural networks. Int. J. Appl. Math. Comput. Sci. **16**(1), 85–99 (2006)
56. Witczak, M.: Modelling and Estimation Strategies for Fault Diagnosis of Non-linear Systems. Springer, Berlin (2007)
57. Witczak, M.: Fault Diagnosis and Fault-Tolerant Control Strategies for Non-Linear Systems: Analytical and Soft Computing Approaches. Springer International Publishing, Heidelberg, Germany (2014)
58. Witczak, M., Buciakowski, M., Aubrun, C.: Predictive actuator fault-tolerant control under ellipsoidal bounding. Int. J. Adapt. Control Signal Process **30**(2), 375–392 (2015)
59. Witczak, M., Buciakowski, M., Puig, V., Rotondo, D., Nejjari, F.: An LMI approach to robust fault estimation for a class of nonlinear systems. Int. J. Robust Nonlinear Control **26**(7), 1530–1548 (2015)
60. Witczak, M., Puig, V., de Oca, S.: A fault-tolerant control strategy for non-linear discrete-time systems: application to the twin-rotor system. Int. J. Control **86**(10), 1788–1799 (2013)
61. Witczak, P., Witczak, M., Korbicz, J., Aubrun, C.: A robust predictive actuator fault-tolerant control scheme for Takagi-Sugeno fuzzy systems. Bull. Pol. Acad. Sci. Tech. Sci. **63**(4), 93–102 (2015)
62. Zhang, X., Polycarpou, M., Parisini, T.: Fault diagnosis of a class of nonlinear uncertain systems with Lipschitz nonlinearities using adaptive estimation. Automatica **46**(2), 290–299 (2010)
63. Zhang, Y., Jiang, J.: Bibliographical review on reconfigurable fault-tolerant control systems. In: IFAC Symposium Fault Detection Supervision and Safety of Technical Processes, SAFE-PROCESS, pp. 265–276. Washington, D.C., USA (2003)
64. Zhang, Y., Jiang, J.: Bibliographical review on reconfigurable fault-tolerant control systems. Annu. Rev. Control **32**(2), 229–252 (2008)

Systems Approach in Complex Problems of Decision-Making and Decision-Support

Jerzy Józefczyk and Maciej Hojda

Abstract The chapter introduces a unified representation of decision-making and decision-support in the form of a decision system with a distinct decision-making plant and a distinct decision-making algorithm. The used systems approach is exemplified by such complex problems of decision-making or decision-support where the sought decisions are interconnected. The investigated cases concern the determination of optimal decisions in technological systems. The joint problem of scheduling the spatially deployed tasks, together with a collision-free control of executors to minimize the total execution time is the one most extensively presented. Proactive and reactive decision-making is considered with the use of an uncertain approach for the former case. The connection between this problem and that of multi-robot task allocation is indicated. Integrated planning of production as well as raw material and product transportation in a supply chain to minimize the total cost serves as an example of a complex decision-support problem. Joint admission control and rate allocation in computer networks is included as the last example of complex decision-making problems. Numerical examples appended to all of the considered problems confirm the advantage of the systems approach over methods assuming separate determination of component decisions.

1 Introduction

The issue of decision-making is extensive, and examples of adequate topics with possible different names can be found in many scientific disciplines and relevant applications. Decision-making has a strong interdisciplinary character. However, it

J. Józefczyk (✉) · M. Hojda
Department of Computer Science and Systems Engineering, Wroclaw University of Science and Technology, Łukasiewicz St. 5, 50-370 Wrocław, Poland
e-mail: jerzy.jozefczyk@pwr.edu.pl

M. Hojda
e-mail: maciej.hojda@pwr.edu.pl

P. Kulczycki et al. (eds.), *Automatic Control, Robotics, and Information Processing*, Studies in Systems, Decision and Control 296,
https://doi.org/10.1007/978-3-030-48587-0_19

is not the intention of this chapter to present a comprehensive discussion on decision-making from the perspective of different scientific disciplines. Considerations are limited to and focused on selected aspects of decision-making. Understanding the decision as information processing for a planned activity together with considering only normative decision-making are the two factors constituting the scope of this chapter. The latter factor assumes access to all the necessary information concerning decision-making. We chose not to examine many vital aspects of decision-making investigated in different research disciplines such as psychology or cognitive science. The connection of the notion of 'decision' with information processing inherently places further considerations in the area of information science and engineering as well as information systems.

The chapter presents a unified plant-based approach for the description of a decision process. As a consequence, the notion of a 'decision-making system' is provided, and it comprises both a decision-making subject and its object, that is a decision-maker. From here on, the considerations are limited to a case with a specific information system, rather than a human as the decision-maker. Such a system can decide as a result of processing the input information on a decision-making plant and on its environment. From an information system point of view, it is crucial to distinguish between the decision-making system and decision-support system. The latter is a vital class of information systems present in a wide range of real-world applications. Many individual activities have to be carried out to design both decision systems. Among these, formulating a decision-making plant's formal representation in the form of a mathematical model and designing the decision-maker's functioning are the most important. To obtain a specific information system, it is necessary to involve software engineers, and, in the case of decision-support system, to properly include the crucial role of users. The current work focuses on the second aforementioned activity, i.e. on designing a functioning decision-maker, so on methods and algorithms of decision-making. In the considerations, we omit the discussion on implementation issues, and representation of a decision-making plant in the form of a mathematical model is taken into account only as far as it is necessary to understand the essence of the presentation.

Operations research, control theory, and optimization are substantial sources of methods and algorithms for the outlined scope of considerations. We will refer to and use selected results coming from the aforementioned research areas while presenting selected decision problems. In the chapter, we present several selected problems of decision-making for different applications. Those problems have been investigated, in recent years, by the Division of Intelligent Decision Support Systems at the Wroclaw University of Science and Technology. The decision problem has been selected based on its significant complexity, which means that more than a single decision should be made, and that a simple decomposition might result in substantial worsening of the outcome. Systems paradigm was chosen as the basis of further considerations, which means that taking into account, in a decision-making process, connections among individual decisions adds a specific value which improves the quality of a jointly made complex decision. The systems paradigm seems to be self-evident. For instance, a similar opinion can be encountered among the statements of

Aristotle, who said that "the whole is greater than the sum of its parts." However, a paradigm of reductionism has also been present in the scientific research up to and including the current times. It postulates the examination of a whole via its decomposition into individual parts and their independent consideration. Knowledge about the whole is then the sum of knowledge of the individual parts.

The presentation in the chapter is composed as follows. We explain the preliminaries in the second sub-chapter. The selected problem of task scheduling on unrelated executors combined with the control of their movement is the content of the next sub-chapter. The determination of optimal decision requires solving a difficult discrete-continuous optimization problem. Different versions of the joint decision-making are investigated depending on available information about the decision-making plant. Apart from reactive decision-making, the proactive case which entails handling some parameters as non-deterministic variables is also considered. The more straightforward joint problem of the combined raw material allocation along with the transportation of raw material and the resulting product is presented in the fourth sub-chapter. It is an example of decision-support as decisions are subject to the approval of a user. The problem represents real-world applications in supply chains in the alimentary industry. The third problem, presented in the fifth sub-chapter, deals with decision-making in modern computer networks where some management activities can be performed automatically. The presented approach for the selected problem refers to the analogous works in the area of economic systems. Conclusions drawn from the results together with an indication of other engaging cases are included in the sixth sub-chapter.

Presented considerations, in general, and the notation used, in particular, refer to earlier works which have been performed independently. Therefore, some notations might be repeated for different discussed problems unless it could cause misunderstandings. The complete unification of the notation would obstruct referring the presented issues to referenced and more extensively cited works.

2 Preliminaries

Let us start the formulation of the notions 'decision-making system' (DMS) and 'decision-support system' (DSS) with a decision-making subject which is considered as an input-output plant and called a decision-making plant (DMP) with separate inputs and outputs, Fig. 1. A decision is denoted by a vector $d = [d_1, d_2, \ldots, d_j, \ldots, d_n]^T$, consisting of n component decisions d_j and belonging to a set of admissible decisions $\mathbf{D}$, i.e. $d \in \mathbf{D}$. A decision-making plant is also characterized by a vector of information $w = [w_1, w_2, \ldots, w_k, \ldots, w_l]^T, w \in \mathbf{W}$ which describes the plant and is substantial in the process of decision-making, and where $\mathbf{W}$ is a set of admissible information. An output vector $q = [q_1, q_2, \ldots, q_i, \ldots, q_m]^T, q \in \mathbf{Q}$ serves as a criterion, i.e. the evaluation of decision d, where q_i, m are the current criterion and the number of criteria, respectively, and $\mathbf{Q}$ is a set of admissible vector evaluations. For $n = 1$ and (or) $m = 1$, it

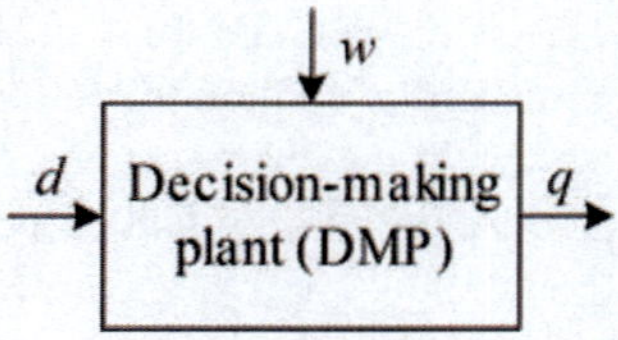

Fig. 1 Decision-making plant

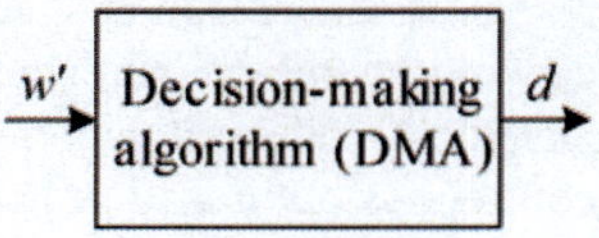

Fig. 2 Decision-making algorithm

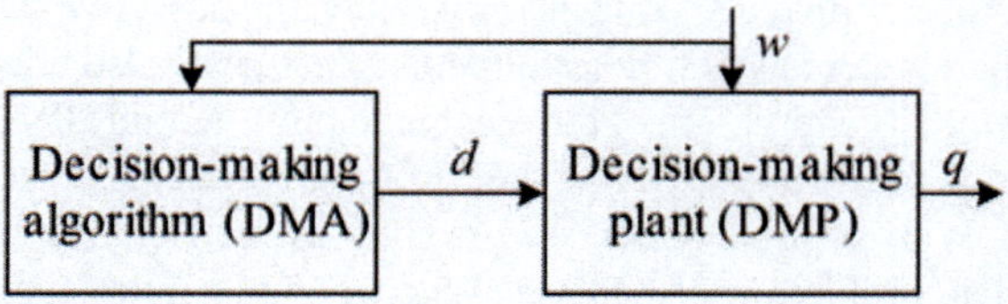

Fig. 3 Decision-making system using information on w

is enough to make a single decision, and (or) a single criterion evaluates this decision. Any mathematical expression, e.g. function, relation, logic expression, which establishes the relationship between inputs d, w and output q, can be used as a model of DMP. A decision-making algorithm (DMA), acting as a decision-maker in information systems, can also be presented in the form of an input-output plant with a decision d as the output and the vector w' as the input which is used for making a decision. The latter vector can be an equivalent of w or its part. It can also contain values of previous decisions or their evaluation in the case of multiple decision-making, Fig. 2. Both introduced input-output plants constitute a decision-making system (DMS) of different structures depending on information used by DMA. The most typical structure of DMS is shown in Fig. 3 where $w' \equiv w$.

Let us mention two other well-known structures: the open structure without full information on w as well as the closed structure representing sequenced decision-making where information on the results of previous decisions is employed. Single decision-making processes like designing, planning, or management are carried out in open structures. Closed structures are typical for control, which can be considered as a particular case of decision-making where vector w or a part of w represents disturbances affecting the control plant.

It is worth noting that a decision made by DMA at control systems is directly implementable without a consultation with a user of such a system. In this case, the user's role is limited to stating the problem, preparing the data, and designing the DMA. Decision-support systems permit a more active role of their users, allowing for acceptance or rejection of decisions determined by DMA. The implementation of a decision is dependent on the user's agreement. Lack of the corresponding agreement results in waiving the activity or redetermination of DMA, possibly for other conditions, defined by the user anew. The aforementioned activities of designing, planning, and management are usually implemented as decision-making. Descriptions of the considered problems will refer to the block-scheme shown in Fig. 3. The complexity, the usage of systems thinking, and the connection with operations research are the main factors behind the choice of problems presented in the subsequent sub-chapters.

3 Joint Task Scheduling and Executor Movement Control

Content of this sub-chapter adopts and expands individual results presented in [18, 27, 29, 33].

3.1 Two-Level Decision-Making Algorithm

In classic deterministic task scheduling problems, task execution times (as performed by executors, machines, or processors) constitute the basic information used in decision-making (determination of schedules) which is assumed to be given a priori. In the considered complex decision-making problem, this assumption does not hold when task execution times are a result of other decisions. Such a situation takes place when tasks are executed in different but known locations (workstations) to which the executors are required to drive-up to (move) to perform the specified task. Ergo, each task consists of two parts: driving up to the workstation (executor movement) and job execution. Furthermore, drive-up time is a result of the control carried out over the executor's movement mechanism.

In the considered case, the decision—the schedule—which for each task determines the corresponding executor and the starting moment (time) of the execution, is equivalent to determining separate and independent routes for the executors. Routes together with the aforementioned movement control mechanisms for individual executors constitute the decision variable set. It is imperative to emphasize the relation between the elements of that set. Firstly, routes determine the starting and ending positions of the executor movement control. Secondly, drive-up times, treated as intermediate results of movement control, are the data for the route determination problem. The object of decision-making is the group of moving (mobile) executors, e.g. mobile robots or autonomous vehicles (the applications will be further expanded near the end of this sub-chapter).

We introduce the notation to formalize the decision-making problem. Fundamentally, this complex problem is related to classic scheduling of non-preemptive tasks and independent executors to minimize the makespan, as presented in [44] and expanded upon in [25, 49]. Let $\mathbf{J} = \{J_1, J_2, \ldots, J_j, \ldots, J_n\}$ be the set of n independent and non-preemptive tasks J_j, $j = 1, 2, \ldots, n$, where j is the index of the current task. Unrelated parallel executors M_i make up the set $\mathbf{M} = \{M_1, M_2, \ldots, M_i, \ldots, M_m\}$ where i and m are the index of executor M_i and their number, respectively. Tasks are executed at the workstations—a single task at a single workstation. To label the workstations, we use the set $\mathbf{J}$. The depot, which is the beginning and the end of the route for each executor, is separate from $\mathbf{J}$. We denote as $\bar{\mathbf{J}} = \mathbf{J} \cup J_{n+1}$ the set of all workstations and the depot, where $n+1$ is the index of the depot. Time p_{ikj} of the execution of task J_j by the executor M_i after driving up from the workstation J_k is an element of the matrix of all task execution times $p = [p_{ikj}]_{\substack{i=\overline{1,m} \\ k,j=\overline{1,n+1}}}$. It is the sum of $\hat{p}_{ikj}$ which is the drive-up time of the executor M_i to the workstation J_j from the workstation J_k, as well as time $\bar{p}_{ij}$ of the execution of the job at the workstation. Consequently, $p_{ikj} = \hat{p}_{ikj} + \bar{p}_{ij}$. Routes $R_i = (r_i(l))_{l=\overline{1,\rho_i}}$, along which the executors move, are ρ_i-long sequences of workstations (their indexes) $r_i(l) \in \{1, 2, \ldots, n\}$ which are joined in a matrix $\gamma = [\gamma_{ikj}]_{\substack{i=\overline{1,m} \\ k,j=\overline{1,n+1}}}$ where $\gamma_{ikj} = 1(0)$ if executor M_i performs a job J_j after driving up from the workstation J_k (if otherwise). Then the routes R_i can be calculated recursively from γ in the following way

$$\gamma_{i,r_i(l-1),r_i(l)} = 1 \rightarrow \exists!_{r_i(l+1)\in\bar{\mathbf{J}}}\gamma_{i,r_i(l),r_i(l+1)} = 1 \text{ and } r_i(0) = n+1. \tag{1}$$

Constraints on the matrix γ guarantee that the solutions are feasible: that all tasks are executed and all routes, which start and end in the depot, are connected. Those constraints define the set $\mathbf{\Gamma}$ of feasible matrices γ, so $\gamma \in \mathbf{\Gamma}$, [27]. It is important to note that the drive-up times depend on the decisions made regarding the control of the movement mechanisms, which are $p_{ikj}(u(t))$ where $u(t) = [u_1(t), \ldots, u_i(t), \ldots, u_m(t)]^{\mathrm{T}}$ is the control vector for all the movement mechanisms for all the executors, and that $t \in [0, C_{\max}]$, where $C_{\max}$ is the total execution time. To calculate $u(t)$, one needs to solve the problem of movement control for all the executors jointly and those executors move within a single workspace. The sub-problem is explained in detail for various movement mechanisms in [18, 27]. Here, we present only the general idea.

The movement mechanism, which is a subject of control (decision-making), is described in the state space defined by vectors of generalized positions and velocities. Let $x_i(t) = [y_i^{\mathrm{T}}(t), \dot{y}_i^{\mathrm{T}}]^{\mathrm{T}}$ be the state vector, and let its components be λ-element vectors of generalized positions and velocities. Then the model in the state space is described as follows

$$\dot{x}_i^{(\nu)}(t) = f_i^{\nu}(x_i(t), u_i(t)), \ \nu = 1, 2, \ldots, 2\lambda, \tag{2}$$

where $u_i(t) = [u_i^{(1)}(t), \ldots, u_i^{(\nu)}(t), \ldots, u_i^{(\lambda)}(t)]^{\mathrm{T}}$ is an λ-element control vector. This vector, the form of which depends on the selected type of model, belongs to the set of feasible controls $\mathbf{D}_u(t)$ defined by:

- interval constraints on the state variables $x_i^{(\nu)}(t) \in \mathbf{X}_i^{\nu} = [\underline{x}_i^{(\nu)}, \bar{x}_i^{(\nu)}]$,
- interval constraints on the control variables $u_i^{(\nu)}(t) \in \mathbf{U}_i^{\nu} = [\underline{u}_i^{(\nu)}, \bar{u}_i^{(\nu)}]$,
- areas $\mathbf{D}_i(t)$ i $\bar{\mathbf{D}}_j$ taken up by the executors and depots (the only stationary obstacles), respectively.

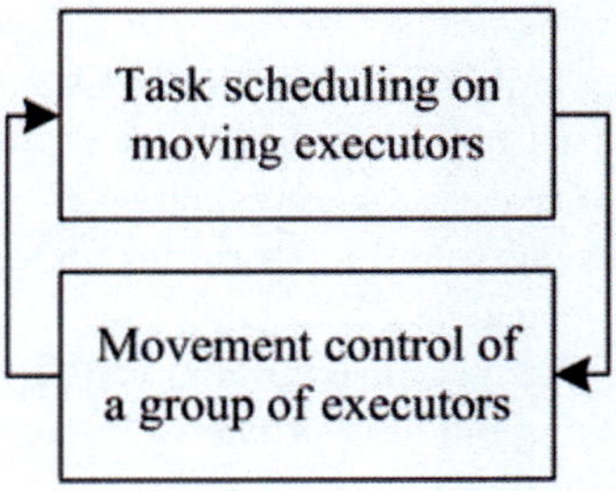

Fig. 4 Structure of the joint problem of route determination and movement control of a group of executors

Movement without collisions requires satisfying two conditions for each executor M_i: $\mathbf{D}_i(t) \cap \mathbf{D}_g(t) = \emptyset,\ g \neq i$ and $\bar{\mathbf{D}}_j \cap \mathbf{D}_i(t) = \emptyset,\ j \in \bar{\mathbf{J}}$ at any given moment t. Collision-free movement control is one of the two interconnected sub-problems, the other one being the route determination formulated as task scheduling. Together, those sub-problems make up the complex joint problem of route determination and movement control of a group of executors, Fig. 4.

Formal dependence between both sub-problems is such that movement control has an impact on the drive-up times $\hat{p}_{ijk}$. On the other hand, the control of the movement of executors $u(t; \gamma)$ depends on the routes. More specifically, the order of visiting workstations determined by γ has an impact on the movement control parameters, that is on starting, ending, and intermediate points of the route as well as the forbidden areas $\mathbf{D}_i(t)$ (so also on $\mathbf{D}_u(t; \gamma)$). In consequence, solving one of the sub-problems changes the data for the other.

We are consequently treating the problem as a generalization of the classic task scheduling problem with a makespan criterion. The following form of the makespan serves as the criterion for the considered complex problem

$$C_{\max}(\gamma, u(t)) = \max_{i=\overline{1,m}} \sum_{j=1}^{n+1} \sum_{k=1}^{n+1} \gamma_{ikj}(\hat{p}_{ikj}(u(t)) + \bar{p}_{ij}),$$

where the argument of the maximum operator is the executor's M_i operating time. Finally, the complex decision-making problem described here consists in determining a pair of decisions $(\gamma, u) \in \mathbf{D}_{\gamma u}(t)$ which minimizes $C_{\max}(\gamma, u(t))$, where $\mathbf{D}_{\gamma u}(t) = (\mathbf{\Gamma} \times \mathbf{D}_u(t)) \cap \bar{\mathbf{D}}_{\gamma u}(t)$. Set $\bar{\mathbf{D}}_{\gamma u}(t)$ consists of such pairs of decisions for which the following dependency holds $\gamma_{i,r_i(l),r_i(l+1)} = 1 \rightarrow \bar{\sigma}_i(x(t)) = \bar{y}_{r_i(l)}$, $t \in [\bar{t}_{r_i(l)}, \underline{t}_{r_i(l+1)}]$. The mapping $\bar{\sigma}_i(x(t))$ yields the location of the executor M_i at any given moment t. When the executor is at a workstation $r_i(l)$, its location is limited to $\bar{y}_{r_i(l)}$. This ensures the continuity of routes when the executors travel between subsequent workstations.

Let us assume that given are: the set of executors $\mathbf{M}$, the set of workstations (tasks) $\bar{\mathbf{J}}$ and their locations $\bar{y}_j$ on a surface, execution times $\bar{p}_{ij}$ for the jobs preformed at workstations, and the models of the movement mechanism (2).

Papers [18, 26, 28, 30, 33] presented various algorithms for solving the sub-problems of route determination and movement control of a group of executors. Let us note that the first sub-problem is an NP-hard combinatorial problem and the second problem is typically solved in robotics. In the second case, the literature presents solution algorithms for various movement mechanisms as well as an analysis of special cases.

The fundamental goal of the performed work was designing the algorithms for solving the joint problem. We present two such algorithms here: the first one using a simple decomposition and sequentially solving the resulting sub-problems (AD) and the second, iterative improvement algorithm (AI).

Algorithm AD

1: Set the movement times $\hat{p}^0_{ikj}$.
2: Solve the route determination sub-problem for $\hat{p}_{ikl} = \hat{p}^0_{ikj}$ in order to calculate γ^0.
3: Solve the problem of collision-free movement control of a group of executors – calculate $u^0(\gamma^0)$.

Driving times $\hat{p}^0_{ikj}$ are calculated for each executor M_i separately and for each pair of workstations (k, j). This is done under the assumption that no collisions can occur. In the case of collisions, drive-up times can change significantly $\hat{p}^0_{ikj}$, which is why the values calculated in the step AD can differ from the final values. Makespan $C^{AD}_{\max}(\gamma^0, u^0(\gamma^0))$ is an intermediate result of the algorithm.

In the algorithm AI, both sub-problems are solved iteratively, multiple times, one after the other, until the stop condition is satisfied. The algorithm stops if there are no improvements to the solution in a given number of iterations η. Similarly as in AD, the makespan $C^{AI}_{\max}(\gamma(\theta - \eta), u(\theta - \eta))$ is an intermediate result of the algorithm. Movement control of a group of executors is performed in steps 4-9. Algorithm AOK is explained in more detail in [18]. It stops the movement of executors until they no longer collide with other executors on the route to the next workstation.

Algorithm AI

1: Set $\theta := 0$.
2: Calculate control $u(\theta)$ and movement times $\hat{p}(\theta)$ in a collision-free system where $\hat{p}(\theta)$ is the matrix of all movement times in the iteration θ.
3: Calculate $\gamma(\theta)$ which is a matrix representing the routes for $\hat{p}(\theta)$. Set $\theta := \theta + 1$.
4: Verify if any executors should leave the workstation. If not, then go to Step 8.
5: Calculate control $u(\theta)$ for the next fragment of routes.
6: Verify if there are no collisions. If not, go to Step 8.
7: Apply the collision avoidance algorithm AOK. Modify $u(\theta)$.
8: Perform the control for the next step of the route.
9: Check if the executors finished their movement. If not, go to Step 4.
10: Check if the stop criterion is met. If not, go to Step 3.
11: Let the algorithm stop with the result $(\gamma(\theta - \eta),\ u(\theta - \eta))$ where the matrix $\gamma(\theta - \eta)$ determines the routes of executors as in (1).

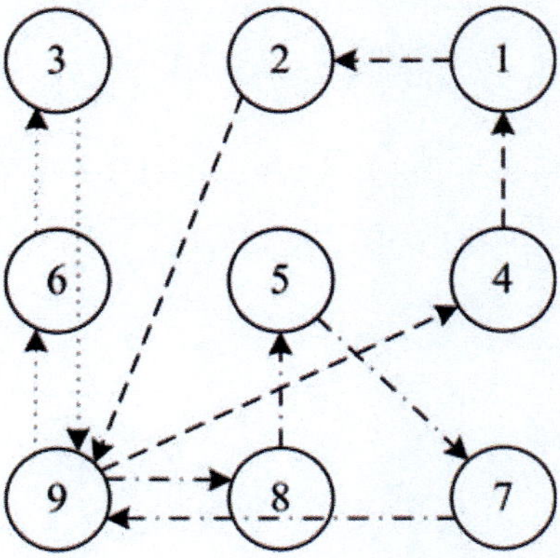

Fig. 5 Results of the AD algorithm

Let us compare both algorithms on a simple example of eight workstations ($n = 8$) and a depot, all located on a grid with $0.5m$ spacing [18]. Tasks are executed by three autonomous E-puck robots ($m = 3$) with two-wheeled movement mechanisms. The following kinematic model was adopted

$$\dot{x}_i^{(1)}(t) = 0, 5(u_i^{(1)}(t) + u_i^{(2)}(t)) \cos(x_i^{(3)}(t)),$$

$$\dot{x}_i^{(2)}(t) = 0, 5(u_i^{(1)}(t) + u_i^{(2)}(t)) \sin(x_i^{(3)}(t)),$$

$$\dot{x}_i^{(3)}(t) = u_i^{(1)}(t) - u_i^{(2)}(t)$$

where $x_i^{(1)}$, $x_i^{(2)}$ are the coordinates of a selected (inner) point of the ith executor in a Cartesian coordinate system, and $x_i^{(3)}$ is the angle between the axis of the robot and the abscissa. Moreover, $u_i^{(1)}$ and $u_i^{(2)}$ are the angular velocities of the right and the left wheel, respectively. Assumed were uniform constraints on control $\bar{u}_i^{(\rho)} = -\underline{u}_i^{(\rho)} = 1$, $\rho = 1, 2$, zeroed starting states $x_i = \bar{0}$, and equal job execution times $\bar{p}_{ij} = 10s$. The routes $R_1 = (9,\ 6,\ 3,\ 9)$, $R_2 = (9,\ 4,\ 1,\ 2,\ 9)$, $R_3 = (9,\ 8,\ 5,\ 7,\ 9)$ and $C_{\max}^{\text{AD}} = 172\ s$ for AD, as well as $R_1 = (9,\ 6,\ 3,\ 2,\ 9)$, $R_2 = (9,\ 5,\ 1,\ 9)$, $R_3 = (9,\ 8,\ 7,\ 4,\ 9)$ and $C_{\max}^{\text{AI}} = 136\ s$ for AI (see Figs. 5 and 6) are the results of both algorithms.

Jointly solving both sub-problems shortened the makespan by over 20%.

3.2 *Reactive Decision-Making*

Algorithms AD and AI work offline, so the decision pair (γ, u) is calculated before the executors start driving, and the decision-making system has an open structure Fig. 7. The information about collisions is encoded in w. The main disadvantage of both algorithms is the lack of information regarding the ongoing state of the

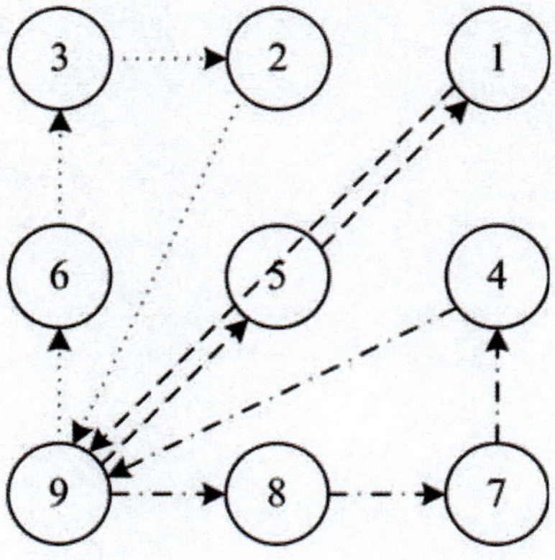

Fig. 6 Results of the AI algorithm

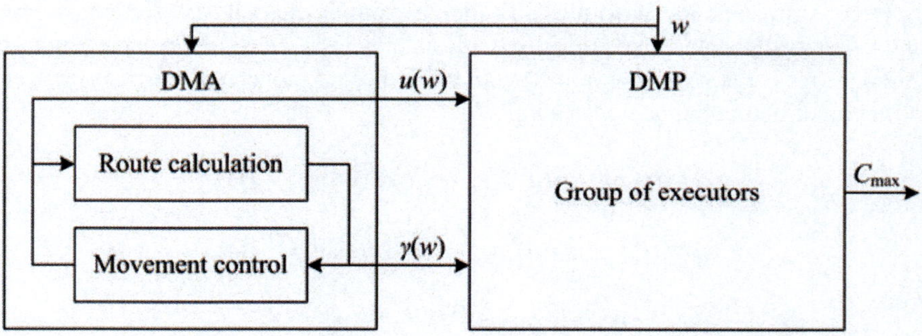

Fig. 7 Open-loop decision-making system

decision-making plant. A closed-loop structure was designed to eliminate that flaw, Fig. 8. A significant advantage of such a structure is its ability to benefit from the information about the DMP during the execution of DMA. Two types of feedback are considered. The first one is the location $y^{\mu}_{r_i(l)}$ of the ith executor moving along the lth fragment of its route. It permits the modification of the movement control which can be necessary due to inaccuracies of the models and disturbances other than collisions. The top index μ is the number of the control iteration. Each iteration has the same fixed length. Clearly, the corrections do not affect the current solution $(\gamma^{\chi}, u^{\chi})$.

The drive-up times of the executors are the second type of feedback. The top index χ in variables in Fig. 8 corresponds to the current event, which is understood as finishing at least one task. In the moment of the event, the factual task execution times are compared to the execution times calculated in Step $\chi - 1$ (starting moments are calculated based on $\hat{p}^0$ in the step AI.2). If those times are equal, the solution procedure is continued without change. Otherwise, the solution $(\gamma^{\chi}, u^{\chi})$ is subject to a corrective procedure. Firstly, the task allocation on moving executors is performed

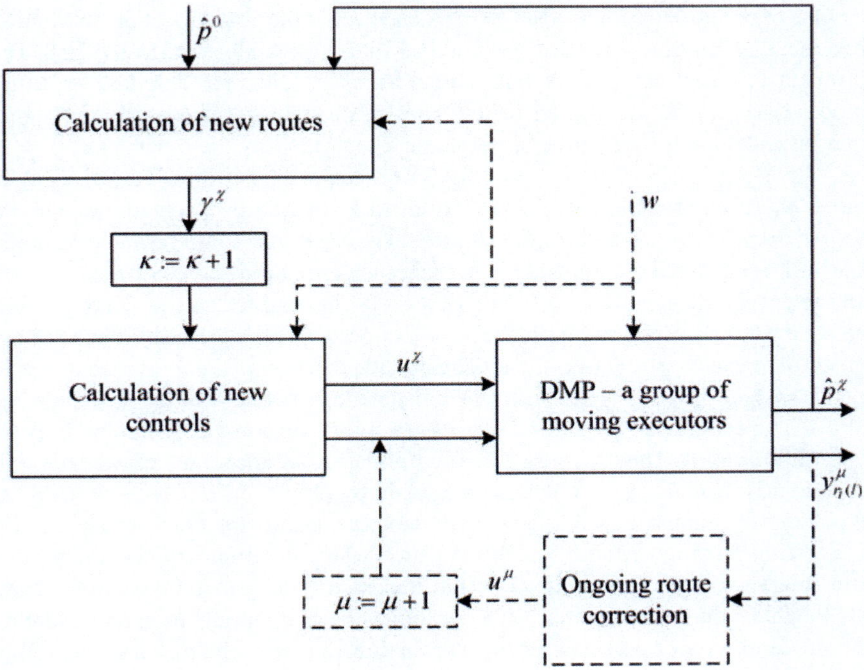

Fig. 8 Closed-loop decision-making system

only for the tasks that have not been started yet. Then the movement control problem is solved and new control values are used.

This approach has been tested in [29, 32]. This closed-loop schedule correction of the schedule γ is often (for other applications) called the reactive rescheduling.

3.3 Decision-Making for Multi-robot Systems

The considered complex decision-making problems also exist in applications of autonomous vehicles, including mobile robots, e.g. in production systems or environments hazardous to humans [8, 11, 22]. A crucial class of those applications consists of the so-called Multi-Robot Task Allocation (MRTA) problems [7, 16, 38]. In the MRTA class, the main division on specific problems is done by considering either Single Task (ST) robots or Multi-Task (MT) robots, Single-Robot (SR) tasks or Multi-Robot (MR) tasks. The former division determines how many tasks can a robot perform at a time. The latter division determines whether tasks are divisible between multiple robots. Finally, one can consider either Instantaneous Assignment (IA) if all the data is available prior to decision-making or Time-extended Assignment (TA) when the data arrives as tasks are under execution.

One of the problems in the MT-MR-IA category is the task allocation on multiple robots under multiple modes of execution – Multi-Task Allocation (MTA) [19]. This problem assumes that there exist executors **M**, tasks **J**, and multiple task execution modes $\mathbf{K} = \{K_1, K_2, \ldots, K_\kappa, \ldots, K_l\}$. The decision whether an executor is assigned to a task along with the mode of execution is provided by the binary matrix $\gamma' = [\gamma'_{ij\kappa}]_{i=\overline{1,m}, j=\overline{1,n}, \kappa=\overline{1,l}}$ where $\gamma'_{ij\kappa} = 1(0)$ if the executor M_i performs the task J_j in the mode K_κ (if otherwise). If $\gamma'_{ij\kappa} = 1$ then the task is partially executed, i.e. with the degree of $\eta_{ij\kappa} \geq 0$. Then the executor uses up $e_{ij\kappa} \geq 0$ of a non-renewable resource. The task is considered completed if it reaches the execution degree of $E > 0$ across all the executors. The limit of resources $F > 0$, separate for each executor, cannot be exceeded. Furthermore, each executor can perform a single task in exactly one mode (if it executes the task at all). The quality criterion is the total resource used $Q(\gamma') = \sum_{i=\overline{1,m}, j=\overline{1,n}, \kappa=\overline{1,l}} \gamma'_{ij\kappa} e_{ij\kappa}$. We denote the optimal solution as y^*.

Such a problem is NP-hard even in its feasibility version, i.e. searching only for feasible solutions. The expected difficulty in finding solutions in a reasonable time encouraged formulating and solving a Substitutive Multi-Task Allocation (SMTA) problem where some leeway is given regarding constraint violation. In the paradigm of decision-making systems, this is a situation where the decision-maker has a certain reserve of resources and can permit their use as long as this extension has a predictable, bounded character. An algorithm using an approximation procedure for the single-executor case was used to solve the substitutive problem. This Single Task Allocation (STA) problem is NP-hard as well. However, it has a 2-approximate solution algorithm which can be used for each task J_j separately. It is denoted as 2RTA(j) and described in detail in [19]. This subroutine is used in the AMR algorithm for the SMTA problem.

Algorithm AMR

1: Let $\gamma' = \bar{0}$, $j = 1$.

2: Calculate $\gamma'_j = [\gamma'_{ij\kappa}]_{i=\overline{1,m}, j=\overline{1,n}, \kappa=\overline{1,l}}$ using the solution for the executor M_j obtained with the use of 2RTA(j).

3: If $j < n$ then set $j := j + 1$ and go to Step 2. Otherwise, end the algorithm.

The solution γ' has the following property. If the solution to the MTA problem exists, then γ' is a feasible solution to the SMTA problem and $Q(\gamma') \leq 2Q(\gamma^*)$ as well as $Q(\gamma') \leq 2mF$.

Let us consider the following computational example. Let $m = 3$, $n = 5$, $l = 2$, $E = 9$, $F = 16$, $\eta_{ij1} = 2j$, $\eta_{ij2} = 3j$, $e_{ij1} = i$, $e_{ij2} = 5i$. The algorithm AMR yields the solution

$$\gamma'_1 = \begin{bmatrix} 0\,1 \\ 0\,1 \\ 0\,1 \end{bmatrix}, \gamma'_2 = \begin{bmatrix} 1\,0 \\ 1\,0 \\ 1\,0 \end{bmatrix}, \gamma'_3 = \gamma'_4 = \begin{bmatrix} 1\,0 \\ 1\,0 \\ 0\,0 \end{bmatrix}, \gamma'_5 = \begin{bmatrix} 1\,0 \\ 0\,0 \\ 0\,0 \end{bmatrix} \text{ and } Q(\gamma') = 43.$$

In the paper [20], the MTA problem is combined with the traveling salesman problem. This is done to ensure communication between spatially distributed workstations. The proposed solution algorithm uses decomposition to simplify the problem and solves the resulting sub-problems in a fixed order.

3.4 Related Problems

So far, the considered task allocation and movement control problems were such that the sub-problem from the upper level of the system depended on the results obtained on the lower level and the generated solution was a consequence of the results obtained at the lower level, Fig. 4. This approach can be called reactive since new routes are a reaction to the change in the drive-up times of the executors. Another concept is to solve those complex decision-making problems using the proactive approach, where it is assumed a-priori that for the determination of routes, the movement times can vary and the whole problem is solved under parametric uncertainty. Introduction of uncertainty eliminates the need for considering the problem at the lower level. The structure is simpler; however, uncertainty complicates the upper-level problem.

When considering decision-making systems with parametric uncertainty, firstly, one has to select the method of representation for the uncertain parameters. Secondly, a new quality criterion is needed—deterministic criteria can no longer be applied directly. Of the many possible approaches to uncertainty, we assume that the so-called interval uncertainty is used. More to the point, we assume that the drive-up times can take values from intervals bordering on values $\hat{p}^-_{ikj}$ and $\hat{p}^+_{ikj}$, so $\hat{p}^-_{ikj} \leq \hat{p}^+_{ikj}$, where $\hat{p}_{ikj} \in [\hat{p}^-_{ikj}, \hat{p}^+_{ikj}]$. Using this approach, we do not need any additional information regarding the uncertain parameters, such as probability distributions or membership functions. In consequence, there is no difference between possible values of the parameters. It is clear that the lack of exact information about the values of parameters forbids using deterministic quality criteria. The most common approach to evaluation is then determinization with respect to unknown parameters.

The determinization procedure used here consists of the calculation of criterion for the worst case of possible uncertain parameter values, and it is performed by the operator at the maximum values. Moreover, the difference $C_{\max}(\gamma; \hat{p}) - C'_{\max}(\hat{p})$, not the criterion for the deterministic case is the argument of the operator. Such a difference is called regret. The expression $C_{\max}(\gamma; \hat{p}) = \max_{i=\overline{1,m}} \sum_{j=1}^{n+1} \sum_{k=1}^{n+1} \gamma_{ikj}(\hat{p}_{ikj} + \bar{p}_{ij})$ and $C'_{\max}(\hat{p}) = \min_{\gamma \in \Gamma} C(\gamma; \hat{p})$ are the makespan and its optimal value for the fixed parameters $\hat{p}_{ikj}$ from the intervals $[\hat{p}^-_{ikj}, \hat{p}^+_{ikj}]$, respectively. The matrix $\hat{p}$ of the values $\hat{p}_{ikj}$ is referred to as a set of all possible scenarios $\hat{\mathbf{P}}$ and is a Cartesian product of drive-up times $\hat{p}_{ikj}$, so $\hat{p} \in \hat{\mathbf{P}}$. Regret, calculated for each $\hat{p}$ and γ, is the difference between the current and the optimal value of the makespan. Then for a fixed decision γ, regret can be calculated for each scenario $\hat{p}$. Using the robust approach, so considering the worst case, consists in determining the highest value of regret, i.e. $z(\gamma) = \max_{\hat{p}\in\hat{\mathbf{P}}}[C_{\max}(\gamma;\, \hat{p}) - \min_{\sigma\in\Gamma} C_{\max}(\sigma;\, \hat{p})]$. This approach leads to a deterministic quality criterion, which is the way in which γ is evaluated.

To obtain the optimal decision γ^*, one has to minimize $z(\gamma)$ regarding $\gamma \in \mathbf{\Gamma}$. As can be seen, calculating γ^* requires solving three nested optimization problems. Just solving the deterministic problem, which is the internal minimization, is an NP-hard combinatorial problem. Except for trivial cases, obtaining solutions in a reasonable time requires using polynomial approximation algorithms or heuristics. For the latter case, Tabu Search and Simulated Annealing were presented in [35, 41]. In the paper [47], and more in-depth in the Ph.D. thesis [46], analytical properties were provided for an analogous problem without drive-up times. Also, an algorithm based on Scatter Search was described. It was shown that, amongst others, there are no approximation algorithms, which is also the case for the problem with moving executors considered here.

To summarize the considerations of this sub-chapter, we point out that the decision-making sub-problem from the upper level of the whole system in the deterministic version also exists as a self-standing problem under the name "routing-scheduling," which is a problem of determining the routes for executors joined with task scheduling. The optimal decision-making presented here applies only to a special case of the "routing-scheduling" problem in which not only other quality criteria (also multiple criteria) are considered but also the executors can be understood more generally, as a device equipped with limited resources or a person with certain skills. For such problems, many applications can be found, not just in complex technical systems but also in healthcare, forestry, or management of planes at airports. More examples can be found in the survey papers [4–6, 17, 39, 42].

4 Joint Allocation and Transportation in a Supply Chain

In the previous sub-chapter, the defining element of the complex decision-making problem was the movement of executors performing different application-dependent tasks. In this sub-chapter, we consider movement but understood as transportation of goods or other resources necessary to perform production or service tasks. Decision-making regarding transportation is intertwined with decisions regarding production and services. We focus on a fragment of the so-called supply chains, which consist of J production units working in parallel, I raw material inventories, and K product inventories. The considered supply chain includes the tasks: transportation of raw material to production units, the production from raw material and the transportation of the product to the inventories (see Fig. 9).

In a supply chain, there are three main interconnected decision-making sub-problems.

– Transportation of raw material amounting to w_i from each inventory i, $i \in \{1, 2, \ldots, I\}$ to each production unit j, $j \in \{1, 2, \ldots, J\}$. The decision is the

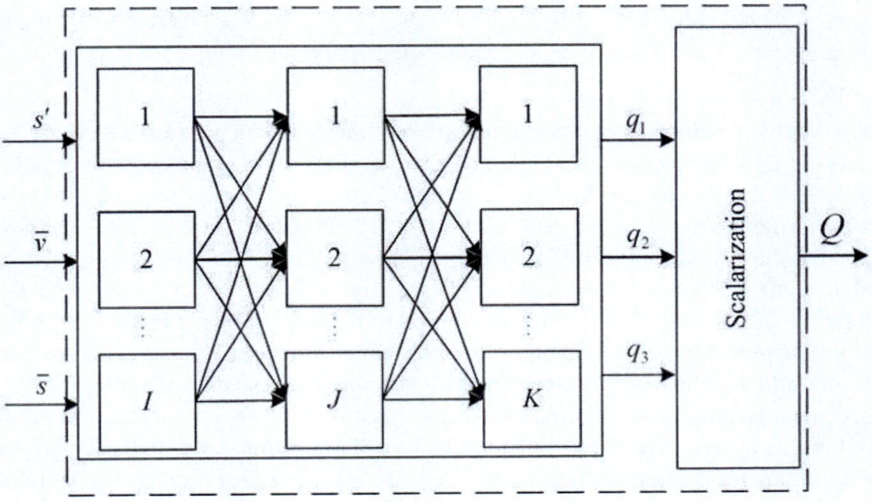

Fig. 9 Supply chain as an input-output decision-making plant

transportation plan presented as the matrix $s' = [s'_{ij}]_{\substack{i=\overline{1,I}\\ j=\overline{1,J}}}$, where $s'_{ij} \in \mathbf{R}_+ \cup \{0\}$ is the non-negative amount of the raw material transported between the ith inventory and the jth production unit.

- Allocation of a global amount of raw material V located in inventories to production units so that $\sum_{j=1}^{J} v_j \leq V$, where v_j is the amount of the raw material provided to the jth production unit. Further, we assume that the current production unit j can, according to its productivity e_j, manufacture the product in the amount $\bar{v}_j = v_j/e_j$ from the raw material provided in the amount of v_j. The decision variable is the distribution of the raw material $v = [v_1,\ v_2,\ \ldots,\ v_J]^{\mathrm{T}}$ or equivalently the amount of product $\bar{v} = [\bar{v}_1,\ \bar{v}_2,\ \ldots,\ \bar{v}_J]^{\mathrm{T}}$.
- Transportation of the product from each production unit j to the inventories of the product $k,\ k \in \{1,\ 2,\ \ldots,\ K\}$ with requirements $\bar{w}_k$. In this sub-problem, the decision variable is the transportation plan of the product represented by the matrix $\bar{s} = [\bar{s}_{jk}]_{\substack{j=\overline{1,J}\\ k=\overline{1,K}}}$, where $\bar{s}_{jk} \in \mathbf{R}_+ \cup \{0\}$ is the non-negative amount of product transported between the jth production unit and the kth inventory of the product.

Execution of the decisions from each sub-problem generates costs which are then used to evaluate those decisions.

- Raw material transportation cost $q_1(\bar{v}, s') = \sum_{i=1}^{I}\sum_{j=1}^{J} c'_{ij}s'_{ij}$, where c'_{ij} is the unit cost of transportation between the ith inventory and the jth production unit.
- Production cost $q_2(\bar{v}) = \pi \max\limits_{j=\overline{1,J}} T_j(\bar{v}_j)$ where $T_j(\bar{v}_j) = T_j(v_j/e_j)$ and π are the production time of the jth production unit processing a raw material in the amount of v_j and the cost incurred each unit of time that the production is carried out. We assume that the production cost is proportional to the production time.

– Product transportation costs $q_3(\bar{v}, \bar{s}) = \sum_{j=1}^{J} \sum_{k=1}^{K} \bar{c}_{jk}\bar{s}_{jk}$, where $\bar{c}_{jk}$ is the unit transportation cost between jth production unit and the kth inventory of the product.

Since the allocation vector v, and consequently, the production size vector $\bar{v}$, are the data for the transportation sub-problems, then solving those sub-problems is possible only after the allocation is known. Sequential solution of the sub-problems, denoted as algorithm A1, is simpler than jointly solving all of the sub-problems. The first and the third sub-problem are classic transportation problems with known solution algorithms. The second sub-problem can be analytically solved for a wide variety of time models T_j, e.g. [27]. Jointly solving all of the sub-problems is done using the algorithm denoted as A2. This algorithm is based on the observation that the "worse" solution to the allocation sub-problem might imply data for the transportation sub-problems yielding lower transportation costs than in the case of the optimal solution of the allocation sub-problem. Such a reduction of transportation costs can be greater than the increase of production costs. Thus, we join the transportation and solution costs into a single, scalarized criterion in the form

$$Q(\bar{v}, s', \bar{s}) = q_1(\bar{v}, s') + q_2(\bar{v}) + q_3(\bar{v}, \bar{s}).$$

The taken model of the production process in which the production cost is proportional to the production time, describes such a type of production when the goal is to minimize its time. This can take place e.g. in food processing of quickly decaying food, such as fruits.

The possibility of adequately solving the problem of minimization of $Q(\bar{v}, s', \bar{s})$ with regards to $\bar{v}, s', \bar{s}$ depends on the form of the functions T_j which express the dependence of the production time on the allocated raw material and the number of production units J. In general, it is only required that T_j are strictly increasing. Exact algorithms were determined for linear functions $T_j = \alpha_j \bar{v}_j = \alpha_j v_j / e_j$, where $\alpha_j \in R_+$, in the case of two production units ($J = 2$) or for larger number of such production units which have identical productivity ($e_j = e$). For the general case, a heuristic evolutionary algorithm was developed [12].

Without characterizing the details of the algorithms, we demonstrate their use on a simple computational example. Let $w_1 = 275$, $w_2 = 225$, $\bar{w}_1 = 100$, $\bar{w}_2 = 80$, $\bar{w}_3 = 70, e_1 = 1/2, e_2 = 2/3, \pi = 1/2000, \gamma_1(v_1) = 0,25v_1^3, \gamma_2(v_2) = v_2, c' = \begin{bmatrix} 3 & 5 \\ 4 & 3 \end{bmatrix}, \bar{c} = \begin{bmatrix} 8 & 7 & 9 \\ 15 & 2 & 9 \end{bmatrix}$. Results of both algorithms are provided in Table 1. Note that using A2 algorithm, which is based on the systems approach, lowered the total cost of transportation and production by 12668 units which make up 18% of the cost obtained with the use of A1. This makes A2 the recommended algorithm.

The considered joint decision-making algorithm was implemented in a decision-support system. The general concept of that computer system is presented in Fig. 10, [12, 13]. The system implemented various exact, approximate, and heuristic solution algorithms. The system is divided into two parts. The first part is used for the one-time calculation of the solution to the joint allocation and transportation sub-problems for

Table 1 Results for the computational example comparing the sequential and the joint algorithm

	A1	A2
$\bar{v}_1$	32.83	53.23
$\bar{v}_2$	217.17	196.77
$s'_{1,1}$	65.66	106.46
$s'_{1,2}$	100.76	70.16
$s'_{2,1}$	0	0
$s'_{2,2}$	225.00	225.00
$\bar{s}_{1,1}$	32.83	53.23
$\bar{s}_{1,2}$	0	0
$\bar{s}_{1,3}$	0	0
$\bar{s}_{2,1}$	67.17	46.77
$\bar{s}_{2,2}$	80.00	80.00
$\bar{s}_{2,3}$	70.00	70.00
$Q(\bar{v}, s', \bar{s})$	70747	58078

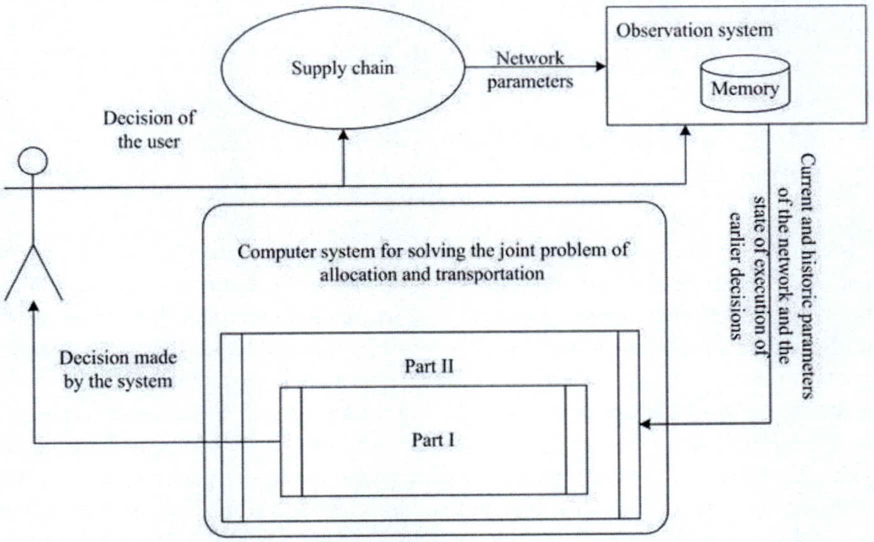

Fig. 10 Architecture of the computer decision-support system for the joint problem of allocation and transportation

the supply networks provided by the user. The user can select whether desired is the optimal, approximate, or a heuristic solution. The first option is assumed by default and if it proves to be unobtainable, then an approximate one with a given approximation coefficient is attempted next. The heuristic solution is only calculated when the search for the optimal or the approximate solution fails. Alternatively, the user can declare that the speed of obtaining the solution should be valued above its quality. In this case, the heuristic solution is selected. If so desired, both, the approximate and the heuristic solutions can be calculated and juxtaposed with regards to the quality of the solution obtained. This first part of the system is used for solving the non-stationary transportation network (stationary in a special case).

The second part of the system is used to calculate an ongoing decision for the non-stationary supply chain based on the current parameters of the network, on the previous decision (regarding transportation), and on its stage of execution. It can be said that the second part of the system expands the first one by considering non-stationary transportation networks. This part of the system can be used as a module in a larger application that also consists of a data processing module which observes the changes in parameters of the transportation network and communicates them to the solution subsystem. The implementation was designed to be a part of an SCM or ERP application.

5 Admission Control and Rate Allocation in Computer Networks

Recently, we can observe the development of existing computer networks and the rise of new network technologies, e.g. content delivery networks, software-defined networks, autonomic networks as well as cloud and edge computing. These modern solutions enable the automation of activities traditionally performed by network operators and providers, which, for example, concern the current network structure, to better address customers' needs. The enormous growth of network customers is undoubtedly the impulse for the development of computer network technologies. The literature contains many works devoted to solving decision-making and decision-support problems for the application in the mentioned types of computer networks, e.g. [37, 48, 50]. We describe in this sub-chapter selected traditional decision-making problems significant for computer networks for which we have applied the systems approach. This approach has been presented for the first time in [14] and then developed in other works conducted in the Division of Intelligent Decision Support Systems at the Wroclaw University of Science and Technology, e.g. [9, 15]. The mentioned problems are admission control (AC) and rate allocation (RA), which will further be referred to as sub-problems of the joint problem of admission control and rate allocation, the joint problem in brief. Both adequately solved connected sub-problems prevent the congestion in computer networks, i.e. the state of the network when the amount of sent data might be higher than the avail-

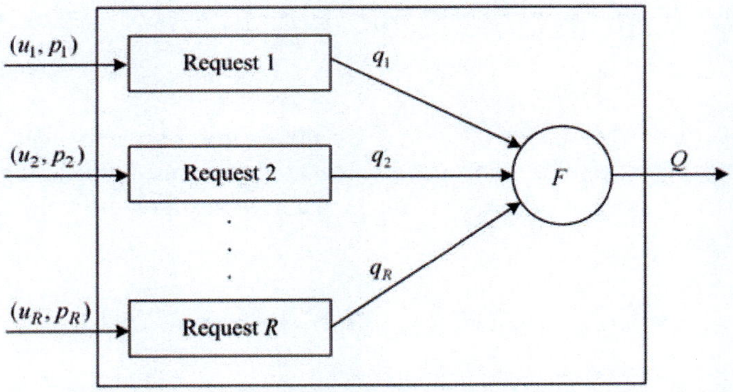

Fig. 11 Computer network as an input-output plant

able capacity. Computer networks are usually required to guarantee the quality of service (QoS) which means the minimum rate control or the minimum amount of data possible to be sent in a time unit. The necessity to hold the defined QoS in the presence of a limited network capacity may result in rejecting some waiting requests for the data transmission. We assume that every request has an assigned route in a computer network which is the result of previously solved routing problem. A class of the network flow is arbitrary. The input-output plant can represent a computer network, Fig. 11.

Let us assume that there are R requests in a computer network constituting a set of requests $\mathbf{R}$. We define two decision variables p_r and u_r for every request $r \in \mathbf{R}$. They form decision vectors $p = [p_1, p_2, \ldots, p_r, \ldots, p_R]^T$ and $u = [u_1, u_2, \ldots, u_r, \ldots, u_R]^T$. The binary variables p_r take the value of 1(0) if the request is accepted for the transmission (if otherwise).The value of u_r denotes the rate of request r and has to satisfy the QoS requirement that $u_r \geq u_{r,\min}$. We also assume that the maximization of utility is the purpose of the network. Local utilities q_r of individual requests constitute the global utility $\bar{Q}(q_1, q_2, \ldots, q_r, \ldots, q_R)$ of the network as a whole. They are expressed by formulas $q_r = f_r(p_r, u_r; a_r)$, where

$$f_r(p_r, u_r; a_r) = \begin{cases} \varphi_r(u_r; a_r), & \text{dla } p_r = 1, \\ -a_{r,0}, & \text{dla } p_r = 0, \end{cases}$$

and a_r, $a_{r,0}$ are given parameters characterizing the network. Additionally, parameter a_r can be a vector. The form of the function $\varphi_r(u_r; a_r)$ depends on the so-called class of network flow, e.g. for the enhanced inelastic flow

$$\varphi_r(u_r; a_r) = \begin{cases} \ln(u_r + 1), & \text{dla } u_r \geq u_{r,\min}, \\ 0, & \text{dla } u_r < u_{r,\min}. \end{cases}$$

The global utility is the function of the pair of decisions (p, u), i.e.

$$\bar{Q}(q_1, q_2, \ldots, q_R) = F(f_1(p_1, u_1; a_1), f_2(p_2, u_2; a_2), \ldots, f_R(p_R, u_R; a_R)) \triangleq Q(p, u; a, x),$$

where $x = [x_1, x_2, \ldots, x_r, \ldots, x_R]^{\mathrm{T}}$ is the vector of parameters x_r which describe, among others, the number of elementary flows constituting a single request as well as the priority of the request. The global utility is often the weighted sum of local utilities, so,

$$Q(p, u; a, x) = \sum\nolimits_{r=1}^{R} z_r(p_r; x_r) f_r(p_r, u_r; a_r), \tag{3}$$

where $z_r(p_r; x_r)$ is the weight parameter of the rth request. This parameter is dependent on the decision whether the request is accepted or rejected and on the parameter x_r.

The set of links $\mathbf{L}$, the route for every request r as the subset of links $\mathbf{L}_r$, available link capacities $U_l, l \in \mathbf{L}$ are the data provided for the joint problem of admission control and rate allocation considered therein. We also assume the knowledge of bounds $u_{r,\min}$ and $u_{r,\max}$ of intervals comprising feasible rates for requests. Then, the decision-making problem deals with the determination of the pair $(p^*, u^*) \in \mathbf{D}$ maximizing (3), where

$$\mathbf{D} = \{(p, u) : (\forall_{l \in \mathbf{L}} \sum\nolimits_{r \in \mathbf{R}_l} p_r u_r \le U_l) \wedge (\forall_{r \in \mathbf{R}_l} u_r \in [u_{r,\min}, u_{r,\max}])\}$$

is the set of feasible solutions of the joint problem, and $\mathbf{R}_l$ is the set of requests using the link l for the transmission.

We can distinguish both sub-problems AC and RC in the maximization of (3), namely

$$\max_{(p,u) \in \mathbf{D}} Q(p, u; a, x) = \max_{p \in \mathbf{D}_p} \max_{u \in \mathbf{D}_u(p)} Q(p, u; a, x) = \max_{p \in \mathbf{D}_p} Q(p, u^*(p); a, x) \triangleq \max_{p \in \mathbf{D}_p} Q_{\mathrm{AC}}(p; a, x),$$

where

$$\mathbf{D}_p = \{p : \forall_{l \in \mathbf{L}} \sum\nolimits_{r \in \mathbf{R}_l} p_r u_{r,\min} \le U_l\},$$

and

$$\mathbf{D}_u(p) = \{u : (\forall_{l \in \mathbf{L}} \sum\nolimits_{r \in \mathbf{R}_l} p_r u_r \le U_l) \wedge (\forall_{r \in \mathbf{R}_l} u_r \in [p_r u_{r,\min}, p_r u_{r,\max}])\}.$$

The internal maximization $\max_{u \in \mathbf{D}_u(p)} Q(p, u; a, x)$ stands for the sub-problem RA, and the maximization of the criterion $Q_{\mathrm{AC}}(p; a, x)$ expresses the sub-problem AC. We obtain the latter criterion after replacing u with $u^*(p)$ in $Q(p, u; a, x)$. Let us notice that the value of $u^*(p)$ depends on the requests accepted for the transmission.

In consequence, the exact algorithm AO solving the joint problem consists of two steps. The pair $(p^*, \ u^*) \triangleq (p^*, \ u^*(p^*))$ is the optimal solution.

Algorithm AO

1: Solve the sub-problem RA, i.e. $\max_{u \in \mathbf{D}_u(p)} Q(p, u; a, x) \triangleq Q_{\mathrm{AC}}(p; \ a, x)$ for every $p \in \mathbf{D}_p$, and let $u^*(p) = \arg \max_{u \in \mathbf{D}_u(p)} Q(p, u; a, x)$.

2: Select $p^* \in \mathbf{D}_p$ such that $\max_{p \in \mathbf{D}_p} Q_{\mathrm{AC}}(p; a, x) \triangleq Q_{\mathrm{AC}}(p^*; a, x)$.

It is worth noting that the utility functions $\varphi_r(u_r; a_r)$ are additive, monotonic, and differentiable. The global utility function is additive as well $Q(p, u; a, x)$, while the constraints defining $\mathbf{D}$ are linear. Such properties of the optimization problem allow us to use the Kuhn-Tucker conditions. For the considered problem, they are necessary and sufficient conditions for the existence of maximum of $Q(p, u; a, x)$ for the pair $(p, \ u)$ fulfilling those conditions. In consequence, it is enough to solve $L + 3R$ equations where L is the number of links in a computer network. Details and other cases are available in [14].

The comparison of the proposed systems approach with the sequential approach has been conducted to evaluate the pertinence of the former approach. The latter approach starts with solving the sub-problem AC. Then, for the given set of accepted requests, the rate control is determined in the framework of the sub-problem RA. Block-schemes of both approaches are provided in Figs. 12 and 13. It is understandable that the sequential solving of the sub-problems usually leads to worse outcomes. For the considered problem, such property is justified by the fact that the exact solution of the sub-problem RA is derived only for a single vector p.

Let us consider the following computational example. The computer network consists of a single link with the capacity $U_1 = 10$ Mb /s. There are three requests ($R = 3$) which need the same minimum transmission rate $u_{r,\min} = 1$ Mb/s, $r = 1, \ 2, \ 3$. Local utility functions $f_r(p_r, u_r; a_r)$ have a logarithmic form $\varphi_r(u_r; a_r) = \ln u_r$ and take values 0 for $p_r = 0$. Moreover, $x_1 = 5$, $x_2 = 3$, $x_3 = 2$. It is easy to obtain the exact solution $p^* = [1, \ 1, \ 0]^{\mathrm{T}}$, $u^* = [6.25, \ 3.75, \ 0]^{\mathrm{T}}$ and the corresponding global utility $Q_{\mathrm{AC}}(p^*, u^*; a, x) = 13.13$. On the other hand, in the sequential approach, when $p = [1, \ 1, \ 1]^{\mathrm{T}}$, the vector of rates $u = [5, \ 3, \ 2]^{\mathrm{T}}$ results in the maximum value of the global utility function $Q_{\mathrm{AC}}(p, u; a, x) = 12.73$.

There are cases when both approaches provide the same solution, as indicated in [14]. For example, the following condition of equivalence is valid

$$\forall_{r \in \mathbf{R}} \forall_{\substack{s \in \mathbf{R} \\ s \neq r}} \ c_r > x_r \ [f_s(1, \ u_{s,\min} + u_{r,\min}; \ a_r) - f_s(1, \ u_{s,\min}; \ a_r)]$$

if vector p maximizing the criterion $\tilde{Q}(p; c) = \sum_{r \in \mathbf{R}} c_r p_r$ is the solution of AC, where c_r stands for weights of individual requests.

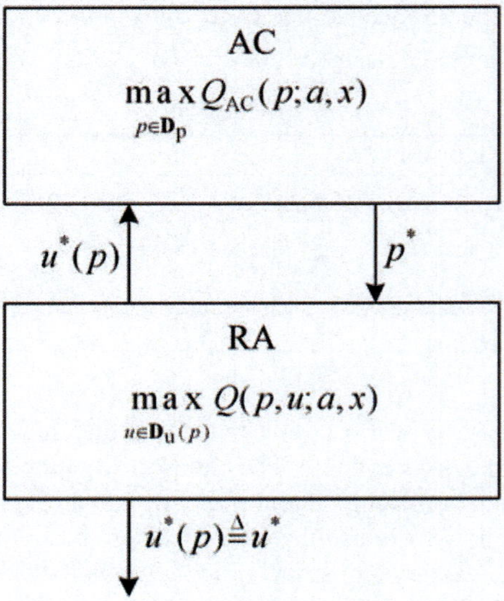

Fig. 12 Two-level structure of the joint problem of admission control (AC) and rate allocation (RA)

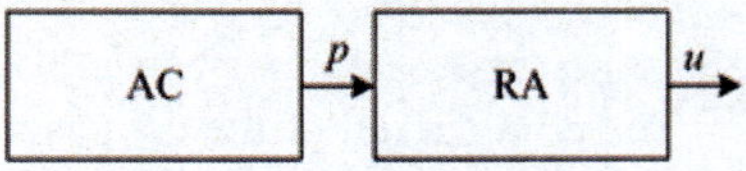

Fig. 13 Structure of the sequential solving of sub-problems AC and RA

The presented deterministic case is somewhat idealized. Namely, it assumes both the specific forms of utility functions and the exact values of parameters a_r, x_r which are hardly achievable for working computer networks. Therefore, a significant part of the research dealt with cases without full information on parameters. In particular, it concerned the parameters x_r, especially challenging to determine. The formalism of uncertain variables [3] was applied in [14].

Finally, it is worth noting the universality of the applied approach using the utility as the evaluation of made decisions. The notion 'utility', used for computer networks as an example of the technological system, has been taken from investigations of economic systems, where it was considerably earlier applied to the determination of resource allocation and strategy selection for companies to maximize their profit under the conditions of competitiveness. This issue has been developed in [31], where the analogy between decision-making for computer networks and economic systems was discussed in detail.

6 Other Problems and Conclusions

Three selected complex decision-making and decision-support problems investigated in the Division of Intelligent Decision Support Systems at the Wroclaw University of Science and Technology have been presented in depth. There are other such problems. Let us now characterize some of them.

The first one concerns the intensively developed topic of decision-making for contemporary computer networks, which deals with joint data deployment in a network of servers and assignment of clients to selected replicas of data. A wide area network, together with local networks connected to it via access routers, is considered and regarded as the decision-making plant (DMP). Caching of data at servers and assignment of clients to servers are decision variables. Clients can download required data from assigned servers. Initially, both problems have been solved separately and sequentially with a cost criterion. The systems approach, which has been described in detail in [10], assumes a joint determination of both decisions. The sum of the three costs, namely the cost of data transmission among servers, the cost of request processing at cache servers, and the cost of storing data at servers, is used to evaluate the decisions. As a consequence, the quadratic binary optimization problem has been solved using two decision-making algorithms basing on the Lagrange decomposition and the random rounding. Many particular cases or related problems have been investigated in the Ph.D. thesis [9]. Other approaches to a similar problem are available in [1, 2].

The next investigated problem deals with decision-making for wireless sensor networks. Sensors are deployed at predefined locations within an area of interest, perform measurements, and provide results of measurements to sinks where the measurement data are stored. The wireless sensor network, along with sinks located at a given measurement area, is now the decision-making plant (DMP). Deployment of sensors and sinks in the measurement area, determination of sensors' ranges to cover all measurement points in the measurement area (coverage) as well as routing of all of the data from the measurement directly to sinks or to sinks via other sensors, are crucial decisions which should be made. According to the literature, those three decisions are predominantly made independently, starting from the deployment of sensors and sinks, and with the use of different criteria. Among those, the consumption of limited amount of energy, sensors' working time, investment cost, and data acquisition time are the most important ones. The three partial problems, i.e. the deployment of sensors and sinks, coverage, and routing have been jointly solved in [23] to minimize the sum of sensor and sink deployment costs, measurement cost as well as processing and transmission costs at sensors. The resultant NP-hard optimization problem has been solved by an original heuristic algorithm based on an algorithm solving the known circulation problem [21].

Development of Internet technology implies the necessity to optimize different transactions performed with the use of this medium. Internet shopping is an example of such an important transaction. An analysis of this application shows new decision-making problems, significant not only for retailers but also for clients. Solving those problems can facilitate the management of corresponding enterprises as well as minimize purchase costs and improve the comfort of shopping [24, 40]. The particular Internet shopping optimization problem with discounts and variable delivery costs has been investigated in [34]. The decision-making plant (DMP) is composed of a client's list of purchases and a set of shops together with their characteristics. A matrix whose entries inform about a number (an amount) of products purchased at individual shops serves as the decision. Unlike the previous cases considered in the literature, the complexity and the systemic feature of the problem investigated in [34] consists in taking into account in the evaluation of the decision, i.e. in the criterion, different factors affecting the purchases made in every shop. The considered factors are the purchase cost, the delivery cost, and the discounts, which are dependent on the client's total payment. The original criterion combining the mentioned factors was proposed, and a strongly NP-hard optimization problem was formulated. Two heuristic algorithms, one based on the Tabu Search metaheuristic, the other based on the Simulated Annealing metaheuristic, were proposed as a solution tool.

The systemic approach can also be applied for the generalized versions of traditional problems from the scope of operations research, which are closer to real-world applications of these problems and refer to some atypical but practical situations. Let us mention, as an illustration of such a situation, the generalization of the traditional task scheduling problem with non-preemptive, independent tasks, parallel machines, and the sum of completion times as a criterion, i.e. $\mathrm{P}|r_j|\sum C_j$ according to Graham's notation [44]. The generalization deals with choosing the machines deployment inside a given area [45]. Tasks are moved towards known machines' locations to be executed. Times required for the movement of tasks towards machines imply different release times, i.e. times when tasks are ready for execution. However, every such time is not given a priori, and it depends on the decision on the deployment of the machines. The sets of tasks and machines, together with their corresponding characteristics, constitute the decision-making plant (DMP). There are two decisions: the schedule, so the determination of starting time and the corresponding machine for every task as well as the deployment of machines. Consequently, problem $\mathrm{P}|r_j(x)|\sum C_j$ needs a solution where matrix x represents the deployment of machines. As pointed out and proven in [43], this problem, referred to as ScheLoc in the literature [36], is strongly NP-hard. The evolutionary solution algorithm and the comparison with exact results generated for small instances of the problem by the solver CPLEX CP Optimizer are provided in [43].

The presented examples of the application of systems approach to solving decision-making and decision-support problems, which means the joint and not the separate determination of interconnected partial decisions, univocally confirm its advantage in comparison with the reductionist approach based on a simple decomposition. Solving these problems according to the systems approach usually results in severe optimization problems. However, the increase in opportunities for implementing decision-making algorithms, which has been made possible due to the advances in information technologies, is the additional reason for the use of the systems approach in difficult, complex decision-making or decision-support problems with essential real-world applications.

References

1. Bateni M., Hajiaghayi M.: Assignment problem in content distribution networks: unsplittable hard-capacitated facility location. In: 20th Annual ACM-SIAM Symposium on Discrete Algorithms, pp. 805–814. Society for Industrial and Applied Mathematics, Philadelphia (2009)
2. Bektas, T., Cordeau, J., Erkut, E., Laporte, G.: Exact algorithms for the joint object placement and request routing problem in content distribution networks. Comput. Oper. Res. **35**(12), 3860–3884 (2008)
3. Bubnicki, Z.: Analysis and Decision Making in Uncertain Systems. Springer, Berlin, London, New York (2004)
4. Cappanera, P., Scutella, M.G.: Joint assignment, scheduling, and routing models to home care optimization: a pattern-based approach. Transp. Sci. **49**(4), 830–852 (2015)
5. Castillo-Salazar, J.A., Landa-Silva, D., Qu, R.: Workforce scheduling and routing problem: literature survey and computational study. Ann. Oper. Res. **239**(1), 36–67 (2014)
6. Chen, X., Thomas, B.W., Hewit, M.: The technician routing problem with experience-based service times. Omega **61**, 49–61 (2016)
7. Chvatal, V.: A formal analysis and taxonomy of task allocation in multi-robot systems. Math. Oper. Res. **4**, 233–235 (1979)
8. Correll, N., Martinoli, A.: Multirobot inspection of industrial machinery. IEEE Robot. Autom. Mag. **16**, 103–112
9. Drwal, M.: Algorithms for data placement problem: decentralization and competitiveness. Ph.D. thesis, Wroclaw University of Science and Technology, Faculty of Computer Science and Management (2013)
10. Drwal, M., Jozefczyk, J.: Decomposition algorithms for data placement problem based on Lagrangian relaxation and randomized rounding. Ann. Oper. Res. **222**(1), 261–277 (2014)
11. Fields, M., Haas, E., Hill, S., Stachowiak, C., Barnes, L.: Effective robot team control methodologies for battlefield applications. In: Proceedings of the IEEE International Conference on Intelligent Robots and Systems, pp. 5862–5867 (2009)
12. Filcek, G.: Algorithms for joint raw material allocation and transportation in production systems (in Polish). Ph.D. thesis, Wroclaw University of Science and Technology, Faculty of Computer Science and Management (2011)
13. Filcek, G., Józefczyk, J.: Heuristic algorithm for integrated allocation and transportation in three-stage supply network. Procedia Soc. Behav. Sci. **54**(4), 1298–1307 (2012)
14. Gąsior, D.: Congestion avoidance algorithms using the formalism of uncertain variables (in Polish). Ph.D. thesis, Wroclaw University of Science and Technology, Faculty of Computer Science and Management. http://www.dbc.wroc.pl/publication/4288 (2010)
15. Gąsior, D., Drwal, M.: Pareto-optimal Nash equilibrium in capacity allocation game for self-managed networks. Comput. Netw. **57**(14), 2817–2832 (2013)
16. Gerkey, B., Mataric, M.: A formal analysis and taxonomy of task allocation in multi-robot systems. Int. J. Robot. Res. **23**(9), 939–954 (2004)
17. Grangier, P., Gendrau, M., Lahuede, F., Rousseau L.-M.: The vehicle routing problem with cross-docking and resource constraints. Technical report, CIR-RELT-2017-23, Universite de Montreal (2017)
18. Hojda, M.: Algorithms of task rescheduling with the use of control results (in Polish). Ph.D. thesis, Wroclaw University of Science and Technology, Faculty of Computer Science and Management (2011)
19. Hojda, M.: Algorithms of task allocation in multi-robot systems with energy constraints (in Polish). In: Automation of Discrete Processes, vol. 2, pp. 105–113. Jacek Skalmierski's Computer Studio Press (2014)
20. Hojda, M.: Task allocation in robot systems with multi-modal capabilities. In: INCOM 2015, Ottawa, Canada, pp. 2109–2114 (2015)
21. Hoffman, A.J.: Some recent applications of the theory of linear inequalities to extremal combinatorial analysis. In: Symposium in Applied Mathematics, pp. 113–127. American Mathematical Society (1960)

22. Hoshino, S., Seki, H., Naka, Y.: Development of a flexible and agile multi-robot manufacturing system. In: Proceedings of 17th World Congress of the International Federation of Automatic Control, pp. 15786–15791 (2008)
23. Jagusiak, S., Józefczyk, J.: An algorithm for joint location, coverage and routing in wireless sensor networks. J. Theor. Appl. Comput. Sci. **6**(2), 1–11 (2012)
24. Jiang, Y., Liu, Y., Wang, H., Shang, J., Ding, S.: Online pricing with bundling and coupon discounts. Int. J. Prod. Res. **56**(5), 1773–1788 (2018)
25. Józefczyk, J.: Scheduling tasks on moving executors to minimize the maximum lateness. Eur. J. Oper. Res. **131**, 171–187 (2001)
26. Józefczyk, J.: Task Scheduling on Moving Executors in Complex Operation Systems (in Polish). Wroclaw University of Technology Press, Wrocaw (1996)
27. Józefczyk, J.: Selected Decision-making Problems in Complex Operation Systems (in Polish). Wroclaw University of Technology Press, Wrocław (2001)
28. Józefczyk J.: Knowledge-based two-level control of a group of mobile executors. Integr. Comput. Aided Eng. **10**(2), 191–201 (2003)
29. Józefczyk, J.: Decision-making algorithms in two-level complex operation system. Decis. Support Syst. **38**, 171–182 (2004)
30. Józefczyk, J.: Hybrid solution algorithms for task scheduling problem with moving executors. Eng. Appl. Artif. Intell. **37**(16), 181–186 (2004)
31. Józefczyk, J., Gąsior, D.: Utility-based models and decision making problems for selected network processes. Kybernetes **44**(6/7), 1107–1121 (2015)
32. Józefczyk, J., Hojda, M.: Decision making for autonomous vehicles in the presence of uncertain information. In: Conference on Control and Fault-Tolerant Systems (SysTol'10), Nicea, France. http://www.ieeex-plore.ieee.org/iel5/5672506/5675889/05676032.pdf?arnumber (2010)
33. Józefczyk, J., Hojda, M.: Algorithms of joint task scheduling and movement control on moving executors (in Polish). In: Advances of Automatics and Robotics, vol. 1, pp. 34–49. Kielce University of Technology Press (2011)
34. Józefczyk, J., Ławrynowicz, M.: Heuristic algorithms for the internet shopping optimization problem with price sensitivity discounts. Kybernetes **47**(4), 831–852 (2018)
35. Józefczyk, J., Salwach, T.: An uncertain routing-scheduling problem in a class of manufacturing systems. In: 8th International Conference on the Modern Information Technology in the Innovation Processes of the Industrial Enterprises (MITIP 2006), Budapest, Hungary, pp. 547–552 (2006)
36. Kalsch, M.T., Drezner, Z.: Solving scheduling and location problem in the plane simultaneously. Comput. Oper. Res. **37**(2), 256–264 (2010)
37. Karakus, M., Durresi, A.: Quality of service (QoS) in software defined networking (SDN): a survey. J. Netw. Comput. Appl. **80**, 200–218 (2017)
38. Khamis, A., Hussein, A., Elmogy, A.: Multi-robot Task Allocation: A Review of the State-of-the-Art, vol. 604, pp. 31–51. Springer (2015)
39. Kim, B.I., Koo, J., Park, J.: The combined manpower-vehicle routing problem for multi-staged services. Expert Syst. Appl. **37**, 8424–8431 (2010)
40. Lopez-Loces, M.C., Musiał, J., Pecero, J.E., Fraire-Huacuja, J., Błażewicz, J., Bouvry, J.: Exact and heuristic approaches to solve the Internet shopping optimization problem with delivery costs. Int. J. Appl. Math. Comput. Sci. **26**(2), 391–406 (2016)
41. Markowski, M., Józefczyk, J.: Heuristic algorithms for solving uncertain routing-scheduling problem. In: Artificial Intelligence and Soft Computing, ICAISC. Springer Lecture Notes in Computer Science, vol. 5097, pp. 1052–1063 (2008)
42. Paraskevopoulos, D.C., Laporte, G., Repoussis, P.P.: Resource constrained routing and scheduling: review and research. Eur. J. Oper. Res. **263**, 737–754 (2017)
43. Piasecki, B., Józefczyk, J.: Evolutionary algorithm for joint task scheduling and deployment of executors. In: Automation of Discrete Processes. Theory and Applications, vol. 1, pp. 169–178 (2018)
44. Pinedo, M.L.: Scheduling: Theory, Algorithms, and Systems. Springer, New York (2012)

45. Rajabzadeh, M., Ziaee, M., Bozorgi-Amiri, A.: Integrated approach in solving parallel machine scheduling and location (ScheLoc) problem. Int. J. Ind. Eng. Comput. **7**(4), 573–584 (2016)
46. Siepak, M.: Robust task scheduling algorithms with interval uncertainty (in Polish). Ph.D. thesis, Wroclaw University of Science and Technology, Faculty of Computer Science and Management (2013)
47. Siepak, M., Józefczyk, J.: Minmax regret algorithms for uncertain $P||C_{\max}$problem with interval processing times. In: 21st International Conference on Systems Engineering, Las Vegas, USA, pp. 115–119 (2011)
48. Singh, S., Chana, I.: QoS-aware autonomic resource management in cloud computing: a systematic review. ACM Comput. Surv. (CSUR) **48**(3), 42:1–42:46 (2016)
49. Thomas, W.: Task scheduling on moving executors in computer control systems to minimize the average flowtime (in Polish). Ph.D. thesis, Wroclaw University of Science and Technology, Faculty of Computer Science and Management (2006)
50. Zehra, U., Shah M.A.: A survey on resource allocation in software defined networks (SDN). In: 23rd International Conference on Automation and Computing (ICAC), Huddersfield, pp. 1–6 (2017)

Advanced Ship Control Methods

Roman Śmierzchalski and Anna Witkowska

Abstract The chapter presents two main streams of research in vessel control at sea: dynamic positioning (DP) of the vessel and decision support in case of collision at sea. The control structure and basic requirements for the DP system are defined. Selected issues of automatic control of a dynamically positioned vessel are discussed. A review of advanced methods of controlling a DP ship is carried out, taking into account the tasks of particular subsystems. In the scope of decision support, the issue of collision avoidance at sea is discussed. This problem is defined as a dynamic multi-criteria optimization task which consists in looking for effective solutions that meet specific criteria within a set of acceptable solutions. To solve this problem, the evolutionary method of path planning is used.

1 Introduction

Sea-going vessels perform a number of tasks at sea: transport of cargoes and/or passengers, extraction of marine resources, fishing, etc. These tasks are carried out with the use of control systems, where the basic criterion is to ensure safety at sea. In most cases, these systems allow automatic control, in which the function of the operator is to supervise and intervene in emergency or critical states. In the case of ship navigation, automated processes include: course stabilization, control of ship motion along the route of the voyage, and dynamic positioning of the ship. Taking into account factors such as unpredictability of navigational situations at sea and hydrometeorological phenomena in the marine environment, the complexity of

R. Śmierzchalski (✉) · A. Witkowska
Faculty of Electrical and Control Engineering, Gdansk University of Technology, 11/12 Gabriela Narutowicza Street, 80-233 Gdansk, Poland
e-mail: roman.smierzchalski@pg.edu.pl

A. Witkowska
e-mail: anna.witkowska@pg.edu.pl

P. Kulczycki et al. (eds.), *Automatic Control, Robotics, and Information Processing*, Studies in Systems, Decision and Control 296,
https://doi.org/10.1007/978-3-030-48587-0_20

systems used on board and the increasing amount of available information make control and decision making by the operator difficult in many situations. Therefore, automatic control and decision support systems are used to support the operator's work on board. In this way, it is possible to steer a ship with a certain accuracy and the decision support systems allow to make a proper choice of acceptable solutions. The basic function of decision support systems is to protect the operator against erroneous decisions. This includes cruise route planning and collision avoidance at sea. According to the regulations in force, the final decision is always made by the human operator, the navigator in case of potential accident.

This chapter deals with advanced ship control and decision support systems, which differ from the existing systems in land-based solutions due to specific tasks performed by the vessel in the marine environment and the resulting different control objectives. The main purpose of control is to move the vessel according to the chosen route along the trajectory, or to stabilize the vessel position and heading in the case of a dynamically positioned vessel. The control criteria are minimisation of operating costs and safety of maritime traffic in the presence of unpredictable hydrodynamic disturbances of the marine environment affecting the vessel motion. The maritime safety includes the safety of ships, people, cargoes, and the protection of the marine environment. The decomposition of tasks, along with the multi-layered structure of ship control divided into systems, allows the achievement of the assumed goal. In navigation systems, in the ship steering layer where automatic control and decision support systems are used, the primary objective is to optimize the trajectory of the ship with simultaneous fulfilment of the condition of safety at sea. Automation of other ship propulsion and power systems enables to reduce the number of crew members, fuel consumption, and total cost of ship operation, and to increase the reliability of equipment operation. The paper presents two main issues of ship control related to the development of autonomous ships, which are dynamic positioning and decision support in a collision situation. The issue of dynamic positioning concerns stabilization of vessel's position and heading determined by its course, and vessel control at low manoeuvring speeds. Depending on the DP class, specific requirements and control modes are defined. The structure and advanced algorithms of dynamically positioned ship control are presented. In the scope of decision support systems for navigators, the method of modelling collision situations at sea and the method and algorithms of their avoidance are presented. In order to avoid collisions, an evolutionary method of planning passageways is used to solve the task of dynamic multi-criteria optimization which allows to avoid collisions at sea.

2 Selected Problems of Steering a Dynamically Positioned Vessel

A dynamically positioned vessel, as defined by DNV (Det Norsk Veritas), automatically maintains a fixed position and course, and tracks a specific trajectory with a slow- motion speed (<2–3 knots) using only the thrust forces generated by its own propulsion system. Dynamically positioned vessels are used at sea to support the drilling and exploration industries. They require systems that ensure accurate positioning and heading-keeping, and meet the requirements for safety and control accuracy. Modern offshore vessels are equipped with Dynamic Positioning (DP) systems and modern propulsion systems to perform many additional functions, such as dynamic mooring, automatic tracking of a set trajectory, etc. The main types of vessels that can be distinguished in this area are:

- DS (Drillship)—drilling vessel;
- PVC (Platform Cable Vessel)—vessel for laying cables on the seabed;
- Platform Supply Vessel (PSV)—vessel for transportation of goods, tools, equipment, and personnel between a port and an offshore installation;
- RV (Research Vessel)—research vessel;
- AHTS (Anchor Handling Tug & Supply) Vessel—platform support unit designed to support platform anchors;
- ERRV (Emergency Response Rescue Vessel)—rescue vessel supporting a platform in emergency situations, designed for rapid response.

Depending on the type and intended use of the vessel, the most important tasks of the DP system include:

- maintaining a position within a defined area, due to the limited length and flexibility of the drilling line (drilling vessels);
- maintaining an appropriate angle (heading) to the quay to allow efficient unloading of cargo by a crane (container ships);
- maintaining a position within a defined area to maintain a safe distance from the oil platform (supply vessels);
- slow speed tracking on a specific trajectory (seabed research vessels).

According to the definition by the DNV organization, a dynamic positioning system is a complete installation on the ship that is necessary for the implementation of dynamic positioning. This installation consists of the following subsystems: power supply system, drive system, and DP control system. The DP system component which is responsible for control, navigation, and steering is the Dynamic Control System (DCS). Depending on the type of vessel and operational tasks, this system allows the following modes of operation: simultaneous control of heading and position in automatic mode in normal (beyond extreme) conditions of interference with the marine environment; control of position and heading in joystick mode or with automatic heading control; manual control of the drive in start/stop mode; automatic control of position with preprogrammed change—tracking of the target trajectory. The DP system should ensure that the ship's position and heading are automatically maintained at a predetermined accuracy, and the ship is manoeuvred within a limited operating range at a speed not exceeding 2–3 knots, while operating only on active rudder and propulsion propulsors to minimise energy losses.

DP systems are designed to perform these tasks independently of:

- the impact of environmental disturbances acting on the ship: counteracting wind and/or current and wave forces (control algorithms resistant to disturbance changes);
- changes in the ship's loading status, e.g. during unloading operations (algorithms to adapt to changes in the ship's load and algorithms to stabilize the side list);
- system failures, e.g. when one of the propulsors is lost (fault-tolerant algorithms).

Vessels of AHTS and PSV type are among the most specialized vessels. Due to the variety of tasks performed and the safety of work, they require the most sophisticated control systems that meet the requirements of dynamic positioning class DP3. The most recent offshore units are also equipped with modern electric propulsion systems, characterized by the best fulfilment of safety requirements, the ability to conduct operations in the sea area, as well as economic efficiency and flexibility in the operational profile for transit, position maintenance, manoeuvring, and loading of the vessel. These complex systems enable the use of fully integrated computer-controlled systems on board ships to operate safely and reduce costs. For economic and safety reasons, the redundancy, i.e. duplication of equipment and DP systems, is applied. If one unit fails, the replacement unit is automatically switched on. DP vessels are divided into 3 classes due to their reliability, considered in relation to the failure of on-board equipment (e.g. executive devices, electrical installations). The rules for dynamically operated ships described as class DP1, class DP2, and class DP3 are defined by the Classification Societies on the basis of the International Maritime Organization (IMO). In DP1 systems, the course departure occurs at the occurrence of 1 failure, while DP2 has some redundancy, as a result of which a single failure does not cause the course departure. DP3 is the highest class and to meet it, the equipment used for system construction is to be fireproof, waterproof, and waterproof. In these systems the course loss does not occur.

2.1 DP Vessel Model

When describing the movement of a dynamically positioned vessel, it is assumed that the vessel moves on the water surface in a horizontal plane. The vessel's movement on the water surface is due to the interaction of propulsion forces and additional environmental disturbances such as wind, wave, and currents. A vessel moving horizontally has three degrees of freedom of movement: surge, sway, and yaw. Assuming additionally that the vessel moves at low speed, the remaining degrees of freedom of its movement, associated with rolling, pitching, and heaving, are neglected [1].

Considering the movement of a DP vessel, wave interference is dominant at low speeds. It generates forces consisting of two components with different low and high frequencies. The second-order, zero-mean frequency wave interference causes slow-changing ship movements, while the first-order, high-frequency wave interference causes fast-changing movements. Due to the nature of wave interference, ship

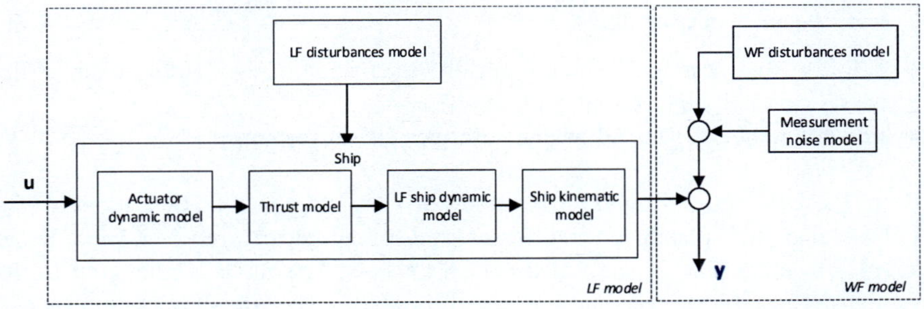

Fig. 1 Low frequency (LF) and wave frequency (WF) ship model

dynamics can be modelled using the superposition rule [2]. According to this principle, the total movement of a vessel is the superposition of low and high frequency movements. The measured output signal y of the position and heading of the vessel is interpreted as the superposition of the low frequency movement $y_{LF} \in R^3$ and the high frequency movement $y_{WF} \in R^3$, extended by Gauss' measurement noise $w \in R^3$ with zero mean value:

$$y = y_{LF} + y_{WF} + w \tag{1}$$

The dynamics of the ship can therefore be considered separately as the low frequency (LF) dynamics and the wave frequency (WF) dynamics. Since wave frequency motions cannot be compensated by the propulsion system, the low frequency dynamic model is used to synthesize the control law which will only compensate for low frequency disturbances, while the wave frequency dynamic model is used for state observer synthesis and wave filtration (Fig. 1). The DP controller compensates only for low frequency disturbances, assuming that wave frequency disturbances are properly filtered out in the state observer.

Figure 1 presents a simplified diagram of the ship model, including its division into low and wave frequency motion models. Detailed descriptions of mathematical models of a DP ship can be found in [3–5].

The LF vessel model includes:

- LF disturbance model, which describes forces caused by the interaction of average disturbances with wind and ocean currents, 2-nd order of wave disturbances;
- LF ship dynamics model, which describes the movement of the vessel submerged in water, caused by hydrodynamic forces and low frequency environmental disturbances, and compensated by the thrust of propulsors;
- thrust model, which describes the force generated by the propulsors;
- actuator dynamics model, which describes physical limitations of the propulsors resulting from saturation and rate of changes of control signals;
- ship kinematics model, which describes the relation between ship's velocities and position.

The WF vessel model includes:

- WF disturbance model, which gives the description of the displacement of the vessel caused by the action of a first-order wave;
- measurement noise model, which describes the sensor noise.

From the DP control system design point of view, the mathematical models of a DP vessel are formulated at two different levels of complexity, required in nonlinear simulation models or in simplified control models. The models belonging to the former group take into account nonlinearities associated with ship kinematics and dynamics, as well as stochastic environmental interference. Nonlinearities in kinematics result from trigonometric transformations of the rotation matrix of coordinates describing the vessel motion with respect to the coordinate system associated with a fixed earth point and that associated with the moving vessel. Nonlinear dynamics is associated with the Coriolis forces and nonlinear damping. Environmental disturbances, due to their stochastic nature, are difficult to model and predict. A detailed overview of the wave, wind, and current models used in ship control systems can be found in the work by [6]. In addition, the wave profile contains a nonlinear description which is modelled using frequency spectra. Pierson-Moscowitz, Jonswap, Torsethaugen, or Bretschneider spectra are most commonly used to describe an irregular wave. Additional simplifying assumptions can be made [3] when keeping the ship in a preset position in the presence of very small displacements ($|v| < 2$–3 m/s).

Simplified models intended for the synthesis of a control system usually contain linear models of slow-moving dynamics and linear wave spectrum approximation. Some authors [7] describe the method of taking into account current disturbances in the DP ship model by determining the vector of total forces generated by current disturbances and damping forces. Simplified mathematical models are burdened with measurement errors, process noise, and parametric uncertainties, which result from changes in environmental conditions, water depth, loading of the vessel, and/or losses of thrust forces caused by damage to propellers and rudders. Improving the quality of the positioning process is possible by the use of effective algorithms that offer advanced nonlinear, adaptive, and resilient control and observation techniques.

2.2 Structure of the DP System

Automatic access and keeping the object on the target position and heading in the presence of external disturbances is possible thanks to the operation of complex computer systems for measurement, estimation, and regulation, as well as the use of executive and driving devices. The control takes place at low speeds of object movement and requires defining the coordinates of the target points and the direction in which the object should move to reach individual target points. The structure of the DP system is schematically shown in Fig. 2. The elements of this structure were described in detail by [8].

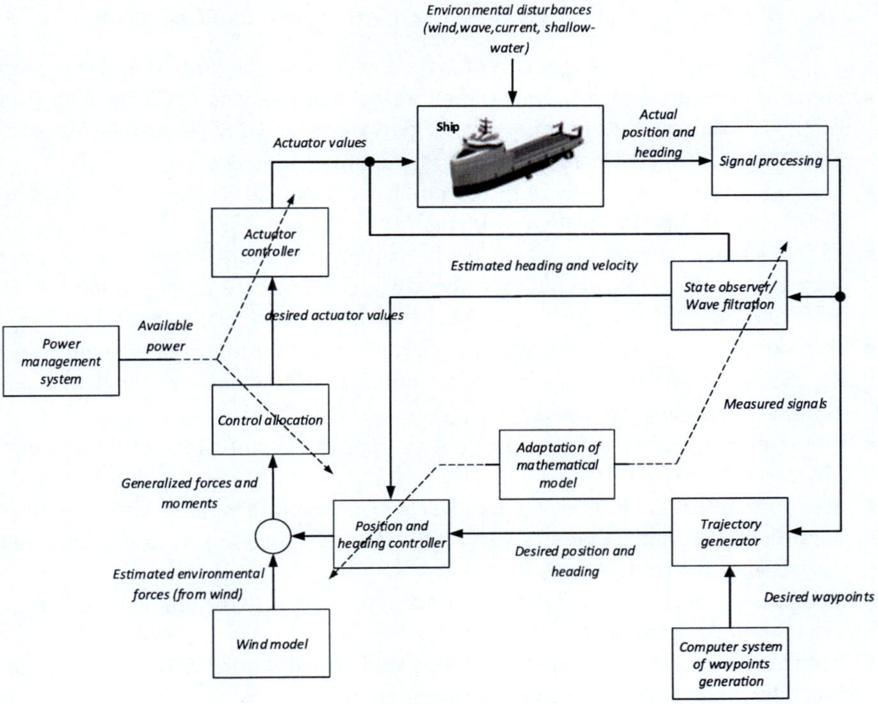

Fig. 2 Structure of the DP vessel dynamic positioning system

The set position and heading values are entered into the system via the user interface. Based on the navigational reference system and the states observer data, an estimation of the current vessel position and speed is obtained. The position error, which is the difference between the preset position and the estimated current position, is transmitted to the DP controller. The DP controller of ship's position and heading controls independently the ship's movement in three degrees of freedom by defining the desired force and torque vector. The generalised force and torque vector is transmitted to the allocation control system, where the force and torque setpoints are distributed to the propulsor setting control signals: rotational speeds of the main propeller and azimuthal and tunnel rudders, and angles of the rudder blade, determined in such a way as to minimise the amount of energy required to perform the control task.

In the DP control system, due to the uncertainty present in the model, an additional block is introduced which defines the mechanism of adaptation of the object model. The adaptation mechanism defines variable coefficients of resistance and masses accompanying the ship on the basis of object's movements and reactions depending on the forces and moments of operation.

The DP control system shown in Fig. 2 consists of the following blocks:

- ship—hydrodynamic description of kinetic and kinematic properties of the ship;
- signal processing—measurement, monitoring, analysis, and fusion of data from reference and measurement systems: gyrocompass, GPS (Global Positioning Systems), DGPS (Differential Global Positioning Systems);
- computer system for generating waypoints - a computer system for supporting the decision of the DP system operator;
- ship trajectory generator—determines the parameters of ship's movement between waypoints, taking into account the ship dynamics and determining smooth characteristics of reference signals and their derivatives (velocities and accelerations);
- ship position and heading controller—a high-level controller which determines the required force and torque vector to minimize the deviation of the actual ship position from the set position;
- power management system—controls and ensures a sufficient level of energy power on the vessel;
- actuator control—a low-level controller which determines the desired vector of controls, taking into account their amplitude and speed limits, to minimize the deviation of the actual value;
- adaptation of mathematical model—makes corrections to the model based on the setpoint and current position errors;
- control allocation—determines the required thrust forces and angles of the propellers to minimize the energy consumption.

When analysing the operating diagram of the DP system shown in Fig. 2, the basic problems in this task are: signal filtration and estimation of unmeasured quantities, selection of the DP control method, and allocation of forces. The most important requirements of the designed DP ship control systems include their robustness, proper compensation of environmental interference, and optimal distribution of forces to the redundant set of actuators.

2.3 *Review of Selected Methods of Dynamic Positioning of the Vessel*

The development of methods related to the dynamic positioning of ships dates back to the 1960s. A description of early research activities in this field can be found in the paper by [9]. The review of modern control methods and selected most important research achievements is summarized in the works by [10, 11]. The research on the development of effective methods of dynamic positioning of the ship has developed in several directions. The basic division includes: controllers designed with the use of PID algorithms, linear and nonlinear methods of mathematical modelling, and artificial intelligence algorithms. The first DP systems were designed with the use of conventional PID controllers in a cascade arrangement with low-pass and notch filters. Classic PID controllers ensure that the position is maintained if the object is linear and constant over time, and the interference is constant. The PID-AFB (PID with Acceleration Feedback) type controller proposed by [12] contains an additional integrating component in the feedback, associated with the measured acceleration of the ship. In the paper by [13], the concept of a nonlinear PID controller was applied using kinematic motion equations in the design. This approach made the system less sensitive to external interference. In the following years, some authors, [14] as one of the first, proposed a multidimensional optimal controller with Kalman Filter (KF) for linear discrete models of the DP ship, assuming that the measurement is subject to error in Gauss' decomposition. The development of Kalman filters initiated a new direction of research in DP systems, which concerned state vector estimation and fast variable interference filtration. The first papers [15, 16] included Kalman filters in combination with conventional Linear Quadratic Gaussian (LQG) optimal controllers. Currently, DP systems make a wide use of various types of DP control systems with Kalman filters and position and heading controllers in the feedback. In terms of Kalman filter design, they differ primarily in the algorithm of parameter selection. Problems related to KF tuning were considered using the amplification scheduling method, Bayesian methods, Monte Carlo methods, as well as covariance matrix estimation and iterative artificial intelligence methods.

Systems with nonlinear dynamics require model linearization. In this case, the extended Kalman filter EKF [17] is widely accepted in DP systems. An alternative to the EKF filter is the next in the family of Kalman filters—the unscented Kalman filter (UKF) [18]. The UKF filter performs estimations in a recursive way, using statistical linearization and so-called unscented transformation (UT). In most cases, DP systems are described by nonlinear models with non-Gauss noise. In the paper by [19], the application of a particle filter (PF) in positioning systems was analysed in terms of both theory and application. The general PF algorithm was compared with classical solutions, including an extended Kalman filter [20]. Unlike Kalman filters, the particle filters do not need additional assumptions about the probability density function and can provide optimal estimation in state models when the noise is of no-Gaussian type.

Simultaneously, nonlinear estimation methods in DP systems were developed in scientific centres. In the 1990s, a new approach to the problem of designing a state observer was developed in the robotics environment [21]. This approach utilised a design concept based on the passive theory and the strategy of constructing a controller and an observer taking into account their dynamics. The nonlinear passive observer proposed by [2] used a similar technique for marine vehicles. Based on the position measurement, the observer estimates the remaining states, along with slowly changing disturbances assuming that these disturbances are only distorted by white noise. The results of simulation studies of a multi-dimensional DP vessel motion control system with different types of observers were presented in the paper by [22, 23]. The systems were investigated using observers built on the basis of Kalman filter, extended Kalman filter, Kalman-Bucy continuous filter, and nonlinear observer.

The nonlinear control of a DP ship with uncertainties is still largely an unexplored field. Two main approaches are currently being used to deal with system uncertainties. These include: robust and adaptive control. Conventional methods of robust control, such as LQG control, sliding mode control, and H∞ robust control, rely on a fixed-structure controller that provides adequate performance across the full range of process changes, while adaptive control methods, such as the adaptive backstepping control [24–26] and their modifications such as dynamic surface control (DSC) [27], make it possible to design a dynamically changing feedback loop. These methods usually assume that the unknown disturbances are limited. Another active direct surface control (ADSC) method is a modification of the backstepping and DSC methods. Robust adaptive controllers take into account disturbances with unknown limits [28], which are estimated by modifying the adaptation laws and introducing an additional estimated parameter σ (σ—modification leakage) to prevent uncontrolled changes in the estimated parameters. When using a first order low-pass filter, there is no need for analytical differentiation of the virtual control law. The active disturbance rejection control (ADRC) method considered by [29] is a solution which avoids complex disturbance modelling and is an alternative to dynamic positioning systems.

A separate group of methods of control in DP systems are artificial intelligence methods such as genetic algorithms [30], artificial neural networks [31], and fuzzy algorithms [32]. The paper by [33] presents a robust adaptive controller being a modification of the vectorial backstepping method. This controller uses RBF neural networks to approximate the uncertainty of model ship dynamics and the sensitivity of the system to varying parameters [34, 35]. Taking into account the incomplete information on ship dynamics [36], the authors designed a nonlinear controller for the DP system, based on the nonlinear model predictive control (MPC) algorithm which neglects environmental disturbances. The article by [37] presents dynamic matrix control (DMC) for a dynamic positioning system with uncertain ship dynamics and disturbances.

An integral part of the DP control structure are control allocation (CA) systems. They are responsible for distribution of forces and moments from the DP controller to the redundant set of actuators for rudders and propellers. The control allocation algorithms should take into account physical limitations of the rudder and propulsion, their location on the ship, as well as limitations resulting from the amount of electrical and mechanical energy currently available on the ship. Most of the presented solutions are based on classical optimization methods, such as the Lagrange method or the least squares method, to calculate the pseudo-inverse control matrix. Typically, CA methods take the form of a quadratic optimization task with a target function that minimizes the activity of actuators while taking into account the saturation constraints. Such solutions have been considered using the method of penalty function, direct allocation (DA), redistributed pseudo-inverse (RPI), and cascaded generalized inverse. Classical solutions assume that for a DP vessel, the efficiency matrix of the propulsors is known and constant regardless of how this matrix changes over time for different states and control inputs of the system. The problem of allocation control is usually considered as a static or quasi-dynamic optimization problem and solved by taking into account non-adaptive linear models, while neglecting the dynamics of the actuators. The control allocation task is solved using model predictive control allocation (MPCA) algorithms, in which linear dynamics of actuators can be taken into account at the level of allocation control system. An important goal of the CA system is to provide robustness to actuator damage and compensate for uncertainties in the efficiency matrix. These uncertainties may result from thrust losses, which depend on environmental conditions, water density, ship speed, and propulsion failure. To deal with damage information, the thrusters' efficiency matrix is typically estimated by fault detection and diagnosis (FDD) systems, which are strongly associated with the fault tolerant (FT) control strategy. The fault-tolerant algorithms dedicated to the DP control system were considered by [38]. In addition, the authors [39] proposed robust adaptive control algorithms to estimate unknown errors due to rudder failures and unknown environmental disturbances. The limitation of FT algorithms is that they require for proper system operation that the fault detection and isolation mechanism is incorporated directly into the control law formula, which means that they are taken into account when designing the control law. Another way to reduce the influence of the uncertainty from thrust losses is to use adaptive dynamic allocation control. This approach to DP system design has been presented in the literature by [40–42] and (Witkowska and Śmierzchalski [25]. In these articles, a dynamic approach based on updating allocation rules by static optimization and defining adaptation laws based on Lagrange function and Lapunow function theory was considered.

The group of modern methods of dynamic positioning includes hybrid algorithms based on the concept of switching logic with supervisory switching logic (SSL). The articles by [43, 44] discussed the advantages of controllers with SSL over the traditional adaptive controller. Such a structure may lead to frequent switching of the control system, which in turn may cause destabilization of the whole system. In marine applications, the controller with SSL proposed by [45], which switched between different operational tasks (dynamic positioning, stopping manoeuvre, trajectory tracking) corresponding with ship's speed, is also worth mentioning. [46] presented a hybrid DP control system which automatically switched according to the peak frequency of the wave, determined on the basis of spectral analysis of vessel position measurements.

Currently, DP systems are developing towards adaptive noise observer-based control which allows online estimation of unknown disturbances. In the DP system, the adaptive observer is combined in a feedback loop with vector backstepping control [47] and DSC control [48]. Data fusion algorithms are also extensively developed [49] to combine position signals from several GPS (Global Positioning System) systems that utilize properties of the Kalman filter and the particle filter in a cascade system.

3 Supporting Decisions in Maritime Collision Situations

3.1 Avoiding Collisions at Sea

A collision avoidance manoeuvre consists in steering the ship by the navigator in such a way as to avoid a situation of excessively close approach of the ship to another object. A collision situation may occur when the point of intersection of trajectories of two or more ships is reached by each of them at the same or similar time. These points are defined as potential points of collision (PPC) [8]. The basic action performed by the own vessel to avoid a collision is a course change. At the same time, it is assumed that other ships do not change their speed, and that there can be an infinite number of potential collision points, depending on the change in ship's course [50]. A development of the collision avoidance method which initially determined only the collision avoidance manoeuvre is the determination of the entire safe ship's trajectory. Trajectory determination is significantly different from the manoeuvre determination procedure. The main difference concerns the way of generating control signals of the ship taking action. The collision avoidance manoeuvre determines only one manoeuvre to avoid a collision in a given navigational situation, while the safe trajectory determination refers to the definition of a function describing how to avoid a time-dependent collision, which in practice, when reduced to a discrete form, can be described by a sequence of manoeuvres or segmented trajectories assigned to a specific moment in time. Avoiding collisions at sea, in terms of determining a safe trajectory, is included in the task of dynamic optimization with constraints. Designing control systems intended to act in collision situations, along with simulation models taking into account the dynamic properties of the ship for these processes. Due to the dynamics of the process, its model can be described in the form of differential equations. Initially, analytical methods to determine safe trajectories were based on optimization methods, including dynamic programming, maximum principles [51], and methods based on game theory [52]. By analysing individual analytical methods, it is possible to solve selected continuous tasks for simple models of ship traffic. Taking into account practical aspect, a more convenient approach to solving the collision avoidance task is to replace the continuous dynamic task with a discrete form, described by difference equations. The discrete model allows to describe the state of the system for each discrete moment.

3.2 Navigator Decision Support in a Collision Situation

The navigator encounters continuously changing and unforeseeable factors that are important for the safety of navigation while navigating the ship. These factors have the form of changes in the marine environment, hydrometeorological conditions at sea, and the threat to navigation resulting from meeting other ships in motion. This threat occurs when the courses of the approaching ships cross the course of the own ship and may create a situation of excessive closeness. A critical state occurs when the vessel close-up parameters are exceeded and, additionally, the approaching vessel violates International Regulations for Preventing Collisions at Sea (COLREGs). When making a decision, the navigator chooses from among the acceptable decisions a decision that avoids collision with minimal loss of road. Decision making in a collision situation is normally divided into stages: once the radar information has been collected and processed, the navigation situation is analysed and assessed. Then, on this basis a decision is made and action is taken to avoid a collision. At present, the initial stages of decision making are carried out by the Automatic Radar Plotting Aid (ARPA) systems used on most ships. ARPA systems automatically process radar signals and on this basis the situation is visualised. An additional function is to simulate a manoeuvre on the radar screen. The visualised development of the navigational situation allows to assess the collision risk. In accordance with the International Convention for the Safety of Life at Sea (SOLAS), ARPA systems have only an advisory function and support the navigator in avoiding a collision. The final decision and selection of the way in which the manoeuvre will be executed is done by the navigator, after predicting the behaviour of other ships and, consequently, the development of future events. Based on analysing the situation and assessing the risk of collision, the navigator constantly changes the trajectory. Avoiding a collision is reduced to correcting the set trajectory and defining a new trajectory which will enable safe passing of other ships - moving objects. When determining the new trajectory, the navigator takes into account hydrometeorological conditions in a given navigation area, as well as navigation constraints determining the acceptable area of vessel manoeuvring. As previously indicated, on the basis of the analysis of disasters at sea, the weakest element in making decisions in emergency situations such as a collision at sea is a human being, i.e. the navigator in this case. As it has been stated, in normal shipping conditions the navigator conducts the navigation in accordance with his experience and maritime practice. However, in states of threat of collision, i.e. in situations which do not occur very often, the navigator's behaviour during the decision-making process may be subjective and lead to erroneous decisions that will result in collision. The navigator's behaviour in critical states depends on individual predispositions, skills, experience, and maritime practice, as well as cognitive abilities. When designing a particularly intelligent collision avoidance algorithm, cognitive behaviour that influences the decision-making process is especially important. Therefore, the cognitive behaviour of the navigator is widely researched. In order to limit subjective and misinterpreted decisions, the IMO has introduced the International Regulations for Preventing Collisions at Sea. The scale of effectiveness of the application of regulations in maritime navigation is significant. However, ambiguous interpretation of the COLREGs regulations by a navigator who bases his actions on his own maritime practice may lead to a wrong decision and consequently to a collision.

In order to prevent wrong decisions from being made, in addition to adequate training for navigators, modern control methods allow the use of navigator decision-support systems to navigate the ship in a safe manner. Taking into account the SOLAS Safety Convention and the provisions on the operation of ARPA systems, it would be possible to extend the functionality of these systems with additional modules to enhance decision-making. These modules, in terms of decision-making, would make it possible to operate on the basis of implemented algorithms. The solution to a collision situation would consist in automatic generation of the solution in the form of a proposed safe trajectory. It is important that these modules do not interfere with basic functions of ARPA systems. These additional functions would significantly extend the scope of operation of the conventional ARPA collision avoidance system. The input data for calculation procedures, being the necessary information for making decisions such as location of permanent navigational obstacles, positions and directions of movement and proximity parameters of ships, will be taken from ECDIS (Electronic Chart Display and Information System), ARPA, and AIS (Automatic Identification System) systems. The classical approach to ship decision making and control is based on models and analytical control methods. At the same time, the aim is to make models, through their verification, similar to real objects and processes. In order to eliminate navigator's subjectivity in decision making and possible misinterpretation of regulations and maritime situations, modern trends in ship steering aim at introducing on board Decision Support Systems (DSS). These systems, equipped with additional computing functions, are designed as advisory systems. The functions of the additional system must be taken into account in the multi-layered structure of ship control. The collision avoidance system is linked to the steering and navigation layer. The system has the ability to identify alternatives in a given navigational situation and assess them against the established criteria in the event of a collision. The system shall propose effective solutions selected from the set of acceptable alternatives for a given navigational situation. These solutions constitute a compromise between maritime safety and economic costs related to collision avoidance.

In many cases, however, it is difficult to get the right decision because it depends on many factors. This difficulty results from: the adopted decision-making model, the way the optimization task is formulated, the area of solutions, the adopted contradictory criteria and limitations, and the level of credibility of the data on the navigation environment obtained from the navigation system. Decision support is usually reduced to the task of multi-criteria optimization, and its solution consists in determining a number of best possible solutions in a given situation in the form of decision variables. The proposed values of decision variables allow for the achievement of the assumed objectives, defined by criteria that are usually contradictory. The solutions proposed by the system are effective solutions in the set of acceptable solutions. In a further procedure, the system or the operator will choose from among the effective solutions one compromise solution with respect to the specified criteria.

When designing a decision support system in the event of a collision, a model of this process is to be defined. This model should describe the relations between decision variables and process variables. In solving collision avoidance tasks, two approaches are used: (1) knowing the approximate model of the process, precise solutions are determined or (2) for the exact model, approximate solutions are determined. The first approach is to determine the exact control for a simplified approximate ship model, while the second approach is to determine the approximate control for an exact ship model, usually described by nonlinear equations. For a simplified, approximate form of the model, the decision support system is used to locate the solution area, analyse the sensitivity of the solution due to changes in parameters, and determine the extent to which the model is compatible with the actual process and the extent to which the solution of the task can be used in the actual control processes. However, for the methods of determining the collision manoeuvre and the safe trajectory, the model data and the solutions are not homogeneous. The collision avoidance manoeuvre is determined using an accurate model of ship motion generated by the ARPA system, which determines parameters of object movement on the basis of filtered radar signals, assuming that the tracked ships maintain their direction and speed of movement. Such a model makes it possible to use algebraic dependencies to obtain an accurate solution describing the collision manoeuvre. On the other hand, when determining the safe trajectory and using the vessel traffic model, the solution of the discrete task of dynamic optimization by approximate methods will also be an approximate solution in the form of a trajectory. The determined safe trajectory determines points of turn using geographical coordinates and time moments of reaching these points. Avoiding a collision by executing a safe trajectory requires control that will allow the ship to reach its own individual points of turn on the trajectory at specific moments of time. In this case, the condition of physical feasibility of the collision avoidance manoeuvre related to the dynamic properties of the vessel must be met.

3.3 *Intelligent Ship Control and Decision Support Methods*

Most of the work relating to analytical methods of ship control are based on control theory, e.g. (Burns and others, [53]; Zhang and others, [4, 54]. These methods require the use of ship control models that accurately describe dynamic properties of the ship and take into account the influence of hydrometeorological conditions on ship's movement. Inaccuracies in modelling dynamic properties of a ship and complexity of the process of determining ship control in marine environment are the reasons why the analytical methods have limited ability to determine proper control and only for selected conditions. Another approach to this task is based on the intelligent control methods developed in recent years [55]. Also, the application of mechatronics [56] allows for a different approach to this type of tasks. Techniques used in ship control based on artificial intelligence methods are alternatives to analytical methods, and are increasingly widely used in practice [57–63]. The use of artificial intelligence methods in ship control, especially in navigation, is discussed in the papers by [55, 57, 64]. In the automation of navigation, artificial intelligence was used in control of ship movement along a given trajectory, in situations of collision avoidance, and in cruise route mapping. Control along a given trajectory is considered in case of external hydrometeorological disturbances. It is noteworthy that in the scope of artificial intelligence methods used in ship control, evolutionary algorithms, fuzzy logic, neural networks, and expert systems are used. A frequent method of creating computational systems is to combine these techniques into hybrid systems. Artificial intelligence systems can support the operator's decision by determining the collision avoidance trajectory that will meet the safety and optimal control requirements. These systems can also be used to control autonomous surface vehicles. Intelligent control systems which are currently under development are capable of independently steering a surface vehicle in limited waters and ports without human intervention [65]. A number of works have been carried out on intelligent collision avoidance systems. For example, in the works by [66, 67], the use of the CBR (Case Based Reasoning) based learning system to avoid collisions is presented. The assessment of the risk of collision was based on fuzzy sets of information combined from various post-mortem devices. An agent system was used to make a decision. The collision avoidance agent system was tested for various collision scenarios on a specialized multi-computer simulation platform [68]. The analysis of collision avoidance for the situation of a meeting of two ships using ANFIS (Adaptive-Network-Based Fuzzy Inference Systems) is presented in the article by [69]. The decision-making process was based on the fuzzy Takagi-Sugeno system of inference. The issue of simultaneous consideration of COLREG rules and intelligent collision avoidance is presented in the article by [70]. The authors focus on the formulation of an appropriate decision-making model. Decisions are determined by fuzzy logic modules that operate in parallel tracks, where subsequent choices are formulated using the Bayesian network. The use of a hybrid system in which the neural network is connected to fuzzy sets is presented in the paper by [59]. The intelligent ship management system operates on the basis of a predictive model of avoiding damage, built on a neural network and fuzzy sets. The information about the parameters of the encountered obstacles is taken from the ship's navigation devices and is subject to data fusion. The predictive neural-fuzzy model determines the next waypoints in order to avoid obstacles. The ship motion between successive waypoints is controlled on the basis of LOS (Line of Sight). Tests with selected scenarios have been carried out in conditions close to real conditions, with the ESSO model ship as the control object.

3.4 Planning of Collision Avoidance Paths in Evolutionary Terms

Due to the fact that the process of collision avoidance at sea concerns control in a short time horizon, the aim is to develop an effective method of determining the optimal path for a moving object in the environment with static and dynamically varying constraints. The concept of an effective method consists in solving the problem in such a way that it allows to determine an optimal pathway that will bypass all static and dynamically variable constraints in near-real time. The input data for the path optimization task is the information about the analysed environment and about the currently identified static and dynamic constraints. The correctness of the obtained solution depends on the accuracy of the adopted process model and the applied optimum search method. Proper model definition and its verification with the real process is a prerequisite for obtaining a reliable solution. On the other hand, achieving a globally optimal solution for the task of determining the path is closely related to the choice of the optimization method. Taking into account the specificity of the task, the effectiveness of algorithms, it can be stated that evolutionary algorithms are an effective tool for solving this problem in near-real time [71–74]. They have been permanently included in the group of optimization algorithms and their usefulness has been confirmed by numerous practical applications [71]. They are characterized by a specific scheme of operation and the possibility of easy adaptation to different areas of the problem being solved (continuous and discreet space, with and without constraints).

Initial applications of evolutionary algorithms to avoid obstacles by robots were related to the issue of path planning in stationary environment. This planning was shown in a number of works, where mainly schematics of evolutionary algorithms in off-line mode were presented, and their parameters and capabilities discussed. In most works, simple, standard, evolutionary algorithms were used [74, 75], without taking into account specific knowledge about the nature of the path and the environment. Moreover, in most cases, the environment was discretised, which forced the generation of path nodes at designated points, and that significantly limited the space of solutions. Significant progress in this area was made by introducing planning of the path of changes into the standard evolutionary algorithm. This planning consisted in taking into account the current environment, parallelisation of calculations, and creating a variety of paths. The researchers also experimented with evolutionary path planning methods, using dynamic chromosome structures and a modified crossing operator to optimize actual processes [76]. However, most approaches assumed the existence of the environment containing discretised path nodes [77]. As a result, the above methods did not meet the condition of solving the problem in near-real time. Real-time control can be achieved by using methods that refer to potential fields known in physics, where the trajectory is determined by the "repulsive" effect of constraints on a moving object [78]. A breakthrough in this area came when heuristics representing specific knowledge of the task was used in evolutionary calculations to determine the path strategy in near-real time [79, 80]. Attempts have also been made to use classification systems and the genetic programming paradigm for planning the pathway [75].

The results of the research on the solution of the specific problem of path planning in collision situation confirm the necessity of introducing heuristic knowledge about the process into the evolutionary algorithm. Moreover, the space of solutions should not be limited to searching the paths in a fixed map of nodes, but in the whole space of acceptable solutions. Additionally, the path planning system should be able to adapt to the changing diversity of the environment, at the same time achieving the objectives of optimization in the assumed environment in near-real time.

3.5 Evolutionary Method of Avoiding Collisions at Sea

The first works that presented the use of evolutionary algorithms to solve the task of avoiding collisions at sea concerned many aspects, such as: defining the task, the population structure, genetic operators, domain modelling, and the ship management system [818283]. The starting point for the works on the method used to avoid collisions at sea was the EP/N (Evolution Planner/Navigator) system [79, 80]. It allowed for evolutionary search for the robot's path in a stationary environment. The idea of applying the evolutionary mechanism to solve the collision avoidance problem was also studied by [84–86]. In these papers, a genetic algorithm was used to optimize the automatic collision avoidance system. This algorithm contained a single gene being a combination of parameters of ship's position, latitude, longitude, and speed, as well as parameters of external factors: wave, force, and direction of wind, etc. The genetic algorithm was used to optimize the collision avoidance system. A detailed description of the coding method for planning a secure trajectory was given by [85]. The presented solutions concern: introduction of marine environment and weather disturbances into the chromosome model, the way to determine the potential risk of collision using a probability density map, and the proposed algorithm testing methodology and collision avoidance system.

Also noteworthy are the works by, presenting the version of ESoSST (Evolutionary Sets of Safe Ship Trajectories). They make use of evolutionary algorithms and some elements of game theory to solve ship collision situations. By defining individual positions and vessel traffic parameters, the method determines approximate safe trajectories for all ships participating in the meeting. The trajectories are presented as a sequence of manoeuvres in the event of a meeting with a larger number of objects in confined waters, taking into account the collision risk, and using defined domains referring to a specific navigational situation and a zoom factor. The paper tests different versions of adaptation functions, specialized operators, and selection methods.

After developing the EP/N (Evolutionary Planner/Navigator) system in the works [81, 82, 87], an extended version of this system was developed. The new version bears the name of vEP/N++, where the symbol v indicates change in the speed parameter of the own vessel, and the symbol ++ indicates the new version of the system. This system makes it possible to take into account time variable limitations and speed changes of the object moving along individual path segments. The task of collision avoidance is reduced to path adaptation to the modelled marine environment. Initially, a population of paths is created in the evolutionary system. Then, the algorithm seeks the best solution from the population. The solution is the path that is best adapted to the environment. The scale of path adaptation is determined by the survival function. Appropriately shaped areas of danger—domains of the met objects, take into account the COLREGs. The algorithm also takes into account the dynamic aspect of the environment by recording current changes in the strategy of the approaching objects and the speed of the own ship. While constructing the algorithm, the initial population and chromosomes of individuals, paths in this case, are defined, along with specialized genetic operators. In the initial phase, the input data is entered, parameterizing the work of the algorithm and describing the environment, and random population initialization is started. The principle of the algorithm is based on an iterative process of creating new individuals (paths) in the population. A genetic operator randomly selected from the set of specialized operators modifies a randomly selected path or paths. Then, after evaluating the adaptation quality of all paths in the modelled environment, paths that are better adapted than those currently in existence are introduced into the new population, while the worst paths are removed. The population is a set of solutions—a set of paths, which are evaluated in the environment with constraints in terms of the quality of their adaptation, determined by the target function. A single path—an individual in the population—represents a specific solution to the problem. The model of the environment in the vE/PN+++ system consists of static and dynamic constraints [88], which represent static and moving obstacles. In vE/PN+++, the dynamic constraints can be mobile vessels or surface vehicles, while the static obstacles are land, islands, shallows, traffic separation schemes, etc. All these obstacles are simplified to polygons. In cases of obstacles with extensive coastline, genetic operators may introduce a large number of turns into the path. In order to limit their number, a sequence of lines is created to simplify the shape of the prohibited area. For safety reasons, a redundant zone approximating the area is created around each prohibited area. At the same time, the basic assumption of the path planning task is that the starting and target points, or the mission area, are not within the fixed limits increased by the redundant zone. In the algorithm, in the initiation phase, an individual representing a single path S from the starting point s_0 to the end point—target point s_e is defined (Fig. 3).

The path S consists of elementary line segments s_i, where $i = 1, \ldots, n-1$, and n is the number of waypoints, connected at waypoints (y_i, x_i). Including waypoints to the construction of an individual allows them to be easily removed and added to the existing path. In order to increase the effectiveness of the assessment of path adaptation to the environment, the information about the coordinates (y_i, x_i) of the waypoints is complemented by the data about the ship speed along the segment between the

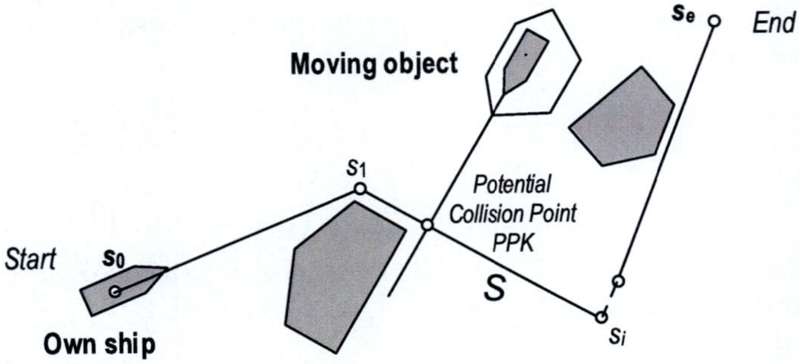

Fig. 3 Path S from starting point s_0 to target point s_e

current and the next waypoint, and a marker specifying the type of waypoint, and the angle and length of the section between two path segments, with the top at a given waypoint. In the initiation phase, the initial population is generated randomly, the coordinates of the waypoints (y_i, x_i) in elementary individuals are freely selected from the admissible space defined by the constraints of the environment. Each individual in the population contains the starting and the target point, and randomly selected waypoints. The population contains a defined number of individuals, modified in the reproduction and creation phase by genetic operators. These operators are the basic tool for creating individuals with new characteristics to improve their adaptation. This mechanism allows the algorithm to find better and better paths in the examined environment, relative to a given adaptation function determined by the total cost of path, *Total_Cost(S)*. The adaptation function in the navigation environment is based on opposing criteria related to maritime safety and traffic economics, which tend to minimize, on the one hand, the risk of collision with fixed and movable objects, and, on the other hand, the cost of changing the ship's route determined by the set path. The total cost of path, *Total_Cost(S)*, takes into account both the safety criterion, *Safe_Cost(S)*, and the traffic economics, *Econ_Cost(S)*, of the vessel moving along the path, according to the following equation [64, 83, 89]:

$$Total\_Cost(S) = Safe\_Cost(S) + Econ\_Cost(S) \tag{2}$$

The safety criterion *Safe_Cost*(*S*) is met if the path does not exceed static constraints and moving areas of danger, defined as domains representing moving objects [64, 88–91]. This criterion is frequently referred to as "path cleanliness" and defined as the maximum quality of segment's waypoints in terms of their distance from constraints:

$$Safe\_Cost(S) = w_c \cdot clear(S) \tag{3}$$

where $clear(S) = \max\{c_i\}$, for $i = 1, \ldots, n-1$, n is the number of waypoints, c_i is the difference between the distance from the waypoint of segment s_i closest to the obstacle and the assumed distance d, and w_c is the weight factor.

The criterion *Econ_Cost(S)*, evaluating the path from the economic point of view, consists of three weighted values: the length *dist(S)* of the path *S*, defined as the sum of the lengths of sections s_i $(i = 1, \ldots, n-1)$:

$$dist(S) = \sum_{i=1}^{n-1} dist(s_i) \tag{4}$$

the value *smooth*(*S*), defined as the maximum turning angle between successive path segments at the waypoints of section s_i:

$$smooth(S) = \max\{smooth(s_i)\}, \quad dla \quad i = 1, 2, \ldots, n-1 \tag{5}$$

and the time needed to complete the path *S:*

$$time(S) = \sum_{i=1}^{n-1} time(s_i) \quad (6)$$

By taking into account the weight coefficients w_d, w_s and w_t, the sum of the above individual criteria reduces *Econ_Cost(S)* to the following single-criterion form:

$$Econ\_Cost(S) = w_d \cdot dist(S) + w_s \cdot smooth(S) + w_t \cdot time(S) \quad (7)$$

After taking into account the formulated criteria of path safety and non-violation of static and dynamic constraints, as well as the economic criteria evaluating the transition in terms of energy expenditure, determined by the turning angles of path segments and the length and time to complete the determined path, the problem is reduced to the dynamic optimization task. The solution to this task is searched in the set of acceptable paths that meet the condition of avoiding static and dynamic constraints. The solution to the task with respect to criterion (1) will be a safe transition path, configured in the constrained environment using a sequence of segments defined by the coordinates (y_i, x_i) of their points of turn and the time t needed to reach these points.

In summary, the task of driving a ship or a surface vehicle, including collision avoidance, is understood as an evolutionary process of searching for safe passageways. It consists in moving safely the maritime object from its starting point to a given destination point or mission area, while avoiding stationary obstacles and other moving objects. To this end, an evolutionary optimal trajectory of the ship in relation to a specific criterion should be determined. Along the given trajectory, other disturbances (e.g. hydrometeorological disturbances) may act on the ship, causing deviations from this trajectory. These deviations are to be corrected by the control system, which generates signals to the executive elements—rudders and ship's propulsions. The structure of control of a moving object was presented by [8]. This structure contains two systems. The first system generates passage routes in the control layer, where the preset trajectory is determined and corrected in case of potential collision with another object, as well as the ship motion control along this trajectory is performed. The task of the second system is measuring and tracking other moving objects (navigation layer).

Additional disturbances, which make correcting the previously determined trajectory necessary as a result of trajectory and/or speed changes of other moving objects, are treated as dynamic constraints. Tracking the moving objects allows to observe changes in their motion parameters and introduce these changes to the control system. Then, based on the new parameters, determined online, the trajectory is corrected by the system using an evolutionary algorithm. This task is done by the superior object management system, which adapts the ship's trajectory to new environmental conditions. The *on-line* mode of system operation in the case of changes in motion parameters of the encountered objects (changes in their strategy: location, direction, and speed) was presented by [92]. The review of collision avoidance methods, understood as determination of the collision manoeuvre and safe trajectory, and the comparison of solutions obtained with the use of evolutionary algorithms, were presented in the papers by [89]. The paper by [93] included a number of results of experiments on the use of evolutionary collision avoidance methods. Also, works on a new version of the competing evolutionary algorithm were carried out. The main goal of the application of this algorithm was to obtain solutions close to globally optimal solutions [94]. Starting from the same starting point, two parallel evolutionary algorithms compete on the basis of a ranking, and despite the increase in calculation time, they are able to search the space of solutions more effectively. The result with the maximum value of the adaptation function is the final solution. Two types of evolutionary algorithms were used. The first type, a so-called single-population evolutionary algorithm, was repeatedly run and the solutions from the previous run were passed on to the next run. The second algorithm was a two-population algorithm, which solved the problem on the basis of competition between two populations: the base population and the elite population [95]. The vEP/N++ system was applied to the multilayer integrated ship control system [63, 96]. It was also used for linking evolutionary algorithms with a fuzzy regulator to control a ship on the trajectory [97].

3.6 Summary

The summary covers two issues of dynamic positioning and collision avoidance at sea. Referring to dynamic positioning, modern control methods enable complex low-speed manoeuvring operations at sea, in inland and port areas. The multitasking of the DP ship forces the use of more and more advanced control techniques which combine different algorithms in search of the best solutions in terms of resistance to model errors and interference, control accuracy, and optimal power distribution. Research in the field of algorithms resistant to drive damage can reduce material and economic losses caused by interruption of operation and loss of stability of the unit. Data fusion algorithms, on the other hand, enable accurate determination of ship's position from several sensors, which increases the accuracy of estimation and control in difficult conditions of the marine environment. In terms of decision support and collision avoidance with the use of evolutionary algorithms, they allow to obtain solutions in near-real time. Evolutionary mechanism of self-adaptation allows to determine the solution in case of dynamic changes of the environment, or change of the strategy of another object. In addition to collision situations, these methods can be used to control an autonomous surface vehicle, optimize the route of the voyage, and/or avoid moving areas of danger. The proposed methods will increase the level of safety at sea and, through route optimization, will increase the economic effects of maritime transport. In addition to maritime applications, evolutionary methods can also be used to control autonomous air and land vehicles performing both civil and military missions. Optimization of the control of these types of vehicles concerns the accuracy of mission execution and energy savings, which will increase the scope and duration of facility's operation.

References

1. Price, W.G., Bishop, R.E.D.: Probabilistic Theory of Ship Dynamics. Chapman and Hall, London, UK (1974)
2. Fossen, T.I., Strand, J.P.: Passive nonlinear observer design for ships using Lyapunov methods: full-scale experiments with a supply vessel. Automatica **35**(1), 3–16 (1999)
3. Fossen, T.I.: Guidance and Control of Ocean Vehicles. Wiley, First edition (1994)
4. Fossen, T.I.: Handbook of Marine Craft Hydrodynamics and Motion Control. Wiley, Ltd, Chichester, UK (2011)
5. Perez, T.: Ship Motion Control. Springer (2005)
6. Koyama, T.: On the optimum automatic steering systems of ships at sea. J. Soc. Naval Architects **122**, 18–35 (1967)
7. Fossen, T.I., Perez, T.: Kalman filtering for positioning and heading control of ships and offshore rigs: estimating the effects of waves, wind, and current. IEEE Control Syst. **29**(6), 32–46 (2009)
8. Śmierzchalski, R.: Ship automation and control. Volume 18—The Committee of Automation and Robotics of the Polish Academy of Sciences. Gdansk University of Technology Publishing House). Gdańsk (In Polish) (2013)
9. Fay, H.: Dynamic Positioning Systems, Principles, p. 1989. Design and Applications, Editions Technip, Paris, France (1989)
10. Skjong, E., Volden, R., Rodskar, E., Molinas, M., Johansen, T.A., Cunningham, J.: Past, present, and future challenges of the marine vessel's electrical power system. IEEE Trans. Transp. Electrification **2**, 522–537 (2016)
11. Sørensen, A.J.: A survey of dynamic positioning control systems. Ann. Rev. Control **35**, 123–136 (2011)
12. Fossen, T.I., Skjetne, R., and Lindegaard, K.P.W.: Inertia shaping techniques for marine vessels using acceleration feedback. In: Proceedings of the 15th IFAC World Congress. Barcelona, Spain (2002)
13. Fossen, T.I.: Marine Control Systems. Marine Cybernetics (2002)
14. Balchen, J.G., Jenssen, N.A., Mathisen, E., Sælid, S.: A Dynamic Positioning System Based on Kalman Filtering and Optimal Control. Modeling IFAC/IFIP Symposium on Automation in Offshore Oil Field Operation, Amsterdam, The Netherlands (1976)
15. Fossen, T.I., Sagatun, S.I., Sørensen, A.J.: Identification of dynamically positioned ships. Control Eng. Pract., CEP **4**(3), 369–376 (1996)
16. Sørensen, A.J., Sagatun, S.I., Fossen, T.I.: Design of a Dynamic positioning system using model-based control. IFAC J. Control Eng. Pract. **4**(3), 359–368 (1996)
17. Grimble, M., Patton, R., Wise, D.: The design of dynamic ship positioning control systems using extended Kalman filtering techniques. OCEANS' **79**, 488–497 (1979)
18. Shin, E.-H., El-Sheimy, N.: An unscented kalman filter for in-motion alignment of low-cost IMUs. In: Plans, pp. 273–279 (2004)
19. Gustafsson, F.: Particle filter theory and practice with positioning applications. IEEE Aerosp. Electron. Syst. Mag. **25**(7 PART 2), 53–81 (2010)
20. Jaroś, K., Witkowska, A., Śmierzchalski, R.: Designing particle kalman filter for dynamic positioning. In: Kościelny, J., Syfert, M., Sztyber, A. (eds.) Advanced Solutions in Diagnostics and Fault Tolerant Control. DPS 2017. Advances in Intelligent Systems and Computing, p. 635. Springer, Cham (2018)
21. Berghuis, H., Nijmeijer, H.: A passivity approach to controller-observer design for robots. IEEE Trans. Robot. Autom. **9**(6), 740–754 (1993)

22. Tomera, M.: Nonlinear observers design for multivariable ship motion control. Pol. Marit. Res., Issue Special **19**(74), 50–56 (2012)
23. Tomera, M.: Dynamic positioning system for "Blue Lady". Simulation tests. Pol. Marit. Res., Issue Special **19**(74), 57–65 (2012)
24. Fossen, T.I., Grøvlen, A.: Nonlinear output feedback control of dynamically positioned ships using vectorial observer backstepping. IEEE Trans. Control Syst. Technol. **6**(1), 121–128 (1998)
25. Witkowska, A., Śmierzchalski, R.: Adaptive dynamic control allocation for dynamic positioning of marine vessel based on backstepping method and sequential quadratic programming. Ocean Eng. **163**, 570–582 (2018)
26. Witkowska, A., Śmierzchalski, R.: Adaptive backstepping tracking control for an over–actuated DP marine vessel with inertia uncertainties. Int. J. Appl. Math. Comput. Sci. **28**(4), 679–693 (2018). https://doi.org/10.2478/amcs-2018-0052
27. Swaroop, D., Hedrick, J.K., Yip, P.P., Gerdes, J.C.: Dynamic surface control for a class of nonlinear systems. IEEE Trans. Autom. Control **45**(10), 1893–1899 (2000)
28. Du, J., Abraham, A., Yu, S., Zhao, J.: Adaptive dynamic surface control with Nussbaum gain for course-keeping of ships. Eng. Appl. Artif. Intell. **27**(1), 236–240 (2014)
29. Han, J.: From PID to active disturbance rejection control. IEEE Trans. Industr. Electron. **56**(3), 900–906 (2009)
30. McGookin, E.W., Murray-Smith, D.J., Li, Y., Fossen, T.I.: Ship steering control system optimization using genetic algorithms. Control Eng. Pract. **8**(4), 429–443 (2000)
31. Lee, T.H., Cao, Y., Lin, Y.M.: Application of an on-line training predictor/controller to dynamic positioning of floating structures. Tamkang J. Sci. Eng. **4**(3), 141–154 (2001)
32. Lee, T.H., Cao, Y., Lin, Y.M.: Dynamic positioning of drilling vessels with a fuzzy logic controller. Int. J. Syst. Sci. **33**(12), 979–993 (2002)
33. Du, J.L., Yang, Y., Wang, D.H., Guo, C.: A robust adaptive neural networks controller for maritime dynamic positioning system. Neurocomputing **110**, 128–136 (2013)
34. Witkowska, A., Niksa-Rynkiewicz, T.: Dynamically positioned ship steering making use of backstepping method and artificial neural networks. Pol. Marit. Res. **4**, **25**(100), 5–12 (2018). https://doi.org/10.2478/pomr-2018-0126
35. Niksa-Rynkiewicz, T., Witkowska, A.: Analysis of impact of ship model parameters on changes of control quality index in ship dynamic positioning system. Pol. Marit. Res. **1**, **26**(101), 6–14 (2019). https://doi.org/10.2478/pomr-2019-0001
36. Wang, Y., Sui, Y., Wu, J., Jiao, J.: Research on nonlinear model predictive control technology for ship dynamic positioning system. In: IEEE International Conference on Automation and Logistics, pp. 348–351 (2012)
37. Hu, X., Du, J., Shi, J.: Adaptive fuzzy controller design for dynamic positioning system of vessels. Appl. Ocean Res. **53**, 46–53 (2015)
38. Benetazzo, F., Ippoliti, G., Longhi, S., Raspa, P.: Advanced control for fault-tolerant dynamic positioning of an offshore supply vessel. Ocean Eng. **106**, 472–484 (2015)
39. Zhang, J.X., Yang, G.H.: Robust adaptive fault-tolerant control for a class of unknown nonlinear systems. IEEE Trans. Industr. Electron. **64**(1), 585–594 (2017)
40. Tjønnås, J., Johansen, T.A.: Adaptive optimizing nonlinear control allocation. In: Proceedings of the 16th IFAC World Congress, vol. 38, no. 1, pp. 1160–1165 (2005)
41. Tjønnås, J., Johansen, T.A.: Optimizing adaptive control allocation with actuator dynamics. In: Proceedings of 46th IEEE Conference on Decision and Control, pp. 3780–3785 (2007)
42. Tjønnås, J., Johansen, T.A.: Adaptive control allocation. Automatica **44**, 2754–2766 (2008)
43. Hespanha, J.: Tutorial on supervisory control. In: Lecture Notes for the Workshop Control Using Logic and Switching for the 40th Conference on Decision and Control, pp. 1–46 (2001)
44. Hespanha, J.P., Liberzon, D., Morse, A.S.: Overcoming the limitations of adaptive control by means of logic-based switching. Syst. Control Lett. **49**(1), 49–65 (2003)
45. Tomera, M.: Hybrid switching controller design for the maneuvering and transit of a training ship. Int. J. Appl. Math. Comput. Sci. **27**(1), 63–77 (2017)

46. Nguyen, T.D., Sørensen, A.J., Tong Quek, S.: Design of hybrid controller for dynamic positioning from calm to extreme sea conditions. Automatica **43**(5), 768–785 (2007)
47. Hu, X., Du, J., Sun, Y.: Robust adaptive control for dynamic positioning of ships. IEEE J. Oceanic Eng. **42**(4), 826–883 (2017)
48. Du, J.L., Hu, X., Krstic, M., Sun, Y.Q.: Robust dynamic positioning of ships with disturbances under input saturation. Automatica **73**, 207–214 (2016)
49. Jwo, D.J., Chung, F., Weng, T. Adaptive kalman filter for navigation sensor fusion. In: Sensor Fusion and Its Applications, pp. 65–90 (2001)
50. Zhao-Lin, W.: Quantification of action to avoid collision. J. Navig. **37**, 420–430 (1984)
51. Zwierzewicz, Z.: On a computer oriented analytical method for solving optimal control problems via maximum principle-zermello navigational problem. In: Międzyzdroje: 3rd International Symposium on Methods and Models in Automation and Robotics MMAR, Międzyzdroje (1996)
52. Lisowski, J.: The dynamic game models of safe navigation. Int. J. Marine Navig. Saf. Sea Transp. **1**(1), 11–18 (2007)
53. Burns, R.S., Blackwell, G., Calvert, S.: An automatic guidance, navigation and collision avoidance system for ships at sea. IEE Colloquium on Control in the Marine Industry, pp. 3/1–3/3 (1988)
54. Zhang, X.K., Jia, X.L., Wang, X.C.: Robust track-keeping autopilot for ships. In: Proceedings of the 3rd IFAC Workshop on Control Applications in Marine Systems, pp. 138–142 (1995)
55. Statheros, T., Howells, G., McDonald-Maier, K.: Autonomous ship collision avoidance navigation concepts, technologies and techniques. J. Navigation **61**, 129–142 (2008). The Royal Institute of Navigation
56. Sutton, R., Roberts, G.N.: Martronics: a philosophy for marine systems design. In: Proceedings of the IMechE Part M: J. Engineering for the Maritime Environment, vol. 223, pp. 160–174 (2009)
57. Burns, R.S.: The application of artificial intelligence techniques to modelling and control of surface ships. In: Eleventh Ship Control Systems Symposium, Computational Mechanics Publications, Southampton, Boston, vol. 1, pp. 77–83 (1995)
58. Enab, Y.M.: Intelligent controller design for the ship steering problem. IEE Proc. Control Theory Appl. **143**(1), 17–24 (1996)
59. Hong, X., Harris, C.J., Wilson, P.A.: Autonomous ship collision free trajectory navigation and control algorithms. In: Proceedings of the 7th IEEE International Conference on Emerging Technologies and Factory Automation, ETFA '99, vol. 2, pp. 923–929 (1999)
60. Hwang, C.N.: The integrated design of fuzzy collision-avoidance and hinf-autopilots on ships. J. Navig. **2002**(55), 117–136 (2002)
61. Lee, S.B., Kim, I.: A ship automatic navigation in dynamic environment using artificial intelligence. In: Proceedings of the Fourth International Conference on Knowledge-Based Intelligent Engineering Systems and Allied Technologies, vol. 1, pp. 418–422 (2000)
62. Stamenkovich, M.: An application of artificial neural networks for autonomous ship navigation through a channel. In: Position Location and Navigation Symposium, '500 Years After Columbus—Navigation Challenges of Tomorrow', IEEE PLANS '92, pp. 346–352 (1992)
63. Śmierzchalski, R.: Ship as intelligent machine. In: X International Scientific and Technical Conference on Marine Traffic Engineering, Szczecin 2003, Szczecin, pp. 239–248 (2003)
64. Śmierzchalski, R.: Intelligent marine control systems. In: Enhanced Methods in Computer Security, Biometric and Artificial Intelligence Systems, Międzyzdroje, pp. 311–319 (2005)
65. Campbell, S., Naeem, W., Irwin, G.: A review on improving the autonomy of unmanned surface vehicles through intelligent collision avoidance manoeuvres. Ann. Rev. Control **36**(2), 267–283 (2012)
66. Liu, Y., Liu, H.: Case learning base on evaluation system for vessel collision avoidance. In: Proceedings of the Fifth International Conference on Machine Learning and Cybernetics, pp. 2064–2069 (2006)
67. Liu, Y., Yang, C., Yang, Y., Lin, F., Du, X., Ito, T.: Case learning for CBR-based collision avoidance systems avoidance systems. Appl. Intell. **36**(2), 308–319 (2012)

68. Yang, S., Lina Li, L., Suo, Y., Chen, G.: Study on construction of simulation platform for vessel automatic anti-collision and its test method. In: Proceedings of IEEE International Conference on Automation and Logistics, pp. 2414–2420 (2007)
69. Zhuo, Y., Hearn, G.E.: A ship based intelligent anti-collision decision-making support system utilizing trial manoeuvres. In: Proceedings of Chinese Control and Decision Conference, pp. 3982–3987 (2008)
70. Perera, L.P., Carvalho, J.P., Guedes Soares, C.: Intelligent ocean navigation and fuzzy-bayesian decision-action formulation. IEEE J. Oceanic Eng. **37**(2), 204–219 (2012)
71. Branke, J.: Evolutionary Optimization in Dynamic Environments. Kulwer Academic Publisher (2002)
72. Kurata, J., Grattan, K., Uchiyama, H.: Path planning for a mobile robot by integrating mapped information. In: 3rd IFAC Conference IAV Intelligent Autonomous Vehicle, Madrid, pp. 319–323 (1998)
73. Pongpunwattana, A., Rysdyk, R.: Evolution-based dynamic path planning for autonomous vehicles. In: Innovations in Intelligent Machines, pp. 113–145 (2007)
74. Shibata, T., Fukuda, T.: robot motion planning by genetic algorithm with fuzzy critic. In: Proceedings of the 8th IEEE International Symposium on Intelligent Control, Chicago (1993)
75. Handley, S.: The genetic planner: the automatic generation of plans for a mobile robot via genetic programming. In: Proceedings of the 8th IEE International Symposium on Intelligent Control (1997)
76. Zhuang, H., Du, S., Wu, T.: On-line real-time path planning of mobile robots in dynamic uncertain environment. J. Zhejiang Univ. Sci. A **7**(4), 516–524 (2006)
77. Yanualailai, J., Ha, J.H., Nakagiri, S.I.: A solution to the two-dimensional find path problem. In: 3rd IFAC Conference IAV Intelligent Autonomous Vehicle, Madrid, Spain, pp. 313–318 (1998)
78. Lee, S.M., Kwon, K.Y., Joh, J.: A fuzzy logic for autonomous navigation of marine vehicles satisfying COLREG guidelines. Int. J. Control Autom. Syst. **2**(2), 171–182 (2004)
79. Michalewicz, Z., Xiao, J.: Evaluation of paths in evolutionary planner/navigator. In: Proceedings of the International Workshop on Biologically Inspired Evolutionary Systems, Tokyo, pp. 45–52 (1995)
80. Xiao, J., Michalewicz, Z., Zhang, L., Trojanowski, K.: Adaptive evolutionary planner/navigator for mobile robots. IEEE Trans. Evol. Comput. **1**(1), 18–28 (1997)
81. Śmierzchalski, R.: Evolutionary guidance system for ship in collisions situation at sea. In: 3rd IFAC Conference IAV Intelligent Autonomous Vehicle, Madrid, pp. 391–396 (1998)
82. Śmierzchalski, R.: Speed mutation evolutionary ship trajectory planning in navigation traffic areas. In: IFAC Conference Control Applications in Marine Systems, Fukuoka (1998)
83. Śmierzchalski, R., Michalewicz, Z.: Adaptive modeling of a ship trajectory in collision situations. In: 2nd IEEE World Congress on Computational Intelligence, ICEC'98, Alaska, USA, pp. 342–347 (1998)
84. Ito, M., Zhnng, F., Yoshida, N.: Collision avoidance control of ship with genetic algorithm. In: Proceedings of the IEEE International Conference on Control Applications, vol. 2, pp. 1791–1796 (1999)
85. Zeng, X.: Evolution of the safe path for ship navigation. Appl. Artif. Intell. **17**(2), 87–104 (2003)
86. Zeng, X., Ito, M.: Planning a collision avoidance model for ship using genetic algorithm. In: IEEE International Conference on Systems, Man, and Cybernetics, vol. 4, pp. 2355–2360 (2001)
87. Śmierzchalski, R.: Ship's domains in evolutionary guidance system. In: 5th International Symposium on Methods and Models in Automation and Robotics, Międzyzdroje, pp. 381–386 (1998)
88. Śmierzchalski, R.: Static and dynamic constrains in evolutionary problem of ship's trajectory planning at sea. In: 3rd World Multiconference on Systems, Cybernetics and Informatics (SCI'99), USA, Orlando, pp. 1124–1130 (1999)

89. Śmierzchalski, R., Michalewicz, Z.: Modeling of a ship trajectory in collision situations at sea by evolutionary algorithm. IEEE Trans. Evol. Comput. **4**(3), 227–241 (2000)
90. Śmierzchalski, R., Michalewicz, Z.: Path planning in dynamic environments. In: Innovations in Machine Intelligence and Robot Perception, Springer, pp. 135–154 (2005)
91. Śmierzchalski, R., Weintrit, A.: Domains of navigation objects as an aid in planning the trajectory of a ship in a collision situation at sea (in Polish) (Volume I). In: III Navigation Symposium Gdynia, vol. I, Gdynia, pp. 265–279 (1999)
92. Śmierzchalski, R.: Experiments in on-line mode trajectory planning at sea by evolutionary algorithm, Warszawa 2002. Zeszyty Naukowe serii „Elektronika", pp. 137–147 (2002)
93. Śmierzchalski, R.: Simulation of ship's trajectory planning by evolutionary computation. Bezpieczeństwo na morzu, Politechnika Gdańska, Gdańsk (2001)
94. Dziedzicki, K., Łebkowski, A., Śmierzchalski, R., Tobiasz, M., Gierusz, W.: A multi-population, competitive evolutionary algorithm for ship trajectory planning. In: Computational Intelligence: Methods and Applications, pp. 158–165 (2008)
95. Martikainen, J.: Methods for improving reliability of evolutionary computation algorithms and accelerating problem solving. Helsinki: Ph.D. work, Helsinki University of Technology, Department of Electrical and Communications Engineering, Espoo (2006)
96. Łebkowski, A., Śmierzchalski, R.: System of safe ship steering in a collision situation at sea (Tom 2). In: IEEE International Conference on Methods and Models in Automation and Robotics MMAR, Międzyzdroje, vol. 2, pp. 1177–1182 (2004)
97. Gierusz, W., Dziedzicki, K., Śmierzchalski, R. and Łebkowski, A.: Intelligent ship steering system with fuzzy logic and robust controllers governed by evolutionary algorithm (Tom 16). Polish J. Environ. Stud. **16**(4b), 47–51 (2007)

On-line Diagnostics of Large-Scale Industrial Processes

Jan Maciej Kościelny

Abstract Basic problems of on-line diagnostics of large-scale industrial processes have been characterized, as well as the methods aimed at solving them, focusing on the authors results obtained in the area of fault isolation. Particular attention was given to fault distinguishability, multiple faults isolation, inference methods of uncertain signals, decomposition of the system and diagnostics in decentralized structures, as well as the application of graph models in diagnostic systems design for industrial processes. Effective and robust diagnostic algorithms of complex dynamic large-scale systems and their implementation in the realized diagnostics systems are also described. The summary emphasizes the significance of on-line diagnostics in ensuring process safety.

1 Introduction: State of the Art Versus Practice of On-line Diagnostics of Complex Industrial Processes

Over the past 30 years, research has been conducted around the world concerning fault detection and isolation derived from modeling and identification, as well as artificial intelligence techniques. The results have been presented in [1–12] and in numerous scientific papers, including survey papers such as [13–22]. Various scientific conferences are also a significant source of knowledge in this area, particularly cyclical specialist conferences such as: IFAC Symposium on Fault Detection, Supervision and Safety for Technical Processes—SAFEPROCESS, International Conference on Diagnostics of Processes and Systems—DPS, International Workshop on Principles of Diagnosis DX, Workshop on Advanced Control and Diagnosis, Conference on Control and Fault-Tolerant Systems—SysTol.

J. M. Kościelny (✉)
Faculty of Mechatronics, Warsaw University of Technology, ul. Św. A. Boboli 8, Warsaw 02-525, Poland
e-mail: jm.koscielny@mchtr.pw.edu.pl

P. Kulczycki et al. (eds.), *Automatic Control, Robotics, and Information Processing*, Studies in Systems, Decision and Control 296,
https://doi.org/10.1007/978-3-030-48587-0_21

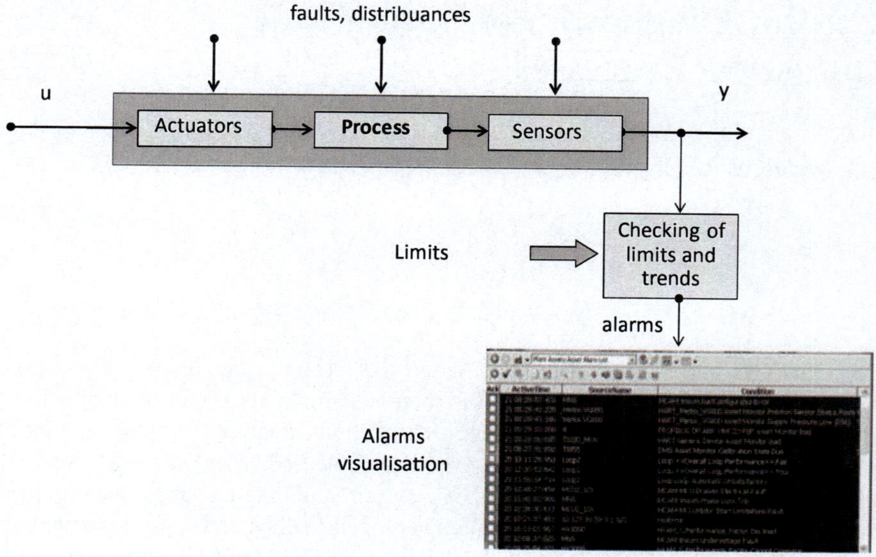

Fig. 1 Scheme of alarm-based diagnostics

Two approaches to industrial processes diagnostics can be distinguished: classic and advanced. For the first one (Fig. 1), which still dominates in industrial practice, mainly simple methods of limits control are used for fault detection. Fault isolation is usually performed by the operator. In some cases, the operator is supported by the system inferring on the causes of the alarms on the basis of rules and logical functions. Advanced diagnostics, in turn, is based on the use of process models to detect faults, and automatic inference or classification methods are used for fault isolation.

In classic solutions, the role of a diagnostic system of a process is played by an alarm system, which is an integral part of the contemporary process control and monitoring systems (DCS, SCADA). Alarm systems have multiple disadvantages [23], the main of them being a very large number of alarms in an state of emergency. It is typical for alarms to exceed 100 or even 500 per minute. EEMUA (The Engineering Equipment and Materials Users' Association) data shows that the average daily number of alarms in the petrochemical industry is approximately 1500, and in the power industry 2000, whereas according to recommendations it should not exceed 144. The excess of alarms is a huge problem to operators and was even a cause of the explosion at the Texaco Refinery, Milford Haven in 1994. Excessive number of alarms causes information overload and stress as a consequence. This may lead to further mistakes of the operator staff that accumulates with previous faults causing serious failures [24]. Other serious defects of alarm systems its inability to detect some of the faults due to masking their symptoms by control systems and detection delays.

Despite these obvious disadvantages, a classic approach is commonly used in practice. Alarm systems are constantly upgraded with various mechanisms of alarm management (especially in DCS class control systems), but the main cause of their weakness: the use of limit control for fault detection, is still the same.

Diagnostic system (DS) for industrial processes are based on advanced methods of fault detection and isolation utilizing different types of models. Automatic diagnostics with the use of partial models of a process and advanced location algorithms (Fig. 2) allows to precisely indicate the occurring faults and significantly reduce the time to detect and isolate the failure when compared to the diagnostics realized by the alarm system and the operator. However, despite its advantages, DS are not commonly used yet.

Known applications of the nature of pilot implementations and industrial research, were mainly realized by research centres. Leading companies in the automation industry do not yet offer their own DS solutions for industrial processes. This is caused by the sensitivity of the models to non-stationarity of the controlled system, as well as effects of renovation works causing changes in the values of the system's parameters. It is then necessary to tune in the models impeding DS exploitation. Another cause for not implementing DS in industry is insufficient robustness of the

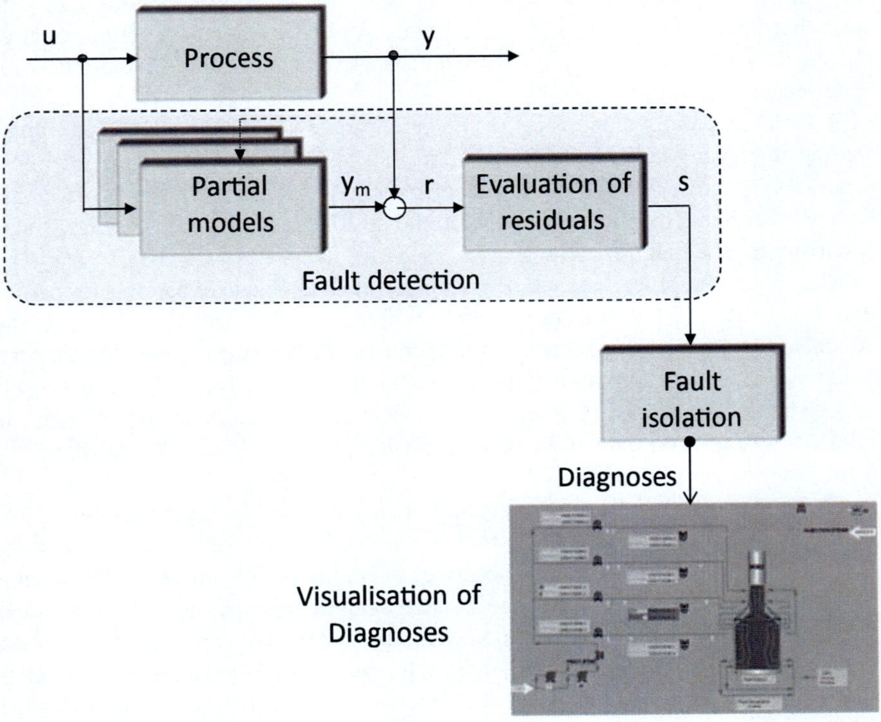

Fig. 2 Scheme of diagnostics based on partial models

applied fault Isolation method to various types of changes occurring in the diagnosed system [25]. There are also not enough experts employed in the industry.

It can be stated that between the theoretical level of diagnostics of dynamic systems and state of the systems, a huge gap is to be filled in the years to come due to safety reasons.

2 Fundamental Problems of On-line Diagnostics of Complex Processes

The Institute of Automatic Control and Robotics at Warsaw University of Technology (IAiR PW) has been conducting research in the field of on-line diagnostics of industrial processes for about 30 years. Its scientific goal is to develop methods of solving basic problems of fault detection in complex dynamic systems, as well as effective and robust algorithms for real-time automatic diagnostics of industrial processes.

Problems of on-line diagnostics of complex processes concern not only methods of designing diagnostic systems for complex systems, but also real-time diagnostic algorithms. They are mainly related to the scale of the diagnosed system and are specialised, with numerous technical solutions applied in extremely complex petrochemical, chemical, power and other installations. This makes it difficult to transfer experiences to new systems.

A factor impeding diagnostics of industrial processes is the lack of data representing emergency states. Obtaining training data for emergency states (e.g. chemical processes) is virtually impossible, and even more so for states with multiple faults. On the other hand, diagnostic systems should be able to recognize emergency states occurring for the first time.

This makes classification methods requiring the knowledge of patterns for particular states useless for detecting states with faults. Fault isolation can only be conducted by automatic inference methods and determining the relationship between faults and diagnostic signals values is possible only on the basis of expert knowledge. The construction of mathematical models of the process acknowledging the influence of faults on the values of residuals is very difficult and expensive, and in some cases even impossible.

In complex systems diagnostics, partial models are used for fault detection representing the normal state of a given part of a process [23]. Due to the non-linearity of industrial processes and changeability of operating points, linear models cannot be used for detection. The preference is to use analytical models describing physical phenomena occurring in the process. Such models reflect the properties of the system in the entire range of operating points of the system. However, developing models based on the description of physical phenomena is extremely hard or even impossible for many systems, because of the nature of phenomena occurring during the industrial processes is not entirely known. Due to these difficulties and limitations, approximation models are often created on the basis of measuring data and expert knowledge of the model's structure. These are different types of neural models [9]), TSK or Wang and Mendel type fuzzy models [9, 10]) and additive models [26]. Usability range of these types of models is limited to the range of input and output signals variability, on the basis of which the model was trained. Fault detection robustness is an important issue, which has been the subject of numerous publications [8] and thus will not be considered here.

The main problems of diagnosing large-scale process are related to, among others, ensuring high fault distinguishability, multiple faults isolation, inference of the uncertain symptoms and fault signatures, inference with system structure and diagnostic system variability, and delays in symptoms occurrence.

The main design problems are: the choice of the set of instrumentation for diagnostics (particularly for new installations), the choice of the set of the models for fault detection being the basis for diagnostic tests, determining fault-diagnostic signals values relation, and decomposition of the system into parts diagnosed by separate diagnostic units.

Fault detection and isolation methods should ensure the robustness of the diagnostic system, so that no false diagnoses are generated as a result of these problems. This chapter characterizes the abovementioned research issues the highlighting the authors achievements and publications from this area. Fault isolation was the main focus.

2.1 Faults Distinguishability

The aim of diagnostics is to detect early and precisely (accurately) indicate the fault. The accuracy of the obtained diagnoses is determined by the number of faults indicated in the diagnosis [10]. The smaller the number, the more accurate the diagnosis. The accuracy of diagnosis is then dependent on the obtained fault distinguishability (isolability). Fault distinguishability is defined as an ability to distinguish (isolate) a fault as a result of tests and inference. Undistinguishable faults are characterized by the fact that the interpreted symptoms are the same for them. Two faults are distinguishable (unconditionally) if the values of at least one of the symptoms are different for both faults. Fault distinguishability is then dependent on the symptoms taken into consideration during diagnostics.

It is usually desirable to obtain isolability of all faults, however it is not always necessary. It seems that it is enough to isolate only faults whose effects may be particularly dangerous. Another requirement is to isolate faults demanding various procedures providing safety to the system.

The obtained fault detectability and isolability is mainly affected by the set of the measured process variables, thus a set of performed tests. The more signals are measured, the more models can be constructed for particular tests. Methods of increasing fault distinguishability were the subject of numerous works conducted in the IAiR PW.

In the initial works [24, 27], quotient sets were used for the analysis of fault isolability in the binary assessment of residues. In [10] the possibility of increasing fault isolability by using multivalued residual assessment was described. Conducting an analysis is enabled by the proposed Fault Isolation System (FIS), which is an adaptation of the information system defined by Pawlak [28, 29]. FIS is a generalization of the binary diagnostic matrix. The significant extensions of FIS to the binary diagnostic matrix are as follows:

(a) each diagnostic signal can have an individual (finite) set of its values,
(b) any FIS element may have one diagnostic signal value or their subset.

In multivalued classifications of diagnostic signals, it is not possible to unambiguously determine isolability. It depends on the combination of the obtained test results. In [10, 30, 31] unconditional and conditional fault distinguishability in FIS is defined.

Faults $f_k, f_m \in F, k, m \in [1, K]; k \neq m$ are undistinguishable (unconditionally undistinguishable, i.e. remaining in $f_k R_{WN} f_m$ relation) in FIS due to diagnostic signals $s_j \in S$ if and only if their signatures are unique

$$f_k R_N f_m \Leftrightarrow \forall_{s_j \in S} V_{j,k} = V_{j,m}. \tag{1}$$

The signatures of unconditionally undistinguishable faults are unique.

Faults $f_k, f_m \in F$ are conditionally undistinguishable (remaining in $f_k R_{WN} f_m$ relation) in FIS due to diagnostic signals $s_j \in S$ if and only if for each signal, the subsets of its values corresponding to faults f_k and f_m have a common part and these faults are not unconditionally undistinguishable:

$$f_k R_{WN} f_m \Leftrightarrow \forall_{s_j \in S} V_{j,k} \cap V_{j,m} \neq \varnothing \wedge \exists_{s_j \in S} V_{j,k} \neq V_{j,m}. \tag{2}$$

Conditional fault undistinguishability means that v_j values of the diagnostic signals fulfilling the following condition may occur:

$$\forall_{s_j \in S} v_j \in V_{j,k} \cap V_{j,m}, \tag{3}$$

at which two given faults are undistinguishable. However, other values of diagnostic signals are possible, which means the same faults are isolable. The following condition is then met:

$$\exists_{s_j \in S} \left[v_j \in V_{j,k} \wedge v_j \notin V_{j,m} \right] \vee \left[v_j \notin V_{j,k} \wedge v_j \in V_{j,m} \right]. \tag{4}$$

Faults $f_k, f_m \in F$ are (unconditionally) distinguishable in FIS if there is a diagnostic signal for which the subsets of the values corresponding to these faults are disjoined:

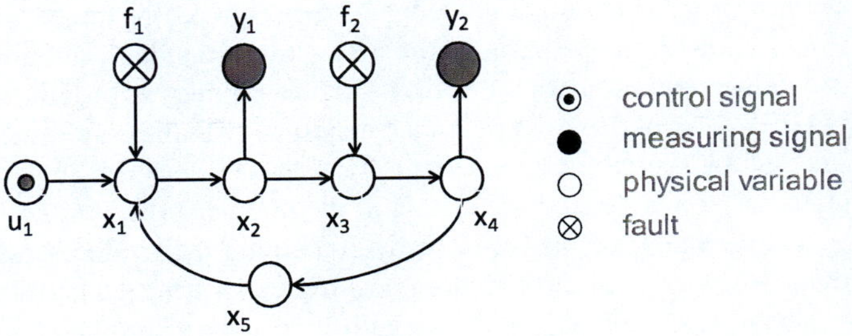

Fig. 3 Graph of a process illustrating fault isolability obtained on the basis of dissimilarity of the symptoms sequence [36]

$$f_k R_R f_m \Leftrightarrow \exists_{s_j \in S} V_{j,k} \cap V_{j,m} = \varnothing. \tag{5}$$

During inference, the values of diagnostic signals are often interpreted as the symptoms of faults, but the sequences of the occurring symptoms can also be taken into consideration. The diagnosed system is a dynamic system therefore the same fault is detected after different periods of time by different diagnostic tests. The order of symptoms occurrence is important information that has been used in diagnostic processes to increase fault isolability [32, 33], assuming that only the computational form of the residues is known.

Let's denote by $es_{j,p}(f_k)$ the elementary sequence, i.e. a sequence of two symptoms j and p for f_k fault. To distinguish any pair of faults, it is sufficient that the order of any pair of symptoms for these faults is different:

$$es_{j,p}(f_k) = \langle s_j, s_p \rangle, es_{j,p}(f_n) = \langle s_p, s_j \rangle. \tag{6}$$

The source of knowledge on the sequence of the symptoms may be the knowledge of a qualitative model of the process in the form of a cause-and-effect graph GP [34]. This will be illustrated by the example of the GP graph shown in Fig. 3.

Two models can be used for fault detection: $y_1 = g_1(u_1)$, $y_2 = g_2(u_1)$. They allow for calculating two residuals: $r_1 = y_1 - g_1(u_1)$, $r_2 = y_2 - g_2(u_1)$. They are sensitive to both faults f_1 and f_2. If f_1 occurs, $s_1 \neq 0$ symptom will occur first and then a $s_2 \neq 0$ symptom. The occurrence of f_2 will cause the reverse sequence of the symptoms, symptom $s_2 \neq 0$ will occur at first, and then $s_1 \neq 0$. Therefore: $s_{1,2}(f_1) = s_1, s_2, es_{1,2}(f_2) = s_2, s_1$, which ensures fault distinguishability.

The aforementioned example shows that in feedback systems, it is possible to design residues characterized by different sequences of symptoms occurrence for the same faults, when the choice of measurements is correct. The sequences of some pairs of symptoms can be determined on the basis of expert knowledge, without modelling the influence of faults on the system outputs [33, 35].

In [36] a connection between the applied forms of faults-diagnostic signals values relation and the obtained fault isolability is described. Formal conditions are provided for fault unisolability and isolability on the basis of a binary diagnostic matrix, fault directions in the residue region, knowledge of the sequence of the occurring symptoms, FIS system and on the basis of the areas in the residue region. Generalization of isolability definition was introduced by combining appropriate conditions concerning the values of diagnostic signals (residues) with the conditions describing the sequence of the symptoms.

For industrial processes, the greatest fault distinguishability with a given set of tests (diagnostic signals) can be achieved by combining multivalued residue assessment with the knowledge of the symptom occurrence sequence. The following fault undistinguishability/distinguishability conditions can be formulated for this case [36]:

Faults $f_k, f_m \in F$ are distinguishability (unconditionally distinguishable) on the basis of FIS and the sequence of the symptoms if and only if the signatures and the sequences of these faults are unique:

$$f_k R_N f_m \Leftrightarrow \left[\forall_{s_j \in S} V_{j,k} = V_{j,m}\right] \wedge \left[\forall_{s_j,s_p} es_{j,p}(f_k) = es_{j,p}(f_m)\right] \tag{7}$$

Faults $f_k, f_m \in F$ are (unconditionally) distinguishable on the basis of FIS and the sequence of the symptoms if there is a diagnostic signal for which subsets of the values corresponding to these faults are disjoint or have different elementary sequences of symptoms:

$$f_k R_R f_m \Leftrightarrow \left[\exists_{s_j \in S} V_{j,k} \cap V_{j,m} = \emptyset\right] \vee \{\left[\exists es_{j,p}(f_k) =< s_j, s_p >\right] \wedge \left[\exists es_{j,p}(f_m) =< s_p, s_j >\right]\} \tag{8}$$

In other cases, i.e. when one of the aforementioned conditions is not fulfilled, conditional distinguishability is obtained. The provided definitions unambiguously show that utilizing knowledge (usually incomplete) on the sequence of the symptoms allows us to achieve greater fault isolability in comparison to an inference based only on the signatures determined by the standard values of diagnostic signals.

For the case of knowing the internal form of the residuals (i.e. their fault-dependence) a method for designing secondary sequential residuals was developed [36, 37] allowing us to obtain isolating sequences for residuals unisolable by primary residues, provided that they are detected by at least two residuals. For certain pairs of faults, the sequences of the symptoms with the required properties can be obtained, such as simultaneous symptoms, symptoms in any order, shifted toward each other by a desired delay. A significant advantage of the designed pairs of related residuals is the fact that it allows for calculating the size of fault. The procedure then leads not only to a location, but also enables identification of single faults.

In [38], an original method is provided for isolating double faults for linear systems with the use of a directional residues method. Double faults location is based on the assessment of coplanarity of the residue vector with planes determined by directional vectors for particular pairs of faults. The conditions for isolability of a particular pair of faults from other faults and other pairs of faults was formulated. Isolability can already be tested at the diagnostic system design stage. A method of residual structuring has also been developed, i.e. designing secondary residues in such a way as to increase isolability or robustness of the double faults isolability.

However, both of these approaches have limited application in the diagnostics of complex systems. Due to the high costs of the design of the models acknowledging fault influence, they can only be used for critical processes, when design costs are negligible in comparison to the costs of losses during states of emergency.

The summary of the results of works in the field of fault distinguishability is presented in the monograph [35].

2.2 *Multiple Faults*

In large-scale industrial processes, the number of possible faults is very large (thousands—to tens of thousands). For this reason, simultaneous existence of many faulty elements of the process is a typical situation, and multiple faults may be a serious problem. Therefore, diagnostic algorithms for industrial processes should be able to detect not only single, but also multiple faults. However, the majority of faults occurs sequentially, at various time intervals. With on-line diagnostics, DGN diagnosis in n time instance should indicate a subset of faults, that has occurred since generating the previous diagnosis [10, 24]. The probability of more than one fault occurring in a short time period is low.

In [24, 39] the method of dynamic decomposition of the diagnosed system was proposed to be used in complex process diagnostic algorithms. It involves dynamic isolation of a sub-system where the fault is searched for, based on the first observed symptom since the previous diagnosis. Each sub-system is described by a subset of faults, subset of tests and the diagnostic relationship is denoted a Cartesian product of these subsets. Assuming no false values of diagnostic signals during binary assessment of the residuals values, it has been shown that the use of the dynamic decomposition of the system method leads to correct diagnoses generated, assuming the occurrence of single faults, when:

(a) faults occur sequentially at intervals greater than the time to formulate subsequent diagnoses;
(b) faults occur at the same time and dynamically isolated sub-systems corresponding to them are disjoint (in this case, faults are isolated in parallel in separate inference processes conducted assuming single faults).

Nevertheless, it should be emphasized that after each diagnosis, the set of the available diagnostic signals should be reduced by those which are sensitive to the detected fault. They can be included in the set of the available diagnostic signals again after bringing back up the state of the faulty element [10, 24, 39, 40].

Diagnostic inference assuming single faults is unreliable only if two or more faults occur in a shorter period than the time of formulating the diagnosis and the isolated sub-systems are not separate. For this case, an effective algorithm with two parts:

coarse and accurate isolation of multiple faults is provided [40]. A coarse diagnosis indicates a subset of possible faults containing faults for which all the symptoms characterizing them have been observed. Precise diagnosis indicates states with multiple faults for which the signature matches the obtained values of diagnostic signals. Diagnostic algorithms for multiple faults described in [40] significantly reduce computational complexity for diagnostics. Combining both of these algorithms gives an effective inference mechanism for double faults, as well as faults with a greater multiplicity.

A different approach to multiple faults diagnostics is presented in [41]. The extension of the basic, one-level model based on a binary diagnostic matrix into a two-level one was proposed. An approach to multiple fault diagnosis was added based on the inconsistence analysis.

2.3 *Diagnosis Uncertainties*

During diagnostic inference, some uncertainties occur that may lead to generating false diagnoses. The majority of data used in practice for diagnostics are uncertain [9, 42]. The causes of uncertainties are disturbances and measuring noise, model inaccuracy and difficulties in determining threshold values in decision algorithms. They lead to the uncertainty of diagnostic signals constituting the outputs of fault detection algorithms and inputs of the fault isolation algorithm.

Diagnoses generated in fault isolation algorithms are also affected by the uncertainty of the faults-symptoms relationship, defined by experts at the diagnostic system design stage. What is more, one can never be really sure if any possible fault has been included in the design phase. The occurrence of the omitted faults leads either to indicating other faults (unisolable with the omitted) or to a combination of test results different from the signatures of faults included in the knowledge base, which results in the inability to formulate a diagnosis.

Known methods for acknowledging uncertainty and lack of precision in diagnostic inference are: Bayes theory [43], fuzzy logic [44], Dempster-Shafer theory [45, 46], rough sets [29] and certainty factor [47, 48].

Bayes theory allows us to include uncertainty related to the disability to formulate validity/falsity of diagnosis. In Bayes theory we formulate diagnoses as hypotheses on the state of the system based on observations (diagnostic signals values) taking into consideration a'priori probability of particular states and conditional probability of test results in particular states of the system. This corresponds to the *IF* < observation > *THEN* < hypothesis > inference rule. In chapter 3.7 of the monograph [10], the principles of diagnoses formulation were specified on the basis of this theory. It was assumed that diagnostic inference is conducted on the basis of binary diagnostic signals generated by the threshold assessment of the absolute residual values.

Probability of particular states of the system z_i under condition of occurrence of the observed binary values of diagnostic signals V are dependent on conditional probabilities $P(V/z_i)$ of binary diagnostic signals in particular states of the diagnosed system and a'priori probability $P(z_i)$ of these states:

$$P(z_i/V) = P(z_i)P(V/z_i) \Big/ \sum_{i:z_i \in Z} P(z_i)P(V/z_i) \tag{9}$$

In diagnostics, we infer on the state of the system on the basis of numerous observations (diagnostic signals). Combining observations is a significant problem when inferring the basis of Bayes theory. The problem is simplified if the assumption of the independence of the observation is correct. In practice, this is only the case when Bayes theorem can be successfully used. In fact, this assumption is not strictly met, because the subsets of measurements used to calculate particular diagnostic signals are not always disjoint, and in such cases, they are dependent on the same disturbances and measuring of noise.

Knowledge of the above probabilities makes it possible to formulate a probabilistic diagnosis. The diagnosis indicates pairs: state of the system and the probability of its occurrence on the condition of obtaining the values of diagnostic signals for which the probability exceeds the accepted threshold value.

$$DGN(P) = \{z_i, P(z_i/V) : P(z_i/V) \geq K\}. \tag{10}$$

The advantage of the Bayes approach is the use of a well-established probability theory and knowledge of a'priori probabilities of particular faults. Knowledge about fault intensity of particular elements is increasingly available i.e. it is necessary for risk assessment in the safety analysis of technical systems, especially high and increased risk systems. It allows us to use reliability data for the elements of the system for diagnostic inference. The drawback of this way of inference, usually, is not fulfilling the assumption the independence of observation and difficulties with determining conditional probabilities $P(V/z_i)$ experimentally or theoretically. To determine them, it is necessary to have either the frequency of occurrence of 0 and 1 values of diagnostic signals in particular states of the system or distribution of the density of residue values probability. In practice, it is impossible to get such experimental data for industrial processes.

There is one more limitation on the use of formula (9) for making diagnoses. In industrial processes, the number of possible faults K is large, which leads to such a huge number of states of the system $|Z| = 2^K$, that it is impossible to take all of them into consideration during diagnostic inference. If the set of the system states is limited to the full up state and single faults states, the number of the analysed states is K + 1 and the diagnostic procedure gets significantly simplified. It is not purposeful to eliminate the full up state from this set even if faults symptoms have been observed, as we assume the possibility of false symptoms occurrence. Even in such a case, theoretical and experimental determination of probabilities $P(V/z_i)$ even for single faults is in fact impossible. The only way is to arbitrarily accept these values including threshold values for particular residues.

In large-scale systems, Bayesian networks cannot be used for fault isolation, which are useful for diagnostics of relatively simple systems. The reason is combinatorial explosion related to the analysis of all possible states of the network's inputs [49] and the difficulties in determining conditional probabilities described above.

Fuzzy logic is an effective tool for considering the lack of precision of variable values used during inference. Theoretical grounds of fuzzy sets can be found in [50, 51], while the examples of their diagnostic application are described in [10, 42, 52–55].

Fuzzy assessment of residual values allows for taking imprecise values of diagnostic signals into consideration. Diagnosis is specified by determining the firing threshold of the rules on faults (alternatively states of the system with multiple faults) on the basis of the degree of compatibility of the values of diagnostic signals with the values specified in the rules' premises. Firing thresholds of the rules are interpreted as a degree of certainty of the occurrence of a given fault or a state by a specified subset of faults. When utilizing fuzzy sets, we take into account the lack of observation precision leading to uncertainty of the symptoms and the frequency of fault occurrence is ignored.

Imprecise values of diagnostic signals during classic logic-based inference translates into uncertain diagnosis which can be true or false. Fuzzy logic allows us to quantify uncertainties of particular fault occurrence.

The application of fuzzy logic in diagnostic inference is presented in [10, 52, 53]. It is based on a combination of fuzzy residues assessment and fuzzy inference, performed assuming single faults on the basis of rules derived from a binary diagnostic matrix or FIS [9, 10, 53]. This approach is presented in Fig. 4.

In this case, diagnosis indicates a fault f_k along with a certainty factor δ_k. They take the following form:

$$DGN(\delta) = \{\langle \delta_k, f_k \rangle : \delta_k \geq G\}. \tag{11}$$

Wherein δ_k is the degree of activation of the rule of the occurrence of f_k fault and G is a certain threshold value of a rule activation degree at which the fault is

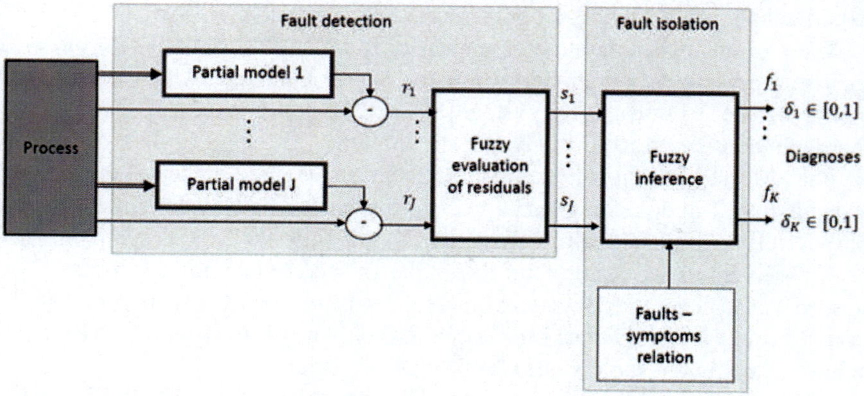

Fig. 4 Diagnosing with the use of fuzzy logic [8]

indicated in the diagnosis (e.g. G = 0, 1). It is characteristic that the fuzzy system in this case covers only the stages of blurring and inference and there is no stage.

Development of the method is presented in [42], where the ways of including the uncertainty of faults-symptoms relations defined by a human at the DS design stage are also described. The uncertainty can be defined in a few ways by the expert—designer of the diagnostic system. The simplest way is to assign each signature with a signature certainty factor. Another solution is defining the certainty factor of the knowledge on each pair of the fault-diagnostic signals value relationship in FIS. A fuzzy Fault Information System is created this way [10, 56]. The extended form of this notation is presented in [42, 57].

The rule base that is used in a certain inference process is usually incomplete. It does not include rules for all possible combinations of diagnostic signal values, because in practice not all such combinations are possible. However, one can never be sure that a specified fault set contains all possible faults. Therefore, there may also be no rules corresponding to these faults.

Hence, it is advisable to calculate in the inference algorithm not only certainty of occurrence of particular faults, but also certainty of the diagnosis. This is possible when PROD operator is used during inference and unisolable faults are combined into elementary blocks [10, 42]. In this case, there are no contradictory rules in the rule base, i.e. those with different conclusions under the same conditions).

The value μ_S of the sum of degrees of activation of all rules in the rule base calculated with the use of a PROD operator is 1, if the rule base is complete [50], and belongs to the range [0, 1] in the case of an incomplete base. The difference $\mu_{US} = 1 - \mu_S$ is the measure of uncertainty of the obtained diagnosis. The closer the value of the sum is to 0, the more certain the diagnosis. The high value of this factor can mean that not all faults are included in the base or that false values of diagnostic signals are created. The value of the factor μ_{US} is then a measure of belief in the occurrence of another, unknown state of the system.

A new method of diagnosis formulation in uncertainty conditions, which is a fusion of the Dempster-Shafer theory-based inference and fuzzy logic is described in [58]. An approach combining fuzzy inference with Bayes law is also being developed [59]. The proposed algorithm is a fusion of Bayes theorem as a subjective interpretation of conditional probabilities and fuzzy inference. It utilizes the knowledge of a'priori probabilities of the up state and faults, and instead of conditional probabilities of observations, takes into account the degrees of fulfilment of the conditions occurring in the fuzzy inference rules.

2.4 *Decomposition of the System and Diagnostics in Decentralized Structures*

When diagnosing complex industrial installations, there is a need to decompose such process into smaller parts diagnosed at the same time by separate diagnostic units.

The problems of decomposition were addressed by Siljak [60]. Decomposition of the system and decentralized diagnostics have a lot of advantages [10, 24, 61–66], such as parallelization of diagnostic tasks, reduction of computational complexity, better adaptation of diagnostic information for the needs of different users, and a possibility to start the diagnostic system in stages in order of a particular subsystem.

Decomposition of the system can be realized on-line or off-line at the diagnostic system design stage. On-line decomposition is used in all fault location algorithms intended for large-scale processes developed in IAiR PW. Below, the issues of decomposition of complex processes occurring at the diagnostic system design stage are addressed.

Decomposition of the system includes separation of the parts of the process to be diagnosed by dedicated computer units and assigning diagnostic tasks to those units. Solving the problem of a diagnostic system decomposition for a large-scale process involves finding the answer to the following questions:

- which parts (areas) of the process should be covered by particular diagnostic sub-systems?
- how to assign subsets of diagnostic tests to individual sub-systems, bearing in mind the relationships between these sub-systems?

Separation of completely independent parts of the process is usually impossible. Each division should be carried out in an optimal or at least rational way. In the case of optimal divisions, the index of relationship between particular sub-systems should be minimized. The size of sub-systems usually has to be reduced due to the computing power of diagnostic units. Rational division consists of separating sub-systems corresponding to the technological nodes of the process. This approach does not guarantee minimization of the relationship between the sub-systems, but is convenient for process maintenance, and often leads to low values of the indices of relationships between the separated parts. However, in large technological nodes, their further division is needed. The size of the sub-systems is determined by the number of the processed variables, i.e. the number of the measuring signals and control signals used in diagnostics, as well as the number of possible faults in a sub-system.

To perform decomposition, one has to know the model of the process and diagnostic system. There are two main groups of approaches to decomposition of the system and diagnostic system. The former is based on the use of analytical models of the process [61], the latter utilizes different forms of graph description of a diagnostic system [10, 52, 64, 66, 67]. In large-scale processes, obtaining a quantitative model is very hard and expensive, therefore, the Author's work focused on the use of qualitative models.

For the decomposition of the diagnosed system purposes, a network was applied representing the relationship between the elements of the set of diagnostic signals S [10, 24, 64]. The network is defined by a G_S graph combining the elements of the set of diagnostic signals S and function ψ defined on a set of graph arcs, assigning to each graph a number equal to the number of faults detected jointly by a given pair of diagnostic signals. The principles of separating independent parts of the system, the way of isolating sub-systems corresponding to technological nodes of the process, were provided and the problem of the division minimizing the indices the relationship between the sub-systems has been formulated. The following solutions were applied to this problem [24, 64], heuristic algorithm [68] and a genetic algorithm [66].

In [67] a model of the process in the form of a GP graph was applied for the division of the system into a particular number of sub-systems with a limited size (defined by the number of process variables and faults). On the other hand, to assign the isolated diagnostic sub-systems to sub-sets of tests, thus subsets of diagnostic signals in such a way as to minimize the relationships between them, a triple graph of a system GSD was applied. The vertices of the graph correspond to the set of process variables Z, diagnostic signals S and faults F, whereas the arcs represent the relationships determined on the Cartesian products of sets $Z \times S$ and $S \times F$. The degree of connection between two subsystems with indices m and n, where $m \neq n$ depends on the number of arcs in GSD graph connecting these sub-systems. Figure 5 presents the idea of the proposed decomposition.

Decomposition ensures minimization of the interdependence between the sub-systems, e.g. reduces the need for mutual exchange of information between them.

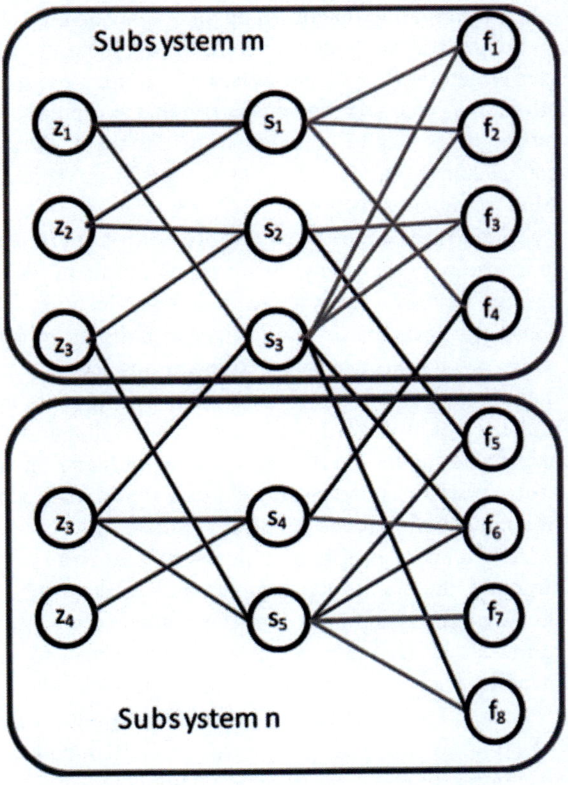

Fig. 5 An example of decomposition of a diagnostic system determined by GSD graph into two sub-systems [67]

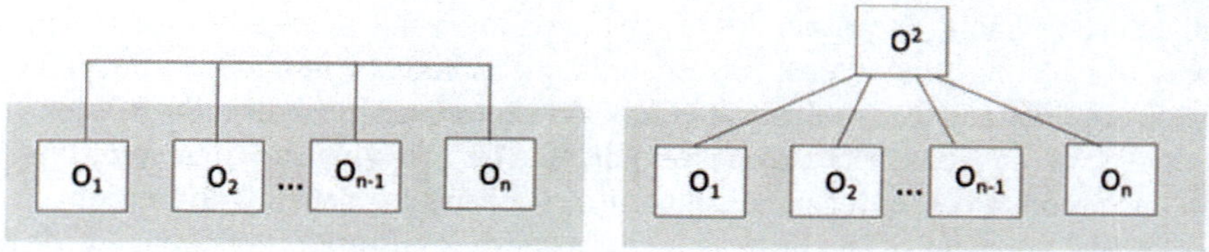

Fig. 6 The structures of decentralized diagnostics: one-level and two-level [70, 71]

The fundamental difference of this solution when compared to the previous ones is the fact that graph of the process is subject to decomposition, whereas, in previous solutions, it was a suitably transformed graph of a diagnostic system. The drawback of earlier solutions was the need to design a diagnostic system for the entire complex process, and then its decomposition.

The structures of contemporary automation systems are decentralized and spatially distributed. It is advisable for diagnostics, that control tasks and protecting actions also be realized in decentralized structures. The main problem of diagnostic inference in this structure is connected to the relationships between the sub-systems. The symptoms of faults that emerged in one subsystem may also be observed in another subsystem. Therefore, there is a need to take such symptoms into account when formulating diagnoses by all diagnostic units.

Diagnostic methods on one-level and hierarchical (mostly two-level) decentralized structures (Fig. 6) using classic logic are presented in [10, 24, 63, 69]. A one-level structure is characterized by the fact that particular parts of the system are diagnosed by computer diagnostic units assigned to them. These units are network coupled and can exchange data to include symptoms observable in other subsystems. There is no superior diagnostic unit.

Diagnostics can also be performed in a hierarchical structure. First level units diagnose only the subsystems assigned to them. With proper decomposition of the system, first level subsystems are independent, i.e. subsets of the detected faults and subsets of tests are disjoint. Superior units in the system detect and isolate faults, whose symptoms are observed in various lower-level sub-systems. On the basis of diagnoses of the first level unit associated to them and locally realized test results, higher-level units formulate more general diagnoses covering all of the subsystem assigned to them. In [10, 24] a method of hierarchical description of complex diagnostic systems is presented used to diagnostics in a hierarchical structure, the rules of diagnostic inference are also provided.

A special case of hierarchical structure is a two-level structure. In a two-level structure, the first level computer units diagnose the subsystem assigned to them not including the symptoms occurring in other subsystems [69]. All tests detecting faults in more than one sub-system are realized by a superior unit, which formulates its own diagnosis and specifies diagnoses made at a lower level.

Generalization of inference methods in decentralized structures on the basis of classic logic are fuzzy inference methods intended for diagnostics in a one-level structure [70] and a two-level structure [71].

2.5 *The Use of Graph Models in the Design of Industrial Process Diagnostic Systems*

To address the problems described in Sect. 2, original methods have been developed using a graph model of a process (GP graph) and a diagnostic system graph (GDS). GP described e.g. in [34, 72] is a qualitative model describing the cause-and-effect relationships between variables in the process, including fault influence.

A directed GP graph is an extension of the known SDG—Signed Directed Graph group which are used to represent cause-and-effect relationships between the variables or alarms in technological installation. Extension which have been proposed enable incorporating faults in GP graph.

Figure 7 presents exemplary graph for a simple laboratory set-up—a set of 3 serially connected tanks [34].

An important application of a GP graph when designing diagnostic systems for complex processes, is using it for automatic generation of all structures of the model for fault detection, assuming that the set of measuring devices is known. The structure of the model is understood as a pair: a modelled variable and a set of variables affecting it. This approach is dedicated for designing the models aimed at fault detection with the use of registered measuring and control data. These may be fuzzy, neural, neural-fuzzy, additive or parametric models. Each structure of the model defines a sub-graph in a GP. It contains the nodes of all faults, for which a residual calculated from the model is sensitive. For these faults, there are directed paths from fault to a modelled variable. It is then possible to determine a binary diagnostic matrix necessary for fault isolation on the basis of a GP.

The obtained set of the model structures allows for the assessment of the obtained fault detectability and isolability, as well as designing the set of models (residuals) ensuring maximal detectability and isolability at a given set of control and measured signals [34, 73]. In [73], necessary and sufficient conditions have been formulated which a set of measuring devices has to meet in order to ensure fault detectability and isolability. A GP graph can be used to determine which measurements need to be performed in the system in order to achieve the assumed fault isolability and

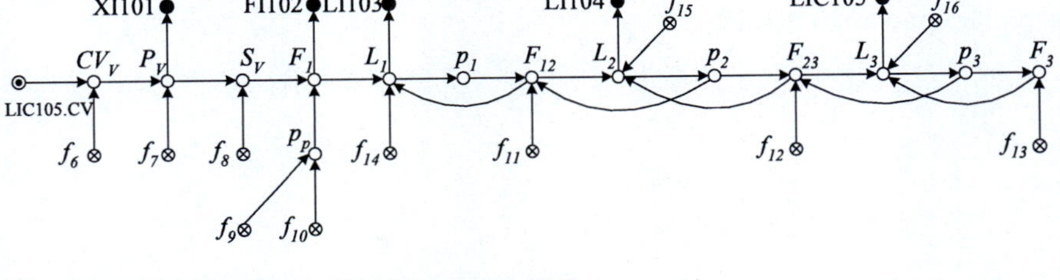

Fig. 7 Exemplary graph of a part of the process [34]

detectability. A GP graph is also used for the decomposition of a diagnostic system for large-scale processes [67]. This problem is described in point 2.4.

The application of a GP graph is not limited to process diagnostics. It can also be used as the first stage of designing a process simulator, as well as for the analysis and reduction of a set of alarms in control systems [74] and to support safety analysis. In [75] a new approach was presented for safety analysis by a HAZOP method with the use a GP graph. In the HAZOP method, a methodology of providing the completeness of specified threats is not defined. The work proves that the completeness of such an analysis may be increased by the application of a qualitative model for the process in the form of a GP graph. A GP graph is used to determine the causes of the parameters' deviation.

In a classic HAZOP analysis, relationships between the separated nodes may not be noticed, especially in the case of internal feedback in the process. Application of a GP graph increases the likelihood of including all of the threats due to their systematic modelling in the graph and explicit relationships between the separated nodes. In the case of HAZOP analysis support, it is possible to reveal a sub-graph covering all possible threats (faults) for a given deviation of the process parameter.

The proposed procedure of supporting HAZOP analysis with the use of a GP graph allows for:

- visualisation of cause-and-effect relationships occurring in the process with the use of a graph,
- complementing the analysis by identification of potential dependencies between the nodes,
- increasing the level of completeness of the analysis.

A qualitative model of a process in the form of a GP graph has many advantages. There is no need to have a mathematical model of the diagnosed system, expert knowledge is enough. The influence of faults is analysed qualitatively, thus it is not necessary to develop a complex analytical description of the influence of faults on the modelled variables, which is almost impossible in the case of complex processes. This model allows us to solve basic diagnostic systems design problems concerning the choice of measuring signals, a set of detection algorithms and the analysis of fault detectability and isolability. According to the author, the proposed methods are alternative, but at the same time a more intuitive approach in comparison to other methods utilizing qualitative models: SDG graphs [76–78], bond graphs [79, 80] and structural model [3, 81, 82].

3 Diagnostic Methods and Systems for Large-Scale Processes

Solving the aforementioned diagnostic problems of dynamic process allowed to develop effective and robust diagnostic algorithms for complex dynamic large-scale systems and their implementation in the realized diagnostic systems.

The basis of all on-line diagnostic algorithms of complex industrial processes, presented in [9, 10, 24, 32, 39, 53, 83], is the use of the DDS method—a method for dynamic decomposition of a diagnostic system. Decomposition means that, after detecting a symptom, an appropriate subsystem is isolated, defined by subsets of possible faults and diagnostic signals detecting these faults. Further diagnostics are then performed for such a subsystem.

The application of a DDS method significantly reduces the computational burden for making diagnoses for single and multiple faults. It protects against inference errors under assumption of single faults in the cases of multiple faults occurrence [40]. Multiple faults occurring at the same time or in short time intervals are correctly isolated, assuming single faults, if the dynamically determined subsets of diagnostic signals used for their detection are disjoint.

Probably the first diagnostic method described in publications aimed at real-time diagnostics of large-scale industrial processes was DTS—Dynamiczne Tablice Stanu (Dynamic State Tables) method [24, 39, 83]. This method combined inference rules corresponding to the rows of a binary diagnostic matrix, classic logic and the concept

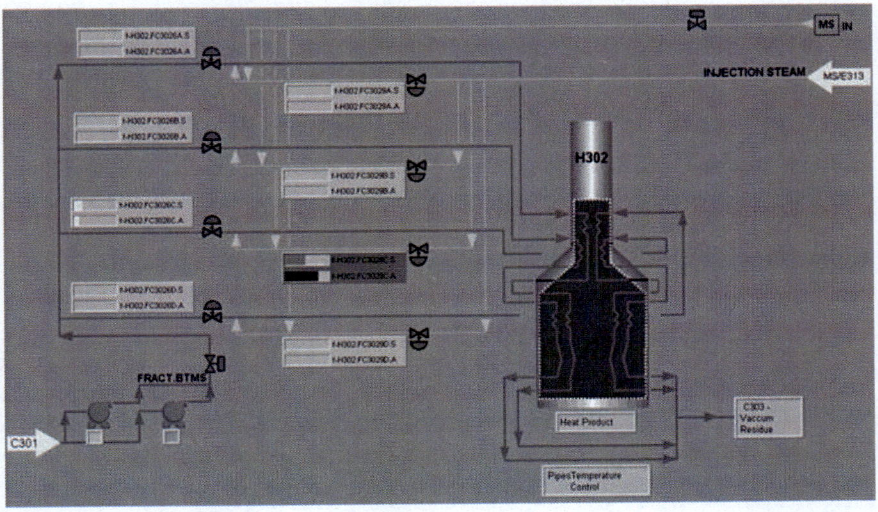

Fig. 8 Visualisation of the faults of control valves and measuring paths of the flow of mazout and steam in a stove. Synoptic diagram show the distribution of the indices corresponding to particular faults, where the value of the degree of certainty of a given fault from the range 0–1 is projected. The colour of the bar on the index (yellow, purple, red), which is not visible in the picture, depends on the degree of certainty of a fault [87]

of dynamic decomposition of a diagnosed system. This method was implemented in the OSA process monitoring system. The OSA system developed in 1986 was the first in the country and one of the first in the world to perform advanced diagnostics in real time. It was implemented in the Lublin Sugar Factory [24, 84].

In subsequent versions of the methods, the extensions were:

- the use of FIS-resulting rules to the notation of faults-symptoms relation, introducing multivalued fuzzy residue evaluation and the use of fuzzy logic for F-DTS method inference [53]
- including the knowledge on the delays in symptoms occurrence to protect against false diagnoses and to increase the achieved fault isolability—T-DTS method [9, 10, 32].

All variants of a DTS method were also presented in the monographs [9, 10]. These and further developments of diagnostic algorithms for industrial processes were implemented in subsequent, increasingly advanced diagnostic systems: DIAG [9], AMandD [30, 31, 85, 86, 87] and DiaSter [8, 88].

These systems have undergone industrial research and were implemented as pilot projects in, among others, Lublin Sugar Factory, Zakłady Azotowe Puławy, EC Siekierki, PKN Orlen and in the laboratories of Universite' des Sciences et Technologies de Lille. An example of visualisation of diagnoses in AMandD system is presented in Fig. 8.

4 Summary—The Significance of On-line Diagnostics in Ensuring Process Safety

One of the research fields currently being developed is in the area of industrial automation related to safety and cyber security, as well as reliability of ICS—Industrial Control Systems. Research topics are fault detection, isolation and identification, systems of automatic diagnostics of industrial processes, fault tolerant control systems, process modelling techniques including emergency scenarios for the needs of constructing process simulators for operators training, cyberattacks detection methods, risk assessment and reduction methods etc. These issues are becoming increasingly important, especially for large-scale processes taking place in the power, petrochemical, chemical, iron and steel or food industry, where many are critical systems. In the case of such installations, all failures cause not only huge economic losses, but can also pose a threat to human life and cause environmental polution.

Technical safety is considered as 'safety'—the problem of counteracting serious industrial failures caused by unreliability of the components of technical installations (e.g. pipeline rupture), damages to the control system elements and human errors. Another aspect is technical safety interpreted as 'security'—protection against intentional hostile attacks from outside (e.g. hacker attacks on control systems) and sabotage operations carried out from the inside of the facility. Despite various causes, the effects of serious faults, human errors and attacks can be the same, i.e. fire, explosion, environmental contamination, destruction of the installation, and shut-down of the process.

This work describes the problems of on-line diagnostics of large-scale industrial processes, realized by a control system or integrated expert system. Automatic diagnostics of technological systems, measuring devices and actuators directly affects safety, as well as reliability indices. It increases such factors as DC (Diagnostics Coverage) and SFF (Safe Failure Fraction) defined by standards on functional safety (PN-EN 61508, PN-EN 61511), which have a direct impact on risk reduction [89]. Reducing diagnostic time also decreases the MTTR (Mean Time to Repair) index, thus increasing the availability of the system.

The requirements of ensuring a proper security level are defined by existing legal regulations and technical standards. A basic way of risk reduction to the acceptable level is using SIS (Safety Instrumented Systems) implementing the so-called safety instrumented functions, i.e. lock and automatic shut-down algorithms. SIS operations are related to shut-down the entire process or its part resulting in economic losses. On-line diagnostics ensures early fault detection. Diagnoses support process operators (Fig. 9) which allows them to take appropriate measures. They are also used by FTC (Fault Tolerant Control Systems) to realize automatic re-configuration of the system structure in states of emergency. In such cases, SIS does not operate, and the process

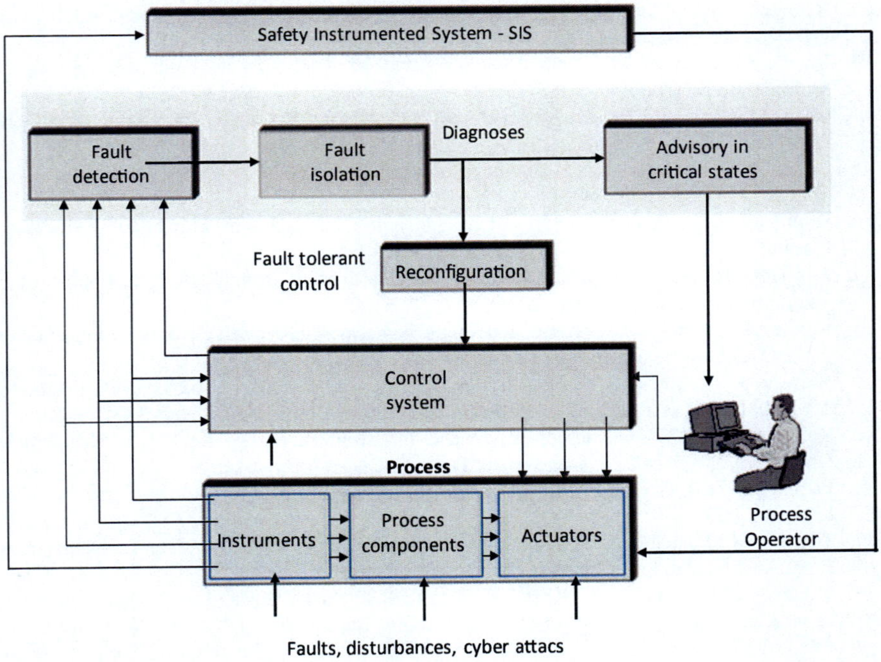

Fig. 9 Diagnostic system of the structure of control and safety of the process system [90]

is not shut-down, which translates into the reduction of economic losses emergency states. A new process protective layer [90] was proposed to tackle this problem.

Another goal of advanced on-line diagnostics is conducting a modern maintenance strategy based on the assessment of the technical conditions of a given technical installation. In addition to abrupt faults in the apparatus of technological processes, there are often slow-changing destructive phenomena modifying their characteristics and deteriorating operational properties. The source of these changes is material corrosion, and deposition of various substances in the device components etc. The rational course of action is to replace periodic inspections and repairs with the strategy of carrying out repairs on the basis of current assessments of the technical condition of the system and estimating the time to a critical state. This strategy of maintenance is known as Predictive Maintenance.

Advanced on-line diagnostics is an effective way to detect not only faults, but also cyberattacks [74]. It allows for recognizing cyberattacks in a situation where other protective layers against cyber threats have turned out to be ineffective.

References

1. Bartyś, M.: Chosen Issues of Fault Isolation, Theory, Practice and Applications. PWN, Warszawa (2014)
2. Basseville, M., Nikiforov, I.V.: Detection of Abrupt Changes—Theory and Application. Prentice-Hall, Englewood Cliffs, NJ (1993)
3. Blanke, M., Kinnaert, M., Lunze, J., Staroswiecki, M.: Diagnosis and Fault-Tolerant Control. Springer, Berlin (2004)
4. Chen, J., Patton, R.: Robust Model-Based Fault Diagnosis for Dynamic Systems. Kluwer Akademic Publishers, Boston (1999)
5. Gertler, J.: Fault Detection and Diagnosis in Engineering Systems. Marcel Dekker, Inc., New York-Basel-Hong Kong (1998)
6. Himmelblau, D.: Fault Detection and Diagnosis in Chemical and Petrochemical Processes. Elsevier, Amsterdam (1978)
7. Isermann, R.: Fault Diagnosis Systems. An Introduction from Fault Detection to Fault Tolerance. Springer, New York (2006)
8. Korbicz, J., Kościelny, J.M. (eds.): Modeling, Diagnostics and Process Control. Implementation in the DiaSter System. Springer, Berlin, Heidelberg (2010)
9. Korbicz, J., Kościelny, J.M., Kowalczuk, Z., Cholewa, W. (eds.): Fault Diagnosis: Models, Artificial Intelligence Methods, Applications. Springer, Berlin, Heidelberg (2004)
10. Kościelny, J.M.: Diagnostics of Automated Industrial Processes, (in Polish. Akademicka Oficyna Wydawnicza Exit, Warsaw (2001)
11. Patton, R., Frank, P., Clark, R. (eds.): Issues of Fault Diagnosis for Dynamic Systems. Springer, London (2000)
12. Witczak, M.: Modelling and Estimation Strategies for Fault Diagnosis of Non-Linear Systems. From Analytical to Soft Computing Approaches. Springer, Berlin (2007)
13. Calado, J.M.F., Korbicz, J., Patan, K., Patton, R.J., Sá da Costa, J.M.G.: Soft computing approaches to fault diagnosis for dynamic systems. Eur. J. Control **7**(2–3), 248–286 (2001)
14. Frank, P.M.: Fault diagnosis in dynamic systems via state estimations methods. A survey. In: Tzafestas S.G., et al. (eds.) System Fault Diagnostics, Reliability and Related Knowledge Based Approaches, vol. 2, D. Reidl Publishing Company, Dordrecht (1987)

15. Frank, P.M.: Fault diagnosis in dynamic systems using analytical and knowledge-based redundancy. Automatica **26**, 459–474 (1990)
16. Isermann, R.: Model-based fault-detection and diagnosis—status and applications. Annu. Rev. Control **29**(2005), 71–85 (2005)
17. Isermann, R., Ballé, P.: Trends in the application of model-based fault detection and diagnosis of technical process. Control Eng. Pract. **5**(5), 709–719 (1997)
18. Korbicz, J.: Robust fault detection using analytical and soft computing methods. Bull. Polish Acad. Sci. Tech. Sci. **54**(1), 75–88 (2006)
19. Leonhardt, S., Ayoubi, M.: Methods of fault diagnosis. Control Eng. Pract. **5**(5), 683–692 (1997)
20. Venkatasubramanian, V., Rengaswamy, R., Yin, K., Kavuri, S.N.: A review of process fault detection and diagnosis, Part I: quantitative model-based methods. Comput. Chem. Eng. **27**, 293–311 (2003)
21. Venkatasubramanian, V., Rengaswamy, R., Kavuri, S.N.: A review of process fault detection and diagnosis, Part II: qualitative model and search strategies. Comput. Chem. Eng. **27**, 313–326 (2003)
22. Venkatasubramanian, V., Rengaswamy, R., Kavuri, S.N., Yin, K.: A review of process fault detection and diagnosis, Part III: process history based methods. Comput. Chem. Eng. **27**, 327–346 (2003)
23. Kościelny, J.M., Syfert, M.: Application properties of methods for fault detection and isolation in the diagnosis of complex large-scale processes. Bull. Polish Acad. Sci. Tech. Sci. **62**(3), 571–582 (2014)
24. Kościelny, J.M.: Diagnostics of Continuous Automatized Industrial Processes by Dynamic Table of States Method (in Polish). In: Scientific Papers of Warsaw Technical University, Series: Electronics, vol. 95. Warszawa (1991)
25. Kościelny, J.M., Syfert, M.: Robust diagnostics of complex chemical processes: main problems and possible solutions. Chem. Process Eng. **39**(2), 1–19 (2018)
26. Łabęda-Grudziak, Z.: Zastosowanie addytywnego modelu regresji do generacji residuów dla potrzeb detekcji uszkodzeń. Ph.D. Dissertation, PW, Wydział Mechatroniki (2011)
27. Kościelny, J.M.: Recognition of fault in the diagnosing process. Appl. Math. Comput. Sci. **3**(3), 559–572 (1993)
28. Pawlak, Z.: Systemy Informacyjne. Podstawy teoretyczne, WNT, Warszawa (1983)
29. Pawlak, Z.: Rough Sets. Theoretical Aspects of Reasoning About Data. Springer, Netherlands (1991)
30. Kościelny, J.M., Bartyś, M., Rzepiejewski, P., Sá da Costa, J.: Actuator fault distinguishability study. Control Eng. Pract. **14**, 645–652 (2006)
31. Kościelny, J., Syfert, M., Wnuk, P.: Advanced monitoring and diagnostic system 'AMandD'. IFAC Proc. Volumes **39**(13), 635–640 (2006)
32. Kościelny, J.M., Zakroczymski, K.: Fault isolation algorithm that takes dynamics of symptoms appearances into account. Bull. Polish Acad. Sci. Tech. Sci. **49**(2), 323–336 (2001)
33. Syfert, M., Kościelny, J.M.: Diagnostic reasoning based on symptom forming sequence. IFAC Proc. Volumes **42**(8), 89–94 (2009)
34. Sztyber, A., Ostasz, A., Kościelny, J.M.: Graph of a process—a new tool for finding model's structures in model based diagnosis. IEEE Trans. Syst., Man, Cybern.: Syst. **45**(7), 1004–1017 (2015)
35. Kościelny, J.M., Syfert, M., Sztyber, A.: Fault distinguishability in diagnostic of complex dynamical systems (in Polish). Akademicka Oficyna Wydawnicza EXIT (2018)
36. Kościelny, J.M., Rostek, K., Syfert, M., Sztyber, A.: Fault isolability with different forms of faults-symptoms relation. Int. J. Appl. Math. Comput. Sci. **26**(4), 815–826 (2016)
37. Kościelny, J.M., Syfert, M., Tabor, Ł.: Application of knowledge about residuals dynamics for fault isolation and identification. In: Conference on Control and Fault-Tolerant Systems. SysTol′13, Nice, IEEE Xplore Digital Library, pp. 275–280 (2013)
38. Kościelny, J.M., Łabęda-Grudziak, Z.: Double fault distinguishability in linear systems. Int. J. Appl. Math. Comput. Sci. **23**(2), 395–406 (2013)

39. Kościelny, J.M.: Fault isolation in industrial processes by dynamic table of states method. Automatica **31**(5), 747–753 (1995)
40. Kościelny, J.M., Bartyś, M., Syfert, M.: Method of multiple fault isolation in large scale systems. IEEE Trans. Control Syst. Technol. **20**(5), 1302–1310 (2012)
41. Ligęza, A., Kościelny, J.M.: A new approach to multiple fault diagnosis. Combination of diagnostic matrices, graphs, algebraic and rule-based models. The case of two-layer models. Int. J. Appl. Math. Comput. Sci. **18**(4), 465–476 (2008)
42. Kościelny, J.M., Syfert, M.: Fuzzy diagnostic reasoning that takes into account the uncertainty of the faults-symptoms relation. Int. J. Appl. Math. Comput. Sci. **16**(1), 27–35 (2006)
43. Pearl, J.: Probabilistic Reasoning in Intelligent Systems. Networks of Plausible Inference. Morgan Kaufmann Publishers Inc., San Francisco, CA, USA (1988)
44. Zadeh, L.: Fuzzy sets. Inf. Control **8**(3), 338–353 (1965)
45. Dempster, A.P.: Upper and lower probabilities induced by a multivalued mapping. Ann. Math. Stat. **38**(2), 325–339 (1967)
46. Shafer, G.: A Mathematical Theory of Evidence. Princeton University Press (1976)
47. Buchanan, B.G., Shortliffe, E.H.: Rule Based Expert Systems: The Mycin Experiments of the Stanford Heuristic Programming Project. Addison-Wesley Longman Publishing Co., Inc., Boston, MA, USA (1984)
48. Shortliffe, E.H., Buchanan, B.G.: A Model of Inexact Reasoning in Medicine. Readings in Uncertain Reasoning, pp. 259–275. Morgan Kaufmann Publishers Inc., San Francisco, CA, USA (1990)
49. Cooper, G.F.: The computational complexity of probabilistic inference using Bayesian belief networks (research note). Artif. Intell. **42**(2–3), 393–405 (1990)
50. Piegat, A.: Fuzzy Modelling and Control. Springer, Berlin (2001)
51. Yager, R.R., Filev, D.P.: Essentials of Fuzzy Modelling and Control. Wiley (1994)
52. Kościelny, J.: Application of fuzzy logic fault isolation in a three-tank system. IFAC Proc. Volumes (14th World Congress IFAC), **32**(2), 7754–7759 (1999)
53. Kościelny, J., Sędziak, D., Zakroczymski, K.: Fuzzy logic fault isolation in large scale systems. Int. J. Appl. Math. Comput. Sci. **9**(3), 637–652 (1999)
54. Kościelny, J., Syfert, M.: Current diagnostics of power boiler system with use of fuzzy logic. IFAC Proc. Volumes **33**(11), 669–674 (2000)
55. Kościelny, J.M., Syfert, M.: Fuzzy logic application to diagnostic of industrial processes. IFAC Proc. Volumes **36**(5), 711–716 (2003)
56. Sędziak, D.: Methods of fault isolation in industrial processes (in Polish). Ph. D. Thesis, Warsaw University of Technology, Department of Mechatronics (2001)
57. Syfert, M.: Diagnostics of industrial processes with the use of partial models and fuzzy logic. Ph. D. Thesis, Warsaw University of Technology, Department of Mechatronics (2003)
58. Sztyber, A., Kościelny, J.M.: Diagnostic reasoning framework combining fuzzy logic and dempster-shafer theory. In: IEEE Xplore Digital Library. IEEE International Conference on Prognostics and Health Management (ICPHM), Ottawa, Ontario (2016)
59. Kościelny, J.M., Sztyber, A.: Metoda wnioskowania diagnostycznego uwzględniająca niepewności symptomów na drodze fuzji teorii Bayes'a i logiki rozmytej. In: Kowalczuk, W.Z., Domżalski, M. (eds.) Advanced Systems for Automation and Diagnostics, Pomorskie Wydawnictwo Naukowo-Techniczne, pp. 117-128 (2015)
60. Siljak, D.: Large-Scale Dynamic Systems: Stability and Structure. North Holland, New York (1978)
61. Boem, F., Ferrari, R.M.G., Parisini, T., Polycarpou, M.M.: Distributed fault diagnosis for nonlinear systems. In: Preprints of the 8th IFAC Symposium on Fault Detection, Supervision and Safety of Technical Processes, pp. 1089–1094 (2012)
62. Indra, S., Travé-Massuyés, L., Chanthery, E.: Decentralized diagnosis with isolation on request for spacecraft. In: Preprints of the 8th IFAC Symposium on Fault Detection, Supervision and Safety of Technical Processes, vol. 8, no. PART 1, pp. 283–288 (2012)
63. Kościelny, J.M.: Diagnostics of processes in decentralized structures. Arch. Control Sci. **7**(3/4), 181–202 (1998)

64. Kościelny, J.M., Sikora, A.I.: Dekompozycja złożonych obiektów diagnozowania. Archiwum Automatyki i Telemechaniki **1**(1991), 115–131 (1991)
65. Pulido, B., Zamarre, J.M.: Using structural decomposition methods to design gray-box models for fault diagnosis of complex industrial systems: a beet sugar factory case study. In: First European Conference of the Prognostics and Health Management Society (2012)
66. Wnuk, P., Kościelny, J.M.: Diagnostic system decomposition with genetic optimization. Pomiary Automatyka Kontrola, Nr **6**(2011), 641–647 (2011)
67. Kościelny, J.M., Sztyber, A.: Decomposition of complex diagnostic systems. In: Preprints of the 10th IFAC Symposium on Fault Detection, Supervision and Safety of Technical Processes, pp. 755–762 (2018)
68. Kernighan, B.W., Lin, S.: An efficient heuristic procedure for partitioning graphs. Bell Syst. Tech. J. **49** (1970)
69. Kościelny, J.M., Bartyś, M., Syfert, M.: Diagnostics of industrial processes in decentralised structures with application of fuzzy logic. IFAC Proc. Volumes (17th World Congress of IFAC), **41**(2), 6944–6949 (2008)
70. Syfert, M., Kościelny, J.M., Bartyś, M.: A fuzzy inference approach to fault diagnosis refinement in decentralized diagnostics. In: Advanced Solutions in Diagnostics and Fault Tolerant Control, pp. 143–154. Springer (2017)
71. Syfert, M., Bartyś, M., Kościelny, J.M.: Refinement of fuzzy diagnosis in decentralized two-level diagnostic structure. In: Preprints of the 10th IFAC Symposium on Fault Detection, Supervision and Safety of Technical Processes, pp. 160–167 (2018)
72. Kościelny, J.M., Ostasz, A.: Application of causal graph GP for description of diagnosed process. IFAC Proc. Volumes **36**(5), 801–806 (2003)
73. Sztyber, A.: Sensor placement for industrial process diagnosis using graph of a process. Ph. D. Thesis, Warsaw University of Technology, Department of Mechatronics (2015)
74. Kościelny, J.M., Syfert, M., Wnuk, P.: The idea of on-line diagnostics as a method of cyberattack. In: Advanced Solutions in Diagnostics and Fault Tolerant Control, Springer, pp. 449–457 (2017)
75. Kościelny, J.M., Sztyber, A., Syfert, M.: Graph description of the process and its applications. In: Trends in Advanced Intelligent Control, Optimization and Automation, pp. 550–559. Springer (2017)
76. Chang, C.C., Yu, C.C.: On-line fault diagnosis using the signed directed graph. Ind. Eng. Chem. Res. **29**(7), 1290–1299 (1990)
77. Kramer, B.L., Palowitch, M.A.: A rule-based approach to fault diagnosis using the signed directed graph. AIChE J. **33**(7), 1067–1078 (1987)
78. Shiozaki, J., Matsuyama, H., Tano, K., O'Shima, E.: Fault diagnosis of chemical processes by the use of signed directed graphs: extension to five-range patterns of abnormality. Int. Chem. Eng. **37**(4), 651–659 (1985)
79. Bouamama, B.O., Harabia, R.E., Abdelkrimb, M., Gayedb, M.B.: Bond graphs for the diagnosis of chemical processes. Comput. Chem. Eng. **36**, 301–324 (2012)
80. Samantaray, A., Medjaherb, K., Bouamama, B.O., Staroswiecki, M., Dauphin-Tanguy, G.: Diagnostic bond graphs for online fault detection and isolation. Simul. Model. Pract. Theory **14**(3), 237–262 (2006)
81. de Flaugergues, V., Cocquempot, V., Bayart, M., Pengov, M.: Structural analysis for FDI: a modified, invertibility-based canonical decomposition. In: 20th International Workshop on Principles of Diagnosis, pp. 59–66 (2009)
82. Düstegör, D., Frisk, E., Cocquempot, V., Krysander, M., Staroswiecki, M.: Structural analysis of fault isolability in the damadics benchmark. Control Eng. Pract. **14**, 597–608 (2006)
83. Kościelny, J.M.: Diagnostyka on-line aparatury technologicznej i układów automatyki metodą DTS—dynamicznych tablic stanu. Archiwum Automatyki i Telemechaniki, 4/1988, pp. 593–606 (1988)
84. Kościelny, J.M., Pieniążek, A.: Algorithm of fault detection and isolation applied for evaporation unit in sugar factory. Control Eng. Pract. **2**(4), 649–657 (1994)

85. Syfert, M., Rzepiejewski, P., Wnuk, P., Kościelny, J.M.: Current diagnostics of the evaporation station. IFAC Proc. Volumes **38**(1), 365–370 (2005)
86. Kościelny, J.M., Leszczyński, M., Gąsecki, A., Syfert, M.: Monitoring of the degree of coking in H-Oil plant. IFAC Proc. Volumes **42**(8), 1258–1263 (2009)
87. Kościelny, J.M., Syfert, M., Leszczyński, M., Gąsecki, A.: Pilot tests of the advanced system of process diagnostics in PKN ORLEN. Pomiary Automatyka Kontrola **55**(3/2009), 136–140 (2009)
88. Syfert, M., Wnuk, P., Kościelny, J.M.: DiaSter—intelligent system for diagnostics and automatic control support of industrial processes. JAMRIS—J. Autom., Mob. Rob. Intell. Syst. **5**(4), 41–46 (2011)
89. Pawlak, M., Kościelny, J.M., Wasiewicz, P.: Method of increasing the reliability and safety of the processes through the use of fault tolerant control systems. Eksploatacja i Niezawodność—Maintenance and Reliability **17**(3), 398–407 (2015)
90. Kościelny, J.M., Bartyś, M.: The requirements for a new layer in the industrial safety systems. In: 9th IFAC Symposium on Fault Detection, Supervision and Safety of Technical Processes, Safeprocess 2015, pp. 1333–1338. Arts de Métiers ParisTech, Paris, France (2015)
91. Kościelny, J., Syfert, M., Fajdek, B., Kozak, A.: The application of a graph of a process in HAZOP analysis in accident prevention system. J. Loss Prev. Process Ind. **50**(2017), 55–66 (2017)

Applications of Computational Intelligence Methods for Control and Diagnostics

Jacek Kluska, Tomasz Żabiński, and Tomasz Mączka

Abstract This chapter discusses examples of computational intelligence applications in automatics, i.e. methods of designing adaptive single-loop fuzzy control systems in a continuous case and the idea of monitoring and diagnostics of the condition of selected machine parts and technological processes. The first part discusses the concept of adaptive control. It assumes a dynamic plant to be linear, however, not fully known, and shows the methods of designing a fuzzy adaptive controller based on frequency domain stability conditions. Considerations involve direct control systems with a reference model in the system with feedback from output as well as a fuzzy controller with the properties of a non-linear PID algorithm. The second part describes methods of using classifiers for real-time diagnostics for machine part conditions and the implementation of technological processes used i.e. in the aviation industry, including the detection of anomalies. Methods for diagnosing a cold forging process and the degree of mechanical imbalance of a CNC milling machine head are given as examples. The discussed subject is in line with the idea of "Industry 4.0".

1 Introduction

This chapter presents applications of computational intelligence methods in automatics that use fuzzy logic and machine learning algorithms. It gives examples of applications of the mentioned methods in the adaptive control of continuous

J. Kluska (✉) · T. Żabiński
Politechnika Rzeszowska, Al. Powstańców Warszawy 12, 35-959 Rzeszów, Poland
e-mail: jacklu@prz.edu.pl

T. Żabiński
e-mail: tomz@prz.edu.pl

T. Mączka
Żbik Sp. z o.o., ul. Emilii Plater 7, 35-079 Rzeszów, Poland
e-mail: t.maczka@zbiksoft.pl

P. Kulczycki et al. (eds.), *Automatic Control, Robotics, and Information Processing*, Studies in Systems, Decision and Control 296,
https://doi.org/10.1007/978-3-030-48587-0_22

single-loop systems and diagnostics of the machine condition. Methods of designing a fuzzy adaptive controller with output feedback in a continuous case are discussed. Procedures are based on the results of absolute stability criterion and analytical methods of fuzzy systems modeling. As a result of their application, easily interpretable fuzzy controller rules that minimize a given quality index are obtained. The controller design process can be frequently fully automated.

Applications of some machine learning methods are discussed, mainly classifiers for real time diagnostics of the machine parts condition and the implementation of technological processes used i.e. in the aerospace industry, including the detection of anomalies. The discussed subject generally refers to the idea of "Industry 4.0" [39, 40, 48, 56, 58, 62, 80, 85, 86, 88, 95, 100]. Preliminary results are promising. However, data obtained in industrial conditions (e.g. for machines used in the aerospace industry) are usually difficult, i.e. they come en masse in real time, are big, often imbalanced, heavily disturbed, drifting, etc. [24, 45], which raises new problems, not yet satisfactorily solved [10, 17–20, 41, 42, 45, 57, 68, 69, 72, 96, 98, 99].

This chapter is divided into two parts. The first part is devoted to fuzzy adaptive control of one input and one output systems and includes Sects. 2–8. In Sect. 2, a brief description of tests for adaptive fuzzy control with a reference model is given. Section 3 presents notations and assumptions for the control system architecture containing a PID-like fuzzy controller, called PID-FC. In Sect. 4, absolute stability conditions for a continuous closed-loop system containing the PID-FC are formulated. Section 5 discusses the conditions under which the fuzzy controller is non-linearity in the bounded sector, assuming that it is treated as the so-called P1-TS system, referring to the terminology taken from the analytical theory of fuzzy systems [29, 31]. In Sect. 6 the idea of an adaptation procedure is presented. Section 7 gives a step-by-step procedure for constructing an adaptive PID-FC. In Sect. 8, conclusions regarding the procedure for an adaptive PID-FC design are formulated.

The second part of the chapter shows the application of classifiers (neural, genetic, based on support vectors and decision trees methods, minimal-distance methods etc.) for the diagnostics of a cold forging process and the degree of imbalance of the CNC milling machine tool head.

2 Characteristics of Research on Adaptive Fuzzy Control

The results of various techniques of stability analysis that are commonly used in non-linear control systems, both in time and frequency domain (small-gain theorem, Lyapunov theory, Nyquist, Popov or circle criteria), enable the design of fuzzy adaptive systems [1, 7, 22, 36, 71, 77]. When designing a fuzzy controller, attention should be paid to ensuring the interpretability of its rules [5, 31], otherwise other methods of designing non-linear non-fuzzy controllers may turn out to be more attractive [11, 14, 23]. Fuzzy control was initially proposed as a control method that does not require any knowledge of a controlled plant, but in this case it is difficult to guarantee stability and desired control quality. In this chapter, the assumption is that there is a limited knowledge of the plant's parameters.

Regarding the stability of fuzzy control systems, two issues are worth noting. First of all, most of the work focuses on designing an adaptive fuzzy controller based on the Lyapunov's methods [3, 61, 83]. Secondly, all stability criteria for continuous non-linear systems provide only sufficient conditions, except for simple exceptions, where conditions are both necessary and sufficient [26].

Numerous publications are devoted to fuzzy control of non-linear plants, e.g., [6, 9, 13, 25, 37, 44, 46, 47, 60, 73, 83, 101], but absolute stability results were used with all necessary assumptions in only a few publications e.g. regarding the requirements for reference signals [1, 27, 38]. In this chapter a linear time invariant plant (LTI) is considered, with the assumption that there is a limited knowledge of some of the plants parameters. Direct adaptive control has been applied, in which the adaptation mechanism uses signals present in the control system to modify the controller's parameters. The reference model determines the desired quality of a closed-loop system. A PID-like fuzzy controller is used. A wide variety of fuzzy PID-like controllers have been developed [8, 15, 43, 50, 52, 55, 59, 79, 87, 91]. However, the concept of PID-FC considered in this chapter is similar to that described in [16], and the adaptation mechanism extends the one presented in paper [89]. PID-FC is assumed to satisfy a specific sector condition.

Sector bounds and restrictions for reference signals result from stability conditions that are different from the classic circle criterion [2, 65, 66]. The design methodology presented in this chapter is a significant extension of the methods proposed in similar papers, in which the classic circle criterion for the fuzzy type P controller was used [89], and the number of controller inputs is less than three [90]. This chapter proves that the stability conditions for the PID-FC control system guarantee that the control error signal tends to zero for some classes of reference signals during the adaptation process. Less restrictive and more general conditions for the non-linear fuzzy controller were formulated and the gradient descent method was generalized using analytical theory of the so-called P1-TS fuzzy system [31]. Moreover, the process of designing an adaptive fuzzy controller was automated for the most laborious steps of the design procedure.

3 The Control System Containing a PID-FC Controller or Its Variants

In paper [89] an interesting concept of adaptation with the use of a fuzzy controller was given. In addition to a number of advantages, the method also has some drawbacks, which include the possibility of a non-zero steady-state error in the control system and the lack of noise suppression, as well as the fact that the procedure for selecting the reference system is based on a trial and error method. In an attempt to avoid these drawbacks an extended approach to the problem of suitable adaptation with a non-linear reference model will be presented.

Firstly, the feedback system containing the PID-FC (or its variants) as in Fig. 1 will be considered. A plant transfer function is $\hat{g}_0(s) = [\hat{u}(s)]^{-1}\hat{y}(s)$, where $\hat{v}(s) = \mathscr{L}[v(t)]$ is the Laplace transform of the function $v : [0, \infty] \to \mathbb{R}$. Controllers with two or three inputs will be mainly investigated, i.e. PID-FC or PI-FC or PD-FC. The controller type is determined by the appropriate positions of the switches $p_1, \ldots, p_5$. A polynomial $\hat{N}(s)$ with coefficients $q_1, q_2 \geq 0$ and a constant I are also associated with each controller type:

- for PID-FC we assume $\hat{N}(s) = 1 + q_1 s + q_2 s^2$, $I = 1$,
- for PI-FC we assume $\hat{N}(s) = 1 + q_1 s$, $q_2 = 0$, $I = 1$,
- for PD-FC we assume $\hat{N}(s) = 1 + q_1 s$, $q_2 = 0$, $I = 0$.

The signal $\varphi(t)$ is the controller output; $\varphi(t) = u(t)$ in the case of PD-FC and $\varphi(t) = du/dt$ for PI-FC. The original system model (Fig. 1) can be described by

$$e(t) = z(t) - (g * \varphi)(t), \tag{1}$$

where $e(t) = w_0(t) - y(t)$ is the control error, $w_0(t)$ is reference input signal, $y(t)$—plant output, $z(t)$ is the difference between w_0 and the zero input response, i.e. $z(t) = w_0(t) - y_0(t)$, and $(g * \varphi)(t) = \int_0^t g(t-\tau)\varphi(\tau)\,d\tau$. All cases of the systems containing PI-FC, PD-FC, or PID-FC controllers can be described by (1), where the artificial impulse response is given by [27]

$$g(t) = (1 - I)g_0(t) + I \int_0^t g_0(\zeta)d\zeta, \quad I \in \{0, 1\}. \tag{2}$$

Definition 1 The fuzzy controller with the inputs $e(t)$, $\dot{e}(t)$ and $\ddot{e}(t)$ with the output $\varphi(t)$ is non-linearity in a bounded sector $[\beta, \beta + K] \subseteq [0, \infty)$, if for $t \geq 0$ there exist

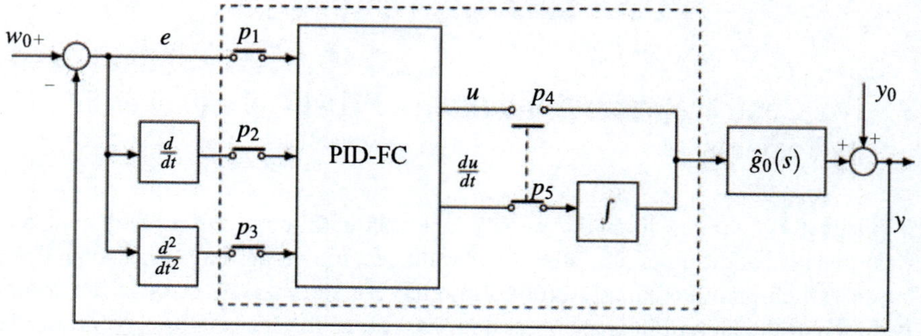

Fig. 1 Closed-loop control system containing PID-FC controller

four nonnegative constants β, K, q_1 and q_2 such that the inequality

$$(Kh + \beta h - \varphi)\,(\varphi - \beta h) \geq 0, \quad 0 \leq \beta, \ K < \infty\,, \tag{3}$$

is satisfied, where the artificial signal

$$h(t) = e + q_1\dot{e} + q_2\ddot{e}. \tag{4}$$

Note that the signal $h(t)$ does not appear in the system shown in Fig. 1.

4 Absolute Stability Conditions

Below, the closed-loop system containing a linear plant and PID-FC (or its variants) is investigated, and it is shown how to guarantee the control error to be bounded and zero in the steady state. For simplicity, the plant transfer function is assumed to be $\hat{g}_0(s) = [\hat{A}(s)]^{-1}\hat{B}(s)$, where $\hat{A}(s)$ and $\hat{B}(s)$ are polynomials that have no common zeros. If m and n are the degrees of $\hat{B}(s)$ and $\hat{A}(s)$, respectively, then $m < n$. One can check that the original closed-loop system (1)–(4) can be written in the following equivalent form

$$h(t) = z_1(t) - (g_1 * \varphi_1)\,(t), \tag{5}$$

where $\varphi_1(t) = \varphi(t)/K - \beta h(t)/K$ satisfies the sector inequality

$$(h(t) - \varphi_1(t))\,\varphi_1(t) \geq 0, \tag{6}$$

i.e. $\varphi_1(t)$ belongs to the sector [0, 1], and

$$z_1(t) = \mathscr{L}^{-1}\left[\frac{\hat{N}(s)\hat{z}(s)}{1 + \beta\hat{N}(s)\hat{g}(s)}\right], \tag{7}$$

$$g_1(t) = \mathscr{L}^{-1}\left[\frac{K\hat{N}(s)\hat{g}(s)}{1 + \beta\hat{N}(s)\hat{g}(s)}\right], \tag{8}$$

where $\mathscr{L}^{-1}$ denotes the inverse Laplace transform and $\hat{N}(s)$ is defined in a general form

$$\hat{N}(s) = 1 + q_1 s + q_2 s^2, \quad 0 < q_1 < \infty, \quad 0 \leq q_2 < \infty. \tag{9}$$

The transformation of the original system to the model (5)-(9) is correct if

$$\inf_{\mathrm{Re}\, s \geq 0} \left| 1 + \beta \hat{N}(s) \hat{g}(s) \right| > 0 \tag{10}$$

is satisfied. It is assumed that the system satisfies a generalized Hurwitz condition (GHC) [53, 64, 84], i.e.

$$1 + r\hat{N}(s)\hat{g}(s) = 0 \;\Rightarrow\; \mathrm{Re}\, s < 0, \;\; \forall\, r \in [\beta, \beta + K]. \tag{11}$$

Note that fulfillment of (11) implies that (10) is true.

Theorem 1 *Consider the system containing PID-FC or PI-FC or PD-FC as in (1)–(4), where* $\hat{N}(s) = 1 + q_1 s + q_2 s^2$, *and* $I = 1$, $0 < q_1, q_2 < \infty$ *for PID-FC,* $I = 1$, $0 < q_1 < \infty$, $q_2 = 0$ *for PI-FC,* $I = 0$, $0 < q_1 < \infty$ *and* $q_2 = 0$ *for PD-FC. In addition, the following conditions are met*
(C1) GHC as in (11),
(C2) $g_1 \in L^1(0, \infty) \cap L^2(0, \infty)$, $var_{[0,\infty)} g_1(t) < \infty$, *where* $g_1(t)$ *as in (8),*
(C3) $z_1 \in L^2(0, \infty)$, $\exists M : |dz_1/dt| < M$, $\lim_{t \to \infty} z_1(t) = 0$, *where* $z_1(t)$ *as in (7),*
(C4) frequency condition

$$\inf_{\omega} \mathrm{Re} \left(\hat{g}_1(j\omega) + 1 \right) > 0. \tag{12}$$

Then $e \in L^2(0, \infty)$, $|e| < \infty$ *and* $\lim_{t \to \infty} e(t) = 0$.

Proof The proof is given in [28].

Condition (12) can be readily interpreted on an appropriate complex plane. In the simplest case, for P-FC, the result coincides with the circle criterion [2, 21]. Note also, that for the fulfillment of condition (C2) of Theorem 1, it is necessary to guarantee to be $n = n_1 \geq m + 2$ in case of PD-FC or PID-FC, or $n = n_1 \geq m + 1$ in case of PI-FC. If any of these conditions and GHC are satisfied, then condition (C2) of Theorem 1 holds.

5 The Fuzzy Controller as a Rule-Based System of P1-TS Type

Our goal is to provide conditions that are not conservative and easy to implement, in which inequality (3) is satisfied. The results will be directly used to design a PID-FC adaptive controller. The controller will be treated as a so-called P1-TS system introduced in [30] in the framework of the analytical theory of fuzzy-rule-based systems [31]. P1-TS systems were implemented as a high-speed and low-cost hardware device in FPGA technology [33, 34] and used for the synthesis of control or decision support systems [32, 35, 74–76, 82].

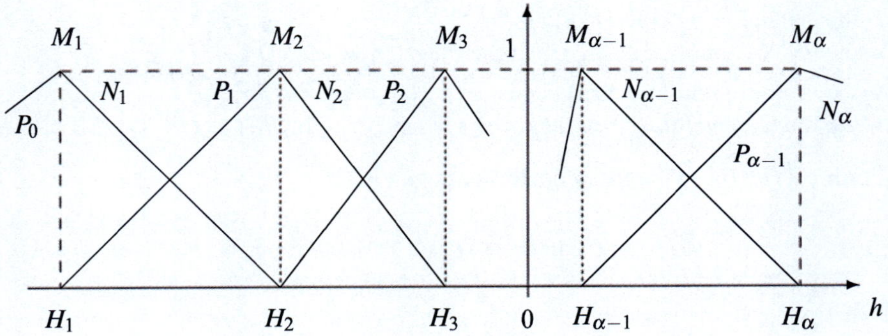

Fig. 2 Strong fuzzy partition of the universe implemented with triangular membership functions for the fuzzy controller as a one-input P1-TS system with the input $h \in D_h = [H_1, H_\alpha]$

Before considering PID-FC, the simplest fuzzy controller with one input h and output φ are investigated. The input universe has the form (Fig. 2)

$$D_h = \bigcup_{i=1}^{\alpha-1} \left[H_i, H_{i+1}\right], \tag{13}$$

where $2 \leq \alpha < \infty$, $H_0 = 2H_1 < 0$, and $H_{\alpha+1} = 2H_\alpha > 0$ is assumed. The fuzzy rules can be expressed as

$$\text{If } h \text{ is } M_i, \text{ then } \varphi = Q_i, \tag{14}$$

where $Q_i \in \mathbb{R}$, membership functions $M_i : D_h \to [0, 1]$ for $i = 1, \ldots, \alpha$, are triangular and make strong partition as shown in Fig. 2.

They can be defined as

$$M_i\,(h) = 0 \vee \left[P_{i-1}\,(h) \wedge N_i\,(h)\right],\ h \in D_h, \tag{15}$$

for $i = 1, \ldots, \alpha$, where $\vee = \max$, $\wedge = \min$, and [31]

$$N_i\,(h) = \frac{H_{i+1} - h}{H_{i+1} - H_i},\ P_i\,(h) = 1 - N_i\,(h), \tag{16}$$

for $h \in \left[H_i, H_{i+1}\right]$, $i = 1, \ldots, \alpha$. The controller output is

$$\varphi\,(h) = \frac{\sum_{i=1}^{\alpha} M_i(h)\,Q_i}{\sum_{i=1}^{\alpha} M_i(h)},\quad h \in D_h. \tag{17}$$

Using the notations typical for the analytical theory of fuzzy rule-based systems [31], for $h \in \left[H_i, H_{i+1}\right]$ in the P1-TS system we will denote

- the generator: $\mathbf{g}(h) = [1, h]^T \in \mathbb{R}^2$,
- the real vector of consequents of the rules: $\mathbf{Q}_i = \left[Q_i, Q_{i+1}\right]^T \in \mathbb{R}^2$,
- the vector of constant coefficients: $\Theta_i = \left[\theta_{i,0}, \theta_{i,1}\right]^T \in \mathbb{R}^2$,
- the fundamental matrix of the fuzzy system: $\Omega_i = \left[\mathbf{g}(H_i), \mathbf{g}(H_{i+1})\right] \in \mathbb{R}^{2\times 2}$.

For $h \in \left[H_i, H_{i+1}\right]$ the fuzzy controller output is

$$\varphi_i(h) = \Theta_i^T \mathbf{g}(h) = \mathbf{Q}_i^T \Omega_i^{-1} \mathbf{g}(h) = \theta_{i,0} + \theta_{i,1} h, \tag{18}$$

where

$$\theta_{i,0} = \frac{Q_i H_{i+1} - H_i Q_{i+1}}{H_{i+1} - H_i}$$

and

$$\theta_{i,1} = \frac{Q_{i+1} - Q_i}{H_{i+1} - H_i}.$$

It is required that φ is sector bounded non-linearity.

It can be demonstrated that the controller with the input $h \in D_h$ as in (13) and membership functions as in Fig. 2, can be unambiguously defined by the following "input–output" pairs of ordered points:

$$S = \{(H_i, Q_i) \mid i = 1, \ldots, \alpha\}. \tag{19}$$

where $H_1 < \cdots < 0 < \cdots < H_\alpha$. Suppose β and K are given. Let us define

$$\lambda_i = \frac{1}{K}\left(\frac{Q_i}{H_i} - \beta\right), \quad i = 1, \ldots, \alpha. \tag{20}$$

It can be proved that with simple and natural assumptions, regardless of what α is (even or odd), if $\lambda_i \in [0, 1]$ for all $i = 1, \ldots, \alpha$, then the controller is non-linearity in a bounded sector (we omit the proof).

If a set of points is given H_i: $H_1 < H_2 < \cdots < H_\alpha$, $(H_1 < 0 < H_\alpha)$ which makes a partition of the universe D_h and there is a collection of coefficients λ_i such that $(\lambda_1, \ldots, \lambda_\alpha) \in [0, 1]^\alpha$, then a one-input controller given by

$$S = \{(H_i, \beta H_i + \lambda_i K H_i) \mid i = 1, \ldots, \alpha\}, \tag{21}$$

is non-linearity in the sector $[\beta, \beta + K] \subseteq [0, \infty)$. Condition (21) is convenient for the adaptive fuzzy controller design. In contrast to [89], neither odd number α of rules nor the controller function to be symmetric one is required at the moment. Thus, we have greater freedom in choosing a non-linear function of the (fuzzy adaptive) controller.

Next, PID-FC is considered as the most universal, since the results for PI-FC and PD-FC are easy to obtain. Take the following notation for the universes for the control error e and its derivatives: $e \in \bigcup_{i=1}^{a-1} [A_i, A_{i+1}] = [A_1, A_a], \dot{e} \in \bigcup_{j=1}^{b-1} [B_j, B_{j+1}] = [B_1, B_b], \ddot{e} \in \bigcup_{k=1}^{c-1} [C_k, C_{k+1}] = [C_1, C_c]$, where $2 \leq a, b, c < \infty$. The input vector of PID-FC is $(e, \dot{e}, \ddot{e}) \in D = [A_1, A_a] \times [B_1, B_b] \times [C_1, C_c]$, and the cuboid D can be expressed as the union of subcuboids $D_{i,j,k}$

$$D = \bigcup_{i=1}^{a-1} \bigcup_{j=1}^{b-1} \bigcup_{k=1}^{c-1} D_{i,j,k} \,, \tag{22}$$

where $D_{i,j,k} = [A_i, A_{i+1}] \times [B_j, B_{j+1}] \times [C_k, C_{k+1}]$ for $i = 1, \ldots, a-1,\ j = 1, \ldots, b-1$ and $k = 1, \ldots, c-1$. From now on, the behavior of PID-FC in the subcuboid $D_{i,j,k}$ is considered. For any $(e, \dot{e}, \ddot{e}) \in D_{i,j,k}$, the fuzzy system can be described by the rule

$$\text{If } e \text{ is } M_i \text{ and } \dot{e} \text{ is } M_j \text{ and } \ddot{e} \text{ is } M_k, \text{ then } \varphi = Q_{i,j,k} \,, \tag{23}$$

where the fuzzy sets M_i, M_j and M_k have triangular membership functions and make a strong fuzzy partition in the appropriate universes, $(i = 1, \ldots, a,\ j = 1, \ldots, b,\ k = 1, \ldots, c)$. The membership functions can be defined as follows

$$M_i(e) = 0 \vee [P_{i-1}(e) \wedge N_i(e)], \quad i = 1, \ldots, a, \tag{24}$$

$$M_j(\dot{e}) = 0 \vee [P'_{j-1}(\dot{e}) \wedge N'_j(\dot{e})], \quad j = 1, \ldots, b, \tag{25}$$

$$M_k(\ddot{e}) = 0 \vee [P''_{k-1}(\ddot{e}) \wedge N''_k(\ddot{e})], \quad k = 1, \ldots, c, \tag{26}$$

where the linear functions N_i, N'_j, N''_k, P_i, P'_j, and P''_k are defined as follows

$$N_i(e) = \frac{-e + A_i}{A_i - A_{i-1}}, \quad A_i > A_{i-1}, \tag{27}$$

$$N'_j(\dot{e}) = \frac{-\dot{e} + B_j}{B_j - B_{j-1}}, \quad B_j > B_{j-1}, \tag{28}$$

$$N''_k(\ddot{e}) = \frac{-\ddot{e} + C_k}{C_k - C_{k-1}}, \quad C_k > C_{k-1}, \tag{29}$$

$P_i = 1 - N_i$, $P'_j = 1 - N'_j$, $P''_k = 1 - N''_k$, $(\,i = 1, \ldots, a+1,\ \ j = 1, \ldots, b+1,\ k = 1, \ldots, c+1)$, and additionally $A_0 = 2A_1 < 0$, $A_{a+1} = 2A_a > 0$, $B_0 = 2B_1 < 0$, $B_{b+1} = 2B_b > 0$, $C_0 = 2C_1 < 0$, and $C_{c+1} = 2C_c > 0$. Every rule in the form of (23) can be written equivalently by means of 8 rules

$$\text{If } e \text{ is } N_i \text{ and } \dot{e} \text{ is } N'_j \text{ and } \ddot{e} \text{ is } N''_k, \text{ then } \varphi = Q_{i,j,k,1},$$

$$\text{If } e \text{ is } P_i \text{ and } \dot{e} \text{ is } N'_j \text{ and } \ddot{e} \text{ is } N''_k, \text{ then } \varphi = Q_{i,j,k,2},$$
$$\text{If } e \text{ is } N_i \text{ and } \dot{e} \text{ is } P'_j \text{ and } \ddot{e} \text{ is } N''_k, \text{ then } \varphi = Q_{i,j,k,3},$$
$$\text{If } e \text{ is } P_i \text{ and } \dot{e} \text{ is } P'_j \text{ and } \ddot{e} \text{ is } N''_k, \text{ then } \varphi = Q_{i,j,k,4}, \tag{30}$$
$$\text{If } e \text{ is } N_i \text{ and } \dot{e} \text{ is } N'_j \text{ and } \ddot{e} \text{ is } P''_k, \text{ then } \varphi = Q_{i,j,k,5},$$
$$\text{If } e \text{ is } P_i \text{ and } \dot{e} \text{ is } N'_j \text{ and } \ddot{e} \text{ is } P''_k, \text{ then } \varphi = Q_{i,j,k,6},$$
$$\text{If } e \text{ is } N_i \text{ and } \dot{e} \text{ is } P'_j \text{ and } \ddot{e} \text{ is } P''_k, \text{ then } \varphi = Q_{i,j,k,7},$$
$$\text{If } e \text{ is } P_i \text{ and } \dot{e} \text{ is } P'_j \text{ and } \ddot{e} \text{ is } P''_k, \text{ then } \varphi = Q_{i,j,k,8}.$$

Using the terminology from [31], for $(e, \dot{e}, \ddot{e}) \in D_{i,j,k}$ in the P1-TS system, let us denote the following:

- the generator: $\mathbf{g}(e, \dot{e}, \ddot{e}) = [1, e, \dot{e}, e\dot{e}, \ddot{e}, e\ddot{e}, \dot{e}\ddot{e}, e\dot{e}\ddot{e}]^T \in \mathbb{R}^8$,
- the real vector of consequents of the rules: $\mathbf{Q}_{i,j,k} = \left[Q_{i,j,k,1}, \ldots, Q_{i,j,k,8}\right]^T \in \mathbb{R}^8$,
- the vector of constant coefficients:
  $\Theta_{i,j,k} = [\theta_{000}, \theta_{100}, \theta_{010}, \theta_{110}, \theta_{001}, \theta_{101}, \theta_{011}, \theta_{111}]^T \in \mathbb{R}^8$,
- the fundamental matrix of the fuzzy system
  $\Omega_{i,j,k} = \left[\mathbf{g}\left(\mathbf{v}_{i,j,k,1}\right), \ldots, \mathbf{g}\left(\mathbf{v}_{i,j,k,8}\right)\right] \in \mathbb{R}^{8\times 8}$,

where the vertices of the subcuboid $D_{i,j,k}$ are
$\mathbf{v}_{i,j,k,1} = \left(A_i, B_j, C_k\right)$, $\mathbf{v}_{i,j,k,2} = \left(A_{i+1}, B_j, C_k\right)$, $\mathbf{v}_{i,j,k,3} = \left(A_i, B_{j+1}, C_k\right)$,
$\mathbf{v}_{i,j,k,4} = \left(A_{i+1}, B_{j+1}, C_k\right)$, $\mathbf{v}_{i,j,k,5} = \left(A_i, B_j, C_{k+1}\right)$, $\mathbf{v}_{i,j,k,6} = \left(A_{i+1}, B_j, C_{k+1}\right)$,
$\mathbf{v}_{i,j,k,7} = \left(A_i, B_{j+1}, C_{k+1}\right)$, $\mathbf{v}_{i,j,k,8} = \left(A_{i+1}, B_{j+1}, C_{k+1}\right)$.
The controller output for $(e, \dot{e}, \ddot{e}) \in D_{i,j,k}$ is given by the function

$$f_{i,j,k}(e, \dot{e}, \ddot{e}) = \Theta^T_{i,j,k}\mathbf{g}(e, \dot{e}, \ddot{e}) = \mathbf{Q}^T_{i,j,k}\Omega^{-1}_{i,j,k}\mathbf{g}(e, \dot{e}, \ddot{e}). \tag{31}$$

The whole system of rules (23) in cuboid D can be equivalently written in the form of $8(a-1)(b-1)(c-1)$ fuzzy rules, as in (30) and the conditions for PID-FC to be non-linearity in the bounded sector are much more complicated for the controller with three inputs than for one or two inputs. Therefore, it is important to provide relatively simple conditions on the basis of which it can be determined whether the considered fuzzy controller is non-linearity in the bounded sector. Therefore, the following theorem can be formulated

Theorem 2 *Suppose the vector of consequents of the rules (30) of the PID-FC in each subcuboid $D_{i,j,k}$ is given by*

$$\mathbf{Q}_{i,j,k} = r_{i,j,k}\left[\mathbf{v}^T_{i,j,k,1}, \ldots, \mathbf{v}^T_{i,j,k,8}\right]^T [1, q_1, q_2]^T \tag{32}$$

where $i = 1, \ldots, a$, $j = 1, \ldots, b$, $k = 1, \ldots, c$. *If* $r_{i,j,k} \in [\beta, \beta + K]$ *for all* i, j, *and* k, *then the considered PID-FC belongs to the sector* $[\beta, \beta + K]$ *for all* $(e, \dot{e}, \ddot{e}) \in D$.

Proof Consider the subcuboid $D_{i,j,k}$. To simplify the notation, the indices i, j, and k in some parts of the proof will be omitted. For the vector $\Theta^T = [0, r, rq_1, 0, rq_2, 0, 0, 0]$, with (31), the vector of consequents of the rules is equal

$$\mathbf{Q} = \Omega^T \Theta = \Omega^T r\, [0, 1, q_1, 0, q_2, 0, 0, 0]^T .$$

Hence, $\mathbf{Q} = r \left[\mathbf{v}_{i,j,k,1}^T, \ldots, \mathbf{v}_{i,j,k,8}^T\right]^T [1, q_1, q_2]^T$ and the crisp fuzzy controller output is $f(e, \dot{e}, \ddot{e}) = r_{i,j,k}(e + q_1\dot{e} + q_2\ddot{e})$ for $(e, \dot{e}, \ddot{e}) \in D_{i,j,k}$. This completes the proof of Theorem 2.

Theorem 2 gives sufficient conditions. The number of checks in the form of (32) to be performed is equal to $(a-1)(b-1)(c-1)$. Notice that in the literature, the conditions for the fuzzy controllers with triangular membership functions to be non-linearity in a bounded sector were formulated only for the controllers with one or two inputs, and an odd number and evenly distributed fuzzy sets (see [90], p. 502). None of these restrictions appears in Theorem 2, which makes the fuzzy controller design process more flexible. According to Theorem 2, it is easy to guarantee the sector condition (3) be satisfied, if we decompose the PID-FC into two serially connected fuzzy controllers FC-1 and FC-2, where e.g. FC-1 performs the function (4) and FC-2 is defined by (21). This simple idea will be used further for the three-input PID-FC. In the case of PI-FC or PD-FC, the FC-2 can be assumed to be linear, whereas FC-1—non-linear, and then the conditions from [90] can be applied, checking beforehand whether all assumptions of the stability theorem are met.

As illustrated in Theorem 2, consider the following data: $q_1 = 12$, $q_2 = 36$, $[\beta, \beta + K] = [1, 5]$, $e \in [-9, 12]$, $\dot{e} \in [-4, 5]$, $\ddot{e} \in [-1, 1]$, and points that establish the fuzzy partition: $(A_1, A_2, A_3, A_4) = (-9, -3, 6, 12)$, $(B_1, B_2, B_3, B_4, B_5) = (-4, -2, 1, 3, 5)$, and $(C_1, C_2) = (-1, 1)$. In this case, there are $8 \times 3 \times 4 \times 1 = 96$ fuzzy rules in the form of implications (30). Instead of a large number of fuzzy rules in the whole universe D, the behavior of the controller function in a sub-cuboid $D_{2,2,1} = [-3, 6] \times [-2, 1] \times [-1, 1]$ is checked (as one of 12 conditions to be checked). For $(i, j, k) = (2, 2, 1)$ the fundamental matrix is

$$\Omega_{2,2,1} = \begin{bmatrix} 1 & 1 & 1 & 1 & 1 & 1 & 1 & 1 \\ -3 & 6 & -3 & 6 & -3 & 6 & -3 & 6 \\ -2 & -2 & 1 & 1 & -2 & -2 & 1 & 1 \\ 6 & -12 & -3 & 6 & 6 & -12 & -3 & 6 \\ -1 & -1 & -1 & -1 & 1 & 1 & 1 & 1 \\ 3 & -6 & 3 & -6 & -3 & 6 & -3 & 6 \\ 2 & 2 & -1 & -1 & -2 & -2 & 1 & 1 \\ -6 & 12 & 3 & -6 & 6 & -12 & -3 & 6 \end{bmatrix} .$$

Table 1 The inputs and outputs of the adaptive closed-loop control system

#	Component	Input signals	Output signals
1.	Reference model	$w_0(t)$	$y_m(t)$
2a.	FC-1	$e(t)$, $\dot{e}(t)$ $\ddot{e}(t)$	$h(t)$
2b.	FC-2	$h(t)$	$\varphi(t)$
3.	Integral block	$\varphi(t)$	$u(t)$
4.	Plant	$u(t)$	$y(t)$
5.	Adaptation procedure	$\varepsilon(t)$, $\dot{\varepsilon}(t)$ $\ddot{\varepsilon}(t)$	$Q_1(t), \ldots, Q_\alpha(t)$

8 fuzzy rules in the subcuboid $D_{2,2,1}$ can be easily written, where the vector of consequents of the rules is as follows

$$\mathbf{Q}_{2,2,1} = r\,[-63, -54, -27, -18, 9, 18, 45, 54]^T .$$

One can check that

$$f_{2,2,1}(e, \dot{e}, \ddot{e}) = \mathbf{Q}_{2,2,1}^T \Omega_{2,2,1}^{-1} \mathbf{g}(e, \dot{e}, \ddot{e}) = r\,(e + 12\dot{e} + 36\ddot{e}) .$$

Thus, for $r \in [1, 5]$ it can be deduced, that in $D_{2,2,1}$ the fuzzy controller is non-linearity in the sector $[1, 5]$.

6 The Idea of Adaptation

The adaptive fuzzy control system includes the following components:

1. The non-linear reference model,
2. PID-FC which consists of two parts:

   2a. FC-1,
   2b. FC-2,

3. The integral block,
4. The adaptation procedure,
5. The plant.

The inputs and outputs of the above components are given in Table. 1 where the control error $e(t)$ and the adaptation error $\varepsilon(t)$ are defined as follows

$$e(t) = w_0(t) - y(t), \tag{33}$$

$$\varepsilon(t) = y_m(t) - y(t), \tag{34}$$

and the values of $Q_1(t), \ldots, Q_\alpha(t)$ are consequents of the fuzzy rules resulting from the adaptation procedure. The adaptive PID-like fuzzy controller is decomposed into two serially connected fuzzy controllers FC-1 and FC-2 as described in Sect. 4, where FC-1 performs the function (4) and FC-2 is defined by the set of pairs (19)–(21).

As the reference model is non-linear, it is possible to obtain the closed-loop system with better performance parameters than in the case of the linear one, e.g., shorter settling time and no or short overshoot. This justifies the use of PID-FC instead of the conventional one, because the conventional PID controller is a special case of PID-FC. The minimized quality index is defined as follows

$$J = \frac{1}{2}\eta_1\varepsilon^2 + \frac{1}{2}\eta_2\dot{\varepsilon}^2 + \frac{1}{2}\eta_3\ddot{\varepsilon}^2, \quad \eta_1, \eta_2, \eta_3 \geq 0.$$

Supposing that the consequents of the rules for FC-2 will be changed according to the adaptation rule

$$\frac{dQ_i}{dt} = -\eta\frac{\partial J}{\partial Q_i}, \quad \eta > 0, \quad i = 1, \ldots, \alpha, \tag{35}$$

the following adaptation law is obtained

$$\begin{aligned}\frac{\partial J}{\partial Q_i} &= \frac{\partial J}{\partial \varepsilon}\frac{\partial \varepsilon}{\partial y}\frac{\partial y}{\partial \varphi}\frac{\partial \varphi}{\partial Q_i} + \frac{\partial J}{\partial \dot{\varepsilon}}\frac{\partial \dot{\varepsilon}}{\partial \dot{y}}\frac{\partial \dot{y}}{\partial \varphi}\frac{\partial \varphi}{\partial Q_i} + \frac{\partial J}{\partial \ddot{\varepsilon}}\frac{\partial \ddot{\varepsilon}}{\partial \ddot{y}}\frac{\partial \ddot{y}}{\partial \varphi}\frac{\partial \varphi}{\partial Q_i} \\ &= -\left(w_1\varepsilon + w_2\dot{\varepsilon} + w_3\ddot{\varepsilon}\right) M_i(h),\end{aligned} \tag{36}$$

where $M_i(h)$ is the membership degree defined in (24) and the coefficients are $w_1 = \eta_1 \partial y/\partial\varphi$, $w_2 = \eta_2 \partial\dot{y}/\partial\varphi$, as well as $w_3 = \eta_3 \partial\ddot{y}/\partial\varphi$. The partial derivatives $\partial y/\partial\varphi$, $\partial\dot{y}/\partial\varphi$ and $\partial\ddot{y}/\partial\varphi$ do not have to be known. It is sufficient to know their sign, which is assumed not to change in time and be positive. The adaptation rule is as follows

$$\frac{dQ_i}{dt} = \begin{cases} \left(w_1\varepsilon + w_2\dot{\varepsilon} + w_3\ddot{\varepsilon}\right) M_i(h) & \text{if } |h| \geq \delta \\ 0 & \text{if } |h| < \delta \end{cases}, \tag{37}$$

for $i = 1, \ldots, \alpha$, where $\delta \geq 0$ is the dead zone. The values of w_1, w_2, and w_3 are defined as design constant.

Let $Q_i^L = \beta H_i$ and $Q_i^H = (\beta + K) H_i$, where Q_i^L is the minimum value of the vector of the rule consequents and Q_i^H—the maximum value, which result from the sector condition. If after adaptation process of the i-th fuzzy rule, its consequent $Q_i(k)$ of this rule in the discrete time k is not included in the interval $\left[Q_i^L, Q_i^H\right]$, then

$$Q_i(k+1) = \begin{cases} Q_i^L + \sigma_1\left(Q_i^H - Q_i^L\right) & \text{if } Q_i(k) < Q_i^L \\ Q_i^H - \sigma_2\left(Q_i^H - Q_i^L\right) & \text{if } Q_i(k) > Q_i^H \end{cases}, \tag{38}$$

for $i = 1, \ldots, \alpha$, where the values $\sigma_1 \geq 0$ and $\sigma_2 \geq 0$ are given by the designer. The fact that only two rules are active simultaneously was used in software implementation.

7 Procedure for Adaptive PID-FC Design

Below, a step-by-step procedure for designing an adaptive PID-FC is provided. It follows from the assumptions and results obtained in previous sections. Let $\hat{g}_0(s, p)$ denote the plant transfer function with an unknown parameter $p \in [P_1, P_2]$, where P_1 and P_2 are known and p_n denotes the nominal value of p.

Step 1. Designing a linear PID controller.

For a given plant model with transfer function $\hat{g}_0(s, p_n)$ and required closed-loop system response parameters, the linear PID controller

$$\hat{g}_{PID}(s) = k_p \left(1 + (T_i s)^{-1} + T_d s\right) = k_r s^{-1} \hat{N}(s), \tag{39}$$

where $k_r = k_p T_i^{-1}$, the polynomial $\hat{N}(s)$ is given by (9), $q_1 = T_i$, $q_2 = T_i T_d$, is designed by a chosen classical method, e.g. root-locus.

Result: FC-1 performing the function (4) for $p = p_n$ and FC-2 adjusted temporarily for $\varphi = k_r h$. In addition, closed-loop system response parameters are achieved, i.e. settling time and overshoot.

Step 2. Determining the absolute stability sector (ASS).

In the first phase of this step, the GHC sector (GHS for short) is calculated. This sector is defined as

$$GHS = \bigcap_{p \in [P_1, P_2]} [\beta(p), \beta(p) + K(p)] = [\beta, \beta + K], \tag{40}$$

where $\beta(p)$ and $\beta(p) + K(p)$ are the sector bounds, for which the condition (11) is satisfied.

In the second stage of this step, for every $p \in [P_1, P_2]$ a family of intervals $ASS_{\beta_1}, \ldots, ASS_{\beta_k}$ is determined in which the lower sector bound $\beta_1, \ldots, \beta_k \in GHS$, whereas the upper sector bound K depends on β_i and p, i.e. $K = K(\beta_i, p)$ is computed using the frequency condition (C4) from Theorem 1. Formally, such intervals can be expressed by

$$ASS_{\beta_i} = \bigcap_{(C4\,\text{holds})} \; \bigcap_{p \in [P_1, P_2]} [\beta_i, \beta_i + K(\beta_i, p)], \tag{41}$$

for $i = 1, \ldots, k$. During the experiments it was observed that in order to design a non-linear PID-FC with better achievements of quality indices than those obtained for a linear PID, the obtained sector ASS must include k_r value obtained in *Step 1*. Due to this requirement, such a final sector ASS is suggested to be chosen that includes k_r with some top and bottom margin $\delta_1 > 0$ and $\delta_2 > 0$, respectively

$$ASS = [k_r - \delta_1, k_r + \delta_2] \in \left\{ASS_{\beta_1}, \ldots, ASS_{\beta_k}\right\}, \tag{42}$$

where $\beta_1, \ldots, \beta_k \in GHS$.

Result: $ASS = [\beta, \beta + K]$.

Step 3. Designing non-linear PID-FC for the reference model.

The reference model has a standard structure of a closed-loop control system. It consists of a non-linear PID-FC system and the plant transfer function for $p = p_n$. During this step, the non-linear part of PID-FC, i.e. FC-2, is obtained. At the same time, it provides closed-loop system response parameters better than those achievable for the linear PID controller tuned in *Step 1*. This can be done manually using the trial and error method as in [89] or automatically. An automatic numerical procedure was developed in MATLAB [51] to obtain a non-linear PID-FC, which provides a closed-loop system step response with no overshoot and settling time shorter than achievable for a linear PID. Having obtained the ASS, piecewise linear controller functions candidates, which are non-linearities in the ASS, are automatically generated. In the automatic procedure, the partition D_h as in (13), and the number of consequents candidates Q_i for each H_i value in FC-2 specified in (19), are defined by the designer. The candidates Q_i are then obtained by the use of a formula (21) in which λ_i is calculated automatically based on the designers specifications, e.g. for three values Q_i: $\lambda_{i,1} = 0 \Rightarrow Q_{i,1} = H_i\beta$, $\lambda_{i,2} = (K - \beta)/(2K) \Rightarrow Q_{i,2} = H_i(\beta + K)/2$, $\lambda_{i,3} = 1 \Rightarrow Q_{i,3} = H_i(\beta + K)$. All possible combinations of piecewise linear functions are automatically generated. When the particular fuzzy controller is created, the closed-loop system is simulated, and the structure of the controller that fulfills the designer requirements is saved that meets the design requirements. It is possible that more than one non-linear controller fulfills the requirements. In such cases, the final controller is chosen by the designer on the basis of the report created by the procedure. The final fuzzy system FC-2 can be described by fuzzy rules (14). In addition, a reference signal must be selected so that condition (C3) of Theorem 1 is satisfied.

Result: The reference model with the non-linear PID-FC, which provides better closed-loop system response parameters than achievable for linear PID.

Step 4. Setting parameters of the adaptive fuzzy system.

In order to track the reference model output by the closed-loop system, the fuzzy controller with identical universe partition, as defined in *Step 3*, is used in the main system loop. The initial values of the consequents are set to obtain a fuzzy controller equivalent to the linear PID one tuned in *Step 1*.

Result: The closed-loop system with PID-FC equivalent to the linear PID one tuned in *Step 1*.

Step 5. Closed-loop system simulations.

In order to perform the adaptation, the reference signal $w_0(t)$, adaptation parameters, i.e. w_1, w_2, w_3, dead zones δ, σ_1 and σ_2 are to be chosen by the designer. It is recommended that the reference signal should provide sufficient excitation of the system. The adaptation parameters w_1, w_2 and w_3 are to be chosen by a trial-and-error method. The dead zone δ in (37) can be set to zero and, depending on the need, it can be increased in order to decrease the influence of very small fluctuations of the error $\varepsilon(t)$ on modifications of consequents of the rules (Q_i).

The parameters σ_1 and σ_2 can be set to zero and, depending on the need, they can be increased to prevent reaching the limits by Q_i values too fast.

In order to calculate signal h, as well as well as the adaptation signal (the right-hand side of the Eq. (37)), the first and second derivatives of the errors e and ε must be calculated.

8 Conclusions Regarding Adaptive PID-FC Design

A systematic design procedure of a non-linear PID-like adaptive fuzzy controller for a closed-loop single-input-single-output control system and a plant given by transfer function with limited knowledge of some model parameters is given, assuming a limited knowledge of the controlled plant transfer function. The procedure is based on an absolute stability criterion to guarantee the closed-loop control system stability and achievement of system performance indices better than reachable for a linear (conventional) PID controller. The adaptive control system works in a direct mode and consists of a non-linear dynamic reference model, PID-FC, and an adaptation procedure, which tunes consequents of the controller rules. The adaptation procedure guarantees that the non-linear controller function remains in a bounded sector that ensures stability of the closed-loop control system. The conditions under which the PID-FC is a non-linearity in a bounded sector were developed by the use of the so-called P1-TS analytical theory. The absolute stability conditions for PID-FCs and all its variants were formulated. An important aspect of the results presented in this chapter is the design procedure of a stable closed-loop system with a fuzzy controller equipped with an integral part. The controller's integral part has a significant practical relevance, due to its ability for plant input step disturbance compensation. The authors performed experiments with a disturbance signal for the system described in [89], for which a nonzero steady-state error remains.

The procedure for PID-like adaptive fuzzy controller design consists of five steps and can be easily implemented with software. On the basis of the given absolute stability criteria, the process for designing a PID-like non-linear fuzzy controller, which provides better closed-loop system response parameters than achievable for a linear PID, can be automated. This is a significant advantage of the proposed method as compared to the trial and error approach usually applied in practice or in literature. The software tools for performing and testing the proposed procedure were developed in the Matlab/Simulink environment (some in C language).

9 A System of Intelligent Diagnostics of Machines and Technological Processes in Real Time

In this part, the concept of a system for diagnostics of machines and technological processes as well as the implementation of an IT hardware and software platform designed to perform monitoring, diagnosing, and supervision functions are discussed. Next, the results obtained for some processes will be presented. Matlab/Simulink packages (also in External mode) and TwinCAT 3 [81] are used to develop software and diagnostic methods due to the possibility of rapid prototyping. Other tools and programming languages such as Python, Java, Visual Studio, C#, ST (for PAC/PLC controllers) are also used to develop "system" software. Weka [12], See5 [67], Statistica [78], DTREG [70] are used also (to select, teach and test classifiers), as well as FPGA to build interfaces to accelerate the operation of the system in real time mode.

The procedure for building the real time diagnostic system consists of many steps that include (Fig. 3):

1. Data acquisition and saving in files of appropriate format. This data comprises of signals coming from sensors (accelerometers, acoustic or temperature sensor, etc.). Numerous experiments on real plant using the appropriate software is required (Simulink, TwinCAT 3).
2. Selection of signal processing methods and their attributes (features) and initial off-line testing of classifiers. The time window needs to be selected, data should be cleaned and transformed (e.g. using an FFT or continuous wavelet transform), attributes determined and a set of classifiers selected, their learning and initial testing conducted.
3. Real time implementation of the classifier. This requires prior determination of signal pre-processing algorithms, features extraction and classifier selection.
4. Launching the implementation in real time on the target equipment, which will be mounted on the production site, is the last stage of the procedure.

As a part of the work carried out, a real time prototype hardware and software platform for the intelligent machine diagnostics system was built [92]. It enables intelligent on-line diagnostics in real time with the idea of Industry 4.0 . The important thing is that the design, testing and implementation processes take place on the same platform constructed using the Beckhoff industrial automation system. The platform enables comprehensive communication and can be easily expanded. It was used, i.e. to predict damage in a cold forging process and to diagnose the degree of mechanical imbalance of a CNC milling machine tool head. The works were carried out in cooperation with companies from the Aviation Valley Cluster in southeastern Poland.

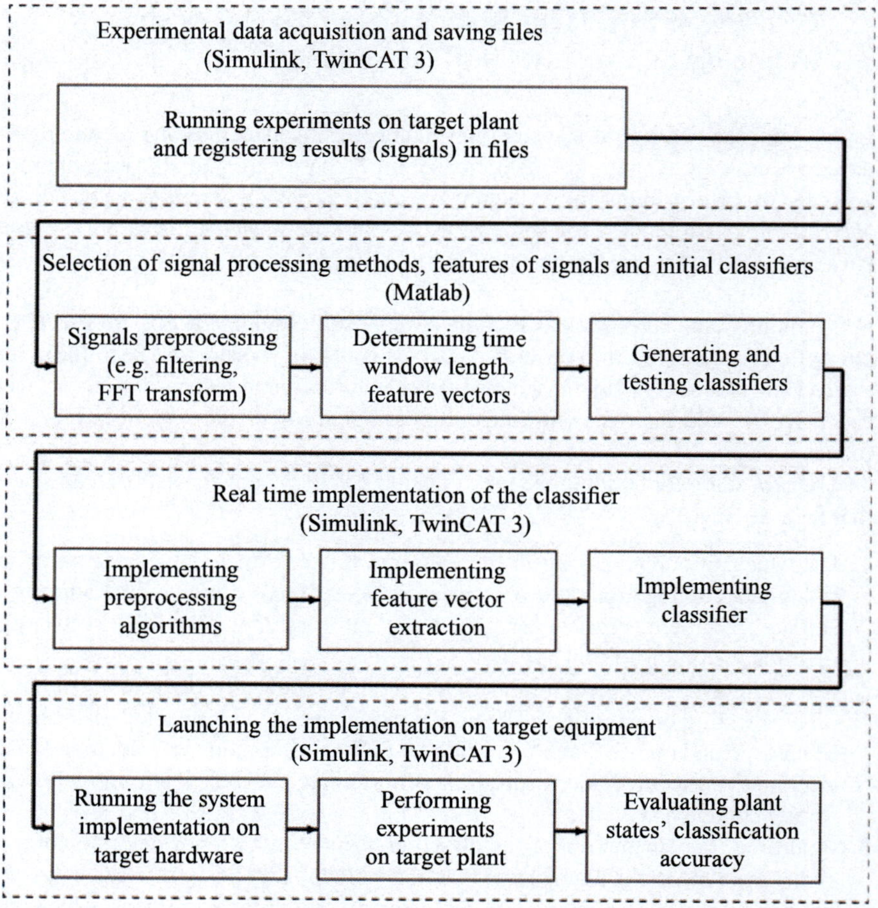

Fig. 3 Procedure for the construction of the real-time diagnostics system of a technological process

9.1 Failure Prediction in the Cold Forging Process

As the first example, the prediction of failure in a cold forging technological process is briefly discussed and the results obtained are presented. In the cold forging machine, producing fasteners, the following main phases are distinguished: wire feed, cut off, upsetting, cone punch change, forging, punch change, cut-off knife return [49, 94]. The in-die piezoelectric force sensor sends slightly noisy waveforms lasting 180 milliseconds, sampled at 10 μs, shown in Fig. 4, where 4 selected classes of waveforms determining the state of the cold forging process are distinguished.

Therefore, the data collected are records of 4 classes, which should be detected early enough by a real time classifier. Three quality indicators were taken to assess classifiers: accuracy (Acc), sensitivity (Sen) and specificity (Spe). For example,

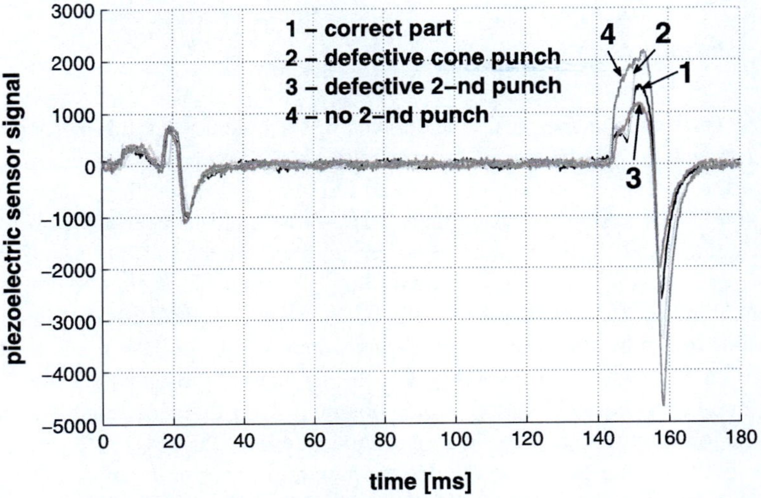

Fig. 4 Four classes of signals received from a piezoelectric sensor placed in a machine that produces fasteners

for 133 records of every class and the use of learning/off-line testing with 10-fold cross-validation and 3 selected attributes in the frequency domain, as a result of the classification algorithms: single decision tree (SDT), probabilistic neural network (PNN), support vector machines (SVM), multi-layer perceptron (MLP), linear discriminant analysis (LDA) and k-means methods, the following results were obtained:

- for **class 1**: *Acc*—PNN (99.06%), *Sen*—LDA (99.25%), *Spe*—PNN and SVM (99.24%);
- for **class 2**: *Acc*—PNN, SVM and K-Means (99.06%), *Sen*—SDT (99.24%), *Spe*—PNN, SVM, LDA and K-Means (99.25%);
- for **class 3**: *Acc*—PNN (99.62%), *Sen*—PNN (99.24%), *Spe* – SDT, PNN and SVM (99.75%);
- for **class 4**: *Acc*—PNN (99.25%), *Sen*—PNN and SVM (97.78%), *Spe*—PNN (99.75%).

These are very good results, which according to the authors may be the basis for the implementation of the classifier, which will require systematic "training" on the real plant due to the non-stationarity of the process and concept drift phenomenon.

9.2 CNC Milling Tool Head Mechanical Imbalance Diagnosis

As a second example, the process of diagnosing the mechanical imbalance of a CNC milling machine tool head will be briefly discussed, which is particularly important in the aerospace industry [97]. Head imbalance is a condition that occurs in the spindle-tool system when vibrations are transmitted to bearings as a result of centrifugal forces. In order to conduct preliminary off-line tests of classifiers, the masses of cutting inserts placed on the milling head were changed. The problem was the prediction of the imbalance of the head. According to ISO 1940-1: 2003 norm, the following imbalance grades are distinguished: small (G 0.4), medium (G 2.5), large (G 6.3) and very large (G 40) [93]. In the example in question, the raw data record contains the acceleration values obtained from the accelerometer mounted near the lower spindle bearing in subsequent time intervals (every 40 μs). 8 signal attributes in the time domain were selected: maximum, minimum, mean value, kurtosis, skewness, 3rd order moment, RMS, standard deviation. In the frequency domain 6 attributes were selected: RMS in the band 190–210 Hz (F1), RMS in the band 390 -410 Hz (F2), RMS in the band 10–500 Hz (F3), RMS in the band 500–1000 Hz, field under the amplitude spectrum in the band 10–500 Hz and field as above in the 500–1000 Hz band. 27334 data records and 14 attributes were selected using a specific method of selecting features. The idea of the attribute selection method was based on the fact that for the selected attribute, the same set of values was randomly transferred to different records and the accuracy of the obtained prediction was evaluated. A significant deterioration in the quality of the prediction means that the variable is indeed significant, whereas a slight deterioration in the quality of the prediction means that the variable is irrelevant. As a result of the selection, it turned out that the most important indicator is F1, then F2 and F3 subsequently. It is worth noting here that the result obtained in this way has a convincing physical interpretation, because the machine rotational speed was 12 000 rpm, which corresponds to the frequency of 200 Hz. 13 classifiers were taught and tested using a 10-fold cross-validation. The method of fuzzy c-means (FC-Means), probabilistic neural network (PNN), single decision tree (SDT), boosted decision trees (BDT), decision tree forest (DTF), multi-layer perceptron (MLP), radial basis functions neural network (RBFNN), support vector machines (SVM), k-nearest neighbors method (kNN), group method of data handling (GMDH) and several others.

In the assessment of the classification quality, Acc, Sen, Spe and additionally the area under the ROC curve (Auc) was taken into account. The best results were obtained for FC–Means: $Acc, Sen, Spe, Auc \approx 99 \ldots 100\%$.

Even simple examples show that having a high accuracy classifier is not acceptable due to the lack of balanced data. In practice, most of the collected data concerns the normal state (i.e. the normal state of the machine or the correctly executed process), while not much—the abnormal one, hence the available data is imbalanced [63]. Therefore, the records from the majority class can be deleted, duplicated from a minority class or artificially generated, but the authors do not prefer such methods.

Table 2 The results of the G 0.4 normal state classifier for 577 learning records

	$\#Attr.$	Acc	Sen	Fpr	Auc	t_{mc} (ms)	t_{src} (ms)
AANN	7	0.9899	1.00	0.1956	1.00	1.745	0.042
LOF	2	1.00	1.00	0.00	1.00	19.253	7.597
OC-SVM	6	1.00	1.00	0.00	1.00	0.044	<0.001

A serious problem arises: "How to train, test and evaluate a predictive system for imbalanced data?" It seems that none of the methods proposed in the literature so far are satisfactory. One-class classifiers are proposed, that is, those that detect abnormal states (anomalies). A real assumption was applied here that available data regard only the normal state and at least one abnormal. This assumption was applied in [54], where dedicated algorithms were used: autoencoder (AANN), one-class support vector machine (OC-SVM) and the local outlier factor (LOF) method. The following data partition has been made

$$Data\ set = \underset{\substack{\text{normal}\\ \text{state}}}{\{s_0\}} \cup \underset{\substack{\text{known}\\ \text{abnormal states}}}{\{s_1, \ldots, s_k\}} \cup \underset{\substack{\text{unknown}\\ \text{abnormal state}}}{\{s_*\}} \tag{43}$$

and classification errors have been defined for the normal state (Err_0), i-th of the known abnormal state (Err_i, $i = 1, \ldots, k$), and a state that has never occurred (Err_*). In addition, a global quality indicator for a one-class classifier has been defined

$$Err = 0.1\,(1 - Acc) + 0.3\,(1 - Sen) + 0.3Fpr + 0.3\,(1 - Auc)\,, \tag{44}$$

where Fpr is a false positive ratio also known as the false alarm ratio. The global indicator seems sensible, since it takes the error $(1 - Acc)$ with a weight of 10%, insensitivity $(1 - Sen)$—with a weight of 30%, false alarm ratio (Fpr)—with a weight of 30%, and the complement to the area under the receiver operating characteristic curve $(1 - Auc)$—with a weight of 30%. In addition, the mean time of the model construction should be taken into account (t_{mc}), as well as the mean time of a single record classification (t_{src}),which is very important for constructing the classifier as a real time system. It should be emphasized that for the construction of the classifier, only records for the normal state are used. This classifier qualifies the new input record either as a normal or abnormal state, without indicating the type of difference. Sample results of the normal state classifier are shown in Table 2.

It is worth noting that despite LOF being the best method, OC-SVM was selected, because both the mean time of the model construction and the mean time of a single record classification were short and are suitable for real time applications.

10 Conclusions on the Diagnostics of Machines and Technological Processes in Real Time

A hardware and software platform for an industrial IT system was built and a real time learning system was constructed for predicting some of the failures in the cold forging process and detecting the degree of mechanical imbalance of the CNC milling machine tool head. The development of intelligent diagnostic and predictive systems is a key element enabling the implementation of Industry 4.0 in practice. The results of research in the field of building a system for diagnostics of machine parts working in the metal and aerospace industries are promising. Other systems for classification were also built, including a system for detecting the blunting of the drill, which is not described here. These systems work correctly for a certain time, but they must be re-trained or "continuously trained", like a well-known advanced neural network with the possibility of self-organization—Fuzzy ARTMAP [4]. That is why the Fuzzy ARTMAP network has been implemented in the free WEKA software and in the programming language of the Matlab package and work is underway on its application in the platform. However, attention should be paid to difficult problems that are waiting for a solution, as the results published so far in the literature seem to be still unsatisfactory. They include:

- taking into account the non-stationarity of the process, especially a concept drift,
- more efficient processing of huge data streams flowing in real time,
- selection of the best algorithms for certain types of data ("no free lunch"),
- statistical evaluation of predictive systems for imbalanced, drifting and partly contradictory data.

Acknowledgements This work is financed by Polish Ministry of Science and Higher Education under the program "Regional Initiative of Excellence" in 2019–2022. Project number 027/RID/2018/19, funding amount 11 999 900 PLN.

References

1. Aracil, J., Gordillo, F. (eds.): Stability Issues in Fuzzy Control. Studies in Fuzziness and Soft Computing. Physica-Verlag, Heidelberg, New York (2000). http://www.springer.com/us/book/9783790818550#aboutBook
2. Ban, X., Gao, X.Z., Huang, X., Yin, H.: Stability analysis of the simplest Takagi-Sugeno fuzzy control system using circle criterion. In: 2006 IEEE International Conference on Fuzzy Systems, pp. 2056–2061 (2006). https://doi.org/10.1109/FUZZY.2006.1681985
3. Birdwell, J.D., Wang, Y.: Lyapunov stability analysis of systems using the fuzzy-PID controller. In: Proceedings of 1994 American Control Conference—ACC '94, vol. 1, pp. 966–970 (1994). https://doi.org/10.1109/ACC.1994.751888
4. Carpenter, G.A., Grossberg, S., Markuzon, N., Reynolds, J.H., Rosen, D.B.: Fuzzy ARTMAP: a neural network architecture for incremental supervised learning of analog multidimensional maps. IEEE Trans. Neural Netw. **3**(5), 698–713 (1992). https://doi.org/10.1109/72.159059

5. Casillas, J., Cordon, O., Herrera, T.F., Magdalena, L. (eds.): Interpretability Issues in Fuzzy Modeling. Studies in Fuzziness and Soft Computing. Springer, Berlin, Heidelberg (2003). http://www.springer.com/us/book/9783540029328
6. Diao, Y., Passino, K.M.: Adaptive neural/fuzzy control for interpolated nonlinear systems. IEEE Trans. Fuzzy Syst. **10**(5), 583–595 (2002)
7. Ding, Y., Ying, H., Shao, S.: Typical Takagi-Sugeno PI and PD fuzzy controllers: analytical structures and stability analysis. Inf. Sci. **151**, 245–262 (2003). https://doi.org/10.1016/S0020-0255(02)00302-X
8. Duan, X., Deng, H., Li, H.: A saturation-based tuning method for fuzzy PID controller. IEEE Trans. Ind. Electron. **60**(11), 5177–5185 (2013). https://doi.org/10.1109/TIE.2012.2222858
9. Er, M.J., Mandal, S.: A survey of adaptive fuzzy controllers: nonlinearities and classifications. IEEE Trans. Fuzzy Syst. **24**(5), 1095–1107 (2016)
10. Feng, Z., Chen, X.: Adaptive iterative generalized demodulation for nonstationary complex signal analysis: Principle and application in rotating machinery fault diagnosis. Mech. Syst. Signal Process. **110**, 1–27 (2018). https://doi.org/10.1016/j.ymssp.2018.03.004. http://www.sciencedirect.com/science/article/pii/S0888327018301237
11. Fliess, M., Join, C.: Model-free control. Int. J. Control **86**(12), 2228–2252 (2013). https://doi.org/10.1080/00207179.2013.810345
12. Frank, E., Hall, M.A., Witten, I.H.: The WEKA workbench. In: Online Appendix for "Data Mining: Practical Machine Learning Tools and Techniques", 4th edn. Morgan Kaufmann (2016)
13. Hamzaoui, A., Essounbouli, N., Benmahammed, K., Zaytoon, J.: State observer based robust adaptive fuzzy controller for nonlinear uncertain and perturbed systems. IEEE Trans. Syst. Man Cybern. Part B (Cybern.) **34**(2), 942–950 (2004). https://doi.org/10.1109/TSMCB.2003.818562
14. Han, S.I., Ha, H., Lee, J.: Barrier Lyapunov function-based model-free constraint position control for mechanical systems. J. Mech. Sci. Technol. **30**(7), 3331–3338 (2016). https://doi.org/10.1007/s12206-016-0642-3
15. Hu, B., Mann, G.K.I., Gosine, R.G.: New methodology for analytical and optimal design of fuzzy PID controllers. IEEE Trans. Fuzzy Syst. **7**(5), 521–539 (1999). https://doi.org/10.1109/91.797977
16. Hu, B.G., Mann, G.K., Gosine, R.G.: A systematic study of fuzzy PID controllers-function-based evaluation approach. IEEE Trans. Fuzzy Syst. **9**(5), 699–712 (2001)
17. Jardine, A.K., Lin, D., Banjevic, D.: A review on machinery diagnostics and prognostics implementing condition-based maintenance. Mech. Syst. Signal Process. **20**(7), 1483–1510 (2006). https://doi.org/10.1016/j.ymssp.2005.09.012. http://www.sciencedirect.com/science/article/pii/S0888327005001512
18. Javed, K., Gouriveau, R., Li, X., Zerhouni, N.: Tool wear monitoring and prognostics challenges: a comparison of connectionist methods toward an adaptive ensemble model. J. Intell. Manuf. **29**(8), 1873–1890 (2018). https://doi.org/10.1007/s10845-016-1221-2
19. Javed, K., Gouriveau, R., Zerhouni, N.: A new multivariate approach for prognostics based on extreme learning machine and fuzzy clustering. IEEE Trans. Cybern. **45**(12), 2626–2639 (2015). https://doi.org/10.1109/TCYB.2014.2378056
20. Jia, F., Lei, Y., Guo, L., Lin, J., Xing, S.: A neural network constructed by deep learning technique and its application to intelligent fault diagnosis of machines. Neurocomputing **272**, 619–628 (2018). https://doi.org/10.1016/j.neucom.2017.07.032. http://www.sciencedirect.com/science/article/pii/S0925231217312894
21. Kaczorek, T.: Linear Control Systems: Analysis of Multivariable Systems. Wiley, New York, NY, USA (1992)
22. Keel, L.H., Bhattacharyya, S.P.: Controller synthesis free of analytical models: three term controllers. IEEE Trans. Autom. Control **53**(6), 1353–1369 (2008)
23. Khadraoui, S., Nounou, H.N., Nounou, M.N., Datta, A., Bhattacharyya, S.P.: A model-free design of reduced-order controllers and application to a DC servomotor. Automatica **50**, 2142–2149 (2014)

24. Khan, S., Yairi, T.: A review on the application of deep learning in system health management. Mech. Syst. Signal Process. **107**, 241–265 (2018). https://doi.org/10.1016/j.ymssp.2017.11.024. http://www.sciencedirect.com/science/article/pii/S0888327017306064
25. Khanesar, M.A., Kaynak, O., Teshnehlab, M.: Direct model reference Takagi-Sugeno fuzzy control of SISO nonlinear systems. IEEE Trans. Fuzzy Syst. **19**(5), 914–924 (2011)
26. Kluska, J.: Fuzzy logic control (in Polish). Sci. Lett. (OW Rzeszow University of Technology, Rzeszow) (1992)
27. Kluska, J.: Stability analysis of SISO, MISO and MIMO continuous fuzzy control systems based on theorems of Kudrewicz. Arch. Control Sci. **5**, 5–38 (1996)
28. Kluska, J.: Absolute stability of continuous fuzzy control systems. In: Aracil, J., Gordillo, F. (eds.) Stability Issues in Fuzzy Control, vol. 44, pp. 15–45. Physica-Verlag, Heidelberg, New York (2000)
29. Kluska, J.: New results in analytical modeling using Takagi-Sugeno expert system. In: Hryniewicz, O., Kacprzyk, J., Koronacki, J., Wierzchoń, S.T. (eds.) Issues in Intelligent Systems Paradigms, pp. 79–94. Exit, Warsaw (2005)
30. Kluska, J.: Transformation lemma on analytical modeling via Takagi-Sugeno fuzzy system and its applications. In: Rutkowski, L., Tadeusiewicz, R., Zadeh, L.A., Żurada, J.M. (eds.) Artificial Intelligence and Soft Computing—ICAISC 2006, pp. 230–239. Springer, Berlin, Heidelberg (2006)
31. Kluska, J.: Analytical Methods in Fuzzy Modeling and Control. Studies in Fuzziness and Soft Computing. Springer, Berlin, Heidelberg (2009). https://link.springer.com/book/10.1007/978-3-540-89927-3
32. Kluska, J.: Selected applications of P1-TS fuzzy rule-based systems. In: Rutkowski, L., Korytkowski, M., Scherer, R., Tadeusiewicz, R., Zadeh, L.A., Zurada, J.M. (eds.) Artificial Intelligence and Soft Computing, pp. 195–206. Springer International Publishing, Cham (2015)
33. Kluska, J., Hajduk, Z.: Digital implementation of fuzzy Petri net based on asynchronous fuzzy RS flip-flop. In: Rutkowski, L., Siekmann, J.H., Tadeusiewicz, R., Zadeh, L.A. (eds.) Artificial Intelligence and Soft Computing—ICAISC 2004, pp. 314–319. Springer, Berlin, Heidelberg (2004)
34. Kluska, J., Hajduk, Z.: Hardware implementation of P1-TS fuzzy rule-based systems on FPGA. In: Rutkowski, L., Korytkowski, M., Scherer, R., Tadeusiewicz, R., Zadeh, L.A., Żurada, J.M. (eds.) Artificial Intelligence and Soft Computing, pp. 282–293. Springer, Berlin, Heidelberg (2013)
35. Kluska, J., Kusy, M., Obrzut, B.: The classifier for prediction of peri-operative complications in cervical cancer treatment. In: Rutkowski, L., Korytkowski, M., Scherer, R., Tadeusiewicz, R., Zadeh, L.A., Zurada, J.M. (eds.) Artificial Intelligence and Soft Computing, pp. 143–154. Springer International Publishing, Cham (2014)
36. Kluska, J., Wiktorowicz, K.: Integral evaluations for a class of multivariable fuzzy control systems. IFAC Proc. Vol. **35**(1), 97–102 (2002)
37. Koo, T.J.: Stable model reference adaptive fuzzy control of a class of nonlinear systems. IEEE Trans. Fuzzy Syst. **9**(4), 624–636 (2001)
38. Kudrewicz, J.: Frequency-Domain Methods in the Theory of Nonlinear Dynamical Systems (in Polish). WNT, Warsaw (1970)
39. Lasi, H., Fettke, P., Kemper, H.G., Feld, T., Hoffmann, M.: Industry 4.0. Bus. Inf. Syst. Eng. **6**(4), 239–242 (2014). https://doi.org/10.1007/s12599-014-0334-4
40. Lee, J., Kao, H.A., Yang, S.: Service innovation and smart analytics for industry 4.0 and big data environment. Procedia CIRP **16**, 3–8 (2014). Product Services Systems and Value Creation. Proceedings of the 6th CIRP Conference on Industrial Product-Service Systems. https://doi.org/10.1016/j.procir.2014.02.001. http://www.sciencedirect.com/science/article/pii/S2212827114000857
41. Lee, J., Wu, F., Zhao, W., Ghaffari, M., Liao, L., Siegel, D.: Prognostics and health management design for rotary machinery systems. reviews, methodology and applications. Mech. Syst. Signal Process. **42**(1), 314–334 (2014). https://doi.org/10.1016/j.ymssp.2013.06.004. http://www.sciencedirect.com/science/article/pii/S0888327013002860

42. Lei, Y., Li, N., Guo, L., Li, N., Yan, T., Lin, J.: Machinery health prognostics: a systematic review from data acquisition to RUL prediction. Mech. Syst. Signal Process. **104**, 799–834 (2018). https://doi.org/10.1016/j.ymssp.2017.11.016. http://www.sciencedirect.com/science/article/pii/S0888327017305988
43. Li, H.-X., Zhang, L., Cai, K.-Y., Chen, G.: An improved robust fuzzy-PID controller with optimal fuzzy reasoning. IEEE Trans. Syst. Man Cybern. Part B (Cybern.) **35**(6), 1283–1294 (2005). https://doi.org/10.1109/TSMCB.2005.851538
44. Lin, C.M., Li, H.Y.: TSK fuzzy CMAC-based robust adaptive backstepping control for uncertain nonlinear systems. IEEE Trans. Fuzzy Syst. **20**(6), 1147–1154 (2012)
45. Liu, R., Yang, B., Zio, E., Chen, X.: Artificial intelligence for fault diagnosis of rotating machinery: a review. Mech. Syst. Signal Process. **108**, 33–47 (2018). https://doi.org/10.1016/j.ymssp.2018.02.016. http://www.sciencedirect.com/science/article/pii/S0888327018300748
46. Liu, Y.J., Tong, S.: Adaptive fuzzy control for a class of unknown nonlinear dynamical systems. Fuzzy Sets Syst. **263**, 49–70 (2015)
47. Liu, Y.J., Tong, S.C., Wang, W.: Adaptive fuzzy output tracking control for a class of uncertain nonlinear systems. Fuzzy Sets Syst. **160**(19), 2727–2754 (2009)
48. Lu, Y.: Industry 4.0: a survey on technologies, applications and open research issues. J. Ind. Inf. Integr. **6**, 1–10 (2017). https://doi.org/10.1016/j.jii.2017.04.005. http://www.sciencedirect.com/science/article/pii/S2452414X17300043
49. Mączka, T., Żabiński, T., Kluska, J.: Discovering rules for production speed prediction in fasteners manufacturing. In: Korbicz, J., Kowal, M. (eds.) Intelligent Systems in Technical and Medical Diagnostics, pp. 199–209. Springer, Berlin, Heidelberg (2014)
50. Mann, G.K.I., Hu, B.-G., Gosine, R.G.: Analysis of direct action fuzzy PID controller structures. IEEE Trans. Syst. Man Cybern. Part B (Cybern.) **29**(3), 371–388 (1999). https://doi.org/10.1109/3477.764871
51. Matlab: Version 9.6.0 (R2019a). The MathWorks Inc., Natick, Massachusetts (2019)
52. Misir, D., Malki, H.A., Chen, G.: Design and analysis of a fuzzy proportional-integral-derivative controller. Fuzzy Sets Syst. **79**(3), 297–314 (1996). https://doi.org/10.1016/0165-0114(95)00149-2. http://www.sciencedirect.com/science/article/pii/0165011495001492
53. Mitkowski, W.: Stabilization of Dynamic Systems. WNT, Warszawa (1991)
54. Mączka, T.: Application of computational intelligence methods and decision support in production systems (in Polish). Ph.D. thesis, Rzeszow (2016)
55. Mohan, B.M., Sinha, A.: Analytical structures for fuzzy PID controllers? IEEE Trans. Fuzzy Syst. **16**(1), 52–60 (2008). https://doi.org/10.1109/TFUZZ.2007.894974
56. Muhuri, P.K., Shukla, A.K., Abraham, A.: Industry 4.0: a bibliometric analysis and detailed overview. Eng. Appl. Artif. Intell. **78**, 218–235 (2019). https://doi.org/10.1016/j.engappai.2018.11.007. http://www.sciencedirect.com/science/article/pii/S0952197618302458
57. Noman, M.A., Nasr, E.S.A., Al-Shayea, A., Kaid, H.: Overview of predictive condition based maintenance research using bibliometric indicators. J. King Saud Univ. Eng. Sci. (2018). https://doi.org/10.1016/j.jksues.2018.02.003. http://www.sciencedirect.com/science/article/pii/S1018363917303720
58. Oesterreich, T.D., Teuteberg, F.: Understanding the implications of digitisation and automation in the context of industry 4.0: a triangulation approach and elements of a research agenda for the construction industry. Comput. Ind. **83**, 121–139 (2016).https://doi.org/10.1016/j.compind.2016.09.006. http://www.sciencedirect.com/science/article/pii/S0166361516301944
59. Patel, A.V., Mohan, B.: Analytical structures and analysis of the simplest fuzzy PI controllers. Automatica **38**(6), 981–993 (2002). https://doi.org/10.1016/S0005-1098(01)00297-7. http://www.sciencedirect.com/science/article/pii/S0005109801002977
60. Precup, R., David, R., Petriu, E.M.: Grey wolf optimizer algorithm-based tuning of fuzzy control systems with reduced parametric sensitivity. IEEE Trans. Ind. Electron. **64**(1), 527–534 (2017). https://doi.org/10.1109/TIE.2016.2607698
61. Qi, R., Brdyś, M.A.: Indirect adaptive controller based on a self-structuring fuzzy system for nonlinear modeling and control. Appl. Math. Comput. Sci. **19**, 619–630 (2009). https://doi.org/10.2478/v10006-009-0049-8

62. Qin, J., Liu, Y., Grosvenor, R.: A categorical framework of manufacturing for industry 4.0 and beyond. Procedia CIRP **52**, 173–178 (2016). The Sixth International Conference on Changeable, Agile, Reconfigurable and Virtual Production (CARV2016). https://doi.org/10.1016/j.procir.2016.08.005. http://www.sciencedirect.com/science/article/pii/S221282711630854X
63. Rafajłowicz, E.: Image-driven, model-free control of repetitive processes based on machine learning. In: 2017 10th International Workshop on Multidimensional (nD) Systems (nDS), pp. 1–6 (2017)
64. Rasvan, V.: Absolute Stability of Control Systems with Time-Delay. Nauka, Moskva (1983) (in Russian)
65. Ray, K.S., Ananda, S.G., Ghosh, A.M., Majumder, D.D.: L_2-stability and the related design concept for SISO linear system associated with fuzzy logic controller. IEEE Trans. Syst. Man Cybern. **SMC-14**(6), 932–939 (1984). https://doi.org/10.1109/TSMC.1984.6313322
66. Sandberg, I.W.: A frequency-domain condition for the stability of feedback systems containing a single time-varying nonlinear element. Bell Syst. Tech. J. **43**(4), 1601–1608 (1964). https://doi.org/10.1002/j.1538-7305.1964.tb04100.x
67. See5: Data mining tools See5 and C5.0 (an informal tutorial) (2019). http://www.rulequest.com/see5-info.html
68. Shao, H., Jiang, H., Zhang, H., Duan, W., Liang, T., Wu, S.: Rolling bearing fault feature learning using improved convolutional deep belief network with compressed sensing. Mech. Syst. Signal Process. **100**, 743–765 (2018). https://doi.org/10.1016/j.ymssp.2017.08.002. http://www.sciencedirect.com/science/article/pii/S0888327017304260
69. Shao, H., Jiang, H., Zhao, H., Wang, F.: A novel deep autoencoder feature learning method for rotating machinery fault diagnosis. Mech. Syst. Signal Process. **95**, 187–204 (2017). https://doi.org/10.1016/j.ymssp.2017.03.034. http://www.sciencedirect.com/science/article/pii/S0888327017301607
70. Sherrod, P.H.: DTREG predictive modelling software (2019). http://www.dtreg.com. Accessed 18 Aug 2019
71. Siddique, N.: Stability Analysis of Intelligent Controllers, pp. 243–267. Springer International Publishing, Cham (2014). https://doi.org/10.1007/978-3-319-02135-5_9
72. Sikorska, J., Hodkiewicz, M., Ma, L.: Prognostic modelling options for remaining useful life estimation by industry. Mech. Syst. Signal Process. **25**(5), 1803–1836 (2011). https://doi.org/10.1016/j.ymssp.2010.11.018. http://www.sciencedirect.com/science/article/pii/S0888327010004218
73. Škrjanc, I., Blažič, S., Matko, D.: Model-reference fuzzy adaptive control as a framework for nonlinear system control. J. Intell. Robot. Syst. **36**(3), 331–347 (2003)
74. Smoczek, J.: P1-TS fuzzy scheduling control system design using local pole placement and interval analysis. Bull. Pol. Acad. Sci. Tech. Sci. **62**(3), 455–464 (2014)
75. Smoczek, J.: Experimental verification of a GPC-LPV method with RLS and P1-TS fuzzy-based estimation for limiting the transient and residual vibration of a crane system. Mech. Syst. Signal Process. **62**, 324–340 (2015)
76. Smoczek, J., Szpytko, J.: Robust predictive control of an overhead crane. J. KONES Powertrain Transp. **24**(2), 231–238 (2017)
77. Spooner, J.T., Ordonez, R., Maggiore, M., Passino, K.M.: Stable Adaptive Control and Estimation for Nonlinear Systems: Neural and Fuzzy Approximation Techniques. Wiley, New York, NY, USA (2002)
78. Statistica: Dell Statistica (data analysis software system), version 13.1. Dell Inc. (2016). http://software.dell.com
79. Tang, K.S., Man, K.F., Chen, G., Kwong, S.: An optimal fuzzy PID controller. IEEE Trans. Ind. Electron. **48**(4), 757–765 (2001). https://doi.org/10.1109/41.937407
80. Trstenjak, M., Cosic, P.: Process planning in industry 4.0 environment. Procedia Manuf. **11**, 1744–1750 (2017). 27th International Conference on Flexible Automation and Intelligent Manufacturing, FAIM2017, 27–30 June 2017, Modena, Italy. https://doi.org/10.1016/j.promfg.2017.07.303. http://www.sciencedirect.com/science/article/pii/S2351978917305115

81. TwinCAT: TwinCAT 3, Version 1.7. Beckhoff Automation GmbH and Co. KG, Germany (2019). http://download.beckhoff.com/download/document/automation/twincat3/TC1300_C_EN.pdf. Accessed 17 Jan 2019
82. Ulu, C.: Exact analytical inverse mapping of decomposable TS fuzzy systems with singleton and linear consequents. Appl. Soft Comput. **23**, 202–214 (2014)
83. Van, M.: An enhanced robust fault tolerant control based on an adaptive fuzzy PID-nonsingular fast terminal sliding mode control for uncertain nonlinear systems. IEEE/ASME Trans. Mechatron. **23**(3), 1362–1371 (2018). https://doi.org/10.1109/TMECH.2018.2812244
84. Vidyasagar, M.: Nonlinear Systems Analysis. Prentice Hall, Englewood Cliffs, New Jersey (1993)
85. Wagner, T., Herrmann, C., Thiede, S.: Industry 4.0 impacts on lean production systems. Procedia CIRP **63**, 125–131 (2017). Manufacturing Systems 4.0 Proceedings of the 50th CIRP Conference on Manufacturing Systems. https://doi.org/10.1016/j.procir.2017.02.041. http://www.sciencedirect.com/science/article/pii/S2212827117301385
86. Wan, J., Tang, S., Shu, Z., Li, D., Wang, S., Imran, M., Vasilakos, A.V.: Software-defined industrial internet of things in the context of industry 4.0. IEEE Sens. J. **16**(20), 7373–7380 (2016). https://doi.org/10.1109/JSEN.2016.2565621
87. Wang, N., Meng, X.: Analytical structure of three-dimensional fuzzy controller. In: 2007 IEEE International Conference on Control and Automation, pp. 2349–2354 (2007). https://doi.org/10.1109/ICCA.2007.4376781
88. Wang, S., Wan, J., Zhang, D., Li, D., Zhang, C.: Towards smart factory for industry 4.0: a self-organized multi-agent system with big data based feedback and coordination. Comput. Netw. **101**, 158–168 (2016). https://doi.org/10.1016/j.comnet.2015.12.017. http://www.sciencedirect.com/science/article/pii/S1389128615005046
89. Wiktorowicz, K.: Output feedback direct adaptive fuzzy controller based on frequency-domain methods. IEEE Trans. Fuzzy Syst. **24**(3), 622–634 (2016)
90. Wiktorowicz, K.: Design of state feedback adaptive fuzzy controllers for second-order systems using a frequency stability criterion. IEEE Trans. Fuzzy Syst. **25**(3), 499–510 (2017)
91. Ying, H.: Fuzzy Control and Modeling: Analytical Foundations and Applications, 1st edn. Wiley-IEEE Press (2000)
92. Żabiński, T., Mączka, T., Kluska, J.: Industrial platform for rapid prototyping of intelligent diagnostic systems. In: Mitkowski, W., Kacprzyk, J., Oprzędkiewicz, K., Skruch, P. (eds.) Trends in Advanced Intelligent Control, Optimization and Automation, pp. 712–721. Springer International Publishing, Cham (2017)
93. Żabiński, T., Mączka, T., Kluska, J., Kusy, M., Gierlak, P., Hanus, R., Prucnal, S., Sęp, J.: CNC milling tool head imbalance prediction using computational intelligence methods. In: Rutkowski, L., Korytkowski, M., Scherer, R., Tadeusiewicz, R., Zadeh, L.A., Zurada, J.M. (eds.) Artificial Intelligence and Soft Computing, pp. 503–514. Springer International Publishing, Cham (2015)
94. Żabiński, T., Mączka, T., Kluska, J., Kusy, M., Hajduk, Z., Prucnal, S.: Failures prediction in the cold forging process using machine learning methods. In: Rutkowski, L., Korytkowski, M., Scherer, R., Tadeusiewicz, R., Zadeh, L.A., Zurada, J.M. (eds.) Artificial Intelligence and Soft Computing, pp. 622–633. Springer International Publishing, Cham (2014)
95. Żabiński, T., Mączka, T., Kluska, J., Madera, M., Sęp, J.: Condition monitoring in industry 4.0 production systems—the idea of computational intelligence methods application. Procedia CIRP **79**, 63–67 (2019). 12th CIRP Conference on Intelligent Computation in Manufacturing Engineering, 18–20 July 2018, Gulf of Naples, Italy. https://doi.org/10.1016/j.procir.2019.02.012. http://www.sciencedirect.com/science/article/pii/S221282711930126X
96. Zhang, J., Wang, P., Yan, R., Gao, R.X.: Long short-term memory for machine remaining life prediction. J. Manuf. Syst. **48**, 78–86 (2018). Special Issue on Smart Manufacturing. https://doi.org/10.1016/j.jmsy.2018.05.011. http://www.sciencedirect.com/science/article/pii/S0278612518300803
97. Zhang, L., Gao, R.X., Lee, K.B.: Spindle health diagnosis based on analytic wavelet enveloping. IEEE Trans. Instrum. Meas. **55**(5), 1850–1858 (2006). https://doi.org/10.1109/TIM.2006.880261

98. Zhang, W., Li, C., Peng, G., Chen, Y., Zhang, Z.: A deep convolutional neural network with new training methods for bearing fault diagnosis under noisy environment and different working load. Mech. Syst. Signal Process. **100**, 439–453 (2018). https://doi.org/10.1016/j.ymssp.2017.06.022. http://www.sciencedirect.com/science/article/pii/S0888327017303369
99. Zhao, R., Yan, R., Chen, Z., Mao, K., Wang, P., Gao, R.X.: Deep learning and its applications to machine health monitoring. Mech. Syst. Signal Process. **115**, 213–237 (2019). https://doi.org/10.1016/j.ymssp.2018.05.050. http://www.sciencedirect.com/science/article/pii/S0888327018303108
100. Zhong, R.Y., Xu, X., Klotz, E., Newman, S.T.: Intelligent manufacturing in the context of industry 4.0: a review. Engineering **3**(5), 616–630 (2017). https://doi.org/10.1016/J.ENG.2017.05.015. http://www.sciencedirect.com/science/article/pii/S2095809917307130
101. Zhou, Q., Shi, P., Lu, J., Xu, S.: Adaptive output-feedback fuzzy tracking control for a class of nonlinear systems. IEEE Trans. Fuzzy Syst. **19**(5), 972–982 (2011)

Applications

Consequences and Modeling Challenges Connected with Atmospheric Pollution

Zbigniew Nahorski and Piotr Holnicki

Abstract Emissions of pollutants into the atmosphere is a major problem affecting the further development of humanity. Many shorter living pollutants are a source of risk to human health. This is, among others, connected with a huge increase in the population of cities, where the concentration of harmful compounds in the air is usually much higher. On the other hand, in the long-term perspective, excessive greenhouse gas emissions may be a reason for climatic changes covering the whole globe, with difficult to predict effects. Complicated dependencies between emissions and effects require a systemic approach, involving modeling and computer aided decisions on rationally reducing the pollutant emissions, both at local, national, international, and even global levels. This chapter contains an overview of related issues, with particular emphasis on the problems studied at the Systems Research Institute, Polish Academy of Sciences. They are related to all elements of integrated atmospheric modeling: human activity, the emissions of pollutants and their dispersion in the atmosphere, the assessment of pollution impact on population health risk, as well as computer supported decision making to reduce the emissions harmful effects.

Z. Nahorski (✉) · P. Holnicki
Systems Research Institute, Polish Academy of Sciences, Warszawa, Newelska 6, 01-447 Warsaw, Poland
e-mail: nahorski@ibspan.waw.pl; nahorski@wit.edu.pl

P. Holnicki
e-mail: holnicki@ibspan.waw.pl

Z. Nahorski
Warsaw School of Information Technology, Warszawa, Newelska 6, 01-447 Warsaw, Poland

P. Kulczycki et al. (eds.), *Automatic Control, Robotics, and Information Processing*, Studies in Systems, Decision and Control 296,
https://doi.org/10.1007/978-3-030-48587-0_23

1 Introduction

Human activity changes the Earth. The development of technology was a necessary factor for the population to grow, however, as a side effect created various types of waste. Among those are wastes that pollute the atmosphere and cause, alter alia, such well-known effects as the ozone hole, climate change, acid rain or smog. These issues fall within the scope of many sciences, such as geophysics, environmental protection, physics and chemistry, biology, medicine and health sciences, engineering and technical sciences, and social sciences, with possible breakdowns into finer disciplines and interdisciplinary researches. To study atmospheric pollution issues, it is necessary to consider various spatial and temporal scales. In order to describe the complex relationships between variables affecting the evolution of environmental processes, many studies use mathematical models, whose large size and interdisciplinary nature involves interest in systemic approaches. Automation and automatic control have a long modeling tradition that includes large scale interdisciplinary systems. Among them are environmental systems, as evidenced by the creation of the TC8.3 *Modeling and Control of Environmental Systems* technical committee within the *International Federation of Automatic Control* (IFAC). Research problems related to various aspects of this complex research problem are also within the scope of other IFAC Committees belonging to the IFAC Coordination Groups 8 and 9. Important activities of the *International Institute for Applied Systems Analysis* (IIASA) founded in 1972 in Laxenburg, Austria, were paramount to the development and promotion of a systemic approach to this area, followed later by the *Institute for Environment and Sustainability* founded by the European Union in Ispra, Italy, as one of the *Joint Research Center* institutes of the European Commission, as well as many institutions acting in different countries.

This chapter presents some modeling studies that relate to various aspects of atmospheric pollution considered at the Institute of Systems Research, Polish Academy of Sciences (SRI PAS), in particular in its Computer Modeling Department (CDM). They include topics from all major issues connected with integrated atmospheric modeling, such as modeling human activity that leads to pollution emissions, dispersion of the pollutants in the atmosphere, their impacts on the human health risk, and emission reduction along with its economic aspects.

2 Air Pollution Influence Ranges

Atmospheric air quality covers a very wide spectrum of issues, both due to the type of pollutants involved (gases, dusts, aerosols, smog, deposition of acid chemical compounds), the range of the associated processes (from the local effects to the entire globe influence), and also due to environmental effects (health risk for people, soil acidification, contamination of surface and underground water, destruction of forest areas and agricultural crops, climate change, and degradation of the stratospheric ozone layer).

Pollutants are emitted to the atmosphere both from natural sources (e.g. from volcanic eruptions, through emissions by the oceans, from organic pollution), but also in direct connection with human activity in the energy, industry, municipal services, transport networks, agriculture, forest management, land use change, etc., and can occur in various forms, for example gaseous, such as SO_2, CO_2, NO_x, NH_4, O_3, two-phase, as particulate matter, microorganisms, or three-phase, as aerosols. Some primary pollutants that are emitted directly by pollution sources, undergo chemical reactions in the atmosphere, which result in the formation of secondary pollutants, which may be even more dangerous for the environment [67, 90, 93, 94].

The gaseous secondary pollutants include e.g. NO_2, arising from the primary NO and the tropospheric ozone O_3, generated as a result of photochemical reactions. Particulate matter secondary pollutants are often formed by chemical or photochemical transformations of gaseous primary pollutants, for example transformations of sulfur dioxide (SO_2) into sulfate aerosol ($SO_2^{=}$), or nitrogen dioxide (NO_2) into nitrate aerosol (NO_3^-), and also transformation of organic compounds into the organic particles.

A large part of the pollutants reaching the atmosphere is of anthropogenic nature. These emissions and their dispersion in the atmosphere are the main goal of most studies aimed at assessing human activity effects. However, the biosphere and geophysical processes, such as the absorption of gases by vegetation or oceanic waters, influence substantially the concentration of certain substances in the atmosphere, e.g. CO_2.

An important parameter characterizing all the substances in the atmosphere is their lifetime, i.e. their average residence time in the atmosphere, that is generally accepted as a period, in which the emitted mass of a pollutant drops in the atmosphere to the level of 1/e of the initial value due to atmospheric processes [89, 90, 94]. The measure of lifetime is an important factor deciding on the pollutant impact on the environment. The lifetime of a pollutant (its temporal range) is connected directly with a spatial range of its essential environmental impact that is illustrated in Fig. 1a. The most often distinguished spatial ranges are (Fig. 1b): the local (the microscale), the regional (the mesoscale), the continental (the synoptic scale), and the global (the planetary scale) ones.

In most of the implementations, the local scale models operate in the ranges from 20 km up to over 100 km, while the mesoscale models from about 20 km up to a few thousand kilometers [15, 27, 29, 90]. In practice, the mesoscale models usually cover both of the above categories. The temporal activity of the main pollutants in this case is between one day to about a week. This is the period, in which the pollutant atmospheric transport, the chemical transformations and the deposition on the earth surface are the main processes determining their environmental impact.

The spatial range of regional and continental models is not precisely defined in literature. The global range models include atmospheric phenomena occurring across whole continents or even the globe. Extremely slowly reacting gases (cf. Fig. 1), in particular emitted as a part of human activity, such as carbon dioxide (CO_2), methane

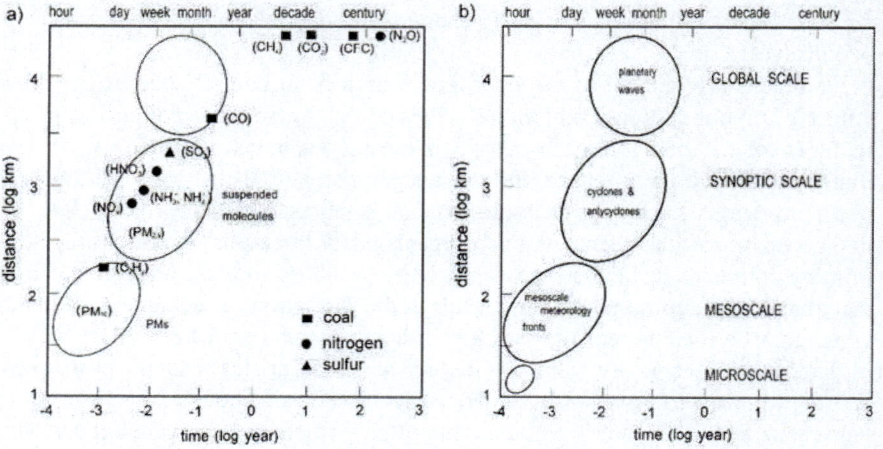

Fig. 1 **a** Space-time ranges of the selected pollutant impacts, and **b** atmospheric processes associated with their propagation

(CH_4) and nitrous oxide (N_2O), as well as, gases with very long lifetimes, expressed in thousands or even tens of thousands of years, such as chlorofluorocarbons, e.g. freons (CFCl) and halons as well as other industrial gases (HFC, PHC, SF_6), can be transported over very large distances and can migrate to the stratospheric layer before they are removed from the atmosphere. The long lifetime of these gases and the relatively slow transformation processes in relation to the tropospheric pollutant mixing cause that simple models are often utilized to describe their concentration changes, in which the atmosphere or troposphere and stratosphere are described as reservoirs with appropriate inflows and outflows.

Time-space range is an important reference in classification of pollutant propagation models. The mathematical formulas applied there describe the processes and phenomena occurring in the atmosphere, which correspond, among others, to the pollutant propagation and transformation.

Analysis and modeling of the air quality in the mesoscale includes a segment that is very important from the point of view of environmental quality protection, and namely, dispersion of the short-term pollution in the urban and regional range. The sulfur and nitrogen pollutants (primary and secondary), ammonia NH_3, particulates (mainly PM_{10} and $PM_{2.5}$), typical transport contaminants (including C_6H_6 and CO), and some organic gases, are involved in these processes. The most important products of the atmospheric transformations are ozone, sulfate aerosol, nitrogen and ammonium ions, as well as secondary particles. These pollutants are the main source of environment acidification. In Poland, the impact of polycyclic aromatic hydrocarbons (PAHs) is very important, especially of the highly carcinogenic benzo(a)pyrene (BaP) coming primarily from coal combustion (often of a very low quality) in the municipal sector. Mesoscale models have recently been used more and more frequently to illustrate the behavior of slowly reactive substances. The reason for this is to link the local emissions to the local measurements, and to help estimating the biogenic emissions.

Formation of smog is quite characteristic for the urban atmosphere. Depending on the formation mechanisms and the dominant components two variants are distinguished. The so-called black smog or London smog usually arises in longer periods of the meteorological stagnation, temperature inversion and the fog. Sulfur dioxide and fine dust are absorbed by the droplets of mist where they can react with other substances. Smoke and sulfuric acid molecules also act as water vapor condensation centers which further increases the smog formation intensity.

Photochemical smog is typical for many large cities with intensive industrial production and high car traffic. Also known as Los Angeles smog, it is a mixture of many gaseous and material substances. The predominant components are: ozone, carbon monoxide, sulfur dioxide, nitrogen oxides, PAN (PeroxyAcetylNitrate), sulfuric and nitric acid ions, formaldehyde, ammonia. Some of these substances, emitted as primary pollutants (NO, CO, SO_2), enter photochemical reactions with active organic gases during periods of prolonged insolation, resulting in accumulation of ozone that causes its high concentrations, which further intensifies the whole process. The phenomenon is particularly dangerous during periods of meteorological stagnation, combined with temperature inversion.

For long living gases, such as greenhouse gases, the environmental damage is generally global, mainly related to the gradual increase of the global average annual temperature, which, among others, leads to excessive melting of the glaciers and rising the sea levels. Despite its global nature, the damage caused by the effects of increased emissions is unevenly distributed and certain regions of the globe are more vulnerable than others to unstable meteorological conditions. The greenhouse gases impact on warming is expressed by the GWP (*Global Warning Potential*) factor, which determines the retention of energy in the atmosphere by a given gas within a certain time, in relation to the energy captured by CO_2. The GWP coefficient is connected with the gas lifetime and hence depends on the adopted retention time, which is usually taken as 100 years. The *Intergovernmental Panel on Climate Change* (IPCC) guidelines assume that 100-year GWP of CH_4 is equal to 25, of N_2O is equal to 298, and of SF_6 is 22,800. By using this coefficient, emissions can be reduced to one equivalent value, usually expressed in terms of CO_2eq.

Another global effect, observed and documented during several dozen recent years (associated with the impact of the above-mentioned CFCs and halons), is the destruction of the stratospheric ozone layer above the Arctic regions, mainly in the Antarctic area. This is the result of specific meteorological conditions prevailing in the stratosphere in the winter-spring period and the effect caused by chlorites and bromines, which are formed from CFCs and halons and play the role of the catalysts in the chemical degradation of the ozone layer (cf. [90, 92]). The *Montreal Protocol* (signed in 1987) and its continuations practically eliminated the emission of these gases, nevertheless the negative effects continue because of their very long lifetimes.

3 Integrated Atmospheric Modeling

Besides the cognitive goals, atmospheric modeling is primarily used to help in elaborating appropriate steps to mitigate harmful effects of atmospheric pollution. Integrated modeling includes all stages of the pollutants formation, their propagation, their harmful impact, and the selection of actions leading to the improvement of the air quality [2, 3, 17, 69, 75, 79, 80, 88]. The *European Environmental Agency* (EEA) has introduced a DPSIR framework methodology for this purpose. It specifies the following components: activities (Drivers), emissions (Pressures), air quality (State), effects (Impact), and corrective actions (Responses), as presented in Fig. 2. In the above specifications, the Drivers are all human activities that evoke emissions, in particular from energy production, industry, transport, agriculture, and forestry, while the Pressures are pollutant emissions into the atmosphere as a result of these activities. The State, in term, is expressed by concentrations of harmful substances in the atmosphere, the Impacts are, in particular, deterioration of human population health, decline of plant or animal forms, destruction of buildings, etc. In the long-term they also cause structural changes. The last of the components, i.e. the Responses, include all measures to improve air quality, which also involves costs.

A better understanding of the processes related to atmospheric pollution involves a significant development of measurement techniques, starting from local measurements, such as measurements of gas emissions from arable land within a submeter range, gas exchange in the forest areas within a range of several to several dozen

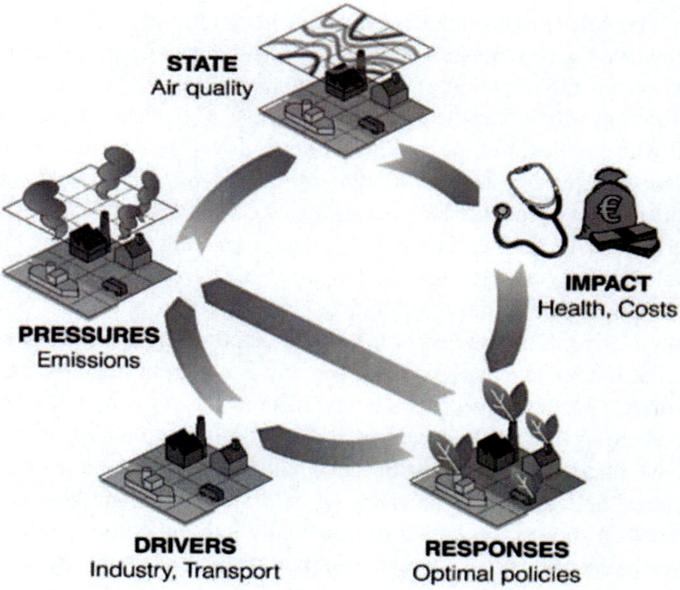

Fig. 2 DPSIR methodology in the integrated atmospheric modeling (as presented in FP7-ENV project APPRAISAL at http://appraisal-fp7.terraria.com/site/index.php)

meters, ground-based measurements of the atmospheric concentrations at different altitudes, corresponding to various spatial ranges, measurements performed on flying devices (drones, balloons), to satellite measurements covering the entire globe, see e.g. [64]. Integrated atmosphere modeling gives an opportunity to synthesize knowledge from these and other sources and to develop coherent tools to help in understanding, analyzing, anticipating and making decisions in this complex system, where correct actions are essential for humanity fate.

Taking into account the harmful effects of mesoscale pollution, the European Union in 2008 issued a Directive, known under the acronym CAFE (*Clean Air for Europe*). A review of the solutions in order to improve the air quality in the European Union countries was carried out in EU FP7 project, acronym APPRAISAL (Air Pollution Policies foR Assessment of Integrated Strategies At regional and Local scales) and can be found in book [88], in particular in chapters [2, 3] and articles [69, 75].

In order to reduce greenhouse gas emissions, the *Framework Convention on Climate Change* (UNFCCC) was adopted at the United Nations Conference in Rio de Janeiro in 1992. It was supplemented by a treaty called the Kyoto Protocol adopted in 1997, in which the signatories pledged to limit emissions from their territories. As its continuation, the Paris Agreement was adopted in 2015 after long negotiations.

4 Prediction of Emission Levels

Assessment of the pollutant emission level from a certain area consists in an inventory of all activities (all sources at this area) and then summing up the emissions generated by them. The identification of the emission sources depends to a large extent on the size of the area. However, after their determination, the emissions are usually calculated according to the following mathematically simple relationship

$$E_i = EP_i \cdot A_i, \tag{1}$$

where i is the source number, A_i is its activity expressed in physical quantities, for example in tones of burnt fuel, and EP_i is the pollutant emission factor, i.e. its quantity emitted for a unit activity.

For example, for an entire country taken as the inventory area, a source may be the combustion of the gasoline 95 sold in the country during a year. Assuming (with a good approximation) that all gasoline sold was actually consumed, it is possible to determine from the stoichiometric equations how much CO_2 was emitted into the atmosphere. However, to calculate emissions with higher spatial resolution, the emission sources have to be specified more precisely. They are generally divided into three classes:

- point sources, i.e. emissions from energy or industrial installations stacks,
- line sources, from transport sector,
- area sources, being the sum of smaller emissions from a certain separated area.

Division into aforementioned classes is often arbitrary. The selection of point sources is generally based on the emission amount and the stack height, but for different scales and availability of the source data, the criteria may change. Similarly, only roads with sufficiently large traffic may be classified as line sources. All smaller point or line sources are then included in the emission of the area sources.

To model dispersion of the air pollutants in smaller areas, for example in cities, estimation of the emissions with high spatial resolution is required, even in grids with a cell dimensions much smaller than 1 km. In particular, to model traffic pollution in narrow and densely built-up streets (of a canyon type) it is necessary to estimate the emission in the cells of a few to several dozen meters. Such spatial accuracy requires disaggregation of data available for larger areas, which of course introduces an additional uncertainty. Typically, the disaggregation is carried out using proxy physical quantities associated with emissions and known with a high spatial and/or temporal resolution. A typical quantity of this kind is, for example, population density or car traffic intensity.

As a part of the FP7-MC-IRSES project *Geoinformation technologies, spatio-temporal approaches, and full carbon account for improving accuracy of GHG inventories*, acronym GESAPU, a method of compiling emissions with high spatial resolution depending only on the resolution of the digital maps used, was developed. This method was used to estimate greenhouse gas emissions from Poland for 2012, an example of which is shown in Fig. 3. Its more detailed description and presentation of results in color maps can be found in [4] published in Ukrainian, and in a more compact way in [5, 7, 8].

The emission estimation follows the scheme developed by the IPCC. It is performed separately for each substance, taking into account point, line, and area sources. These sources are divided into elementary objects. An elementary object is any point source with a sufficiently high stack and emission quantity or a part of a linear or area source with homogeneous parameters. For example, the roads of a fixed class (motorway, expressway, …) are divided into sections of a constant speed limit, while the area sources are divided into parts with dense residential buildings, farms, meadows, or arable lands. The elementary area objects are approximated with polygons, which allows processing them in the GIS system. The most detailed description is obtained from inspection of each object using all the publicly available information: from Google Maps to determine the exact coordinates, land-use maps, the local database provided by the Central Statistical Office, the data provided on the websites of enterprises, municipalities, counties, and voivodships. This helps, inter alia, in avoiding locational errors, resulting in large inaccuracies. For example, in the global CARMA database of large point sources, which is often used in inventories, large urban sources, like the cogeneration plants, are often located in city centers. This may in practice mean an error of up to several kilometers in relation to their actual locations, significantly exceeding the dimensions of sub-kilometer grids (see Fig. 4 and examples in [5, 7]). Also some examples of incorrect allocation of the source to the city were found, for example, location of a source in Krąków instead of Kraków, which means an error of more than two hundred kilometers [57].

Fig. 3 Total emissions from coal combustion in Poland in 2009, spatial resolution 2 × 2 km [tons CO_2eq./ km^2] compiled in the GESAPU- FP7-MC-IRSES project

To reduce the disaggregation errors, data from the lowest possible level were used. Another useful solution consists in maintaining the source emission data and information about objects in GIS databases. They are then used to compile the final inventory in the desired regular or irregular grids, such as administrative areas, which helps in avoiding the typical rounding errors, associated with the conversion of the inventories created in one grid to another grid, e.g. to an irregular administrative one.

An additional improvement in the disaggregation accuracy of the most inaccurate area emissions was obtained using conditional auto-regression (CAR) modeling, taking into account the spatial correlation of the area variables. In this method [30, 31] it is assumed, that the value of a certain variable y_i in the i-th grid cell (e.g. the emission of a pollutant in this cell), depending on variables x_k is described by the normal distribution with the expected value of the form

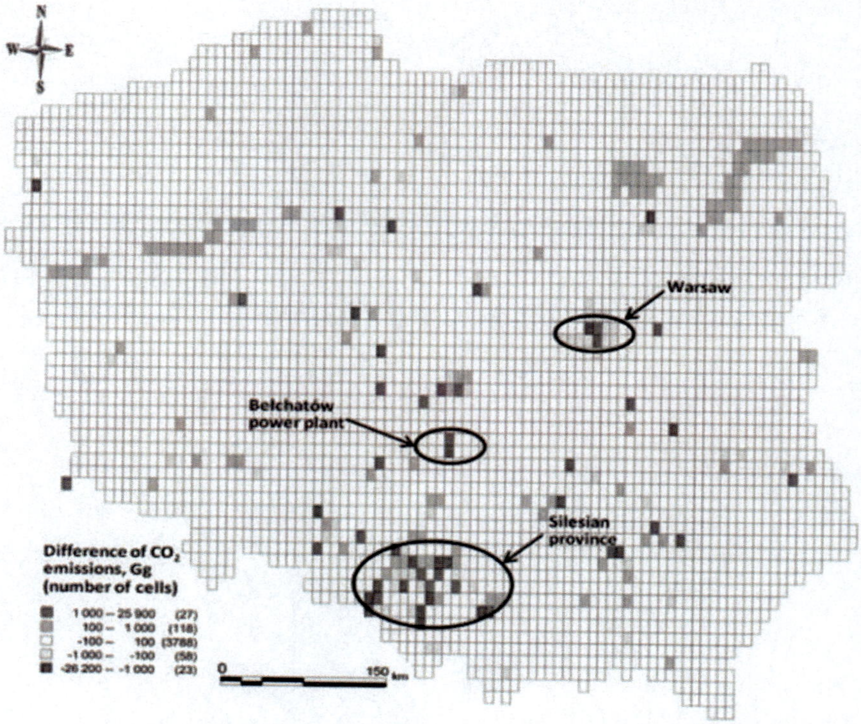

Fig. 4 Differences between the inventories of emission from fossil fuels in the EDGAR base [85] and that carried out in the GESAPU project (after aggregating to the EDGAR grid, approximately 7 × 11 km). Dark fields denote the erroneous location of the point emission sources (graph taken from [5])

$$\mu_i = \sum_{k=1}^{K} \beta_k x_k + \theta_i, \tag{2}$$

where x_k are some proxy variables (covariates), β_k are coefficients, and θ_i is a spatially correlated variable with the variance σ_y^2 constant in the whole considered area (independent on i). For example, y_i may denote ammonia emissions from the soil after fertilization, in tons per year, and x_k may be the land-use data, specified in the CORINE Land Cover Maps created by the *European Environmental Agency*. This latter data may consist of agricultural areas not irrigated artificially, the orchards, pastures, complex crops, and natural vegetation agricultural areas, all expressed in the area units [31]. In another example, y_i can express dairy cattle numbers in the commune, and x_k the data on the cultivation area in a municipality, the pasture area or other agricultural area [8].

Model (2) has a structure typical for linear regression, it is however assumed that the error θ_i depends on the corresponding errors in the neighboring cells, i.e. that it has a conditional probability distribution, assumed to be normal with the expected value of $\sum_{j \neq i} w_{ij}\theta_j$, where the index j specifies neighboring cells and w_{ij} are appropriately selected weights, and with the variance equal to τ^2/J which depends on the number of the neighboring cells, denoted by J.

In disaggregation, the basic problem of the parameter estimation is in the fact that the value of y_i in the fine grid is not measured and has to be estimated. We only know the values in the coarse grid, whose cells consist of a number of the finer grid cells. The method presented in [30, 31] allows us to obtain maximum likelihood estimates of the parameters in (2), from which a value of μ_i can be calculated. In consequence, the disaggregated value estimates in the fine grid can be obtained. The improvement of the model fit depends on the force of the spatial correlations. For example, taking into account the spatial correlation in (2) for the aforementioned livestock case, and comparing it with independent data on the available number of horses in the municipalities showed that the mean square error of the residuals dropped by over 45% and the mean absolute error by almost 30% when the spatial correlation was considered in disaggregation. It is worth mentioning that, in the example analyzed in [32] the CAR method proved to work better, as it gave more accurate results, compared to the kriging method (see Fig. 5). The variance of the values estimated by both methods can be of a great help in assessing the uncertainty of the final emission prediction in the fine grid.

The main advantage of compiling the spatial inventory is the availability of local parameters that can be used to predict emissions. These parameters are related e.g. to the type of the soil or the accepted local farming managements in the agriculture, or the raw materials used in local industry production. On the other hand, when estimating the spatial emission with a high resolution the statistical data averaging effect, which reduces errors in large grids, is usually much smaller in the fine grids. Moreover, the statistical errors due to necessary data disaggregation increase there. The use of all the methods described above mitigates these errors and hence improves the inventory accuracy. Another method elaborated in CDM, takes into consideration various data of different origin, even linguistic ones, using fuzzy logic reasoning to increase accuracy of a gridded inventory [70–74].

The estimation of emissions requires a detailed knowledge of many types of data. While the data in the large energy or industrial installations are usually recorded in quite good detail, data from municipal and housing sector are hardly available and usually only appear in some aggregate forms, often only for larger territorial units. Some emissions, in particular the emissions of N_2O from agriculture, are poorly understood and their estimation accuracy is rather dubious. As a result, the estimation of pollutant emissions is characterized by various uncertainty levels, depending on both the type of a source and the type of pollution (see Table 1). For example, the emissions from large industrial installations can be estimated within a few percent accuracy, e.g. 2–4%, while the N_2O emissions with accuracy even greater than 100%.

However, the total national emissions mostly do not exceed 10%, and are often smaller due to prevailing share of the large installation emissions which are assessed more accurately, and also due to the statistical averaging. The level of uncertainty depends on the emission source structure, but also on their estimation accuracy that

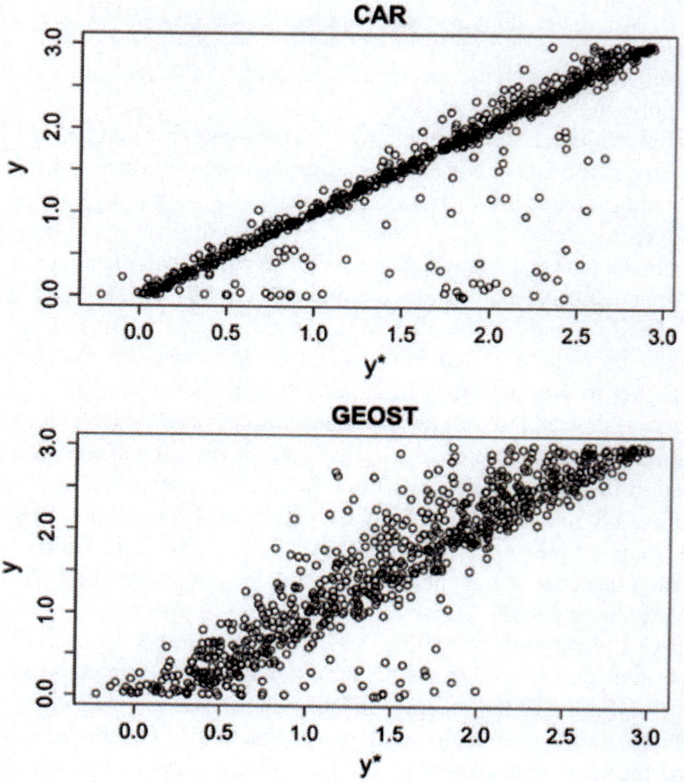

Fig. 5 Scatter plots of the real y, and the estimated y^* quantities obtained by the CAR (upper panel) and kriging (bottom panel) methods. Graphs taken from [32]

Table 1 Estimation of the GHG emission uncertainty from selected sources, according to [104]. The uncertainty classes presented in the left panel are defined in the right one

Source	Uncertainty class	Uncertainty class	Uncertainty range
Fossil fuel combustion, CO_2	1	1	< 10%
Fugitive fuel emission	1–5	2	10–25%
Chemical industry	1–3	3	25–50%
Agriculture, forestry, land use	2–4	4	50–100%
Manure management fertilizers, CH_4	1–5	5	100% <
Fuel combustion in the energy sector N_2O	1–5		

is connected with the expenditures on the estimation improvement. The emission uncertainty has been the subject of many studies, see e.g. [37, 39–41, 44, 45, 57–59, 76].

In [49] an independent method to estimate the variance of emission data compiled in the national inventory reports (NIR) was proposed. It was based on the smoothing splines. Despite the short data series, the estimates obtained by this method were quite similar to those provided by a few countries at that time.

The countries reporting emissions in the terms of the NIRs also correct the submitted earlier inventories, according to the newly acquired knowledge. In [96] the question was asked whether these corrections, called revisions, can be used to assess the temporal evolution of the NIR accuracy. In the methods presented in [34, 36, 43] the revisions are analyzed either as time series or as sets of revised values in a given year. In [35] a combined method is suggested, in which it is assumed that the estimated emission error contains two components: the one arising from the emission inventory compiling, and the other from their correction. The inventory compilation error is connected with the compilation year, and the correction error with the revision year, which gives

$$S_y^t = H_y + V^t, \tag{3}$$

where S_y^t is the finally estimated emission error in the year y after its revision in the year $t > y$, H_y is the compilation error in the year y, and V^t is the correction error due to the revision in the year t.

The method is a nonparametric one, and works also on an incomplete data set, provided that it meets some mild existence conditions, see [35]. Under these conditions the uniqueness of a solution to the problem is proved and the algorithm to find it is developed.

The time evolution of the errors calculated using that method is depicted in Fig. 6 for a few selected EU countries. The series exhibit the features which are traced for all countries analyzed. The inventory compilation errors are irregular and random. The correction errors are sometimes slightly irregular, but generally converge, at a slower or faster rate, to the final value of zero. This indicates that the countries generally try to improve the methodology for estimating emissions, i.e., the learning effect is observed there. The correction errors are usually higher in the initial periods of time and are a significant part of the total error at that time.

Nevertheless, the emission inventories are currently the most accurate methods of assessing pollutant emissions. The measurements of the pollutants concentrations in the atmosphere bring valuable information, but provide only the values in discrete points of the space that may be very sensitive to local emission sources (e.g. communication or close furnace gases emissions). Moreover, they are hardly usable to assess the actual anthropogenic pollution emissions, as the measurements include the biogenic pollutions that are difficult to assess. The methods to assess the emissions from atmospheric measurements have nevertheless been developed [82].

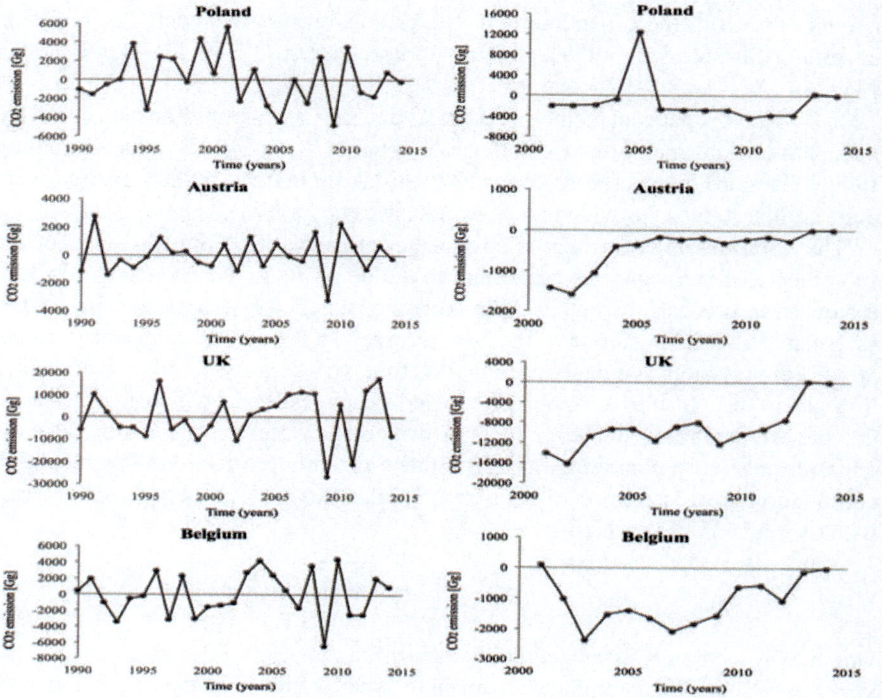

Fig. 6 The estimated time series of the inventory compilation errors (left) and their correction errors (right) for a few selected countries. Graphs taken from [35]

Modeling of pollution transport is now mainly used for estimating the emissions of short-living pollutions. Increasingly, such mesoscale modeling is also used for long-living pollutions, for example to assess the biogenic emissions [99]. As the Paris Agreement stipulates the demonstration of greenhouse gas reduction in atmospheric concentrations, the mesoscale modeling seems to be the main tool that can be used for this purpose.

5 Mesoscale Models

The aim of atmospheric pollution models is both to illustrate and explain the propagation process dynamics and to present the possibilities of using this knowledge to support decisions and invent effective methods of air quality control. This decides the choice of the model and other tools necessary to support environmental quality management. In deterministic pollution transport modeling it is usually assumed that the processes can be modeled using a system of advection-diffusion (transport) equations [27, 29, 89, 90, 94]. They also describe the pollutants physical and chemical changes, their depositions on the earth or water surface, and their washout by precipitation. The basic transport equations have the following general form [27, 89, 90, 94]:

$$\frac{\partial c}{\partial t} + \upsilon \nabla c - K_h \Delta c + \gamma c = Q, \quad (4)$$

along with the relevant boundary and initial conditions, where c is the concentration of the substance modeled, $\boldsymbol{v}$ is the wind field vector, K_h is the horizontal diffusion coefficient, γ is the aggregated deposition rate (dry/wet/absorbed by the surface waters), and Q is the total emission field.

Such models can be quite complex, especially if they include a time-varying, three-dimensional description of many physical and chemical processes. For this reason, in numerical algorithms, a decomposition of both the spatial variables and individual processes, is often used. The vertical profile is approximated by introducing a layered structure. Then, in each layer, the model is divided into separate modules describing the emission field, the horizontal transport, the horizontal diffusion, the deposition, and the precipitation washout, as well as the chemical transformations. The exchange of the components between the adjacent layers is most often described by an appropriate parameterization [15, 89, 94].

The role of the module significantly depends on the model scale. In local scale models, the spatial processes impact range is limited and ranges from 1 to about 20 km. They can be applied to model the town districts or the cities that are characterized by largely focused and diversified pollution sources in a limited area. This results in a small pollution dispersion time between the source and the receptor, i.e. the point of measuring/calculating the substances concentration in the air. A relatively short pollutant transport time enables disregarding the slowly occurring chemical transformations and formation of many secondary pollutants. On the other hand, a large number of the reacting compounds and their high concentrations can result in forming some secondary pollutants (such as ozone) in this scale. The mathematical descriptions may be quite complicated in this case [89, 94], especially when taking into account the terrain features and coverage, and the local meteorological phenomena and their influence on the wind field.

In the regional scale models (the range from 20 to over 100 km), where the development and coverage of the terrain, as well as the local thermal conditions have a significant impact on the pollutant spread, it is generally necessary to consider the vertical pollutant distribution. In the continental scale models (well over 100 km), a single averaged mixing layer is often considered. This approach was used, among others, in the operational EMEP and RAINS models, analyzing the spread of the pollutant energy sources in Europe [79, 80]. The higher scale modeling results are generally used as the boundary conditions in the lower scale models.

From the modeling point of view, there are significant differences between the description of phenomena on a smaller spatial scale (the local and urban models) and a large one (the large-scale regional and global models). In the former case, the assumption of the dispersion homogeneity and stationarity is often justified. In the larger scales these assumptions are no longer valid (in particular, the meteorological conditions change in space and time), so the whole process has to be considered non-stationary. Due to a long time horizon, factors such as the depositions or some chemical changes that are often omitted in the small scale models, must also be taken into account. On the other hand, the pollutant horizontal diffusion is generally omitted in the large scale operational models, especially the global ones. This is due to a larger aggregation of the emission field, as well as to a coarser discretization of the calculated area. The diffusion process becomes then a negligible subscale process, when compared to the averaging errors and the so-called numerical diffusion.

The Gaussian models are often applied for the numerical solution of the local scale propagation equations in short- or medium-term problems [90, 94]. The modeling is based there on the analytical solution of Eq. (4), obtained under the simplifying assumption of the meteorological and emission fields stationarity. In the larger scale models, two basic approaches used differ primarily in fixing the reference coordinate system, in relation to which the movement of the air particles is analyzed.

In the Eulerian models, the natural coordinate system is used with the origin fixed in a point located within the modeling area. This type of modeling is often used to numerically solve complex equations used in broad applications, from the local to the global scale (in the latter case the Cartesian coordinates are usually replaced with the spherical ones). Spatial and temporal variability of meteorological fields and complex characteristics of the emission field are routinely taken into account there. With fine discretization of the area, a large system of complex nonlinear equations arises whose solution creates a heavy computing burden.

In Lagrangian models a mobile coordinate system is used that moves along the particle trajectory in the wind field. The temporal and spatial discretization used in the computational algorithm involves the division of the transported pollutant mass into the computational elements, whose movement along the wind field lines is analyzed independently in conformity with the advection equation solution. Such computational algorithms are often used to determine the share of the individual emission sources observed in the pollution receptors in the form of the so-called transfer matrices. The transfer matrices allow for quick calculations of the pollutant concentrations for known emissions by solving a set of linear equations. As this approach requires a much smaller computational burden for calculating the total pollution caused by a given emission field, it is often applied for analysis of emission reduction scenarios in air quality management decision-making projects [17, 91, 94].

In the emission source description, not only the emission intensity, but also the technical characteristics and the emission fluctuations in time may play an important role, especially if the pollutant dispersion evolution in time is taken into account. In the larger scale models, the basic problem is to accurately determine the pollutant trajectory, since even small inaccuracies in its calculation can generate a large final error, due to the large receptor-to-source distances.

6 Some Mesoscale Model Applications

The atmospheric pollution dispersion models are commonly used in the short- or long-term forecasting of pollutant concentration distributions, their cumulative deposition values, or for monitoring the exceedance of the established limits (e.g. the acceptable concentration levels or the critical loads). Since the basic input data are the meteorological and emission parameters, the simulation studies enable comparison of the various emission reduction strategies and their environmental impact. Such comparisons are usually of a qualitative character as their goal is to help in selection of practically relevant solutions.

6.1 Distributions of Air Pollution Concentrations in a City

The results presented in this subsection pertain to an air quality analysis in the Warsaw metropolitan area (Poland), for the emission and meteorological data in 2012, see [23–26, 66]. The regional scale CALMET/CALPUFF modeling system [100] is applied to forecast pollution. The main goal is to obtain the mean annual pollutants concentrations maps to identify the areas, where the permissible pollution levels are exceeded, and to indicate the main emission sources responsible for those exceeding.

The basic pollutants, significantly influencing the ambient air quality decline in the city were as follows: sulfur and nitrogen oxides, PM_{10} and $PM_{2.5}$ dusts (i.e. particular matters with the diameters up to 10 or 2.5 μm, respectively), CO, C_6H_6, the aerosols, heavy metals and carcinogenic benzo(a)pyrene (BaP). The following categories of emission sources were distinguished: high point sources (mainly energy sector), other point sources (industry), line sources (transportation), and area sources (residential sector). Moreover, the inflow of pollutants from the neighboring communes/cities, having significant influence on the pollutants total concentrations, were taken into account. That inflow was implemented as boundary conditions.

The line and area sources prevail in forming the main pollutants concentrations in Warsaw. The influence of energy sources, in spite of their high production rates of PMs, SO_2 and NO_x, is relatively low, mainly due to efficient filters, desulphurization systems, and high stacks (150–300 m), what causes a significant pollutants blow-out affecting mainly the areas beyond Warsaw. Figure 7 presents the distributions of the mean annual concentrations of PM_{10} and NO_x. The areas, where the permissible concentration levels are exceeded, are marked dark. Nitrogen oxides, which are the typical transportation pollutants, exceed the levels in the center and along the main streets. The residential sector (the area sources) is mainly responsible for PM pollution. The higher PM_{10} concentrations in the western districts are the results of the distinct majority of the S-W winds in the year considered and of the significant pollutant trans-boundary inflow from this direction [23, 24].

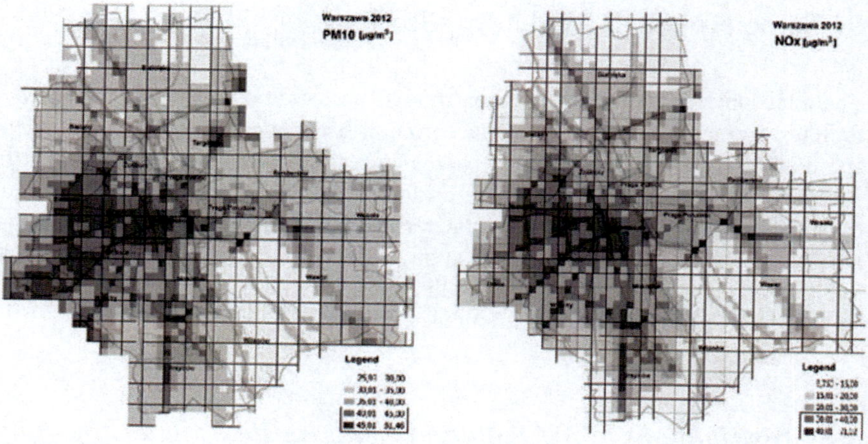

Fig. 7 Distributions of PM_{10} and NO_x mean annual concentrations in Warsaw. The colors circled in the legends are those for which the permissible concentration level is exceeded

The pollutant concentrations and the contributions of the emission sources depends to much extent on the receptor locations. The diagrams presented in Fig. 8 illustrate this on the examples of the PM_{10} and $PM_{2.5}$ concentrations. The transportation-related pollutants, as PM_{10}, dominate in the vicinity of a street crossing (Fig. 8a), while the area sources have a greater share in the residential district (area)

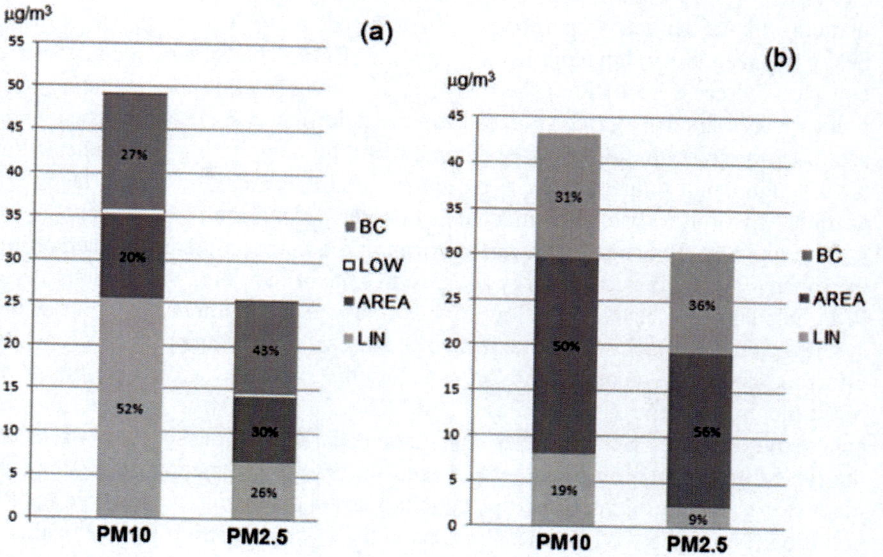

Fig. 8 The share of the emission sources depending on the receptor location **a** a street crossing, **b** a residential district (BC—the boundary conditions, LOW—the low point sources, Area—the area sources, LIN—the line sources)

PM concentrations (Fig. 8b). Also the $PM_{10}/PM_{2.5}$ ratio depends on the receptor location, due to the changing relative share of the emission categories affecting the concentrations. As seen in Fig. 8, in both cases the PM concentrations strongly depend on the transboundary inflow from the sources located out of Warsaw. In the effect, in both cases the mean annual concentrations exceed the limit values, which are of 40 $\mu g/m^3$ for PM_{10} and 25 $\mu g/m^3$ for $PM_{2.5}$. The dominating share of the line sources with typical transportation pollutants, like NO_x, CO, and C_6H_6, is well visible around the main streets. For example, in the vicinity of the main crossroads, the share of the line emissions of the NO_x concentration may be over 90% [22, 23].

As shown in [22, 28], the mean annual BaP concentrations are above the threshold of 1 ng/m^3 in the entire city area, that reflects the general situation in Poland, connected with this pollution. High concentrations are mainly observed in the peripheral agglomeration districts, as a result of coal combustion in residential individual installations. In Warsaw, it relates to low buildings, mainly in the peripheral districts or in the neighboring cities.

6.2 Evaluation of Air Pollution Dispersion Models

The measurements of emissions and the meteorological conditions that are used in the simulations, are highly uncertain, which is mainly due to a large spatial volatility of these phenomena. In consequence, the achieved model results may differ a lot from the measurements, especially for the so-called episodes, i.e. the short-term periods of usually high pollutant concentrations in the air. This is why evaluation of the dispersion models is difficult and justified mainly for large data sets, where statistical averaging takes place.

A few essential aspects are usually taken into account [90, 94] in the qualitative model evaluation. The correctness and adequacy of a mathematical description of the processes analyzed and the equations applied, as well as the adopted simplifications and parameterizations, are reviewed. The sensitivity analysis enables evaluating the impact of the uncertainty of the input data or the model parameters on the uncertainty of the modeled concentrations [6, 18, 26, 102]. The statistical inference allows for a quantitative evaluation of the model results and the goodness of fit. A lot of statistical indicators of that fit were elaborated, like FB—*the fractional bias*, MG—*the geometric mean*, MSE—*the mean square error*, SD—*the standard deviation*, r—*the correlation coefficient*, FAC2—*the factor of 2 index*, and many others [90, 94, 103].

The factor of 2 index is defined as $FAC2 = C_m/C_p$, where C_m denotes the concentration computed by the model and C_p is the real measurement value. It compares in a simple way the model results with the available observation data. In general the model tested is considered sufficiently good, when for all or at least almost all considered points it holds

$$0.5 \leq FAC2 \leq 2. \tag{5}$$

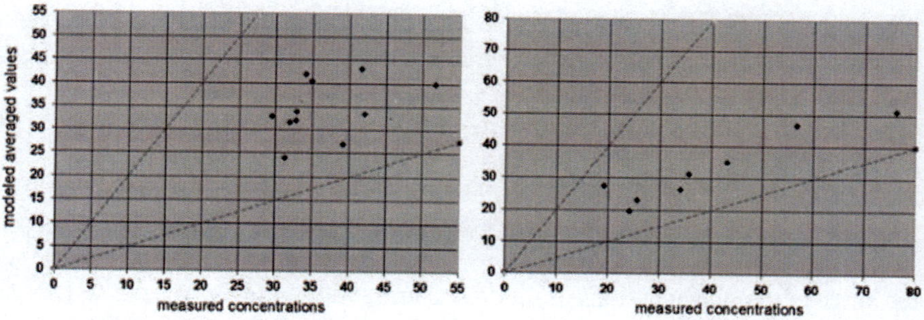

Fig. 9 FAC2 index, i.e. the ratio of modeled vs. measured annual mean concentrations [μg/m^3] for the PM_{10} (left) and NO_x (right) values in the receptor points associated with the pollution measurement stations in Warsaw

This index was used in [24] to evaluate the simulation result for the computed annual mean concentrations of the main pollutants in Warsaw in the year 2012. Figure 9 presents the FAC2 values for PM_{10} and NO_x concentrations, when the annual average values were computed by the CALPUFF model with the spatial resolution step of 0.5 km. Each dot in a scatter plot represents a receptor point connected with a real measurement station in which the mean annual concentrations were recorded. In both cases (PM_{10} and NO_x) the FAC2 values are within the required range given by (5).

The uncertainty analysis of the model prediction connected with the input data imprecision that usually constitutes the main source of uncertainty in this kind of modeling, is discussed in [26]. The normal distribution of each emission category input data was assumed and the Monte Carlo algorithm was applied for assessing uncertainty of the concentration forecasts. The scatter plots in Fig. 10 present distributions of the standard deviation (Fig. 10a) and the relative uncertainty (Fig. 10b) of the NO_x concentrations computed in 563 receptors covering the Warsaw area (each dot represents a receptor point). The relative concentration dispersion at the receptor

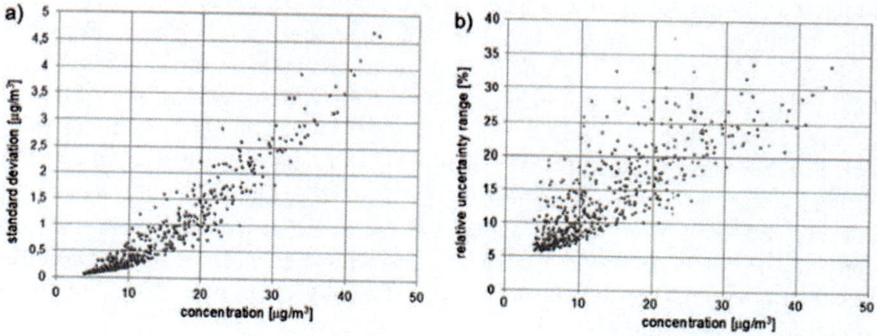

Fig. 10 **a** The standard deviation and **b** the relative dispersion of the predicted NO_x concentrations, due to the input emission data uncertainty

is calculated as $(c_{97.5} - c_{2.5})/c_M$, where $c_{2.5}$ and $c_{97.5}$ denote 2.5- and 97.5- percentile concentrations, respectively, and c_M is the mean value.

7 Assessment of Air Pollution Health Effects

Among the harmful effects of short-term polluting concentrations, the most important are the air pollution impacts on human health. The estimation of the pollutants concentration values and their comparison with the permissible level for a given compound gives only a rough idea on the overall atmospheric air quality in a city/region. Due to the global trend of fast growth in the number of urban residents and the simultaneous increase in industrialization and transportation, the air quality problem becomes more and more serious. Hence, the ultimate pollutants impact on human health is crucial [67, 92–94, 102]. The actual pollutants impact on population health requires a complex analysis of their concentrations and the size of the population with which they interact, in order to implement effective air quality control strategies, limiting the degradation of the environment.

There are many indicators of the air pollution impact on the population health risk, see e.g. [65, 67]. One such indicators is the inhabitants exposure (E) to a given pollutant [20, 28]. The exposure is usually calculated for a specific pollution k. It can be assigned to a selected emission source, but can also be aggregated to an entire emission category. In the first case, the exposure $E_{i,k}$ [μg/m^3] is calculated according to the formula

$$E_{i,k} = \frac{1}{Pop} \sum_j C_{i,j,k} Pop_j, \tag{6}$$

where $C_{i,j,k}$ denotes the concentration [μg/m^3] of compound k in grid cell j (represented by the receptor point in this cell), coming from source i, Pop_j is the population in the grid cell with receptor j [person], and Pop is the total population in the area studied. Hence, this indicator reflects the average exposure of a person to compound k, emitted by source i and expressed in the same units as the concentration. Its value depends on the population spatial distribution and the pollution concentration distribution in the air. To calculate the aggregated value, an additional summation over all given category sources is performed

$$E_k = \frac{1}{Pop} \sum_j Pop_j \sum_i C_{i,j,k}. \tag{7}$$

An intake fraction (iF) is, in turn, an index characterizing the source—exposure relationship. It can refer to a single source or a whole emission category, similar to the exposure. It is defined as a fraction of a pollutant emitted by a given source, which is inhaled by a population within the range of the source influence. For a single emission source, this indicator is calculated according to the following expression

$$(iF)_{i,k} = \frac{BR}{Q_{i,k}} \sum_j C_{i,k,j} \cdot Pop_j = \frac{BR * Pop}{Q_{i,k}} \quad (8)$$

where $Q_{i,k}$ [g/s] is the source emission and BR [m^3/s/person] is the pollution inhalation factor (~0.00021) [21, 66]. The intake fraction aggregated for the selected emission source category is calculated as follows

$$(iF)_k = \frac{BR}{Q_k} \sum_j Pop_j \sum_i C_{i,j,k} = \frac{Pop \cdot BR}{Q_k} E_k, \quad (9)$$

where Q_k [g/s] is the total emission of all given category sources.

It can be seen from (8) and (9), that *iF* is a measure of the source impact on the residents within its reach. It depends on such factors as the location of the source, the emission stack effective height, the population within its range, and meteorological conditions. On the other hand, *iF* is a measure of the sensitivity of the inhabitants exposure to source emission changes. This is key information in searching for the optimal strategy to reduce emissions in a city/region.

The calculation of the *iF* index estimates for Warsaw, using the mean annual concentration distributions of the most important pollutants discussed in Sect. 6.1, are presented in [20, 21]. In addition to the emission sources located within the borders of Warsaw, the city surrounding sources (in about 20 km wide belt), were also considered in simulations. A homogenous spatial resolution grid was used, with a step of 0.5 km inside the city and 1 km in the surroundings.

The *iF* estimates, aggregated for the source categories, are presented in Table 2. The *iF* values are rather low for the point sources, while dominant influence of the surface and line sources is clearly visible. The aggregated *iF* values in the case when the Warsaw surrounding area are taken into account (the upper part of the table) are further reduced by inclusion of the remote sources whose impact on the city area is relatively low. If only the emissions from the Warsaw area are taken into account

Table 2 The aggregated *iF* values for chosen emission categories for Warsaw, Poland, in 2012

	SO_2	NO_x	PM_{10}	PM_{25}	CO	C_6H_6	BaP	Pb
	Warsaw + vicinity							
High point.	0.7	0.6	1.2	1.9	0.8	0.6	1.4	1.3
Low point	2.0	3.8	4.1	4.3	2.6	2.7	2.3	5.4
Area	8.4	8.8	9.4	9.2	8.7	10.1	8.4	10.0
Line	15.2	13.9	22.1	23.7	24.6	24.5	13.2	22.2
	Warsaw (within the administrative border)							
Area	23.5	23.6	24.7	24.5	23.7	26.0	23.1	25.9
Line	26.0	24.4	36.9	38.8	34.7	34.7	21.3	33.0

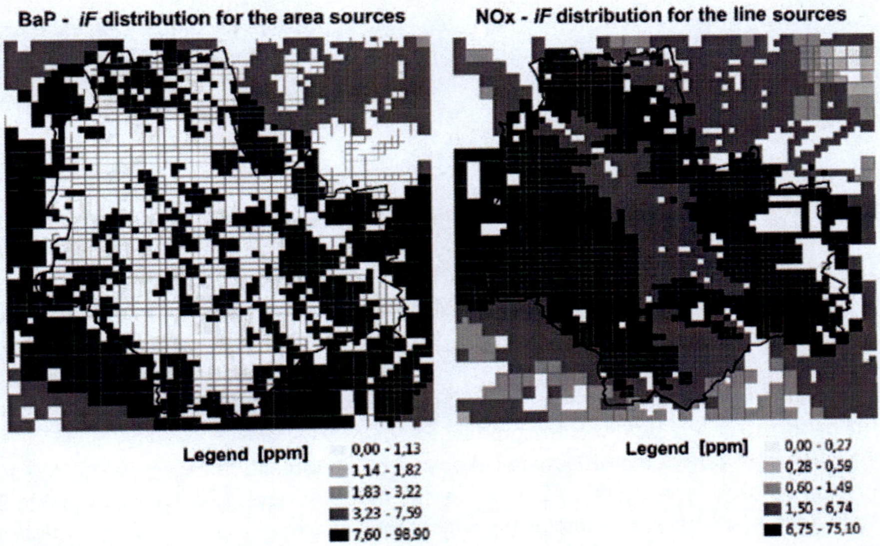

Fig. 11 The *iF* spatial distributions for the area and line emission categories

(the bottom part of the table), the values of *iF* grow strongly. They then correspond to the similar estimates [81], obtained for cities of a comparable size, both in terms of the population and area, see [21].

Figure 11 shows the maps of two selected *iF* spatial distributions for the area and line sources. The distributions provided are representative also for other pollutants from the same emission categories, since the *iF* value depends primarily on the source location and its parameters, the population density, and the meteorology that are the same during the simulation and hence the same for all pollutants, and all the emission sizes/types. The influence of the peripheral districts sources and those located in the immediate Warsaw vicinity dominate within the area emission (municipal sector) category. The impact of the central part of the city is small due to the central heating system operating there. Among the line emission (communication) sources, those located within the Warsaw borders dominate, as a dense street network coincides there with a high population density.

Indices (6)–(9) can be used to estimate health effects. For this, the exposure—response relationships, which are largely based on earlier studies, like the cohort ones, are used. From them, the relative risk can be calculated as follows

$$RR = \frac{P_1}{P_0}, \tag{10}$$

where P_1 is the frequency (the empirical probability) of the health effects occurrence at a given exposure, and P_0 is the frequency of the same effects occurrence among participants not exposed to the pollutant considered. This can, in turn, be used to estimate the premature death rate due to exposure. It can be calculated from the expression $M = M_b \cdot RR$, where M_b is the basic population mortality. However, this index is usually made dependent on changes in exposure, according to the following expression

$$\Delta M = M_b \exp(\beta \Delta E), \tag{11}$$

where $\beta = \ln(RR/\Delta E)$. This expression can also be approximated by the formula $\Delta M = M_b(1 + \beta\Delta E)$, since the values in the exponent are small. This allows an interpretation of β, as a mortality increase rate per the exposure unit increase.

Another health index is the reduction of expected life time, assessed on the basis of life expectancy tables. The death risk is calculated from the formula $H_b = m/Pop$, where m is the number of the people who died in a certain period of time, for example in a year, and Pop is the number of the population studied. The risk of death due to the exposure to a pollutant is the product $H_b \cdot RR$.

Now, the value of the inhabitants exposure to a pollutant, as expressed in (6) and (7) can be used to estimate its impact on the mortality level [28, 67, 102]. Table 3 shows the mortalities calculated this way among Warsaw residents, in dependence on the pollution type. It can be noticed there that, the fine particulate matter fractions are mainly responsible for mortality. Their most harmful effects are also indicated in most other publications. Among the emission categories, the influence of the surface and linear sources is the most significant. Together, they are responsible for a little more than half of the estimated mortality cases. Next are the external pollutants that inflow from outside of Warsaw whose share is close to a half.

In [67], the iF estimates derived from the anthropogenic emission of $PM_{2.5}$ in Poland, in 2000 and influencing other European countries, were calculated. The

Table 3 The annual mortality in Warsaw depending on the pollution type

Sources	Point		Line	Area	Inflow	Sum	%
	High	Low					
$PM_{2.5}$: mortality	7.9	24.6	823	1466	2566	4887	90.9
NO_x: mortality	7.1	7.8	379.8	34.9	26.9	456.5	8.5
SO_2: lung cancer	0.7	0.3	1.8	3.3	1.4	7.6	0.1
BaP: lung cancer	0.0	0.0	0.3	1.6	1.0	2.9	0.1
Cd: cancer	0.0	0.0	0.0	0.7	0.0	0.8	0.0
Ni: cancer	0.0	0.0	0.0	0.0	0.0	0.0	0.0
Pb: cardiologic diseases—adult	0.0	0.1	6.8	4.1	0.5	11.4	0.2
As: lung cancer	0.0	0.0	0.0	0.0	0.0	0.0	0.0
CO: ischemic disease	0.0	0.0	6.5	0.2	3.9	10.7	0.2
C_6H_6: leukemia	0.0	0.0	0.1	0.1	0.0	0.3	0.0
Total	15.8	32.9	1218	1510	2600	5377	100
%	0	1	23	28	48	100	

estimate for Poland is certainly the largest of them, with the value of 1.23, while Ukraine (0.18) and Germany (0.12) are ranked next. However, in terms of the exposure, Poland (0.78) is ranked first, followed by Slovakia (0.20), the Czech Republic (0.16), Belarus (0.15), and Lithuania (0.15). And the rank according to the shares in shortening the life expectancy includes Poland (49%), the Czech Republic (10%), Germany (7%) and France (5%). Hence, the ranks vary considerably, mainly because different variables were used in calculating estimates for particular indicators.

A weighted index called QALY (*the quality adjusted life years*) is also often used. In this case the life quality with the values between 0 (death) to 1 (full health), is taken into account. In the DALY (*disability adjusted life years*) index the weights are associated with functionality loss. The weights in both the above indices are determined on the basis of population surveys and expert evaluations.

The above methodology makes it also possible to take into account other factors in health risk estimation. For example, physical effort significantly contributes to lowering of health risk [13, 48, 68, 78].

8 Emission Mitigation

8.1 Emission Reduction Policies

A popular way to enforce reducing emissions has been taxing emissions. In optimization theory, tax is a penalty that the emitter must take into account while minimizing costs. Another, more flexible way is the emission permit trading. This is considered in the sequel. The permit is an allowable emission granted to an emitter. In the so-called cap-and-trade system each emitter is granted a limited number of permits that are, however, tradable. A simpler case of long-lived pollutants, greenhouse gases, is discussed in this section, as actually their total inflow to the atmosphere is of the main interest in climate change predictions and hence pollution transport in the atmosphere need not be considered.

The central planner considers minimization of the following cost criterion

$$J = \sum_n c^{(n)}\left(x^{(n)}\right) \tag{12}$$

where n is the number of the emitters, $x^{(n)}$ is the n-th emitter emission, and $c^{(n)}$ is the reduction cost, and the marginal cost is usually assumed. The marginal cost is 0, if the emissions are not reduced and increases when $x^{(n)}$ is decreased. In addition, the following condition must be met

$$\sum_n x^{(n)} \leq L, \tag{13}$$

where L is the total emission limit, imposed by the central planner. The solution of this problem can be found easily, if all the marginal cost functions $c^{(n)}$ are known to the central planner. However, the emitters are not interested in disclosing this information due to business competitive reasons.

In the end, the problem can be solved by emissions trading, as proposed in [84]. A two-level approach is applied. At the upper level, the overall limit is divided between the emitters $\sum_n L^{(n)} = L$, where $L^{(n)}$ is the emission permit limit for the n-th emitter. The permits are tradeable and it is assumed that $e^{(n)} < 0$ denotes the permits sold, and $e^n > 0$ the bought ones, and then the following condition $\sum_n e^{(n)} = 0$ is satisfied. This condition can be included in the problem (12)–(13) by minimizing the Lagrange function

$$J_\lambda = \sum_n \left[c^{(n)}\left(x^{(n)}\right) - \lambda e^{(n)}\right], \tag{14}$$

with the constraint

$$\sum_n \left[x^{(n)} - e^{(n)} - L^{(n)}\right] \leq 0. \tag{15}$$

The above problem can be easily decomposed, letting the emitters solve, at the lower level, the following problem of minimizing their own costs

$$J_\lambda^{(n)} = c^{(n)}\left(x^{(n)}\right) - \lambda e^{(n)} \tag{16}$$

subject to

$$x^{(n)} - e^{(n)} \leq L^{(n)} \tag{17}$$

where $\sum_n e^{(n)} = 0$. The multiplier λ can be interpreted as the marginal (shadow) price of the emission unit at the global optimal point.

The above solution is derived, assuming that the emission values are known exactly. In practice, however, the actual emissions are estimated with uncertainty which can be large. Hence, the bound (17) is of a random nature. Discussions of methods that can be used to check the compliance under uncertainty in this case, can be found in [33, 38]. A popular method used in optimization with the constraints of a random nature is to admit that the constraint may be met only with a certain (a sufficiently large) probability, i.e. the adoption of a constraint of the following form

$$P\left\{\hat{x}^{(n)} - \hat{e}^{(n)} - L^{(n)} \geq q_{1-\alpha}\right\} \leq \alpha, \quad 0 < \alpha < 1 \tag{18}$$

where $q_{1-\alpha}$ is the $(1-\alpha)$th quantile of the probability distribution. The hat sign over variables emphasizes that these are the uncertain estimates. This form of the constraint allows writing the emitters bound as

Table 4 The proposed organization of the emission permit markets for uncertain emissions, divided into two classes (A and B) presented in two separate columns

A. Grouping of the emitters according to their emission uncertainty	B. Direct consideration of the emission uncertainties in the trade
1. Only the emitters with sufficiently low uncertainties trade. This solution is adopted in the present markets. They cover primary the emissions from the large energy/industrial installations	Modification of the limits according to rule (19) and then normal emission trading. The emitters are punished here for the emission uncertainty by decreasing the emission limits [1, 9, 87]
2. Creation of several parallel markets, depending on the participant emission uncertainty [97]	Conversion of the traded emissions to keep the total emission uncertainty constant [86]
	Conversion of the traded emissions, so the buyer satisfies condition (19), taking additionally into account the uncertainty that is transferred with the purchased emissions [50–52, 54]
	As above, but the transferred uncertainty modifies the buyer's total uncertainty after the transaction [53]

$$\hat{x}^{(n)} - \hat{e}^{(n)} + q_{1-\alpha} \leq L^{(n)} \tag{19}$$

The quantile $q_{1-\alpha}$ can be interpreted here as the emission uncertainty measure, referred to in the literature as "unaccounted emission'" [86] or "risk charge" [95]. The above compliance checking rule was proposed independently in [50, 87].

Emission uncertainty also affects trade. A question arises there of how to compare the value of an almost certain emission unit with those burdened with, say, 60% uncertainty. This issue has been discussed in literature, but due to the article length limitation, the proposed solutions are only mentioned in Table 4, without going into details.

In [56, 62, 77] a computer system is described, that simulates the market operations using programmable agents as emitters, which use electronic auction methods to set prices. A sample result is presented in Fig. 12 for the bilateral auction. Results for 16 countries, several auction methods, and several negotiation strategies can be found in [62].

8.2 *Programs and Projects to Reduce Emissions*

To mitigate the harmful effects of atmospheric emissions, it is necessary to develop programs of reducing them at various levels, starting with enterprises, through administrative units, and ending with national authorities or even transnational structures, such as the European Union. The planning of emission reduction projects requires consideration of the impacts on economic and social effects, like population health

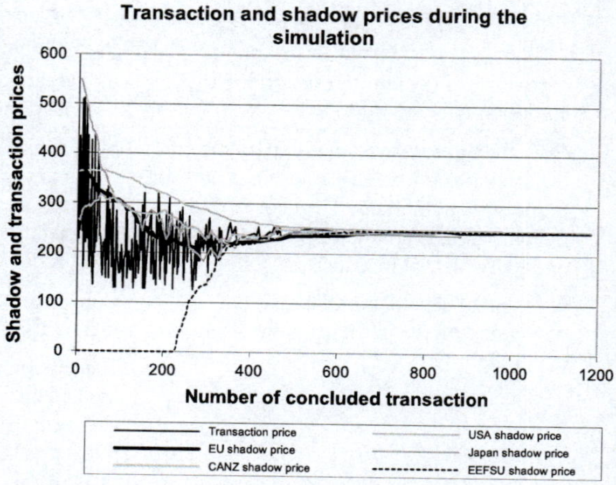

Fig. 12 Transaction prices and marginal costs for an emission permit market with 5 countries/regions using the bilateral auctions (taken from [56]). EU—European Union (EU12), CANZ—Canada and New Zealand, EEFSU—Eastern Europe and Former Soviet Union Republics)

risk or the labor market, as well as the impacts on nature or climate change. Reduction can usually be obtained in many ways, even in a single emitter, not to mention more complex structures, like joint emissions from the production and supply chain, or air quality improvement programs, created by administrative units of various levels. The optimization in such cases must usually consider many indicators and then has to be of the multi-criteria type, giving the Pareto set consisting of the points in the optimized indicators (criteria) space, in which no indicator can be improved without worsening another one.

Complicated relations between the variables and phenomena occurring in these problems require using a computer support [14, 46] whose basic goal is to solve the optimization tasks. Even simple minimization of pollutant concentration in a selected area involves an extensive calculation [16, 17, 19], since the dispersion equation system (4) has to be solved in each subsequent minimization iteration. If many indicators are considered, for example related to climate change, energy security, air quality, and health effects, that should be optimized concurrently, the computational burden substantially growths as presented in [42, 47]. This is the result of multi-criteria tasks, where optimization must be repeated at least so many times, as many points we want to find in the Pareto set. For grids consisting of a large number of cells it forms a very computationally demanding task, even for supercomputers. For this reason, attempts are being made to replace the partial differential equations model (4) by simpler ones, with small computational requirements, such as neural network models, used in the RIAT system [83].

Emission reduction projects can be of a technological, logistical or management nature. An example of an energy-saving solution, with energy management in a renewable energy research and training center can be found in [60, 61, 63]. The center uses renewable energy sources: wind turbines, photovoltaic panels, biogas or natural gas fueled micro-turbines working in the heat and electricity cogeneration, and a small hydroelectric plant. The energy generation inside the center varies a lot during days and seasons, depending on the meteorological variables (wind, sunshine, precipitation). On the other hand, the energy consumption is also very volatile. In addition to the normal day/night or working-day/holiday fluctuations, a large power consumption happens during experiments carried out in the center laboratories or during training. The center is equipped with energy storage devices, like hot water tanks, battery and flywheels sets, that help in energy management. But their capacity is too small to take over or provide sufficient energy, according to needs. There is also a reciprocating engine generation device, so that the center can operate to some extent in an island mode. During normal operation it is however connected to a medium voltage power grid, with which it can exchange energy. For this reason, the management of the center energy network, which uses cost-effectively all the practically free renewable energy coming from the sun, wind, and water, is complicated.

The management system developed consists of two basic modules. One of them helps in the long-term energy supply planning on the basis of different time scales energy demand predictions and subscriptions. The planning may extend from one hour to a year, depending on the arrangements with the medium voltage distribution network operator. This includes, inter alia, trading on the long-term and medium-term energy markets, as well as heat management, due to the systems considerable inertia. It is mostly a decision support module, as there is usually enough time that the final decisions can be made by a human. The other module, which is mainly discussed in the above cited articles, manages deviations from the rolling energy supply plans, given by the planning module. For this purpose, an agent system was designed, in which the software agents, Fig. 13, located on each generating or receiving energy device, negotiate the energy exchange in small time intervals (the simulations showed that considering the processing speed, even sub-second intervals can be applied). The negotiations are based on electronic auctions, and in particular, on the bilateral auction, in which a solution is found in a few interactions between the auction participants, saving negotiation time.

It is easy to notice, that the solution presented has a good automatic control theory interpretation. It has a hierarchical bi-level structure. In the upper level, the planning

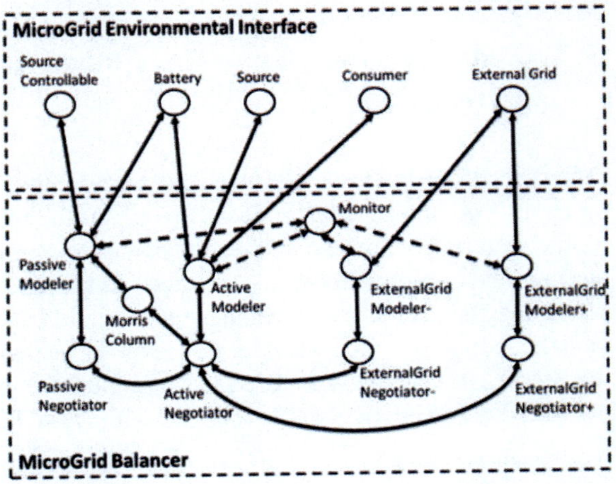

Fig. 13 The agent cooperation scheme of the deviation management module (taken from [63])

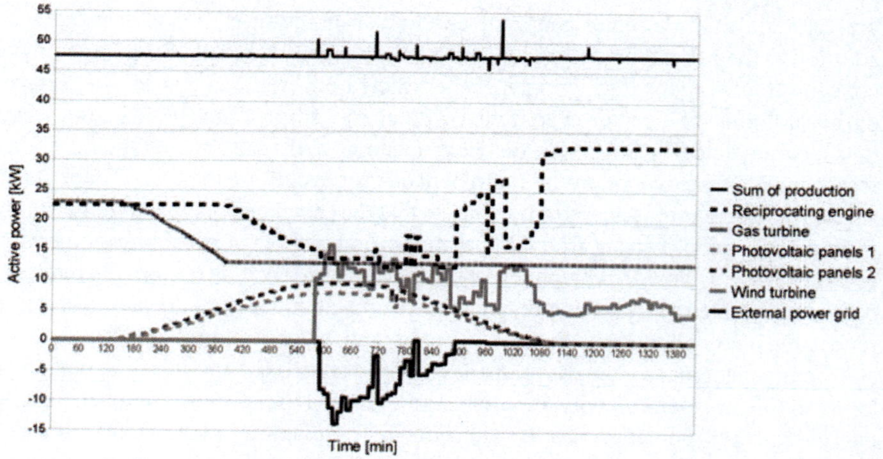

Fig. 14 Simulated energy generation curves by different devices in a research and development center during a day, in order to obtain a constant energy production in the center (taken from [63]). An agent system is used to coordinate and optimize the production

module uses a rolling predictive optimization of the energy production curve. In the lower level, the deviation management module is a tracking system, that adjusts production on-line to the actual energy demand. In both cases, the cost criterion is minimized under the condition of the produced and consumed energy equality (and at such small time intervals in the latter module, with a good approximation of the power equality). However, the criterion may also contain other components. Figure 14 presents an example of the simulated energy generation by several devices, that cooperate to obtain a constant electric energy production at a minimum cost. No storing energy device is used, hence very short spikes in the energy produced occur because of switching production between devices.

8.3 Reducing Greenhouse Gas Emissions and Economic Development

The costs of the intensive and extensive emission reductions affect economic development. A complex model is needed to estimate the emission reduction costs. In addition to natural phenomena and technological constraints, it is necessary to take into account the economic relations and social processes. The early complicated models assessing global effects of greenhouse gas emission reduction and the resulting losses, and the mitigation costs, were developed by the teams led by Nordhaus [98] and later by Stern [101]. These types of models were later developed in many projects, such as ADAM, AMPERE, CLIMSAVEC, EMF-27, or RECIPE, with the participation of large international research groups. They are mainly applied to predict the global or regional climate changes, under assumed scenarios.

The smaller scale models, like national ones, usually use neoclassical production functions to model economic development, see e.g. [55]. The macroeconomic sectoral models of the Polish economy considered in [10–12] are used to optimize the paths to achieve greenhouse gas emission reductions in the year 2050. The following economy sectors are specified: investment goods production, consumption good production, materials and raw materials production, and, also, energy production. It is assumed that, there are two (or three) generalized technologies in each sector: the old one, emitting a lot of greenhouse gases, the new one, emitting much less, and possibly the third one, that is intermediate. The final model is obtained by establishing the relationships between the sectorial variables and the balances of the cross-sectoral and foreign exchanges. It is then used in the bi-criteria optimization, where the total domestic consumption in the period from 2005 to 2050 is maximized and the greenhouse gas emission in 2050 is minimized. The decision variables used were: the investments in particular technologies in each sector, the gross products, and the foreign exchange balances in each year. The solution of this task required multiple solving of quite a large linear optimization task (3500 variables, 5500 constraints).

The simulation results revealed, that every percent of the emission reduction in 2050 implicates a reduction of the total consumption in the considered period from 0.56 to 0.74%, when compared to the business-as-usual scenario, depending on the required emission to be achieved in 2050 (from 43 to 70%). However, the solution path includes crisis years, due to necessary intensive technology conversion. As direct foreign investment and European funds are not considered in the model, the actual economic results may be more optimistic.

9 Summary

Air pollution affects various areas, important for human beings. They extend to very wide spatial and temporal scales. Many processes of pollution emission and absorption are not fully understood, nor can they be assessed accurately. The meteorology, especially in local environments, may differ considerably from the general weather forecasts prepared for larger areas. The measurements are scarce in space and, in some cases, also made for a too short time period. Hence, the modeling of pollution concentrations is still a difficult scientific task, in particular when they have to be predicted with sufficiently high confidence in small space areas and in small time periods. However, the modeling helps a better understanding of the processes, connected with the overall area pollution evolution and enables obtaining quantitative estimates, which agree quite well with the averaged measured values.

The modeling of pollution impacts is also challenging, due to their complicated and insufficiently investigated nature and a lack of data.

In this complicated and large system, many partial models are created. This paper attempts to outline this wide subject. It also presents some models developed with participation of the employees from CMD, SRI PAS and obtained simulation results. These papers do not exhaust all modeling issues, connected with the atmospheric pollution problems, nor a full review of the model solutions, or even all studies performed by the Department employees. To keep a reasonable volume of the chapter, some of these studies were only noted by referring to the cited publications, and some, especially the older ones, were completely omitted.

Many researchers investigate the issues related to the widely understood atmospheric pollution processes. This is a challenging area, where creation of new methods and detection of new research directions, along with a high interest in the results from the international community is a serious stimulus for researchers dealing with these and related topics.

Acknowledgements The authors would like to acknowledge the co-authors of the joint publications that were used to prepare this chapter. Some parts of the chapter were created in effect of the following projects: EU projects FP7-MC-IRSES acronym GESAPU and FP7-ENV-2012 acronym APPRAISAL, Coca Cola Foundation project acronym TAPAS, the Polish Ministry and Higher Education project NCN N N519 580238, China-Poland Scientific and Technological Cooperation projects in 2008–2010 and 2015–2016, and Polish-Austrian Scientific and Technological Cooperation project in 2009–2010.

References

References With SRI PAS Affiliations (Researchers Associated With CMD Are Underlined)

1. Bartoszczuk, P., Horabik, J.: Tradable permit systems: considering uncertainty in emission estimates. Water Air Soil Pollut. Focus **7**, 573–579 (2007)
2. Belis, C., Baldasano, J., Blond, N., Bouland, C., Buekers, J., Carnevale, C., Cherubini, A., Clappier, A., De Saeger, E., Douros, J., Finzi, G., Fragkou, E., Gama, C., Graff, A., Guariso, G., Janssen, S., Juda-Rezler, K., Karvosenoja, N., Maffeis, G., Martilli, A., Mills, S., Miranda, A.I., Moussiopoulos, N., Nahorski, Z., Pisoni, E., Ponche, J.-L., Rasoloharimahefa, M., Real, E., Reizer, M., Relvas, H., Roncolato, D., Tainio, M., Thunis, P., Viaene, P., Vlachokostas, C., Volta, M., White, L.: Current European AQ planning at regional and local scale. In: Guariso, G., Volta, M. (eds.) Air Quality Integrated Assessment. A European Perspective, pp. 37–68. Springer (2017)
3. Belis, C., Blond, N., Bouland, C., Carnevale, C., Clappier, A., Douros, J., Fragkou, E., Guariso, G., Miranda, A.I., Nahorski, Z., Pisoni, E., Ponche, J.-L., Thunis, P., Viaene, P., Volta, M.: Strengths and weaknesses of the current EU situation. In: Guariso, G., Volta, M. (eds.) Air Quality Integrated Assessment. A European Perspective, pp. 69–83. Springer (2017)
4. Бунь Р.А., Чарковська Н.В., Данило О.Я., Топилко П.І., Галущак М.О., Нахорскі З., Хорабік-Пизел І.: *Просторове моделювання емісії парникових газів від стаціонарних джерел* (Bun, R., Charkovska, N., Danylo, O., Topylko, P., Halushchak, M., Nahorski, Z., Horabik-Pyzel, J.: Spatial modeling of greenhouse gas emissions from stationary sources). PP Soroka T.B., Lviv 2016
5. Bun, R., Nahorski, Z., Horabik-Pyzel, J., Danylo, O., See, L., Charkovska, N., Topylko, P., Halushchak, M., Lesiv, M., Valakh, M., Striamets, O., Kinakh, V.: Development of a high resolution spatial inventory of GHG emissions for Poland from stationary and mobile sources.

Mitig. Adapt. Strateg. Glob. Change **24**(6), 853–880 (2018). https://doi.org/10.1007/s11027-018-9791-2
6. Carnevale, C., Douros, J., Finzi, G., Graff, A., Guariso, G., Nahorski, Z., Pisoni, E., Ponche, J.-L., Real, E., Turrini, E., Vlachokostas, Ch.: Uncertainty evaluation in air quality planning decisions: a case study for Northern Italy. Environ. Sci. Policy **65**, 39–47 (2016)
7. Charkovska, N., Halushchak, M., Bun, R., Nahorski, Z., Oda, T., Jonas, M., Topylko, P.: A high-definition spatially explicit inventory of greenhouse gas (GHG) emissions from the industrial processes. Mitig. Adapt. Strateg Glob. Change **24**(6), 907–939 (2019). https://doi.org/10.1007/s11027-018-9836-6
8. Charkovska, N., Horabik-Pyzel, J., Bun, R., Danylo, O., Nahorski, Z., Jonas, M., Xu, X.: High resolution spatial distribution and associated uncertainties of greenhouse gas emissions from the agricultural sector. Mitig. Adapt. Strateg Glob. Change **24**, 881–905 (2018). https://doi.org/10.1007/s11027-017-9779-3
9. Ermoliev, Y., Ermolieva, T., Fisher, G., Jonas, M., Makowski, M.: Cost effective and environmentally safe emission trading under uncertainty. In: Marti, K., Ermoliev, Y., Makowski, M. (eds.) Coping with Uncertainty. Robust Solutions. Lecture Notes in Economics and Mathematical System, vol. 633, pp. 79–99. Springer, Berlin, Heidelberg (2010)
10. Gadomski, J., Nahorski, Z.: Kyoto Protocol and the growth of a small country open economy: a simple mathematical model approach. Roczniki Kolegium Analiz Ekonomicznych SGH **15**(2006), 139–150 (2006)
11. Gadomski, J.,, Kruś, L., Nahorski, Z.: A multicriteria approach for analysis of the impact of GHG limiting policies on economic growth in Poland. In: E3S Web of Sciences. Energy and Fuels 2016, vol. 14(01002), pp. 1–13 (2017)
12. Gadomski, J., Kruś, L., Nahorski, Z.: A multicriteria model for analysis of the impact of EU GHG limiting policies on economic growth: the case of Poland. Control Cyber. **47**(1), 55–83 (2018) (w druku)
13. Götschi, T., Tainio, M., Maizlish, N., Schwanen, T., Goodman, A., Woodcock, J.: Contrasts in active transport behaviour across four countries: how do they translate into public health benefits? Prev. Med. **74**, 42–48 (2015). https://doi.org/10.1016/j.ypmed.2015.02.009
14. Holnicki, P.: Evaluation of environmental impact of air pollution sources. Cybern. Syst. **35**, 595–605 (2004)
15. Holnicki, P.: Modelowanie propagacji zanieczyszczeń atmosferycznych w zastosowaniu do kontroli i sterowania jakością środowiska. EXIT, Warszawa (2006)
16. Holnicki, P.: On the real-time emission control—case study application. Control Cybern. **35**, 351–367 (2006)
17. Holnicki, P.: Some aspects of the integrated approach to air quality management based on optimization techniques. Arch. Environ. Protect. **36**(1), 145–159 (2010)
18. Holnicki, P.: Uncertainty in integrated modeling of air quality. In: Nejadkoorki, F. (ed.) Advanced Air Pollution, pp. 239–260. InTech, Rijeka (2011)
19. Holnicki, P., Kałuszko, A.: Decision support for optimal emission reduction. Manag. Environ. Qual. Int. J. **15**, 250–257 (2004)
20. Holnicki, P., Nahorski, Z., Kałuszko, A.: Intake Fraction (*iF*) assessment in an urban area. IFAC Pap. Online **51**(5), 79–84 (2018). https://doi.org/10.1016/j.ifacol.2018.06.214
21. Holnicki, P., Kałuszko, A., Nahorski, Z., Tainio, M.: Intra-urban variability of the intake fraction from multiple emission sources. Atmos. Pollut. Res. **9**, 1184–1193 (2018). https://doi.org/10.1016/j.apr.2018.05.003
22. Holnicki, P., Kałuszko, A., Nahorski, Z., Stankiewicz, K., Trapp, W.: Air quality modeling for Warsaw agglomeration. Arch. Environ. Prot. **43**(1), 48–64 (2017)
23. Holnicki, P., Kałuszko, A., Stankiewicz, K.: Particulate matter air pollution in an urban area. A case study. Oper. Res. Decis. **3**, 43–56 (2016)
24. Holnicki, P., Kałuszko, A., Trapp, W.: An urban scale application and validation of the CALPUFF model. Atmos. Pollut. Res. **7**, 393–402 (2016)
25. Holnicki, P., Nahorski, Z.: Air quality modeling in Warsaw Metropolitan Area. J. Theor. Appl. Comput. Sci. **7**(1), 56–69 (2013)

26. Holnicki, P., Nahorski, Z.: Emission data uncertainty in urban air quality modeling—case study. Environ. Model. Assess. **20**(6), 583–597 (2015)
27. Holnicki, P., Nahorski, Z., Żochowski, A.: Modelling of Environmental Processes. WSISiZ Monographs, Warszawa (2000)
28. Holnicki, P., Tainio, M., Kałuszko, A., Nahorski, Z.: Burden of diseases due to multiple air pollutants emitted from local and external sources in Warsaw, Poland. Int. J. Environ. Res. Public Health **14**(11), 1359 (2017). https://doi.org/10.3390/ijerph14111359
29. Holnicki, P., Żochowski, A.: Wybrane metody matematyczne analizy jakości powietrza atmosferycznego. PWN, Warszawa (1990)
30. Horabik, J., Nahorski, Z.: A statistical model for spatial inventory data: a case study of N_2O emissions in municipalities of Southern Norway. Clim. Change **103**(1–2), 263–276 (2010). https://doi.org/10.1007/s10584-010-9913-7
31. Horabik, J., Nahorski, Z.: Improving resolution of spatial inventory with a statistical inference approach. Clim. Change **124**(3), 575–589 (2014). https://doi.org/10.1007/s10584-013-1029-4
32. Horabik-Pyzel, J., Nahorski, Z.: Uncertainty of spatial disaggregation procedures: conditional autoregressive versus geostatistical models. Ann. Comput. Sci. Inf. Syst. **8**, 449–457 (2016). https://doi.org/10.15439/2016F539
33. Hryniewicz, O., Nahorski, Z., Verstraete, J., Horabik, J., Jonas, M.: Compliance for uncertain inventories via probabilistic/fuzzy comparison of alternatives. Clim. Change **124**(3), 519–534 (2014)
34. Jarnicka, J., Nahorski, Z.: Estimation of temporal uncertainty structure of GHG inventories for selected EU countries. Ann. Comput. Sci. Inf. Syst. **8**, 459–465 (2016). https://doi.org/10.15439/2016f318
35. Jarnicka, J., Nahorski, Z.: Estimation of means in a bivariate discrete-time process. In: Atanassov, K.T., Kacprzyk, J., Kałuszko, A., Krawczak, M., Owsiński, J., Sotirov, S., Sotirova, E., Szmidt, E., Zadrożny, S. (eds.) Uncertainty and Imprecision in Decision Making and Decision Support: Cross Fertilization, New Models and Applications. Advances in Intelligent Systems and Computing, vol. 559, pp. 3–11. Springer (2018). https://doi.org/10.1007/978-3-319-65545-1_1
36. Jarnicka, J., Żebrowski, P.: Learning in greenhouse gas emission inventories in terms of uncertainty improvement over time. Mitig. Adapt. Strateg. Glob. Change **24**(6), 1143–1168 (2019). https://doi.org/10.1007/s11027-019-09866-5
37. Jonas, M., Bun, R., Nahorski, Z., Marland, G., Gusti, M., Danylo, O.: Quantifying greenhouse gas emissions. Mitig. Adapt. Strateg. Glob. Change **24**(6), 839–852 (2019). https://doi.org/10.1007/s11027-019-09867-4
38. Jonas, M., Gusti, M., Jęda, W., Nahorski, Z., Nilsson, S.: Comparison of preparatory signal analysis techniques for consideration in the (post-)Kyoto policy process. Clim. Change **103**(1–2), 175–213 (2010)
39. Jonas, M., Krey, V., Wagner, F., Marland, G., Nahorski, Z.: Uncertainty in an emissions constrained world. Clim. Change **124**(3), 459–476 (2014). https://doi.org/10.1007/s10584-014-1103-6
40. Jonas, M., Marland, G., Winiwarter, W., White, T., Nahorski, Z., Bun, R., Nilsson, S.: Benefits of dealing with uncertainty in greenhouse gas inventories. Clim. Change **103**(1–2), 3–18 (2010)
41. Jonas, M., White, T., Marland, G., Lieberman, D., Nahorski, Z., Nilsson, S.: Dealing with uncertainty in GHG inventories: how to go about it? In: Marti, K., Ermoliev, Y., Makowski, M. (eds.) Coping with Uncertainty, Robust Decisions. Lecture Notes Series on Economics and Mathematical Systems (LNEMS), vol. 633, pp. 229–245. Springer (2010)
42. Lehtveer, M., Makowski, M., Hedenus, F., McCollum, D., Strubegger, M.: Multi-criteria analysis of nuclear power in the global energy system: assessing trade-offs between simultaneously attainable economic, environmental and social goals. Energy Strategy Rev. **8**, 45–55 (2015). https://doi.org/10.1016/j.esr.2015.09.004
43. Lesiv, M., Bun, A., Jonas, M.: Analysis of change in relative uncertainty in GHG emissions from stationary sources for the EU 15. Clim. Change **124**(3), 505–518 (2014). https://doi.org/10.1007/s10584-014-1075-6

44. Lieberman, D., Jonas, M., Nahorski, Z., Nilsson, S. (eds.): Accounting for Climate Change: Uncertainty in Greenhouse Gas Inventories—Verification, Compliance, and Trading. Springer, Dordrecht (2007)
45. Lieberman, D., Jonas, M., Winiwarter, W., Nahorski, Z., Nilsson, S.: Accounting for climate change: introduction. Water Air Soil Pollut. Focus **7**(4–5), 421–424 (2007)
46. Makowski, M.: Advances in modeling methodology for supporting environmental policy-making. Arch. Environ. Prot. **36**(1), 129–143 (2010)
47. McCollum, D.L., Krey, V., Riahi, K., Kolp, P., Grubler, A., Makowski, M., Nakicenovic, N.: Climate policies can help resolve energy security and air pollution challenges. Clim. Change **119**(2), 479–494 (2013). https://doi.org/10.1007/s10584-013-0710-y
48. Mytton, O.T., Tainio, M., Ogilvie, D., Panter, J., Cobiac, L., Woodcock, J.: The modelled impact of increases in physical activity: the effect of both increased survival and reduced incidence of disease. Eur. J. Epidemiol. **32**(3), 235–2502 (2017). https://doi.org/10.1007/s10654-017-0235-1
49. Nahorski, Z., Jęda, W.: Processing national CO_2 inventory emission data and their total uncertainty estimates. Water Air Soil Pollut. Focus **7**(4–5), 513–527 (2007)
50. Nahorski, Z., Jęda, W., Jonas, M.: Coping with uncertainty in verification of the Kyoto obligations. In: Studziński, J., Drelichowski, L., Hryniewicz, O. (red.) Zastosowanie informatyki i analizy systemowej w zarządzaniu, pp. 305–317. IBS PAN, Warszawa (2003)
51. Nahorski, Z., Horabik, J.: Greenhouse gas emission permit trading with different uncertainties in emission sources. J. Energy Eng. **134**(2), 47–52 (2008)
52. Nahorski, Z., Horabik, J.: Compliance and emission trading rules for asymmetric emission uncertainty estimates. Clim. Change **103**(1–2), 303–325 (2010)
53. Nahorski, Z., Horabik, J.: A market for pollution emission permits with low accuracy of emission estimates. In: Kaleta, M., Traczyk, T. (eds.) Modeling Multi-Commodity Trade: Information Exchange Methods, pp. 151–165. Springer (2012)
54. Nahorski, Z., Horabik, J., Jonas, M.: Compliance and emission trading under the Kyoto protocol: rules for uncertain inventories. Water Air Soil Pollut. Focus **7**(4–5), 539–558 (2007)
55. Nahorski, Z., Ravn, H.: Review of mathematical models in economic environmental problems. Ann. Oper. Res. **97**, 165–201 (2000)
56. Nahorski, Z., Stańczak, J., Pałka, P.: Simulation of an uncertain emission market for greenhouse gases using agent-based methods. Clim. Change **124**(3), 647–662 (2014)
57. Oda, T., Bun, R., Kinakh, V., Topylko, P., Halushchak, M., Marland, G., Lauvaux, T., Jonas, M., Maksyutov, S., Nahorski, Z., Lesiv, M., Danylo, O., Horabik-Pyzel, J.: Errors and uncertainties in a gridded carbon dioxide emission inventory. Mitig. Adapt. Strateg. Glob. Change **24**(6), 1007–1050 (2019). https://doi.org/10.1007/s11027-019-09877-2
58. Ometto, J.P., Bun, R., Jonas, M., Nahorski, Z. (eds.): Uncertainties in Greenhouse Gas Inventories: Expanding Our Perspective. Springer (2015)
59. Ometto, J.P., Bun, R., Jonas, M., Nahorski, Z., Gusti, M.: Uncertainties in greenhouse gases inventory—expanding our perspective. Clim. Change **124**(3), 451–458 (2014). https://doi.org/10.1007/s10584-014-1149-5
60. Pałka, P., Radziszewska, W., Nahorski, Z.: Application of an auction algorithm in an agent-based power balancing system. In: Pechenizkiy, M., Wojciechowski, M. (eds.) New Trends in Databases and Information Systems, pp. 231–240. Springer, Berlin, Heidelberg (2012)
61. Pałka, P., Radziszewska, W., Nahorski, Z.: Balancing electric power in a microgrid via programmable agents auctions. Control Cybern. **41**(4), 777–797 (2012)
62. Pałka, P., Stańczak, J., Radziszewska, W., Nahorski, Z.: Multi-agent auction simulation of the GHG international emission permit trading. In: Proceedings of the 4th International Workshop on Uncertainty in Atmospheric Emissions, pp. 159–176. SRI PAS, Warszawa (2015). http://www.ibspan.waw.pl/unws2015/index.php?go=publications
63. Radziszewska, W.: An intelligent distributed system for flexible management of variable energy supply and demand in microgrids. Ph.D. thesis. SRI PAS, Warszawa (2015)
64. See, L., Schepaschenko, D., Lesiv, M., McCallum, I., Fritz, S., Comber, A., Perger, C., Schill, C., Zhao, Y., Maus, V., Siraj, M., Albrecht, F., Cipriani, A., Vakolyuk, M., Garcia, A., Rabia,

A., Singha, K., Marcarini, A., Kttenborn, T., Hazarika, R., Schepaschenko, M., van der Velde, M., Kraxner, F., Obersteiner, M.: Building a hybrid land cover map with crowdsourcing and geographically weighted regression. ISPRS J. Photogramm. Remote Sens. **103**, 48–56 (2015). https://doi.org/10.1016/j.isprsjprs.2014.06.016

65. Tainio, M.: Burden of disease caused by local transport in Warsaw, Poland. J. Transp. Health **2**(3), 423–433 (2015). https://doi.org/10.1016/j.jth.2015.06.005
66. Tainio, M., Holnicki, P., Loh, M., Nahorski, Z.: Intake fraction variability between air pollution emission sources inside the urban area. Risk Anal. **34**, 2021–2034 (2014)
67. Tainio, M., Kukkonen, J., Nahorski, Z.: Impact of airborne particulate matter on human health: An assessment framework to estimate exposure and adverse health effects in Poland. Arch. Environ. Prot. **36**(1), 95–115 (2010)
68. Tainio, M., Monsivais, P., Jones, N.R.V., Brand, C., Woodcock, J.: Mortality, greenhouse gas emissions and consumer cost impacts of combined diet and physical activity scenarios: a health impact assessment study. BMJ Open **7**, e014199 (2017). https://doi.org/10.1136/bmjopen-2016-014199
69. Thunis, P., Miranda, A., Baldasano, J.M., Blond, N., Douros, J., Graff, A., Janssen, S., Juda-Rezler, K., Karvosenoja, N., Maffeis, G., Martilli, A., Rasoloharimahefa, M., Real, E., Viaene, P., Volta, M., White, L.: Overview of current regional and local scale air quality modelling practices: assessment and planning tools in the EU. Environ. Sci. Policy **65**, 13–21 (2016). https://doi.org/10.1016/j.envsci.2016.03.013
70. Verstraete, J.: The spatial disaggregation problem: simulating reasoning using a fuzzy inference system. IEEE Trans. Fuzzy Syst. **25**(3), 627–641 (2003). https://doi.org/10.1109/TFUZZ.2016.2567452
71. Verstraete, J.: Solving the map overlay problem with a fuzzy approach. Clim. Change **124**(3), 591–604 (2014)
72. Verstraete, J.: Fuzzy quality assessment of gridded approximations. Appl. Soft Comput. **55**, 319–330 (2017)
73. Verstraete, J.: Solving the general map overlay problem using a fuzzy inference system designed for spatial disaggregation. Mitig. Adapt. Strateg. Glob. Change **24**(6), 1101–1122 (2018). https://doi.org/10.1007/s11027-018-9823-y
74. Verstraete, J.: Artificial Intelligent Methods for Handling Spatial Data. Fuzzy Rulebase Systems and Gridded Data Problems. Studies in Fuzziness and Soft Computing, vol. 370. Springer Nature (2019). https://doi.org/10.1007/978-3-039-00238-1_3
75. Viaene, P., Belis, C.A., Blond, N., Bouland, C., Juda-Rezler, K., Karvosenoja, N., Martilli, A., Miranda, A., Pisoni, E., Volta, M.: Air quality integrated assessment modelling in the context of EU policy: A way forward. Environ. Sci. Policy **65**, 22–28 (2016). https://doi.org/10.1016/j.envsci.2016.05.024
76. White, T., Jonas, M., Nahorski, Z., Nilsson, S. (eds.): Greenhouse Gas Inventories: Dealing with Uncertainty. Springer, Dordrecht (2011)
77. Xu, X., Horabik, J., Nahorski, Z.: Pricing of uncertain certified emission reductions in a China coal mine methane project with an extended Rubinstein-Ståhl model. Clim. Change **124**(3), 617–632 (2014). https://doi.org/10.1007/s10584-014-1057-8
78. Zapata-Diomedi, B., Knibbs, L.D., Ware, R.S., Heesch, K.C., Tainio, M., Woodcock, J., Veerman, J.L.: A shift from motorised travel to active transport: What are the potential health gains for an Australian city? Plos One **12**(10), e0184799 (2017)

References Without SRI PAS Affiliations

79. Amann, M., Cofala, J., Klaasen, G.: The SO_2 control cost module in the RAINS 7.0 model. Working Paper, IIASA, Laxenburg (1994)

80. Amann, M., Cofala, J., Klimont, Z., Schöpp, W.: RAINS review 2004—Modelling of emission control potentials and cost. IIASA, Laxenburg (2004)
81. Apte, J.S., Bombrun, E., Marshall, J.D., Nazaroff, W.W.: Global intraurban intake fractions for primary air pollutants from vehicles and other distributed sources. Environ. Sci. Technol. **46**, 3415–3423 (2012)
82. Belis, C.A., Karagulian, F., Larsen, B.R., Hopke, P.K.: Critical review and meta-analysis of ambient particulate matter source apportionment using receptor models in Europe. Atmos. Environ. **69**, 94–108 (2013). https://doi.org/10.1016/j.atmosenv.2012.11.009
83. Carnevale, C., Finzi, G., Pisoni, E., Volta, M., Guariso, G., Gianfreda, R., Maffeis, G., Thunis, P., White, L., Triacchini, G.: An integrated assessment tool to define effective air quality policies at regional scale. Environ. Model. Softw. **38**, 306–315 (2012). https://doi.org/10.1016/j.envsoft.2012.07.004
84. Coase, R.: The problem of social cost. J. Law Econ. **3**, 1–44 (1960)
85. Emissions Database for Global Atmospheric Research (Joint Research Centre). http://edgar.jrc.ec.europa.eu/. Cited 03 Aug 2017
86. Gillenwater, M., Sussman, F., Cohen, J.: Practical policy applications of uncertainty analysis for national greenhouse gas inventories. Water Air Soil Pollut. Focus **7**, 451–474 (2007)
87. Godal, O., Ermoliev, Y., Klaassen, G., Obersteiner, M.: Carbon trading with imperfectly observable emissions. Environ. Resour. Econ. **25**, 151–169 (2003)
88. Guariso, G., Volta, M. (eds.): Air Quality Integrated Assessment. A European Perspective. Springer Nature, Cham (2017)
89. Jacobson, M.Z.: Fundamentals of Atmospheric Modeling. Cambridge University Press, Cambridge (2005)
90. Juda-Rezler, K.: Oddziaływanie zanieczyszczeń powietrza na środowisko. Oficyna Wydawnicza PW, Warszawa (2000)
91. Juda-Rezler, K.: Uniwersalna funkcja celu dla zintegrowanych modeli oceny wpływu zanieczyszczeń powietrza na środowisko. Oficyna Wydawnicza PW, Warszawa (2004)
92. Juda-Rezler, K.: New challenges in air quality and climate modeling. Arch. Environ. Prot. **36**(1), 3–28 (2010)
93. Juda-Rezler, K., Toczko, B. (red.): Pyły drobne w atmosferze. Główny Inspektorat Ochrony Środowiska, Warszawa (2016)
94. Markiewicz, M.T.: Podstawy modelowania rozprzestrzeniania się zanieczyszczeń w powietrzu atmosferycznym. Oficyna Wydawnicza PW, Warszawa (2000)
95. Marland, E., Cantrell, J., Kiser, K., Marland, G., Shirley, K.: Valuing uncertainty. Part I: the impact of uncertainty in GHG accounting. Carbon Manag. **5**(1), 35–42 (2014)
96. Marland, G., Hamal, K., Jonas, M.: How uncertain are estimates of CO_2 emissions? J. Ind. Ecol. **13**(1), 4–7 (2009). https://doi.org/10.1111/j.1530-9290.2009.00108.x
97. Monni, S., Syri, S., Pipatti, R., Savolainen, I.: Extension of EU Emissions Trading Scheme to other sectors and gases: consequences for uncertainty of total tradable amount. Water Air Soil Pollut. Focus **7**, 529–538 (2007)
98. Nordhaus, W.D.: Managing the Global Commons: The Economics of Climate Change. MIT Press, Cambridge MA (1994)
99. Rivier, L., Peylin, P., Ciais, P., Gloor, M., Rödenbeck, C., Geels, C., Karstens, U., Bousquet, P., Brandt, J., Heimann, M., Aerocarb experimentalists: European CO_2 fluxes from atmospheric inversions using regional and global transport models. Clim. Change **103**(1–2), 93–115 (2010). https://doi.org/10.1007/s10584-010-9908-4
100. Scire, J.S., Strimaitis, D.G., Yamartino, R.J.: A User's Guide for the CALPUFF Dispersion Model. Earth Technology Inc. (2000)
101. Stern, N.: The Economics of Climate Change: The Stern Review. Cambridge University Press, Cambridge, UK (2007)

102. Tainio, M.: Methods and uncertainties in the assessment of the health effects of fine particulate matter ($PM_{2.5}$) air pollution. Ph. D. dissertation. National Institute for Health and Welfare, Kuoppio (2009)
103. Thunis, P., Pederzoli, A., Pernigotti, D.: Performance criteria to evaluate air quality modeling applications. Atmos. Environ. **59**, 476–482 (2012)
104. Verifying Greenhouse Gas Emissions. Methods to Support International Climate Agreements. National Academy of Sciences, Washington D.C. (2010)

System with Switchings as Models of Regulatory Modules in Genomic Cell Systems

Andrzej Świerniak, Magdalena Ochab, and Krzysztof Puszyński

Abstract In recent times there has been an increasing interest in switching control systems due to their ability to display simultaneously several types of dynamics in different modes of the analyzed system. Such models are used for many issues, both theoretical and practical, e.g. in control of mechanical systems and electric drives, in the automotive industry, in aircraft and air traffic control. In this paper, we present the application of switching systems in biological models of intercellular signal pathways. According to the type of biological system, switches in the model may be dependent on the state of the system or on time. The properties of the biological system can be partially characterized by searching for an analytical solution for the subsystems of the full model. Besides, transition graphs can describe the qualitative behaviour of the system by presenting jumps between subsystems. Moreover, in the systems with switching, we can find stationary points, which reflect specific system behaviour: stabilization or oscillation.

1 Introduction

A characteristic feature of a hybrid system is the capability to maintain simultaneously different types of dynamics, for example continuous and discrete in different modes or subsystems [11, 43]. As examples of such a system we can mention power converters [21], computer disk drivers [14], stepper motors [9], constrained robotic

A. Świerniak (✉) · M. Ochab · K. Puszyński
Faculty of Automatic Control, Electronics and Computer Science, Silesian University of Technology, Akademicka Street 16, 44-100 Gliwice, Poland
e-mail: andrzej.swierniak@polsl.pl

M. Ochab
e-mail: magdalena.ochab@polsl.pl

K. Puszyński
e-mail: krzysztof.puszynski@polsl.pl

P. Kulczycki et al. (eds.), *Automatic Control, Robotics, and Information Processing*, Studies in Systems, Decision and Control 296,
https://doi.org/10.1007/978-3-030-48587-0_24

systems [5], intelligent vehicles [46],sampled-data systems [16], discrete event systems [35] and many others.

System control theory has got a growing interest in many new fields of science, such as system biology, biotechnology and biomedical engineering. Real biological systems have very complicated dynamics, so discovering their structure and features are constrained by the capability of biological experiments. Mathematical modelling and analytical system analysis can be used as support for such research.

Mathematical models can be created for biological systems for a different level of complexity, starting from epidemiological models, which describe the behaviour of groups of people or animals [48], going through models of functioning one organism (such as human [3]) to models of intercellular pathways [18, 40]. On all levels of complexity, biological models are characterized by very complicated structures [44]. With the present work, we consider models of intercellular interplay between molecules, which include simultaneously continuous elements (eg. change of protein levels) and discrete elements (step switch of gene state). Additionally, in biological systems, we can distinguish signals dependent on the state of the system, time or external forces. Modelling such systems is a great challenge because it enforces finding compromise between the complexity of the mathematical model and its accuracy. The vast majority of biological models are continuous, highly nonlinear models, which on the one side allow describing complicated dynamics but on the other side are very difficult to analyze, and their solution can be found only by numerical simulations [18, 19].

Due to specific features of biological systems, there is a need to adjust the mathematical tool to real systems. One of the possible solutions is to use hybrid systems, which are characterized by different types of behaviour in different subsystems. Specific subsystems can maintain a relatively simple structure because switches between them enable modelling of very complicated dynamics of the whole system. Moreover, hybrid systems by definition, include two types of dynamics: continuous and discrete.

One of the types of hybrid systems is a system described by Liberzon in [23], which includes two types of components. The discrete component (σ) is the set of rules, which define switches between subsystems in a specific time (τ) and state of the system. The continuous part of the system is denoted by x and consists of a set of difference or differential equations, which depend on the state of the system ($x(t)$) and the type of the subsystem (σ). The system can be described by the following formulas:

$$\begin{aligned} \dot{x}(t) &= f_{\sigma(t)}(x(t)), \qquad x \in R^n, \\ \sigma(t) &= \lim_{\tau \to t^-} \phi(x(\tau), \sigma(\tau)), \qquad \sigma \in \mathcal{N}, \end{aligned} \tag{1}$$

where $\mathcal{N}$ is the set of natural numbers defining the number of existing subsystems.

Among hybrid systems, we can exclude a class of switched systems [20]. They are continuous systems with discrete switchings, which can result in a step change of the system structure, parameters or state [23].

Systems with switchings can be used in modelling biological systems, because a large part of the intercellular processes, such as production, degradation or transport, can be described by a simple, linear function. The complexity of the system increases, when we take into account the regulation of such processes, especially the kinetics of the enzymatic reactions which is highly nonlinear. Enzymatic reactions can be activated only by a specific enzyme and their rate is limited by biological cell properties [2]. In hybrid systems of biological processes, subsystems can reflect the state of the cell with or without an enzyme, where the rate of the process is completely different. Moreover, by the switch between subsystems, we can model external impacts, such as the rapid change of the environment (eg. increase or decrease of temperature, emergence of toxins) or drug applications, which results in a step change of the structure or parameters of the system [15].

2 Systems with Swichings

The main classification of systems with switching can be made in respect with rules of the switchings. In dependence on the type of switching, systems are divided into state-dependent and time-dependent [23].

2.1 Systems with Switchings Dependent on the State

Systems with switchings dependent on the state include a set of switching surfaces, which determines thresholds between subsystems. In this case, subsystems are also called regions or domains. Continuous dynamics of the system inside domains is described by differential equations (linear or nonlinear). When the trajectory reaches a threshold, there is a step switch between domains, resulting in a change of the structure, parameters or state of the system as well as two or three of them simultaneously. If, during the switch the state of the system does not change, trajectories are continuous and the only switch is in differential equations describing the system.

In models of protein production, which takes into account gene state, activation and deactivation of the allele can be modelled as a step change of the system state: usually, the number of active alleles can be equal 0, 1 or 2 [2]. Whereas in proteins interactions models, switches between different domains can model the change of the process reaction rates by enzymes. In such models during the change of the subsystem, the system state is steady and only a structure or values of parameters are changing [1, 34].

2.2 Systems with Switchings Dependent on the Time

In systems with switchings dependent on time there exists a switching signal, which is piece-wise steady, right-side continuous function $\sigma : [0, \infty) \to \mathcal{N}$. The switching function σ has a finite number of discontinuities and has obtained a steady values in intervals between times of the switch. The function of the switching signal defines, which subsystem describes system behaviour in a specific time. The general equation describing systems with the time-dependent switch is the following:

$$\dot{x}(t) = f_{\sigma(t)}(x(t)), \tag{2}$$

where x is system variable, f is a function describing the model in the specific subsystem, depending on the switching signal σ and state of the system.

Systems with switchings dependent on the time can be used in modelling biological processes, which properties are changing with time. One of the examples can be a system with cyclic changes, such as cell cycle or circadian rhythm, where process activation is tightly dependent on the time. Another example is modelling pharmacological therapy, where drug application—single or multiple—induces step change of the state or structure of the system [30].

3 Intercellular Pathways

Proteins are basic elements of all intercellular pathways. There are a big variety of proteins, which have varied and important functions, for example structural, enzymatic, transport, storage and many others [2]. Protein production is a highly complicated process with many intermediate stages, correctness of which is essential for normal cell behaviour. Information about protein structure is included in genetic material, precisely in the gene. Generally, genes are present in the cell in two copies, alleles, which usually are in an inactive state [49]. Gene activation is a very complicated process, which is regulated by many different enzymes and transcription factors. An active gene is transcripted to a molecule called mRNA (messenger RNA). mRNA after post-transcriptional modification and transport to the cytoplasm is used as the matrix for protein synthesis in the process named translation. The newly created protein molecule undergoes a series of processes, such as 3D folding and addition of some functional groups, which ends the process or creation of a fully functional protein [41].

3.1 Feedback Loops

In intercellular systems, autoregulation of protein level is a very common phenomenon. In the cell there exist both types of autoregulation: negative and positive. Negative feedback loops are more common than positive ones. Negative autoregulation enables to maintain the dynamic state of equilibrium in the cell, called homeostasis. Its role is especially important in the case of external stress or noise [36]. On the contrary, positive feedback loops enable signal propagation in the cell and introduces the bistability which allows the cell to choose its fate. For example, they are the main part of the signalling cascade, where they enhance cell response for a single stimuli [12]. The feedback loop can exist on a different stage of the proteins pathway by induction of gene activation or deactivation, enhancing of transcription or translation but also by regulation of mRNA or protein degradation. Usually, feedback is not a direct action, but works through several other molecules, such as enzymes [2].

3.2 Cell Regulation and Its Modelling

The majority of intercellular processes are not spontaneous but are strictly regulated by regulatory factors such as enzymes. Moreover, there is a nonlinear dependence between the number of regulatory factors and rate of the reaction: if the number of factors is negligible, the reaction is not processed, with the increase of the factor quantity, the rate of the process increases until maximum level. Saturation of the process results from biological cell limitations. One of the most common functions to describe enzymatic reaction dynamics is the Michaelis-Menten function [27]:

$$v = v_{\max} \frac{[S]}{K_M + [S]}, \tag{3}$$

where $[S]$ is enzyme concentration, $v_{\max}$ is maximum reaction rate and K_M is constant, numerically equal to the substrate concentration at which the reaction rate is half of $v_{\max}$.

Another function used to model enzymatic reaction is the Hill function [17], which has sigmoidal shape depending on the Hill parameter n:

$$v = v_{\max} \frac{[S]^n}{K_M^n + [S]^n}. \tag{4}$$

With the increase of the Hill parameter n there increases the steepness of the function. In some biological processes change of the rate is very rapid, which enforces using high values od n parameters. For example, the transcription rate is tightly connected with the number of active alleles: after the alleles activation reaction increases to the maximum level of its mRNA production. Steep Hill function (with high n value)

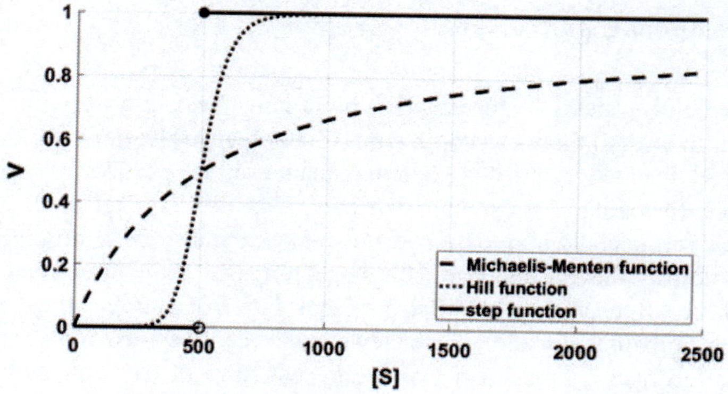

Fig. 1 Reaction rate (v) in dependence on the enzyme concentration ($[S]$) described by: Michaelis-Menten function, Hill function and step function

may be approximated by the step function (Fig. 1). As a result, the continuous model with nonlinear function is converted into a piece-wise continuous model with linear functions.

3.3 *Protein P53 and Its Function*

Protein p53 is called the guardian of the genome because of its role in maintaining cell correctness by preventing genome mutation [2]. The main p53 function is regulation of the cell response to a variety of harmful factors, both extracellular and intercellular. According to the level of the cell damage, protein p53 activates different processes, especially DNA damage repair. Moreover, it induces a temporary cell cycle blockade, until the DNA damages are repaired to prevent duplication of the abnormal cell. In the case of major DNA damages, which can not be effectively repaired, protein p53 activates the process of programmed cell death to prevent tumorigenesis [50].

The main role of p53 is regulation of gene transcription. As transcription factor p53 activates genes of proteins responsible for a variety of intercellular processes, both connected with autoregulation of p53 level and cell response to damages. Proper functioning and regulation of p53 level are very important for the proper performance of the whole organism. In normal cells p53 is maintained on a low level, which is rapidly increased in the case of intercellular damages [47]. P53 level is mainly regulated by two feedback loops: one positive and one negative. Main factors included in the negative feedback loop are p53 and MDM2. In the nucleus, p53 activates expression of MDM2, which induces the degradation of p53. This mechanism is responsible for maintaining a low level of p53 in the normal state. The positive feedback loop enables reaching and maintaining a high level of p53 in case of major DNA damages [8]. The positive feedback loop contains a set of different proteins, which transduce information. One of these proteins is PTEN, which transcription is activated by p53. PTEN activates other proteins, which finally enhance blockade of the MDM2 transport from cytoplasm to the nucleus. As a result MDM2 level in the nucleus is small, so the degradation of p53 is constrained, and the accumulation of p53 is possible. The first step of cell response to DNA damages is increased degradation of MDM2, which decreases protein p53 degradation, and enables p53-dependent gene activation (eg. MDM2 and PTEN). Depending on the level of damage, different strengths of responses are observed [39].

4 Analysis of the System with Switchings

4.1 *Stability of the System with Switchings*

Stability is a very important feature of any analyzed system. In the case of biological systems, they maintain steady internal conditions, so usually they are stable - levels of all the molecules are limited by natural cell properties. Unstable behaviour of a biological system can occur for example in cell death models. However many intercellular systems are characterized by regular oscillations. Stability analysis of such systems can be valuable for the better characteristic of a given biological system.

Stability analysis of the systems with switching is not an easy task, as we see in many reports [4, 23, 42]. If the subsystems are linear, it is quite easy to determine their stability, however, it is not sufficient to determine the stability of the whole system. Switchings among stable systems can result in destabilization of the whole system. On the other hand, switchings between unstable subsystems can lead to the stabilization or regular oscillations [24, 25].

Another characteristic behaviour of the system with switchings is sliding modes [7, 45]. This phenomenon can occur in systems with state-dependent switchings if trajectories on both sides of the threshold are directed to this threshold. After reaching the threshold, the trajectory can not leave the given subsystem and is then sliding on the threshold [22, 37].

4.2 *Analytical Solution*

In most cases finding an analytical solution for the whole piece-wise linear system is impossible. However, finding the analytical solution only for the subsystems can be very useful in system analysis. The biggest difficulty in determining the stability of the whole system is finding the time of the switch. In a system with a time-dependent switch, the time of the switch is given, so we can easily find a full analytical solution. In the system with a state-dependent switch, finding the time of the switch is much more complicated. After determining the solution in the given subsystem, we have to find the time for which state of the system will reach the threshold value. This is possible only in simple subsystems.

Table 1 Meaning and values of parameters of the protein production model

Parameter	Description	Values	Unity
q_a	Allele activation rate	2.78×10^{-4}	1/s
q_d	Allele deactivation rate	2.78×10^{-4}	1/s
p_1	mRNA production rate	0.05	Molecules/s
t_1	Protein production rate	0.1	1/s
d_1	mRNA degradation rate	1.5×10^{-4}	1/s
d_2	Protein degradation rate	2.0822×10^{-4}	1/s
d_3	Drug-induced protein degradation rate	3×10^{-7}	1/s
$DRUG$	Drug molecules number	1.4×10^{3}	Molecules
N_A	Number of genes alleles	2	Molecules

Example 1 To present the benefits of determining an analytical solution we use a model of protein production. The presented model is a modified version of the model we published in [29]. The protein production model contains three variables: gene (G), mRNA (M) and protein (P). The gene equation includes two parts: the first describes allele activation, second—allele deactivation:

$$\frac{dG}{dt} = q_a \cdot (N_A - G(t)) - q_d \cdot G(t), \tag{5}$$

where N_A is the number of alleles in the cell. The equation describing mRNA level change consists of production (first part) and degradation (second part):

$$\frac{dM}{dt} = p_1 \cdot G(t) - d_1 \cdot M(t). \tag{6}$$

Similarly in the equation describing protein level changes, there are two parts responsible for protein production (first one) and degradation (second one):

$$\frac{dP}{dt} = t_1 \cdot M(t) - d_2 \cdot P(t), \tag{7}$$

Parameters are denoted by small letters and are presented in the Table 1.

In the analysed system we assume that we want to decrease the number of proteins under a given therapeutic target. Drug application happens at a specified time, so can be modelled by time dependent switching. Once the system has switched, meaning drug application, the protein equation changes to the following formula:

$$\frac{dP}{dt} = t_1 \cdot M(t) - (d_2 + d_3 \cdot DRUG) \cdot P(t), \tag{8}$$

where $DRUG$ is drug number and d_3 is a rate of the protein degradation induced by drug.

Due to the relatively simple, linear structure of the system we can find analytical solutions for both subsytems. An analytical solution for the number of active alleles is following:

$$G(t) = \frac{N_A\, q_a - N_A\, q_a\, e^{-t(q_a+q_d)}}{q_a + q_d} + G_0\, e^{-t(q_a+q_d)}. \tag{9}$$

Analytical solution for mRNA level has more complicated form:

$$\begin{aligned} M(t) = \Big[& e^{-t d_1}\Big(M_0\, d_1\, q_a^2 - M_0\, d_1^2\, q_a + M_0\, d_1\, q_d^2 \\ & -M_0\, d_1^2\, q_d - N_A\, p_1\, q_a^2 + G_0\, d_1\, p_1\, q_a \\ & +G_0\, d_1\, p_1\, q_d + 2\, M_0\, d_1\, q_a\, q_d - N_A\, p_1\, q_a\, q_d\Big) + \\ & +e^{-t(q_a+q_d)}\Big(-G_0\, d_1\, p_1\, q_a - G_0\, d_1\, p_1\, q_d \\ & +N_A\, d_1\, p_1\, q_a\Big) + N_A\, p_1\, q_a\, q_d - N_A\, d_1\, p_1\, q_a \\ & +N_A\, p_1\, q_a^2\Big] \cdot \Big[\frac{1}{d_1\, (q_a+q_d)\, (q_a - d_1 + q_d)}\Big]. \end{aligned} \tag{10}$$

An analytical solution for protein level is different in both subsytems, due to an additional part describing drug-induced degradation. However, the general structure of the equations is the same in both equation, so the only difference is in the protein degradation formula: before drug application degradation rate is equal to d_2 and after the drug application it is equal to $d_2 + d_3 \cdot DRUG$. In the subsystem before drug application, the analytical solution is given by:

$$\begin{aligned} P(t) = & \frac{e^{-t\,(q_a+q_d)}\, f_1}{g_1} - \frac{e^{-t\,d_1}\, f_2}{g_2} + \frac{e^{-t\,d_2}\, f_3}{g_3} \\ & + \frac{f_4\,(g_1 - g_2 - g_3)}{g_1\, g_2\, g_3}, \end{aligned} \tag{11}$$

where

$$\begin{aligned} f_1 &= t_1\, (G_0\, q_a + G_0\, q_d - N_A\, q_a) \\ f_2 &= G_0\, d_1\, p_1\, t_1 - N_A\, p_1\, q_a\, t_1 - M_0\, d_1^2\, t_1 \\ & \quad +M_0\, d_1\, q_a\, t_1 + M_0\, d_1\, q_d\, t_1 \\ f_3 &= P_0\, d_2^3 - P_0\, d_1\, d_2^2 - P_0\, d_2^2\, q_a - P_0\, d_2^2\, q_d \\ & \quad -M_0\, d_2^2\, t_1 - N_A\, p_1\, q_a\, t_1 + P_0\, d_1\, d_2\, q_a \\ & \quad +P_0\, d_1\, d_2\, q_d + G_0\, d_2\, p_1\, t_1 + M_0\, d_2\, q_a\, t_1 + M_0\, d_2\, q_d\, t_1 \\ f_4 &= q_a\, p_1\, t_1\, N_A \end{aligned}$$

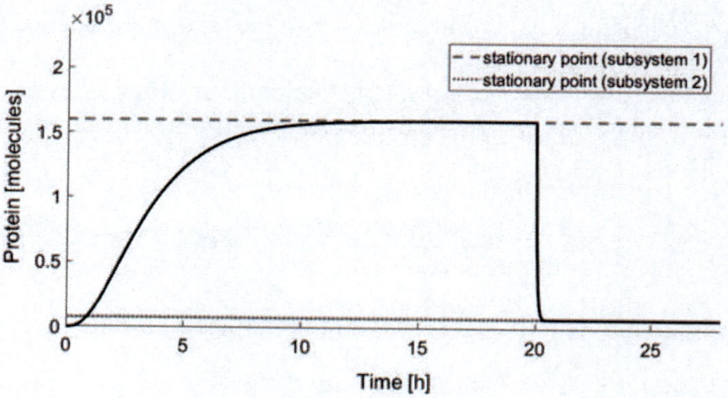

Fig. 2 Time course of protein level in system with time-dependent switch. Drug application 20 hours after start of the simulation

$$
\begin{aligned}
g_1 &= (q_a + q_d)\,(q_a - d_1 + q_d)\,(q_a - d_2 + q_d) \\
g_2 &= d_1\,(d_1 - d_2)\,(q_a - d_1 + q_d) \\
g_3 &= d_2\,(d_1 - d_2)\,(q_a - d_2 + q_d)\,.
\end{aligned}
\tag{12}
$$

After drug application all the parameters d_2 should be replaced by the formula $(d_2 + d_3 \cdot DRUG)$. In the system with time-dependent switching we can determine an analytical solution for the whole system. It will be consisted of two parts: first describing behaviour before drug application, and second—after the switch. Both subsystems are stable and we can easily determine their stationary points:

$$\lim_{t\to\infty} G(t) = N_A \frac{q_a}{q_a + q_d}, \tag{13}$$

$$\lim_{t\to\infty} M(t) = N_A \frac{p_1\, q_a}{d_1 (q_a + q_d)}, \tag{14}$$

$$\lim_{t\to\infty} P(t) = N_A \frac{t_1\, p_1\, q_a}{d_2\, d_1\, (q_a + q_d)}. \tag{15}$$

And for the system after switch:

$$\lim_{t\to\infty} P(t) = N_A \frac{t_1\, p_1\, q_a}{(d_2 + d_3 \cdot DRUG)\, d_1\, (q_a + q_d)}. \tag{16}$$

Independently at the time of the switch and parameter values, the given system is always stable. An exemplary time course is presented on the Fig. 2.

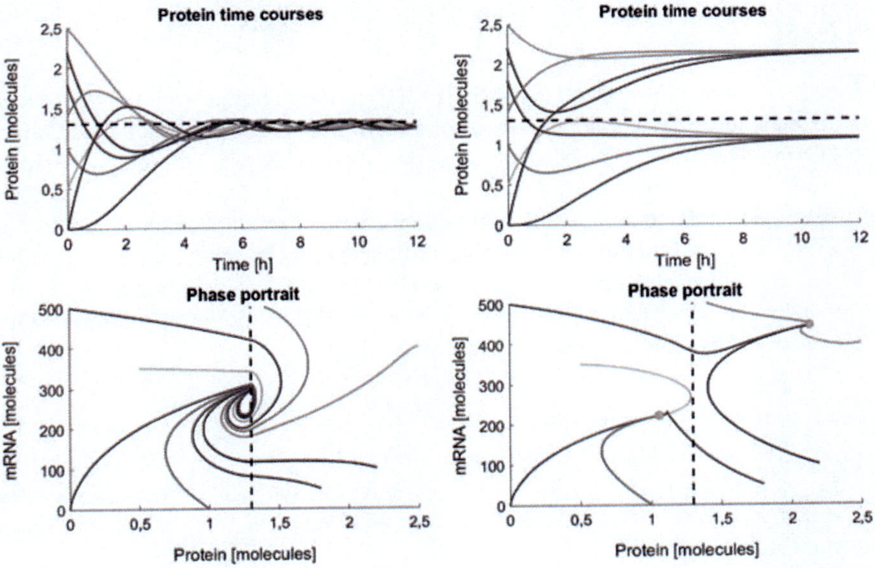

Fig. 3 Response of systems with autoregulation. Left columns: system with negative feedback. Right column: system with positive feedback

Another example of a biological system with switchings can be a system with autoregulation. In such systems switching depends on the state, after reaching the threshold value of variables, the process is activated and the parameter values are changed. In the created system autoregulation works by protein influence on its gene state. In the system with a negative feedback loop the increase of protein (P) over threshold value ($\theta = 130\,000$) causes increase of the allele deactivation rate from q_d to $5 \cdot q_d$. For some specific values of parameters in the system there exists a limit cycle, which results in protein level oscillation. In both subsystems its stationary point is on the other side of threshold (Fig. 3 left column), so trajectories circulate between subsystems.

In a system with a positive feedback loop autoregulation works in the opposite direction: if the protein is below the threshold value, gene activation rate is low ($q_a/2$), but if the protein level is higher then the threshold, gene activation rate is high ($q_a \cdot 2$). In both subsystems there is a stationary point, so depending on the initial conditions, there can be two different final protein level (Fig. 3 right column).

Equations describing the model and the analytical solution are the same as in the model depending on time, but with different parameter values. However the analytical solution for the protein has got a quite complex formula (Eq. 13), so we can not find the analytical formula for time of the switch. Nevertheless, an analytical solution can be used to determine time courses and stationary points in subsystems.

4.3 Analysis of the Piece-Wise Linear Systems

Piece-wise linear differential equation (PLDE) models are one group of the systems with switchings. Such systems consist of a set of linear models, which are relatively easy to analysis. Switching between subsystems is state-dependent, so they can maintain very complex dynamics. Such a system can be intuitively used for biological systems. Basic processes, such as production, degradation and transport, can be modelled by a linear function with little loss of accuracy. Complex regulation, such as activation and deactivation of the processes by enzymes, can be modelled by switchings between subsystems. The basic equation in PLDE model of the intercellular process can have the following formula:

$$\frac{dx_i}{dt} = \alpha_i(X) - \gamma_i(X)x_i, \quad i = 1, \ldots, n, \tag{17}$$

where x_i is a protein level, α_i is a protein production rate, γ_i is a protein degradation rate and X is a set of switching functions (with values 0 or 1) dependent on the system state x.

The state space is divided into regulatory domains by threshold values, denoted by θ_{ij}, where i stand for the variable and j is the number of the threshold for the given variable. The model in each subsystem is described by a set of linear (affine) functions. On the thresholds, the system structure or parameter values are changed, so the system is not continuously differentiable. To simplify the system description, there are introduced switching variables (Z_{ij}), which are boolean. Switching variables inform if variable i is above or below the given threshold (θ_{ij}).

4.3.1 Transition Graph

Transitions between domains can be presented by transition graphs. The transition graph is made of nodes (reflecting domains) and arches (arrows presenting transitions between domains). For piece-wise linear models, transition graphs can be easily created, because they use information about localization of focal points in each domain [38]. For a simple model with two variables, and two threshold values, the transition graph can have a form presented on Fig. 4. On the graph there are presented transitions from all domains to domain {11}, where it is the high level for both variables. Based on this graph we can assume, that in the domain {11} the stationary point is located.

4.3.2 Stationary Points in PLDE Systems

In the PLDE system there can exist two types of stationary points: regular stationary points (RSPs) and singular stationary points (SSPs) [26]. The regular stationary

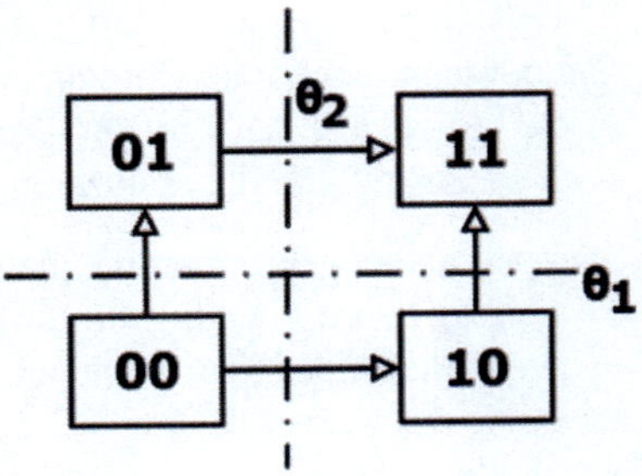

Fig. 4 Transition graph for simple model. Threshold values θ_1 and θ_2 are presented by dashed lines

point is located in the given domain. It can be calculated by solving the condition for equilibrium state for all domains:

$$\frac{dx_i}{dt} = 0; \quad i = 1, \ldots, N. \tag{18}$$

The final point of trajectory in each domain is called a focal point and is equal to:

$$\Phi_i = \alpha_i / \gamma_i. \tag{19}$$

If the focal point is located inside the given domain, it is the asymptotically stable regular stationary point, and trajectories in the close neighbourhood are going to reach it. We need to keep in mind, that attraction pool is not equal to the domain, so in some domains with RSP, trajectories can move to another domain and reach another stationary point.

Singular stationary points are much more difficult to determine, due to their localization on the threshold or the crossing on a few thresholds. Using the method proposed by Mestl and coworkers [10, 26, 38] there is the possibility to localize SSP in the PLDE systems. In the simplest case, SSP is located on the one threshold value. This is possible if the given variable is directly dependent on its threshold value. Precisely, in the system, SSP exists on the threshold θ_{ij} for variable $x_i = \theta_{ij}$ if derivative $\partial F_i / \partial Z_{ij}$ is bigger than 0, where $F_i = dx_i/dt$ and Z_{ij} is the switching variable.

The PLDE system is not continuous, so if we want to calculate a derivative, we need to replace the step function by its continuous, sigmoidal approximation with values from interval [0,1]. Thus, we can define switching function $Z_{ij} = Z(x_i, \theta_{ij}, \delta)$ as:

$$Z_{ij} = \begin{cases} 0 & \text{dla } x_i \leq \theta_{ij} - \delta \\ \text{increase monotonically from 0 to 1} & \text{for } x_i \in \langle \theta_{ij} - \delta, \theta_{ij} + \delta \rangle \\ 1 & \text{dla } x_i \geq \theta_{ij} + \delta, \end{cases}$$

where parameter δ is distance from the threshold value. For δ close to zero, function Z_{ij} will be a good approximation of the step function.

If the linear models inside domains contain not only simple functions of production and degradation, but also production based on another variable, transport or change of the molecules, the algorithm for preliminary regions selections will not work. In such a case, we should analyze all the regions, with the following algorithm:

- create a set of differential equations describing given regions, by replacing variables by appropriate threshold values.
- determine values (0 or 1) of switching variables not related to analyzed threshold.
- equate derivatives to 0,
- calculate values of switching variables,
- check if values of switching variables are included in interval [0, 1],
- calculate values of remaining variables and check if they are contained in the assumed interval,

If in the system there exists a SSP on the crossing of two or more threshold values, limit cycle can exist. However, it is impossible to determine if the oscillations around SSP are stable because it depends on the localization of other threshold values. In some cases, the oscillatory trajectory can hit the threshold and jump to another domain, with a strong attraction to another stationary point.

Example 2 We create a simple system of protein regulation to present analysis of the PLDE models. The system consists of two proteins x_1 and x_2, which reciprocally activates their production:

$$\frac{dx_1}{dt} = \alpha_1 + \alpha_{12} \cdot Z_2 - \gamma_1 x_1, \tag{20}$$

$$\frac{dx_2}{dt} = \alpha_2 + \alpha_{21} \cdot Z_1 - \gamma_2 x_2, \tag{21}$$

where Z_1 and Z_2 are switching variables, defining levels of variables x_1 and x_2 in relation to their thresholds, θ_1 and θ_2 respectively. Parameter values are presented in the Table 2.

Table 2 Parameter values in PLDE model of two proteins

Parameter	Description	Values	Unit
α_1	x_1 basic production rate	100	1/s
α_{12}	x_1 induced production rate	100	1/s
α_2	x_2 basic production rate	50	1/s
α_{21}	x_2 induced production rate	80	1/s
γ_1	x_1 degradation rate	3	1/s
γ_2	x_2 degradation rate	1.5	1/s
θ_1	x_1 threshold value	50	Molecules
θ_2	x_2 threshold value	50	Molecules

Table 3 Focal points in the system of two proteins

Starting domains	x_1	x_2	Ending domain
{00}	33.3	25	{00}
{01}	66.7	25	{10}
{10}	33.3	65	{01}
{11}	66.7	65	{11}

To localize RSP we have to equate derivatives to 0 and calculate focal points for all domains (Table 3). In two domains, focal points are localized inside domain, so in the system there are two RSPs. Trajectories of the two remaining domains are moving towards threshold values, where there is a change of the parameter values.

Looking for singular stationary points, we need to analyse both thresholds and the crossing of them. Variable x_1 is not dependent directly on the switching variable Z_1, so on the threshold θ_1 is not SSP. Similarly, on the threshold θ_2 is not SSP, because variable x_2 is not dependent on Z_2. To check if SSP is located on the crossing of the threshold θ_1 and θ_2 we follow up the presented algorithm. Firstly we form equations, describing analysed region:

$$0 = \alpha_1 + \alpha_{12} \cdot Z_2 - \gamma_1 \cdot \theta_1, \tag{22}$$

$$0 = \alpha_2 + \alpha_{21} \cdot Z_1 - \gamma_2 \cdot \theta_2, \tag{23}$$

thus the numbers are:

$$0 = 100 + 100 \cdot Z_2 - 3 \cdot 50, \tag{24}$$

$$0 = 50 + 80 \cdot Z_1 - 1.5 \cdot 50. \tag{25}$$

From the above equation we can calculate values of switching variables: $Z_2 = -0.5$ oraz $Z_1 = -0.3125$. Both switching variables are not included in interval [0,1], so in this region SSP does not exist. On the phase space portrait there are denoted RSPs, and the exemplary trajectories which are reaching them, depending to initial conditions (Fig. 5).

4.3.3 Analysis PLDE System

Application of PLDE models to biological systems is reasonable, if the rate of processes are changing very rapidly, dependently on the number of regulatory factors. One example of such system is the p53 regulatory module (Sect. 3.3). We can create PLDE model of the simplified system of p53 level regulation. This model and its stability analysis were presented in our earlier publications [31, 33].

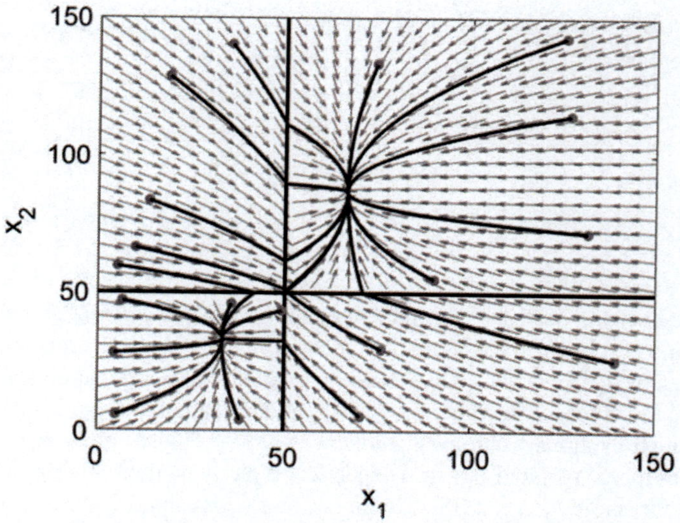

Fig. 5 Localization of RSP and exemplary trajectories for system od 2 protein regulation (model from example 2)

The Model of p53 regulatory module contains only crucial proteins, that enables us to describe basic system dynamics with a quite simple model. The model contains 4 variables:

- P—protein p53,
- C—protein MDM2 in cytoplasm;
- N—protein MDM2 in nucleus;
- T—protein PTEN;

and 4 threshold values, which divide state space into 12 regulatory domains. To check system response to different extortion, we introduce external stress R, which causes cell damages and increases degradation of MDM2 (both cytoplasmic and nuclear). In each domain, the system is described by a set of linear differential equations with different parameter values. We model the influence of one variable on the other by changing values of process rate when protein level is crossing through the threshold. A parameter dependent on the system state is denoted by a star (*). The general model has the following form. Changes of the protein p53 level are dependent on its production and degradation, which is induced by nuclear MDM2:

$$\frac{dP(t)}{dt} = p_1 - d_1^*(N(t))\ P(t). \tag{26}$$

The equation describing changes in the MDM2 level consists of three following parts: production (dependent on the p53 level), transport to nucleus (dependent on the PTEN level) and degradation (increased by stress).

$$\frac{dC(t)}{dt} = p_2^*(P(t)) - k_1^*(T(t))\ C(t) - d_2\ (1+R)\ C(t). \tag{27}$$

The equation describing MDM2 level changes in nucleus consists of transport from cytoplasm and stress-dependent degradation:

$$\frac{dN(t)}{dt} = k_1^*(T(t))\ C(t) - d_2\ (1+R)\ N(t), \tag{28}$$

Protein PTEN changes are dependent on the production (induced by p53) and spontaneous degradation:

$$\frac{dT(t)}{dt} = p_3^*(P(t)) - d_3\ T(t). \tag{29}$$

In the model there exist four threshold values:

- one threshold value for protein $p53$ (θ_P), which reflects an activation of production MDM2 (C) and PTEN (T)
- one threshold value for PTEN (θ_T) separating system for regions with normal - fast and limited transport of MDM2 to the nucleus.
- two threshold variables for nuclear MDM2 (θ_{N1} and θ_{N2}), which differentiate 3 p53 degradation rates: slow, medium and fast.

As a result of the existing four thresholds, there exist four switching variables, to define if the given variable is over or below its threshold (Z_P, Z_T, Z_{N1}, Z_{N2}). Each domain can be described by 3-values vector: $B = \{B_P, B_N, B_T\}$, where B_P, B_N and B_T has values determining if their level in relation to thresholds (0 or 1 for B_P and B_T and 0, 1 or 2 for B_N). For example domain {011} denoting subspace, where $P < \theta_P$, $\theta_{N1} < N < \theta_{N2}$ and $T \geq \theta_T$.

Values of parameters dependent on the state space can be described as:

$$p_2^* = p_{20} + p_{21} Z_P, \tag{30}$$
$$p_3^* = p_{30} + p_{31} Z_P, \tag{31}$$
$$d_1^* = d_{10} + d_{11} Z_{N1} + d_{12} Z_{N2}, \tag{32}$$
$$k_1^* = k_{10} - k_{11} Z_T, \tag{33}$$

where

$$Z_P = \begin{cases} 0 & \text{for} \quad P < \theta_P \\ 1 & \text{for} \quad P \geq \theta_P, \end{cases} \tag{34}$$

$$Z_{N1} = \begin{cases} 0 & \text{for} \quad N < \theta_{N1} \\ 1 & \text{for} \quad N \geq \theta_{N1}, \end{cases} \tag{35}$$

Table 4 Parameter values in p53 regulatory module model

Parameter	Description	Value	Unit
p_1	Basic production rate of P	8.8	1/s
p_{20}	Basic production rate of C	2.4	1/s
p_{21}	P-induced production rate of C	21.6	1/s
p_{30}	Basic production rate of T	0.5172	1/s
p_{31}	P-induced production rate of T	3.6204	1/s
d_{10}	Basic degradation rate of P	9.8395×10^{-5}	1/s
d_{11}	First N-induced degradation rate of P	6.5435×10^{-5}	1/s
d_{12}	Second N-induced degradation rate of P	1.6283×10^{-4}	1/s
d_2	Degradation rate of C i N	1.375×10^{-5}	1/s
d_3	Degradation rate of T	3×10^{-5}	1/s
k_{10}	Basic transport rate of N	1.5×10^{-4}	1/s
k_{11}	T-blocked transport rate of N	1.4713×10^{-4}	1/s

Table 5 Threshold values in p53 regulatory module model

Threshold	Description	Value	Unit
θ_P	Threshold for P	4.5×10^4	Molecules
θ_{N1}	1. threshold for N	4×10^4	Molecules
θ_{N2}	2. threshold for N	8×10^4	Molecules
θ_T	Threshold for T	10^5	Molecules

$$Z_{N2} = \begin{cases} 0 & \text{for} \quad N < \theta_{N2} \\ 1 & \text{for} \quad N \geq \theta_{N2}, \end{cases} \tag{36}$$

$$Z_T = \begin{cases} 0 & \text{for} \quad T < \theta_T \\ 1 & \text{for} \quad T \geq \theta_T. \end{cases} \tag{37}$$

We can examine the influence of different stress levels for cell response by changing values of parameter R. Values of parameters in the presented model are given in Table 4 and threshold values in Table 5. Values of parameters are fitted based on biological experiments (e.g. [18]) and our previous works [28].

Biological results show, that for different levels of DNA damage different cell responses are activated. Results for the presented model can imitate different behaviour for stress values from 0 to infinity. In system without external stress ($R = 0\,a.u.$) there exists one regular stationary point in domain {020}. In this domain, protein p53 level and PTEN level are low, and protein MDM2 level is high, that is consistent with biological observations. The transition graph for such a configuration shows that trajectories reach domain {020}. There are also closed sequences of transitions between domains: {021}, {011}, {111} and {121} (Fig. 6). To check

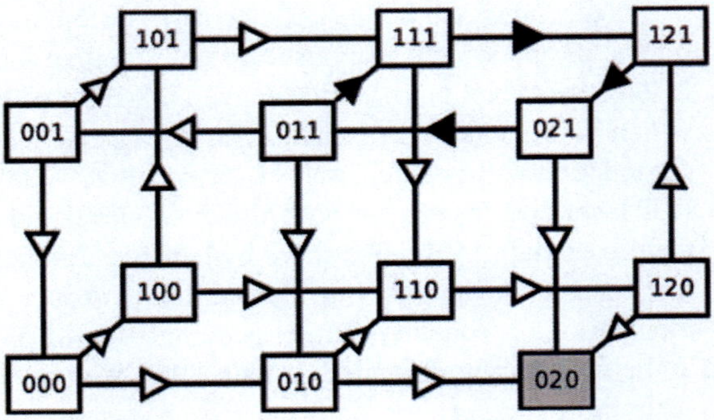

Fig. 6 Transition graph for p53 model without stress ($R = 0\,a.u.$) [31]

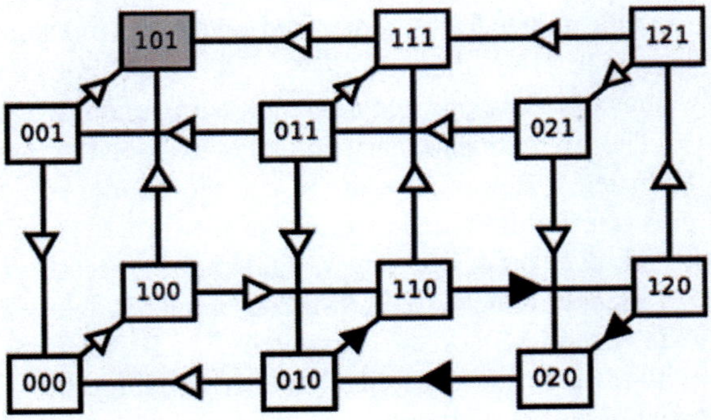

Fig. 7 Transition graph for p53 model with stress $R = 5\,a.u.$ [31]

Table 6 Variable p53 values in RSP

Stress (R [a.u.])	Domain	P_s
0–0.8635	[020]	3.3559×10^4
1.0322–1.9154	[111]	5.3714×10^4
1.9154–$+\infty$	[101]	8.9435×10^4

if there can exist a stable oscillation we need to determine if there is SSP. In the case of medium stress ($R = 5\,a.u.$), the transitions between domains are completely different. Trajectories are moving to domain {101} and there is a closed sequence of domains for low PTEN level: {020}, {010}, {110} i {120} (Fig. 7).

Complex analysis for a whole range of stress parameter values shows which intersection of stress there can exist a stationary point: RSP or SSP. In the analyzed system RSP can exist in 3 different domains depending on the stress:

- in domain {020} for $R \in [0, 0.8635\rangle$,
- in domain {111} for $R \in \langle 1.0322, 1.9154\rangle$,
- in domain {101} for $R \in \langle 1.9154, +\infty\rangle$.

Precise values of protein p53 number in RSP are presented in Table 6.

For stress $R \in [0.8635, 1.0322\rangle$ in the system there are not any regular stationary points, so the system response can not stabilize, but it oscillates in a limit cycle . If the stress level is low (smaller than 0.8635 a.u.) RSP exist in the domain {020}, which corresponds to normal, unstressed cells. The MDM2 level is high, so the degradation p53 is quick, and its level is low. With the increase of the stress dose, RSP is moving to domains with higher p53 level and smaller MDM2 level: {111} and then to domain {101}. For stress higher than 1.9154 a.u. nuclear MDM2 is quickly degraded, so the activity of p53 is sufficient to accumulate PTEN and block MDM2 transport. Consequently, further accumulation of p53 is possible, and the p53 level in the stationary point in the domain {101} is very high.

In biological experiments, there is an increase of p53 level after external stress. Low stress causes low DNA damages, so an increase of p53 is relatively small and the cell induce DNA repair and cell cycle blockade. With a stress dose increase, increased DNA damage, and the p53 level can reach very high values. For high values of stress, p53 accumulation is big enough to induce cell death.

The singular stationary point can exist on the threshold between domains. In the analyzed system, SSP can exist in the middle of a threshold, on the crossing of 2 or 3 thresholds. The derivatives of the time derivative of all variables $\partial F_i/\partial Z_{ij}$ are equal to zero because any variable is not directly regulated by its threshold value. As a result in the system, SSP does not exist on a single threshold. To determine the existence of SSP on the crossing of 2 or 3 thresholds, we analyse all existed regions with the algorithm described in Sect. 4.3.2. For the whole range of stress R, SSP can exist only in 2 regions.

SSP 1: $[\theta_P, \theta_{N2}, T < \theta_T]$

The first singular stationary point exists on the crossing of threshold values for p53 θ_P and nuclear MDM2 θ_{N2}. The PTEN level must be lower than θ_T. Equations describing this state are following:

$$0 = p_1 - (d_{10} + d_{12} Z_{N2})\theta_P, \tag{38}$$
$$0 = p_{20} + p_{21} Z_P - (k_{10} + d_2(1 + R))C, \tag{39}$$
$$0 = k_{10} C - d_2(1 + R)\theta_{N2}, \tag{40}$$
$$0 = p_{30} + p_{31} Z_P - d_3\ T. \tag{41}$$

Values of switching variables Z_P i Z_{N2} are included in the range (0, 1) and $T < \theta_T$ for stress $R \in \langle 0.8635, 7.7038\rangle$ $a.u.$.

SSP 2: $[\theta_P, \theta_{N2}, T > \theta_T]$

The second singular stationary point exists for P equal θ_P, N equal θ_{N2} and T bigger than θ_T. Equations describing this region are presented below:

Table 7 Values of variables in SSPs

R	P	N	T
0.8635–7.7038	θ_P	θ_{N2}	$T < \theta_T$
0.7059–1.0322	θ_P	θ_{N2}	$T > \theta_T$

$$0 = p_1 - (d_{10} + d_{12} Z_{N2})\theta_P, \tag{42}$$
$$0 = p_{20} + p_{21} Z_P - (k_{10} - k_{11} + d_2(1 + R))C, \tag{43}$$
$$0 = (k_{10} - k_{11})C - d_2(1 + R)\theta_{N2}, \tag{44}$$
$$0 = p_{30} + p_{31} Z_P - d_3\ T. \tag{45}$$

Switching variables Z_P and Z_{N2} are in intersection (0, 1) and variable T is bigger than θ_T for stress $R \in \langle 0.7059, 1.0322\rangle\ a.u.$. This SSP exists only for a very small range of stress and its influence on the system dynamics is negligible. Values of variables for the SSPs are presented in the Table 7.

The last possibility is localization on the crossing of three thresholds. In the analysed system there are two regions: $[\theta_P, \theta_{N1}, \theta_T]$ and $[\theta_P, \theta_{N2}, \theta_T]$. In the region on the θ_{N2}, based on calculations, we can assume that the SSP exists but is unstable, and the oscillations are not observed.

Due to the existence of two SSPs on the crossing of the thresholds in the system two types of oscillations can exist. The SSP in region $[\theta_P, \theta_{N2}, T < \theta_T]$ exists for a much bigger range of stresses and have a bigger attraction pool. Oscillations around this stationary point are between medium and low p53 levels during when the PTEN level is low. Oscillations result from the negative loop between p53 and MDM2. Because of the oscillations of p53 after the stress, the final, apoptotic decision can be postponed in time. The stress response can be activated even by a minor increase of p53 level, which induces transcription of the protein responsible for DNA repair, cell cycle blockade and autoregulation (such as MDM2 and PTEN). If the damages are small, the oscillation of p53 and MDM2 levels enables cell repair and return to the normal state. In the case of significant cell damages, in the first stage of cell response, there are also p53 oscillations. The cell tries to repair the DNA, however, if it is not possible, the mean p53 level is high enough to induce accumulation of PTEN, which results in a MDM2 transport blockade. A substantial decrease of MDM2 level in the nucleus results in the steady, high level of p53, which enables an activation of the program of the cell death [6, 13]. The presented model does not include a variable for DNA damages, so it can not present a return to the normal state, and the p53 oscillation after small damages are undumped. Nevertheless, the crucial behaviour after the stress occurrence is similar to results of biological experiments.

The second singular stationary point stands for the oscillations between p53 and MDM2 in the high PTEN level. This SSP exists only for a very narrow range of external stress R. Moreover, the PTEN level should be high (over threshold value θ_T) for low p53 level, which is not observed in regular cells. Practically, PTEN

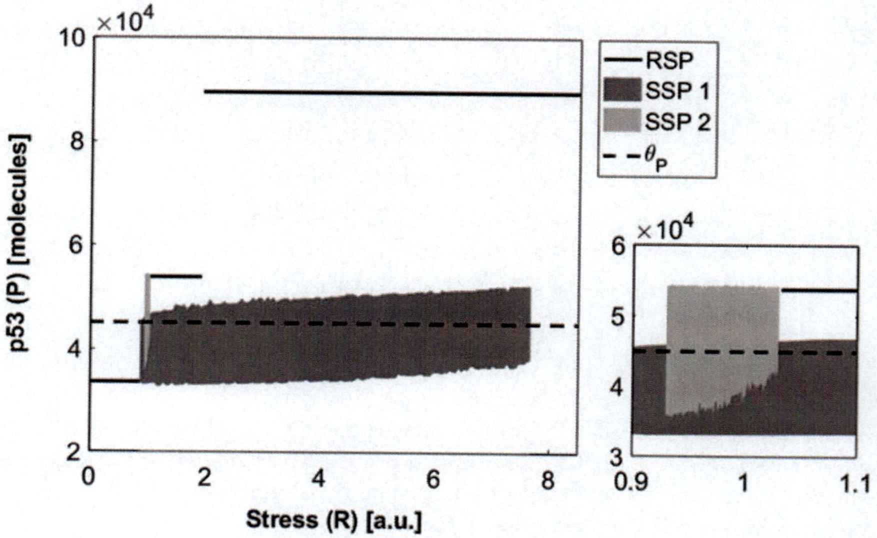

Fig. 8 P53 level (P) in dependency on the stress value R. On the right magnification for $R \in \langle 0.9, 1.1 \rangle$

increases after gene activation by p53, so these theoretical results do not correspond to normal cell behaviour.

Bifurcation diagrams

Depending on the stress R stationary point RSP and SSP are located in different localizations in the state space. For small values of stress, in the system exists only one RSP in the domain {020}, characterized by low p53 and PTEN levels and high MDM2 level. With an increase of R, in the system there appears a second point: SSP. Depending on the initial conditions, trajectories can stabilize in the RSP or oscillate around the SSP. With the further increase of R, the RSP disappears and the only possible response is characterized by undumped oscillations. For the narrow range of R in the system there exist two SSPs, so there are two different limit cycles for different protein levels. With a further increase of R, there appears RSP in the domain {111}, which is characterized by the increased p53 level. For stress R bigger than 1.9154, RSP moves from domain {111} to domain {101}, where p53 is very high. SSP exists in the system only for R smaller than 7.7038, which means that for a very big stress, the only possible response is high p53 increase, which is a determinant of apoptosis. The protein p53 level in dependence on the stress is presented in Fig. 8, where grey areas present a range of oscillations around both SSPs.

With the increase of R, other variables also change their level in the stationary point and the range of the oscillation. In Fig. 9, dependency between PTEN level and

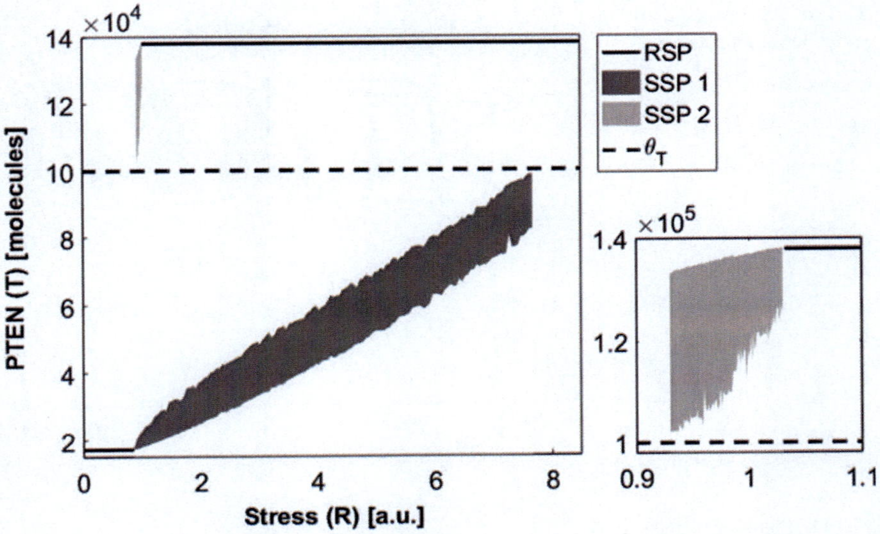

Fig. 9 PTEN level (T) in dependency of stress value R. On the right magnification for $R \in \langle 0.9, 1.1\rangle$

the stress R is presented. In the RSPs the level of PTEN is steady. The range of the PTEN oscillations around SSPs is increasing at R. Moreover, oscillations around the second SSP are very close to the threshold value θ_T.

On the bifurcation diagrams we can see a lack of oscillations around the second SSP for $R \in \langle 0.8635, 0.95\rangle$ (Figs. 8 and 9). The lack of oscillations is a result of close localization of SSP to the threshold θ_T. Trajectories are moving around in a limit cycle, hit the threshold θ_T and jump to a domain with a very strong attraction to another stationary point. With an increase of R, the second SSP moves to higher PTEN values and for R bigger than 0.93, trajectories do not cross θ_T so we can observe stable oscillations.

Calculation of localization of stationary points is not sufficient to determine if there are oscillations around the given SSP. For piece-wise linear models, analytical determination of the attraction pools for SSP is usually impossible. Numerical simulations help us to find a possible oscillatory response for the given system. In some cases attraction pools are very small, like for SSP in region $[\theta_P, \theta_{N2}, T > \theta_T]$.

Numerical results

To illustrate the results of analytical model analysis, we made numerical investigations of a cell response for two different Rs and different initial conditions. For $R = 1\,[a.u.]$ there exist two SSPs and consequently, we can observe two types of oscillations around them. It is worth noticing, that both SSPs are located on the same

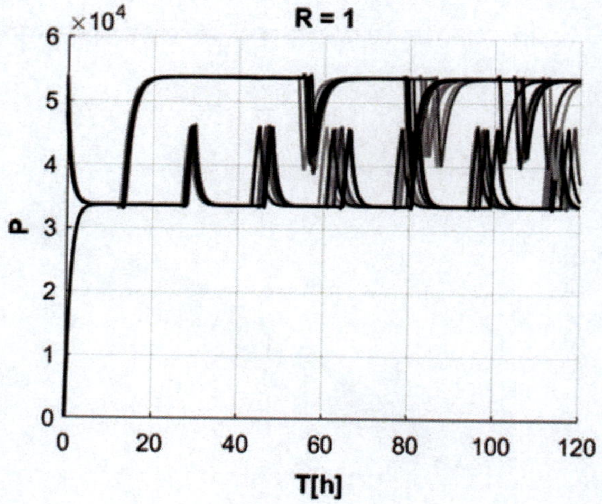

Fig. 10 P53 time courses for different initial conditions and for $R = 1$ [33]

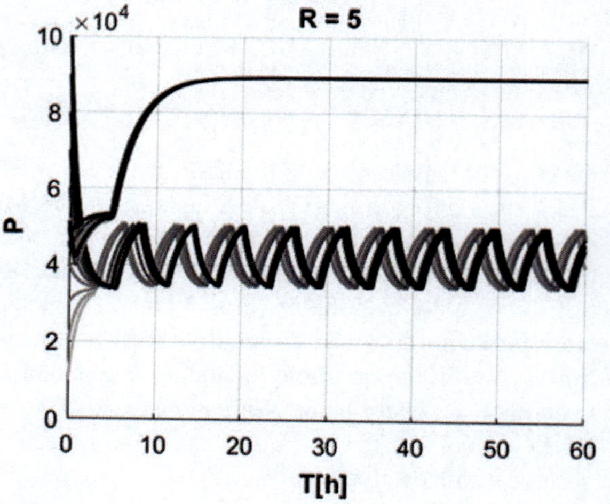

Fig. 11 P53 time courses for different initial conditions and for $R = 5$ [33]

threshold crossing: θ_P and θ_{N2} but one is above θ_T and the second is below. As a result, limit cycles have a different localization that we can see on the p53 time courses (Fig. 10).

For higher values of R (e.g. $R = 5\,a.u.$) in the system there exist one RSP and one SSP. Depending on the initial conditions trajectories stabilize in the domain $\{101\}$ or oscillate in a limit cycle (Fig. 11). In one of our earlier works [32], we present division of the population into subpopulations with different types of response. In biological results, usually, cell response is not the same but there are cell subpopulations with different behaviours depending on the cell properties.

5 Conclusions

Recently a variety of real systems, have been described and analysed using mathematical tools used for systems with switchings. In the present work, we extend this catalogue by the biological systems, more precisely, the models of genomic intercellular systems.

Biological systems have a highly nonlinear structure and some of their reaction rates are dependent on the level of regulatory proteins, such as enzymes. Because of these properties, intercellular systems are usually modelled by the Michaelis-Menten function or the Hill function, which, in many cases, can be approximated by the step function. Genetic networks are attractive objects of the application of the presented methodology, which allows the replacement of complex, highly non-linear functions describing the interplay between molecules in the cell by simple, usually linear subsystems with the switchings between them.

As we shown in the present work, the proposed methodology allows, among others, to get analytical solutions for subsystems of the analysed system, finding stationary points and indicating oscillation trajectories. The obtained information can be used not only for better understanding of complex control systems, like genetic-cellular, networks but also can be a base for elaboration of new hypothesis on the difference in the behaviour of normal and damaged (tumour) cells, which can lead to new, explorative therapies and improvement of existing ones.

It is worth noticing, that presented methods of analysis of genomic-cellular network as systems with switchings do not solve the problem, but should be treated as a contribution for future work. This approach enables modelling complex biological systems, while maintaining the simplicity of their analysis. For example, in the presented models, we assume that all cells in a population behave in the same way. They have common threshold values and times of switching. In reality, cell population are often highly heterogeneous, which creates a need for differentiation of threshold values or parameter values in specific subsystems. Similarly, parameter values, which illustrate conditions in the cell, such as temperature, enzyme concentration, etc., can differentiate for analysed cells. Including the above modifications in PLDE models require modifications of presented method analysis or even the invention of completely new methods.

The next issue that can be successfully solved by the application of system with switchings, is the problem of the synthesis of the complex systems into one big model. An example of such models can be a detailed ATM-p53-NFκB model or an NFκB-HSF model. In both cases standard analysis leads to various problems of fitting the nonlinear dynamics of subsystems, to get the correct, complete system. Simplifications of the components to systems with switchings can significantly help with this synthesis. This is a challenge for our future works.

Acknowledgements Research presented in this study were partially financed by National Science Center in Poland by grants number: UMO-2016/23/B/ST6/03455 (KP) and DEC-2016/21/B/ST7/02241 (AS) and Ministry of Science and Higher Education funds under the project 02/010/BK_19/0143 (MO).

References

1. Acary, V., De Jong, H., Brogliato, B.: Numerical simulation of piecewise-linear models of gene regulatory networks using complementarity systems. Physica D: Nonlinear Phenomena **269**, 103–119 (2014)
2. Alberts, B., Bray, D., Hopkin, K., Johnson, A., Lewis, J., Raff, M., Roberts, K., Walter, P.: Essential Cell Biology. Garland Science (2013)
3. Aloulou, I., Brun, J.-F., Mercier, J.: Evaluation of insulin sensitivity and glucose effectiveness during a standardized breakfast test: comparison with the minimal model analysis of an intravenous glucose tolerance test. Metabolism **55**(5), 676–690 (2006)
4. Bacciotti, A., Ceragioli, F.: Stability and stabilization of discontinuous systems and nonsmooth lyapunov functions. ESAIM: Control, Optimisation Calc. Var. **4**, 361–376 (1999)
5. Back, A., Guckenheimer, J., Myers, M.: A dynamical simulation facility for hybrid systems. In: Hybrid Systems, pp. 255–267. Springer (1993)
6. Bar-Or, R.L., Maya, R., Segel, L.A., Alon, U., Levine, A.J., Oren, M.: Generation of oscillations by the p53-mdm2 feedback loop: a theoretical and experimental study. Proc. Natl. Acad. Sci. **97**(21), 11250–11255 (2000)
7. Bartoszewicz, A., Patton, R.J.: Sliding mode control. Int. J. Adapt. Control Signal Process. **21**(8–9), 635–637 (2007)
8. Batchelor, E., Loewer, A., Mock, C., Lahav, G.: Stimulus-dependent dynamics of p53 in single cells. Mol. Syst. Biol. **7**(1), 488 (2011)
9. Brockett, R.W.: Hybrid models for motion control systems. In: Essays on Control, pp. 29–53. Springer (1993)
10. Casey, R., De Jong, H., Gouzé, J.-L.: Piecewise-linear models of genetic regulatory networks: equilibria and their stability. J. Math. Biol. **52**(1), 27–56 (2006)
11. Cortes, J.: Discontinuous dynamical systems. IEEE Control Syst. **28**(3), (2008)
12. Ferrell Jr., J.E.: Self-perpetuating states in signal transduction: positive feedback, double-negative feedback and bistability. Curr. Opin. Cell Biol. **14**(2), 140–148 (2002)
13. Geva-Zatorsky, N., Rosenfeld, N., Itzkovitz, S., Milo, R., Sigal, A., Dekel, E., Yarnitzky, T., Liron, Y., Polak, P., Lahav, G. et al.: Oscillations and variability in the p53 system. Mol. Syst. Biol. **2**(1), 2006.0033 (2006)
14. Gollu, A., Varaiya, P.: Hybrid dynamical systems. In: Proceedings of the 28th IEEE Conference on Decision and Control, pp. 2708–2712. IEEE (1989)
15. Gouzé, J.-L., Sari, T.: A class of piecewise linear differential equations arising in biological models. Dyn. Syst. **17**(4), 299–316 (2002)
16. Hagiwara, T., Araki, M.: Design of a stable state feedback controller based on the multirate sampling of the plant output. IEEE Trans. Autom. Control **33**, 812–819 (1988)
17. Hill, A.V.: The possible effects of the aggregation of the molecules of haemoglobin on its dissociation curves. J. Physiol. **40**, 4–7 (1910)
18. Jonak, K., Kurpas, M., Szoltysek, K., Janus, P., Abramowicz, A., Puszynski, K.: A novel mathematical model of atm/p53/nf-κ b pathways points to the importance of the ddr switch-off mechanisms. BMC Syst. Biol. **10**(1), 75 (2016)
19. Kardyńska, M., Paszek, A., Śmieja, J., Spiller, D., Widłak, W., White, M.R., Paszek, P., Kimmel, M.: Quantitative analysis reveals crosstalk mechanisms of heat shock-induced attenuation of nf-κb signaling at the single cell level. PLoS Comput. Biol. **14**(4), e1006130 (2018)
20. Klamka, J., Czornik, A., Niezabitowski, M.: Stability and controllability of switched systems. Bull. Polish Acad. Sci.: Tech. Sci. **61**(3), 547–555 (2013)
21. Kluk, P., Kogut, P., Milewski, A.: Automatycznie sterowany ultradźwiękowy generator mocy. Elektronika: konstrukcje, technologie, zastosowania **52**(9), 198–201 (2011)
22. Levant, A.: Sliding order and sliding accuracy in sliding mode control. Int. J. Control **58**(6), 1247–1263 (1993)
23. Liberzon, D.: Switching in Systems and Control. Springer Science & Business Media (2003)
24. Liberzon, D., Hespanha, J.P., Morse, A.S.: Stability of switched systems: a lie-algebraic condition. Syst. Control Lett. **37**(3), 117–122 (1999)

25. Liberzon, D., Morse, A.S.: Basic problems in stability and design of switched systems. IEEE Control Syst. **19**(5), 59–70 (1999)
26. Mestl, T., Plahte, E., Omholt, S.W.: A mathematical framework for describing and analysing gene regulatory networks. J. Theor. Biol. **176**(2), 291–300 (1995)
27. Michaelis, L., Menten, M.L.: Die kinetik der invertinwirkung. Universitätsbibliothek Johann Christian Senckenberg (2007)
28. Ochab, M., Puszynski, K., Swierniak, A.: Application of the piece-wise linear models for description of nonlinear biological systems based on p53 regulatory unit. In: Proceedings of the XII National Conference on Applications of Mathematics in Biology and Medicine, pp. 85–90 (2016)
29. Ochab, M., Puszynski, K., Swierniak, A.: Reachability of the therapeutic target in the systems with parameters switch. In: Proceedings of the International Conference on Bioinformatics and Biomedical Engineering, pp. 573–584. Springer (2016)
30. Ochab, M., Puszynski, K., Swierniak, A.: Influence of parameter perturbations on the reachability of therapeutic target in systems with switchings. Biomed. Eng. Online **16**(1), 77 (2017)
31. Ochab, M., Puszynski, K., Swierniak, A., Klamka, J.: Variety behavior in the piece-wise linear model of the p53-regulatory module. In: Proceedings of the International Conference on Bioinformatics and Biomedical Engineering, pp. 208–219. Springer (2017)
32. Ochab, M., Swierniak, A., Klamka, J., Puszynski, K.: Influence of the stochasticity in threshold localization on cell fate in the plde-model of the p53 module. In: Proceedings of the Polish Conference on Biocybernetics and Biomedical Engineering, pp. 205–217. Springer (2017)
33. Ochab, M., Swierniak, A., Klamka, J., Puszynski, K.: Apoptotic regulatory module as switched control system—analysis of asymptotic properties. In: Proceedings of the 9th International Conference on Bioinformatics Models, Methods and Algorithms, pp. 119–126, (2018)
34. Ochab, M., Swierniak, A., Puszynski, K.: Analysis of the bifurcation behavior of simple biological production-degradation system with switchings. In: Proceedings of the XXIII National Conference on Applications of Mathematics in Biology and Medicine, pp. 131–136 (2017)
35. Passino, K.M., Michel, A.N., Antsaklis, P.J.: Lyapunov stability of a class of discrete event systems. IEEE Trans. Autom. Control **39**(2), 269–279 (1994)
36. Perrimon, N., McMahon, A.P.: Negative feedback mechanisms and their roles during pattern formation. Cell **97**(1), 13–16 (1999)
37. Perruquetti, W., Barbot, J.P.: Sliding Mode Control in Engineering, vol. 11. M. Dekker (2002)
38. Plahte, E., Mestl, T., Omholt, S.W.: Global analysis of steady points for systems of differential equations with sigmoid interactions. Dyn. Stab. Syst. **9**(4), 275–291 (1994)
39. Puszyński, K., Hat, B., Lipniacki, T.: Oscillations and bistability in the stochastic model of p53 regulation. J. Theor. Biol. **254**(2), 452–465 (2008)
40. Puszyński, K., Jaksik, R., Świerniak, A.: Regulation of p53 by sirna in radiation treated cells: Simulation studies. Int. J. Appl. Math. Comput. Sci. **22**(4), 1011–1018 (2012)
41. Salomon, E.P., Berg, L., Martin, D., Ville, C.: Biologia. MULTICO, Warszawa (1996)
42. Sontag, E.D.: Stability and stabilization: discontinuities and the effect of disturbances. In: Nonlinear Analysis, Differential Equations and Control, pp. 551–598. Springer (1999)
43. Sun, Z.: Switched Linear Systems: Control and Design. Springer Science & Business Media (2006)
44. Swierniak, A., Kimmel, M., Smieja, J., Puszynski, K., Psiuk-Maksymowicz, K.: System engineering approach to planning anticancer therapies. Springer (2016)
45. Utkin, V.I.: Sliding mode control design principles and applications to electric drives. IEEE Trans. Industr. Electron. **40**(1), 23–36 (1993)
46. Varaiya, P.: Smart cars on smart roads: problems of control. IEEE Trans. Autom. Control **38**(2), 195–207 (1993)
47. Vousden, K.H., Lu, X.: Live or let die: the cell's response to p53. Nat. Rev. Cancer **2**(8), 594–604 (2002)
48. Wang, Z., Bauch, C.T., Bhattacharyya, S., d'Onofrio, A., Manfredi, P., Perc, M., Perra, N., Salathé, M., Zhao, D.: Statistical physics of vaccination. Phys. Rep. **664**, 1–113 (2016)

49. Zhu, J., Adli, M., Zou, J.Y., Verstappen, G., Coyne, M., Zhang, X., Durham, T., Miri, M., Deshpande, V., De Jager, P.L., et al.: Genome-wide chromatin state transitions associated with developmental and environmental cues. Cell **152**(3), 642–654 (2013)
50. Zilfou, J.T., Lowe, S.W.: Tumor suppressive functions of p53. Cold Spring Harb. Perspect. Biol. **00**(a001883), 1–13 (2009)

Modelling and Control of Heat Conduction Processes

Wojciech Mitkowski and Krzysztof Oprzędkiewicz

Abstract This chapter presents a few examples of heating problems described by a one-dimensional partial parabolic equation. Two examples concern the heating of a metal bar using two methods: with an electric heating element and by electric current flowing through such a bar. The next example concerns the solidification process and describes possibilities of controlling the phase transition front. The next subchapter is dedicated to a practical process of heating thin slabs in a heating furnace. A control algorithm is proposed which has been partially verified in practice. The chapter summary indicates systems of fractional order used for modelling of the heat processes. Other technical applications have also been mentioned.

1 Introduction

Heating problems are one of the most important issues in the technical sciences. In industrial processes production costs can be minimized by reducing energy consumption [8, 9, 26, 31, 66], which is mainly converted to heat. Parabolic partial differential equations with various boundary conditions are used for mathematical modelling of heat processes [18, 19, 71]. In practice, when designing appropriate temperature control systems it should be adequate to use linear, dimensional stationary models, e.g. in the form of adequate transfer function [50, 54, 67, 68]. Another class of model used for modelling of heat processes are systems with delay [56, 68, p. 71]. At the next design stage, for the chosen model class an automation engineer makes a synthesis of an appropriate type of controller, often a PID controller, also in a version discrete at the time allowing for a digital control [28, 68, p. 79]. An another idea is to use dynamic feedback control system, as it is presented in [33, 38].

W. Mitkowski (✉) · K. Oprzędkiewicz
Faculty of Electrical Engineering, Automatics, Computer Science and Biomedical Engineering, Department of Automatic Control and Robotics, AGH University of Technology, Kraków, Poland
e-mail: wojciech.mitkowski@agh.edu.pl

K. Oprzędkiewicz
e-mail: kop@agh.edu.pl

P. Kulczycki et al. (eds.), *Automatic Control, Robotics, and Information Processing*, Studies in Systems, Decision and Control 296,
https://doi.org/10.1007/978-3-030-48587-0_25

Temperature changes cause changes of properties of the heated material and can cause phase transitions. For a more detailed observation of the occurring phenomena more complex mathematical models are required, namely models with parameters spread over time in the form of partial differential equations of various types [19], also of fractional order in time or space [36, 49, 55]. The properties of the material depend on the method of heating it [35, 37, 73]. For instance, the type of obtained steel depends on the technological process at set temperature intervals. In some temperature ranges, the materials with memory return to their previously set shape [73], which can be used e.g. in medicine or during orthopaedic surgeries.

This chapter presents few examples of heating, considering a one-dimensional (spatially) parabolic equation. Previous results obtained by the authors have been used [43, 44]. Firstly, two examples from the paper [29] were discussed: the first concerns heating a metal bar with an electric heating element, the other heating by electric current flowing through the bar (electric conductor). Flowing through the resistive conductor of a finite length, the electric current causes heat emission and heats up the conductor, allowing appropriate plastometric measurements to be taken [29, 32, p. 184].

The next example [29, 37] concerns the solidification process (omitting the mass concentrations of individual components of the liquid and solid phase). This example indicates the possibilities of controlling the transition front [35, 45]. The problem of solidification (problem of changing the state of aggregation) can be modelled using parabolic partial differential equations with a moving limit. The alloy properties (here an alloy with memory [73]) depend on the crystallization front speed and on the final distribution of admixture in the crystal [16]. Only a simplified mathematical model of a one-dimensional solidification process is presented in this chapter [18, p. 178, 20] (see also [13, 71, p. 233]). An effective solidification process control system has been presented using a finitely dimensional approximation.

A practical process of heating thin slabs in a furnace has also been discussed and an appropriate control algorithm has been proposed [28, p. 49, 70].

This chapter summary indicates fractional order systems [17, 64] used for modelling of heat processes [25, 54]. Other technical applications have also been indicated. They are described in papers [2, 3, 21–24, 27, 31, 46, 69, 70].

2 Bar Heating

Let us consider a laboratory system for heating a thin bar [29, 32, p. 182, 52] which the schematic diagram is shown in Fig. 1. The laboratory system was built (and is still being developed) by K. Oprzędkiewicz at the AGH's Department of Automatics and Robotics.

A copper bar of length $l = 260$ mm and diameter 2 mm is heated on one end with an electric heating element (spiral wound on a ceramic insulator covering the bar at 1/13 of its length (for a detailed description of the laboratory rig see e.g. [40, 41, 43, 44]). The bar is not thermally insulated and the heat transfer takes place along

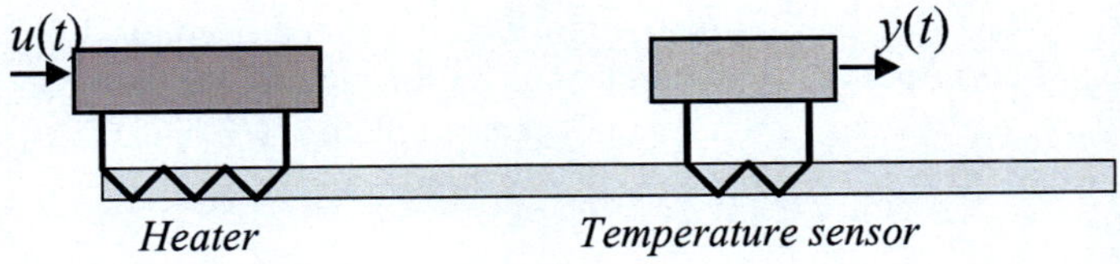

Fig. 1 Bar heating system

its entire length. The bar temperature is measured using a resistance temperature sensor wound at the interval [25*l*/52, 27*l*/52] of the bar length. The change of sensor resistance is read in a bridge system and converted to a current signal by an APU-11 transducer. The heating element is also controlled by a current signal, forming a current input u and current output y (see Fig. 1) in a 0–5 [mA] normalized range. The identification of the model parameters [15, 52] was performed using the least squares method (e.g. [11]).

The temperature distribution $x(z, t)$ in the bar can be described using the following equations:

$$\frac{\partial x(z,t)}{\partial t} = a\frac{\partial^2 x(z,t)}{\partial z^2} - R_a x(z,t) + b(z)u(t), \quad t \ge 0, \quad z \in [0,1],$$

$$\left.\frac{\partial x(z,t)}{\partial z}\right|_{z=0} = \left.\frac{\partial x(z,t)}{\partial z}\right|_{z=1} = 0, \quad t \ge 0,$$

$$x(z,0) = 0, \quad z \in (0,1), \; y(t) = \int_0^1 c(z)x(z,t)\mathrm{d}z. \tag{1}$$

The symbol $x(z, t)$ denotes the temperature at moment t in the point z. The characteristic functions $b(z)$ and $c(z)$ are determined by the construction of the system. These functions are as follows:

$$b(z) = \begin{cases} 1 & \text{for} \quad 0 \le z \le z_0 \\ 0 & \text{for} \quad z_0 < z \le 1, \end{cases} \tag{2}$$

$$c(z) = \begin{cases} \overline{c} & \text{for} \quad z_1 \le z \le z_2 \\ 0 & \text{for} \quad 0 \le z < z_1 \quad \text{and} \quad z_2 < z \le 1, \end{cases} \tag{3}$$

where $z_0 = \frac{1}{13}$, $z_1 = \frac{25}{52}$, $z_2 = \frac{27}{52}$, $\overline{c} = 25{,}7922$.

The boundary value problem (1) can be interpreted as an abstract problem (see e.g. [28, 34, pp. 175, 176]) in the Hilbert space X with a scalar input $u(t)$ and a scalar output $y(t)$

$$\dot{x}(t) = Ax(t) + Bu(t), \quad y(t) = Cx(t), \quad X = L^2(0,1;R), \quad t \ge 0. \tag{4}$$

Using the fact [34, p. 177] that appropriately defined operator A has a discrete spectrum and corresponding eigenvectors (eigenfunctions) constituting in X the basis of this space, the system parameters (2) and its solution $x(z, t)$ can be presented in the following form:

$$x(z,t) = \sum_{i=0}^{\infty} x_i(t) h_i(z). \tag{5}$$

In formula (4), it should be assumed that

$$A = \mathrm{diag}(\lambda_0, \lambda_1, \lambda_2, \ldots), \quad B = [b_0\ b_1\ b_2\ \ldots]^{\mathrm{T}},$$

$$C = [c_0\ c_1\ c_2\ \ldots], \quad x(t) = [x_0(t)\ x_1(t)\ x_3(t)\ldots]^{\mathrm{T}}, \tag{6}$$

where

$$b_i = \int_0^1 b(z)h_i(z)dz, \quad c_i = \int_0^1 c(z)h_i(z)\mathrm{d}z, \tag{7}$$

$$\lambda_i = -i^2\pi^2 a - R_a, \quad i = 0, 1, 2, \ldots, \tag{8}$$

and $a = 0.000945$, $R_a = 0.0271$. The eigenvectors of the operator A have the form as follows:

$$h_i(z) = \begin{cases} 1 & \text{for} \quad i = 0, \\ \sqrt{2}\cos(i\pi z) & \text{for} \quad i = 1, 2, 3, \ldots \end{cases} \tag{9}$$

For the mathematical model of the form (4) with parameters (6), it can be easily proved [34, pp. 232–236, 241] that the bar heating process (1) can be stabilized with the use of a finite dimensional feedback (continuous or discrete over time). The parameters of the finite dimensional stabilizer can be determined using the finite dimensional techniques.

An example of the bar heating control system design is schematically presented in Fig. 2.

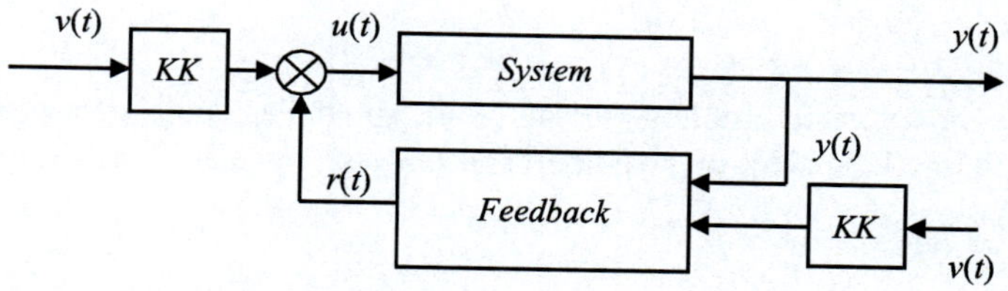

Fig. 2 Closed-loop control system with dynamic feedback

After elementary calculations we can see that the asymptotically stable control system shown in Fig. 2 has the following property:

$$\text{when} \quad v(t) = \overline{v} = \text{const.}, \quad \text{then} \quad y(t) \to \overline{v} \quad \text{for} \quad t \to \infty. \tag{10}$$

Without going into details (discrete version see [44]), the dynamically stabilizing feedback depends on two parameters K_1 and G_1. For example, for KK = 1.2310, K_1 = –0.3003, G_1 = 9.8058 the stabilizing feedback is as follows:

$$\begin{bmatrix} \dot{w}_1(t) \\ \dot{w}_2(t) \end{bmatrix} = \begin{bmatrix} -10.0231 & 0 \\ -0.0324 & -0.0358 \end{bmatrix} \begin{bmatrix} w_1(t) \\ w_2(t) \end{bmatrix} + \begin{bmatrix} 9.8058 & 0.0769 \\ 0 & 0.1077 \end{bmatrix} \begin{bmatrix} y(t) \\ v(t) \end{bmatrix},$$

$$r(t) = -0.3003 w_1(t). \tag{11}$$

and guarantees that the condition (10) is satisfied. The constant KK is determined from the appropriate static characteristics of the system. Appropriate simulation experiments were run for $v(t) = 1(t)$. The results are shown in Fig. 3. Circle denotes the time diagram $y(t)$ for an open system (without feedback) at KK = 0.9344, asterisk denotes the time diagram $y(t)$ for the system at KK = 1.2310, $K_1 = -0.3003$, $G_1 =$ 9.8058 and cross denotes the time diagram $y(t)$ for the system at KK = 1.8730, K_1 = –0.9503, G_1 = 9.8058.

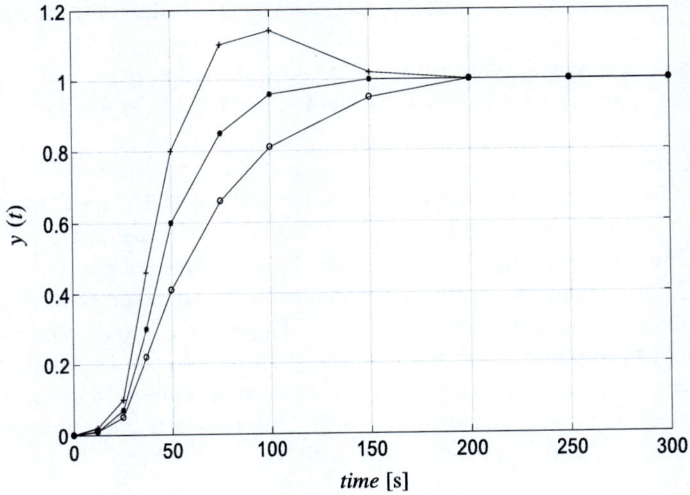

Fig. 3 The step response of the system for three sets of gain values

3 Conductor Heating with Electric Current

Consider heating with a current $I(t)$ flowing through a resistive conductor (see [29, 32, p. 183]) of length L. The system scheme is shown in Fig. 4. The amount of heat [kcal] generated as a result of the current flow of intensity I [A] through a heating conductor of resistance R [Ω] in time t [h] is calculated according to the formula $Q = 860RI^2t$. The conductor resistance changes depending on the temperature change x according to the relationship $R(x) = R_0(1 + \alpha x)$, where α is the resistance temperature coefficient, R_0—resistance at temperature 0 °C. The temperature distribution $x(z, t)$ in the conductor can be described by the following equation (z is a spatial variable, t is time):

$$a_p \frac{\partial x(z,t)}{\partial t} = \frac{\partial^2 x(z,t)}{\partial z^2} + Q(z,t),$$

$$x(z,0) = x_p, \quad t \geq 0, x(0,t) = x_p, \quad x(L,t) = x_p, \quad 0 \leq z \leq L. \tag{12}$$

The density of heat sources is a nonlinear function described as follows [29, 32, p. 184]:

$$Q(z,t) = 860b(z)R_0[1 + \alpha x(z,t)]I(t)^2. \tag{13}$$

where $b(z)$ is an appropriate function described by (2) with a compact carrier. Function $b(z)$ was determined experimentally in order to strengthen the heating in the middle of the conductor (for $z_s = L/2$). Also, function $b(z)$ allows accounting for complex thermal effects caused by the nonlinearity of the thermal conductivity coefficient.

Numerical calculations were made using an appropriate differential scheme obtained after discretization of spatial variable z and time t. The results were remarkably good and were confirmed by practical experiments. The details are described in [29]. The relevant computer simulation results are presented in Figs. 5, 6 and 7. Figure 5 shows temperature trends $x(z, t_k)$ along the conductor at the final moment of control $t_k = 2$ and at the initial moment t_0. The control was effected by changing the intensity of the current $I(t)$ flowing through the conductor. Figure 6 presents the temperature diagram $x(L/2, t)$, and Fig. 7—the control current diagram $I(t)$.

All calculations were made for $w = 2$. This parameter determines the width (carrier) of nonlinear function $b(z)$, and function $b(z)$—the impact of the nonlinearity

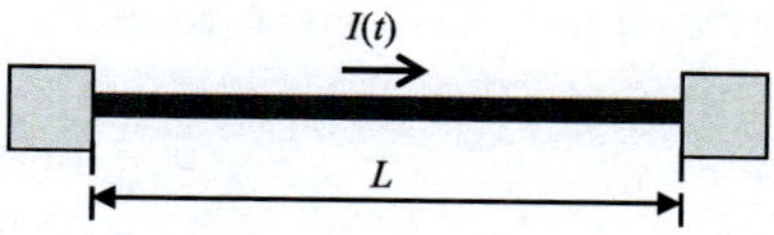

Fig. 4 Heating a thick conductor with electric current

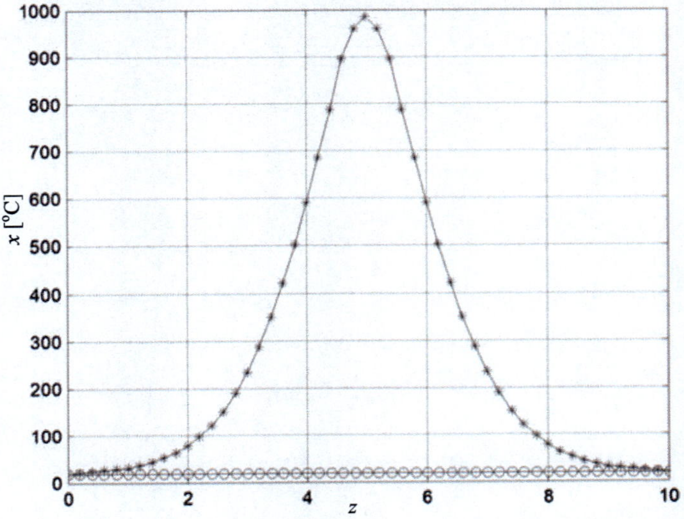

Fig. 5 Initial (circle) and final (asterisk) temperature distribution in the electrically heated conductor

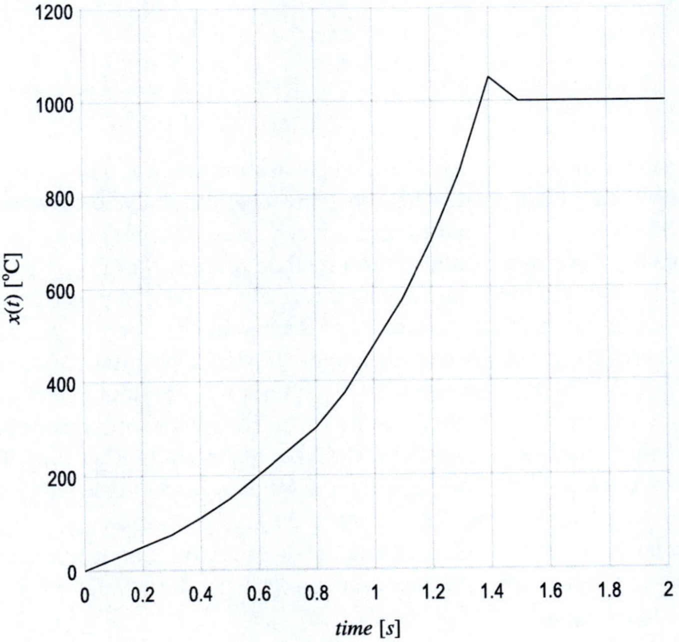

Fig. 6 Temperature trend in the middle point of the resistive conductor

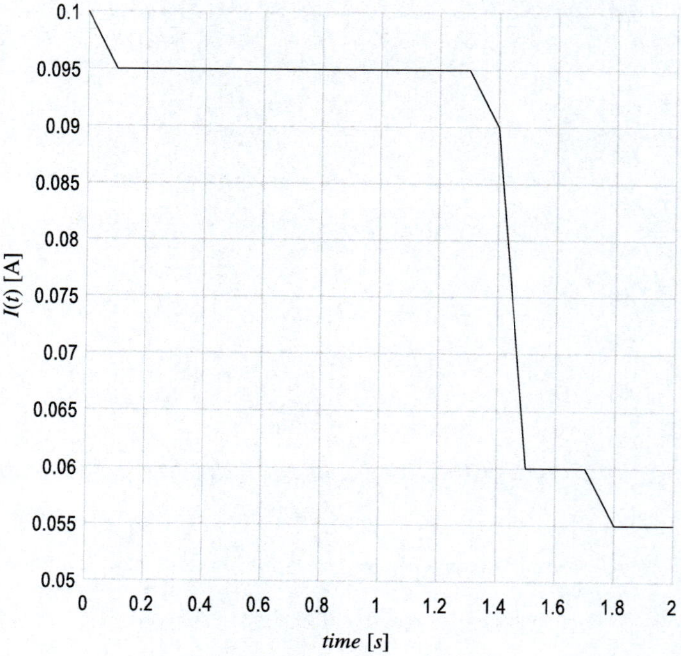

Fig. 7 Temperature stabilizing control $I(t)$ in the middle of the conductor at 1000 °C

of the thermal conductivity coefficient on the heating process. The coefficient w was identified with the least squares method using measurements on a real object.

Mathematical optimum control methods can be used for searching for the control algorithms [4]. An experimental control method presented in Fig. 7 was proposed, and the results were temperature changes as in Fig. 5. Then, a preset temperature of 1000 °C was maintained in the middle of the conductor (see Fig. 6). Temperature stabilization in the middle of the bar length allowed appropriate plastometric tests to be conducted. The temperature of both bar ends was constant (room temperature, about 20 °C), hence the supplied energy caused a significant temperature increase in the middle of the bar, leading the material to its fusibility limit [46]. The heating process was controlled with a current $I(t)$ in order to increase the bars plasticity but not to cause its total melting. The temperature stabilization at $z = L/2$ guaranteed that the conductor will not melt as a result of an excessive temperature. These works have been continuing with the AGH metallurgists in the laboratory (e.g. with M. Pietrzyk and M. Suliga in 2003, see also [72]).

4 Controlling the Phase Transition Front

Let us consider the following task (single-layer Stefan problem with free boundary [20, p. 178, 29, 30]): we are looking for the function of time $g(t)$, $g(0) = z_N$ (where z_N is the initial position of the phase transition front $g(t)$) and the function of time and spatial variable $x(z, t)$ describing the temperature, so that the following conditions are satisfied in the area $0 \le z \le g(t)$, $t \ge 0$ and in its boundary:

$$\frac{\partial x(z,t)}{\partial t} = a^2 \frac{\partial^2 x(z,t)}{\partial z^2}, \quad z \in [0,\ g(t)], \quad t \ge 0, \quad g(0) = z_N,$$

$$\lambda \frac{\partial x(0,t)}{\partial z} = -\alpha\big[u(t) - \gamma x(0,t)\big], \quad x(z,0) = \varphi(z), \quad z \in [0,\ z_N],$$

$$k \frac{\partial x(z,t)}{\partial z} |z = g(t) = \rho q \frac{\mathrm{d}g(t)}{\mathrm{d}t}, \quad x(g(t), t) = 0, \tag{14}$$

where $a^2 = k/(c\rho)$ is the thermometric conductivity coefficient, λ—heat exchange coefficient, α—external conductivity coefficient, ρ and q are the phase transition heat coefficient and density of the solidifying substance respectively, c is specific heat, k—thermal conductivity. The auxiliary coefficient γ is either 1 or 0, depending on the considered boundary condition.

If $\varphi(z)$ and $u(t) - x(0, t) < 0$, the conditions (14) can describe a simplified solidification process of a layer of initial thickness z_N at constant phase transition temperature equal to zero.

In order to perform the numerical calculations, let us discretize the spatial variable with a constant step $h = z_N/N$, where N is a preset natural number. The appropriate situation is shown in Fig. 8. Further forming of the appropriate differential scheme is described in detail in [29]. The initial position of the phase transition front $g(t)$ (e.g. between the solid and liquid phase) is designated z_N. Let us introduce designations $z_i = ih$ and let the discrete time moment t_j designate such value t for which the solidification front $g(t_j) = z_{N+j}$. Further, let $x[i, j] = x(ih, t_j) = x[j]$, where $t_j = t_{j-1} + \tau(j-1)$, $t_0 = 0$, $j = 1, 2, \ldots, NN$ after rearranging [29] from (14),

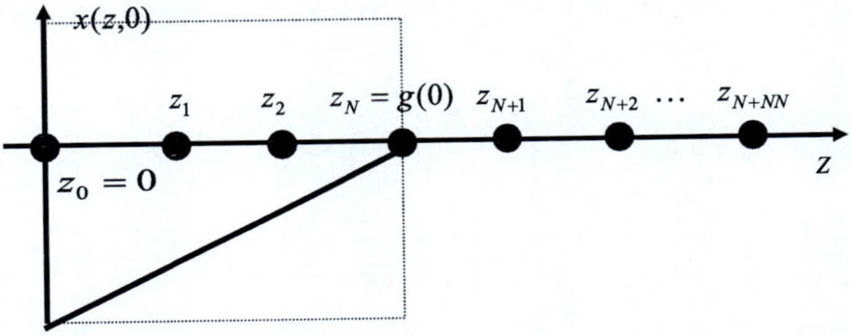

Fig. 8 Initial temperature distribution

the following differential diagrams were obtained:

$$x[j] = [I - \tau(j-1)A]^{-1}[x[j-1] + \tau(j-1)Bu[j]], \tag{15}$$

$$\tau(j-1) = -\frac{\rho q h^2/k}{x[N+j-2, j-1]}, \tag{16}$$

where $j = 1, 2, \ldots, NN, x[0]$ is set by the initial distribution $\varphi(z) < 0$. For example, for $N + j - 1 = 7,\ j = 4,\ N = 4$ we have

$$A = \frac{a^2}{h^2}\begin{bmatrix} F-2 & 1 & 0 & 0 & 0 & 0 & 0 \\ 1 & -2 & 1 & 0 & 0 & 0 & 0 \\ 0 & 1 & -2 & 1 & 0 & 0 & 0 \\ 0 & 0 & 1 & -2 & 0 & 0 & 0 \\ 0 & 0 & 0 & 0 & 0 & 0 & 0 \\ 0 & 0 & 0 & 0 & 0 & 0 & 0 \\ 0 & 0 & 0 & 0 & 0 & 0 & 0 \end{bmatrix}, B = \frac{a^2}{h^2}\begin{bmatrix} G \\ 0 \\ 0 \\ 0 \\ 0 \\ 0 \\ 0 \end{bmatrix}, \tag{17}$$

where

$$F = \frac{\lambda}{\lambda + \alpha h \gamma}, \quad G = \frac{\alpha h}{\lambda + \alpha h \gamma}. \tag{18}$$

It is also possible to control the phase transition front by an appropriate change of temperature $u(t)$. Let us consider a single-layer phase transition problem described by conditions (14). The layer of initial thickness z_N is cooled on the left-hand side with control $u(t) < 0$. The control system shown in Fig. 9 can be used to control the phase transition. Such a control system for temperature stabilization along the bar was proposed in [63]. It is worth mentioning that the proposed control system was obtained using the method of Lyapunov functionals. In our case the controller has the following form:

$$u(t) = Kx(0, t), \quad K > 0. \tag{19}$$

The examples of numerical calculations were made and are presented below for $a = 1, \lambda = 1, \alpha = 1, k = 1, \rho = 1, q = 1, z_N = 1$, where $h = z_N/N$. In the

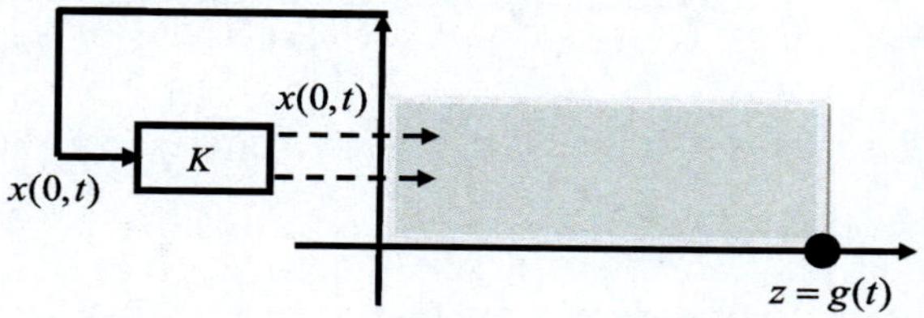

Fig. 9 Crystallization front control

system (14) the constant $\gamma = 1$ and $u(t) = Kx(0, t)$. Figures 10 and 11 present the results of front position simulations received from the differential scheme (15), (16) and (19) at $N = 5$, $NN = 20$, for $K = 1.5$ and $K = 1.8$, respectively.

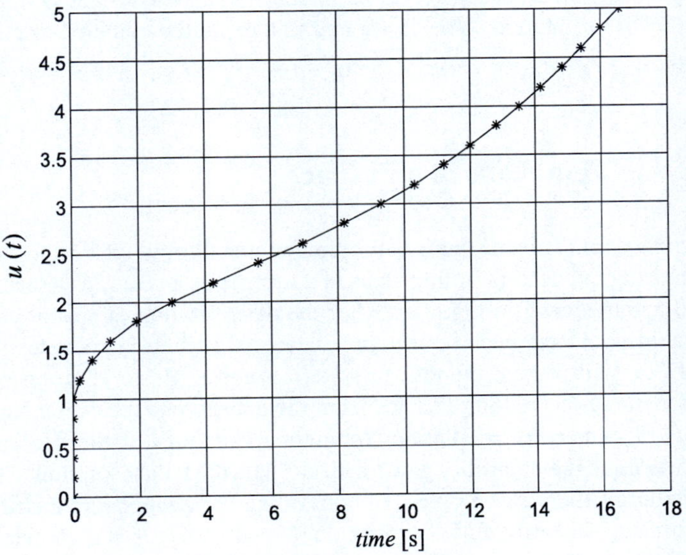

Fig. 10 Crystallization front as a function of time for $K = 1.8$

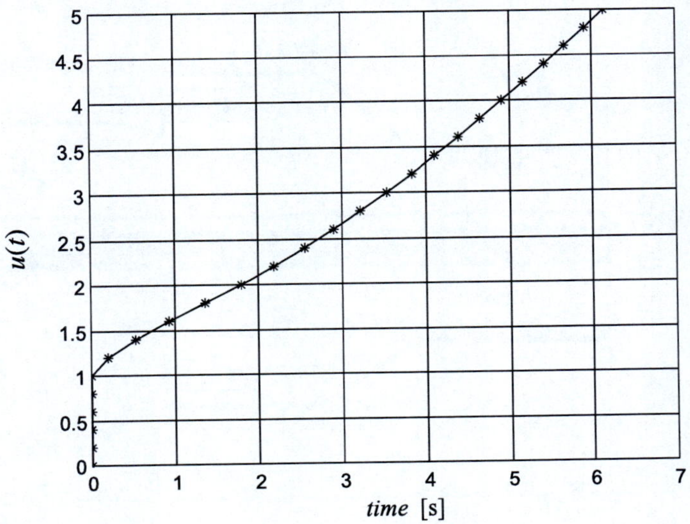

Fig. 11 Crystallization front as a function of time for $K = 1.5$

The results shown in Figs. 10 and 11 indicate that the solidification front speed depends on the gain coefficient of the controller K. The changes of solidification front speed have a fundamental impact on structure defects of the solidifying material.

More detailed simulations showed that the correct results can be obtained only in a certain range of K changes. Further studies are needed because the numerical scheme (16) is unstable in some cases, and as a result the obtained calculations are erroneous.

5 Heating Thin Slabs in a Furnace

The practical application of the L-Q problem in metallurgy [28, 35, pp. 49–61] will be shown using the heating of thin slabs in a furnace as an example (see the heating furnace diagram shown in Fig. 12). After the discretization of spatial variable z, a finite-dimensional model will be used to describe a distributed parameter object (L-Q problem). We will present the task of state restoration (restoration of temperature distribution along the heating furnace from the appropriate measurement data) and the results of numerical calculations (numerical solutions of the Riccati equation used to determine the controller gain) with the use of Matlab/Simulink.

In metallurgy, there is a problem of controlling the temperature distribution in a heating furnace. A heating furnace example [5, p. 432, 6, p. 66] is shown in Fig. 12. At the inlet to the furnace are slabs fed from e.g. continuous steel casting system. Such systems have been installed in recent years in many steel works (e.g. Sendzimir Steelworks in Krakow, Poland), allowing slabs of identical dimensions to be obtained

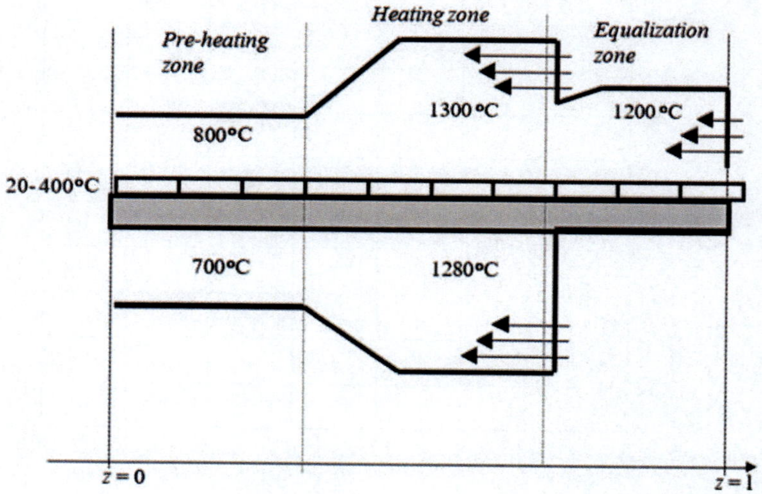

Fig. 12 Heating furnace diagram

which facilitates a further heating process. We are interested in temperature stabilization $x(l, t)$ at the outlet from the heating furnace (e.g. furnace of length $l = 27$ m, width 20 m and slab dimensions $6\,\text{m} \times 0.6\,\text{m} \times 0.15\,\text{m}$). In case of "thin" slabs, the following partial differential equation ("transport" equation) is a good mathematical model of the process [5, pp. 14, 15, 6, p. 64, [12, pp. 283]:

$$b(z,t)\frac{\partial x(z,t)}{\partial t} + b(z,t)v(t)\frac{\partial x(z,t)}{\partial z} + x(z,t) = u(z,t),$$
$$x(0,t) = x^p(t), \quad x(z,0) = x^0(z), \tag{20}$$

where z is a one-dimensional spatial variable, t is time moment, $b(z, t)$—appropriate thermal conductivity coefficient, $v(t)$—slab transfer speed through the furnace, u—temperature distribution control $x(z, t)$ in space and time obtained by adjustment of flame temperature of gas burners inside the furnace. The Biot number Bi $= \alpha S/\lambda$ is used to determine the slab thickness, where α is the convective heat-transfer coefficient, S—slab thickness, a λ—thermal conductivity coefficient. The slab can be called thin if Bi ≤ 0.25 [5, p. 15, 6, p. 64].

In book [5] there are proposed more complex mathematical models of slab heating in a furnace, accounting for e.g. three spatial variables. At least two approaches are used in the applications: build a very accurate mathematical model or build a model as simple as possible. The use of accurate models is now possible as a result of the huge computing power of computers. A simplified model, which does not lose the important properties of the real object, is interesting cognitively (insight into the essence of things) and allows us to more effective use control algorithms in real time.

The goal of control is to stabilize the temperature inside the furnace and at its outlet according to the present temperature distribution. Let g be an approximate desirable temperature distribution (at selected points along the furnace length). The diagram of the control system achieving our goal is presented in Fig. 13. The synthetizing function has the following form [14, pp. 150,160, 164, 171, 175]:

$$u(t) = -R(t)^{-1}B(t)^{\mathrm{T}}\{K(t)[x(t) - \beta(t)] + k(t)\}, \tag{21}$$

and parameters $K(t)$, $k(t)$ and $\beta(t)$ of feedback (21) will be specified below. Other parameters $f(t)$, $g(t)$ and e of the control system shown in Fig. 13 will be specified below too.

In expression (21), $K(t)$ is the solution of the Riccati differential equation:

$$\dot{K}(t) = K(t)B(t)R(t)^{-1}B(t)^{\mathrm{T}}K(t) - A(t)^{\mathrm{T}}K(t) - K(t)A(t) - C(t)^{\mathrm{T}}C(t),$$

$$K(T_k) = E^{\mathrm{T}}E, \quad t \in [0, T_k], \quad K(t)^{\mathrm{T}} = K(t) \geq 0. \tag{22}$$

The time function $k(t)$ present in equality (21) is specified by the equation

$$\dot{k}(t) = [K(t)B(t)R(t)^{-1}B(t)^{\mathrm{T}} - A(t)^{\mathrm{T}}]k(t) - C(t)^{\mathrm{T}}[g(t) + C(t)\beta(t)],$$

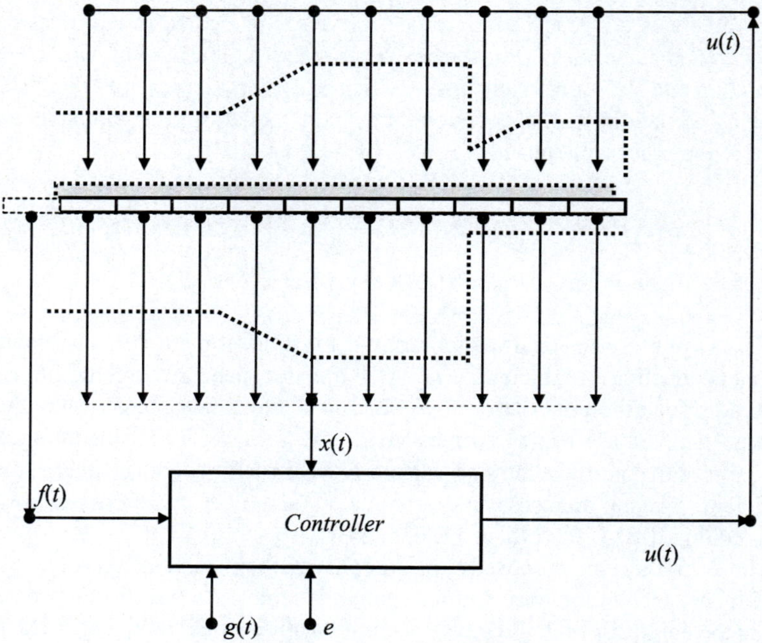

Fig. 13 Control system diagram

$$k(T_k) = E^{\mathrm{T}}[E\beta(T_k) + e], \tag{23}$$

and the function $\beta(t)$ is the solution of the following differential equation:

$$\dot{\beta}(t) = A(t)\beta(t) + f(t), \quad \beta(0) = 0. \tag{24}$$

The equality (21) defines an affine, non-stationary optimal controller in the sense of minimization (e.g. [14, pp. 179–182]) of the following cost function:

$$J(u; T_k) = E \cdot x(T_k) + e^2 + \int_0^{T_k} [C(t)x(t) + g(t)^2 + \sqrt{R(t)}u(t)^2]\mathrm{d}t, \tag{25}$$

at limitations constituting a finitely dimensional approximation of the model (20) in the form of the following linear differential equation:

$$\dot{x}(t) = A(t)x(t) + B(t)u(t) + f(t). \tag{26}$$

It is intuitively understood that our control goal will be reached by minimizing the cost function (25). Depending on the method of approximation of Eq. (20), we can obtain various finitely dimensional approximate models of the form (26). In further discussion, the magnitude $x_i(t) = x(ih, t)$ can be approximately considered as the slab temperature inside the furnace in point $z = ih, h > 0,\ i = 0, 1, 2, \ldots, n,\ h = l/n$ and at time t. Thus, we have $x(t) = [x_1(t) \ldots x_n(t)]^{\mathrm{T}},\ u_i(t) = u(ih, t),\ u(t) = [u_1(t) \ldots u_n(t)]^{\mathrm{T}}$. The function $f(t)$ in the approximated model (26) depends on the slab temperature $x^p(t)$ at the inlet to the furnace. At set $x^p,\ x^O$ and u, the Eq. (26) has an unequivocally determined solution x. In equality (25), defining the cost function, T_k is the final control time, real matrix $R(t) > 0$ (positive definite matrix) is a symmetric, regular matrix of time, E and $C(t)$ are matrices of appropriate dimensions, e is a vector, g is a vector function, e.g. continuous in time.

For instance, for $n = 3$ (which does not limit the generality of discussion, but simplifies the equations) it can be assumed that:

$$E = C(t) = \begin{bmatrix} 0 & 0 & 0 \\ 0 & 0 & 0 \\ 0 & 0 & -\gamma \end{bmatrix}, \quad g(t) = \begin{bmatrix} 0 \\ 0 \\ x^{\bullet}(t) \end{bmatrix}, \quad e = g(T_k), \tag{27}$$

where $x^{\bullet}$ in the present temperature curve at the outlet from the furnace. In equalities (27), for physical reasons, it can also be assumed that $C(t) = \begin{bmatrix} 0 & 0 & -\gamma \end{bmatrix}$, but generally it is easier to use a square matrix (larger possibility of interpreting the indicator (25)). Similarly, for $n = 3$ in equality (26), using appropriate difference quotients, it can be assumed [24, 28]:

$$A(t) = \begin{bmatrix} a_1(t) & 0 & 0 \\ c(t) & a_2(t) & 0 \\ 0 & c(t) & a_3(t) \end{bmatrix}, \quad B(t) = \begin{bmatrix} b_1(t)^{-1} & 0 & 0 \\ 0 & b_2(t)^{-1} & 0 \\ 0 & 0 & b_3(t)^{-1} \end{bmatrix}, \tag{28}$$

where

$$a_i(t) = -\frac{1}{b_i(t)} - c(t), \quad c(t) = \frac{v(t)}{h}, \quad b_i(t) = b(ih, t), \tag{29}$$

and

$$f(t) = \begin{bmatrix} c(t) & 0 & 0 & 0 & 0 \end{bmatrix}^{\mathrm{T}} x^p(t). \tag{30}$$

Based on the function $f(t)$ dependent on the temperature measured at the outlet from the furnace, the measurement of temperature $x(t)$ along the furnace and based on the known, preset desirable temperature distribution $g(t)$ and the set temperature at the outlet from the furnace e, the controller (21) generates the control $u(t)$ specifying the temperature of the flames of the burners placed along the furnace length. The optimal control system diagram is shown in Fig. 14.

The temperature measurement $x(t)$ along the whole furnace is not possible in practice. The temperature is measured only in selected spatial points. The system output is available for measurements, e.g. $y(t) = C(t)x(t) = x_n(t)$. In such a case,

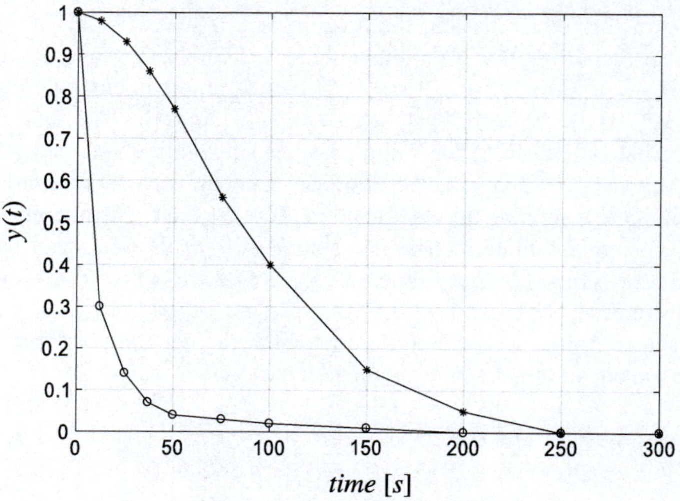

Fig. 14 Output $y(t) = x_n(t)$ from an open-loop (asterisk) and closed-loop system (circle)

the system state can be relatively easily reconstructed asymptotically using dynamic systems called Luenberger observers (e.g. [34, p. 88]). Based on the measurement of control $u(t)$ and output $y(t) = C(t)x(t)$, the observer generates a state estimate $\hat{x}(t)$, where $\hat{x}(t) \to x(t)$, when $t \to \infty$. In this case, the controller (21) can be replaced with the following feedback:

$$u(t) = -R(t)^{-1}B(t)^{\mathrm{T}}\{K(t)[\hat{x}(t) - \beta(t)] + k(t)\}. \tag{31}$$

Very simple results are obtained in the stationary case when $T_k \to \infty$. If in (25) and (26) relevant matrices and vectors $A, B, C, R, E = 0, g, e = 0$ are real and constant and the pair $(A; B)$ is stabilizable and the pair $(C; A)$ is detectable, then [14, pp. 179–182]:

$$K(t) \to K = \text{const}, \quad \text{when} \quad T_k \to +\infty, \tag{32}$$

where $K = \text{const}$ is the only solution of the Riccati algebraic equation:

$$KBR^{-1}B^{\mathrm{T}}K - A^{\mathrm{T}}K - KA - C^{\mathrm{T}}C = 0 \quad \text{so that } K = K^T \geq 0, \tag{33}$$

and the state matrix of a closed system is exponentially stable (the eigenvalues of the state matrix of a closed system have negative real parts), so

$$\mathrm{Re}\lambda\left[A - BR^{-1}B^TK\right] < 0. \tag{34}$$

Examples of numerical calculations are shown for a stationary case. The numerical solution of the algebraic Riccati equation (33) was obtained with the use of

Matlab/Simulink. The calculation assumptions were $E = 0, \gamma = -1, l = 1, b = 100, v = 0, 1, R = 1$ (see (25)–(30)) at the appropriate, following constant values

$$a_i(t) = a = -\frac{1}{b} - c, \quad c = \frac{v}{h}, \quad h = l/n, x_i(0) = 1. \tag{35}$$

As a result of the numerical solution of the algebraic Riccati equation (33) for $n = 3$ was obtained. The K matrix takes the following form:

$$K = \begin{bmatrix} 0.0351 & 0.0427 & 0.0281 \\ 0.0427 & 0.0924 & 0.1506 \\ 0.0281 & 0.1506 & 0.7257 \end{bmatrix}, \quad K = 0.7615, \tag{36}$$

and the eigenvalues of matrix $A - BR^{-1}B^TK$ of the closed-loop system are as follows: $\{-1.0035, -0.3899 + 0.1066j, -0.3899 - 0.1066j\}$. These eigenvalues have negative real parts, so the closed-loop system is asymptotically stable.

Figure 14 presents the temperature diagrams $y(t) = x_n(t)$ at the furnace outlet, respectively for $u(t) = 0$ (asterisk) and for optimal control (circle) in the sense of cost function (25). The control signal takes the form: $u_o(t) = -R^{-1}B^TKx_o(t)$ for $x^{\bullet}(t) = 0$. The cost function (25) is equal: $J(0) = 6.3607$ and $J(u_o) = 0.6478$, respectively. As we can see, the applied control system significantly improves (reduces) the cost function (25) with appropriate coefficients.

6 Fractional Order Models

Thermal processes are still a focus of interest for the researchers using the results of applied sciences. The development of IT tools allows to employ new classes of mathematical models that describe the thermal and other processes, e.g. related to diffusion or diffusion and turbulence. A recent trend involves a return to the models in the form of various types of differential equations of fractional order. The integral and differential calculus of fractional order was most likely introduced and used as early as 1695–1822 by de l'Hospital (1661–1704), Leibniz (1646–1716), Newton (1643–1727) and Euler (1707–1783) and Laplace (1749–1827). The further development of the theory and applications of fractional order systems took place in the 19th and 20th centuries. As a result of development of IT tools, many researchers in recent years have taken interest in the differential equations of fractional order (e.g. [17, 36, 39, 47, 49–51, 55, 58, 60–62, 64]) which appear in various applications, such as transport of mass, various types of flows, temperature distributions in materials, diffusion, dispersion, supercapacitors [36] and many others [27]. Algorithms of numerical solution of fractional order systems have also been developed recently and they are generally available, e.g. in the Matlab/Simulink package. Various modelling methods are used for distributed-parameter systems [1, 12, 18, 19, 65], particularly e.g. electric chain systems are used for thermal systems modelling [25, 47]. Selected examples of diffusion equations discussed in the literature are given in [39, 49].

Fractional advection-dispersion equation is used for e.g. liquid flow modelling in porous materials. It has the following form:

$$\frac{\partial u(x,t)}{\partial t} = K_\alpha \frac{\partial^\alpha u(x,t)}{\partial |x|^\alpha} + K_\beta \frac{\partial^\beta u(x,t)}{\partial |x|^\beta},$$
$$0 \le t \le T, \quad 0 < x < L, \quad 1 < \alpha < 2, \quad 0 < \beta < 1, \quad K_\alpha \ne 0, \quad K_\beta \ne 0. \quad (37)$$

In partial differential equations of type (37), initial conditions $u(x, 0) = g(x)$ are additionally specified, so are the boundary conditions, e.g. $u(0, t) = u(L, t) = 0$. The designations are as follows: $u(x, t)$—dissolved substance concentration, K_α—coefficient of dispersion, K_β—average fluid velocity, x is a spatial variable, t is time.

The Riesz fractional equation is often used during the description of diffusion processes (see [49], similarly earlier [48], where experimental studies with the use of a thermal camera were described in detail):

$$\frac{\partial u(x,t)}{\partial t} = K_\alpha \frac{\partial^\alpha u(x,t)}{\partial |x|^\alpha},$$
$$0 \le t \le T, \quad 0 < x < L, \quad 1 < \alpha < 2, \quad K_\alpha \ne 0. \quad (38)$$

In addition, the Eq. (38) has the initial conditions $u(x, 0) = g(x)$ and the boundary conditions, e.g. $u(0, t) = u(L, t) = 0$. When the appropriate magnitudes are measured, using the identification methods we can determine e.g. the fractional order α of the considered system. Various modifications of the least squares method are usually applied.

There are studies (e.g. [55, 57, 59]) in which the mathematical model (1) is replaced with the following boundary problem of fractional orders α, β:

$$\frac{\partial^\alpha x(z,t)}{\partial t^\alpha} = a \frac{\partial^\beta x(z,t)}{\partial z^\beta} - R_a x(z,t) + b(z)u(t), \quad t \ge 0, \quad z \in [0,1],$$

$$\left.\frac{\partial x(z,t)}{\partial z}\right|_{z=0} = \left.\frac{\partial x(z,t)}{\partial z}\right|_{z=1} = 0, \quad t \ge 0,$$

$$x(z,0) = 0, \quad z \in (0,1),$$

$$y(t) = \int_0^1 c(z)x(z,t)\mathrm{d}z. \quad (39)$$

Fractional calculus can be also used to modify classic control systems using a PID controller. Replacing integer orders of integral and derivative actions by their

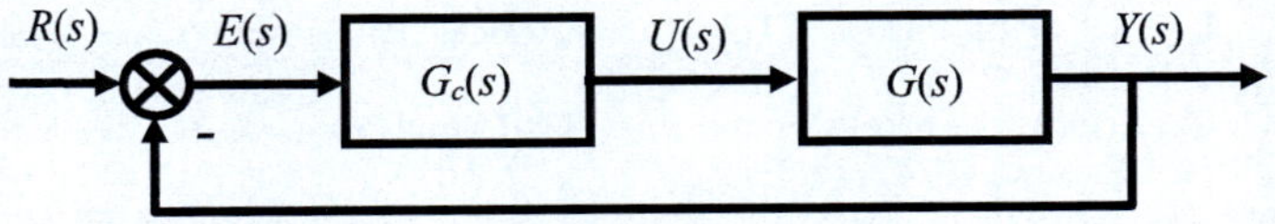

Fig. 15 Closed-loop control system with FOPID controller

fractional order equivalents we obtain the fractional order PID controller (FOPID). This allows us to improve the control quality due to the possibility to use two additional parameters to precisely tune a controller. Parameters of a FOPID controller can be estimated using biologically inspired optimization algorithms (see for example [53]). An example of such a control system is shown in Fig. 15. A more detailed system description is given in [50, 51], and discrete feedback in [41, 42].

In Fig. 15 $R(s)$, $E(s)$, $U(s)$ and $Y(s)$ are the Laplace transforms of reference value, error, control signal and process value respectively, $G(s)$ is the transfer function of a plant, $G_c(s)$ is the transfer function of FOPID controller.

The problem of designing the control system from Fig. 15 consists in determining the parameters (particularly the fractional order α and the fractional order β) of the controller transfer function:

$$G_c(s) = k_p + k_\alpha s^{-\alpha} + k_\beta s^\beta. \tag{40}$$

7 Final Remarks

The studies on the control of thermal processes in residential buildings are also interesting, e.g. [2, 7, 10]. Chain systems of various types, described e.g. in [7, 8, 10, 25, 36] are used to model the temperature distribution in buildings.

Acknowledgements This chapter is sponsored by AGH project 16.16.120.773.

References

1. Auer, A.: Analog Modeling of Distributed Parameters Processes. PWN, Warszawa (1976). (in Polish)
2. Baranowski, J., Długosz, M., Ganobis, M., Skruch, P., Mitkowski, W.: Applications of mathematics in selected control and decision processes. Mat. Stosow. Numer. Spec. **12**(53), 65–90 (2011)
3. Baranowski, J., Długosz, M., Mitkowski, W.: Remarks about DC motor control. Arch. Control Sci. **18**(LIV)(3), 289–322 (2008)
4. Bołtianski, W.G.: Mathematical Methods of Optimal Control. WNT, Warszawa (1971). (in Polish)

5. Butkowski, A.G.: Optimal Control Theory of Distributed Parameter Systems. Nauka, Moskwa (1965). (in Russian)
6. Ciepiela, A., Kozioł, R.: Complex Automation of Industrial Processes at the Example of Rolling Mill. WNT, Warszawa (1978). (in Polish)
7. Długosz, M.: Agregation of state variables in an RC model. Build. Ser. Eng. Res. Technol. **39**(1), 66–80 (2018)
8. Długosz, M., Chronowski, J., Baranowski, J., Piątek, P., Mitkowski, W., Skruch, P.: Wireless home automation system using Z-Wave standard. Pomiary Autom. Robot. **7–8**, 100–106 (2013). (in Polish)
9. Długosz, M., Mitkowski, W., Pietrzyk, W.: Modelowanie i sterowanie tlenowym procesem konwertorowym. Modeling and control of oxygen convertor process. Inform. Technol. Mater. **3**(2), 61–71 (2003). (in Polish)
10. Długosz, M., Skruch, P.: The application of fractional-order models for thermal process modelling inside buildings. J. Build. Phys. **39**(5), 440–451 (2016)
11. Eykhoff, P.: System Identification. Wiley, London (1974)
12. Friedly, J.C.: Dynamic Behavior of Processes. Prentice-Hall, New Jersey (1972)
13. Gosiewski, A., Niezgódka, M.: Mathematical models of a controlled crystallization process. Arch. Autom. Telekomun. **23**(3), 255–265 (1978). (in Polish)
14. Górecki, H., Fuksa, S., Korytowski, A.: Mitkowski W Optimal Control of Linear Systems with Square Performance Criterion. PWN, Warsaw (1983). (in Polish)
15. Grabowski, P.: Example of identification of the parabolic system. In: Mitkowski, W. (ed.) Models with Point Control and Observation. Prace z automatyki, pp. 181–186. Wydawnictwo AGH, Kraków (1997). (in Polish)
16. Holly, K., Danielewski, M.: Interdiffusion and free-boundary problem for r-component ($r \geq 2$) one-dimensional mixtures showing constant concentration. Phys. Rev. B **50**(18), 13336–13346 (1994)
17. Kaczorek, T.: Selected Problems in Fractional Systems Theory. Springer-Verlag, Berlin (2011)
18. Kącki, E.: Partial Differential Equations in Physics and Technics. WNT, Warszawa (1992). (in Polish)
19. Lions, J.L.: Selected Methods of Solution of Nonlinear Boundary Problems. Mir, Moskwa (1972). (in Russian)
20. Michlin, S.G., Smolicki, C.L.: Approximated Methods of Solution of Differential and Integral Equations. PWN, Warszawa (1970). (in Polish)
21. Mitkowski, P., Mitkowski, W.: Homogenization of one dimensional elliptic system. Comput. Methods Mater. Sci. (Inform. Technol. Mater.) **8**(3), 160–164 (2008)
22. Mitkowski, P.J., Mitkowski, W.: Ergodic theory approach to chaos: remarks and computational aspects. Int. J. Appl. Math. Comput. Sci. **22**(2), 259–267 (2012)
23. Mitkowski, P.J., Mitkowski, W.: Stepped basic function in asymptotic homogenization of elliptic system. Automatyka/Automatics **16**(1), 45–57 (2012)
24. Mitkowski, W.: Dynamic properties of chain systems with applications to biological models. Arch. Control Sci. **9**(XLV)(1–2), 123–131 (1999)
25. Mitkowski, W.: Finite-dimensional approximations of distributed RC networks. Bull. Polish Acad. Sci. Tech. Sci. **62**(2), 263–269 (2014)
26. Mitkowski, W.: Cost minimization of electrical energy deliver to resistance heating device. In: Gutenbaum, J. (ed.) Automatics, Control Management, pp. 269–282. IBS-PAN, Warszawa (2002). (in Polish)
27. Mitkowski, W.: Time-based process models. Hut. Wiad. Hut. **85**(1), 2–5 (2018). (in Polish)
28. Mitkowski, W.: Heating of thin slabs in heating stove. Automatyka **2**(2), 195–204 (1998). (in Polish)
29. Mitkowski, W.: Heating and phase transitions. In: Bańka, S. (ed.) Recent Techniques of Control and Measurements and Control Methods, pp. 149–160. Komitet AiR PAN and Wydawnictwo Politechniki Szczecińskiej, Szczecin (2006). (in Polish)
30. Mitkowski, W.: Numerical solution of single layer phase transition problem. In: Bubnicki, Z., Grzech, A. (eds.) Knowledge Engineering and Expert Systems, vol. 2, pp. 223–228. Oficyna Wydawnicza Politechniki Wrocławskiej, Wrocław (2000). (in Polish)

31. Mitkowski, W.: Remarks about energy transfer in an RC ladder network. Int. J. Appl. Math. Comput. Sci. **13**(2), 193–198 (2003)
32. Mitkowski, W.: Matrix Equations and Its Applications, wyd. 3. Wydawnictwo AGH, Kraków (2012). (in Polish)
33. Mitkowski, W.: Stabilization of linear infinite dimensional systems with the use of dynamic feedback. Arch. Autom. Telemech. **33**(4), 515–528 (1998). (in Polish)
34. Mitkowski, W.: Stabilization of Dynamic Systems. WNT, Warszawa (1991). (in Polish)
35. Mitkowski, W.: Dynamic Systems. Supplementary Materials to Lecture. Wydawnictwa Wydziału Elektrotechniki, Automatyki, Informatyki i Elektroniki AGH, Kraków (2000). (in Polish)
36. Mitkowski, W.: Non integer order systems and their selected applications. Hut. Wiad. Hut. **81**(1), 39–43 (2014). (in Polish)
37. Mitkowski, W.: Remarks about growth dynamic of ferrite grain in austenite. Prz. Mech. **LXII**(12), 29–31 (2003). (in Polish)
38. Mitkowski, W., Bauer, W., Zagórowska, M.: Discrete-time feedback stabilization. Arch. Control Sci. **27**(2), 309–322 (2017)
39. Mitkowski, W., Obrączka, A.: Simple identification of fractional differential equation. Solid State Phenom. **180**, 331–338 (2012)
40. Mitkowski, W., Oprzędkiewicz, K.: A sample time assign for a discrete interval parabolic system with the two-dimensional uncertain parameter space. In: Bubnicki, Z., Grzech, A. (eds.) Proceedings of the 15th International Conference on System Science, vol. II, pp. 322–330. Oficyna Wydawnicza Politechniki Wrocławskiej, Wrocław (2004) (see also: Syst. Sci. **30**(1) (2004))
41. Mitkowski, W., Oprzędkiewicz, K.: A sample time optimization problem in a digital control system. In: Korytowski, A., Malanowski, K., Mitkowski, W., Szymkat, M. (eds.) 23rd IFIP TC7 Conference on System Modeling and Optimization, Cracow, Poland, July 2007, Revised Selected Papers, pp. 382–396. Springer, Berlin (2009)
42. Mitkowski, W., Oprzędkiewicz, K.: Fractional-order P2D$^{\beta}$ controller for uncertain parameter DC motor. In: Mitkowski, W., Kacprzyk, J., Baranowski, J. (eds.) Advances in the Theory and Applications of Non-integer Order Systems; Proceedings of 5th Conference on Non-integer Order Calculus and Its Applications, Cracow, Poland; Lecture Notes in Electrical Engineering, vol. 257, pp. 249–259. Springer International Publishing (2013). https://doi.org/10.1007/978-3-319-00933-9
43. Mitkowski, W., Oprzędkiewicz, K.: Control of heating process with the use of fuzzy controller. Automatyka **5**(1/2), 429–437 (2001). (in Polish)
44. Mitkowski, W., Oprzędkiewicz, K.: Control of heating process. Automatyka **1**(1), 301–310 (1997). (in Polish)
45. Mitkowski, W., Pietrzyk, M.: The issue of moving border applied to modeling transformation austenite-ferrite. Prz. Mech. **LX**(4), 22–25 (2001). (in Polish)
46. Mitkowski, W., Skruch, P.: Control system of a rotating Timoshenko beam. Arch. Control Sci. **13**(XLIX)(3), 281–288 (2003)
47. Mitkowski, W., Skruch, P.: Fractional-order models of the supercapacitors in the form of RC ladder networks. Bull. Polish Acad. Sci. Tech. Sci. **61**(3), 581–587 (2013)
48. Obrączka, A., Kowalski, J.: Modeling of heat distribution in ceramic materials using non integer order differential equations. In: Materiały XV PPEEm 2012, Arch. PTETiS, vol. 32, pp. 133–135. Gliwice (2012). (in Polish)
49. Obrączka, A., Mitkowski, W.: The comparison of parameter identification methods for fractional, partial differential equation. Solid State Phenom. **210**, 265–270 (2014)
50. Oprzędkiewicz, K.: Approximation method for fractional order transfer function with zero and pole. Arch. Control Sci. **24**(LX)(4), 409–425 (2014)
51. Oprzędkiewicz, K.: Non integer order, state space model of heat transfer process using Caputo-Fabrizio operator. Bull. Polish Acad. Sci. Tech. Sci. **66**(3), 249–255 (2018)
52. Oprzędkiewicz, K.: Example of identification of the parabolic system. Elektrotechnika **16**(2), 99–106 (1997). (in Polish)

53. Oprzędkiewicz, K., Dziedzic, K.: A tuning of a fractional order PID controller with the use of particle swarm optimization method. In: Proceedings of 16th International Conference Artificial Intelligence and Soft Computing, ICAISC 2017 Zakopane, Poland, pp. 394–407 (2017)
54. Oprzędkiewicz, K., Gawin, E.: Non integer order, state space model for one dimensional heat transfer process. Arch. Control Sci. **26**(2), 261–275 (2016)
55. Oprzędkiewicz, K., Gawin, E., Mitkowski, W.: Modeling heat distribution with the use of a non-integer order, state space model. Int. J. Appl. Math. Comput. Sci. **26**(4), 749–756 (2016)
56. Oprzędkiewicz, K., Mitkowski, W.: Accuracy analysis for fractional order transfer function models with delay. In: Babiarz, A., Czornik, A., Klamka, J., Niezabitowski, M.' (eds.) Theory and Applications of Non-integer Order Systems; 8th Conference on Non-integer Order Calculus and Its Applications. Lecture Notes in Electrical Engineering, vol. 407, pp. 253–263. Springer International Publishing (2017). https://doi.org/10.1007/978-3-319-45474-0
57. Oprzędkiewicz, K., Mitkowski, W.: Memory-efficient, non integer order, discrete, state space model of heat transfer process. Int. J. Appl. Math. Comput. Sci. (AMCS) **28**(4), 649–659 (2018)
58. Oprzędkiewicz, K., Mitkowski, W,. Gawin, E.: Application of fractional order transfer functions to modeling of high-order systems. In: Proceedings of MMAR 2015, pp. 1169–1174 (2015)
59. Oprzędkiewicz, K., Mitkowski, W., Gawin, E.: An accuracy estimation for a non integer order, discrete, state space model of heat transfer process. In: Szewczyk, R., Zieliński, C., Kaliczyńska, M. (eds.) Innovations in Automation, Robotics and Measurement Techniques, Proceedings of Automation 2017, Warsaw, Poland, pp. 86–98. Springer (2017)
60. Oprzędkiewicz, K., Mitkowski, W., Gawin, E.: An estimation of accuracy of Oustaloup approximation. In: Szewczyk, R., Zieliński, C., Kaliczyńska, M. (eds.) Challenges in Automation, Robotics and Measurement Techniques. Proceedings of Automation 2016, Warsaw, Poland, pp. 299–307. Springer (2016)
61. Oprzędkiewicz, K., Mitkowski, W., Gawin, E.: The PLC implementation of fractional-order operator using CFE approximation. In: Szewczyk, R., Zieliński, C., Kaliczyńska, M. (eds.) Innovations in Automation, Robotics and Measurement Techniques, Proceedings of Automation 2017, Warsaw, Poland, pp. 22–33. Springer (2017)
62. Oprzędkiewicz, K., Stanisławski, R., Gawin, E., Mitkowski, W.: A new algorithm for CEF-approximated solution of a discrete-time noninteger-order state equation. Bull. Polish Acad. Sci. Tech. Sci. **65**(4), 429–437 (2017)
63. Parks, P.C., Pritchard, A.J.: On the construction and use of Liapunov functionals. In: Congress IFAC, Stability, Technical Session, vol. 20, pp. 59–76. Warszawa (1969)
64. Podlubny, I.: Fractional Differential Equations. Holt, Rinehart, and Winston San Diego, USA (1999)
65. Roszkiewicz, J.: Distributed Parameter RC Systems. WKiŁ, Warszawa (1972). (in Polish)
66. Sirazetdinow, T.K.: Optimization of Distributed-Parameter Systems. Nauka, Moskwa (1977). (in Russian)
67. Skoczowski, S.: Deterministic identification of transfer function models and its application. In: Gutenbaum, J. (ed.) Automatics, Control, Management, pp. 339–356. IBS-PAN, Warszawa (2002). (in Polish)
68. Skoczowski, S.: Technique of Temperature Control. PAK, Warszawa (2000). (in Polish)
69. Skruch, P., Mitkowski, W.: Optimum design of shapes using the pontryagin principle of maximum. Automatyka **13**(1), 65–78 (2009)
70. Swietlichnyj, D., Pietrzyk, M., Mitkowski, W.: Optimization of hot working parameters assuring desired microstructure using control theory. In: Palmiere, E.J., Mahfouf, M., Pinna, C. (eds.) Proceedings of the International Conference on Thermomechanical Processing: Mechanics, Microstructure and Control, pp. 453–460. The University of Sheffild, England (2002)

71. Tichonow, A.N., Samarski, A.A.: Equations of Mathematical Physics. PWN, Warszawa (1963). (in Polish)
72. Trebacz, L., Mitkowski, W., Pietrzyk, M.: Parameter sensitivity of the SICO test. In: Owen, D.R.J., Onate, E., Suarez, B. (eds.) Computational Plasticity. Proceedings of Fundamentals and Applications/COMPLAS VIII, CIMNE, Barcelona, vol. 2, pp. 995–998 (2005)
73. Żochowski, A.: Mathematical Problems in Shape Optimization and Shape Memory Materials. Peter Lang, Frankfurt (1992)

Active Suppression of Nonstationary Narrowband Acoustic Disturbances

Maciej Niedźwiecki and Michał Meller

Abstract In this chapter, a new approach to active narrowband noise control is presented. Narrowband acoustic noise may be generated, among others, by rotating parts of electro-mechanical devices, such as motors, turbines, compressors, or fans. Active noise control involves the generation of "antinoise", i.e., the generation of a sound that has the same amplitude, but the opposite phase, as the unwanted noise, which causes them to interfere destructively, rather than constructively. In the range of low frequencies (below 1 kHz), the active approach is more effective than passive methods that employ dampers, barriers, absorbers, and other forms of acoustic isolation.

1 Introduction

One of the negative aspects of the growth of technology is the increase of various risk factors, such as acoustic noise. Long-term exposure to noise may cause fatigue, loss of focus and insomnia, or, in case of high levels of the noise, partial or total hearing loss.

Traditional methods of reducing acoustic noise rely on measures such as hearing protections, attenuators, dampeners, or acoustic isolators, among others. Such solutions, which can be referred to as passive, are most effective in the ranges of medium and high frequencies. Unfortunately, at low frequencies, the volume and mass of passive solutions grows substantially, which is a major drawback of this group of methods.

M. Niedźwiecki (✉) · M. Meller
Department of Automatic Control, Gdańsk University of Technology, Faculty of Electronics, Telecommunications and Informatics, Narutowicza 11/12, 80-233 Gdańsk, Poland
e-mail: maciekn@eti.pg.edu.pl

M. Meller
e-mail: michal.meller@eti.pg.edu.pl

P. Kulczycki et al. (eds.), *Automatic Control, Robotics, and Information Processing*, Studies in Systems, Decision and Control 296,
https://doi.org/10.1007/978-3-030-48587-0_26

Recently, there has been a considerable growth in interest for active noise control (ANC). Active noise control methods rely on the phenomenon of destructive interference of acoustic waves. Without going into details, an ANC system generates an "antisound", i.e., an acoustic wave whose superposition with the undesired acoustic field results in the, typically local, cancellation of the latter. In contrast to passive methods, active noise control is most effective in the low-frequency range. This property stems from the fact that the volume of the space in which the cancellation takes place increases with the wavelengths of the interfering acoustic fields [1].

Figure 1 depicts three basic configurations of ANC systems in the classical application of noise cancellation in an acoustic duct. The differences between the configurations include, among others, the controller's principle of operation, the number of sensors required in the system, and their placement in the duct. The elements common to all configurations are the noise source, the acoustic duct, a loudspeaker, and the measurement microphone, placed at the end of the duct, which senses the residual noise.

In the ANC systems literature, the undesired acoustic noise is typically referred to as the primary noise, and the fragment of the duct between the noise source and the measurement microphone is called the primary path. The loudspeaker that is placed inside the duct is the canceling loudspeaker. Its purpose is to generate an acoustic wave, called a secondary noise, that will cancel the primary noise. Consequently, the fragment of the duct between the loudspeaker and the measurement microphone is called the secondary path. Finally, the role of the measurement microphone is to sense the control results and to enable the computation of adjustments to the control signal.

Most ANC systems employ the so-called feedforward approach (Fig. 1a). Feedforward systems take advantage of the fact that sound travels at speeds that are many orders of magnitude slower than the propagation velocity of electrical signals. Using an additional microphone, typically placed close to the noise source, the so-called reference signal is collected. The reference signal carries the information about the primary noise ahead of time, in the sense that it reaches the controller well before the primary noise acoustic wave reaches the end of the duct. It follows that, by taking into account the transfer functions of the primary and the secondary paths, one may compute how to excite the loudspeaker to cancel the primary noise. An important feature of feedforward ANC systems is their capability to cancel wideband noise. On the other hand, to work effectively, feedforward systems require that the transfer delay of the primary path is longer than the time needed by the controller to compute the control signal. Moreover, the performance of feedforward ANC systems depends heavily on the quality of the reference signal or, more precisely, on the level of correlation between the reference signal and the primary noise [2].

The feedback configuration, depicted in Fig. 1b, is free of the above limitations. In this approach, the control signal is generated solely using the signal from the measurement microphone. Feedback systems, however, are capable of eliminating narrowband noise only, which stems from the fact that the computation of the control signal relies on predictions of the primary noise, which are accurate only for narrowband signals [3].

Fig. 1 Typical configurations of active noise control systems in acoustic ducts

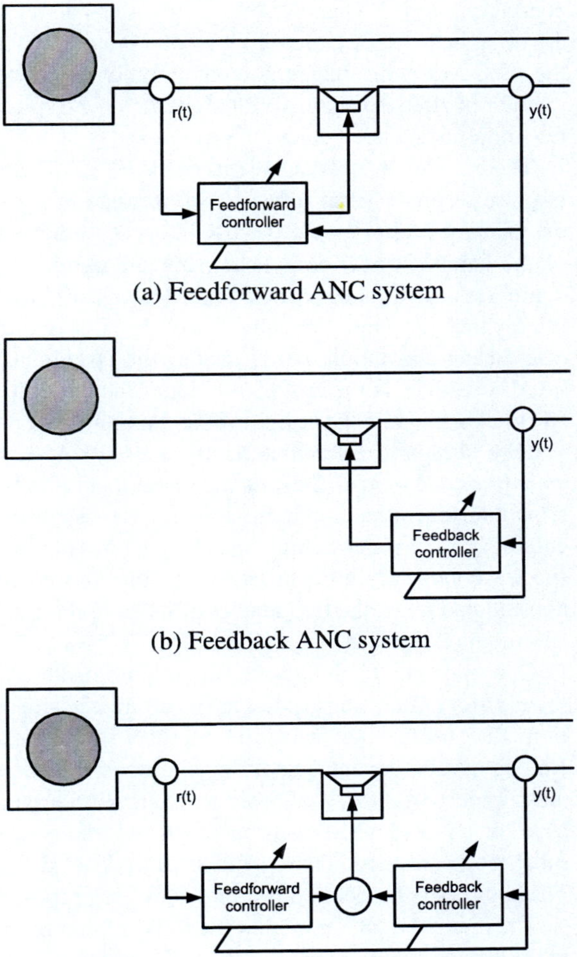

(a) Feedforward ANC system

(b) Feedback ANC system

(c) Hybrid ANC system that combine feedforward and feedback approaches

The hybrid configuration, shown in Fig. 1c, attempts to combine the advantages and, at the same time, to avoid the disadvantages of the above two approaches. This configuration employs both the feedforward and the feedback controllers, whose roles are to cancel the wideband and the narrowband components of the primary noise, respectively.

The most popular control algorithm in active noise control is the FX-LMS filter [4]. The FX-LMS controller employs an adaptive finite impulse response (FIR) filter, whose coefficients are adjusted using a version of the least mean square (LMS) algorithm, modified so as to take into account the influence of the secondary path. The FX-LMS algorithm is suitable both for the feedforward [2] and the feedback configurations [1]. The relatively easy tuning and the variety of modifications, which are often heuristic [2], are only two of its numerous advantages. On the other hand, the FX-LMS controller is computationally complex, which stems from the large number of adjusted adaptive filter coefficients (typically between few hundreds and few thousands tapped delay line weights).

There exists several equivalents of the FX-LMS algorithm that employ the infinite impulse response (IIR) filters [5–7]. Remarkably, in active noise control systems, the primary reason to use the IIR filters is, rather unexpectedly, not the reduction of the computational complexity, but the elimination of the risk of the adaptive controller destabilizing, related to the leakage of the canceling signal to the reference microphone [2]. However, the poles of the IIR filter might temporarily leave the unit disc during adaptation, which makes the issue of setting the initial values of the filter's coefficients the critical problem. The difficulty of ensuring safe operation of IIR filters is, likely, the reason of the dominance of the FX-LMS algorithm.

Several algorithms dedicated to the use of ANC systems that employ feedback were proposed as well. Despite their apparent differences, the basic principle underlying their operation is the same in all cases. Rejection of disturbances in feedback control systems requires high open loop gain at the frequencies that correspond to the spectral content of the disturbance [8]. Since the rejected disturbance signal might be nonstationary, i.e., the frequencies of its components might vary, the use of adaptive controllers is advised.

One may satisfy the above two requirements by combining a straightforward second order filter, whose poles are located at the unit circle, with a suitable frequency estimation mechanism, whose output serves to adapt the locations of the filter's poles. Among several advantages of this approach, its flexibility stands out as one of the most important. One is allowed to employ both parametric frequency estimators, based on a model of the disturbance [9–11], and nonparametric ones, such as FFT or subspace-based methods (e.g. classical MUSIC [12] and ESPRIT [13] algorithms). The controllers based on this approach were proposed in [14–16], among others.

The internal model principle (IMP) is another powerful framework for constructing feedback ANC controllers. The IMP principle states that a two degrees of freedom controller with transfer function $R(\zeta)/S(\zeta)$ (depending on the time domain, the symbol ζ corresponds to the complex variable s or z) ensures asymptotic cancellation of a disturbance $N(\zeta)/D(\zeta)$ if, and only if, a subset of the poles of the controller replicates the roots of the polynomial $D(\zeta)$, i.e., when $S(\zeta) = S'(\zeta)D(\zeta)$. IMP adaptive controllers may employ both indirect and direct adaptation mechanisms [17, 18]. In the first case, a model of the disturbance is estimated explicitly and the controller is synthesized by solving a Diophantine equation that includes the disturbance model [18]. In the direct approach, the controller is adapted without any intermediate steps, e.g., using the Youla-Kucera parametrization [19].

Note that each of the above referenced solutions requires the knowledge of the transfer function of the secondary path. There exist many applications where the plant may be regarded as stationary and its identification may be carried out once, typically in the open loop configuration. However, when the plant is subject to unpredictable changes, it is necessary to equip the controller with means to perform the identification on-line, without interrupting the operation of the ANC system.

The most popular solution of the problem of identifying a secondary path is the auxiliary noise approach, proposed in [20]. The method relies on introducing an additional signal, in the form of white noise, that excites the plant, to the control loop. The main disadvantage of such an approach is that the overall level of disturbances in the system increases, which conflicts with the control objective. Moreover, in its basic form, the auxiliary noise method converges slowly, and the modifications that eliminate this problem are known to increase the complexity and computational cost of the controller considerably [2].

The most important commercial application of ANC systems are active headsets. Due to their small size, such systems typically employ feedback, and the high-frequency noise is canceled passively [21–23]. Closely related to active headsets are the active noise control motorcycle helmets [24–27], which were demonstrated to offer a considerable reduction of noise. Another interesting application is ANC anti-snore system, which is very challenging due to severely limited freedom of placing the system's components and the variability of the secondary path, caused by movements of the sleeping person's head [28–30].

A highly promising area for ANC systems is the cancellation of acoustic noise generated by medical apparatus . The systems belonging to this category might aim at improving a patient's comfort or at reducing the long-term impact of the noise to medical staff. In the first group, one may point to an infant incubator ANC systems, where the purpose of the ANC system is to improve the infant's sleep quality and to eliminate the risk of hearing damage or loss [31–33]. In the second group, the active noise control of MRI noise stands out as the most important application [34–38].

Several other examples of interesting applications of ANC systems can be found in the survey paper [39].

2 Problem Formulation

To simplify mathematical analysis, we will assume that all signals considered below, such as control signal, reference signal, narrowband disturbance and measurement noise are complex-valued. In Sect. 5.2 we will describe a simple approach which allows one to use the obtained results when all signals are real-valued (which is a typical situation in practice.

Consider the problem of the reduction of a narrowband disturbance observed at the output of a discrete-time system (in acoustic applications this role is played by the secondary path) governed by the equation

$$y(t) = K_p(q^{-1})u(t-1) + d(t) + v(t) \tag{1}$$

where $t \in \mathbb{Z} = \{\ldots, -1, 0, 1, \ldots\}$ denotes normalized discrete time, $y(t)$ denotes the corrupted complex-valued output signal, q^{-1} is the backward shift operator, the signal

$$d(t) = a(t)e^{j\phi(t)}, \quad \phi(t) = \sum_{i=1}^{t} \omega(i) \tag{2}$$

is a complex-valued sinusoidal disturbance (cisoid) with a slowly-varying amplitude $a(t)$ and slowly-varying angular frequency $\omega(t) \in (-\pi, \pi]$, $v(t)$ denotes complex-valued wideband noise obeying the assumption

(A1) $\{v(t)\}$ is a sequence of independent complex-valued random variables with zero mean and variance σ_v^2; real and imaginary components of $v(t)$ are mutually independent,

and finally, $K_p(q^{-1})$ denotes transfer function of a stable, linear, single-input single output system obeying the condition

(A2) $K_p(e^{-j\omega}) \neq 0, \quad \forall \omega \in (-\pi, \pi].$

Our goal will be to design a minimum-variance controller, i.e., to find such control rule which minimizes the signal observed at the output of the ANC system in the mean squared sense

$$u(t): \quad \mathrm{E}\left[|y(t)|^2\right] \longrightarrow \min .$$

We will consider both purely feedback control strategies and hybrid strategies which combine feedback control and feedforward control (provided that the reference signal is available).

3 Feedback Control Strategies

3.1 Control in the Case of Full Prior Knowledge of the System and Disturbance

Suppose that the transfer function of the system $K_p(q^{-1})$ is known and that disturbance has the form $d(t) = a_0 e^{j\omega_0 t}$ [which is equivalent to assuming that $a(t) \equiv a_0$ and $\omega(t) \equiv \omega_0$] and is measurable. To compensate sinusoidal disturbance of this form, one should generate such a sinusoidal control signal $u(t)$ for which the system output, also sinusoidal due to system linearity, would have the same amplitude as $d(t)$ but opposite polarity. Since linear systems basically scale and shift sinusoidal inputs, the narrowband character of $u(t)$ justifies the following steady state approximation

$$K_p(q^{-1})u(t-1) = k_p u(t-1) \tag{3}$$

where $k_p = K_p(e^{-j\omega_0}) \in \mathbb{C}$ denotes the complex-valued system gain/attenuation at the frequency ω_0. Combining relationships (3) and (1), one obtains a simple formula $y(t) = k_p u(t-1) + d(t) + v(t)$, leading to the following form of control signal

$$u(t) = -\frac{d(t+1)}{k_p} \tag{4}$$

The steady state signal observed at the output of the open-loop system governed by (4) has the form $y(t) = v(t)$, which means that its variance takes the smallest achievable value equal to $\lim_{t\to\infty} \mathrm{E}[|y(t)|^2] = \sigma_v^2$.

3.2 *Control in the Case of Full Prior Knowledge of the System and Partial Prior Knowledge of the Disturbance (Known Frequency)*

Suppose that the amplitude of the disturbance with constant and known frequency ω_0 changes slowly with time according to the random walk model

$$a(t+1) = a(t) + \eta(t+1) \tag{5}$$

where the sequence of one-step amplitude changes $\{\eta(t)\}$ obeys the assumption

(A3) $\{\eta(t)\}$ is a sequence of independent random variables with zero mean and variance σ_η^2; the sequence $\{\eta(t)\}$ is independent of $\{v(t)\}$.

It is straightforward to check that under assumptions made above, the disturbance signal can be expressed in a recursive form

$$d(t+1) = e^{j\omega_0} d(t) + \widetilde{\eta}(t+1) \tag{6}$$

where $\{\widetilde{\eta}(t)\}$, $\widetilde{\eta}(t) = e^{j\omega_0}\eta(t)$, is a sequence of independent complex-valued random variables with zero mean and variance $\widetilde{\sigma}_\eta^2 = \sigma_\eta^2$.

Using the approximation $K_p(q^{-1})u(t-1) \cong k_p u(t-1)$, one can show that the optimal, minimum-variance control rule has the following steady state form [40]

$$\begin{aligned} \widehat{d}(t+1|t) &= e^{j\omega_0}[\widehat{d}(t|t-1) + \mu_a y(t)] \\ u(t) &= -\frac{\widehat{d}(t+1|t)}{k_p} \end{aligned} \tag{7}$$

where $\widehat{d}(t+1|t)$ denotes the predicted value of the disturbance and $\mu_a \in \mathbb{R}$ denotes the adaptation gain given by $\mu_a = -\xi/2 + \sqrt{\xi^2/4 + \xi}$, $\xi = \sigma_\eta^2/\sigma_v^2$. Under control rule (7), system output is given by $y(t) \cong c(t) + v(t)$, where

$$c(t) = d(t) - \widehat{d}(t|t-1) = e^{j\omega_0}(1-\mu_a)c(t-1) + \widetilde{\eta}(t) - \mu_a e^{j\omega_0} v(t+1) \quad (8)$$

is the quantity which will be further referred to as cancellation error. Based on this relationship it is easy to show that the steady state variance of the output signal is given by $\lim_{t\to\infty} \mathrm{var}[y(t)] = \sigma_c^2 + \sigma_v^2$, where $\sigma_c^2 = \lim_{t\to\infty} \mathrm{var}[c(t)] = \big[\sigma_\eta^2 + \sqrt{\sigma_\eta^4 + 4\sigma_\eta^2\sigma_v^2}\big]/2$.

3.3 Control in the Case of Partial Prior Knowledge of the System and Partial Prior Knowledge of the Disturbance (Known Frequency)

In practical situations the quantities k_p and μ_a not only are unknown but can also change over time due to the nonstationarity of the system and/or disturbance. Prior to designing an adaptive version of the controller, capable of automatic tuning of its settings, we will analyze properties of the "realistic" version of the controller (7)

$$\begin{aligned} \widehat{d}(t+1|t) &= e^{j\omega_0}[\widehat{d}(t|t-1) + \mu y(t)] \\ u(t) &= -\frac{\widehat{d}(t+1|t)}{k_n} \end{aligned} \quad (9)$$

obtained when the true system gain k_p is replaced with its "nominal" gain $k_n \neq k_p$, and when the optimal adaptation gain μ_a is replaced with an arbitrarily chosen gain $\mu \neq \mu_a$.

Denote by $\beta \in \mathbb{C}$, $\beta = k_p/k_n \neq 1$ the system gain modeling error. Similarly as in the optimal regulator case (7), the output signal can be expressed in the form $y(t) = c(t) + v(t)$ where now

$$c(t) = d(t) - \beta\widehat{d}(t|t-1) = e^{j\omega_0}(1-\mu\beta)c(t-1) + \widetilde{\eta}(t) - \mu\beta e^{j\omega_0} v(t+1) \quad (10)$$

Comparison of (8) and (10) leads to an interesting conclusion—when the adaptation gain coefficient μ is chosen so as to fulfill the condition

$$\mu\beta = \mu_a \quad (11)$$

the cancellation error (10) becomes identical with that yielded by the optimal controller (7). This means that by choosing the value of μ appropriately one can "compensate" the modeling error which is the consequence of adopting the inappropriate value of the system gain $k_n \neq k_p$. It is worth noticing that when $\mathrm{Im}[\beta] \neq 0$, the compensation mentioned above is possible provided that the adopted adaptation gain is complex-valued $\mu \in \mathbb{C}$. Having this in mind, we will design an adaptive algorithm

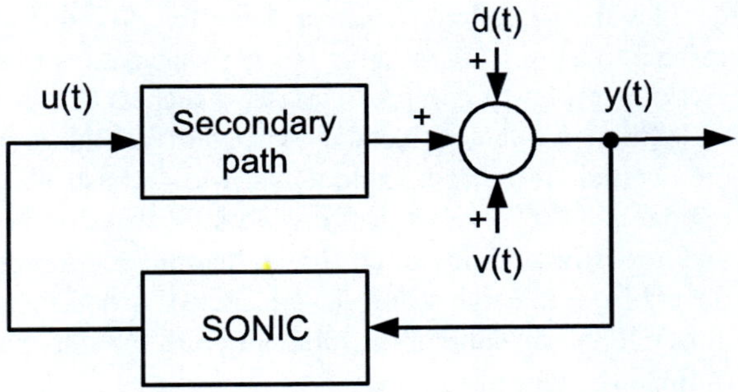

Fig. 2 Block diagram of an ANC system incorporating SONIC controller

for on-line tuning of a complex-valued adaptation gain μ. We will adjust μ recursively by minimizing the following local measure of fit, made up of exponentially weighted squares of system outputs

$$V(t;\mu) = \sum_{\tau=1}^{t} \rho^{t-\tau} |y(\tau;\mu)|^2 \tag{12}$$

where $\rho \in [0.999, 0.9999]$ denotes the so-called forgetting constant, determining the effective width of the local analysis window [in steady state equal to $1/(1-\rho)$ sampling intervals].

Using the method of recursive prediction error (RPE) [41], the following adaptive control algorithm, further referred to as SONIC (Self-Optimizing Narrowband Interference Canceller), was derived in [40]—see Fig. 2

$$\begin{aligned} z(t) &= e^{j\omega_0}\left[(1-c_\mu)z(t-1) - \frac{c_\mu}{\widehat{\mu}(t-1)}\, y(t-1) \right] \\ r(t) &= \rho r(t-1) + |z(t)|^2 \\ \widehat{\mu}(t) &= \widehat{\mu}(t-1) - \frac{z^*(t)y(t)}{r(t)} \\ \widehat{d}(t+1|t) &= e^{j\omega_0}[\, \widehat{d}(t|t-1) + \widehat{\mu}(t)y(t)\,] \\ u(t) &= -\frac{\widehat{d}(t+1|t)}{k_{\text{n}}} \end{aligned} \tag{13}$$

where $c_\mu \in [0.005, 0.05]$ denotes a small positive constant, and $z^*(t)$ denotes a complex conjugate of $z(t)$. The variable $z(t)$ is given by the formula

$$z(t) = \frac{\partial y(t;\mu)}{\partial \mu}$$

where

$$\frac{\partial}{\partial \mu} = \frac{1}{2}\left[\frac{\partial}{\partial \text{Re}[\mu]} - j\,\frac{\partial}{\partial \text{Im}[\mu]} \right]$$

denotes the so-called Wirtinger derivative [42–45]—symbolic differentiation with respect to a complex variable [Wirtinger calculus facilitate minimization of nonanalytic functions of complex variables, such as (12)].

Without explaining details of the derivation, we will remark that it is based on the coefficient fixing technique, used to "robustify" self-tuning minimum-variance regulators [17]. Within this framework the value of one of system parameters is fixed and not estimated. It can be shown that this modeling error (the fixed value usually differs from the true value) is "corrected" when identification is carried out in the closed loop—in spite of the modeling bias the self-tuning regulator converges to the optimal regulator [46].

The self-optimization loop, made up of the first three recursions of SONIC, allows not only for compensation of the system modeling error ($\beta \neq 1$), but also for optimization of the controller. As shown in [40], under assumptions (A1)–(A3) it holds that

$$\lim_{t\to\infty} \mathrm{E}[\widehat{\mu}(t)] \cong \frac{\mu_a}{\beta} \tag{14}$$

which means that the SONIC controller converges, in the mean sense, to the optimal controller (7). Even more importantly, since optimization of μ is performed using the RPE approach, the controller (13) will attempt to minimize the cancellation error also in cases where the assumptions (A1)–(A3)—not very realistic from the practical viewpoint—are not fulfilled, i.e., it should respond adequately to the changes in plant dynamics, changes of the signal-to-noise ratio ξ, etc.

The robustness of SONIC can be demonstrated using a simple simulation experiment, carried out for a system with switched dynamics—all details are given in Table 1. The narrowband disturbance and wideband noise were generated using the following settings: $\sigma_v = 0.1$, $\sigma_e = 0.001$, $\omega_0 = 0.1$, $d(0) = 1$.

While the first two changes [from $K_1(q^{-1})$ to $K_2(q^{-1})$ at instant $t = 15000$ and from $K_2(q^{-1})$ to $K_3(q^{-1})$ at instant $t = 30000$] were confined to plant parameters, the last change was more substantial: at instant $t = 45000$, the first-order inertial system $K_3(q^{-1})$ with a single real pole was switched to the second-order nonminimum phase

Table 1 System switching schedule in the transient behavior experiment and the corresponding modeling errors.

Time interval	System	$\lvert\beta\rvert$	Argβ[°]
$0 < t < 15000$	$K_1(z) = \dfrac{0.0952}{1 - 0.9048z^{-1}}$	0.708	−47.9
$15000 \le t < 30000$	$K_2(z) = \dfrac{0.0238}{1 - 0.9762z^{-1}}$	0.234	−79.3
$30000 \le t < 45000$	$K_3(z) = \dfrac{0.2}{1 - 0.8z^{-1}}$	0.913	−27.1
$45000 \le t \le 60000$	$K_4(z) = \dfrac{+0.1 - 0.14z^{-1}}{1 - 1.8391z^{-1} + 0.8649z^{-2}}$	1.960	121.1

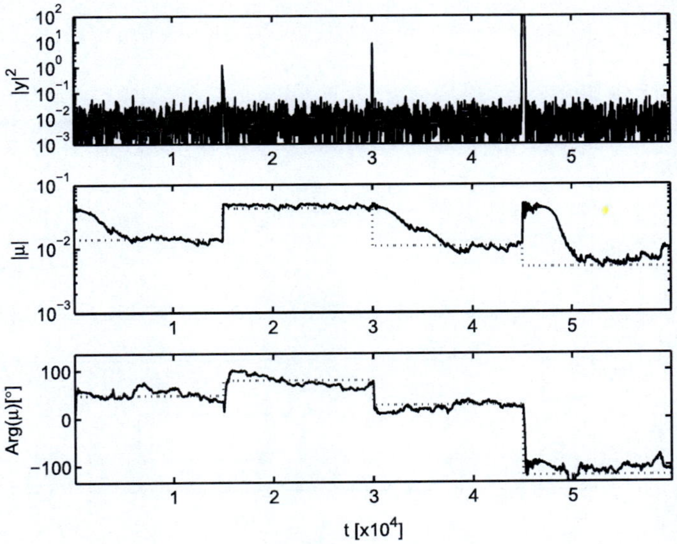

Fig. 3 Typical simulation results obtained using SONIC. Solid lines—the estimated quantities, dotted lines—optimal steady state values

system $K_4(q^{-1})$ with a pair of complex poles. Since the phase shift introduced by $K_4(q^{-1})$ at the frequency ω_0 differs from the analogous shift of $K_3(q^{-1})$ by more than $\pi/2$, the last change causes temporal instability of the closed-loop system, making the task of disturbance rejection even harder.

Figure 3 (illustrating typical behavior) and Fig. 4 (illustrating mean behavior) show results obtained for the SONIC algorithm with the following settings: $c_\mu = 0.005$, $\rho = 0.9995$, $k_n = e^{j\omega_0}$. Additionally, the algorithm was equipped with safety jacketing mechanisms which will be described in Sect. 5.1 [with $\mu_{\max} = 0.05$, $\mu_{\min} = 0$, $\Delta\mu_{\max}(t) = \widehat{\mu}(t-1)/50$, $r_{\max} = 1600$, $r_{\min} = 0$].

The adaptation process was started at instant $t = 1$ using the following initial conditions: $\widehat{d}(0) = e^{j\omega_0}$, $r(0) = 100$, $z(0) = 0$, $\widehat{\mu}(0) = 0.02$. The algorithm copes favorably with both the initial convergence problem and with abrupt plant changes. When the experiment is started or when a change to the plant dynamics occurs, the magnitude of the adaptation gain $\widehat{\mu}(t)$ temporarily increases to quickly compensate large initial modeling errors; later on, it gradually decays to settle down around its optimal steady state value. Note the very quick response to phase errors and usually much slower response to magnitude errors—the effect caused by diverse sensitivity of system output to two types of modeling errors.

The simulation experiment shows also that the proposed control scheme has a self-stabilization property. When instability occurs at the instant $t = 45000$ [which is unavoidable since, due to the sign mismatch, the stabilizing gain $\widehat{\mu}$ for $K_3(q^{-1})$ does not stabilize $K_4(q^{-1})$], it causes rapid growth of the output signal $y(t)$, which

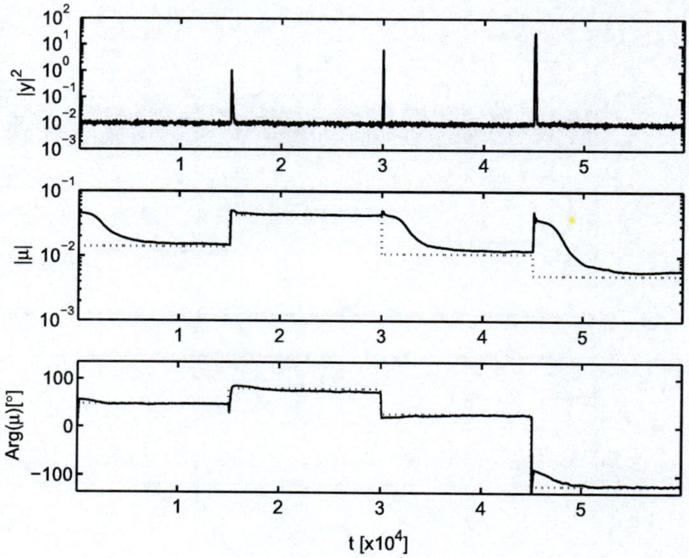

Fig. 4 Simulation results obtained using SONIC, averaged over 100 process realizations. Solid lines—the estimated quantities, dotted lines—optimal steady state values

in turn speeds up convergence of $\widehat{\mu}$ to a new stabilizing value. In this way, after a burst observed at the system output, the closed-loop stability is regained.

3.4 Control in the Case of Partial Prior Knowledge of the System and Partial Prior Knowledge of the Disturbance (Unknown Frequency)

Suppose that the frequency $\omega(t)$ of the narrowband disturbance is not known and slowly varies with time. Prior to designing the fully adaptive minimum-variance regulator, we will analyze properties of the modified algorithm (9), obtained by means of replacing the known frequency ω_0 with the current estimate of the instantaneous frequency

$$\begin{aligned} \widehat{d}(t+1|t) &= e^{j\widehat{\omega}(t|t-1)}[\widehat{d}(t|t-1) + \mu y(t)] \\ \widehat{\omega}(t+1|t) &= g[\mathscr{Y}(t)] \\ u(t) &= -\frac{\widehat{d}(t+1|t)}{k_n} \end{aligned} \tag{15}$$

where $\widehat{\omega}(t+1|t)$ denotes the one-step-ahead prediction of the instantaneous frequency $\omega(t+1)$ based on observations $\mathscr{Y}(t) = \{y(i), i \le t\}$ gathered up to the instant t.

Estimation of the instantaneous frequency can be carried out using a simple gradient algorithm [47]

$$\widehat{\omega}(t+1|t) = \widehat{\omega}(t|t-1) + \gamma \,\mathrm{Arg}\left[\frac{\widehat{d}(t+1|t)e^{-j\widehat{\omega}(t|t-1)}}{\widehat{d}(t|t-1)}\right] \tag{16}$$

where $\gamma, 0 < \gamma \ll 1$ denotes adaptation gain in the frequency estimation loop, and $\mathrm{Arg}(\cdot) \in (-\pi, \pi]$ denotes the principal argument of a complex number. In the case of slowly time-varying disturbance signals, one can use the following approximation

$$\mathrm{Arg}\left[\frac{\widehat{d}(t+1|\,t)e^{-j\widehat{\omega}(t|\,t-1)}}{\widehat{d}(t|\,t-1)}\right] = \mathrm{Im}\left\{\log\left[1 + \frac{\mu y(t)}{\widehat{d}(t|\,t-1)}\right]\right\} \cong \mathrm{Im}\left[\frac{\mu y(t)}{\widehat{d}(t|\,t-1)}\right]$$

which leads to the following control algorithm

$$\begin{aligned}\widehat{d}(t+1|t) &= e^{j\widehat{\omega}(t|t-1)}[\widehat{d}(t|t-1) + \mu y(t)]\\ \widehat{\omega}(t+1|t) &= \widehat{\omega}(t|t-1) + \gamma \mathrm{Im}\left[\frac{\mu y(t)}{\widehat{d}(t|\,t-1)}\right]\\ u(t) &= -\frac{\widehat{d}(t+1|t)}{k_n}\end{aligned} \tag{17}$$

Analysis of tracking properties of the algorithm (17) can be performed using the method of local averaging [48] and the approximating linear filtering (ALF) technique [11, 49], developed for the purpose of studying adaptive notch filters (ANF). Consider a local analysis window $T = [t_1, t_2]$ of width $n = t_2 - t_1 + 1$ obeying the condition $n \gg 2\pi/\omega(t)$, $\forall t \in T$. Assuming that $K_p(e^{-j\omega})$ is a smooth function of ω, that the amplitude of the disturbance is constant and equal to a_0, and that the instantaneous frequency $\omega(t)$ varies slowly with time, the output of a system excited by a narrowband input signal can be approximately expressed in the form $K_p(q^{-1})u(t-1) \cong k_T u(t-1)$, $t \in T$, where $k_T = \sum_{t\in T} K_{\mathrm{p}}(e^{-j\omega(t)})/n$ denotes the mean gain of the system in the time interval T. Using this approximation, the output signal can be expressed in the form $y(t) \cong d(t) - \beta\widehat{d}(t|t-1) + v(t), t \in T$, where $\beta = k_T/k_n$. Denote by $w(t)$ the one-step change of the instantaneous frequency

$$\omega(t) = \omega(t-1) + w(t).$$

The ALF technique amounts to studying dependence between the cancellation error $c(t) = d(t) - \beta\widehat{d}(t|t-1)$ and frequency tracking error $\Delta\widehat{\omega}(t) = \omega(t) - \widehat{\omega}(t|t-1)$, and "exciting" signals $v(t)$ i $w(t)$, while neglecting all moments of these quantities of order higher than 1, as well as all "cross-terms". Using this approach, one arrives at the following linear relationships

$$\begin{aligned}\Delta\widehat{x}(t) &= (1-\mu\beta)\Delta\widehat{x}(t-1) + ja_0^2\Delta\widehat{\omega}(t-1)\mu\beta s(t-1)\\ \Delta\widehat{\omega}(t+1) &= \Delta\widehat{\omega}(t) - \frac{\gamma}{a_0^2}\,\mathrm{Im}[\mu\beta\Delta\widehat{x}(t)] - \frac{\gamma}{a_0^2}\,\mathrm{Im}[\mu\beta z(t)] + w(t+1)\end{aligned} \tag{18}$$

where $\Delta x(t) = c(t)d^*(t)$ and $s(t) = v(t)d^*(t)$.

Equations of the linear approximating filter (18) allow one to quantify frequency tracking errors. Note that—similar to (10)—both ALF equations incorporate terms that depend on $\mu\beta$, and neither of them incorporates terms dependent exclusively on μ or β. This means that, also in this case, system modeling errors can be compensated by a suitable choice of a complex-valued gain μ.

Interesting analytical results can be obtained in the case where the instantaneous frequency evolves according to the random walk model, i.e., when

(A4) $\{w(t)\}$ is a sequence of zero-mean independent random variables with variance σ_w^2; the sequence $\{w(t)\}$ is independent of $\{v(t)\}$ and $\{\eta(t)\}$.

Assume, for simplicity, that the true system gain is known ($\beta = 1$), which means that there is no need to adopt complex-valued gain μ, i.e., $\mu \in \mathbb{R}$. Let $s_I(t) = \mathrm{Im}[s(t)]$. Under the restrictions mentioned above the following result can be obtained after solving ALF equations

$$\Delta\widehat{\omega}(t) \cong H_1(q^{-1})s_I(t) + H_2(q^{-1})w(t)$$

where

$$\begin{aligned}H_1(q^{-1}) &= -\frac{\eta\mu(1-q^{-1})q^{-1}}{a^2[1-(1+\lambda)q^{-1}+(\lambda+\eta\mu)q^{-2}]}\\ H_2(q^{-1}) &= \frac{1-\lambda q^{-1}}{1-(1+\lambda)q^{-1}+(\lambda+\eta\mu)q^{-2}}\end{aligned}$$

and $\lambda = 1-\mu$. Linear filters $H_1(q^{-1})$ and $H_2(q^{-1})$ are asymptotically stable for all values $\mu, \gamma \in (0, 1)$. Since under the assumption (A4) the processes $\{s_I(t)\}$ and $\{w(t)\}$ are orthogonal, the steady state expression for the variance of the frequency tracking error takes the form

$$\mathrm{E}\{[\Delta\widehat{\omega}(t)]^2\} = I[H_1(z^{-1})]\mathrm{E}[s_I^2(t)] + I[H_2(z^{-1})]\mathrm{E}[w^2(t)]$$

where

$$I[X(z^{-1})] = \frac{1}{2\pi j}\oint X(z)X(z^{-1})\frac{dz}{z}$$

is an integral evaluated along the unit circle in the $\mathscr{Z}$-plane, and $X(z^{-1})$ denotes any stable proper rational transfer function. Using the method of residue calculus [50] and taking into account the facts that $\mathrm{E}[s_1^2(t)] = a_0^2\sigma_v^2/2$ and $\mathrm{E}[w^2(t)] = \sigma_v^2$, one arrives at

$$\mathrm{E}[(\Delta\widehat{\omega}(t))^2] \cong \frac{\gamma^2\mu}{4a_0^2}\sigma_v^2 + \left[\frac{1}{2\gamma} + \frac{1}{2\mu}\right]\sigma_w^2 . \tag{19}$$

Denote by μ_ω and γ_ω the coefficients that minimize the mean squared error (19). Straightforward calculations lead to

$$\mu_\omega = \sqrt[4]{8\zeta}, \quad \gamma_\omega = \sqrt[4]{\zeta/2}, \quad \zeta = \frac{a_0^2\sigma_w^2}{\sigma_v^2}$$

and

$$\mathrm{E}[(\Delta\widehat{\omega}(t))^2 | \mu_\omega, \gamma_\omega] \cong \sigma_w^2\sqrt[4]{2\zeta^{-1}} . \tag{20}$$

The mean squared frequency estimation error cannot be smaller than the so-called *a posteriori* Craméra-Rao lower (PCRB)[1] As shown in [11], when the variables $v(t)$ and $w(t)$ are normally distributed, PCRB takes the form

$$\mathrm{PCRB} \cong \sigma_w^2\sqrt[4]{2\zeta^{-1}} . \tag{21}$$

Comparison of relationships (20) and (21) leads to the conclusion that the optimally tuned algorithm (17) is—in spite of its very simple form—statistically efficient, i.e., it guarantees the highest achievable estimation accuracy when the instantaneous frequency $\omega(t)$ drifts in a random way. More than that, as shown in the paper [51], under more realistic frequency variation scenarios the simple gradient algorithm (17) yields results that are comparable with those obtained by means of applying the high-resolution (and computationally much more demanding) frequency estimation algorithm known as ESPRIT (Estimation of Signal Parameters via Rotational Invariance Techniques) [52].

Combining all adaptation mechanisms described so far, one obtains the following extended version of the SONIC algorithm [47]

$$self - optimization$$

$$\begin{aligned}
z(t) &= e^{j\widehat{\omega}(t|t-1)}\left[(1-c_\mu)z(t-1) - \frac{c_\mu}{\widehat{\mu}(t-1)}y(t-1)\right] \\
p(t) &= \rho r(t-1) + |z(t)|^2 \\
\widehat{\mu}(t) &= \widehat{\mu}(t-1) - \frac{y(t)z^*(t)}{p(t)}
\end{aligned} \tag{22}$$

[1]Since in the case considered $\omega(t)$ is a random variable (rather than an unknown deterministic constant) the classical Craméra-Rao bound does not apply.

predictive control

$$\widehat{d}(t+1|t) = e^{j\widehat{\omega}(t|t-1)}[\widehat{d}(t|t-1) + \widehat{\mu}(t)y(t)]$$

$$u(t) = -\frac{\widehat{d}(t+1|t)}{k_{\mathrm{n}}[\widehat{\omega}(t|t-1)]} \tag{23}$$

frequency estimation

$$\widehat{\omega}(t+1|t) = \widehat{\omega}(t|t-1) + \gamma\,\mathrm{Im}\left[\frac{\widehat{\mu}(t)y(t)}{\widehat{d}(t|t-1)}\right] \tag{24}$$

where c_μ, ρ, γ and $k_n[\widehat{\omega}(t|t-1)] = K_n(e^{j\widehat{\omega}(t|t-1)})$ are previously defined regulator settings, and $K_n(q^{-1})$ denotes the "nominal" transfer function of the system. If no prior knowledge about the system is available, one can set $K_n(q^{-1}) \equiv 1$.

Figure 5 shows typical results obtained for a real-valued system. The algorithm (22)–(24), after modifications described in Sects. 5.1 and 5.2, was implemented on a regular PC. The sampling rate was equal to 1 kHz. The left loudspeaker was used to generate disturbance, and the right one—to generate the antisound. The error microphone was placed 15 cm away from the right loudspeaker. The instantaneous frequency of the disturbance was slowly varying, in a sinusoidal way, between 241 Hz and 250 Hz with a period equal to 20 sec. The nominal gain of the system was set to $k_n = 1$ and the remaining settings of SONIC were equal to: $c_\mu = 0.01$, $\mu_{\max} = 0.05$, $\gamma = 0.01$. Application of the extended SONIC led to attenuation of the harmonic noise which, depending on the instantaneous frequency, ranged between 15 and 20 dB.

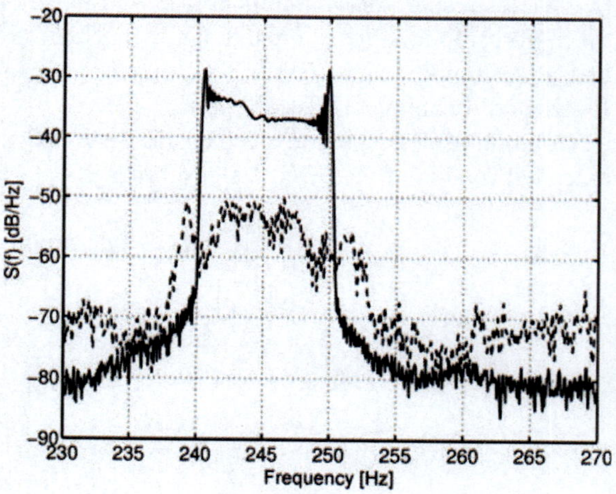

Fig. 5 Comparison of spectral density functions of disturbances observed at the output of a real acoustic system: without control (solid line) and under control of the extended SONIC algorithm (broken line)

4 Control Incorporating both Feedback and Feedforward Mechanisms

An obvious advantage of SONIC, typical of all feedback ANC systems, is due to the fact that it does not require deployment of a reference sensor. Such a sensor may be expensive and/or difficult to mount. Additionally, it may introduce acoustic feedback, which deteriorates performance of the ANC system. However, this advantage comes at a price: without access to a reference signal, SONIC needs to learn the properties of the disturbance, such as its instantaneous frequency $\omega(t)$, by observing the error signal $y(t)$, i.e., the very signal it is trying to cancel. Such an internal "conflict of interests" (things that are good for identification are bad for control and *vice versa*) is an inherent limitation of many adaptive control systems. Under nonstationary conditions, this may result in episodes of turbulent, or even bursting, behavior, not acceptable from a practical viewpoint. Additionally, fluctuations of the attenuation rate (since periods of higher attenuation are interleaved with periods of lower attenuation), even if in a small range, make the residual noise less comfortable for a human listener than a constant-intensity noise.

The drawbacks described above can be eliminated by equipping the control system with a reference sensor (microphone, accelerator), placed close to the source of the unwanted sound, and measuring the reference signal

$$r(t) = d_*(t) + v_*(t) \tag{25}$$

where $d_*(t)$ denotes the reference disturbance, strongly correlated with $d(t)$, and $v_*(t)$ denotes sensor noise, independent of $v(t)$. The proposed solution is obtained by replacing in SONIC the estimate $\widehat{\omega}(t|t-1)$ yielded by the feedback frequency estimation loop, with a suitably modified (smoothed or simply delayed) estimate of the instantaneous frequency $\omega_*(t)$ of the signal $d_*(t)$, obtained by means of processing the reference signal $r(t)$. Such a hybrid feedforward/feedback solution, depicted in Fig. 6, has several advantages over a purely feedback design. First, the reference signal is a non-vanishing source of information about the instantaneous frequency of the disturbance. Secondly, even if the ANC system is switched off, the signal-to-noise ratio is usually much higher at the reference point than at the cancellation point. Finally, since the reference signal is measured ahead of time, estimation of the instantaneous frequency of $d(t)$ can be based not only on the past, but also on a certain number of "future" (relative to the local time of the controller) samples of the disturbance. Such noncausal estimates, which incorporate smoothing, are more accurate than their causal counterparts.

The instantaneous frequency $\omega_*(t)$ of the signal $d_*(t)$ can be estimated using a slightly modified version of the algorithm used to track the signal $d(t)$

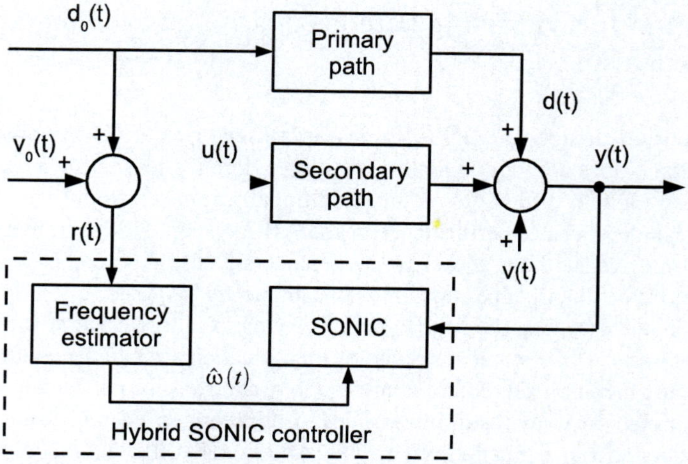

Fig. 6 Block diagram of the ANC system incorporating hybrid SONIC controller

$$
\begin{aligned}
\varepsilon(t) &= r(t) - \widehat{d}_*(t|t-1) \\
\widehat{d}_*(t+1|t) &= e^{j\widehat{\omega}_*(t)}[\widehat{d}_*(t|t-1) + \mu_*\varepsilon(t)] \\
\widehat{\omega}_*(t+1) &= \widehat{\omega}_*(t) + \gamma_*\mu_* \operatorname{Im}\left[\frac{\varepsilon(t)}{\widehat{d}_*(t|t-1)}\right]
\end{aligned}
\tag{26}
$$

where μ_* and γ_*, $0 < \mu_*, \gamma_* \ll 1$, denote small adaptation gains. Unlike the coefficient μ in the algorithm (24), the coefficient μ_* in (26) is constant and real-valued.

Using the ALF approach, one can show that if the frequency $\omega_*(t)$ changes slowly, the quantity $\widehat{\omega}_*(t)$ can be viewed as an approximately unbiased estimate of $\omega_*(t-\tau_{\text{est}})$, where $\tau_{\text{est}} = \text{int}[t_\omega]$, $t_\omega = 1/\gamma_*$, denotes the so-called estimation delay [53] introduced by the algorithm (26). This is a straightforward consequence of the fact that the mean trajectory of the estimates $\widehat{\omega}_*(t)$ is delayed with respect to the true trajectory $\omega_*(t)$ by (approximately) τ_{est} sampling intervals [54]. Due to the causality principle, in a majority of adaptive control applications estimation of unknown system/signal coefficients can be based exclusively on past data samples. In the case of the ANC system depicted in Fig. 6 the situation is different. Since the acoustic delay, i.e., delay with which the sound wave emitted by the source of disturbance reaches the point at which it is supposed to be canceled, is considerably longer than the electrical delay with which reference measurements are transmitted to the control unit, the controller has the advantage of knowing the disturbance (or, more precisely, of knowing the signal correlated with the disturbance) before it reaches the cancellation point. Denote by τ_{el} the electrical delay, by $\tau_{\text{ak}} \gg \tau_{\text{el}}$—the acoustic delay, and by τ_{prz}—the time needed to complete one cycle of computations using the algorithm (26). Since it holds approximately that $\omega(t) \cong \omega_*(t-\tau_{\text{ak}})$, when the processing unit is sufficiently fast the control algorithm has at its disposal the estimate of the instantaneous frequency of $d(t)$ with a time advance equal to $\tau_0 = \tau_{\text{ak}} - \tau_{\text{el}} - \tau_{\text{prz}}$

sampling intervals. Hence, after accounting for the estimation delay, as an estimate of the instantaneous frequency $\omega(t)$, one can use the following quantity

$$\widehat{\omega}(t) = \widehat{\omega}_*(t - \tau_d) \tag{27}$$

where $\tau_d = \max\{\tau_0 - \tau_{est}, 0\}$. When $\tau_0 < \tau_{est}$ only a partial correction of the bias can be achieved.

The hybrid SONIC algorithm incorporating both feedback and feedforward mechanisms has the form

$$\begin{aligned} z(t) &= e^{j\widehat{\omega}_*(t-\tau_d)}\left[(1-c_\mu)z(t-1) - \frac{c_\mu}{\widehat{\mu}(t-1)}y(t-1) \right] \\ p(t) &= \rho r(t-1) + |\, z(t)\,|^2 \\ \widehat{\mu}(t) &= \widehat{\mu}(t-1) - \frac{y(t)z^*(t)}{p(t)} \\ \widehat{d}(t+1|t) &= e^{j\widehat{\omega}_*(t-\tau_d)}[\widehat{d}(t|t-1) + \widehat{\mu}(t)y(t)] \\ u(t) &= -\frac{\widehat{d}(t+1|t)}{k_n[\widehat{\omega}_*(t-\tau_d)]} \end{aligned} \tag{28}$$

where $\widehat{\omega}_*(t)$ denotes the estimate of the instantaneous frequency of the reference signal $d_*(t)$ obtained using the algorithm (26).

Unlike most of the existing hybrid schemes, hybrid SONIC is not made up of two controllers—the reference signal is used only to extract information about the instantaneous frequency of the disturbance, rather than to form the reference-dependent control (compensation) signal. Therefore it can be characterized as a feedback ANC with an external (feedforward) frequency adjustment mechanism. Since the reference signal is usually a more reliable source of information about the instantaneous frequency of the disturbance than the error signal (which is minimized by the controller), hybrid SONIC has better tracking and robustness properties than its original, purely feedback version.

Time shift is the simplest form of noncausal estimation of the frequency $\omega_*(t)$ [55]. The estimation accuracy can be further improved by means of application of the fixed-delay adaptive notch smoothing (ANS) algorithm

$$\widehat{\omega}(t) = \widehat{\omega}_*(t - \tau_d|t) \tag{29}$$

where $\widehat{\omega}_*(t-\tau_d|t)$ denotes the estimate of the frequency at the instant $t-\tau_d$ based on the data collected up to the instant t, i.e., data incorporating τ_d "future" measurements of the reference signal. The ANS algorithms of this form were presented in [56–59].

The properties of the hybrid SONIC algorithm were checked in a simulation experiment incorporating a realistic model of the controlled system obtained via

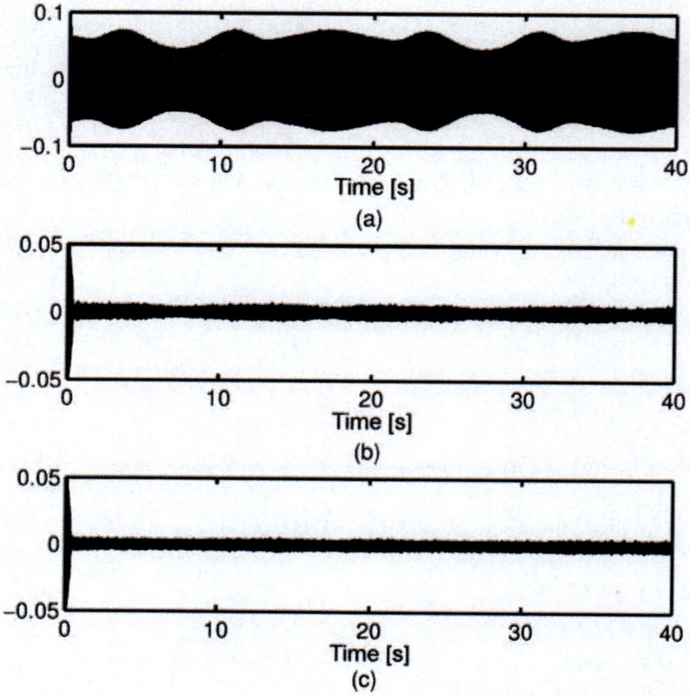

Fig. 7 Comparison of signals observed at the output of the simulated acoustic system. **a** Without active noise cancellation. **b** Obtained using the extended SONIC controller **c** Obtained using the hybrid SONIC controller

identification. The aim of this experiment was to compare the algorithms (22)–(24) and (28). The simulated sampling rate was equal to 8 kHz. The assumed transportation delays introduced by the primary path and secondary path were equal to 100 and 60 sampling intervals, respectively. The instantaneous frequency of the disturbance was varying between 190 and 210 Hz. Finally, the nominal gain of the secondary path was set to its true gain measured at the frequency 200 Hz.

The results of the experiment are shown in Fig. 7. Note fluctuations of the amplitude of the residual output signal in the case which can be observed when the frequency of the disturbance is estimated in the feedback loop—the effect caused by periodic deterioration of identifiability conditions caused by operation of the control loop. Note also that no such effect appears when the hybrid algorithm is used, which results in the improvement of cancellation efficiency by 5–15 dB.

5 Extensions

5.1 Recommended Safety Guards

To robustify the SONIC controller against abrupt changes of plant and/or disturbances (e.g. the changes of the nonstationary level of the disturbance, measurement noise variance, impulsive disturbances), and to improve its behavior during the initial transient phase, one may employ several heuristic modifications.

Excessive variability of the adapted parameters may be prevented by upper-bounding the magnitude of the complex gain $|\widehat{\mu}(t)|$, its one-step changes $|\widehat{\mu}(t) - \widehat{\mu}(t-1)|$, and the scaling variable $r(t)$. These modifications can be regarded as safety guards typical to adaptive control. Additionally, to prevent the shutdown of the algorithm, it is recommended to introduce additional lower bounds on the variables $\widehat{\mu}(t)$ and $r(t)$.

Denote by clip(x, a, b), $x \in \mathrm{C}$, $\mathrm{a, b} \in \mathrm{R}_+$ the complex clip function

$$\mathrm{clip}(x, a, b) = \begin{cases} a\frac{x}{|x|} & \text{dla} \quad |x| < a \\ x & \text{dla} \ \ a \le |x| \le b \\ b\frac{x}{|x|} & \text{dla} \quad |x| > b \end{cases} .$$

The modified algorithm (13) has the form

$$\begin{aligned} z(t) &= e^{j\omega_0}\left[(1-c_\mu)z(t-1) - \frac{c_\mu}{\widehat{\mu}(t-1)}y(t-1)\right] \\ r(t) &= \mathrm{clip}(\rho r(t-1) + |z(t)|^2, r_{\min}, r_{\max}) \\ \Delta\widehat{\mu}(t) &= \mathrm{clip}(\frac{z^*(t)y(t)}{r(t)}, 0, \Delta\mu_{\max}) \\ \widehat{\mu}(t) &= \mathrm{clip}(\widehat{\mu}(t-1) - \Delta\widehat{\mu}(t), \mu_{\min}, \mu_{\max}) \\ \widehat{d}(t+1|t) &= e^{j\omega_0}[\widehat{d}(t|t-1) + \widehat{\mu}(t)y(t)] \\ u(t) &= -\frac{\widehat{d}(t+1|t)}{k_n}. \end{aligned} \tag{30}$$

5.2 Modifying the SONIC Controller to Work with Real-Valued Signals and Systems

A variant of the SONIC controller designed to work with real-valued signals was derived in [60]. It was assumed that the controlled plant is governed by Eq. (1), and the assumptions regarding the disturbance signals were modified. The measurement noise $v(t)$ was modeled as real-valued Gaussian-distributed zero-mean white noise with variance σ_v^2, and the narrowband disturbance $d(t)$ as governed by

$$\begin{aligned} d(t) &= \alpha^{\mathrm{T}}(t)\mathbf{f}(t) \\ \boldsymbol{\alpha}(t) &= [\alpha_1(t), \alpha_2(t)]^{\mathrm{T}}, \;\; \mathbf{f}(t) = [\sin(\omega_0 t), \cos(\omega_0 t)] \;, \end{aligned} \tag{31}$$

where the vector $\boldsymbol{\alpha}(t)$ is a two-dimensional random walk process

$$\begin{aligned} \boldsymbol{\alpha}(t+1) &= \boldsymbol{\alpha}(t) + \mathbf{e}(t) \\ \mathbf{e}(t) &= [e_1(t)\; e_2(t)], \;\; \mathbf{e}(t) \sim \mathcal{N}(\mathbf{0}, \sigma_e^2 \mathbf{I}) \;, \end{aligned}$$

which can be regarded as a real-valued counterpart of the complex-valued disturbance governed by (2) and (5).

The inner control loop proposed in [60] takes the form

$$\begin{aligned} \widehat{\boldsymbol{\alpha}}(t+1|t) &= \widehat{\boldsymbol{\alpha}}(t|t-1) + \hat{\mathbf{M}}(t)\mathbf{f}(t)y(t) \\ u(t) &= -\widehat{\boldsymbol{\alpha}}^{\mathrm{T}}(t)\mathbf{K}_n\mathbf{f}(t) \;, \end{aligned} \tag{32}$$

where $\widehat{\alpha}(t+1|t)$ is the one-step prediction of the vector $\alpha(t)$,

$$\mathbf{K}_n = \begin{bmatrix} \mathrm{Re}[K_n(e^{j\omega_0})] & \mathrm{Im}[K_n(e^{j\omega_0})] \\ -\mathrm{Im}[K_n(e^{j\omega_0})] & \mathrm{Re}[K_n(e^{j\omega_0})] \end{bmatrix}$$

denotes the nominal gain matrix of the plant at frequency ω_0, and

$$\hat{\mathbf{M}}(t) = \begin{bmatrix} \mathrm{Re}[\widehat{\mu}(t)] & -\mathrm{Im}[\widehat{\mu}(t)] \\ \mathrm{Im}[\widehat{\mu}(t)] & \mathrm{Re}[\widehat{\mu}(t)] \end{bmatrix}$$

where $\widehat{\mu}(t)$ is the complex-valued gain coefficient, whose value is adjusted in the outer self-optimization loop.

The self-optimization loop reads

$$\begin{aligned} z_y(t) &= -c_\mu \mathbf{f}^{\mathrm{T}}(t)\hat{\mathbf{M}}^{-1}(t-1)\mathbf{z}_\alpha(t-1) \\ \mathbf{z}_\alpha(t) &= \mathbf{z}_\alpha(t-1) + \hat{\mathbf{M}}(t-1)\mathbf{f}(t)z_y(t) + \mathbf{H}\mathbf{f}(t)y(t) \\ r(t) &= \rho r(t) + |z_y(t)|^2 \\ \widehat{\mu}(t) &= \widehat{\mu}(t-1) - \frac{z_y^*(t)y(t)}{r(t)} \;, \end{aligned}$$

where $c_\mu > 0$ is a small constant and

$$\mathbf{H} = \frac{\partial \mathbf{M}}{\partial \mu} = \frac{1}{2}\begin{bmatrix} 1 & j \\ -j & 1 \end{bmatrix}, \quad \mathbf{M} = \begin{bmatrix} \mathrm{Re}[\mu] & -\mathrm{Im}[\mu] \\ \mathrm{Im}[\mu] & \mathrm{Re}[\mu] \end{bmatrix},$$

The above equations were derived using the approach similar to the complex-valued case, i.e., using the Wirtinger calculus and the coefficient fixing technique.

Remarkably, one may obtain almost exactly the same control results using algorithm (13), modified by projecting the complex control signal on the real axis [61]

$$
\begin{aligned}
z(t) &= e^{j\omega_0}\left[(1-c_\mu)z(t-1) - \frac{c_\mu}{\widehat{\mu}(t-1)}y(t-1)\right] \\
r(t) &= \rho r(t-1) + |z(t)|^2 \\
\widehat{\mu}(t) &= \widehat{\mu}(t-1) - \frac{z^*(t)y(t)}{r(t)} \\
\widehat{d}(t+1|t) &= e^{j\omega_0}[\widehat{d}(t|t-1) + \widehat{\mu}(t)y(t)] \\
u(t) &= -\mathrm{Re}\left[\frac{\widehat{\mathrm{d}}(\mathrm{t}+1|\mathrm{t})}{\mathrm{k_n}}\right] .
\end{aligned}
\tag{33}
$$

Taking into account the effort required to develop real-valued versions of algorithms (22)–(24), (26)–(28) from scratch, one is allowed to conclude that the projection method is the recommended approach to the implementation of the SONIC-family controllers in a real-world systems.

5.3 Application of Improved Estimators of Instantaneous Frequency

In many applications, the instantaneous frequency of disturbances varies in, approximately, a piecewise linear way. In such cases, the estimates obtained using the algorithm (16) will be biased, which will affect the cancellation performance of the controller adversely. To mitigate this issue, an improved version of the frequency estimator, which equips it with the capability to track the local trend in the frequency trajectory, was proposed in [62]. The extended algorithm has the form

$$
\begin{aligned}
\widehat{\alpha}(t+1|t) &= \widehat{\alpha}(t|t-1) + \gamma_\alpha g(t) \\
\widehat{\omega}(t+1|t) &= \widehat{\omega}(t|t-1) + \widehat{\alpha}(t+1|t) + \gamma_\omega g(t) \\
g(t) &= \mathrm{Arg}\left[\frac{\widehat{d}(t+1|t)}{\widehat{d}(t|t-1)e^{j\widehat{\omega}(t|t-1)}}\right] ,
\end{aligned}
\tag{34}
$$

where $\widehat{\alpha}(t+1|t)$ and $\widehat{\omega}(t+1|t)$ denote the one-step predictions of frequency rate and frequency, respectively, while γ_α and γ_ω are small gains that satisfy the condition $0 \le \gamma_\alpha \ll \gamma_\omega$. Note that, for $\gamma_\alpha = 0$ and zero initial conditions, the algorithm (34) reduces to (16).

Behavior of the algorithm (34) was analyzed using the approximating linear filter approach, under the assumption that the instantaneous frequency is governed by the following model

$$\omega(t+1) = \omega(t) + \alpha(t+1)$$
$$\alpha(t+1) = \alpha(t) + w_\alpha(t+1)\,, \tag{35}$$

where $\omega(t)$ and $\alpha(t)$ denote the frequency and frequency rate, respectively, and $\{w_\alpha(t+1)\}$ is the sequence of one-step changes of the process $\alpha(t)$. The analysis demonstrated that the proposed improved estimator is statistically efficient. Moreover, simulation results presented in [62] demonstrate that the application of the extended frequency estimator can result in 5–15 dB of improvement in the cancellation performance of the SONIC controller.

5.4 Extension to the Multifrequency Case

The control algorithms presented so far are capably only of canceling sinusoidal disturbances. In many applications, however, one may encounter the need to cancel nonsinusoidal disturbances, among whom the most important group consists of the so-called pseudoperiodic signals. Pseudoperiodic disturbances may be represented as sums of sinusoidal components, whose frequencies are multiples of some fundamental frequency $\omega_0(t)$

$$s(t) = \sum_{k=1}^{K} a_k(t) e^{j \sum_{\tau=0}^{t} \omega_k(\tau)}$$
$$\omega_k(t) = m_k \omega_0(t), \quad k = 1, 2, \ldots, K. \tag{36}$$

In signal processing, the problem of tracking signals with such a structure is often referred to as the comb filtering problem.

Cancellation of signals governed by (36) may be accomplished using a parallel structure, which consists of multiple SONIC controllers, each tracking and canceling a single harmonic component [63]. Unfortunately, such a solution is hardly efficient, because it does not exploit the relationships between the frequencies of all components. As a result, the subcontrollers in the parallel structure are prone to loosing track of weaker components or switching to different—typically stronger—components.

A novel solution of this problem was presented in papers [64, 65], where a new adaptive comb filtering algorithm was proposed, and its application in active noise control was explained. The new algorithm employs a centralized frequency estimation mechanism, which optimally combines the information carried by all components of the tracked signal. It was shown that the new estimator is a statistically efficient estimator of the fundamental frequency and, via Eq. (36), of the frequencies of all harmonic components.

In the case of application in ANC systems, the new algorithm allows one to improve the cancellation performance considerably. Figure 8 depicts the comparison of power spectral densities of the residual noise from an MRI scanner, resulting

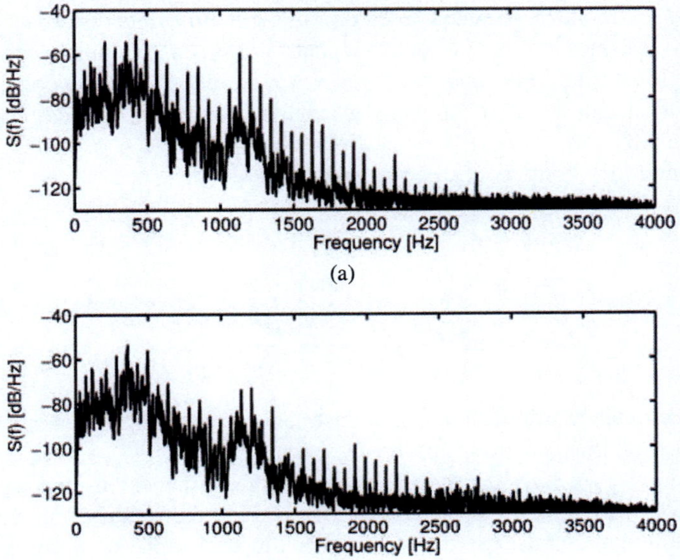

(b)

Fig. 8 Comparison of residual MRI noise power spectral densities: **a** Basic SONIC controller. **b** Comb SONIC controller

from the application of two ANC schemes: standard SONIC algorithm combined with the parallel structure, and the comb variant of the SONIC controller [64]. Poor results obtained using the parallel structure are caused by multiple subcontrollers switching to neighboring, stronger harmonic components. In contrast, the application of the centralized frequency estimation mechanism eliminates this phenomenon completely, which results in unquestionably better controller performance.

5.5 *Extension to the Multivariate Case*

The multivariate SONIC controller was proposed in papers [66, 67]. The structure of the controller is close to its univariate version, given in (13). It consists of the inner control loop, whose task is to track and cancel a disturbance, and the outer self-optimization loop, which performs the optimization of the controller's gains. The inner loop has the form

$$\begin{aligned} \widehat{\mathbf{d}}(t+1|t) &= e^{j\omega_0}[\widehat{\mathbf{d}}(t|t-1) + \widehat{\mathbf{M}}(t)\mathbf{y}(t)] \\ \mathbf{u}(t) &= -\mathbf{K}_n^{-1}\widehat{\mathbf{d}}(t+1|t)\ , \end{aligned} \tag{37}$$

where $\widehat{\mathbf{d}}(t+1|t)$ is the one-step prediction of the multivariate disturbance, $\mathbf{u}(t)$ and $\mathbf{y}(t)$ are N-element vectors of control signals and system outputs, respectively, $\mathbf{K}_n$ is the nominal plant gain matrix at frequency ω_0, and $\hat{\mathbf{M}}(t)$ is a complex matrix of adaptation gains, i.e., a multivariate counterpart of the complex gain $\widehat{\mu}(t)$ in algorithm (13).

The outer loop consists of three recursions

$$\begin{aligned}\tilde{\mathbf{z}}(t) &= e^{j\omega_0}\left[(1-c_\mu)\tilde{\mathbf{z}}(t-1) - c_\mu \mathbf{y}(t-1)\right]\\ \tilde{r}(t) &= \rho\tilde{r}(t-1) + |\tilde{\mathbf{z}}(t)|^2\\ \hat{\mathbf{M}}(t) &= \hat{\mathbf{M}}(t-1)\left[\mathbf{I} - \frac{\mathbf{y}(t)\tilde{\mathbf{z}}^H(t)}{\tilde{r}(t)}\right],\end{aligned} \tag{38}$$

which, due to the multiplicative gain adjustment mechanism, are a nontrivial extension of the first three equations of the algorithm (13).

The multivariate SONIC controller may undergo further extensions, e.g. to cancel disturbances with unknown frequency, or to employ the hybrid configuration. See [67] for simulated and experimental results illustrating the applications of these concepts in the multivariate case.

5.6 *Robustification Against Impulsive Noise*

The robustification of the SONIC controller against the adverse effect of impulsive noise was considered in papers [68, 69]. Impulsive noise is typically modeled as heavy-tailed white noise or using α-stable distributions. One may robustifiy the SONIC controller to such noises by replacing the cost criterion (12) with one that employs the L_1 norm

$$V(t;\mu) = \sum_{\tau=1}^{t} \rho^{t-\tau}|y(\tau;\mu)|\,. \tag{39}$$

The study presented in paper [68] shows that the application of the modified cost criterion improves the controller performance in the presence of α-stable measurement noise, $\alpha \in (1,2)$. However, it was observed that the basic variant of the SONIC controller is also quite robust to this kind of disturbances, in the sense that its performance was completely satisfactory from the practical standpoint.

A much more significant challenge than α-stable noises are the disturbances that result from events such as shocks in the walls of an acoustic duct. The presence of such artifacts may cause the SONIC controller to diverge from the optimal settings, or even to destabilize. A suitable modification of the controller, which includes a mechanism that detects such events and freezes the adaptation process until they decay, was proposed in [69]. Typical control results, obtained in real-world acoustic duct using the robust version of the algorithm, are shown in Fig. 9.

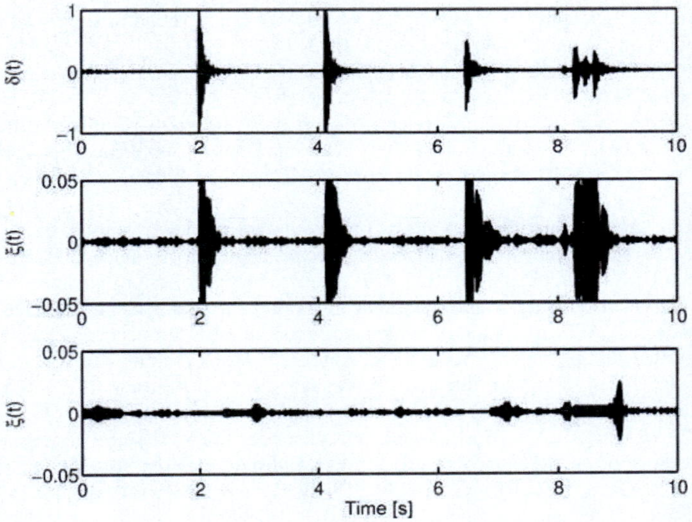

Fig. 9 Comparison of control results obtained using the basic SONIC controller (middle plot) and the robust SONIC controller (bottom plot) in the presence of disturbances caused by mechanical shocks to the acoustic duct (top plot)

References

1. Nelson, P.A., Elliot, S.J.: Active Control of Sound. Academic Press, London (1993)
2. Kuo, S.M., Morgan, D.: Active Noise Control Systems: Algorithms and DSP Implementations. Wiley, New York (1995)
3. van Trees. H.L., Bell K.L., Tian, Z.: Detection, Estimation and Modulation Theory, Part I: Detection, Estimation, and Filtering Theory. Wiley, New York (2013)
4. Morgan, D.R.: An analysis of multiple correlation cancellation loops with a filter in the auxiliary path. IEEE Trans. Acoust. Speech Signal Process. **28**, 454–467 (1980)
5. Crawford, D.H., Stewart, R.W.: Adaptive filtered-V algorithm for active noise control. J. Acoust. Soc. Am. **101**, 2097–2103 (1997)
6. Eriksson, L.J., Allie, M.C., Greiner, R.A.: The selection and application of an IIR adaptive filter for use in active sound attenuation. IEEE Trans. Acoust. Speech Signal Process. **35**, 433–437 (1987)
7. Eriksson, L.J.: Development of the filtered-U algorithm for active noise control. J. Acoust. Soc. Am. **89**, 257–267 (1991)
8. Skogestad, S., Postlethwaite, I.: Multivariate Feedback Control: Analysis and Design. Wiley, New York (2005)
9. Nehorai, A.: A minimal parameter adaptive notch filter with constrained poles and zeros. IEEE Trans. Acoust, Speech Signal Process. **33**, 983–996 (1985)
10. Regalia, P.A.: An improved lattice-based adaptive IIR notch filter. IEEE Trans. Signal Process. **39**, 2124–2128 (1991)
11. Tichavský, P., Händel, P.: Two algorithms for adaptive retrieval of slowly time-varying multiple cisoids in noise. IEEE Trans. Signal Process. **43**, 1116–1127 (1995)
12. Schimidt, R.O.: Multiple Emiter Location and Signal Parameter Estimation. IEEE Trans. Antenn. Propag. **34**, 276–280 (1986)
13. Roy, R., Kailath, T.: ESPRIT - Estimation of Signal Parameters via Rotational Invariance Techniques. IEEE Trans. Acoust. Speech Signal Process. **37**, 984–995 (1989)

14. Bodson, M., Douglas, S.C.: Adaptive algorithms for the rejection of sinusoidal disturbances with unknown frequency. Automatica **33**, 2213–2221 (1997)
15. Bodson, M.: Rejection of periodic disturbances of unknown and time-varying frequency. Int. J. Adapt. Control Signal Process. **19**, 67–88 (2005)
16. Guo, X., Bodson, M.: Adaptive rejection of multiple sinusoids of unknown frequency. In: EUSIPCO 2007, Proceedings of the 2007 European Control Conference, pp. 121–128 (2007)
17. Åström, K.J., et al.: Theory and applications of self-tuning regulators. Automatica **13**, 457–476 (1977)
18. Landau, L.D., Constantinescu, A., Rey, D.: Adaptive narrow band disturbance rejection applied to an active suspension—an internal model principle approach. Automatica **41**, 563–574 (2005)
19. Silva, A.C., Landau, I.D., Airimitoaie, T.-B.: Direct adaptive rejection of unknown time-varying narrow band disturbances applied to a benchmark problem. Eur. J. Control **19**, 326–336 (2013)
20. Eriksson, L.J., Allie, M.C.: Use of random noise for on-line transducer modeling in an adaptive active attenuation system. J. Acoust. Soc. Am. **85**, 797–802 (1989)
21. Gan, W.S., Kuo, S.M.: An integrated active noise control headsets. IEEE Trans. Consum. Electron. **48**, 242–247 (2002)
22. Gan, W.S., Mitra, S., Kuo, S.M.: Adaptive feedback active noise control headset: implementation, evaluation and its extensions. IEEE Trans. Consum. Electron. **51**, 975–982 (2005)
23. Kuo, S.M., Chen, Y.-R., Chang, Ch-Y, Lai, Ch-W: Development and evaluation of light-weight active noise cancellation earphones. Appl. Sci. **8**, 1178 (2018)
24. Castae-Selga, R., Pea, R.: Active noise hybrid time-varying control for motorcycle helmets. IEEE Trans. Control Syst. Technol. **18**, 602–612 (2010)
25. Liu, L.C., Kuo, S.M., Raghuathan, K.: An audio integrated motorcycle helmet. J. Low Freq. Noise Vib. Active Control **29**, 161–170 (2010)
26. Raghunathan, K., Kuo, S.M., Gan, W.S.: Active noise control for motorcycle helmets. In: Proceedings of the 2010 International Conference on Green Circuits and Systems, pp. 170–174. Shanghai, China (2010)
27. Wang, T.W., Gan, W.S., Kuo, S.M.: Subband-based active noise equalizer for motorcycle helmets. In: Proceedings of the 2010 ASIPA ASC 2010, pp. 555–559. Singapore (2010)
28. Chakrovorthy, S.R.: Active snore noise control systems. M.S. Thesis, Northern Illinois University (2005)
29. Chakravarthy, S., Kuo, S.M.: Application of active noise control for reducing snore. In: ICASSP 2006, Proceedings of the 2006 International Conference on Acoustics Speech and Signal Procesing, pp. 305–308. Toulose, France (2006)
30. Kuo, S.M., Gireddy, R.: Real-time experiment of snore active noise control. In: Proceedings of the 2007 International Conference on Control Applications, pp. 1085–1092. Singapore (2007)
31. Kuo, S.M., Liu, L., Gujjula, S.: Development and application of audio-integrated active noise control system for infant incubators. J. Noise Control Eng. **58**, 163–175 (2010)
32. Liu, L., Du, L., Kolla, A.: Wireless communication integrated hybrid active noise control system for infant incubators. In: Proceedings of the 2016 IEEE Signal Processing in Medicine and Biology Symposium, pp. 1–6. Philadelphia, USA (2016)
33. Yu, X., Gujjula, S., Kuo, S.M.:Active noise control for infant incubators. In: Proceedings of the 2009 Annual International Conference of the IEEE Engineering in Medicine and Biology Society, pp. 2531–2534. Minneapolis, USA (2009)
34. Chambers, J., Bullock, D., Kahana, Y., Kots, A., Palmer, A.: Developments in active noise control sound systems for magnetic resonance imaging. Appl. Acoust. **68**, 281–295 (2007)
35. Chen, C.K., Chiueh, T.T., Chen, J.H.: Active cancellation system of acoustic noise in MR imaging. IEEE Trans. Biomed. Eng. **46**, 186–191 (1999)
36. Kannan, G., Milani, A.A., Panahi, I., Briggs, R.W.: An efficient feedback active noise control algorithm based on reduced-order linear predictive modeling of fMRI acoustic noise. IEEE Trans. Biomed. Eng. **58**, 3303–3309 (2011)
37. Panahi, I.M.S.: Development of ANC methods for fMRI rooms: Design challenges and algorithms. In: Proceedings of the 2013 International Symposium on Image and Signal Processing and Analysis, pp. 655–660. Trieste, Italy (2013)

38. Takar, M.S., Kumar Sharma, M., Pal, R.: A review on evolution of acoustic noise reduction in MRI. In: Proceedings of the 2017 Recent Developments in Control, Automation Power Engineering, pp. 235–240. Noida, India (2017)
39. Kajikawa, Y., Gan, W.S., Kuo, S.M.: Recent advances on active noise control: open issues and innovative applications. APSIPA Trans Signal Inf. Process. **1**, 1–21 (2012)
40. Niedźwiecki, M., Meller, M.: A new a pproach to active noise and vibration control—Part I: the known frequency case. IEEE Trans. Signal Process. **57**, 3373–3386 (2009)
41. Söderström, T., Stoica, P.: System Identification. Prentice Hall, Englewood Cliffs, New Jersey (1988)
42. Van den Bos, A.: Complex gradient and Hessian. IEE Proc. Image Signal Process. **141**, 380–382 (1994)
43. Brandwood, D.H.: A complex gradient operator and its application in array theory. Proc. Inst. Elect. Eng. **130**, 11–16 (1983)
44. Haykin, S.: Adaptive Filter Theory. Prentice Hall, Englewood Cliffs (1996)
45. Wirtinger, W.: Zur formalen theorie der funktionen von mehr komplexen veränderlichen. Math. Ann. **97**, 357–375 (1927)
46. Niedźwiecki, M.: Steady state and parameter tracking properties of self-tuning minimum variance regulators. Automatica **25**, 597–602 (1989)
47. Niedźwiecki, M., Meller, M.: A new a pproach to active noise and vibration control—Part II: the unknown frequency case. IEEE Trans. Signal Process. **57**, 3387–3398 (2009)
48. Bai, E.-W., Fu, L.-C., Sastry, S.: Averaging analysis for discrete time and sampled data adaptive systems. IEEE Trans. Circ. Syst. **35**, 137–148 (1988)
49. Niedźwiecki, M., Kaczmarek, P.: Tracking analysis of a generalized adaptive notch filter. IEEE Trans. Signal Process. **54**, 304–314 (2006)
50. Jury, M.: Theory and Application of the $\mathscr{Z}$-transform Method. Wiley, New York (1964)
51. Niedźwiecki, M., Meller, M.: SONIC—Self-optimizing narrowband interference canceler: comparison of two frequency tracking strategies. In: Proceedings of the 8th IEEE International Conference on Control and Automation, pp. 1892–1896. Xiamen, China (2010)
52. Manolakis, D.G., Ingle, V.K., Kogon, S.M.: Statistical and Adaptive Signal Processing: Spectral Estimation, Signal Modeling, Adaptive Filtering, and Array Processing. McGraw-Hill, Boston (2000)
53. Niedźwiecki, M.: Identification of Time-Varying Processes. Wiley, New York (2000)
54. Niedźwiecki, M., Meller, M.: Hybrid SONIC: joint feedforward-feedback narrowband interference canceller. Int. Journ. Adaptive Contr. and Signal Process. **27**, 1048–1064 (2013)
55. Niedźwiecki, M., Sobociński, A.: A simple way of increasing estimation accuracy of generalized adaptive notch filters. IEEE Signal Process. Lett. **14**, 217–220 (2007)
56. Niedźwiecki, M., Kaczmarek, P.: Generalized adaptive notch and comb filters for identification of quasi-periodically varying system. IEEE Trans. Signal Process. **53**, 4599–4609 (2005)
57. Niedźwiecki, M.: Adaptive notch smoothing revisited. In: EUSIPCO 2008, Proceedings of the 16th European Signal Processing Conference, pp. 1–5. Lausanne, Switzerland (2008)
58. Niedźwiecki, M.: Generalized adaptive notch smoothing revisited. IEEE Trans. Signal Process. **58**, 1565–1576 (2010)
59. Niedźwiecki, M., Meller, M.: New algorithms for adaptive notch smoothing. IEEE Trans. Signal Process. **59**, 2024–2037 (2011)
60. Niedźwiecki, M., Meller, M.: Self-optimizing adaptive vibration controller. IEEE Trans. Autom. Control **54**, 2087–2099 (2009)
61. Meller, M.: Samooptymalizujące adaptacyjne tłumienie zakłóceń wąskopasmowych. Ph.D. Dissertation, Politechnika Gdańska, Wydział Elektroniki, Telekomunikacji i Informatyki (2010)
62. Niedźwiecki, M., Meller, M.: An improved frequency estimator for an adaptive active noise control scheme. In: EUSIPCO 2010, Proceedings of the 18th European Signal Processing Conference, pp. 353–357. Aalborg, Denmark (2010)
63. Niedźwiecki, M., Meller, M.: Multifrequency self-optimizing narrowband interference canceller. In: IWAENC 2010, Proceedings of the 18th International Workshop on Acoustic Echo and Noise Control, pp. 1–4. Tel Aviv, Israel (2010)

64. Niedźwiecki M., Meller, M., Kajikawa, Y., Łukwiński, D.: Estimation of nonstationary harmonic signals and its application to active control of MRI noise. In: CASSP 2013, Proceedings of the 2013 IEEE International Conference on Acoustics, Speech and Signal Processing, pp. 5661–5665. Vancouver, Canada (2013)
65. Niedźwiecki, M., Meller, M.: Generalized adaptive comb filters/smoothers and their application to the identification of quasi-periodically varying systems and signals. Automatica **49**, 1601–1613 (2013)
66. Meller M., Niedźwiecki, M.: Multiple-channel frequency-adaptive active vibration control using SONIC. In: ISPA 2013, Proceedings of the 8th International Symposium on Image and Signal Processing and Analysis, pp. 620–625. Trieste, Italy (2013)
67. Meller, M., Niedźwiecki, M.: Multichannel self-optimizing narrowband interference canceller. Signal Process. **98**, 396–409 (2014)
68. Niedźwiecki M., Meller, M.: Robust algorithm for active feedback control of narrowband noise. In: EUSIPCO 2015, Proceedings of the 23rd European Signal Processing Conference, pp. 275–279. Nice, France (2015)
69. Niedźwiecki M., Meller, M.: Active feedback noise control in the presence of impulsive disturbances. In: ICASSP 2015, Proceedings of the 2015 IEEE International Conference on Acoustics, Speech and Signal Processing, pp. 659–663. Brisbane, Australia (2015)

Methods of Device Noise Control

Marek Pawełczyk, Stanisław Wrona, and Krzysztof Mazur

Abstract Technological development has a high impact on the everyday life of humans and their working conditions. Noise generated by devices is an important problem in numerous environments. Such a problem is explored in this chapter. A classical protection solution is to apply passive sound insulating materials. However, passive barriers are often ineffective, especially at low frequencies. They may also be inapplicable due to technical reasons. An alternative way is to use active control methods. If a device generating noise is surrounded by a thin-walled casing, or if it can be enclosed in an additional casing, control inputs can be applied directly to the structure, and as a whole the casing can be used as an active barrier enhancing acoustic isolation of the device. If appropriately implemented, it results in a global noise reduction instead of local zones of quiet. The aim of the authors is to present the theoretical basis for developing active device casings. The chapter presents exemplary plants, discusses the principles of building a mathematical model and using it to build optimal noise reduction systems, passive, semi-active and active. Examples of results confirming the development of a new concept of global noise reduction for machines and devices are given.

M. Pawełczyk (✉) · S. Wrona · K. Mazur
Silesian University of Technology, Department of Measurements and Control Systems, Akademicka 16, 44-100 Gliwice, Poland
e-mail: marek.pawelczyk@polsl.pl

S. Wrona
e-mail: stanislaw.wrona@polsl.pl

K. Mazur
e-mail: krzysztof.jan.mazur@polsl.pl

P. Kulczycki et al. (eds.), *Automatic Control, Robotics, and Information Processing*, Studies in Systems, Decision and Control 296,
https://doi.org/10.1007/978-3-030-48587-0_27

1 Introduction

Sound accompanies humans in their everyday life. It is used for communication, during leisure or for many medical and technical applications. However, sometimes the type of sound is unpleasant or unwanted. This is an informal definition of noise. It may reduce concentration and annoy, or even impair the hearing system, when people are prolongly exposed to a loud noise. Besides, it also has a negative impact on other basic human systems. It can cause stress reactions, lead to pathological alterations in the myocardium and the vascular walls, and even deteriorate vision acuity. Also, in households, there are many everyday devices including PCs, hoovers, food processors, heaters with fans, washing machines, etc., which generate annoying noise. Passive noise barriers are in many cases ineffective or cannot be applied [1], because they increase the size and weight of the devices or result in overheating, which causes the breaking of the devices. An innovative solution is to modify properties of the structure by appropriately arranging local masses and ribs, so that the structure is more resistant to the frequencies of noise concentration.

An alternative is to use active control methods [2–4]. In classical active noise control (ANC) an additional secondary sound source is used to cancel the noise from the original primary source [5–8]. The physical justification is given by Young's principle of destructive interference. The secondary source can also change the radiation acoustic impedance thereby reducing the sound power radiated or absorb the primary sound power [9]. In practice there are many problems related to physical aspects of the cancellation phenomenon as well as related to control. In a diffuse acoustic field, the global active noise control performed in that way is practically unfeasible for an entire enclosure [10]. The solution is thus the local control in a particular area or some areas and creation of the so-called "zones of quiet". Actually, the control is performed at a given point in space and the attenuation propagates from this point in the form of a zone. Such a zone can be created, e.g. around the user's head. This, however, is related to the imperfection of the entire system in the case of user movements and non-stationary noise. The authors, after gaining experience in the field of classical noise reduction, decided to look for another solution that would guarantee global reduction.

The sound source is one of the most crucial components of each ANC system. Its properties should be adequate for the given specific environmental conditions. For some applications, the sound source is subject to dust, high temperature, high humidity or precipitation. If feasible, it should then be moved to a less harsh area. However, in most situations, such a solution does not apply. The idea of using vibrating plates as secondary sound sources in the ANC is well known. However, vibrating plates are harder to control than loudspeakers because of nonlinearity and high variations of amplitude response [11, 12]. Radiation from vibrating plates is still of scientific interest [13–15]. For many applications, multiple actuators are mounted on a single plate, which complicates the control problem further [16]. The system becomes multichannel, and commonly used algorithms for noise control require several actual secondary path models to operate. If the noise source is isolated from the area where the reduction is required by a wall (plate, window, casing, etc.) such a wall can be made to vibrate in order to protect the noise against passing through [17, 18]. This approach is known as active structural acoustic control (ASAC). Vibration of the plates can be generated by piezoelectric patches or stacks [19], or by different types of shakers [20]. Dependent on the control strategy, they can be used to reduce sound transmission and emission by the plate. The vibrating plates can also play the role of a sound source and can be used instead of loudspeakers for ANC to create zones of quiet. The idea can be extended to multi-plate structures, although, it significantly complicates the control problem. The number of actuators needed to control the whole structure increases. The model of such a structure is much more complex.

In this chapter, the theoretical basis for developing active device casings is outlined. All types of casing design solutions that are relevant to the operation of a given device are allowed, including reinforcements, fasteners, ventilation openings, and complex shapes. The general concept is to enclose the noise generating device by a thin-wall casing or use its own casing, and make its walls vibrate by using piezoelectric or electrodynamic exciters. The control system should provide signals to drive the exciters in order to acoustically isolate the device inside (block the noise going out, according to the ASAC idea). The use of active control for the whole device casing was initially proposed by Fuller et al. [21], but no real implementation, as well as plant and control system analysis, have been provided. This task was undertaken by the authors of this chapter, who, together with a team of collaborators, completed a number of projects aimed at comprehensively developing the concept and using it on convincing real-life examples [22].

Due to the high complexity of the plant caused by the multidimensionality and interactions of vibrating walls on each other affecting the boundary conditions, the presented research was split into three stages. Firstly, the casing of a rigid frame structure in which every wall is made of a separate panel/panels was considered [23–25]. In this design, the interaction of forced vibrations of individual walls on the other is reduced. It does not protect, however, against the complex phenomena of interference of noise generated by the vibration of each panel. The configuration of double-panel walls was also considered. In the second stage, a light frameless flexible casing was examined, in which the control of the wall vibration greatly stimulates the other walls. The vibrating wall changes the boundary conditions of neighbouring walls. Additional interference is generated through the acoustic field as well [26, 27]. Finally, a regular device made of thin walls and available on the market was considered. It was a washing machine with the whole associated complexity, vents for heat removal, brackets for the engine, fold of the structure, and non-stationary noise.

Physical models of separate single and double-panel walls as well as models of the whole casing were developed [28–31]. They include various vibroacoustic interactions. The wall models were verified by laboratory measurements and by numerical analysis with a multiphysics package [32]. They were used for optimization of distribution of sensors, actuators and passive elements to increase acoustic isolation [33].

A number of control structures and algorithms, and data exchange procedures were designed. A global reduction of over 10 dB is achieved in the entire room (in some zones it was even over 20 dB) for tonal or multitonal noises generated inside the casing. The results are confirmed by a formal metrological analysis based on measurements with a calibrated sound analyser and in accordance with current standards [34, 35]. Moreover, thanks to the virtual microphone control approach the active casing constitutes an autonomous solution, which does not involve employment of any microphones around the casing in the room, and it uses a structural point or surface sensors bonded to the walls only, preferably from the inside [36]. Additionally, it guarantees stable operation for changes in the room around the casing caused by moving people or rearrangement of room equipment. In the case of complex noise, e.g. generated by the washing machine during fast tumbling and spinning, the reduction is still global in the entire room, although the reduction level is lower.

Although this chapter is mainly focused on active control methods, it is worth mentioning that an ability to reduce device and machinery noise by controlling vibration of its casings was also examined for problems where the crucial restriction is the system complexity and cost. Semi-active control was considered, where piezoelectric patches bonded to the casing walls transform the mechanical energy into electrical energy, which is then dissipated by an electric circuit [37]. Automatic control is responsible only to assure appropriate switching in the circuit in this solution. Additionally, a passive method improving acoustic isolation of the walls for the chosen frequency bands was developed [32, 33]. It does not employ any control system and it does not consume any energy during operation. For this purpose, a method for shaping frequency response of the vibration panels according to precisely defined demands was developed. It is based on mounting several additional ribs and masses to the panel surface at locations followed on from an optimization process. It can also be used to improve properties of the casing as a plant for active structural acoustic control. Then, the location of actuators on the walls can also be included in the overall optimization process.

The chapter is organized as follows. In Sect. 2 the considered plants are introduced, emphasizing important features related to active control and mathematical modeling. The employed mathematical model of the plant with its simplifying assumptions is briefly described. In Sect. 3, the model is employed to optimize the actuators and sensors placement, which strongly influences the final performance of the control system. In Sect. 4, a passive method improving acoustic isolation of the walls for chosen frequency bands is outlined. In Sect. 5, a semi-active control with piezoelectric patches bonded to the casing walls is considered. In Sect. 6, an adaptive control algorithm is shown. Several control approaches and techniques are presented and discussed. Selected experimental results are also provided to better visualize the performance of the active casing approach. The chapter is summarized with concluding remarks and directions for future research.

2 The Plant and Its Mathematical Modeling

Three types of active casings are considered in the presented research. The first type is a rigid casing (presented in Fig. 1a), in which each wall is a separate panel or double panel. The panels of the casing are mounted to a heavy frame. It results in limited vibrational couplings between the walls, facilitating initial attempts to control their vibrations individually. However, it does not protect against complex acoustic interference of the noise generated by the vibration of each panel and couplings between walls through the acoustic field, especially for low-frequency noise. The second type of casing is a light-weight casing, which in contrast to the previous structure is made without an explicit frame (visible in Fig. 1b). Such a structure results in greater vibrational couplings between individual walls, in addition to the couplings through the acoustic field inside and, to a lesser extent, outside the casing. The casing is made of metal plates bolted directly together. The third type is a casing of a real device (presented in Fig. 1c). Similar to a light-weight casing, it exhibits strong couplings between the walls. In addition, it has a number of irregularities (embossing, bendings, internal mountings) that complicate the mathematical modelling of the structure.

To obtain a more reproducible environment for a control system evaluation, a loudspeaker placed inside the casings was frequently used to generate the primary noise. However, active control during real device operations (e.g. spinning cycle of a washing machine, etc.) is also being considered by the authors through ongoing research.

In front of each controlled casing wall, an *error microphone* is placed at a distance of 500 mm. These microphones are used strictly for the control purpose. Additionally, to evaluate the noise reduction performance, multiple microphones are placed at greater distances from the casing, corresponding to potential locations of the user (they are referred to as the *room microphones*). To control vibrations of the device casing, inertial mass actuators EX-1 are used. They are light-weight (115 g) actuators of small dimensions (diameter 70 mm), compared to the size of the casing. For the washing machine, to facilitate laboratory control experiments, they are mounted on the casing walls from the outside, although, in final applications they can be attached to the inner side. Depending on the given casing and followed from optimization, different numbers of actuators have been used: 15 for the rigid casing, 21 for the light-weight casing, and 11 for the washing machine casing.

2.1 *Mathematical Modelling*

The response of a physical plant excited to vibrations is determined by its mechanical structure. Even slight modifications of the structure may strongly affect its frequency characteristics. On the other hand, the performance of an active noise/vibration control system is highly dependent on both the plant structure itself and an arrangement of actuators and sensors applied to it, affecting the observability and controllability

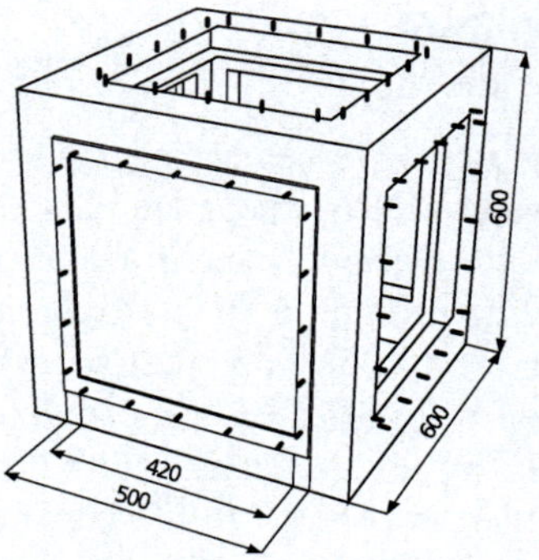

(a) A schematic representation of the rigid casing.

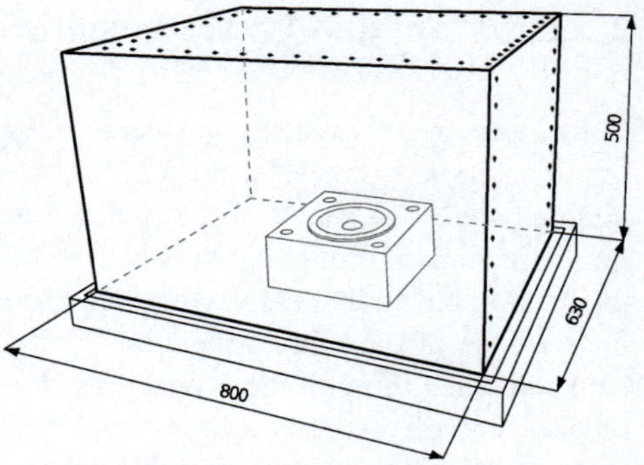

(b) A schematic representation of the light-weight casing.

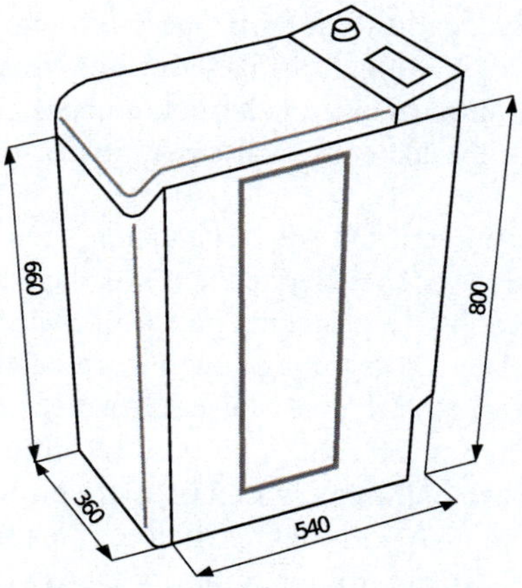

(c) A schematic representation of the real device — a washing machine.

Fig. 1 Schematic representations of active casings. All dimensions are given in (mm)

of the system [38]. Errors in this respect can significantly impede or prevent noise reduction [39]. Therefore, the active control of an object like the considered device casing should be preceded by a thorough investigation of its mechanical structure. Firstly, to maximize the susceptibility of the plant to active control by plausible modifications (e.g. by mounting some additional passive elements on the casing walls to appropriately shape the frequency response [32, 33]). Secondly, to efficiently apply actuators and sensors, so the maximum advantage can be taken from them (e.g. by maximizing measures of controllability and observability of the obtained system).

The efficiency of the selected arrangement of actuators and sensors determines the control properties of the system, and consequently, the synthesis of appropriate control strategies depends on it. It is known that an inadequate arrangement of actuators and sensors will make it impossible to meet certain control objectives,

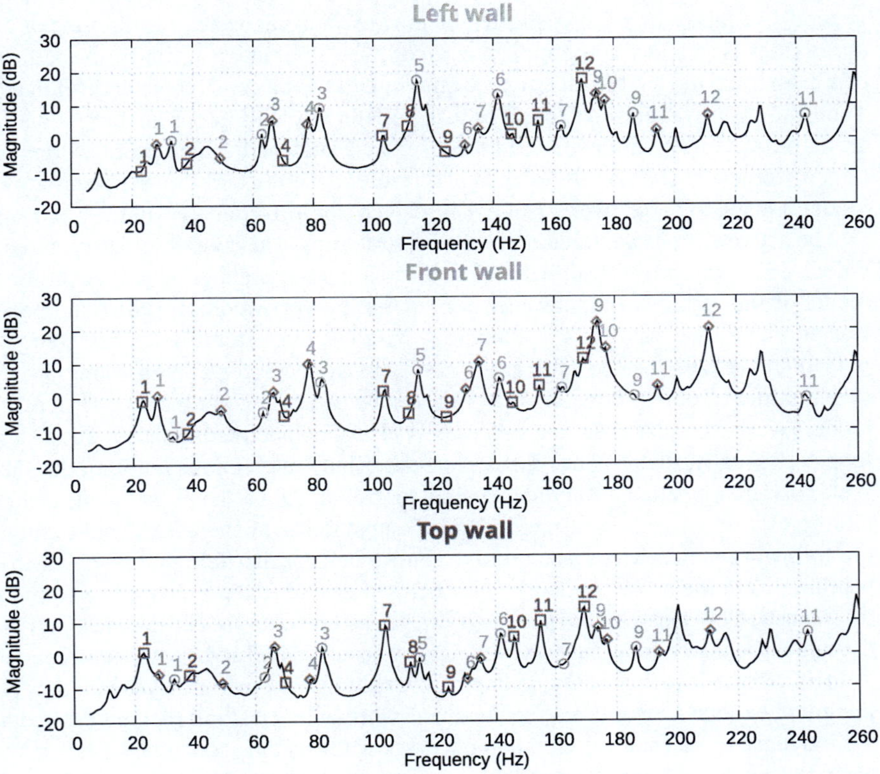

Fig. 2 Spatially averaged frequency responses of light-weight casing walls. Two pairs of walls are symmetrical (left and right, front and back), hence only one of each pair is presented in the figure. Initial 12 eigenmodes originating at each wall are marked: eigenmodes originating at left wall are marked with a circle, at front wall with a diamond, at top wall with a square

regardless of the applied control strategy [39]. This is directly related to the loss of the controllability and observability properties. For this reason, the results obtained from the analysis of controllability and observability provide valuable guidelines for the proper arrangement of actuators and sensors. However, the analysis of the conditions of controllability and observability, due to the qualitative nature of these concepts, is insufficient to determine the most favorable distribution of actuators and sensors from the point of view of control objectives [38]. Therefore, quantitative measures of the observability and controllability of the system were formulated and used for the appropriate definition of the optimization problem [32]. The coordinates of additional elements attached to the structure are the main optimization variables (the optimization process is further considered in Sect. 3).

Sometimes an intuition, expert knowledge, or simply the trial and error method can be used to improve the properties of a vibrating structure and/or its active control system. But it is limited to the simplest cases. A complicated scenario would generally require a more sophisticated method to obtain expected results. One of the approaches is to apply an optimization algorithm. To employ it, firstly, in the form of a theoretical model, the investigated structure has to be constructed and validated. Then, the objectives and constraints need to be defined in an appropriate form. Finally, an optimization algorithm should be chosen and launched utilizing the aforementioned elements. The quality of obtained results depends both on the accuracy of modelling and the effectiveness of the employed optimization algorithm.

The active casings are three-dimensional structures. The couplings between individual walls, of both vibrational and acoustical nature, are significant. A physical model of the whole casing has been formulated by Wyrwal et al. [28]. The model was developed according to a methodology based on three steps. In the first step, the objective was to extract the dynamical subsystems for each considered casing, which was important from the point of view of the casing behaviour. For the second step, the models were formulated for the subsystems identified and distinguished during the first step. Finally, the physical phenomena describing interactions between the subsystems were modelled, providing the final mathematical model of the whole device casing. During extraction of the dynamical subsystems of the considered casings, the following factors were taken into account: the acoustic field inside the casing, resulting from the noise emitted by the enclosed device (the primary noise source); the acoustic field between the panels (if double-panel walls are considered); and the dynamical behaviour of casing walls excited by acoustic field and by actuators bonded to their surface. During the modelling of interactions between subsystems, the following phenomena were taken into account: vibro-acoustical interactions related to the influence of mechanical vibration of casing walls on the acoustic field inside the casing; acousto-vibrational interactions reflecting an impact of acoustic field inside the casing on the vibrations of the walls, including cross-coupling between casing walls.

The complexity of the model formulated in such a way is very high, which makes it difficult to implement and employ such model in an optimization process of actuators arrangement. During optimization, the model has to be solved many thousands of times, which would be very time consuming. However, noteworthy, is an analysis of spatially averaged (over the area of each wall) frequency responses of casing walls. Such responses for light-weight casing walls are presented in Fig. 2 (left, front and top walls; right and back walls are omitted as they are symmetrical to the left and front walls, respectively). Resonances that originate on one wall, are clearly visible on the others. Moreover, all of the resonances visible as peaks in the given frequency range can be assigned to one of the walls where they originate. Such assignment of the eigenmodes is consistent with the mathematical model of an individual plate developed in details and validated in [32].

It leads to a conclusion that the observed natural frequencies and modeshapes of the whole structure are a consequence of superposition of resonances of each wall excited individually (but as a part of the structure). Therefore, it is justified to analyse the walls separately for the purpose of optimization of actuators locations, considering only eigenmodes due to the given wall (if the resonance is controlled with actuators at the wall where it originates, it will be reduced for the whole casing).

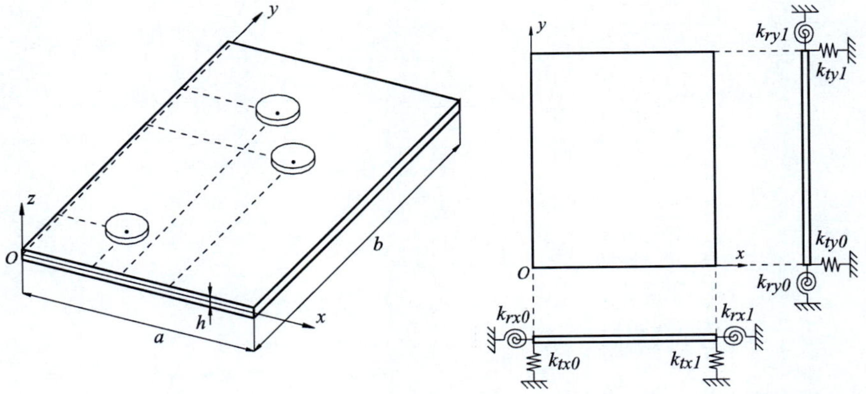

(a) A visualization in an isometric projection.

(b) A multiview orthographic projection of the plate with boundary conditions represented as rotational and translational springs.

Fig. 3 Rectangular plate with actuators bonded to its surface

The mathematical model employed for analysis of individual walls is based on the Mindlin plate theory. It includes the impact of actuators' masses. The model is solved utilising the Rayleigh-Ritz assumed mode-shape method. Characteristic orthogonal polynomials having the property of Timoshenko beam functions are used as eigenfunctions satisfy edge constraints. The model is presented schematically in Fig. 3. Due to real (and hence imperfect) mountings of each of the casing walls, boundary conditions elastically restrained against both rotation and translation are employed. It is noteworthy that the spring constants describing boundary conditions of the walls cannot be measured or calculated directly. Therefore, for the purpose of fitting the model to the behaviour of a real vibrating structure, an optimization algorithm is used to identify them. The process of identification based on experimental data is described in detail in [40].

3 Optimization of Actuators Arrangement

The mathematical model introduced in the previous section can be employed for the process of optimization of actuators arrangement on the active device casing. Each wall is evaluated in this process separately. The search space following on from the actuators arrangement problem is very complicated. An efficient algorithm needs to be chosen to find a solution satisfying defined demands. Evolutionary algorithms have proven to be a versatile and effective technique for solving nonlinear optimization problems with multiple optima [41]. However, they usually require evaluation of

Wall	Act.	$x_{a,i}$ (m)	$y_{a,i}$ (m)
Top	1	0.100	0.066
	2	0.322	0.445
	3	0.389	0.564
	4	0.610	0.091
	5	0.695	0.247
Left	1	0.077	0.455
	2	0.424	0.152
	3	0.542	0.222
	4	0.565	0.304
Front	1	0.134	0.065
	2	0.307	0.164
	3	0.674	0.445
	4	0.735	0.263

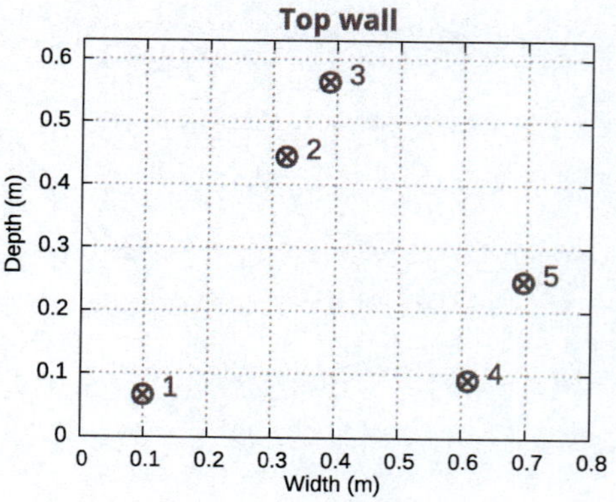

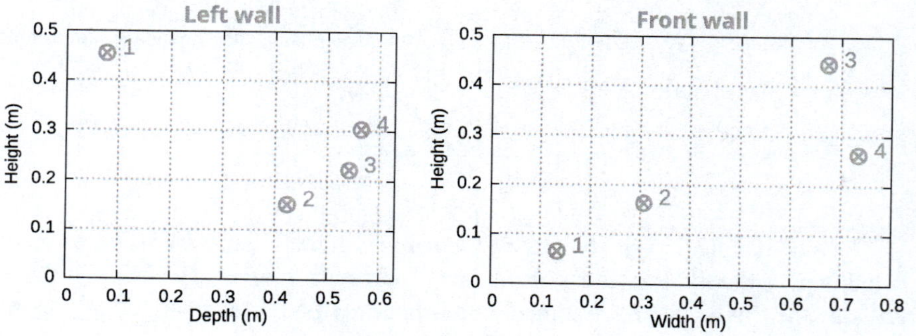

Fig. 4 Arrangement of actuators on the light-weight casing walls. Two pairs of walls are symmetrical (left and right, front and back), hence only one of each pair is presented

numerous solutions resulting in high computational cost. To mitigate this drawback, memetic algorithms can be utilised, which are hybrid forms of a population-based approach coupled with separate individual learning [42, 43]. Memetic algorithms combine advantages of a global search, as for evolutionary algorithms, and local refinement procedures, which enhance convergence to the local optima [44]. Because of complementary properties, they are particularly useful in solving complex multi-parameter optimization problems. Hence, a memetic algorith can be employed to perform the optimization.

The optimization variables are the coordinates of actuators on the plate surface. The optimization index J is a measure of controllability of the least controllable mode:

$$J = \min_i \lambda_{c,i} \, , \qquad i \in \{1, 2, \ldots, N_{\text{mod}}\} \, , \tag{1}$$

where $\lambda_{c,i}$ is the i-th diagonal element of the controllability Gramian matrix, corresponding to the i-th eigenmode [32].

To better clarify the process, light-weight casing is used as an example. The goal in the considered example is to maximize the controllability of eigenmodes in the frequency range up to 300 Hz. Hence, depending on the number of eigenmodes included in the given frequency range, a different number of eigenmodes was considered for each wall of the casing: $N_{\mathrm{mod}} = 25$ for top wall, $N_{\mathrm{mod}} = 21$ for front and back wall, $N_{\mathrm{mod}} = 17$ for left and right wall. Five actuators locations have been optimized for the top wall and four actuators locations for the remaining walls.

As the two pairs of walls are symmetrical (left and right, front and back), the same configuration has been calculated for a given pair. The obtained arrangement of actuators is presented in Fig. 4. It follows from analysis of many optimization scenarios that if the number of considered modes is high, the optimal trade-off locations of actuators ensuring controllability for each mode are impossible to intuitively determine without a mathematical model and optimization tools. Especially, this is the case when the actuators themselves load the walls and thus change their modeshapes to irregular forms.

4 Passive Control

In applications where the external source of energy is completely unavailable, passive control methods can be applied to reduce noise and vibrations of a device casing. One of the approaches well explored in the literature is to add passive components to the system (dampers, springs, passive shunt circuits, etc.) to increase the damping in the system. The authors propose an alternative method, which is based on appropriately located passive elements, which can be used to alter the frequency response of the vibrating structure, thus improving its sound-insulation properties [32, 33].

For device casings, it means mounting several additional ribs and masses to the walls surface at locations followed on from an optimization process. General rules are known: additional masses lower the natural frequencies of the plate, whereas stiffeners elevate them. However, the presence of additional masses and stiffeners has not been analysed and used together, especially for shaping the frequency response according to precisely defined demands. The shaping of the frequency response can provide high passive attenuation for certain frequency bands. Moreover, if active control is considered, the sensors and actuators arrangement can be optimized together with passive masses and ribs, improving the structure control susceptibility. The limits are related mainly to the maximum dimensions and mass of the created structure. Visualization in an isometric projection of a rectangular plate with additional elements has been given in Fig. 5.

The optimization process fundamentals resemble the process described in the previous section. However, there are two main extensions that have to be included. Firstly, the mathematical model of the plate has to be developed to include the mass loading and stiffening effect of the additional elements, such as additional masses and ribs. Derivations of such model in detail has been done in [32]. A state space model of a rectangular orthotropic plate with additional masses and stiffeners bonded to its surface is then developed. Presence of sensors and actuators is also modelled as additional masses. Plate edges are considered elastically restrained against both

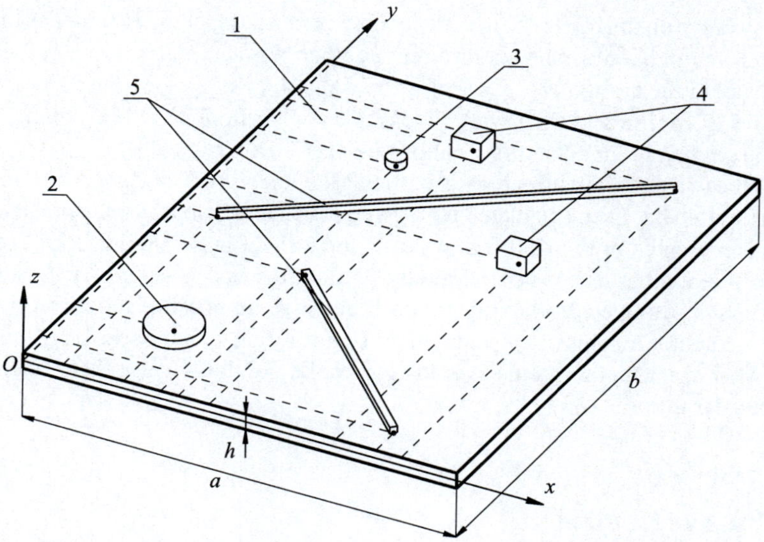

Fig. 5 Visualization of an isometric projection of a rectangular plate (1) with actuators (2), sensors (3), additional masses (4) and stiffeners (5) bonded to its surface

translation and rotation. The Mindlin plate theory, which includes the effect of shear deformation and rotational inertia is used, along with the Engesser theory associated with the consideration of torsion for the stiffeners.

The second extension is that the optimization problem needs to be reformulated, incorporating new optimization variables. For additional masses, assuming their are of small size, the most important aspect is their location. However, for additional ribs, there are more parameters that can be optimized. These include not only location, but also orientation, cross-section shape and dimensions, length, etc. Hence, the search space followed from the frequency response shaping problem discussed above is very complicated. The number of its dimensions often reaches several dozen. For such an optimization task, the employment of the memetic algorithm is even more justified [33].

The adopted cost function should reflect the discrepancy between the desired and actual frequency responses of the plate. Various approaches can be utilized to define the cost function. It can be defined on the basis of natural frequencies, ω_i, and magnitudes of the response, where i stands for the eigenmode number. On the other hand, overall transmission in a whole bandwidth can also be specified to be either enhanced or attenuated, depending whether the role of the plate is to act as a structural noise source or a noise barrier, respectively. Moreover, if an active control application is considered, measures of controllability and observability of the system expressed in the form of diagonal elements of the controllability and observability

Gramian matrices, $\lambda_{c,i}$ and $\lambda_{o,i}$, respectively, can also be introduced in the cost function. Therefore, in the general form the cost function can be presented as:

$$J = f(\omega_i, \lambda_{c,i}, \lambda_{o,i}, \Theta), \tag{2}$$

where Θ is a vector representing parameters of the elements to be mounted to the plate, including their shapes and locations. More detailed forms of particular cost functions are presented in [33], where optimization results for different scenarios are presented and discussed.

5 Semi-active Control

Noise passing through a barrier can be reduced by controlling the barrier's vibrations [11, 46]. For problems where the system complexity and the cost is substantially limited, alternatively to the ASAC system, semi-active control can be applied to reduce vibrations, thus limiting the emitted noise. The semi-active system requirement for external energy is usually very low and it is used only to favourably modify in real-time the characteristics of damping elements embedded in the system. In general, a semi-active system dissipates the mechanical energy already present in the system, thus they are inherently stable (e.g. piezoelectric element attached to a vibrating structure transforms the mechanical energy into electric energy, which is then dissipated in a passive electric shunt circuit). The efficiency of such a system lies in the adaptation of its properties to the current (or predicted) state of the plant, external dis-

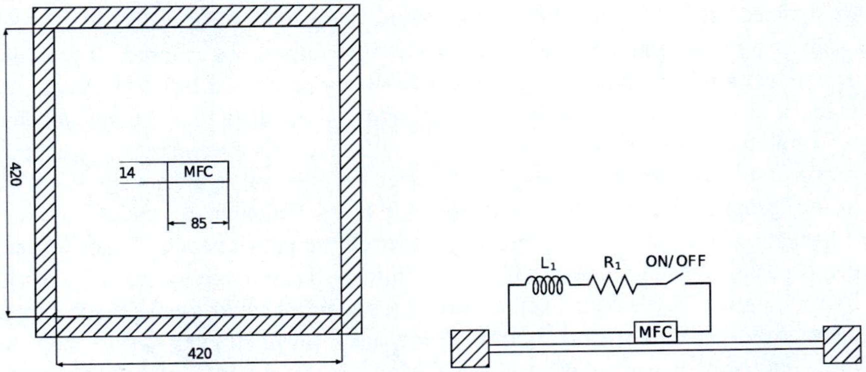

(a) A Pulse-switching shunt circuit mounted on a single vibrating plate (L_1 and R_1 are electric inductance and resistance, respectively).

(b) Shunt MFC bonded to the rigid casing wall.

Fig. 6 Schematic representations of an exemplary shunt MFC circuit attached to casing wall urface [45]. All dimensions are given in (mm)

turbances etc., so the introduced additional damping is optimal. Such adaptation can be obtained, e.g. by appropriately switching (keying) passive electric shunt circuits. The MFC of PZT patches can be successfully used to build the vibration absorber [47]. The structure can be arranged into a number of configurations, allowing different behaviour depending on the application.

A schematic representation of an exemplary shunt MFC circuit attached to a casing wall surface is presented in Fig. 6 [45]. It is a pulse-switching shunt circuit, which is a modified version of a state-switching circuit [47]. The circuit consists of additional electric LR elements, which combined with the natural capacitance of piezoceramic creates a LRC resonant circuit, thus increasing the quality factor. For the sake of brevity, the semi-active control approach for device casing is only outlined in this section, however, more detailed considerations can be found in [45].

6 Active Control

There are two basic active approaches to reduce the noise transmitted by a casing: reduction of noise measured by error microphones placed around the casing or reduction of the casing vibrations measured by structural sensors and thus reducing noise emissions indirectly. During previous research, both approaches were investigated by the authors. The second approach is satisfactory only for some frequencies and cannot be effectively used to reduce noise in most applications. The approach with error microphones provides much better results, even for global noise reduction. For applications where the employment of microphones around the device is impossible, the Virtual Microphone Control approach can be employed [36]. It allows us to estimate and minimize the sound pressure levels at specified locations in space, based on the measured vibration of the structure and its device casing walls.

In active noise and vibration control systems, a normalized Filtered-x Least Mean Square (FxLMS) feedforward algorithm is widely employed [48–51]. A reference signal can be obtained, e.g. by a microphone placed close to the noise source or by an attached vibration sensor. However, a useful reference signal is sometimes impossible to acquire due to, e.g. device structure preventing placement of a sensor at an appropriate location, hence the reference sensor is unable to provide in-advance information about the primary noise. As a result the performance of such a system can be degraded. Fortunately, often the dominant noise components are generated by rotating device elements, and thus they are tonal or multi-tonal. In such a case, alternatively to feedforward control, a feedback control strategy can be employed [52]. For the considered casing, Internal Model Control (IMC) architecture can be used, which is based on the error signal only [24]. It allows to directly apply techniques well developed for feedforward control, including the FxLMS algorithm. For the active casing it has been found sufficient to use a scalar reference signal $x(n)$ [24], which is an estimate of the primary noise at one selected j-th sensor obtained by subtracting from the selected error signal $e_j(n)$ the contribution of control signals

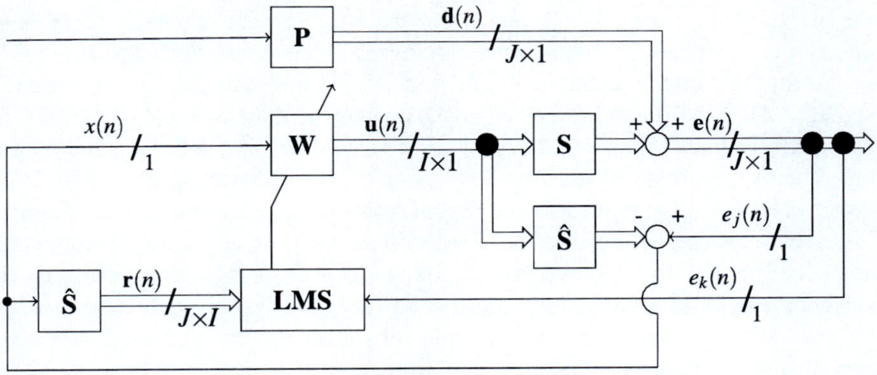

Fig. 7 Multi-channel feedback IMC control system with the FxLMS algorithm

$\mathbf{u}_i(n)$ from all actuators using models of respective secondary paths $\hat{\mathbf{s}}_{ij}(n)$ as shown in Fig. 7 and in the following equation:

$$x(n) = e_j(n) - \sum_{i=1}^{I} \hat{\mathbf{s}}_{ij}(n)^T \mathbf{u}_i(n). \tag{3}$$

For the washing machine casing, the vibrational and acoustic cross couplings are of significant magnitude. It results in a requirement imposed on the control system to take into account all paths between actuators and sensors (neglecting some of the paths may result in instability). However, such a multi-input multi-output (MIMO) control system represents a very high computational demand (in the considered system there are 11 inputs and 4 outputs). To respond to this feature, a switched-error modification is introduced to the control algorithm [53]. The modification consists in adaptation of all control filters according to only one error signal $e_k(n)$ (k-th out of J error signals), and cyclically changing the sensor which provides the signal (cyclically changing the k in the range from 1 to J). Such an algorithm results in approximately J times reduced computation complexity, but it is less vulnerable to cross couplings of the structure compared to a control system where some of the paths are neglected (e.g. as in separated control systems for individual walls). However, as expected, the cost for it is the convergence speed. This is due to the fact that at one time, control filters are adapting to only one of the error signals. It is approximately $2J$ times slower, but it is still a matter of only tens of seconds to converge, which is in accordance to predefined practical requirements. Further details of switched-error modification are provided in [53].

In the schematic representation of the resulting control system given in Fig. 7, symbol **W** is the adaptive control filters vector (of dimension $(I \times 1)$, where I is the number of actuators), **P** is the primary paths vector (of dimension $(J \times 1)$, where J is the number of error sensors), defined between the reference and error sensors. The symbol **S** stands for the secondary paths matrix of dimension $(J \times I)$ defined between the inputs of the actuators and outputs of the error sensors. These paths include electronics necessary for signal conditioning and data conversion. The symbol $\hat{\mathbf{S}}$ stands for the secondary path model. In turn, $x(n)$ is the estimated scalar reference signal, $\mathbf{r}(n)$ is the filtered-reference signals matrix of dimension $(J \times I)$, $\mathbf{u}(n)$ is the control signals vector of dimension $(I \times 1)$. Further, signals $\mathbf{d}(n)$ and $\mathbf{e}(n)$ are the primary disturbances vector and the error signals vector, respectively, both of dimension $(J \times 1)$, at positions of the error sensors where noise reduction is desired. Signal $e_j(n)$ is the j-th selected error signal for estimation of primary disturbance, as defined for the IMC architecture. Signal $e_k(n)$ is the k-th selected error signal currently employed for adaption and cyclically changed as previously described. In the control system configured for the light-weight casing, the number of actuators $I = 21$. On the other side, the number of error sensors (outer microphones) $J = 5$.

The coefficients of i-th control filter at the $(n+1)$-st sample, $w_i(n+1)$, are updated for the error signals $e_k(n)$ according to the formula:

$$\mathbf{w}_i(n+1) = \alpha\mathbf{w}_i(n) - \mu(n)\mathbf{r}_{ik}(n)e_k(n), \tag{4}$$

where $\mathbf{w}_i(n) = \left[w_{i,0}(n), w_{i,1}(n), \ldots, w_{i,N-1}(n)\right]^T$ is the vector of coefficients of the i-th adaptive Finite Impulse Response (FIR) control filter at sample n, and N is the filter order. Symbol $\mathbf{r}_{ik}(n) = [r_{ik}(n), r_{ik}(n-1), \ldots, r_{ik}(n-(N-1))]^T$ is a vector of regressors of the ik-th filtered-reference signal, $\mu(n)$ is a step-size, and $0 \ll \alpha < 1$ is the leakage coefficient.

The i-th control filter is used to calculate i-th control signal, $u_i(n+1)$, obtained as follows:

$$u_i(n+1) = \mathbf{w}_i(n)^T\mathbf{x}_u(n), \tag{5}$$

where $\mathbf{x}_u(n) = [x(n), x(n-1), \ldots, x(n-(N-1))]^T$ is the vector of regressors of the reference signal. The ik-th filtered-reference signal is calculated as:

$$r_{ik}(n) = \hat{\mathbf{s}}_{ik}(n)^T\mathbf{x}_r(n), \tag{6}$$

where $\hat{\mathbf{s}}_{ik}(n) = \left[\hat{s}_{ik,0}(n), \hat{s}_{ik,1}(n), \ldots, \hat{s}_{ik,M-1}(n)\right]^T$ is the vector of coefficients of the M-th order FIR model of the ik-th secondary path and $\mathbf{x}_r(n) = [x(n), x(n-1), \ldots, x(n-(M-1))]^T$ is a vector of regressors of the reference signal. It is justified to consider further reduction of the computational burden, as discussed in [54].

6.1 Experimental Results

To better visualize the performance of the active casing approach, selected experimental results are given in this subsection. Results obtained for a light-weight device casing are presented. A reference microphone placed next to the loudspeaker is used

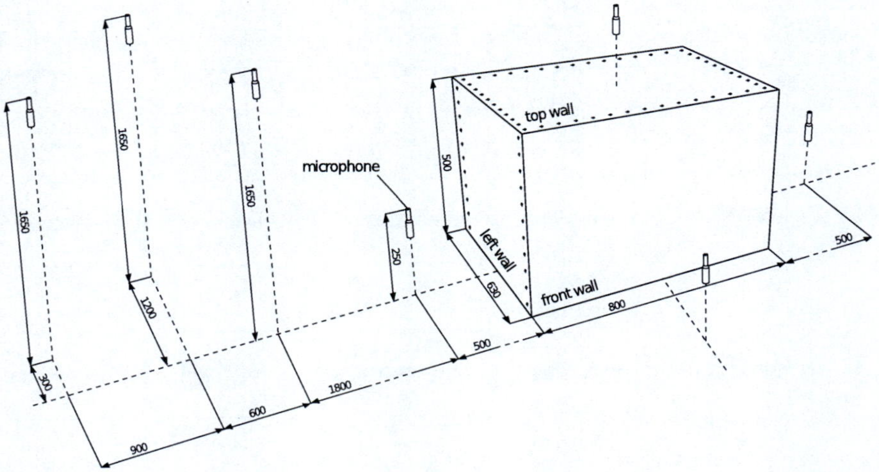

Fig. 8 A schematic representation of the laboratory setup with light-weight active casing. All dimensions are given in (mm)

to obtain the reference signal (in the presented experiment, a regular feedforward architecture has been used for the control algorithm). In front of each casing wall, a microphone is placed at the distance of 500 mm. The laboratory setup is presented schematically in Fig. 8.

Inertial exciters NXT EX-1 are used to control vibrations of the casing walls. They are mounted on the walls of the inner side. The number of actuators depends on the particular wall; four actuators are mounted to the front, right, back and left wall, and five actuators to the top wall. Their placement has been optimized using a method that maximizes a measure of the controllability of the system.

The error signal is obtained by the error microphones. The primary disturbance is generated by a loudspeaker enclosed in the casing as a tonal signal of frequency incremented by 4 Hz in the range from 1 to 500 Hz. The considered frequency range includes the low frequencies where the speaker starts to transmit sound.

The control performance is evaluated by noise reduction level observed by the room microphones. For each frequency of the primary disturbance, a 60 s experiment was performed. During its initial 5 s the active control was off, and variance of the signal acquired by different sensors was estimated. Then, active control was turned on. When the control algorithm converged, the final 5 s of the experiment were used to estimate the variance of the signal acquired by corresponding sensors.

Results in the time domain of an exemplary experiment for the frequency of primary disturbance equal 129 Hz are presented in Fig. 9. The first five rows present control signals, where the convergence rate can be observed. In the sixth row, signals measured by error microphones used in this experiment as error sensors are shown. In the seventh row of the Figure, signals measured by three room microphones are presented. The reference microphone measurement is also shown for completeness.

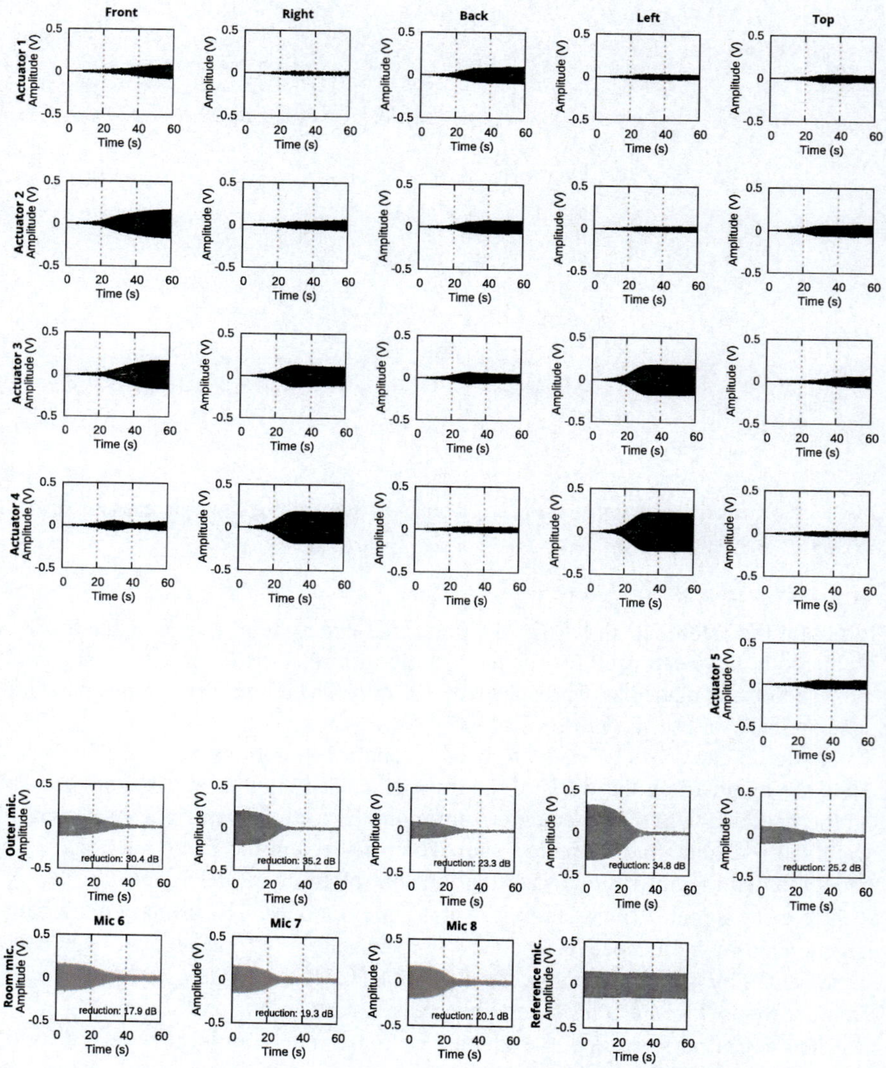

Fig. 9 Time plots for the experiment performed for primary disturbance of 129 Hz and light-weight casing with ASAC-SE algorithm. The outer microphones were used as error sensors

In Fig. 10 frequency characteristics of the experiments performed are presented. In the last rows of these figures, the mean reduction obtained by the room microphones is shown. This is considered as the main point for evaluation of active control performance. Remaining plots present variances in dB scale of signals acquired by error sensors and individual room microphones, without (grey) and with (black) control. Additionally, bellow each individual frequency characteristic, a reduction

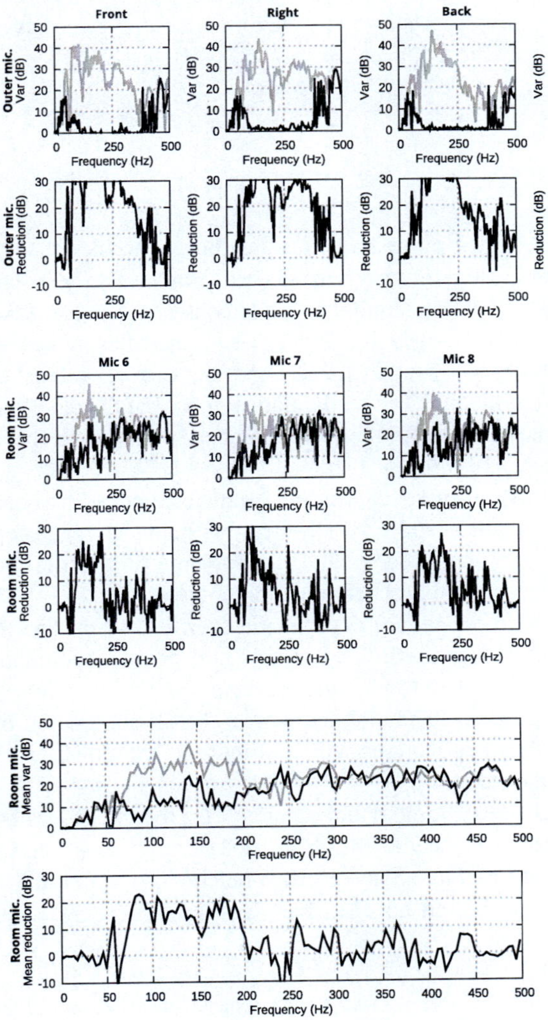

Fig. 10 Frequency characteristics for the experiment performed for light-weight casing with ASAC-SE algorithm. The outer microphones were used as error sensors

characteristic is also presented, calculated as a difference between noise level without and with control (reduction is marked with a black colour).

Taking into account the error signals, the control algorithm performed very well for frequencies up to approximately 400 Hz. Beside lowest frequencies, where the inertial actuators were lacking power, the noise at the error microphones was reduced to the noise floor level. Above the frequency of 400 Hz the noise reduction was weaker, however, noise enhancement at error microphones practically never occurred. It is also noteworthy to assess noise reduction observed by the room microphones. A high level of noise reduction has been achieved, even exceeding 20 dB.

7 Conclusions

In this chapter the problem of device noise reduction has been considered. The authors propose to use original casings, if they are made of thin walls, or to put the devices into additional thin-wall casings, and then try to control vibration of the casings. The theoretical basis for developing active device casings has been presented. An exemplary plant has been introduced, emphasizing important features related to active control and mathematical modeling. The mathematical model of the plant has been briefly described. The employment of the model for plant optimization has been introduced.

A wide variety of approaches to improve device casing sound insulation have been outlined, including passive, semi-active and active control methods. The decision for the means of reduction which are most beneficial depends on the application and restrictions imposed on the system, mainly related to the complexity and cost.

The chapter is summarized with exemplary experimental results, where one of the aforementioned techniques have been used. They confirm high practical potential of casing vibration control method applied to globally enhance acoustic isolation of industrial devices and home appliances. The authors believe that dissemination of this idea and following commercial applications will contribute to the reduction of noise pollution in the human environment.

This area also has very high scientific potential in many disciplines. It can be explored in various directions. For instance, more sophisticated ribs can be optimized to enhance passive isolation of the structure, many electric branches can be optimally switched in the semi-active approach, and novel active algorithms reducing noise emission globally can be developed.

Acknowledgements Support from the National Science Centre, Poland, decision no. DEC-2017/25/B/ST7/02236, is greatly appreciated.

References

1. Kuo, S.M., Morgan, D.: Active Noise Control Systems: Algorithms and DSP Implementations. Wiley (1996)
2. Lorente, J., Ferrer, M., de Diego, M., Gonzalez, A.: The frequency partitioned block modified filtered-x NLMS with orthogonal correction factors for multichannel active noise control. Digit. Signal Process. **43**, 47–58 (2015)
3. Ferrer, M., Gonzalez, A., de Diego, M., Piñero, G.: Distributed affine projection algorithm over acoustically coupled sensor networks. IEEE Trans. Signal Process. **65**(24), 6423–6434 (2017)
4. Morzyński, L., Szczepański, G.: System aktywnej redukcji hałasu do zastosowania w hełmach strażackich-propozycja rozwiazania. *Bezpieczeństwo Pracy: nauka i praktyka* (2018)

5. Lam, B., Elliott, S., Cheer, J., Gan, W.-S.: Physical limits on the performance of active noise control through open windows. Appl. Acoust. **137**, 9–17 (2018)
6. Jung, W., Elliott, S.J., Cheer, J.: Estimation of the pressure at a listener's ears in an active headrest system using the remote microphone technique. J. Acoust. Soc. Am. **143**(5), 2858–2869 (2018)
7. Murao, T., Shi, C., Gan, W.-S., Nishimura, M.: Mixed-error approach for multi-channel active noise control of open windows. Appl. Acoust. **127**, 305–315 (2017)
8. Wang, S., Tao, J., Qiu, X., Pan, J.: Mechanisms of active control of sound radiation from an opening with boundary installed secondary sources. J. Acoust. Soc. Am. **143**(6), 3345–3351 (2018)
9. Pawelczyk, M.: Application-Oriented Design of Active Noise Control Systems. Academic Publishing House Exit (2013)
10. Elliott, S., Nelson, P.: The active control of sound. Electron. Commun. Eng. J. **2**(4), 127–136 (1990)
11. Christopher, C.: Fuller, Sharon Elliott, and Philip A. Academic Press, Nelson. Active control of vibration (1996)
12. Pietrzko, S.: Contributions to Noise and Vibration Control Technology. AGH University of Science and Technology (2009)
13. Zawieska, W.M., Rdzanek, W.P., Rdzanek, W.J., Engel, Z.: Low frequency estimation for the sound radiation efficiency of some simply supported flat plates. Acta Acustica united with Acustica **93**(3), 353–363 (2007)
14. Szemela, K., Rdzanek, W.P., Żyłka, W.: The radiation efficiency measurements of real system of a thin circular plate embedded into a thick square baffle. Arch. Acoust. **43**(3), 413–423 (2018)
15. Szemela, K., Rdzanek, W.P., Pieczonka, D.: The total acoustic power of a clamped circular plate located at the boundary of three-wall corner region. Acta Physica Polonica A **119**(6A), 1050–1060 (2011)
16. Keir, J., Kessissoglou, N., Norwood, C.: Active control of connected plates using single and multiple actuators and error sensors. J. Sound Vibr. **281**(1–2), 73–97 (2005)
17. Morzyński, L.: System aktywnej redukcji hałasu przenikajacego przez przegrody szklane, wykorzystujacy algorytm rotacyjny. *Bezpieczeństwo Pracy: nauka i praktyka*, pp. 21–25 (2017)
18. Morzyński, L., Krukowicz, T.: Dwupłytowa struktura aktywna do ograniczania transmisji hałasu przez obudowy i przegrody dźwiekoizolacyjne. *Bezpieczeństwo Pracy: nauka i praktyka*, pp. 24–28 (2017)
19. Tuma, J., Suranek, P., Mahdal, M., Wagnerova, R.: Linear piezoactuators in systems of the active vibration control. In: Proceedings of the 23rd International Congress on Sound and Vibration, Athens, Greece (2016)
20. Camperi, S., Tehrani, M.G., Elliott, S.J.: Parametric study on the optimal tuning of an inertial actuator for vibration control of a plate: theory and experiments. J. Sound Vibr. **435**, 1–22 (2018)
21. Fuller, C., McLoughlin, M., Hildebrand, S.: Active acoustic transmission loss box. WO Patent App. PCT/US1992/008,401 (1994)
22. Pawelczyk, M.: Final report for the project entitled „Active reduction of device acoustic noise by controlling vibration of the device casing", financed by the National Science Centre, Poland, decision no. DEC-2012/07/B/ST7/01408. Silesian University of Technology (2016)
23. Mazur, K., Pawelczyk, M.: Active control of noise emitted from a device casing. In: Proceedings of the 22nd International Congress of Sound and Vibration, Florence, Italy (2015)
24. Wrona, S., Pawelczyk, M.: Active reduction of device narrowband noise by controlling vibration of its casing based on structural sensors. In: Proceedings of the 22nd International Congress on Sound and Vibration, Florence, Italy (2015)
25. Mazur, K., Wrona, S., Pawelczyk, M.: Design and implementation of multichannel global active structural acoustic control for a device casing. Mech. Syst. Signal Process. **98**, 877–889 (2018)

26. Mazur, K., Pawelczyk, M.: Internal model control for a light-weight active noise-reducing casing. Arch. Acoust. **41**(2), 315–322 (2016)
27. Wrona, S., Pawelczyk, M.: Feedforward control of a light-weight device casing for active noise reduction. Arch. Acoust. **41**(3), 499–505 (2016)
28. Wyrwal, J., Zawiski, R., Pawelczyk, M., Klamka, J.: Modelling of coupled vibro-acoustic interactions in an active casing for the purpose of control. Appl. Math. Model. **50**, 219–236 (2017)
29. Klamka, J., Wyrwal, J., Zawiski, R.: On controllability of second order dynamical systems—a survey. Bull. Polish Acad. Sci.—Techn. Sci. **65**(3), 279–295 (2017)
30. Wyrwal, J.: Approximate controllability of infinite dimensional system with internal damping dependent on fractional powers of system operator. IET Control Theor. Appl. **10**(18), 2370–2377 (2016)
31. Klamka, J., Wyrwal, J.: Approximate controllability of stochastic nonlinear infinite dimensional systems. A short survey. In: Proceedings of the 21st International Conference on Methods and Models in Automation and Robotics, pp. 505–510 (2016)
32. Wrona, S., Pawelczyk, M.: Shaping frequency response of a vibrating plate for passive and active control applications by simultaneous optimization of arrangement of additional masses and ribs. Part I: Modeling. Mech. Syst. Signal Process. **70–71**, 682–698 (2016)
33. Wrona, S., Pawelczyk, M.: Shaping frequency response of a vibrating plate for passive and active control applications by simultaneous optimization of arrangement of additional masses and ribs. Part II: Optimization. Mech. Syst. Signal Process. **70–71**, 699–713 (2016)
34. Wiora, J., Wrona, S., Pawelczyk, M.: Evaluation of measurement value and uncertainty of sound pressure level difference obtained by active device noise reduction. Measurement **96**, 67–75 (2017)
35. Wiora, J., Kozyra, A., Wiora, A.: A weighted method for reducing measurement uncertainty below that which results from maximum permissible error. Measur. Sci. Technol. **27**(3), 035007 (2016)
36. Mazur, K., Pawelczyk, M.: Virtual microphone control for an active noise-cancelling casing. Solid State Phenom. **248**, 57–66 (2016)
37. Kurczyk, S., Pawelczyk, M.: Nonlinear structural acoustic control with shunt circuit governed by a soft-computing algorithm. Arch. Acoust. **43**(3), 397–402 (2018)
38. Williams, T., Cheng, X.: Degrees of controllability and observability for close modes of flexible space structures. IEEE Trans. Autom. Control **44**(9), 1791–1795 (1999)
39. El Jai, A., Pritchard, A.: Sensors and Controls in the Analysis of Distributed Systems. Halsted Press (1988)
40. Wrona, S., Pawelczyk, M.: Identification of elastic boundary conditions of light-weight device casing walls using experimental data. In: Proceedings of the 21st International Conference on Methods and Models in Automation and Robotics, pp. 212–217 (2016)
41. Goldberg, D.E.: Genetic Algorithms in Search Optimization and Machine Learning, vol. 412. Addison-wesley Reading Menlo Park (1989)
42. Nalepa, J., Blocho, M.: Adaptive cooperation in parallel memetic algorithms for rich vehicle routing problems. Int. J. Grid Util. Comput. **9**(2), 179–192 (2018)
43. Nalepa, J., Czech, Z.: New selection schemes in a memetic algorithm for the vehicle routing problem with time windows. In: International Conference on Adaptive and Natural Computing Algorithms, pp. 396–405. Springer (2013)
44. Neri, F., Cotta, C., Moscato, P.: Handbook of Memetic Algorithms, vol. 379. Springer (2012)
45. Kurczyk, S.: Soft computing algorithms for noise and vibration control. PhD thesis, Silesian University of Technology, Gliwice, Poland (2018)
46. Kurczyk, S., Pawełczyk, M.: Nonlinear structural acoustic control with shunt circuit governed by a soft-computing algorithm. Arch. Acoust. (2018)
47. Mao, Q., Pietrzko, S.: Control of Noise and Structural Vibration. Springer (2013)
48. Hansen, C., Snyder, S., Qiu, X., Brooks, L., Moreau, D.: Active Control of Noise and Vibration. CRC Press (2012)

49. Bismor, D., Pawelczyk, M.: Stability conditions for the leaky LMS algorithm based on control theory analysis. Arch. Acoust. **41**(4), 731–739 (2016)
50. Bismor, D.: Variable step size partial update LMS algorithm for fast convergence. In: Proceedings of the 22nd International Congress of Sound and Vibration, Florence, Italy (2015)
51. Lifu, W., Qiu, X., Guo, Y.: A generalized leaky FxLMS algorithm for tuning the waterbed effect of feedback active noise control systems. Mech. Syst. Signal Process. **106**, 13–23 (2018)
52. Niedźwiecki, M., Meller, M., Gajdzica, M.: Robust algorithm for active feedback control of narrowband noise. In: Proceedings of the 23rd European Signal Processing Conference (EUSIPCO), pp. 275–279. IEEE (2015)
53. Mazur, K., Pawelczyk, M.: Multiple-error adaptive control of an active noise-reducing casing. In: Progress of Acoustics, pp. 701–712. Polish Acoustical Society (2015)
54. Bismor, D., Czyz, K., Ogonowski, Z.: Review and comparison of variable step-size LMS algorithms. Int. J. Acoust. Vibr. **21**(1), 24–39 (2016)